The
ENCYCLOPEDIA
of
FIELD AND
GENERAL GEOLOGY

ENCYCLOPEDIA OF EARTH SCIENCES SERIES

Series Editor: Rhodes W. Fairbridge

Volume

I   THE ENCYCLOPEDIA OF OCEANOGRAPHY/*Rhodes W. Fairbridge*

II   THE ENCYCLOPEDIA OF ATMOSPHERIC SCIENCES AND
ASTROGEOLOGY/*Rhodes W. Fairbridge*

III   THE ENCYCLOPEDIA OF GEOMORPHOLOGY/*Rhodes W. Fairbridge*

IVA   THE ENCYCLOPEDIA OF GEOCHEMISTRY AND
ENVIRONMENTAL SCIENCES/*Rhodes W. Fairbridge*

IVB   THE ENCYCLOPEDIA OF MINERALOGY/*Keith Frye*

VI   THE ENCYCLOPEDIA OF SEDIMENTOLOGY/*Rhodes W. Fairbridge
and Joanne Bourgeois*

VII   THE ENCYCLOPEDIA OF PALEONTOLOGY/*Rhodes W. Fairbridge
and David Jablonski*

VIII   THE ENCYCLOPEDIA OF WORLD REGIONAL GEOLOGY, PART 1:
Western Hemisphere (Including Antarctica and Australia)/
Rhodes W. Fairbridge

X   THE ENCYCLOPEDIA OF STRUCTURAL GEOLOGY AND PLATE
TECTONICS/*Carl K. Seyfert*

XI   THE ENCYCLOPEDIA OF CLIMATOLOGY/*John E. Oliver
and Rhodes W. Fairbridge*

XII   THE ENCYCLOPEDIA OF SOIL SCIENCE, PART 1: Physics,
Chemistry, Biology, Fertility, and Technology/*Rhodes W. Fairbridge
and Charles W. Finkl, Jnr.*

XIII   THE ENCYCLOPEDIA OF APPLIED GEOLOGY/*Charles W. Finkl, Jnr.*

XIV   THE ENCYCLOPEDIA OF FIELD AND GENERAL GEOLOGY/
Charles W. Finkl, Jnr.

XV   THE ENCYCLOPEDIA OF BEACHES AND COASTAL
ENVIRONMENTS/*Maurice L. Schwartz*

ENCYCLOPEDIA OF EARTH SCIENCES, VOLUME XIV

The ENCYCLOPEDIA of FIELD AND GENERAL GEOLOGY

EDITED BY

Charles W. Finkl, Jnr.

Coastal Education and Research Foundation, Inc.
and
Florida Atlantic University

VNR VAN NOSTRAND REINHOLD COMPANY
New York

Copyright © 1988 by **Van Nostrand Reinhold Company Inc.**
Library of Congress Catalog Card Number: 87-21618
ISBN: 0-442-22499-0

Printed in the United States of America

Van Nostrand Reinhold Company Inc.
115 Fifth Avenue
New York, New York 10003

Van Nostrand Reinhold Company Limited
Molly Millars Lane
Wokingham, Berkshire RG11 2PY, England

Van Nostrand Reinhold
480 La Trobe Street
Melbourne, Victoria 3000, Australia

Macmillan of Canada
Division of Canada Publishing Corporation
164 Commander Boulevard
Agincourt, Ontario, M1S 3C7, Canada

16 15 14 13 12 11 10 9 8 7 6 5 4 3 2 1

Library of Congress Cataloging-in-Publication Data
The Encyclopedia of field and general geology.
  (Encyclopedia of earth sciences; v. 14)
  Bibliography: p.
  Includes index.
  1. Geology—Dictionaries. I. Finkl, Charles W., 1941-   . II. Series.
QE5.E515 1988   550′.3′21   87-21618
ISBN 0-442-22499-0

PREFACE

Fieldwork, supplemented by laboratory studies, is a cornerstone for the geological sciences. Even with the advent of modern remote-sensing techniques that enable geoscientists to interpret a great many features using high-altitude or space photography, they must still visit the field for the purposes of ground truthing. Because geology, the study of the form, nature, and history of the Earth, is one of the so-called field sciences, it is essential that geologists become familiar with field methods in their special area of interest. It is also important for those working in laboratories to have at least rudimentary knowledge of field conditions and sampling procedures because extant environmental conditions often affect, either directly or indirectly, interpretations of lab data. Although new, highly sophisticated field equipment lessens many tasks that were previously burdensome, it is still essential for geologists to know the field relationships of rocks and minerals.

Field geology, according to the *Glossary of Geology* (American Geological Institute, 1980), is geology as practiced by direct observation in the field. Original, primary reconnaissance—i.e., fieldwork sensu stricto—provides much of our present knowledge of the Earth's complicated history. The purpose of this volume is to provide an introduction to general fieldwork by way of selected topics that illustrate some specific technique or methodology. Because the scope of the subject is so great and because the length of the volume is restricted for practical purposes, this compendium is an eclectic collection. Although we attempted to represent as many aspects of field geology as possible, we gave only cursory mention to some very specialized subject areas and omitted others because of limited interest and space limitations. For the topics that are included here, it is noted that, because fieldwork presents great diversity and all conditions cannot be foreseen, readers using this encyclopedia will likely have to modify methods or concepts to suit their particular needs.

Coverage and Scope

This volume, *Field and General Geology* (Vol. XIV in the Encyclopedia of Earth Science Series), perhaps more than any other in this geoscience collection, is geared to the practical geologist. Field geology should also be of interest to professional earth scientists who teach or are involved with administrative duties but occasionally make field inspections or conduct specialized field surveys. Most topics deal with aspects of exploration surveys, geotechnical engineering, field techniques, mapping, prospecting, and mining. The general topic of surveys, e.g., is broken down into numerous specialized entries that detail important aspects of primary field endeavors *viz. Acoustic Surveys, Aerial Surveys, Harbor Surveys,* and *Sea Surveys.* Special efforts were made to include topics that consider aspects of environmental geology, in particular those subjects that involve field inspections related, e.g., to the placement of artificial fills, sediment control in canals and waterways, the geologic effects of cities, or the importance of expansive soils to environmental management and engineering. The potpourri that follows a general theme is provided as background to descriptions of the field techniques. Some widely ranging topics dealing with legal affairs, geological methodology, the scope and organization of geology, report writing, and other concepts such as those related to plate tectonics and continental drift provide a necessary perspective to the arena of field geology. Although this volume contains topics that describe specific techniques or procedures—*viz.* alidade, block diagrams, field notes, plane table mapping, profile construction, or sampling—this is not a how-to book on field geology. For detailed procedures readers are referred to specialized governmental and proprietary manuals or to the following list of further readings.

Because this volume contains many topics that are linked by the common thread of geology in the service of human beings, it is viewed as a companion to *Applied Geology,* Vol. XIII in the series. These two volumes (XIII and XIV) are also closely interrelated because they were originally designed as a single integrated work. Page limits and marketability dictated that a range of subjects be removed from *Applied Geology,* which are treated in the present volume and are rounded out by additional topics that were specially prepared for this volume to support the field sciences.

Literature

The study area of field geology is so great that, as a discrete discipline, it almost defies definition in a pragmatic sense. Developments in cognate areas of interest constitute a great literature in themselves, and many geologists have difficulty keeping abreast of advances in related fields. So diverse and pervasive is the subject of field geology that no English-language journal is devoted strictly to field topics. Field techniques and methodologies are described in a plethora of journals too numerous to list here. There are, however, several good texts, field manuals, and guides that provide introductions to field geology. Some reading sources of particular interest, including classical works, are listed in the following selection.

Ahmed, F., and D. C. Almond, 1983, *Field Mapping for Geology Students.* London: Allen & Unwin, 72p.

Anderson, J. G. C., 1983, *Field Geology in the British Isles.* Oxford: Pergamon, 324p.

Anon., 1968, *Field Training in Geography.* Washington, D.C.: Association of American Geographers, 69p.

Armstrong, A. T., 1975, *Handbook on Quarrying.* Adelaide, South Australia: Department of Mines, 238p.

Barnes, J. W., 1981, *Basic Geological Mapping.* New York: Wiley, 112p.

Bates, D. E. B., and J. F. Kirkaldy, 1977, *Field Geology.* New York: Arco, 215p.

Beckmann, H., 1976, *Geological Prospecting of Petroleum.* New York: Wiley, 183p.

Berkman, D. A., and W. R. Ryall, 1976, *Field Geologists' Manual.* Parkville, Victoria: Australasian Institute of Mining and Metallurgy, 295p.

Boyer, R. E., 1971, *Field Guide to Rock Weathering.* Boston: Houghton Mifflin, 37p.

Branagan, D. F., and G. H. Packham, 1967, *Field Geology of New South Wales.* Sydney: Science Press, 191p.

Chief of Engineers, 1941, *Basic Field Manual: Conventional Signs, Military Symbols, and Abbreviations.* Washington, D.C.: U.S. Government Printing Office (War Dept. FM 21-30), 66p.

Compton, R. R., 1962, *Manual of Field Geology.* New York: Wiley, 378p.

Dietrich, R. V., J. T. Dutro, and R. M. Foose, 1982, *AGI Data Sheets for Geology in the Field, Laboratory, and Office.* Falls Church, Va.: American Geological Institute.

Freeman, T., 1971, *Field Guide to Layered Rocks.* Boston: Houghton Mifflin, 44p.

Fry, N., 1984, *The Field Description of Metamorphic Rocks.* New York: Wiley, 110p.

Gardiner, V., and R. Dackombe, 1983, *Geomorphological Field Manual.* London: Allen & Unwin, 254p.

Geike, A., 1879, *Outlines of Field Geology.* London: Macmillan, 222p.

Hamilton, W. R., A. R. Woolley, and A. C. Bishop, 1980, *The Larousse Guide to Minerals, Rocks and Fossils.* New York: Larousse, 320p.

Hayes, C. W., 1921, *Handbook for Field Geologists.* New York: Wiley, 166p.

Hawkes, H. E., and J. S. Webb, 1962, *Geochemistry in Mineral Exploration.* New York: Harper & Row, 415p.

Hobbs, B. E., D. M. Winthrop, and P. F. Williams, 1976, *An Outline of Structural Geology.* New York: Wiley, 571p.

Joyce, M. D., 1982, *Site Investigation Practice.* New York: E. & F. N. Spon, 369p.

Knill, 1978, *Industrial Geology.* Oxford: Oxford University Press, 344p.

Lahee, F. H., 1961, *Field Geology.* New York: McGraw-Hill, 926p.

Langstaff, C. S., and D. Morrill, 1981, *Geologic Cross Sections.* Boston: International Human Resources Development Corp., 108p.

Lattman, L. H., and R. G. Ray, 1965, *Aerial Photographs in Field Geology.* New York: Holt, Rinehart & Winston, 221p.

Levinson, A. A., 1974, *Introduction to Exploration Geochemistry.* Calgary, Alberta: Applied, 612p.

Lindgren, W., 1933, *Mineral Deposits.* New York: McGraw-Hill, 929p.

Moseley, F., 1981, *Methods in Field Geology.* Oxford: Freeman, 211p.

Murray, J. W., 1981, *A Guide to Classification in Geology.* New York: Wiley, 112p.

Peters, W. C., 1978, *Exploration and Mining Geology.* New York: Wiley, 696p.

Pettijohn, F. J., 1957, *Sedimentary Rocks.* New York: Harper & Row, 718p.

Pfleider, E. P., 1972, *Surface Mining.* New York: American Institute of Mining, Metallurgical, and Petroleum Engineers, 1,061p.

Platt, J. L., and J. Challinor, 1980, *Simple Geological Structures.* London: Allen & Unwin, 56p.

Pough, F. H., 1960, *A Field Guide to Rocks and Minerals.* Boston: Houghton Mifflin, 349p.

Rapp, G., Jr., 1971, *Color of Minerals.* Boston: Houghton Mifflin, 30p.

Roberts, J. L., 1982, *Introduction to Geological Maps and Structures.* Oxford: Pergamon, 332p.

Rock-Color Chart Committee, 1980, *Rock-Color Chart.* Boulder, Colo.: Geological Society of America.

Romey, W. D., 1971, *Field Guide to Plutonic and Metamorphic Rocks.* Boston: Houghton Mifflin, 53p.

Siegel, F. R., 1974, *Applied Geochemistry.* New York: Wiley, 353p.

Spock, L. E., 1962, *Guide to the Study of Rocks.* New York: Harper & Row, 298p.

Thomas, J. A. G., 1979, *An Introduction to Geological Maps.* London: Allen & Unwin, 67p.

Tucker, M. E., 1982, *The Field Description of Sedimentary Rocks.* New York: Wiley, 112p.

How to Use this Encyclopedia

As in most encyclopedic works, topics are arranged in alphabetical order. The organizational levels of entries in this series—reviews of major disciplines, component building blocks, and elements and phenomena—permit direct access to primary subject matter. The wide range of subjects, however, often necessitates grouping specific concepts or applications under general headings or according to key words or phrases. As an aid to the reader, entries are cross-referenced both in the body of the text and at the end of each article. (*Biogeochemistry,* e.g., appears under that title, is listed in the *B*s, and is cross-referenced at the end of the *Geochemistry* entry.) The abbreviation q.v. (quode vide), as in "cross-section (q.v.)," means that an entry with that title appears elsewhere in this volume. Cross-references to related subjects in other volumes in the Encyclopedia of Earth Science Series are also included. For a list of volumes in this series see Table 1.

If a subject does not appear as a main entry or a cross-reference or in-text cross-reference, the reader should consult the comprehensive subject index. An author citation index is also provided because some researchers find it useful to look up specific subjects by author or wish to find publications by the experts quoted here.

References to cited works follow each entry. These bibliographic citations are designed to lead the interested reader to additional, more detailed information or to other summaries that contain extensive reference lists. Efforts were made to ensure that complete bibliographic information was included in each citation to facilitate easy retrieval in libraries, through reference services, or from computerized databanks. Journal titles are abbreviated according to the style recommended by *Serial Publications Commonly Cited in Technical Bibliographies of the United States Geological Survey*. Citations for books follow an abbreviated format that gives the main titles and a key word for the publisher; thus, the Van Nostrand Reinhold Publishing Company becomes Van Nostrand Reinhold and John Wiley & Sons, Inc. becomes Wiley & Sons.

Abbreviations, Units, and Symbols

Standard abbreviations are used for measures of area, length, volume, mass, pressure, and temperature. The Systéme International d'Unités (S.I.), as established by the International Organization for Standardization in 1960, is used throughout this volume. Equivalent units are often given in parentheses, however, because some tables, figures, and maps retain units of the English system (Customary Units) or because it is convenient to retain older units for practical purposes. Even though the pascal, e.g., is the S.I. unit of pressure, the millibar still finds applications in many practices.

With such a wide range of topics in field geology, it is impossible to list here all the units used in this volume. Most units are defined locally as they are used in each article, but for those readers wishing to brush up on refinements of S.I., the following tables may be useful. Table 2 gives values and symbols for basic S.I. units. Table 3 provides prefixes, symbols, and multiplying factors that can be applied to basic units. These prefixes, which indicate fractions or multiples of basic or derived S.I. units, are useful because some primary S.I. units have been found to be of inconvenient size for practical applications. Some units that are not part of S.I. but that are widely used by specialists are listed in Table 4. Although most nations use S.I. units in scientific and industrial applications, the United States, Brunei, Burma, and North and South Yemen remain the last officially to make the transition from Customary Units to S.I. Because the momentum of the national conversion process has been temporarily stifled in the United States, both kinds of units continue to be used by geoscientists working in the field or in other applied situations. Table 5 facilitates work in both systems. It lists a number of traditional units that can be converted from

---

TABLE 1. Published Volumes in the Encyclopedia of Earth Science Series

Volume	
I	THE ENCYCLOPEDIA OF OCEANOGRAPHY/*Rhodes W. Fairbridge*
II	THE ENCYCLOPEDIA OF ATMOSPHERIC SCIENCES AND ASTROGEOLOGY/ *Rhodes W. Fairbridge*
III	THE ENCYCLOPEDIA OF GEOMORPHOLOGY/*Rhodes W. Fairbridge*
IVA	THE ENCYCLOPEDIA OF GEOCHEMISTRY AND ENVIRONMENTAL SCIENCES/ *Rhodes W. Fairbridge*
IV B	THE ENCYCLOPEDIA OF MINERALOGY/*Keith Frye*
VI	THE ENCYCLOPEDIA OF SEDIMENTOLOGY/*Rhodes W. Fairbridge and Joanne Bourgeois*
VII	THE ENCYCLOPEDIA OF PALEONTOLOGY/*Rhodes W. Fairbridge and David Jablonski*
VIII	THE ENCYCLOPEDIA OF WORLD REGIONAL GEOLOGY, PART 1: Western Hemisphere (Including Antarctica and Australia)/*Rhodes W. Fairbridge*
X	THE ENCYCLOPEDIA OF STRUCTURAL GEOLOGY AND PLATE TECTONICS/ *Carl K. Seyfert*
XI	THE ENCYCLOPEDIA OF CLIMATOLOGY/*John E. Oliver and Rhodes W. Fairbridge*
XII	THE ENCYCLOPEDIA OF SOIL SCIENCE, PART 1: Physics, Chemistry, Biology, Fertility, and Technology/*Rhodes W. Fairbridge and Charles W. Finkl, Jnr.*
XIII	THE ENCYCLOPEDIA OF APPLIED GEOLOGY/*Charles W. Finkl, Jnr.*
XIV	THE ENCYCLOPEDIA OF FIELD AND GENERAL GEOLOGY/*Charles W. Finkl, Jnr.*
XV	THE ENCYCLOPEDIA OF BEACHES AND COASTAL ENVIRONMENTS/*Maurice L. Schwartz*

TABLE 2. The International System of Units (S.I.).

Physical Quantity	Name of Unit	Value	Symbol
Length	meter	base unit	m
	millimeter	0.001 m	mm
	centimeter	0.01 m	cm
	kilometer	1,000 m	km
	international nautical mile (navigation)	1,852 m	n mi
Mass (commonly called "weight")	kilogram	base unit (1,000 g)	kg
	gram	0.001 kg	g
	tonne	1,000 kg	t
Time interval	second	base unit	s
Area	square meter	S.I. unit	m^2
	square millimeter	$0.000001\ m^2$	mm^2
	square centimeter	$0.0001\ m^2$	cm^2
	hectare	$10,000\ m^2$	ha
Volume	cubic meter	S.I. unit	m^3
	cubic millimeter	$10^{-9}\ m^3$	mm^3
	cubic centimeter	$0.000001\ m^3$	cm^3
	cubic decimeter	$0.001\ m^3$	dm^3
Volume (fluids only)	liter	$0.001\ m^3$	l
	milliliter	$0.001\ l$	ml
	kiloliter	$1,000\ l\ (1\ m^3)$	kl
Velocity and speed	meter per second	S.I. unit	m/s or $m\ s^{-1}$
	kilometer per hour	0.27 m/s	km/h or $km\ h^{-1}$
	knot (navigation)	1 n mi/h or 0.514 m/s	kn
Force	newton[a]	S.I. unit	N
Energy	joule[a]	S.I. unit	J
Power	watt[a]	S.I. unit	W
Density	kilogram per cubic meter	S.I. unit	kg/m^3 or $kg\ m^{-3}$
	tonne per cubic meter	$1,000\ kg/m^3$	t/m^3 or tm^{-3}
	gram per cubic centimeter	$1,000\ kg/m^3$	g/cm^3 or $g\ cm^{-3}$
Density (fluids only)	kilogram per liter	$1,000\ kg/m^3$	kg/l or kgl^{-1}
	gram per milliliter	$1,000\ kg/m^3$	g/ml or $g\ ml^{-1}$
Pressure	pascal	S.I. unit (N/m^2)	Pa
Pressure (meteorology)	bar	100,000 Pa	bar
	millibar	100 Pa	mbar
Electric current	ampere[b]	base unit	A
Potential difference or electromotive force	volt [a, b]	S.I. unit	V
Electrical resistance	ohm[a, b]	S.I. unit	
Frequency	hertz[a]	S.I. unit	Hz
	revolution per minute	$\frac{1}{60}$ Hz	rpm or rev/min
Temperature	kelvin	base unit	K
	degree Celsius[c]	K	°C
Plane angle	radian	S.I. unit	rad
	milliradian	0.001 rad	mrad
	degree	$\pi/180$ rad	°
	minute	$\frac{1°}{60}$	′
	second	$\frac{1'}{60}$	″
Amount of substance	mole	base unit	mol

[a]Decimal multiples commonly associated with this unit are *kilo* ($\times$ 1,000), *mega* ($\times$ 1,000,000), and *giga* ($\times$ 1,000,000,000).

[b]Decimal submultiples associated with this unit are *milli* ($\times$ 0.001) and *micro* ($\times$ 0.0000001).

[c]The units of temperature on the Celsius scale (°C) and the thermodynamic scale (K) are equal. A temperature t on the Celsius scale is related to a temperature T on the thermodynamic scale by the relationship $t = T - 273.15$.

Source: After Berkman, D. A., 1976. *Field Geologists' Manual.* Parkville, Victoria: Australasian Institute of Mining and Metallurgy, pp. 275–276.

TABLE 3. Prefixes, Symbols, and Multiplying Factors

Multiplying Factor	Prefix	Symbol
$1{,}000{,}000{,}000{,}000 = 10^{12}$	tera	T
$1{,}000{,}000{,}000 = 10^{9}$	giga	G
$1{,}000{,}000 = 10^{6}$	mega	M
$1{,}000 = 10^{3}$	kilo	k
$100 = 10^{2}$	hecto	h
$10 = 10^{1}$	deca	da
$0.1 = 10^{-1}$	deci	d
$0.01 = 10^{-2}$	centi	c
$0.001 = 10^{-3}$	milli	m
$0.000001 = 10^{-6}$	micro	μ
$0.000000001 = 10^{-9}$	nano	n
$0.000000000001 = 10^{-12}$	pico	p
$0.000000000000001 = 10^{-15}$	femto	f
$0.000000000000000001 = 10^{-18}$	atto	a

TABLE 4. Some Non-S.I. Units with S.I.

Condition of Use	Unit	Symbol	Value in S.I. Units
Permissable Universally with S.I.	minute	min	1 min = 60 s
	hour	h	1 h = 3,600 s
	day	d	1 d = 86,400 s
	year	a	$1 \text{ a} = 3.1536 \times 10^{7}$ s
	degree (of arc)	°	$1° = (\pi/180)$ rad
	minute (of arc)	'	$1' = (\pi/10{,}800)$ rad
	second (of arc)	''	$1'' = (\pi/648{,}000)$ rad
	liter[a]	l	$1\ l = 1 \text{dm}^3$
	tonne[b]	t	$1 \text{ t} = 10^{3}$ kg
	degree Celius	°C	[f]
	revolution	r	$1 \text{ r} = 2\pi$ rad
Permissible in Specialized fields	electronvolt	eV	$1 \text{ eV} = 1.60219 \times 10^{-19}$ J
	unit of atomic mass	u	$1 \text{ u} = 1.66053 \times 10^{-27}$ kg
	astronomical unit[c]		1 AU = 149.600 Gm
	parsec	pc	1 pc = .30857 Tm
Permissible for a Limited Time	nautical mile		1 nautical mile = 1,852 m
	knot[d]		1 nautical mile per hour = (1,852/3,600) m/s
	angstrom	Å	$1 \text{ Å} = 0.1 \text{ nm} = 10^{-10}$ m
	hectare[e]	ha	$1 \text{ ha} = 10^{4} \text{m}^2$
	bar	bar	1 bar = 100kPa
	standard atmosphere	atm	1 atm = 101.325 kPa

[a]The word liter standing alone must be typed in full unless the typewriter is equipped with a special looped "ell."
[b]Care must be taken in the interpretation of this word when it occurs in French text of Canadian origin where the implication may be a "ton of 2,000 lb."
[c]This unit does not have an international symbol. The abbreviations used are AU in French and UA in French.
[d]There is no internationally recognized symbol for knot (1 nautical mile per hour), but kn is frequently used.
[e]The "acre" and "hectare," by international agreement, can be used for a limited period of time as needed in the agriculture and surveying sectors.
[f]The S.I. unit of temperature is the kelvin. The Celsius temperature scale (previously called Centigrade) is the commonly used scale for temperature measurements, except for some scientific work where a thermodynamic scale is used.

Source: After Kenting Limited, n. d. *The International System of Units (S.I.) for the Petroleum Industry in Canada.* Calgary, Alberta, Canada.

TABLE 5. Alphabetical List of Traditional Units Showing the Conversion Factor to Selected Metric (S.I.) Units

Traditional Unit	Multiply Value in Traditional Units by Factor × 10E to Obtain Value in Metric Units[a] Factor E	Selected Metric (S.I.) Unit Name	Symbol
acre	4.046856 E + 03	square meter	m^2
acre	4.046856 E − 01	hectare	ha
acre-foot	1.233482 E + 03	cubic meter	m^3
ampere hour	3.6* E + 00	kilocoulomb	kC
API gravity at 60°F	ASTM D1250 Table 3 × 1000	kilogram per cubic meter at 15°C	kg/m^3
angstrom unit	1.0* E − 01	nanometer	nm
arpent	3.418894 E − 01	hectare	ha
atmosphere (atm)	1.01325* E + 02	kilopascal	kPa
atmosphere technical (at or atm)	9.80665* E + 01	kilopascal	kPa
bar	1.0* E + 02	kilopascal	kPa
barrel (42 U.S. gal)	1.589873 E − 01	cubic meter	m^3
barrel per foot	5.216118 E − 01	cubic meter per meter	m^3/m
barrel per inch	6.25934 E − 01	cubic meter per centimeter	m^3/cm
barrel per acre-foot	1.28893 E − 04	cubic meter per cubic meter	m^3/m^3
barrel per cubic mile	3.81431 E − 02	cubic meter per cubic kilometer	m^3/km^3
barrel per long ton (U.K.)	1.564763 E − 01	cubic meter per tonne	m^3/t
barrel per short ton (U.S.)	1.752535 E − 01	cubic meter per tonne	m^3/t
barrel per day	6.624471 E − 03	cubic meter per hour	m^3/h
barrel per day psi	2.305916 E − 02	cubic meter per day kilopascal	$m^3/(d \cdot kPa)$
BCF (billion cubic feet: 60°F 1 atm)	2.826231 E + 07	cubic meter (API)	m^3(API)
	1.195307 E + 00	gigamole	Gmol
Btu (International Table)	1.055056 E + 00	kilojoule	kJ
Btu per barrel	6.636102 E + 00	kilojoule per cubic meter	kJ/m^3
Btu per brake horsepower hour	3.930148 E − 01	watt per kilowatt	W/kW
Btu per square foot second	1.135653 E + 01	kilowatt per square meter	kW/m^2
Btu inch per square foot second °F	5.192204 E − 01	kilowatt per meter degree Celsius	$kW/(M \cdot °C)$
Btu per square foot hour °F	5.678263 E − 03	kilowatt per square meter degree Celsius	$kW/(m^2 \cdot °C)$
Btu per square foot second °F	2.044175 E +01	kilowatt per square meter degree Celsius	$kW/(m^2 \cdot °C)$
Btu per cubic foot	3.725895 E + 01	kilojoule per cubic meter	kJ/m^3
Btu per standard cubic foot (60°F − 1 atm)	8.826705 E − 01	kilojoule per mole	kJ/mol
Btu per gallon	2.320808 E − 01	kilojoule per liter	kJ/l
Btu per gallon (U.S.)	2.787163 E − 01	kilojoule per liter	kJ/l
Btu per hour	2.930711 E − 04	kilowatt	kW
Btu per minute	1.758427 E − 02	kilowatt	kW
Btu per pound	2.326010 E + 00	kilojoule per kilogram	kJ/kg
Btu per pound °F	4.1868 E + 00	kilojoule per kilogram degree Celsius	$kJ/(kg \cdot °C)$
Btu per pound mole	2.326000 E + 00	joule per mole	J/mol
Btu per second	1.055056 E + 00	kilowatt	kW
Btu foot per square foot hour °F	1.73073 E + 00	watt per meter degree Celsius	$W/(m \cdot °C)$
calorie (International Table)	4.1868* E + 00	joule	J
calorie (thermochemical)	4.184* E + 00	joule	J

*The conversion factor is exact.
[a]To go from metric to customary units, divide the conversion factor instead of multiplying.

TABLE 5. (*continued*)

Traditional Unit	Multiply Value in Traditional Units by Factor × 10E to Obtain Value in Metric Units[a] Factor E	Selected Metric (S.I.) Unit Name	 Symbol
calorie per centimeter °C second	4.1868* E + 02	watt per meter degree Celsius	W/m · °C)
calorie (IT) per cubic centimeter second	4.1868* E + 12	microwatt per cubic meter	μW/m^3
calorie (IT) per square centimeter second	4.1868* E + 07	milliwatt per square meter	mW/m^2
calorie (IT) per gram °C	4.1868* E + 00	kilojoule per kilogram degree celsius	kJ/(kg · °C)
calorie (thermochemical) per pound	9.224141 E + 00	joule per kilogram	J/kg
centigrade (= Celsius)			
centipoise	1.0* E + 00	millipascal second	mPa · s
centistoke	1.0* E + 00	square millimeter per second	mm^2/s
chain	2.01168* E + 01	meter	m
cubem (cubic mile)	4.168182 E + 00	cubic kilometer	km^3
cubic inch	1.638706 E − 02	cubic decimeter	dm^3
cubic foot (see standard cu. ft.)	2.831685 E − 02	cubic meter	m^3
cubic foot gas (60°F 1 atm) per acre foot	9.690510 E − 04 2.291262 E − 05	mole per cubic meter cubic meter API per cubic meter	mol/m^3 m^{3}API/m^3
cubic foot of gas (60°F − 1 atm) per barrel	1.777646 E − 01	cubic meter API per cubic meter	m^{3}API/m^3
cubic foot of gas per barrel of oil	7.518255 E + 00	mole per cubic meter	mol/m^3
cubic foot per foot	9.290304* E − 02	cubic meter per meter	m^3/m
cubic foot per pound	6.242797 E + 01	cubic decimeter per kilogram	dm^3/kg
cubic yard	7.645549 E − 01	cubic meter	m^3
cycle per second	1.0* E + 00	hertz	Hz
"cc"	1.0* E + 00	cubic centimeter	cm^3
CV (cheval vapeur)	7.354990 E − 01	kilowatt	kW
darcy	9.869233 E − 01	square micrometer	μm^2
decibel	0.1 × log$_{10}$ (ratio of two intensities, e.g., watts)		dB
degree (angle) per foot	5.726145 E − 02	radian per meter	rad/m
degree (angle)	1.745329 E − 02	radian	rad
degree Centigrade (= Celsius)			
degree Fahrenheit	(°F − 32) 5/9* E + 00	degree Celsius	°C
degree Fahrenheit as interval	5/9* E + 00	degree Celsius	°C
degree Rankine	5/9* E + 00	kelvin	K
degree F per hundred foot	1.822689 E − 02	degree Celsius per meter	°C/m
dyne	1.0* E − 05	newton	N
dyne per centimeter	1.0* E + 00	millinewton per meter	mN/m
dyne per square centimeter	1.0* E − 01	pascal	Pa
erg	1.0* E − 07	joule	J
erg per square centimeter	1.0* E + 00	millijoule per square meter	mJ/m^2
erg per year	3.170979 E − 15	watt	W
fathom	1.8288* E + 00	meter	m
foot	3.048* E − 01	meter	m
foot-candle	1.076391 E + 01	lux	lx

*The conversion factor is exact.
[a]To go from metric to customary units, divide the conversion factor instead of multiplying.

(*continued*)

TABLE 5. (*continued*)

Traditional Unit	Multiply Value in Traditional Units by Factor × 10E to Obtain Value in Metric Units[a] Factor E	Selected Metric (S.I.) Unit Name	Symbol
foot per barrel	1.917134 E + 00	meter per cubic meter	m/m^3
foot per cubic foot	1.07639 E − 01	meter per cubic meter	m/m^3
foot per °F	5.4864* E − 01	meter per kelvin	m/K
foot per gallon (U.S.)	8.051964 E + 01	meter per cubic meter	m/m^3
foot per gallon	6.7046 E + 01	meter per cubic meter	m/m^3
foot per mile	1.893939 E − 01	meter per kilometer	m/km
foot pound-force	1.355818 E + 00	joule	J
foot pound-force per minute	2.259697 E − 02	watt	W
gal (see milligal)			
gallon (Cdn. & new U.K.)	4.54609* E + 00	liter	*l*
gallon (old U.K.)	4.546092 E + 00	liter	*l*
gallon (U.S.)	3.785412 E + 00	liter	*l*
gallon (U.S.) per foot	1.241933 E − 02	cubic meter per meter	m^3/m
gallon per foot	1.4914 E − 02	cubic meter per meter	m^3/m
gallon per horsepower hour	1.693466 E + 00	liter per megajoule	*l*/MJ
gallon per mile	2.824809 E + 02	liter per 100 kilometer	*l*/100km
gallon (U.S.) per mile	2.352146 E + 02	liter per 100 kilometer	*l*/100km
gallon per pound	1.002241 E + 01	liter per kilogram	*l*/kg
gallon (U.S.) per pound	8.345405 E + 00	liter per kilogram	*l*/kg
gallon (U.S.) per short ton	4.172702 E + 00	liter per tonne	*l*/t
gallon (U.S.) per long ton	3.725627 E + 00	liter per tonne	*l*/t
gamma (magnetic flux density)	1.0* E + 00	nanotesla	nT
gas constant: value	8.31432 E + 00	joule per mole kelvin	J/(mol · K)
gas gravity (density relative to air)	2.896 E + 01	gram per mole	g/mol
grain	6.479891* E + 01	milligram	mg
grain per 100 SCF	2.292768 E + 01	milligram per cubic meter API	mg/m^3
	5.421110 E − 01	milligram per mole	mg/mol
grain per gallon	1.42538 E − 02	gram per liter	g/*l*
gram mole	1.0* E + 00	mole	mol
horsepower (boiler)	9.80950 E + 00	kilowatt	kW
horsepower (550 ft-lb/s)	7.456999 E − 01	kilowatt	kW
horsepower (electric)	7.46* E − 01	kilowatt	kW
horsepower (hydraulic)	7.46043 E − 01	kilowatt	kW
horsepower (metric)	7.35499 E − 01	kilowatt	kW
horsepower (U.K.) & "indicated" or "brake"	7.457 E − 01	kilowatt	kW
hundredweight	4.535924 E + 01	kilogram	kg
inch	2.54* E + 00	centimeter	cm
inch to the fourth power	4.162314 E + 05	millimeter to the fourth power	mm^4
inch of mercury (Hg) at 0°C	3.386389 E + 00	kilopascal	kPa
inch of mercury (Hg) at 60°C	3.37685 E + 00	kilopascal	kPa
inch of water (H_2O) at 60°F	2.48843 E − 01	kilopascal	kPa
kilogram-force (kgf)	9.80665* E + 00	newton	N
kilogram-force per square centimeter	9.80665* E + 01	kilopascal	kPa
kilogram-force per square millimeter	9.80665* E + 00	megapascal	MPa

*The conversion factor is exact.
[a]To go from metric to customary units, divide the conversion factor instead of multiplying.

TABLE 5. (*continued*)

Traditional Unit	Multiply Value in Traditional Units by Factor × 10E to Obtain Value in Metric Units[a] Factor E	Selected Metric (S.I.) Unit Name	Symbol
kilopond (kp)	9.80665* E + 00	newton	N
kilowatt hour	3.6* E + 02	kilojoule	kJ
kip	4.448222 E + 00	kilonewton	kN
knot (international)	5.144444 E − 01	meter per second	m/s
link	2.01168 E − 01	meter	m
magnetic permeability (cgs e.m.u.)	1.256637 E + 00	microhenry per meter	μH/m
magnetic susceptibility (cgs e.m.u.)	1.579137 E + 01	microhenry per meter	μH/m
MCF (thousand cubic foot 60°F−1 atm)	2.826231 E + 01 1.195307 E + 00	cubic meter (API) kilomole	m³API kmol
MCF per acre foot (60°F−1 atm)	9.690510 E − 01 2.291262 E − 02	mole per cubic meter cubic meter API per cubic meter	mol/m³ m³API/m³
microcalorie per square centimeter second	4.1868* E + 01	milliwatt per square meter	mW/m²
micron	1.0* E + 00	micrometer	μm
microsecond per foot	3.280840 E + 00	microsecond per meter	μs/m
mil	2.54* E + 01	micrometer	μm
mile (U.S. and Canada)	1.609344* E + 00	kilometer	km
mile per gallon	3.540060 E − 01	kilometer per liter	km/l
mile per U.S. gallon	4.251437 E − 01	kilometer per liter	km/l
mile (international nautical)	1.852* E + 00	kilometer	km
millicalorie per second centimeter °C	4.1868* E + 02	milliwatt per meter degree Celsius	mW/(m · °C)
millidarcy	9.869233 E − 04	square micrometer	μm²
milligal	1.0* E + 01	micrometer per second squared	μm/s²
millimeter of mercury (Hg) 0°C	1.333222 E − 01	kilopascal	kPa
millimho	1.0* E + 00	millisiemens	mS
millimicron	1.0* E + 00	nanometer	nm
millimicrosecond	1.0* E + 00	nanosecond	ns
MMCF (million cubic foot 60°F − 1 atm	2.826231 E + 04 1.195307 E + 00	cubic meter (API) megamole	m³API Mmol
million years	1.0* E + 00	megayear	Ma
millisecond per foot	3.289474 E + 00	millisecond per meter	ms/m
neper per foot	3.777207 E − 01	decibel per meter	dB/m
oersted	7.957747 E + 01	ampere per meter	A/m
ounce (avdp)	2.834952 E + 01	gram	g
ounce (fluid UK)	2.841308 E + 01	cubic centimeter	cm³
parts per billion (mass basis)	1.0* E + 00	microgram per kilogram	μg/kg
parts per million (ppm) (mass basis)	1.0* E + 00	milligram per kilogram	mg/kg
parts per million (ppm) (by volume)	1.0* E + 00 multiply parts by density in kg/m³	cubic meter per liter milligram per cubic meter	m³/l mg/m³
parts per thousand (0/00) (mass basis)	1.0 E + 00	gram per kilogram	g/kg
parts per thousand (0/00) (by volume)	1.0* E + 00	cubic centimeter per liter	cm³/l
pint	5.68261 E − 01	liter	l
pound-force	4.448222 E + 00	newton	N
pound-force foot (see foot pound-force)			

*The conversion factor is exact.
[a]To go from metric to customary units, divide the conversion factor instead of multiplying.

(*continued*)

TABLE 5. (continued)

Traditional Unit	Multiply Value in Traditional Units by Factor × 10E to Obtain Value in Metric Units[a] Factor E	Selected Metric (S.I.) Unit Name	Symbol
poundal	1.382550 E − 01	newton	N
pound-force per 100 square foot	4.788026 E − 01	pascal	Pa
pound-force per square foot	4.788026 E + 01	pascal	Pa
pound-force per square inch (psi)	6.894757 E + 00	kilopascal	kPa
pound-force second per square foot	4.788026 E + 01	pascal second	Pa · s
pound-mass (avdp)	4.535924 E − 01	kilogram	kg
pound-mass per horsepower hour	1.689659 E + 02	milligram per kilojoule	mg/kJ
pound-mass per barrel	2.853010 E + 00	kilogram per cubic meter	kg/m³
pound-mass per foot	1.488164 E + 00	kilogram per meter	kg/m
pound-mass foot per second	1.352549 E − 01	kilogram meter per second	kg · m/s
pound-mass per cubic foot	1.601846 E + 01	kilogram per cubic meter	kg/m³
pound-mass per gallon	9.97763 E + 01	kilogram per cubic meter	kg/m³
pound-mass per gallon (U.S.)	1.198264 E + 02	kilogram per cubic meter	kg/m³
pound-mass per cubic inch	2.767990 E + 04	kilogram per cubic meter	kg/m³
pound-mass per thousand cubic foot	1.601846 E − 02	kilogram per cubic meter	kg/m³
pound-mass per square foot	4.882428 E + 00	kilogram per square meter	kg/m²
pound mole	4.535924 E − 01	mole	mol
psi (pound-force per square inch)	6.894757 E + 00	kilopascal	kPa
psi per foot	2.262059 E + 01	kilopascal per meter	kPa/m
quart	1.136522 E + 00	liter	l
quart (U.S.)	9.463529 E − 01	liter	l
quarter section (160 acres)	6.474970 E + 01	hectare	ha
RPM	1.0* E + 00	revolution per minute	r/min
second per quart (U.S.)	1.056882 E + 00	second per liter	s/l
section (640 acres)	2.589988 E + 02	hectare	ha
square inch	6.4516* E + 00	square centimeter	cm²
square foot	9.290304* E − 02	square meter	m²
square mile	2.589988 E + 00	square kilometer	km²
square yard	8.361274 E − 01	square meter	m²
standard cubic foot (60°F 1 atm − ideal gas)	2.826231 E − 02	cubic meter API	m³API
	1.195307 E + 00	mole	mol
TCF (trillion cubic foot 60°F 1 atm)	1.195307 E + 00	teramole	Tmol
	2.826231 E + 10	cubic meter (API)	m³API
"thou"	2.54* E + 00	micrometer	μm
thirty-second of an inch	7.93750 E − 01	millimeter	mm
ton (U.S. short−2,000 lb)	9.071847 E − 01	tonne	t
ton (U.K. long−2,240 lb)	1.016047 E + 00	tonne	t
ton-mile	1.431744 E + 01	megajoule	MJ
ton-mile per foot	4.697322 E + 01	megajoule per meter	MJ/m
ton (metric)	1.0* E + 00	tonne	t
yard	9.144* E − 01	meter	m

Note: Although the conversion factors have been calculated for the S.I. unit judged to be the most frequently required, for any particular application reference should be made to other tables of recommended units.

*The conversion factor is exact.

[a]To go from metric to customary units, divide the conversion factor instead of multiplying.

Source: From Kenting Limited, n. d. The International System of Units (S.I.) for the Petroleum Industry in Canada. Calgary, Alberta, Canada.

TABLE 6. S.I. Units Commonly Used

Item	Name	Symbol	Printer
Drilling, Cementing, and Formation Testing			
Linear	meter	m	M
(tool dimensions—always)	millimeter	mm	MILLIM
Area	square meter	m^2	M2
	hectare	ha	HECTARE
Volume and capacity	cubic meter	m^3	M3
	liter	l	LITRE
Mass	kilogram	kg	KG
	tonne	t	TONNE
Other			
Time	second	s	S
	minute	min	MIN
	hour	h	HR
	day	day	D
	week	wk	WEEK
	month	mo	MONTH
	year	yr	ANN
General Geology, Geophysics, and Reservoir Engineering			
Dip, gradient	meter per kilometer	m/km	M/KILOM
	degree	°	DEG
Geographical coordinates	degree or decimal degree	°	DEG
	minute	′	MNT
	second	″	S
Universal transverse mercator coordinates	meter	m	M
Distance	kilometer	km	KILOM
Elevation	meter	m	M
Depth	meter	m	M
Thickness of formations	meter	m	M
Area	square meter	m^2	M2
	hectare	ha	HECTARE
Volume of sediment in a basin	cubic kilometer	km^3	KILOM3
Geological age	megayear	Ma	MEGAANN
Reservoir Geology and Engineering			
Volume of reservoir or fluid	cubic meter	m^3	M3
Volume of pore space or fluid per volume of sediment	cubic meter per cubic meter	m^3/m^3	M3/M3
Permeability	square micrometer	μm^2	MICROM2
Formation pressure	megapascal	MPa	MEGAPA
Capillary pressure	pascal	Pa	PA
Head	meter	m	M
Pressure gradient	kilopascal per meter	kPa/m	KILOPA/M
Gas-oil ratio	cubic meter API per cubic	m^3API/m^3	M3/M3
	meter kilomole per cubic meter	$kmol/m^3$	KILOMOL/M3
Productivity index	cubic meter per day kilopascal	$m^3/(d \cdot kPa)$	M3/(D.KILOPA)
Pipeline Operations			
Flow	liter per second	l/s	LITRE/S
	cubic meter per second	m^3/s	M3/S
	cubic meter per hour	m^3/hr	M3/HR
Pressure	pascal	Pa	PA
Force	newton	N	N
Energy, work			
—quantity of heat	joule	J	J
Compressor rating:			
—compressor heads	kilopascal	kPa	KILOPA
—flow	cubic meter per second	m^3/s	M3/S
Pumping rating:			
—dynamic head	meter	m	M
—flow	liter per second	l/s	LITRE/S
—conversion factor	kilopascal per meter	kPa/m	KILOPA/M

(*continued*)

TABLE 6. (*continued*)

Item	Name	Symbol	Printer
Gradient:			
—pressure	kilopascal	kPa/km	KILOPA/KILOM
—slope	meter per kilometer	m/km	M/KILOM
Viscosity:			
—dynamic (gas)	micropascal second	μPa · s	MICROPA.S
—kinematic	square millimeter per second	mm^2/s	MILLIM2/S
Rotational frequency	revolution per minute	r/min	R/MIN
	radian per second	rad/s	RAD/S
Concentration	mole per cubic meter	mol/m^3	MOL/M3
	gram per cubic meter	g/m^3	G/M3
	milligram per kilogram	mg/kg	MILLIG/KG
	cubic centimeter per cubic meter	cm^3/m^3	CENTIM3/M3
Density—gas	kilogram per cubic meter	kg/m^3	KG/M3
	gram per mole	g/mol	G/MOL
Gravity—liquid density	kilogram per cubic meter	kg/m^3	KG/M3
Velocity	meter per second	m/s	M/S
	kilometer per hour	km/hr	KILOM/HR
Sound intensity	watt per square meter	W/m^2	W/M2
	decibel	dB	DECIBEL
	Seismological Investigations and Survey		
Amount of explosive	kilogram	kg	KG
Attenuation	decibel	dB	DECIBEL
Energy of source	megajoule	MJ	MEGAJ
Frequency	hertz	Hz	HZ
Pressure of shock wave	gigapascal	GPa	GIGAPA
Travel time	second	s	S
Velocity	meter per second	m/s	M/S
Wavelength	meter	m	M
	Gravity and Magnetic Surveys		
Gravitational variation	micrometer per second squared	μm/s^2	MICROM/S2
Density	kilogram per cubic meter	kg/m^3	KG/M3
Magnetic flux density (intensity)	nanotesia	nT	NANOT
Magnetic permeability	microhenry per meter	μH/m	MICROH/M
Magnetic susceptibility	microhenry per meter	μH/m	MICROH/M
	Rock Mechanics		
Strength, stress,	megapascal	MPa	MEGAPA
bulk modulus, elastic constant	gigapascal	GPa	GIGAPA
Elastic modulus	gigapascal	GPa	GIGAPA
Viscosity	pascal second	Pa · s	PA.S

Source: From Kenting Limited, n. d. *The International System of Units (S.I.) for the Petroleum Industry in Canada.* Calgary, Alberta, Canada.

Customary to metric units using a multiplying factor. Other specialized units are given in Table 6 for drilling, cementing, and formation testing; general geology; reservoir geology and engineering; pipeline operations; seismological investigations; gravity and magnetic survey; and applications in rock mechanics. A more detailed discussion of abbreviations, units, and symbolization may be found in the entry entitled *Units, Numbers, Constants, and Symbols* (Vol. II in this series).

Acknowledgments

The preparation of a work of this magnitude and duration involves the generous support and cooperation of contributors (acknowledged in the section, "Contributors") and many other helpers who worked behind the scenes to get the job done. I am especially indebted to our contributors for their patience and perseverance during the preparation of this volume. Those contributors who originally submitted papers in 1974 are, in a way, the real heroes of this volume. For their efforts in this geoscience series they have had to update their résumés patiently with "in preparation" footnotes for over a decade, surviving changes of publishers three times under my editorship. I am also most grateful to colleagues at Florida Atlantic University and the Coastal Education and Research Foundation (CERF), who critically read papers that were submitted for inclusion and who made suggestions for improvement both in content and style.

To those contributors who volunteered to write additional articles, to other correspondents who indicated topics or concepts that should be incorporated into the volume, and to those who suggested names of other experts as potential contributors, a very special word of thanks is due: Y. E. Abdelhady (Egypt), S. E. Allen (England), G. D. Aitchison (Australia), Frederick Betz, Jr., (United States), T. Boldizsar (Hungary), R. Bowen (Zambia), Y. Chatelin (France), George V. Chilingar (United States), F. De Connick (Belgium), J. Demek (Czechoslovakia), Donald R. Coates (United States), Robert G. Font (United States), Keith Frye (United States), D. R. Fussell (England), A. A. Geodekyan (USSR), Robert R. Gilkes (Australia), Keith Grant (Australia), C. P. Gravenor (Canada), N. M. Harrison (South Africa), C. B. Hunt (United States), F. MacIntyre (Greece), J. D. Mather (England), Christopher C. Mathewson (United States), M. Hamid Metwali (Poland), P. E. La Moreaux (United States), S. O. Nielsen (Denmark), Nikola Prokopovich (United States), W. W. Ristow (United States), Robert L. Schuster (United States), J. H. Tatsch (United States), and I. Valeton (West Germany).

A special word of thanks for inspiration and encouragement goes to my series editor, Rhodes W. Fairbridge. His interest and cooperation in the development of this volume is warmly appreciated. Thanks to Rhodes is especially deserved because he proved to be a real *fons et origo* of information for many subjects. Having spent a decade as volume editor, putting several of these works together, it was always a pleasure working under Rhodes, one of the grand scholars of geology.

The efforts of Pamela A. Matlack, my research assistant at CERF, are appreciated for her attention to details in the filing of manuscripts, assistance in the preparation of cross-references, and redrafting of figures.

Last but not least I thank the publisher for continued support and interest throughout this lengthy project. Of special note are the dedicated efforts of Bernice Pettinato, production and managing editor, who deserves a special word of thanks for enhancing the production of this book.

CHARLES W. FINKL, JNR.

MAIN ENTRIES

Acoustic Surveys, Marine
Aerial Surveys
Alidade
Alluvial Systems Modeling
Artificial Deposits and Modified Land
Atmogeochemical Prospecting
Augers, Augering (Soil)

Biogeochemistry
Blasting and Related Technology
Block Diagram
Blowpipe Analysis
Borehole Drilling
Borehole Mining

Canals and Waterways, Sediment Control
Cartography, General
Cat Clays
Cities, Geologic Effects of
Coal Mining
Compass Traverse
Computers in Geology
Cratering
Cross-Sections

Dispersive Clays

Environmental Engineering
Environmental Management
Expansive Soils, Engineering Geology
Exploration Geochemistry
Exploration Geophysics
Exposures, Examination of

Field Geology
Field Notes, Notebooks
Floating Structures in Waves
Fossils and Fossilization
Fractures, Fracture Structures

Geoanthropology
Geoarchaeology
Geobotanical Prospecting
Geohistory, American Founding Fathers

Geological Cataloguing
Geological Fieldwork, Codes
Geological Highway Maps
Geological Methodology
Geological Survey and Mapping
Geological Surveys, State and Federal
Geological Writing
Geologic Reports
Geologic Systems, Energy Factors
Geology, Philosophy of
Geology, Scope and Classification
Geomagnetism and Paleomagnetism
Geomicrobiology
Geomythology
Geonomy
Geophilately
Geostatistics
Groundwater Exploration

Harbor Surveys
Hardrock versus Softrock Geology
Hydrochemical Prospecting

Igneous Rocks, Field Relations
Indicator Elements

Lake Sediment Geochemistry
Land Capability Analysis
Landslide Control
Legal Affairs
Lichenometry
Lithogeochemical Prospecting

Map Abbreviations, Ciphers,
  and Mnemonicons
Map and Chart Depositories
Map Colors, Coloring
Maps, Environmental Geology
Maps, Logic of
Maps, Physical Properties
Map Series and Scales
Map Symbols
Marine Exploration Geochemistry
Marine Magnetic Surveys

Marine Mining
Mathematical Geology
Mineral Identification, Classical Field
  Methods
Mineraloids
Minerals and Mineralogy
Mine Subsidence Control
Mining Preplanning

Offshore Nuclear Plant Protection
Open Space

Pedogeochemical Prospecting
Petroleum Exploration Geochemistry
Petroleum Geology
Photogeology
Photointerpretation
Placer Mining
Plane Table Mapping
Plate Tectonics and Continental Drift
Plate Tectonics and Mineral Exploration
Popular Geology
Professional Geologists' Associations
Profile Construction
Prospecting
Punch Cards, Geologic Referencing

Remote Sensing, General
Remote Sensing and Photogrammetry,
  Societies and Periodicals

Rock-Color Chart
Rock Particles, Fragments

Samples, Sampling
Saprolite, Regolith, and Soil
Satellite Geodesy and Geodynamics
Scientific Method (Scientific Methodology)
Sea Surveys
Sedimentary Rocks, Field Relations
Serendipity
Slope Stability Analysis
Soil Fabric
Soil Sampling
Soils and Weathered Materials, Field
  Methods and Survey
Surveying, Electronic Distance
  Measurement
Surveying, General

Tephrochronology
Terrain Evaluation Systems

Uniformitarianism

Vegetation Mapping
VLF Electromagnetic Prospecting

Well Logging

CONTRIBUTORS

ALLEN F. AGNEW, 1435 Menlo Drive, Corvallis, Oregon 97330. *Geological Surveys, State and Federal.*

FREDERIK P. AGTERBERG, Geomathematics Section, Geological Survey of Canada, 601 Booth Street, Ottawa, Ontario K1A 0E8, Canada. *Mathematical Geology.*

CLAUDE C. ALBRITTON, JR., 3436 University Boulevard, Dallas, Texas 75205. *Geology, Philosophy of.*

J. W. AMBROSE, deceased. *Profile Construction.*

N. T. ARNDT, Max-Planck-Institut für Chemie, Postfach 30 60, D-6500 Mainz, West Germany. *Igneous Rocks, Field Relations.*

JAMES M. BARKER, Duval Corporation, Industrial Mineral Exploration, 5357 East Pima Street, Tucson, Arizona 85712. *Geoanthropology.*

JAMES P. BARKER, 1116 East 17th Street, Tucson, Arizona 85719. *Geoanthropology.*

C. E. G. BENNETT, C.S.I.R.O. Division of Mineralogy, Private Bag, Post Office, Wembley, Western Australia, 6014, Australia. *Borehole Mining.*

ANN STARCHER BENTLEY, Charles Ryan Associates, Charleston, West Virginia 16802. *Coal Mining.*

RAMESWAR BHATTACHARYYA, Department of Ocean Engineering, U.S. Naval Academy, Annapolis, Maryland 21402. *Floating Structures in Waves.*

K. BLOOMFIELD, British Geological Survey, Overseas Directorate, Keyworth, Nottingham, England NG12 5GG. *Tephrochronology.*

R. W. BOYLE, Earth Sciences, Geological Survey of Canada, 601 Booth Street, Ottawa, Ontario K1A 0E8, Canada. *Indicator Elements.*

ROBERT R. BROOKS, Department of Chemistry and Biochemistry, Massey University, Palmerston North, New Zealand. *Biogeochemistry; Geobotanical Prospecting.*

PAUL L. BROUGHTON, Husky Oil Operations Ltd., Box 6525, Postal Station "D," Calgary, Alberta T2P 3G7, Canada. *Borehole Drilling.*

RICHARD E. CHAPMAN, Department of Geology and Mineralogy, University of Queensland, St. Lucia, Queensland 4067, Australia. *Petroleum Geology.*

RAYMUNDO J. CHICO, 110 Inverness Circle East, Unit A, Englewood, Colorado 80112. *Rock-Color Chart.*

GEORGE V. CHILINGAR, Petroleum Engineering Department, University of Southern California, Los Angeles, California 90089-1211
(with F. L. Margot, L. G. Adamson, R. A. Armstrong, W. H. Fertil, I. Ershaghi, and N. Hashem)
Well Logging.

DONALD R. COATES, Department of Geological Sciences and Environmental Studies, State University of New York, Binghamton, New York 13901. *Environmental Management; Groundwater Exploration; Legal Affairs; Open Space.*

MICHAEL J. CRUICKSHANK, Conservation Division, U.S. Geological Survey, Menlo Park, California 94025. *Marine Mining.*

JOHN G. DENNIS, Department of Geological Sciences, California State University, Long Beach, 1250 Bellflower Boulevard, Long Beach, California 90840. *Plate Tectonics and Continental Drift.*

WALTER S. DIX, 3636 16th Street N.W., B430, Washington, D.C., 20010. *Surveying, General.*

ROBERT H. DOTT, SR., Geological Highway Map Editor, American Association of Petroleum Geologists, P.O. Box 979, Tulsa, Oklahoma 74101. *Geological Highway Maps.*

CHARLES H. DOWDING, Department of Civil Engineering, Northwestern University, Evanston, Illinois 60201. *Fractures, Fracture Structures.*

R. DRAZNIOWSKY, The AGS Collection of the University of Wisconsin, Milwaukee Library, P.O. Box 399, Milwaukee, Wisconsin, 53201. *Cartography, General.*

LARRY D. DYKE, Department of Geology, Texas A&M University, College Station, Texas 77843. *Landslide Control.*

TUNCER B. EDIL, Department of Civil and Environmental Engineering, Engineering Building, The University of Wisconsin, 1415 Johnson Drive, Madison, Wisconsin, 53706. *Soil Fabric; Soil Sampling.*

RHODES W. FAIRBRIDGE, Department of Geological Sciences, Columbia University, New York, New York 10027. *Geological Writing; Geology, Scope and Classification; Professional Geologists' Associations; Prospecting; Serendipity; Uniformitarianism.*

ROBERT J. FERENS, Branch of Cartography, Publications Division, U.S. Geological Survey, National Center, Reston, Virginia 22092. *Map Series and Scales.*

CHARLES W. FINKL, JNR., Center for Coastal Research, P.O. Box 8068, Char-

lottesville, Virginia 22906. *Alidade; Block Diagram; Field Geology; Geological Fieldwork, Codes; Map Abbreviations, Ciphers, and Mnemonicons; Map and Chart Depositories; Map Colors, Coloring; Map Symbols; Professional Geologists' Associations; Saprolite, Regolith, and Soil; Soils and Weathered Materials, Field Methods and Survey.*

IRVING FISHER, 9 Ricker Park, Portland, Maine 04101. *Blowpipe Analysis; Mineral Identification, Classical Field Methods.*

JOHN J. FISHER, Department of Geology, University of Rhode Island, Kingston, Rhode Island 02881. *Punch Cards, Geologic Referencing.*

ROBERT W. FLEMING, U.S. Geological Survey, MS 903 KAE, Box 25046, Denver, Colorado 80225. *Cratering; Slope Stability Analysis.*

G. H. W. FRIEDRICH, Institut für Mineralogie und Lagerstättenlehre der RWTH, Wüllnerstrasse 2, D-5100 Aachen, West Germany. *Pedogeochemical Prospecting.*

KEITH FRYE, Department of Geophysical Science, Old Dominion University, Norfolk, Virginia 23508. *Mineraloids; Minerals and Mineralogy.*

JAMES C. GAMBLE, Pacific Gas & Electric Co., 77 Beale Street, San Francisco, California 94106. *Mine Subsidence Control.*

A. A. GEODEKYAN, VNIIGEOINFORM-SYSTEM, Varshavskoye shosse 8, Moscow, USSR. *Petroleum Exploration Geochemistry.*

CHARLES P. GIAMMONA, Civil Engineering Department, Texas A&M University, College Station, Texas 77843. *Sea Surveys.*

ALAN C. GRANT, Atlantic Geoscience Centre, Geological Survey of Canada, Bedford Institute of Oceanography Box 1006, Dartmouth, Nova Scotia B2Y 4A2. *Harbor Surveys.*

K. GRANT, 14 Pokana Circuit, Kaleen, A.C.T. 2617, Australia. *Terrain Evaluation Systems.*

RICHARD E. GRAY, G.A.I. Consultants, Inc., 570 Beatty Road, Pittsburgh, Pennsylvania 15146. *Mine Subsidence Control.*

JOHN C. GRIFFITHS, Department of Geochemistry and Mineralogy, Pennsylvania State University, University Park, Pennsylvania 16802. *Geostatistics.*

W. R. GRIFFITHS, Branch of Exploration Research, U.S. Geological Survey, MS 973, Box 25046, Federal Center, Denver, Colorado 80225-0046. *Placer Mining.*

M. GRANT GROSS, Chesapeake Institute, Johns Hopkins University, Baltimore, Maryland 21218. *Cities, Geologic Effects of.*

ROBERT B. HALL, 11280 West 20th Avenue, Apartment 32, Lakewood, Colorado 80215. *Hardrock versus Softrock Geology.*

FEKRI HASSAN, Department of Anthropology, Washington State University, Pullman, Washington 99163. *Geoarchaeology.*

ALLEN W. HATHEWAY, Department of Geological Engineering, 125 Mining Building, University of Missouri-Rolla, Rolla, Missouri, 65401-0249. *Geologic Reports.*

ROBERT M. HAZEN, Department of Mineralogy and Petrology, University of Cambridge, Downing Place, Cambridge CB2 3EW, England. *Geohistory, American Founding Fathers.*

R. HERMANN, Institut für Mineralogie und Lagerstättenlehre der RWTH, Wüllnerstrasse 2, D-5100 Aachen, West Germany. *Pedogeochemical Prospecting.*

STAN HOFFMAN, BP Canada, Selco Division, Suite #700, 890 West Pender Street, Vancouver, B.C. V6C 1K5, Canada. *Lake Sediment Geochemistry.*

PAUL M. HOPKINS, deceased. *Placer Mining.*

ROY E. HUNT, 149 Richard Street, Bricktown, New Jersey 08724. *Remote Sensing, General.*

J. A. JACOBS, Department of Geology and Geophysics, Cambridge University, Madingley Road, Cambridge CB3 0EZ, England. *Geonomy.*

A. S. JOYCE, Department of Applied Geology, South Australian Institute of Technology, North Terrace, Adelaide, South Australia 5000, Australia. *Atmogeochemical Prospecting; Exploration Geochemistry; Hydrochemical Prospecting.*

W. D. KELLER, Department of Geology, University of Missouri, Columbia, Missouri 65211. *Geologic Systems, Energy Factors.*

L. E. KOCH, deceased. *Geological Methodology; Scientific Method (Scientific Methodology).*

W. E. KRUMBEIN, Universität Oldenberg, Ammerländer Heerstrasse 67-99, Postfach 943, 29 Oldenberg, West Germany. *Geomicrobiology.*

STEVEN L. KRUPA, Department of Ocean Engineering, Florida Atlantic University, Boca Raton, Florida 33431. *Samples, Sampling.*

A. W. KÜCHLER, Department of Geography, The University of Kansas, Lawrence, Kansas 66045-2121. *Vegetation Mapping.*

PETER F. LAGASSE, Resource Consultants, Inc., 402 West Mountain Avenue, P.O. Box Q, Fort Collins, Colorado, 80522. *Alluvial Systems Modeling.*

P. F. F. LANCASTER-JONES, deceased. *Compass Traverse; Plane Table Mapping.*

SIMO H. LAURILA, P.O. Box 99, Jamestown, Colorado 80455. *Surveying, Electronic Distance Measurement.*

ROY LEMON, Department of Geology, Florida Atlantic University, Boca Raton, Florida 33431. *Exposures, Examination of; Field Notes, Notebooks; Samples, Sampling.*

ROY C. LINDHOLM, Department of Geology, George Washington University, Washington, D.C. 20052. *Sedimentary Rocks, Field Relations.*

KENNETH V. LUZA, Oklahoma Geological Survey, 830 Van Vleet Oval, Room 163, University of Oklahoma, Norman, Oklahoma 73019. *Maps, Physical Properties.*

TERRY S. MALEY, USDI, Bureau of Land Management, 3380 American Terrace, Boise, Idaho 83706. *Acoustic Surveys, Marine; Marine Magnetic Surveys.*

CHRISTOPHER C. MATHEWSON, Department of Geology, Texas A&M University, College Station, Texas 77843. *Expansive Soils, Engineering Geology; Landslide Control; Mining Preplanning.*

J. F. M. MEKEL, International Institute for Aerial Survey and Earth Sciences (ITC), 3 Kanalweg, Delft, The Netherlands. *Aerial Surveys; Photogeology.*

ANTHONY M. MELONE, Klohn Leonoff Ltd., 10180 Shellbridge Way, Richmonf, British Columbia V6X 2W7, Canada. *Canals and Waterways, Sediment Control.*

DANIEL F. MERRIAM, Department of Geology, P.O. Box 27, The Wichita State University, Wichita, Kansas 67208-1595. *Computers in Gelogy.*

LARRY L. MINTER, Department of Geology, Texas A&M University, College Station, Texas 77840. *Mining Preplanning.*

FRANK MOSELEY, Department of Geological Sciences, The University of Birmingham, P.O. Box 363, Birmingham, B15 2TT, England. *Geological Survey and Mapping.*

JAN M. MUTMANSKY, Department of Mineral Engineering, 104 Mineral Sciences Building, The Pennsylvania State University, University Park, Pennsylvania 16802. *Coal Mining.*

L. DAVID NEALEY, Geological Division, U.S. Geological Survey, 601 East Cedar Avenue, Flagstaff, Arizona 86001. *Remote Sensing and Photogrammetry, Societies and Periodicals.*

R. L. NICHOLS, Anchorage W-22, 15 Pleasant Street, Harwich Port, Maine 02646. *Geophilately.*

GERALD OSBORN, Department of Geology and Geophysics, The University of Calgary, Calgary, Alberta T2N 1N4, Canada. *Lichenometry.*

KURT OTHBERG, Idaho Geological Survey, Room 332, Morrill Hall, University of Idaho, Moscow, Idaho 83843
(with Pamela Palmer)
Maps, Environmental Geology.

THEODOULOS M. PANTAZIS, Geological Survey Department, Ministry of Agriculture and Natural Resources, Nicosia 107, Cyprus. *Lithogeochemical Prospecting.*

T. C. PARKS, C.S.I.R.O. Minerals Research Laboratories, Private Bag, Wembley, Western Australia 6014, Australia. *Borehole Mining.*

EDWARD B. PERRY, U.S. Army Corps of Engineers, Waterways Experiment Station, P.O. Box 631, Vicksburg, Mississippi 39180-0631. *Dispersive Clays.*

EDWARD J. PETUCH, Department of Geology, Florida Atlantic University, Boca Raton, Florida 33431. *Fossils and Fossilization.*

NIKOLA P. PROKOPOVICH, U.S. Bureau of Reclamation, MD230 N.P.P., 2800 Cottage Way, Sacramento, California 95825. *Cat Clays.*

BRUCE B. REDPATH, URS/Blume Engineers, 2855 Telegraph Avenue, Berkeley, California 94705. *Cratering.*

D. ANDREW ROBERTS, The Museum Documentation Association, Building 0, 347 Cherry Hinton Road, Cambridge CB1 4DH, England. *Geological Cataloguing.*

ROBERT L. SCHUSTER, Engineering Geology Branch, U.S. Geological Survey, MS966, Box 25046, Federal Center, Denver, Colorado 80225-0046. *Slope Stability Analysis.*

ROBERT R. SHARP, JR., Los Alamos Scientific Laboratory, P.O. Box 1663, Los Alamos, New Mexico 87545. *Blasting and Related Technology.*

FREDERICK R. SIEGEL, Department of Geology, George Washington University, Washington, D.C. 20052. *Marine Exploration Geochemistry.*

HOWARD E. SIMPSON, 2020 Washington Avenue, Golden, Colorado 80401. *Artificial Deposits and Modified Land.*

THOMAS A. SIMPSON, Department of Mineral Engineering, University of Alabama, Mineral Industries Building, University, Alabama 35486. *Environmental Engineering.*

MICHAEL SOLOMON, Department of Geology, University of Tasmania, Hobart, Tasmania 7001, Australia. *Plate Tectonics and Mineral Exploration.*

DANN J. SPARIOSU, Department of Geology, University of Georgia, Athens, Georgia 30602. *Geomagnetism and Paleomagnetism.*

RONALD D. STEIGLITZ, College of Environmental Sciences, University of Wisconsin-Green Bay, Green Bay, Wisconsin 54301-7001. *Cross-Sections; Land Capability Analysis; Rock Particles, Fragments.*

V. A. STROGONOV, Ul. Profsoyusnaya, 48-4-48, Moscow, USSR. *Petroleum Exploration Geochemistry.*

DAVID V. THIEL, School of Science, Griffith University, Narthan, Brisbane, Queensland, Australia, 4111. *VLF Electromagnetic Prospecting.*

P. VANICEK, Department of Surveying Engineering, University of New Brunswick, P.O. Box 4400, Fredericton, N.B., Canada E3B 5A3. *Satellite Geodesy and Geodynamics.*

DAVID J. VARNES, U.S. Geological Survey, Federal Center, Box 25046, Stop 903KAE, Denver, Colorado 80225-0046. *Maps, Logic of.*

H. Th. VERSTAPPEN, International Institute for Aerial Survey and Earth Sciences (ITC), 144 Boulevard 1945, P.O. Box 6, Enschede, The Netherlands. *Photointerpretation.*

DOROTHY B. VITALIANO, 1114 Brooks Drive, Bloomington, Indiana 47401. *Geomythology.*

WILLIAM M. VOORHIS, 11613 Maple Glen Court, St. Louis, Missouri 63146. *Augers, Augering (Soil).*

IAN WATSON, Department of Geology, Florida Atlantic University, Boca Raton, Florida 33431. *Exposures, Examination of; Field Notes, Notebooks; Offshore Nuclear Plant Protection; Samples, Sampling.*

A. EASTON WREN, 82 Eagle Ridge Drive S.W., Calgary, Alberta T2V 2U4, Canada. *Exploration Geophysics.*

The
ENCYCLOPEDIA
of
FIELD AND
GENERAL GEOLOGY

A

ACOUSTIC SURVEYS, MARINE

Marine acoustic surveys represent one of several geophysical techniques commonly used to delineate and study the character and structure of the ocean bottom and subbottom. In acoustic surveys (see Vol. I, *Acoustics, Underwater),* sound energy is transmitted through seawater to the ocean bottom or subbottom where it is reflected off the sea bottom and received by the ship through listening devices called *hydrophones.* The time period required for the sound to travel to the reflecting surface and back to the ship is a function of the distance to the reflecting surface. Because most acoustic survey systems provide continuous operation, surveys are normally operated at a ship speed of 10–12 nautical miles per hour (1 nautical mi = 1,852 m), and a continuous profile of the reflecting surface is produced on board the ship by a *graphic recorder* (Fig. 1).

There are two basic types of acoustic surveys, both of which operate on the described principle. In one type, commonly referred to as the *echo sounder system,* the sound pulse is reflected at the water-sediment interface. This system has provided our present state of knowledge on the topography of the sea bottom (e.g., Maley et al., 1974). The second type of acoustic survey system is referred to as a *continuous seismic reflection system.* In *reflection profiling,* the sound pulse not only reflects at the water-sediment interface but also penetrates the sea bottom sediments and reflects off subbottom layers or structures that offer the necessary acoustic impedance (Ewing and Ewing, 1970).

Seismic refraction techniques are similar to *seismic reflection* techniques because both provide for penetration of the subbottom by sound energy. However, seismic refraction methods are not commonly used at sea because they are much more expensive and cumbersome to use than the seismic reflection systems (see *Seismological Methods).*

Navigation

On a marine geophysical survey, there is no system more essential than the ship's positioning system. If the ship's position cannot be plotted accurately, the depth or subbottom information supplied by the acoustic system or any other geophysical data collected will be of questionable value. Most *survey ships* are now equipped with one or more of the numerous *electronic positioning systems* now available. A well-equipped survey ship will have a satellite navigation system aboard, primarily for calibrating the ship's electronic positioning system.

There are two basic types of electronic positioning systems. In the *circular, or ranging, system,* a master station is placed aboard the ship and two slave stations are placed on shore. With this system the lane width is constant, and the position of the ship is determined by the intersection of two circular arcs measuring the distance from each slave station to the ship. Because this system typically has a short range and high degree of accuracy, it is best suited for nearshore or coastal surveys.

The other type of electronic positioning system is called a *hyperbolic system.* In this system a master station and two slave stations are all on shore. For a fix, it is necessary to get an intersection of two hyperbolic lines on the plotting sheet. This system typically has a long–range capability of hundreds of nautical miles and is accurate to within several hundred meters. The type of electronic positioning system used will depend on variables such as distance from shore, accuracy required, mobility of operation, and costs (see *Surveying, Electronic).*

FIGURE 1. Depth recorder with echogram displaying continuous profile of the sea bottom. The horizontal lines on the echogram represent 20 fathom depth intervals with a total of 400 fathoms for each phase; vertical lines are time lines spaced approximately 15 min apart.

1

Survey Design

Before starting an acoustic survey, the ship's proposed trackline should be constructed on the plotting sheet. Existing bathymetric charts should be studied so that the tracklines are laid out at right angles to the general trend of the topography. Trackline spacing depends on time available, quality of navigation, scale of the plotting sheet, and type of information needed. Most surveys are run at a line spacing ranging from several hundred meters to 10 nautical mi. A 10 nautical mi line spacing would be considered reconnaissance level, whereas a trackline spacing of less than 1 or 2 nautical mi would be considered detailed. Cross-check lines should be run periodically during the survey to determine if the acoustic values correspond where the cross-check lines intersect the regular tracklines. If the data do not agree, the navigation system is usually at fault.

Survey Operations

Survey operations are almost always maintained on an around-the-clock basis. The acoustic instrumentation, like all geophysical equipment, must be monitored constantly. The graphic recorders should be annotated at least every 15 min with information such as time, date, time zone, ship's name, and depth phase. The ship's position is plotted every 3–5 min on the plotting sheet that displays the hyperbolic or circular navigational lines. A log is maintained of the ship's speed, course changes, sea state, and wind speed and direction. This log is used to smooth the ship's track to adjust for erratic fluctuations in the track that result from instrument or plotting error.

Echo Sounder

Echo sounders provide a continuous profile of the ocean bottom. Every 24 hr this profile is removed from the graphic recorder and is scaled for depth values, typically at 20 fathom intervals, including topographic highs and lows. The depth values are recorded in a log book with their respective time values, and then the soundings are plotted on the smoothed trackline.

History The echo sounder is a relatively recent development in the history of depth sounding. Sailors first used a marked staff to negotiate ships around shoal areas that might represent a hazard to navigation (see Vol. I, *Sounding*). The main purpose of depth sounding was to ensure that the ship did not run aground. A later method of depth sounding provided for greater depth measurement by use of a weighted line to sound the bottom. Lead was used as a weight, and the line consisted of either hemp rope or piano wire. The weighted line could be operated to depths of 100 ft by hand; if deeper soundings were desired,

a winch was used to release and retrieve the line. Two significant problems were associated with weighted line soundings; it was very time consuming, usually requiring several hours to measure a single depth, and it was extremely difficult to keep the line vertical because of ship drift and subsurface currents (Cohen, 1970).

In 1922, the first bathymetric survey utilizing a continuous echo sounder was conducted (see Vol. I, *Bathymetry*). The echo sounder took advantage of the fact that water is an excellent conductor of sound and that the vertical velocity of sound in seawater is a function of the depth, temperature, and salinity variations. Depth sounding instruments are adjusted for a constant sounding velocity of about 800–820 fathoms per second. The early echo sounder was used as a navigation aid in water depths of less than 100 fathoms because it was not very effective at greater depths.

Transducer The *transducer* is the portion of the echo sounding system that converts electrical energy into acoustic energy and projects or transmits the sound pulse to the ocean bottom. The sound pulse is reflected when it strikes a medium that has an acoustic impedance different than that of water. A hydrophone receives the reflected sound pulse or echo and converts it back to electrical energy.

Transducers may be attached to the ship's hull or towed at the aft or side of the ship. Hull-mounted transducers are mounted in a dome on the hull of the ship; the transducer must, however, be stabilized because the ship's movement may severely deteriorate the return signal. Towed transducers are hydrodynamically designed in the shape of a fish (Fig. 2). These transducers may be towed just beneath the surface or suspended from a line to reduce the effect of the ship's movement. In deep-towed methods, the transducer is towed as close to the bottom as possible. This method avoids inaccuracies by allowing a very narrow beam to sound the bottom, therefore avoiding the side echo problem.

Wide versus Narrow Beam Until recently, most survey ships utilized wide-beam sonar systems with beams as wide as 55°. Such sonar systems do not require stabilization because at least a part of the wide beam will reach bottom and return a signal to the ship. In deep ocean surveying, a wide beam covers such a large area of the seafloor (7.44 square nautical mi at 3,000 fathoms for a 55° beam) that, if the topography is irregular, the resulting echogram profile will show only an average bottom depth. This problem prompted the development of narrow-beam systems so that researchers could obtain more accurate representations of the seafloor.

The width of the sound beam depends on the diameter of the transducer and the frequency of the acoustic energy. On most bathymetric sur-

FIGURE 2. Towed transducer, hydrodynamically designed like a fish to reduce the effect of the transducer's movement through water.

veys, wide- and narrow-beam transducers are used simultaneously by operating at two different frequencies and keeping the transducers separated. The greatest advantage of the narrow–beam system is that it provides the most accurate profile of the sea bottom. Wide–beam systems are more useful for exploratory or reconnaissance surveys where an overview is required of a large area. The major disadvantage of the wide-beam transducer is that ocean bottom features appear as distorted shapes. For example, sea–mounts are conical in shape but show up on the graphic recorder as hyperbolic shapes.

Narrow–beam transducers must be stabilized in heavy seas or the movement of the ship will deteriorate the return signal. Stabilized beams keep the sound beam in a vertical orientation even though the ship may roll 40° or 50° from vertical. Stabilization is achieved by utilizing gimbal mounts controlled by a gyroscope.

Multiple-Beam Sonar System An example of a highly directional, multiple-beam sonar system is the Harris Model 853 D, which is manufactured by the Harris ASW Division of the General Instrument Corporation. Twenty projector transducers are arranged fore and aft along the keel of the ship; this produces a pitch-compensated beam 54° wide athwartship and 2 2/3° wide fore and aft. Forty hydrophones, electronically reduced to 16 preformed beams (effectively 16 hydrophones), are arranged athwartship, each with a roll-compensated receiving beam 2 2/3° wide athwartship by 28° wide fore and aft. The primary received echo comes from that area of the sea bottom covered by the intersection of the transmitted beam with the received beam. Each of the 16 pos-

sible intersections produces a beam with a cross-section of 2 2/3° by 2 2/3° as shown in Fig. 3. Hence, if all 15 beams are operating simultaneously, they cover a swath on the seafloor that is 2 2/3° fore and aft and 40° athwartship. The vertical beam may be operated simultaneously with a corresponding pair of side beams; alternatively, all 15 beams may be operated at the same time and displayed on a single graphic recorder. Figure 3 illustrates the arrangement of the 15 beams, the ship, and the seafloor. During the survey, two graphic recorders provide a visual display of the water depth (Fig. 4); one records only the vertical depth, and the other records both a port-side and a starboard-side beam (Maley, 1973).

Graphic Recorder The graphic recorder provides a continuous profile of the ocean bottom by amplifying and displaying the return signal to the ship. Depending on the type of recorder used, wet or dry paper on rolls 48 cm wide is maintained in the recorder. The sea bottom is displayed on the paper with a vertical exaggeration ranging from 6:1 to 30:1. The amount of vertical exaggeration depends on the ship's speed and the paper speed. The recorder controls the frequency of sound sent through the water; it also gates to eliminate noise or superfluous marks on the paper. Most survey-quality recorders have a scale of 0–400 fathoms for general operation and another scale of 0–4,000 fathoms that is used to check the phase. While operating over bottom topography of high relief, it is possible to lose track of the phases (Fig. 5). If this should happen, the proper phase may be determined by switching the recorder over to the 0–4,000 fathom phase (Fig. 6).

False Bottom Multiple returns of the bottom are likely to occur if the reflective surface consists of gravel and the depth is shallow. Such returns result when the sound pulse rebounds one or more times before it is received by the ship's hydrophone (Fig. 7). A false bottom can also be caused by phenomenon known as the deep-sea scattering layer (Fig. 8), which is due to microscopic marine organisms that produce a reflecting surface of sufficient density to obscure the true bottom.

Automatic Data Systems Computers are now available for linking navigation systems, sounding equipment, and other geophysical instrumentation so that sounding data may be displayed with digital plotting equipment. The computer and plotter can, e.g., prepare a plotting sheet at any desired scale and then make a continuous plot of the ship's track. As the plotter marks the ship's position, the sounding data are plotted simultaneously on the sheet.

Continuous Acoustic Reflection

The instrumentation and operating principles used in the seismic profiler are almost identical to

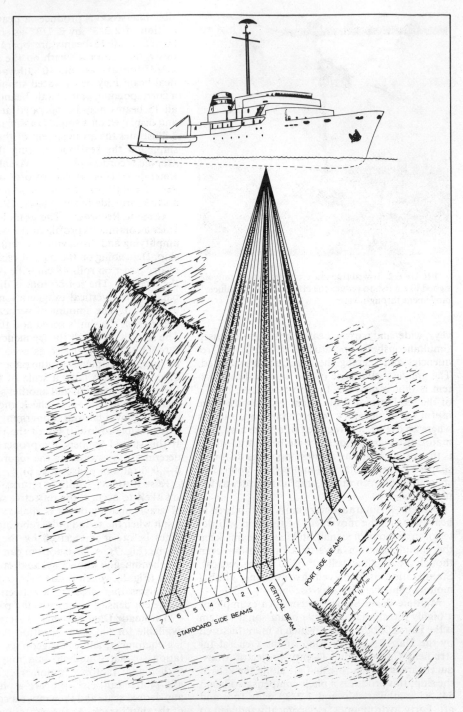

FIGURE 3. Coverage of the seafloor by multiple-beam sonar system in which fifteen independent 2 2/3° beams each covers a small square area of the seafloor.

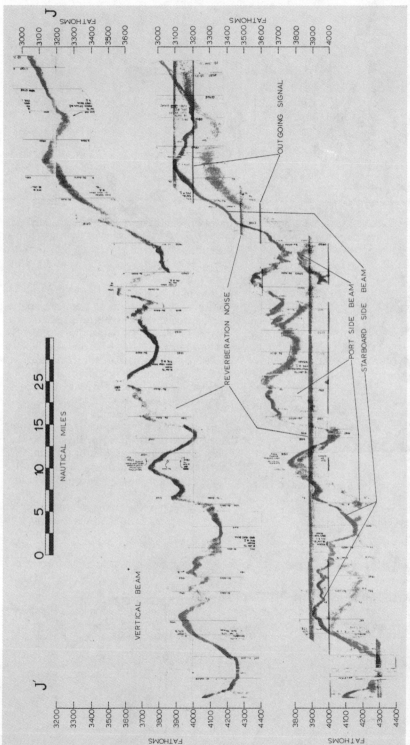

FIGURE 4. Echograms collected during an east-west traverse along the north wall of the Puerto Rico Trench. The profile on the top record was sounded by the vertical beam; simultaneously, the two profiles on the bottom echogram were sounded by the no. 6 port-side and the no. 6 starboard-side beams. Depths shown by side beams are not corrected for slant angle. Note that the echo trace for the upslope, port-side beam is darker than the trace for the downslope, starboard-side beam.

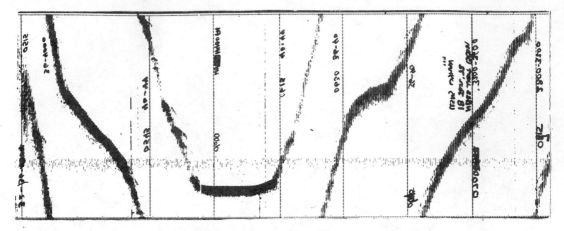

FIGURE 5. Echogram collected during a north-south traverse over the Puerto Rico Trench utilizing a narrow-beam echo sounder. In steep terrain like this, the echogram phases every few minutes and must be monitored constantly so investigators do not lose track of the correct phase.

FIGURE 6. A phase check is made periodically on the depth recorder by switching to the 0–4,000 fathom scale and listening through the headphones for the return echo. The return echo should occur simultaneously with the mark on the echogram.

those in the echo sounder. The seismic profiler, however, utilizes a lower-frequency pulse that is able to penetrate the bottom sediments to show traces of certain subbottom interfaces. Because a sound pulse is reflected from interfaces between layers of different acoustic impedance, the seismic profiler will show the bottom and each successive layer or feature with a different acoustic impedance (see Vol. I, *Seismic Reflection Profiling at Sea*). If the velocity at which sound travels through the subbottom material is known, then it is possible to calculate the distance to the reflecting surface. In a continuous seismic reflection system, there are three major units: (1) a sound source, (2) a receiving hydrophone, and (3) a recorder.

Sound Sources Some of the more common sound sources include direct electrical discharge

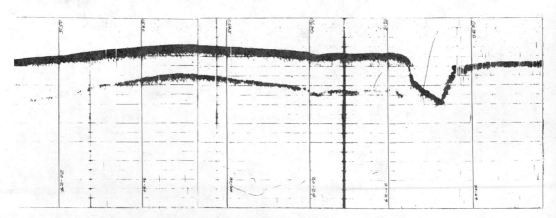

FIGURE 7. Echogram collected during a traverse across the lower Hudson Canyon, utilizing a narrow-beam echo sounder. The false bottom, which lies below the actual sea bottom, is caused by a multiple return of the echo.

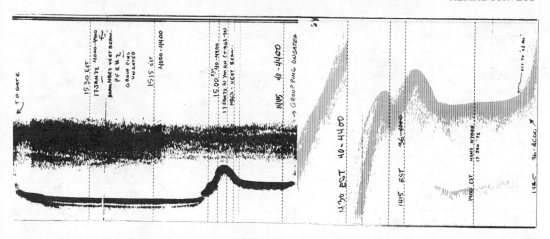

FIGURE 8. Echogram collected during a north-south traverse across the Puerto Rico Trench. The echogram displays well-defined subbottom penetration into the sediments of the abyssal plain. Also well displayed above the abyssal plain is a deep-sea scattering layer.

(sparker), direct discharge of compressed air or gas explosion (air gun), explosives, and electromagnetic transducers. Sparkers and air guns are the most commonly used sound sources at sea; explosives are rarely used during marine seismic exploration because they present safety problems and are inconvenient. Different problems require different equipment and different operating frequencies. High frequencies give high resolution and low penetration, whereas low frequencies give low resolution and high penetration. Thus, low-frequency sound penetrates deeper but gives relatively poor delineation of structural features.

Hydrophones The hydrophone arrays are the receiving portion of the seismic reflection system. The arrays are placed in a neutrally buoyant liquid-filled tube to reduce the noise level caused from the motion of the hydrophones as they are drawn through the water and are usually towed 150–300 m behind the ship. It is important to maintain a satisfactory signal-to-noise ratio to produce a readable graphic record.

Recorders The recorder translates variation in signal level into variations in printing density. As the paper is advanced by the recorder at a constant rate, a stylus produces a vertical line so that reflecting surfaces are indicated by a greater printing density. The line or darkened band on the graphic recorder represents the boundary between two contrasting materials; the greater the acoustic contrast between the two materials, the darker the interface will appear on the profile. The vertical exaggeration of the profile, normally 30:1, is controlled by ship's speed and the rate of paper advance in the recorder.

TERRY S. MALEY

References

Cohen, P. M., 1970, *Bathymetric Navigation and Charting.* Annapolis, Md.: U.S. Naval Institute, 138p.

Ewing, J., and M. Ewing, 1970, Seismic reflection, in A. E. Maxwell, ed., *The Sea,* vol. 4. New York: Wiley-Interscience, 1–57.

Maley, T. S., 1973, Multiple-beam sonar system to determine the axis of a linear submarine feature, *EOS Trans., Am. Geophys. Union Abstracts* **11,** 1110.

Maley, T. S., F. Siebert, and G. L. Johnson, 1974, Topography and structure of the Western Puerto Rico Trench, *Geol. Soc. America Bull.* **85,** 513–518.

Cross-references: *Harbor Surveys; Marine Magnetic Surveys; Surveying, Electronic; Seismological Methods.* Vol. I: *Acoustics (Underwater); Bathymetry; Seismic Reflection Profiling at Sea; Sounding.*

AERIAL PHOTOGRAPHY—See AERIAL SURVEYS; PHOTOGEOLOGY; PHOTOINTERPRETATION. Vol. XIII: PHOTOGRAMMETRY; REMOTE SENSING, ENGINEERING GEOLOGY.

AERIAL SURVEYS

Initial Stages

In 1855 Nadar made the first aerial photographs from a captive balloon at the modest height of 80 m near Petit Bicetre in the neighborhood of Paris. During 1856 he followed with a dozen good quality photographs of Paris. Similar pictures over Boston were made in 1860 by Samuel A. King and J. W. Black. Photographs from a free balloon over Paris by Triboulet date from 1879. In 1898 Albert Heim, a well-known Swiss geologist, made his famous flight over the Alps and the Juras, taking aerial photographs for geological purposes. Wilbur Wright, flying over Centocelli in Italy in 1909, was the first to take photographs from an airplane. World War I, when air navigation developed quickly, also gave a strong impetus to aerial reconnaissance. In 1915 Lt. Col. J. R. L. Moore Brabazon, later chief of the photographic section of the RAF, designed the first aerial camera. The development of conventional aerial photography resulted in the application of these techniques to geology and other disciplines in the 1920s and early 1930s (Colwell, 1960; Reeves, 1975).

Further Developments

World War II and space flight programs, with attendant developments in electronics and computer techniques, gave rise to refinements and new techniques for viewing the Earth. The Mercury and Gemini projects included photographic experiments with hand-held cameras, showing the value of broad synoptic views of the Earth's surface. Concurrently, remotely sensed data from unoccupied satellites and the development of sidelooking airborne radar systems were viewed with interest for various Earth resources applications. The first satellite system specially dedicated to the sensing of Earth resources, known as the Earth Resources Technology Satellite (ERTS-1), was placed in orbit in June 1972. The Skylab mission followed in May 1973. Both were equipped with cameras and multispectral scanner systems. LANDSAT-2 and LANDSAT-3 were launched in 1975 and 1978 respectively. SEASAT-1 was put in orbit in 1978.

Sensors

Sensors carried aboard air- or spacecraft offer a synoptic and comprehensive view of terrain not afforded by ground survey methods. Photographic systems are still the most important sensors, and when flown in a stereoscopic mode, they allow for three-dimensional viewing (see *Photo Interpretation*). The assortment of available camera and film types makes photography the most adaptable remote sensing (q.v.) system for the visible and near infrared parts of the spectrum.

Photogrammetric cameras belong to the so-called *framing cameras,* which in contrast to panoramic cameras, expose the full film format instantaneously. The most common provide 9 × 9 in image frames on rolls of aerial film. The image scale depends on the flight altitude and the camera's focal length. Stereoscopic photography is achieved by choosing exposure intervals that overlap about 60%.

Black and white aerial photographs cover the visible spectrum (400–750 nanometers, nm). These proven mapping tools are difficult to surpass in availability and spatial resolution. *Color photographs* have the same spectral range (400–750 nm) as conventional black–and–white photographs. Modern color films have a high resolution, and it is visually easier to distinguish between subtle color differences than between shades of black and white. Such photographs can therefore be interpreted with greater accuracy. *Black–and–white infrared photographs* cover the range between 400 nm and 1,100 nm, recording radiant energy beyond the visible part of the spectrum. Such photographs are often used because infrared rays penetrate areas covered by haze. Furthermore, they are especially useful in delineating areas with contrasting absorption for the reflective infrared. They thus serve in mapping coastal features and drainage patterns in regions covered by heavy tropical forest (see *Photogeology*). They can also be helpful in soil moisture investigations. *Color infrared photographs* are also known as *false color photographs* because the colors recorded appear different than they are in nature. Such film is sensitive to the green, red, and infrared bands, which are rendered in blue, green, and red respectively on the photograph. These photographs make it possible to distinguish painted objects from living vegetation. Therefore, the first applications were in military reconnaissance for the detection of camouflage. Healthy deciduous vegetation is recorded in magenta; healthy evergreens, which have a lower infrared reflectivity, in bluish-purple; and dead or dying leaves, as green (evergreens) or yellow-white (deciduous). Due to the relationship between type and condition of the vegetation and its infrared reflectivity, false color photographs are especially useful in *vegetation surveys* (see *Vegetation Mapping*). In geology they may give false information when there is a direct relationship between the rock or soil material and vegetation (Jerie, 1970).

In *multiband photography* a number of photographs are taken simultaneously, each with a different film-filter combination so that specific parts of spectral reflections are recorded separately. This makes it possible to define better the spectral signatures of rock and soil materials and of vegetation cover. At present there are two types of cameras with either four or nine lens/filter combinations. Quantitative information on the spectral signature of a feature in the various bands can be obtained by *densitometry.* Various enhancement and color-coding techniques are employed for visual interpretation.

Sidelooking airborne radar (SLAR) has been a routinely operational system since the 1970s. Radar, having its own source of energy, is not limited by adverse light or weather conditions and operates even at night. Pulses of electromagnetic energy are transmitted and reflected back as signals from terrain targets. The target returns for each energy transmission are converted into light, which is imaged as a single line on photographic film. Moving the film at a velocity proportional to the speed of the aircraft, an image is built up and is recorded as a continuous strip map. A radar image contains information on the morphology and roughness of a given terrain and on the relative reflectivity of its constituent materials. It has become common practice to take parallel and overlapping strips, scanned from the same direction. In this way a stereoscopic effect can be obtained that considerably improves possibilities from interpretation of radar imagery. SLAR evolved from systems designed for military reconnaissance during World War II, the so-called Planned Position Radar (PPR), which had a rotating antenna underneath the aircraft. In the earliest SLAR equipment, such rotating aerials were locked in a sidelooking position. Notwithstanding the use of relatively long antennas, the swath width of SLAR systems is limited because of the increasing loss of resolution from near to far range. In synthetic aperture radar (SAR) it is possible to obtain a good resolution in all ranges and to use a short antenna.

Thermal scanners record thermal energy on magnetic tape or, after transforming it to light, on photographic film. Whereas tonal differences in conventional photography are a function of light reflectance, tonal variations on thermal infrared images result from thermal emission differences between adjacent features. Because of strong atmospheric absorption, the only bands that can be used are from 3,000 nm to 5,000 nm and from 8,000 nm to 13,000 nm. The radiation that is measured is emitted, not reflected as in near infrared photography. Thermal information is useful for locating forest fires and thermal pollution of lakes and streams, helps chart the movement of water bodies with contrasting temperatures in the oceans, and is useful in hydrology and related fields (e.g., karst phenomena) for detecting faulting with groundwater–controlled temperature effects. Sequential observation in volcanic regions may give clues to increased activity.

Multispectral scanners (MSS) provide data in a multiband mode similar to that obtained from multiband camera systems. The scanners are used in both air- and spacecraft. The received energy is optically separated into discrete wavelength bands that are recorded on magnetic tape. This information can be used to produce separate images for each band or can be computer processed to determine the spectral signature of various terrain targets. Apart from cameras and other equipment, both the ERTS (now LANDSAT) and the Skylab satellites have been equipped with multiband scanner systems; four bands (green, red, and two in infrared) and thirteen bands respectively. Due to a failure, however, the Skylab multispectral scanners never functioned. LANDSAT-2 and 3 were equipped with scanner systems similar to those of LANDSAT-1. SEASAT-1 was designed especially for the observation of ocean currents. Apart from other equipment it carried a SAR and thermal scanners. After functioning for 99 days, a power failure silenced SEASAT-1. LANDSAT-D has been planned to carry a thermatic mapper. Some observers, however, argue that it would be more useful to develop a satellite based on *multispectral linear arrays*. Such a system allows for an arrangement of the scanners that will ensure a considerable overlap of successive images so they can be viewed stereoscopically. Satellites afford sequential observation, which is important in the study of rapidly evolving processes (e.g., crops, floods, coastal erosion, volcanic activity). Thus, in agriculture a crop can be watched from sowing till reaping, diseases and pests can be spotted at an early stage, and a reliable estimate of the yield may be available well before harvest time. LANDSAT and Skylab data can contribute to an understanding of megatectonics and regional relationships. In areas where geological maps are poor or lacking, they can be profitably used in the reconnaissance stage provided such areas are well exposed and favorable atmospheric conditions prevail (see *Photogeology*). Automatic thermatic mapping from multiband information may be possible for relatively uncomplicated conditions found in land use and crop types. It seems feasible in more complicated situations where geomorphological expression is obscure and where visual stereoscopic interpretation cannot be considered by computer.

Geophysical Sensors

Sensors used in airborne geophysical measurements record a single quantity at the location of the sensor instead of imaging electromagnetic radiation. The sampled physical quantity and its measuring device may also be one or more of the following.

Local Magnetic Field The intensity of the Earth's magnetic field normally lies between 30,000 gammas and 70,000 gammas (1 gamma = 10^{-9} tesla). In airborne surveys the intensity of the field is measured without regard to its direction. The relative or absolute variations are measured by magnetometers. The flux-gate magnetometer, constructed on the principle of flux saturation of magnetically permeable Fe-Ni alloy cores, has been in use since 1946 and has a precision of a few gammas. The *proton free-precision magnetometer* is based on the principle of nuclear resonance of the gyromagnetic spin of protons (present in water or liquid hydrocarbons). This type of instrument came into use around 1954 and is accurate to within a few gammas. The measuring sequence takes about a second. The *optically pumped magnetometer,* also called an alkali-vapor magnetometer, utilizes the nuclear resonance of the gyromagnetic spin of cesium or rubidium atoms that is present in their vapor. In use since about 1962, this instrument can measure a few hundredths of a gamma.

Local Level of Radioactivity Radioactivity results from naturally decaying elements such as thorium, uranium, and potassium. The radioactivity is measured by *scintillation detectors,* which consist of a large (several cubic centimeters) single crystal of NaJ, with traces of T1 as activator. The scintillations may, in addition, be analyzed in a *spectrometer* to discriminate different energy levels in the radiation, which may be used to obtain information on radioactive elements.

Electromagnetic Field Electromagnetic fields are produced by distant thunderstorms [frequency band below 1,000 Hz, audiofrequency magnetics (AFMAG)] or by artificial distant communications transmitters [frequency band between 10,000 Hz and 30,000 Hz, known as very low frequencies (VLF) (see *VLF Electromagnetic Surveying*)]. In both cases the orientation in space is measured without regard to intensity. The measuring devices consist of two receiving coils, aligned perpendicularly to each other. The ratio of the induced signals equals the tangent of the angle that the field vector subtends with one coil-plane. In an electromagnetic system it is also possible to transmit a field locally by either a local ground transmitter (TUEAIR) or an airborne transmitter flown in either the receiving plane or in a separate plane flying nearby. In both cases the intensity of the received field and its phase relation to the transmitted field are measured.

Photogrammetry

Photogrammetry (q.v.) may be defined as a system of measuring data recorded on reliable geometric measurements (Hallert, 1960; Moffit, 1967). It is an important aid in terrain mapping, replacing a considerable amount of control surveying by photogrammetric methods. As early as 1850, Laussedat used photographs for measuring purposes. Photography from a ground station (*terrestrial photogrammetry*) predominated at the beginning, but today *aerial photogrammetry* is more common. A relatively new branch is *space photogrammetry,* which is related to all aspects of *extra-terrestrial photography.* The main task of photogrammetry can be defined as a production

of an orthogonal projection (map) on a certain plane and scale by means of one or two central projections (photographs) of the object. Two photographs are usually required to produce an orthogonal projection, the process being called *stereo photogrammetry.* In particular cases, if the object is more or less a place, an orthogonal projection can be obtained from a single photograph. During aerial photography, the area is covered by several runs of photographs taken from different air stations along parallel flight lines. These photographs overlap each other in such a way that all points of the object appear in at least two photographs (about 60% or more longitudinal and 20% or more lateral overlap). If the photographs are developed and placed in projectors that are congruent with the taking camera, by illuminating the pictures from above it is possible to reconstruct the bundles of rays that formed the photographs. Further, following a certain routine known as orientation, the corresponding rays will intersect and form a three–dimensional model geometrically similar to the photographed object. Taking advantage of stereoscopic vision, measurements can be performed in this model instead of in the object.

Forest Surveys

The increasing demand for forest products requires efficient management of world forests. A rapid assessment of the resources is obtained through aerial survey because it provides essential information on logging planning, damage assessment, flood control, wildlife, and recreation planning (Weber, 1971). This information, which is important for management planning, may be mapped or recorded otherwise. In the tropical regions, where panchromatic film is widely used, photoscales ranging from 1:20,000 to 1:50,000 are applied for forest-type mapping (Stellingwerf, 1975). Because identification of tree species is difficult, even on large scale photographs, the best time for taking aerial photographs is at the end of the rainy season when the sky is free of clouds (Kenneweg, 1972). In areas frequently under cloud cover, such as the tropical rain forest of the Amazon drainage basin, use may also be made of radar. In the temperate regions, photoscales between 1:5,000 and 1:15,000 are suitable for forest stand classification and tree species identification (Fig. 1). In forest areas with mixtures of coniferous and broad leaved (deciduous) species, infrared black-and–white film is preferred over panchromatic film to distinguish these two important species groups. The best time for aerial photography of forest areas is early summer. Color infrared film is used for the registration of forest damage caused by diseases, insects, fungi, and environmental pollution. Forest fires can be

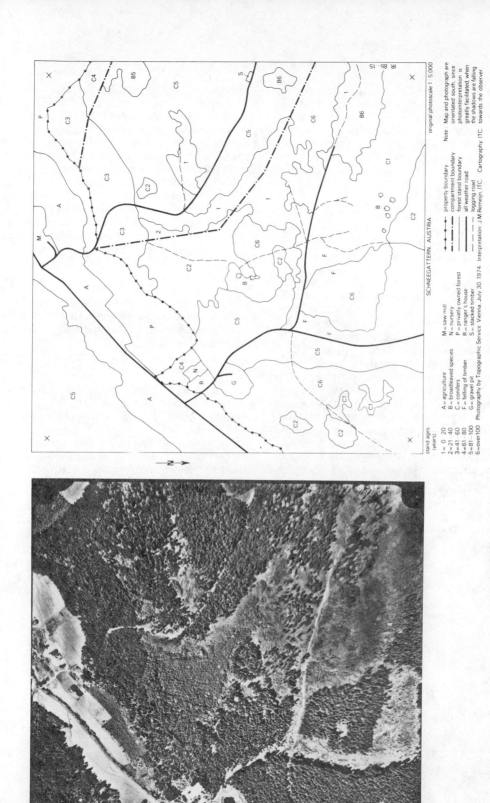

FIGURE 1. Forest survey with aerial photographs, Austria.

detected and mapped, even through dense smoke and during the night by heat sensors such as the *thermal infrared line scanner.* Timber volumes are estimated from aerial photographs by measuring stand area, tree height, tree crown diameter, stand age, stocking density, site class, merchantable volume, and timber volume growth (see *Vegetation Mapping*).

Geological Surveys

Geological maps form the basis for exploration and civil engineering. Aerial photographs and other remote sensing imagery or data provide essential information in this respect that has to be checked and supplemented by fieldwork. Remote sensing should be employed in stages. Small-scale information, as provided by LANDSAT and Skylab, can be important in the reconnaissance stage, in poorly known areas, and where regional relationships are the main concern.

For civil engineering projects, very detailed maps may be needed (see *Remote Sensing*). This implies the use of large scale (1:5,000 or so) photographs. The geologist is interested in two main aspects, the lithology, or composition, of the rocks and their structural arrangements. Both find expression in surface morphology, and geological interpretation from aerial photographs largely depends on an understanding and analysis of phy-

siography (Miller, 1961; Mekel, 1978). Combined with field observations, such geological interpretations may yield important information in mineral exploration, hydrogeology, and civil engineering. Mineral exploration is concerned with the search for specific environments where concentrations of mineral substances, either fuel or ore, occur. *Hydrogeological* investigations focus on surface drainage basins, groundwater basins, the extent of hydrological units, and the presence of aquifers (Fig. 2). These should be supplemented with data on the amount of precipitation and evaporation (Meyerink, 1974). *Civil* engineering requires a land classification that gives information about the stability of the terrain and its resistance to erosion. The location and quantity of available construction materials are also of interest. For dam construction and airfield and highway projects, estimates of cut-and-fill volumes are important.

Geophysical Surveys

Airborne geophysical surveys are conducted with three main objectives. It is customary to accompany regional geological mapping by an *airborne magnetic survey.* The flight lines are spaced 1–2 km apart and flown with a terrain clearance of a few hundred meters. The object is to obtain structural geological information (Telford et al.,

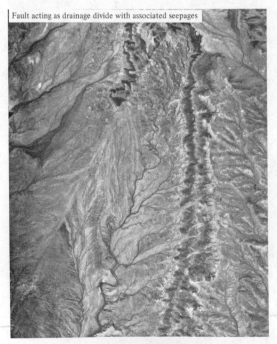
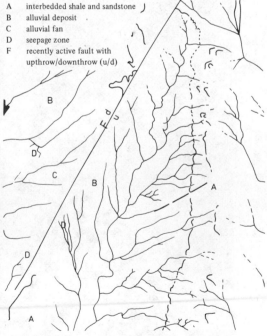

A   interbedded shale and sandstone
B   alluvial deposit
C   alluvial fan
D   seepage zone
F   recently active fault with upthrow/downthrow (u/d)

Fault acting as drainage divide with associated seepages

FIGURE 2. Hydrogeological survey with aerial photographs, Utah, United States.

1978). In some cases—e.g., in offshore oil exploration—this is the best method for rapidly gaining the desired information. Detailed information on mineralization, especially sulfide mineralization, is sought by flying an *electromagnetic transmitting and receiving system* that registers changes in the electrical conductivity of the ground (see *Exploration Geophysics*). The line spacing is a few hundred meters and terrain clearance as low as flight regulations allow. The airborne search for radioactive minerals is done by flying a *scintillometer detector and spectrometer*. This search is frequently combined with other surveys (see *Exploration Geochemistry*).

Soil Surveys

Soil surveys record the properties of soils and their distribution patterns. The resulting soil maps give important information for agriculture and engineering. It is especially in semidetailed and smaller-scale surveys that aerial photography and other remote sensing imagery are currently used to locate boundaries between different soils (Fig. 3). Surface features related to those boundaries are visible on various types of imagery and are called *elements*. Examples of elements include relief, drainage pattern, land use, and vegetation. Some elements, like relief, are consistently related to certain kinds of soils and can be used for mapping. Physiographic analysis also studies elements to establish landscape processes. A knowledge of those processes and their related soil distribution patterns then leads to appropriate delineations of the soils (Buringh, 1960; Goosen, 1967). The photo-interpretation map shows different units that are studied in the field. Image interpretation leads to a considerable increase of surveyed area per time unit and to a corresponding reduction (up to five to six times) in mapping cost per hectare.

Urban Surveys

In urban areas significant information concerning geography, urban planning, civil engi-

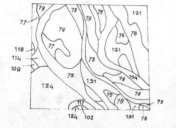

A Section of the Soil Map of the Netherlands, scale 1 : 200,000, in the Maas valley near Well and an assembly of the standard 1 : 20,000 aerial photographs of the same area are shown.
Nearly all the boundaries on the map, which was made without aerial photo-interpretation, can be readily seen on the photographs. The lines on the photographs are the result of extensive photo-interpretation. Comparison shows that even a small scale map can be upgraded by the use of aerial photographs. Boundaries which do not appear on the photo-interpretation are usually of minor importance on a small scale map.
(Example by H.F. Gelens, ITC Soils Department).

FIGURE 3. Soil survey with aerial photographs, Meuse Valley, Netherlands.

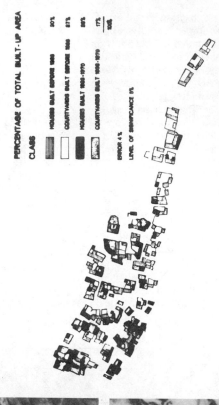

FIGURE 4. Urban survey with aerial photographs, Tunis City, Tunesia. The area represented in the photographs is a part of the city of Tunis. It shows a spontaneous development along the lake in the southwest. By comparison with aerial photographs of two different years it was possible to detect the changes in the physical structure over an 8 yr period; it appears that a considerable increase in built-up and enclosed spaces for residential use took place. This is a clear example of monitoring urban dynamics by the use of aerial photography. Using these techniques the planner and administrator can be provided quickly and efficiently with basic information about type, location, and speed of physical changes in urban areas.

neering, and architecture may be gained from aerial photographs. Because there is a high information content per square kilometer, large-scale photographs (1:10,000) are usually preferred. The resulting data are presented in *thematic maps,* tables, diagrams, and reports and can be used for urban administration, management, and planning (Branch, 1971; Bruijn, 1979). The information obtained can be subdivided into two classes, with special reference to population and its activities and use of the available space and the material outfit or artifacts. Because *artifacts* can be seen and measured on aerial photographs, information about the population and its activities and use of space is derived in an indirect way. Whether the emphasis lies on photo-interpretation or fieldwork depends largely on the object of the survey. In the case of features with a large areal extent and a repetitious character, like residential areas, the emphasis is on photo-interpretation; fieldwork becomes dominant when the artifacts give little or no information on activities and space use, a situation commonly found in city centers. Processes of change (occupation of new land, reconstruction of existing urban areas) can be studied efficiently by sequential photography at 2 yr or 3 yr intervals. Measurement of the speed of change makes prediction possible (Fig. 4).

J. F. M. MEKEL

References

Bodechtel, J., and H. G. Gierloff, 1974, *The Earth from Space,* Newton Abbot, England: David and Charles Emden, Ltd., 176p.

Branch, M. C., 1971, *City Planning and Aerial Information.* Cambridge, Mass.: Harvard University Press, 283p.

Bruijn, C. A. de, 1979, Input of air-photo and remote sensing data into urban information systems, *Seventh European Symposium on Urban Data Management.* The Hague, 10p.

Buringh, P., 1960, The application of aerial photographs in soil surveys, in *Manual of Photographic Interpretation.* Washington, D.C.: American Society of Photogrammetry, 63–66.

Colwell, R. N., ed., 1960, *Manual of Photographic Interpretation.* Washington, D.C.: American Society of Photogrammetry, 868p.

Estes, J. E., and L. W. Senger, 1974, *Remote Sensing Techniques for Environmental Analyses.* Santa Barbara, Calif.: Hamilton Publishing Co., 340p.

Goosen, D., 1967, Aerial photo-interpretation in soil survey, *FAO Soils Bull. No. 6,* 55p.

Hallert, B., 1960, *Photogrammetry.* New York: McGraw-Hill, 340p.

Hofstee, P., 1978, Aerial photography as a data source for traffic management, *Proceedings of the International Symposium on Remote Sensing.* Freiburg: International Archives of Photogrammetry, 959–967.

Jerie, H. G., 1970, Photogrammetry for natural resources surveys, *International Symposium of ISP, Commission VII.* Dresden, September 1970, 895–922.

Kenneweg, H., 1972, Zur frage der erkennung und abgrenzung von rauchschaden aus luftbildern, *Mitteilungen der Fortlichen Bundes-Versuchsanstalt* **97,** 295–305.

Mekel, J. F. M., 1978, The use of aerial photographs and other images in geological mapping, in *ITC Textbook of Interpretation,* 2nd. ed., vol. 1, 206p; vol. 2, 190p.

Meyerink, A. M. J., 1974, *Photo-Hydrological Reconnaissance Surveys.* Enschede: ITC Publications, 371p.

Miller, V. C., 1961, *Photo-Geology.* New York: McGraw-Hill, 248p.

Moffit, F. H., 1967, *Photogrammetry.* Scranton, Pa.: International Textbook Company, 540p.

Parasnis, D. S., 1973, *Mining Geophysics.* Amsterdam: Elsevier, 395p.

Reeves, R. G., ed., 1975, *Manual of Remote Sensing,* vol. 1: *Theory, Instruments and Techniques;* vol. 2: *Interpretation and Application.* Falls Church, Va.: American Society of Photogrammetry, 2144p.

Smith, J. T., 1968, *Manual of Color Aerial Photography,* Falls Church, Va.: American Society of Photogrammetry, 550p.

Soil Survey Staff, 1966, *Aerial Photo-Interpretation in Classifying and Mapping Soils.* Washington, D.C.: Soil Conservation Service (USDA Handbook 294), 89p.

Stellingwerf, D. A., 1975, The tropical zone, in R. G. Reeves, ed., *Manual of Remote Sensing.* Falls Church, Va.: American Society of Photogrammetry, 101–128.

Strandberg, C. H., 1967, *Aerial Discovery Manual.* New York: Wiley, 249p.

Telford, W. A., L. P. Geldart, R. E. Sherif, and D. A. Keys, 1978, *Applied Geophysics.* Cambridge: Cambridge University Press, 860p.

Vink, A. P. A., 1964, Aerial photographs and the soil sciences, *Proceedings of the Toulouse Conference on Aerial Surveys and Integrated Studies,* 81–141.

Weber, F. P., 1971, Applications of airborne remote sensing in forestry, in Joint Report of the International Union of Forest Research Organization, Section 25, *Application of Remote Sensors in Forestry.* Freiburg, Germany: Rombach and Co., 75–88.

Cross-references: *Exploration Geophysics; Magnetic Susceptibility, Earth Materials; Photogeology; Photo Interpretation; Remote Sensing, Engineering Geology; Remote Sensing, General; Sea Surveys; Surveying, Electronic; Terrain Evaluation Systems; Urban Geology; Vegetation Mapping; VLF Electromagnetic Prospecting.*

AIRBORNE GEOPHYSICS—See ATMOGEOCHEMICAL PROSPECTING; EXPLORATION GEOCHEMISTRY; EXPLORATION GEOPHYSICS.

ALIDADE

The alidade is possibly one of the oldest known mapping instruments (Low, 1952). Its current function, however, is largely incorporated into modern surveying instruments (Ahmed and Almond, 1983) (see also *Surveying, Electronic; Surveying, General*). The term *alidade* (from the Arabic) originally referred only to a rule or straightedge but now commonly identifies the entire instrument, which incorporates simple telescopic or other sighting devices with index reading or recording accessories (Breed and Hosmer, 1945; Lahee, 1961; Bates and Jackson, 1980). The alidade is used for determining the directions of objects and is commonly deployed in detailed survey (q.v.), especially plane table, mapping (q.v.). Modern telescopic alidades, like the one shown in Fig. 1, include a focusing telescope, a striding level, and vernier scale for measuring vertical angles; a magnetic needle, stadia hairs in the eyepiece for determining distances read on a stadia rod; and commonly a Beaman arc for rapid conversion of stadia readings to true horizontal-vertical distances. Regardless of the form of the sighting device, the basic principles of operation from plane surveying (Forbes, 1955) are identical.

To use an alidade, a flat surface containing a map sheet is required. This surface must be capable of being leveled (made perpendicular to the direction of gravity) and rotated in a horizontal plane. A drawing board mounted on a tripod that permits leveling and rotation is known as a *plane table* (see *Plane Table Mapping*). Locating points to make a map is accomplished by one or more of three methods: radiation, intersection, and resection [see, e.g., Tracey (1914), Greenly and Williams (1930), and Bouchard and Moffit (1959) for classic descriptions of this technique].

Radiation

The plane table is set up near the center of the area to be mapped, then is leveled and securely clamped to prevent rotation. A point on the map sheet is chosen arbitrarily to represent the point occupied by the plane table. This point does not have to be at or near the center of the sheet. The scale of the map and extent of the area to be mapped govern the location of this point.

The beveled edge (fiducial) of the alidade is placed alongside this point and rotated until a feature to be mapped is observed in the sighting device. With the distance to the feature determined by taping or by stadia measurement, it is plotted to scale along the line drawn against the edge of the alidade. There is, however, no need to draw this line because it would clutter the map. The location of the feature (e.g., fence post, edge of road, tree, outcrop) is marked with a sharp pencil or a Rapidograph-type pen or pricked with a plotting needle on the map sheet and marked.

Other features are similarly located at the ends of lines radiating from the point on the map sheet designating the plane table location. If the area to be mapped is small and the entire terrain is visible from the plane table location, the entire map can be made by this simple method. For larger areas and where visibility is limited, radiation is still a

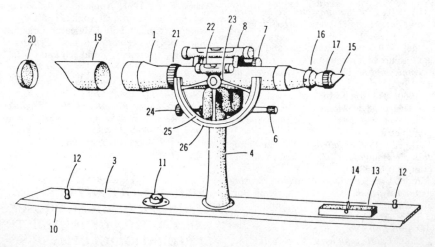

FIGURE 1. A typical telescopic alidade, commonly called a *high-standard* or *topographic* model. The component parts are as follows: (1) telescope, (3) blade, (4) pedestal, (6) axis clamp screw, (7) tangent screw, (8) striding level, (10) fiducial edge, (11) bull's eye level, (12) knobs, (13) compass box, (14) lifting lever, (15) ordinary or prism eyepiece, (16) cross and stadia hairs, (17) knurled ring, (19) sunshade, (20) protective lens, (21) knurled retaining ring, (22) vernier scale, (23) frame, (24) tangent screw, (25) index lines, (26) calibrated arc. The *explorer's* alidade is similar except that it is 5–7 cm shorter than the topographic model, and its stand is attached directly to the blade. (Note: Accessory component parts numbered 2, 5, 9, and 18 are not shown in this partial diagram, or they are hidden from view in the drawing.) (From Compton, 1962).

practical method. A series of points, each located by radiation from the preceding point, can be established, and at each location in turn, the terrain visible from that point can be mapped by radiation. Such a procedure is, however, seldom practical, and other methods (to be described) may be used to advantage.

Intersection

Two points, A and B, are established at convenient locations on the ground, and the distance between the points is determined by taping or stadia. This baseline is then drawn to scale on the map sheet. The location of the baseline on the map sheet is not critical provided the area to be mapped will fit on the map sheet in proper relation to the baseline.

The plane table is then set up over point A and leveled. The edge of the alidade is placed along the line AB on the map sheet, and the table is rotated until the line of sight intersects point B. The table is now oriented, i.e., lines on the map are parallel to corresponding lines on the ground. The alidade is then pivoted about point A toward C. The distance AC is not determined.

Then the plane table is set up at B, oriented as before, and the ray BC is drawn after C is sighted. The intersection of the rays locates point C. Point C may now be used as a point from which detail can be located by radiation, or it may be one of a series of inaccessible points that have been plotted to provide detail for the map.

Each time the plane table is moved from one point to another, it is necessary to orient it before continuing. Orientation is a simple matter provided that the terrain point over which the plane table is placed is located on the map sheet and that a second terrain point located on the map sheet is visible from the setup.

Resection

Assume that in the example given for intersection the topographer had not drawn the ray from A to C and did not discover this omission until he or she reached point B and had drawn the ray from B to C. It is still possible for the typographer to locate point C on the map without going back to into A. He or she occupies point C and orients using the ray from B in the direction of C. Then, if the alidade edge is placed on A and is rotated about A until A is sighted, the back ray, from A to C, intersects the ray from B toward C at the map position of C.

This example assumes that the ray from A to C was not drawn due to oversight. Actually, in many cases the location on the ground of point C may not have been chosen when point A was occupied because the topographer or surveyor needed a view of the terrain from B before de-

termining where C should be located. Resection is also used to locate on the map a ground point that is occupied by the plane table but to which no ray has been drawn. Such a point may be required to fill in detail by radiation, and its exact location is not determined until the surveyor actually occupies it and sees that it will be satisfactory.

To obtain the map location in this case, the surveyor needs three terrain points. They must be visible from the occupied station and plotted on the map. This is the classic three–point problem. There are several methods for solving it. The most practical solution is known as *Lehmann's method.*

Lehmann's Method

The problem is one of orientation because if the plane table were properly oriented, back rays from the three plotted stations would establish the occupied point by resection; the third ray would provide a check. Because it is impossible to orient the table exactly, the rays intersect to form a *triangle of error.* By following two rules, the surveyor may establish the exact location by successive approximations; experienced surveyors usually establish the point of the first try. The first rule is that the point sought is either to the left or the right of the three rays (it falls within the triangle of error only when the point sought falls within the triangle formed by the three signals). The second rule is that the distance on the map of the point sought from each of the rays is proportional to the distance from the point sought to the signals. The method fails if the three signals and the point sought fall on a circle. The three points should be chosen so this situation does not occur. Because the location of the occupied point can be closely approximated, this is not a difficult matter.

Orientation

The *declinator,* a compass in a rectangular housing, is often used for preliminary orientation. The compass establishes magnetic north, but this does not limit its usefulness in orientation. Assume, e.g., that the direction of line AB, described in the section on intersection, has been established with respect to true north. With the plane table set up at A and oriented on B, the declinator is placed at any convenient place on the table; the box is moved until the needle is centered; and a line is drawn along the side of the box. At any subsequent station, approximate orientation can be established by placing the declinator along the line on the table and rotating the table until the needle is centered. The accuracy of such orientation is seldom better than 1° because it is difficult to center the needle exactly. Where local attraction exists, the accuracy may be reduced.

Maps are usually oriented with respect to true north (see *Surveying, General*). However, because of the convergence of the meridians, true meridians (north-south lines) at the edges of a map sheet are not parallel to the meridian at the center of the sheet. This effect is magnified when a large area is mapped and the direction of all sheets is referred to a central meridian on one sheet. This direction is referred to as *grid north,* and it coincides with true north only in the center of the central sheet. Most all map sheets show the relationship of true, grid, and magnetic north on a sketch in the margin. The direction in which the magnetic needle points, at a given station, varies with time. *Isogonic* charts, published by the U.S. Coast and Geodetic Survey, show the deviation of the magnetic needle from true north for a specific year. Annual variation in the declination is also indicated.

Stadia

The telescopic alidade has two stadia hairs on the reticule, which carries the vertical and horizontal cross hairs. The stadia hairs are parallel to and equidistant from the horizontal hair. The distance between the stadia hairs is such that the angle they form is 1 in 100. Thus, with the line of sight horizontal, the intercept between the hairs on the vertical rod 100 m from the alidade is 1 m. When the line of sight is not horizontal, the horizontal distance and difference in elevation of the sighted point can be determined by formulas. Pocket calculators, charts, and diagrams make such computation a simple operation.

Various reticule patterns are in common use. All have the two hairs with the 1:100 ratio. Additional hairs establish other ratios, 1:200, 1:400, etc., to permit the determination of distances greater than the length of the rod multiplied by 100.

CHARLES W. FINKL, JNR.

References

Ahmed, F., and D. C. Almond, 1983, *Field Mapping for Geology Students.* London: Allen and Unwin, 72p.
Bates, R. L., and J. A. Jackson, 1980, *Glossary of Geology.* Falls Church, Va.: American Geological Institute, 715p.
Breed, C. B., and G. L. Hosmer, 1945, *The Principles and Practice of Surveying.* New York: Wiley, 327p.
Bouchard, H., and F. H. Moffitt, 1959, *Surveying.* Scranton, Pa.: International Textbook Co., 664p.
Compton, R. R., 1962, *Manual of Field Geology.* New York: Wiley, 378p.
Greenly, E., and H. Williams, 1930, *Methods in Geological Surveying.* London: Thomas Murby, 420p.
Lahee, F. H., 1961, *Field Geology.* New York: McGraw-Hill, 926p.
Low, J. W., 1952, *Plane Table Mapping.* New York: Wiley, 365p.
Tracey, J. C., 1914, *Plane Surveying.* New York: Wiley.

Cross-references: *Compass Traverse; Cross Section; Field Geology; Geological Survey and Mapping; Plane Table Mapping; Profile Construction.*

ALLUVIAL SYSTEMS MODELING

Methods of predicting river response include the use of both physical and mathematical models. Engineers have long used small-scale *hydraulic models* to assist them in anticipating the effect of altering conditions in a stretch of river. With proper awareness of the scale effects that exist, the results of hydraulic model testing can be extremely useful for this purpose. A more recent and perhaps more elegant method of predicting short- and long-term changes in rivers involves the use of mathematical models. To study a transient phenomenon in natural alluvial channels, the equations of motion and continuity for sediment-laden water and the continuity equation for sediment can be used. These equations are powerful analytical tools for the study of unsteady flow problems. However, because of mathematical difficulties, practical solutions are usually obtained by numerical analysis using iteration procedures and digital computers.

Physical Models

There are many problems in *hydraulic engineering* for which the basic equations are known but that are geometrically so complicated that the direct application of the equations is impossible. Many such problems can be solved by the use of models that duplicate this complicated geometry and in which the resulting flow patterns can be observed directly. These models are physical models—i.e., small-scale physical replications of the prototype that are use to test the performance of a design or to study the details of a phenomenon. The performance tests or proposed structures can be made at moderate costs and small risks on small-scale physical models. Similarly, the interaction of structure and the river environment can be studied in detail.

Natural phenomena are described by appropriate sets of governing equations. If these equations can be integrated, a given phenomenon in time and space domains can be predicted mathematically. In some cases related to river engineering, all the governing equations are not known. Also, the known equations cannot be directly treated mathematically for the complex geometries involved. In such cases, models are used to integrate physically the governing equations.

Rigid Boundary Models

Similitude between a prototype and a model implies that the relative importance of the parameters governing the flow is the same from model to prototypes. While the scaling of characteristic lengths can be easily visualized, determining the relative importance of other parameters such as flow velocity of fluid density presents a more difficult problem (see Vol. VI, *Sediment Transport, Initiation*).

To satisfy the conditions of similitude in clear water, geometric, kinematic, and dynamic similarity must exist between the prototype and the model. *Geometric similarity* refers to the similarity of form between the prototype and its model. *Kinematic similarity* refers to similarity of motion, and *dynamic similarity* is a scaling of masses and forces. For kinematic similarity, patterns or paths of motion between the model and the prototype should be geometrically similar. If similarity of flow is maintained between the model and the prototype, mathematical equations of motion will be identical for the two. Considering the equations of motion, the dimensionless ratio of $V\sqrt{gy}$ (*Froude number*) and Vy/v (*Reynolds number*) are both significant parameters in models of rigid boundary clear water open channel flow (see Vol. VI, *Reynolds and Froude Numbers*). Here, V represents the flow velocity, y represents channel depth, g is the gravitational acceleration, and v is the kinematic viscosity of the fluid flowing.

It is seldom possible to achieve kinematic, dynamic, and geometric similarity at the same time in a model. For instance, in open channel flow, gravitational forces predominate, and hence, the effects of the Froude number are more important than those of the Reynolds number. Therefore, the Froude criterion is used to determine the geometric scales but only with the knowledge that some scale effects—i.e., departure from strict similarity—exist in the model.

Ratios (or scales) of velocity, time, force, and other characteristics of flow for the two systems (model and prototype) are determined by equating the appropriate dimensionless parameter. If the two systems are denoted by the subscript m for model and p for prototype, then the ratio of corresponding quantities in the two systems can be defined. The subscript r is used to designate the ratio of the model quantity to the prototype quantity. For example, the length ratio is given by

$$L_r = \frac{x_m}{x_p} = \frac{y_m}{y_p} = \frac{z_m}{z_p} \quad (1)$$

for the coordinate directions x, y, and z. Eq. (1) assumes a condition of exact geometric similarity in all coordinate directions.

Open channel models are frequently distorted. A model is said to be *distorted* if variables exist that have the same dimension but are modeled by different scale ratios. Thus, geometrically distorted models can have different scales in horizontal (x, y) and vertical (z) directions, and two equations are necessary to define the length ratios in this case:

$$L_r = \frac{x_m}{x_p} = \frac{y_m}{y_p} \quad \text{and} \quad z_r = \frac{z_m}{z_p} \quad (2)$$

In free surface flows, the length ratio is often selected arbitrarily but with certain limitations kept in mind. The Froude number is used as a scaling criterion because gravity has a predominant effect. However, if a small length ratio is used—i.e., the water depths are very shallow—then surface tension forces may become important and complicate the interpretations of results of the model. The length scale is made as large as possible so that the Reynolds number is sufficiently large that friction becomes a function of the boundary roughness and essentially independent of the Reynolds number. A large length scale also ensures that the flow is turbulent in the model as it usually is in the prototype.

Boundary roughness can be characterized by *Manning's roughness coefficient, n,* in free surface flow. Analysis of Manning's equation and substitution of the appropriate length ratios, based on the Froude criterion, result in an expression for the ratio of the roughness that is given by

$$n_r = L_r^{1/6}. \quad (3)$$

It is not always possible to achieve boundary roughness in a model and prototype that correspond to that required by Eq. (3), and additional measures, such as adjustment of the slope, may be necessary to offset disproportionately high resistance in the model.

Mobile Bed Models

In modeling response to river development works, three-dimensional mobile bed models are often used. In these models the bed and sides are molded of materials that can be moved by the model flows. Similitude in mobile bed models implies that the model reproduces fluvial processes such as bed scour, bed deposition, lateral channel migration, and varying boundary roughness (see Vol. VI, *Fluvial Sediment Transport*). It has not been considered possible to simulate faithfully all these processes simultaneously on scale models.

Distortions of various parameters are often made in such models.

Two approaches are available to design mobile bed models. One is the analytical derivation of distortions explained by Einstein and Chien (1956), and the other is based on hydraulic geometry relationships given by Lacey, Blench, and others (Mahmood and Shen, 1971). In both these approaches, a first approximation of the model scales and distortions can be obtained by numerical computations. The model is built to these scales and then verified for past information obtained from the prototype. In general, the model scales need adjusting during the verification stage.

Model verification consists of the reproduction of observed prototype behavior under a given set of conditions. This reproduction is specifically directed to one or more alluvial processes of interest. For example, a model may be verified for bed-level changes over a certain reach of the river. The predictive use of the model should be restricted to the aspects for which the model has been verified. This use is based on the premise that if the model has successfully reproduced the phenomenon of interest over a given hydrograph as observed on the prototype, it will also reproduce the future response of the river over a similar range of conditions.

Mobile bed models are more difficult to design, and their theory is extremely complicated as compared to clear water rigid bed models. However, many successful examples of their use are available. In general, all important river training and control works are studied on physical models. The interpretation of results from a mobile bed model requires a basic understanding of the fluvial processes and some experience with such models. In many cases, it is possible to obtain only qualitative information from mobile bed models. However, this information is of great help in comparing the performance of different designs.

Mathematical Models

A physical scale model is a means for extracting information from some source other than the prototype. As Gessler (1971) points out, a physical model can be looked on as an analog computer since there is a high degree of analogy between prototype and model. With a distorted physical model, the geometric analogy is weakened considerably, but still, under most conditions, the analogy of the overall behavior is strong.

Once one gets used to the idea of looking at a model as an analog computer, the next logical step would be to model the process under study on a digital computer in numerical form. This step clearly requires a complete set of governing equations (some of them differential equations). Such equations would include basic flow equations, the differential equation of nonuniform and unsteady flow, the sediment transport equation, the differential equation formulating continuity of sediment transport, and criterion to predict the bed deformations, just to mention the most obvious equations involved. It is clear that the interaction among these equations is extremely complex. This, after all, is the reason for attempting to model these processes physically. But with the availability of high speed digital computers, it becomes entirely feasible to study some of the characteristics of a river system numerically. Of course, the results cannot be better than the basic equations used in the analysis, and most equations available are for one- or two-dimensional flow fields only. But when an overall river system is considered, a river can be viewed as a highly two-dimensional system, and with certain simplifying assumptions, a river can be modeled as a one-dimensional system. Only when one starts looking into the details do three-dimensional processes become important (Gessler, 1971).

The increasing use of numerical mathematical models for flood and sediment routing, degradation, and aggradation and long-term channel development studies is indicative of their potential for contributing to the complex problems of river system development and response.

PETER F. LAGASSE

References

Einstein, H. A., and N. Chien, 1956, Similarity of distorted river models with movable beds, *Am. Soc. Civil Engineers Trans.* **121,** 440–457.

Gessler, J., 1971, Modeling of fluvial processes, in H. W. Shen, ed., *River Mechanics.* Fort Collins, Colo.: Water Resources Publ., 20p.

Graf, W. H., 1971, *Hydraulics of Sediment Transport.* New York: McGraw-Hill, 513p.

Henderson, F. M., 1966, *Open Channel Flow.* New York: Macmillan, 522p.

Mahmood, K., and H. W. Shen, 1971, Regime concept of sediment-transporting canals and rivers, in H. W. Shen, ed., *River Mechanics.* Fort Collins, Colo.: Water Resources Publ., 39p.

Shen, H. W., ed., 1971, *River Mechanics.* Fort Collins, Colo.: Water Resources Publ., 1322p.

Simons, D. B., P. F. Lagasse, Y. H. Chen, and S. A. Schumm, 1975, *The River Environment.* Reference document prepared for U.S. Department of the Interior, Fish and Wildlife Service. Fort Collins: Colorado State University, 520p.

Cross-references: *Alluvial Plains, Engineering Geology; Deltaic Plains, Engineering Geology; Hydromechanics; Marine Sediments, Geotechnical Properties; River Engineering.* Vol. VI: *Alluvial Sediments; Fluvial Sediment Transport; Reynolds and Froude Numbers; Sediment Transport—Initiation and Energetics.*

ANTHROPOLOGICAL GEOLOGY—
See GEOANTHROPOLOGY;
GEOARCHAEOLOGY.

APPLIED GEOCHEMISTRY—See
BIOGEOCHEMISTRY; BLOWPIPE
ANALYSIS; EXPLORATION
GEOCHEMISTRY; HYDROCHEMICAL
PROSPECTING; INDICATOR
ELEMENTS; LAKE SEDIMENT
GEOCHEMISTRY;
LITHOGEOCHEMICAL
PROSPECTING; MARINE
EXPLORATION GEOCHEMISTRY;
MINERAL IDENTIFICATION,
CLASSICAL FIELD METHODS;
PEDOGEOCHEMICAL PROSPECTING;
PETROLEUM EXPLORATION
GEOCHEMISTRY; PROSPECTING.

APPLIED GEOLOGY—See Vol. XIII:
GEOLOGY, APPLIED.

APPLIED GEOMORPHOLOGY—See
Vol. XIII: GEOMORPHOLOGY,
APPLIED.

APPLIED MINERALOGY—See
BLOWPIPE ANALYSIS; MINERAL
IDENTIFICATION, CLASSICAL FIELD
METHODS; MINERALS AND
MINERALOGY; PLACER MINING;
PROSPECTING; ROCK PARTICLES,
FRAGMENTS; TEPHROCHRONOLOGY.
Vol. XIII: MINERAGRAPHY; MINERAL
ECONOMICS.

APPLIED OCEANOGRAPHY—See
ACOUSTIC SURVEYS, MARINE;
FLOATING STRUCTURES; HARBOR
SURVEYS; MARINE MAGNETIC
SURVEYS; OFFSHORE NUCLEAR
PLANT PROTECTION. Vol. XIII:
COASTAL ENGINEERING; COASTAL
INLETS, ENGINEERING GEOLOGY;
OCEANOGRAPHY, APPLIED;
SUBMERSIBLES.

ARCHAEOLOGICAL GEOLOGY—See
GEOANTHROPOLOGY;
GEOARCHAEOLOGY.

ARTIFICIAL DEPOSITS AND MODIFIED LAND

Artificial deposits and modified land are con-structed of and in part cut from rock and other earth materials at or on the earth's surface and so can be represented appropriately on a geologic map. Such land alterations are a result of human economics, aesthetics, or convenience; they are designed and constructed using widely varying de-grees of care, depending mainly on expected use or need. Although the materials involved are geo-logic, the design, construction, and testing tech-niques are primarily those of civil engineering, particularly the specialty of soil mechanics (q.v.). These techniques share an area of overlap with the geologic specialty of engineering geology. For more technical information on the engineering as-pects of land alteration, see Sowers and Sowers (1970) and Terzaghi and Peck (1967).

Characteristics of Modified Land

Land that has been altered by human activity can be classified in several ways. One, a strictly generic basis, differentiates *excavated* areas from *filled* areas, but this classification presents a problem in that the two are closely interrelated and both commonly occur on the same construc-tion site. An alternate basis differentiates land al-terations in which excavation plays an essential role from those in which its role is minor. The role of filled areas is similar in character in either case. The following definitions may be helpful.

Artificial deposits are land altered mainly by placement of earth, broken rock, sand and gravel, or some combination thereof, locally mixed with trash or garbage in varying amounts, with minor preliminary excavation necessary for proper placement.

Modified land is land altered mainly by exca-vation with minor associated landfilling or by in-termingled excavation and landfilling, with land-fills built of the material excavated.

A *strip mine, quarry, pit, or bank* is a surface working for the extraction of rock, ore, sand and gravel, or other earth material that constitutes one aspect of modified land.

A *landfill* is an accumulation of rock, other earth materials, trash and garbage, or some com-bination thereof, typically broader than high, and commonly intended for commercial, residential, or recreational development or for waste dis-posal; it includes some deposits of human origin and the filled parts of artificially modified land.

An *embankment* is an accumulation of rock, earth materials, or some combination thereof that is typically of linear form and commonly is intended to retain water or support a transportation facility. It includes many human-made deposits. An *earthwork* is excavation, landfill, or embankment that is constructed by grading.

Perhaps any geologic material can be used in the construction of a landfill or embankment provided the deposit is properly designed and the finished product meets design specifications. The point of concern is whether the engineering behavior of the deposit meets the requirements of the intended use. If the engineering behavior is unsatisfactory, the shortcoming can perhaps be offset by change in either the design of the structural foundation or the style of the building, by a change in use, or perhaps by additional mechanical compaction. The farther the engineering behavior of a deposit is from a norm for a given use, the greater can be the cost of such modification. The decision to use or not to use is ultimately made by the owner or the developer and may be based on a cost-benefit ratio.

Rock and earth materials, especially those that consist of more or less discrete physical components, have been classified in various ways. One method, the Unified Soil Classification (q.v.) (U.S. Army Corps of Engineers, 1953; U.S. Bureau of Reclamation, 1963), has been widely adopted. Categories of this classification are based mainly on particle size and partly on the amounts by weight of each size category together with the behavioral characteristics of any very fine particles present. Each category possesses reasonably distinct engineering behavior characteristics that permit field estimation of how the material will function and the design of a landfill that will meet requirements.

Field classification of a soil is not highly technical. Thus, the desirability of one soil material relative to another can be approximated for a general type of use. Aspects of special interest include the very fine particles (clay and silt sizes) that are too small to be distinguished by the unaided eye (less than 0.07 mm, or about 0.003 in) and the character of the *grading*. If there is even distribution by weight of all particle sizes from largest to smallest, the material is well graded. A rough approximation can be made of the kinds of compaction equipment that would be useful, the potential internal drainage characteristics, and the response of the deposit to frost action. Tables containing such information can be found in various soil engineering publications (e.g., U.S. Bureau of Reclamation, 1963, Fig. 8; The Asphalt Institute, 1969, Table VI-2).

Some geologic materials are more desirable than others for the construction of landfills and embankments having maximum physical integrity. Desired characteristics can include the following, but not all may be available in any one material:

1. The constituent particles are hard and tough, are of angular to subangular shape, and resist disintegration and weathering well.
2. Stones larger than about 25.6 cm (10 in) in diameter (boulders) are absent or rare and, if undesired, can be cast aside readily.
3. Stones from about 25.6 cm to 7.5 cm (10–3 in) in diameter (cobbles) are rare to few and, if undesired, can be cast aside readily.
4. Stones from about 7.5 cm (3 in) to 4.7 mm (0.18 in) in diameter (pebbles) constitute roughly 10% by weight.
5. Particles between 4.7 mm and 0.07 mm (about 0.18 in and 0.003 in) in diameter (sand) are well graded.
6. Particles less than 0.07 mm (about 0.003 in) in diameter (silt and clay size) constitute less than roughly 20% by weight.
7. Silt- and clay-size material is of low plasticity and nonswelling when wetted and does not shrink when dried.
8. Organic material constitutes less than roughly 10% by weight.
9. Flaky and needlelike material, or some combination thereof, constitutes less than roughly 5% by weight.
10. Trash and garbage, although not geologic, can constitute a serious threat to the internal integrity of a deposit (see Vol. XVI, *Sanitary Landfill*).

Artificial deposits and modified land are subject to the same environmental hazards as similar features of natural origin. The kind and degree of response by altered land is greatly influenced by factors such as the steepness of cut slopes, the character of the material used in landfills and deposits, the design and methods used in their construction, and subsequent changes in physical conditions both within a deposit and surrounding it. Those hazards most likely to cause damage are landsliding, earthquakes, compaction and compression, and expensive clays. Other potentially significant hazards include flooding, tsunamis, radioactivity of the fill material, and pollution of both surface and groundwater.

Characteristics of Artificial Deposits

There are three general kinds of essentially freestanding artificial deposits. They can be classified by the method of emplacement as (1) *dumped,* with possible associated spreading but little or no mechanical compaction, (2) *layered* and *compacted* according to specifications, and (3) *hydraulically spread* (dredged), commonly without compaction. Dumping and layering are commonly used; hydraulic spreading is used locally for large nearshore or riverbank deposits.

A *dumped landfill* commonly is emplaced by spilling, as from the tailgates of the hauling trucks and, for this discussion, includes dropping from the end of a conveyor belt. In some cases some compaction is performed by the passage of the trucks. A *layered and compacted* landfill is sometimes called an *engineered landfill*. Materials of undesirable composition are excluded, fragment size is restricted, and the material used is placed in layers (*lifts*) as much as 1 m thick. Each lift is compacted in a specified manner and to a specified degree. Thus, the characteristics of the finished deposit are determined by an engineer. Landfills that are *hydraulically spread* normally consist of material taken from the bottom or bank of a water body. Placement involves two distinct sedimentary processes: (1) deposition in open water and (2), where above water, current (fluvial) deposition on a sloping surface of previously deposited material.

Compaction of a landfill by wheeled equipment is least effective at or near a very steep confining face or along an unconfined face. Large power equipment cannot compact effectively closer than 0.5–1 m from the former, and the use of hand compactors may be necessary. Large compactors can work to the edge of an unconfined face, but much of their effectiveness is lost within 1 m of that edge. In that situation a landfill larger than is actually needed is often emplaced, and the loose material of the free face is trimmed back to properly compacted material.

Awareness of the method employed for emplacement of a landfill permits a rough estimation of the general behavior of the landfill relative to that of landfills emplaced by other methods. Some aspects of that character include potential for surface settlement, landsliding of side slopes, amplification of ground motion and erosion by running water. A relative evaluation of these characteristics for artificial deposits emplaced by the three principal methods is shown in Table 1. The best criterion for evaluation, however, remains whether or not a specific deposit meets the need of the planned use.

Local regulations commonly specify that earthworks be so constructed or protected that they do not endanger life or property and define those conditions under which permits, plans, specifications, and a report by an engineering geologist or a soils engineer are required. Many of the grading regulations are based on the *Uniform Building Code* published by the International Conference of Building Officials (1970) and do not specifically consider the character of local geologic materials or conditions. Some are incomplete or outdated. A preliminary investigation of the characteristics of an earthwork before purchase, use, or change in use is prudent and may warrant a more technical study.

Characteristics of Modified Land

The two principal kinds of modified land are identified most readily by the nature of the terrain occupied. Strip mines, rock quarries, sand, gravel,

TABLE 1. Generalized Relative Potential for Structural Failure of Artificial Deposits

Potential for Failure or Damage	Emplacement Method		
	Dumped, with or without Spreading, Little or No Mechanical Compaction	Layered and Mechanically Compacted to Optimum Specifications	Hydraulically Spread, Little or No Mechanical Compaction
Surface settlement (by subsequent compaction/consolidation, by earthquake shaking, or by change of water table)	High[b]	Low[b]	Intermediate
Landsliding of side slopes (by subsequent change of internal and external stresses and by earthquake shaking)	High	Low	Low
Amplification of ground motion (by earthquake shaking)	High	Low	Intermediate
Surface erosion (by runoff)	High	Intermediate	Low

[a]Failure or damage by natural processes that do not normally affect the supportive capacity of a deposit are not included here (e.g., failure by erosion or by landsliding caused by a stream in flood).

[b]These terms are used in a relative sense only.

and other borrow pits or banks typically are associated with piles of sized aggregate on the excavation floor, and scattered piles of ungraded, unsorted waste nearby. Areas for residential, commercial, industrial, and public grounds are characterized by intermingled cutting and filling of varying depths to yield a gently sloping or terraced surface. Short broad ridges of stored topsoil may be present.

The expectable behavior of undisturbed geologic materials in either kind of modified land is not changed except for some increase in landslide potential in the higher, steeper cut faces. Behavior of the artificial deposits will be similar to that of those artificial deposits described above and in Table 1.

Land-use Aspects

Artificial deposits and modified land are distinguished from natural surficial units by the fact that the altered land is the direct result of human activity rather than natural processes. Alteration of the land can introduce land-utilization problems, initiate water supply pollution, and affect property value. Landfills containing trash or garbage have commonly caused problems in all three ways.

Among the land-utilization problems introduced, the most common are (1) natural compaction (consolidation), which can lead to settlement damage from improperly constructed landfills and deposits; (2) poor drainage, which can lead to cross-lot flooding, ponding, wet basements, and eroded and debris-mantled ground; (3) reduction of slope stability from improper construction or placement of material on underlying slopes that are too steep, that are mantled with vegetation, or that are thus overloaded; (4) exposure of unsuitable foundation material in the areas excavated; and (5) loss of topsoil by stripping for sale.

The potential for pollution of surface-water supplies with sediment because of increased erosion of bare earth is significantly greater in areas of newly modified land and artificial deposits. The sediment thus produced has been known to fill small lakes in development areas. The rate of erosion decreases, however, as the bare area becomes paved or covered with vegetation. The pollution of shallow groundwater supplies in alluvium or other surficial deposits is likely to be most serious where the landfill components of an altered area contain trash, garbage, or other wastes. Groundwater supplies in subsurface layers that are separated from the altered land by nearly impermeable layers of rock are least likely to become polluted.

The value of property that contains altered land usually increases in value, but adjacent property can either increase or decrease, reflecting the use of the altered land. The value of land adjacent to a pit or quarry may decrease because of operational unsightliness, dust or noise pollution, and heavier traffic in the vicinity. It can increase, however, if some desirable future sequential development is planned or if the potential for future resource production is increased.

Landfills that contain trash are common in urban areas; most are located in small valleys or abandoned pits and quarries, and some are placed in excavations prepared for the purpose. Many older deposits consist simply of dumped wastes covered with a layer of earth; these are classified as nonsanitary landfills. They constitute poor foundation sites for construction because of potential settlement and can be a serious source of water pollution (see Vol. IVA, *Enviromental Pollution*).

Many newer landfills are carefully planned, situated, and managed to meet the specific standards for sanitary landfills. Such standards include the daily or more frequent addition of a specified thickness of nearly impermeable earth and a thick terminal cover of similar material. In some cases a pit is lined with impermeable earth or an artificial membrane. In either case, the hazard of groundwater pollution is sharply reduced.

Geologic Mapping of Altered Land

In areal geologic mapping (maps at a scale of 1:24,000, or even less detailed), artificial deposits have received more attention from geologists than modified land. For economic reasons, quarries, like mines, have long been symbolized on areal bedrock maps, but the symbolization of pits and banks is perhaps more recent. Most other excavations have been largely ignored because some, like roadcuts, are seemingly permanent, and foundation excavations are soon backfilled. In a few cases the latter have been indicated because they yielded significant geologic data. The advent of areal maps of surficial geology, mainly in the late 1940s, was followed in the 1950s by the representation on some of those maps of an excavation margin, together with an indication of the physical character or stratigraphic sequence of the material exposed and its thickness. Other modified land, which consists of intermixed cut-and-fill (graded) areas, rarely is shown on areal maps, probably because of the time required to delineate complex boundaries. The detailed representation of excavations and differentiated graded areas commonly is made, however, on site-specific maps (usually at a scale of 1:12,000 or even more detailed) used by planners and civil engineers (see *Maps, Enviromental Geology*).

Artificial deposits received little or no recognition for many years in areal mapping, but significant representation closely followed the introduction of areal surficial geologic maps and preceded the areal mapping of modified land. Now artificial deposits are represented on surficial geologic maps as well as on maps showing both surficial and bedrock information.

In the United States, recognition of the special needs of urban areas for detailed geologic information led in the early 1950s to inclusion of a unit for artificial fill on some maps of natural surficial deposits (e.g., Jahns, 1951). Simpson (1959) later tried a three-part classification that was based in part on the relative trash content of deposits, but one class included areas of undifferentiated cut-and-fill. Subsequently, Simpson (1973) used a classification that distinguished primarily between artificial deposits and modified land and secondarily between uses. This map was oriented toward land-use planners and the general public.

Map units used to represent various classifications of artificial deposits and modified land thus record areas of differing geologic character. Such a record can become important because the area may be forgotten with time; a map, then, can serve as a key to a source of groundwater pollution or to a cause of damaging surface settlement beneath a home. In addition, these maps are used as an unbiased reference in local land-use planning and zoning regulations and in locating sources of earth fill for reuse. Because a map shows only those features that existed at the time that the mapping was done, it is soon outdated in an area of dynamic urban growth.

Identification of altered land depends mainly on recognition of incongruous topographic characteristics, the disturbed nature of constituent earth materials, unusual vegetative cover, and evidence from old maps and records. The principal factors that contribute to concealment are low height, irregular plan, variable profile, surface-form reflection of surrounding terrain, use of local earth materials, natural character of vegetal cover, and minor to moderate alteration of former contours, drainage routes, or shorelines. Old maps are found in a great variety of places, including libraries, federal and state map-making agencies, and state, county, township, and city administrative departments. Old records may be found in county, township, and city assessor offices, and development plans may be found in offices of corporations and homeowners' associations, consultants' offices, and newspaper files.

Geologic mapping of land that has been or may be altered is most useful to land-use planners, civil engineers, and architects if the geologic materials involved are described in terms readily understandable to members of those professions. The popular Unified Soil Classification System is the basis for the classification of earth material on an engineering geologic map by Gardner et al. (1971).

HOWARD E. SIMPSON

References

Asphalt Institute, 1969, *Soils Manual,* 2nd ed. College Park, Md., 269p.
Gardner, M. E., H. E. Simpson, and S. S. Hart, 1971, Preliminary engineering geologic map of the Golden quadrangle, Jefferson County, Colorado, *U.S. Geol. Survey Misc. Field Studies Map MF-308* (1972).
International Conference of Building Officials, 1970, *Uniform Building Code,* vol. 1. Whittier, Calif., 412–423, 598–608.
Jahns, R. H., 1951, Surficial geology of the Mount Toby quadrangle, Massachusetts, *U.S. Geol. Survey Geol. Quad. Map GQ-9.*
Simpson, H. E., 1959, Surficial geology of the New Britain quadrangle, Connecticut, *U.S. Geol. Survey Geol. Quad. Map GQ-119.*
Simpson, H. E., 1973, Map showing man-modified land and man-made deposits in the Golden quadrangle, Jefferson County, Colorado, *U.S. Geol. Survey Misc. Geol. Inv. Map I-761-E.*
Sowers, G. B., and G. F. Sowers, 1970, *Introductory Soil Mechanics and Foundations,* 3rd ed. New York: Macmillan, 556p.
Terzaghi, K., and R. B. Peck, 1967, *Soil Mechanics in Engineering Practice,* 2nd ed. New York: Wiley, 729p.
U.S. Army Corps of Engineers, 1953, *The Unified Soil Classification System.* Vicksburg, Miss.: Waterways Experimental Station (Tech. Mem. 3-357), 30p.
U.S. Bureau of Reclamation, 1963, *Earth Manual—A Guide to the Use of Soils as Foundations and as Construction Materials for Hydraulic Structures,* rev. ed. Denver, Colo., 783p.

Cross-references: *Environmental Engineering; Geotechnical Engineering; Land Capability Analysis; Maps, Environmental Geology; Medical Geology; Mine Subsidence Control; Unified Soil Classification.* Vol. XII, Pt. 1: *Pollution; Sludge Disposal; Soils, Nonagricultural Uses.* Vol. XII, Pt. 2: *Minesoils.*

ATMOGEOCHEMICAL PROSPECTING

This aspect of exploration geochemistry (q.v.) may be defined as prospecting on the basis of pathfinder atmophile elements (q.v. in Vol. IVA) or compounds. Atmophile substances exist as gases under conditions of pressure and temperature that prevail near the Earth's surface and, therefore, seek to enter the atmosphere.

Because of the speed and economy with which aircraft or ground vehicles can sample the atmosphere, explorers have always hoped that atmogeochemical surveying procedures might be developed to search for ore deposits. Particular attention has centered on the use of Hg in sulfide prospecting, Rn in uranium prospecting, and hydrocarbons in petroleum prospecting. Useful results have been obtained for these gases using soil gas, gas dissolved in water, and gas entrapped in soil or rock. Limited technical success has resulted from the use of airborne or vehicle-borne detectors, but experiments are continuing. In addition to the desire for an airborne regional geochemical technique, there is a second important attraction in atmogeochemical prospecting: the high mobility of gases makes it possible to prospect for deeply buried (blind) ore deposits, some of which may not even have entered the weathering zoning.

Mercury Vapor Detection

Publications on *mercury vapor detectors* have been prominent since about the mid–1960s (Ling, 1958; James and Webb, 1964; Barringer, 1966; Hadeishi and McLaughlin, 1966; Vaughn, 1967; Hatch and Ott, 1968; McCarthy et al., 1969; Robbins, 1973; McNerney and Buseck, 1973). Robbins (1973) has described the design and operation of a portable mercury spectrometer that utilizes the *Zeeman effect* to provide a dual–beam method of atomic absorption analysis that overcomes the problem of interfering substances, such as SO_2 and NO_2, which limited the use of some early spectrometers. Field tests of the instrument in various parts of the world have produced variable results. Low–level atmospheric tests were carried out in North America and Australia using vehicle–mounted instruments with air intakes about 1.5 m and 0.3 m respectively above the ground. The North American tests were disappointing: seven Cu mines and twenty Pb-Zn-Ag and Au mines produced atmospheric concentrations no higher than about the detection limit of the spectrometer (5 ng/m³). Similar results were recorded in the Sudbury, Kirkland Lake, and Noranda areas of Canada, but 10–12 ng/m³ were recorded at Cobalt, Ontario. Higher levels were recorded over two Hg mines northeast of Phoenix, but it was apparent that anomalous concentrations fluctuate rapidly. Results in the Pilbara region of Western Australia using a more sensitive spectrometer (1 ng/m³) and an air intake closer to the ground registered anomalies of 2-4 ng/m³ over Cu-Zn mineralization, but results were sensitive to temperature, wind, and moisture. These results seem to preclude airborne mercury sniffing and severely limit the potential of even ground vehicle detectors.

In contrast, experiments with Hg contained in *soil gas* have shown much more promise, results being less affected by short–term meteorological changes. One approach is to insert an inlet tube into a soil auger hole, plug the hole (e.g., with an inflatable bladder), then pump soil gas through a portable spectrometer for immediate analysis. Improved sensitivity is obtained if, instead of performing field analyses, a silver or gold foil is inserted in the pumping line to collect Hg by amalgamation from a relatively large volume of soil gas; the collected Hg is subsequently evaporated from the collector and analyzed under laboratory conditions. This approach has been tested under laboratory conditions and recommended by McNerney and Buseck (1973). A third, and similar, approach is to omit pumping soil gas by leaving a gold or silver foil collector for several hours or days suspended in an auger hole or placed in plastic hemisphere above the soil. The collector methods improve sensitivity and precision but reduce the speed of operation offered by field analysis. Consequently, the main attraction of Hg sniffing relies on the potential for discovery of deeply buried blind deposits (cf. Sakrison, 1971) rather than speed of operation. There has been a significant shift in attention from analysis of Hg in gas samples at the reconnaissance stage to that of transient Hg in bedrock samples at the detailed follow-up stage.

Alpha Scintillometry

Field analysis of Rn^{222} in soil gas using α scintillometry has received a good deal of attention in U prospecting. Other radiogenic gases, He, Kr, and Xe, have received less attention, mainly because they have very short half-lives. The simplest method of U prospecting is *gamma-ray surveying* (using the daughter product Bi^{214}), but much less than a meter of soil cover is sufficient to shield such gamma radiation emanating from a U deposit. Rn, being a gaseous daughter product of U, has the potential to migrate through overburden and outline hidden deposits. For example, Andrews and Wood (1972) have calculated that Rn^{222} with a half-life of about 4 days, is theoretically capable of diffusing 6.5 m through soil to produce a discernible anomaly; in practice, they have detected U deposits concealed by 6-9 m of overburden. Caneer and Saum (1974) have claimed even greater actual penetration in describing a discovery of U in sandstone 100–120 m below the surface.

A possible explanation of the discrepancy between the theoretical diffusion distance and the apparent diffusion distance is that several daughter products precede Rn in the decay series from U, including Ra, which has a long half-life and is moderately mobile. Stevens et al. (1971) provided details of three other case histories, including both successful and unsuccessful soil gas surveys and an attempted, but inconclusive, atmospheric gas sampling survey. Measurement of Rn in soil gas seems well suited to prospecting for U in arid regions, especially during enhanced emission periods in dry weather following rain, which temporarily traps Rn in the soil (Andrews and Wood, 1972). In damp regions Rn emission into soil gas or the atmosphere is inhibited.

Leakage Halos

Prospecting for oil and gas on the basis of *leakage halos* of hydrocarbons is another form of atmogeochemical prospecting (McCrossan and Ball, 1971; Poll, 1975; Devine and Sears, 1975). Samples are collected typically from deep horizons in soil or from boreholes in bedrock. Gas may be collected directly or evolved artificially from soil or rock samples in the laboratory for analysis by gas chromatography to identify the abundance and character of hydrocarbon compounds contained.

Gaseous Dispersion Patterns

Other gases of potential use in geochemical exploration include I_2, SO_2, H_2S, CO_2, N_2, H_2, He, and Ar. I_2 accompanies some sulfide deposits (e.g., porphyry copper), and it can be detected in atmospheric or other gas samples using an *atomic absorption* approach (Barringer, 1971). No significant results appear to have been published.

Urban pollution problems have led to the development of mobile, sensitive SO_2 detectors that may prove useful in locating oxidizing sulfide deposits in regions remote from industrial pollution. SO_2 can be detected also by dogs trained to sniff out oxidizing sulfides, and promising experiments have been conducted in Finland, Sweden, Canada, and Norway.

CO_2 determinations may be a suitable approach to detecting sulfide deposits that oxidize within carbonate host rocks, produce acids, and thereby, decompose carbonates. Sulfides in noncarbonate host rocks may still be detectable by gas surveys based on the ratio of O_2 (consumed by oxidation) to CO_2 (unaffected).

The gases He, Ar, N_2, Ne, H_2, and CO_2 have been the subject of Russian attempts to use gaseous dispersion patterns in prospecting (Ovchinnikov et al., 1973; Eremeev et al., 1973). In some cases the gases are used as direct pathfinders, but in other cases they are used to seek deep structural weaknesses thought to constitute favorable zones for mineralization. Actual sampling seems to be conducted mainly in bedrock using deep boreholes.

The most recent development in atmogeochemical prospecting appears to be airborne detection of organometallic compounds of Hg, Cu, Pb, Zn, Ag, Ni, and Co using *correlation spectroscopy* (Levinson, 1974). The potential of this approach is highlighted further by results obtained by Curtin et al. (1974), who detected 29 metals in exudates from coniferous trees. These metals would enter the atmosphere as aerosols or organometallic vapors.

Finally, two atmospheric, but not strictly atmogeochemical, prospecting techniques deserve comment in this section because of their similar aims in providing regional airborne prospecting procedures. The first is *airborne gamma-ray spectrometry* (e.g., Schwarzer and Adams, 1973), which plots the surface distribution of U, K, and Th; the second is airborne sampling of aerosols and particulate matter entering the atmosphere from soils, vegetation, water, and biological activities (Weiss, 1971).

A. S. JOYCE

References

Andrews, J. N., and D. F. Wood, 1972, Mechanism of radon release in rock matrices and entry into groundwaters, *Inst. Mining and Metallurgy Trans.* **81**, B206–209.

Barringer, A. R., 1966, Interference-free spectrometer for high-sensitivity mercury analyses of soils, rocks and air, *Inst. Mining and Metallurgy Trans.* **75B**, 120–124.

Barringer, A. R., 1971, Optical detection of geochemical anomalies in the atmosphere, *Canadian Inst. Mining and Metallurgy Spec. Vol.* **11**, 474.

Caneer, W. T., and N. M. Saum, 1974, Radon emanometry in uranium exploration, *Mining Eng.* May, 26–29.

Curtin, G. C., H. D. King, and E. L. Mosier, 1974, Movement of elements into the atmosphere from coniferous trees and subalpine forests of Colorado and Idaho, *Jour. Geochem. Explor.* **3**, 245–263.

Devine, S. B. and H. W. Sears, 1975, Experiment in soil geochemical prospecting for petroleum, Della Gas Field, Cooper Basin, *APEA Jour.* 103–110.

Eremeev, A. N., V. A. Sokolov, A. P. Solovov, and I. N. Yanitskii, 1973, Application of helium surveying to structural mapping and ore deposit forecasting, in M. J. Jones, ed., *Geochemical Exploration, 1972.* 183–192.

Hadeishi, T. and R. D. McLaughlin, 1966, Hyperfine Zeeman effect atomic absorption spectrometer for mercury, *Science* **174**, 404–407.

Hatch, W. R. and W. L. Ott, 1968, Determination of sub-microgram quantities of mercury by atomic absorption spectrophotometry, *Anal. Chemistry* **40**, 2085–2087.

James, C. H., and J. S. Webb, 1964, Sensitive mercury vapour meter for use in geochemical prospecting, *Inst. Mining and Metallurgy Trans.* **73**, 633–641.

Levinson, A. A., 1974, *Introduction to Exploration Geochemistry.* Calgary: Applied Publ. Ltd., 612p.

Ling, C., 1958, Portable atomic absorption photometer for determining nanogram quantities of mercury in the presence of interfering substances, *Anal. Chemistry* **40**, 1876–1878.

McCarthy, J. H., et al., 1969, Mercury in soil gas and air—a potential tool in mineral exploration, *U.S. Geol. Survey Circ. 609*, 16p.

McCrossan, R. G., and N. L. Ball, 1971, An evaluation of surface geochemical prospecting for petroleum,

Caroline area, Alberta, *Canadian Inst. Mining and Metallurgy Spec. Vol.* 11, 529–536.

McNerney, J. J., and P. R. Buseck, 1973, Geochemical exploration using mercury vapor, *Econ. Geology* 68, 1313–1320.

Ovchinnikov, L. N., V. A. Sokolov, A. I. Fridman, and Yanitskii, I. N., 1973, Gaseous geochemical methods in structural mapping and prospecting for ore deposits, in M. J. Jones, ed., *Geochemical Exploration 1972,* 177–182.

Poll, J. J. K., 1975, Onshore Gippsland geochemical survey: a test case for Australia, *APEA Jour.* pp. 93–101.

Robbins, J. C., 1973, Zeeman spectrometer for measurement of atmospheric mercury vapour, in M. J. Jones, ed., *Geochemical Exploration 1972.* London: Institute Mining and Metallurgy.

Sakrison, H. C., 1971, Rock geochemistry—its current usefulness on the Canadian Shield, *Canadian Inst. Mining Metallurgy Bull.* 64, 28–31.

Schwarzer, T. F., and J. A. S. Adams, 1973, Rock and soil discrimination by low altitude airborne gamma-ray spectrometry in Payne County, Oklahoma, *Econ. Geology* 68, 1297–1312.

Stevens, D. N., G. E. Rouse, and R. H. De Voto, 1971, Radon-222 in soil gas: three uranium exploration case histories in the western United States, *Canadian Inst. Mining Metallurgy Spec. Vol.* 11, 258–264.

Vaughn, W. W., 1967, A simple mercury vapor detector for geochemical prospecting, *U.S. Geol. Surv. Circ. 540,* 8p.

Weiss, O., 1971, Airborne geochemical prospecting, *Canadian Inst. Mining and Metallurgy Spec. Vol.* 11, 502–514.

Cross-references: *Exploration Geochemistry; Exploration Geophysics; Indicator Elements.* Vol. XII, Pt. 1: *Aeration, Respiration and Atmosphere; Diffusion Processes.*

AUGERS, AUGERING (SOIL)

A common and relatively inexpensive method of obtaining disturbed soil samples is by auger borings. The depth to which auger borings may be taken is controlled by the depth to which the soil will not collapse and close an uncased hole or by the power limitations of the auger used (Knill, 1978; Peters, 1978). Auger holes are impossible to bore without casing in flowing sands or highly plastic soils below the water table.

Hand Augers

Hand augering is the simplest, most portable method of obtaining subsoil samples. Since the auger must be turned into and removed from the ground by hand, these borings are usually limited to 2 m or 3 m in depth, though on occasion, hand auger borings are made to depths of 5 m or more.

Power Augers

There are two general types, the two-person portable power auger, which is a simple extension of the hand auger and is limited to much the same depths as hand augering, and the truck- or trailer-mounted power auger. Truck- or trailer-mounted augers, while being much less portable, have the necessary power to go through much more compact soils and to much greater depths.

There are two main types of power augers. They are a single-flight cutter and the continuous flight. *Single flight cutters* have a cutter head mounted on the spindle of the powering device. As the machine rotates, the cutter head is lowered into the ground, penetrates into the soil for 10–30 cm, and is raised back out of the ground. After removing the soil accumulated atop the cutter assembly, the head is lowered again into the hole, repeating the cycle until the desired depth has been attained.

The fastest method of auger boring is *continuous-flight augering.* The cutter head is attached to the leading section of flights. This head cuts a hole about 5 cm larger in diameter than the flights that follow. The flights, acting as a screw conveyor, bring the cuttings to the surface where they are cleared away from the hole.

This method is fast and has the advantages of speed and the ability to go deeper than a single-flight cutter since there is no problem of a collapsing hole. The main drawback in continuous-flight augering is the accurate determination of sample depth because of the soil movement up the flight during rotation and soil falling on the flight off the open sidewalls.

An important variation of the continuous-flight auger is the *hollow stem auger.* As the name implies, the flights have a hollow stem and a point that can be pulled. This allows accurate disturbed or undisturbed samples to be taken at any desired depth.

With the exception of the hollow stem auger, all sampling from conventional auger holes results in disturbed samples (Waterman and Hazen, 1968). Because of the limitations of auger borings, they are normally used for preliminary surveys or in conjunction with a program of other types of borings.

Auger borings cannot be made through soils containing boulders or through hard rock. Continuous-flight borings, which can go deeper than any other type of auger boring, are limited to depths of 50 m or less.

When more detailed information or greater depths are required or when soil conditions do not permit the use of augers, subsurface investigations are conducted by cased borings or rotary borings (see *Borehole Drilling*). In borings of these types it is possible to obtain undisturbed samples by several methods, most commonly with a thin walled tube sampler.

WILLIAM M. VOORHIS

References

Compton, R. R., 1962, *Manual of Field Geology.* New York: Wiley, 378p.

Jumikis, A. R., 1962, *Soil Mechanics.* New York: Van Nostrand, 791p.

Knill, J. L., 1978, *Industrial Geology.* London: Oxford University Press, 344p.

McKinstry, H. E., 1948, *Mining Geology.* Englewood Cliffs, N.J.: Prentice–Hall 680p.

Peters, W. C., 1978, *Exploration and Mining Geology.* New York: Wiley, 696p.

Waterman, G. C., and S. Hazen, 1968, Development drilling and bulk sampling, in E. P. Pfeider, ed., *Surface Mining.* New York: American Institute of Mining, Metallurgical and Petroleum Engineers, 89.

Cross-references: *Borehole Drilling; Exploration Geology; Foundation Engineering; Pipeline Corridor Evaluation; Soil Sampling; Urban Engineering Geology.*

B

BANK STABILIZATION—See CANALS AND WATERWAYS, SEDIMENT CONTROL.

BATHYMETRIC SURVEY—See ACOUSTIC SURVEYS, MARINE; HARBOR SURVEYS; MARINE MAGNETIC SURVEYS; SEA SURVEYS. Vol. XIII: COASTAL ENGINEERING; OFFSHORE NUCLEAR PLANT PROTECTION.

BIOGEOCHEMICAL PROSPECTING— See BIOGEOCHEMISTRY.

BIOGEOCHEMISTRY

Botanical methods of prospecting involve the use of vegetation in searching for ore deposits. Although these methods have been used for several centuries, there is much confusion about terminology because there are two distinct methods of botanical prospecting. *Geobotanical methods* (see *Geobotanical Prospecting*) are visual and rely mainly on interpretation of the vegetative cover to detect mineralization. *Biogeochemical methods* are entirely different in scope and involve chemical analysis of the vegetation. This entry considers this latter method.

Biogeochemical methods were first used just before World War II when Tkalich (1938) found that vegetation could be used to delineate a Siberian iron ore body. Since then, much biogeochemical work has bee carried out in the USSR (Kovalevsky, 1987; Malyuga, 1964), Canada (Warren and Delavault, 1950), the United States (Cannon and Starrett, 1956), and Australasia (Cole et al. 1968; Brooks, 1983).

The Biogeochemical Cycle

Analysis of the accumulation of elements in vegetation and the upper humic layer of soils is the basis of biogeochemical prospecting. The mechanisms whereby plants accumulate trace elements are extremely complicated but, in essence, involve uptake via the root system, the passage of the elements through the aerial parts of the plants into organs such as the leaves and flowers, and finally, a return of these elements to the upper layers of the soil when the leaves or flowers wither and fall. The elements are then leached (q.v. in Vol. XII, Pt. 1) through the various soil horizons and are reaccumulated by vegetation in a series of events known as the *biogeochemical cycle* (q.v. in Vol. XII, Pt. 1, individual elemental cycles).

Trees and shrubs with extensive root systems can penetrate beneath a nonmineralized overburden and under favorable conditions can indicate the existence of minerals at depth. This possibility is the greatest advantage of the biogeochemical method, but it is worthless unless accumulation of trace elements from depth is achieved in a *reproducible* manner and to a degree that is *proportional* to the concentration of the element or elements that are sought. A large number of factors can affect the accumulation of elements by plants, and these are summarized in Table 1 together with ways of reducing the consequent adverse effect on the reliability of the biogeochemical method.

Chemical Analysis of Vegetation

Advances in biogeochemical prospecting have of necessity always been linked with the progress of chemical analysis. For this reason, the method was developed much later than the geobotanical technique. Four main methods of analysis are now used: colorimetry, emission spectrography, atomic absorption spectrophotometry, X-ray fluorescence.

Colorimetry depends on the visual or spectrophotometric determination of colors developed with reagents specific for certain elements. One of the most successful reagents is dithizone, which unfortunately forms colored complexes with at least twenty elements. Adjustment of the pH with buffers makes the procedure much more specific, but serious interferences can result unless the analytic conditions are standardized exactly. Although colorimetry with dithizone has now been replaced almost entirely by atomic absorption and ICP spectrometry, reagents exist that are still used frequently for some elements such as molybdenum and tungsten that are not easily determined by other methods. An example of this is the use

TABLE 1. Factors Affecting Elemental Uptake by Plants and Methods of Reducing Their Effect

Factor	Relative Importance	Methods of Reducing Effect
Type of plant	Very great	Selection of plants by orientation survey
Organ sampled	Great	Selection of barrier–free* organs by orientation survey
Age of organ	Significant	Selection by orientation survey
Root depth	Significant	Use of ratios of two elemental concentrations in same sample
pH	Fairly significant	Selection of elemental ratios of pair of elements of some availability over pH range encountered
Health of plant	Fairly significant	Selection of health specimens only
Drainage	Fairly significant	Avoidance of poorly drained areas where possible, also use of elemental ratios
Availability of element	Fairly significant	Use of elemental ratios with two elements of same availability
Antagonism of other elements	Minor	None
Rainfall	Minor	Carrying out work over short period
Variable shading	Minor	Avoidance of shady sites if possible or use of elemental ratios
Temperature of soil	Minor	Carrying out work over short period

Source: After Brooks, 1983.
*See Kovalevsky, 1987.

of dithiol for the colorimetry determination of tungsten and molybdenum. Colorimetric procedures are still useful for simple field tests and will continue to be used in the foreseeable future for specific cases.

Atomic absorption spectrophotometry represents one of the greatest modern advances in chemical analysis. It affords a steady, accurate, sensitive, and inexpensive way of analyzing a wide range of elements in many types of samples. It does, however, suffer from the disadvantages that not all elements may be determined by its use and, moreover, the necessity of bringing the sample into solution. Nevertheless, there can be little doubt that this technique is by far the most extensively used procedure for chemical analysis of vegetation and will remain so for many years.

Emission spectrography which involves either arcing a solid sample in a suitable type of electrical discharge, or nebulizing a solution into a plasma source (ICP) followed by analysis of the radiation via a spectrometer with prism or grating optics, is also used extensively for the analysis of vegetation. Although optical instruments cannot handle samples at the same speed as atomic absorption units, this disadvantage has been offset by the development of direct–reading plasma or dc–arc instruments that can analyze a solid or liquid sample for up to 25 elements simultaneously and represent the ultimate in speed of analysis. Unfortunately, such instruments are extremely expensive, and their use is scarcely justified when only a few elements are to be analyzed.

X-ray fluorescence is a nondestructive procedure that can be used for the simultaneous analysis of plant material for many elements. The procedure is not especially sensitive but is particularly useful if many elements are to be determined.

Sample Preparation and Analysis by Atomic Absorption or Plasma Emission (ICP) Spectrometry

Since in the majority of cases chemical analysis of plant material is carried out by atomic absorption or ICP spectrophotometry, the following method of preparation of plant material for analysis is designed to suit the latter instrumental techniques.

About 50 g of plant material is washed and dried overnight at 110°C in a drying oven. The dry weight is recorded and the dried material is ground in a small mill (e.g., hammer mill) and thoroughly mixed. About 1 g of material is placed in a 25 ml Pyrex squat beaker and ashed at just under 500°C. About 2–3 hours will be required for this operation. There has been some discussion about the relative merits of dry ashing and wet ashing of plant material (Brooks, 1983; Mitchell, 1964). Wet ashing has the disadvantage of using highly explosive perchloric acid and the further danger of contamination from the added mineral acids. Dry ashing is much simpler and cleaner, although the loss of volatile elements such as lead and cadmium can occur. If ashing temperatures do not exceed 500°C and provided that there is adequate air during the ignition stage, losses of these and most other elements will not be significant. If determination of very volatile elements such as arsenic or selenium is to be performed, then wet ashing must be used.

After ignition, 1 g of ash is weighed into a small test tube and digested with 10 ml 2M hydrochloric acid (prepared from redistilled, constantly boiling 6M acid). The tubes should be placed for 5 min in a rack immersed in hot water. This will ensure complete dissolution of the ashed vegetation, and the samples are then ready for chemical analysis by atomic absorption spectrophotometry. If the data are to be reported on an ash weight rather than dry weight basis, the vegetation can be ashed immediately without the necessity of predrying and recording dry weights.

Essential and Nonessential Elements in Biogeochemical Prospecting

From the very earliest days, biogeochemists had noticed that certain ore elements were difficult to detect by the biogeochemical method. Whereas investigators had often been successful in detecting elements such as uranium and nickel, other elements such as copper and zinc posed problems. This difficulty was interpreted by Timperley et al. (1970), who showed that plants have a regulatory mechanism that ensures a constant level of elements essential to plant growth and yet allows a more unrestricted entry of nonessential elements. Concentrations of nonessential elements in plants are thus more likely to reflect the nature of the substrate than in the case of essential elements such as copper and zinc. Kovalevsky (1987) has, however, addressed the problem by selecting plants or plant organs that have no barrier (i.e., barrier-free) to uptake of certain elements. These observations do not necessarily indicate that the biogeochemical method will always be a failure for the essential elements, but they do indicate that researchers can expect a higher failure rate than in the case of the nonessential elements.

Biogeochemical Orientation and Exploration Surveys

Like most geochemical methods, the biogeochemical technique usually requires a preliminary orientation survey in an area of known mineralization before it can be applied to an unknown area (see *Exploration Geochemistry*). The ideal situation is where bedrock mineralization is masked by an overburden of transported soil or where a geochemical barrier such as siliceous hardpan results in soil samples that are not indicative of the true nature of the bedrock. Ideally, bedrock samples in a region of known geology should be compared with plant and soil elemental concentrations by use of a suitable statistical procedure such as correlation analysis. If vegetation

data agree more closely with bedrock data than do the soil data, the plant may then be sampled in an area of unknown geology.

It is preferable to select two or three species since it is most unlikely that any one species will be found at each point on a preselected uniform grid pattern of sampling sites. When several plants are used, levels can be standardized by dividing each elemental concentration in a plant by the appropriate plant–soil absorption coefficient (PSC), which is the concentration of an element in the plant divided by the concentration of the same element in the soil (Brooks, 1973; Kovalevsky, 1987). This formula then gives a predicted value for the appropriate element in the bedrock at that particular site. The biogeochemical data can now be used to plot isoconcentration contours or subjected to trend surface analysis (Agterberg, 1964) or other statistical procedures in the same way as other geochemical data.

Multielement Analysis and Biogeochemical Prospecting

The potential scope of the biogeochemical method has been greatly extended in recent years by two simultaneous developments in instrumentation. The first of these is the appearance of sophisticated direct-reading dc–arc and ICP spectrometers that can analyze simultaneously over a score of different elements in a given sample. The interpretation of these multielement data has been rendered possible by modern computers that can use some form of discriminant analysis for this interpretation.

If plant material is analyzed for only one element, the species concerned may not always indicate the presence of mineralization in the substrate because of the natural variation of this element from specimen to specimen. If, however, several elements can be determined simultaneously and if these are either ore elements or at least suitable pathfinder elements for the mineralization that is sought, then the prospects for success with the biogeochemical method are enhanced.

The work of Nielsen et al. (1973) illustrates this situation. Nine elements were analyzed in the bark of *Eucalyptus lesouefii* growing over two types of substrates (amphibolites and ultrabasic rocks) in Western Australia. Although *mean* concentrations of chromium and nickel were appreciably higher over ultrabasics than over amphibolites, several individual specimens did not show this pattern for both elements. Investigators used a computer to carry out discriminant analysis of the biogeochemical data. The area was divided into 63 quadrats (see *Geobotany*) of which 34 were amphibolites and 29 were ultrabasics. For each rock type, a regression equation of the following form was calculated

$$Y_a = l_1x_1 + l_2x_2 + l_3x_3 \ldots l_9x_9 + C_a.$$

The variables x_1–x_9 were the concentrations of each of the nine elements for specimens of *E. lesouefii* (in this case growing over amphibolites), and l_1–l_9 were coefficients selected by the computer to maximize the difference in values of Y_a (amphibolites) and Y_b (ultrabasics). The extent of this difference can be measured by the so-called Mahalanobis D^2 statistic (Mahalanobis, 1936). This procedure was also used for discrimination of geobotanical data from the same area (see *Geobotany*). The data are shown in Table 2.

Note that using multielement analysis, it was possible to predict the identity of up to 29 of 34 amphibolite sites and 18 of 29 ultrabasics. The method was not particularly useful for prediction of the nature of ultrabasic quadrats since it should be possible to predict *by chance* 14 or 15 quadrats in 29. Note also (see *Geobotany*) that the geobotanical method was much more successful in this area because of the greater number of variables (36) used.

Advantages and Disadvantages of Biogeochemical Prospecting Methods

Advantages

1. Large trees can sometimes penetrate thick overburden to give evidence of mineralization at depth. An effective depth of at least 30 m (100 ft) has been reported for the juniper in prospecting for uranium in the Colorado Plateau.
2. Plants with extensive root systems can effectively sample a greater soil volume than the size of samples normally collected in soil surveys.
3. In densely overgrown areas with a deep humus layer in the soil, plant sampling is easier and quicker than soil sampling.
4. Plant samples are usually appreciably lighter than soil samples, a practical factor that is useful in fieldwork.
5. Chemical analysis of plant ash is sometimes freer from interference problems since the concentrations of elements such as iron, which frequently cause problems in chemical procedures, are substantially less in plant ash than in rocks or soils.
6. In cases where effective soil sampling depends on the selection of a particular horizon, plant sampling can be much simpler since it depends only on the selection of a particular easily available and identifiable plant organ.
7. In cases where the soil is either almost nonexistent or complicated by the presence of features such as a siliceous hardpan, biogeochemical prospecting can be more advantageous than soil sampling.
8. The method is particularly useful in areas of permafrost or in cold regions frozen for a large part of the year.

TABLE 2. Use of Discriminant Analysis of Biogeochemical Data to Predict the Nature of the Substrate

Variables Used (Elemental Concentrations in Bark of *E. Lesouefii*)	D^2	Amphibolites (34 Quadrats)	Ultrabasics (29 Quadrats)
Ca	0.122	17	17
Co	1.66	23	15
Cr	9.36	24	14
Cu	2.18	24	15
Mg	0.789	24	13
Mn	0.165	21	10
Ni	3.15	24	14
Pb	1.22	20	13
Zn	3.76	26	14
Ca, Co, Cr, Cu, Mg, Mn, Ni, Pb, Zn	16.76	29	17
Cr, Ni	10.14	27	15
Cr, Ni, Zn	10.62	25	14
Co, Cr, Ni	13.49	26	16
Cu, Ni	3.56	23	16
Cr, Mn, Ni	10.15	27	14
Ca, Cr, Ni	11.93	27	13
Cr, Mg, Ni	13.92	25	18
Cr, Ni, Pb	10.51	27	16

Source: After Brooks, 1983.

Disadvantages

1. Plant sampling demands more skill and experience from the fieldworker, particularly as far as recognition of species is concerned.
2. Unknown factors such as pH, drainage, aspect, and age of organ can greatly affect the reliability of the method unless they can be controlled or allowed for.
3. The method depends on an even distribution of the selected species in the exploration area or at least the availability of more than one species for prospecting.
4. Biogeochemical prospecting methods usually require an orientation survey over a known anomaly before they can be used with confidence in a new area. The same is admittedly true for soil sampling where a decision has to be made as to which horizon should be sampled. Such soil surveys are, however, simpler than those involving vegetation.
5. Sometimes the method can be applied only at certain seasons of the year (e.g., during the growing period, if sampling of deciduous leaves is involved).
6. Biogeochemical methods are not invariably applicable, whereas soil sampling methods usually are.
7. Plant samples are much more subject to contamination than are soils.
8. Elements essential in plant nutrition can sometimes cause difficulties in biogeochemical prospecting.

This listing is far from complete, but we may conclude that the greatest single advantage of this technique is its penetrating power and that its greatest disadvantage is its variability caused by factors difficult to control.

Latest Developments in the Field

A new development in biogeochemical prospecting involves the analysis of herbarium material. Geobotanical use of herbarium material has been done by Persson (1956) and is described in this volume (in *Geobotanical Prospecting*). Until recently, however, chemical analysis of herbarium material had not been feasible because of the size of sample needed for classical methods of analysis (5–10 g of plant material). With atomic absorption and plasma emission (ICP) spectrophotometry, however, it is now possible to analyze a small leaf sample (1 ml or less, weight about 0.02 g) for several elements. Brooks et al. (1977), e.g., used herbarium material to analyze over half of all 400 species of the genera *Hybanthus* and *Homalium* and identified several new *hyperac-cumulators* of nickel (>0.1% nickel in dried leaves), all of which were associated with nickeliferous substrates. The results showed that it was possible to identify most of the world's major ultrabasic areas within the tropical and warm-temperate zones by means of elevated nickel concentrations in the plants. This work has now been extended to a biogeochemical survey of eastern Indonesia using herbarium specimens and resulted in the identification of a new hyperaccumulator of nickel, *Rinorea bengalensis*. Further specimens of this species collected from throughout southeast Asia were then analyzed and resulted in the identification of a previously unknown ultrabasic area in West Irian (Brooks and Wither, 1977). This latter specimen had been collected by two Japanese botanists in 1940, though its significance remained unknown for over 25 yr. Other herbarium work on species from central Africa (Brooks, 1977) has resulted in the identification of a hyperaccumulator of cobalt (*Haumaniastrum robertii*), which is only the second known hyperaccumulator of this element—i.e., in addition to the previously known *Crotalaria cobalticola* (Duvigneaud, 1959).

There is an inherent paradox in herbarium work. The more successful it becomes, the greater the demands that will be placed on herbaria and the greater the resistance of curators to furnishing further material. This reluctance to disturb specimens can be answered to some extent by use of progressively smaller specimens (2–3 mg), but then the problem arises of whether such a small sample is representative of the whole.

An exciting new biogeochemical development is the use of plant exudates for mineral prospecting. Curtin et al. (1974) showed that condensed exudates from common conifers such as *Pinus contorta*, *Picea engelmannii* and *Pseudosuga menziesii* showed the presence of a large number of trace elements transported for the substrate. The authors suggested air sampling of these exudates as a tool in mineral exploration. Further work in this field was carried out by Beauford et al. (1975), who also suggested the possibility of airborne sampling programs.

Because new ore deposits are becoming progressively more difficult to find, and because such deposits are likely to be found in the more inaccessible (often well-vegetated) regions of the Earth, it is clear that the biogeochemical method will continue to have a place in future prospecting operations. It is not likely that it will be used by itself, but if used judiciously in combination with other methods, it should continue to be a useful component of the exploration geochemist's armory of techniques.

ROBERT R. BROOKS

References

Agterberg, R. P., 1964, Computer techniques in geology, *Earth-Sci. Rev.* **3**, 47-77.

Beauford, W., J. Barber, and A. R. Barringer, 1975, Heavy metal release from plants into the atmosphere, *Nature* **256**, 35-37.

Brooks, R. R., 1973, Biogeochemical parameters and their significance for biogeochemical prospecting, *J. Appl. Ecol.*, **10**, 825-836.

Brooks, R. R., 1977, Cobalt and copper uptake by the genus *Haumaniastrum, Plants and Soil* **48**, 541-545.

Brooks, R. R., 1983, *Biological Methods of Prospecting for Minerals.* New York: Wiley, 322p.

Brooks, R. R., and E. D. Wither, 1977, Nickel accumulation by *Rinorea bengalensis* Wall. O.K., *Jour. Geochem. Explor.* **7**, 295-300.

Brooks, R. R., F. Lee, R. D. Reeves, and T. Jaffre, 1977, Detection of nickeliferous rocks by analysis of herbarium specimens of indicator plants, *Jour. Geochem. Explor.* **7**, 49-57.

Cannon, H. L. and W. H. Starrett, 1956, Botanical prospecting for uranium on La Ventura Mesa, Sandoval County, New Mexico, *U.S. Geol. Surv. Bull.* **1009M**, 391-407.

Cole, M. M., D. M. J. Provan, and J. S. Tooms, 1968, Geobotany, biogeochemistry and geochemistry in the Bulman-Waimuna Springs area, Northern Territory, Australia, *Inst. Mining and Metallurgy Trans. Sec B*, **77**, 81-104.

Curtin, G. C., H. D. King, and E. L. Mosier, 1974, Movement of elements from coniferous trees in subalpine forests of Colorado, *Jour. Geochem. Explor.* **3**, 245-263.

Duvigneaud, P., 1959, Plant cobaltophytes in Upper Katanga (in Fr.), *Royal Bot. Soc. Belgium Bull.* **91**, 111-134.

Kovalevsky, A. L., 1987, *Biogeochemical Exploration for Mineral Deposits.* Utrecht: VNU Science Press, 220p.

Mahalanobis, P. C., 1936, On the generalized distance in statistics, *India Natl. Inst. Sci. Proc.* **12**, 49-55.

Malyuga, D. P., 1964, *Biogeochemical Methods of Prospecting,* New York: Consultants Bureau, 205p.

Mitchell, R. L., 1964, The spectrochemical analysis of soils, plants and related materials, *Commonwealth Bur. Soil Sci. Tech. Commun.* **44A**, 225p.

Nielsen, J. S., R. R. Brooks, C. R. Boswell, and N. J. Marshall, 1973, Statistical evaluation of geobotanical and biogeochemical data by discriminant analysis, *Jour. Appl. Ecology* **10**, 251-258.

Persson, H., 1956, Studies of the so-called "copper mosses," *Hattori Bot. Lab. Jour.* **17**, 1-18.

Timperley, M. H., R. R. Brooks, and P. J. Peterson, 1970, The significance of essential and non-essential trace elements in plants in relation to biogeochemical prospecting, *Jour. Appl. Ecology* **7**, 429-439.

Tkalich, S. M., 1938, Experience in the investigation of plants as indicators in geological exploration and prospecting (in Russian), *Akad. Nauk SSSR Fil., Vest. Dal'nevost.* **32**, 3-25.

Warren, H. V., and R. E. Delavault, 1950, History of biogeochemical prospecting in British Columbia, *Canadian Inst. Mining and Metallurgy Trans.* **53**, 236-242.

Cross-references: *Atmogeochemical Prospecting; Exploration Geochemistry; Geobotanical Prospecting; Hydrogeochemical Prospecting; Lithogeochemical Prospecting; Pedogeochemical Prospecting; Prospecting; Vegetation Mapping.*

BIOGEOLOGY—See GEOMICROBIOLOGY; GEOBOTANICAL PROSPECTING; LICHENOMETRY.

BLASTING, EXPLOSIVES—See BLASTING AND RELATED TECHNOLOGY; CRATERING.

BLASTING AND RELATED TECHNOLOGY

Uses of Explosives

Military uses of explosives are legion. In civil projects, they are utilized for fragmentation or displacement of geologic materials in surface, underground, and underwater construction (Langefors and Kihlstrom, 1978); in quarrying and mining (Lewis and Clark, 1964); in the stimulation of water (Walton, 1970), oil, and gas wells (McLamore, 1973); and in the preparation of ore for in situ leaching (Navarro, 1974) or of oil shale for in situ retorting (Miller et al., 1974). For these applications, nuclear as well as conventional blasting has been proposed (Teller et al., 1968). Other uses are in controlling avalanches (LaChapelle, 1966); breaking ice and log jams; removing sunken vessels from shipping channels; cutting timbers, steel plate, and structural shapes; demolishing brick, concrete, or steel structures, heavy industrial equipment, and castings; and preparing land for agricultural use (du Pont, 1969).

Capable of rapidly drilling holes and removing rock, *shaped charges* are routinely used for furnace tapping in the steel industry, for oil well casing perforation (Clark et al., 1971), for submarine trenching without drilling blast holes, and for severing submerged piles (Gregory, 1973). An array of industrial fabrication and metallurgical processes are facilitated by *explosive forming* or metal working (Ezra, 1973), and explosive-actuated devices and fasteners are regularly used in contemporary mechanical and architectural designs (Manon, 1977). Generally, the grander the scale of civil use for explosives, the greater the value to *applied geology.*

Types of Explosives

The basic types of explosives are *mechanical,* exemplified by a contained gas that explodes when heated, types of which find use in subsurface coal mining; *chemical,* which comprise most dry-packaged commercial types and liquid-solid mixes used in construction and mining; and *atomic,* either of the nuclear fission or thermonuclear fusion variety. The chemical types, of principal interest here, are classified as *detonating* high explosives, characterized by internal development of a shock wave propagated by energy of the chemical reaction, or *deflagrating* low explosives, which burn rapidly enough to generate large volumes of violently expanding gas but fall short of generating a shock (Cook, 1958).

High explosives are further categorized as *primary* or *secondary.* Primary explosives can be detonated through simple ignition by any adequate source of heat, whereas secondary explosives require a primary *detonator* and sometimes a *booster* as well (Manon, 1977). Some *blasting agents* that incorporate ammonium nitrate (AN) are categorized apart from explosives because they are not *cap sensitive* (Dick, 1972). They are, nevertheless, secondary explosives as classified by Cook (1958) and simply require booster charges, or *primers.* But almost all combustible gases and even dusts (Nagy et al., 1968), though not normally used in blasting, are explosive when mixed with air in certain proportions. Consequently, explosives are here limited simply to substances purposely used for blasting.

Development of Explosives Used in Conventional Blasting

The first explosive used was black powder, a 75:13:12 mixture of saltpeter (potassium nitrate), charcoal (carbon), and sulfur. Known before 1260 A.D., its origins are obscure. Though early utilized as a propellant in guns, not until 1613 did Weigel, in Saxony, suggest its application to mining. Even then, it was 1627 before Weindl, in Hungary, used it in a blast and 1689 before it came to the great mining district of Cornwall (du Pont, 1969).

The invention in 1831 of the *safety fuse* by Bickford in England, the discoveries in 1846 of nitroglycerine (NG) by Sobrero in Italy and nitrocellulose (guncotton) by Schonbein in Switzerland, and the substitution in 1857 of less expensive sodium nitrate for potassium nitrate in black powder by du Pont in the United States brought notable developments in explosives. Among these were inventions by Nobel, in Sweden, such as the mercury fulminate *blasting cap,* or detonator in 1864 that, with a safety fuse, facilitated the use of NG and other secondary explosives; the mixture of NG (an oily liquid) in a 3:1 ratio with in-

active diatomaceous earth (kieselguhr) as an absorbent in 1867, to form solid *guhr dynamite;* the replacement of diatomaceous earth with an active absorbent of NG in 1869 to produce stronger *straight* dynamite; and the formation of 92–93% NG with collodion cotton, in 1875, to form plastic, conveniently handled, *blasting gelatin,* which is less sensitive than liquid NG (Cook, 1958).

The Swedish chemists Ohlsson and Norrbein patented the first explosive based on AN and AN mixtures in 1867. This development introduced the world's cheapest source of conventional explosive energy, a main ingredient in *permissible explosives* as now allowed in gassy and dusty coal mines in the United States and the forerunner of modern blasting agents. In 1935, the first actual blasting agent, a mixture of AN and a carbonaceous *sensitizer* in a metal container, was patented in the United States. By 1955 the precursors of field-mixed ANFO recipes utilizing fertilizer grade AN prills and fuel oil (FO), so popular today because of their economy, had been patented (though instead of FO, they included 1–12% carbon black).

Most recently, *slurries,* or *water gels*—mixtures of AN and explosive (e.g., TNT) or nonexplosive (e.g., aluminum, paraffin, sugar, carbon) sensitizers, or both, in an aqueous medium with thickeners and waterproofing gelatinization agents—have come into wide use. Tested by Cook in 1956 and first used in the Mesabi Iron Range in 1959, they are more powerful than ANFO and, though more expensive, do not suffer from its lack of water resistance and low density (Dick, 1972). Finally, numerous concurrent developments such as instantaneous, milli-, and multisecond delay electric blasting caps; improved *permissible practices* as well as explosives; and small–diameter, flexible *detonating cord* (with a cloth- or plastic-bound core of the high explosive, PETN) have further increased the safety and effectiveness of blasting.

Explosive Properties Important in Blast Design

Along with overall cost, factors recognized as most pertinent in blast design are the parameters of the explosive to be used, the charge distribution in space and time, and the natural properties of the material to be broken or moved (Atchison, 1968). While the relative *sensitivity* of an explosive is of great importance to operational safety, its *bulk density,* ρ_e and *detonation velocity,* V_e, are among the principal parameters determining its performance. This is because the *brisance,* or shattering action of an explosive, can be identified with these two properties and its detonation pressure, and detonation pressure and available *energy per unit weight,* also of considerable import, are both roughly proportional to $\rho_e V_e$. A more detailed account of explosives, their properties, commercially available types, and civil applications is provided by Manon (1977).

The efficiency of energy transfer from an explosive to the material being blasted depends on how closely its *detonation impedence, $\rho_e V_e$,* matches the *acoustic impedance, $\rho_m V_m$* (density × sonic velocity) of the material (Clark, 1968). Another unit explosive parameter of importance, especially in blasting low–strength, fractured, or unconsolidated materials, is the *volume of gas* generated, since this relates to performance in expansion of cracks and mass movement of fragments. The *fume class* (relating to production of poisonous gases) and *water resistance* can also be critical (Dick, 1968).

Blast Effects and Their Prediction

Early attempts to understand blast effects on Earth materials quantitatively showed these effects to be so complex as to require empirical treatment and to differ, depending on whether the explosion is at the surface (Robinson, 1944), underground (Lampson, 1946), underwater (Cole, 1948), or elsewhere. A major part of the energy of an explosion in air goes into *airblast*, characterized by a spherically diverging *shock* or steep pressure rise and outward air flow, decaying to a negative phase with a reversal of flow. The pressure history or peak airblast *overpressure* at any point depends on its separation distance, or *range;* the shape, type, and weight of the *explosive charge;* and position of each relative to the ground surface (Swisdak, 1975).

An explosion in water is similar to one in air, except the shock is transmitted much more efficiently, and inertial effects cause the *gas bubble* to undergo as many as ten or more repeated cycles of expansion and contraction. In underwater blasting, the effect on surroundings of shock waves transmitted through the water or water-bearing deposits, the hydrostatic pressure, the difficulties of working in and beneath water, and the need for waterproof blasting components must be considered. *Hydrostatic confinement* sometimes makes it possible to shatter a bottom formation by *dobying*, the detonation of charges lowered and placed at regular intervals over the underwater surface.

In Earth solids, explosion effects are notably different in snow (Livingston, 1968), ice (Livingston, 1960), soils (Lampson, 1946), alluvium (Murphey, 1961), and various rocks [e.g., salt (Nicholls and Hooker, 1962), coal (du Pont, 1969), tuff (Short, 1961), limestone (Atchison and Pugliese, 1964), marble (Atchison and Roth, 1961), granite (Nicholls and Hooker, 1965), and taconite (Gnirk and Pfeider (1968)]. The differences are related to the predominant type of failure (compressive, shear, tensile, or *spallation*) ex-perienced by the material and depend largely on whether it tends toward fluid, plastic, or brittle behavior (Fig. 1).

The spatial distribution and timing of individually detonated charges in a *blasting round* or composite blast are important because they collectively control the *burden,* or volume to be broken or moved by each charge; *degree of material confinement,* or number, extent, and relative location of free faces allowing displacement; *fragmentation* (Fig. 2), or extent to which the material will be broken; *coupling,* or extent to which the charges are in direct contact with the material being blasted; and *blasting vibrations* or *particle motions* that may be detrimental to nearby structures. In practice, the allowable burden for a particular explosive in a particular material is best determined empirically, and one of the simpler methods is by *cratering experiments.* Here, constant concentric charges are *stemmed* and detonate at the bottom of increasingly deeper vertical holes until the *optimum depth, D_o,* providing a crater of maximum volume, or the *critical depth, D_c,* giving fragmentation to the surface but no appreciable crater, is determined.

Using *scaling law* (Coates and Gyenge, 1973), where the *burden distance* or minimum range to the nearest *free face, D,* is taken proportional to the cube root of the charge weight, *W,* one has $D = KW^{1/3}$ for any particular idealized crater shape or fragmentation zone. The empirical *scaling factor, K,* is characteristic of the combined explosive and medium properties (Ash, 1968). When determined in the field for the cases where $D = D_o$ or D_c, the scaling factor allows a reasonable estimate of an effective burden distance for any desired charge weight (or vice versa), providing the shot geometry, explosive, and material properties remain the same.

However, to utilize such a scaling factor in arriving at a burden distance or hole spacing for a multiple-charge *bench blast* at the surface, a *tunneling round* underground, or most any practical application, account must be take of all changes in *similitude,* and these are always considerable. In fact, the difficulties are such that the combined effects of various explosives, *hole patterns, charge distributions, firing sequences,* and *stemming* or *decoupling* methods are sometimes determined on a first approximation by *model experiments* prior to field tests. This is true whether effective rock fragmentation and reduced ground vibrations are the primary aim, as in some pit or quarry operations (Bergman et al., 1974), or smooth-finished walls, obtained by *cushion blasting* (with the charges decoupled in stemming) or *presplitting* (with a lightly loaded outer row of initially fired blast holes), are most important, as in some construction and subsurface mine development work (Fennel et al., 1968).

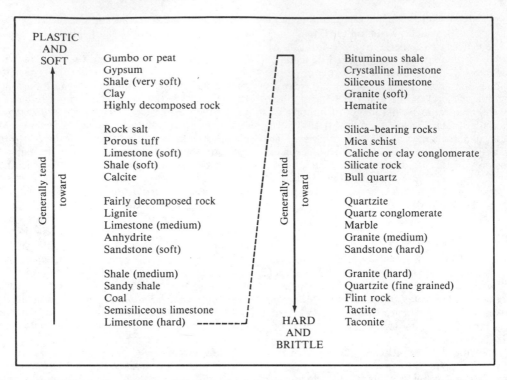

FIGURE 1. Classification of Earth materials for blasting (based on general characteristics).

Importance of Geotechnical Properties and Applied Geology in Blast Design

Stronger, brittler, and unbroken rocks require an explosive with higher energy, detonation velocity, and explosion pressure and, for a given blast configuration, demand a greater *powder factor,* or ratio of explosive to burden. Also, for a constant input of explosive energy, the difficulty of blasting certain types of rocks has been qualitatively shown to increase as their density, sonic velocity, Young's modulus, and tensile or compressive strengths (Gnirk and Pfeider, 1968) increase. However, the pertinent mechanisms and material parameters are complicated enough that no direct quantitative relationships have been found between the measurable properties of explosives and those of a particular rock that will define its *blastability* (Clark et al., 1969).

In large part, the complexities in rock blasting result from the mineralogical, textural, and structural variations, including incipient discontinuities and subtle *anisotropies,* ubiquitously found in *rock masses* but not always apparent in their *substance.* Further complications arise from the fact that rock strength increases variability with both increased confinement and *rate of dynamic loading,* while there is a general increase in efficiency of explosive *energy transmission* but a decrease in

rock strength caused by moisture content (itself variable) or saturation. Likewise, more energy is absorbed by less dense, porous rocks, and *preferred fracture orientations* and difficulties in *block caving* can result from abnormal *in situ stresses,* be they *gravitational, residual,* or *tectonic.*

Among the heterogeneities in all types of rock masses that can be important controls in blasting are microstructures, ore-waste or other contacts, *facies* changes, alteration zones, and variations in grain size, porosity, and hardness (Belland, 1966). Obvious discontinuities such as *faults, joint systems,* and other fractures or even visibly unapparent potential *planes of weakness* commonly control the *burden geometry* in a blast as well as the shape of the individual fragments (Gnirk and Pfeider, 1968), while joint spacings often determine fragment size (Atchison, 1968). In sedimentary rocks, *bedding planes* or *laminae, ripple surfaces,* or any type of stratigraphic parting or break may have low tensile strength and control fragmentation by blasting, whereas the *cleat* in coal, the *foliation* or *schistosity* in metamorphics, the *rift* or *grain* in many intrusive igneous rocks, and the *flow banding* in some extrusives can do likewise.

Recognition of such features and their importance in blasting can greatly aid in blast design

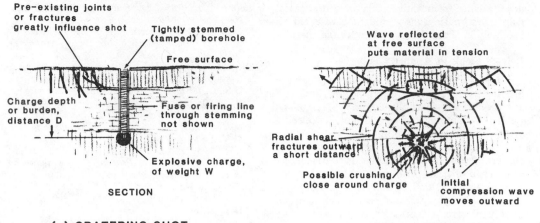

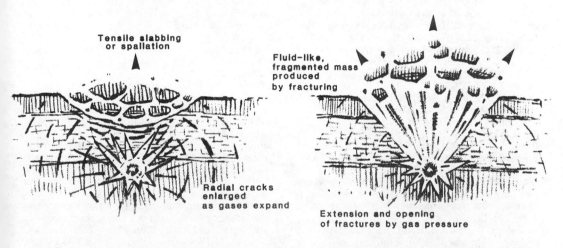

FIGURE 2. Stages in fragmentation by blasting.

(Cohen, 1980). For instance, close jointing can be an advantage if blasting is done to a free face parallel rather than normal to a predominantly developed vertical joint. Where joints are more widely spaced, the *blast holes* should be drilled in the solid blocks to alleviate early escape of gases without fracturing. In bedded or schistose deposits, blasting should be done perpendicular to these features and in the direction of their dip. Once a problem in obtaining the required fragmentation or desired effect in blasting a particular rock mass for a specific purpose becomes apparent, it can usually be overcome by an improved *explosive distribution* that takes account of geology, though this may require the use of smaller–diameter, more closely spaced blast holes (Dick and Olsen, 1972).

Quite often, then, a close appraisal of the geologic feature of a rock mass can ascertain its likely *stability* while it undergoes a particular type of blasting, predict any serious problems that might arise (such as from water influx or natural gases), help in estimating costs, and otherwise aid appreciably in economic and effective blasting.

Related Fields of Applied Geology

Geological knowledge is equally applicable to blasting overpressure, ground motion, and damage predictions (Nicholls et al., 1971), related fields that can control blast design; site selection and design of surface and underground structures to resist earthquakes and explosions, matters of concern in tunnel (Duke and Leeds, 1961) and hydroelectric projects (Campbell and Dodd, 1968), as well as in general civilian and military construction engineering; and peaceful uses of nuclear explosives (IAEA, 1974) or underground nuclear testing (Sharp, 1972), special areas of blasting technology.

ROBERT R. SHARP, JR.

References

Ash, R. L., 1968, The design off blasting rounds, in *Surface Mining*. New York: AIME, 373–397.

Atchison, T. C., 1968, Fragmentation principles, in *Surface Mining*. New York: AIME, 355–372.

Atchison, T. C., and J. M. Pugliese, 1964, *Comparative Studies of Explosives in Limestone*. Washington, D.C.: U.S. Department of the Interior, Bureau of Mines (RI 6395), 25p.

Atchison, T. C., and J. Roth, 1961, *Comparative Studies of Explosives in Marble*. Washington, D.C.: U.S. Department of the Interior, Bureau of Mines (RI 5797), 20p.

Belland, J. M., 1966, Structure as a control in rock fragmentation, *Canadian Inst. Mining and Metallurgy Bull.* **59**(647), 323–328.

Bergman, O. R., F. C. Wu, and J. W. Edl, 1974, Model rock blasting measures: effect of delays and hole patterns on rock fragmentation, *Eng. Mining Jour.*, June, 124–127.

Campbell, R. B., and J. S. Dodd, 1968, Estimated rock stresses at Morrow Point underground power plant from earthquakes and underground nuclear blasts, *9th Symp. Rock Mech., AIME Proc.*, 84–114.

Clark, G. B., 1968, Explosives, in *Surface Mining*. New York: AIME, 341–354.

Clark, G. B., C. J. Haas, J. W. Brown, and D. A. Summers, 1969, *Rock Properties Related to Rapid Excavation*. Springfield, Va.: National Technical Information Service (PB 184767), 188–207.

Clark, G. B., R. R. Rollins, J. W. Brown, and H. N. Kalia, 1971, Rock penetration by jets from lined cavity explosive charges, *12th Symp. Rock Mech., AIME Proc.*, 621–651.

Coates, D. F., and M. Gyenge, 1973, Blasting, in *Incremental Design in Rock Mechanics*. Ottawa: Information Canada (Mines Bureau Mon. 880, 3-1–3-7.

Cohen, C. J., 1980, Bench blasting related to joint design, *World Mining* **33**(4), 49.

Cole, R. H., 1948, *Underwater Explosions*. Princeton, N.J.: Princeton University Press, 437p.

Cook, M. A., 1958, *The Science of High Explosives*. New York: Reinhold Publ. Co., 440p.

Dick, R. A., 1968, Selecting the proper explosive—factors that should influence your choice, *Mining Eng.* **20**(11), 37.

Dick, R. A., 1972, *The Impact of Blasting Agents and Slurries on Explosives Technology*. Washington, D.C.: U.S. Department of the Interior, Bureau of Mines, (IC 8560), 58p.

Dick, R. A., and J. J. Olsen, 1972, Choosing the proper borehole size for bench blasting, *Mining Eng.* **24**(3), 41–45.

Duke, C. M., and D. J. Leeds, 1961, Effects of earthquakes on tunnels, in *Protective Construction in a Nuclear Age*, vol. 1. New York: Macmillan, 302–328.

du Pont de Nemours and Co., E. I., 1969, *Blasters' Handbook*, 15th ed. Wilmington, Dela.: Explosives Dept., 516p.

Ezra, A. A., 1973, *Principles and Practice of Explosive Metalworking*. London: Industrial Newspapers Ltd., 270p.

Fennel, M. H., R. P. Plewman, and A. N. Brown, 1968, Smooth blasting and presplitting, in *Papers and Discussions, 1966–1967*. Association of Mine Managers of South Africa, 811–863.

Gnirk, P. R., and E. P. Pfeider, 1968, On the correlation between explosive crater formation and rock properties, *9th Symp. Rock Mech., AIME Proc.*, 321–345.

Gregory, C. E., 1973, *Explosives for North American Engineers*. Clausthal, Germany: Trans. Tech. Publications, 276p.

International Atomic Energy Agency, 1974, *Peaceful Nuclear Explosions III—Applications, Characteristics and Effects*. New York: UNIPUB, Inc., 488p.

La Chapelle, E. R., 1966, The control of snow avalanches, *Sci. American* **214**(2), 92–101.

Lampson, C. W., 1946, *Final Report on Effects of Underground Explosions*, Princeton, N.J.: Princeton University, National Defense Research Committee of Office of Scientific Research and Development (NDRC No. A-479/OSRD No. 6645), 83p.

Langefors, U., and Kihlstrom, B., 1978, *The Modern Technique of Rock Blasting*. New York: Wiley, 38p.

Lewis, R. S., and G. B. Clark, 1964, Explosives and blasting, in *Elements of Mining*. New York: Wiley, 103–173.

Livingston, C. W., 1960, *Explosions in Ice*. Wilmette, Ill.: U.S. Army Snow, Ice and Permafrost Research Establishment, Corps of Engineers, (Tech. Rept. 75), 49p, Apps. A–D.

Livingston, C. W., 1968, *Explosions in Snow*. Hanover, N.H.: U.S. Army Cold Regions Research and Engineering Laboratory, Material Command (Tech. Rept. 86), 124 pp.

Manon, J. J., 1977, *Explosives: Their Classification and Characteristics; Chemistry and Physics; Mechanics of Detonation; Commercial Applications; and Selection for Specific Job*. New York: McGraw-Hill, 29p.

McLamore, R. T., 1973, *Development of Non-nuclear Explosive Fracturing Technology to Increase Domestic Energy Supplies and Reserves*. Redmond, Wash.: Petroleum Tech. Corp. (Rept. 73-H-014-R2), 20p.

Miller, J. S., C. J. Walker, and J. L. Eakin, 1974, *Fracturing Oil Shale with Explosives for in Situ Recovery*.

Washington, D.C.: U.S. Department of the Interior, Bureau of Mines (RI 7874), 100p.

Murphey, B. R., 1961, *High Explosive Crater Studies: Desert Alluvium.* Albuquerque, N.M.: Sandia Corp. [SC–4614(RR)], 8p.

Nagy, J., A. R. Cooper, and H. G. Dorsett, Jr., 1968, *Explosibility of Miscellaneous Dusts.* Washington, D.C.: U.S. Department of the Interior, Bureau of Mines (RI 7208), 31p.

Navarro, R., 1974, *Structural and Free Field Seismic Measurements on the Zonia Mine Blast, Arizona.* Springfield, Va.: National Technical Information Service (USGS/AEC Report NVO-472-2), 20p. 2 apps.

Nicholls, H. R., and V. E. Hooker, 1962, *Comparative Studies of Explosives in Salt.* Washington, D.C.: U.S. Department of the Interior, Bureau of Mines (RI 6041), 46p.

Nicholls, H. R., and V. E. Hooker, 1965, *Comparative Studies of Explosives in Granite: III.* Washington, D.C.: U.S. Department of the Interior, Bureau of Mines (RI 66993), 46p.

Nicholls, H. R., C. F. Johnson, and W. I. Duvall, 1971, *Blasting Vibrations and Their Effects on Structures,* Washington, D.C.: U.S. Department of the Interior, Bureau of Mines (Bull. 656), 105p.

Robinson, C. S., 1944, *Explosions—Their Anatomy and Destructiveness.* New York: McGraw–Hill, 88p.

Sharp, R. R., Jr., 1972, Requirements and engineering parameters of contained nuclear tests, in A Geological Engineering Evaluation of an Underground Nuclear Test Site. Ph.D. dissertation, University of Arizona, 357–404.

Short, N. M., 1961, Excavation of contained TNT explosions in tuff, University Park, Pa., *4th Symp. Rock Mech., AIME Proc.,* 171–178.

Swisdak, M. M., 1975, *Explosion Effects and Properties: Part I—Explosion Effects in Air.* Silver Spring, Md.: Naval Surface Weapons Center, White Oak Laboratory, (Rept. NSWC/WOL/TL 75–116) 139p.

Teller, E., W. K. Talley, G. H. Higgins, and G. W. Johnson, 1968, *The Constructive Uses of Nuclear Explosives.* New York: McGraw–Hill, 320p.

Walton, W. C., 1970, Shooting production wells to increase yield, in *Groundwater Resource Evaluation.* New York: McGraw–Hill, 353–359.

Cross-references: *Borehole Drilling; Borehole Mining; Cratering; Exploration Geology; Shaft Sinking; Tunnels, Tunneling.*

BLOCK DIAGRAM

A block diagram is a sketch of a relief model—in particular, a representation of a landscape in a perspective projection. Lobeck (1958), perfecting the art of block diagram construction, defined these illustrations as plane figures that represent an imaginary rectangular block of the Earth's crust in what appears as a three-dimensional perspective. The top of the block gives a bird's-eye view of the ground surface, and the side gives the underlying geologic structure. These three-dimen-

sional landscape models, when suitably cut and placed, enable examination of the surface and two of the lateral faces. Different types of block diagrams have been developed (Monkhouse and Wilkinson, 1971), but most are based on the simple isometric diagram that is prepared from a series of profiles. The need for these pseudo–three-dimensional diagrams is, according to Lawrence (1971), often encountered in the interpretation of landforms. They also provide useful summaries of field conditions for a specified or hypothetical block of the Earth's crust.

In a one-point perspective (Fig. 1), on the one hand, the nearest face is drawn not in perspective but identical with the plane section (Schou, 1962). A two-point perspective, on the other hand, contains two visible faces drawn in perspective (Fig. 2). The isometric block, a third form, is constructed with opposite faces of equal length that are parallel. This three-dimensional model is imagined as being viewed from an infinite distance. The isometric block resembles a photograph that was taken using a telephoto lens. The diagram is similarly distorted. Most blocks, however, are normally viewed at distances corresponding to natural vision.

The value of a block diagram as an illustration depends on the combination of methods used in its construction. Because the block diagram coordinates features of two-dimensional topographic and geologic maps—i.e., the diagram de-

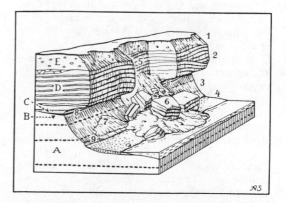

FIGURE 1. Block diagram in one-point perspective, horizontal and vertical scales are equal. (A) Senonian limestone with concretions of black flint, (B) fish clay, (C) Ceritium limestone, (D) Danian limestone with continuous beds of gray flint, and (E) boulder clay. (1) Slope of boulder clay, (2) precipice of overhanging limestone, (3) undercutting of cliff by wave action, (4) beach with flint pebbles, (5) scar due to fall of limestone, (6) fallen limestone and boulder clay, (7) scree washed away below the high-water mark, (8) bed of fish clay, and (9) wave-cut scar. (From Schou, 1962)

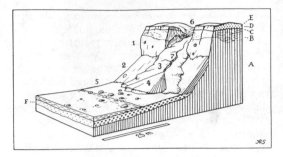

FIGURE 2. Block diagram in two-point perspective; horizontal and vertical scales are equal. (A) Boulder clay, (B) diluvial sand, (C) humus stratum, (D) blown sand, (E) recent humus stratum, and (F) marine sediments. (1) Precipice in clay, (2) scree, (3) landslide debris, (4) scree cone, (5) beach, (6) landslide scar, and (7) rain gully. (From Schou, 1962)

tails the relationship between subsurface structure and surface morphology—it provides immediate visual representation not otherwise easily obtainable. Thus, the advantage of a block diagram is that many phenomena of a three-dimensional nature, difficult to show on a two-dimensional map, become directly observable in a graphic form that appears to have three dimensions. The block diagram thus becomes a valuable instructional aid for those who find it difficult to abstract three-dimensional features from two-dimensional drawings. Maps (see *Geological Survey and Mapping; Maps, Environmental Geology; Maps, Physical Properties*) and sectional constructions (see *Cross Sections*), e.g., require considerable ability to select and interpret if their content is to be coordinated effectively.

In addition to providing valuable illustrative material, the block diagram is also important for research. Schou (1962), e.g., stresses that the process of construction compels constant observation of the subject matter of the map so that attention is directed to problems not readily evident by other methods of study. The block diagram clearly shows the limitations of map material, helping researchers to avoid specious conclusions. The method also suggests new ideas and solutions, especially in problems of the morphological nature. When constructed from measurements and observations in the field (see *Geological Survey and Mapping*), these diagrams are frequently used to illustrate details and associations of landscape features.

Robinson (1960) reports that many geologists first illustrated their reports with cross–sections to show structural relationships. The top line of a cross-section (q.v.) is a profile of the land surface, and some pictorial sketching was occasionally added to make a more realistic appearance. The next advancement involved cutting out

a block of the Earth's crust and tilting it upward and sideways so that two sides as well as the top became visible. W. M. Davis, D. Johnson, and A. K. Lobeck, renowned geomorphologists and field researchers, notably advanced the technique to where it is now a standard form of graphic expression.

Techniques of Block Diagram Construction

The principles of block diagram construction follow rules for perspective drawing, especially those that deal with linear perspective. Because the point of sight is constant, there is theoretically only one point from which a picture should be viewed to obtain the optimal spatial effect. Perspective pictures ideally should be viewed with one eye from a distance equal to that from the point of sight to the plane of the picture. For a photograph, this distance corresponds to the focal length of the lens. A strong spatial illusion is obtained in this manner. For an enlargement, the distance should correspond to the focal length times the amount of enlargement. An intensified spatial effect is often seen in enlargements because the viewing distance corresponds to the normal visual distance.

Construction of a simple block is illustrated in Fig. 3, which shows methods for two-point and one-point perspectives. The development of a two-point perspective involves the following steps (see Fig. 3, left side). Draw a horizon line V_1V_2. Select two points on this line as vanishing points V_1 and V_2 and another point, A, below the horizon. Point A represents the nearest part of the block surface and the uppermost point of the foremost edge. From A draw rays to V_1 and V_2, marking the edges of the block. From A, plot any lengths along these lines, marking off the edges AB and AD. From B, draw a line to V_2; from D, draw a line to V_1. Their point of intersection marks the block's fourth corner, C. Now plot the block thickness A–E from A along a line at right angles to the horizon. The lower edges of the block are constructed by drawing lines from E to the vanishing points and from B and D dropping perpendicular lines to intersect them at K and G.

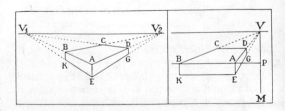

FIGURE 3. Construction of a simple block diagram in two-point perspective (left side) and one-point perspective (right side). (From Schou, 1962)

Block construction in one-point perspective is less complicated (see Fig. 3, right side). Draw a horizon line, selecting one vanishing point, V, thereon. From this point V, drop a vertical line VM which will serve as the main vertical. Choose a point P on this line, and from it draw a line parallel to the horizon. On that line plot the length of the front edge of the block A–B. The line AB can be placed to the right or left of the main vertical VM, depending on which edge of the block is to be visible. The block should not, however, be drawn too far to one side of the main vertical. From A and B draw lines to V, and on one of them, e.g., AV, choose a point D as a corner of the block. Through D draw a line parallel to the horizon. The intersection of this line with BV determines the fourth corner of the block. Construct the vertical edges of the block and its lower limits as described for the two-perspective block.

These instructions are useful for sketching block diagrams. They are, however, not precise enough for construction from maps. Consult Schou (1962) for detailed instructions that show methods for block tilting, transferring map pictures to block surfaces in one-point perspective, and applying longitudinal measurements, elevations, and angles on the block sides. These techniques are briefly summarized in the following sections.

Block Tilting

Figure 4 shows three blocks drawn with the horizon at the same height but with different distances between their vanishing points. The vertical plane is the plane of the drawing paper; the horizontal plane is the horizon at the level of the eye. It is assumed that the block to be drawn lies below the horizon plane and behind the drawing plane. Figure 5 shows a horizontal plane that is intersected by a vertical plane—e.g., a sheet of

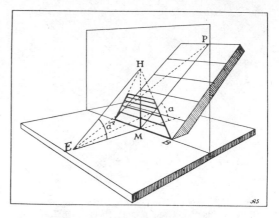

FIGURE 5. Relationship between block tilting, height of horizon, and viewing distance. In this drawing the height of horizon and viewing distance is the agent of the angle of tilt—that is, HM/EM = tan $a°$. The edges of the block, represented by parallel lines on the block surface, must converge toward the same point in the perspective picture. (From Schou, 1962)

glass placed vertical to a table top. Behind the glass is a block with its surface leaning at $a°$ to the table top. Point E is the point of sight. EM is at right angles to the drawing plane, and the block is so placed that its medial line touches M. To determine the position of the horizon and the positions of the vanishing points for parallel line systems on the block surface, sight from E along a line parallel with the median line of the block surface. The intersection with this line, H, with the drawing plane determines the height of the horizon.

Transfer of Map Picture to a Two-Point Perspective Block

Figure 6 illustrates the construction of a two-point perspective block diagram from a map sheet. In most cases it is necessary to make the block drawing on an enlargement of the map grid because a block constructed from a map sheet is on a much smaller scale than the map. The principles involved in drawing the map in the plane of the drawing paper are analogous to those already explained. The description for the procedure, after Schou (1962), is as follows:

1. Draw a horizon line HH.
2. At right angles to the horizon draw a line MM. The intersection of this line with HH is called F.
3. Well down on MM, select a point of sight E. Successively lower positions of E will determine larger sizes of the block to be constructed.
4. Between the point of sight and the horizon, determine a point, A, on MM where Fa/AE

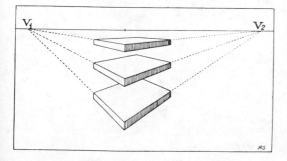

FIGURE 4. Relationship between the tilting of blocks and the distance to the front corner below the horizon. Note that as the block is tilted increasingly forward more surface area becomes visible. Tilting can be used to enhance the visual perception of surface features. (From Schou, 1962)

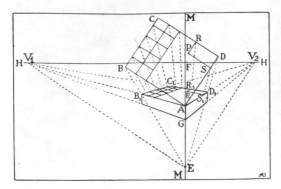

FIGURE 6. Construction of a block diagram in two-point perspective from a map sheet (ABCD). (From Schou, 1962)

+ tan 0, where O is the desired block tilt. It is suggested to select O between 30° and 40°, tan 35° = 0.7.

5. Place the map sheet or map section so that the corner that will be the foremost corner of the block lies at point A. The direct use of the map may be eliminated by superimposing a square grid that is transferred to the block surface. Map details can then be sketched in, using the squares as proportional guides.
6. From E draw lines parallel with the map edges AB and AD. Where these lines intersect the horizon, mark the vanishing points V_1 and V_2.
7. From A draw lines to V_1 and V_2. These lines show the directions of the block sides.
8. Determine the length of the sides by drawing lines from E to the map corners B and D. From B_1 and D_1 draw lines to V_2 and V_1 respectively to form the back edges of the block. The intersection C_1 forms the position of the fourth corner.
9. To transfer a point P from the map to the block surface, first determine its rectangular coordinates, R and S, on the map edges, and then transfer these points to the block edges. This is accomplished by drawing lines from E to R and S. The intersection of these lines with the block edges establishes the positions of the points R_1 and S_1 on them. Lines on the block surface at right angles to the edges must have the direction of V_1 and V_2. Draw a line on the block surface from R_1 in the direction of V_2, and one from S_1 sighted on V_1. The intersection of these lines, P_1, is the position of point P in the perspective picture.
10. From A, plot the required block thickness AG down the line MM. Construct the lower edges of the block by drawing lines from G to V, B_1 and V_2. From block corners B_1 and D_1 draw lines parallel to AG.

Transfer of Map Picture to a One-Point Perspective Block

A corresponding construction in one-point perspective is given in Fig. 7. In the following brief description, the numbers refer to the preceding more detailed instructions for two-point construction.

1. Draw the horizon line HH.
2. Draw the line VE at right angles to HH.
3. Mark E, the point of sight, or eye point.
4. Determine M thus: MV/ME = tan 0. Through M draw a line parallel to the horizon.
5. On this line lay the map sheet ABCD or similar figure.
6. From A and D draw lines to V.
7. From E draw lines to one of the other map corners, B or C. This will determine B_1 and C_1, and the back edge of the block can be drawn parallel to the horizon.
8. Transfer points as in two-point perspective construction.

Relief Drawing

Relief models are three-dimensional representations of the physical features or relief of an area. They may be constructed in any size or medium and are frequently not to true scale because the vertical scale is exaggerated to accentuate the relief. Many techniques are employed in the preparation of relief drawings—viz., contour lines, ground hachures, hill shading (shaded relief), layer tinting, pictorial symbols (physiographic diagram), or combinations of these methods and photography. All these techniques require some degree of skill. A useful approach to the problem of relief drawing is provided by the so–called section–line method of graphic construction as already described.

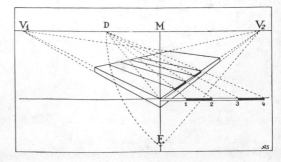

FIGURE 7. Method for inserting exact measurements in a one-point perspective diagram. (From Schou, 1962)

Section-Line Drawing Section (cross-section) lines (see *Cross–Section*) are lines on the block surface formed by a number of vertical planes parallel to one of the block sides. It is helpful to imagine a block sliced by a series of parallel cuts in an analogous way that a slab of bacon is cut by a slicing machine. Surface relief produced by means of section lines is exact but rather time consuming. Each line is constructed by determining its position on the block surface and inserting the elevations, obtained from the appropriate position on the map, along these lines. Creation of a three-dimensional illustration using section lines depends on physiological and mental peculiarities in human visual faculties. Thus, one perceives a three-dimensional impression that does not exist in reality.

Shading Shade includes the following categories: unlit surfaces, sources where light does not penetrate, shadows on illuminated surfaces, and some surfaces that lie in half-shade. For most purposes in block diagrams, exact shading is unnecessary. The spatial illusion may be enhanced by limiting the shading to unlit surfaces, providing the shading is uniform.

Unilluminated surfaces can be toned dark either continuously by watercolor wash, drawing ink, or press-on film or screen overlays or discontinuously by hachuring, stippling, or using screens in the printing process. Hachuring is perhaps the easiest and most common method. A light source must be assumed so as many slopes as possible are in the shade. In applying the hachuring, the perspective effect must be maintained. The direction of the section lines at some points must be determined and the hachuring completed in conformity with these directions. The method is detailed by Schou (1962, pp. 27–29) through the use of informative illustrations and comprehensive instructions. The three-dimensional effect may be enhanced by shading the block. The unlighted side needs to be hatched vertically because the section lines run parallel to the block edges. Symbols may also be drawn on the lighted and unlighted sides in the same form, preferably with different line thickness (see Figs. 1 and 2).

Ground Hachuring This method is also based on the three-dimensional illusions obtained by section-line perspective. Here the surface is covered wholly or partly with dense hachuring. Each line must be drawn in conformity with the section lines of the area (Fig. 8). Note that different thicknesses of the ground hachures produce perceptions of depth and steepness of slope.

Applications

When the final product is in the form of a block diagram that is keyed to geologic, landform, soil, and vegetation features, a great deal of infor-

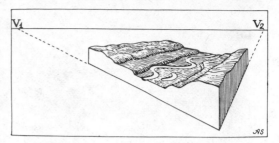

FIGURE 8. Example of ground hachuring in a block diagram. Note that a light source was assumed between V_1 and V_2 near the top left of the drawing. This enhanced the shadow effect along the stream and scarps cut into the terraces. (From Schou, 1962)

mation can be summarized in graphic form. The CSIRO (Australia) Land Research Series (Speck et al., 1964) makes extensive use of block diagrams to help illustrate different kinds of land systems. The block diagram in Fig. 9, e.g., makes use of hachures and shading to illustrate surface relief. Geologic structures are shown on the block sides and provide visual expression of relationships between relief and structure. The accompanying table is keyed to the block diagram with numbers that identify land units with their associated landforms, soils, and vegetation. Such representations that combine graphic and tabular summaries provide a convenient means for presenting field relationships in a format that is easy to remember.

CHARLES W. FINKL, JNR.

References

Lawrence, G. R. P., 1971, *Cartographic Methods.* London: Methuen, 153p.
Lobeck, A. K., 1958, *Block Diagrams and Other Graphic Methods Used in Geology and Geography.* Amherst, Mass.: Emerson-Trussell Book Company, 206p.
Monkhouse, R. J. and H. R. Wilkinson, 1971, *Maps and Diagrams.* London: Methuen, 330p.
Robinson, A. H., 1960, *Elements of Cartography.* New York: Wiley, 343p.
Schou, A., 1962, *The Construction and Drawing of Block Diagrams.* London: Nelson, 33p.
Speck, N. H., R. L. Wright, G. K. Rutherford, K. Fitzgerald, F. Thomas, J. M. Arnold, J. J. Basinski, E. A. Fitzpatrick, M. Lazarides, and R. A. Perry, 1964. *General Report on the Lands of the West Kimberley Area, W.A.,* CSIRO (Australia) Land Research Series No. 9, Melbourne, Victoria, 219p.

Cross-references: *Cartography, General; Field Notes, Notebooks; Geological Survey and Mapping; Map Symbols; Profile Construction.*

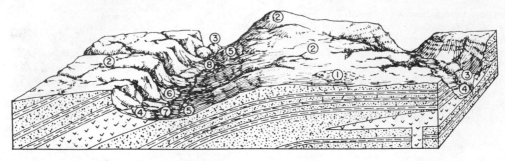

QUARTZITE   SHALE   BASALT

Unit	Approx. Area (%)	Land Forms	Soils	Vegetation
1	2	Summit remnants: up to 2 miles in extent; pebble-strewn slopes mainly less than 1%	Yellowish sandy soils high in laterite gravels: Tableland family (5)	Low open woodland with scattered shrubs and *Plectrachne pungens*. *E. brevifolia–E. collina* and *E. collina–E. dichromophloia* communities (4, 6)
2	74	Quartzite plateaux and mountain summits: plateaux and dip slopes typically less than 5%, with indented escarpments up to 1000 ft high comprising vertical upper walls and benched steep slopes; mountain slopes up to 50%; basal screes and boulder fans sloping up to 35%	Mainly outcrop with some sandy skeletal soil (24)	Low, very open woodland, scattered or patchy shrubs and *Plectrachne pungens*. *E. brevifolia* community (1*d*)
3	5	Basalt and dolerite hills: up to 250 ft high; rounded crests up to 100 yd wide and benched slopes up to 60%, with boulder mantles	Outcrop with red basaltic soil: Walsh family (4)	Very open, grassy woodland, with scattered shrubs and ground storeys of combinations of *Sehima nervosum, Sorghum* spp., *Dichanthium fecundum*, and *Plectrachne pungens*. *E. tectifica* alliance (14*a*, 14*c*, 15)
4	4	Lower slopes on basalt and dolerite: concave, up to 5%, and up to ¼ mile long, colluvial mantles and local outcrop	Moderate to deep, reddish sandy to loamy basaltic soils: Frayne family (3)	Similar to unit 2
5	4	Lower slopes on quartzite: concave, up to 10%, and up to ¼ mile long; colluvial mantles and local outcrop	Some outcrop with yellowish sandy soils of variable depth: Tableland family (5)	Open woodland with moderately dense shrub layer and *Plectrachne pungens*. *E. brevifolia* community (1*d*); locally 28
6	2	Colluvial aprons and fans: up to 500 yd long with gradients 1 in 20 to 1 in 60 and gullied up to 10 ft	Mainly yellowish sandy soils: Tableland family (5)	Spinifex with scattered trees and shrubs and patches of open woodland. *Plectrachne pungens* grassland (54) and *E. collina–E. dichromophloia* community (6)
7	4	Drainage floors: up to ¼ mile wide, gradients 1 in 200 to 1 in 500; marginal slopes up to 1%	Variable soils but mainly greyish sands over tough loamy subsoils: Tarraji family (18), commonly mottled	Grassy woodland, sparse to moderate shrubs and *Aristida hygrometrica*. *E. papuana* community (22*b*); also 25
8	5	Channels: up to 300 ft wide and 30 ft deep	Channels, bed-loads range from sand to boulders. Banks, brownish loamy alluvial soils: Robinson family (21)	Fringing woodlands. *E. camaldulensis–Terminalia platyphylla* community (42)

Comparable with the rocky quartzite plateaux and mountain ranges of Buldiva land system, North Kimberley area.

FIGURE 9. Block diagram illustrating surface and subsurface features associated with the Precipice Land System in northwestern Western Australia. Descriptions of landforms, soils, and vegetation are keyed to the block diagram by circled numbers that identify specific land units. Note the effective use of ground hachuring, shading, and geologic structure on the sides of the block. Numbers in parentheses in the tabular portion refer to descriptions of soils and vegetation given in other chapters of Speck et al., 1964. (From Speck et al., 1964).

BLOWPIPE ANALYSIS

Blowpipe analysis refers to the series of physical and chemical tests made using the flame of a *blowpipe*. The blowpipe is a metal tube about 25 cm long, tapering from a diameter of about 8 mm to an orifice of about 0.3 mm at the tip. It is usually made of brass and bent at the right angle near the tip so the operator can observe the flame (Berzelius, 1845; Fletcher, 1894). The operator places the tip of the blowpipe in the luminous flame of a Bunsen burner, alcohol lamp, or broad wick candle and blows a jet of air into the flame. This produces an elongated flame (Fig. 1) composed of a low-temperature inner blue cone of un-burned gases, a visible flame in which gases are rapidly oxidized (the reducing flame), and an in-visible outer zone of high temperature in which no gas combustion takes place (the oxidizing flame) (Butler, 1910).

A mineral fragment held in the reducing flame will be heated in the effective absence of oxygen and, if oxygen containing, will be reduced. A fragment in the oxidizing flame reaches high temperature in air and tends to oxidize. The maximum temperature of the flame is just beyond the tip of the reducing flame and has been estimated to reach 1,500°C, depending on the fuel. The following sections describe briefly some typical blowpipe tests. (see Vol. IVA: *Chemical Mineralogy; Geochemistry: Testing for Elements*).

Fusibility

Small fragments are held in the flame, and their ease of melting is observed. This ease of fusion is a function of the size of the fragments, the time in the flame, and the steadiness of the flame, but results can be very useful. For example, within the garnet group, almandine can be distinguished from pyrope by fusibility.

Flame Color

A mineral powder moistened with acid or in some cases merely sprinkled through the flame will give characteristic flame colors. Color filters may be used to cancel interfering colors of over-powering flame-coloring elements such as sodium.

Tests of Charcoal

A charcoal block provides a useful support on which minerals can be held for testing. Minerals such as orpiment composed entirely of volatile elements can be completely removed by oxidizing flames but leave a ring sublimate, in the case of orpiment, of Sb_2O_4 and Sb_2O_3, plus the strong odor of SO_2. In some cases a plaster of Paris block is substituted for the charcoal, and sublimate

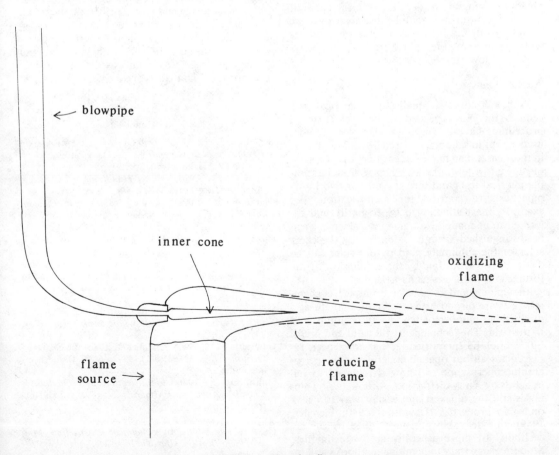

FIGURE 1. Blowpipe flame.

colors such as blues, which do not show up well on charcoal, are distinct on the white background. The charcoal block serves as a miniature smelter for sulfide minerals. Roasting in the oxidizing flame removes sulfur and any arsenic or antimony. The oxidized powder is mixed with charcoal and a flux and is heated in the reducing flame, producing a tiny globule of metal.

Bead Tests

Borax and other fluxes are melted and held in a 2–3 mm diameter circular loop of platinum wire. The molten flux will dissolve minerals and, on cooling, produce a bead of distinctive color. Bead tests are indicative for elements, and the color can be checked hot and cold, from the oxidizing or reducing flame, and in various fluxes.

Tube Tests

A mineral powder held in a high–melting–point glass tube closed at one end and heated is, in effect, heated in the absence of oxygen. The decomposition products volatilize and deposit along the tube walls in a temperature fractionation process. Heating a mineral powder placed in the curve of a slightly bent open glass tube draws air over the sample, oxidizing it, and fractionation products will deposit along the walls of the tube in the sequence of lowering temperature.

Applications

Although complete qualitative analytical procedures for blowpipe and simple wet chemical procedures have long existed, they were seldom used except in schools. A practicing mineralogist, in most cases, can narrow identification of a sample down to a few minerals by inspection and simple physical tests and then use one or two blowpipe tests to confirm the identification. For example, the distinction between barite and celestite can be made by a flame test, arsenides can be distinguished from antimonides by the character of the sublimate, and pyrophyllite can be distinguished from talc by the exfoliation of the former in a flame, as its name states.

Blowpiping and its associated simple wet chemical tests are disappearing techniques taught at a diminishing number of schools and seldom used in the field. The technique is still valid and useful but cannot be quantified. It has long been replaced for all but opaque minerals by the petrographic microscope and now is replaced for all minerals by X-ray diffraction. Although blowpiping is still the quickest and easiest way to check on some properties (Davidson, 1940; Fletcher, 1949), this desirability is outweighed by the accessibility of more generally useful techniques. Modern elementary mineralogy textbooks still include some data on blowpiping, but if the technique has any future, it probably lies with hobbyists or prospectors who do not have available the expensive equipment of sophisticated techniques. Smith's (1946) reference text, *Identification and Qualitative Chemical Analysis of Minerals,* is the most complete as well as the last textbook that considers blowpiping in detail.

Historically, blowpiping appears to have been derived from the primitive smelting and assaying techniques whose origins are lost in antiquity. Agricola, in *De Re Metallica,* published in 1556, discusses small forced draft furnaces used in assaying but makes no mention of blowpipes. The first manual of blowpipe techniques was written by the Swedish mineralogist Engstrom in 1770, and the techniques were apparently in use in the Swedish mining industry. Berzelius (1845), the great Swedish chemist, developed the technique and made it an almost indispensible tool for mineral and chemical analysis. Mineralogy books written since 1800 almost invariably contain a section on blowpiping, and the technique seems to have reached its peak of use at about 1900. The long persistence of what now appears to be a primitive technique is due to three factors: (1) The tests were accurate and reproducible; (2) they dealt with minerals in a solid state rather than necessitating dissolution as required by wet chemistry, which is commonly difficult with minerals; and (3) the equipment was readily available, inexpensive, portable, and easily cleaned.

IRVING FISHER

References

Berzelius, J. J., 1845, *The Use of the Blowpipe in Chemistry and Mineralogy,* J. D. Whitney, trans. Boston: W. Ticknor and Co., 237p.

Butler, G. M., 1910, *Pocket Handbook of Blowpipe Analysis.* New York: Wiley, 80p.

Clarke, E., R. T. Prider, and C. Teicher, 1946. *Elementary Practical Geology.* Nedlands: University of Western Australia, 170p.

Davidson, E. H., 1940, *Field Tests for Minerals,* 2nd ed. London: Butler & Tanner, 60p.

Elderhorst, W., 1874, in H. B. Nason and C. F. Chandler, eds., *Manual of Qualitative Blow-Pipe Analysis and Determinative Mineralogy,* 6th ed. New York: Zell, 312p.

Fletcher, E. F., 1949, *Practical Instructions in Quantitative Assaying with the Blowpipe.* New York: Wiley, 142p.

Galbraith, F. W., 1963, Chemical tests for the elements, *Geotimes* 7(8), 35–36; 8(1), 35–36 (AGI Data Sheets 42a and 42b).

Smith, O. C., 1946, *Identification and Qualitative Chemical Analysis of Minerals.* New York: Van Nostrand, 385p.

Cross-references: *Lithogeochemical Prospecting; Mineral Identification, Classical Field Methods.* Vol. IVA: *Chemical Mineralogy; Flame Spectroscopy; Geochemistry; Testing for Elements; Mineralogy.*

BOREHOLE DRILLING

Borehole penetration of rock strata in the shallow subsurface is designed to recover representative samples for chemical and physical tests, to provide an access conduit for sophisticated indirect testing with geophysical sondes, or to provide a conduit to a specific subsurface horizon for extraction purposes. Variations and combinations of these drilling logics may be dependent on drilling depths. Shallow subsurface borehole drilling is usually considered within 300 m of the surface. It is usually directed toward exploration and production in sedimentary terrains within the disciplines of groundwater hydrology, Pleistocene geology, coal and oil shale exploration and delineation of industrial mineral deposits, and corresponding delineation of metallic ore bodies in igneous and metamorphic regions.

The major categories of subsurface drilling techniques are normal rotary, rotary coring, and reverse circulation. Each methodology is designed to provide optimum benefits within several geologic and economic constraints. Rotary drilling has maintained its place of leadership as the most versatile and economical boring method. Nonrotary methods, including percussion air hammers and explosive charge bits, are largely limited to specialized applications and experimental studies in the shallow subsurface.

Normal Rotary Drilling

Normal rotary drilling is the rotation of a cutting bit on the end of a length of drill stem in which the downward pressure behind the cutting edge abrades the rock face. The dislodged rock fragments and powder, termed the *cuttings,* are carried to the surface between the drill pipe and the borehole wall (Fig. 1).

The initial developments in rotary drilling techniques were designed for use in boring water wells. The basic principles of rotary drilling were applied to the search for oil as early as 1866. The extensive application of rotary drilling methodology to oil well exploration dates back to the exploration and discovery of the Corsicana field in northern Texas in October 1895. Two brothers, M. C. and C. E. Baker, had perfected the rotary methodology for drilling water wells in the Dakota Territory in about 1882. They moved to the Corsicana field with their equipment and established the American Well and Prospecting Company to build and develop rotary drilling rigs.

Water was used as the first circulating medium in the rotary rigs. Rotary mud was used first in drilling during January 1901 on the Lucas gusher discovery well, Spindletop field, Texas.

Normal rotary drilling is an inexpensive means of sampling in terms of unit cost per length, but

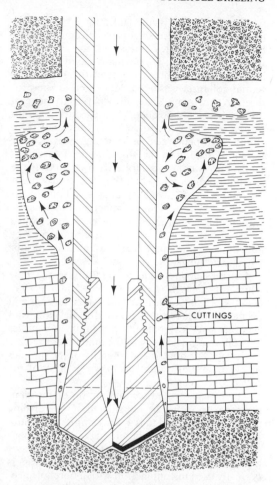

FIGURE 1. Conventional rotary drilling system. Rock cuttings are conveyed to the surface between the drill pipe stem and the borehole wall.

some quantity and quality constraints limit its usefulness. The samples carried toward the surface in the annulus between the drill pipes and the wall tend to mix with chips of wall rock or to be lost in formational fractures. Wall rock and face sample mixing becomes more acute when unconsolidated to semiconsolidated sediments have been penetrated. These conditions are widespread in the shallow subsurface terrains normally associated with both soft coal basins and groundwater exploration in surficial deposits. Buildup of a sufficient mud cake on the borehole wall lessens the influence of this contamination problem. Nevertheless, persistent slumping of unconsolidated sand or gravel horizons into the borehole continues to be a major contamination source and may ultimately force specialized treatment of the borehole or abandonment.

Sidewall contamination problems may be controlled by casing or by additives to the circulation medium. *Casing* is the technique of drilling through an oversized steel pipe driven in tandem behind the drill bit. Shallow oil, gas, and water wells in production are normally cased the entire length of the borehole to protect them from potential collapse. Casing is perforated at the production horizons. The upper ten to several hundred feet are frequently cased during exploration drilling to protect the borehole from slumping of penetrated sands and gravels. Rotary borehole drilling in the shallow subsurface normally uses water to generate its own mud as the circulation medium. When the specific gravity of the mud is insufficient to retain the stability of the borehole wall, various commercial additives—e.g., barite preparations—may be added. Allied preparations such as swelling bentonites may be added to seal ground fractures and unconsolidated sand horizons.

Cuttings do not tend to intermix in the hole when turbulent flow is maintained in the fluid column. Mixing of heavier and lighter cuttings will occur if laminar flow is allowed to develop, and accurate samples would not be obtained. Samples are collected on a screen in the mud pit at the base of the rig. The quality and quantity of the samples recovered are also influenced by the mesh of the collection screen.

The efficiency and contamination of the sample recovery must be considered if the drilling program is designed to recover samples to be subjected to chemical or related analyses. These considerations are weighed against the economics of greater speed in rotary drilling. Rotary drilling is typically used in conjunction with a borehole geophysics program. Radiation and electric geophysical logs have the highest clarity and meaningful delineations when applied to rotary boreholes with sufficient mud cake. Borehole drilling with other techniques are considered to yield more reliable samples but have the disadvantage of logging through drill stems and casing.

Rotary Core Drilling

Rotary core drilling employs a core barrel attached to the end of the drill stem. The core is cut by a tungsten carbide or diamond–studded drill bit and collected in the core barrel. It is normally retrieved by pulling the drill rods (Fig. 2). Coring is employed in conjunction with rotary drilling of the lithologies overlying the target horizon. The large diameter of the core barrel requires a proportionally larger–diameter borehole with a slow penetration rate that increases coring costs. Since coring usually involves sampling a specific zone of a horizon, the exact depth interval of the target must be obtained from an adjacent normal rotary test hole drilled prior to coring. This step also increases the cost of a coring program.

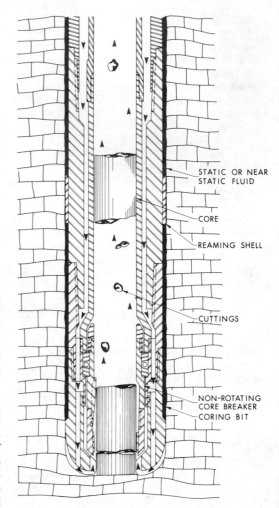

STATIC OR NEAR STATIC FLUID

CORE

REAMING SHELL

CUTTINGS

NON-ROTATING CORE BREAKER

CORING BIT

FIGURE 2. Continuous rotary coring system.

Coring is frequently utilized for evaluation of shallow coal seams. It can commence 1 ft or 2 ft above the seam and continues below the coal to ensure as complete a recovery as possible. The quality of coring results depends entirely on the percentage of recovery. One would theoretically have the best sample of coal or another lithology obtainable short of an outcrop exposure if 100% recovery is achieved. However, determination of core recovery and, if losses occur, the assignment of those losses to a specific stratigraphic interval are not readily obtainable unless meticulous measurements are made during the coring process. There is a general expansion of core length on removal from the barrel related to fracturing of the material during the drilling. It is common to recover in excess of 100% of the cored interval. If soft sediment, such as unconsolidated sand, occurs within a cored coal seam, it is likely to be lost through washing. Its presence or position may not be determinable if one still has an apparent 100% recovery due to expansion. Geophysical density logging (see *Well Logging*) of the borehole is essential to clarify these stratigraphic ambiguities. When the cored stratigraphic intervals are relatively shallow, the core barrel containing the cut core is retrieved by pulling the entire length of the drill stem. If the stratigraphic target interval is deep or if a considerable number of drill stem rods are involved, then a wire–line system is favored whenever feasible. The inner core barrel is removed from the bottom of the hole and retrieved without having to take the rods from the hole. The success of the wire–line method depends a on relatively long bit life. Its use to date in Precambrian Shield areas has been limited compared to its use in softer sedimentary terrains.

The rubber sleeve core barrel was developed to improve core recovery of unconsolidated sands under conditions unfavorable for obtaining conventional barrels. This modification of the basic core barrel is finding current application in the shallow subsurface sampling of bituminous (tar) sands. The barrel operates on the bottom of a conventional drill string, and as the bit advances into the formation, cutting the core, the central rod attached to the upper barrel remains stationary to support the upper end of the rubber sleeve. Advancement of the lower end of the nonrotating barrel forces the rubber sleeve to shrink over the wall of the core. The rubber sleeve is smaller in diameter than the core to form a tightly stretched skin holding the contents of the core in place. The rubber sleeve is displaced into the nonrotating barrel as coring progresses. The core that is being cut is not in compression from the core above it to prevent compressive column failure (see Hildebrandt et al., 1957, for additional details on this methodology).

Reverse Circulation Drilling

Reverse circulation drilling, a relatively new modification of the rotary drilling technique, employs dual wall pipe (Fig. 3). The drilling medium, usually air, is forced down the annulus between the inner and outer pipes and over the drill bit before returning to surface up the center. The distance between the bit stub and the cutting edges of the bit is approximately 4 in (102 mm) when applied to hollow subsurface applications. The sample has only marginal exposure to contamination. The outer diameter of the drill pipe is usually 4 5/8 in (117.5 mm) and the drill bit is 4 7/8 in (124 mm), allowing 1/8 in (3.3 mm) clearance between the outside of the pipe and the wall of the hole. The bit commonly employed is a tricone to produce rock or coal chips of 1/4 in (6.4 mm) maximum dimension. When this drilling tech-

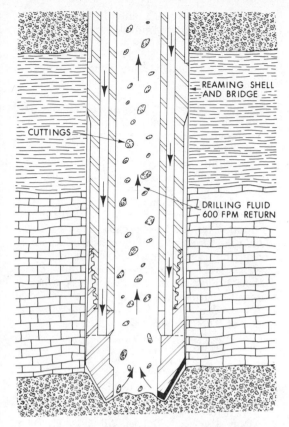

FIGURE 3. Reverse circulation drilling method with dual wall pipe and tungsten carbide insert bit.

nique is applied to shallow-depth soft coal, the cuttings have a high percentage of minus 28 mesh. This percentage normally exceeds 50% of the sample. A drill rig with a diamond coring bit is capable of recovering a continuous core in up to 5 in (127 mm) lengths from the more indurated rock zones but is highly inefficient in the soft sediment horizons. Core sections are 2 1/8 in (53 mm) diameter and return along with all the rock cuttings up the center of the pipe.

Rotary reverse circulation drilling costs are a function of formation hardness. A higher percentage of coarse indurated sandstone and conglomerate lithologies within the zone of penetration will slow drilling and thereby increase costs but will not reduce the effectiveness of the method.

The reverse circulation method is very applicable to drilling unconsolidated and friable lithologies and coal seams. Fractured formations prone to lost circulation with this drilling method usually pose few problems since the method resembles drilling a cased hole. Sample mixing or contamination from the borehole wall is negligible. Sample chips and powder are recovered by means of a cyclone at the surface. A cyclone is a device for removing small or powdered solids from the gas or liquid circulation medium by centrifugal forces. Reverse circulation drilling has the distinct advantage over normal rotary drilling in the shallow surface for providing relatively complete and uncontaminated samples. The method is considerably slower and more expensive than normal rotary drilling. The major disadvantage of the reverse circulation method is the application of geophysical logging techniques that may be desirable for stratigraphic correlation purposes. Logging tools must be run through casing pipe, which is severely detrimental to the instrumentation results. The borehole lacks a suitable mud cake for stability so the pipestem cannot be readily pulled prior to geophysical logging.

There are few quantitative comparative evaluations of these drilling techniques to recover statistically reliable samples. The reader is referred to Irvine et al. (1974) for a representative evaluation of these methods.

PAUL L. BROUGHTON

References

Hildebrant, A. B., H. C. Bridwell, and J. M. Kellner, 1957, Development and field testing of a core barrel for recovering unconsolidated oil sands, in W. C. Goins et al., eds., *Drilling.* AIME Petrol. Trans. (Rept. Ser. No. 6), 271–273.
Irvine, J. A., S. H. Whitaker, P. L. Broughton, and T. E. Tibbets, 1974, Analysis of borehole sampling methods in lignite seam evaluation, in G. R. Parslow, ed., *Fuels: A Geological Appraisal.* Saskatchewan Geological Society (Spec. Publ. No. 2), 203–228.

Cross-references: *Augers, Augering; Borehole Mining; Blasting and Related Technology.* Vol. XIII: *Pipeline Corridor Evaluation; Rapid Excavation and Tunneling; Shaft Sinking; Tunnels, Tunneling; Well Logging.*

BOREHOLE MINING

Borehole mining is a general term for any process in which the wanted mineral or metal values of a deposit are *fluidized* in place by some means and removed from the deposit via borehole. Researchers have proposed (Shock and Conley, 1974; Anon., 1974a) that it should supplant the accepted, but narrower, term *chemical mining,* which implies that the wanted values are dissolved in a reagent that is circulated through the deposit, the *permeability* of which may be artificially increased for the purpose. Thus, *borehole mining* encompasses *slurry mining,* which has been proposed for the recovery of insoluble min-

erals—e.g., coal, phosphate rock—as suspended solids. *Borehole mining* is also perhaps more apt for the long-established *Frasch process* (see Vol. IVA: *Sulfur: Element and Geochemistry*), in which superheated water is injected into sulfur deposits and the resulting sulfur melt is pumped to the surface (Frasch, 1976), and for the *in situ* (in-place) *retorting* of petroleum from oil shales and tar sands, by burning some of the hydrocarbons in injected oxygen (Anon., 1974b, c; Maugh, 1977a, b).

Within the field of chemical mining, Hunkin (1971) and Shock and Conley (1974) distinguish *complete solution mining,* i.e., the dissolution of massive deposits such as evaporites, leaving behind cavernous voids, from *leach mining* or *in situ leaching,* in which a disseminated ore mineral is selectively dissolved, leaving behind a porous gangue. In contrast, *direct leaching* is not a variety of borehole mining and merely describes the leaching of conventionally mined ore without prior activation processes such as oxidation, chlorination, or roasting; it is sometimes used to imply also that gangue has not been separated from the ore mineral before leaching.

Advantages that have been attributed to chemical mining are as follows (Hunkin, 1971; Shock and Conley, 1974; Lackey, 1975):

Low capital investment,
In most cases, a short preproduction period and high initial recovery rates,
Minimum disturbance of the environment and minimal waste disposal problems,
Low labor intensity,
Low safety hazards,
Applicability to otherwise inaccessible deposits—i.e., to deposits too deep or too faulted for conventional mining or to grades too low to be economic by conventional mining. Also, where markets are limited, leach mining may be applied to low-grade deposits on a smaller scale than conventional methods would require to be economic.

Conditions required for in situ leaching are as follows:
1. The ore mineral must be capable of selective leaching by an economically and technically feasible reagent. Reagents used to date have been inexpensive so that recycling was unnecessary—e.g., water, oxygen, dilute sulfuric acid, or acid ferric sulfate (the latter being a by-product in the oxidation of many base metal sulfide ores, a process which is accelerated by bacterial action and is the basis of *bioextractive metallurgy*) (Anon., 1974a; Trudinger, 1971, Murr, 1980).
2. The ore body must possess adequate *permeability* and pore structure. An ideal ore body

would contain pores of only one size and type, large enough for the flow of leachant under a moderate pressure gradient yet numerous enough to intersect the disseminated mineral grains to be leached; in principle, a single passage of leachant would then suffice for complete extraction. More likely, the leachant will circulate through the ore body via relatively few large channels, while the extraction depends on diffusion along finer, stagnant pores. The fine pores are a vital prerequisite and must not be blocked by any side effects of the leaching reaction. The larger channels may be created by prior fracturing with chemical (Porter and Carlevato, 1974; Anon., 1974a) or nuclear explosives (Lewis et al., 1974, Lewis and Braun, 1973) or by hydraulic fracturing (Daneshy, 1974; Anon., 1974a); the leaching process then has analogies with *heap* (or *dump*) *leaching*.

3. After passing through the ore body, the leachant must be collected without leakage to the environment. The following examples illustrate various circumstance under which this collection may be achieved.

4. The mineral values must be recovered from the pregnant leach liquor, which ideally is recycled. This may involve solvent extraction, ion exchange, cementation, electrolysis, hydrogen reduction, or any other of the techniques that make up *hydrometallurgy*.

5. The process must comply with existing legal requirements—in particular, the controls that have been imposed on the discharge of possible pollutants in wells (Rouse, 1974; Anon., 1974a).

Examples of in-situ leaching

For details of developments in the field, refer to Aplan et al. *Proceedings of the Solution Mining Symposium* (1974) and the review by Lackey (1975); only a brief overview is given here. Deposits may be classified according to whether the deposit is permeable or impermeable and the depth of burial.

Permeable Deposits Sedimentary rocks are more likely than other rock types to possess a pervasive *permeability* that permits intimate contact between ore mineral and flowing leachant and hence a closer approach to so–called single-pass extraction. The techniques applied, therefore, have some analogies with those of petroleum recovery (which is, however, distinguishable by the involvement of immiscible fluids). *Hydraulic fracturing* of the ore by pressurizing the boreholes can be regarded in this situation as a means of increasing the effective hole diameter, whereas explosive fragmentation would short circuit the natural permeability and thereby depart from the ideal single-pass situation. The containment of leachant requires the mineralized zone to be at least underlain by an impermeable stratum; fur-

ther confinement by impermeable rock above and around the deposit is desirable, otherwise the deposit should lie beneath the static water table to prevent the unrestricted sideways loss of injected fluids. The flow of the leachant in an *aquifer* may be controlled by injecting water at various rates down surrounding boreholes (additional to the normal inlet or collector holes). Account may thus be taken of prevailing groundwater movements (provided they are slight) and anisotropies in permeability. Sandstone uranium operations are examples of this (Hunkin, 1971; Anon., 1974a).

Kennecott Copper Corporation has conducted pilot scale in situ leaching at a depth of 100 m in a hydraulically fractured low-grade copper deposit at Ruth, Nevada (Anon., 1972b). The same report describes plans by Dowell and Asarco to leach an Arizona copper deposit in situ at a depth of 400 m after hydraulic fracturing. The lithology of neither deposit was specified in the report, but since explosive fragmentation seems to have been unnecessary, we include them as examples of permeable deposits.

Impermeable Deposits Crystalline rocks are characterized by a low porosity so that joint and fracture systems are expected to be the only avenues for leakage of leachant from the ore body. A grout curtain has been suggested for preventing such leakage where it occurs (Livingston, 1957). The ore body usually requires explosive fragmentation. In some low-grade ores, mineralization is intensified along previous fractures and cleavages, which are easily reopened during fragmentation, thus providing easy access to leachant (Braun et al., 1974; Bennett and Parks, 1974). Some low-grade deposits host a smaller high-grade body for which conventional mining is appropriate, and the resulting shafts and crosscuts provide a convenient means of fragmenting the remaining ore by blasting and subsidence. At the Old Reliable Copper Mine, situated in a hillside in Arizona, the remnant low-grade ore was fragmented by explosives stacked directly in abandoned tunnels (termed a *coyote* blast); leachant was sprinkled over terraces on the hillside and was collected as it drained from the base of the deposit (Anon., 1972a; Longwell, 1974). At the Big Mike open pit copper mine, in Nevada, drill hole explosives blasted the low-grade walls into the pit for percolation leaching (Ward, 1974).

Leaching at an intermediate depth is being tested at the Agnew Lake Mine, Ontario, where uraniferous metaconglomerate has been fragmented in a stope between drives at the 260 m and 300 m levels. Leachant percolates from the upper level and is collected by a dam at the lower level (Lang and Morrey, 1976).

When impermeable deposits are deep and abandoned mine workings are absent, it is impractical to use chemical explosives in drill holes.

The use of downhole nuclear blasting, in preparation for in situ leaching, is the subject of several feasibility studies (Lewis and Braun, 1973; Lewis et al., 1974). A deep confined chimney of fragmented ore offers the possibility of leaching at elevated temperatures and pressures and with volatile reagents such as ammonia. Also there is growing interest in the extraction of geothermal energy from hot dry rock at depth, by downhole fracturing and subsequent circulation of water over the rock (Laughlin, 1975). Technological advances in this field could also apply to deep in situ leaching.

C. E. G. BENNETT
T. C. PARKS

References

Anon., 1972a, Ranchers' big blast shatters copper orebody for *in-situ* leaching, *Eng. and Mining J.* **173**(4), 98–100.

Anon., 1972b, Asarco and Dow Chemical to leach deep copper ore body *in-situ* in Arizona, *Eng. and Mining J.* **173**(6), 19.

Anon., 1974a, Solution mining opening new reserves, *Eng. and Mining J.* **175**(7), 62–71.

Anon., 1974b, Oil from shale through new Occidental method, *Eng. and Mining J.* **175**(2), 25–26.

Anon., 1974c, *In-situ* tests on Utah tar sands underway by USBM, *Eng. and Mining J.* **175**(11), 200.

Aplan, F. F., W. A. McKinney, and A. D. Pernichele, eds., 1974, *Proceedings of the Solution Mining Symposium.* New York: American Institute of Mining Enginners, 469p.

Bennett, C. E. G., and T. C. Parks, 1974, Leaching of low-grade nickel sulfide serpentinite ores: A way to chemical mining?, in *Proceedings of the Australia Institute of Mining and Metallurgy Annual Conference,* Melbourne, 401–416.

Braun, R. L., A. E. Lewis, and M. E. Wadsworth, 1974, In-place leaching of primary sulfide ores: Laboratory leaching data and kinetics model, *Metallurgy Trans.* **5**, 1717–1726.

Daneshy, A. G., 1974, Principles of hydraulic fracturing, in F. F. Aplan, W. A. McKinney, and A. D. Pernichele, eds., *Proceedings of the Solution Mining Symposium.* New York: American Institute of Mining, Metallurgy, and Petroleum Enginneers, 15–32.

Frasch, G., 1976, Perkin medal, address of acceptance, *Chemtech.* **6**, 99–105.

Hunkin, G. G., 1971, A revive of *in-situ* leaching, *Soc. Mining Engineers Trans.* Preprint 71-AS-88, 28p.

Lackey, J. A., 1975, Solution mining (*in-situ* leaching)—A literature survey, *Amdel Bull. No. 19*, 40–61.

Lang, L. C. and W. B. Morrey, 1976, Scientific blasting for *in-situ* leaching proves successful at Agnew Lake Mines, *Eng. and Mining Jour.* **177**(1), 100–102.

Laughlin, A. W., 1975, Hot dry rock tested for geothermal energy, *Geotimes* **20**(3), 20–21.

Lewis, A. E., and R. L. Braun, 1973, Nuclear chemical mining of primary copper sulfides, *Soc. Mining Engineers. Trans.* **254**, 217–228.

Lewis, A. E., R. L. Braun, C. Sisemore, and R. G. Mallon, 1974, Nuclear solution mining—Breaking and leaching considerations, in F. F. Aplan, W. A.

McKinney, and A. D. Pernichele, eds., *Proceedings of the Solution Mining Symposium.* New York: American Institute of Mining, Metallurgy, and Petroleum Engineers, 56–75.

Livingston, C. W., 1957, Method of mining ores *in-situ* by leaching, U.S. Patent 2,818,240.

Longwell, R. L., 1974, In-place leaching of a mixed copper ore body, in F. F. Aplan, W. A. McKinney, and A. D. Pernichele, eds., *Proceedings of the Solution Mining Symposium.* New York: American Institute of Mining, Metallurgy, and Petroleum Engineers, 233–242.

Maugh, T. H. II, 1977a, Oil shale: Prospects on the upswing . . . again, *Science* **198**, 1023–1027.

Maugh, T. H. II, 1977b, Underground gasification: An alternate way to exploit coal, *Science* **198**, 1132–1134.

Murr, L. E., 1980, Theory and practice of copper sulphide leaching in dumps and *in situ, Minerals Sci. Engineering* **12**, 121–189.

Porter, D. D., and H. G. Carlevato, 1974, *In-situ* leaching: A new blasting challenge, in F. F. Aplan, W. A. McKinney, and A. D. Pernichele, eds., *Proceedings of the Solution Mining Symposium.* New York: American Institute of Mining, Metallurgy, and Petroleum Engineers, 33–43.

Rouse, J. V., 1974, Environmental aspects of *in-situ* mining and dump leaching, in F. F. Aplan, W. A. McKinney, and A. D. Pernichele, eds., *Proceedings of the Solution Mining Symposium.* New York: American Institute of Mining, Metallurgy, and Petroleum Engineers, 3–14.

Shock, D. A. and F. R. Conley, 1974, Solution mining—Its promise and its problems, in F. F. Aplan, W. A. McKinney, and A. D. Pernichele, eds., *Proceedings of the Solution Mining Symposium.* New York: American Institute of Mining, Metallurgy, and Petroleum Engineers, 79–97.

Trudinger, P. A., 1971, Microbes, metals and minerals, *Minerals Sci. Engineering* **3**(4), 13–25.

Ward, M. H., 1974, Surface blasting followed by *in-situ* leaching, Big Mike Mine, in F. F. Aplan, W. A. McKinney, and A. D. Pernichele, eds., *Proceedings of the Solution Mining Symposium.* New York: American Institute of Mining, Metallurgy, and Petroleum Engineers, 243–251.

Cross-references: *Blasting and Related Technology; Borehole Drilling; Economic Geology.*

BORE LOG DATA—See SAMPLES, SAMPLING; WELL LOGGING. Vol. XIII: ELECTROKINETICS; WELL DATA SYSTEMS.

BORING AND DRILLING—See AUGERS, AUGERING; BOREHOLE DRILLING; BOREHOLE MINING. Vol. XIII: RAPID EXCAVATION AND TUNNELING; SHAFT SINKING; TUNNELS AND TUNNELING.

C

CANALS AND WATERWAYS, SEDIMENT CONTROL

Sediment control in canals and waterways is normally achieved by the complementary application of various sediment exclusion and ejection techniques. *Sediment exclusion* applies to methods by which sediment is deflected away from canal headworks and is prevented from entering the canal with the diversion water. *Sediment ejection* is the removal of sediment from the canal after the diverted water has transported the sediment into the canal. Numerous sediment exclusion and ejection schemes have been reported in the literature. This entry references only the general comprehensive publications by the ASCE (1972, 1975) and Melone et al. (1975), any of which contains a thorough bibliography of individual reports.

Sediment Control in Perspective

Both the expansion of cities and an increase in the demand for food throughout the world have made it necessary to divert more water from rivers into constructed canals for municipal uses and for irrigation. The removal of sediment from canals is a major economic consideration in the operation and maintenance of these canals. For some large projects elaborate and costly *desilting works* have been built; for smaller projects costly desilting works cannot be economically justified. Therefore, simple inexpensive methods for removing or excluding sediment from canals must be utilized.

When the sediment concentration in diverted water is quite substantial, sediment may accumulate in the canals and interfere with the operation of the water conveyance system. Some problems often encountered are (1) the deposition of sediment in some reaches of the canal, thereby reducing the channel's conveyance capacity and making frequent cleaning necessary; (2) excess sediment's entering power turbines; (3) municipalities requiring settling ponds or water treatment facilities for removing sediment from their water supply; and (4) sediment carried in irrigation canals being detrimental to or in some instances increasing the soil productivity on farms where the water is used.

A great number of variables and a complex interaction of these variables are involved in river mechanics and sediment-related problems (see *Alluvial Systems Modeling*). As a result, analytical design criteria do not presently exist for most river and canal structures. It is therefore imperative that each field situation be individually analyzed and that any necessary sediment control works be designed for the specific field conditions.

Rational Approach to Sediment Control

The selection and design of sediment control works proceed in two distinct steps. The first is to analyze the field conditions; the second is to select a general sediment control method and make the appropriate modifications to fit the specific field conditions. Before detailed design of sediment control works can proceed, a program for field data collection must be established. At proposed diversion sites, analysis of river flow and sediment data is essential so the design can proceed beyond conceptual drawings. The interaction of water and sediment is treated in detail in texts by Graf (1971), Yalin (1972), Bogardi (1974), and Shen (1971, 1972). The initial field investigation may indicate that certain sediment control techniques are totally unworkable and can thus be eliminated from consideration.

During the preliminary design stage, the present state of knowledge in sedimentation engineering is adequate to the point of selecting a sediment control scheme. Final design normally requires a hydraulic model study so that aspects of sediment and water interaction that are not analytically understood can be physically modeled and observed. Having selected a general type of sediment control, a physical model provides a means by which refinements and modifications can be investigated so that compatibility with specific field conditions is ensured.

Sediment Exclusion

Canal Headworks General guidelines for diverting water from a main channel into a canal system can be major factors in minimizing the degree to which sediment control works are needed. Figure 1 shows the vertical distribution of suspended sediment for different particle size groups from one sediment sample. Note that the coarse sediment sizes are more heavily concentrated in

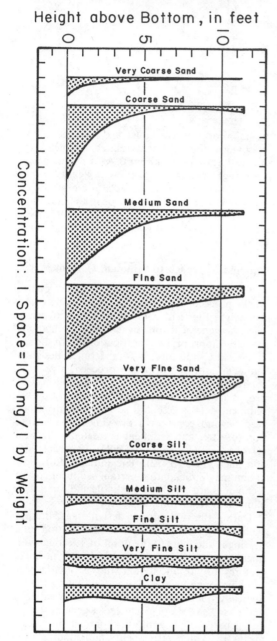

Height above Bottom, in feet

Concentration : I Space = 100 mg/l by Weight

Very Coarse Sand

Coarse Sand

Medium Sand

Fine Sand

Very Fine Sand

Coarse Silt

Medium Silt

Fine Silt

Very Fine Silt

Clay

FIGURE 1. Discharge-weighted concentration of suspended sediment for different particle size groups at a sampling vertical in the Missouri River at Kansas City, Missouri. (From Guy, 1970)

the lower part of the flow, whereas the finer material is more uniformly distributed throughout the vertical. Therefore, by attempting to draw only surface water from the river into the canal, the sediment problem can be significantly reduced.

The sediment inflow to the canal can be further

reduced by the proper selection of the diversion site. The outside of a river bend is a favorable location for canal diversion headworks. Briefly, when a river flows around a bend, the higher-velocity surface water moves along the outside of the bend while that portion of flow carrying the major part of the bed load is deflected inward (Fig. 2). In India and Pakistan extensive *training works* have been constructed to create artificial channel curvature at canal intakes and hence benefit from this flow phenomenon. In one instance, the Kotri Dam design (see reference in ASCE, 1975) utilized a natural island to provide concave curvature at canal headgates located on opposite banks of the river. Caution must be observed when locating a stable bend because flow conditions vary with discharge and, therefore, what may appear to be favorable curvature at one river flow rate may be unfavorable at a higher flow rate. Also, river meanders tend to migrate downstream, which may eventually shift the favorable flow path.

A typical river bend cross–section is shown in Fig. 3. The river thalweg, at the outside of the bend, is near the right bank; point bar formation is evidenced by the shallower depths at the inside of the bend along the left bank. Note that sediment concentrations are much less near the water surface than near the river bed and that the concentrations are less at the outside of the bend than along the point bar at the inside of the bend.

No conclusive data are available to determine quantitatively the influence of the off-take orientation on the exclusion of sediment. The *headworks* orientation for optimum exclusion is probably a function of discharge ratio and must be determined from experiment. One study has shown qualitatively that an outlet at 90° draws

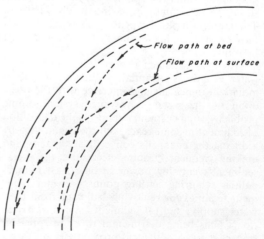

Flow path at bed

Flow path at surface

FIGURE 2. Schematic diagram of flow in curved channel.

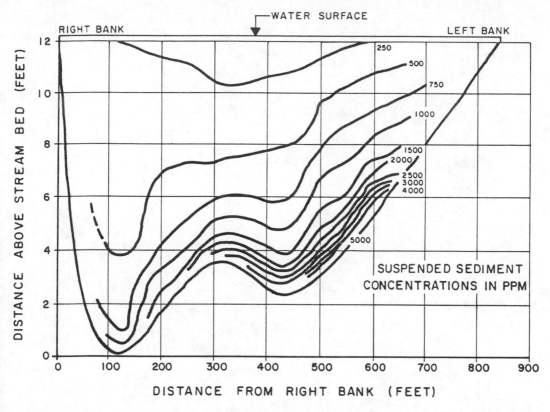

FIGURE 3. Distribution of suspended sand (mg/l), Missouri River near Omaha, Nebraska. (From ASCE, 1975)

more sediment than an outlet at 45° to the direction of flow in the main channel. An explanation is that the higher–velocity surface water is more likely to shoot past a perpendicular off-take than one at 45°. Consequently, the more slowly moving water near the bed that carries the greater concentration of sediment tends to flow into the diversion channel.

Thus, the point of diversion, angle of diversion, and canal headworks design can be selected to benefit from natural flow phenomena. Each of these considerations serves to minimize the sediment problem in the canal system and hence may reduce both the number and size of necessary sediment control structures.

Guide Walls In addition to selecting a diversion site with a desirable natural flow pattern, local *guide walls* have proven beneficial in further inducing a favorable curvature of flow. The aim of guide walls is to create a local channel bend immediately opposite the canal intake so that the major portion of the sediment load is deflected toward the inside of the bend and away from the canal headworks.

The U.S. Bureau of Reclamation has successfully combined guide walls with continuous *sluic-ing operations* to increase the sediment exclusion efficiency of many canal intake structures. Two typical designs resulting from extensive model studies are shown in Figs. 4 and 5. Table 1 presents model data indicating the pronounced effect of the guide walls. Note that in Table 1, C_s is defined as the sediment concentration in the sluiceway, and C_H is the sediment concentration passing through the diversion headworks. The subscript P refers to the preliminary design; the subscript GW refers to the design with a guide wall included. A high C_s/C_H is desirable. For example, if $C_s/C_H = 6$, then the sediment concentration through the sluiceway is six times the concentration passing through the headworks into the canal.

Guide walls have typically been designed by a trial and error procedure using physical models. Although the final design almost certainly requires a hydraulic model, conceptual sketches can be considered by utilizing various empirical equations that interrelate equilibrium channel width and meander length and radius of curvature for a given discharge. One set of these relations was developed by Schumm (cited in Shen, 1971) based on data collected at 36 stable alluvial river reaches.

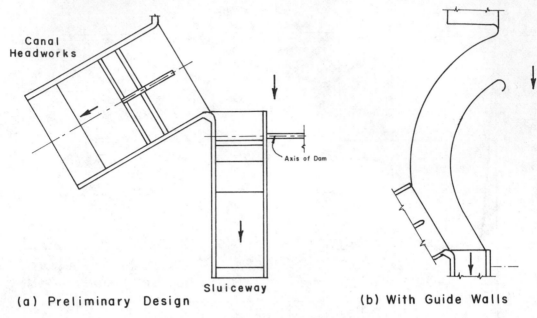

FIGURE 4. Bartley Headworks. (From Enger and Dodge, 1954)

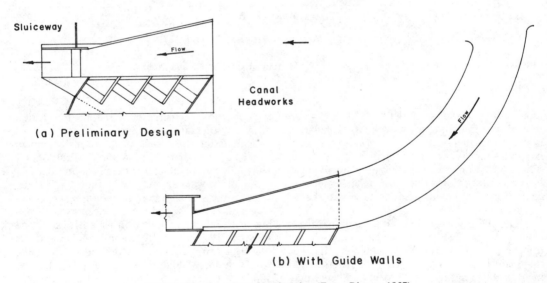

FIGURE 5. Angostura Headworks. (From Rhone, 1957)

TABLE 1. Effect of Guide Walls on Sediment Exclusion

Diversion Headworks	Discharge (cfs)		Preliminary Design $(C_s/C_H)_P$	With Guide Walls $(C_s/C_H)_{GW}$	Improvement Ratio $(C_s/C_H)_P/(C_s/C_H)_{GW}$
	Headworks	Sluiceway			
Bartley	59	38	0.41	6.69	16.3
Angostura	500	150	0.17	6.79	39.9

Source: From Melone, Richardson, and Simons, 1975.

Guide Walls and Stilling Ponds   In India and Pakistan flow rates up to 625 m³/s are diverted from major rivers. A common practice on these large river systems is to construct a guide wall parallel to the canal headworks, extending upstream from a head regulation structure across the river (Fig. 6). The guide wall produces a pocket into which sediment is deposited instead of passing into the canal.

Depending on the regulation of undersluices, the pocket is classified either as a *still pond* or a *semiopen* system. A still pond exists when the undersluices are kept closed. Although this method is effective at reducing sediment entry into the canal, it necessitates repeated canal closures to pass the settled sediment through the undersluices. A semiopen flow system exists when the undersluices are kept continuously open. This system is less effective in reducing sediment inflow to the canal since the velocity upstream of the undersluices is greater than the velocity in a still pond. The chief advantage of the semiopen system is that canal closure is not necessary. The exclusion efficiency of both the still pond and semiopen systems is further enhanced by the operating procedure of the regulation structure outside of the pocket. Undersluices included in the weir outside the pocket maintain a well-defined channel in front of them and can be operated to create a flow curvature that deflects sediment away from the pocket. The best location for undersluices that will work in conjunction with the pocket varies with each project.

Guide Vanes.   Both bottom and surface vanes have been installed near canal intakes to create favorable secondary currents that reduce sediment inflow to canals. Bottom vanes are located on the stream bed and deflect the heavier sedi-ment concentration along the bed away from the intake structure. Surface vanes are set near the water surface at an angle that directs that portion of the flow transporting a relatively light sediment load toward the canal intake.

The U.S. Bureau of Reclamation (Enger and Carlson, 1962) investigated guide vanes in conjunction with the design of the headworks of the Socorro Main Canal, located at the San Acacia Diversion Dam. The bureau conducted 37 tests to determine the effects of vane spacing, angle, length, elevation, cross-section, number, and location with respect to the canal headworks. The flow patterns created by guide vanes are depicted in Fig. 7. The model study indicated that, for a river discharge of 250 m³/s and diversion discharge of 5 m³/s, the vane characteristics were not critical to the performance of the vanes. The bottom and surface vanes produced approximately equivalent results; the concentration ratio (canal concentration/river concentration) was reduced from 2.38 with no vanes to an average of 0.1 with the guide vanes.

In a series of papers, King (see reference in Melone et al., 1975) discussed the design of curved bottom vanes (Fig. 8). The so-called King's vanes excluded practically the whole of the bed load when a single discharge was run, though their efficiency depended greatly on the vanes' being in line with the oncoming current. This alignment cannot be assured if there is a range of river discharges or if a widely varying proportion of the river discharge is diverted into the canal. Then the vanes perform inefficiently due to sand banks forming downstream of the vanes, and where vanes are not parallel to the river flow, coarse material is thrown into suspension. When discharge conditions are relatively steady, curved bottom

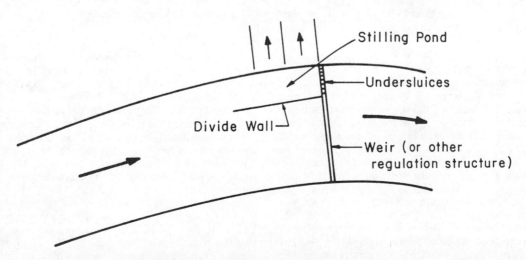

FIGURE 6. Definition sketch for divide wall and stilling pond.

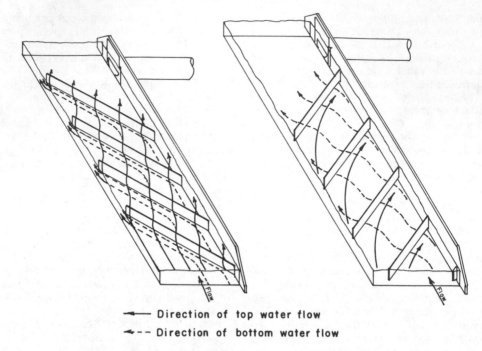

→ Direction of top water flow

◄-- Direction of bottom water flow

FIGURE 7. Flow patterns past guide vanes. (From Carlson and Enger, 1962)

PLAN

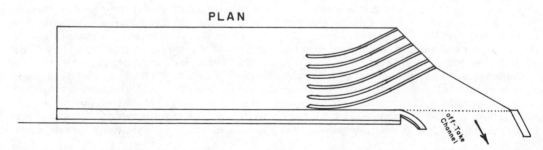

FIGURE 8. Typical design sketch for King's curved bottom guide vanes.

guide vanes perform efficiently, especially if complete exclusion is required, but under varying flow conditions they should be avoided.

Tunnel Diverters Tunnel diverters are designed to benefit from the vertical distribution of sediment; that is, the more highly concentrated and coarser sediment is transported along the bed of a river, while the upper portion of the flow has a much lower sediment content. The tunnels provide a lower chamber that splits the flow horizontally; the lower portion of the flow passes into the tunnel and is directed back to the river, leaving the upper portion of the flow to be diverted into the canal. Tunnel diverters were originally developed for large diversions in India and Pakistan although with sufficient model testing, they should function well for small diversions.

The main design consideration is that sufficient velocity be maintained through a tunnel so that no sediment settles and clogs the tunnel. This consideration causes each tunnel diverter design to be very site specific. For instance, the required velocity is related to the sediment size distribution being transported; the ease with which a compatible tunnel can be designed is then related to the discharge and head available. A hydraulic model study is usually undertaken to determine an optimum design.

Early model studies have provided the considerations in the design of tunnel diverters that must be investigated for each site. Studies have shown that the percentage of flow necessary for efficient sluicing through the tunnel varies with each design, that the position of the tunnel opening is more important than the number of tunnels, that the opening of the tunnel should be at the upstream end only because side openings decreased the efficiency, that the zone of suction of each tunnel extends up to two times the width of the tunnel, that a restriction or bell-mouthed entrance increases the efficiency, and that the outfall channel must have sufficient capacity to transport the sediment back to the river.

Sediment Ejection

Sediment ejectors downstream from the canal headworks are beneficial for two reasons: (1) to remove sediment that was not excluded from the canal and cannot be transported through the canal system and (2) to provide a factor of safety should exclusion devices fail to perform according to design. Sediment ejectors should always be included immediately downstream from the canal headworks. In some instances ejectors will be needed throughout the canal system. With each turnout and corresponding withdrawal of water, the main canal will have a reduced capacity to transport sediment. Therefore, sediment that was in suspension in the upper reaches of the canal may still settle to the bed in the lower reaches of the canal. The function of sediment ejectors is, first, to complement the performance of the sediment exclusion structures and, second, to serve as sediment control devices along the entire canal system.

Settling Basins Such basins are a popular method for the removal of both bed load and sus-pended load at canal headworks. The underlying principle is simply to provide a section wide and long enough so that the resulting reduced flow velocity will allow the sediment to settle out. A recommendation is to widen the initial section of every canal just downstream from the headworks, even for projects where the employed exclusion techniques are considered to be adequate. Then, if the exclusion devices fail or excess sediment enters the canal, maintenance need be performed at only one section of the canal as opposed to long lengths along the canal system.

Settling basins can be considered to be of two types. One type is termed *off-line*. The All-American Canal Desilting Works in southwestern United States is of this type. Water is diverted from the Colorado River and routed directly to the desilting installation from which the clear water is then routed to the canal. The diversion into the All-American Canal (Fig. 9) is about 345 m^3/s, which carries a sediment load of 90,000 ton/day. The All-American Desilting Works is unique in that it consists of three basins, each containing 24 rotary scrapers, each 40 m in diameter. Due to its size and cost of operation it is unlikely similar installations would be economically feasible in the future for sediment removal from irrigation water. The off–line desilting basin is considered only in special cases.

A second type of settling basin is termed *in-line*. An in-line settling basin is simply a widened or deepened section of the canal. The enlarged cross section will cause a lower flow velocity, allowing the sediment to settle out. The Fort Lar-

FIGURE 9. Imperial Dam and Desilting Works, Colorado River, Arizona, United States (looking upstream), with the All-American Canal Desilting Basin (left) and Gila Valley Main Canal (right). (From U.S. Bureau of Reclamation)

amie Canal Desilting Basin in Wyoming is of this type (Fig. 10). One major drawback of in-line basins, which exclude sediment by sluicing, is that flow to the canal must be periodically cut off in order to perform the sluicing operation.

For both off-line and in-line settling basins, various physical and economic constraints need to be considered. Each type of settling basin must be provided with some means of removing the sediment after settling out. For in-line basins sluicing is a common method, though caution must be observed because, depending on the percentage of flow diverted, the river may have a greatly reduced capacity to transport the sediment returned to the river. Therefore, in some instances, if sediment is routed back to the river, the equilibrium of the river may be upset and aggradation can take place downstream from the diversion dam. Other settling basins require mechanical means such as hydraulic dredges, draglines, and scrapers for the removal of sediment. In each case an expense is involved in the operation, storage, and maintenance of equipment. Not only is cost a factor, but also the physical constraint of a disposal area must be considered. The sediment must be dumped and spread over the disposal area; sometimes double handling is necessary because the sediment must be hauled by trucks some distance to the disposal area. Sediment removal by sluicing and mechanical methods are the only means available. Sluic-

ing is by far the cheaper, but the design engineer must be familiar with the stability of the river system.

Vortex Tube The vortex tube sediment ejector (Fig. 11), like many hydraulic devices, has no one proven design that is suitable for universal application. Each vortex tube now in use was specially designed to meet the field conditions for each project. Properly designed vortex tubes have proven to be extremely effective in the removal of coarse bed material up to the gravel and cobblestone size range.

The actual hydraulics of the flow of the water-sediment mixture through the vortex tube has not been determined; hence, a conceptual approach and design is not possible. The main feature of the vortex tube sand trap is a tube with an opening along the top side. As water flows over the tube, a shearing action across the open portion sets up a vortex motion within and along the tube. This whirling action catches the bed load as it passes over the lip of the opening and carries the sediment to the outlet at the downstream end of the vortex tube. Due to uncertainties in the design, a downstream valve is sometimes included to adjust the tube outflow for maximum efficiency. Investigations have provided qualitative guidelines and cautions, but no concrete information is available that would allow a design engineer, knowing the field conditions, to design a vortex tube ejector and confidently predict its performance. The design engineer has at his or her disposal only experience with existing vortex tubes, comparison with experimental studies, and various guidelines that have been suggested. A summary of design guidelines is given by Melone et al. (1975).

Tunnel-Guide Vane Ejector Another sediment ejector in wide use in the lower reaches of canal systems can be found in the Upper Bari Doab Canal at Madhopur Headworks. The ejector system consists of three or four tunnels directed into

FIGURE 10. Fort Laramie Canal Desilting Basin, looking downstream (a) and upstream (b). (From Carlson, 1951)

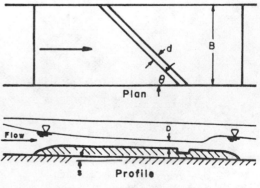

FIGURE 11. Vortex tube sediment ejector.

the flow and covered by a roof that extends beyond the tunnel mouth for about 45 cm. The roof serves to bifurcate the flow, allowing for that portion of the flow carrying the coarser sediment along the canal bed to be diverted out of the canal to an outfall channel.

To achieve high efficiency, each tunnel–guide vane ejector must be designed for the specific field conditions. Experience has shown that a velocity of 3 m/s with a 75 cm head differential is sufficient for maintaining the ejector operation. The major drawback to this system is that up to 25% of the canal flow is used in the ejector operation. This drawback requires an original headworks diversion from the river significantly above the nominal canal requirements. In areas where water is scarce, the tunnel–guide vane ejector may not be acceptable.

ANTHONY M. MELONE

References

ASCE, 1972, Sediment control methods: Control of sediment in canals, *Hydraulics Jour.* **98**(HY9): 1647–1689.

ASCE, 1975, *Sedimentation Engineering.* New York, 745p.

Bogardi, J., 1974, *Sediment Transport in Alluvial Streams.* Budapest, Hungary: Akademiai Kiado, 650p.

Carlson, E. J., 1951, *Hydraulic Model Studies of Fort Laramie Canal Desilting Basin, North Platte Project.* Denver, Colo.: Bureau of Reclamation, U.S. Department of the Interior, (Hydraulic Laboratory Report No. HYD 313.

Carlson, E. G., and P. F. Enger, 1962, *Hydraulic Model Tests of Bottom and Surface Vanes to Control Sediment Inflow into Canal Headworks.* Denver, Colo.: Bureau of Reclamation, U.S. Department of the Interior, August. (Hydraulic Laboratory Report No. HYD 499).

Enger, P. F., and E. J. Carlson, 1962, *Hydraulic Model Study to Determine a Sediment Control Arrangement for Socorro Main Canal Headworks, San Acacia Diversion Dam, Middle Rio Grande Project, New Mexico.* Denver, Colo.: Bureau of Reclamation, U.S. Department of the Interior (Hydraulic Laboratory Report No. HYD 479).

Enger, P. F., and R. A. Dodge, 1954, *Hydraulic Model Studies of Bartley Diversion Dam—Progress Report No. 3 on General Studies of Headworks and Sluiceway Structures, Missouri River Basin Project, Nebraska.* Denver, Colo.: Bureau of Reclamation, U.S. Department of the Interior, February (Hydraulic Laboratory Report No. HYD 384).

Graf, W. H., 1971, *Hydraulics of Sediment Transport.* New York: McGraw-Hill, 513p.

Guy, H. P., 1970, *Fluvial Sediment Concepts, Book 3.*

Lepold, L. B., M. G. Wolman, and J. P. Miller, 1964, *Fluvial Processes in Geomorphology.* San Francisco: W. H. Freeman, 522p.

Melone, A. M., E. V. Richardson, and D. B. Simons, 1975, *Exclusion and Ejection of Sediment from Canals.* Fort Collins: Department of Civil Engineering, Colorado State University [No. CID–211 (d)], 192p.

Mercer, A. G., 1971, Diversion structures, in H. W. Shen, ed., *River Mechanics.* Fort Collins, Colo.: H. W. Shen.

Rhone, T. J., 1957, *Hydraulic Model Studies to Develop a Sediment Control Arrangement for Angostura, Diversion, Middle Rio Grande Project, New Mexico.* Denver, Colo.: Bureau of Reclamation, U.S. Department of the Interior, February (Hydraulic Laboratory Report No. HYD 419).

Shen, H. W., ed., 1971, *River Mechanics,* vols. 1 and 2. Fort Collins, Colo.: H. W. Shen, 1162p.

Shen, H. W., ed., 1972, *Sedimentation.* Fort Collins, Colo.: H. W. Shen, 798p.

Uppal, H. L., 1951, Sediment excluders and extractors, in *Proceedings of the 4th General Meeting of the International Association for Hydraulic Research, Bombay, India,* 261–316.

Yalin, M. S., 1972, *Mechanics of Sediment Transport.* New York: Pergamon Press, 290p.

Cross-references: *Alluvial Systems Modeling.* Vol. XIII: *Channelization; Coastal Inlets, Engineering Geology; Dams, Engineering Geology; Hydrology; Land Drainage; River Engineering.*

CARTOGRAPHY, GENERAL

Cartography is the art, science, and technology of making maps, together with their study as scientific documents and works of art. In this context, maps may be regarded as including all types of maps, plans, charts and sections, three-dimensional models, and globes representing the Earth or any heavenly body at any scale. This definition of cartography was adopted by the International Cartographic Association, October 26–27, 1967, in Bad Godesberg, Germany. According to Raisz (1938), "The surveyor measures the land, the cartographer collects the measurements and renders them on a map, and the geographer interprets the facts thus displayed."

Early Cartography

It is difficult to establish a date for the beginning of cartography. There are indications that some sort of cartography was undertaken 5,000 yr ago. A clay tablet showing an estate in a Babylonian town, which flourished in 2500 B.C., could be taken as evidence of this. Quite a number of Babylonian city plans from the seventh century B.C. have survived. It is known, too, that Babylonians established the *sexigesimal system,* dividing the circle into 360° and the degrees into minutes and seconds, as well as dividing the day into hours, minutes, and seconds Andree (1877). The invention of the *gnomon,* a device for determining latitude from the angle of the sum, could be credited to the Babylonians. Although such discoveries were essential to establish the basis for relating the motion of the Earth to the sky, they were not a great contribution to cartography.

The first scientific approach to cartography was made by the Greeks. Eratosthenes developed the *system of parallels and meridians.* He also undertook to measure the Earth. The result was remarkable because he came within 14% of the true size. Claudius Ptolemy, however, contributed the most to the development of early cartography. His influence was so strong that cartography was dominated by the Ptolemy teaching for more than 1,500 yr. His most important works were the eight volumes of *Geographia,* in which he presented detailed cartographic instructions. He also constructed the first world map on a modified conic projection (see Vol. I: *Oceanography*).

The growing political power of the Romans replaced the Greeks also in the field of cartography. Unfortunately, their original contributions were not so great, and not many maps of that time survived. We know about their existence only from descriptions or references to them. One of the most important maps preserved is the *Peutinger Table.* It is a map showing roads in parts of the Roman Empire at about A.D. 250. The map is rather peculiar in shape and size, for if assembled it would be 3 m long and only 25 cm high; it was not drawn to scale and was without a geographical grid.

After the Roman Empire collapsed, the Arabs became active in science. Once again Ptolemy's *Geographia,* translated into Arabic, served as a basis for cartography. One of the most noted Arab geographers, Idrisi, described and calculated the circumference of the Earth and compiled a map of the world.

Despite the fact that exploration and traveling in the medieval period was intensified, expanded geographical knowledge was not reflected in maps. Surviving copies can be divided into three distinct groups. Maps showing the world were known as *Mappae Mundi.* They offered relatively little of scientific value, and as a common feature they were centered on Jerusalem. Second, there were sketches of individual regions that supplemented descriptions, being more accurate and obviously of more value. However, the most important cartographic achievements of that time were the first nautical, charts known as *protolans;* the Pison chart ca. 1300 and the famous *Catalan Atlas* of 1375 are excellent examples.

Improved shipbuilding and experience gained in navigation awakened further interest in cartography. The fifteenth and sixteenth centuries witnessed some tremendous advances. Not only were individuals trained in the preparation and production of maps, but also special schools were established to teach the arts of navigation and map production. The Portuguese at that time were leading in this field, and Prince Henry V ("the Navigator") especially promoted exploration. For economic as well as political reasons, maps were considered confidential, classified material. The situation has not changed much even to the present time (detailed nautical charts and topographic maps of many areas are still classified) (Harvey, 1980).

The great change in maps between the medieval period and the fifteenth and sixteenth centuries can be seen on the *Fra Mauro map of the world,* 1457–1459. The circle form still resembled the medieval *Mappae Mundi,* but the content was much different. The Fra Mauro map also marked the beginning of the Italian contribution to the revival of cartography. Mercantile centers such as the city republics of Venice, Bologna, Rome, and others also became cartographic centers. The art of cartography eventually spread throughout Europe, and as a result, cartographic centers such as Munster, Koln, and Nurnberg were able to produce maps of superior quality. Martin Waldseemuller produced his famous world map in 1507, on which for the first time the name *America* appears; it is a very fine example of such cartography.

A great example of English cartography of that time (1583) would be a map of England and Wales, *Britannia Insularum in Oceano Maxima,* which is the masterpiece of the famous cartographer William Saxton. For a long time, this map was used by many cartographers as the basis for maps of England.

During the sixteenth and especially the seventeenth centuries, cartography was dominated by the Dutch. Noted cartographers and publishers such as Gerhard Mercator, his friends Ortelius, Blaeu, and Hondius broke away completely from the Ptolemaic tradition and compiled maps according to the new information collected from the latest explorations. One of the important achievements of Mercator was the construction of the now famous projection known as the *Mercator projection,* which is used extensively even today. Among many other maps produced by him, the map of Flanders, which he personally surveyed, compiled, drafted, and engraved, deserves special attention. He also used, for the first time, on his map of the world in 1538 the designations *North America* and *South America.* His greatest work, however, *Atlas sive Cosmographicae,* was published after his death.

The development of eighteenth century cartography was strongly influenced by French scientific experiments, sponsored by the French Academy Royal, regarding the size and shape of the Earth, the measuring of longitude, etc. New data influenced the production of more accurate maps. For example, Jean Domenique Cassini prepared a world map that was incorporated in the floor of the Paris Observatory in 1682; it had an equidistant projection centered on the North Pole. Jacques Cassini began the triangulation of France in 1733 that was accomplished by his son, Caeser François Cassini, in 1746. This work resulted in the survey of France and the preparation of a detailed map, the famous *Cassini Carte Géometrique de la France,* on a scale of 1:86,400, published in 1798. Similar government mapping was initiated in Belgium in 1770; Great Britain, 1784 (actually work started much later); the United States, 1879.

The development of American cartography, private as well as government, was very active from the beginning. The publication of Tanner's *New American Atlas* in 1823 could easily compete with European atlases of that time. The increased exploration of the American West created demands for more detailed maps. The mapping responsibility was taken over by the newly reorganized Army Topographical Bureau. The first of the four great western surveys was conducted from 1867 through 1879. Due to the rapid growth of mapping responsibilities in 1879, the U.S. Geological Survey was established, and all government mapping was entrusted to this organization.

Twentieth Century Cartography

The demands for more and better maps, as well as the rapid development in technology, have changed twentieth century cartography. One such change is the use of photogrammetry (see *Photogeology*), the process of making maps from photographs, especially aerial photographs. It is possible to use a multiple choice of colors in map printing, enabling the production of physical as well as thematic colored maps at low cost and at greater speed (see *Maps, Logic of*). The 1980s witnessed the increased application of computers, especially in the compilation of maps and projections.

Automation in cartography shows very fine results, bringing forth another important aspect that is characteristic of twentieth century cartography—international cooperation. The first attempt in this direction was made by the German geographer, Albrecht Penck. He proposed the preparation of a world map on the scale of 1:1,000,000, using uniform symbols, colors, size of sheets (see *Map Symbols*), etc. Despite two world wars the publication of this fine work was not stopped, which indicates a recognition of the importance of having a complete map coverage of the world for manifold economic and research purposes.

One of the most remarkable achievements in this field was the publication of the *Map of Hispanic America,* in 107 sheets, on the scale of 1:1,000,000, prepared and published by the American Geographical Society. Mapping cooperation among countries—e.g., those of the Pan-

American Union consisting of Latin American countries and the United States—has proved to be one of the more important means for improving the economic as well as the political situation. Some fine results are also achieved through the United Nations with the publication of the *Geological Map of Asia and the Far East,* the *Population Map of the World,* and others.

R. DRAZNIOWSKY

References

Andree, R., 1877, Die Anfänge der Kartographie, *Globus* **31**, 24–27, 33–43.

Bagrow, I., 1964, *History of Cartography.* Cambridge, Mass.: Harvard University Press, 312p.

Campbell, E. M. J., 1962, Landmarks in British cartography: The beginnings of the characteristic sheet to English maps, *Geog. Jour.* **128**, 411–415.

Colvocoresses, A. P., 1975, Evaluation of the cartographic application of ERTS-i imagery, *Am. Cartographer* **2**(1), 5–18.

Crone, G. R., 1953, *Maps and Their Makers, An Introduction to the History of Cartography.* London: Hutchinson, 181p.

Cuff, D. J., and M. T. Mattson, 1981. *Design and Production of Thematic Maps.* New York: Methuen, 160p.

Harvey, P. D. A., 1980, *The History of Topographical Maps.* New York: Thames and Hudson, 199p.

Lawrence, G. R. P., 1979, *Cartographic Methods.* New York: Methuen, 154p.

McEwen, R. B., and A. E. Elassal, 1978, USGS digital cartographic data acquisition, *U.S. Geol. Survey Open File Report,* 21p.

Peucker, K., 1901, Zur Kartographischen Darstellung der dritten Dimension, *Zeitschr. Geog.* **7**, 22–41.

Raisz, E. J., 1938. *General Cartography.* New York: McGraw-Hill, 370p.

Reeves, R. G., ed., 1975, *Manual of Remote Sensing.* Falls Church, Va.: American Society of Photogrammetry, 86p.

Robinson, R. B., and R. D. Sale, 1969, *Elements of Cartography.* New York: Wiley, 415p.

Ristow, W. W., 1943, Maps—How to make them and read them, A bibliography of general and specialized works on cartography, *New York Public Library Bull.* **47**(6), 381–386.

Tooley, R. V., 1952, *Maps and Map-makers.* London: Batsford, 140p.

Cross-references: *Geological Highway Maps; Map Abbreviations, Ciphers, and Mnemonicons; Map Series and Scales; Map Symbols; Maps, Physical Properties; Maps, Environmental Geology; Maps, Logic of; Photogeology; Remote Sensing, Societies and Periodicals; Vegetation Mapping.*

CAT CLAYS

The terms *cat clay* or *cat clay phenomenon* are well known to pedologists but unfortunately are not familiar to many geologists, though the phenomenon is an important and widespread geo-chemical and biogeochemical process. An engineering geologist should be able to predict and warn of the possibility of formation of cat clay in a developing area because of its low agricultural productivity, acidity, and corrosiveness.

A more scientific synonym for cat clay is *acid sulphate soils*. The English term *cat clay* is a translation of the original Dutch name *Katteclei-gronden*. The Dutch name was established in about the seventeenth century and was given to soils of some reclaimed areas that became grad-

ually highly acidic and developed prominent yellowish mottling and crusts composed of jarosite and related sulfates, which resembled cat excreta. In northern Germany similar clays were called *Maibolt*, a combination of two words: *hay field* and *Kobolt*, 'an evil ghost.' The first scientific description of the phenomenon was made by Linnaeus in 1735 who described cat clays as *argilla vitriolacea*.

Typical cat clay develops after drainage and aeration of reduced marsh, lake, or marine sediments containing notable amounts of pyrite and/

FIGURE 1. Unreclaimed (A) and reclaimed (B) tidal marshes on the northern shore of San Pablo Bay, north of the city of San Francisco, California. The lowest pH values of reclaimed soils ranged from 3.0 to 3.5 (January, 1985)

or other sulfides (see Vol. XV: *Coastal Soils*). Their oxidation produces soluble, highly acidic sulfates and leads to a partial decomposition of clays and other alumosilicates (Finkl, 1982). Soils of such reclaimed areas gradually become acidic, with pH values of about 2–4, and rich in soluble toxic aluminum ions derived from decomposed aluminosilicates. Geologically, such cat clays are only ephemeral features. Gradual leaching of soils eventually decreases their acidity and improves their agricultural properties (see Vol. XII, Pt. 1: *Leaching*). Typical cat clays are, therefore, essentially restricted to relatively newly reclaimed areas in the coastal zone, usually underlain by Holocene sediments. Intense land development in recent years has led to a spreading of cat clay phenomena on all continents except Antarctica. Several localities with cat clays are recorded on the Atlantic coast of the United States. Probably one of the largest deposit areas of cat clays on the Pacific coast is on the north shore of San Pablo Bay, north of San Francisco, California (Fig. 1).

Improvement of cat clays for agricultural purposes is an expensive procedure. For example, several methods used in the Netherlands involved application of carbonates from barges prior to reclamation to neutralize acidity, spreading of carbonate after reclamation, deep plowing of cat soils underlain by marl, leaching by flushing of cat soil, and even importation of new topsoil. The feasibility of such improvements should be carefully evaluated prior to reclamation of a potential cat clay deposit (see Vol. XII, Pt. 1: *Management of Soils*).

A comprehensive bibliography of early literature on cat clays was given by Van der Spek (1950). More recent literature is summarized in numerous papers presented at the International Symposium on Acid Sulphate soils (Dent, 1986; Dost, 1973a).

A phenomenon geochemically resembling cat clay is not restricted to recently reclaimed lands. Development of an acidic medium with iron sulfates may be observed in air-exposed natural and artificial cuts and mining dumps exposing sulfide-bearing formations to air and moisture—e.g., mines of sulfide ore and lignite strip mining (see Vol. XII, Pt. 2: *Minesoils*). An extremely low pH value of 0.1 was recorded, e.g., by the author in a small dump at a molybdenum-sulfide prospect at the southern end of Pyramid Lake, Nevada (Fig. 2). Another good example of acidic deposits associated with past mining and smelter operations occurs in the Spring Creek basin in northern California (Prokopovich, 1965). Toxic, acidic smelter fumes denuded local vegetation and acidified top soils (pH 3–5) in the vicinity of the old smelter, while erosion of old mine dumps and acidic and toxic surface runoff resulted in a rapid deposition of highly acidic and toxic alluvium

FIGURE 2. Foster's Camp prospect on the southwest bank of the desert of Pyramid Lake, Nevada. The prospect was initially developed in 1920 for gold and silver exploration, but in 1930 it was used for the study of copper–molybdenum mineralization. Some dump materials at the prospect had field pH values of 0.1. (July 1975)

(Fig. 3). The lowest pH of some saturated soil pastes of old tailings was 0.7, while water seeps from some old mines had a pH of 1.4–1.6 and less.

Similar acidity and mineralization was reported at some seepages of natural gas associated with hydrogen sulfide. The hydrogen sulfide (H_2S) associated with methane (CH_4) is generated frequently by a bacterial reaction between CH_4 and SO_4 ions in groundwater. Following reactions between the hydrogen sulfide and soil create iron sulfides while oxidation of hydrogen sulfide and other sulfides creates native sulfur, sulfates, and acidic media. An interesting example of such type of alteration of sediments was described in human excavations on the western margin of the San Joaquin Valley, California (Prokopovich et al., 1971). Pyrite, marcasite, probably colloidal (?) iron sulfate pigment, native sulfur, alunite, jarosite, and gypsum were associated here with hydrogen-sulfide-bearing methane seeping from the Upper Cretaceous beds into Cenozoic fluvial deposits and artificial fill. The pH values of sediments frequently ranged from 1.5 to 4.5 and locally were as low as 0.5.

An inspection of buried discharge pipes of the O'Neill Pumping Plant (Fig. 4) constructed in this vicinity revealed several through holes (Fig. 5) in the 1-cm thick walls of the steel pipes. The corrosion started from the outside surface of pipes and was caused by the acid associated with gas seepages. Proposed remedial action included removal of clayey, gypsiferous backfill; patching the holes; and backfilling excavated areas with clean gravel. This gravely nongypsiferous fill will per-

FIGURE 3. Downstream view of the Spring Creek delta (now excavated) developed in an arm of the Keswick Reservoir on the upper Sacramento River, California. The deltaic alluvium accumulated here during a 10–year interval after flooding of the reservoir was over 13 m thick locally and had some pH values as low as 2.2–2.3. The alluvium contained large amounts of copper, zinc, and iron and originated from an erosion of mine dumps in the watershed.

FIGURE 4. Aerial photograph of the O'Neill Pumping Plant of the Federal San Luis Division of the Central Valley Project in California. The plant is located at the toe of Coast Ranges in the west–central portion of the arid San Joaquin Valley. The gas bearing Upper Cretaceous marine bedrock is capped by clayey gypsiferous colluvium and alluvium. X marks covered portions of six discharged pipes partially air exposed at the pumping plant.

O   1   2

FIGURE 5. Close up view of a corroded outside surface of a buried steel discharge pipe of the O'Neill Pumping Plant. Note one through hole and some shallow cavities.

mit an easy escape of natural gas without generating of hydrogen sulfide.

NIKOLA P. PROKOPOVICH

References

Dent, D., 1986, *Acid Sulphate Soils: A Baseline for Research and Development,* Publication No. 39. Wageningen, The Netherlands: International Institute for Land Reclamation and Improvement (ILRI), p. 204.

Dost, H., ed., 1973a, Acid sulphate soils, in *Proceedings of the International Symposium on Acid Sulphate Soils,* Publication No. 18, vol. 1. Wageningen, The Netherlands: International Institute for Land Reclamation and Improvement (ILRI), p. 296.

Dost, H., 1973b, in *Proceedings of the International Symposium on Acid Sulphate Soils,* Publication No. 18, vol. 2. Wageningen, The Netherlands: International Institute for Land Reclamation and Improvement (ILRI), p. 406.

Finkl, C. W., Jnr., 1982, Soils, in M. Schwartz, ed., *Encyclopedia of Beaches and Coastal Environments.* Stroudsburg, Pa.: Hutchinson Ross, pp. 759-763.

Prokopovich, N. P., 1965, Siltation and pollution problems in Spring Creek, Shasta County, California, *Am. Water Works Assoc. Jour.* 57(8):986-995.

Prokopovich, N. P., C. R. Cole, and C. K. Nishi, 1971, Alteration of sediments by natural gases in western Merced County, California, *Am. Assoc. Petroleum Geologists Bull.* 55(6):826-832.

Van der Spek, J., 1950, Katteklei, *Verslagen Landbouw. Onderz.* 56(2):1-4.

Cross-references: Vol. XII, Pt. 1: *Soil Mineralogy; Sulfur Transformations.* Vol. XIII: *Alluvial Plains, Engineering Geology; Clay, Engineering Geology; Coastal Engineering; Coastal Inlets, Engineering Geology; Deltaic Plains, Engineering Geology; Lime Stabilization.* Vol. XV: *Coastal Zone Management; Polder; Soils.*

CHEMICAL MINING—See BOREHOLE MINING.

CITIES, GEOLOGIC EFFECTS OF

Since 1800, cities have increased in size and number. In the process of building (or rebuilding) and providing needed services to urban populations, cities have emerged as geologically important factors, altering shorelines, creating sediment deposits, and changing river discharges. These effects can be demonstrated in virtually any city. The magnitude and geological significance of such effects and their causes are illustrated in this entry using the New York metropolitan region.

In every city, urban effects were initially small and affected only the area near the city center. With the steam engine and the introduction of large-scale dredging in the late 1880s, however, the changes have grown in size and are particularly evident in coastal cities. Now these changes are large enough to leave an imprint on entire estuarine systems and adjoining continental shelves.

Landfill Effects

Cities are prolific waste producers. These wastes are usually collected and buried in landfills beyond the city limits or deposited in waterways or somewhere in the city. Rates of accumulation range from 20 cm to 200 cm/100 yr. The magnitude of the landfill operation along the lower Hudson River in the New York region is shown in Fig. 1. Much of the coastal landfill around

Lower Manhattan occurred before the 1850s. Between 1688 and 1862, the waterfront advanced 191 m into the East River. Most of the filling-in along the New Jersey shore and in Newark Bay has occurred since 1900. In 1966, about 20% of New York City was built on filled lands; about half of that consisted of former garbage dump or other waste disposal sites.

Sediment Sources

Sediment sources in urban areas are affected by human activities. Erosion and sediment transport are usually greatly accelerated; new sediment sources, such as sewers, become significant; and depositional areas are altered by dredging and waste disposal operations. Sediments deposited in the New York Harbor come from several sources: riverborne sediment and particulate wastes; littoral drift of sands along beaches; dumping of waste solids; and deposition of wastes discharged by sewers and shoreside industries.

Rivers in the New York–New Jersey region

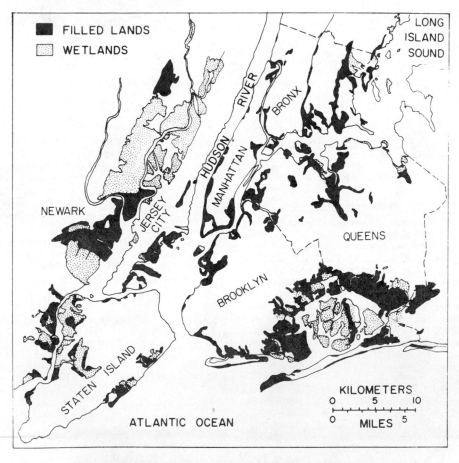

FIGURE 1. Areas of landfill in the New York metropolitan region.

carry sediments derived from erosion of their drainage basins that is often accelerated by human activities. The Hudson River sediment load was estimated at approximately 800,000 tonne/yr in the early 1960s. Sediment loads are estimated at about 70,000 tonne and 95,000 tonne/yr for the Raritan and Passaic rivers respectively, both discharging into New York Harbor. Beach sands moving along the Long Island–New Jersey coasts also deposit in the harbor mouth. This littoral drift of beach sand has been estimated at approximately 1.1 million tonne/yr.

Another large but poorly known source of sediment is sewers and sewage treatment plants that discharge municipal sewage and industrial wastes, either untreated (raw) or after various levels of treatment. The volume of sewage and treated effluents is large, approximately 10% of the annual discharge of the Hudson River at its mouth. Untreated sewage is discharged into the Hudson Estuary from Manhattan Island, parts of Brooklyn, and other localities. In addition, untreated sewage is discharged by both sanitary and storm sewers after heavy rains. Sewage plant effluents typically contain solid concentrations of 50 ppm. Solids removed by sewage treatment are usually larger than 100 μ. Consequently, it seems reasonable to assume that even treated sewage provides appreciable quantities of solids in the fine sand to silt size range. The annual discharge to New York waterways has been estimated at approximately 300,000 tonne.

In summary, littoral drift is the largest known contributor of sediment to the estuary, depositing about 1.1 million tonne/yr of dry solids. The Hudson, Raritan, and other rivers contribute about 1 million tonne/yr. Sewage solids amount to nearly 0.3 million tonne/yr. To these should be added an unknown quantity of waste solids that are discharged directly to the estuary. Thus, the annual contributions of sediment from all sources (natural and human controlled) to the Hudson Estuary are about 2.4 million tonne of solids, on a dry weight basis.

Dredging

Maintenance of navigation facilities is an important activity in port cities. For instance, the increased size and draft of ships made it necessary in the 1870s to dredge navigation channels and anchorages within the harbor. By the mid–1960s, there were 110 km (66 mi) of federally authorized channel projects in the harbor, 53 km (32 mi) of channels in the harbor entrance, and numerous anchorage for vessels throughout the harbor complex. In all, more than 67 km² (26 mi²) of harbor bottom were involved in these projects. Some projects require maintenance dredging every year to maintain adequate depths, while others need little or no dredging. Many channels, being much deeper than the adjacent harbor bottom, act as effective sediment traps.

It seems highly probable that the larger channels at the harbor entrance have increased exchange of waters. The best evidence of increased water exchange comes from slight exchange in annual tidal ranges at the Narrows, where the tidal range changed from 1.37 m (4.49 ft) in 1893–1895 to 1.46 m (4.8 ft) in 1903–1907. Dredging the Hudson River had even more dramatic effects on tides in the river. Cutting a 120 m (400 ft) wide, 8.2 m (27 ft) deep channel through rock and sandbars in the river caused a greatly increased tidal range. After dredging, the tidal range at Albany increased from 0.722 m (2.37 ft) in 1926 to 1.22 m (4.01 ft) in 1930.

To maintain the various channels and anchorages in the harbor, dredging is required each year. Between 1930 and 1970, the average annual dredging in New York Harbor removed about 2.2 million tonne, with 1 million tonne coming from the entrance channels alone. In short, the annual dredging is approximately equal to the entire sediment deposition in the region from all sources (see *Canals and Waterways, Sediment Control*). In addition to the removal of bottom materials to provide needed water depths for navigation, extensive dredging for sand and some gravel production has taken place in Lower New York Bay. These sands have been used primarily for *hydraulic fill.*

The volume of sand produced commercially is poorly known. Tax records indicate that the total volume of sand removed from Lower New York Bay during the late 1960s may well have exceeded 38 million m³ (50 million yd³). For comparison, this figure is about equal to the total amount of materials dredged from navigation channels in New York Harbor during the 1960s. Because of dredging for navigation projects and sand and gravel production, Lower New York Bay is apparently being deepened at a fairly rapid rate.

Sediment and Waste Deposits

Sediment distributions in the Hudson Estuary have been greatly altered by dredging and waste disposal. The complicated geometry of the Hudson Estuary and the New York Bight make it difficult to delineate sediment deposits satisfactorily. Nonetheless, it is possible to indicate general types of sediments on the harbor bottom and in New York Bight and to provide a generalized picture of their distribution (Fig. 2).

Inner portions of New York Harbor are covered by fine sands and silts. These fine-grained deposits have high total contents (5–10%); they also contain anomalously high concentrations of copper, lead, and silver. These fine-grained de-

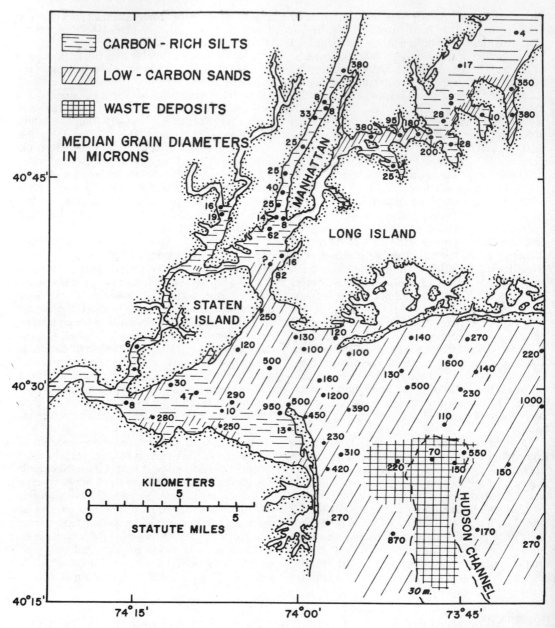

FIGURE 2. Distribution of sediment and waste deposits in the New York metropolitan region and adjoining waters.

posits are apparently mixtures of river-borne sediment and wastes, combined with wastes from sewers and industrial plants around the estuary. These waste deposits cover about 160 km² in the harbor, which is about 41% of its area.

Sediment distributions in the waste disposal area of the continental shelf adjoining New York Harbor are more difficult to characterize. Before human activities changed them, there were prob-

ably two different sediment types in this area. On the Continental Shelf there were relatively coarse-grained sands and some gravels. In Hudson Channel, the sediments were probably fine-grained sand and silts. The distribution of sediment and waste deposits is not well known but general outlines are indicated in Fig. 2. Carbon-rich, metal-rich waste deposits cover at least 150 km² in New York Bight; much of that area was apparently originally a hard sand or gravel bottom. So the waste deposits have locally altered bottom characteristics and caused substantial changes in benthic populations.

Waste deposits have also substantially modified bottom topography in areas long used for waste disposal. Approximately 410×10^6 m³ of liquid and solid wastes were dumped near the New York Harbor entrance between 1888 and 1975. Small hills were formed; one is 1.3 km in diameter and about 9 m high. The head of the Hudson Channel has been filled with approximately 200×10^6 m³ of debris. Sewage sludges apparently do not build up deposits in the disposal areas; they may be moved to other areas or decomposed after release.

Water Supply

The growth of cities and suburbs has significantly altered hydrologic regimes; in some cases the effects extend hundreds of kilometers from the urban area. As urban populations have expanded, individual water use has increased, and as industrial water consumption has expanded, cities have gone farther and farther from the city center to tap unused water supplies. In some areas urban water supply systems have extended across drainage divides so that water has been transferred from one basin to another.

The process is well illustrated by the development and expansion of the New York City water supply system. The city depended originally on wells and ponds. But as these were contaminated, surface water supplies were developed. The first major system was put into service in 1842, drawing its supply from surface waters in the Croton reservoir just north of the city. To satisfy increased demands, the system was expanded and in the 1970s drew water from several reservoirs in the Hudson and Delaware River basins, more than 160 km from the city center.

In addition to taking water from a wide area, cities change river conditions by constructing sewer systems that discharge relatively large amounts of water in a few locations. The New York metropolitan region discharged 110 m³/s (2.97 billion gal/day, 11.2 million m³) per day in 1971 (Table 1). For comparison, the average combined flow of the Hudson and other rivers in the region was about 640 m³/s, or the dry weather sewage discharge was 16.9% of the average river flow. In addition, the discharge of cooling and industrial process waters in the region was 570 m³/s, exceeding average river flow. Furthermore, the storm sewers discharge the runoff from city streets and rooftops before it evaporates or flows into the ground. These discharges were estimated at 83 m³/s in 1964, 12.7% of the total river discharge and precipitation. In all, the amount of water discharged by city sewers, industries, and power

TABLE 1. River and Waste Water Discharges in the New York Metropolitan Region

Type of Discharge	m³/s
Direct precipitation (on harbor)	13
Rivers	
Hudson	560
Passaic	31
Raritan	47
Subtotal, rivers and precipitation	651
Sewers, dry weather flow (1971)	110
Storm water runoff	83
Industrial	~ 100
Subtotal, waste discharges	~ 290
Power plants, cooling waters	570

plants clearly exceeds the average river flow into the harbor.

Future Trends

The effects of urban activities on local waterways and harbors has continued despite increasing efforts (and expenditures) to limit or change the extent of influence. With continued growth of cities and their suburbs as well as the requirements for large industrial facilities, electrical power generating plants, and fuel–processing facilities, it seems likely that the future influence of cities will be felt on the adjacent continental shelf. New facilities, such as offshore installations (see *Floating Structures; Offshore Nuclear Plant Protection*), will cause changes in current patterns and locally influence sediment transport patterns. And the disposal of wastes from increasingly larger populations and greater industrial production can possibly cause pronounced changes in water characteristics that are felt over areas of hundreds to thousands of square kilometers. Some wastes produced in only small quantities and prohibited for ocean disposal have been handled by other disposal options such as landfills. But large volumes of wastes will probably be widely disposed in coastal waters for decades, especially in developing countries.

As consumption of water use for industrial processes and electrical power generation increases, the discharge of coastal rivers will likely be markedly reduced in quantity, especially during periods of low flow. And the timing of peak discharges may be significantly altered.

M. GRANT GROSS

References

Blake, N. M., 1956, *Water for the Cities.* Syracuse, N.Y.: Syracuse University Press, 341p.
Gross, M. G., 1972, Geologic aspects of waste solids and marine waste deposits, New York Metropolitan Region, *Geol. Soc. America Bull.* **83**, 3163–3176.

Gunnerson, C. G., 1973, Debris accumulation in ancient and modern cities, *Jour. Environmental Engineering Division* **99**, 229–243.

Klimm, L. E., 1956, Man's ports and channels, in W. L. Thomas, Jr., ed., *Man's Role in Changing the Face of the Earth.* Chicago, Il.: University of Chicago Press, 522–541.

Legget, R. F., 1973, *Cities and Geology.* New York: McGraw-Hill, 624p.

Cross-references: *Canals and Waterways, Sediment Control.* Vol. XIII: *Channelization and Bank Stabilization; Coastal Engineering; Coastal Inlets, Engineering Geology; Hydrology; Nuclear Plant Siting, Offshore; Urban Engineering Geology; Urban Geomorphology; Urban Hydrology.*

CLAY MINERALOGY—See CAT CLAYS; DISPERSIVE CLAYS; EXPANSIVE SOILS; MINERALS AND MINERALOGY. Vol. XII, Part 1: CLAY MINERALOGY; CLAY MINERALS, SILICATES. Vol. XIII: CLAY, ENGINEERING GEOLOGY

CLAYS—See CAT CLAYS; DISPERSIVE CLAYS; EXPANSIVE SOILS; SAPROLITE, REGOLITH, AND SOIL

COAL MINING

Coal Mining in an Energy-Conscious Society

While oil and gas presently supply about three-fourths of the energy needs in the United States, coal is a more important resource for the future since it has a known recoverable reserve of 4,500 quadrillion BTUs, or about 80% of the recoverable energy in our mineral fuels. These figures indicate why coal is presently considered to be one of the most important factors in combatting the present U.S. energy crisis and why coal mining is an active and growing industry.

The convenience of oil and gas projects and the promise of inexpensive nuclear power have long held back coal production, but the Arab oil embargo and the failure to develop nuclear capacity to meet U.S. energy needs have increased the need for coal in the past few years. Tonnage figures for 1984 show that coal production has risen to 896 million ton, up from about 600 million ton in 1974, while the demand has drastically increased the value of coal mined from over $9 billion in 1974 to about $23 billion in 1984 (Anon., 1984).

Coal mining is carried out by a variety of methods, which can be classified into two broad categories, *surface mining* and *underground mining.*

The surface mining methods produced about 54.4% of the coal mined in the United States during 1974, while underground mining methods accounted for the remainder (Anon., 1975). The surface mining of coal has been increasing in percentage of tonnage ever since it came into being. But the largest portion of the remaining coal reserves are deeper coals that can be mined only underground, and hence, the surface mining of coal is likely to decrease in percentage in the future.

Surface Mining Methods

Of the common surface mining methods, three are similar in nature: area stripping, contour stripping, and open-pit mining. A fourth method, known as augering, is quite different and accounts for only about a few percent of all the coal mined in the United States. At present, these four methods are responsible for nearly all the coal that is produced from surface mines.

Area Stripping *Area stripping,* or *area strip mining,* is the practice of removing *overburden* from a relatively shallow and nearly horizontal coal seam in parallel strips, exposing a narrow width of the coal seam as each strip of overburden is removed from above the coal. The overall operation may look like the one shown in Fig. 1. This method of mining is practiced in the midwest and the far west where the coal seams are relatively shallow and where the topography is relatively flat or gently rolling.

An area stripping operation begins with the clearing of the land surface. This is followed by the drilling of the overburden by large rotary drills (see *Borehole Drilling*). The blast holes produced are loaded with ammonium nitrate and fuel oil mixtures, and these blasting agents (see *Blasting and Related Technology*) are detonated to break and loosen the overburden mixture for the subsequent removal. The overburden material is usually handled by large stripping shovels or by draglines. A *stripping shovel* is used on top of the coal seam as shown in Fig. 2 and transfers the overburden from atop the coal to the area where the coal has been removed, creating a spoil pile in the process. A *dragline* is a similar device, but it is normally utilized from a position on top of the overburden or, alternately, from a position on the closest spoil pile. This device is illustrated in Fig. 3, where Big Muskie, the world's largest dragline, is shown excavating overburden from a position on the highwall. The magnitude of this mechanical excavator can be appreciated by a look at Fig. 4 where a 90-piece marching band is positioned in the 220 yd³ (170 m³) bucket of the machine. Draglines provide much greater flexibility than stripping shovels and thus are becoming more popular for use in stripping. Large *bucket-wheel excavators* have been used successfully in the excavation

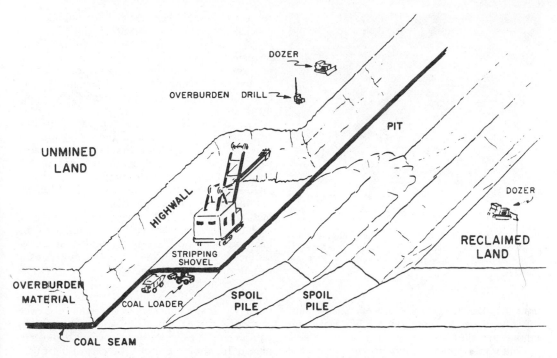

FIGURE 1. Typical area stripping operation. (Drawing by J. P. Gerkin, formerly of Mining Extension Service, West Virginia University)

FIGURE 2. The Silver Spade, a large stripping shovel, illustrates extraction of overburden on an Ohio stripping operation. (Photo courtesy of Bucyrus-Erie Company)

FIGURE 3. Big Muskie, the world's largest dragline, is shown removing overburden from its position on the highwall. (Photo courtesy of Bucyrus-Erie Company)

of the overburden in some countries but have not been applied widely in the United States because the overburden consists primarily of rocky sediments that are not easily excavated by a bucket wheel.

After the excavating machine has exposed the coal seam, the coal is drilled and blasted, if nec-essary, and loaded by loading shovels or front-end loaders into trucks for haulage out of the strip mine to the processing plant or marketplace. The land is restored by dozing to a smooth contour, by replacing the topsoil, and by planting with grasses or trees. The ability to produce a suitable final topography is quite good with area strip-

FIGURE 4. This 90-piece marching band graphically illustrates the huge size of the 220 yd³ (170 m³) bucket of Big Muskie. (Photo courtesy of Bucyrus-Erie Company)

contours around the hills and mountains. The general procedure is illustrated in Fig. 5, which shows a contour mining operation using a stripping shovel. The overburden can be moved in one or more strips, but the total width of mining is limited to an economic depth determined by the coal value or physical limitations of the equipment.

The overburden removed from above the coal seam can be placed either on the outslope of the mine or hauled back along the contour and placed back into the strip pit where the coal has already been removed. If the latter option is chosen, the method is normally referred to as the *haulback method*. While the variations of contour stripping methods are practiced in the more mountainous coal fields of the Appalachian region, the haulback method is utilized in the hilliest of topographies because it allows a more complete restoration of the original landscape.

The sequence of operations in the contour stripping method is similar to that in area stripping. The strip mine area is cleared and the overburden is drilled, normally with rotary drilling equipment. The overburden is then blasted to prepare it for the excavator. The excavating of the overburden material can be accomplished by a number of types of equipment. As in area stripping, the overburden can be removed using strip-

ping, but the revegetation process can be quite difficult where the climate is dry, like that in the mining states of the far west.

Contour Stripping *Contour stripping* is somewhat similar to area stripping but is practiced in rolling hills or mountainous terrain. The method obtains its name from the procedure of following

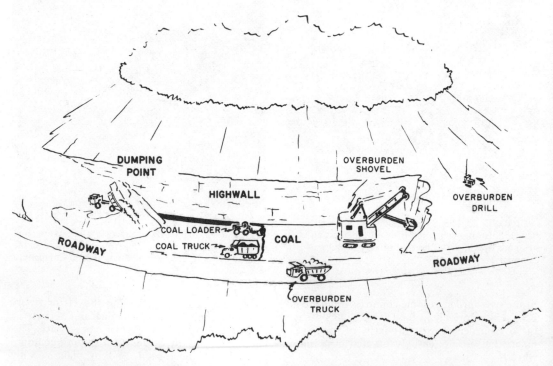

FIGURE 5. Typical contour stripping operation. (Drawing by J. P. Gerkin, formerly of Mining Extension Service, West Virginia University)

ping shovels or draglines. However, the equipment used in the average contour stripping operation is likely to be considerably smaller in size than that used in area stripping. In addition to shovels and draglines, other equipment types can be applied in a contour mine, including front-end loaders, wheel scrapers, and dozers. The front-end loader is now widely used in contour mining, especially where the overburden is to be loaded into trucks and hauled back along the outcrop and placed in the previous pit area. Scrapers are capable of doing the same job although they are not as widely used. Dozers are used in a somewhat different manner since they are capable of moving material only by a dozing action and are unable to lift the material for transport or for stacking purposes.

A great proportion of the overburden excavation is done by shovels, draglines, and loaders. To excavate the coal, the seam may be drilled and blasted or simply excavated. The excavation is performed by smaller power shovels or, as is becoming more common, with rubber-tired front-end loaders. The coal is hauled by trucks normally capable of travel on both rough strip mine roads and more developed highways.

Because of the steepness of the terrain, the reclamation of a contour mine is often quite difficult. Unless the haulback method is used, contour stripping leaves a highwall above the original mine site. The spoil material is a great problem. The spoil area is normally regraded and planted with grasses and/or trees. However, because of the steep topography that often exists, problems with the erosion of the regraded spoil banks and the stability of the spoil materials often occur unless the reclamation has been done with great care.

Open-Pit Mining The *open-pit mining* technique differs from the previous two methods. The overburden is excavated and then hauled out of the pit to a dumping area adjacent to the mine. The open-pit method is applied to thick or steeply dipping coal seams that occur in the far west and in the anthracite beds of Pennsylvania or to sediments that contain multiple beds of coal. The open-pit method often creates a very deep excavation. This excavation presents a serious drawback that makes the method rather undesirable for general use.

In open-pit mining, the overburden can be removed in a variety of ways. Scrapers or loaders can be used, especially for the overburden close to the surface. Power shovels and draglines can also be applied, but the reach of such excavators limits their application. Consequently, they are used only for overburden close to the surface. For overburden well below the surface and for harder materials, the material is often blasted and loaded by power shovels or front-end loaders into large off-highway trucks for transport out of the mine.

Reclaiming the site of an open-pit mine is naturally very difficult since complete restoration can be accomplished only by refilling the completed pit. For this reason, open-pit mining is not likely to be a very popular method of producing coal in the future.

Augering A method that is quite different in concept from the previous three but that is practiced on the surface is known as *augering* (q.v.). Though often classified as a separate method from the surface mining methods, it is normally practiced on the surface and usually performed in conjunction with contour strip mining. Augering, or *auger mining,* is the process of drilling into a coal seam with large-diameter drills called augers at the point where the seam meets the ground surface. The drilling or augering is normally done in a contour mine after the overburden has been removed from the coal seam and before the pit has been backfilled. The auger used for this purpose normally has one to three boring heads and operates at the foot of the highwall to drill the holes in a parallel fashion up to 70 m into the coal seam. The hole diameters are ordinarily about two-thirds the thickness of the coal seam to provide a margin of error and to accommodate variations of coal seam thickness. A pillar of coal that is half a meter or so in width is left to support the highwall. The augering method can also be applied to coal seams in hilly terrain without previously stripping the outcrop.

The primary advantage of augering is that the procedure can recover coal that is too deep to be economically stripped or too poor in quality or too dangerous to be mined by underground methods. In addition, the method produces coal at a lower cost per ton than any other mining method. The principal disadvantage is the fact that the recovery of the coal is ordinarily well below 50%, with the remaining coal being lost for the foreseeable future.

Underground Mining Methods

An *underground mine* is a complex system involving skilled labor, complicated machinery, and an extensive layout of openings that are provided to produce coal in a planned and coordinated fashion. Underground mines can be classified as *shaft mines* (those where access is made through vertical openings called shafts), *slope mines* (those where access is made through slopes that are used primarily with conveyor belt haulage), and *drift mines* (those that are accessible through outcrop openings).

The layout of an underground mine is often related to the geologic fracture system that exists in most coal seams. The vertical cleavages apparent in most coal seams are referred to as *cleats*. The two cleavages that normally exist in coal seams are nearly perpendicular. The most prominent cleavage is known as the *face cleat*. The direction of the face cleat is usually east-west in the coals of the eastern United States. The secondary cleat system is known as the *butt cleat*, and it normally runs in a north-south direction.

To take advantage of the natural planes of weakness in the coal, many mines are laid out so that the opening will be parallel to the cleat system. One such layout is shown in Fig. 6, where each mine opening is represented by a solid line. The openings of a typical mine are arranged in a hierarchy as shown in the drawing. The primary openings that are driven from the outcrop are known as the main entries. A group of these openings is opened up parallel to each other and connected together by crosscuts that are necessary to provide adequate ventilation during the driving of the main entries. The driving of *entries* and *crosscuts* results in blocks of coal known as *pillars* that are left in place to support the mine openings during the subsequent mining activity.

The second hierarchy of openings is known as *submain entries*, and these are usually driven perpendicular to the main entries. When the submain entries are driven perpendicular to the butt cleat, as in Fig. 6, they are referred to as *butt entries*. The development of the butt, or submain, entries enlarges the mine and provides more working places. The third hierarchy of entries, known as *section entries*, is provided for similar reasons, and these are normally driven perpendicular to the submain entries.

The final set of entries is known as *room*, or *panel, entries*, and these are the openings that lead to the working places that are usually laid out in a panel. The working places can be in the form of parallel openings called rooms, as illustrated in Fig. 6. When rooms are utilized for production, the mining method is known as *room-and-pillar mining*. Room-and-pillar mining can be further classified into two subcategories known as *continuous mining* and *conventional mining*. Alternately, the production from the panel can be obtained by means of a *longwall face*, and the method is then called *longwall mining*. The most recent authoritative figures, which are from 1983 (Anon., 1984), indicate that continuous mining accounted for 69%, conventional mining accounts for 10%, and longwalling provided for 20% of the total tonnage of coal mined underground. Many variations of these three methods exist, which is evidenced by study of the standard references on coal mining (Woodruff, 1966; Cummins and Given, 1973; Cassidy, 1973). However, this section describes only a few of the more common mining layouts used.

Conventional Mining *Conventional mining* is

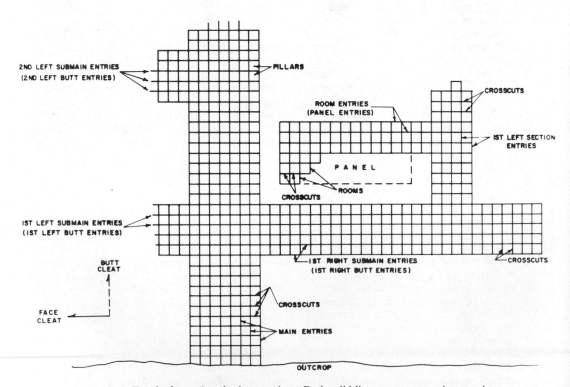

FIGURE 6. Terminology of coal mine openings. Each solid line represents a mine opening.

FIGURE 7. A coal miner moves his cutting machine to the face to begin the undercut. (Photo courtesy of Joy Manufacturing Company)

a method of mining coal using explosives to break the coal in a cycle of operations that requires each piece of equipment to travel from face to face to do its work. This mining method has been practiced for many decades and has been improved over the years mainly by the development of faster, more powerful, and more convenient machines to perform each step in the mining cycle. The conventional mining method is presently used primarily in coal seams that possess soft bottom conditions, in seams that have significant partings (bands of noncoal materials), and in thin seams where other mining methods are difficult to use. The cycle of operations in conventional mining is as follows.

1. The coal face is undercut (i.e., an undercut, or kerf, is cut at the bottom of the coal seam) by a device known as a *cutting machine* that is very similar in action to a huge chain saw. A typical cutting machine used for this purpose is shown in Fig. 7.
2. The face is drilled by a *coal drill* to provide a place for the explosives that are used to break the coal from the seam. This operation is ordinarily performed by a rubber-tired drill similar to the one illustrated in Fig. 8.
3. The coal broken from the face is loaded by a gathering-arm loader that is used to load the coal into shuttle cars for haulage from the face.

FIGURE 8. A mobile coal drill is used to drill the coal face for the subsequent blast. (Photo courtesy of Joy Manufacturing Company)

Figure 9 shows the type of equipment used for this purpose.
4. The roof is bolted by drilling into it with a *roof-bolting machine*. To support the roof, long metal bolts that have an anchoring device at the end are implanted to pin the roof rock in place using the overlying strong sediments. The type of roof–bolting machine used most often is shown in Fig. 10.

FIGURE 9. A miner uses a gathering–arm loader to load a shuttle car with freshly mined coal. (Photo courtesy of Joy Manufacturing Company)

FIGURE 10. A miner uses a roof bolter from a position under the protective canopy. (Photo courtesy of FMC Corporation)

The initial transport of the coal is normally performed by rubber-tired vehicles called *shuttle cars,* which the miners often call "buggies." The shuttle cars are similar in design to a truck but are designed with a conveyor in the bed for loading and unloading purposes. They operate on electrical power supplied by a cable. Figure 11 illustrates a shuttle car being used to haul coal on a mining section. After the shuttle car has moved the coal from the face to the section dump point, the coal is then hauled out of the mine by a belt conveyor or rail haulage system.

The conventional mining process is suitable to both development and production mining. *Development* is the process of opening entries and crosscuts that are necessary to get to the production sections of the mine. In Fig. 12, the development process consists of driving the section and room entries and the related crosscuts. To perform this operation using the conventional mining method, it is necessary to have a number of openings of faces to mine simultaneously. For example, if the conventional mining method were used to develop the section entries shown in Fig. 12, there would be a number of working faces at which to distribute the equipment. Each piece of equipment moves in a definite pattern. For example, the cutting machine can cut the face in Entry 1, then move to Entry 2, then Entry 3, etc. The drill follows the cutting machine, performing its job on the same sequence of faces. The loader and roof bolter follow in turn, each performing its job on all faces in the entry set.

Production mining is the process of opening rooms (production openings) and the related crosscuts and of mining the remaining pillars. This process can be envisioned by again referring to Fig. 12. The room entries are developed by drilling away from the section entries. Then the production process begins by first opening the rooms and the related crosscuts. In an optional process, the pillars are mined, allowing the roof to cave in, creating a caved area that is sometimes known as the *gob.* When the process of mining or extracting the pillars is initiated (called the *pillaring operation*), the work is done along a line of pillars known as a *pillar line.* This line of pillars is often kept at about a 45° angle with the room entries.

The production process using conventional mining is conducted in a similar manner to development mining using a number of faces to utilize all the equipment properly. On the pillar line, this results in mining a number of pillars along the pillar line simultaneously. This process requires a carefully coordinated series of operations to extract the pillars in a safe and orderly manner.

Continuous Mining *Continuous mining,* the most widely used underground mining method, first saw significant use about 1950 and has grown in use since. The method is characterized by large mechanical machines called *continuous miners* that dig or gouge the coal out of the coal seam without the use of explosives. Typical continuous miners of the rotating-drum type are shown in Figs. 13 and 14. The continuous mining method derives its name from the way the coal flows from

FIGURE 11. A shuttle car operator drives or trams his shuttle car (buggy) through a mine opening on his way to a dump point. (Photo courtesy of Joy Manufacturing Company)

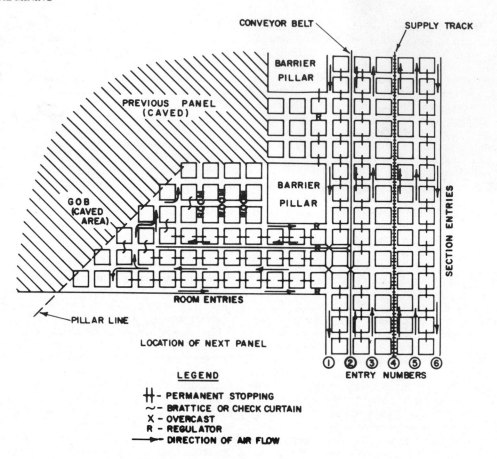

FIGURE 12. Typical room-and-pillar mining layout.

the machine in an essentially continuous fashion. The method is widely used in coal seams that are free of extensive partings and soft bottom (floor), and this allows for a rather wide area of application. The steps in room-and-pillar mining by the continuous mining method are as follows:

1. The coal is ripped from the coal seam by the continuous mining machine. The coal cut from the seam is placed by the continuous miner in shuttle cars.
2. Shuttle cars, ordinarily working in pairs, move the coal from the working place to the section transportation device.
3. The coal is transported out of the section by conveyor belts or rail cars pulled by direct current (DC) electric locomotives. The main transportation system, consisting of conveyor belts or rail, then completes the transportation out of the mine.

The roof control process with continuous mining is handled by roof-bolting devices either at the time of mining the face or after the mining takes place. If bolting takes place after the mining has been performed, it is necessary to have more than one working place so that the continuous miner can move to an alternate working place while the roof-bolting machine is doing its work.

Continuous mining layouts are often quite similar to those of conventional mines. In Fig. 12, e.g., it is quite possible to produce coal from the layout shown by either the conventional or the continuous mining method. The development process (driving the section and room entries) will take place in a similar fashion except that the continuous miner will not require the number of moves needed in the conventional mining method. The pillaring operation is likely to be more variable than the other operations. In continuous mining, the continuous miner will ordinarily work on a single pillar and will extract the entire pillar before moving to another pillar.

The extraction of pillars by continuous miners is performed in several different ways (Fig. 15.) The *open-ending method* is the simplest proce-

FIGURE 13. The working end of a rotating-drum continuous miner is shown in an underground opening. (Photo courtesy of Consolidation Coal Company)

FIGURE 14. A continuous miner operator attacks the coal face with a continuous mining machine. (Photo courtesy of Joy Manufacturing Company)

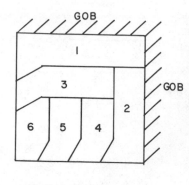

OPEN – ENDING

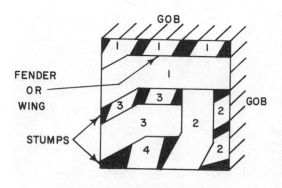

FENDER
OR
WING

STUMPS

POCKET– AND –WING METHOD

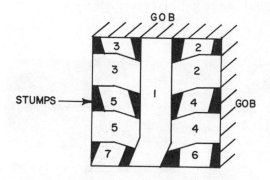

STUMPS

SPLIT– AND– FENDER METHOD

FIGURE 15. Common pillaring methods. The numbers refer to the cut sequence used in the pillaring methods.

dure, but it requires a considerable amount of support in the mining area to protect the miner operator. The *pocket-and-wing* and the *split-and-fender methods* are similar in nature and offer the distinct advantage that the fenders and stumps of coal left in place during the extraction help to support the roof and provide a safer working place. As in conventional mining practice, the extraction of the pillars is ordinarily done in a systematic manner and is often carried out so that the pillar line is extracted on a 45° angle with the room entries.

Longwall Mining *Longwall mining* is an old mining method that has only recently been used with any frequency in the United States. The implementation of large self-propelled hydraulic jacks has made the method more attractive in the United States, and it has thus been used with increasing frequency. It is applied in coal seams that are large in area and that are continuous and relatively uniform in thickness. It is most efficient in deep mines where the roof conditions make other methods less productive.

Longwall mining is the process of mining panels of coal along a long working face that is usually in the range of 150 m to 200 m in length. The panel or block of coal being mined is normally several kilometers in length. The longwall panel is normally developed using continuous mining methods. These consist of driving the panel entries that border the longwall panel and the entries adjacent to the barrier pillars that are necessary for the setup and dismantling of the longwall equipment. These development openings are shown in Fig. 16. The supports used on the longwall face are then put in place to support the roof during the mining process. These supports consist of large self-advancing hydraulic jacks (sometimes called *chocks* or *shields*) that protect the men and equipment at the face and prevent the roof from falling immediately along the longwall face. The breakage of the coal at the face is done by either a *shearer* or *plow*. The shearer or the plow rides along the face and cuts a slice of coal from the entire length of the longwall face. The coal falls into a chain conveyor that carries it to the section belt conveyor, which is placed in the headgate entry system. From there, the coal is transported out of the section to the main haulage system.

After each slice of coal has been removed, the hydraulic jacks are advanced so that they are again against the longwall face. The longwall process continues until the coal in the panel is completely extracted and the roof has been caved. The equipment is then moved to the adjacent solid block of coal or panel, and the process is initiated on the new panel. The primary advantages of this method are the safety of the miners and the ease of ventilation of the working face.

Mining Legislation

Surface Mining In the past, the practice of surface mining for coal left numerous huge scars on the Earth's surface, provided a source of acid

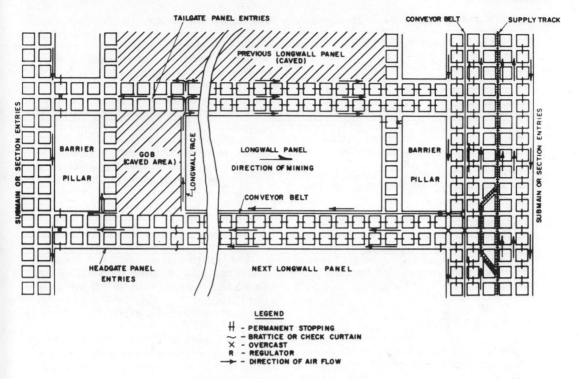

FIGURE 16. Typical longwall mining layout.

mine drainage that polluted streams, and was responsible for plant damage and siltation in the drainage system. Such mining practices resulted because mining companies were not subject to legislative regulation of their practices. Naturally, the least expensive methods were used, calling for little or no reclamation to achieve the least cost. Indeed, the company that did reclaim its mined land was in effect penalized because its profits were reduced by the expense of reclamation.

Legislation of surface mining began in earnest during the 1960s and has continued to affect the mining and reclamation requirements. The many state laws governing surface mining have been summarized by Grim and Hill (1974) and illustrate the progress made in this area. In addition to the state governments, the federal government has been in the act of legislating surface mining since 1977.

The impact of surface mining regulations has been significant. Surface methods now result in much less overall damage to the environment. The regrading and revegetation requirements in most states have significantly reduced topographic damage, and the protection given to streams has improved to the point where minimal problems are occurring in comparison with past results. Strip-mined land is now being reclaimed for use-

ful functions such as agricultural and recreational pursuits, whereas the same land was unfit for these purposes before.

Underground Mining The underground mining of coal in the United States has changed drastically since the passage of the 1969 Federal Coal Mine Health and Safety Law. The law was designed to eliminate the hazards that exist in the underground coal mines and to help prevent injuries and fatalities during such mining. It has accomplished its purpose so far, and the safety record of the U.S. coal industry has improved significantly since the enactment of the legislation. It has, however, also increased the cost of producing coal underground because of the many expensive elements of compliance not required in previous laws.

JAN M. MUTMANSKY
ANN STARCHER BENTLEY

References

Anon., 1974–1975, *Coal Facts*. Washington, D.C.: National Coal Association, 95p.

Anon., 1984, *Coal Production 1984,* Report DOE/EIA-0118(84), Washington, D.C.: Energy Information Administration, U.S. Department of Energy, 144p.

Cassidy, S. M., ed., 1973, *Elements of Practical Coal Mining*. New York: Society of Mining Engineers of the American Institute of Mining, Metallurgical and Petroleum Engineers, 614p.

Cummins, A. G., and I. A. Given, eds., 1973, *SME Mining Engineering Handbook*. New York: Society of Mining Engineers of the American Institute of Mining, Metallurgical and Petroleum Engineers, pp. 12-1–12-262.

Grim, E. C., and R. D. Hill, October 1974, *Environmental Protection in Surface Mining of Coal*. Cincinnati, Ohio: U.S. Environmental Protection Agency, 276p.

Woodruff, S. D., 1966, *Method of Working Coal and Metal Mines*, vol. 3. New York: Pergamon Press, 571p.

Cross-references: *Augers, Augering; Blasting and Related Technology; Borehole Drilling; Mine Subsidence Control; Mining Preplanning*. Vol. XIII: *Cavity Utilization; Economic Geology; Mineral Economics; Rapid Excavation and Tunneling; Shaft Sinking.*

COMPASS TRAVERSE

A *compass traverse* is a method of filling in detail on a topographic, geologic, or other map. It is a method of surveying a *route* such as a stream, path, or the edge of an outcrop by means of a series of *traverse legs,* of each of which the bearing is observed by compass and the length measured directly in the field. It may be described as a zig-zag. The traverse may be *closed* or *open* depending on whether or not it returns to the starting point. An area may be surveyed by means of a grid of traverse lines.

History

For many centuries mariners have navigated by a series of courses, dictated usually by the wind, the bearing of each course being given by the compass and the length from an estimate of speed and elapsed time. The resulting dead reckoning was checked at suitable intervals by astronomical observation or from bearings to known points. Many coastlines were surveyed in this manner, but although there is no exact record when the method was first used in land surveying, it seems that the road maps of John Ogilby (London, 1675) were made by compass traverse, which may also have been used for Der Rom Weg of Erhard Etzlaub (Nuremberg, 1500).

Measurement of Bearing

Equipment A hand-bearing compass with a dial at least 2 in (50 mm) in diameter and graduated in 360° (or 400 grad), equipped with sighting vanes and preferably a prismatic lens to permit simultaneous sighting and bearing observation, is adequate. The compass may be steadied at eye level on a staff or mounted on a tripod for more accurate reading.

Compass Deviation All magnetic compasses are affected by the following:

Magnetic variation. Since the compass needle points to the north magnetic pole, which wanders, it points east or west of true north according to whether the observer is west or east of the agonic line (of zero variation), which in North America extends roughly from Lake Superior to the Bahamas. The variation and the annual rate of change are normally shown on published survey maps or agonic charts;

Local external factors such as buried cables, iron fences, magnetic ore bodies, metal objects on the person, and local annual and diurnal variations from the published data;

Individual compass error due to age or maladjustment;

Personal observational errors.

Before starting a compass traverse, the observer should verify the cumulative errors affecting his or her compass by observing a known bearing or checking by solar or stellar observation. Compass bearings are converted to true bearings by deducting westerly deviation or adding easterly deviation. Some compasses have an adjustable dial to enable true bearings to be read directly.

Measurement of Distance

Methods Several methods of distance measurement are used, each with different equipment. In ascending order of accuracy these are as follows.

Estimation. The distance of objects off the traverse line may be estimated by eye, and when traversing a river by boat this may be the only way to measure each leg.

Measure time elapsed and estimate rate of walking, riding, cycling, etc. Check rate over a known distance.

By car, using the odometer; these are usually calibrated to 0.1 km (or 0.1 mi) and quarter divisions can be estimated. Suitable for rapid road surveys.

A bicycle with a cyclometer graduated as for an odometer. Four actuators instead of the usual one allow reading to 11 yd (6 1/4 m), which is adequate for many purposes. An assembly of a cycle wheel, front fork, and handlebars is an acceptable pedestrian substitute.

Pacing, either human or animal (horse, camel, etc.). Paces are counted mentally or by pedometer, and pace length is checked over a known distance.

A surveyor's measuring wheel (velometer), usually calibrated in yards or meters. A sledgemeter is a similar wheel towed behind a sledge.

Chaining.

Taping.

The last two methods are suitable for short distance only and need two operators. Over long distances compass errors do not ensure accuracy.

Slope Correction Field measurements of distance must be corrected for slope before plotting. The slope of each traverse leg may be estimated, measured by some sort of clinometer, or deduced from heights given by a pocket aneroid after allowing for diurnal variation by comparison with observed readings at a fixed point.

From the observed slope angle the horizontal distance may be deduced from the field measurement with sufficient accuracy by using Table 1.

Alternatively, use published tables or charts (e.g., Forrester, 1966, 209–213) or the simple formula:

Horizontal distance = slope distance
× Cosine slope angle.

Traversing Techniques

At the *start point* (SP) observer selects the first *Turning Point* (TP) and, if necessary, sends out an assistant to mark it. Note that a compass has an accuracy of perhaps 1:250, so if sights are long, broad aiming marks are acceptable. Each leg should be as long as practicable.

Observer then reads the forward bearing of the traverse leg, preferably taking the mean of two or more readings, checks the slope, and notes the reading of any measuring device.

Observer proceeds along the leg, noting the distance to each item of interest reached; if the item is off the direct line, read the distance along the leg to a point opposite the item and the distance of the item from the leg (*the offset*). The offset

TABLE 1. Slope Correction

Slope in Degrees	Gradient	Percentage reduction
5	1:12	0.5
7½	1:8	1
10	1:6	1.5
12	1:5	2
14	1:4	3
16	1:3½	4
18	1:3	5

Source: After Debenham, 1940. Reproduced by kind permission of The Blackie Publishing Group.

is frequently measured by a less accurate means than that used in measuring the traverse (see section on methods for measuring distance). If items of interest are too far from the traverse to measure easily, they can be fixed by two or more bearings from points on the traverse.

On arrival at the first TP, observer reads the back bearing along the first leg; the mean of two or more readings should not differ from the reciprocal of the forward bearing by more than 1°. Observer also notes the final length of the first leg.

This process is then repeated for the second leg and so on until the final leg returns to SP (closed traverse) or reaches a final point whose position is known from other data (open traverse). No compass traverse is complete unless it starts and ends at known points.

Areas can be surveyed by a series of parallel traverse lines, separated by anything from a few hundred yards to a mile or more, originating from a baseline, which may be central or along one edge of an area, and ending at a natural stream line, track, or cut line at the far edge. Each traverse line is driven as straight as possible and, if forced to deviate by obstacles, should return to the original bearing when practicable. Any tracks, outcrops, streams, etc. intersected are traversed to a point about midway between the lines. Traverses along the area margins connect the marked ends of the traverse lines. Finally, traversing at closer centers can be done in an area of special interest.

Observation Notebook

Observations are entered neatly in a notebook, usually on a separate sheet for each leg. Distances are customarily plotted upward in a central column with offsets, etc. on the relevant side. Details are sketched in with no attempt at scale drawing.

Plotting the Traverse

Trial Plot Using the field notes plot the traverse to the final scale, correcting each distance for slope (see "Slope Corrections"). The mean of the forward and back bearing (reciprocal) of each leg is used unless they differ by more than 1° when so-called local iron may be present. In that case, the back bearing is ignored and the correct forward bearing for the next leg is deduced from the *change in bearing* measured at the suspect TP. Any closing error between the final leg and the SP (closed traverse) or known end point (open traverse) is distributed between the TPs by graphic or mathematical means.

Graphic Adjustment, Closed Traverse The traverse is drawn as a straight line to any suitable scale, with the TPs marked; the closing error is then plotted *to the scale of the final plan* at right

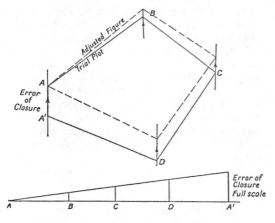

FIGURE 1. Graphic adjustment, closed traverse. (After Debenham, 1940. Reproduced by kind permission of The Blackie Publishing Group.)

Mathematical Adjustment by Coordinates To obtain the true bearing, correct the mean magnetic bearing of each leg by adding or subtracting the compass deviation (see "Compass Deviation"). Convert the true bearing to the reduced, or quadrant, bearing (shortest way to the meridian)—e.g.,

True Bearing	Reduced Bearing
080	N80E
160	S20E
240	S60W
320	N40W

Using lengths corrected for slope (see "Slope Correction"), determine the *latitude* (northing or southing) and *departure* (easting or westing) or each leg from the formulas:

$$\text{Departure} = \text{Horizontal distance} \times \text{Sine reduced angle}$$

$$\text{Latitude} = \text{Horizontal distance} \times \text{Cosine reduced angle}$$

angles at the end of the line (Fig. 1), and the triangle is completed. Lines drawn normal to the base through each TP show the scale displacement of each TP in the trial plot, and these are marked on lines parallel to the closing error on that plot to give the adjusted position of each TP.

Graphic Adjustment, Open Traverse From the trial plot determine the apparent bearing and distance of the end point from the SP. From known data, plot the correct position of the end point and join these two positions to determine the closing error. Draw rays from the SP to the real and apparent end points. Draw lines parallel to the closing error through each TP (Fig. 2). The intercepts of these lines between the two rays drawn from the SP determine the adjustment, parallel to the closing error, of each TP. Note that if the true position of the end point is not known, an open traverse cannot be adjusted.

Traverse tables or pocket calculators simplify the work. Northings and eastings are positive, southings and westings negative. Tabulate these.

For a closed traverse the arithmetic sum of the latitudes and departures should be zero, and for an open traverse it should equal the known difference between the coordinates of start and finish. Any difference, being the closing error, is added or subtracted to each leg, proportionately to the distances, and the adjusted latitudes and departures tabulated (Table 2). The corrected coordinates of each TP can then be determined.

The corrected TPs having been plotted, the details of the traverse can be filled in on the final plot. Suitable scales and offsets are available commercially to simplify plotting.

P. F. F. LANCASTER-JONES

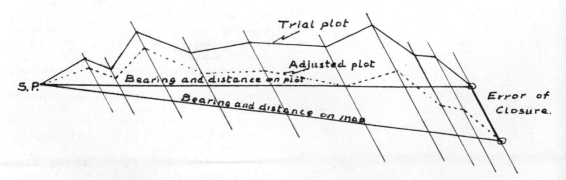

FIGURE 2. Graphic adjustment, open traverse. (After Debenham, 1940. Reproduced by kind permission of The Blackie Publishing Group.)

TABLE 2. Correction by Coordinates

Line	Bearing	Reduced Bearing	Distance	LATITUDES		DEPARTURES		Total Co-ordinates		Station
				Northings	Southings	Eastings	Westings	N+, S-	E+, W-	
AB	97	S 83 E	620		75·5	615·3 626·3 +11·0		−75·5	+626·3	B
BC	176	S 4 E	440		438·9 439·4 +·5	30·7		−439·4 −514·9	+30·7 +657·0	C
CD	270½	N 89½ W	660	5·6			659·8 648·1 −11·7	+5·6 −509·3	−648·1 +8·9	D
DA	359	N 1 W	510	509·8 509·3 −·5			8·9	+509·3 0·0	−8·9 0·0	A
				515·4	514·4	646·0	668·7			
				Diff 1·0		22·7				

N.B. Corrections in Heavy Type   Final figures in heavy type

Working in Full		LAT.	DEP.
S 83 E	600	73·1	595·5
	20	2·4	19·8
	620	75·5	615·3
S 4 E	400	399·0	27·9
	40	39·9	2·8
	440	438·9	30·7
N 89½ W	600	5·1	599·9
	60	·5	59·9
	660	5·6	659·8
N 1 W	500	499·9	8·7
	10	9·9	·2
	510	509·8	8·9

The discrepancies of 1·0 and 22·7 in Latitudes and Departures are distributed by adding half to the lesser column and subtracting half from the greater. The apportioning of the amount to the different co-ordinates is done approximately in proportion to their size. Thus in the Departures the difference of 22·7 is split up into 11·0 to be added to Eastings and 11·7 to be subtracted from Westings. The whole of the 11·0 is added on to the longest co-ordinate, 615·3, but 10·0 might have been added there and 1·0 to the co-ordinate 30·7. It must be realised that more accurate apportioning than this is useless since the assumption itself is only an approximation.

Source: After Debenham, 1940. Reproduced by kind permission of The Blackie Publishing Group.

References

Debenham, F., 1940, *Map Making.* London: Blackie and Son, 73–94.

Forrester, J. D., 1966, *Principles of Field and Mining Geology.* New York: Wiley, 200–273.

Cross-references: *Field Geology; Geological Survey and Mapping; Geomagnetism; Plane Table Mapping; Surveying, General.*

COMPUTER MAPPING—See LAND CAPABILITY ANALYSIS; SATELLITE GEODESY AND GEODYNAMICS; SURVEYING, ELECTRONIC.

COMPUTERS IN GEOLOGY

Several major conceptual revolutions have occurred in geology that were accompanied or followed by technological advances of the time. The progressive technologies allowed for rapid development of science. The first revolution was the result of Copernicus's discovery that the sun, not the Earth, was the center of the solar system. The second resulted from the work of James Hutton and Charles Lyell, founders of modern geology (see *Geohistory, American Founding Fathers*), in their elucidation of time and processes (the present is the key to the past; uniformitarianism). The third revolution was the integrated theory of plate tectonics.

In the seventeenth century, Galileo utilized the telescope, invented in the previous century, to make observations that led to his verification of Copernicus's ideas. The invention and development of the steam engine in the later part of the seventeenth and early part of the eighteenth centuries allowed the massive changes in manufacturing and transportation in the nineteenth century that were instrumental in the creation of leisure time so necessary for scientific pursuits. The next major technological advance was the rapid acceptance and adoption of the computer and computer methods. The ability to acquire, manipulate, and analyze massive amounts of data facilitated the acceptance of plate tectonics (see *Plate Tectonics and Continental Drift*) in a matter of just a few years. In each of these instances, the technological product was an extension of a human faculty—the telescope: sight; the steam engine: muscle; and the computer: the mind. It is with this last technological advance, computers, as applied to geology, that we are concerned here.

Roots of Ideas

Although the roots of quantitative approaches in geology go back to the beginning of modern geology (Merriam, 1981a), the use of computers by geologists is relatively new. The use of computers is linked closely with developments in and the availability of hardware and software. The computer age in geology usually is dated as having started with the publication of a geologically oriented computer program in a recognized journal in 1958 by W. C. Krumbein, father of computer geology, and his co-worker, L. L. Sloss. Advances in many aspects of geology since then have depended on the utilization of computers (Agterberg, 1974; Merriam, 1982).

The origins of modern geology and the computer both date back to the early part of the nineteenth century where, in London, an amazing group of farsighted scientists lived. Included in this group were Charles Lyell (geology), Charles Babbage (computers), Charles Darwin (biology), Humphrey Davy and Michael Faraday (chemistry), and John Herschel (astronomy). These men, following the Scientific Revolution of the seventeenth century, took an active part in the Industrial Revolution and contributed to a scientific golden age, laying down the foundations for modern science.

Lyell, formulator of the principle of uniformitarianism and the first to use statistics in geology, and Babbage, mathematician and creator of the difference and analytical engines, were good friends. They were acquainted professionally through the Royal Society and the Geological Society of London. It is not unlikely that they shared ideas and problems of work as both were ingenious and inquisitive researchers. Although Lyell's work forms the basis for modern geology and he was recognized for his contributions, Babbage's calculating machines were destined for failure for lack of technology, and therefore he was not given his due recognition at the time (Merriam, 1983).

Not much happened in the next 100 yr from this modest beginning. Ada Augusta, Lady Lovelace and daughter of Lord Byron, wrote the first computer program, which was a series of steps to compute Bernoulli numbers using the analytical engine (Table 1). This program was patterned after the series of card instructions used to control the

TABLE 1. Important Events in Computer Applications in Geology

1642	Blaise Pascal devised a calculating machine.
1694	A machine was built by Gottfried Wilhelm Leibniz to multiply and divide directly.
1804	Joseph Marie Jacquard used holes punched into cards to control and operate a loom.
1812	Charles Babbage got the idea of calculating machines.
1822	The first working model of Babbage's difference engine was built.
1834	Babbage started work on his analytical engine, precursor of the modern computer.
1842	Ada Augusta, Lady Lovelace, wrote the first program, a step-by-step description for computing Bernoulli numbers with the analytical engine.
1890	A punch card system was developed by Herman Hollerith for the U.S. Census Bureau.
1941	Z3, the first electronic computer, was made in Germany.
1944	Mark I, the decimal electromechanical calculator, was put into operation at Harvard.
1945	The idea that operating instructions and data could be stored in computer memory was developed by John von Neumann.
1946	ENIAC was built at the University of Pennsylvania.
1949	The first stored-program, digital computer, the EDSAC, was built.
1951	UNIVAC, the first commercial computer, became available.
1952	Digital plotters were introduced.
1953	The first FORTRAN compiler was written.
1954	The IBM 650, the first mass-produced computer, was introduced.
	Core memory was first used in the MIT Whirlwind I computer.
1958	W. C. Krumbein and L. L. Sloss published the first geologically oriented computer program in a major geology journal.
	Transistorized second-generation computers were introduced.
	The ALGOL language was introduced in several countries.
1961	Geo.Ref, the geology bibliographic database, was established.
	The sumposium series *Computer Applications in the Mineral Industries* by the University of Arizona began.
1963	Third-generation microcircuit computers were announced.
	The first regular publication of geological computer programs as *Special Distribution Publications* of the Kansas Geological Survey started.
1964	This was the first year more than 100 papers were published on computer applications in geology.
	A time-sharing system was successfully used at Dartmouth College. BASIC was introduced.
1966	The first series of geological publications concerned exclusively with computer programs was established by Kansas Geological Survey.
	Computer Applications in the Earth Sciences, the first of eight colloquiums sponsored by the Kansas Geological Survey, was held.
	The position of associate editor for computer applications for the *American Association of Petroleum Geologists Bulletin* was established.

TABLE 1. (*continued*)

1967	The American Association of Petroleum Geologists Committee on Electronic Data Storage and Retrieval was formed. COGEODATA (IUGS Committee on Storage, Automatic Processing, and Retrieval of Geologic Data) was formed. The *IUGS Geological Newsletter* published the first international attempt to standardize description of mineral deposits in computer–processible form.
1968	The International Association for Mathematical Geology (IAMG) was founded in Prague at the IGC.
1969	The first issue of the *Journal of Mathematical Geology* of the IAMG was published. *GEOCOM Bulletin,* an international current-awareness publication, was initiated. The U.S. Geological Survey initiated its *Computer Contribution* series. The first book in a series on computer applications in the earth sciences was published by Plenum Publishing Corporation.
1970	An informal research group on computer technology was formed by the Society of Economic Paleontologists and Mineralogists. The first fourth–generation machines, utilizing virtual memory, became available.
1971	The first *GEOCOM Program* was published by Geosystems in London.
1972	Syracuse University established a series of geochautauquas.
1973	Geo.Ref goes on–line on SDC ORBIT.
1974	The U.S. Geological Survey established a computer advisory committee.
1975	A journal, *Computers & Geosciences,* was started by Pergamon Press, New York.
1976	A new book series by Pergamon Press, *Computers and Geology,* was instituted.
1977	The Apple II microcomputer, one of the true home computers, was introduced.
1979	Supercomputers, Cray-1, Cyber 205, and BSP, became available for special scientific problems.
1981	Mathematical Geologists of the United States (MGUS) held their first meeting. Announcement was made of fifth–generation computers utilizing massively parallel concepts and incorporating artificial intelligence functions.
1982	Van Nostrand Reinhold Company initiated a new series, *Computer Methods in the Geosciences.*
1983	COGS (Computer Oriented Geologists Society) formed in Denver. GeoTech 83 took place, the first in a series of meetings sponsored by COGS on microcomputers in geology.
1985	The first issue of *Geobyte* was published, a new publication by AAPG to focus on computer applications in exploration and development of petroleum and energy minerals. COGS membership surpassed 1,000.

Note: Some information is taken from Merriam, 1975.

weaving patterns on Jacquard looms. Later, Herman Hollerith utilized the idea of punch cards for the U.S. Census Bureau. Punch cards also were used in precomputer days for routinely sorting bibliographic and other large datasets; it is hardly surprising then that punch cards were used for input/output (I/O) in the first computers.

Quantification of geology was slow. Geologists were not oriented quantitatively, and between 1830 and 1958 few applications can be cited in the literature. Some applications of trigonometry and geometry were made in crystallography and computations made on age determinations and heat flow in the Earth, e.g. statistics were applied to sedimentological and paleontological problems. Geophysicists, geochemists, engineering geologists, and hydrologists required mathematics to solve their problems. Observations and data had been collecting for 100 yr. The introduction of the computer then ushered in the automated era where complex problems could be solved, large amounts of data manipulated, and data acquisition automated; it was the harbinger of the development of French geostatistics, the application of sophis-

ticated techniques to geological problems, the use of large realistic datasets, and the development of simulation and model studies, especially those involving time (Merriam, 1981*a*).

Early Developments

The conceptual stage of computer applications in geology took place during the 1950s. During that time geologists recognized the potential of the new tool as an extension of the mind just as the microscope had extended sight (Merriam, 1981*b*). Early applications were mainly calculations that had been done previously by hand. Geophysicists, geochemists, and others who were quantitatively inclined simply exchanged their calculators and slide rules for computers where computation was speeded up and fewer errors were made. Many papers containing suggestions of possibilities were published, and the literature was long on ideas but short on applications.

Part of the problem in the early days was simply the limitations of the machines; the number of data that could be processed was limited by

TABLE 2. Progression of Computer Development to Present

Time	Technology	Software	Orientation
1950s	First–generation machines: vacuum tubes	Machine language	Cards and paper tape
1960s	Second–generation machines: transistors	Low–level symbolic languages—e.g., SOAP, FORTRAN, ALGOL, COBOL	Cards, RJEs, magnetic tape, batch
1970s	Third–generation machines: integrated circuits, minis	High–level symbolic languages—e.g., macro–assembler; BASIC, PASCAL, USP; Database management—e.g., ADABAS; Database languages—e.g.,IMS	Interactive time sharing, disks, CRTs
1980s	Fourth–generation machines: VLSI circuits, micros, super-computers	Virtual memory; special high–level languages; user friendly; spread sheets—e.g., VISA-CAC; query languages	Networking, floppy disks, smart terminals, interactive and color graphics

storage space and run time. Only the simplest things could be done and those only with difficulty. Programming was extremely awkward and tedious, e.g., in machine language (Table 2); cards or paper tape had to be punched; run time was long and many errors occurred in the processing; and the vision of applications was limited mainly to analytical technique. In addition, most geologists were cautious and even suspicious of computers that no doubt dated from the time of Darwin and his warning to beware of mathematicians (and by extension, computers).

A Decade of Rapid Development

Toward the end of the 1950s, second generation computers were introduced. With that, programming became easier, machines were made accessible, and computing became more economical. As a result, geologists branched out into modifying statistical techniques and mathematical procedures to solving their problem; algorithms were borrowed from other disciplines; and many papers were published that demonstrated the use of different techniques (Fig. 1). Multivariate statistical techniques of trend analysis, time-series analysis, classification, and correlation analyses were applied. Trend analysis became popular because it was a technique that geologists could understand and whose results they could interpret because they had been using the technique by hand.

In analyzing the subject matter of these early papers, it is interesting to note that certain nationalistic trends occur. For example, geologists in Germany, France, Canada, and Czechoslovakia were concerned with data and their collection and treatment. Geologists in other countries were more concerned with application—in the United Kingdom, e.g., to sedimentological and petrol-

ogical problems. Much work was done on structural and tectonic problem by the Germans, hydrology by the French, and mineral exploration by the Canadians, South Africans, and Czechs. Trend analysis was popular in the United Kingdom, India, and Australia; factor analysis in France; power-spectrum studies in Rumania; and simulation in the United States. These generalities reflected to some extent the availability of hardware, workable software, local problems, and areas of interest of those geologists working with computers (Merriam, 1974a).

Rapid growth and interest in computer use occurred during the 1960s as many researchers saw the potential of this powerful tool. Geo.Ref, the leading bibliographic database, was established, the APCOM (Computer Applications in the Mineral Industries), a series of meetings in the mineral industries and the Kansas colloquia, were established, the Kansas Computer Contributions made their debut, and the International Association for Mathematical Geology (IAMG) was founded. The IAMG, affiliated both with the International Union of Geological Sciences (IUGS) and the International Statistical Institute (ISI), in a few years founded two international journals and a newsletter, sponsored numerous meetings, and fostered and facilitated an exchange of ideas on a worldwide basis (Merriam, 1978).

Pervasion of Computers in Geology

By the 1970s computers had become available and were economical. High-level symbolic languages were available and easy to use with the third-generation machines. Interactive systems were used, widely forecasting the demise of cards and paper tape. Terminals were everywhere. Sequential analysis and spatial analysis were being used extensively, and dimension-free methods

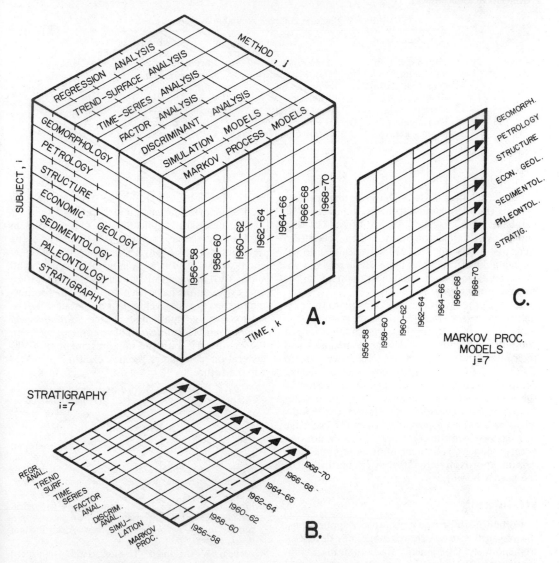

FIGURE 1. (*A*) Subset of geological subjects, methods of data analysis, and time in 2 yr periods. (*B*) Horizontal slice to show time of entry of computer applications in stratigraphy. Dashed lines indicate precomputer applications of methods shown. (*C*) Vertical slice cut to show spread of marked models into various geological fields by year. (After Krumbein, 1969).

were gaining popularity (Table 3). Simulation was introduced and the modeling of geological processes was realistic. Geo.Ref and GeoArchives both went on–line. Additional meetings were held, of which the Syracuse Geochautauquas were most effective for disseminating information on computers; some of the topics were the impact of quantification, data analysis, mineral resources, Computer–Assisted Instruction (CAI), software quantitative stratigraphic correlation, modeling, and subsurface techniques.

Databases proliferated and were accessible through time–sharing networks. Different countries, especially Canada, Rumania, and Czechoslovakia, worked toward developing of archival files; others, toward special files for mapping, mineral resources, geochemistry, paleontology, well information, etc. (Merriam, 1974*b*). GEO-MAP was developed in Sweden for field mapping data; GRENVILLE, in Canada. CRIB contained the mineral-resource data of the U.S. Geological Survey and RASS, their geochemical data. CO-GEODATA (a committee of the IUGS) looked after standards and quality control for the inter-

TABLE 3. Some Techniques Available to Geologists

Sequential	Spatial	Dimension–free
Power spectrum	Autocorrelation	Correlation coefficients
Variograms	2D power spectrums	Regression analysis
Cross–correlation	Kriging	Cluster analysis
Cross–association	Bicubic splines	Principal components
Markov chains	Trend analysis	Factor analysis
Fourier analysis	2D Fourier analysis	Canonical correlation
Moving averages	3D trend analysis	Discriminant functions

national exchange of data. Database management systems (DBMS) came into being to handle these databases. G-EXEC, GIPSY, SASFRAS, GRASP, and CLAIR (all acronyms for their specialties) were just a few of the DBMS that came into existence at this time. The large databases and sophisticated programs available set the stage for the supercomputers.

In 1975, the IAMG founded the quarterly journal, *Computers and Geosciences* (C&G). The journal is devoted to the rapid publication of computer programs in widely used languages and their applications. C&G was the successor to *GEOCOM Bulletin,* which took over publishing computer programs in 1971 at the conclusion of the Kansas Geological Survey's series of successful *Computer Contributions*. The aim of the journal is to serve as a public medium for exchange of ideas between the geological and computer sciences, a concise statement of an interdisciplinary venture approximately 150 yr after constitution of the parent bodies.

Present Status

In the fourth decade of computer applications, the physical size of the machines has decreased tremendously, the capacity for computing has increased, the speed has increased enormously, and computers have become user-friendly. These attributes are due to advances of technology (McIntyre, 1981). Developments such as VM (virtual memory), bubble memory, memory chips, optical fibers, LSI and VLSI (large–scale integrated and very large–scale integrated) circuits have revolutionized the hardware. Microcomputers are ubiquitous. The personal computer (PC), because of this low cost and user orientation, has been accepted almost instantly.

Software is being affected by fast algorithms such as the FFT (fast Fourier transform), user-friendly languages, and telecommunications. Networking is common. Although geostatistiques, a development of the French school of geostatistics, are the only techniques expressly for solving geological problems, the invention of a metalanguage for solving geological problems has been pro-

posed (Griffiths, 1982). Synthesizers will be used to hear, speak, and even understand.

The successful application of mathematics in geology via the computer in the 1960s rivals the development of geophysics in the 1940s and geochemistry in the 1950s. Again, the practical (and successful) applications in numerical exploration and exploitation have been foremost in promoting interest in and development of the subject. The subjects of geomathematics (see *Mathematical Geology* and *Geostatistics*) have come of age with the computer. With the announcement of fifth-generation computers, comes exciting promises for the future.

DANIEL F. MERRIAM

References

Agterberg, F. P., 1974, *Geomathematics.* Amsterdam: Elsevier, 596p.

Griffiths, J. C., 1982, Mathematics in geology: New languages, new perspectives, *Nature and Resources,* **17**(4), 14–17.

Krumbein, W. C., 1969, The computer in geological perspective, in D. F. Merriam, ed., *Computer Applications in the Earth Sciences.* New York: Plenum Press, 251–275.

Krumbein, W. C., and L. L. Sloss, 1958, High-speed digital computers in stratigraphic and facies analysis, *Am. Assoc. Petroleum Geologists Bull.* **42**(11), 2650–2669.

McIntyre, D. B., 1981, Developments at the man-machine interface, in D. F. Merriam, ed., *Computer Applications in the Earth Sciences, An Update of the 70's.* New York: Plenum Press, 23–42.

Merriam, D. F., 1974*a,* Introduction to the impact of quantification on geology, *Syracuse Univ. Geology Contr. 2,* pp. 1–3.

Merriam, D. F., 1974*b,* Resource and environmental data analysis, in *Earth Science in the Public Service.* Washington, D.C.: U.S. Geological Survey Prof. (Paper 921), 37–45.

Merriam, D. F., 1975, *Computer Fundamentals for Geologists: COMPUTe.* Hanover, N.H.: Dartmouth University, 284p.

Merriam, D. F., 1978, The international association for mathematical geology—a brief history and record of accomplishments, *Syracuse Univ. Geology Contr. 5,* pp. 1–6.

CRATERING

Merriam, D. F. 1981b, A forecast for use of computers by geologists in the coming decade of the 80's, in D. F. Merriam, ed., *Computer Applications in the Earth Sciences, An Update of the 70's.* New York: Plenum Press, 369–380.

Merriam, D. G. 1981a, Roots of quantitative geology, *Syracuse Univ. Geology Contr. 8,* pp. 1–15.

Merriam, D. F., 1982, Development, significance, and influence of geomathematics: Observations of one geologist, *Jour. Math. Geology* **14**(1), 1–10.

Merriam, D. F., 1983, The geological contributions of Charles Babbage, *The Compass* **62**,(1), 31–38.

Morrison, P., and E. Morrison, 1959, History of punch cards, in *Charles Babbage and His Calculating Engines.* New York: Dover, xxxiii–xxxv.

Cross-references: *Geostatistics; Mathematical Geology.* Vol. XIII: *Computerized Resources Information Bank.*

CRATERING

A *crater* is formed by the ejection of material, which is caused by a rapid release of energy near or below the surface of the ground. The energy source may be the high-velocity impact of a projectile or the detonation of an explosive charge.

Early interest in the form and structure of craters was probably precipitated by the large craters on the moon's surface and the associated speculation as to their origin—i.e., whether they were created by volcanic activity or were the result of impacts. The presence of meteor-impact craters on the Earth's surface has allowed some direct observation of their characteristics, and in recent years, various satellite missions have resolved some of the uncertainties about lunar craters.

An *impact crater* is formed by ejection of material when a particle, such as a meteor or swarm of closely spaced particles, strikes the ground at a velocity above a critical lower limit, thought to be in the range of 1 km to 2 km/s. The configuration of the crater is governed by the surface geology and the mass, density, and velocity of the impacting material. McCunn (1974) has simulated most forms of lunar craters in a series of model experiments in which he introduced drops of water falling onto the surface of heated silt. The drops of water become superheated on impact and vaporize explosively to produce forms that appear to have direct analogies on the surface of the moon.

Short (1964) has reviewed the features of impact and explosion craters and defined the similarities and differences between them in terms of morphology, mineralogy, and structure. On the surface of the Earth, meteor-impact craters most closely resemble craters formed by explosions at relatively shallow depths. Figure 1 is a sketch of a cross-section of the New Quebec Crater, an impact crater in granite; the Sedan Crater, a 100 kiloton (kt) nuclear crater in alluvium; and an estimation of a crater produced by a 10,000 kt explosion in hard rock.

In recent decades many investigations of *explosion craters* have been conducted using chemical explosives, and several nuclear experiments have also been conducted in the United States and the USSR. Much of this experimental work was prompted by the early promises of the Plowshare Program for peaceful, constructive applications of nuclear explosives. A corresponding interest in the potential military applications of craters continues to stimulate empirical and analytical studies of their shapes and properties. The nomenclature applying to explosion craters has been

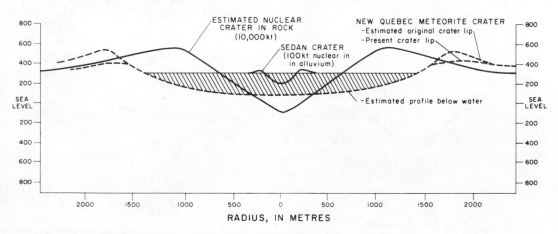

FIGURE 1. Cross–sections of an impact crater and two nuclear craters. (Drawn from information provided by the University of California, Lawrence Livermore Laboratory, and the U.S. Energy Research and Development Administration)

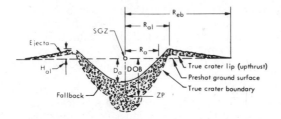

Ra - Radius of apparent crater measured at original ground surface datum

Ral - Radius of apparent lip crest

Reb - Radius of outer boundary of continuous ejecta

Da - Maximum depth of apparent crater below and normal to original ground surface

Hal - Apparent crater lip crest height above original ground surface

Va - Volume of apparent crater below original ground surface

Val - Volume of apparent lip

Vt - Volume of true crater below original ground surface

DOB - Depth of burst

ZP - Zero Point—effective center of explosion energy

SGZ - Surface Ground Zero (point on surface vertically above ZP)

FIGURE 2. Cross-section of an explosion crater. (Adapted from Hughes, 1968)

standardized over the years, at least in the United States. A cross-section of a crater, illustrating the most important dimensional parameters, is shown in Fig. 2. The surface separating fragmented and

dislocated material from relatively intact material defines the true crater. The fallback debris that blankets this surface and that reduces the volume defines the apparent crater.

Phenomenology

The position of the explosive charge above or beneath the ground surface (height of burst, HOB, or depth of burst, DOB) has a great influence on the size and configuration of the resulting crater. The analogy in the case of impact craters would be the velocity, mass, and corresponding rate of penetration of a projectile into the ground. Figure 3 schematically shows, for a given amount of explosive energy, the effects on the shape of the crater of increasing the DOB. As the explosive is buried progressively deeper beneath the surface, crater dimensions will increase to some maximum and then decrease until eventually the material above the charge is barely broken and bulked, resulting in a mound of rubble. This phenomenon is graphically illustrated by the curves in Fig. 4, which are referred to as *cratering curves*. The depth at which an explosive charge forms the crater having the greatest volume is called the *optimum depth of burial*. A more quantitative discussion of crater dimensions is given in the next section.

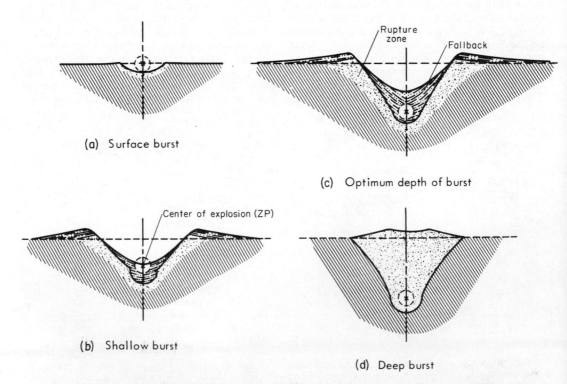

FIGURE 3. Changes in crater cross-section with increasing depth of burial. (Adapted from Hughes, 1968)

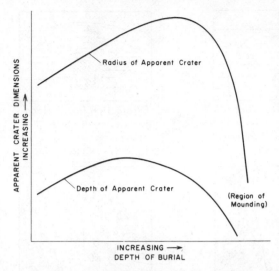

FIGURE 4. Variation of apparent crater dimensions with increasing depth of charge.

The mechanics of *crater formation* are well understood in the case of explosively formed craters; the mechanics are similar but more complex for impact craters. As explained by Teller et al. (1968), the two primary agents acting on the ground in the case of a subsurface explosion are the initial shock wave from the detonation and the subsequent expansion of the explosion gases or by products. The relative importance of these two mechanisms is dependent on the properties of the cratered material and the DOB of the explosive.

The general sequence of events occurring in the formation of an explosion crater is shown schematically in Fig. 5. In the case of a relatively strong, competent material such as rock, the rock is crushed and pulverized close to the explosion. As the shock front expands and its intensity is dissipated by spherical divergence and energy losses to the medium, the effects progressively diminish from crushing, fracturing, and cracking to a point where all motions are elastic. When the *shock front* reaches the ground surface, it is reflected downward as a tensile wave and spalling of the surface material occurs. When the downward-traveling tensile wave meets the boundary of the spherically expanding, gas-filled cavity, the tensile wave relieves some of the compressional stresses, and the top of the cavity expands preferentially toward the surface. Eventually, the expanding cavity and the mound of broken material rising above it will dissociate, venting the explosion gases to the atmosphere. From this point on, the material travels ballistically; i.e., it is in free fall, subject to the forces of gravity and air drag. Some of the debris is ejected beyond the boundaries of

the crater to form the lips, and the rest falls back into the true crater as fallback.

The shock and spall mechanisms in weaker, soillike materials are usually subordinate to the heaving action of the expanding explosion gases. The energy of the shock wave is rapidly dissipated by crushing and compacting of the soils—i.e., closing up pore spaces. As the water saturation of a soil increases, open pore space is diminished and transmission of the shock becomes more efficient, with a consequent increase in crater dimensions. As shown by the hydrodynamic calculations of Terhune et al. (1970), the degree of water saturation is the single most important factor governing crater dimensions in rock, and the same importance has also been demonstrated empirically for soils. Gautier (1971) presented an analysis of the influence of material properties on crater dimensions.

Crater Dimensions and Scaling

The dimensions of explosion craters should scale with the energy yield of the explosion (expressed as weight of TNT), according to the rules of dimensional analysis. However, complications arise when gravity is considered. Although crater dimensions theoretically cannot be scaled over a wide range of explosive energy when gravity is invariant, empirically observed scaling exponents have found wide use in predicting crater dimensions to acceptable levels of accuracy.

The dimensions of a known crater can be scaled up or down to predict the crater dimensions produced by a greater or lesser amount of explosives by means of an empirical scaling formula such as:

$$D = \left(\frac{W}{w} \right)^p d,$$

where D is the dimension (radius, depth, etc.) being scaled, d is the reference dimension of a known crater, W is any explosive energy (weight), w is the explosive energy that produced the reference crater, and p is the scaling exponent. When scaling a crater dimension from one charge weight to another, it is implicit that the scaled DOB stay constant—i.e., that one stays at the same relative point on a cratering curve.

If gravity were not a factor, then dimensions should scale as the cube root of the energy. Chabai (1965) suggested that the scaling exponent, p, should approach 1/4 for large explosions in the presence of a gravity field. Scaling exponents obtained from analysis of experimental data fall between 1/3 and 1/4. Nordyke (1961) arrived at an exponent of 1/3.4 on the basis of buried chemical and nuclear explosions in desert alluvium, and this factor has seen common use since. For practical purposes, a scaling exponent of 1/3 is adequate

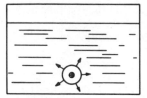

(a) Detonation of explosive

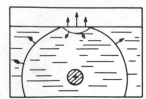

(b) Shock wave reaches surface causing spalling; simultaneous spherical cavity growth

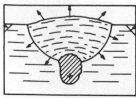

(c) Rarefaction from surface reaches cavity and asymmetrical growth towards surface begins

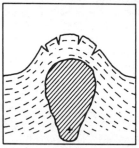

(d) Mound grows and begins to dissociate, vapor filters through broken material

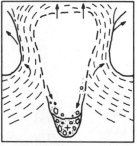

(e) Maximum development of mound, some material slumps from cavity walls, major venting

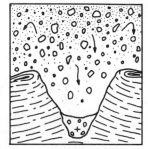

(f) Complete dissociation of mound, ejection and fall-back of material to form apparent crater

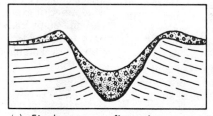

(g) Final crater configuration

FIGURE 5. Crater formation history. (Adapted from Hughes, 1968)

for surface bursts and 3/10 for buried explosions. In any one series of experimental cratering detonations in a given material, the random scatter of crater dimensions will probably exceed the uncertainties of scaling.

It has become common practice to use the energy equivalent of 1,000 tons of TNT (1 kt; one ton of TNT has been defined as 10^9 calories, 1 kt is then equivalent to 10^{12} calories; in SI units 1 kt is equivalent to 4.187 terajoules, tj) as the reference explosive yield when discussing the dimensions of nuclear craters. When dimensions are expressed in meters, scale dimensions are usually expressed in the units of $m/kt^{3/10}$. One ton of TNT has been frequently used as a convenient reference energy when discussing craters produced by high explosives.

The cratering efficiencies of nuclear explosives and conventional explosives do not appear to be equivalent for equivalent releases of energy (expressed in tons of TNT). Some of the energy produced by a nuclear device is expended in radiation and in vaporizing the surrounding material. There are indications that nuclear explosives have only about half the cratering effectiveness of chemical explosives, at least in soillike materials.

Specific examples of scaled dimensions in the form of cratering curves are shown in Figs. 6 and 7 for chemical explosives in soils and for nuclear explosives in dry rock respectively. The large influence of a soil's water content on crater dimensions is depicted graphically in Fig. 6.

As an example of the use of a scaled cratering curve such as Fig. 6, which is based on relatively small chemical explosions in soil, assume that we wish to estimate the radius of the Sedan Crater, formed by a 100 kt nuclear explosion in dry desert alluvium at the Nevada Test Site, a photograph

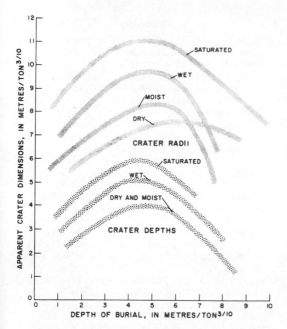

FIGURE 6. Typical scaled crater dimensions for chemical explosions in soils showing influence of degree of saturation.

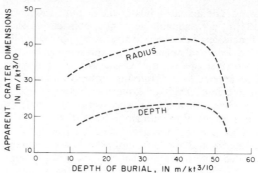

FIGURE 7. Scaled crater dimensions for nuclear craters in dry strong rock.

of which is shown in Fig. 8. The Sedan nuclear explosive was buried at 194 m, and assuming that 100 kt of nuclear explosive energy is equivalent to 50 kt of chemical explosives, the scaled burial depth is then $194/50,000^{3/10}$, or 7.55 m/ton$^{3/10}$. Figure 6 indicates that the scaled crater radius for this DOB in dry soil is about 7.4 m/ton$^{3/10}$, which is then equivalent to 190 m, still assuming that we have the equivalent of 50 kt of chemical explo-

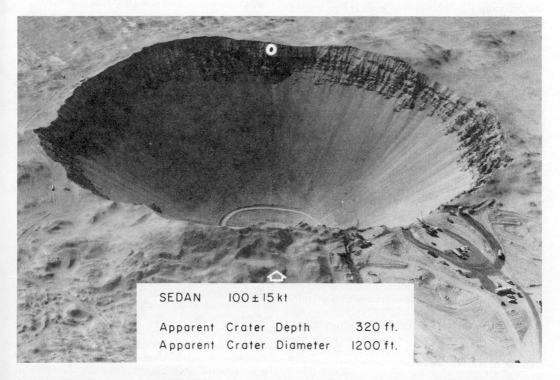

FIGURE 8. The 100 kt Sedan Crater at the Nevada Test Site. (Photograph courtesy of the University of California, Lawrence Livermore Laboratory, and the U.S. Energy Research and Development Administration)

sives. The actual radius of the Sedan Crater was 185 m, a reasonable agreement considering that dimensions have been scaled over approximately five orders of magnitude in explosive yield.

Interest in potential engineering or construction application of cratering prompted applied research into arrays of explosive charges, particularly linear arrays for the excavation of canals. A linear excavation with smooth walls will result if charges are appropriately scaled in a row and detonated simultaneously. Johnson (1971) demonstrated experimentally that charge spacings of approximately 1.4 times the radius of the optimum single crater will produce a smooth-walled crater whose width and depth are about equal to the diameter and depth of the optimum single crater. Spacings greater than 1.4 radii will result in cusping. Spacings of less than 1.4 radii will produce progressively wider and deeper craters. In general, the width and depth of a row crater vary as the square root of the linear charge density; i.e.,

$$\text{width } (W_a), \quad \text{depth } (D_{ar}) \propto \left(\frac{W}{S} \right)^{1/2},$$

where W is the individual charge weight and S is the spacing between charges. Specifically (Johnson, 1971), if the dimensions of a row crater are normalized to the corresponding dimensions of the optimum single crater that would be produced by one of the charges in the row, then

$$\frac{W_a}{2R_a} = \frac{D_{ar}}{D_a} \approx \left(\frac{1.4}{S/R_a} \right)^{1/2}.$$

W_a and D_{ar} are the width and depth of the row crater, R_a and D_a are the radius and depth of the optimum single crater, and S is the charge spacing. The foregoing relationship is valid only if the charges are buried progressively deeper as the spacing is shortened; the optimum DOB of the charges in the row (DOB_{row}) can be expressed in terms of the optimum single-charge DOB (DOB) as:

$$DOB_{\text{row}} = DOB \left(\frac{1.4}{S/R_a} \right)^{1/2}.$$

Much work has been done in the area of computer-assisted, numerical calculation of crater dimensions using large, hydrodynamic, finite-difference codes (Terhune et al., 1970). Because the codes apply only to part of the cratering mechanism and judgment must be used to estimate the final crater configuration, at present they offer no significant improvement over empirical scaling.

BRUCE B. REDPATH
ROBERT W. FLEMING

References

Chabai, A. J., 1965, On scaling dimensions of craters produced by buried explosives, *Jour. Geophys. Research* **70**, 5075–5098.

Gautier, J., 1971, *Etudes Experimentale et de Synthese de l'Influence des Proprietes des Explosifs et des Caracteristiques Mecaniques des Sols et des Roches sur les Dimensions des Crateres.* Paris: Commisariat à l'Energie Atomique (CEA-R-4242), 209p.

Hughes, B. C., ed., 1968, *Nuclear Construction Engineering Technology.* Livermore, Calif.: U.S. Army Engineer Nuclear Cratering Group, Lawrence Radiation Laboratory (NCG-TR-2), 182p.

Johnson, S. M., ed., 1971, *Explosive Excavation Technology.* Livermore, Calif.: U.S. Army Engineer Nuclear Cratering Group, Lawrence Radiation Laboratory (NCG-TR-21), 232p.

McCunn, H. J., 1974, Model study of morphologies caused by exploding super-heated vapor and possible lunar analogies, *Gulf Coast Assoc. Geol. Socs. Trans.* **24**, 211–222.

Nordyke, M. D., 1961, *On Cratering: A Brief History, Analysis, and Theory of Cratering.* Livermore, Calif.: U.S. Army Engineer Nuclear Cratering Group, Lawrence Radiation Laboratory (UCRL-6578), 72p.

Short, N. M., 1964, *Nuclear Explosion Craters, Astroblemes, and Cryptoexplosion Structures.* Livermore, Calif.: U.S. Army Engineer Nuclear Cratering Group, Lawrence Radiation Laboratory (UCRL-7787), 75p.

Teller, E., W. K. Talley, G. H. Higgins, and G. W. Johnson, 1968, *The Constructive Uses of Nuclear Explosives.* New York: McGraw-Hill, 320p.

Terhune, R. W., T. F. Stubbs, and J. T. Cherry, 1970, Nuclear cratering on a digital computer, in *Proceedings: Engineering with Nuclear Explosives.* Las Vegas, Nev.: American Nuclear Society, 334–359.

Cross-references: *Blasting and Related Technology.* Vol. XIII: *Rapid Excavation and Tunneling; Rocks, Engineering Properties; Soil Mechanics.*

CROSS-SECTIONS

A geological cross section is a diagram that displays geological relationships in a vertical plane extending into the Earth. Such diagrams present the appearance of exposing a slice of the Earth's interior to view. Cross-sections may be constructed entirely from either surface or subsurface data, or they may include both. Subsurface information can be derived from written logs of wells, samples, and cores or from geophysical techniques such as electric logging and seismic surveys (see *Well Logging*). Cross-sections are important tools that aid in the interpretation and presentation of geological data, however; they provide only a two-dimensional picture and must be interpreted accordingly. They are most effectively used in conjunction with other illustrative techniques. Langstaff and Morril (1981) and Ragan (1985) provide a thorough discussion of cross-section construction and use.

Types of Cross-Sections

The two basic types of cross-sections are *structural* and *stratigraphic*. The difference between them is significant, and the choice of which to use for a particular problem is critical. Each is constructed differently and portrays the geometrical relationships among geological features in a unique way.

Cross-sections are constructed by first choosing a specific elevation or stratigraphic horizon to serve as a reference plane to which the features of each data point are related. The reference plane, or datum plane, shown as a straight horizontal line, is usually sea level for structural cross-sections (Fig. 1). Such cross-sections therefore show folds, faults, and rock units as they presently exist in the subsurface. This type of section is useful to suggest petroleum traps, the location of faults at depth, or the depth of a layer such as a coal seam. Because structural cross-sections show features in their present orientation, they can be used to depict the relationships of geology to surface topographic features and may be drawn with a profile of the surface (see *Profile Construction*).

Stratigraphic cross-sections are constructed with a stratigraphic horizon, such as an unconformity or formation boundary, as the datum (Fig. 2). The datum is drawn as a horizontal line in spite of topographic or structural irregularities. The choice of an appropriate datum will result in a cross-section that displays depositional relationships such as facies changes, lateral thickness variations, and paleostructure. Such cross-sections are useful to decipher the geological development of an area and depositional conditions.

Construction of Cross-Sections

Cross-sections are drawn between two end points either directly or through intermediate data points. When three or more data points are used they may fall along a straight line, but more likely they will be connected by an irregular line. When the line of profile is not straight, features shown on the cross-section will be somewhat distorted and must be interpreted accordingly.

Before the section is drawn, its purpose must be decided and the line of profile chosen. The purpose will determine the type of datum selected. The location of the profile line determines which features are crossed, their orientation on

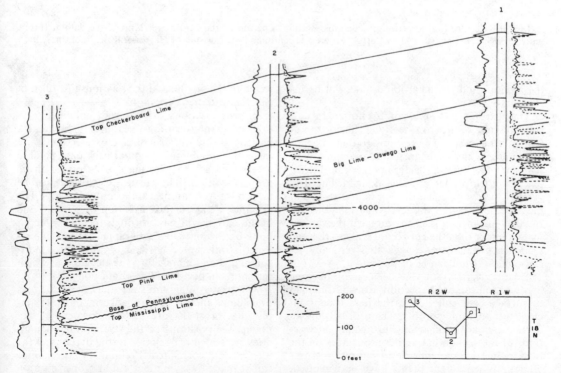

FIGURE 1. Electrical log structural cross-section. Sea level is the datum. (From Moore, 1963; copyright © 1963 by Carl A. Moore and reprinted by permission of Harper & Row, Publishers, Inc.)

101

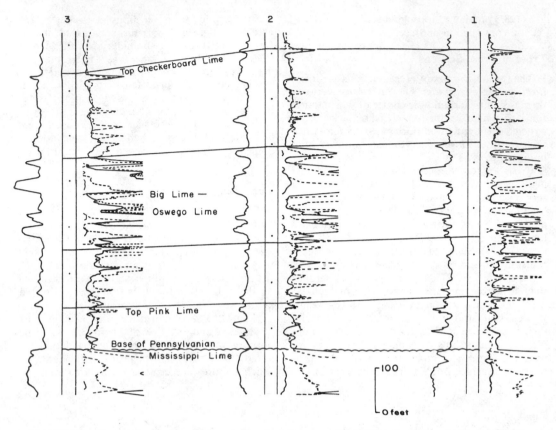

FIGURE 2. Electrical log stratigraphic cross-section. The top of the Mississippi Lime is the datum. (From Moore, 1963; copyright © 1963 by Carl A. Moore and reprinted by permission of Harper & Row, Publishers, Inc.)

the section, and the data points that will be included.

The next step is to decide on the horizontal and vertical scales of the cross-section. The horizontal scale is often taken to correspond to that of available maps of the study area, but any convenient scale may be chosen. The data points can then be placed at their proper locations along the line. The vertical scale may equal the horizontal scale and the section drawn at a one-to-one relationship. In many instances it is an advantage if the vertical scale is larger than the horizontal scale so that the section is expanded or exaggerated vertically because more detail can be displayed.

Once the vertical scale has been established, the construction can begin. Sections drawn from map or surface data collected in the field usually include a topographic profile. The profile is drawn by the standard technique of projecting intersections of the section line and contour lines on the map to the corresponding elevations on the cross-section. When all the points have been plotted, the profile is completed and then the locations of contact lines or other geological features are pro-

jected to the profile and the relationships filled in (see *Plane Table Mapping*).

In cross-sections drawn from columnar sections or geophysical logs, sea level is drawn as a horizontal line on structural cross-sections, and the corresponding elevation at each data point is determined so that the log can be placed in its proper position. The surface elevation must be known for well samples and geophysical logs. For example, if the surface elevation of a well is 380 m the datum will pass through the 380 m depth point on the log. On stratigraphic sections the selected stratigraphic horizon is drawn as a horizontal line and matched to the elevation at which that horizon occurs at each data point.

The stratigraphic units are marked on the logs or columns and correlated between points. If simple lines or sticks are used to locate the data points, the positions of the stratigraphic boundaries are scaled off relative to the datum and the intervals are correlated. Preliminary correlations can be made by laying logs out in proper position on a large table or by pinning them to a wall or bulletin board. Logs can be added, removed, or adjusted vertically until a final cross-section is set. The diagram can then be drafted in finished form and colors or patterns applied. It is good practice to place a bar graph of both vertical and horizontal scales on the cross-section because such a scale changes in direct proportion to the diagram regardless of the size at which it is reproduced.

Cross-sections drawn through intensely folded strata are more difficult to construct. In the case of parallel folding, the lengths of layers in the cross-section must remain the same to preserve true volume relationships, and therefore, the radius of curvature must change toward the centers of folds. In cases where that is insufficient to produce an accurate cross-section, additional features such as a décollement, a change in fold morphology, or plastic flow of incompetent beds must be considered. A detailed discussion of the construction of these so-called balanced cross-sections can be found in Roberts (1982).

Interpretation of Cross-Sections

Several factors must be kept in mind to interpret a geologic cross-section accurately. The first consideration is, as mentioned, the type of cross-section—i.e., whether it is a structural or stratigraphic section.

The second factor is the relationship of the vertical and horizontal scales. If they are not the same, the section is vertically exaggerated and the amount of exaggeration is defined by $VE = $ Vertical scale/Horizontal scale. Expanding the vertical scale is useful to enhance detail, but it introduces distortion into the section. Dips in an exaggerated cross-section appear steeper than true dips. The relationship is given by: tan exaggerated dip $= VE$ tan true dip, where VE is the vertical exaggeration defined earlier. Exaggerated sections are changed in appearance because of steepened dips. Also because the change introduced is not uniform for all dip values, the angular relationship of features with different dips is altered as well.

Vertical exaggeration also alters apparent bed thickness. Since horizontal distances are not changed, vertically dipping beds are shown at true thickness, whereas horizontal beds appear thickened by the factor of the exaggeration. This means that beds will appear thick in areas of gentle dip and seem thin markedly in areas of steep dip.

The final consideration is that of the orientation of the line of section to the strike of the beds or faults. Unexaggerated sections drawn perpendicular to strike will show the true dip of the features. Sections oriented other than normal to the strike will display dips less than the true dips or somewhat distorted structural relationships.

Cross-sections are useful to develop a concept of the geology of an area. Preliminary sections help geologists to understand better the geological and structural complexities present. Cross-sections are also used to present and illustrate a fully developed interpretation in final form. Cross-sections are extremely powerful tools for geological exploration and interpretation. However, they are most effective when used in conjunction with geological and facies maps. Maps and cross-sections provide complementary information that result in a complete view of features and their relationships. Diagrams known as *fence diagrams* are also valuable. On such diagrams the data points are located on a map base in their appropriate relative positions. A network of sections is then drawn between the points, starting from the front of the diagram. This technique produces a three-dimensional picture.

Cross-sections, fence diagrams, and maps can be constructed by computers. Several sophisticated systems are in use by industry and universities. The advantages of data storage and retrieval and speed of construction are self-evident but a geologist still must provide the proper information to interpret correctly the cross-section.

RONALD D. STIEGLITZ

References

Barnes, J. W., 1981, *Basic Geological Mapping.* New York: Halsted Press (Geological Society of London Handbook Series No. 1), New York, 112p.

Bishop, M. S., 1960, *Subsurface Mapping.* New York: Wiley, 198p.

Langstaff, C. S. and D. Morril, 1981, *Geologic Cross-sections.* Boston: International Human Resources Development Corporation, 108p.

Moore, C. A., 1963, *Handbook of Subsurface Geology.* New York: Harper & Row, 235p.

Roberts, J. L., 1982, *Geologic Maps and Structures.* New York: Pergamon Press, 332p.

Ragan, D., 1985, *Structural Geology: An Introduction to Geometrical Techniques,* 3rd ed. New York: Wiley, 393p.

Simpson, B., 1968, *Geological Maps.* New York: Pergamon Press, 98p.

Cross-References: *Block Diagrams; Cartography, General; Compass Traverse; Field Geology; Geological Survey and Mapping; Map Symbols; Plane Table Mapping.*

D

**DATABASE MANAGEMENT—See
COMPUTERS IN GEOLOGY. Vol XIII:
CRIB.**

**DATA CODING—See MAP
ABBREVIATIONS, CIPHERS, AND
MNEMONICONS; PUNCH CARDS,
GEOLOGIC REFERENCING; WELL
LOGGING.**

**DIAGRAMS, GEOLOGICAL
ILLUSTRATION—See BLOCK
DIAGRAMS; CROSS-SECTIONS; MAP
SYMBOLS; PROFILE CONSTRUCTION.**

DISPERSIVE CLAYS

Dispersive clays are a particular type of soil material in which the clay fraction erodes in the presence of water by a process of *deflocculation* (q.v. in Vol. XII, Pt. 1). This deflocculation occurs when the interparticle forces of repulsion exceed those of attraction so that the clay particles go into *suspension,* and if the water is flowing, as in a crack in an earth embankment, the detached particles are carried away and *piping* occurs. The possibility of piping failure in earth dams due to dispersive clay behavior was initially recognized in Australia (Cole and Lewis, 1960). In addition to possibilities of piping failure, the slopes of earth embankments (see *Slope Stability Analysis*) constructed with dispersive clays are susceptible to rainfall erosion. In natural soil deposits of dispersive clay, erosion due to wave action may occur along shorelines of reservoirs, and erosion due to wave and current action may occur along channels constructed in dispersive clay (Sherard et al., 1976; Mitchell, 1976).

Geologic Origin and Topography

The geologic origins of dispersive clays are not well understood. Most of the dispersive soils encountered to date developed in floodplain deposits, slope wash, lake bed deposits, and weathered loess. Clays derived from in situ weathering of ig-

neous and metamorphic rocks, as well as soils derived from limestones, are normally nondispersive. Dispersive clays have been recognized in the United States, Australia, Israel, Ghana, Venezuela, Mexico, Brazil, Trinidad, South Africa, Thailand, and Vietnam.

Dispersive clays are usually not present in the topsoil due to the process of *eluviation* (movement of clay particles downward in the soil profile) (see Vol. XII, Pt. 1: *Leaching*). Therefore, natural soil deposits may show little or no evidence from surface appearance that the underlying soil may be dispersive. The distribution of dispersive clays in a soil profile may vary in plane as well as with depth. When subjected to rainfall, excavated slopes in dispersive clays will exhibit rill erosion, surface cracking, and vertical erosion tunnels resembling a karst topography (Fig. 1). Embankments constructed of dispersive clays, with excellent vegetative cover, also develop rainfall erosion tunnels, commonly called *cave-ins* or *jugs.* The number of rainfall erosion tunnels that develop may decrease with time due to an amelioration process that changes the upper portion of the embankment from a dispersive soil to a nondispersive soil (Perry, 1979).

Identification

Soils with the fraction finer than 0.005 mm ≤ 12% and with a plasticity index ≤ 4 usually do not contain sufficient *colloids* to support dispersive erosion. Such soils are known to have low resistance to erosion, however, and the dispersion characteristics would add little to the known field performance of the soils. Positive identification of dispersive clays is by observed performance of the soil in the field.

Dispersive clays cannot be identified by conventional engineering *index tests* such as particle size distribution, Atterberg limits, and compaction characteristics. Sherard et al. (1976) have obtained a relationship between dispersion and soil pore–water chemistry based on *pinhole erosion tests* and observed dispersive erosion in nature (Fig. 2). The soil pore water correlation has about 85% reliance in predicting dispersive performance. The pinhole erosion test (Fig. 3) is the most reliable test for identifying dispersive clays. In conducting the test, distilled water under a low hydraulic head is caused to flow through a small-

FIGURE 1. Erosion pattern of excavated slope in dispersive clay. *Left:* General view of slope showing rills. *Right:* Close-up view of slope showing vertical erosion tunnels and surface cracking resembling a karst topography.

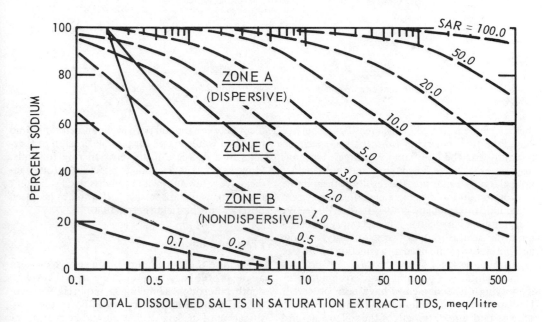

FIGURE 2. Relationship between dispersion and soil pore–water chemistry based on pinhole erosion tests and experiences with erosion in nature. (After Sherard et al., 1976)

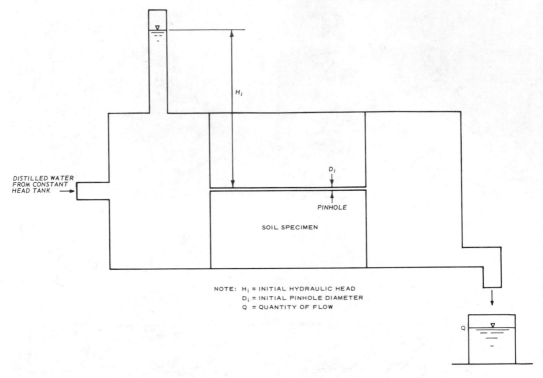

NOTE: H_i = INITIAL HYDRAULIC HEAD
D_i = INITIAL PINHOLE DIAMETER
Q = QUANTITY OF FLOW

FIGURE 3. Schematic representation of pinhole erosion apparatus. (After Perry, 1978)

diameter hole in the soil specimen. For dispersive clays, the flow emerging from the soil specimen is cloudy and the hole rapidly enlarges. For nondispersive clays, the flow is clear and the hole does not enlarge. To interpret the results of the pinhole erosion test and to develop a classification system containing intermediate grades between dispersive and nondispersive clays, it is necessary to determine the relationship between quantity of flow and initial hydraulic head as a function of the size of the pinhole (Fig. 4).

Two limitations of the pinhole erosion test for identifying dispersive clays have been observed. Undisturbed soil samples of high sensitivity (ratio of peak undrained strength of the soil in a natural state to the peak undrained strength after it has been remolded without change in water content) may be classified as dispersive from the pinhole erosion test, while in nature the soil may be resistant to erosion. The natural structure of the soil is apparently destroyed by punching the pinhole in the undisturbed soil specimen, and a reaction analogous to dispersion is obtained in the pinhole erosion test. Soils with high sodium (>80%) and low total dissolved solids (<0.4 meq/l) in the soil pore water may show nondispersive performance in the pinhole erosion test, while the soil may exhibit dispersive performance in the field. This may occur because a decrease in the concentration gra-

dient between the soil pore water and eroding fluid (distilled water meq/l for pinhole erosion test) results in a decrease in the erosion rate for soils. However, available data from case histories indicate very few soils with total dissolved solids ≤1 meq/l for which dispersive performance has been observed in the field (Perry, 1980).

Practical Considerations

Because the distribution of dispersive clays in a soil profile may vary in plane as well as with depth, thorough sampling is required for geotechnical projects. Laboratory tests for the identification of dispersive clays should be part of the subsurface soils investigation for earth dams, channels, and streambank protection. Dispersion characteristics of soil from joint solution cavities in the foundation and abutment area of earth dams should be determined. When dispersive clays are identified, field trials should be considered where excavated slopes and test fills (both untreated and protected with sand and gravel and lime-modified blankets) are exposed to the elements to observe the susceptibility to rainfall erosion (Sherard and Decker, 1977).

When dispersive clays are utilized as embankment material, laboratory filter tests should be conducted to evaluate the adequacy of the pro-

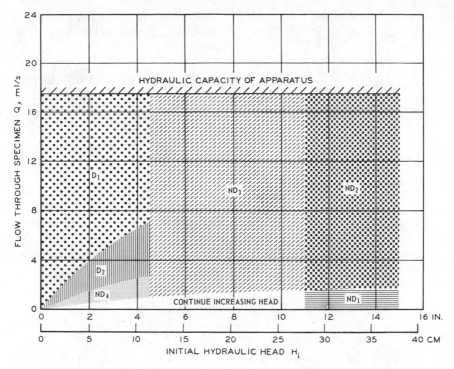

FIGURE 4. Graphic illustration of classification of test results for pinhole erosion apparatus. (After Perry, 1978)

posed filter material concerning internal stability and effectiveness in protecting the base soil (Bourdeaux and Imaizumi, 1975). In zoning earth dams, dispersive clays should not be used in critical locations that might develop concentrated leaks, such as around conduits, adjacent to rigid structures, and in sections of the dam subject to tensile cracking. Dispersive clays should not be placed directly on fractured rock foundations. A lime-modified zone may be required between a dispersive clay core and the rock foundation (McDaniel, 1978; Logani et al., 1979). The possibility of piping failure in homogeneous *earth dams* constructed of dispersive clay, resulting from a decrease in the total ionic concentration of the reservoir water, should be investigated when changing the source of water for the reservoir would be complicated (Ingles and Wood, 1964).

EDWARD B. PERRY

References

Bourdeaux, G., and H. Imaizumi, 1975, Technological and design studies for Sobradinho earth dam concerning the dispersive characteristics of the clayey soils, *Proc. 5th Panamerican Conf. on Soil Mech. and Foundation Eng.* **2**, 99–120.

Cole, D. C. H., and J. G. Lewis, 1960, Piping failure of earthen dams built of plastic materials in arid climates, *Proc. 3rd Australian–New Zealand Conf. on Soil Mech. and Foundation Eng.* 93–99, 234.

Ingles, O. G., and C. C. Wood, 1964, The contribution of soil and water cations to deflocculation phenomena in earth dams, *Proc. 37th Cong. of Australian–New Zealand Assoc. Adv. of Sci.* 1–5.

Logani, K. L., M. Hector, and H. Shex, 1979. Techniques developed during foundation treatment of the Ullum Dam constructed on dispersive soils, *Proc. 13th Internat. Cong. Large Dams* **1**, 729–748.

McDaniel, T. M., 1978, Dispersive problems at Los Esterros Dam. Paper presented at American Society of Civil Engineers convention, Chicago, IL, 1–21.

Mitchell, J. K., 1976, *Fundamentals of Soil Behavior.* New York: Wiley & Sons, 422p.

Perry, E. B., 1979, Susceptibility of clay at Grenada Dam, Mississippi, to piping and rainfall erosion, *U.S. Army Corps Engineers Waterways Expt. Sta. Tech. Rept. No. GL-79-14,* 111p.

Perry, E. B.,1980, Pinhole erosion test for identification of dispersive clays, in *Laboratory Soils Testing.* Washington, D.C.: Office, Chief of Engineers (EM 1110–2–1906), XIII-1–XIII-31.

Sherard, J. L., and R. S. Decker, 1977, Summary—Evaluation of symposium on dispersive clays, in *Dispersive Clays, Related Piping, and Erosion on Geotechnical Projects.* Philadelphia: American Society for Testing and Materials (Spec. Tech. Publ. 623), 467–479.

Sherard, J. L., L. P. Dunnigan, and R. S. Decker, 1976, Identification and nature of dispersive soils, *Jour. Geotech. Eng. Div. (Am. Soc. Civil Engineers)* **102**,(GT4), 287–301.

Sherard, J. L., L. P. Dunnigan, and R. S. Decker, 1977, Some engineering problems with dispersive clays, *Dispersive Clays, Related Piping, and Erosion on Geotechnical Projects*. Philadelphia: American Society for Testing and Materials (Spec. Tech. Publ. 623), 3–12.

Cross-references: *Dams; Slope Stability.* Vol. XIII: *Atterberg Limits and Indices; Clays, Engineering Geology; Consolidation, Soil; Engineering Geology; Engineering Soil Science; Rheology.*

DRAINAGE SURVEYS—See ALLUVIAL SYSTEMS MODELING; CANALS AND WATERWAYS, SEDIMENT CONTROL;

HYDROCHEMICAL PROSPECTING. Vol. XII, Part 1: SOIL DRAINAGE.

DRILL HOLE INFORMATION—See FIELD NOTES, NOTEBOOKS; GEOLOGIC REPORTS; WELL LOGGING. Vol. XIII: WELL DATA SYSTEMS.

DRILLING TECHNOLOGY—See AUGERS, AUGERING; BOREHOLE DRILLING; BOREHOLE MINING. Vol XIII: RAPID EXCAVATION AND TUNNELING; TUNNELS, TUNNELING.

E

ECHO SOUNDING—See ACOUSTIC
SURVEYS, MARINE

ELECTRICAL, ELECTROMAGNETIC
SURVEYS—See MARINE MAGNETIC
SURVEYS; VLF ELECTROMAGNETIC
PROSPECTING; WELL LOGGING.
Vol. XIII: ELECTROKINETICS.

ENGINEERING INDEX PROPERTIES—
See OFFSHORE NUCLEAR PLANT
PROTECTION.

ENGINEERING MAPS—See MAPS,
ENVIRONMENTAL GEOLOGY; MAPS,
PHYSICAL PROPERTIES.

ENVIRONMENTAL ENGINEERING

Environmental engineering lends itself to the application of sound engineering principles to the solution of problems that humans face in their finite environment. The human life environment embraces several systems—chiefly, spheres in which the life scene occurs: (1) the atmosphere, (2) the hydrosphere, (3) the lithosphere, and (4) the biosphere and the interfaces between these sphere boundaries (Strahler and Strahler, 1973). Engineers, assigned a given task, accomplish that task by utilizing the tools of the basic sciences and, through their innovativeness and technical skill, design and solve problems to the satisfaction of the society of which they are a part.

Environmental engineering is a relatively new discipline brought about by the realization that there is no such thing as a superabundance of mineral raw materials, pure water, clean air, and wide open land. Now, it is a question of human survival. The Earth has been around for billions of years, and in a very short time span, humans have taken their toll on the natural systems that occur on the planet Earth. The depletion of natural resources and the modifications humans have imposed on the Earth's ecosystems have resulted in a scramble to undo or correct these situations.

As technological skills have increased, so has the ability of humans to modify their living environment (Liptak, 1974).

Applied geology has been the chief concern of the engineering geologist or geological engineer over the years. The technologies associated with exploration, subsurface investigation, soil mechanics, and/or geotechnics have steadily progressed. New drilling techniques (see *Borehole Drilling*) and methodologies have been pouring forth, providing new approaches to age-old, problems and as near surface mineral deposit bonanzas have become worked out, the searching teams probe deeper into the Earth's crust and have gone to the ocean depths in the never-ending search for minerals to supply the needs of humankind. These geoscientists will continue in their role to fulfill the pressing need for mineral raw materials for construction, industry, and sources of energy. What of the environmental engineer and his or her role in applied geology? Perhaps the major contributions lie in solving problems of reducing or eliminating *environmental hazards*.

Deep-Seated Earth Activity

Forces generated in the Earth's crust create mountains, cause stresses to form rocks, and cause the rocks to flow and fracture, and earthquakes to develop. These activities create environmental pollution and geological hazards. Much research activity has been and is being directed toward a better understanding of what occurs deep within the bowels of the Earth (Tank, 1976).

The elements within the molten core and the reversals of the Earth's magnetic field that affect the human environment are matters of great concern. A better understanding of these phenomena, which are out of the observational field, may result in our ability to predict the occurrence of an earthquake and enable us to determine its frequency and intensity (National Science Board, 1972).

Continental Drift and Seafloor Spreading

For over fifty years, American scientists scoffed at the concept of drifting continents, but since the late 1960s, revolutionary ideas have confirmed the fact that the Earth's crust consists of plates. The pattern of earthquakes on the globe and their distribution provide clues that the earthquake regions are the active boundaries of plates in contact. Offsets of major faults on the land surface provide a means of measurement to indicate that related offsets occur at other places along plate boundaries (National Science Board, 1972).

The orientation of magnetic minerals in igneous rocks provides further clues of plate movement. Each change of magnetic polarity resulting from new rocks formed at the trailing edge of the plates is recorded, and the magnetic record can be converted to give an age-of-rock record (Blyth and deFreitas, 1974).

As geoscientists continue their work on the land and ocean, they accumulate new facts related to the composition of the continental crust and the structure and composition of the upper mantle. Insight into tectonism and the role it plays in shaping the Earth's surface have offered new clues for the explorationists in the search for mineral deposits and hydrocarbons (see *Exploration* and prospecting topics, *Petroleum Geology*). Seismology has produced facts to substantiate and elucidate detail on earthquake zones of the California continental margins as well as showing the thicknesses of crustal portions of the continental interiors and borderlands (Tank, 1976).

The progress of geoscientists and engineers since the 1960s has shown that much more needs to be done to solve some of the pressing problems of environmental pollution and change brought about by natural Earth processes. Environmental engineering (q.v.) in the area of applied geology is greatly aided by the knowledge gained from the geological studies of the deep interior of the Earth and the processes at work along the continental margin and ocean floors. These data provide base levels of natural pollution for projection into time and space for comparison and are a means of alleviating or abating the unnatural condition imposed by humans on their life environment.

THOMAS A. SIMPSON

References

Blyth, F. G. H., and M. H. deFreitas, 1974, *A Geology for Engineers.* New York: Crane, Russak & Company, Inc., 557p.

Liptak, B. G., ed., 1974, *Environmental Engineers' Handbook. Vol. 3: Land Pollution.* Radnor, Pa.: Chilton Book Co., 1,130p.

National Science Foundation, 1972, *Patterns and Perspectives in Environmental Science* (Report prepared for the National Science Board,) Washington, D.C.: U.S. Government Printing Office, 21–48.

Strahler, A. N., and A. H. Strahler, 1973, *Environmental Geoscience: Interaction between Natural Systems and Man.* Santa Barbara, Calif.: Hamilton Publishing Company, 511p.

Tank, R., ed., 1976, *Focus on Environmental Geology,* 2nd ed. New York: Oxford University Press, 538p.

Cross-references: *Petroleum Geology.* Vol. XIII: *Earthquake Engineering; Geotechnical Engineering; Medical Geology; Oceanography, Applied; Urban Engineering Geology.*

ENVIRONMENTAL MANAGEMENT

Environmental management is in the eyes of the beholder. What may appear to be a managed environment to some, may to others be unmanaged or mismanaged. *Environmental management* is interpreted to mean in this entry the deliberate planning and administration of policies that determine the status of the biosphere in which humans live. It covers a broad spectrum of activities that involve the air, water, and land of the planet Earth. Management ranges from the local to the national and international, from policies instituted by private persons to those by industry and governmental bodies. Our concern here involves general policy that has developed from legal and mandated directives with the concurrence of societal objectives and aspirations. It is assumed that decisions reached by environmental managers have been made after careful examination of a pertinent database that was established for this purpose.

Before any type of environmental management is orchestrated, organizers must recognize that something is in need of being managed, that a problem already exists, or that one is likely to arise in the future. Thus, awareness and perception of how humans are affecting the world and how nature influences humankind are prerequisites to environmental programs and schemes for management. Not only is a political and societal support base necessary for effective environmental management, but also the objectives must be consistent with what can be accomplished realistically with the funds that are available.

Objectives

The objectives adopted in programs of environmental management are a reflection of many factors that include assignment of priorities, decisions on constraints, and character of accommodation to the physical systems. Priorities that are established provide the controlling guidelines of how and for what purpose the environment will be managed. The three most common and competing viewpoints are:

1. *The utilitarian-developmental ethic:* Here humankind is supreme and nature is always sub-

servient and should be used and controlled for every human desire.

2. *The conservation ethic:* This view believes that nature best serves humans when resources are carefully administered and prorated so that maximum benefits will occur through time.

3. *The preservation ethic:* Here nature has equal footing with humankind, and natural systems should be protected, left in as virginal a state as possible, and be unmolested by human activities.

Thus, the management of environmental affairs can be largely dependent on the type of ethic that is adopted for nature's materials and processes.

As in all programs, there are constraints that operate and that provide the boundary conditions for the system. The size of the area to be managed is of vital importance. Political subdivisions sometimes constitute unnatural barriers for complete integration of the system. For example, a watershed comprises a natural entity, but it may not be possible to obtain rights to administer it as a coordinated parcel. Feedback mechanisms that operate in the process systems provide unwanted effects when the entire region is not managed. An illustration of this principle is when a dam is constructed. The downstream channel area is subjected to erosion because of sediment loss impounded behind the dam. Thus, management of this land-water ecosystem should also consider the side effects that occur at other locales, instead of being concerned only with geologic conditions at the dam foundation site.

A second constraint for environmental management concerns the time frame (of course, money is also involved) that will be considered for construction and administrative purposes. A common planning preference used for floodplain development is the 100 yr flood, whereas many highway culverts are designed to carry waters for the 25 yr or 50 yr event. The rate of resource depletion can also place a limit on developments, and the degree that materials and processes should be conserved for future generations also needs consideration.

Another factor that influences objectives of how an environment will be managed relates to the sophistication and understandings about the components that constitute the system. Most environments are complex, interrelated, and possess thresholds that, when exceeded, may result in irreversible changes. Interdisciplinary efforts and skills are needed for complete comprehension and management.

As in all human endeavors, decisions are made on the basis of some type of benefit-cost analysis. An environmental program will not be undertaken if the predicted benefits are judged to be less than the costs. Thus, the manipulation of strategies concerned with how to determine true benefits and costs often decides whether or not a program will be approved and financed.

Environmental management is a very broad field, but we are mainly concerned here with geologic considerations. Therefore, this discussion covers only those aspects of the geology of the land-water ecosystems that affect human health, safety, and welfare. The legislative parts of such management appear in the *Legal Affairs* entry.

Nonrenewable Resources

Nonrenewable resources are those for which there will be no second crop for present generations of the human race. The management of materials such as iron, copper, petroleum, and numerous other geologic resources requires different procedures and guidelines than those used for other resources. Nonrenewable resources need careful monitoring throughout all elements of their existence—in mining, in use, and in ultimate demise. For example a management policy that would opt for their utilization as quickly as possible, without consideration for alternative programs or substitutes, would indeed be a bankrupt philosophy.

A wise liaison between government and industry is necessary to achieve maximum long-range benefits from the use of the geologic resources. Some methods that might be adopted for the short range can for the long range prove very wasteful. For example, stockholders are typically interested in large and quick profits, and if this one goal is completely incorporated only the highest grade and most accessible ores will be mined first. This type of management is called *highgrading* and is indulged to obtain early and excessive royalties and rapid investment returns. Such a practice can be very unproductive in the long run, however, and can lead to excessive consumer costs. Proper management in such a situation could alleviate the uneven flow of goods and produce more uniform supply and costs throughout the period of mining.

Another management stratagem is needed to prevent *cultural nullification* of mining and the necessary flow of Earth resources. Appropriate public relations and education programs need to be instituted to prevent repressive societal opinions that would lead to prohibitive mining legislation. Management must convince policymakers that noise, sediment, and air pollution can be prevented or minimized and land properly reclaimed; otherwise, the mines may be closed by an incensed and aroused public. Loss of geologic resources may also occur in and near urban areas where the deposits are paved over before they can be properly mined and extracted. For example, Alfors et al. (1973) report the potential loss of such resources in California would be $17 billion during the period 1970–2000 if nothing is done to remedy current urbanization practices.

A hotly debated issue for many years in the United States has been over strip mining legislation. The mining of coal has produced the greatest amount of land deterioration of any type of mining in the United States (see *Coal Mining*). Although several states had enacted legislation to encourage reclamation, not until 1977 was federal policy set in the Surface Mining Control and Reclamation Act. Thus, the environmental management of coal fields now requires a certain level of land reclamation. Appropriate land reclamation measures include the stabilization of slopes (see *Slope Stability Analysis*) during mining, maintenance of proper drainage of construction and haulage sites, covering of tailings, development of settling basins and containment of sediment and other pollutants, and vegetation planting and slope reduction. An associated problem receiving much discussion concerns what environmental management posture should be adopted for the possible development of western oil shales. Would their use be the ruination of the west, or are they an important consideration toward an energy independence?

In many states sand and gravel is the most important nonrenewable resource. In recognition of this fact, New York requires all quarry operators to submit geology source reports if the material is used on state roads. They must also file reclamation plans to show what will be done when the resource is mined out. Because of their bulk, sand and gravel deposits are cost-sensitive to long-distance haulage, so they are mined as close to the consumption point as possible. Problems in transport occur during increased urbanization and the creation of haulage barriers that restrict their easy mobility (Fakundiny, 1976).

Living Resources

Careful logical engineering of terrain and soil that support living resources—crops, grazing lands, and timbering—is vital to human survival. The mismanagement of such lands can result in the ruination of soils and deleterious erosion and sedimentation, which cause losses in the resource base. For about 50 yr the Soil Conservation Service (U.S. Department of Agriculture) has been involved with programs dealing with conservation measures to inhibit the destruction of soils. They have produced a wide educational base that instructs in the use of check dams, terraces, hillside ditching, crop rotation, and other techniques. Such procedures have greatly reduced sediment losses due to farming in the United States.

A different type of environmental management for agricultural lands occurs when the decision is reached to reclaim lands that were formerly wetlands. About one-third of the Netherlands, the prime example, are lands that have been reclaimed from the coastal water zone. In the United States, however, residents are now looking on them as areas that should be retained in the natural state. Laws for protection of wetlands have been passed in many states—e.g., Massachusetts, Connecticut, and New York—where environmentalists effectively argued their importance in the food chain of life, as regulators of water balance, and as a buffer against storm damages.

The management of forest lands has been a major battleground in the United States for more than 75 yr. It was the site of the classic confrontation between the conservation-utilitarian views of Gifford Pinchot and the preservation leader, John Muir. The current compromise was reached in 1960 Congress passed the Multiple Use–Sustained Yield Act, which mandates that forest lands have five functions: (1) timber resource, (2) recreation, (3) wildlife, (4) forage for grazing, and (5) protection for rivers. The geology side of forest management arises when improper uses cause accelerated soil loss and produce river imbalances, increase the landslide hazard potential, and deplete nutrients in soil. Two conflicts currently unresolved regarding forest management involve clearcutting and fires. Clearcutting is complete removal of all trees (rather than selective cutting), and the fire issue concerns whether the government should attempt to put out those forest fires that occur naturally if lives and property are not in danger. The pros and cons of these issues each have their adherents.

Water Resources

There is no other resource area where the various disciplines interact more fully than they do in the management of water. Water is the basic building block of civilization (Wittfogel, 1956) and a vital ingredient in agriculture and industry and for domestic and municipal use. The transportation, use, and disposal of water in industrialized countries require extensive engineering schemes and management policies. Numerous laws have been written and court cases held about the use of water (see *Legal Affairs*).

In the United States the two largest water management programs are those established by New York City and by the California Water Plan. Extensive litigation in New York, Pennsylvania, and Delaware was necessary to allow New York City to tap Delaware River water in the Catskill Mountains and transport it 160 km to the metropolitan area. The California Water Plan was even more dramatic because water from the northern part of the state is transported 1,000 km to the south, requiring 21 dams and reservoirs, 22 pumping stations, extensive tunnels and pipelines, and construction of the highest dam in the United States. The Australian Snowy Mountain Hydroelectric Project represents another massive environmental management project where water flow to eastern Australia is diverted to the drier western part of the country for irrigation.

Management of groundwater resources is linked to certain laws and to strategies for recharge of the resource. In New York, Long Island has been a leader in trying to conserve the groundwater resource. When excessive pumping in the 1930–1940 period caused lowering of the water table and infiltration of seawater into freshwater aquifers, laws were passed that required large water producers to drill recharge wells so that water could be pumped back into the ground. Another innovation was started in the 1950s with the establishment of more than 2,500 recharge pits. The purpose of these excavated pits is to trap surface water before it flows into the ocean. Percolation of the water in the pits acts as a recharge source for shallow aquifers. In the West, California pioneered in water-spreading techniques to induce surface-water infiltration into the ground where it would not be readily lost by evaporation or by phreatophytes. Another water management scheme was first used in the Raymond Basin where overpumping of groundwater threatened the economic life of the Pasadena–Alhambra area (see *Legal Affairs*). Through court action the State of California was empowered to manage the water through the new technique of prescriptive rights, an apportionment of groundwater according to prior use and need for the water.

Environmental Hazards

Society has typically looked on government as their protector, not only from foreign enemies but also from the geologic hazards associated with earthquakes, volcanic eruptions, hurricanes, floods, and landslides. It is not enough to provide various types of early warning systems, but the government is obliged to disseminate information about the dangers of building on or near terrain that is liable to be subjected to geologic disasters. However, even when adequate information is available, people will build in hazardous places unless they are prohibited by ordinances and other regulations. Thus, the environmental management of developing communities must contain legislation that requires geologic mapping and evaluation of terrain and rock structures, allows for extensive dissemination of the investigation results, and prohibits development in areas of high

risk. For newly planned communities, the environmental manager has three options on how to handle the danger element: (1) avoid the area, (2) develop a program of financial remuneration if disaster does occur, and (3) undertake some type of structural engineering solution. These matters are discussed elsewhere in this volume (see *Landslide Control*). For an established community the management plan usually consists of structures such as dams, levees, etc. to control flooding. More recently a wide range of land use management strategies has been instituted that include stronger building codes, insurance programs, tax relief, and other forms of subsidization.

Service Engineering

This aspect of environmental management has become increasingly important as humans continue to expand urban areas and as the needs for increased energy and services grow faster than the population. The National Environmental Policy Act of 1970 was passed in recognition of the continuing construction endeavors of government and the likelihood of accelerating deterioration of the air, land, and water. This act ushered in a new era of environmental management methods. It mandated that large governmental construction projects had to assess the types of impacts that would result and must present data why other alternatives were not selected for action. These ideas, when coupled with benefit-cost analysis, provide important criteria on which to judge a given project. Service engineering thus involves those construction activities that maintain the free flow of foods, materials, energy, and services necessary for civilization. In the field of service engineering, countless decisions and policies are needed to determine the location of dams, roads, pipelines, and communications systems. The building of the $6.5 billion Alaska Pipeline and the construction and implementation of the 42,000 mi (67,000 km) interstate highway system in the United States provide goods and services that required extensive geological engineering to accomplish.

The environmental management of communities relies heavily on the earth sciences if they are to be designed to cause minimum damage to natural systems. The placement of new towns, green belts, open space, and recreation facilities requires knowledge of the land-water ecosystem. Unless placed in complete wilderness, even recreational resource localities require service engineering know how for their proper utilization and maintenance. Several controversial issues typify the types of environmental conflicts that arise in the management of areas such as the National Seashore. One debate concerns the extent to which the National Park Service (U.S. Department of

Interior) should restore areas that are damaged by storms. Another management problem arises regarding whether artificial sand dunes should continue to be constructed and their growth encouraged. Pioneer studies by Dolan (1973) and Godfrey and Godfrey (1973) showed that the North Carolina barrier islands that contained artificial dunes were altered more and subjected to greater storm damage than other barrier islands that had remained unchanged by humans. Such research has prompted an entire reassessment of management policies in these fragile environments.

The newest combination of threats to humankind's environmental skills is the close relationship of population growth, food, and energy. Geology will be an important ingredient in the solution equation because the common base rests with earth resources and the engineering and technology to make them available. This in turn depends on the husbandry practices of the land. Therefore, on both the national and international scenes, it is important that many different disciplines be involved in those policies that seek resolution of problems that cause environmental deterioration and lower the quality of life. The search for this goal in environmental management should in the words of Ian McHarg, be associated with a "design with nature."

DONALD R. COATES

References

Alfors, J. T., J. L. Burnett, and T. E. Gay, Jr. 1973, Urban geology master plan for California, *California Div. Mines and Geology Bull. 198,* 112p.
Dolan, R., 1973, Barrier islands: Natural and controlled, in D. R. Coates, ed., *Coastal Geomorphology.* Binghamton: State University of New York, 263–278.
Fakundiny, R. H., 1976, Forecasting the effect of land-use plans on the regional market conditions of the sand and gravel business, in D. R. Coates, ed. *Geomorphology and Engineering.* Stroudsburg, Pa.: Dowden, Hutchinson & Ross, 223–242.
Godfrey, P. J., and M. M. Godfrey, 1973, Comparison of ecological and geomorphic interactions between altered and unaltered barrier island systems in North Carolina, in D. R. Coates, ed., *Coastal Geomorphology.* Binghamton: State University of New York, 239–258.
Wittfogel, K. A., 1956, The hydraulic civilizations, in W. L. Thomas, ed., *Man's Role in Changing the Face of the Earth.* Chicago: Chicago University Press, 152–164.

Cross-references: *Environmental Engineering; Terrain Evaluation Systems.* Vol. XIII: *Arid Lands, Engineering Geology; Coastal Zone Management; Geomorphology, Applied; Permafrost, Engineering Geology; River Engineering; Urban Engineering Geology; Urban Geology.*

EQUIPMENT MINING—See BOREHOLE DRILLING; BOREHOLE MINING; COAL MINING; MARINE MINING; PLACER MINING.

EXCAVATIONS—See BLASTING AND RELATED TECHNOLOGY; CRATERING. Vol. XIII: CAVITY UTILIZATION; PIPELINE CORRIDOR EVALUATION; RAPID EXCAVATION AND TUNNELING; SHAFT SINKING; TUNNELS, TUNNELING; URBAN TUNNELS AND SUBWAYS.

EXPANSIVE SOILS, ENGINEERING GEOLOGY

For a soil to be labeled *expansive,* it must contain significant amounts of smectite (montmorillonitic clay minerals) (see Vol. IVB: *Soil Mineralogy*). The expansive soil becomes a problem, however, when variations in the ambient environment produce changes in the soil moisture content that in turn cause a volume change in the soil profile.

Expansive soils are a worldwide problem. Donaldson (1969), e.g., cites nineteen countries where swelling and shrinking soils cause serious engineering problems. Throughout the United States, expansive soils are responsible for $2.3 billion damage annually (Jones and Holtz, 1973), with over $1 billion damage to highways and streets. Expansive soils are, in fact, the single most costly natural disaster; the average yearly loss from earthquakes, hurricanes, tornadoes, and floods combined amounts to only half that due to expansive soils.

A typical approach to the design of light structures on expansive soils is to construct a foundation that resists soil movement. These design procedures, which are based mainly on engineering properties, give little consideration to other environmental parameters.

Most people who have lived on stable foundation materials are horrified by movements due to expansive soil and by unconcerned attitudes often displayed by those who lived with the expansive soil problem. Thus, there are several levels of acceptance of the problem, including not only those who live with it but also, unfortunately, those who build with it.

Building on expansive soils is similar to the risks incurred by building in hurricane or flood zones because of the uncertainty involved in predicting the severity of movement and consequent damage to structures. The engineer assumes a design level of severity based on previous experience, local damage records, and the risk the owner is prepared to take. Only then can he or she predict the response of expansive soils to expected changes in ambient environments and design economical means of combatting the anticipated structural distortion. Factors that must be taken into account when predicting soil moisture-volume changes include climate, soil moisture active zone, soil characteristics, vegetation, topography, drainage, time, site control, and the quality of construction.

Factors Involved in Expansive Soils

Climate is the single most important environmental factor affecting expansive soils. The depth of the *water table* and the soil moisture active zone is largely controlled by climatic conditions. Problems associated with expansive soils may be nonexistent in humid climatic regions where the water table lies near the ground surface. In contrast, the water table in arid regions may be at great depths and exert little influence on surface soil moisture conditions. The *soil moisture active zone* is that portion of the profile that experiences major variations in seasonal moisture content. The depth of this zone is important to foundation engineers because it represents a minimum depth for placement of piers and because it is the critical depth considered in calculations of swell potential.

The thickness of the *active zone* is usually determined through long-term measurement of heave to depths of about 5 m below the surface.

Glenn (1931), using cased steel rods set at depths of 1/2 to 5 m, studied the active zone in central Texas over a 9 mo period. Movements of the rods, which were referenced to a benchmark sunk to a depth of about 6 m, indicated that volume changes below that depth were negligible.

Comparison of moisture equilibrium conditions in an expansive soil profile before and after the placement of a foundation slab is of interest to the engineer because it indicates the magnitude of possible soil movement. In an attempt to correlate equilibrium moisture content with soil index properties, Russam (1965) suggested that after an expansive soil has been covered with an impervious layer, such as a foundation or road, the ratio of natural soil moisture content (W_N) to the plastic limit (*PL*) will reach a constant value. Other authors, including Loxton et al. (1953) and Livneh et al. (1969), report similar findings. Livneh et al., however, point out that the use of soil suction would be preferred to the use of W_N/PL for a clay profile because the latter increases slightly for large *PL* values. The *PL* of a soil is closely correlated with the equilibrium suction value for the soil, and the ratio of W_N/PL observed beneath the active zone, carried to the surface, represents the equilibrium profile that would eventually develop beneath a slab (Fig. 1).

Several climatic ratings attempt to quantify rainfall, evapotranspiration, and soil drainage characteristics (Thornthwaite, 1948; Prescott, 1949). An additional study by the Building Research Advisory Board (BRAB, 1968) produced a climatic rating system for the United States that gives the probability of drought not occurring.

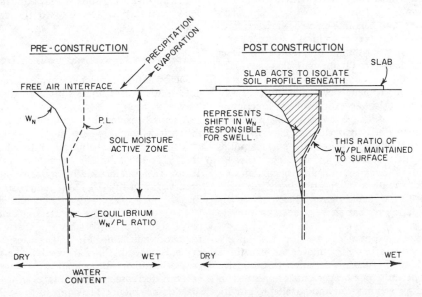

FIGURE 1. Development of an equilibrium moisture profile after a slab is placed.

These ratings, along with a unified soil classification designation, constitute the Federal Housing Administration (FHA) method for selecting the proper foundation.

The *stratigraphy* of the soil profile is a critical factor in the determination of its expansive potential. In soil profiles where an expandable clay is sandwiched between two low-plastic soils, two wetting fronts exist and the expansive potential of the soil profile is significantly greater than the case where a thicker high–plastic clay is exposed to one wetting front. The depth of the active zone, whether it is controlled by the stratigraphy or the water table, also plays a significant role in establishing the expansive potential of the soil profile (Mathewson and Dobson, 1982).

Vegetation, especially large trees, in close proximity to slab foundations may cause large soil volume changes because of the withdrawal of soil water. During relatively wet periods the drawdown effect may be somewhat muted, but during drier seasons the vegetative demand for soil water may result in pronounced, but localized, decreases in volume. Altmeyer (1956), Barber (1956), and Bozozuk and Burn (1960), among others, have studied the deleterious effects of soil volume changes around light structures due to the effects of vegetation. Felt (1953) showed that structures placed over areas with viable soil moisture contents were likely to fail, even after the removal of vegetation, because of volume changes during the rewetting phase. Hammer and Thompson (1966) report that severe structural failures were related to the planting of trees at the same time of construction. This study showed that the most serious failures occurred where large elm, poplar, and willow trees were planted within 12.2 m (40 ft) of a structure. Trees with 25 cm diameters or less and shallow–rooted varieties produced significantly fewer failures. These authors suggested that trees should be planted no closer to the foundation than one-half their expected mature height.

Whether the problem is cyclic, as in the case of stresses imposed by vegetation, or unidirectional, as in the case of heave provided by placing a structure over a heavily desiccated clay, *time* is a crucial factor. Most clays have low permeabilities that are further decreased by swelling; thus, ultimate volume changes in expansive soils and attendant deformation of the overlying structure may involve a time scale measured in years—a design consideration that the engineer cannot ignore.

Topography, especially steep slopes, is an important consideration for building on expansive soils. A form of creep is attributed to expansive soil on slopes in which movements are normal to the slope during expansion and parallel to gravity during shrinkage. As a result, a lateral downslope component is generated. Lytton et al. (1980) have determined that downhill creep in expansive soils can be modeled using a Kelvin model. *Poor drainage,* inherited from the original topography, or poor site finishing can result in ponding around structures. Unless corrected, this condition may result in localized swelling and foundation movement.

Site control is important in controlling the damage due to expansive soil movement. Maintenance or improvement of lot drainage, the installation of gutters, landscaping, and yard maintenance all play a complex role in causing volume changes in expansive soils. In some cases, homeowners who are aware of the expansive soil problem have watered the perimeter of the home's foundation and have experienced few cracks while others have seen little reduction in cracking. Furthermore, actions by homeowners, who are sometimes ignorant of their impact, may result in increased damages.

Quality of construction is another significant factor. Cracking in the concrete foundation will occur when the tensile strength or shear strength is exceeded. The care that is taken in mixing and placing the concrete is at least as important as the amount of steel reinforcing that is embedded in the concrete. Contractors who are unaware or unconcerned about the potential damage that may occur place greater emphasis on rapid completion of a job rather than on assuring proper construction.

The effect of *unloading during construction* must also be considered in any expansive soil problem. Meehan et al. (1975), e.g., report that removing the overburden for a roadcut or building site above an expansive claystone in the Menlo Park, California, area caused the claystone to heave.

Deformation Mechanisms Causing Damage

Deformational mechanisms and the mode of deformation of foundations on expansive soils have not been extensively studied. *Slab deformation* is considered to be large-scale *doming* of the entire soil mass below the foundation (Fig. 2). In a study of home foundations by Mathewson et al. (1975, 1980), in the humid region of central Texas, two distinct deformational models were observed: end lift (EL) and center lift (CL) (Fig. 3).

Large–scale doming of the soil below an impermeable slab is probably the characteristic deformation mechanism in arid regions. Wetting the dry soil beneath the slab results in a volume increase, while drying around the edges results in a volume decrease, thereby generating the large-scale dome. Edge effects appear to be the primary mechanism operating in temperate regions. Wet-

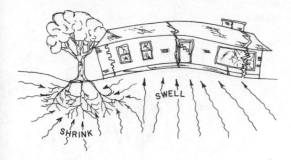

FIGURE 2. Schematic diagram of large-scale dom*ing of the soil mass below a slab foundation.

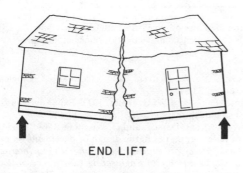

END LIFT

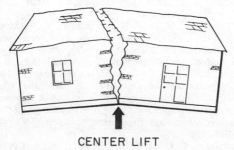

CENTER LIFT

FIGURE 3. Schematic diagram of an end lift (EL) and center lift (CL) deformation of a slab foundation.

ting of dry soil around the foundation edges yields an EL failure, and drying of wet soil around the foundation edge yields the CL failure, resulting in an edge drop movement.

Preventive Considerations

Even though the damaging effects of expansive soils are well known, methods for reducing such deleterious impacts on structures have not been thoroughly investigated. There are, however, several practical precautions that can be implemented in an effort to reduce the risk of foundation failure. The following techniques, which alter the local environmental conditions, have been found useful in this regard:

1. All dwellings and related structures should include a complete system of gutters and downspouts.
2. Gutter downspouts should be connected to drain pipes that carry runoff at least 10 ft (3.2 m) away from the slab.
3. Final lot grades should place a home or similar structure on a mound that has a minimum slope of about 10 cm/m (1 in/ft) for a distance of at least 3 m (10 ft) from the slab or a minimum of 25 cm (10 in) vertical elevation above grade at the property line if it lies less than 3 m (10 ft) from the foundation.
4. Structures should be sited as far away from large trees as possible.
5. Large trees should not be planted near the foundation.
6. Grounds maintenance should strive to maintain the moisture regime that existed when the slab was placed.
7. The floor plan should avoid wing walls or other simple extensions of the slab foundation.
8. Construction should incorporate frame expansion sections at doors and windows or be of flexible frame construction if acceptable.
9. Bearing walls for basements should be supported on piles placed below the active zone or designed to withstand differential movements below the footing (Hart, 1974).
10. Basement slabs should be isolated from the expansive soil by at least 15 cm (6 in) or allowed to float freely (Hart, 1974).
11. If floating floor slabs are used, it is imperative that interior finishings allow for the free movement of the slab.
12. Areas with a high groundwater table or where permeable layers crop out in the excavation should incorporate a peripheral drain that removes excess seepage water (FHA, 1966).
13. Where expansive soils or claystone are interlayered with nonexpansive soils, structures should be founded entirely on one unit or the other (Fig. 4).
14. When excavation is required, the heave of expansive soils in response to unloading should be allowed to occur prior to the construction of slab-on-ground foundations.
15. Walks and streets should be composed of flexible materials (asphalt, loose gravel, stepping stones, timber), or the soil should be stabilized below rigid pavements.

The expansive soil problem results from a complex series of interactions between the structure, topography, soil, and environmental factors that act to cause changes in the soil moisture conditions. The severity of expansive–soil–related damage can be reduced by modifying any three components in this series. Structures can be designed

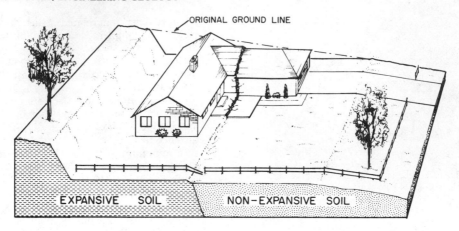

FIGURE 4. Undesirable location of a structure, founded on both expansive and nonexpansive soils. Expansion and heave of the expansive soil leads to differential movement of the foundation with a corresponding failure.

as flexible structures, the soil can be artificially stabilized, or the soil moisture conditions can be maintained at a fixed level. The problem, however, cannot be solved through a simplistic approach.

CHRISTOPHER C. MATHEWSON

References

Altmeyer, W. T., 1956, Discussion of engineering properties of expansive clays, *Am. Soc. Civil Engineers Trans.* **121**, 666-669.

Barber, E. W., 1956, Discussion of engineering properties of expansive clays, *Am. Soc. Civil Engineers Trans.* **121**, 669-673.

Bozozuk, M., and K. N. Burn, 1960, Vertical ground movements near elm trees, *Geotechnique* **10**, 19-32.

BRAB, 1968, *Criteria for Selection and Design of Residential Slabs-on-Ground.* Washington, D.C.: National Academy of Sciences (Rept. No. 33), 288p.

Casagrande, A., 1932, Research on the Atterberg limits of soils, *Public Roads* **13**(8), 121-130.

Donaldson, G. W., 1969, The occurrence of problems of heave and the factors affecting its nature, in *Proceedings of the Second International Research and Engineering Conference on Expansive Soils.* College Station: Texas A&M University Press, 25-36.

Felt, E. G., 1953, Influence of vegetation on soil moisture contents and resulting soil volume change, *Proc. 3rd Internat. Conf. on Soil Mech. and Foundation Eng.* **1**, 24-27.

FHA, 1966, *Minimum Property Standards for One and Two Living Units.* Washington, D.C., 300, 315.

Glenn, E. W., 1931, A Study of the Foundation Soils at College Station, Texas. Ph.D. dissertation, Texas A&M University.

Hammer, M. J., and O. B. Thompson, 1966, Foundation clay shrinkage caused by large trees, *Am. Soc. Civil Engineers Proc.,* SM 6, Paper 4956, **92**, 1-17.

Hart, S. S., 1974, *Potentially Swelling Soil and Rock in the Front Range Urban Corridor, Colorado,* Envir.
Geol. No. 7. Denver, Colo.: Colorado Geological Survey, 23p.

Jones, E. C., and W. G. Holtz, 1973, Expansive soils—The hidden disaster, *Civil Eng.* **8**, 49-51.

Livneh, M., G. Dassiff, and G. Wiseman, 1969, The use of index properties in the design of pavements on expansive clay, in *Proceedings of the Second International Research and Engineering Conference on Expansive Clay Soils.* College Station: Texas A&M University Press, 22p.

Loxton, H. T., N. D. McNicholl, and H. C. Williams, 1953, Soil moisture and density measurements at Australian aerodromes, *Proc. 3rd Internat. Conf. on Soil Mech. and Foundation Eng.* **2**, 112-118.

Lytton, R. L., L. D. Dyke, and C. C. Mathewson, 1980, Creep Damage to Structures on Expansive Clay Slopes. Unpublished report, Texas A&M Research Foundation, prepared for U.S. Army Corps of Engineers Waterways Experiment Station, Vicksburg, Miss., 66p.

Mathewson, C. C., J. P. Castlebery II, and R. L. Lytton, 1975, Analysis and modeling of the performance of home foundations on expansive soils in central Texas, *Assoc. of Eng. Geologists Bull.* **12**(4), 275-302.

Mathewson, C. C., and B. M. Dobson, 1982, The influence of geology on the expansive potential of soil profiles, *Geol. Soc. America Bull.* **93**(7), 565-571.

Mathewson, C. C., B. M. Dobson, L. D. Dyke, and R. L. Lytton, 1980, System interaction of expansive soils with light foundations, *Assoc. of Eng. Geologists Bull.* **17**(2), 55-94.

Meehan, R. L., M. T. Dukes, and P. O. Shires, 1975, A case history of expansive claystone damage, *Am. Soc. Civil Engineers Proc. Jour. Geotech. Eng. Div.* **101**(GT9), 933-948.

Prescott, J. A., 1949, A climatic index for the leaching factor in soils, *Jour. Soil Sci.* **1**, 9-19.

Russam, K., 1965, *The Prediction of Subgrade Moisture Conditions for Design Purposes: Moisture Equilibria and Moisture Changes in Soils Beneath Covered Areas—A Symposium in Print.* Sydney, Australia: Butterworth Scientific Publication, 233-236.

Thornthwaite, C. W., 1948, An approach toward a rational classification of climate, *Geog. Rev.* **38**, 55-94.

Cross-references: Vol. XII, Pt. I: *Electroosmosis; Soil Stabilization.* Vol. XIII: *Atterberg Limits and Indices; Clay, Engineering; Geology Clays, Strength of; Consolidation, Soil; Dispersive Clays; Engineering Geochemistry; Engineering Soil Science; Foundation Engineering; Land Drainage; Permafrost, Engineering Geology; Soil Mechanics.*

EXPERIMENTAL GEOLOGY—See ALLUVIAL SYSTEMS MODELING; GEOLOGICAL METHODOLOGY.

EXPLORATION GEOCHEMISTRY

Geochemistry can be defined as the measurement of the relative and absolute abundance of the elements and isotopes in various parts of the Earth with the object of discovering the principles governing their distribution and migration throughout the geological cycle. *Exploration geochemistry* concentrates particularly on the abundance, distribution, and migration of ore elements, or elements closely associated with ore, with the object of detecting ore deposits. This distinction is only one of emphasis since ores are natural, but not abundant, products of the overall rock-forming cycle.

Motive

As continents become explored more thoroughly, the chances of discovering ore mineral deposits by direct observation decrease. The present land surface, however, presents only a two-dimensional sample of the geology of the Earth's crust; this sample is reduced even more by obscuring layers of soil, vegetation, and water. Because it is feasible to exploit some ore deposits that are obscured by as much as 100–150 m of barren overburden, the probability of discovering a new ore body is greatly increased if indirect methods can be used to detect mineralization below the land surface. Such detection methods may be divided broadly into geophysical and geochemical; the former employ physical characteristics that may persist in an ore body naturally (e.g., magnetic field) or may be induced (e.g., induced potential) (see *Exploration Geophysics*), and the latter make use of often inconspicuous chemical patterns in the vicinity of an ore body. Some of these chemical patterns develop during formation of the ore body, and others develop only after the ore body has been subjected to weathering and erosion of the land surface.

Methods

The methods of exploration geochemistry are essentially refinements of placer prospecting techniques (see *Placer Mining;* Vol. IVB: *Placer Deposits*) that were developed when humans first began to exploit metals. The approach of locating *dispersion trains* of mineralized float in soil and alluvial gravels and tracing them upgradient to an outcropping source has been used by prospectors throughout history (see *Prospecting*). This use of *mechanical dispersion patterns* is readily applicable to the search for mineral species that are relatively stable under conditions prevailing at the Earth's surface (e.g., gold, platinum, cassiterite, chromite, rare earth minerals, etc.). It is also applicable to a wider range of minerals in areas where climatic conditions restrict chemical weathering.

A second basic method relies on recognition of *chemical dispersion patterns* rather than mechanically produced dispersion patterns. These patterns may be produced with or without erosion of the ore deposit and with or without weathering. In addition, chemical patterns are frequently less conspicuous than mechanical dispersion patterns because the element forming the dispersion patterns may

Have a different mineralogy to the ore deposit (e.g., cerussite dispersing from a galena deposit);

Be in solution (e.g., Cu^{2+} ions in groundwater passing out from a chalcopyrite deposit);

Be concealed in another mineral (e.g., Ni in serpentines and clay adjacent to a pentlandite deposit);

Be adsorbed (e.g., Cu adsorbed on clay or organic matter in streams fed by groundwater that has passed through a chalcopyrite deposit);

Be incorporated in organic mater (e.g., Cu in plants or animals).

Although the use of chemical dispersion patterns is regarded as a modern development, it is interesting to note that ancient writings refer to the use of mineral encrustations around springs and acid-tasting water as prospecting guides.

The main refinement, introduced by modern geochemical prospecting, is that by relying on chemical analysis of samples it is possible to detect much weaker dispersion patterns than the old-style prospectors; to utilize elements that do not produce readily recognizable minerals in their dispersion patterns or that are adsorbed, coprecipitated, or concealed in other minerals; and to prospect larger areas more rapidly by using semiskilled personnel to collect reliable data for subsequent interpretation by a skilled prospector.

Historical Developments

Significant developments in exploration geochemistry commenced in the USSR and Scandinavia in the 1930s and 1940s but did not achieve widespread prominence until about 1960. Many of the early applications involved biogeochemical and soil samples. By 1960 Hawkes, of the U.S. Geological Survey, and Webb, at Imperial College in London, had contributed materially to convincing the majority of mineral explorers outside the USSR that geochemistry is a valuable exploration tool (Hawkes, 1957; Hawkes and Webb, 1962). Emphasis shifted from biogeochemical sampling to stream sediment and soil sampling. Chemical analysis of the samples relied heavily on colorimetric techniques often performed in the field after cold extraction of weakly bound metal ions by dilute or weak acids. The limitations of precision, sensitivity, and range of elements imposed by cold extraction colorimetry were overcome in the mid-1960s by the widespread use of optical emission spectrometry on solid samples and atomic absorption spectrometry on solutions prepared by strong acid attack under laboratory conditions.

The availability of inexpensive and precise analysis led to such a boom in geochemical exploration that by 1973 about 28 million geochemical samples were being collected annually throughout the world (Webb, 1973). Since about 1970 X-ray fluorescence spectrometry has become widely available in commercial geochemical laboratories as an additional analytical technique that provides valuable support in analyzing some elements that cannot be dealt with satisfactorily by atomic absorption or optical emission spectrometry (Joyce, 1974). Most analytical procedures now seek to analyze the total content of any given indicator element, but recent literature suggests some revival of interest in partial extraction techniques related to earlier cold extraction procedures. Modern partial extraction techniques differ in seeking more selective reagents and using better controlled conditions.

Other major advances have occurred in the field of interpreting geochemical data. Early interpretation techniques relied entirely on subjective approaches or on very simple statistics based on establishing for each survey area a single threshold value for each element that would be exceeded by a predetermined percentage of arbitrarily defined anomalous samples. Partly because of improved understanding and partly because of the great volume of data that has accumulated so rapidly, more appropriate statistical approaches, usually computer based, have been developed to aid in sorting the data for interpretation. Cumulative frequency analysis, moving average analysis, trend surface analysis, discriminant analysis, cluster analysis, and multiple regression analysis have all been used with varying justification and success. While these techniques are valuable aids to interpretation, especially when dealing with difficult

sampling media, final interpretation still requires subjective skill and experience.

Improved interpretation techniques, coupled with relatively inexpensive, sensitive, and precise analytical techniques, have led to an increasing amount of interest in geochemical surveys involving difficult sampling media such as gas, aerosols, water, and organic muck. Commercial laboratories have very recently commenced offering atomic absorption analysis using a carbon furnace (carbon rod) atomizer that improves the sensitivity of the method to such an extent that many elements can be determined at concentration in the parts per billion range rather than above a few parts per million. This innovation paves the way for improved analysis of natural water samples (and facilitates analysis of gold and platinoids in more conventional samples). Developments in anodic stripping voltametry may also pay dividends in future water–sampling surveys. Although exploration geochemistry originates as a technique to search for metalliferous ore deposits, recent adaptations (Levinson, 1974) include prospecting for gas and petroleum and applications to agriculture, pollution studies, and medicine.

Types of Sampling Media

Stream sediment sampling (see *Hydrochemical Prospecting*) is widely used in regional geochemical reconnaissance in areas with well-defined drainage. The usual technique consists of sampling fine material from the moving bottom load of streams, then analyzing some particular size fraction or heavy mineral fraction. Sample spacing is normally variable, with emphasis placed on sampling minor tributaries entering the main stream being traversed since these are subject to less complex dilution than the major stream.

Sediment sampling is most directly related to early alluvial prospecting techniques used in the search for so-called mother reefs that give rise to trails of resistant detrital minerals. The main difference is that attention is not restricted to immobile elements occurring in resistant primary minerals; i.e., many metals released by weathering of sulfides can be scavenged under favorable conditions by fine-grained secondary minerals and then transported as detritus. Mobile metals of this type (e.g., Cu and Zn) have constituted either the pathfinder or target elements of many geochemical surveys (see *Indicator Elements*). Anomalous concentrations of elements detected in sediment surveys may be displaced appreciably downgradient from their source so that ultimate discovery of a mineral deposit usually involves other geochemical, geophysical or geological follow-up.

Soil-sampling surveys (see *Pedogeochemical Prospecting*) involve chemical analysis of samples collected from a particular soil horizon, then sieved to retain a suitable size fraction or pulverized in bulk. Many surveys have used B-horizon samples to take advantage of the abundance of clays and iron oxides that scavenge many metallic indicator elements. A- and C-horizon samples can be used equally well and may have important advantages in complicated profiles or in situations where a residual character for the soil is uncertain. In flattish terrain, samples are usually collected on a regular grid pattern or along widely spaced traverse lines crossing regional geological trends. In rugged terrain, sampling is more practical as a series of traverses along ridges and spurs.

The simplest results are obtained on residual soils, but colluvium surveys can prove useful. Lateritic soils pose serious problems. Soil surveys are not as amenable to regional reconnaissance as stream sediment surveys because the anomalies sought form smaller targets, thereby necessitating an uneconomical number of samples. The situations in which soil-sampling programs are employed include follow-up surveys in anomalous areas located by stream sediment sampling; follow-up surveys in anomalous areas located by regional geophysical prospecting; follow-up surveys in the vicinity of gossanous outcrops of uncertain origin; delineating drilling targets in the vicinity of known mineralization; and reconnaissance surveys in areas where other forms of prospecting are unsuitable.

Sampling of glaciogene detritus is an approach adapted successfully for geochemical exploration of certain parts of Canada and other glaciated countries. *Mechanical dispersion patterns* are utilized, and the samples have some of the characteristics of soils (e.g., wide distribution) and some of stream sediment (e.g., unidirectional displacement and homogenization to produce a composite sample of the source area).

Lake sediment sampling (see *Lake Sediment Geochemistry*) has received attention in Canada as a probable regional geochemical reconnaissance technique. The choice of lake sediment is conditioned by the absence of satisfactory stream patterns coupled with the advantage of good landing sites for helicopters or light aircraft. Both lake sediment and water tend to be analyzed, and the reliability of the technique is still subject to experimentation.

Estuarine and ocean sediments have been the subject of several surveys by academic and government institutions. The main motive of these surveys has been to study the chemical dispersion patterns surrounding presently forming metal accumulations such as Zn-rich brines, submarine volcanogenic sulfides, and marine phosphorites. A knowledge of these patterns can then be applied in the search for older deposits occurring on the continents. Future exploitation of marine ore deposits will lead to developments in marine geochemistry (see *Marine Exploration Geochemistry*).

Bedrock sampling (see *Lithogeochemical Prospecting*) programs may be executed to provide information on background abundances of elements to facilitate interpretation of soil or sediment geochemical surveys. They are also undertaken in their own right, however, in attempts to detect primary dispersion patterns of anomalous chemistry in the vicinity of blind ore deposits. An impressive example of the potential of bedrock sampling is provided by Sakrison's (1971) description of a mercury anomaly detected in surface rocks about 300 m above a blind copper-zinc deposit at Noranda in Canada. Another impressive example is provided by Lambert and Scott (1973), who examined shales hosting the McArthur River lead-zinc deposits in northern Australia and found zinc anomalies persisting laterally for a distance of at least 20 km from the deposit. Other bedrock-sampling programs are aimed at recognizing favorable host rocks for mineralization. For example, attempts have been made to establish the chemical characteristics of granites likely to host cassiterite deposits, of ultramafic rocks likely to host nickel deposits, and of volcanic rocks likely to host copper-zinc deposits.

Gossan surveys involve routine analysis of ironstone outcrops in an attempt to distinguish true gossans (formed by weathering of sulfide deposits) from false gossans (laterite or ferruginous weathering products of diverse barren rock types) (see Vol. XIII: *Mineragraphy*). In addition to sampling outcrops, some companies now sample ironstone pebbles in conjunction with soil and sediment surveys in the hope of recognizing eroded gossan fragments. Interpretation of gossan analyses is difficult because scavenging in false gossans can produce similar concentrations of elements to those produced by leaching in true gossans. Consequently, interpretation must be based either on immobile pathfinder elements or on multivariate statistical interpretation conditioned by control samples of known parentage.

Biogeochemical surveys (see *Biogeochemistry*) involve the analysis of growing plant material, and operate on the philosophy that suitably deep-rooted plants reach below the surface soil to draw nutrients from weathered bedrock. Surveys of this type were popular in the pioneering stages of exploration geochemistry, but their use declined when it was found that simpler techniques of soil sampling could be substituted satisfactorily in many cases. The main situations in which biogeochemical surveys are necessary or viable alternatives include areas of transported overburden, swampy areas, and regions of very dense vegetation. Brundin and Nairis (1972) drew attention to the usefulness of organic muck samples collected from partly decayed planty matter in streams where conventional sediment sampling is impractical.

Water sampling (see *Hydrochemical Prospecting*) from streams, springs, lakes, swamps, wells, and bores can be used in both reconnaissance and detailed geochemical exploration. Low concentrations, coupled with rapid fluctuations in abundance related to weather variations, have hindered the development of water surveys, but recent improvements in analytical and interpretive techniques have widened the applications. In glaciated terrains (e.g., Sweden and Canada) attention centers mainly on lake and stream waters, but in the arid regions (e.g., Australia) it centers on bore waters. Past applications have made use of U and Rn dissolved in water, but recent developments have added the interest in F, Zn, and SO_4^{--}

Gas sampling (see *Atmogeochemical Prospecting*) is a technique that has two attractions: first, dispersion patterns formed by highly mobile gases may permit detection of deeply buried deposits, and second, developing mobile detectors suitable for rapid regional reconnaissance may be possible. Radiogenic gases such as Rn, He, Kr, and Xe have been used for some time in prospecting for U and Th. In particular, portable detectors measuring the Rn content of soil gas or the content of Rn dissolved in water samples have contributed significantly to the discovery of a number of uranium deposits. Considerable recent attention has centered on seeking Hg gas haloes in the vicinity of various ore deposits, and diverse portable analytical instruments have been developed. These have been tested mounted in aircraft and ground vehicles, but the most reliable results are obtained by pumping soil gas or by degassing rock samples.

All gas surveys in surficial materials or the atmosphere are sensitive to temperature, barometric pressure, humidity, and rainfall. Gases such as H_2S, SO_2, I_2, CO_2, N_2, and O_2 have potential in geochemical exploration but are largely unexploited at this stage. Some experiments have been conducted using dogs to detect the smell of SO_2 produced by weathering sulfides. Hydrocarbon gas leakage haloes are exploited in the search for petroleum.

Aerosol sampling is a technique that has received some attention. Suitable kitelike collectors towed by aircraft can collect fine dust particles emanating from soil and aerosol particles emanating from vegetation. This approach may be developed as a regional geochemical exploration tool appropriate for some tasks in suitable areas.

Airborne measurement of U, K, and Th by gamma-ray spectrometry is a geophysical/geochemical technique that facilitates the production of regional geochemical maps. It is a marked improvement over conventional radiometric surveys

based on total gamma radiation and provides a reconnaissance tool capable of detecting the zones of potassic alteration featured by a number of ore types.

A. S. JOYCE

References

Brundin, N. H., and B. Nairis, 1972, Alternative sample types in regional geochemical prospecting, *J. Geochem. Exploration* 1, 7–46.
Govett, G. J. A., ed., 1981, *Handbook of Exploration Geochemistry.* Amsterdam: Elsevier, 256p.
Hawkes, H. E., 1957, Principles of geochemical prospecting, *U.S. Geol. Survey Bull. 1000-F,* 225–355.
Hawkes, H. E., and J. S. Webb, 1962, *Geochemistry in Mineral Exploration.* New York: Harper & Row, 415p.
Joyce, A. S., 1974, *Exploration Geochemistry.* Adelaide, South Australia: Techsearch, 157p.
Lambert, I. B. and K. M. Scott, 1973, Implications of geochemical investigations of sedimentary rocks within and around the McArthur zinc-lead-silver deposit, Northern Territory, *Jour. Geochem. Exploration* 2, 307–330.
Levinson, A. A., 1974, *Introduction to Exploration Geochemistry.* Calgary: Applied Publ. Ltd., 612p.
Sakrison, H. C., 1971, Rock geochemistry—Its current usefulness on the Canadian Shield, *Canada Inst. Mining and Metallurgy Bull.* 64, 28–31.
Webb, J. S., 1973, Applied geochemistry and the community, *Inst. Mining and Metallurgy London Trans.* 82, 33–38.

Cross-references: *Biogeochemistry; Blowpipe Analysis; Exploration Geophysics; Hydrochemical Prospecting; Indicator Elements; Lake Sediment Geochemistry; Lithogeochemical Prospecting; Marine Exploration Geochemistry; Mineral Identification, Classical Field Methods; Pedogeochemical Prospecting; Prospecting.* Vol. XIII: *Autogeochemical Prospecting; Geochemistry, Applied; Hydrogeochemical Prospecting.*

EXPLORATION GEOPHYSICS

Within the broad spectrum of the contemporary earth sciences, there is a continuing trend to overlap and interrelate. *Exploration geophysics* plays a major role in the areas of mineral and petroleum exploration, marine geology, groundwater exploration, and engineering construction, wherein its primary aim is to add an extra dimension to geological information. In its basic essentials it is a method of *geological exploration,* using instruments whose function is to record changes in the physical properties of rocks in the subsurface. It therefore involves the application of principles of several physical sciences and relates to the measurement of rock properties such as density, velocity, susceptibility, and resistivity and involves the drawing of deductions about the rock types and their geological configurations. There is also some prediction involved in terms of inferring the probability of the presence of mineral deposits, hydrocarbons, and groundwater.

As a general definition, *geophysics* constitutes the application of the principles of physics to the study of the Earth and includes geomagnetism, heat flow, meteorology, volcanology, and the Earth's gravity field. In this context, geophysics has a major influence as a pure science where the objective is simply the advancement of scientific knowledge. The main commercial application, *exploration geophysics,* lies in resource exploration—i.e., prospecting of the shallow, relatively small-scale features of the Earth's crust. Such geological features include structures like anticlines, salt domes, faults, unconformities, reefs, and possibly mineral or hydrocarbon deposits.

Exploration geophysics is one of the most recently established branches of applied science and is essentially an offshoot of several disciplines. Most techniques evolved from early academic studies of the Earth's large-scale features. These techniques are based on a number of fundamental principles of physics such as the laws of gravitation and magnetic attraction and Snell's laws of optics, which govern reflection and refraction seismology and the elements of electromagnetic theory. Although in most cases the principles are quite simple, it is not always a simple matter to apply them to the study of these properties in rocks in the Earth's subsurface since rocks are seldom homogeneous.

Historical Aspects

The search for treasures of the Earth has probably entranced the minds of humans for centuries, certainly since before recorded history. Mining of ores and minerals has been an activity for thousands of years. As Parasnis (1966) points out:

Scholastic controversies as to the origin of ores probably did not seriously distract the old prospectors in their search. Their chief guiding principle was analogy, a valuable principle of ore prospecting to this day. The search for ore should be conducted in those areas where it has already been found and in other areas with natural conditions similar to those in the known case.

The earliest use of applied geophysics was probably in the search for magnetic minerals, particularly iron ore, in the mid-seventeenth century, following the publication of "De magnete" in 1600 by William Gilbert, physician to Queen Elizabeth I. In his treatise, Gilbert described the geomagnetic field and suggested its precise orientation at every point on the Earth's surface. The geophysical prospecting technique was based on observations of local variations (anomalies) in the orientation of the geomagnetic field and represents the first recorded utilization of a physical property for locating specific small-scale features of the Earth's crust. In 1870, Thalen and Tiberg designed the *Swedish mining compass* for prospecting. This instrument was widely used until the early twentieth century when, in 1915, Adolf Schmidt developed a precision magnetometer. In the 1920s a more sensitive type, the *Hotchkiss superdip,* appeared. Electrical prospecting activity since 1915 has been confined mainly to the resistivity, spontaneous potential, telluric currents, and electromagnetic methods (Dobrin, 1960). The multielectrode techniques for resistivity measurements developed by Schlumberger and Wenner are still in use. In the mid-1950s the electromagnetic methods were adopted for airborne use.

The seismic and gravity methods, the most commonly used in petroleum exploration, were originally developed for quite different purposes. The earliest application of gravity was aimed at determining the shape of the Earth. In 1887, Von Sterneck introduced a portable pendulum suitable for measuring the Earth's gravity in geodetic surveys. Later, Baron Roland von Eotvos developed the torsion balance in Hungary. He used the device to map the subsurface extension of the Jura Mountains. In 1915, Hugo de Boeckh suggested the application of the torsion balance to map domes and anticlines where the rock of the central core might be lighter or heavier than the flanking formations. The first successful torsion balance survey was conducted in Czechoslovakia in 1915. About 1917, pendulums were applied to the exploration for salt domes along the Gulf Coast of the United States. By 1924, torsion balance surveys had been conducted in Texas and California, and the first successful torsion balance location, the Nash Dome, had been discovered. The first *gravity meter,* or *gravimeter* (i.e., giving a direct reading of gravity differences), was used in 1935 and quickly replaced both the pendulum and torsion balance.

The earliest *seismograph* was used to record earthquakes. In 1848, Robert Mallet suggested that exploding gunpowder would generate artificial earthquakes suitable for surveying subsurface formations on land and the bottom of the great ocean. In 1850, he constructed the prototype seismograph system consisting of a bowl of mercury with a telescope and applied it to the measurement of sound in granite. The seismic refraction technique was used in the early twentieth century to locate salt domes, and in 1919, Ludger Mintrop in Germany applied for a patent covering the application of the refraction method to determine the nature and depth of geological formations. In 1922, the first *refraction surveys* for oil exploration were conducted in Mexico and the Gulf Coast area. Through the 1920s, the use of the seismic refraction technique became widespread, particularly after discoveries such as that of Orchard Dome. The seismic reflection technique was first applied by Reginald Fessenden in his sonic sounder for determining water depths and locating icebergs. The commercial application of this technique was not fully implemented until 1927, but since that time, it has been the most widely used geophysical method (see *Acoustic Surveys, Marine*). In the late 1940s, the seismic method was adapted for marine use and initiated the geophysical exploration of the continental shelves around the world (see *Sea Surveys*).

Classification of Geophysical Methods

Many geophysical methods have significant versatility in that they can be adapted for use on land, sea, or in the air and can even be modified for use in boreholes (see *Well Logging*). Other diagnostic characteristics might include the simplicity of the operation or ease of data handling and interpretation, but such attempts at classification usually prove inadequate. The most appropriate classification scheme so far derived is that of Parasnis (1962, 1966), who draws distinctions among three types of methods. The *static methods* are based on detecting and measuring accurately the distortions produced in a stationary field of force by inhomogeneities (e.g., ore bodies) in the Earth's crust. The essential feature is that the fields concerned, whether natural or artificial, do not vary with time and include the gravity and magnetic fields of the Earth. The field due to an electric current injected into the ground is an example of an artificial stationary field. *Dynamic methods,* in contrast, measure fields that are not stationary but time variant. These exploit both natural and artificial phenomena and include the seismic and electromagnetic methods. The third group is the class of *relaxation methods,* which lies somewhere intermediate to the other two. They are electrical methods that include induced polarization.

The most important single criterion that separates the natural field methods from the artificial field methods is that in the latter case, the depth of exploration can be controlled within certain limits by judicious positioning of the sources(s) and detector(s) of the field. This is not possible with natural field methods.

Relationship of Exploration Geophysics to Geology

Regardless of classification schemes, no single method is ideal since each has its own range, scope, limitations, advantages, and disadvantages and, in most exploration situations, the most reliable results are derived by combining two or more methods. The reliability of the results is fundamental to the primary role of the exploration geophysicist, which is to interpret the instrument observations in geological terms. Thus, the cardinal aim of exploration geophysics is to add an extra dimension to geological maps (Grant and West, 1965). Measurements of geophysical phenomena are usually taken at or near the surface of the Earth, and this aspect is closely controlled by well-established physical and mathematical laws. While the deductions may often be ambiguous, the nature of the ambiguity is at least well understood. The next stage is interpretive and speculative and involves translating the deductions into reasonable geological situations. This stage is surrounded with some air of mystery as a creative art, and like other arts, success depends on a proper appreciation and balancing of all the relevant factors, which in this case are ecological and physical (Grant and West, 1965). In this context, the geophysicist has to rely on his or her experience and that of others and the process is called *interpretation,* which aptly describe its indeterminate nature. Only when the data are properly interpreted can the geophysical information be used to predict the nature of the subsurface geology.

The exploration for resources in the Earth is directed largely in search of new minerals and fuel and water and for the planning and construction of human facilities. This can often be achieved with a purely geological investigation. If the area is known to contain certain resources and the geological environment in which they occur is understood, then the geologist may be in a position to predict other occurrences, assuming some surface geological conditions exist that simulate the known occurrence situation. However, if the area is not well known or if the geological environment for the resources is not fully understood, then the geological framework can be ascertained only through geophysical investigations.

As Parasnis (1966) points out, the success of a geophysical survey is not measured in terms of anomalies found. The basic criteria for success include the estimated savings in costs in eliminating barren ground. In oil exploration, the locating of a structural geological feature by geophysical methods is considered a success whether the drill proves the structure to be barren or productive.

Exploration Geophysics in the Petroleum Industry

In the petroleum industry, geophysical methods are utilized to determine subsurface structural and stratigraphic conditions, conducive to the trapping of hydrocarbons (see *Petroleum Geology*). The three principal methods used in exploring for petroleum are (1) seismic, (2) gravity, and (3) magnetic surveys. The seismic method and its associated data processing account for about

95% of the total expenditures for petroleum exploration geophysics. The remaining 5% is divided between gravity and magnetic methods. This does not mean that the latter make a proportionately small contribution to the overall exploration effort, but since they are much cheaper than seismic methods, they can therefore be used to obtain coverage over much larger areas (Nettleton, 1971).

Geophysical exploration in the petroleum industry has utilized three basic physical principles—i.e., (1) the measurement of small variations in the magnetic field, (2) the measurement of small variations in the gravitational field, and (3) the propagation of elastic waves through the Earth. These principles form the basis for all geophysical work to the present.

The *seismic method* is more direct in relation to geology than the *potential methods* (gravity and magnetic). In many cases though, the correlation with geology may be uncertain, and in such cases, gravity and magnetic data may contribute additional diagnostic information. As Nettleton (1971) suggests, the dominance of the reflection seismic method may be a function of its basic simplicity in contrast to the potential method, where the action is at a distance.

Exploration Geophysics in the Mining Industry

Applied *mining geophysics* relies on a relatively few physical properties and electrical or electromagnetic phenomena in the search for ore bodies. The gravity and magnetic methods, discussed in the previous section, have an application in mining exploration together with electrical and electromagnetic methods. The electrical conductivity of the Earth varies from place to place more than any other physical property, and this property may be employed in any one of a variety of electrical methods including resistivity and inductive electromagnetic methods (Seigel, 1968). The latter may employ artificial fields, low–frequency radio signals, or natural time-varying fields (Afmag, magnetotelluric). Electrochemical phenomena, either natural or induced, give rise to the methods of *spontaneous polarization* and *induced polarization* (IP). The range of techniques currently available is summarized in Table 1. The principal geophysical techniques are described in some detail on an individual basis later in this entry.

Collection and Presentation of Geophysical Data Survey Operations and Planning

The choice of area for a geophysical survey operation is determined largely by geological considerations. Geophysics can eliminate those areas without significant anomalies and highlight areas where credible anomalies indicate subsurface geological variation. Natural resources are relatively rare and there is therefore much less potential area

than barren, and geophysics can save time, money, and effort in evaluating these areas (Parasnis, 1966). Preliminary planning for the prospect area involves both geological and geophysical factors. The exploration objective is determined from geology and defined in geophysical terms (Landes, 1968).

In any kind of geophysical survey the approach is to use a grid system of traverse lines. The size of the grid mesh is related to the size of the expected or desired target. Any feature smaller than the mesh will slip through the net and not be detected unless it happens to fall on a grid intersection. The large features will be detected by the survey. The desired target size is usually a function of basic economics and the likely return on the exploration and development investment. *Ground surveys* for most geophysical methods suffer from variations in topography and climatic changes. The development of most techniques for airborne and shipborne acquisition has led to large volumes of geophysical data in remote areas. In the air and water environments, it is possible with the aid of modern satellite navigation systems to acquire data from closely controlled traverse grids (see *Remote Sensing, General*).

Computers in the Geophysical Industry

The geophysical industry has not been slow in adapting to the computer age. The advent of digital computers accelerated the transition from analog to digital data recording in all branches of exploration geophysics. Geophysical data-recording instruments are highly sophisticated and have undergone rapid development. Computers are used to process recorded data in a number of ways including editing, applying basic connections, manipulation in terms of improving signal quality, and for display purposes. Figure 1 is an example of a seismic cross-sectional display that is generated from the computer. The general use of computers is largely focused on bulk data handling, display, and automation of interpretation (see *Geostatistics*).

Most geophysical data are displayed in map form. Since the data are usually acquired from an area using a traverse grid, the variation in the particular geophysical method response can be represented on profiles or as a three-dimensional contour map. Computer-generated contour maps are commonplace, while current computer investigations involve interactive modeling techniques via cathode ray terminals.

Since 1955, a vast literature on the application of mathematical surfaces to the interpretation of data in the earth sciences has appeared. Their main use has been to simplify and analyze data so that trends that are not particularly obvious in the data have been revealed (Wren, 1973). Most

TABLE 1. Principal Geophysical Methods and their Application in Mineral Exploration

METHOD (d) : dynamic (s) : static G : Ground V : Airborne W : Marine ● : logging available			PARAMETER		CHARACTERISTIC PHYSICAL PROPERTY
			DESIGNATION	UNIT	
ELECTRICAL	**RESISTIVITY** (d) ●	G W	APPARENT RESISTIVITY	ohm.m	RESISTIVITY CONDUCTIVITY
	INDUCED POLARIZATION (d) ●	G	TIME DOMAIN | CHARGEABILITY POLARIZABILITY FREQUENCY DOMAIN FREQUENCY EFFECT	ms % % %	IONIC - ELECTRONIC OVER VOLTAGE
	SELF POTENTIAL (s) ●	G	NATURAL POTENTIAL	mV	CONDUCTIVITY OXYDABILITY
	MISE A LA MASSE (d) ●	G	APPLIED POTENTIAL	mV	CONDUCTIVITY
	TELLURIC (s)	G	RELATIVE ELLIPSE AREA		HORIZONTAL CONDUCTANCE
	MAGNETOTELLURIC (s)	G	APPARENT IMPEDANCE (RESISTIVITY AND PHASE)	ohm.m radian	RESISTIVITY CONDUCTIVITY
	ELECTROMAGNETIC (d) ●	G W V	PHASE DIFFERENCE TILT ANGLE AMPLITUDE RATIO INPUT : SAMPLING OF DECAY CURVE INDUCED IN RECEIVING COIL BY EDDY CURRENTS IN PHASE OUT OF PHASE COMPONENTS	degrees μV	ELECTRICAL CONDUCTIVITY
	MAGNETIC (s) ●	G W V	EARTH MAGNETIC FIELD : VERTICAL COMPONENT Z, TOTAL INTENSITY EARTH MAGNETIC FIELD : TOTAL INTENSITY, VERTICAL GRADIENT	$1\Upsilon = 10^{-5}$ gauss	MAGNETIC SUSCEPTIBILITY
	GRAVITY (s)	G W	GRAVITY FIELD	milligal (1 gal = 1 cm/s^2)	DENSITY
	RADIOACTIVITY (s) ●	G V	INTENSITY OF GAMMA RAYS	μ/Roentgen or C.P.S.	RADIOACTIVITY
SEISMIC	**REFRACTION** (d) ●	G	TRAVELLING TIME OF SEISMIC WAVES	milliseconds	SEISMIC WAVE VELOCITY DYNAMIC MODULUS
	REFLECTION (d)	W			
	THERMOMETRY ● (s)	G W V	TEMPERATURE	°C	GEOTHERMAL GRADIENT OR TEMPERATURE SURFICIAL THERMIC ANOMALY

Source: From Compagnie Généràle de Géophysique/Géodigit.

MAIN CAUSES OF ANOMALIES	APPLICATIONS			
	DIRECT INVESTIGATION		**INDIRECT INVESTIGATION**	
	TARGET	CASE HISTORIES	TARGET	CASE HISTORIES
CONDUCTIVE VEINS, ORE BODIES, SEDIMENTARY LAYERS (CLAY, MARL...) RESISTIVE LAYERS (LIMESTONES, SALT DOMES, VOLCANIC INTRUSIONS...) SHEAR ZONES, FAULTS, WEATHERINGS, HOT WATERS	MASSIVE SULFIDES QUARTZ, CALCITE, OIL SHALES SPECIAL CLAYS ROCK - SALT GEOTHERMAL RESERVOIRS	FRANCE ITALY AFRICA,...	BULK MATERIAL .... BASE METALS PHOSPHATES .... POTASH .... URANIUM .... COAL .... NATURAL STEAM .... DETAILED TECTONICS	MANY COUNTRIES SENEGAL, MOROCCO FRANCE, GABON FRANCE, USA FRANCE, MOROCCO ITALY, JAPAN MANY COUNTRIES
CONDUCTIVE MINERALIZATIONS: DISSEMINATED OR MASSIVE (GRAPHITE, SULFIDES)	CONDUCTIVE { SULFIDES / OXYDES / ARSENIDES / Mn OXYDES	CANADA AUSTRALIA SENEGAL, CONGO, MOROCCO ITALY, GREECE, FRANCE,...	ASSOCIATED MINERALS (ZINC, Mo, GOLD, SILVER)	CANADA IRAN AUSTRALIA
MASSIVE CONDUCTIVE ORES GRAPHITE ELECTRO - FILTRATION	SULFIDES { PYRITE / PYRRHOTITE / Cu / ... Mn ORE	CANADA AFRICA YUGOSLAVIA ITALY FRANCE,...	ASSOCIATED MINERALS (LEAD, GOLD, SILVER, ZINC, Ni...)	CANADA KATANGA
EXTENSION OF PREVIOUSLY LOCATED CONDUCTIVE ORE BODIES	CONDUCTIVE ORES	SPAIN, MOROCCO SAUDI ARABIA FRANCE	IRON	MAURITANIA AUSTRALIA
CONDUCTANCE OF SEDIMENTARY SERIES SALT DOMES	SALT	FRANCE	POTASH REGIONAL EXPLORATION	GABON
CONDUCTIVE VEINS, ORE BODIES, SEDIMENTARY LAYERS (CLAY, MARL...) SHEAR ZONES, FAULTS, WEATHERINGS, RESISTIVE BASEMENTS	MASSIVE SULFIDES CLAYS NATURAL STEAM	FRANCE ITALY	SHEAR ZONES GENERAL TECTONICS	TUNISIA MOROCCO ITALY FRANCE
CONDUCTIVE MINERALIZATIONS SURFICIAL CONDUCTORS SHEAR ZONES	CONDUCTIVE { SULFIDES / OXYDES Mn OXYDES CONDUCTIVE ORE BODIES: SULFIDES, MAGNETITE, GRAPHITE	CANADA CHILE, IRAN SPAIN, SAUDI ARABIA GREECE, IVORY COAST SENEGAL USA CANADA SAUDI ARABIA SOUTH AFRICA BOTSWANA	KIMBERLITES ASSOCIATED MINERALS GROUND FOLLOW-UP (LEAD, Ni...) SHEAR ZONES WEATHERED ZONES (YELLOW AND BLUE GROUND PIPES) CONDUCTIVITY MAPS	IVORY COAST SPAIN CANADA USA NICARAGUA USA CANADA SOUTH AFRICA ZAMBIA AUSTRALIA
CONTRASTS OF MAGNETIZATION MAGNETITE CONTENT OF THE MATERIALS	MAGNETITE PYRRHOTITE TITANO - MAGNETITE	SWEDEN, SPAIN MOROCCO, CHILE FRANCE NORWAY LIBERIA IVORY COAST CANADA MAURITANIA SPAIN SENEGAL FRANCE UPPER VOLTA AUSTRALIA	IRON ORE .... CHROMITE .... COPPER ORE .... KIMBERLITES .... GEOLOGICAL MAPPING IN TERMS OF MAGNETIC CHANGES (BASIC ROCKS...) AND/OR DISCONTINUITIES INVENTORY OF MINERAL RESOURCES	FRANCE TURKEY MAURITANIA MALI AUSTRALIA CANADA AFRICA FRANCE
DEPOSITS OF HEAVY ORES SALT DOMES (LIGHT) BASEMENT ROCKS	CHROMITE PYRITE CHALCOPYRITE	MALAGASY CANADA PORTUGAL	PLACER CONFIGURATION KARSTIC CAVITIES	AFRICA FRANCE, AUSTRALIA
RADIOACTIVE ELEMENTS URANIUM - THORIUM - K_{40}	URANIUM, THORIUM COAL, LIGNITE MONAZITE RADIOACTIVE MINERALS PHOSPHATES	FRANCE AFRICA SOUTH AFRICA AUSTRALIA FRANCE CANADA	GROUND FOLLOW-UP GEOLOGICAL STRUCTURAL MAPPING (DIFFERENCIATION IN GRANITES)	SENEGAL IVORY COAST MAURITANIA IVORY COAST ALGERIA UPPER VOLTA FRANCE AUSTRALIA
CONTRASTS OF VELOCITY: MARKERS AT VARIABLE DEPTH FISSURED ROCKS	BURIED CHANNELS FAULTS MORPHOLOGICAL TRAPS GENERAL TECTONICS SAND, GRAVEL DEPOSITS HEAVY MINERALS	FRANCE, ITALY AFRICA FRANCE FRANCE	TIN, DIAMONDS .... HEAVY MINERALS .... NATURAL STEAM .... URANIUM .... COAL TIN PLACERS	AUSTRALIA MALAYSIA ITALY CANADA, AUSTRALIA FRANCE INDONESIA
ABNORMAL FLUX OF HEAT THERMAL INERTIA OF ROCKS	THERMAL SPRINGS THERMAL SPRINGS	GERMANY	NATURAL STEAM BORON, SULPHUR DRAINAGE ZONES SUBTERRANEAN VOLCANISM	 FRANCE GREAT BRITAIN

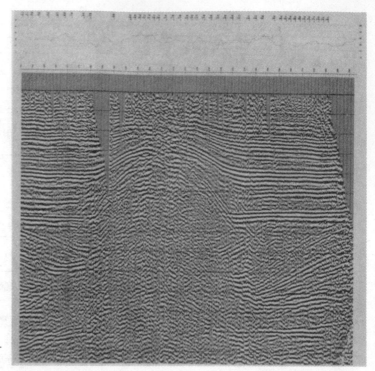

FIGURE 1. A typical computer-generated seismic cross-section.

geophysical datasets are readily adaptable to such statistical treatment on the computer.

Principal Geophysical Methods

Gravity Method The gravity method detects and measures lateral variations in the Earth's gravitational field that are associated with near-surface changes in force density. The objective in exploration work is to associate these changes with variations in rock types. Occasionally the whole field is measured (with a pendulum) or derivatives of the field (with a torsion balance), but usually the difference between the gravity field at two points is measured (with a gravity meter) (Sheriff, 1973).

Geological structures of interest to the explorationist usually give rise to disturbances in the normal density distribution within the Earth, causing anomalies in the field. These anomalies are usually very small in amplitude in contrast to the regional value of the field, and extremely sensitive gravimeters are necessary for adequate resolution. Most contemporary *gravimeters* are of the unstable type where the gravitational force on a mass within the meter is balanced by a spring arrangement, and a third force is provided that acts when the system is not in equilibrium. The Worden meter incorporates temperature-compensating features (Fig. 2) (Sheriff, 1973).

The gravimeter measures the vertical component of attraction, Δg_z, caused by all sources. The basic unit is the *gal* (after Galileo), which is 1 cm/sc^2. Exploration-type anomalies are usually expressed in *milligals* (mgal = 10^{-3} gal).

A typical *gravity survey,* land, marine, or airborne, results in traverse data, which after the ap-

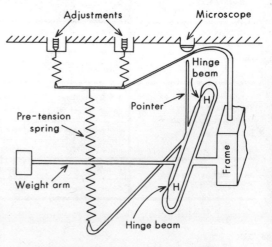

FIGURE 2. Schematic diagram illustrating operation of Worden gravimeter. (After Dobrin, 1960)

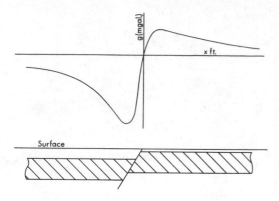

FIGURE 3. Gravity effect of a faulted thick horizontal bed.

plication of various corrections (elevation, terrain, latitude, and Earth tide), is ready for interpretation as profiles or mapped data. Geological structures will produce variations in the gravity values. The example in Fig. 3 shows the effect of a normal fault with a vertical displacement.

Magnetic Method The magnetic method is the oldest form of geophysical prospecting. The real momentum to its current worldwide application came with the development of the airborne magnetometer during World War II, which provided a quick and inexpensive method for surveying inaccessible areas.

Measurements of the magnetic field of the Earth are taken over an area of interest using a *magnetometer,* with the objective of locating concentrations of magnetic materials or of determining the depth to the magnetic (or Precambrian) basement. Differences in the *normal field* are attributed to variations in the distribution of materials having different susceptibility (see Vol. XIII: *Magnetic Susceptibility, Earth Materials*) and perhaps also *remnant magnetization* (Sheriff, 1973). Like the gravity method, the magnetic method is also an application of a *potential field* but has a considerably more complex field mathematically because of variations of the magnetic field with latitude and since different instruments measure different components of the field (see *VLF Electromagnetic Prospecting*). Both methods suffer from the fact that there is no unique solution to the Laplace equation, so it is impossible to separate the mass from the density in the gravity case and the magnetization from the susceptibility in the magnetic case.

Measurements of the magnetic field can be made in the ground, in the air, and at sea using magnetometers. The early Schmidt-type magnetometers could be used to measure either the vertical or horizontal component of the field with an accuracy of one *gamma* (the gamma is the unit of magnetic field intensity and is equal to 10^{-5} *coersted*). They do not measure absolute fields but respond to change in the field component from station to station. These instruments are not readily adaptable for moving ships or aircraft and have largely been replaced by a new generation known as the flux gate, proton, and optically pumped magnetometers, which have an accuracy better than 0.1 gamma and can record digitally at sample increments of 1 s or less. Each of these instruments gives an absolute measurement of the Earth's total magnetic field. In addition, a pair of magnetometers may be operated in tandem to measure differences in the total field—i.e., gradients. The gradient measurement is unaffected by diurnal variations of the field (Reford, 1975).

Like the gravity method, data from a magnetic survey undergo corrections and are displayed as traverse profiles or as datums on a map. The interpretation involves the deducing from the profiles or contoured map the nature of the subsurface geology. Zones of magnetic mineral concentrations produce pronounced anomalies; faults generate linear anomalies or breaks in the magnetic contour patterns. The profile example in Fig. 4 shows the total field response of a common geological situation.

Seismic Method Although the seismic method has some application in mining geophysics, it is applied most universally in petroleum exploration. Also, it gives results that are most readily translated into geological terms. Recent developments in the seismic method have given encouraging results in terms of defining lithologies, porosities, and the presence of hydrocarbons. The complication of the seismic method is that it requires the generation of energy sources in the Earth, whereas with gravity and magnetics, it is necessary to measure only existing fields.

The two individual seismic methods, *reflection* and *refraction,* both yield information on the elastic properties of rocks. The elastic constants of rocks determine their seismic velocity, and these changes in velocity (related to changes in lithology) dictate the behavior of seismic waves in the subsurface. It is these changes of seismic that the method seeks to detect (Dobrin, 1960).

When artificial elastic waves are generated in the near surface, using dynamite or some other acoustic source, the waves pass through the layered rocks and set up oscillatory movement within the medium and at its surface. Seismic waves returning to the surface cause harmonic motion of the ground that can be measured with detector instruments known as *seismometers* or *geophones.*

When a seismic pulse (or energy) propagates through the subsurface, it is controlled by the distribution of rock velocities through which it passes. The geometry of its travel path obeys *Fer-*

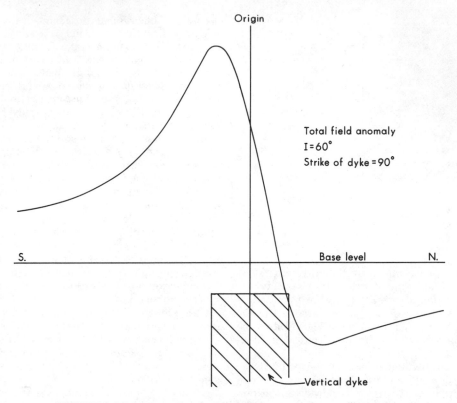

FIGURE 4. Model curve showing total field response from a dike model.

mat's principle, which states that the energy traveling through materials of differing velocities will take the minimum-time travel path. The geometry also obeys *Snell's laws* of optics involving reflection and refraction.

In Fig. 5 we see in the simplest case that a raypath from A to B to C is reflected at B, such that the angle of incidence (i) is equal to the angle of reflection (r), and the travel paths AB and AC are of equal duration. If B lies on the interface be-

tween two layers such that layer 1 has a velocity V_1 and density ρ_1 and layer 2 has a velocity V_2 and density ρ_2, then the amount of energy reflected in this case depends on the contrast in the density-velocity product (acoustic impedance) on the opposite sides of the interface. The reflection coefficient, C_R, is given by the simple relation:

$$C_R = \frac{\rho_2 V_2 - \rho_1 V_1}{\rho_2 V_2 + \rho_1 V_1}.$$

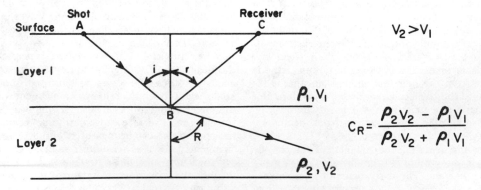

FIGURE 5. Simple two-layer reflection model.

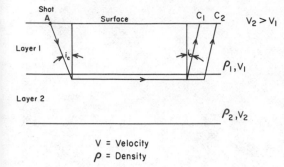

V = Velocity
ρ = Density

FIGURE 6. Simple two-layer refraction model.

Not all the energy is reflected back to the surface. Much of the energy is transmitted across the interface and penetrates deeper layers before it returns to the surface. In Fig. 6 we see the behavior of the transmitted raypath from layer 1 to layer 2 where the lower layer has a higher velocity.

The transmitted ray is bent or refracted away from the vertical such that

$$\frac{\sin i}{\sin R} = \frac{V_1}{V_2}.$$

The maximum possible angle of refraction is $R = 90°$ at which point the refracted ray travels along the interface. The angle of incidence at which this phenomenon occurs is known as the *critical angle, i_c*, which is defined by the relationship,

$$\frac{\sin i_c}{\sin 90°} = \frac{V_1}{V_2}, \quad \text{or } \sin i_c = \frac{V_1}{V_2}.$$

Reflection Method The application of reflection seismology in geophysical exploration represents more than 90% of current geophysical exploration expenditures. The standard method of generating seismic waves is to explode a dynamite charge in a shothole, but numerous other sources exist for both land and marine applications (see *Acoustic Surveys, Marine*). In reflection seismic work, the depth to any interface can be determined by measuring the traveltime of the seismic pulse generated near the surface and reflected back to the surface from the interface. The reflected signals are detected by the seismometers and recorded digitally. In modern seismic recording systems, there are many channels corresponding to the numbers of *geophones* in operation. The response of the geophone is sampled at increments of 1 or 2 ms and the signal may be amplified and filtered prior to being recorded on tape.

The next phase of contemporary reflection seismology consists of data processing and display. Seismic data processing has a twofold ob-

jective—i.e., the organization and application of corrections and data enhancement, which is largely an exercise in filtering. Seismic reflection data are normally presented as a computer-generated section in time (see Fig. 1). The objective for the geophysicist is to interpret the time section in geological-depth-related terms.

In most areas, geological structures can be determined up to depths in excess of 7,000 m. The reflection technique thus provides more reliable structural information than any other method, but it is slower and more expensive. Also, it is somewhat vulnerable to certain areas where reflections are obtained with difficulty.

Refraction Method The main difference between the reflection and refraction techniques lies in the interaction that takes place between the waves and the lithological boundaries they encounter. The geometry of the reflected energy is conceptually simple. The refracted travel path is more complex and, for exploration purposes, utilizes trajectories that lie along the top of high-speed layers. Acquisition techniques vary for the two methods. In structure mapping, the source-to-receiver distance for refraction work is much greater than for reflection, and in general, only the initial arrival of seismic energy is recorded (Dobrin, 1960).

The usual application of refraction is as a reconnaissance tool for regional surveys in relatively unknown areas. It is particularly useful for mapping the Precambrian basement or a continuous geological layer with a velocity higher than the formations that encase it.

The seismic methods have seen rapid technological advancement since the 1940s. Concurrent developments in electronics have simplified some problems peculiar to the seismic technique. The application of filter theory and statistical communication theory to digital data has provided powerful techniques for seismic data processing. *Seismic stratigraphic modeling* originated in the early 1950s with the introduction of the synthetic seismogram. Modeling in this context is essentially a procedure that simulates the seismic response that would be generated from an assumed geological situation. The geological model is usually defined in terms of geometry and velocity/density distribution. The contemporary synthetic seismogram is extremely sophisticated and is an integral part of current technology, correlating observed seismic signal amplitudes with known hydrocarbon reservoirs. This technology is referred to as the *bright-spot* phenomenon, which is now applied to mapping the areal extent and vertical thicknesses of sandstone reservoirs. The important departure from the traditional seismic approach is that in this case there need not be any structural significance. The seismic response is a function of the lithology and the pore-filling me-

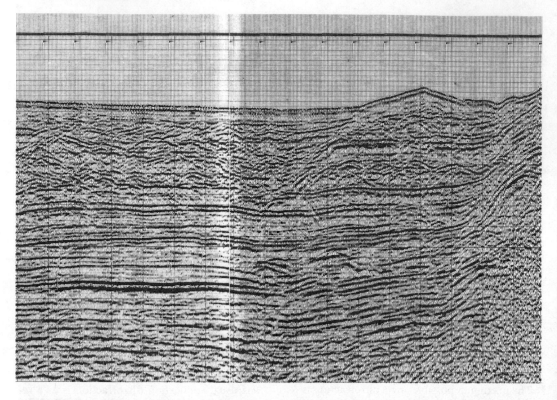

FIGURE 7. A bright spot is a portion of a seismic reflection that appears noticeably stronger than nearby reflections. Note the classic bright spot at the left. At the right is the edge of a piercement–type salt dome. (From Western Geophysical Company)

dium. Figure 7 is an example of a typical bright spot on a seismic section that is diagnostic of a gas-filled sandstone reservoir. The seismic method is therefore capable of predicting the existence of hydrocarbons prior to drilling.

Electrical and Electromagnetic Methods The electrical properties of the subsurface can be explored either electrically or electromagnetically. The three most commonly applied electrical methods are the *self–potential* (SP) (for spontaneous polarization), *resistivity,* and *IP* methods. The SP method depends on a naturally occurring field, whereas the resistivity method requires the introduction of an artificially induced current into the Earth.

Electrical methods are more frequently used in the search for ore bodies than for petroleum. This is largely due to the depth limitation of their applicability, with a maximum depth penetration of approximately 500 m. The SP method has a depth limitation of approximately 30 m. The resistivity method is frequently used to determine depth to bedrock in conjunction with civil engineering projects. The telluric current method uses natural Earth currents as a source and has some application in petroleum exploration.

Electromagnetic methods are often used in mineral exploration when conductive ground connections cannot be made for the common electrical methods. This happens in arid or frozen areas. When an electromagnetic field is produced on the surface of the ground, currents flow in subsurface conductors in accordance with the laws of electromagnetic induction (Parasnis, 1962). These currents give rise to secondary electromagnetic fields that distort the primary field at the surface. The resultant field differs in phase and intensity from the primary field and can reveal the presence of conducting ore bodies. Most electromagnetic prospecting is carried out from aircraft.

Borehole Geophysics (Well Logging)

Well logging is a widely used geophysical technique that has traditionally been the domain of the geologist but has recently been of substantial interest to geophysicists, with the emphasis on bright-spot technology. Well logs (q.v.) have played a significant role in hydrocarbon exploration. The first measurements were made in boreholes in 1927, and by 1933 the Schlumberger brothers had developed a commercial operation for wire line logs. The log is recorded from a sonde that is lowered into the borehole by a cable. Common logs include measurement of resistivity, SP, gamma rays, density, and velocity. Petroleum engineers were quick to recognize the usefulness of these measurements for estimating reservoir parameters such as porosity and hydrocarbon saturation. Geophysicists have implemented log data, particularly sonic and density, in synthetic seismogram modeling. This modeling improves the seismic and log interpretation (Landes, 1968). The recent development of a commercial borehole gravimeter has led to some spectacular hydrocarbon discoveries.

Developments in Exploration Geophysics

Geophysics has gradually changed from an art to a science with modern developments in communication and filter theory. With the seismic method it is now possible to remote sense the subsurface from a stratigraphic and structural standpoint, leading to the generation of pseudosonic logs by inversion of seismic data. The technology of exploration geophysics continues to advance with improvements in marine seismic acquisition and new developments in three-dimensional seismic acquisition and in data processing. Seismic modeling has become very active and is currently being developed from interactive human/computer interfaces. The bright-spot technology has led to a focus on measuring of rock properties such as the effect of fluid saturations on seismic velocity and attenuation.

In mineral exploration, uranium bodies are being located with nuclear logging and IP methods. Superconducting magnetometers have become practical, and computer-enhanced satellite images are being used to locate zones of mineralization (Crook, 1976).

With increasing world demand for mineral, hydrocarbon, and water resources, exploration geophysics continues to develop new technology to define the dwindling occurrences of such resources. Exploration has spread to the frontier areas of the polar regions and continental shelves, and in the face of spiraling economic and often adverse political environments, has persisted and succeeded in locating new deposits of minerals and hydrocarbons and must continue to do so for the foreseeable future.

A. EASTON WREN

References

Crook, T. N., 1976, Exploration geophysics, *Geotimes,* January 1976.
Dix, C. H., 1952, *Seismic Prospecting for Oil.* New York: Harper & Row, 344p.
Dobrin, M., 1960, *Geophysical Prospecting.* New York: McGraw-Hill, 446p.
Grant, F. S. and G. F. West, 1965, *Interpretation Theory in Applied Geophysics.* New York: McGraw-Hill, 584p.
Heiland, C. A., 1940, *Geophysical Exploration.* New York: Prentice-Hall, 344p.
Jakosky, J. J., 1950, *Exploration Geophysics.* Los Angeles: Trija, 346p.
Landes, R. W., 1968, Geosciences in the petroleum industry, in E. R. W. Neale, ed., *The Earth Sciences in Canada.* Ottawa, Ont.: Royal Society of Canada (Spec. Publ. No. 11).
Nettleton, L. L., 1940, *Geophysical Prospecting for Oil.* New York: McGraw-Hill, 286p.
Nettleton, L. L., 1971, *Elementary Gravity and Magnetics for Geologists and Seismologists.* Tulsa, Okla.: Society of Exploration Geophysicists (Mon. Series, No. 1), 121p.
Parasnis, D. S., 1962, *Principles of Applied Geophysics.* London: Methuen, 176p.
Parasnis, D. S., 1966, *Mining Geophysics.* Amsterdam: Elsevier, 356p.
Reford, M. S., 1975, Magnetics, in Petroleum Geophysics: Its Synergistic Elements, symposium sponsored by the Geophysical Society of Houston.
Seigel, H. O., 1968, The changing role of mining geophysics in Canada, in E. R. W. Neale, ed., *The Earth Sciences in Canada.* Royal Society of Canada (Spec. Publ., No. 11).
Sheriff, R. E., 1973, *Encyclopedic Dictionary of Exploration Geophysics.* Tulsa, Okla.: Society of Exploration Geophysicists, 266p.
Telford, W. M., L. P. Geldart, R. E. Sheriff, and D. A. Keys, 1976, *Applied Geophysics.* London: Cambridge University Press, 742p.
Wren, A. E., 1973, Trend surface analysis—A review, *Jour. Canada Soc. Exploration Geophysics* **9,** 39–44.

Cross-references: *Acoustic Surveys, Marine; Marine Magnetic Surveys; Sea Surveys; VLF Electromagnetic Prospecting; Well Logging.* Vol. XIII: *Earthquake Engineering; Electrokinetics, Formation Pressures, Abnormal; Magnetic Susceptibility, Earth Materials; Oceanography, Applied; Seismological Methods.*

EXPLOSIVES, FIELD USE—See BLASTING AND RELATED TECHNOLOGY; CRATERING. Vol. XIII: TUNNELS, TUNNELING.

EXPOSURES, EXAMINATION OF

The *Glossary of Geology* (Gary et al., 1974, p. 245) lists *exposure* [geol] as a "a continuous area in which a rock formation or geologic structure is visible ("hammerable"), either naturally or artificially, and is unobscured by soil, vegetation, water, or the works of man." Natural exposure is most abundant in arid and semiarid climates where predominantly mechanical (rather than chemical) weathering contributes less rapidly to a masking soil profile and where vegetation does not obscure outcrops from view. In cold climates natural exposures are most commonly associated with glaciation. Both Pleistocene and more recent glacial valleys frequently show excellent outcrops (Fig. 1).

In temperate and tropical climates, outcrop, although more limited, again is most widely exposed by active geomorphological process (e.g., streams, waves, and currents). Reconnaissance work in such climates, therefore, is conducted most productively along stream valleys and on coastlines.

The most common artificial exposure is the road cut. Road and railroad cuts often provide the geologist with the best exposures for reconnaissance mapping (see *Field Geology*). Other artificial exposures that might be usefully mapped include unlined tunnels and mine shafts (the geologist should, however, use extreme caution in entering old mine workings).

In mineral exploration and engineering geological site investigation, extensive use is made of pits, trenches, and adits to gain access to artificial exposure in critical areas. The value of these methods over boreholes relates to the greatly superior information that may be gained from in situ measurements, as opposed to cuttings, samples, and core-logging observations.

In mining, all artificial exposure revealed by underground excavation or opencast operations should be mapped as a matter of standard practice. Similarly, the natural materials exposed in the foundation excavations of all critical civil engineering structures such as nuclear power plants (Fig. 2), large bridges, and major dams should be carefully examined and mapped to (1) confirm preconstruction geological observations that set the basis for the foundation (and perhaps even structural) design, (2) provide a basis for design-as-you-go recommendations (e.g., rock bolting), and (3) establish a permanent record of foundation support conditions. The latter becomes an all-important legal document in the case of structural malfunction (e.g., excess differential settlement) or failure.

In major tunnel excavations, modern practice makes use of an exploratory pilot tunnel, the examination and mapping of which sets the basis for a main tunnel design against natural hazards (e.g., advance grouting of faults to prevent excess overbreak and/or groundwater inflow) (see Vol. XIII: *Tunnels, Tunneling*). Furthermore, geological observations in the small-diameter pilot tun-

FIGURE 1. Large exposure in the Animas Canyon, Colorado, is a product of both natural glacial erosion and an artificial railroad cut opened a century ago.

FIGURE 2. The artificial exposures revealed by the foundation excavations for critical civil engineering structures suc as nuclear reactor containment buildings are mapped as a matter of standard practice.

nel are used to select the method of main tunnel excavation (shoot-and-muck versus mechanical mole; lined versus unlined).

Exposures and Mapping

The examination of exposures does not, in most instances, represent an end in itself but one of the early steps in preparing a geological map. Because of the considerable overlap between this and other encyclopedia entries, the reader who wishes to obtain a broad overview on mapping must refer also to the entries on *Geological Survey and Mapping; Field Notes, Notebooks;* and *Samples, Sampling.* In view of this overlap and the availability of some

excellent references in the general area of field mapping (e.g., Barnes, 1981; Lahee, 1961; Moseley, 1981), emphasis in this entry is largely restricted to the special-purpose examination of exposures, an area considered to represent a gap in the literature. However, the following four comments on the procedures relate to both general field mapping and the examination of exposures.

First, before entering an area to commence field (exposure) mapping, the geologist should perform a detailed interpretation of available satellite images and aerial photographs of the project area aimed at identifying the area of exposure, preparing a preliminary (photogeological) map of the area, and planning the most appropriate procedure for the reconnaissance phase of fieldwork. In this regard the photogeological interpretation provides a cost-effective framework of reference into which subsequent, more detailed observations on exposures fit with greater clarity.

Second, before entering the field to commence exposure mapping, the geologist should attempt to obtain base maps and/or stereoscopic coverage of aerial photographs of a scale suited to the level of geological detail required of the project (see Vol. XIII: *Information Centers*). This may range from 1:250,000 for reconnaissance geological maps to 1:25,000 for regional geological maps, to 1:2,500, 1:500, and larger for special-purpose maps (see following section) (see also *Map Series and Scales*).

Third, in examining exposures, the geologist should never lose sight of the big picture. For example, in mapping the geological detail exposed in an adit (see *Field Notes, Notebooks;* Fig. 2), the geologist must remember that the dipping beds he or she observes may represent the limb of a single fold. This fold in turn, in the case history area used, exists in a complexly folded outcrop (Fig. 3) in a mountain fold belt representing part of the tectonically convergent Alpine–Himalayan mountain trend.

Fourth, all exposure observations must be accurately recorded because they are represented as statements of fact on the geological map, and thus, if the map is incorrect, any subsequent interpretation of the map, for whatever purpose, will be incorrect.

Examination of Exposures

The emphasis addressed, nature of the observations, and procedures used for recording exposure examinations vary in response to the purpose of the project under investigation. Because the topic is so broad, the remaining discussion is restricted to exposure examination for civil engineering projects.

Major engineering geological investigations represent some of the most demanding in terms of recording and interpreting what is observed. The importance of the interpretation aspect of exposure examination is emphasized by the fact that the infamous Vaiont Dam disaster (Bolt et al., 1977), where a landslide–generated wave overtopped the dam, killing more than 3,000 people in the town of Longarone in Italy, may be attributed not so much to a lack of geological observation but to a lack of appreciation of the significance of what was observed. Existing landslides (a key indicator of potential future slope failures) were observed in the reservoir area. Mapping showed the river flowing parallel to the axis of a syncline, with the dip of materials into the valley offering a relative configuration of least friction resistance to failure. Observed valley slopes consisting of competent rocks over incompetent rocks represent a geological condition that is highly susceptible to active mass movement (Watson, 1984). Incipient failure in the form of tension cracks and small slides was noted in the area of the great landslide prior to its occurrence. And finally, accelerated slope creep due to heavy rains and associated excess pore water pressures preceded the catastrophic failure.

Examples

Examples of exposure examination are given for two projects, the preconstruction investigation phase of a large arch dam and the foundation construction phase of a nuclear power plant. In considering the case history of the dam, remember that a reconnaissance site feasibility and a regional site selection phase of mapping would already have preceded the preconstruction investigation.

In the preconstruction phase, the main emphasis lies in assessing the engineering significance of geological features associated with the foundations of the dam bypass tunnels, power-plant scheme, and construction material borrow areas. In this connection all natural exposure must be examined and mapped before more costly subsurface investigation techniques are adopted. Although this seems like a statement of the obvious, it is all too often given mere lip service in the rush to mobilize drilling, trenching, and adit subcontractors.

A further point worth emphasizing is that in the preconstruction phase, the geologist is involved in intense communication with the civil engineer regarding the foundation structural interaction of the dam, and that simple geological sketches of exposure are often more meaningful to the engineer than detailed maps. This statement does not infer that mapping should be neglected but that it should be supplemented by other means of data presentation including, if appropriate, physical models.

A good way of recording exposure on a dam abutment is through use of the diagrammatic elevation (see Fig. 3). This is a controlled sketch drawn in the field, using binoculars to assist the eye, and with a cord grid laid over the exposure by tape and compass to impose survey control on the sketch map. The exposure shown in Fig. 3 represents an area of about 20,000 m². Figure 4 shows a photomozaic of the same exposure.

Figure 5 represents a geological map of the dam site (note that the map cross references fold axes shown in the abutment elevation in Fig. 3). Figure 6 shows a developed section along the dam axis. Since, as inferred in the entry *Field Notes, Notebooks,* the performance of abutment rocks under

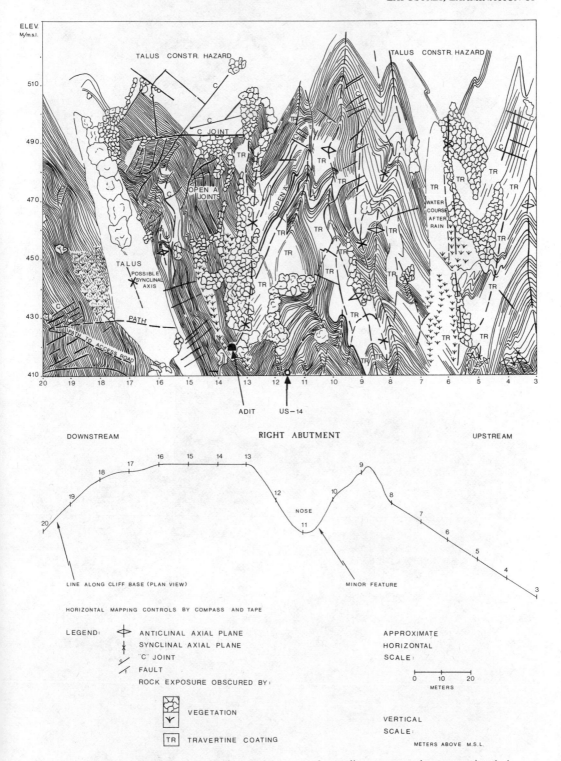

FIGURE 3. The diagrammatic elevation is a useful means of recording exposures in cross-sectional view.

FIGURE 4. A photograph does not highlight structural geologic trends as clearly as a sketch. (Compare photomozaic with the sketch shown in Fig. 3).

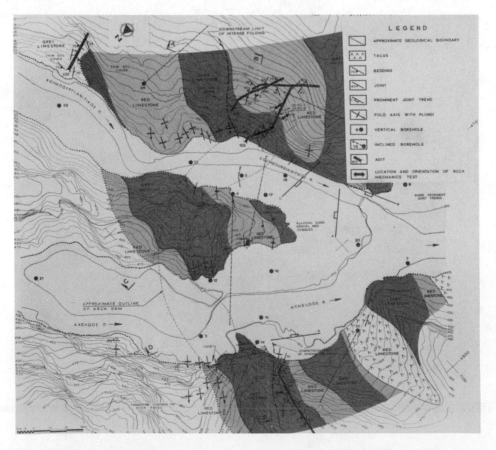

FIGURE 5. Exposure observations are combined with subsurface information to develop a detailed geologic map and cross-section (see Fig. 6) of the project area.

the arch thrust is controlled not so much by the type and strength of the intact rock element as by the flaws (e.g., bedding planes and joints) in the overall rock mass, the recording of structural features is more important than lithology per se, and all records of exposure should reflect this fact. Figure 6, e.g., includes an interpretive step illustrated by stereographic projections. These show discontinuity wedges that need to be further analyzed in terms of shearing resistance to foundation failure. Once engineering geological observations have determined the most favorable location for and type and size of the dam, the civil engineer must finalize its layout in relation to the detailed topography of the site (Fig. 7) and complete its structural (concrete) design.

The case history of the nuclear power plant was selected to demonstrate the fact that exposure examination during construction is essential in that it may reveal new problems (e.g., the fault shown in Figs. 2, 8, and 9) that need to be resolved/designed against before construction can continue. In the case of the selected example, the fault had to be shown to be noncapable (inactive) in terms of Nuclear Regulatory Commission criteria.

Figure 8 shows a relatively large-scale diagrammatic elevation of part of the foundation excavation wall of the nuclear containment building (note the plan view rendering of important features). In spite of the relatively large scale of this exposure sketch, however, it is expanded still further (Fig. 8, Detail A) to show necessary detail in

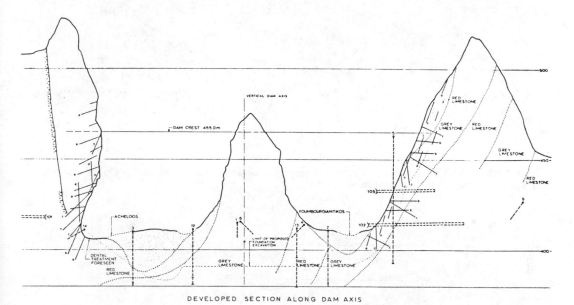

DEVELOPED SECTION ALONG DAM AXIS

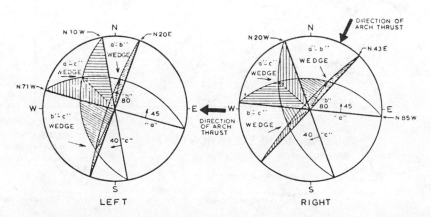

FIGURE 6. Engineering geological observations must be project specific. The cross–section emphasizes the importance of structural features on the design of an arch dam.

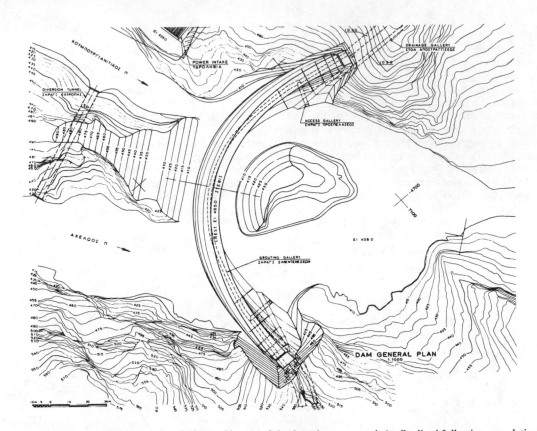

FIGURE 7. The civil engineering design and layout of the dam shown can only be finalized following completion of an engineering geological assessment of the site.

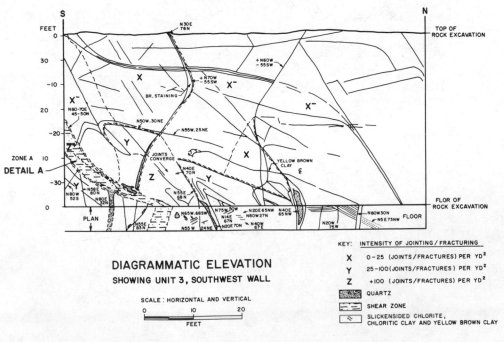

DIAGRAMMATIC ELEVATION

SHOWING UNIT 3, SOUTHWEST WALL

SCALE : HORIZONTAL AND VERTICAL

FIGURE 8. In spite of the large scale of this foundation exposure record, further detail is required in the fault area (Detail A).

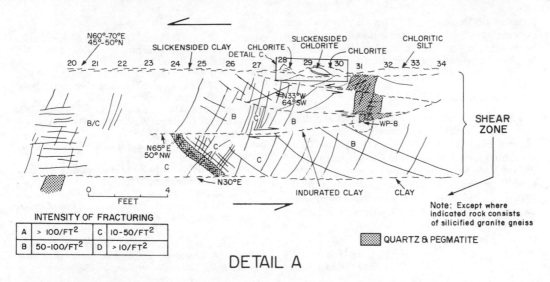

FIGURE 9. Increased detail (compare with Fig. 8) reveals important features of a portion of the fault exposed in the foundation excavation.

the critical fault area (Fig. 9) and yet once again (Fig. 10) to show still further detail. Oriented sketches of petrofabric (microscopic) observations may be used to complete this progression.

All sections of exposed rock on the excavation floor are mapped as a matter of routine procedure. This is not necessarily a time-consuming (costly) operation. For example, the map shown in Fig. 11 was prepared in 3 hr, by two geologists using a tape and (Brunton) compass.

Conclusion

Although the examples cited pertain to engineering geological projects, the overriding philosophy related to the importance of exposure mapping and the techniques outlined have application to the general practice of exposure mapping.

The systematic examination of exposure represents, in most cases, the most important step in the preparation of geological maps, cross–sections, and reports, which, after all, is the main

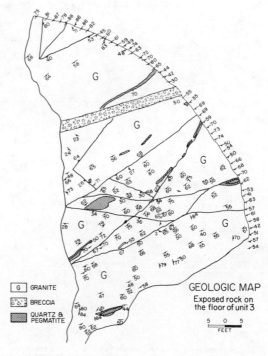

FIGURE 11. The large-scale exposure map is a most economical record in relation to its potential value as a design tool and/or legal document.

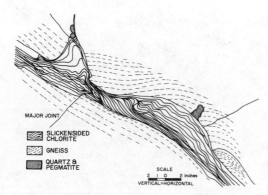

FIGURE 10. Exposure records at the scale shown grade into oriented sketches of petrofabric observations.

141

business of the geologist. The examination and documentation of exposures is a skill that cannot be taught in the classroom but that must be learned, largely through practice, in the field.

Acknowledgments

The authors acknowledge the Coastal Education and Research Foundation; Dames and Moore; Florida Atlantic University; Rio Tinto-Zinc; Surveyer, Nenniger and Chenevert Inc.; The Public Power Corporation of Greece; and Virginia Electric and Power Company. Much of the data published were gathered in employment with, on projects for, or as a result of research support by these institutions.

IAN WATSON
ROY R. LEMON

References

Barnes, J. W., 1981, *Basic Geological Mapping* (Geological Society of London Handbook Series). New York: Halsted Press, 112p.

Bolt, B. A., W. L. Horn, G. A. Macdonald, and R. F. Scott, 1977, *Geological Hazards*. New York: Springer-Verlag, 330p.

Gary, M., R. McAfee Jr., and C. L. Wolf, eds., 1974, *Glossary of Geology*. Washington, D.C.: American Geological Institute, 805p.

Lahee, F. H., 1961, *Field Geology*. New York: McGraw-Hill, 789p.

Moseley, F., 1981, *Methods in Field Geology*. San Francisco: W. H. Freeman, 211p.

Watson, I., 1984, Hydrogeologic control and statistical prediction of active mass movement, *Assoc. Eng. Geologists Bull.* **21**(4), 479–494.

Cross-references: *Field Notes, Notebooks; Fractures, Fracture Structures; Geological Survey and Mapping; Rock Color Chart; Samples, Sampling.*

F

FIELD GEOLOGY

The term *field geology* implies fieldwork, geology as practiced by direct observation of outcrops, exposures, landscapes, and drill cores (cf. Vol. XIII: *Applied Geology*). Those engaged in field geology investigate rocks and rock materials in their natural environment. Field geologists thus try to describe and explain surface features, underground structures, and their interrelationships.

The description of materials in the field may take several forms, depending on the nature of the materials at hand. Soils, e.g., require different considerations than do unconsolidated sediments and involve the use of specialized systems for describing profile morphologies and materials (see *Soil Fabric*), the properties of soil particles (see Vol. XI, Pt. 1: *Particle-Size Distribution*), and sampling methods (see *Soil Sampling*). Descriptions of rock, in constrast, are more familiar to most geologists but may involve rather specific methods for determining, by way of one example, the strength of rock materials (see Vol. XIII: *Rock Mechanics*).

Both the N and L types of hammers have been used by geologists and geomorphologists in assessments of rock strength (Deere and Miller, 1966; Day and Goudie, 1977). As is usual with field practice, numerous precautions must be exercised when using these instruments. Examples of subjective terms and approximate values for rock mass strengths, as applied to geomorphological studies, are given in Table 1. (All tables appear at the end of the entry.) Relationships among strength, hammer orientation, and rebound number are shown in Fig. 1. In addition to field determination of overall rock properties, field geologists are often required to apply knowledge of specific mineral properties such as composition, crystal structure, density, and hardness, among other features (Table 2).

Whatever the nature of the materials being investigated—i.e., whether soils, unconsolidated or loose sediments, or bedrock—it is usually important for most applications to specify the age of the material. Even though age determinations depend on a wide variety of factors that lie beyond the ordinary bounds of field geology per se, the field geologist must have at least rudimentary knowledge of the basic geologic time scales in use. Various schemes applied in Europe and North

America are laid out in Table 3 for the main systems, series, and stages. Because names for chronological units differ from one region to another, it is often important for field geologists to report series/stage names in more than one time scale, especially when preparing summaries of fieldwork for international publications. Still more general knowledge of basic units is often required of field geologists who have occasion to work in different countries that variously employ SI (Système International), c.g.s, and Imperial units. The basic and derived units in the SI and c.g.s systems are given in Table 4 (see also Vol. II: *Units, Numbers, Constants, and Symbols*). Conversion factors for those still working in customary U.S. units are provided in Table 5.

Whatever field methods are used to describe geological features and materials, however sophisticated they may be, Lahee (1961) emphasizes that although field geology is based on observation, many conclusions are predicated on inferences. He states that the "ability to infer and infer correctly is the goal of training in field geology" (p. 4). Proficiency as a geologist is thus largely measured by one's ability to draw reasonable conclusions from observed phenomena and to predict the occurrence of features, conditions, or processes using field experience (see *Geological Methodology*).

Fieldwork, which is critical to advances in knowledge of the geology of the Earth, must be supplemented by laboratory studies. The range of laboratory studies that supports fieldwork is too great to specify because the numbers of methods and techniques are legion in the general endeavors of geochemistry, geophysics, petrology, mineralogy, hydrology, and exploration geology, among others. The determination of the size of sedimentary particles is an example of a common procedure that engages many field geologists at one time or another. Graphs for determining particle sizes in the field, for light- and dark-colored particles, are illustrated in Fig. 2. Place sand grains or rock particles in the innter circle. Compare the size of the particles with those on the graph, using a magnifying glass if necessary. Record the corresponding number (1, 2, 3, 4, 5, 6, 7, 8) in your notebook. For samples with varying sizes, record the most common size first. Comparisons for particle-size groupings obtained in the laboratory using U.S., Tyler, Canadian, British, French, and

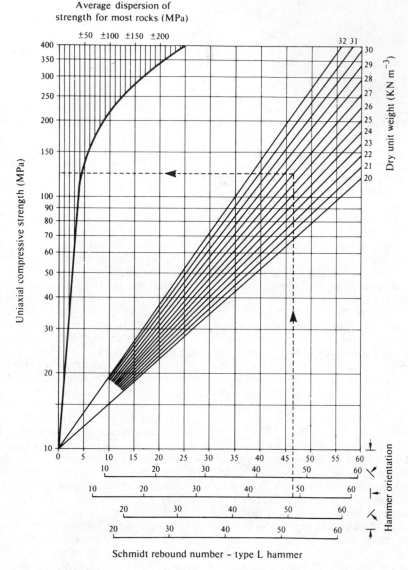

Average dispersion of
strength for most rocks (MPa)

FIGURE 1. Relationships among hammer orientation, Schmidt rebound numbers, and uniaxial compressive strength. (From Gardiner and Dackbombe, 1983; modified after Deere and Miller, 1966)

German standard sieve series are given in Table 6. Grain size classes obtained by field and laboratory methods may be compared to the somewhat different scales used by American geologists, engineers, and soil scientists in Table 7. In this example of grain-size determination, which is more complex than illustrated here (q.v. in Vol. XII, Pt. 1), emphasis is placed on the breadth of knowledge required from what at first are seemingly simple field tasks.

Whether in the acquisition of original data, primary reconnaissance, detailed surveys, geoexploration, or academic pursuits, field geologists are hired by private consulting firms, colleges and universities, mining companies, or national geological surveys. Hypotheses developed in the workroom (see Vol. XIV: *Geology, Philosophy*) also must eventually be tested in the field, as geology is essentially an outdoor science that demands field training.

Much of the fieldwork conducted in the public sector is formalized by geological surveys. As shown in Table 8, almost every country in the world has an official geological survey (see Vol.

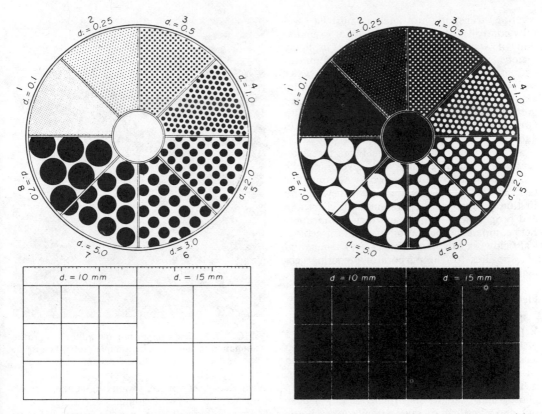

FIGURE 2. Graphs for determining the size of sedimentary particles. Left figure is for dark-colored particles, and right is for light-colored particles. (From Dietrich et al., 1982)

XIII: *Information Centers*), the earliest of which was instituted in the United Kingdom in 1835. Many of these geological surveys are centers of excellence that emphasize aspects of applied geology (q.v. in Vol. XIII) and use a multidisciplinary approach to solve problems. Typical activities of many surveys focus on geological mapping (see *Geological Maps, Mapping*). In North America, state and provincial geological maps provide useful overviews of the general geology and terrain. A list of the principal maps for this part of the world is supplied in Table 9. Although not meant to provide a detailed picture of extant geological conditions at any particular point, the maps provide convenient summaries of salient features. More detailed analyses often require the extraction of data from one map scale to another. To facilitate these kinds of maneuvers or extrapolations, tables have been prepared that give linear (Table 10) and areal (Table 11) equivalents for large range of fractional scales (see also *Map Series and Scales*).

With increasing industrial demands for raw materials and safe sites for urban development, many surveys now concentrate on the assessment of energy resources (q.v. in Vol. XXIV), ground-water supply, and the lessening of risk from earthquakes, slope failures (rockfall, landslip, and landslide), volcanic activity, mining operations, and the disposal of toxic waste.

Disposal of toxic waste is an increasing area of concern for the well-being of human populations, wildlife, and farm animals. Improper disposal of many toxic wastes poses a significant threat to water supplies, both surface and groundwater. The composition of water is determined by physical and chemical means and, usually, by collection of field samples followed by analysis in the laboratory. With increasing availability of portable automatic recording devices and quality analytical instruments, more and more measurements are now performed in the field. The suitability of water for particular uses is dependent on the total concentration of dissolved ions or salts. High levels of lead, arsenic, and fluoride, e.g., render water unsuitable for drinking where calcium, magnesium, iron, and manganese affect the use of water for washing purposes. Sodium and boron in large quantities are toxic to plants. Because many components of water are often present in minute concentrations and are subject to change after sampling, sample collection, preservation, transport, and analysis must be carefully controlled. Standard methods for examination of waters have been prepared in many regions. Some of the more common recommendations for sampling and preservation of water samples are indicated in Table 12. The maximum amounts of permissible impurities in water for human consumption, based on the World Health Organization International Standard, are listed in Table 13.

All these activities involve fieldwork that is essential to understanding and developing the ground on which we live and work. Although the tasks of field geologists are varied, they depend first on background preparation in the classroom and proper field training and then on adequate field equipment, application of codes for geological fieldwork (q.v.), and finally, the examination of exposures for the purpose of preparing geological maps (see *Exposures, Examination of*).

Field Equipment

The field geologist's basic equipment (see Vol. XIII: *Field Geology*) consists of a hammer with a pick or chisel at one end, a pocketknife, a hand lens, a compass and a clinometer, a notebook, and writing utensils. Additional tools include a ruler, meter stick, or tape measure; a scale for close-up photographs, sample bags (paper, cloth, or plastic), and identification tags; a small squeeze bottle of dilute HCl; binoculars; insect repellant, small first aid kit (including an emergency reflective thermal blanket); and a walkie-talkie-type radio (as applicable) (Fig. 3). Miscellaneous items are dictated by terrain and local climatic conditions. A machete or small hand ax, e.g., is often useful in savanna, bush, or forests. In forest country a lightweight string hammock should be considered an acceptable and compact alternative to more conventional camping gear. For many fieldworkers, hammocks are probably easier to handle than a sleeping bag. Fluorescent signal tape may be useful for marking trails; signal flares may be required in emergency or for location by aircraft. On snow or ice fields, specialized equipment such as snow goggles, snowshoes, crampons, etc. are often necessary. It should be remembered that alpine camping equipment, although remarkably compact and lightweight, is designed for short (a day or two), intense operations. Such equipment thus might not be entirely suitable for the usually much more protracted geological field operations of even the fly-camp type. Because there are obvious limitations to what can be carried, traverses should be planned in advance in an effort to limit pack loads as much as possible. Although weight is a primary consideration, bulky clutter can be annoying as well.

The use of firearms should be discouraged.

FIGURE 3. Field equipment required for soil profile examination including spade, sample probe, knapsack, ruler, field guide or manual, and sample bags.

With the exception of venomous snakes, problems caused by wild animals are exaggerated. Most animals will keep clear of humans if they know they are about. If venomous snakes are a nuisance, e.g., around base camp—an effective defense is a .45 caliber handgun with special shot-filled cartridges. In most areas with local guerilla or outlaw activity, firearms might be considered a provocation. These materials will fit into a modest-sized knapsack, but additional packs should be carried if numerous rock samples are to be collected and carried.

A photographic record of sample sites or general shots of exposures are often useful in later reports or slide presentations. Many geologists bring a small camera for black-and-white photographs, which are more suitable than color for reproduction in published reports. Conversion of color slides to black-and-white prints often results in less than satisfactory pictures, so an additional small lightweight camera for taking color photographs is recommended as part of the standard equipment. Some geologists use, in addition, an instant camera so that sample sites can be precisely annotated directly on the picture.

Topographic maps and appropriate aerial photographs are useful for locating sample sites and for preparing field maps. If a pocket stereoscope is available, stereopaired aerial photos will make a convenient base for mapping since mapping units can be delineated directly on the photos or on clear plastic film overlays, which is preferable.

For a more comprehensive discussion of instruments and other equipment used in field mapping, see sections in Lahee (1961), King (1966), and Gardiner and Dackombe (1983) that describe the compass-and-clinometer method, the hand-level method, the barometer method, and the plane table method; techniques of topographic survey; geomorphological mapping; and geophysical methods for subsurface investigation. See also the following entries: *Compass Traverse; Plane Table Mapping; Surveying, General.*

Examination of Exposures

Exposures (q.v.) provide clues to many significant sedimentary, igneous, and metamorphic structural features. Unconformities, joints, veins, faults, folds, bedding planes, graded bedding, chilled margins, and the like seem obvious enough when described in the classroom or illustrated by clear photos or diagrams in textbooks. Such features, whether large or small, are often not obvious in the field, however, especially when quarry walls, road and railway cuts, or natural surfaces are partly masked by vegetative overgrowth, dust or slump material, or the effects of weathering. With experience, the observant field geologist learns to recognize *critical exposures* that provide unmistakable evidence of geological processes, relationships between rock units, or other properties that assist in the recognition and classification of rocks. The ability to read exposures, a skill that is critical to field observations, provides an opportunity for developing working hypotheses to be tested by closer field examination and enables geologists to plan subsequent field and laboratory work effectively. Those hypotheses may then be developed into computer-based theoretical models, but to carry conviction they must be testable and compatible with field evidence, or ground truth, as the remote-sensing specialists call it.

The *scale of observation* is an important consideration. First impressions are afforded by standing back and reading the main features indicated by inclined strata, unconformities, faulting, folding, intercalation of paleosols, color, and other distinctive characteristics. Closer inspection should focus on rock colors, relationships of colors to surfaces of weakness, mineralogical composition, and meso- and microscale structures. Many surface features other than color are of importance for their interpretive values. Lahee (1961), e.g., carefully describes interpretations of smoothed and polished rock surfaces and provides keys for the identification of scratches, grooves, pits, and hollows on rock surfaces. These and other smaller structures are fully described and illustrated by Shrock (1948) and Coneybeare and Crook (1968). The effects of fretwork, or

honeycomb weathering; subsurface corrosion; and phases of soil development should also be meticulously recorded in the field notebook because they may be important considerations in subsequent interpretations.

The field notebook should contain full details of each day's work (see *Field Notes, Notebooks*). Although the types of data recorded in notebooks vary with the project, the following kinds of information should be recorded: date, names of other members in the survey party, location, general characteristics of the area (topography, soil, vegetation, climate-microclimate, nature of outcrops), name of unit or brief rock name, thickness and overall structure of unit, fossils, description of rocks with most abundant type described first (color, induration, grain sizes, grain shapes, fabric, cement, porosity, mineralogy), and nature of contacts (Compton, 1962).

Rock samples are preferably broken directly from the outcrop (talus or scree samples should be avoided if possible), and the exact location should be noted and numbered (see *Samples, Sampling*). Specifications for sample sizes depend on grain size and homogeneity of the rock (size of homogeneous structural units), but samples 8 cm $\times$ 10 cm $\times$ 3 cm are normally adequate for most purposes. When oriented specimens are important, samples should be marked before removal to indicate the top and structural attitude. Fossils require special care in trimming and should be well packed for transport in the knapsack. Each rock or fossil specimen must be marked with a number matching that used in the notes; the number should also correspond to a location on a map or aerial photograph. The number can be inscribed on a piece of adhesive tape and squeezed onto the sample. Ballpoint pen ink is indelible, but felt pen ink is easily smeared or washed away by rain. When several geologists are sampling outcrops in the same area, it is often useful to record the initials of the collector with other numbers. Such efforts often assist in the retrieval of supplementary information.

Geological Mapping

Most primary geological maps (Thomas, 1979) are initially prepared in the field using a topographic map as a base. Some maps are prepared from remotely sensed data using radar or from color-enhanced satellite imagery, but these maps are most useful when checked in the field by a process referred to as *ground truthing* (see *Photogeology; Photo Interpretation;* Vol. XIII: *Photogrammetry; Remote Sensing, Engineering Geology*). Still other kinds of geological maps may contain map units initially interpreted from remotely sensed images that are later correlated with actual conditions after field inspection. Field geologists often combine satellite imagery or aerial photography with ground operations in reconnaissance traverses to produce geological maps of both a general and a specific nature. Peters (1978) discusses techniques of geological mapping as they apply to surface mineral exploration as well as to mapping in underground mines.

Geological maps show the distribution of surficial deposits and solid rocks, sometimes together, depending on the depth and extent of overburden, and sometimes separately. The Institute of Geological Sciences in the United Kingdom, e.g., sometimes publishes two editions for the same area: a solid and drift edition, delineating both surficial materials and exposed bedrock using color, and a solid edition, showing all bedrock by means of color but identifying overlying surficial deposits only by means of symbols. The Geological Quadrangle Maps of the U.S. Geological Survey correspond mostly to the solid and drift edition of British maps (Roberts, 1982). Maps on scales of 1:250,000, 1:1 million, and smaller are almost always solid. The solid versions are very useful for solving structural problems, for mineral exploration, and for mining operations. The drift map is essential for surface engineering work and helpful for Quaternary stratigraphers and geomorphologists (see *Saprolite, Regolith, and Soil*).

Although field maps show the nature of materials as they occur at the ground surface, geologists often distinguish between two slightly different approaches by using the colloquial terms *soft-rock* and *hard-rock* geology (see *Hard versus Soft Rock Geology*). Soft-rock geologists, on the one hand, concentrate on sedimentary materials that either are loose and unconsolidated or may have undergone some lithification and diagenesis but are mostly unmetamorphosed. Hard-rock geologists, on the other hand, deal with igneous rocks as well as beds that have been folded, contorted, and metamorphosed, forming resistant hard-rock units (Bates and Kirkaldy, 1977; Tucker, 1982).

Mapping Sedimentary Rocks The rocks most commonly shown on geological maps are sedimentary in origin. Fieldwork with sedimentary rocks implies a knowledge of stratigraphic-lithological units that are defined by physical characteristics and time-stratigraphic units according to the fossils they contain (see *Sedimentary Rocks, Field Relations*). A range of rock properties important to mapping is normally recorded in relation to textural and compositional characteristics such as grain size, shape and sorting, fabric, porosity, mineralogical composition, and cements (see Vol. VI: *Clastic Sediments and Rocks*). Rock units may be classified into general categories in the field, but if the field geologist is unsure of the category or needs closer refinement, he or she

should clearly number the specimen for later identification in the lab (see *Samples, Sampling*). Maps may be prepared in this manner, provided the various units are consistently identified and their relationships to one another noted. Many field maps may, e.g., simply show rock units A, B, C, D, etc. in the first attempt. Legends are later compiled and classification units correlated with map units. Then come structural interpretations. Inclined and folded strata, faults, and unconformities often pose interesting challenges for the field geologist who must employ a range of techniques to determine sedimentary structure accurately (Freeman, 1971). Primary structures are the details of bedding, such as illustrated by Shrock (1948) and Coneybeare and Crook (1968).

Descriptive terms for basic even, curved, and wavy sedimentary layering accompany diagrams in Fig. 4. Examples of classifications for cross-lamination and cross stratification are shown in Figs. 5 and 6 respectively. The classification proposed by Allen (1963) (Fig. 6) is rather more comprehensive than Jopling and Walker's (1967) system (Fig. 5) but more difficult to use in the field.

Secondary, or tectonic, structures are those caused by diastrophic processes. Platt and Challinor's (1980) booklet dealing with simple geological structures adequately summarizes field methods applicable to completing outcrop maps, determining a bedding plane, finding the vertical thickness and dip of a bed or the displacement of a section due to faulting, and relating topography to a geological structure, among other important field techniques.

Mapping Igneous Rocks Rocks produced by the solidification of magma may take the form of intrusive bodies (Boyer, 1971) that solidify within the Earth's crust or extrusive rocks that are erupted from a volcano or fissure before solidifying at the Earth's surface. Both intrusive and extrusive rocks tend to occur in characteristic suites, as conditioned by their mode of emplacement or extrusion and chemical composition as acid, basic, or intermediate magmas (see *Igneous Rocks, Field Relations*). Igneous rocks are usually depicted on geological maps using special colors and ornamentation. Maps of the U.S. Geological Survey, e.g., use a capital letter to give the stratigraphic age of the intrusion followed by an abbreviated form of the rock name (see *Map Symbols*). Volcanic rocks are mapped in the same way as sedimentary formations, but there are several important distinctions. These rocks are often so heterogeneous that it is not possible to differentiate sequential formations, lava flows and ash falls tend to form lenticular bodies that are erupted from more than one vent, and lava flows and pyroclastic deposits are quickly produced in an instant of geological time (Roberts, 1982). Structures produced by volcanic intrusions such

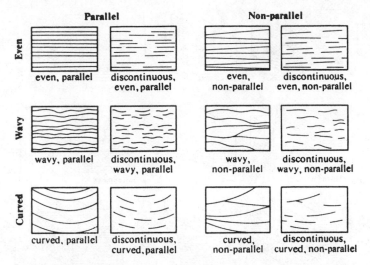

FIGURE 4. Examples of some descriptive terms for sedimentary layering. (From Campbell, 1967, as reproduced in Gardiner and Dackombe, 1983)

as sills and dikes, laccoliths and domes, lopoliths, and necks and plugs often have to be interpreted indirectly. Fieldwork with intrusive igneous rocks can be especially difficult, as Compton (1962) points out, because their interpretation often requires working back in time from the youngest or most obvious features at widely distributed outcrops to relics of former events. Intrusive events and sequences may be understood only after mapping is well advanced. Forms of erupted magma— e.g., ropy (pahoehoe) lava and blocky (aa) flows, ignimbrites, ash-fall deposits, fissure lavas, and pillow lavas—are more easily interpreted as genetic map units because of their continuous exposure in outcrop. Erupted rocks tend to follow

the laws of succession crosscutting relationships (the last crosscut is the youngest).

Mapping Metamorphic Rocks Metamorphic rocks have been modified in a solid state by recrystallization of their minerals by heat (thermal or contact metamorphism), pressure (dynamic metamorphism), or percolation of hot fluids or gases through fractured rocks (hydrothermal metamorphism). Because a given metamorphism is typically the composite of several processes, Compton (1962) advises field geologists to map and classify the premetamorphic lithology, stratigraphic sequence, and structure; determine the amounts and kinds of deformations; and map zones of metamorphic minerals and textures.

Feature	Type A	Type B	Type C	Type S
laminae	preserved on lee side only	continuous from stoss side to lee side	continuous from stoss side to lee side	continuous from stoss side to lee side
sediment	sand and silt, with some fines near base of slip face	sand and silt, fines concentrated near base of slip face and in trough	sand and silt restricted to stoss side, grading laterally into fines in the troughs	no selective concentration of sand, silt, or fines on particular parts of the ripples
ripple morphology	asymmetrical, amplitude variable	asymmetrical, amplitude variable	asymmetrical, amplitude decreases upward through coset	symmetrical, sinusoidal profile
grading of coset	ungraded	ungraded	graded, from fine sand and silt at the base to silt and fines at the top	ungraded
ratio of suspended to traction load	low and steady	intermediate and steady	intermediate and steady	high and steady
current strength stability	steady	steady	decreasing	decreasing

FIGURE 5. Classification of cross-lamination. Note: Types A, B, and S are a transitional series. (From Jopling and Walker, 1967, as reproduced in Gardiner and Dackombe, 1983)

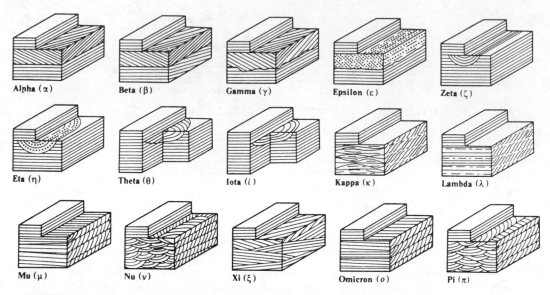

FIGURE 6. Classification of cross-stratification. (From Allen, 1963, as reproduced in Gardiner and Dackombe, 1983)

These procedures are applicable to the two basic kinds of metamorphic units, (1) metamorphosed sedimentary or igneous rocks and (2) metamorphic zones. Fieldwork with metamorphic rocks (Fay, 1984) requires familiarity with metamorphic foliations and lineations, styles of folding, and regional deformation structures. Metamorphic conditions are thus normally mapped as a series of zones based on key minerals, textures, and structures, as described by Compton (1962) and Roberts (1982) for zones of contact metamorphism, mineral zones in regional terrains, structural zones, and regional zones of textural reconstruction.

The understanding and mapping of surficial features and formations are often no less important than getting to the bedrock. Both scientific and economic needs often call for Quaternary stratigraphic mapping and geomorphic cartography (King, 1966).

Geological Illustration

Maps that show the distribution of rocks and structures are, in the broadest sense, geological maps. They take many different forms, and the symbolization used depends on factors such as policies of national geological surveys, nature of the material depicted, and purpose of the map (see *Map Symbols*). Field sketches typically use line symbols for contacts (boundary lines between geological formations), folds, and faults. Solid lines denote accurate locations, whereas dashed and dotted lines indicate approximate and concealed locations respectively. Profile sections (see

Profile Construction), block diagrams (q.v.), joint diagrams, and columnar sections provide useful insight into relationships between rock units. Geological cross-sections (q.v.), panel or fence diagrams, and correlations of lithological or formation logs (see *Well Logging;* Vol. XIII: *Well Data Systems*) are best prepared in the office because they require extensive computations or computerized display (see discussions in Lahee, 1961; Langstaff and Morrill, 1981; Roberts, 1982).

Topographic forms and relationships between geological structure and topography have, on occasion, been expressed in field sketches that border on art form. Notable early European efforts at field sketching include examples from the works of James Hutton (1726–1797), Horace Benedict de Saussure (1740–1799), John Playfair (1747–1819), William Buckland (1784–1856), and Sir Charles Lyell (1797–1875). Albert Heim (1849–1937) and Marcel Bertrand (1847–1907) were the first detailers to illustrate nappes and thrusts respectively. Field sketches by John Wesley Powell (1834–1902) made during advance explorations of the American West, especially his notebook sketches of the Grand Canyon, record remarkable details of structure, topographic form, and stratigraphy. The persuasive field sketches of William Morris Davis (1850–1934) still serve to illustrate the impressions of field relationships between rock units and topography in the interpretation of geomorphic history (Fig. 7). Such artistic efforts of pioneer geologists to record details of field observations represent what today might be termed *geostenography,* a form of data recording through

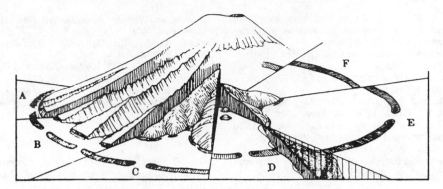

FIGURE 7. One of Davis's freehand illustrations, a deduced sector diagram, showing the upgrowth of a coral reef around a slowly subsiding island. The fringing reef
(A) becomes a barrier reef (B, C, and D). The reef thickness is shown in E, and in F
an atoll has been formed. (From Davis, 1928)

field sketching that is literally almost a lost art.
The camera, the substitute, can also be used to
create art as well as to convey a scientific message,
as illustrated by Shelton's (1966) collections.

Conclusion

Field geology is an important part of geological
endeavor. It is so basic to research and routine
surveys that field training is almost taken for
granted. Most colleges and universities require at
least one field course for those majoring in geological sciences, while many national or state geological surveys and private consulting firms offer
on-the-job training in field techniques. Lahee's
Field Geology (1961) served as the basic handbook for nearly two decades, but now there are
other helpful guides as well (e.g., Berkman and
Ryall, 1976; Bates and Kirkaldy, 1977; Moseley,
1981). The increasing attention given to the proper
training of field geologists shows concern for an
area that was somewhat neglected in the past.
Such interest will ensure a continuous supply of
competent field geologists who are highly qualified to conduct field surveys using sophisticated
equipment as well as master the use of simple tools
that geologists have used for so long.

CHARLES W. FINKL, JNR.

References

Allen, J. R. L., 1963, The classification of cross-strat-
ified units with notes on their origin, *Sedimentology*
2, 93–114.
Bates, D. E. B., and J. F. Kirkaldy, 1977, *Field Geology.*
New York: Arco, 215p.
Bergquist, W. E., E. J. Tinsley, L. Yordy, and R. L.
Miller, 1981, Worldwide directory of national earth-
science agencies and related international organizations, *U.S. Geol. Survey Circ. 834,* 87p.

Berkman, D. A., and W. R. Ryall, 1976, *Field Geolo-
gist's Manual.* Parkville, Victoria: Australian Institute of Mining and Metallurgy, 295p.
Boyer, R. E., 1971, *Field Guide to Plutonic and Meta-
morphic Rocks.* Earth Science Curriculum Project
Pamphlet Series. Boston: Houghton Mifflin, 53p.
Campbell, C. V., 1967, Lamina, laminaset, bed, and
bedset, *Sedimentology* **8,** 7–26.
Compton, R. R., 1962, *Manual of Field Geology.* New
York: Wiley, 378p.
Coneybeare, C. E. B., and K. A. W. Crook, 1968, Man-
ual of sedimentary structures, *Australian Bur. Min-
eral Resources, Geology and Geophysics Bull. 102,*
327p.
Davis, W. M., 1928, *The Coral Reef Problem.* New
York: American Geographical Society (Spec. Publ.
No. 9), 596p.
Day, M. J., and A. S. Goudie, 1977, Field assessment
of rock hardness using the Schmidt test hammer, in
Shorter Technical Methods (II), BGRG Tech. Bull.
No. 18. Norwich, England: Geo Abstracts, p. 19.
Deere, D. U., and R. P. Miller, 1966, Engineering clas-
sification and index properties for intact rock, *U.S.
Air Force Weapons Lab. Tech. Rept. AFWL-TR-65-
116.*
Dietrich, R. V., J. T. Dutro, Jr., and R. M. Foose, 1982,
*AGI Data Sheets for Geology in the Field, Labora-
tory, and Office.* Falls Church, Va.: American Geo-
logical Institute.
Freeman, T., 1971, *Field Guide to Layered Rocks.* Earth
Science Curriculum Project Pamphlet Series. Boston:
Houghton Mifflin, 44p.
Fry, N., 1984, *The Field Description of Metamorphic
Rocks.* Milton Keynes, England: Open University
Press, 110p.
Gardiner, V., and R. Dackombe, 1983, *Geomorphol-
ogical Field Manual.* London: Allen and Unwin, 254p.
Jopling, A. V., and R. G. Walker, 1967, Morphology
and origin of ripple-drift cross lamination with ex-
amples from the Pleistocene of Massachusetts, *Jour.
Sed. Petrology,* **38,** 971–984.
King, C. A. M., 1966, *Techniques in Geomorphology.*
New York: St. Martin's, 342p.
Lahee, F. H., 1961, *Field Geology.* New York: McGraw-
Hill, 926p.

Langstaff, C. S., and D. Morrill, 1981, *Geological Cross Sections*. Boston: International Human Resources Development Corporation, 108p.

Moseley, F., 1981, *Methods in Field Geology*. San Francisco: W. H. Freeman, 211p.

Peters, W. C., 1978, *Exploration and Mining Geology*. New York: Wiley, 696p.

Platt, J. I., and J. Challinor, 1980, *Simple Geological Structures*. London: Pergamon Press, 332p.

Roberts, J. L., 1982, *Introduction to Geological Maps and Structures*. Oxford: Pergamon Press, 332p.

Selby, M. J., 1980, A rock mass strength classification for geomorphic purposes: With tests from Antarctica and New Zealand, *Zeitschr. Geomorphologie* **24**, 31-51.

Shelton, J., 1966, *Geology Illustrated*. San Francisco: W. H. Freeman, 434p.

Shrock, R. R., 1948, *Sequence in Layered Rocks: A Study of Features and Structures Useful for Determining Top and Bottom or Order of Succession in Bedded and Tabular Rock Bodies*. New York: McGraw-Hill, 507p.

Tennent, R. M., ed., 1971, *Science Data Book*. Edinburg: Oliver and Boyd, 347p.

Thomas, J. A. G., 1979, *An Introduction to Geological Maps*. London: Allen and Unwin, 67p.

Tucker, M. E., 1982, *The Field Description of Sedimentary Rocks*. Milton Keynes, England: Open University Press, 112p.

Cross-references: *Block Diagrams; Compass Traverse; Exposures, Examination; Field Maps, Mapping; Fractures, Fracture Structures; Geological Fieldwork, Codes; Geologic Reports; Map Symbols; Plane Table Mapping; Samples, Sampling.* Vol. XIII: *Geology, Applied; Maps, Engineering Purposes; Field Geology; Photogrammetry; Remote Sensing, Engineering Geology.*

TABLE 1. Rock Mass Strength Classification and Ratings

	1	2	3	4	5
Weathering	unweathered r: 10	slightly weathered r: 9	moderately weathered r: 7	highly weathered r: 5	completely weathered r: 3
Spacing of joints	> 3 m r: 30	3-1 m r: 28	1-0.3 m r: 21	300-50 mm r: 15	< 50 mm r: 8
Joint orientations	Very favourable. Steep dips into slope, cross joints interlock r: 20	Favourable. Moderate dips into slope r: 18	Fair. Horizontal dips, or nearly vertical (hard rocks only) r: 14	Unfavourable. Moderate dips out of slope r: 9	Very unfavourable. Steep dips out of slope r: 5
Width of joints	< 0.1 mm r: 7	0.1-1 mm r: 6	1-5 mm r: 5	5-20 mm r: 4	> 20 mm r: 2
Continuity of joints	none continuous r: 7	few continuous r: 6	continuous, no infill r: 5	continuous, thin infill r: 4	continuous, thick infill r: 1
Outflow of groundwater	none r: 6	trace r: 5	slight < 25 l/min/ 10 m^2 r: 4	moderate 25-125 l/min/ 10 m^2 r: 3	great > 125 l/min/ 10 m^2 r: 1
Total rating	100-91	90-71	70-51	50-26	< 26

Source: From Selby, 1980.

Note: r is a rating of rock mass strength as determined in the field and commonly used in engineering classifications.

TABLE 2. Index of Chemical and Physical Properties for Selected Minerals

Name	Composition	Xal. Sys.	D	H	Remarks
Acanthite	Ag_2S	Mon	7.2–7.3	2–2½	Low temp. Ag S, 87% Ag
Achroite	.. ..	.. ..	.. ..	..	Colourless tourmaline
Acmite	$NaFe(SiO_3)_2$	Mon	3.40–3.55	6–6½	A pyroxene
Actinolite	$Ca_2(Mg, Fe)_5(Si_8O_{22})(OH)_2$	Mon	3.0–3.2	5–6	Tremolite with >2% Fe
Adularia	.. ..	.. ..	.. ..	..	Clear orthoclase
Aegirine	.. ..	.. ..	.. ..	..	Impure acmite
Agate	.. ..	.. ..	.. ..	..	Banded chalcedony
Alabandite	MnS	Iso	4.0	3½–4	Black
Alabaster	.. ..	.. ..	.. ..	..	Massive f.gr. gypsum
Albite	$Na(AlSi_3O_8)$	Tric	2.62	6	Na rich plagioclase, Ab_{100} to $Ab_{90}An_{10}$
Alexandrite	.. ..	.. ..	.. ..	..	Gem chrysoberyl
Allanite	$(Ce, Ca, Y)(Al,Fe)_3(SiO_4)_3(OH)$	Mon	3.5–4.2	5½–6	About 28% REO
Allemontite	AsSb	Hex	5.8–6.2	3–4	One cleavage
Allophane	$Al_2O_3.SiO_2.nH_2O$	Amor	1.85–1.89	3	Claylike mineral
Almandite	$Fe_3Al_2(SiO_4)_3$	Iso	4.25	7	A red garnet
Altaite	PbTe	Iso	8.16	3	Tin-white, rare
Alumstone	.. ..	.. ..	.. ..	..	Alunite
Alunite	$K Al_3(SO_4)_2(OH)_6$	Rho	2.6–2.8	4	11.4% K_2O, 37% Al_2O_3
Amazonstone	.. ..	.. ..	.. ..	..	Green microcline
Amblygonite	$(Li,Na)AlPO_4(F, OH)$	Tric	3.0–3.1	6	About 10% Li_2O, 48% P_2O_5
Amethyst	.. ..	.. ..	.. ..	..	Purple quartz
Amosite	.. ..	.. ..	.. ..	..	Anthophyllite asbestos
Amphibole Group	.. ..	.. ..	.. ..	..	See Actinolite, Anthophyllite, Arfvedsonite, Cummingtonite, Glaucophane, Hornblende, Riebeckite, Tremolite
Analcime	$Na(AlSi_2O_6).H_2O$	Iso	2.27	5–5½	A zeolite
Anatase	TiO_2	Tet	3.9	5½–6	Low temp. TiO_2
Anauxite		Mon	2.6	2	Si-rich kaolinite
Andalusite	Al_2SiO_5	Orth	3.16–3.20	7½	Often as square prisms. 63% Al_2O_3
Andesine	$Ab_{70}An_{30}–Ab_{50}An_{50}$	Tric	2.69	6	A plagioclase feldspar
Andradite	$Ca-Fe_2(SiO_4)_3$	Iso	3.75	7	Calcium-iron garnet
Anglesite	$PbSO_4$	Orth	6.2–6.4	3	Secondary, often banded. 68% Pb
Anhydrite	$CaSO_4$	Orth	2.89–2.98	3–3½	41% CaO
Ankerite	$Ca(Fe, Mg, Mn)(CO_3)_2$	Rho	2.95–3	3½	Dolomite with Fe>Mg
Annabergite	$(Ni, Co)_3(AsO_4)_2.8H_2O$	Mon	3.0	2½–3	Nickel bloom. 29% Ni, 25% As
Anorthite	$CaAl_2Si_2O_8$	Tric	2.76	6	Ca-rich plagioclase, An_{100} to $An_{90}Ab_{10}$
Anorthoclase	$(Na, K)AlSi_3O_8$	Tric	2.58	6	Like orthoclase, with Na>K
Anthophyllite	$(Mg, Fe)_7(Si_8O_{22})(OH)_2$	Orth	2.85–3.2	5½–6	Clove brown amphibole var. of asbestos
Antigorite	.. ..	.. ..	.. ..	..	Platy serpentine
Antimony	Sb	Rho	6.7	3	Cl (0001)
Antlerite	$Cu_3SO_4(OH)_4$	Orth	3.9	3½–4	Secondary Cu mineral of arid regions
Apatite	$Ca_5(PO_4, CO_3)_3(F, OH, Cl)$	Hex	3.15–3.20	5	38–42% P_2O_5
Apophyllite	$KCa_4Si_8O_{20}(F, OH).8H_2O$	Tet	2.3–2.4	4½–5	Secondary, in basic lavas
Aquamarine	.. ..	.. ..	.. ..	..	Pale greenish-blue transparent beryl
Aragonite	$CaCO_3$	Orth	2.95	3½–4	Cl (010), (110). 56% CaO
Arfvedsonite	$Na_{2-3}(Fe, Mg, Al)_5Si_8O_{22}(OH)_2$	Mon	3.45	6	Na amphibole
Argentite	Ag_2S	Iso	7.3	2–2½	Sectile, 87% Ag
Arsenic	As	Rho	5.7	3½	Cl (0001)
Arsenopyrite	FeAsS	Mon	5.9–6.2	5½–6	Pseudo-orth. 46% As
Asbestos	.. ..	.. ..	.. ..	..	See Amosite, Anthophyllite, Chrysotile, Crocidolite, Tremolite

(continued)

TABLE 2. (*continued*)

Name	Composition	Xal. Sys.	D	H	Remarks
Asbolite	Cobaltian wad	Amor	2.9–4.3		To 15% Co
Atacamite	$Cu_2Cl(OH)_3$	Orth	3.75–3.77	$3–3\frac{1}{2}$	Cl (010). 59% Cu
Augite	$(Ca, Na)(Mg, Fe^{2+}, Fe^{3+}, Al)(Si, Al)_2O_6$	Mon	3.2–3.4	5–6	Common pyroxene
Aurichalcite	$(Zn, Cu)_5(CO_3)_2(OH)_6$	Mon	3.2–3.7	2	14–23% Cu, 36–47% Zn
Autunite	$Ca(UO_2)_2(PO_4)_2.10–12H_2O$	Tet	3.1–3.2	$2–2\frac{1}{2}$	Yellow-green 67% U_3O_8
Awaruite	$Fe\,Ni_2$	Iso?	7.7–8.1	5	67–77% Ni
Axinite	$(Ca, Mn, Fe)_3Al_2BSi_4O_{15}(OH)$	Tric	3.27–3.35	$6\frac{1}{2}–7$	Crystal angles acute
Azurite	$Cu_3(CO_3)_2(OH)_2$	Mon	3.77	$3\frac{1}{2}–4$	Always blue. 55% Cu
Balas ruby	.. ..	..	..	..	Red gem spinel
Barite	$BaSO_4$	Orth	4.5	$3–3\frac{1}{2}$	Cl (001), (110). 65.7% BaO
Barytes	.. ..	..	..	..	Barite
Bastnaesite	$(Ce, La)(CO_3)(F, OH)$	Hex	4.9–5.2	$4–4\frac{1}{2}$	75% REO
Bauxite	.. ..	..	..	..	A mixture of aluminium hydroxides
Beidellite	$Al_8(Si_4O_{10})_3(OH)_{12}.12H_2O$	Orth?	2.6	$1\frac{1}{2}$	Al-rich montmorillonite
Bentonite	.. ..	..	..	..	Largely montmorillonite
Beryl	$Be_3Al_2(Si_6O_{18})$	Hex	2.75–2.8	$7\frac{1}{2}–8$	14% BeO
Biotite	$K(Mg, Fe^{2+})_3(Al, Fe^{3+})Si_3O_{10}(OH)_2$	Mon	2.8–3.2	$2\frac{1}{2}–3$	Common black mica
Bismite	Bi_2O_3	Mon	8	$4\frac{1}{2}$	72% Bi
Bismuth	Bi	Rho	9.8	$2–2\frac{1}{2}$	Cl (0001)
Bismuthinite	Bi_2S_3	Orth	6.75–6.81	2	Cl (010). 81% Bi
Bismutite	$(BiO)_2CO_3$	Tet	6.1–7.7	$2\frac{1}{2}–3\frac{1}{2}$	75% Bi
Black Jack	.. .. ..	..	..	..	Sphalerite
Blende	.. ..	..	..	..	Sphalerite
Bloodstone	.. ..	..	..	..	Heliotrope
Blue vitriol	.. ..	..	..	..	Chalcanthite
Boehmite	$AlO(OH)$	Orth	3.01–3.06		In bauxite. 85% Al_2O_3
Boracite	$Mg_3B_7O_{13}Cl$	Orth	2.9–3.0	7	62% B_2O_3
Borax	$Na_2B_4O_7.10H_2O$	Mon	1.7	$2–2\frac{1}{2}$	Purple-blue tarnish. 63.3% Cu
Bornite	Cu_5FeS_4	Iso	5.06–5.08	3	Cl (100). 36.5% B_2O_3
Boulangerite	$Pb_5Sb_4S_{11}$	Orth	6–6.3	$2\frac{1}{2}–3$	55% Pb, 25% Sb
Bournonite	$PbCuSbS_3$	Orth	5.8–5.9	$2\frac{1}{2}–3$	Easily fusible. 13% Cu, 42% Pb, 25% Sb
Brannerite	$(U, Ca, Ce)(Ti, Fe)_2O_6$	?	4.5–5.4	$4\frac{1}{2}$	30–50% U_3O_8
Braunite	$3Mn_2O_3.MnSiO_3$	Tet	4.8	$6–6\frac{1}{2}$	64% Mn
Bravoite	$(Ni, Fe)S_2$	Iso	4.66	$5\frac{1}{2}–6$	Steel gray. 24% Ni
Brazilian emerald	.. ..	..	..	..	Green tourmaline
Brittle mica	.. ..	..	..	..	See Chloritoid, Margarite, Ottrelite
Brochantite	$Cu_4(OH)_6SO_4$	Mon	3.9	$3\frac{1}{2}–4$	Green. 56% Cu
Bromyrite	$Ag(Br, Cl)$ with Br>Cl	Iso	6–6.5	$2\frac{1}{2}$	Sectile. 57–65% Ag
Bronzite	$(Mg, Fe)SiO_3$	Orth	3.1–3.3	$5\frac{1}{2}$	Enstatite with 5–13% FeO
Brookite	TiO_2	Orth	3.9–4.1	$5\frac{1}{2}–6$	Adamantine lustre
Brucite	$Mg(OH)_2$	Rho	2.39	$2\frac{1}{2}$	Cl (0001). 69% MgO
Bytownite	$Ab_{30}An_{70}–Ab_{10}An_{90}$	Tric	2.74	6	A plagioclase feldspar
Cairngorm	.. ..	..	..	..	Smoky to black quartz
Calamine	.. ..	..	..	..	Hemimorphite
Calaverite	$AuTe_2$	Mon	9.35	$2\frac{1}{2}$	Easily fusible. 42% Au
Calcite	$CaCO_3$	Rho	2.72	3	Cl (10$\bar{1}$1). 56% CaO
Californite	.. ..	..	..	..	Gem idocrase
Calomel	Hg_2Cl_2	Tet	7.2	$1\frac{1}{2}$	85% Hg
Cancrinite	$(Na_2, Ca)_4(AlSiO_4)_6CO_3.nH_2O$	Hex	2.45	5–6	A feldspathoid
Capillary pyrites	.. ..	..	..	..	Millerite
Carnallite	$KMgCl_3.6H_2O$	Orth	1.6	1	Deliquescent. 16.8% K_2O, 14.6% MgO
Carnelian	.. ..	..	..	..	Red chalcedony
Carnotite	$K(UO_2)_2(VO_4)_2.3H_2O$	Orth	4.1	Soft	50% U_3O_8, 20% V_2O_5
Cassiterite	SnO_2	Tet	6.8–7.1	6–7	Lustre adamantine. 78.6% Sn

TABLE 2. (*continued*)

Name	Composition	Xal. Sys.	D	H	Remarks
Cat's-eye					Gem var. of chrysoberyl or quartz
Celestite	$SrSO_4$	Orth	3.95–3.97	3–3½	56% SrO
Celsian	$BaAl_2Si_2O_8$	Mon	3.37	6	Feldspar with 41% BaO
Cerargyrite	Ag(Cl, Br) with Cl>Br	Iso	5.5–6	2½	Perfectly sectile. 65–75% Ag
Cerussite	$PbCO_3$	Orth	6.55	3–3½	Effer. in HNO_3. 77% Pb
Cervantite	Sb_2O_4	Orth?	4.0–5.0	4–5	After stibnite. 79% Sb
Chabazite	$Ca(Al_2Si_4O_{12}).6H_2O$	Rho	2.05–2.15	4–5	Cube-like zeolite crystals
Chalcanthite	$CuSO_4.5H_2O$	Tric	2.12–2.30	2½	Soluble in water. 35% Cu
Chalcedony			2.6–2.64		Cryptocryst. quartz
Chalcocite	Cu_2S	Orth	5.5–5.8	2½–3	Imperfectly sectile. 79.8% Cu
Chalcopyrite	$CuFeS_2$	Tet	4.1–4.3	3½–4	Brittle, yellow. 31–34.5% Cu
Chalcotrichite					Fibrous cuprite
Chalk					Fine grained calcite
Chalybite					Siderite
Chert	SiO_2		2.65	7	Cryptocryst. quartz
Chessylite					Azurite
Chiastolite					Andalusite with dark cruciform inclusions
Chloanthite					Nickel skutterudite, 3.5–6.5% Co, 14.5–21.5% Ni, 71.5–73.5% As
Chlorite	$(Mg, Fe^{2+}, Fe^{3+})_6$ $AlSi_3O_{10}(OH)_8$	Mon	2.6–2.9	2–2½	Differentiated by chem. analyses and optical properties into Clinochlore, Penninite and Prochlorite
Chloritoid	$Fe_2Al_2Si_2O_{10}(OH)_4$	Mon	3.5	6–7	Brittle mica
Chondrodite	$(Mg, Fe)_3SiO_4(OH, F)_2$	Mon	3.1–3.2	6–6½	Similar species are Clinohumite, Humite, Norbergite
Chromite	$(Fe, Mg)O.(Fe, Al, Cr)_2O_3$	Iso	4.3–4.6	5½	Lustre submetallic, dark brown streak. 43–68% Cr_2O_3
Chrysoberyl	$BeAl_2O_4$	Orth	3.65–3.8	8½	Crystals tabular. 19.8% BeO
Chrysocolla	$Cu_2H_2(Si_2O_5)(OH)_4$	?	2.0–2.4	2–4	Bluish green. 36% Cu
Chrysolite					Olivine
Chrysoprase					Green chalcedony
Chrysotile					Serpentine asbestos
Cinnabar	HgS	Rho	8.10	2½	Red streak. 86% Hg
Cinnamon stone					Grossularite
Citrine					Pale yellow quartz
Clay					See Kaolin, Montmorillonite, Illite
Cleavelandite					White, platy albite
Cliachite					Very fine grained to colloidal Al hydroxides in bauxite
Clinochlore					Chlorite variety
Clinoclase	$Cu_3(AsO_4)(OH)_3$	Mon	4.38	2½–3	Sec. mineral
Clinoenstatite	$(Mg, Fe)SiO_3$	Mon	3.19	6	Monoclinic form of enstatite
Clinoferrosilite	$(Fe, Mg)SiO_3$	Mon	3.6	6	A pyroxene
Clinohumite	$Mg_9Si_4O_{16}(F, OH)_2$	Mon	3.1–3.2	6	Chondrodite group
Clinozoisite	$Ca_2Al_3Si_3O_{12}(OH)$	Mon	3.25–3.37	6–6½	Crystals striated
Cobaltite	CoAsS	Iso	6.33	5½	In pyritohedrons. 29–35% Co, 43–45% As
Cogwheel ore					Bournonite
Colemanite	$Ca_2B_6O_{11}.5H_2O$	Mon	2.42	4–4½	Cl (010) perfect. 50.9% B_2O_3
Collophane					Massive apatite of rock phosphate deposits
Columbite	$(Fe, Mn)(Nb, Ta)_2O_6$ with Nb>Ta	Orth	5.2–6.7	6	Lustre submetallic. 31–79% Nb_2O_5, max 52% Ta_2O_5 (with Nb = Ta)
Common salt					Halite
Copper	Cu	Iso	8.9	2½–3	Malleable
Copper glance					Chalcocite
Copper nickel					Niccolite

(*continued*)

TABLE 2. (*continued*)

Name	Composition	Xal. Sys.	D	H	Remarks
Copper pyrites	. . . .	. . .	. . .	. . .	Chalcopyrite
Cordierite	$(Mg, Fe)_2Al_4Si_5O_{18}$	Orth	2.60–2.66	7–7½	In m. to high grade metamorphics
Corundum	Al_2O_3	Rho	4.02	9	Rhomb. parting. 52.9% Al
Cotton-balls	. . . .	. . .	. . .	. . .	Ulexite
Covellite	CuS	Hex	4.6–4.76	1½–2	Blue. 66.4% Cu
Cristobalite	SiO_2		2.30	7	High temp.quartz, in volcanic rocks (>1470°C)
Crocidolite	. . . .	. . .	3.2–3.3		Blue asbestos variety of riebeckite
Crocoite	$PbCrO_4$	Mon	5.9–6.1	2½–3	Orange-red streak. 23% Cr_2O_3, 64% Pb
Cryolite	Na_3AlF_6	Mon	2.95–3.0	2½	White. 54.4% F
Cubanite	$CuFe_2S_3$	Orth	4.03–4.18	3½	23% Cu
Cummingtonite	$(Fe, Mg)_7(Si_8O_{22})(OH)_2$	Mon	3.1–3.6	6	An amphibole
Cuprite	Cu_2O	Iso	6.0	3½–4	Brownish red streak. 88.8% Cu
Cyanite	. . . .	. . .	. . .	. . .	Kyanite
Cymophane	. . . .	. . .	. . .	. . .	Chrysoberyl
Danaite	$(Fe, Co)AsS$	Mon	5.9–6.2	5½–6	Cobaltian arsenopyrite, to 12% Co
Danburite	$CaB_2(SiO_4)_2$	Orth	2.97–3.02	7	In crystals. 28.4% B_2O_3
Datolite	$CaB(SiO_4)(OH)$	Mon	2.8–3.0	5–5½	Usually in crystals. 21.8% B_2O_3
Davidite	. . . .	. . .	. . .	. . .	Th brannerite. To 9% U_3O_8
Demantoid	. . . .	. . .	. . .	. . .	Green gem andradite
Diallage	. . . .	. . .	. . .	. . .	Diopside with (100) parting
Diamond	C	Iso	3.5	10	Adamantine lustre
Diaspore	$AlO(OH)$	Orth	3.35–3.45	6½–7	85% Al_2O_3
Diatomite	. . . .	. . .	0.4–0.6 (2.2)	2 (5½–6½)	Siliceous tests of diatoms
Dichroite	. . . .	. . .	. . .	. . .	Cordierite
Dickite	$Al_2Si_2O_5(OH)_4$	Mon	2.6	2–2½	Kaolin group clay mineral
Digenite	Cu_9S_5	Iso	5.6	2½–3	With chalcocite. 75–79% Cu
Diopside	$CaMg(SiO_3)_2$	Mon	3.2–3.3	5–6	A pyroxene
Dioptase	$CuSiO_2(OH)_2$	Rho	3.3	5	Emerald green
Disthene	. . . .	. . .	. . .	. . .	Kyanite
Dolomite	$CaMg(CO_3)_2$	Rho	2.85	3½–4	Cl (10$\bar{1}$1). 30.4% CaO, 21.7% MgO, 54.3% $CaCO_3$
Dry-bone ore	. . . .	. . .	. . .	. . .	Smithsonite
Dumortierite	$(Al, Fe)_7O_3(BO_3)(SiO_4)_3$	Orth	3.26–3.36	7	Radiating fibrous
Edenite	$Ca_2NaMg_5(AlSi_7O_{22})(OH)_2$	Mon	3.0	6	Pale iron-free hornblende
Electrum	Au, Ag	Iso	13.5–17.5	3	Natural Au-Ag alloy with >20% Ag
Eleolite	. . . .	. . .	. . .	. . .	Nepheline
Embolite	$Ag(Cl, Br)$ with $Cl = Br$	Iso	5.6	1–1½	Intermediate between cerargyrite and bromyrite
Emerald	. . . .	. . .	. . .	. . .	Green gem beryl
Emery	. . . .	. . .	. . .	. . .	Corundum with magnetite
Enargite	Cu_3AsS_4	Orth	4.43–4.45	3	Cl (110). 48.3% Cu, 19.1% As
Endlichite	. . . .	. . .	. . .	. . .	Arsenical vanadinite, As replacing V
Enstatite	$MgSiO_3$	Orth	3.2–3.5	5½	A pyroxene
Epidote	$Ca_2(Al, Fe)_3Si_3O_{12}(OH)$	Mon	3.35–3.45	6–7	Cl (001)
Epsomite	$MgSO_4.7H_2O$	Orth	1.75	2–2½	Bitter taste. 16.3% MgO
Epsom salt	. . . .	. . .	. . .	. . .	Epsomite
Erythrite	$Co_3(AsO_4)_2.8H_2O$	Mon	2.95	1½–2½	Pink cobalt bloom. 37% Co
Essonite	. . . .	. . .	. . .	. . .	Grossularite
Euclase	$BeAlSiO_4(OH)$	Mon	3.1	7½	Cl (010). 17% BeO
Eucryptite	$LiAlSiO_4$	Hex	2.67		After spodumene

TABLE 2. (*continued*)

Name	Composition	Xal. Sys.	D	H	Remarks
Euxenite	AB_2O_6, A = Y, Ce, Ca, U, Th; B = Ti, Nb, Ta, Fe	Orth	5–5.9	$5\frac{1}{2}$–$6\frac{1}{2}$	22–30% REO, max. 8% U_3O_8, 30–50% $(Nb_2O_5+Ta_2O_5)$
Fahlore					Tetrahedrite
Fayalite	Fe_2SiO_4	Orth	4.14	$6\frac{1}{2}$	Iron olivine
Feather ore					Jamesonite
Feldspar Group	$MAl(Al, Si)_3O_8$, M = K, Na, Ca, Ba, Rb, Sr, Fe				See Plagioclase, Potassium Feldspar, Celsian
Feldspathoid Group					See Cancrinite, Lazurite, Leucite, Nepheline, Petalite, Sodalite
Ferberite	$FeWO_4$	Mon	7.5	5	Wolframite series, 76.3% WO_3
Fergusonite	$(RE, Fe)(Nb, Ta, Ti)O_4$	Tet	4.2–5.8	$5\frac{1}{2}$–$6\frac{1}{2}$	Max. 46% REO, 10% U_3O_8, 54% Nb_2O_5
Ferrimolybdite	$Fe_2(MoO_4)_3.8H_2O$?	Orth?	3	$1\frac{1}{2}$	39% Mo
Ferrosilite	$FeSiO_3$	Orth	3.6	6	A pyroxene
Fibrolite					Sillimanite
Flint	SiO_2		2.65	7	Cryptocryst. quartz
Flos ferri					Arborescent aragonite
Fluorite	CaF_2	Iso	3.18	4	Cl octahedral, 48.9%F
Formanite					Fergusonite with Ta>Nb
Forsterite	Mg_2SiO_4	Orth	3.2	$6\frac{1}{2}$	Magnesian olivine
Fowlerite					Zinc-bearing rhodonite
Franklinite	$(Fe^{2+}, Zn, Mn^{2+})(Fe^{3+}, Mn^{3+})_2O_4$	Iso	5.15	6	Dark brown streak, 5–19% Zn.
Freibergite					Argentiferous tetrahedrite
Gadolinite	$Be_2FeY_2Si_2O_{10}$	Mon	4.0–4.5	$6\frac{1}{2}$–7	Nom. 48% REO, 10% BeO
Gahnite	$ZnAl_2O_4$	Iso	4.55	$7\frac{1}{2}$–8	Zn spinel, green octahedrons
Galaxite	$MnAl_2O_4$	Iso	4.03	$7\frac{1}{2}$–8	Mn spinel
Galena	PbS	Iso	7.4–7.6	$2\frac{1}{2}$	Cl cubic. 86.6% Pb
Garnet Group	$A_3B_2(SiO_4)_3$ A = Ca, Mg, Fe^{2+}, Mn^{2+}; B = Al, Fe^{3+}, Mn^{3+}, Cr.	Iso	3.5–4.3	$6\frac{1}{2}$–$7\frac{1}{2}$	See Almandite, Andradite, Grossularite, Pyrope, Spessartite, Uvarovite
Garnierite	$(Ni, Mg)_2Si_2O_5(OH)_4$?	Amor	2.2–2.8	2–3	Green. 25–30% Ni
Gaylussite	$Na_2Ca(CO_3)_2.5H_2O$	Mon	1.99	2–3	Easily fusible. 20% Na_2O
Gedrite					Al-rich anthophyllite
Geocronite	$Pb_5(Sb, As)_2S_8$	Orth	6.3–6.5	$2\frac{1}{2}$	69% Pb, 8% Sb, 5% As
Gersdorffite	NiAsS	Iso	5.9	$5\frac{1}{2}$	35% Ni, 45% As
Geyserite					Opal of hot spring deposits
Gibbsite	$Al(OH)_3$	Mon	2.3–2.4	$2\frac{1}{2}$–$3\frac{1}{2}$	65.4% Al_2O_3
Glauberite	$Na_2Ca(SO_4)_2$	Mon	2.70–2.85	$2\frac{1}{2}$–3	22% Na_2O
Glaucodot					Danaite
Glauconite	$(K, Na)(Al, Fe^{3+}, Mg)_2(Al, Si)_4O_{10}(OH)_2$	Mon	2.3	2	Green sand mica of marine sediments
Glaucophane	$Na_2(Mg, Fe^{2+})_3Al_2Si_8O_{22}(OH)_2$	Mon	3.0–3.2	6–$6\frac{1}{2}$	Na amphibole
Gmelinite	$(Na_2, Ca)Al_2Si_4O_{12}.6H_2O$	Rho	2.04–2.17	$4\frac{1}{2}$	Var. of chabazite
Goethite	FeO(OH)	Orth	4.37	5–$5\frac{1}{2}$	Cl (010), 62% Fe
Gold	Au	Iso	15.0–19.3	$2\frac{1}{2}$–3	Yellow, soft
Goslarite	$ZnSO_4.7H_2O$	Orth	1.98	2–$2\frac{1}{2}$	Sol. in water. 22% Zn
Graphite	C	Hex	2.3	1–2	Black, platy
Grey copper					Tetrahedrite
Greenockite	CdS	Hex	4.9	3–$3\frac{1}{2}$	Yellow-orange. 77.8% Cd
Grossularite	$Ca_3Al_2(SiO_4)_3$	Iso	3.53	$6\frac{1}{2}$	A garnet
Gummite	$UO_3.nH_2O$		3.9–6.4	$2\frac{1}{2}$–5	Field name for hydrous U oxides. 60–80% U_3O_8
Gypsum	$CaSO_4.2H_2O$	Mon	2.32	2	Cl (010), (100), (011). 32.5% CaO
Haematite					Hematite
Halite	NaCl	Iso	2.16	$2\frac{1}{2}$	53% Na_2O equiv.

(*continued*)

FIELD GEOLOGY

TABLE 2. (*continued*)

Name	Composition	Xal. Sys.	D	H	Remarks
Halloysite	$Al_2Si_2O_5(OH)_4.nH_2O$	Amor	2.0–2.2	1–2	Clay mineral
Harmotome	$(Ba, K)(Al, Si)_2Si_6O_{16}.6H_2O$	Mon	2.45	4½	A stilbite group zeolite
Hastingsite	$NaCa_2(Fe, Mg)_5Al_2Si_6O_{22}(OH)_2$	Mon	3.2	6	Hornblende series
Hausmannite	$MnMn_2O_4$	Tet	4.84	5½	72% Mn
Hauynite	$(Na, Ca)_{4-8}Al_6Si_6O_{24}.(SO_4, S)_{1-2}$	Iso	2.4–2.5	5½–6	A feldspathoid
Heavy spar					Barite
Hectorite	$(Mg, Li)_3Si_8O_{20}(OH)_4$	Mon	2.5	1–1½	Li montmorillonite
Hedenbergite	$CaFe(Si_2O_6)$	Mon	3.55	5–6	End member of diopside series
Heliotrope					Green and red chalcedony
Helvite	$(Mn, Fe, Zn)_4Be_3(SiO_4)_3S$	Iso	3.16–3.36	6–6½	In pegmatites
Hematite	Fe_2O_3	Rho	5.26	5½–6½	Brownish red streak. 70% Fe
Hemimorphite	$Zn_4(Si_2O_7)(OH)_2.H_2O$	Orth	3.4–3.5	4½–5	Cl (110). 54% Zn
Hercynite	$FeAl_2O_4$	Iso	4.39	7½–8	Iron spinel
Hessite	Ag_2Te	Iso	8.4	2½–3	Grey
Heulandite	$(Na, Ca)_{4-6}Al_6(Al, Si)_4Si_{26}O_{72}.24H_2O$	Mon	2.18–2.20	3½–4	A zeolite
Hiddenite					Green spodumene
Holmquistite					Lithium-bearing glaucophane
Hornblende	$Ca_2Na(Mg, Fe^{2+})_4(Al, Fe^{3+}, Ti)AlSi_8O_{22}(O, OH)_2$	Mon	3.2	5–6	Common amphibole
Horn silver					Cerargyrite
Huebnerite	$MnWO_4$	Mon	7.0	5	Wolframite series. 76.6% WO_3
Humite	$Mg_7(SiO_4)_3(F, OH)_2$	Orth	3.1–3.2	6	Chondrodite group
Hyacinth					Brownish to red-orange zircon
Hyalite					Globular, colourless opal
Hyalophane	$(K, Ba)Al(Al, Si)_3O_8$	Mon	2.8	6	Ba-rich orthoclase
Hydromica					Illite
Hydrozincite	$Zn_5(CO_3)_2(OH)_6$	Mon	3.6–3.8	2–2½	Secondary mineral. 59% Zn
Hypersthene	$(Mg, Fe)SiO_3$	Orth	3.4–3.5	5–6	A pyroxene
Ice	H_2O	Hex	0.917	1½	
Iceland spar					Optically clear calcite
Iddingsite	$H_8Mg_3Fe_2Si_3O_{14}?$	Orth	3.5–3.8	3	After olivine
Idocrase	$Ca_{10}(Mg, Fe)_2Al_4(SiO_4)_5(Si_2O_7)_2(OH)_4$	Tet	3.35–3.45	6½	Prismatic crystals
Illite	Hyd.Al, K, Ca, Mg silicate				Mica-like clay mineral, about 38% Al_2O_3
Ilmenite	$FeTiO_3$	Rho	4.7	5½–6	Slightly magnetic. 52.6% TiO_2
Ilvaite	$CaFe^{2+}_2Fe^{3+}(SiO_4)_2(OH)$	Orth	4.0	5½–6	Black or brown
Indicolite					Dark blue tourmaline
Iodobromite	$Ag(Cl, Br, I)$	Iso	5.7	1–1½	To about 15% I, 60% Ag
Iodyrite	AgI	Hex	5.7	1–1½	Sectile. 45% Ag
Iolite					Cordierite (gem var.)
Iridium	Ir	Iso	22.7	6–7	Platinoid metal
Iridosmine	Ir, Os	Rho	19.3–21.1	6–7	Platinoid. Max. 77% Ir, max. 80% Os
Iron pyrites					Pyrite
Jacinth					Hyacinth
Jacobsite	$(Mn^{2+}, Fe^{2+}, Mg)(Fe^{3+}, Mn^{3+})_2O_4$	Iso	5.1	5½–6½	A spinel
Jade					See nephrite and jadeite
Jadeite	$Na(Al, Fe)Si_2O_6$	Mon	3.3–3.5	6½–7	Green pyroxene jade
Jamesonite	$Pb_4FeSb_6S_{14}$	Mon	5.5–6.0	2–3	50.8% Pb, 29.5% Sb
Jargon					Clear, yellow or smoky zircon
Jarosite	$KFe_3(SO_4)_2(OH)_6$	Rho	2.91–3.26	3	6–9% K_2O
Jasper					Red cryptocryst. quartz
Kainite	$MgSO_4.KCl.3H_2O$	Mon	2.1	3	19% K_2O, 16% MgO
Kalinite					Potash Alum

158

TABLE 2. (*continued*)

Name	Composition	Xal. Sys.	D	H	Remarks
Kaliophilite	$K(AlSiO_4)$	Hex	2.61	6	Dimorph. with kalsilite. 30% K_2O, 32% Al_2O_3
Kalsilite	. . . .	. . . .	. . .	. .	End member of nepheline series
Kaolin Group	. . . .	. . . .	. . .	. .	Family of clay minerals, see Anauxite, Dickite, Kaolinite, Nacrite, with 39.5% Al_2O_3
Kaolinite	$Al_2(Si_2O_5)(OH)_4$	Mon	2.6–2.65	2–2½	Earthy
Kernite	$Na_2B_4O_7.4H_2O$	Mon	1.95	3	22.7% Na_2O, 51% B_2O_3
Krennerite	$AuTe_2$	Orth	8.62	2–3	Basal cleavage
Kunzite	. . . . . .	. . . .	. . .	. .	Pink spodumene
Kyanite	Al_2SiO_5	Tric	3.56–3.66	5–7	Blue, Cl (100) perfect, bladed xals. Marked hardness anisotropy
Labradorite	$Ab_{50}An_{50}$–$Ab_{30}An_{70}$	Tric	2.71	6	A plagioclase feldspar
Langbeinite	$K_2Mg_2(SO_4)_3$	Iso	2.83	2½–3½	22.7% K_2O, or 42% K_2SO_4
Lapis lazuli	. . . . . .	. . . .	. . .	. .	Impure lazurite
Larsenite	$PbZnSiO_4$	Orth	5.9	3	Rare olivine
Laumontite	$(Ca, Na)Al_2Si_4O_{12}.4H_2O$	Mon	2.28	4	A zeolite
Lawsonite	$CaAl_2(Si_2O_7)(OH)_2.H_2O$	Orth	3.09	8	In gneisses and schists
Lazulite	$(Mg, Fe^{3+})Al_2(PO_4)_2(OH)_2$	Mon	3.0–3.1	5–5½	Blue gemstone
Lazurite	$(Na, Ca)_4(AlSiO_4)_3$ (SO_4, S, Cl)	Iso	2.4–2.45	5–5½	A feldspathoid
Lechatelierite	SiO_2	Amor	2.2	6–7	Fused silica
Lepidocrocite	$FeO(OH)$	Orth	4.09	5	With goethite. 62% Fe
Lepidolite	$K(Li, Al)_3(Si, Al)_4O_{10}$ $(F, OH)_2$	Mon	2.8–3.0	2½–4	Lithium mica with about 5% Li_2O
Leucite	$K(AlSi_2O_6)$	Iso?	2.45–2.50	5½–6	A feldspathoid
Libethenite	$Cu_2(PO_4)(OH)$	Orth	4	4	53% Cu, 29% P_2O_5
Limonite	$FeO(OH).nH_2O$	Amor	3.6–4.0	5–5½	Field name for brown amorphous hydrous iron oxides, yellowish brown streak, about 60% Fe
Linarite	$PbCu(SO_4)(OH)_2$	Mon	5.3	2½	Deep blue. 15% Cu, 51% Pb
Linnaeite	Co_3S_4	Iso	4.8	4½–5½	58% Co, to 7% Ni
Lithia mica	. . . .	. . . .	. . .	. .	Lepidolite
Lithiophilite	$Li(Mn^{2+}, Fe^{2+})PO_4$	Orth	3.5	5	End member of triphylite series. 9.5% Li_2O, 45% P_2O_5.
Loellingite	$Fe As_2$	Orth	7.4–7.5	5–5½	72.8% As
Magnesiochromite	$(Mg, Fe)(Cr, Al)_2O_4$	Iso	4.2	5½	End member of chromite series, with 21% MgO, 79% Cr_2O_3
Magnesioferrite	$(Mg, Fe)Fe_2O_4$	Iso	4.5	5½–6½	A spinel. 20% MgO, 56% Fe for $MgFe_2O_4$
Magnesite	$MgCO_3$	Rho	3.0–3.2	3½–5	Commonly massive, 47.6% MgO
Magnetic pyrites	. . . .	. . . .	. . .	. .	Pyrrhotite
Magnetite	$(Fe, Mg)Fe_2O_4$	Iso	5.18	6	Iron spinel, strongly magnetic, blk. streak. 72.4% Fe for Fe_3O_4
Malachite	$Cu_2CO_3(OH)_2$	Mon	3.9–4.03	3½–4	Green. 57.3% Cu
Manganite	$MnO(OH)$	Orth	4.3	4	Prismatic crystals, dark brow streak. 62% Mn
Manganosite	MnO	Iso	5.0–5.4	5½	77% Mn
Manganotantalite	$(Mn, Fe)(Ta, Nb)_2O_6$	Orth	7.3	4½	Tantalite with Mn: Fe :: 3:1, 10% Mn, 84% (Nb_2O_5+ Ta_2O_5)
Marcasite	FeS_2	Orth	4.89	6–6½	White iron pyrites. 46.5% Fe
Margarite	$CaAl_2(Al_2Si_2O_{10})(OH)_2$	Mon	3.0–3.1	3½–5	A brittle mica

(*continued*)

TABLE 2. (*continued*)

Name	Composition	Xal. Sys.	D	H	Remarks
Marialite	$3NaAlSi_3O_8.NaCl$	Tet	2.7	$5\frac{1}{2}$–6	End member of scapolite series
Marmatite	.. .. ..		3.9–4.0		Ferroan sphalerite, 10–20% Fe
Martite	.. ..				Hematite octahedrons after magnetite
Meerschaum					Sepiolite
Meionite	$3CaAl_2Si_2O_8.CaCO_3$	Tet	2.7	$5\frac{1}{2}$–6	End member of scapolite series
Melaconite	.. ..			.. ..	Tenorite
Melanite	.. ..			.. ..	Black andradite
Melanterite	$FeSO_4.7H_2O$	Mon	1.90	2	Green-blue
Melilite	$(Na, Ca)_2(Mg, Al)$ $(Si, Al)_2O_7$	Tet	2.9–3.1	5	
Menaccanite	.. ..				Ilmenite
Menaghinite	$CuPb_{13}Sb_7S_{24}$	Orth	6.36	$2\frac{1}{2}$	Jamesonite family
Mercury	Hg		13.6		Fluid, quicksilver
Miargyrite	$AgSbS_2$	Mon	5.2–5.3	$2\frac{1}{2}$	Cherry red streak. 36% Ag, 41% Sb
Mica Group	.. ..				See Biotite, Brittle Mica, Lepidolite, Muscovite, Phlogopite
Microcline	$K(AlSi_3O_8)$	Tric	2.54–2.57	6	Triclinic K feldspar
Microlite	$(Na, Ca)_2(Ta, Nb)_2O_6$ (O, OH, F)	Iso	6.33	$5\frac{1}{2}$	End member of pyrochlore series, 75–80% $(Nb_2O_5+Ta O_5)$
Microperthite	.. ..			..	Microcline and albite microlayers
Millerite	NiS	Rho	5.3–5.7	3–$3\frac{1}{2}$	Capillary crystals 64.7% Ni
Mimetite	$Pb_5Cl(AsO_4)_3$	Hex	7.0–7.2	$3\frac{1}{2}$	Like pyromorphite. 69% Pb, 15% As
Minium	Pb_3O_4		8.9–9.2	$2\frac{1}{2}$	90% Pb
Mispickel					Arsenopyrite
Molybdenite	MoS_2	Hex	4.62–4.73	1–$1\frac{1}{2}$	Platy. 60% Mo
Molybdite					Ferrimolybdite
Monazite	$(Ce, La, Y, Th)(PO_4, SiO_4)$	Mon	5.0–5.3	5–$5\frac{1}{2}$	Max. 30% ThO_2, max. 65% REO
Monticellite	$CaMgSiO_4$	Orth	3.2	5	Rare olivine
Montmorillonite	$(Al, Mg)_8(Si_4O_{10})_3(OH)_{10}.$ $12H_2O$	Mon	2.5	1–$1\frac{1}{2}$	Clay mineral
Montmorillonite Group	.. ..				Family of clay minerals with 39.5% Al_2O_3, see Beidellite, Hectorite, Montmorillonite, Nontronite and Saponite
Moonstone	.. ..				Opalescent albite or orthoclase
Morganite	.. ..				Rose beryl
Mullite	$Al_2Si_2O_{13}$	Orth	3.23	6–7	Formed by heating andalusite, kyanite or sillimanite
Muscovite	$KAl_2(AlSi_3O_{10})(OH)_2$	Mon	2.76–3.1	2–$2\frac{1}{2}$	Common clear mica
Nacrite	$Al_2(Si_2O_5)(OH)_4$	Mon	2.6	2–$2\frac{1}{2}$	Kaolin group clay mineral
Nagyagite	$Pb_5Au(Te, Sb)_4S_{5-8}$	Mon	7.4	1–$1\frac{1}{2}$	Rare
Natroalunite	.. ..				Alunite with Na>K
Natrolite	$Na_2(Al_2Si_3O_{10}).2H_2O$	Mon	2.25	5–$5\frac{1}{2}$	A zeolite
Nepheline	$(Na, K)AlSiO_4$	Hex	2.55–2.65	$5\frac{1}{2}$–6	A feldspathoid. 22% Na_2O, 36% Al_2O_3
Nephrite	.. ..			.. ..	Jade-like var. of tremolite
Niccolite	NiAs	Hex	7.78	5–$5\frac{1}{2}$	Copper-red. 43.9% Ni
Nickel bloom	.. ..			.. ..	Annabergite
Nickel iron	Ni, Fe	Iso	7.8–8.2	5	In meteorites. 5–15% Ni
Nickel skutterudite	$(Ni, Co, Fe)As_3$	Iso	6.1–6.9	$5\frac{1}{2}$–6	2–6% Co, 12–20% Ni, 73–78% As
Nitre	KNO_3	Orth	2.09–2.14	2	Saltpetre
Nontronite	$Fe(AlSi)_8O_{20}(OH)_4$	Mon	2.5	1–$1\frac{1}{2}$	Montmorillonite group clay mineral
Norbergite	$Mg_3(SiO_4)(F, OH)_2$	Orth	3.1–3.2	6	Chondrodite group
Nosean (Noselite)	$Na_8Al_6Si_6O_{24}.(SO_4)$	Iso	2.25–2.4	6	A feldspathoid

TABLE 2. (*continued*)

Name	Composition	Xal. Sys.	D	H	Remarks
Octahedrite	.. ..		.. ..	.. ..	Anatase
Oligoclase	$Ab_{90}An_{10}–Ab_{70}An_{30}$	Tric	2.65	6	A plagioclase feldspar
Olivine Group	$(Mg, Fe)_2SiO_4$				(Forsterite-Fayalite series), also rarer members Larsenite, Monticellite, Tephroite.
Onyx	.. ..		.. ..	.. ..	Layered chalcedony
Opal	$SiO_2.nH_2O$	Amor	1.9–2.2	5–6	Conchoidal fracture
Orpiment	As_2S_3	Mon	3.49	$1\frac{1}{2}$–2	Yellow. 61% As
Orthite	.. ..		.. ..	.. ..	Allanite
Orthoclase	$K(AlSi_3O_8)$	Mon	2.57	6	Common K feldspar
Osmiridium	.. ..		.. ..	.. ..	Iridosmine
Ottrelite	$(Fe^{2+}, Mn)(Al, Fe^{3+})Si_3O_{10}$ .H_2O	Mon	3.5	6–7	Mn chloritoid
Palladium	Pd	Iso	11.9	$4\frac{1}{2}$–5	With platinum
Paragonite	$NaAl_2(AlSi_3O_{10})(OH)_2$	Mon	2.85	2	Na muscovite
Pargasite	$NaCa_2Mg_4Al_3Si_6O_{22}(OH)_2$	Mon	3–3.5	$5\frac{1}{2}$	Greenish Na hornblende
Patronite	Impure V sulphide				Vanadium ore (Peru)
Peacock ore	.. ..		.. ..	.. ..	Bornite
Pearceite	$(Ag, Cu)_{16}As_2S_{11}$	Mon	6.15	3	Var. of polybasite
Pectolite	$NaCa_2Si_3O_8(OH)$	Tric	2.7–2.8	5	Crystals acicular
Penninite	.. ..		.. ..	.. ..	Chlorite variety
Pentlandite	$(Fe, Ni)_9S_8$	Iso	4.6–5.0	$3\frac{1}{2}$–4	With pyrrhotite. 34–35% Ni
Peridot	.. ..		.. ..	.. ..	Gem olivine
Perovskite	$CaTiO_3$	Iso	4.03	$5\frac{1}{2}$	58% TiO_2, variable REO
Perthite	.. ..		.. ..	.. ..	Microcline and albite intergrowth
Petalite	$Li(AlSi_4O_{10})$	Mon	2.4	6–$6\frac{1}{2}$	A feldspathoid. 5% Li_2O
Petzite	Ag_3AuTe_2	Iso?	8.7–9.0	$2\frac{1}{2}$–3	
Phenacite	Be_2SiO_4	Rho	2.97–3.00	$7\frac{1}{2}$–8	In pegmatites. 45.6% BeO
Phillipsite	$(K_2, Na_2, Ca)Al_2Si_4O_{12}.$ 4.5 H_2O	Mon	2.2	$4\frac{1}{2}$–5	Var. of stilbite
Phlogopite	$K(Mg, Fe)_3AlSi_3O_{10}(OH, F)_2$	Mon	2.86	$2\frac{1}{2}$–3	Brown mica
Phosgenite	$Pb_2Cl_2CO_3$	Tet	6.0–6.3	3	Easily fusible. 75% Pb
Phosphuranylite	$Ca(UO_2)_4(PO_4)_2(OH)_4.7H_2O$	Tet		$2\frac{1}{2}$	Yellow secondary U mineral
Picotite	.. ..		.. ..	.. ..	Cr spinel
Piedmontite	Mn^{2+} epidote	Mon	3.4	$6\frac{1}{2}$	Reddish brown
Pigeonite	$(Ca, Mg, Fe)SiO_3$	Mon	3.2–3.4	5–6	Pyroxene in basic volcanics
Pinite	.. ..		.. ..	.. ..	Muscovite after other minerals
Pitchblende	.. ..		.. ..	.. ..	Uraninite
Plagioclase	$NaAlSi_3O_8$ (albite–$Ab_{100}An_0$) to $CaAl_2Si_2O_8$ (anorthite–Ab_0An_{100})	Tric	2.62–2.76	6	See Albite, Oligoclase, Andesine, Labradorite, Bytownite, Anorthite
Plagionite	$Pb_5Sb_8S_{17}$	Mon	5.56	$2\frac{1}{2}$	Jamesonite series
Platinum	Pt alloy	Iso	14–19	4–$4\frac{1}{2}$	Grains in placers
Pleonaste	.. ..		.. ..	.. ..	Iron spinel
Plumbago	.. ..		.. ..	.. ..	Graphite
Polianite	MnO_2	Tet	5.0	6–$6\frac{1}{2}$	Crystalline pyrolusite
Pollucite	$(Cs, Na)_2Al_2Si_4O_{12}.H_2O$	Iso	2.9	$6\frac{1}{2}$	Colourless > 42% Cs_2O
Polybasite	$(Ag, Cu)_{16}Sb_2S_{11}$	Mon	6.0–6.2	2–3	74% Ag, 10% Sb, to 12% Cu
Polycrase	$AB_2O_6, A = Y, Ce, Ca, U, Th;$ $B = Ti, Nb, Ta, Fe$	Orth	4.7–5.9	$5\frac{1}{2}$–$6\frac{1}{2}$	14–30% REO, max. 13% U_2O_8, max. 26% $(Nb_2O_5 + Ta_2O_5)$
Polyhalite	$K_2Ca_2Mg(SO_4)_4.2H_2O$	Tric	2.78	$2\frac{1}{2}$–3	Bitter taste. 15.6% K_2O
Potash Alum	$KAl(SO_4)_2.11H_2O$	Iso	1.75	2–$2\frac{1}{2}$	6–10% K_2O
Potassium feldspar	$KAlSi_3O_8$		.. ..	.. ..	See Orthoclase, Microcline
Potash mica	.. ..		.. ..	.. ..	Muscovite
Powellite	$CaMoO_4$	Tet	4.23	$3\frac{1}{2}$–4	Fluorescent. 48% Mo
Prase	.. ..		.. ..	.. ..	Dull green jasper
Prehnite	$Ca_2Al_2(Si_3O_{10})(OH)_2$	Orth	2.8–2.95	6–$6\frac{1}{2}$	Tabular crystals
Prochlorite	.. ..		.. ..	.. ..	Chlorite variety
Proustite	Ag_3AsS_3	Rho	5.55	2–$2\frac{1}{2}$	Light ruby silver, red streak. 65.4% Ag, 15.2% As

(*continued*)

TABLE 2. (*continued*)

Name	Composition	Xal. Sys.	D	H	Remarks
Psilomelane	. . . .	. . .	. . .	. . .	Field name for massive, hard manganese minerals. About 50% Mn
Purple copper ore	. . . .	. . .	. . .	. . .	Bornite
Pyrargyrite	Ag_3SbS_3	Rho	5.85	$2\frac{1}{2}$	Dark ruby silver, red streak. 22.3% Sb, 59.9% Ag
Pyrite	FeS_2	Iso	5.02	$6-6\frac{1}{2}$	Crystals striated. 46.5% Fe
Pyrochlore	$(Na, Ca)_2(Nb, Ta)_2O_6(OH, F)$	Iso	4.2–4.5	5	Infusible. 3–6% REO, 56–73% Nb_2O_5
Pyrolusite	MnO_2	Tet	4.75	1–2	Sooty. 63.2% Mn
Pyromorphite	$Pb_5(PO_4)_3Cl$	Hex	6.5–7.1	$3\frac{1}{2}-4$	Adamantine lustre. 49–76% Pb, max. 8% As
Pyrope	$(Mg, Fe)_3Al_2(SiO_4)_3$	Iso	3.51	7	Dark red garnet
Pyrophyllite	$AlSi_2O_5(OH)$	Mon	2.8–2.9	1–2	Resembles talc
Pyroxene Group	. . . .	. . .	. . .	. . .	See Aegerine, Augite, Diopside, Enstatite, Jadeite, Spodumene
Pyrrhotite	$Fe_{1-x}S, x = 0$ to 0.2	Hex	4.58–4.65	4	Magnetic. 58–63.5% Fe
Quartz	SiO_2	Rho	2.65	7	46.7% Si
Rammelsbergite	$NiAs_2$	Orth?	7.1	$5\frac{1}{2}-6$	28% Ni
Rasorite	. . . .	. . .	. . .	. . .	Kernite
Realgar	AsS	Mon	3.48	$1\frac{1}{2}-2$	Red. 70% As
Red copper ore	. . . .	. . .	. . .	. . .	Cuprite
Red ochre	. . . .	. . .	. . .	. . .	Hematite
Rhodochrosite	$MnCO_3$	Rho	3.45–3.6	$3\frac{1}{2}-4\frac{1}{2}$	Pink. 49% Mn
Rhodolite	$3(Mg, Fe)O.Al_2O_3.3SiO_2$	Iso	3.84	7	Pale red or purple garnet
Rhodonite	$MnSiO_3$	Tric	3.58–3.70	$5\frac{1}{2}-6$	Pink. 42% Mn
Riebeckite	$Na_2(Fe, Mg)_5Si_8O_{22}(OH)_2$	Mon	3.44	4	Amphibole, end member of glaucophane series
Rock crystal	. . . .	. . .	. . .	. . .	Euhedral clear quartz
Rock salt	. . . .	. . .	. . .	. . .	Halite
Roscoelite	$K(U, Al, Mg)_3Si_3O_{10}(OH)_2$	Mon	2.97	$2\frac{1}{2}$	Vanadium mica
Rubellite	. . . .	. . .	. . .	. . .	Red or pink tourmaline
Ruby	. . . .	. . .	. . .	. . .	Red gem corundum
Ruby copper	. . . .	. . .	. . .	. . .	Cuprite
Ruby silver	. . . .	. . .	. . .	. . .	Pyrargyrite or proustite
Rutile	TiO_2	Tet	4.18–4.25	$6-6\frac{1}{2}$	Adamantine lustre
Samarskite	$(RE, U, Ca, Fe, Pb, Th)(Nb, Ta, Ti, Sn)_2O_6$	Orth	4.1–6.2	5–6	10–22% REO, 28–46% Nb_2O_5, 2–27% Ta_2O_5, O–12% U_3O_8
Sanidine	. . . .	. . .	. . .	. . .	High temp. orthoclase
Saponite	$(Mg, Al)_6(Si, Al)_8O_{20}(OH)_4$	Mon	2.5	$1-1\frac{1}{2}$	Montmorillonite group clay mineral
Sapphire	. . . .	. . .	. . .	. . .	Blue gem corundum
Satin spar	. . . .	. . .	. . .	. . .	Fibrous gypsum
Scapolite	Marialite—meionite series	Tet	2.65–2.74	5–6	Metamorphic
Scheelite	$CaWO_4$	Tet	5.9–6.1	$4\frac{1}{2}-5$	Fluorescent. 70–80% WO_3
Schorlite	. . . .	. . .	. . .	. . .	Common black tourmaline
Scolecite	$Ca(Al_2Si_3O_{10}).3H_2O$	Mon	2.16–2.4	$5-5\frac{1}{2}$	A zeolite
Scorodite	$FeAsO_4.2H_2O$	Orth	3.1–3.3	$3\frac{1}{2}-4$	Green to brown. About 32% As
Scorzalite	$(Fe, Mg)Al_2(PO_4)_2(OH)_2$	Mon	3.35	$5\frac{1}{2}-6$	End member of lazulite series
Selenite	. . . .	. . .	. . .	. . .	Clear crystalline gypsum
Semseyite	$Pb_9Sb_8S_{21}$	Mon	5.8	$2\frac{1}{2}$	Jamesonite series
Sepiolite	$Mg_4(Si_2O_5)_3(OH)_2.6H_2O$	Mon?	2.0	$2-2\frac{1}{2}$	Meerschaum, light, sec. with serpentine
Sericite	. . . .	. . .	. . .	. . .	Fine-grained muscovite
Serpentine	$(Mg, Fe)_3Si_2O_5(OH)_4$	Mon	2.2	2–5	43% MgO
Siderite	$FeCO_3$	Rho	3.83–3.88	$3\frac{1}{2}-4$	48.2% Fe
Siegenite	$(Co, Ni)_3S_4$	Iso	4.8	$4\frac{1}{2}-5\frac{1}{2}$	Linnaeite series. 29% Co, 29% Ni
Sillimanite	Al_2SiO_5	Orth	3.23	6–7	Cl (010) perfect. 63.2% Al_2O_3
Silver	Ag	Iso	10.5	$2\frac{1}{2}-3$	White, malleable
Silver glance	. . . .	. . .	. . .	. . .	Argentite

TABLE 2. (*continued*)

Name	Composition	Xal. Sys.	D	H	Remarks
Sklodowskite	$Mg(UO_2)_2Si_2O_7.6H_2O$	Orth	3.54	?	64% U_3O_8
Skutterudite	$(Co, Ni, Fe)As_3$	Iso	6.1–6.9	5	11–21% Co, 73–79% As, 0–9% Ni
Smaltite	.. .. ..	.. ..	.. ..	.. ..	Skutterudite variety. 13–24% Co, 63–71% As, 1–15% Ni
Smithsonite	$ZnCO_3$	Rho	4.35–4.40	5	52% Zn
Soapstone	.. ..	.. ..	.. ..	.. ..	Talc
Sodalite	$Na_4Al_3Si_3O_{12}Cl$	Iso	2.15–2.3	5½–6	A feldspathoid
Soda nitre	$NaNO_3$	Rho	2.29	1–2	36.5% Na_2O
Spathic iron	.. ..	.. ..	.. ..	.. ..	Siderite
Specular iron	.. ..	.. ..	.. ..	.. ..	Foliated hematite
Sperrylite	$PtAs_2$	Iso	10.50	6–7	54% Pt
Spessartite	$Mn_3Al_2(SiO_4)_3$	Iso	4.18	7	Brown to red garnet
Sphalerite	$(Zn, Fe)S$	Iso	3.9–4.1	3½–4	38–67% Zn, max. 5% Cd
Sphene	$CaTiO(SiO_4)$	Mon	3.40–3.55	5–5½	Wedge-shaped xals. 40% TiO_2
Spinel Group	$(Mg, Fe, Zn, Mn)Al_2O_4$	Iso	3.6–4.0	8	In octahedrons
Spodumene	$LiAl(Si_2O_6)$	Mon	3.15–3.20	6½–7	A pyroxene. 8% Li_2O
Stannite	Cu_2FeSnS_4	Tet	4.4	4	Easily fusible. 29–31% Cu, 27% Sn, 12–14% Fe
Staurolite	$(Fe, Mg)_2Al_9Si_4O_{23}(OH)$	Orth	3.65–3.75	7–7½	In cruciform twins. 56% Al_2O_3
Steatite	.. ..	.. ..	.. ..	.. ..	Talc
Stephanite	Ag_5SbS_4	Orth	6.2–6.3	2–2½	68.5% Ag, 15.2% Sb
Sternbergite	$AgFe_2S_3$	Orth	4.1–4.2	1–1½	34% Ag, 35% Fe
Stibnite	Sb_2S_3	Orth	4.52–4.62	2	71.7% Sb
Stilbite	$NaCa_2Al_5Si_3O_{36}.14H_2O$	Mon	2.1–2.2	3½–4	A zeolite
Stillwellite	$(Ce, La, Ca)BSiO_5$	Rho	4.57		58% REO, 11% B_2O_3
Stolzite	$PbWO_4$	Tet	8.3–8.4	2½–3	45% Pb, 50% WO_3
Stromeyerite	$(Cu, Ag)S$	Orth	6.2–6.3	2½–3	53% Ag, 31% Cu
Strontianite	$SrCO_3$	Orth	3.7	3½–4	Efferv. in HCl. 90% SrO
Sulphur	S	Orth	2.05–2.09	1½–2½	Burns with blue flame
Sunstone					Brilliant translucent oligoclase
Sylvanite	$(Au, Ag)Te_2$	Mon	8.0–8.2	1½–2	Cl (010) perfect. 25% Au, 15% Ag
Sylvite	KCl	Iso	1.99	2	Cl cubic perfect. 63% K_2O
Talc	$Mg_3(Si_4O_{10})(OH)_2$	Mon	2.7–2.8	1	Greasy feel
Tantalite	$(Fe, Mn)(Ta, Nb)_2O_6$; with $Ta>Nb$	Orth	6.2–8.0	6–6½	52–86% Ta_2O_5, max. 31% Nb_2O_5 (with Ta = Nb)
Tennantite	$(Cu, Fe, Zn, Ag)_{12}As_4S_{13}$	Iso	4.6–5.1	3–4½	Max. 11% Fe, 9% Zn, 14% Ag, 4% Pb, 13% Bi, 1% Co, 30–53% Cu
Tenorite	CuO	Tric	5.8–6.4	3–4	Black. 79.9% Cu
Tephroite	$Mn_2(SiO_4)$	Orth	4.1	6	Rare olivine
Tetrahedrite	$(Cu, Fe, Zn, Ag)_{12}Sb_4S_{13}$	Iso	4.6–5.1	3–4½	In tetrahedrons. Max. 45% Cu, 13% Fe, 8% Zn, 18% Ag, 17% Hg, 16% Pb, 4% Ni, 4% Co, 4% Bi
Thenardite	Na_2SO_4	Orth	2.68	2½	In saline lakes
Thomsonite	$NaCa_2Al_5Si_5O_{20}.6H_2O$	Orth	2.3	5	A zeolite
Thorianite	ThO_2	Iso	9.7	6½	To 17% U_3O_8
Thorite	$Th(SiO_4)$	Tet	5.3	5	Usually hydrated
Thulite	.. ..	.. ..	.. ..	.. ..	Pink-red zoisite
Tiger's-eye	.. ..	.. ..	.. ..	.. ..	Yellow brown quartz after crocidolite
Tin	Sn	Tet	7.3	2	Very rare
Tinstone	.. ..	.. ..	.. ..	.. ..	Cassiterite
Titanic iron ore	.. ..	.. ..	.. ..	.. ..	Ilmenite
Titanite	.. ..	.. ..	.. ..	.. ..	Sphene
Topaz	$Al_2(SiO_4)(F, OH)_2$	Orth	3.4–3.6	8	Cl (001) perfect
Torbernite	$Cu(UO_2)_2(PO_4)_2.8$–$12H_2O$	Tet	3.22	2–2½	Green. 61% U_3O_8, 13.5–15% P_2O_5, 6–7% Cu
Tourmaline	$XY_2Al_6(BO_3)_3(Si_6O_{18})(OH)_4$ X = Na, Ca; Y = Al, Fe, Li, Mg	Rho	3.0–3.25	7–7½	Trigonal section

(*continued*)

TABLE 2. (*continued*)

Name	Composition	Xal. Sys.	D	H	Remarks
Tremolite	$Ca_2Mg_5(Si_8O_{22})(OH)_2$	Mon	3.0–3.3	5–6	Ca amphibole, short fibre asbestos
Tridymite	SiO_2	Orth	2.26	7	In volcanic rocks (870–1470°C)
Triphylite	$Li(Fe, Mn)PO_4$	Orth	3.42–3.56	4½–5	9.5% Li_2O, 45% P_2O_5
Troilite					Pyrrhotite
Trona	$Na_2CO_3.NaHCO_3.2H_2O$	Mon	2.13	3	Alkaline taste. 41% Na_2O
Troostite					Manganiferous willemite
Tungstite	$WO_3.nH_2O$	Orth?	?	2½	Sec. mineral
Turgite	$2Fe_2O_3.nH_2O$	?	4.2–4.6	6¼	With goethite
Turquoise	$CuAl_6(PO_4)_4(OH)_8.5H_2O$	Tric	2.6–2.8	6	Blue-green. 5.5–7.8% Cu, 28–35% P_2O_5
Tyuyamunite	$Ca(UO_2)_2(VO_4)_2.5–8H_2O$	Orth	3.7–4.3	2	Ca analogue of carnotite. About 56% U_3O_8, 20% V_2O_5
Ulexite	$NaCaB_5O_9.8H_2O$	Tric	1.96	1	7.7% Na_2O, 43% B_2O_3
Uralian emerald					Green gem andradite
Uralite					Hornblende after pyroxene
Uraninite	UO_2 to UO_3	Iso	9.0–9.7	5½	Pitchy lustre, nom. U_3O_8
Uranophane	$Ca(UO_2)_2Si_2O_7.6H_2O$	Orth	3.81–3.90	2–3	63% U_3O_8
Uranosphaerite	$Bi_2O_3.2UO_3.3H_2O$	Orth	6.36	2–3	61% U_3O_8, 42% Bi_2O_3
Uvarovite	$Ca_3Cr_2(SiO_4)_3$	Iso	3.45	7½	Green garnet
Vanadinite	$Pb_5(VO_4)_3Cl$	Hex	6.7–7.1	3	19.4% V_2O_5, 68–73% Pb
Variscite	$Al(PO_4).2H_2O$	Orth	2.4–2.6	3½–4½	Green, massive. 43–45% P_2O_5
Verde antique					Variegated serpentine and white marble
Vermiculite		Mon	2.4	1½	Altered biotite
Vesuvianite					Idocrase
Violarite	$Ni_2Fe S_4$	Iso	4.8	4½–5½	34–43% Ni, 15–18% Fe
Vivianite	$Fe_3(PO_4)_2.8H_2O$	Mon	2.58–2.68	1½–2	Cl (010) perfect. 28% P_2O_5
Wad	Hyd. Mn oxides				25–48% Mn
Wavellite	$Al_3(OH)_3(PO_4)_2.5H_2O$	Orth	2.33	3½–4	35% P_2O_5, 38% Al_2O_3
Wernerite					Scapolite
White iron pyrites					Marcasite
White mica					Muscovite
Willemite	Zn_2SiO_4	Rho	3.9–4.2	5½	58.5% Zn
Witherite	$BaCO_3$	Orth	4.3	3½	Efferv. in HCl. 77.7% BaO
Wolframite	$(Fe, Mn)WO_4$	Mon	7.0–7.5	5–5½	About 75% WO_3
Wollastonite	$Ca(SiO_3)$	Tric	2.8–2.9	5–5½	Cl (001), (100)
Wood tin					Cassiterite
Wulfenite	$PbMoO_4$	Tet	6.5–7.5	3	Orange-red. 56% Pb, 26.6% Mo
Wurtzite	$(Zn, Fe)S$	Hex	4.0	4	Max. 67% Zn, 8% Fe, 3.6% Cd
Xenotime	YPO_4	Tet	4.4–5.1	4–5	61.4% REO, 38.6% P_2O_5
Yellow copper ore					Chalcopyrite
Zeolite Group					See Analcime, Chabazite, Heulandite, Natrolite, Stilbite.
Zinc blende					Sphalerite
Zincite	ZnO	Hex	5.68	4–4½	80% Zn, orange-yellow streak
Zinc spinel					Gahnite
Zinkenite	$Pb_6Sb_{14}S_{27}$	Hex	5.3	3–3½	Jamesonite series
Zinnwaldite	Fe, Li mica	Mon	3	2.5–3	About 5% Li_2O
Zircon	$ZrSiO_4$	Tet	4.68	7½	67.2% ZrO_2
Zoisite	$Ca_2Al_3Si_3O_{12}(OH)$	Orth	3.3	6	Orth var. of clinozoisite

Source: From Berkman and Ryall, 1976, who reproduced data, with permission from M. Gary, R. McAfee, Jr., and C. L. Wolf, 1972, *Glossary of Geology* (American Geological Institute: Washington, D.C.); C. S. Hurlburt, 1961, *Dana's Manual of Mineralogy* (Wiley: New York); C. Palache, H. Berman, and C. Frondel, 1944, *Dana's System of Mineralogy,* 7th ed., Vols. 1 and 2 (Wiley: New York).

CAINOZOIC — QUATERNARY STAGES

TABLE 3. Geological Time Scale Showing the Main Stages (Marine and Glacial) for Europe, New Zealand, and North America

FORMAL AGE – SYSTEM/ SUBSYSTEM	RADIO-METRIC AGE	EUROPEAN MARINE SERIES/STAGES	EUROPEAN GLACIAL SERIES/STAGES	CENOZOIC PLANKTONIC FORAM ZONES	NEW ZEALAND MARINE STAGES	GREAT BRITAIN STAGES	NORTH AMERICAN STAGES	NORTH AMERICAN SYSTEM/ SUBSYSTEM
HOLOCENE		Flandrian	Subatlanticum / Subboreal / Atlanticum / Boreal / Preboreal (Postglacial)					HOLOCENE
PLEISTOCENE		Weichselian (Tubantian)	Würm		Oturian	Devensian	Wisconsian	PLEISTOCENE
		Eemian	R-W Interglacial			Ipswichian	Sangamonian	
		Saalian (Drenthian)	Riss			Wolstonian	Illinoian	
		Holsteinian (Needian)	M-R Interglacial		Terangian	Hoxnian	Yarmouthian	
		Elsterian	Mindel			Anglian	Kansan	
		Cromerian (Taxandrian)	G-M Interglacial			Cromerian	Aftonian	
		Menapian	Günz			Beestonian	Nebraskan	
		Waalian	D-G Interglacial		Castlecliffian	Pastonian		
		Eburonian	Danube			Baventian		
		Icenian / Tiglian	B-D Interglacial		Nukumaruan	Antian		
		Pretiglian	Biber			Thurnian	NTH. AM. MAMMAL: Irvingtonian	
		Amstelian				Ludhamian	CALIFORNIAN MARINE: Hallian	
QUATERNARY	1.5?/ 1.8	Villafranchian (Calabrian)		N22		Waltonian		

(*continued*)

CAINOZOIC — TERTIARY STAGES

FORMAL AGE – SYSTEM SERIES	RADIO-METRIC AGE	EUROPEAN MARINE STAGES	CENO ZOIC PLANK TONIC FORAM ZONES	CALIFORNIAN MARINE STAGES	S.E. AUSTRALIAN MARINE STAGES	INDO-PACIFIC STAGES	NEW ZEALAND MARINE STAGES	NORTH AMERICAN MAMMAL STAGES	NORTH AMERICAN SYSTEM/SUBSYSTEM
PLIOCENE UPPER	1.8	"Astian", Piacenzian, Zanclean, Tabianian	N.22, N.21, N.20, N.19, N.18	Wheelerian, Venturian, Repettian	Kalimnan, Cheltenhamian	?, Th	Waitotaran, Opoitian	1.7 Blancan	TERTIARY
PLIOCENE MID. / LOWER	5.5	"Messinian"		Delmontian					4 Hemphillian	UPPER (= NEOGENE)
MIOCENE UPPER	10.5	Tortonian	N.15	Mohnian	Mitchellian	Tg	Kapitean, Tongaporutuan	10 Clarendonian	
MIOCENE MID.	14	Serravallian / Langhian	N.10, N.9, N.8	Luisian 14?, Relizian 15.3	Bairnsdalian	Upper Tf, 12.5 Lower Tf (1.2), 15?	Waiauan, Lillburnian, Clifdenian, Altonian 15.2	12 Barstovian	
MIOCENE LOWER	22.5	Burdigalian / Aquitanian	N.7, N.6, N.5, N.4	Saucesian 22.5	Balcombian, Batesfordian, Longfordian 21.1	Upper Te, Tes 22.5	Awamoan 16.8, Hutchinsonian ?, Otaian	17 Hemingfordian, 21.3 Arikareean	
OLIGOCENE UPPER	30	Chattian	P.23 N.3, N.2, N.1, P.21, P.19/20	Zemorrian ?	Janjukian	Lower Te, Te 1–4	Waitakian, Duntroonian	26 Whitneyan	LOWER (= PALEOGENE)
OLIGOCENE MID. / LOWER	32 / 36	Rupelian / Lattorfian	P.18	Refugian	?	Td	Whaingaroan	Orellan, Chadronian	
EOCENE UPPER	45	Priabonian	P.17, P.16	Narizian	Aldingan	Tc	Runangan	Duchesnian	
EOCENE MID.	49	Lutetian	P.15, P.14, P.10	Ulatisian	Johannian ?	Tb	Kaiatan, Bortonian, Porangan	Uintan	
EOCENE LOWER	53.5	Ypresian	P.9, P.8, P.7, P.6	Penutian		Ta3	Heretaungan, Mangaorapan	Bridgerian, Wasatchian	
PALEOCENE UPPER	60	Thanetian	P.5, P.4, P.3	Bulitian	Wangerripian	Ta2	Waipawan	Clarkforkian, Tiffanian	
PALEOCENE LOWER	65	Montian / Danian	P.2, P.1	Ynezian, "Danian"		Ta1, ?	Teurian	Torrejonian, Dragonian, Puercan	

TERTIARY

MESOZOIC — CRETACEOUS STAGES

FORMAL AGE – SYSTEM/ SUBSYSTEM		RADIO-METRIC AGE	EUROPEAN SERIES/STAGES		AUSTRALIAN STAGES	NEW ZEALAND MARINE STAGES	NORTH AMERICAN STAGES	NORTH AMERICAN SYSTEM/ SUBSYSTEM
CRETACEOUS	UPPER	65	Maastrichtian			Haumurian		UPPER
		70	Campanian	Senonian		Piripauan		
		76	Santonian	Senonian		Teratan		
		82	Coniacian	Senonian		Mangaotanean		
		88	Turonian			Arowhanan		
		94	Cenomanian			Ngaterian		
	LOWER	100	Albian			Motuan		CRETACEOUS
		106	Aptian			Urutawan		LOWER
						Korangan		
		112	Barremian	Neocomian		Mokoiwian		
		118	Hauterivian	Neocomian				
		124	Valanginian	Neocomian				
		130	Berriasian	Neocomian				
		135						

(*continued*)

MESOZOIC — JURASSIC STAGES

FORMAL AGE - SYSTEM/ SUBSYSTEM	RADIO-METRIC AGE	EUROPEAN SERIES/STAGES		GERMANY	AUSTRALIAN STAGES	NEW ZEALAND MARINE STAGES	NORTH AMERICAN STAGES	NORTH AMERICAN SYSTEM/ SUBSYSTEM	
UPPER	-135-	Purbeckian	Tithonian / Volgian	Malm		Puaroan		**UPPER**	**JURASSIC**
	-141-	Portlandian				Ohauan			
	-146-	Kimmeridgian				Heterian			
	-151-	Oxfordian							
	-157-	Callovian		Upper Dogger		Temaikan		**MIDDLE**	
MIDDLE	-162-	Bathonian		Middle Dogger					
	-167-	Bajocian		Lower Dogger					
	-172-	Toarcian		Upper Lias		Ururoan		**LOWER**	
LOWER	-178-	Pliensbachian		Middle Lias					
	-183-	Sinemurian		Lower Lias		Araturan			
	-188-	Hettangian							
	-200-								

JURASSIC

MESOZOIC — TRIASSIC STAGES

FORMAL AGE – SYSTEM/SUBSYSTEM	RADIO-METRIC AGE	EUROPEAN STAGES	GERMANY	AUSTRALIAN STAGES	NEW ZEALAND MARINE STAGES	NORTH AMERICAN STAGES	NORTH AMERICAN SYSTEM SUBSYSTEM
TRIASSIC — UPPER (NEOTRIAS)	200	Rhaetian	Keuper		Otapirian		TRIASSIC — UPPER
		Norian			Warepan		
		Carnian			Otamitan		
					Oretian		
TRIASSIC — MIDDLE (MESOTRIAS)		Landinian	Muschelkalk		Kaihikuan		TRIASSIC — MIDDLE
		Anisian			Etalian		
TRIASSIC — LOWER (PALAEOTRIAS)	240	Scythian — U.S.S.R.: Olenekian, Induan	Bunter		Pre-Etalian	CANADA: Spathian, Smithian, Dienerian, Griesbachian	TRIASSIC — LOWER

PALAEOZOIC — PERMIAN STAGES

FORMAL AGE — SYSTEM/SUBSYSTEM	RADIO-METRIC AGE	EUROPEAN SERIES/STAGES (U.S.S.R.)	GERMANY	FRANCE & BELGIUM	AUSTRALIA	NEW ZEALAND MARINE STAGES	U.S.A. SERIES/STAGES	NORTH AMERICAN SYSTEM/SUBSYSTEM
PERMIAN — UPPER	240	Tatarian	Zechstein		F	Makarewan / Waiitian / Puruhauan	Ochoan	PERMIAN — UPPER
		Kazanian	Zechstein		E	Braxtonian / Mangapirian / Telfordian	Guadalupian	
PERMIAN — LOWER		Kungurian	Rotliegendes		D (2 / 1)		Leonardian	PERMIAN — LOWER
		Artinskian	Rotliegendes		C	?		
		Sakmarian / Sakmarian (S.S.)	Rotliegendes		B			
	290	Asselian		Autunian	A		Wolfcampian	

PALAEOZOIC — CARBONIFEROUS STAGES

NORTH AMERICAN SYSTEM/SUBSYSTEM

PENNSYLVANIAN			MISSISSIPPIAN	
UPPER	MIDDLE	LOWER	UPPER	LOWER

U.S.A.

UPPER		MIDDLE		LOWER	UPPER			LOWER	
Virgilian	Missourian	Des Moinesian	Atokan (= Derryan)	Morrowan	Springeran	Chesterian	Meramecian	Osagean	Kinderhookian

NEW ZEALAND STAGES — (blank)

AUSTRALIAN STAGES — (blank)

EUROPEAN SERIES/STAGES — WESTERN EUROPE

UPPER CARBONIFEROUS		MIDDLE CARBONIFEROUS				LOWER CARBONIFEROUS			
C | B | A		D | C | B | A				C | B | A			
Stephanian	Cantabrian	Westphalian				Namurian (G1 | R | H | E₂ | E₁)		Visean	Tournaisian

EUROPEAN SERIES/STAGES — U.S.S.R.

UPPER CARBONIFEROUS		MIDDLE CARBONIFEROUS			LOWER CARBONIFEROUS	
Gzhelian	Kasimovian	Moscovian	Bashkirian	(1974)	Serpukhovian	

RADIOMETRIC AGE: 290 … 362

FORMAL AGE - SYSTEM/SUBSYSTEM

UPPER (= Silesian)	LOWER (= Dinantian)

CARBONIFEROUS

(continued)

PALAEOZOIC — DEVONIAN STAGES

NORTH AMERICAN SYSTEM/SUBSYSTEM	SERIES/STAGES U.S.A. (Series)	SERIES/STAGES U.S.A. (Stages)	NEW ZEALAND STAGES	AUSTRALIAN STAGES	EUROPEAN SERIES/STAGES FRANCE & BELGIUM	ENGLAND & WALES	FORMAL AGE – SYSTEM/SUBSYSTEM	RADIOMETRIC AGE
DEVONIAN — UPPER	Chautauquan	Conewangoan			Famennian	Famennian	DEVONIAN — UPPER	362
		Cassadagan						
	Senecan	Chemungian			Frasnian	Frasnian		
		Fingerlakesian						
DEVONIAN — MIDDLE	Erian	Taghanican			Givetian (BELGIUM & GERMANY)	Givetian	DEVONIAN — MIDDLE	
		Tioughiogan						
		Cazenovian			Couvinian / Eifelian (CENTRAL EUROPE)	Eifelian		
DEVONIAN — LOWER	Ulsterian	Onesquethawan = Onondagan			Emsian / Zlichovian — ? (Coblenzian)	Breconian	DEVONIAN — LOWER	
		?			Siegenian / Pragian (Coblenzian)			
		Deerparkian = Oriskany				Dittonian		
		?			Gedinnian / Lochkovian — ?			
		Helderbergian						
		Keyseran						413

PALAEOZOIC—SILURIAN STAGES

NORTH AMERICAN SYSTEM/SUBSYSTEM			CENTRAL EUROPE		ESTONIA	ENGLAND & WALES		
SILURIAN					EUROPEAN SERIES STAGES			
UPPER			Budnanian	Pridolian	Ohesaare	Downtonian		Salopian
					Kaugatuma			
MIDDLE				Kopaninnian	Kuressaare	Whitecliffian	Ludlow	
					Paadla	Leintwardinian		
						Bringewoodian		
					Rootsiküla (Kaarma)	Eltonian		
			Litenian	Motolian	Jaagarahu	upper	Wenlock	
						middle		
					Jaani	lower		
LOWER				Zelkovician	Adavere	Telychian	Llandovery	Valentian
						Fronian		
					Raikküla	Idwian		
					Tamsal-Juuru	Rhuddanian		

RADIOMETRIC AGE: 413 … 445

FORMAL AGE-SYSTEM: SILURIAN

(continued)

PALAEOZOIC — ORDOVICIAN STAGES

NORTH AMERICAN SYSTEM SUBSYSTEM	NORTH AMERICAN STAGES 'SERIES' U.S.A.	N. CHINA & MANCHURIA	AUSTRALIAN STAGES	ESTONIA / SWEDEN	ENGLAND & WALES	RADIOMETRIC AGE	FORMAL AGE – SYSTEM SUBSYSTEM
ORDOVICIAN — UPPER (Cincinnatian)	Richmond		Bolindian	Porkuni Fii	Hirnantian (Ashgill)	445	ORDOVICIAN — UPPER
	Maysville			Pirgu Fic	Rawtheyan (Ashgill)		
	Eden			Vormsi Fib	Cautleyan (Ashgill)		
			Eastonian	Nabala Fia (Harjuan)	Pusgillian (Ashgill)		
				Rakvere E	Onnian (Caradoc)		
ORDOVICIAN — MIDDLE (Champlainian)	'Trenton Group'			Oandu Diii (Kurna)	Actonian (Caradoc)		
	Wilderness		Gisbornian	Keila Dii	Marshbrookian (Caradoc)		
	Porterfield			Jõhvi Di	Longvillian (Caradoc)		
	Ashby			Idavere Ciii (Purtse)	Soudleyan (Caradoc)		
	Marmor			Kukruse Cii (Viruan)	Harnagian (Caradoc)		
	'Black River Grp' / 'Chazyan'		Darriwilian	Uhaku Cb/Ck	Costonian (Caradoc)		
	Whiterock			Lasnamägi Cib	Llandeilo		
			Yapeenian	Aseri Cia (Ontikan)			
		N. CHINA & MANCHURIA	Castlemanian	Aluouja / Valaste / Hunderum (Kunda)	Llanvirn		
ORDOVICIAN — LOWER (Canadian)	Cassinian	Toufangian	Chewtonian	Langevoja / 'Limbata Lst' (Volkhov)			ORDOVICIAN — LOWER
	Jeffersonian		Bendigonian	Billingen (Latorp)	Arenig		
	Demingian		Lancefieldian	Hunneberg			
	Gasconadian		Warendian	Ceratopyge (Iruan)			
	Trempealeauan	Wolungo - Wanwanian	Datsonian	Dictyonema / Pakeroni	Tremadoc	500	

PALAEOZOIC — CAMBRIAN STAGES

FORMAL AGE – SYSTEM/ SUBSYSTEM		RADIO-METRIC AGE	ENGLAND & WALES	SWEDEN	AUSTRALIAN STAGES	N. CHINA & MANCHURIA	U.S.A.		NORTH AMERICAN SYSTEM SUBSYSTEM	
C A M B R I A N	UPPER	500	— ? —	Olenid Series	Payntonian	Fengshanian (= Yenchouan) ?	Trempealeauan	St Croixan	UPPER	C A M B R I A N
			Dolgelly				Franconian			
			— ? —			- - - - ? - - - - -				
			Ffestiniog			Daizanian				
			Maentwrog		Idamean	Paishanian	Dresbachian			
		515	?		Mindyallan	Kushanian				
	MIDDLE		Menevian	Paradoxides Series		Changhsian or Tsizuan		Acadian	MIDDLE	
						Hsuchuanian or Tangshihian or Mapanian				
			Solvan		Templetonian					
			?	?	Ordian	Mantuan				
	LOWER	540	'Caerfai'	Holmia Series	Lower Cambrian		Waucoban	LOWER	
		570								

(*continued*)

PRECAMBRIAN STAGES

AGE (m.y.)	AUSTRALIA	Geological Survey of Western Australia	CANADA	U.S.A.	U.S.S.R.	SOUTH AFRICA — Sedimentary Successions	SOUTH AFRICA — Metamorphic Complexes	INDIA
500	(CAMBRIAN)	(CAMBRIAN)	(CAMBRIAN)	(CAMBRIAN)	(CAMBRIAN)	(CAMBRIAN)		(CAMBRIAN)
	ADELAIDEAN — Marinoan; Sturtian; Torrensian; ?–?–?; Willouran	UPPER PROTEROZOIC	HADRYNIAN	PRECAMBRIAN Z — 800	RIPHEAN — Terminal Riphean 570±10; 675±25; Upper Riphean 950±50	Nama 550, 600; Malmesbury/Gariep 700; 850	MOZAMBIQUE BELT 450	Upper — PRECAMBRIAN V 600; PRECAMBRIAN IV 900
1000	—1400?	—880	ca 1000 Grenville		Middle Riphean 1350±50		Namaqualand Metl. Cplx 900; 1200	PROTEROZOIC — Middle
1500	CARPENTARIAN	MIDDLE PROTEROZOIC —1640	HELIKIAN — Neohelikian —1400?; Elsonian; Palaeohelikian; ca 1800	PRECAMBRIAN Y —1600	Lower Riphean —1700±50	Waterberg 1750	NAMAQUALAND–NATAL BELT; Richtersveld Craton 1650	Lower — PRECAMBRIAN III 1600
2000	LOWER PROTEROZOIC ("Nullaginian") —1700±20; 2200–2250	LOWER PROTEROZOIC —2400	Hudsonian; APHEBIAN ca 2560	PRECAMBRIAN X	LOWER PROTEROZOIC (Aphebian)	Transvaal 2200, 2300; Ventersdorp 2400		
2500	PROTEROZOIC / ARCHAEAN	PROTEROZOIC / ARCHAEAN	PROTEROZOIC / ARCHAEAN	PRECAMBRIAN W —2500	—2600	Witwatersrand 2600, 2800	LIMPOPO BELT 2700	ARCHAEAN — Upper — PRECAMBRIAN II 2500
3000					ARCHAEAN	Pongola 3000, 3100		Lower — PRECAMBRIAN I 3000
3500						Swaziland 3250, 3450	Ancient Gneiss Complex	3500

Source: From Berkman and Ryall, 1976.

Notes:

1. Age of Adelaidean doubtful. May be younger than 1000 m.y.
2. Marinoan, Sturtian, Torrensian, and Willouran are local subdivisions used in the Adelaide Geosyncline.
3. Carpentarian–Lower Proterozoic boundary recently determined.
4. Lower Proterozoic boundary is revised age of Hamersley Basin succession.

TABLE 4. Derived Units in the SI and c.g.s. Systems

Quantity and recommended symbol	Dimensions	SI unit	c.g.s. unit	Ratio c.g.s.:SI units
mass, m	M	kilogram (kg)	gram (g)	10^{-3}
length, l	L	metre (m)	centimetre (cm)	10^{-2}
time, t	T	second (s)	second (s)	1
area, A, S	L^2	m^2	cm^2	10^{-4}
volume, V	L^3	m^3	cm^3	10^{-6}
density, ρ	ML^{-3}	$kg\ m^{-3}$	$g\ cm^{-3}$	10^3
velocity, u, v	LT^{-1}	$m\ s^{-1}$	$cm\ s^{-1}$	10^{-2}
acceleration, G	LT^{-2}	$m\ s^{-2}$	gal	10^{-2}
momentum, p	MLT^{-1}	$kg\ m\ s^{-1}$	$g\ cm\ s^{-1}$	10^{-3}
moment of inertia, I, J	ML^2	$kg\ m^2$	$g\ cm^2$	10^{-7}
angular momentum, L	ML^2T^{-1}	$kg\ m^2\ s^{-1}$	$g\ cm^2\ s^{-1}$	10^{-7}
force, F	MLT^{-2}	newton (N)	dyne (dyn)	10^{-5}
energy of work, E, W	ML^2T^{-2}	joule (J)	erg	10^{-7}
power, P	ML^2T^{-3}	watt (W)	$erg\ s^{-1}$	10^{-7}
pressure or stress, p	$ML^{-1}T^{-2}$	pascal (Pa)	$dyn\ cm^{-2}$	10^{-1}
surface tension, γ	MT^{-2}	$N\ m^{-1}$	$dyn\ cm^{-1}$	10^{-3}
viscosity, η	$ML^{-1}T^{-1}$	$kg\ m^{-1}\ s^{-1}$	poise	10^{-1}
frequency, ν, f	T^{-1}	hertz (Hz)	s^{-1}	1

Source: From Tennent, 1971, as reproduced in Gardiner and Dackombe, 1983.

TABLE 5. Conversion Factors for U.S. and Metric Units

To convert A to B multiply by	A	B	To convert B to A multiply by
Length			
2.54×10^{-2}	inch (in)	metre (m)	39.37
0.3048	foot (ft)	metre (m)	3.2468
0.9144	yard (yd)	metre (m)	1.0936
1.8288	fathom (fm)	metre (m)	0.5468
20.1168	chain (ch)	metre (m)	4.97×10^{-2}
201.168	furlong (fl)	metre (m)	4.97×10^{-3}
1609.34	mile (mi)	metre (m)	6.214×10^{-4}
1853.2	nautical mile (UK)	metre (m)	5.3961×10^{-4}
1852.0	nautical mile (Int.)	metre (m)	5.3996×10^{-4}
1853.25	nautical mile (US)	metre (m)	5.3959×10^{-4}
100.0	cable	fathom (fm)	10^{-2}
1.6094	mile (mi)	kilometre (km)	0.6214
8.0	mile (mi)	furlong (fl)	0.125
1760.0	mile (mi)	yards (yd)	5.6818×10^{-4}
5280.0	mile (mi)	foot (ft)	1.8939×10^{-4}
Area			
6.4516×10^{-4}	sq. inch (in^2)	sq. metre (m^2)	1.55×10^3
9.2903×10^{-2}	sq. foot (ft^2)	sq. metre (m^2)	10.764
0.8361	sq. yard (yd^2)	sq. metre (m^2)	1.196
2 589 988.0	sq. mile (mi^2)	sq. metre (m^2)	3.861×10^{-7}
4046.856	acre	sq. metre (m^2)	2.4711×10^{-4}
2.590	sq. mile (mi^2)	sq. kilometre (km^2)	0.3861
0.4047	acre	hectare (ha)	2.471
640.0	sq. mile (mi^2)	acre	1.5625×10^{-3}
10 000	hectare (ha)	sq. metre (m^2)	10^{-4}
3.861×10^{-3}	hectare (ha)	sq. mile (mi^2)	259.0045
247.105	sq. kilometre (km^2)	acre	4.047×10^{-3}

(continued)

TABLE 5. (*continued*)

To convert A to B multiply by	A	B	To convert B to A multiply by
Volume			
1.6387×10^{-5}	cubic inch (in^3)	cubic metre (m^3)	6.1024×10^4
16.387	cubic inch (in^3)	cubic centimetre (cm^3)	6.1024×10^{-2}
2.8317×10^{-2}	cubic foot (ft^3)	cubic metre (m^3)	35.314
28316.8	cubic foot (ft^3)	cubic centimetre (cm^3)	3.5315×10^{-5}
28.3168	cubic foot (ft^3)	litre (l)	3.5315×10^{-2}
0.7646	cubic yard (yd^3)	cubic metre (m^3)	1.3079
3.785×10^{-3}	gallon (US)	cubic metre (m^3)	264.2
4.546×10^{-3}	gallon (UK)	cubic metre (m^3)	219.97
0.21998	litre (l)	gallon (UK)	4.546
0.26418	litre (l)	gallon (US)	3.7853
8.0	bushel	gallon (UK)	0.125
9.608	bushel	gallon (US)	0.104
231.0	gallon (UK)	cubic inch (in^3)	4.329×10^{-3}
1233.482	acre foot	cubic metre (m^3)	8.1071×10^{-4}
0.8326	gallon (US)	gallon (UK)	1.2011
2.8413×10^{-2}	fluid ounce (fl. oz)	litre (l)	35.195
Mass			
2.835×10^{-2}	ounce (oz)	kilogram (kg)	35.273
0.4536	pound (lb)	kilogram (kg)	2.2046
6.3503	stone (st)	kilogram (kg)	0.1575
50.8023	hundredweight (cwt)	kilogram (kg)	1.9684×10^{-2}
1016.04	ton	kilogram (kg)	9.8421×10^{-4}
907.20	short ton	kilogram (kg)	1.1023×10^{-3}
1.016	ton	metric tonne (t)	0.9842
2204.6	metric tonne (t)	pounds (lb)	4.536×10^{-4}
14.5939	slug	kilogram (kg)	6.852×10^{-2}
Weight			
4.448	pound force (lbf)	newton (N)	0.2248
0.1383	poundal (pdl)	newton (N)	7.233
10^5	newton (N)	dyne (dyn)	10^{-5}
32.17	pound-force (lbf)	poundal (pdl)	0.031 08
980.7	gram-force (gf)	dyne (dyn)	1.0197×10^{-3}
Pressure or stress			
15.44×10^6	ton-force per sq. in	pascal (Pa)	6.4767×10^{-8}
157.47	ton-force per sq. in	kilogram-force per sq. cm	6.3504×10^{-3}
107.3×10^3	ton-force per sq. ft	pascal (Pa)	9.3197×10^{-6}
1.0936×10^4	ton-force per sq. ft	kilogram-force per sq. m	9.1441×10^{-5}
6.895×10^3	pound-force per sq. in	pascal (Pa)	1.4503×10^{-4}
7.03×10^{-2}	pound-force per sq. in	kilogram-force per sq. cm	14.225
47.9	pound-force per sq. ft	pascal (Pa)	2.0877×10^{-2}
4.882	pound-force per sq. ft	kilogram-force per sq. m	0.204 83
101.325×10^3	standard atmosphere	pascal (Pa)	9.869×10^{-6}
1.033	standard atmosphere	kilogram-force per sq. m	9.6805
14.697	standard atmosphere	pound-force per sq. in	6.804×10^{-2}
760	standard atmosphere	millimetres of mercury (mm Hg)	1.315×10^{-3}
33.901	standard atmosphere	feet of water (ft H$_2$O)	2.9498×10^{-2}
3.05×10^{-2}	foot of water (ft H$_2$O)	kilogram-force per sq. cm	3.2787×10^{-3}
2.989×10^3	foot of water	pascal (Pa)	3.3456×10^{-4}
10^5	bar	pascal (Pa)	10^{-5}
133.322	millimetre of mercury	pascal (Pa)	7.5×10^{-3}
$9.806\ 65 \times 10^4$	kilogram-force per sq. cm	pascal (Pa)	1.0197×10^{-5}
Energy or work			
1.3558	foot pound-force	joule (J)	0.7376
1.3558×10^7	foot pound-force	erg	7.3757×10^{-8}
0.1383	foot pound-force	metre kilogram force	7.2307
4.2140×10^2	foot poundal	joule (J)	2.373×10^{-3}

TABLE 5. (*continued*)

To convert A to B multiply by	A	B	To convert B to A multiply by
4.2140×10^9	foot poundal	erg	2.373×10^{-10}
1.055×10^3	BTU	joule (J)	9.4787×10^{-4}
10^7	joule	erg	10^{-7}
4.1855	calorie at 15 °C	joule (J)	0.2389

Power

550	horse-power	foot pound-force per second	1.8182×10^{-3}
7.457×10^9	horse-power	erg per second	1.341×10^{-10}
7.457×10^2	horse-power	watt (W)	1.341×10^{-3}
1.3405	kilowatt (kW)	horse-power	0.74599
1.3558	foot pound-force per second	watt (W)	0.7376

Density

16.019	pound per cubic foot	kilogram per cubic metre	6.243×10^{-2}
1.6019×10^{-2}	pound per cubic foot	gram per cubic centimetre	6.243×10^{-3}
1.0012	ounces per cubic foot	gram per litre	0.9988

Unit weight

16.019	pound-force per cubic foot	kilogram-force per cubic metre	6.243×10^{-2}
1.571×10^2	pound-force per cubic foot	newton per cubic metre	6.3654×10^{-3}
27.68	pound-force per cubic inch	gram-force per cubic centimetre	3.613×10^{-2}
271.4×10^3	pound-force per cubic inch	newton per cubic metre	3.6846×10^{-6}

Compressibility

1.45×10^{-4}	sq. in per pound-force	sq. m per newton	6.897×10^3
14.22	sq. in per pound-force	sq. cm per kilogram-force	7.032×10^{-2}
9.324×10^{-6}	sq. foot per ton-force	sq. m per newton	1.0725×10^5
0.914	sq. foot per ton-force	sq. cm per kilogram-force	1.0941

Speed

2.54	inch per second	centimetre per second	0.3937
30.48	foot per second	centimetre per second	3.281×10^2
0.447	mile per hour	metre per second	2.2371
0.5144	knot (Int.)	metre per second	1.944
1.0973	foot per second	kilometre per hour	0.9113
0.618 18	foot per second	mile per hour	1.61765
0.5925	foot per second	knot (Int.)	1.6878
3.6	metre per second	kilometre per hour	0.2778
0.9659×10^{-8}	foot per year	metre per second	1.0353×10^8

Rate of flow and discharge

2832	cubic foot per second	cubic centimetre per second	3.53×10^{-4}
2.832×10^{-2}	cubic foot per second	cubic metre per second	35.311
76464	cubic yard per second	cubic centimetre per second	1.3078×10^{-5}
0.7646	cubic yard per second	cubic metre per second	1.3078
101.941	cubic foot per second	cubic metre per hour	9.8096×10^{-3}
2446.57	cubic foot per second	cubic metre per day	4.0874×10^{-4}
28.3161	cubic foot per second	litres per second	3.5316×10^{-2}
11.573 75	cubic metre per day	litres per second	8.6402×10^{-2}
0.408 735	cubic metre per day	cubic foot per second	2.4466
4.3813×10^{-2}	million gallons (US) per day	cubic metre per second	22.824
5.261×10^{-2}	million gallons (UK) per day	cubic metre per second	19.008

(*continued*)

TABLE 5. (*continued*)

To convert A to B multiply by	A	B	To convert B to A multiply by
0.2713	acre-feet per day	million gallons (UK) per day	3.686
0.3259	acre-feet per day	million gallons (US) per day	3.0684
Yield			
0.699 725	cubic feet per acre	cubic metres per hectare	1.4291
1.120 85	pounds per acre	kilograms per hectare	0.892 18
Coefficient of consolidation			
0.1075	sq. in per minute	sq. cm per second	9.3023
1.075×10^{-5}	sq. in per minute	sq. m per second	9.3023×10^2
2.94×10^{-5}	sq. ft per year	sq. cm per second	3.4014×10^4
2.94×10^{-9}	sq. ft per year	sq. m per second	3.4014×10^8
Concentration			
1.0012×10^{-3}	ounce per cubic foot	gram per cubic centimetre	9.988×10^2
1.0012	ounce per cubic foot	kilogram per cubic metre	0.9988
16.019×10^{-3}	pound per cubic foot	gram per cubic centimetre	62.426
16.019	pound per cubic foot	kilogram per cubic metre	6.2426
Dynamic viscosity			
47.8803	pound-seconds per square foot	newton-seconds per sq. metre	2.0885×10^{-2}
47.8803	slugs per foot-second	newton-seconds per sq. metre	2.0885×10^{-2}
10^{-3}	centipoise (cP)	newton-seconds per sq. metre	10^3
Kinematic viscosity			
$9.290\ 30 \times 10^4$	sq. foot per second	centistoke (cSt)	1.0764×10^{-5}
$9.290\ 30 \times 10^{-2}$	sq. foot per second	sq. metre per second	10.764
10^{-6}	centistoke (cSt)	sq. metre per second	10^6

Source: From Gardiner and Dackombe, 1983.

TABLE 6. Comparison of Particle Sizes for Sieve Analysis Using U.S., Tyler, Canadian, British, French, and German Standard Sieve Series

U.S.A.[1]		TYLER[2]	CANADIAN[3]		BRITISH[4]		FRENCH[5]		GERMAN[6]
*Standard	Alternate	Mesh Designation	Standard	Alternate	Nominal aperture	Nominal Mesh No.	Opening (mm)	No.	Opening
31.5 mm	1¼″		31.5 mm	1¼″					
26.5 mm	1.06″	1.05″	26.5 mm	1.06″					25.0 mm
25.0 mm	1″		25.0 mm	1″					20.0 mm
22.4 mm	⅞″	.883″	22.4 mm	⅞″					
19.0 mm	¾″	.742″	19.0 mm	¾″					
16.0 mm	⅝″	.624″	16.0 mm	⅝″					18.0 mm
13.2 mm	.530″	.525″	13.2 mm	.530″					16.0 mm
12.5 mm	½″		12.5 mm	½″					12.5 mm
11.2 mm	⁷⁄₁₆″	.441″	11.2 mm	⁷⁄₁₆″					
9.5 mm	⅜″	.371″	9.5 mm	⅜″					10.0 mm
8.0 mm	⁵⁄₁₆″	2½	8.0 mm	⁵⁄₁₆″					8.0 mm
6.7 mm	.265″	3	6.7 mm	.265″					
6.3 mm	¼″		6.3 mm	¼″					6.3 mm
5.6 mm	No. 3½	3½	5.6 mm	No. 3½			5.000	38	5.0 mm
4.75 mm	4	4	4.75 mm	4					
4.00 mm	5	5	4.00 mm	5	4.00 mm	4	4.000	37	4.0 mm
3.35 mm	6	6	3.35 mm	6	3.35 mm	5			
2.80 mm	7	7	2.80 mm	7	2.80 mm	6	3.150	36	3.15 mm
2.36 mm	8	8	2.36 mm	8	2.40 mm	7	2.500	35	2.5 mm
2.00 mm	10	9	2.00 mm	10	2.00 mm	8	2.000	34	2.0 mm
1.70 mm	12	10	1.70 mm	12	1.68 mm	10	1.600	33	1.6 mm
1.40 mm	14	12	1.40 mm	14	1.40 mm	12			
1.18 mm	16	14	1.18 mm	16	1.20 mm	14	1.250	32	1.25 mm
1.00 mm	18	16	1.00 mm	18	1.00 mm	16	1.000	31	1.0 mm
850 µm	20	20	850 µm	20	850 µm	18			

(continued)

TABLE 6. (continued)

U.S.A.[1]		TYLER[2]	CANADIAN[3]		BRITISH[4]		FRENCH[5]		GERMAN[6]
*Standard	Alternate	Mesh Designation	Standard	Alternate	Nominal aperture	Nominal Mesh No.	Opening (mm)	No.	Opening
710 μm	25	24	710 μm	25	710 μm	22	0.800	30	800 μm
600 μm 500 μm	30 35	28 32	600 μm 500 μm	30 35	600 μm 500 μm	25 30	0.630 0.500	29 28	630 μm 500 μm
425 μm	40	35	425 μm	40	420 μm	36	0.400	27	400 μm
355 μm	45	42	355 μm	45	355 μm	44	0.315	26	315 μm
300 μm	50	48	300 μm	50	300 μm	52			
250 μm 212 μm	60 70	60 65	250 μm 212 μm	60 70	250 μm 210 μm	60 72	0.250 0.200	25 24	250 μm 200 μm
180 μm	80	80	180 μm	80	180 μm	85	0.160	23	160 μm
150 μm 125 μm 106 μm	100 120 140	100 115 150	150 μm 125 μm 106 μm	100 120 140	150 μm 125 μm 105 μm	100 120 150	0.125	22	125 μm
90 μm	170	170	90 μm	170	90 μm	170	0.100	21	100 μm 90 μm
75 μm	200	200	75 μm	200	75 μm	200	0.080	20	80 μm
63 μm	230	250	63 μm	230	63 μm	240	0.063	19	71 μm 63 μm 56 μm
53 μm	270	270	53 μm	270	53 μm	300	0.050	18	50 μm
45 μm	325	325	45 μm	325	45 μm	350	0.040	17	45 μm
38 μm	400	400	38 μm	400					40 μm

Source: From Berkman and Ryall, 1976.

1. U.S.A. Sieve Series - ASTM Specification E-11-70.
2. Tyler Standard Screen Scale Sieve Series.
3. Canadian Standard Sieve Series 8-GP-1d.
4. British Standards Institution, London BS-410-62.
5. French Standard Specifications, AFNOR X-11-501.
6. German Standard Specification DIN 4188.

*These sieves correspond to those recommended by ISO (International Standards Organization) as an International Standard and this designation should be used when reporting sieve analysis intended for international publication.

TABLE 7. Grain-size Scales used by American Geologists, Engineers, and Soil Scientists

GRAIN-SIZE SCALE USED BY AMERICAN GEOLOGISTS
Modified Wentworth Scale — after Lane, et al., 1947, Trans. American Geophysical Union, v. 28, p. 936-938

phi	mm	mm	inches	U.S. Standard Sieve Series	GRADE NAME		
−12	4096		161.3				
					very large		
−11	2048		80.6				
					large		
−10	1024		40.3			Boulders	
					medium		
−9	512		20.2				
					small		
−8	256		10.1				
					large		
−7	128		5.0			Cobbles	
					small		GRAVEL
−6	64		2.52	63 mm	very coarse		
−5	32		1.26	31.5 mm			
					coarse		
−4	16		0.63	16 mm	medium	Pebbles	
−3	8		0.32	8 mm			
					fine		
−2	4		0.16	No. 5	very fine		
−1	2		0.08	No. 10			
					very coarse		
0	1		0.04	No. 18	coarse		
+1	1/2	0.500		No. 35			
					medium	Sand	SAND
+2	1/4	0.250		No. 60	fine		
+3	1/8	0.125		No. 120			
					very fine		
+4	1/16	0.062		No. 230	coarse		
+5	1/32	0.031					
					medium	Silt	
+6	1/64	0.016			fine		
+7	1/128	0.008					
					very fine		
+8	1/256	0.004			coarse		MUD
+9	1/512	0.002					
					medium	Clay size	
+10	1/1024	0.001			fine		
+11	1/2048	0.0005					
					very fine		
+12	1/4096	0.00025					

AGI-DS-rvd-82

GRAIN-SIZE SCALE USED BY ENGINEERS
(A.S.T.M. Standards D422–63; D643–78)

mm	inches	U.S. Standard Sieve Series	GRADE NAME	
				Boulders
305	12.0			
				Cobbles
76.2	3.0	3.0 in.		
				Gravel
4.75	0.19	No. 4		
			coarse	
2.00	0.08	No. 10		
			medium	Sand
0.425		No. 40		
			fine	
0.074		No. 200		
				Silt
0.005				
				Clay Size

GRAIN-SIZE SCALE USED BY SOILS SCIENTISTS
U.S. Dept. of Agriculture, Agriculture Handbook No. 436 (1975)

mm	inches	U.S. Standard Sieve Series	GRADE NAME	
76.2	3.0	75 mm		
				Gravel
2.0	0.08	No. 10		
			very coarse	
1.0	0.04	No. 18		
			coarse	
0.500		No. 35		
			medium	Sand
0.250		No. 60		
			fine	
0.100		No. 140		
			very fine	
0.050		No. 270		
				Silt
0.002				
				Clay

Source: From Dietrich et al., 1982.

TABLE 8. Worldwide Directory of National Geological Surveys

Afghanistan
  Department of Mines and Geology
  Ministry of Mines and Industries
  Darulaman
  Kabul

Albania
  Ministry of Industry and Mining
  Tirane

Algeria
  Sous-Direction de la Géologie
  Ministère de L'Industrie Lourde
  Rue Ahmed Bey de Constantine
  Algiers

Angola
  Direcção de Servicos de Geologia e Minas
  Caixa Postal 1260-C
  Luanda

Argentina
  Servicio Geologico Nacional
  Secretaria de Estado de Mineria
  Avenida Santa Fe 1548
  Buenos Aires

Australia
  Bureau of Mineral Resources, Geology, and
    Geophysics
  Post Office Box 378
  Canberra City, A.C.T.

Austria
  Geologische Bundesanstalt
  Rasumofskygasse 23
  A-1031 Vienna

Bangladesh
  Geological Survey
  Pioneer Road
  Segun Bagicha
  Dacca

Belgium
  Service Géologique de Belgique
  13 rue Jenner
  1040 Brussels

Benin (formerly Dahomey)
  Direction des Mines, de la Geologie, et des
    Hydrocarbures
  Ministere de l' Equipment
  B.P. 249
  Cotonou

Boliva
  Servicio Geológico de Bolivia (GEOBOL)
  Casilla de Correo 2729
  La Paz

Botswana
  Geological Survey Department
  Ministry of Mineral Resources and Water Affairs
  Private Bag 14
  Lobatsi

Brazil
  Instituto de Pesquisas Especiais (INPE)
  Caixa Postal 515
  São José dos Campos
  São Paulo, SP

Bulgaria
  Geologischeski Institut
  Bulgarska Akademiya na Naukite
  Akademik Bonchev St., Blok 2
  Sofia

Burma
  Department of Geological Survey and Exploration
  Ministry of Mines
  Kanbe Road, Yankin P.O.
  Rangoon

Burundi
  Ministère de la Géologie des Mines
  Boite Postale 1160
  Bujumbura

Cameron
  Direction des Mines et de le Géologie
  Ministry of Mines and Energy
  B.P. 70
  Yaoundé

  Office National de la Recherche Scientifique et
    Technique
    (ONAREST)
  B.P. 1457
  Yaoundé

Canada
  Geological Survey of Canada
  Department of Energy, Mines, and Resources
  601 Booth Street
  Ottawa, Ontario K1A 0GI

Chad
  Direction des Mines et de la Géologie
  B.P. 816
  Fort-Lamy

Chile
  Dirección de Geologia y Minas
  Corporacion Nacional del Cobre de Chile
  Huerfanos 1189, 8 Piso

China
  Academia Sinica
  5th Bureau
  Earth and Environmental Science
  Institute of Geology
  Lanzhou

Columbia
  Instituto Nacional de Investigaciones Geológico-
    Mineras
    (INGEOMINAS)
  Carrera 30, No. 51-59
  Apartado Aéreo 4865
  Bogota

Congo
  Service des Mines et de la Géologie
  B.P. 12
  Brazzaville

TABLE 8. (*continued*)

Costa Rica
Departmento de Geologia
Instituto Costarricense de Electricidad (ICE)
Apartado 10032
San José

Cuba
Instituto de Geologia
Academia de Ciencias de Cuba
Ave. Van-Troi, No. 17203
La Habana

Cyprus
Geological Survey Department
Ministry of Agriculture and Natural Resources
Nicosia

Czechoslovakia
Ústave Geologický
Ceskoslovenská Akadémie Ved
Lysolaje 6
Praha 6

Denmark
Geological Survey of Denmark
Thoravej 31
DK 2400 Kobenhavn N.V.

Ecuador
Dirección General de Geologiá y Minas
Ministerio de Recursos, Naturales y Energéticos
Carrion #1016 y Paez
Quito

Egypt
Geological Survey and Mining Authority
Ministry of Industry
3, Salah Salem Street
Abbasia, Post Office Building
Cairo

El Salvador
Centro de Investigaciones Geotécnicas
Ministerio de Obras Públicas
Avenida Peralta, final, costado Oriente Talleres
El Coro
San Salvador

Ethiopia
Ethiopian Institute of Geological Survey (EIGS)
Ministry of Mines, Energy, and Water Resources
P.O. Box 486
Addis Ababa

Fiji
Ministry of Lands and Mineral Resources
Mineral Resources Department
Private Mail Bag, G.P.O.
Suva

Finland
Geologinen Tutkimuslaitos
Kivimiehentie 1 02150 Espoo 15

France
Bureau des Recherches Géologiques et Minières
(BRGM)
Avenue de Concyr, B.P. 6009
45018 Orleans Cedex

French Guiana
Bureau des Recherches Géologiques et Minières
B.P. 42
Cayenne

German Democratic Republic
Zentrales Geologisches Institut
Invalidenstrasse 44
104 Berlin

Germany, Federal Republic of
Bundesanstalt für Geowissenschaften und Rohstoffe
(BGR)
(Geobund)
Alfred-Bentz-Haus
Postfach 510153
Stillweg 2
3000 Hannover 51

Ghana
Geological Survey of Ghana
P.O. Box M 32
Accra

Greece
Institute of Geological and Mining Research
70 Messoghion Street
Athens 608

Greenland
Grønlands Geologiske Undersøgelser
Ostervoldgade 5-7, Tr. KL
DK- 1350 Copenhagen, K
Denmark

Guadeloupe
Office de la Recherche Scientifique et Technique
Outre-Mer (OSTOM)
B.P. 504
97165 Pointe-A-Pitre
Cedex F 1960

Guatemala
División de Geologia
Instituto Goegráfico Nacional
Avenida las Américas 5:76, Zona 13
Guatemala City

Guinea
Direction Générale des Mines et Géologie
Conarky
Guinea-Bissau (formerly Portuguese Guinea)
Direccao Geral, Geologia e Minas
Comissariado Dos Recursos Naturais
Bissau

Guyana
Geological Surveys and Mines Department
P.O. Box 1028, Brickdam
Georgetown

Haiti
Department of Mines and Energy Resources
Village Willy Lamothe
Delmas 9
P.O. Box 2174
Port-au-Prince

(*continued*)

TABLE 8. (*continued*)

Honduras
 Exploracion Geotérmica
 Empressa Nacional de Energia Electrica (ENEE)
 Apartado 99
 Tegucigalpa

Hong Kong
 Mines Department, Branch Office
 Canton Road Government Offices
 Canton Road
 Kowloon

Hungary
 Központi Földtani Hivatal
 H-1251 Budapest 1
 Iskola u. 13

Iceland
 Division of Geology and Geography
 Museum of Natural History
 Laugavegi 105 and Hverfisgata 116
 P.O Box 5320
 105 Reykjavik

India
 Geological Survey of India
 27 Jawaharal Nehru Road
 Calcutta 700016

Indonesia
 Direktorat Sumber Daya Mineral
 Jalan Diponegoro 57
 Bandung

Iran
 Geological and Mineral Survey of Iran
 Ministry of Industry and Mines
 P.O. Box 1964
 Tehran

Iraq
 Directorate General of Geological Survey and
  Mineral Investigation
 State Organization for Minerals
 P.O. Box 2330
 Baghdad

Ireland
 Geological Survey of Ireland
 14 Hume Street
 Dublin 2

Israel
 Geological Survey of Israel
 30 Malchei Israel Street
 Jerusalem

Italy
 Servizio Geologico d'Italia
 Salita S. Nicoló da Tolentino 1B
 00187 Roma

Ivory Coast
 Direction des Mines et de la Géologie
 Ministère des Mines
 BP. V 28
 Adidjan

Jamaica
 Geological Survey Division
 Ministry of Mining and Natural Resources
 Hope Gardens
 Kingston 6

Japan
 Geological Survey of Japan
 Ministry of International Trade and Industry (MITI)
 3, Higashi, Yatabe-cho 1-chome
 Tsukuba-gun, Ibaraki-Ken 300-21

Jordan
 Department of Geological Resources and Mining (NRA)
 P.O. Box 39
 Amman

Kenya
 Geological Survey of Kenya
 Mines and Geological Department
 Ministry of Natural Resources
 P.O. Box 30009
 Nairobi

Korea (North)
 Geology and Geography Research Institute
 Academy of Sciences
 Mammoon-dong
 Central District
 P'yongyang

Korea (South)
 Korea Research Institute of Geoscience and Mineral
  Resources (KIGAM)
 219-5 Garibong-dong
 Youngdeungpo-gu
 Seoul 150-06

Kuwait
 Kuwait Institute for Scientific Research
 P.O. Box 12009 Shamiah
 Kuwait

Laos
 Departement de Géologie et des Mines
 Ministère de l'Industrie et du Commerce
 Vientiane

Lebanon
 Directorate General of Public Works
 Ministry of Public Works
 Beirut

Lesotho
 Department of Mines and Geology
 P.O. Box 750
 Maseru

Liberia
 Liberian Geological Survey
 Ministry of Lands and Mines
 P.O. Box 9024
 Monrovia

Libya
 Geological Research and Mining Department
 Industrial Research Center
 P.O Box 3633
 Tripoli

TABLE 8. (*continued*)

Liechtenstein
Landesbauamt des Fürstentums Liechtenstein
Städtle 49
9490 Vaduz

Luxembourg
Service Géologique
Ponts et Chaussées
4 bd Roosevelt
Luxembourg

Madagascar
Service Geologique
Ministère des Mines et de l'Economie et du
Commerce
B.P. 322, Ampandrianomby
Antananarivo

Malawi
Geological Survey Department
Ministry of Natural Resources
P.O. Box 27, Liwonde Road
Zomba

Malaysia
Geological Survey Department
Bangunan Ukor
Jalan Gurney
Kuala Lumpur

Mali
Direction Nationale des Mines et de la Géologie
B.P. 223
Loulouba, Bamako

Malta
Public Works Department
Beltissebh

Martinique
Arrondissement Mineralogique de la Guyane
B.P. 458
Fort-au-France

Mauritania
Direction des Mines et de la Géologie
Ministère de l'Industrialisation et des Mines
B.P. 199
Nouakchott

Mauritius
Ministry of Agriculture and Natural Resources and
the Environment
Port Louis

Mexico
Instituto de Geologia
Universidad Nacional Autónoma de México (UNAM)
Cuidad Universitaria
México 20, D.F.

Mongolia
Institute of Geology
Academy of Sciences
UI, Leniadom 2
Ulaanbaatar

Morocco
Division de la Géologie
Ministère de l'Energie et des Mines
Quartier Administratif
Rabat

Mozambique
Direcção Nacional de Geologia é Minas e Defensa
Praca 25 de Junho/Maputo
P.O. Box 217
Maputo

Namibia (Southwest Africa)
Geological Survey
P.O. Box 2168
Windhoek

Nepal
Department of Mines and Geology
Ministry of Industry and Commerce
Lainchaur
Kathmandu

Netherlands
Rijks Geologische Dienst
Spaarne 17, P.O. Box 157
2000 AD Haarlem

New Caledonia
Service des Mines et de la Geologie
Rte. No. 1
B.P. 465
Noumea

New Hebrides
Geological Survey
British Residency
Port Villa

New Zealand
New Zealand Geological Survey
Department of Scientific and Industrial Research
(DSIR)
P.O. Box 30-368
Lower Hutt

Nicaragua
Corporación Nicaraguense de Mines e Hidrocarburos
(CONDEMIA)
Apartado Postal No. 8
Manague, D. N.

Niger
Direction des Mines et de la Géologie
Ministère des Mines et de l'Hydraulique
B.P. 257
Niamey

Nigeria
Geological Survey Department of Nigeria
Ministry of Mines and Power
P.M.B. 2007
Kaduna South, Kaduna State

Norway
Norges Geologiske Undersøkelse
P.B. 3006 Ostmarkneset
Leiv Erikssons Vei 39
7001 Trondheim

Oman
Ministry of Petroleum and Minerals
P.O. Box 551
Muscat

(*continued*)

TABLE 8. (*continued*)

Pakistan
Geological Survey of Pakistan
P.O. Box 15
Quetta

Panama
Corporación de Desarrollo Minero-Cerro Colorado
(CODEMIN)
Apartado 5312
Panama 5

Papua New Guinea
Geological Survey
Department of Minerals and Energy
P.O. Box 2352
Konedobu

Paraguay
Direccion de Recursos Minerales
Ministerio de Obras Publicas y Communicaciones
Calle Alberdi y Oliva
Asuncion

Peru
Instituto Geologico Minero y Metalúrgico
(INGEMMET)
Pablo Bermudez 211
Apartado 211
Lima

Philippines
Bureau of Mines and Geosciences
P.O. Box 1595
Pedro Gill Street
Manila 2801

Poland
Instytut Nauk Geologicznych PAN
Al. Zwirki i Wigury 93
02-089 Warszawa

Portugal
Direcção General de Mines e Serviços Geologicos
Ministèrio de Industria
Rua Antonio Enes, 5
Lisboa 1000

Qatar
Industrial Development Technical Center (IDTC)
P.O. Box 2599
Doha

Reunion
Services des Travaux Publiques
St. Denis

Romania
Institutul de Cerecetari Geologice şi Geofizice
Str. Caransebeş No. 1
Sector 7, Bucharest

Rwanda
Ministère des Ressources Naturelles, des Mines et des
Carrieres
B.P. 413
Kigali

Saudi Arabia
Ministry of Petroleum and Mineral Resources
Directorate General of Mineral Resources
P.O. Box 345
Jiddah

Senegal
Direction des Mines et de la Géologie
Ministère du Developpement Industrial
Route de Ouakam
B.P. 1238
Dakar

Sierre Leone
Geological Survey Division
Ministry of Lands, Mines, and Labor
New England, Freetown

Singapore
Public Works Department
Structural Design and Investigation Branch
Ministry of National Development
National Development Building
Maxwell Road
Singapore 0106

Solomon Islands
Ministry of Natural Resources
Geology Department
P.O. Box G24
Honiara

Somalia
Geological Survey Department
Ministry of Minerals and Water Resources
P.O. Box 744
Mogadishu

South Africa
Geological Survey of South Africa
Department of Mines
233 Visagie Street (Private Bag X112)
Pretoria 0001

Spain
Servicio de Geologia
Ministerio de Obras Públicas y Urbanismo
Avenida de Portugal, 81
Madrid-11

Sri Lanka
Geological Survey Department
48 Sri Jinaratanana Road
Colombo 2

Sudan
Geological and Mineral Resources
Ministry of Energy and Mining
P.O. Box 410
Khartoum

Suriname
Geologisch Mijnbouwkundige Dienst
Klein Wasserstraat 1 (2-6)
Paramaribo

Swaziland
Geological Survey and Mines Department
P.O. Box 9
Mbabane

Sweden
Sveriges Geologiska Undersökning (SGU)
Box 670
S-751 28 Upsala

TABLE 8. (*continued*)

Switzerland
  Geologische Kommission der Schweizerischen
    Naturforschende Gesellschaft
  Bernoullianum
  4056 Basel

Syria
  Directorate of Geological Research and Mineral
    Resources
  Ministry of Petroleum
  Fardos Street
  Damascus

Taiwan
  Institute of Geology
  National Taiwan University
  1 Roosevelt Road, Section 4
  Taipei

Tanzania
  Geology
  Ministry of Water, Energy, and Minerals
  P.O Box 903
  Dodoma

Thailand
  Department of Mineral Resources
  Ministry of Industry
  Rama VI Road
  Bangkok 4

Togo
  Direction des Mines et de la Géologie
  Ministère des Mines, et Resources Hydrauliques
  B.P. 356
  Lome

Tonga
  Ministry of Lands, Survey and Natural Resources
  P.O. Box 5
  Nukúalofa

Trinidad and Tobago
  Ministry of Energy and Energy-Based Industries
  P.O. Box 96
  Port-of-Spain

Tunisia
  Service Géologique de Tunisie
  95 Avenue Mohamed V
  Tunis

Turkey
  Mineral Research and Exploration Institute of Turkey
  Eskisehir Yolu
  Ankara

Uganda
  Geological Survey and Mines Department
  P.O. Box 9
  Entebbe

United Kingdom
  Institute of Geological Sciences
  Exhibition Road
  London SW7 2DE

United States
  U.S. Geological Survey
  National Center
  12201 Sunrise Valley Drive
  Reston, Virginia 22092

Upper Volta
  Direction de la Geologie et des Mines
  B.P. 601
  Ouagadougou

Uruguay
  Instituto Geologico del Uruguay
  Calle J. Herrera y Obes 1239
  Montevideo

U.S.S.R.
  All-Union Scientific Research Geological Institute
    (VSEGEI)
  Sredniy Prospekt 72B
  199026 Leningrad

  Institute of Geology
  Akademiya Nauk USSR
  109017 Moscow ZH-17

Venezuela
  Dirección de Geologia
  Dirección General Sectorial de Mines y Geologia
  Ministerio de Energia y Minas
  Torre Norte, Piso 19
  Centro Simon Bolivar
  Caracas

Vietnam
  Geologic Section
  State Committee of Sciences
  Hanoi

Western Sahara (formerly Spanish Sahara)
  Dirección General de Plazas y Provincias Africanas
  Servicio Minero y Geológico
  Castellana No. 5
  Madrid 1, Spain

Yemen (Aden)
  Overseas Geological Surveys of London, England
  c/o Aden Public Works Department
  Aden

Yemen (San'A)
  Geological Authority
  Yemen Oil and Mineral Resources Corporation
    (YOMICO)
  P.O. Box 81
  Saña

Yugoslavia
  Zavod za Geoloskli i Geofizicka Istrazivanja
  Karadjordjeva 48
  Belgrade

Zaire
  Service Geologique du Zaire
  Ministry of Mines
  B.P. 898
  44 Avenue des Huileries
  Kinshasa

Zambia
  Geological Survey Department
  Ministry of Mines
  P.O. RW 135
  Ridgeway, Lusaka

Zimbabwe
  Geological Survey of Rhodesia
  Ministry of Mines and Lands
  P.O. Box 809, Causeway
  Salisbury

Source: After Bergquist et. al (1981).

TABLE 9. State and Provincial Geological Maps

STATE GEOLOGICAL MAPS

State	Title	Date	Scale	Sheets
Alabama	Geological Map of Alabama	1926	500,000	1
Alaska	1. Preliminary Geological Map of Alaska	1978	2,500,000	2
	2. Surficial Geology of Alaska	1964	1,584,000	2
Arizona	Geologic Map of Arizona	1969	500,000	1
Arkansas	Geologic Map of Arkansas	1976	500,000	1
California	Geologic Map of California	1977	750,000	1
Colorado	Geologic Map of Colorado	1980	500,000	1
Connecticut	1. Preliminary Geologic Map of Connecticut	1956	253,440	1
	2. Glacial Geologic Map of Connecticut	1929	125,000	1
Delaware	Generalized Geologic Map of Delaware	1976	576,000	1
Florida	Geologic Map of Florida	1964	1,900,800	1
Georgia	Geologic Map of Georgia	1978	500,000	1
Hawaii	Geology and Groundwater Resources of the Island of _____ (each island)		various	
Idaho	Geologic Map of Idaho	1978	500,000	1
Illinois	1. Geologic Map of Illinois	1967	500,000	1
	2. Quaternary Deposits of Illinois	1979	500,000	1
Indiana	1. Geologic Map of Indiana (Pub. 112)	1932	250,000	2
	2. Geologic Map of Indiana (Atlas, map 9)	1956	1,000,000	1
	3. Glacial Geology of Indiana (Atlas, map 10)	1963	1,000,000	1
Iowa	1. Geologic Map of Iowa	1969	500,000	1
	2. 'Quaternary Map'—Surficial	1969	1,900,800	1
Kansas	Geologic Map of Kansas (7-1)	1964	500,000	1
Kentucky	1. Geologic Map of Kentucky	1929	500,000	1
	2. 'Geologic Map of Kentucky'	1979	1,000,000	1
Louisiana	'Generalized Surface Geologic Map'	1960		1
Maine	Preliminary Geologic Map of Maine	1966	500,000	1
Maryland	Geologic Map of Maryland	1968	250,000	1
Massachusetts	Bedrock Geologic Map of Massachusetts and Rhode Island	1916	250,000	1
Michigan	1. & 2. Centennial Geologic Map of Southern/Northern Peninsula of Michigan	1936	500,000	1 each
	3. & 4. Map of the Surface Formations of the Southern/Northern Peninsula of Michigan	1955 1957	500,000	1 each
Minnesota	1. Geologic Map of Minnesota	1970	1,000,000	1
	2. Map of Northern Part of Minnesota—	1932	500,000	1 each
	3. Map of Southern Part of Minnesota—showing surficial deposits	1932	500,000	
Mississippi	Geologic Map of Mississippi	1969	500,000	1
Missouri	Geologic Map of Missouri	1979	500,000	1
Montana	Geologic Map of Montana	1955	500,000	2
Nebraska	Geologic Bedrock Map of Nebraska	1969	1,000,000	1
Nevada	Geologic Map of Nevada	1979	500,000	1
New Hampshire	1. Geologic Map of New Hampshire	1955	250,000	1
	2. Surficial Geologic Map	1950	250,000	1
New Jersey	Geologic Map of New Jersey	1912	250,000	1
New Mexico	1. Geologic Map of New Mexico	1965	500,000	2
	2. Surface Deposits of New Mexico	1965	500,000	4
New York	1. Geologic Map of New York, Series 5	1970	250,000	6
	2. Quaternary Geology of New York, Series 28	1977	250,000	5
North Carolina	Geologic Map of North Carolina	1958	500,000	1
North Dakota	1. Bedrock Geologic Map of North Dakota	1969	1,000,000	1
	2. Geologic Highway Map of North Dakota	1977	1,000,000	1
Ohio	1. Geologic Map of Ohio	1947	500,000	1
	2. Glacial Map of Ohio	1961	500,000	1
Oklahoma	Geologic Map of Oklahoma	1954	500,000	1
Oregon	1. Geologic Map, Oregon, West—	1961	500,000	1
	2. Geologic Map, Oregon, East— (of 121st Meridian)	1977	500,000	2
Pennsylvania	Geologic Map of Pennsylvania	1960	250,000	2
Rhode Island	Bedrock Geologic Map of Rhode Island	1971	125,000	1
South Carolina	1. Geologic Map of Crystalline Rocks of South Carolina	1965	250,000	1
	2. Geologic Map of the Cretaceous & Tertiary Rocks of Coastal Plain of South Carolina	1936	500,000	1
South Dakota	1. Geologic Map of South Dakota	1953	500,000	1
	2. Geologic Map of South Dakota	1953	500,000	1
Tennessee	Geologic Map of Tennessee	1966	250,000	4
Texas	1. Geologic Map of Texas	1932	500,000	4
	2. Geologic Highway Map of Texas	n/a	n/a	n/a
Utah	*map in preparation*			
Vermont	1. Centennial Geologic Map of Vermont	1961	250,000	1
	2. Surficial Geologic Map of Vermont	1970	250,000	1
Virginia	Geologic Map of Virginia	1963	500,000	1
Washington	Geological Map of Washington	1961	500,000	2
West Virginia	Geologic Map of West Virginia	1968	250,000	2
Wisconsin	1. Geologic Map of Wisconsin	1949	1,000,000	1
	2. Glacial Deposits of Wisconsin; Sand & Gravel Resource Potential	1976	500,000	1
Wyoming	Geologic Map of Wyoming (Bedrock)	1955	500,000	2
Puerto Rico	Provisional Geologic Map of Puerto Rico & Adjacent Islands. Map I-392	1964	240,000	1

Province	Title	Date	Scale	Sheets
Alberta	Geological Map of Alberta	1972	1,267,000	1
British Columbia	Generalized Geological Map of the Northern Cordillera	1976	2,500,000	1
Manitoba	Geological Map of Manitoba	1981	1,000,000	1
	Mineral Map of Manitoba	1980	1,000,000	1
	Surficial Geological Map of Manitoba	1979	1,000,000	1
New Brunswick	Geological Map of New Brunswick	1979	500,000	1
Newfoundland	Geology, Island of Newfoundland	1967	1,000,000	1
	Geological Map of Labrador	1972	3,000,000	1
Northwest Territories	Geology, Yukon Territory and Northwest Territories	1963	3,000,000	1
Nova Scotia	Geological Map of Nova Scotia	1979	500,000	?
Ontario	West Central	1975	1,013,760	1
	Northeast	1971	1,013,760	1
	Northwest	1971	1,013,760	1
	Southern	1979	1,013,760	1
	East-Central	1979	1,013,760	1
	Explanatory Notes and Sheets	1979	n/a	1
Prince Edward Island	Surficial Deposits of Prince Edward Island	1973	126,720	1
Québec	La carte géologique du Québec	1969	1,013,760	2
Saskatchewan	Geological Map of Saskatchewan	1972	1,267,000	1
	Geological Map of Saskatchewan	1980	1,000,000	2
Yukon Territory	Geology, Yukon Territory and Northwest Territories	1963	3,000,000	1
	Macmillan River	1980	1,000,000	3

Source: From Dietrich et al., 1982.

TABLE 10. Fractional Map Scales and Imperial System Equivalents

Fractional Scale of Map	Miles per Inch	Feet per Inch	Chains per Inch	Metres per Inch	Inches per 1000 Feet
1: 200	0.003	16.667	0.252	5.080	60
1: 240	0.004	20	0.303	6.096	50
1: 250	0.004	20.83	0.316	6.350	48
1: 400	0.006	33.33	0.505	10.160	30
1: 480	0.008	40	0.606	12.192	25
1: 500	0.008	41.667	0.631	12.700	24
1: 600	0.009	50	0.758	15.240	20
1: 1 000	0.016	83.333	1.263	25.400	12
1: 1 200	0.019	100	1.515	30.480	10
1: 1 500	0.024	125	1.894	38.100	8
1: 2 000	0.032	166.667	2.525	50.800	6
1: 2 400	0.038	200	3.030	60.960	5
1: 2 500	0.039	208.333	3.156	63.500	4.800
1: 3 000	0.047	250	3.788	76.200	4
1: 3 600	0.057	300	4.545	91.440	3.333
1: 4 000	0.063	333.333	5.050	101.600	3
1: 4 800	0.076	400	6.061	121.920	2.500
1: 5 000	0.079	416.667	6.313	127	2.400
1: 6 000	0.095	500	7.576	152.400	2
1: 7 000	0.110	583.333	8.838	177.800	1.714
1: 7 200	0.114	600	9.091	182.880	1.667
1: 7 920	0.125	660	10	201.168	1.515
1: 8 000	0.126	666.667	10.100	203.200	1.500
1: 8 400	0.133	700	10.605	213.360	1.429
1: 9 000	0.142	750	11.363	228.600	1.333
1: 9 600	0.152	800	12.121	243.840	1.250
1: 10 000	0.158	833.333	12.626	254	1.200
1: 10 800	0.170	900	13.635	274.321	1.111
1: 12 000	0.189	1 000	15.152	304.801	1
1: 13 200	0.208	1 100	16.666	335.281	0.909
1: 14 400	0.227	1 200	18.181	365.761	0.833
1: 15 000	0.237	1 250	18.938	381.001	0.800
1: 15 600	0.246	1 300	19.695	396.241	0.769
1: 15 840	0.250	1 320	20	402.337	0.758
1: 16 000	0.253	1 333.333	20.202	406.400	0.750
1: 16 800	0.265	1 400	21.210	426.721	0.714
1: 18 000	0.284	1 500	22.725	457.201	0.667
1: 19 200	0.303	1 600	24.240	487.681	0.625
1: 20 000	0.316	1 666.667	25.250	508.002	0.600
1: 20 400	0.322	1 700	25.755	518.161	0.588
1: 21 120	0.333	1 760	26.664	536.449	0.568
1: 21 600	0.341	1 800	27.270	548.641	0.556
1: 22 800	0.360	1 900	28.785	579.121	0.526
1: 24 000	0.379	2 000	30.303	609.601	0.500
1: 25 000	0.395	2 083.333	31.563	635.001	0.480
1: 30 000	0.473	2 500	37.879	762.002	0.400
1: 31 680	0.500	2 640	40	804.674	0.379
1: 40 000	0.631	3 333.333	50.505	1 016	0.300
1: 45 000	0.710	3 750	56.818	1 143	0.267
1: 48 000	0.758	4 000	60.606	1 219.202	0.250

(*continued*)

TABLE 10. (*continued*)

Fractional Scale of Map	Miles per Inch	Feet per Inch	Chains per Inch	Metres per Inch	Inches per 1000 Feet
1: 50 000	0.789	4 166.667	63.131	1 270	0.240
1: 60 000	0.947	5 000	75.758	1 524	0.200
1: 62 500	0.986	5 208.333	78.914	1 587.500	0.192
1: 63 360	1	5 280	80	1 609.300	0.189
1: 80 000	1.263	6 666.667	101.010	2 032	0.150
1: 90 000	1.420	7 500	113.636	2 286	0.133
1: 96 000	1.515	8 000	121.212	2 438.405	0.125
1: 100 000	1.578	8 333.333	126.263	2 540	0.120
1: 125 000	1.973	10 416.667	157.828	3 175	0.096
1: 126 720	2	10 560	160	3 218.7	0.095
1: 200 000	3.157	16 666.667	252.525	5 080	0.060
1: 250 000	3.946	20 833.333	315.657	6 350	0.048
1: 253 440	4	21 120	320	6 437.4	0.0473
1: 380 160	6	31 680	480	9 656.1	0.0316
1: 500 000	7.891	41 666.667	631.313	12 700	0.024
1: 760 320	12	63 360	960	19 312.2	0.0158
1: 1 000 000	15.783	83 333.333	1 262.626	25 400	0.0120
1: 5 000 000	78.914	416 666.667	6 313.131	127 000	0.002
1:10 000 000	157.828	833 333.333	12 626.262	254 000	0.001

Source: From Berkman and Ryall, 1976, who reproduced the data, with permission, from W. R. Moran, 1958, *Handbook for Geologists,* Union Oil Company of California.

TABLE 11. Fractional Map Scales and Unit Plan Areas

Fractional Scale of Map	Inches per Mile	Acres per Square Inch	Square Inches per Acre	Square Miles per Square Inch
1: 200	316.80	0.0064	156.816	0.000 010
1: 240	264.00	0.0092	108.900	0.000 014
1: 250	253.44	0.0100	100.362	0.000 015
1: 400	158.40	0.0225	39.204	0.000 040
1: 480	132.00	0.0367	27.225	0.000 057
1: 500	126.720	0.0399	25.091	0.000 06
1: 600	105.600	0.0574	17.424	0.000 09
1: 1 000	63.360	0.1594	6.273	0.000 25
1: 1 200	52.800	0.2296	4.356	0.000 36
1: 1 500	42.240	0.3587	2.788	0.000 56
1: 2 000	31.680	0.6377	1.568	0.0010
1: 2 400	26.400	0.9183	1.089	0.0014
1: 2 500	25.344	0.9964	1.004	0.0016
1: 3 000	21.120	1.4348	0.697	0.0022
1: 3 600	17.600	2.0661	0.484	0.0032

TABLE 11. (*continued*)

Fractional Scale of Map	Inches per Mile	Acres per Square Inch	Square Inches per Acre	Square Miles per Square Inch
1: 4 000	15.840	2.5508	0.392	0.0040
1: 4 800	13.200	3.6731	0.272	0.0057
1: 5 000	12.672	3.9856	0.251	0.0062
1: 6 000	10.560	5.7392	0.174	0.0090
1: 7 000	9.051	7.8117	0.128	0.0110
1: 7 200	8.800	8.2645	0.121	0.0129
1: 7 920	8	10	0.100	0.0156
1: 8 000	7.920	10.203	0.098	0.0159
1: 8 400	7.543	11.249	0.089	0.0176
1: 9 000	7.040	12.913	0.077	0.0202
1: 9 600	6.600	14.692	0.068	0.0230
1: 10 000	6.336	15.942	0.063	0.0249
1: 10 800	5.867	18.595	0.054	0.0291
1: 12 000	5.280	22.957	0.044	0.0359
1: 13 200	4.800	27.778	0.036	0.0434
1: 14 400	4.400	33.058	0.030	0.0516
1: 15 000	4.224	35.870	0.028	0.0560
1: 15 600	4.062	38.797	0.026	0.0606
1: 15 840	4	40	0.025	0.0625
1: 16 000	3.960	40.812	0.024	0.0638
1: 16 800	3.771	44.995	0.022	0.0703
1: 18 000	3.520	51.653	0.019	0.0807
1: 19 200	3.300	58.770	0.017	0.0918
1: 20 000	3.168	63.769	0.016	0.0996
1: 20 400	3.106	66.345	0.015	0.1037
1: 21 120	3	71.111	0.014	0.1111
1: 21 600	2.933	74.380	0.013	0.1162
1: 22 800	2.779	82.874	0.012	0.1295
1: 24 000	2.640	91.827	0.011	0.1435
1: 25 000	2.534	99.639	0.010	0.1557
1: 30 000	2.112	143.480	0.007	0.2242
1: 31 680	2	160	0.006	0.2500
1: 40 000	1.584	255.076	0.004	0.3985
1: 45 000	1.408	322.830	0.003 1	0.5044
1: 48 000	1.320	367.309	0.002 7	0.5739
1: 50 000	1.267	398.556	0.002 5	0.6227
1: 60 000	1.056	573.921	0.001 7	0.8967
1: 62 500	1.014	622.744	0.001 6	0.9730
1: 63 360	1	640	0.001 6	1
1: 80 000	0.792	1 020.304	0.000 9	1.5942
1: 90 000	0.704	1 291.322	0.000 77	2.0173
1: 96 000	0.660	1 469.240	0.000 68	2.2957
1: 100 000	0.634	1 594.225	0.000 627	2.4909
1: 125 000	0.507	2 490.976	0.000 401	3.8922
1: 126 720	0.500	2 560	0.000 390	4
1: 200 000	0.317	6 376.900	0.000 157	9.9639
1: 250 000	0.253	9 963.906	0.000 100	15.5686
1: 253 440	0.250	10 240	0.000 098	16
1: 380 160	0.167	23 040	0.000 043 4	36
1: 500 000	0.127	39 855.626	0.000 025 0	62.2744
1: 760 320	0.083	92 160	0.000 010 9	144
1: 1 000 000	0.063	159 422.507	0.000 006 3	249.0976
1: 10 000 000	0.0063	15 942 250.70	0.000 000 063	24 909.76

Source: From Berkman and Ryall, 1976, who reproduced the data, with permission, from W. R. Moran, 1958, *Handbook for Geologists,* Union Oil Company of California.

TABLE 12. Recommendations for Acquiring and Preserving Water Samples

Measurement	Volume required (ml)	Container[1]	Preservation	Holding time[2]
Acidity	100	P, G	Cool 4°C	24 hrs
Alkalinity	100	P, G	Cool 4°C	24 hrs
Arsenic	100	P, G	HNO_3 to pH 2	6 mo
Biochemical Oxygen Demand (BOD)	1000	P, G	Cool 4°C	6 hrs
Chloride	50	P, G	None	7 days
Cyanides	500	P, G	Cool 4°C NaOH to pH 12	24 hrs
Dissolved Oxygen	300	G	Det. on site	Nil
Fluoride	300	P, G	Cool 4°C	7 days
Hardness	100	P, G	Cool 4°C	7 days
Iodide	100	P, G	Cool 4°C	24 hrs
Metals				
Dissolved	200	P, G	Filter on site HNO_3 to pH 2	6 mo
Suspended	200	P, G	Filter on site	6 mo
Total	100	P, G	HNO_3 to pH 2	6 mo
Mercury				
Dissolved	100	P, G	Filter HNO_3 to pH 2	38 days (G) 13 days (P)
Total	100	P, G	HNO_3 to pH 2	38 days (G) 13 days (P)
pH	25	P, G	Cool 4°C Det. on site	6 hrs
Selenium	50	P, G	HNO_3 to pH 2	6 mo
Silica	50	P	Cool 4°C	7 days
Specific Conductance	100	P, G	Cool 4°C	24 hrs
Sulphate	50	P, G	Cool 4°C	7 days
Sulphide	50	P, G	2 ml zinc acetate	24 hrs
Temperature	1000	P, G	Det. on site	Nil
Turbidity	100	P, G	Cool 4°C	7 days

Source: From Berkman and Ryall, 1976.
Notes:
1. Plastic or glass.
2. Properly preserved samples may be held for longer periods.

TABLE 13. Maximum Allowable Concentration of Impurities in Water for Human Consumption

Substance	Maximum allowable concentration mg/litre (ppm)
Total soluble salts	1500
Iron	1
Manganese	0.5
Copper	1.5
Zinc	15
Arsenic	0.05
Lead	0.05
Calcium	200
Sulphate	400
Magnesium	150
Chloride	600
Magnesium and sodium sulphates	1000
Nitrate	45
Fluoride	1.5
Cyanide	0.2
pH	6.0–9.2

Source: From Berkman and Ryall, 1976.
Note: World Health Organization Standard

FIELD METHODS—See ALIDADE; BLOWPIPE ANALYSIS; COMPASS TRAVERSE; EXPOSURES, EXAMINATION; FIELD NOTES, NOTEBOOKS; GEOLOGICAL FIELD WORK, CODES; GEOLOGICAL SURVEY AND MAPPING; ROCK-COLOR CHART; SAMPLES, SAMPLING. Vol. XIII: FIELD GEOLOGY.

FIELD TESTS—See BLOWPIPE ANALYSIS; LITHOGEOCHEMICAL PROSPECTING; MINERAL IDENTIFICATION, CLASSICAL FIELD METHODS.

FIELD NOTES, NOTEBOOKS

Field notes record the detailed observations of geologists working in the outdoors. Typical notebook entries include sketches of geomorphological landforms and outcrop features, preliminary maps and cross-sections (q.v.), detailed maps of critical or complex areas (e.g., contacts and faults), stratigraphic sections, tabulated quantitative data (e.g., structural measurements), pit and trench logs (see Vol. XIII: *Pipeline Corridor Evaluation*) lists and descriptions of samples and fossils, and a variety of written notes. The notebook contains the first record of the field geologist's observations and the interpretation of what he or she sees in the field and is a testament to the old Chinese proverb that "the faintest ink is better than the best memory."

The field notebook represents the first link in a long chain of geological data gathering, mapping (see *Geological Survey and Mapping*), interpretation, and presentation. It represents not only the first record but also the most complete record of raw (uninterpreted) data. As such it is a unique record and must be considered of prime importance.

Every care must be taken to preserve the field notebook in possibly rigorous outdoor conditions. At the conclusion of the project assignment or field season, it should be filed in a secure place as part of the permanent record. Except in the case of the independent researcher, the notebook is rightly the property of the company, geological survey department, or agency conducting the field operation and not of the individual geologist. Geologists resuming work on extended projects or in a new field season should take photo or microfilm copies to the field rather than the original notebook.

A basic prerequisite of the field notebook is that it be neat and readily comprehensible to geological colleagues who might need to use it as a primary reference for expanding or extending the mapping of an area. The first page of the notebook should include the name of the geologist who compiled it, the general field area worked, and a short statement of the project objective. Each subsequent page should include a specific location and a date of entry. Other points to emphasize include the following:

1. Field notebooks should never be destroyed.
2. Pages should never be removed.
3. Care should be taken to prevent deterioration in storage (e.g., from mildew, dampness, insects).
4. A single field notebook should not be used by more than one individual or for more than one project or field area.

The Notebook

Field notes should be written only in a bound notebook. Commercialy available field books, commonly referred to as an engineer's notebook, are large pocket size (19 cm by 12 cm), have stiff covers, and contain about 60 pages. The covers are generally bright orange or yellow, a precaution against loss and their stiffness provides a suitable base for writing and sketching. The books typically include mathematical tables, trigonometric formulas, and other data commonly used by geologists, surveyors, and civil engineers.

Many government surveys and commercial companies issue their own books. These frequently include additional information on soil classification systems, rock weathering guidelines, etc. Private-issue books contain mailing information printed on the inside cover, and many companies advertise a cash reward to expedite return of lost books. The individual geologist would be wise to follow this good example.

Notebook pages are alternately ruled and of square-graph format. The latter facilitates field sketching, logging, and mapping (Fig. 1). Graph paper also helps the geologist to tabulate more neatly quantitative data such as strike and dip readings.

It is false economy not to use books with good-quality waterproof paper because notebooks cannot escape some degree of rough usage and exposure to inclement weather. Specially designed leather notebook cases are considered an essential part of the field geologist's equipment (Fig. 2). Cases are designed to carry pens and pencils and have loops for attachment to a stout waist belt (not a shoulder strap). As implied by this discussion, loose–leaf or ring-binder books with soft covers and nonrainproof paper should never be used.

General Note Taking

Many geologists take such pride in their notebooks that all entries are made in waterproof ink,

using a draftsman's pen. However, a medium soft (HB or No. 2) pencil is preferred as the best all-purpose note-taking instrument. It does not run when wet, fades only slightly with time, and does not react chemically with the paper. Pens are not recommended unless it is certain that a proven waterproof, indelible, and durable ink is used. There is an enormous range of inks in the countless commercial brands of ballpoint pens available. Many are not waterproof and smudge easily. Over time, many inferior inks break down, become discolored, etch the paper, or soak through to the back of the page(s).

With regard to the format for recording field notes, procedures vary primarily as a function of the type of project under consideration. For example, the notes taken by a survey geologist compiling a general geological map may be expected to differ from those of an engineering geologist, a hydrogeologist, or a marine geologist.

Even within a specific discipline and for a similar project, the procedures for recording field notes vary slightly as a function of the procedures recommended by the company or agency coordinating the field program. The format for note taking and the content of notes varies also in response to the relative abundance of exposures (see *Exposures, Examination of*), nature of rock types, general geological complexity of the project area, and time available.

At a given geological locality, it is a good rule to assume that only one opportunity will be available for data gathering so that notes and sketches are made as complete as possible. This is a practical consideration because, in many cases, the cost and complex logistics associated with organizing field operations in remote areas virtually preclude their being repeated (the geological fieldwork on the moon during the Apollo missions is an extreme case). The interpretation of field observations may be entered in the field notebook but should be clearly identified as such and kept separate from the record of factual data.

The use of portable tape recorders is not recommended. However, if used, tapes should be transcribed into the field notebook as soon as possible and a notation made as to the origin of the entry. Tapes should not be considered in any way as a substitute for field notebook entries.

Reconnaissance Survey Notes The preliminary observations and reconnaissance traverses that mark the first phase of field operations are more strongly oriented toward note taking than map making. They are therefore briefly covered to supplement further discussion in other sections (see *Exposures, Examinations of*).

On large projects, fieldwork commences with a small-aircraft flyover to confirm satellite and/ or aerial photograph interpretations and to update a strategy for more detailed ground recon-

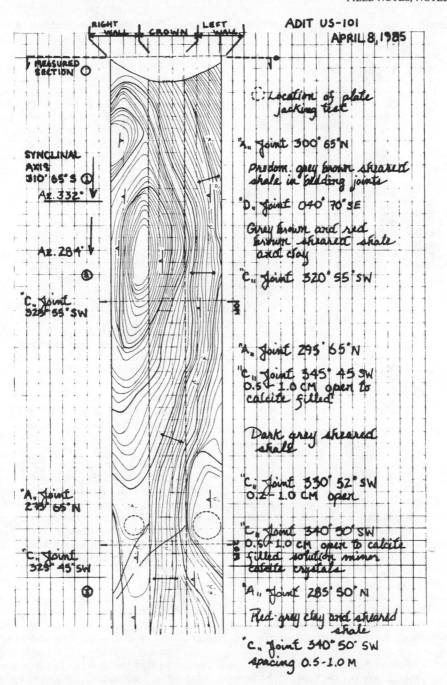

FIGURE 1. Typical field notebook entry, an exploration adit map, and supporting notes on rock types, structural trends, sample locations, and the proposed locations of plate-jacking tests.

naissance. Helicopters offer an added advantage for this task, enabling frequent ground checks; however, their far greater cost is an obvious factor to be considered.

Airborne reconnaissance is normally followed by vehicle-supported note taking along roads and trails or, if an off-road vehicle is available, across open country. All notebook entries made at this time should be cross-referenced to topographic and/or photogeological maps.

FIGURE 2. Geologist Amanda Moore wears a leather case on her belt to carry and protect her field notebook.

In difficult terrain or where exposures are poor, a reconnaissance stage of operation is essential for spotting the best places for more detailed studies. For example, in measuring thick stratigraphic sections of inclined strata in mountainous terrain, it is important to locate the most complete, structurally uninvolved, but at the same time accessible sections and to observe the major lithostratigraphic units (e.g., Fig. 3). A meticulous binocular study of a mountainside from a vantage point on the far side of the valley is an obvious first step in planning operations. Notes and sketches supplemented by Polaroid-type pictures from hilltops and other vantage points may supplement notes taken on regional structures, geomorphology, and the type of rock exposure (see *Field Geology*). These may also be invaluable in finding the selected route when the section is actually measured. Such a procedure may be used to spot the most accessible stream gullies and routes across the tops of talus slopes etc. and may reduce the problem of locating the right route when climbing up to the section from the valley floor through trees.

Outcrop Notes As emphasized in the entries *Exposures, Examinations of* and *Samples, Sampling,* the procedures followed at outcrops are largely a function of the objective of the field survey and the geological complexity of the area under investigation. This subsection stresses some general considerations in note taking at outcrops.

Perhaps the most common misconception of the student geologist is that photographs may be used to replace the controlled sketch and map of the exposure. It is stressed that, although photographs may provide a useful notebook supplement, these should always be accompanied by a sketch (e.g., Fig. 4; see also Figs. 3 and 4 in *Exposures, Examination of*).

A further point of emphasis relates to the importance of notation on the outcrop locality. Each exposure account should begin with a general description of the outcrop locality, indicating, as appropriate, place name, legal subdivision, map location with grid reference, and location of any air photo coverage. Localities may be identified by an appropriate numbering system; a simple approach is to assign three numbers, the first identifying the particular field operation (letter abbreviations are often used instead), a second number identifying the geologist, and a third assigned sequentially to each new location encountered as the field program proceeds. These numbers should be code deciphered on the first page of the notebook and cross-referenced to base maps and aerial photographs. To avoid marking photographs, a small pinhole may be made through the print and the identification numbers written on the back. The importance of properly locating outcrops cannot be overemphasized. More than a few important sections, fossil and mineral localities, have been lost and never rediscovered because these essential note-taking steps were not followed. The remaining discussion summarizes some further points on note taking that are relative to rock types and lithologic environments.

In describing and measuring a succession of sedimentary rocks, the first task is to divide the succession into recognizable lithostratigraphic units, each of which has a certain homogeneity, or lithology, that sets it off from units below it and above it in the section. The procedures to be followed in establishing the boundaries of such units and assigning these a rank within a hierarchy of lithostratigraphic units represent the extension of notetaking to field mapping.

General note-taking detail should include the following:

Generic rock name (sandstone, shale, etc.) and qualifying description (e.g., silty, argillaceous) based on hand-lens examination and field tests (e.g., acid test);

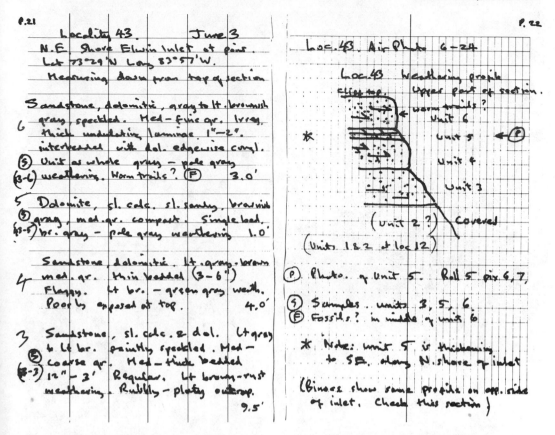

FIGURE 3. Typical notebook entry to describe units in a sedimentary sequence.

Texture (e.g., coarse grained, finely crystalline);

Relative density based on visual examination, tactile sense, hammer ring, and/or Schmidt hammer rebound readings;

Color of fresh and weathered surfaces (use of a color chart is recommended) with sketches of weathering characteristics, as appropriate (see *Rock-Color Chart*);

Internal depositional features (cross-bedding, graded bedding);

Thickness of bedding, dip and strike;

Bedding plane features (e.g., ripple marks, dessication features, ichnofossils, fossils);

Jointing (e.g., attitude, spacing, continuity, roughness, infilling);

Notation relative to samples collected and photographs taken.

There is no hard and fast rule as to whether stratigraphic sections should be measured from bottom to top or top to bottom. In many cases, operational logistics are the controlling factor. For example, where helicopters are used in mountainous terrain, it is often easier to start from the top. Other things being equal, it is probably advisable to traverse up a section first, making reconnaissance observations and noting en route the float or loose material removed by erosion from in situ outcrop, and then to make the detailed measurements and collect samples on the way down. A notation indicating which way the section was measured should be entered in the notebook.

Many of the general features noted in relation to sedimentary rocks, apply equally to igneous and metamorphic rocks (e.g., generic description, weathering, sampling). However, in describing these different rock types, the geologist should modify his or her note-taking format accordingly. Large plutons, e.g., are often remarkably homogeneous, and note-taking emphasis may be concentrated on joints, faults, fractures, and mineralized zones rather than on general lithology. In smaller intrusive bodies, textural and compositional changes and contact relationships with country rock should be noted. In metamorphic rocks, the orientation of foliation and other petrofabric trends are of prime importance.

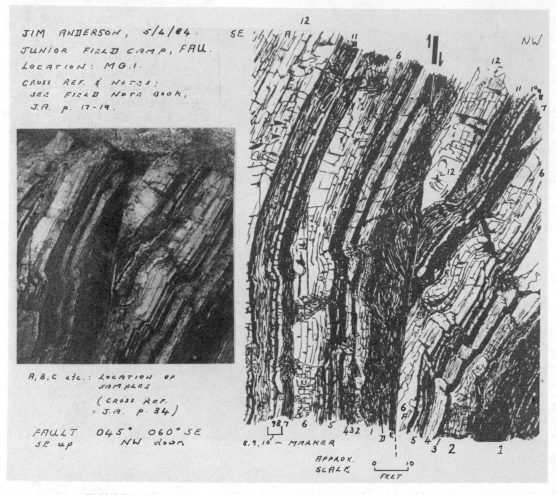

FIGURE 4. Photographs should always be accompanied by an annotated sketch.

Conclusion

Field notes represent the first step in preparing the geological map and report. In geologically complex areas the field notebook contains more information than the map. In all areas, the notebook is as important as the geological map. Furthermore, because the field notebook is central to any subsequent special-purpose mapping and may, in certain cases, be used as a legal document, considerable care should be taken in its neat, accurate, and comprehensive preparation.

ROY R. LEMON
IAN WATSON

Cross-references: *Field Geology; Geological Field-work, Codes; Geological Writing; Geologic Reports; Map Colors, Coloring; Map Series and Scales; Map Symbols; Rock Color Chart; Samples, Sampling; Style, Writing.*

FLAME TESTS—See BLOWPIPE ANALYSIS; MINERAL IDENTIFICATION, CLASSICAL FIELD METHODS.

FLOATING STRUCTURES IN WAVES

Floating structures are built for the purpose of carrying humans, material, and/or weapons on the sea. To accomplish their mission, they must float in a stable upright position and be strong enough to withstand the rigors of heavy weather and wave impact. The success of a design for floating structures depends ultimately on their performance in a seaway.

A floating structure is designed to operate for a long time in various seaway environments. The basic requirements to withstand severe sea conditions are basic stability features and satisfactory seakeeping performance. Offshore drilling began in the United States in the early 1960s, and three types of floating structures—floating, semisubmersible, and self-elevating types—have evolved with time.

Stability

There is a specific requirement for the intact stability of the floating structures by the U.S. Coast Guard, but there are no requirements for damaged stability. Figure 1 illustrates typical plots for the righting arm for all three types of floating platforms. Plot A is for the self-elevating type, which is being towed with the legs raised; plot B indicates the stability characteristic for a unit having a hull form similar to that of a ship or barge, whereas a column-stabilized (i.e., semisubmersible) floating structure at its operating draft may have an intact stability characteristic, as shown by plot C. For large heel angles, a portion of the main superstructure may be immersed, as in the case of semisubmersibles. The dotted line of plot C indicates a modification of the righting arm due to additional buoyancy.

The U.S. Coast Guard requirements for stability are shown in Fig. 2. The area under the curve for righting moment should be at least 1.4 times the area under the heeling moment curve, both areas being measured up to either the angle of lower flooding or the angle of the second crossing of the two curves, whichever gives a smaller angle. For the survival of the structure, a wind speed of 100 knots should be assumed, whereas for operating conditions, a wind speed of 70 knots may be assumed provided that within a specified time period (usually 3 hr) it would be possible to change the loading to fulfill the 100 knot require-

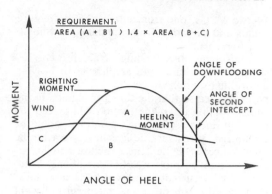

FIGURE 2. Requirements for stability.

ment. Wind–heeling moment may be determined either by direct calculations or by wind tunnel tests.

Floating offshore structures that have ship-shape hulls may require a minimum standard of stability depending on the country of registration. The various classification societies also have published rules for construction and classification of mobile offshore structures. The rules of the American Bureau of Shipping (1973) for offshore mobile structures were first published in 1968, and specifications for both intact and damaged stability were included. For intact stability, the U.S. Coast Guard criteria were adopted, and for damaged stability, the rules state that if a watertight compartment is flooded, the structure shall be able to withstand a 50 knot wind from any horizontal direction without submerging any opening through which lower flooding could occur. For the different types of structures, the rules specify the locations where the damage is to be assumed. In general, these are limited to those compartments that might be damaged by impact from a vessel alongside the unit.

The basic stability is primarily governed by the height of the center of gravity of the unit. This stability, in turn, restricts the amount of weight on top of the platform, i.e., drilling gears, mud, cement, pipes, etc. To determine the stability of the floating structures in various operating conditions, the operating manual for a floating structure normally includes a curve that shows, on the basis of the draft, the maximum allowable vertical center of gravity (VCG) that will comply with the stability requirements. It must be pointed out that the allowable height of the center of gravity position for various drafts of the floating structure is obtained after plotting separate curves, one showing the limiting VCG at each draft for compliance with the intact stability and another for compliance with the damaged stability requirements. For each draft, the ordinate of the allowable VCG curve is obtained from the lower value

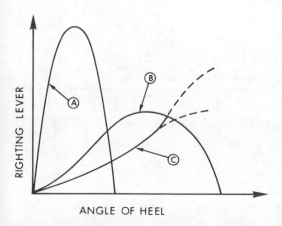

FIGURE 1. Typical plots for the righting arm of three types of floating platforms.

of two VCGs, one from intact stability requirements and the other from the damaged stability
requirements.

Semisubmersible floating structures are designed to have minimum amplitudes of heaving
and rolling motions when they are stationed in a
particular sea so that the time per annum during
which heaving or rolling motion will exceed the
practical limits for drilling operations should be
minimized. Many such designs have proved to be
highly successful; i.e., the structure remains almost upright in most weather conditions. Compliance with current stability criteria may, however, require an operating metacentric height
($\overline{GM}$), which causes increased motion and longer
periods of downtime during severe weather conditions. Sometimes it may be appropriate to reduce the area ratio requirement for intact stability
of semisubmersibles, but the dynamic behavior of
the floating structures must be taken into account
in that case.

Seakeeping

The successful design of a floating structure
such as a ship, offshore platform, or buoy depends on a proper evaluation of the forces acting
on it as well as its motions when floating in a seaway. For example, the midship bending moment
due to wave action may cause excessive stress in
the midship section of a ship. Also because of severe motions, a floating platform may not be able
to perform its mission in a rough sea.

The experimental as well as theoretical work on
seakeeping for ocean platforms and drilling rigs

is carried out in the early stage of design. The seakeeping technology first developed and applied to
ships has influenced considerably the design of
floating structures.

In the majority of experiments, a so-called survival wave condition is required. Considering the
North Sea weather conditions, the survival wave
has become 30 m in height, which is the expected
extreme for North Sea drilling operations. Also,
for proper seakeeping experimentation, both
winds and currents should be simulated.

To study the effects of waves on floating structures, it is necessary to understand sea waves,
which are not regular but highly complex in nature. Statistical means are adopted to study this
irregular behavior of the seaway and to obtain
motion characteristics of floating structures.

A floating structure in waves is almost always
in oscillatory motion. The different kinds of oscillatory motions a floating structure experiences
can be described with help of Fig. 3. As shown in
this figure, there are six kinds of motion; three
are linear and the other three are rotational about
the three principal axes. Accordingly,

a = Surging: Motion backward and forward in
the direction of travel.
b = Swaying: An athwartship motion of the
vehicle.
c = Heaving: Motion vertically up and down.
d = Rolling: Angular motion about the longitudinal axis. When structure rolls its heels
alternately from starboard to port and then
back to starboard.

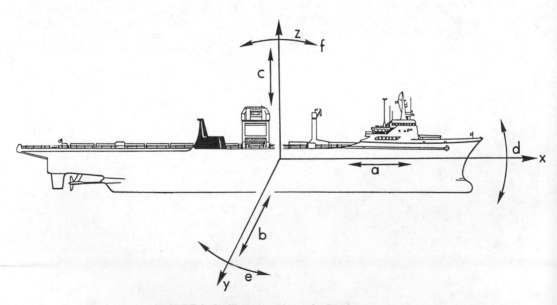

FIGURE 3. Oscillatory motions of a floating structure.

e = Pitching: Angular motion about the transverse axis. When the floating structure pitches, it alternately trims by the bow and by the stern.

f = Yawing: Angular motion about the vertical axis.

Only three kinds of motions—namely, heaving, rolling, and pitching—are purely oscillatory because these motions act under a restoring force or moment when the floating structure is disturbed from its equilibrium position. In the case of surging, swaying, or yawing, the floating structure will not come back to its original position if disturbed from its equilibrium position unless the exciting forces (or moments in case of angular motions), which cause such disturbance themselves, act from the opposite directions alternately.

Even though a floating structure actually experiences all six kinds of motions simultaneously, often only one motion at a time is calculated to simplify the problem. The simplified equation of heaving motion only is expressed as

$$a\ddot{z} + b\dot{z} + cz = F_o \cos \omega_e t \qquad (1)$$

where each term represents a force: e.g., $a\ddot{z}$ = inertia force, where a = virtual mass of the floating structure comprising the actual mass of the structure and the added mass for heaving motion; $b\dot{z}$ = damping force that resists the motion, where b is the damping force coefficient; cz = restoring force that always tends to bring the floating structure back to its equilibrium position, where c is the restoring force coefficient, or commonly known as the spring constant; $F_o \cos \omega_e t$ = encountering force acting on the mass of the ship, where F_o is the amplitude of the encountering force and ω_e is the encountering frequency. $\ddot{z}$, $\dot{z}$, and z are heaving acceleration, heaving velocity, and heaving motion respectively.

For the uncoupled pitching motion the equations are given as

$$a\ddot{\theta} + b\dot{\theta} + c\theta = M_o \cos \omega_e t, \qquad (2)$$

where each term of Eq. (2) represents the moment of force: e.g., $a\ddot{\theta}$ = inertia moment for pitching, a is the virtual mass moment of inertia for pitching; $b\dot{\theta}$ = damping moment for pitching, where b is the damping moment coefficient; $c\theta$ = restoring moment for pitching. Again, $\ddot{\theta}$, $\dot{\theta}$, and θ are the pitching acceleration, pitching velocity, and pitching motion respectively.

Theoretical studies are performed to predict motions, but the mathematics of nonlinear and coupled motions is rather complicated. Relative calculations can be made for the coupled heaving and pitching motions, however, the equations of which are written as

$$a\ddot{z} + b\dot{z} + cz + d\ddot{\theta} + e\dot{\theta} + h\theta \\ = F_o \cos(\omega_e t + \sigma) \qquad (3)$$

$$A\ddot{\theta} + B\dot{\theta} + C\theta + D\ddot{z} + E\dot{z} + Hz \\ = M_o \cos(\theta_e t + \tau)$$

where $a, b, \ldots h$ and $A, B, \ldots H$ are various coefficients, F_o is the exciting force amplitude for heaving, M_o is the exciting moment amplitude for pitching motion, and σ and τ are phase differences for the exciting force and pitching moment respectively relative to wave motion. It should be noted that the coefficients in Eqs. (1), (2), or (3) depend on the geometry of the floating structure's wave length, etc. For the analytical seakeeping studies, linear methods are widely used to determine two-dimensional added mass, damping, and wave excitation. The agreement between the analytical prediction and experimentation with models is quite satisfactory for the seakeeping design.

Motion in an irregular seaway is determined using the following steps:

1. For the particular seaway in which the floating structure is to operate, choose a suitable wave spectrum.
2. Transform the wave spectrum to the encountering wave spectrum where the frequency of encounter is considered instead of the absolute wave frequency. However, the area under the modified spectrum is the same as that under the original spectrum, since total energy remains the same.
3. Obtain a plot in which the ordinates represent the amplitude of motion (either pitch, roll, or heave as the case may be) to a base of encountering frequency distribution. This can be obtained analytically or by experimentation in regular or irregular waves in a tow tank.
4. Modify the diagram obtained in the third step so that the ordinates represent the ratio of the square of the motion amplitude divided by the square of the wave amplitude. This diagram is termed the *response amplitude operator* (RAO), or simply, the *transform spectrum*.
5. Determine the motion amplitude spectrum by multiplying the ordinates of the transformed wave spectrum by the ordinates of the RAOs for the corresponding frequencies of encounter.
6. Finally, determine the area under the motion amplitude spectrum to obtain the necessary motion characteristics—e.g., average of one-third highest or average of one-tenth highest

motion amplitude as in the case of waves using the following formulation:

$$\text{Significant motion amplitude} = 2\sqrt{\text{area under response spectrum}}$$

These steps can be explained with the help of Fig. 4. However, note that this method, used for the response in an irregular seaway, is valid only if the responses are linearly proportional to the wave excitation (i.e., wave amplitude). This is not far from the truth if the seaway is moderate and only moderate responses are expected. Numerous model experiments have shown that in principle the responses of marine vehicles are linear, and so the linear technique for the prediction of motion responses provides a very useful means for sea-keeping design of marine vehicles.

Because severer sea conditions are considered for floating structures than for marine vehicle design, the linear superposition technique used to predict the seakeeping characteristics of marine vehicles in moderate sea conditions described earlier may prove to be erroneous since the response of the platforms will be nonlinear.

Another area of importance for the platform design is the very low-frequency lateral oscillations in an irregular seaway. The low-frequency

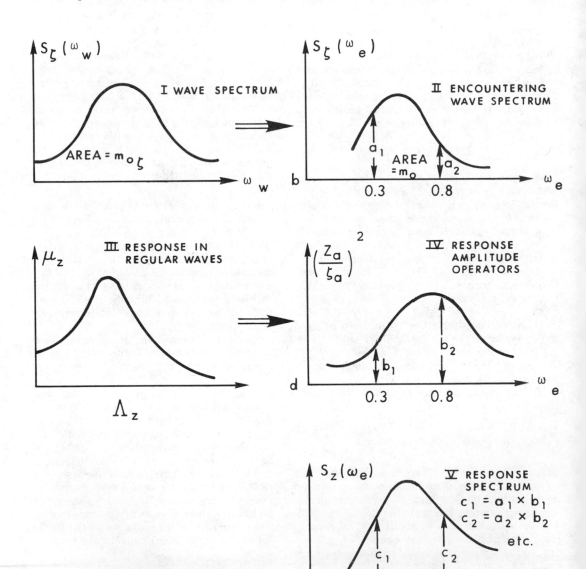

FIGURE 4. Steps for determining motion in an irregular seaway.

oscillations caused by nonlinear interactions between adjacent parts of the wave spectrum will thus exist independently of any mooring system.

It is not always the motion characteristics that are important in studying the behavior of a floating structure in a seaway but the dynamic effects caused by the motions. These effects are shipping of green water on the deck; severe wave impact on the structure, known as *slamming;* and the effects of acceleration due to pitch, heave, roll, or all combined.

Shipping of green water can have a very detrimental effect if watertight integrity is not maintained. Many of the electrical systems could be so damaged that the malfunction of a piece of apparatus on the weather deck may seriously impair the mission of the floating structure. If a floating system experiences repeated slams, this will not only damage the structure and other components but also will have a considerable effect on the personnel operating it. This factor is especially important for the satisfactory operation of naval vessels, the mission of which is to act as a floating platform for weapons systems.

For marine vehicles, speed plays an important role. However, there is a loss of speed while the vehicle is underway in waves, due to the increase in added resistance caused by motions. There is also a loss in propeller efficiency, resulting in higher fuel consumption, thereby limiting the cruising range. The heavier the seaway, the greater the loss of speed. To overcome this loss, it is often necessary either to improve the resistance and propulsion characteristics of the vessel and/or design the machinery plant for adequate reserve power. Model tests can predict with reasonable accuracy the still-water resistance and propulsion performance of a marine vehicle, but the problem of their determination in a seaway is still the subject of research. The maximum speed that can be attained by a marine vehicle is not necessarily governed by the available power but mostly by the acceleration it experiences in a seaway.

To reduce the dynamic effects, various means of motion stabilization are sometimes adopted. For example, damping tanks and fin stabilizers are the most effective means of stabilizing the motions.

The investigation of various problems encountered in motions of floating structures are pursued in four different ways:

Analytically—i.e., on a theoretical basis;
Experimentally—i.e., by means of model tests in controlled environments;
Empirically—i.e., through statistical observations;
Directly—i.e., with the trials of full-scale structures after they are built.

Design Criteria

Before the seakeeping performance of a floating structure in a particular seaway is analyzed in the design office, the following data should be available at the designer's disposal:

Information on existing and new seakeeping test techniques;
Improvement of the theoretical predictions of the behavior of floating structures in waves, including nonlinear effects;
A set of design spectra representing various representative seaways since the response of a floating structure is very much dependent on the spectral distribution of the seaway energy.

Similarly, the statistical approach enables a designer to predict the seakeeping characteristics in numerical terms as follows:

Number of times green water may be shipped over the deck per hundred pitch (or roll) oscillations or during half an hour of operation in a particular seaway;
Number of slams the floating structure may have to endure per hundred pitching oscillations or during any specified amount of time;
The largest vertical acceleration at any particular point on the structure.

Both theoretical and experimental studies help the designers to find out influences of various features of the floating system on seakeeping characteristics that are extremely valuable in designing a floating structure for seaway operation, as well as from the point of view of economic consideration. These studies and full-scale tests are necessary to provide reliable design criteria.

RAMESWAR BHATTACHARYYA

References

Aguirre, J. E., and T. R. Boyce, 1973, Estimation of wind forms on offshore drilling platforms, *Trans. RINA.*

American Bureau of Shipping, 1973, *Rules for Building and Classing Offshore Mobile Drilling Units.*

Beckwith, L., and M. E. Skillman, 1975, Assessment of the stability of floating platforms, *Trans. North East Coast Inst. of Engineers and Shipbuilders,* May.

Bell, A. O., 1974, Service performance of drilling unit, *Trans. RINA.*

Bhattacharyya, R., 1974, On the application of seakeeping research, *Proc. Seakeeping Seminar, Shipbuilding Design and Research Institute* (NIPKIK) (Varna, Bulgaria) (in cooperation with the UNDP and IMCO).

Dalzell, J. F., 1974, Ocean platforms, *Proc. 17th Am. Towing Tank Conf.*

Kim, C. H., and F. Chou, 1973, Motions of a semi-submersible drilling platform in head seas, *Marine Technology.*

McCormick, M. E., 1972, *Ocean Engineering Wave Mechanics.* New York: Wiley Interscience.

Numata, E., and W. H. Michel, 1974, Experimental study of stability limits for semi-submersible drilling platforms, *Proc. Offshore Technology Conf.*

Paulling, J. R., 1974, Elastic response of stable platform structures to wave loading, International Symposium on the Dynamics of Marine Vehicles and Structures in Waves.

Roseman, D., et al., 1968, Vehicles for ocean engineering, *Trans., SNAME.*

St. Denis, M., 1974, On the motions of oceanic platforms, in International Symposium on the Response of Marine Vehicles and Structures in Waves, London.

Suhara, T., F. Tasai, and H. Mitsyasu, 1974, A study of motion and strength of floating marine structures in waves, *Proc. Offshore Technology Conf.*

Cross-references: *Sea Surveys.* Vol. XIII: *Coastal Engineering; Ocean, Oceanographic Engineering, Oceanography, Applied; Submersibles.*

FOSSILS AND FOSSILIZATION

The term *fossil* is derived from the Latin *fossilis,* simply meaning dug-up. As the word implies, fossils are initially buried in sedimentary layers in the earth and must be excavated, either artificially by human mechanisms or naturally by geological processes such as erosion or orogeny. In an area of sedimentary deposition with undisturbed conditions, remains of living organisms are constantly being deposited along with sediments. *Fossil,* then, is a somewhat subjective term, referring to the traces or remnants of animal or plant life that have been naturally preserved within the Earth's crust during some past geologic time. Geologists sometimes refer to unreplaced or unaltered organic remains from the Holocene, or from the past few thousand years, as *subfossils.* All the fossils laid down since the evolution of preservable remains, taken collectively, are referred to as the *fossil record.*

The oldest known fossils, or traces of life, date from the late Precambrian (Proterozoic), between 570 m.y. and 800 m.y. ago. Most of these primitive organisms were microscopic and unicellular and resembled modern bacteria and blue-green algae. Microscopic fossils of this type have been found imbedded in Precambrian chert nodules from Canada. Another of the oldest fossils are *stromatolites.* These are dome-shaped masses composed of layers of sand grains and carbonates and were produced by lamelliform mats of blue-green algae with sand agglutinated onto their mucus-covered surfaces. Layer on layer of sand and algae built up toward the sea surface, with the newest layer growing on top of the previous layer.

This accretionary style of growth eventually formed the characteristic structure and shape of a stromatolite.

The fossil record, then, is biased by the preservability of the organism. This preservability is related to the percentage of the organism that is covered by hard parts or parts that are resistant to immediate decomposition. Abundant soft-bodied animals, such as various types of worms and jellyfish, would not leave any direct record via body parts but could possibly record their existences in the form of burrows, tracks, or impressions in the sediment. Plants, if preserved in a reducing environment that prevents decomposition, such as swamps and bogs, will yield complete fossils, including stems, leaves, flowers, seeds, and pollen. The soft parts of vertebrate animals, conversely, completely decompose at death, leaving behind only the skeleton. In the case of elasmobranchs, only the teeth plates preserve.

Invertebrates with hard exoskeletons or support structures such as mollusks, echinoderms, brachiopods, arthropods, and corals leave behind the most complete fossils. Since sediment deposition is a constant event in aquatic environments, marine sediments and sedimentary rocks are usually far more fossiliferous than those from terrestrial environments. Considering these two points, it is not difficult to explain why marine invertebrates make up, by far, the bulk of the fossil record.

Based on preservability and mode of preservation, fossils fall into two broad categories. The first encompasses the recognizable remains of an organism, either partial or complete (*body fossils,* or *Korperfossilien* in the German literature). The second takes into account the evidence of activity or lifestyle, usually in the form of some trace such as tracks, burrows, excreta, etc. (*trace fossils,* or *Spurenfossilien* of Abel, 1935; Seilacher, 1953; and German literature). Scientists who study these two aspects of the fossil record, either separately or together, are called paleontologists (q.v. in Vol. VII).

In studying the phylogeny, evolution, and ecology of fossil organisms, the paleontologist must be familiar with living forms and their lifestyles. At the same time, the paleontologist must be aware of geological aspects such as tectonism, sedimentology (q.v. in Vol. VI), stratigraphy, geochemistry (q.v. in Vol. IVA), and structural geology. The synthesis of all of these aspects, and more, is a powerful tool for understanding the history and evolution of the Earth. Paleontology, in turn, is further broken up into many subdisciplines, some of which include *invertebrate and vertebrate paleontology* (partitioned by specialty group), *paleobotany* (fossil plants), *paleopalynology* (fossil pollen), *micropaleontology* (microscopic fossil protista and monera), *taphonomy* (the fossilization and preservation process), *ichnology* (trace fossils), *paleobiogeography* (ancient distributions of organisms), *paleoecology* (paleoenvironments and ancient ecosystems), and applied aspects such as *paleoclimatology, geochronology,* and *biostratigraphy.*

Preservation Of Body Fossils

The type and completeness of preservation depend on how long and under what physical conditions an organism is buried. Once encased in sediment and exposed to infiltration and percolation of mineral-rich groundwater, most hard parts undergo some degree of mineralogical replacement. This can vary from total organic replacement by silicates, such as petrified wood or dinosaur bones, to only partial replacement of the protein component of mollusk and brachiopod shells by calcite.

If a shell, e.g., is buried in soft sediment with little groundwater infiltration, it often will preserve completely, retaining delicate spines and shell ornamentation (Fig. 1). Exceptional fossilization conditions with only partial replacement will sometimes allow the preservation of porphyrin color patterns on the shell. Preservation of the original color pattern is a boon to the system-

atic paleontologist and has been found in Paleozoic brachiopods, Mesozoic cephalopods, and more commonly, Cenozoic gastropods such as the Pliocene cone shells shown in Fig. 2. Unlike shells, most vertebrate bones undergo complete remineralization, by ion exchange, within a short period of time.

Besides replacement, aragonite fossils, such as seashells, can be mineralogically altered to calcite by exposure to groundwater. This recalcification from a metastable to stable carbonate variety often obliterates the fine details of the original aragonitic skeleton. This process commonly occurs in fossilized scleractinian corals, and Fig. 3 shows the general appearance of recalcified coral heads. If groundwater infiltration is intense enough, in some cases the entire body fossil is leached away, leaving only a hollow cavity in the sediment or rock. The surfaces of the cavity, referred to as the *mold,* preserve the exterior morphology of the organism. The cavity is often later filled by a mineral deposit, usually calcite or microcrystalline silicates, to produce an exact replica of the original fossil. The secondary, completely replaced replica is called a *cast,* or *pseudomorph* (either calcitic or siliceous). If the replica is produced by the infilling of consolidated mud or fine particulates, it is referred to as a *steinkern.* Steinkerns often pre-

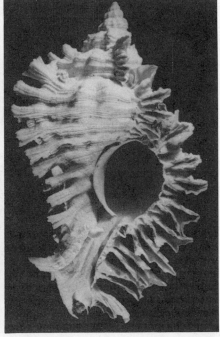

FIGURE 1. Extraordinary preservation and only partial replacement of Pliocene marine gastropods from Sarasota, Florida. Left, *Hystrivasum locklini;* right, *Chicoreus floridanus.*

FIGURE 2. Preservation of color patterns in Pliocene cone shells from Sarasota, Florida. Left, *Conus yaquensis* with brick-and-mortar pattern; right, *Conus vaughanensis* with spotted pattern.

serve only the interior morphology of the organism.

Ichnology And Trace Fossils

The study of trace fossils and evidences of lifestyle activities represents an important subdiscipline of paleontology called *ichnology*. The naming

FIGURE 3. Recalcified heads of the corals *Montastrea annularis* and *Montastrea cavernosa* from a Pleistocene reef deposit, Miami, Florida.

of the trace fossils, or *ichnofossils,* many of which cannot be correlated with any known organisms, embodies yet another subdiscipline called *ichnotaxonomy.* For more information regarding ichnotaxonomy, see Seilacher (1953). A detailed review of ichnology is given by Ager (1963).

The ichnotaxonomic system, like zoological nomenclature, is hierarchically divided into *ichnogenera* and *ichnospecies.* Unlike the phylogenetic relationships of zoological nomenclature, however, the ichnofossils of any ichnotaxon, although similar in appearance, may have been made by completely unrelated organisms. Depending on which group of ichnofossils, every ichnotaxon traditionally has a specific prefix or suffix added to it. For instance, the ichnogenera of dinosaur footprints often end in –*ipus,* signifying a footprint. The American Triassic dinosaur genus *Anchisauripus,* as an example, is known only from its footprints, as are several dinosaurs from Lesotho, southern Africa (Ellenberger and Ellenberger, 1960). Ichnogenera of wormlike organisms often bear the prefix *Helminth* (worm), as seen in the ichnotaxa *Helminthopsis* and *Helminthoida.*

Ichnofossils can be divided into several distinct categories. Some of these may be the most abundant or often the only type of fossil remains present. The following sections describe some of the more commonly encountered of these ichnofossils.

Tracks Tracks are the evidences of the locomotive abilities of an organism that could lift its body off the substrate. Tracks are usually present as individual footprints, such as those laid down by dinosaurs or birds. Very often, as in the case of dinosaurs or large amphibians such as the Pennsylvanian *Eogyrinus,* the impression left from the tail dragging in the mud is also preserved. The position of the tail imprint in relation to the footprints can also indicate what mode of locomotion the animal used, such as bipedal, fishlike slithering, or elevated quadripedal as in crocodilians.

Trails (*Kiechspuren* of Seilacher, 1953, and in the German literature) Marks left by bottom-dwelling organisms as they crawl over unconsolidated sediments are usually placed in this category. Trails can be in the form of corrugated, treadlike patterns produced by multiappendaged organisms such as trilobites, merostomes, onychophorans, or myriapods. They can also be simple smooth-sided meanderings that were produced by polychaetes, holothurians, or other wormlike organisms. Some types of trails are tightly coiled and extremely convoluted and may represent marks made by organisms as they browsed on surficial algae films. Many groups of Recent gastropods, such as *Strombus,* feed in this manner and leave similar trails. These browsing trace fossils are referred to as *Wiedespuren* by Seilacher (1953) and in the German literature.

Burrows In aquatic and intertidal situations, the majority of animal life in soft substrate areas is infaunal, transiently burrowing into the sediments or living inside permanent buried tubes. The constant overturning of the sediments, as the infaunal organisms move through them, leads to a process called *bioturbation*. Bioturbated sediments often lose their stratigraphic identity, with all previously distant depositional layers having been mixed together by the churning of the infauna. Bioturbation, then, is a form of trace fossil.

Two distinct types of burrows are known to be present in the fossil record: dwelling and habitation burrows (*Wohnbauten* of Seilacher, 1953), and feeding structures (*Fressbauten* of Seilacher, 1953). Dwelling burrows are simple in form, often being vertical tubes with both ends open at the sediment surface. In the Recent, burrows of these types are often produced by polychaete worms, sipunculids, and hemichordates like *Balanoglossus*.

Feeding structures are more complex in form than dwelling burrows, often with single large depressions, mounds, or radiating groups of depressions around the openings to the burrow. These depressions result from the ingesting of sediments, with the accompanying edible detritus, by the animal, while associated mounds are produced by the expelled waste sediments. Some organisms, such as the crustacean *Callianassa* or the echinoid *Micraster* (Nichols, 1959), have a combined burrow, with a habitation area, an open vertical shaft to the surface, and lateral feeding burrows for the mining and ingesting of the sediment and for excreta.

Terrestrial burrows are also known in the fossil record. In form, they more often resemble the previously mentioned complex combination burrow and are known to have been produced by rodents, snakes, tortoises, and small insectivores (Ager, 1963). An example of a particularly bizarre type of terrestrial burrow is the elongated and coiled *devil's corkscrew* of the Miocene of Nebraska (Ager, 1963). These large coiled burrows were given the ichnotaxon *Daimohelix* and are now thought to have been produced by a primitive burrowing beaver.

Borings Rock seafloors, shells, bones, and other hard substrates are almost always bored by large numbers of specialized animals. Primary among these are the boring sponges of the genus *Cliona* and rock-boring bivalve mollusks of the genera *Lithophaga, Gastrochaena, Rupellaria,* and *Spengleria*. The boring sponges, which range from the early Silurian to the Recent, honeycomb coral heads, shells, and other hard carbonate substrates, etching them with acids and enzymes. This intense, concentrated boring weakens the substrate and causes it to crumble. Boring bivalves produce long tubular depressions and, along with *Cliona,* are the major bioerosion mechanisms on coral reefs. *Cliona* borings in a gastropod shell are shown in Fig. 4. Other types of borings are produced by sipunculids, echiurins, cirrepeds, polychaetes, and phoronids.

Excreta The fossilized excrement of organisms falls into two categories. If they are produced by large vertebrates, such as reptiles, mammals, or fishes, they are referred to as *coprolites*. This term was first introduced by Buckland (1892) for fossilized fecal remains that are elongated or oval in shape and with a lumpy or convoluted surface. Some ornithischian dinosaur coprolites, presumably from *Stegosaurus,* have been found to exceed 25 cm in length.

Smaller vertebrate and invertebrate groups produce *fecal pellets*. These are usually rounded in shape, occur in clumps, and rarely exceed 20 mm in size. The fecal pellets of gastropod mollusks are often cross shaped or triangular in cross-section. Some types of calcified gastropod fecal pellets form huge accumulations in quiet lakes and ponds and, after lithification, result in *pellet lime-*

FIGURE 4. *Cliona* sponge borings on the ventral side of *Fulguropsis elongatum,* Pliocene of Florida.

stones. The gastropod *Mexipyrgus,* from volcanically heated pools in northern Mexico, produces thick layers of carbonate-rich fecal pellets that eventually produce a *pellet tufa* and a *pellet travertine* (Taylor, 1966).

Gastroliths These rounded, polished stones are commonly associated with herbivorous dinosaurs and probably served as grinding stones in the crop or gizzard. Recent seed-eating and herbivorous birds have a similar grinding arrangement, but on a much smaller scale. Pebbles and grit are swallowed and stored in the gizzard where, through strong muscular contractions, they are used to crush tough vegetable matter. The gizzard stones, through constant grinding and rubbing together, take on a polished appearance much like that produced in lapidary. Gastroliths have been found associated with the rib cages of hadrosaurs, plateosaurs, and some sauripods, indicating that they were associated with the gastrointestinal system.

Drill Holes, Bites, and Breaks Organic remains, particularly molluscan shells, very often preserve evidences of predation or attempted predation by crabs, fishes, or drilling marine gastropods. This latter group, primarily genera of the families Muricidae, Thaididae, and Naticidae, use their tooth-covered, ribbonlike radula within an extendable proboscis to drill a round hole in the shells of their prey. Once the shell is pierced, the victim's flesh is rasped out or digested by injected gastric enzymes. A drill hole of a naticid gastropod in the shell of another marine gastropod is shown in Fig. 5. This drilling process first evolved in the Mesozoic (Vermeij, 1978).

Healed breaks in gastropod and bivalve shells, caused by the cracking action of crabs with specialized claws, are often very common, particularly in Cenozoic mollusks. The crab holds the shell in one claw and attempts to peel back the edge of the shell to reach the retracted animal (Vermeij, 1976). If the shell cannot be broken further and the mollusk is still safe, the crab will abandon the shell and search elsewhere for prey. The broken mollusk will then quickly heal the crab break, leaving a characteristic longitudinal scar on the shell like the one shown in Fig. 6. Toothmarks are also found on mollusk shells. Judging from the characteristic pattern of toothmarks on many ammonite shells, the Cretaceous marine mosasaur, *Tylosaurus,* apparently fed on large pelagic cephalopods.

Eggs Although extremely rare, there are a few cases where vertebrate eggs have been preserved in the fossil record. The most famous of these was the discovery in Mongolia of a fossilized egg clutch of the ceratopsian dinosaur *Protoceratops.* The description of the eggs was given by van Straelen (1925). Neogene bird eggs have also been described (Ladd, 1934).

Commercial Value and Uses Of Fossils

Besides their intrinsic beauty and scientific interest, fossils do have commercial uses and monetary value. Fossil collecting is a big hobby, and there are literally thousands of amateur paleontologists in the United States alone. There is also a wide spectrum of fossil-collecting and paleontological societies, ranging from local fossil clubs

FIGURE 5. Naticid gastropod boring near the spire of *Pyruella rugosicostata,* Pliocene of Florida.

FIGURE 6. Healed crab predation attempt on ventral side of *Conus symmetricus,* Pliocene of the Dominican Republic.

composed of a handful of enlightened amateurs to professional groups like the Paleontological Society or the Society of Economic Paleontologists and Mineralogists. As collection artifacts, some fossils can sell for large sums of money—e.g., from a few dollars for a well-preserved trilobite up to tens of thousands of dollars for complete dinosaur skeletons. Several scientific supply houses carry large selections of fossils, and several more large commercial fossil dealers work through mail order catalogs. These companies advertise in many of the paleontological journals and newsletters. Swap-meets and trading by mail are also good ways to build a fossil collection. Information on these aspects can be obtained from local fossil clubs and museums.

The only true gem-grade fossil used in personal adornment is *amber.* Amber is the permineralized sap of different types of trees, most often evergreens. After burial, the normally gummy sap hardens into a clear, golden lump, which can be polished into beads for jewelry. Before alteration to amber, insects such as ants, flies, and wasps and arachnids such as spiders and mites are very often trapped and surrounded by the oozing sap and form *inclusions.* These trapped insects are perfectly preserved, down to every body hair, and increase the value of the amber. Amber-encased insects are also the main source of a fossil record for paleoentomologists.

Probably the most useful fossil, to humans, is *coal.* The three forms of coal, lignite, bituminous coal, and anthracite, formed from the compacted remains of plants that were buried in the reducing environments of swamps and bogs. Complete plant fossils, usually ferns and horsetails, are frequently found in Carboniferous (Mississippian-Pennsylvanian) coal beds. The coal deposits of the Carboniferous period are the largest and best developed worldwide, but smaller deposits are also known that date from Mesozoic and Cenozoic times.

In certain areas of the world, marine fossils are mined and crushed for roadbeds and landfill, making them a commercial resource. In southern Florida, Pliocene and Pleistocene fossil beds, composed mostly of shells and corals, are mined from large quarries and have been used for the foundations of many large housing subdivisions along the Gulf of Mexico coast. These types of pit mines, or shell pits, form a multi-million dollar industry. For example, a large portion of U.S. Interstate Highway 75 (I-75) along western Florida was built on the crushed and compacted fossils taken from one of these quarries.

To the geologist, the most pragmatic application of fossils to geological problems is the concept of the *index fossil,* or *guide fossil.* These are fossils, usually a genus or species, that can be used to identify and date the layer in which they are found. Ideally, an index fossil should be common, be widespread over a number of environments, and have evolved and become extinct within a known and well-defined period of time. Index fossils can be indicative of any length of time. Therapod dinosaurs, e.g., are index fossils for the Mesozoic era, while one therapod group, the tyrannosaurids, are index fossils for only the Cretaceous period. By using index fossils, a geologist can determine the time of deposition of most fossiliferous strata, and this is the foundation of *relative dating* techniques.

Where To Find Fossils

Since fossils are initially buried, their exposure on the land surface can occur only by a number of catastrophic mechanisms that disturb the original areas of deposition. These mechanisms can be natural processes such as erosion by wind and water, glaciation, and tectonism, or artificial processes. The fossil collector must be able to recognize these areas of highest potential yield and must also know the geology of the area in which the collecting is being undertaken. Geological maps (q.v.) and field guides are available from most state geological departments (q.v. in Vol. XIII) and from most museums and large public libraries.

Erosion is probably the primary way in which potentially rich fossil beds are exposed. Eroded cliffs along lakes, bays, and the open ocean are some of the best places to collect fossils. A classic example is the Calvert Cliffs area of Maryland, along Chesapeake Bay. Here, extremely rich fossil beds of Miocene age are exposed, often yielding complete whale skeletons and giant shark's teeth. The banks of rivers, walls of canyons, and steep ravines are also excellent places to look for fossils.

Eroded anticlines in areas of folded mountains often expose huge fossiliferous deposits. Glaciated areas are often good places to look for fossils, especially in spots where a sheer, clifflike face has been exposed. Large Ordovician coral reef deposits have been exposed in exactly this manner in southern Wisconsin and Iowa. Rich faunas of nautiloids, brachiopods, crinoids, and trilobites are often found in huge slabs exposed right at the surface by Pleistocene glaciers. Tectonically active areas that produce uplift and thrust faulting will also expose large deposits of fossils.

Another good place to look for fossils is in quarries and open pit mines. Some phosphate mines in North Carolina and Florida cut through the Pliocene and well into Miocene sediments and regularly expose large numbers of vertebrate fossils. Some of the more spectacular fossils include whale skeletons, primitive elephants, camels, horses, and giant crocodilians. Spoil banks along

211

deep canals are also good places to find fossils. Some banks along canals in southern Florida yield entire Pliocene coral reefs, while the spoil banks along the Delaware Canal in Delaware contain large numbers of Cretaceous marine fossils, including belemnites, *Exogyra* oysters, and mosasaur bones.

EDWARD J. PETUCH

References

Abel, O., 1935, *Vorzeitliche Lebensspuren*. Jena, East Germany: Gustav Fischer, 344p.

Ager, D. V., 1963, *Principles of Paleoecology*. New York: McGraw-Hill, 371p.

Buckland, W., 1892, On the discovery of coprolites, or fossil faeces, in the Lias at Lyme Regis, and in other formations, *Geol. Soc. London Trans.* **2**(3), 223–236.

Ellenberger, F., and P. Ellenberger, 1960, Sur une nouvelle dalle à pistes de vertebrés, découverte au Basutoland (Afrique de Sud), *Soc. Géol. France Compte Rendus* **9**, 236–238.

Ladd, H. S. 1934, Geology of Vitilevu, Fiji, *Bernice P. Bishop Mus. Bull.* **119**, 1–263.

Nichols, D., 1959, Mode of life and taxonomy of irregular sea-urchins, *Systematics Assoc. Pub.* **3**, 61–80.

Seilacher, A., 1953, Studien zur Palichnologie. 1. Ueber die Methoden der Palichnologie, *Neues Jahrb. Geologie u. Paläontologie* **96**, 421–452.

Straelen, V. van, 1925, The microstructure of the dinosaurian eggshells from the Cretaceous beds of Mongolia, *Am. Mus. Novitates* **173**, 1–62.

Taylor, D. W., 1966, A remarkable snail fauna from Coahuila, Mexico, *Veliger* **9**(2), 152–228.

Vermeij, G. J., 1976, Interoceanic differences in vulnerability of shelled prey to crab predation, *Nature* **260**, 135–136.

Vermeij, G. J., 1978, *Biogeography and Adaptation*. Cambridge: Harvard University Press, 237p.

Cross-references: *Coal Mining; Exposures, Examination of; Field Geology; Geological Cataloguing; Geological Methodology; Geological Survey and Mapping; Geological Surveys, State and Federal; Geology, Scope and Organization; Samples, Sampling.*

FRACTURES, FRACTURE STRUCTURES

Planes of Weakness and Their Patterns

Rock mass behavior, which for engineering purposes can be described by its resistance (strength), deformability, and permeability, is affected by the combined behavior of continuous (or intact) rock and the planes of weakness (discontinuities) (see Vol. XIII: *Residual Stress, Rocks*). Hereafter, planes of weakness will be called *discontinuities*, because such planes represent a rupture of the continuity of intact or solid rock. Discontinuities are planar features (or surfaces) of smaller resistance and higher deformability and permeability than the intact rock. This section describes the practical effects of these discontinuities, their geometric properties and their observation in the field.

Figure 1 shows practical effects of discontinuities in rock masses. Horizontal stress relief joints restrict the size of granite blocks that can be cut from the quarry in Fig. 1*a*. Horizontal bedding plane discontinuities weakened this shale block (Fig. 1*b*) which had to be reinforced with steel rock bolts during excavation. These discontinuities also provided the path for water flow as demonstrated by the horizontal upper limits to the wetted (black) portions of the rock face. Dipping micaceous shear zones in the granitic gneiss in (Fig. 1*c*) provided planes of low friction along which blocks could slide into the excavation. This instability led to moving the construction crane back and reinforcing the slope with rock bolts. The persistent or extensive joint in the metamorphosed sediments shown by the white region in Fig. 1*d* provided a surface along which the slope above the roadway failed.

The importance of discontinuities in the behavior of rock masses can not be underestimated. Since all rock masses are intersected by such discontinuities, no large volume of rock is without them. In fact, in many instances what is most important in large rock volumes is what is not rock.

Terminology

Discontinuities are characterized on an individual basis as well as in the context of the rock mass. The following definitions describe the commonly encountered discontinuity characteristics (Einstein and Dowding, 1981). Individual discontinuities can be described with the following terms that are illustrated in Fig. 2*a*:

Surface: The interface between the wall rock and a feature. The two surfaces of a discontinuity can be in tight (matching) contact, touch each other at a few asperities (defined later) or be completely separated as in a fault.

Wall Rock: The material rockward from the discontinuity surface. The wall rock can vary from sound to completely weathered. It can be described in geologic-petrographic terms, in geotechnical terms (resistance, deformability, permeability), or both. Indexes of geotechnical properties, such as rebound hardness, are frequently employed.

Filler: Material completely or partly filling the space between discontinuity surfaces. The material can be a product of formation or transported into and then deposited or precipitated along the discontinuity. Weathered wall rock does not qualify as filler although, from a mineralogic and geotechnical point of view, the two may be identical.

FIGURE 1. Practical effects of discontinuities in rock masses. (*a*) Stress relief joints limit size of granite blocks. (*b*) Bedding planes transmit water and weaken shale. (*c*) Foliation shear zones control excavation in gneiss. (*d*) Persistent joint and bend failure in metamorphosed sediments.

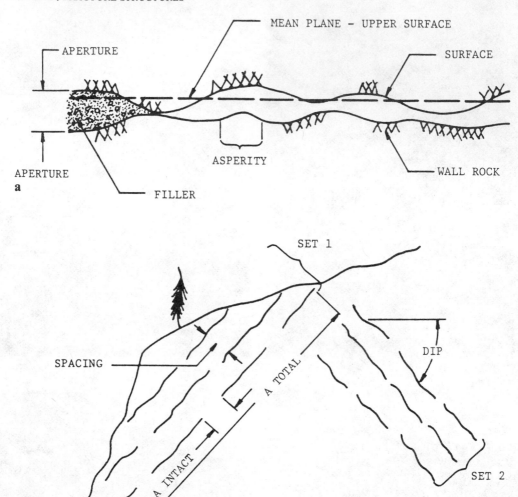

FIGURE 2. Geometrical terminology of discontinuities. (*a*) Individual discontinuity; (*b*) discontinuities in the rock mass. (From Einstein and Dowding, 1981)

Aperture (width): The distance between discontinuity surfaces measured perpendicularly to the mean plane. Aperture is not necessarily the distance between sound rock. For instance, if weathered rock is in tight contact, the aperture is zero. Furthermore the term is somewhat ambiguous if the surfaces touch only at a few positions, as in Fig. 2*a*.

Undulations, Asperities, and Roughness: The deviations of the surface from a mean plane on various scales (undulations are the largest; roughness, the smallest). The scales often overlap and the terms are frequently interchanged. Asperities

significantly affect discontinuity strength (Einstein and Dowding, 1981; Goodman, 1980).

Rock mass discontinuities can be described with the following terms, illustrated in Fig. 2*b*:

Discontinuity Set: Discontinuities at various locations in the rock mass that are approximately parallel to each other.

Attitude: The orientation (i.e., strike and dip) of the discontinuity plane in space. The strike of discontinuity set 2 is perpendicular to the plane of the drawing, and the dip is approximately 50°.

Spacing: The distance between two adjacent discontinuities within a set.

Discontinuity system: Several sets.

Persistence: The proportion of continuity along the mean plane, i.e., $(A_{total} - A_{intact})/A_{total}$ shown for a joint in set 1. The term *continuity* is sometimes used.

Cleft-water pressure: The pressure of water filling discontinuities in a rock mass. The cleft-water pressure may or may not be equal to the pore-water pressure in the intact rock.

Geometric Properties

Geometric characteristics of discontinuities (persistence, attitude, spacing) are of equal and sometimes greater practical importance in design than the mechanical properties of small samples of a discontinuity that can be measured in the laboratory. Persistence is perhaps the most important because the discontinuities can be so much weaker than the intact rock. The persistence of the joint in the metamorphosed sediments in Fig. 1*d* allowed it to dominate the performance of the slope. Unfortunately, there is no direct method to measure persistence physically. It is usually inferred by the frequency of observation of discontinuities with a particular attitude; the most persistent are those that are observed most often. Persistence can also be inferred by thickness, with the thicker discontinuities being most likely to extend the farthest.

Attitude in relation to imposed loads and directions of possible movement is also important. The shear zone in the gneiss in Fig. 1*c* would not lead to instability if it were located on the other side of the excavation. It would dip away from the opening on the other side and therefore could not direct sliding into the opening. In the design of excavations the most frequently observed discontinuities (usually joints) are assumed to be located at the critical locations (i.e., the bottom of an excavation on the side that causes the most difficulty) and to be persistent enough to allow failure.

Discontinuities in rock masses occur in a wide variety of sizes as shown in Table 1 (Goodman, 1980). This table is presented to compare thicknesses. As with most attempts to place numerical limits on natural phenomena, these limits are not exact and may vary from descriptions in other references. However, the relative sense is the same. Faults are the thickest, most persistent, and most

TABLE 2. Descriptive Terminology for Joint Spacing and thickness of Bedding Units

Descriptive Term (joints)	Joint Spacing or Thickness of Beds (Meters)	Descriptive Term (Bedding)
Very close	<0.05	Very thin
Close	0.05–0.3	Very thin
Moderately close	0.3–1	Medium
Wide	1–3	Thick
Very wide	>3	Very thick

Source: Based on data from Deere et al., 1969.

popularized by their association with earthquakes. Shear zones are small faults and/or collections of joints that greatly weaken a rock mass. Joints are usually defined as fractures along which no relative displacement has occurred. However, subsequent weathering from percolation of fluids tends to increase their thickness. Unlike faults and shear zones, which tend to be singular in occurrence, joints occur in semiparallel families or sets. Fissures or fractures can be found as flaws in core-sized intact specimens, or they can be persistent enough to be considered joints. Microfissures are usually found in individual crystals of rock. The remainder of this section deals with joint- and shear-zone-sized discontinuities. The others are treated elsewhere in this encyclopedia.

Table 2 (Deere et al., 1969) presents a quantitative guide for describing the spacing between joints in a particular set or family. This nomenclature also conveys the typicality of the spacing. Joint spacing greater than 1 m is unusual, while that greater than 3 m is very unusual. Of course, spacing is heavily dependent on rock type, with shaley sediments and foliated metamorphic rock tending to have the smallest joint spacing.

Methods of Observation and Reporting

Joints and other larger discontinuities may be mapped from aerial photographs or through direct field observation. An example of the usefulness of aerial photography is shown in Fig. 3, a photograph of faults and shear zones in an igneous-metamorphic complex. The shear zones have been differentially weathered into depressions that are highlighted by trapped snow. Such aerial photography is useful to define linear trends that are persistent over long distances; however, much larger scale photographs are necessary to detect typical joint-sized discontinuities.

Field observations of discontinuities are made at rock outcrops. Unfortunately, rock outcrops are often scarce or not available in critical areas. For instance, even above the tree line in mountainous regions, outcrops of bedrock (excluding talus and other dislodged rock) are often less than

TABLE 1. Discontinuity Thickness

Faults	>20 m
Shear zones	5 mm–20 m
Joints	0.01–5 mm
Fissures (fractures)	0.01–0.1 mm
Microfissures	<0.01 mm

FIGURE 3. Aerial photograph of faults and shear zones highlighted by trapped snow.

15% of the available surface area. Because of the lack of suitable outcrop, one may be tempted to infer discontinuity occurrences from nearby and geologically similar outcrops. However, to do so requires knowledge of the exact stress history relationship of the two areas, which is almost never known.

Figure 4 shows how field observations of average strikes and dips of regional and outcrop discontinuities in a fold can be presented (Billings, 1972). Fig. 4a shows the location of the field observations, and Fig 4b shows the points of tangency of the average bedding plane orientations on the lower hemisphere of a stereonet. Each of the three structurally similar groups of orientations are also grouped on the stereonet.

At each of the locations where the average bedding orientation was determined, two other sets of discontinuities perpendicular to the bedding planes could also have been observed. Hypothetical stereonet representations of the strikes and dips of observed joints at outcrop B are given in Fig. 4c. A particular set of joints will be identified by the tendency of its joints to group about a mean orientation. The variation about the mean

results from the local curvature of the joints and bedding. Figure 4d is an isometric or three-dimensional representation of the mean orientation in each of the three groups in the stereonet in Figure 4c.

CHARLES H. DOWDING

References

Billings, M. P., 1972, *Structural Geology.* Englewood Cliffs, N.J.: Prentice-Hall, 606p.

Deere, D. U., A. H. Merritt, and R. F. Coon, 1969, Engineering classification of in-situ rock, *U.S. Air Force Weapons Lab. Tech. Rept. No. AFWL-TR-67-144* or *NTIS AD-848-798,* 270p.

Einstein, H. H., and C. H. Dowding, 1981, Shearing resistance and deformability of rock discontinuities, in Y. S. Touloukian, ed., *Physical Properties of Rocks and Minerals.* New York: McGraw-Hill, 177–219.

Goodman, R. E., 1980, *Introduction to Rock Mechanics.* New York: Wiley, 478p.

Cross-references: *Exposures, Examination of; Photogeology.* Vol. XIII: *Detailed Mapping; Exploration Geology; Geological Structures; Rocks, Engineering Properties; Structural Geology; Weathering.*

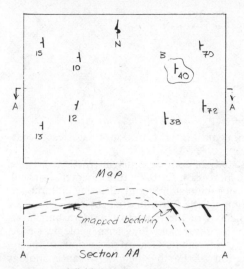

A: Map of field measurements of strike and dip of fold.

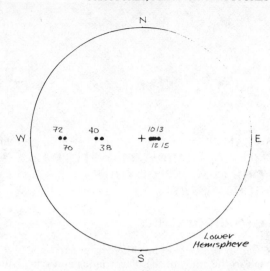

B: Average bedding plane orientations plotted on a stereonet.

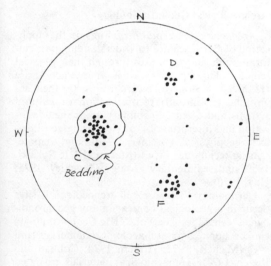

C: All joint observations of outcrop B.

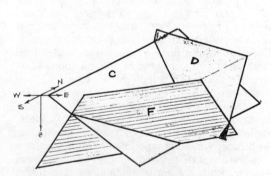

D: Three-dimensional view of average orientation of planes **C, D,** and **F** at outcrop B. **C** = bedding; **D** and **F** = crossbedding.

FIGURE 4. Reporting of regional and outcrop discontinuities. (*A* and *B* after Billings, 1972)

G

GAS SAMPLING—See
ATMOGEOCHEMICAL PROSPECTING; PETROLEUM EXPLORATION GEOCHEMISTRY.

GEOANTHROPOLOGY

Geoanthropology is an interdisciplinary approach to the study of the diachronous and synchronous interaction between Earth processes and humanity (MacNeish, 1967). Anthropology, an eclectic social science, covers all human (hominid) activity—past, present, and future. Earth science is also eclectic and so steeped in mathematics, physics, chemistry, and biology that it is difficult to separate unique Earth science concerns from those of other physical sciences. Just as anthropologists use insights from other disciplines to understand humans, Earth scientists cross disciplinary boundaries to understand Earth processes. Such broad and inclusive disciplines and definitions yield a large number of possible topics (Tuttle, 1975) in geoanthropology.

Anthropology, especially archaeology (see *Geoarchaeology*), and geology have much in common. Both deal with processes, factors, ideas, hypotheses, and theories that cannot commonly be directly viewed or experimented with. *Artifacts* (any evidence of hominid activity) and *fossils* are similar. Both are partial remains of past things and events representing processes and activities of larger spatio-temporal systems.

Anthropologists and geologists deal with relatively large time frames. Geologic time begins with the formation of the Earth about 4.5 b.y. ago. Anthropologic time begins more recently with the appearance of very primitive human ancestors as much as 70 m.y. ago. The most significant period for anthropologists begins about 4.5 m.y. ago (Late Pliocene) with the appearance of the first hominid, a subjective point on an evolving continuum that is difficult to define (Coppens, 1976). Modern humans evolved during the Pleistocene, which dominates human archaeology (Butzer and Isaac, 1975). Cultural anthropology centers on the past 10,000–12,000 yr (Recent) owing to difficulty in deducing culture from an increasingly incomplete record back through time. Since written history covers only the past 5,000 yr, archaeologists rely on Earth scientists for chronological and en-

vironmental insight into prehistorical occurrences.

A study of contemporary interactions between Earth processes, Earth features, humans, and culture aids understanding of differing cultural perceptions of the Earth and its environment. Then predictions of the impact of modern human activity on short- and long-term geological processes can be made. General discussions of the application of Earth sciences to anthropology include Butzer (1971) and Brothwell and Higgs (1970).

Archaeological Geoanthropology

Archaeological geoanthropology is the application of Earth science to understanding past cultural events and processes through their physical remains. Three primary areas of interaction are (1) the chronology of past cultures, (2) their environments, and (3) reconstruction of cultural systems and events. Secondary areas include location and interpretation of sites using geochemical, geophysical (Breiner, 1973), and remote-sensing techniques (q.v.) (Martin, 1971; Solecki, 1960) and sedimentary geology (Shackley, 1975; Hay, 1976).

Chronology Because all archaeological sites occur in a geological context, many geochronological techniques (see Vol. XIII: *Geochronology*) can be applied to archaeological materials and sites (Michels, 1973; Aitken, 1974; Colman and Pierce, 1977). Archaeological materials are dated by establishing contemporaneity with a particular climate, flora, fauna, topographical feature, or geological event or material of known age. Dates may be expressed in actual years (absolute age) or as a sequential relationship (relative age). A recent trend is widespread conversion of relative time sequences into absolute ones by radiometric analysis. Data from geomorphology, stratigraphy, paleontology, palynology, chemical analysis, paleomagnetism (Tite, 1972), paleotemperature (Emiliani, 1970), dendrochronology (Bannister and Robinson, 1975), varves (Flint, 1971), and thermoluminescence (Aitken et al., 1968) are used to date in archaeology.

Geomorphology Archaeological materials can be dated if they correlate to geomorphological features or events (Klein, 1969; Giddings, 1964). Geomorphological events, with glaciation the major exception, affect local areas so geomorphology has limited regional chronological value.

Stratigraphy Correlating artifacts within an enclosing stratum, then ordering the strata within a site are basic to archaeology. Sites can be internally and externally stratified. Internal stratification is the sequencing of artifacts within sites, while external stratification allows regional sequencing of sites. Haury (1950), Jennings (1957), and Aikens (1970) are examples of internal and external stratification that discuss stratigraphy as applied to archaeology.

Palynology Palynology, the study of fossil and living pollen, is usually applied to environmental reconstruction. It can be used in chronology by fitting the pollen spectrum from a site into a known pollen sequence. Theory and methodology are given in Faegri and Iversen (1964). A practical example is discussed by Martin and Mehringer (1965).

Paleontology Fossil animals or plants occur at many archaeological sites and are used for dating (Hester, 1960; Bordes, 1966) by relating them to a regional sequence. Many fossil species are known and often occur at archaeological sites but remain unrecognized, particularly if microscopic in size. Human fossils (Pilbeam, 1972) are the only aspect of paleontology that is largely ignored by geoscientists.

Rapid mammalian evolution during the Pleistocene makes these data particularly useful for archaeological chronology. European Upper Pleistocene fauna (Thenius, 1962) are very well known. Fauna from Africa (Cooke, 1963) and North America (Hibbard et al., 1965) are relatively well synthesized.

Chemical Dating Common chemical-dating techniques are by measuring fluorine replacement of hydroxyl in teeth and bone (Oakley, 1970), obsidian hydration (Erickson, 1975), fission tracks (Fleischer, 1975), and chemical alteration (patination) on the surface of lithic fragments (Goodwin, 1960). Fluorine replacement and patination are used only for establishing contemporaneity of two or more samples from within a site. A famous use of fluorine dating was the exposure of the Piltdown Man hoax by Oakley and Weiner (1955).

Two recent chemical-dating techniques are fluorine diffusion (Taylor, 1975) and amino acid racemization (Bada and Helfman, 1975). Fluorine diffusion, conceptually similar to obsidian hydration, differs from fluorine replacement in that it is based on diffusion of aqueous fluorine 19 into the chipped surfaces of lithic materials.

In amino acid dating, the proportion of L-amino acids to D-amino acids in fossil bone samples is measured (Bada et al, 1974). An equilibrium (racemic) mixture contains equal amounts, but in biological systems, L-amino acids are dominant and new bone may be composed entirely of L-amino acids. Through time, L-amino acids in fossils alter to D-amino acids until a racemic mixture is established. The D- to L-amino acid ratio increases with the age of the sample. Accuracy depends on knowing the average temperature of the sample through time. This problem is minimized in a site by ^{14}C calibration but remains for samples out of context and in regional chronology.

Radiometrics Radiometric dating (Michael and Ralph, 1971) began with ^{14}C dating by Libby (1955) and others from 1946 to 1949. With radiometric-dating techniques, absolute ages are obtained for both archaeological and geological materials. Since 1949, eight isotopes have been used in the radiometric dating of various Quaternary materials and environmental-climatic factors (Broecker, 1965), although ^{14}C remains the most relevant technique.

Environmental Reconstruction Environmental reconstruction (Butzer, 1971) bears on both local and regional environments existing during and following a site's formation (Winter and Wylie, 1974). It is based largely on geomorphology, pedology, sedimentology and petrology, paleontology, and palynology.

Geomorphology Geomorphological data are applied to the reconstruction of topography at the time a site was occupied and to the interpretation of changes affecting a site following abandonment. Geomorphic features and their change through time—both a result of the interaction of structure, relief, topography, substrate, erosion, climate, and hydrology—are valuable environmental indicators. Examples of the use of geomorphology to reconstruct regional and local archaeological environments are Brunet (in Byers, 1967), Judson (1953), Arnold (1957), and Klima et al. (1962).

Pedology The study of soils provides environmental data because solid development depends on climate, geomorphology, biota, bedrock, and time. Soil develops as a result of chemical action on rock. The rate and type of chemical activity (q.v. in Vol. XII, Pt. I) are a product of many factors, the most important of which is climate. Thus, a soil profile records climatic change or stability. Soils and paleosols, particularly by composition, stratification, and sequencing, can be used to infer relative age, climatic and geomorphic history, and general ecology. Cornwall (1958) introduces archaeological soils and Yaalon (1971) provides a general discussion of paleosols.

Sedimentology and Petrology Interpretation of Earth materials containing artifacts is not a well-developed archaeological tool, although the geological methodology exists (see, e.g., Rigby and Hamblin, 1971). Beds of particular significance to anthropologists are surficial or unconsolidated sediments of Quaternary age. Flint

(1970, p. 367) states that Quaternary beds involve "the same range of sediments, the same kinds of problems, and the same techniques as does the stratigraphy of older deposits." In comparison to older deposits, Quaternary deposits are usually less altered and indurated, have clearer indications of their origin and the climate during deposition, and are less disturbed. Factors unimportant in correlating older rocks, such as morphology, weathering, color, induration, and solid development, become useful for correlation of Quaternary rocks and sediments.

Beds containing artifacts must be placed within the context of rocks located above, below, and laterally to determine the main depositional environment. Then each bed can be studied in detail, including microscopically, to determine its relationship to the overall environment and to the artifacts.

Interpretation is facilitated by excavation of archaeological sites in three dimensions. Environments particularly amenable to archaeological study include those associated with caves, lakes, streams, glacial and periglacial terranes, beaches, alluvial fans, and dunes.

Volcanic rocks are useful in chronology and rarely contain artifacts. Metamorphic rocks are of little use in geoanthropology.

Paleontology Floral and faunal remains reflect local and regional climate and ecology for each stratum, and fossil successions indicate changes through time. Heizer (1960) discusses the value of faunal remains in reconstructing environments, and Cornwall (1974) introduces *archaeological paleontology.* Examples are given by Lance (1959), Thenius (1962), Miller (1965), Moreau (1963), Fritts (1971), and Heizer and Napton (1970).

Palynology Palynological studies are probably the most useful botanical technique (Faegri and Iverson, 1964) for environmental reconstruction in archaeology. Pollen preserved in stratified aquatic deposits tend to reflect the regional environment, while pollen from soil samples tends to reflect the local ecology. Clark (1954), Mehringer and Haynes (1965), and Schoeneher and Eddy (1964) provide samples in archaeological contexts.

Cultural Geoanthropology *Cultural Reconstruction* Archaeology is mainly descriptive, relying on cultural anthropology and other social sciences for most theory and data utilized for cultural reconstruction (Thomas, 1974). Culture, used by anthropologists as the basic unifying principle for understanding human behavior, is all nonbiological aspects of human behavior shared by members of a group and transmitted from one generation to the next by social learning. Part of culture is material, but most is ideas, attitudes, knowledge, etc. that leave no material record. Inferring nonmaterial culture from incompletely preserved material aspects of culture is very difficult. Anthropologists have resorted to statistical approaches and to uniformitarianism, by which cultural patterns of existing cultures or well-understood past cultures are used to interpret ancient cultures.

Lithic Analysis Lithic analysis combines knowledge of physical properties of lithic material with their manufacture and use in a culture. The major assumption of lithic analysis is that methods of manufacture and usage are reflected in the edge-wear patterns and general morphology of lithic artifacts, leading to inferences about tool use, manufacture, and general culture (Semenov, 1964; Wilmsen, 1970).

Petrology Petrology is used to identify lithic sources, thus indicating prehistoric trade routes and cultural interactions. Petrological techniques are used in pottery analysis (Williams, 1956; Shepard, 1965). Hester and Heizer (1973) provide a bibliography of petrological analysis in anthropology.

Hydrology Climate can be inferred by estimating the hydrological budget. These estimates can be used to infer archaeological settlement patterns and limits of resource exploitation (Weide, 1974; Byers, 1967; Raikes, 1967).

Other Mineralogy (Arnold, 1971) and geochemistry applied to disease (Cannon and Hopps, 1971; Freedman, 1975) and nutrition (Cannon and Davidson, 1968) are further applications (see Vol. XIII: *Medical Geology*).

Contemporary Cultures

Non-Western Cultures Study of non-Western approaches to Earth science can yield much information about these cultures. Western geologists tend to ignore non-Western geological systems. Work to date is limited to collecting geological mythology (see *Geomythology*) (Vitaliano, 1973) and non-Western solid classifications, geological typologies, and taxonomies (Carter, 1969). Attempts to understand non-Western cultural attitudes toward the Earth are Levi-Strauss (1966) and Haudricourt (1970).

Culture as a Geological Process Throughout most of geologic time change resulted from the interaction of noncultural processes. Since hominids developed culture, they have significantly modified Earth processes. Thus, an understanding of present geological events and the future evolution of the Earth's surface is partly related to the human impact on the environment (Heizer, 1955; Thomas, 1956; Smith, 1976).

Alteration of the environment involves physical, biological, and cultural processes that hominids consciously or unconsciously initiate, inhibit, or deflect. Such alteration is cumulative and is accelerating, particularly since humans developed advanced technology and large populations. Culture affects all levels of the environment, with magnitudes ranging from local changes in solid composition, through regional alteration of the hydrological cycle, to global changes in atmospheric or oceanic composition.

Human-induced extinction of some plants and animals was an early outgrowth of culture that is continuing. This factor will affect the future fossil record and lead to major changes in existing ecological systems. The significance and role of human activity in modern faunal extinctions are discussed in Krantz (1970), and evidence of possible human complicity in Pleistocene and post-Pleistocene extinctions is found in Martin and Wright (1967).

The domestication of plants and animals has significant environmental effects, ranging from alterations in plant communities (Elton, 1958), through increased population, to changes in erosion and evaporation rates. The role of human activity in changing climatic factors is discussed in Charney et al. (1975). Smith (1976) further suggests that cultural activity may cause global changes in the Earth's environment. The recent appearance of artificial chemicals and materials is creating new environmental stresses and ecological niches (Bradbury, 1975).

JAMES P. BARKER
JAMES M. BARKER

References

Aikens, C. M., 1970, Hogup cave, *Utah Univ. Anthropology Papers.*

Aitken, M. J., 1974, *Physics and Archaeology.* Oxford: Clarendon, 291p.

Aitken, M. J., D. W. Zimmerman, and S. J. Fleming, 1968, Thermoluminescent dating of ancient pottery, *Nature* **219,** 442-445.

Arnold, B. A., 1957, Late Pleistocene and Recent changes in landforms, climate, and archaeology in central Baja California, *California Univ. Pubs. Geography* **10,** 201-318.

Arnold, D. E., 1971, Ethnomineralogy of Tecul, Yucatan potters: etics, and emics, *Am. Antiquity* **36,** 20-40.

Bada, J. L., and P. M. Helfman, 1975, Amino acid racemization dating of fossil bones, *World Archaeology* **7,** 160-173.

Bada, J. L., R. A. Schroeder, and G. F. Carter, 1974, New evidence for the antiquity of man in North America deduced from aspartic acid racemization, *Science* **184,** 791-793.

Bannister, B., and W. J. Robinson, 1975, Tree-ring dating in archaeology, *World Archaeology* **7,** 210-226.

Bordes, F., 1966, Observations sur les faunes du Riss et du Wurm, *L'Anthropologie* **69,** 31-46.

Bradbury, J. P., 1975, Diatom stratigraphy and human settlement in Minnesota, *Geol. Soc. America Spec. Paper 171,* 74p.

Breiner, S., 1973. *Applications Manual for Portable Magnetometers.* Sunnyvale, Calif.: Geometrics, 58p.

Broecker, W. S., 1965, Isotope geochemistry and the Pleistocene climatic record, in H. E. Wright and D. G. Frey, eds., *The Quaternary of the United States.* Princeton, N.J.: Princeton University Press, 737-753.

Brothwell, D. R., and E. S. Higgs, eds., 1970, *Science in Archaeology.* London: Thames and Hodson, 720p.

Butzer, K. W., 1971, *Environment and Archaeology; An Ecological Approach to Prehistory.* Chicago: Aldine-Atherton, 703p.

Butzer, K. W., and C. L. Isaac, eds., 1975, *After the Australopithecines: Stratigraphy, Ecology, and Culture Changes in the Middle Pleistocene.* Chicago: Aldine, 911p.

Byers, D. S., 1967, *The Prehistory of the Tehaucan Valley.* Volume I: *Environment and Subsistence.* Austin: University of Texas Press, 332p.

Cannon, H. L., and D. F. Davidson, 1968, Relation of geology and trace elements to nutrition, *Geol. Soc. America Special Paper 90,* 68p.

Cannon, H. L., and H. C. Hopps, 1971, Environmental geochemistry in human health and disease, *Geol. Soc. America Mem. 123,* 230p.

Carter, W. E., 1969. *New Lands and Old Traditions: Kekchi Cultivators in the Guatemalan Lowlands.* Gainesville: University of Florida Press, 153p.

Charney, J., P. H. Stone, and W. J. Quirk, 1975, Drought in the Sahara: A biophysical feedback mechanism, *Science* **187,** 434-435.

Clark, J. G. G., 1954, *Excavations at Star Carr: An Early Mesolithic Site at Seamer, Near Scarborough, Yorkshire.* Cambridge: Cambridge University Press, 200p.

Colman, S. M., and K. L. Pierce, 1977, Summary table of Quaternary dating methods, *U.S. Geol. Survey Map MF 904.*

Cooke, H. B. S., 1963, Pleistocene mammalian faunas of Africa, with particular reference to southern Africa, *Viking Fund Publ. Anthropology* **36,** 65-116.

Coppens, Y., ed., 1976, *Earliest Man and Environments in the Lake Rudolf Basin.* Chicago: University of Chicago Press, 615p.

Cornwall, I. W., 1958, *Soils for the Archaeologist.* London: Phoenix House, 230p.

Cornwall, I. W., 1974, *Bones for the Archaeologist.* London: Dent, 259p.

Elton, C. S., 1958, *The Ecology of Invasions by Animals and Plants.* London: Methuen, 181p.

Emiliani, C., 1970, The significance of deep-sea cores, in D. R. Brothwell and E. S. Higgs, eds., *Science in Archaeology.* London: Thames and Hodson.

Erickson, J. E., 1975, New results in obsidian hydration dating, *World Archaeology* **7,** 151-159.

Faegri, K., and J. Iversen, 1964, *Textbook of Pollen Analysis.* Copenhagen: Munksgaard, 237p.

Fleischer, R. L., 1975, Advances in fission track dating, *World Archaeology,* **7,** 136-150.

Flint, R. F., 1971, *Glacial and Quaternary Geology.* New York: Wiley, 892p.

Freedman, J., ed., 1975, Trace element geochemistry in health and disease. *Geol. Soc. America Spec. Paper 155,* 118p.

Fritts, H. C., 1971, Dendroclimatology and dendroecology, *Quaternary Research* **1,** 419-449.

Giddings, J. L., Jr., 1964, *The Archaeology of Cape*

Denbigh. Providence, R.I.: Brown University Press, 331p.

Goodwin, A. J., 1960, Chemical alteration (patination) of stone, in R. F. Heizer and S. F. Cook, eds., *The Application of Quantitative Methods in Archaeology*. Viking Fund Publ. Anthropology No. 28, 300-312.

Haudricourt, A., 1970, Aspects qualitatifs des civilisations agricoles de la Société de Communaute Primitive, *VII Cong. Internat. des Sciences Anthropologie et Ethnology* **5,** 506-507.

Haury, E. W., 1950, *The Stratigraphy and Archaeology of Ventana Cave, Arizona*. Tucson: University of Arizona Press, 599p.

Hay, R. L., 1976, *Geology of the Olduvai Gorge*. Berkeley: University of California Press, 242p.

Heizer, R. F., 1955, Primitive man as an ecologic factor, *Kroeber Anthropological Soc. Papers* **13,** 1-31.

Heizer, R. F., 1960, Physical analysis of habitation residues, *Viking Fund Publ. Anthropology* **28,** 93-157.

Heizer, R. F., and S. F. Cook, eds., 1960, *The Application of Quantitative Methods in Archaeology*, Viking Fund Publ. Anthropology No. 28, 358p.

Heizer, R. F., and L. K. Napton, 1970, Archaeology and the prehistoric Great Basin lacustrine subsistence regime as seen from Lovelock Cave, Nevada, *University Archaeol. Research Facility Contr. 10*, 202p.

Hester, J. J., 1960, Late Pleistocene extinction and radiocarbon dating, *Am. Antiquity* **26,** 58-77.

Hester, T. R., and R. F. Heizer, 1973, *Bibliography of Archaeology I: Experiments, Lithic Technology, and Petrography*. Reading, Mass.: Addison-Wesley (Module in Anthropology No. 29), 56p.

Hibbard, C. W., D. E. Ray, D. E. Savage, D. W. Taylor, and J. E. Guilday, 1965, Quaternary mammals of North America, in H. E. Wright and D. G. Frey, eds., *The Quaternary of the United States*. Princeton, N.J.: Princeton University Press, 509-526.

Jennings, J. D., 1957, Danger Cave, *Soc. Am. Archeol. Mem. 14*, 328p.

Judson, S., 1953, Geology of the San Jon site, Eastern New Mexico, *Smithsonian Misc. Colln. 121*, 70p.

Klein, R. G., 1969, *Man and Culture in the Late Pleistocene*. San Francisco: Chandler, 259p.

Klima, B., et al., 1962, Stratigraphie des Pleistozäns und Alter des Paläolithischen Rastplates in der Ziegelei von Dolni' Vestonice, *Anthropozoikum (Praha)* **11,** 93-145.

Krantz, G. S., 1970, Human activities and megafaunal extinctions, *Am. Scientist* **28,** 164-170.

Lance, J. F., 1959, Faunal remains from the Lehner mammoth site, *Am. Antiquity* **25,** 35-42.

Levi-Strauss, C., 1966, The scope of anthropology, *Current Anthropology* **7,** 112-123.

Libby, W. F., 1955, *Radiocarbon Dating*, 2nd ed. Chicago: University of Chicago Press, 175p.

MacNeish, R. S., 1967, An interdisciplinary approach to an archaeological problem, in D. S. Byers, ed., *The Prehistory of the Tehaucan Valley*. Vol. I: *Environment and Subsistence*. Austin: University of Texas Press, 14-24.

Martin, A., 1971, Archeological sites—Soils and climate, *Photogramm. Eng.* **37**(4), 353-357.

Martin, P. S., and P. J. Mehringer, 1965, Pleistocene pollen analysis and biogeography of the southwest, in H. E. Wright and D. G. Frey, eds., *The Quaternary of the United States*. Princeton, N.J.: Princeton University Press, 433-452.

Martin, P. S., and H. E. Wright, eds., 1967, *Pleistocene Extinctions: The Search for a Cause*. New Haven: Yale University Press, 433-452.

Mehringer, P. J., and C. V. Haynes, 1965, The pollen evidence for the environment of early man and extinct mammals at the Lehner mammoth site, southeastern Arizona, *Am. Antiquity* **31,** 17-23.

Michael, H. N., and E. K. Ralph, eds., 1971, *Dating Techniques for the Archaeologist*. Cambridge, Mass.: MIT Press, 226p.

Michels, J. W., 1973, *Dating Methods in Archaeology*. New York: Seminar Press, 230p.

Miller, R. R., 1965, Quaternary fresh-water fishes of North America, in H. E. Wright and D. G. Frey, eds., *The Quaternary of the United States*. Princeton, N.J.: Princeton University Press, 569-582.

Moreau, R. E., 1963, The distribution of tropical birds in relation to past climatic changes, in *Viking Fund Publ. Anthropology No. 36*, 28-42.

Oakley, K. P., 1970, *Frameworks for Dating Fossil Man*. London: Weidenfeld, 366p.

Oakley, K. P., and J. S. Weiner, 1955, Piltdown man, *Am. Scientist* **43,** 573-583.

Pilbeam, D. R., 1972, *The Ascent of Man*. New York: Macmillan, 207p.

Raikes, R., 1967, *Water, Weather, and Prehistory*. London: Baker, 208p.

Rigby, J. K., and W. K. Hamblin, eds., 1971, Recognition of ancient sedimentary environments, *Soc. Econ. Paleontologists and Mineralogists Spec. Publ. 16*, 340p.

Sauer, C. O., 1969, *Agricultural Origins and Dispersals: The Domestication of Animals and Foodstuffs*. Cambridge, Mass.: MIT Press, 175p.

Schoeneher, J., and F. W. Eddy, 1964, Alluvial and palynological reconstruction of environments—Navajo Reservoir District, *Mus. New Mexico Papers Anthropology No. 13*, 115p.

Semenov, S. A., 1964, *Prehistoric Technology*. New York: Barnes and Noble, 211p.

Shackley, M. O., 1975, *Archeological Sediments*. New York: Halsted Press, 159p.

Shepard, A. O., 1965, Ceramics for the archaeologist, *Carnegie Inst. Washington Publ. 609*, 414p.

Smith, P. E. L., 1976, *Food Production and Its Consequences*. Menlo Park, Calif.: Cummings, 120p.

Solecki, R. S., 1960, Photointerpretation in archaeology, in D. S. Simonett, ed., *Manual of Photographic Interpretation*. Washington, D.C.: American Society of Photogrammetry, 717-733.

Taylor, R. E., 1975, Fluorine diffusion: A new dating method for chipped lithic materials, *World Archaeology* **7,** 125-135.

Thenius, E., 1962, Die Grossaugetiere des Pleistozans von Mittleeuropa, *Zietschr. Saugerierkunds* **27,** 65-82.

Thomas, D. H., 1974, *Predicting the Past*. New York: Holt, Rinehart, and Winston, 84p.

Thomas, W. L., Jr., ed., 1956, *Man's Role in Changing the Face of the Earth*. Chicago: University of Chicago Press, vol. 1, 390p; vol. 2, 391p.

Tite, M. S., 1972, *Methods of Physical Examination in Archaeology*. New York: Seminar Press, 389p.

Tuttle, R. H., ed., 1975, *Paleoanthropology, Morphology, and Paleoecology.* Chicago: Aldine, 453p.

Vitaliano, D. B., 1973, *Legends of the Earth: Their Geologic Origins.* Bloomington: Indiana University Press, 305p.

Weide, D. L., 1974, Postglacial Geomorphology and Environments of the Warner Valley–Hart Mountain Area, Oregon, Ph.D. dissertation, University of California, Los Angeles.

Williams, H., 1956, Petrographic notes on the tempers of pottery from Chuipicuaro Cerrods., Tetelcate, and Ticoman, Mexico, *Am. Philos. Soc. Trans.* **45,** 576–580.

Wilmsen, E. M., 1970, Lithic analysis and cultural inference: Paleo-Indian case, *Arizona University Anthropology Papers 16,* 87p.

Winter, J. C., and H. G. Wylie, 1974, Paleoecology and diet at Clyde's Cavern, *Am. Antiquity, 39,* 303–315.

Wright, H. E., and D. G. Frey, eds., 1965, *The Quaternary of the United States.* Princeton, N.J.: Princeton University Press, 922p.

Yaalon, D. H., ed., 1971, *Paleopedology: Origin, Nature, and Dating of Paleosols.* Jerusalem: International Society of Soil Science, 350p.

Cross-references: *Cities, Geologic Effects; Geoarchaeology; Geomythology.* Vol. XIII: *Environmental Geomorphology; Geochronology.*

GEOARCHAEOLOGY

Scope

Geoarchaeology deals with the application of the methods and principles of the earth sciences for interpreting the archaeological record. Though the contributions from the earth scientists have often been related to age determination of archaeological finds through stratigraphic studies, a great deal of work is now devoted to paleoenvironmental reconstruction and the study of the dynamic relationship between geological processes and prehistoric cultures. Studies of the geology of formation, modification, and destruction of archaeological sites are of current interest. In addition, petrographic, mineralogic, and geochemical analyses of archaeological artifacts are widely practiced. Photogeologic, geochemical, and geophysical exploration methods are applied to locate and delineate archaeological sites. Some studies are also concerned with the structural and geologic settings of important sites that are threatened by geological hazards.

Historical Perspective

Archaeology shares with geology an interest in the past—geology with the Earth's past, archaeology with that of humankind. Both aim at elucidating historical events and attempt to explain the principles underlying patterns, processes, and

changes. Archaeology has developed out of the antiquarian studies that accompanied the surge of interest in the classical world during the Renaissance. However, it was not until the 1700s that these antiquarian studies provided some insights into the origin of humankind as a result of the discovery of flint implements and human remains in association with extinct animals of prediluvian age. The antiquity of humanity was thus established, and a science of the human remote past, prehistoric archaeology, was born. In 1859 this new science was christened by Charles Lyell (Daniel, 1962), who voiced then his unqualified acceptance of the geological antiquity of the archaeological finds first made by Boucher de Perthes from the Somme gravel at Abbeville, France, and later documentcd by John Prestwich and John Evans.

Dating

The association of hand axes and other flint artifacts with the bones of elephants, rhinoceroses, and hippopotamuses in Europe established not only human antiquity, but also the potential use of historical geology for the determination of the age of archaeological finds. This has been and continues to be one of the main areas of cooperation between geologists and archaeologists. Although today a great deal of emphasis is placed on *chronometric methods* (see Vol. XIII: *Geochronology*) such as K/Ar, $Ar^{40/39}Ar$ radiocarbon, U-series, Th-U, thermoluminescent, ESR, paleomagnetic, and fission track dating (Michael and Ralph, 1971; Michels, 1973; Berger and Suess, 1979; Aitken, 1985; Taylor, 1987). However, regional lithostratigraphy, biostratigraphy, and paleogeomorphology for relative dating, especially when materials suitable for chronometry are not present in all archaeological sites are still widely used. Accordingly, the association of archaeological remains with rock units and certain vertebrate fossil assemblages is also used for correlating sites. Correlation on the basis of the pollen content of archaeological sediments is facilitated in certain regions—e.g., northern Europe—by the frequent changes in vegetation accompanying glacial and postglacial climatic change. Geologists also attempt to attribute certain geomorphic features—e.g., alluvial terraces—to certain climatic events and then tie these events with established geologic-climate units elsewhere. This method was widely applied in the U.S. Southwest by Antevs (1941) and Bryan and Ray (1940). Application of radiocarbon dating reveals that this approach is problematic (Waters, 1986). *Varve chronology* was also used in Europe for dating archaeological sites (de Geer, 1940). The principles of stratigraphy are also applied for interpreting the chronological order of cultural units at archaeological sites (Harris, 1979). Because the thickness of cultural layers is often minute and the changes in facies are vastly complex, the study of the stratification at an archaeological site is referred to as *microstratigraphy*.

Paleoenvironments

Besides dating, the contributions from geology to archaeology have now expanded to those areas that are of the greatest interest to the contemporary archaeologist: the explanation of past cultural events and culture change. The emphasis in contemporary archaeology is on examining the processes and the general principles that could explain those events and those that may elucidate the evolution of human culture. Within this framework, many archaeologists have adopted an ecological perspective on culture and culture change. They have sought to explain the past in terms of the interaction between people and their natural habitat. From the 1920s to the 1940s this approach to prehistory became an established tradition. This included the work by Caton-Thompson (archaeologist) and Gardner (geologist) on paleoclimate and culture changes in palaeolithic Egypt (Butzer, 1971); by Bryan in 1926 and Hack in 1942 on arroyo cutting and prehistoric agriculture in the U.S. Southwest; and Huzayyin in 1939 and Passarge in 1941 on the hydrology of the Nile and the rise of the Egyptian civilization (Hassan, 1981).

Methods Reconstruction of the prehistoric environment is often based on many different methods. Those related to the earth sciences include sedimentology and stratigraphy, geomorphology, pedology, paleontology, palynology, and isotope geochemistry. Sedimentology and stratigraphy are sometimes based on the application of facies models. Special techniques were also designed for paleoenvironmental analysis of archaeological cave sediments (Bordes, 1972; Farrand, 1973). In conjunction with regional stratigraphy, geomorphological investigations are also useful in the construction of paleogeographic settings (Butzer and Hansen, 1968; Kraft et al., 1980). The use of fossil bones, pollen, and other biological remains to reconstruct paleoenvironments in archaeology is in general linked with paleoclimatic studies and the formulation of geologic-climate units (Kurtén, 1968; Wijmstra, 1978). Determination of paleotemperatures from cave sediments or deep-sea cores using oxygen isotopes ($^{18}O/^{16}O$) also contributes to the reconstruction of prehistoric environments (Duplessy, 1978).

Types of Paleoenvironments Archaeologists are interested in different types of paleoenvironments, especially the landscape where sites are lo-

cated because of its possible impact on configuration of spatial food-getting activities and settlement location. They study the implications of climatic change for cultural evolution. Accordingly, Butzer (1977) recognizes three categories of paleoenvironments: (1) microenvironments, or depositional environments (e.g., lacustrine, deltaic, riverine); (2) mesoenvironments, or topographic setting and relief, providing a partial definition of the biotope; and (3) macroenvironments, or zonal types (e.g., periglacial, humid tropical). A similar scheme is proposed by Gladfelter (1977). Hassan (1985) makes a distinction between various environmental systems: climatic-morphogenetic systems (e.g., glacial tropical, semiarid), geomorphic systems (riverine, littoral, deltaic), and depositional systems (e.g., alluvial, eolian, lacustrine, gravitational). These systems are not spatial units. Their relationship to space is like that between rock-stratigraphic units (which are not time units) and time. The spatial expression of these systems are climatic zones (q.v. in Vol. IVA) (e.g., arid region), geomorphic landscapes (q.v. in Vol. III) (e.g., the Sahara), morphotopes (e.g., a playa or ephemeral stream), and submorphotopes (e.g., channel island).

Prehistoric Ecology

Ecological interpretations are at present popular in archaeology. Butzer (1980a) has suggested that this ecological, or contextual, approach ought to guide archaeology and to serve as its all-encompassing paradigm.

The study of prehistoric cultures entails an attempt to understand human adaptation. Human beings must provide food and shelter for themselves, and archaeologists are interested in the implications of environmental potentials and limitations for subsistence (food-getting) activities.

Subsistence Archaeologists often wish to know the spatial configuration of food-getting activities (e.g., which places in the landscape were utilized); the annual scheduling of hunting, fishing, farming, and transport of goods; and the proportional use of food resources (e.g., how much emphasis is placed on hunting gazelle relative to wild cattle or collecting mussels). These questions cannot be answered adequately without an intimate knowledge of the landscape and paleoclimate. The reconstruction of paleoclimate and morphotopes allows for further inferences about vegetation and fauna and makes possible estimates of cost of travel as a function of slope or surface materials (e.g., stony pediment versus mud). The suitability of land for grazing, farming, or some other kind of use can be evaluated, and such an evaluation may provide some insights into the subsistence activities of prehistoric people in certain locations. This approach, called *catchment analysis,* a term borrowed from fluvial geomorphology, was pioneered by Vita-Finzi (geologist) and Higgs (zooarchaeologist) and later elaborated by others (Roper, 1979).

The scheduling of food-getting activities is often attuned to seasonal natural events such as rainfall, flooding, heat waves, and dust storms. These events influence the migration of game animals, the maturation of cultivated plants, the availability of land for special activities and the feasibility of rapid riverine transport (Hassan, 1974, 1985). As may be seen, some of these issues are similar to those dealt with in agricultural geography, applied geography, applied climatology, and environmental geology. These disciplines are also concerned with the relationship between landscapes and settlement.

Settlement The reconstruction of the cultural landscape—the geographic and environmental location of archaeological settlements, the size hierarchy of the settlements, and the spacing of settlements—provides the basis for many archaeological inferences on prehistoric population size, political organization, and social relations among adjacent communities and for documenting changes in settlement in response to environmental changes. Exploration and prospecting for archaeological sites (referred to as *archaeological surveying*) usually constitute the first step in settlement studies. This prospecting often involves the use of topographic and geologic maps, aerial photos, and landsat images for evaluating a region for the most likely places for archaeological sites. Aerial photos are also a powerful tool in locating visual clues for sites such as changes in natural vegetation, soil, moisture, and relief (see *Aerial Surveys*). Geophysical and geochemical methods come next as a tool for determining the aerial extent of archaeological sites and to locate buried features such as cemeteries, ditches, or kilns. The methods commonly used include high-resolution magnetometry, electric resistivity, acoustic sounding, microgravity, thermal infrared, and ground-penetrating radar (see *Exploration Geophysics*) (Dolphin, 1979; Tite, 1972; Weymouth, 1987; Parington, 1983). Geochemical prospecting often consists of determining the phosphate content of surface sediments since human occupations tend to generate organic residues rich in phosphate (see *Pedogeochemical Prospecting*) (Hassan, 1981; Provan, 1971; Eidt, 1984).

An understanding of the erosional/depositional history of a region is indispensible for an evaluation of the number, size, and location of archaeological sites. Sites may be buried under alluvium, enlarged by deflation, or totally removed by runoff erosion. It is thus always necessary not only to reconstruct the environment during the

time of human occupation but also following abandonment of the site.

The adjustment to a landscape may also be reflected in the location of sites in certain bio- or morphotopes. To test the association between sites and certain locational features, *geoekistic maps* (Hassan, 1985) based on the propensity of certain places in the landscape for geological hazards, access to water and other key resources, cost of travel (a function of relief and surface geology), availability of raw materials, and other relevant factors may be prepared. The various morphotopes may be thus ranked and the actual location of sites judged to conform with or vary from the optimal choice (Hassan, 1980). This method may be also a means for generating hypotheses about the differential weight of certain factors. Defensibility, e.g., may be given greater weight than distance to procure water under conditions of intergroup competition and strife.

Culture Change Macromodels of cultural evolution have in many instances been constructed with definite environmental overtones. Archaeologists are interested in the climatic record especially during the periods of major cultural transformations, such as at the time of the emergence of hominids (Isaac and McCown, 1976; Butzer and Isaac, 1975; Coppens et al., 1977), the time of the replacement of the archaic forms of the genus *Homo* with modern humans (Butzer, 1971), and when agriculture superseded hunting and gathering as the dominant model of subsistence (Reed, 1977). The relationship between climate, sea level, and intercontinental migrations has also been a fruitful area of investigation (Hopkins, 1980). The role of geological events in the rise of civilizations and cultural homeostasis is another area where the input from the earth sciences is valuable (Butzer, 1976, 1980b).

Site Formation Processes

Interpretation of archaeological materials is inadequate without a knowledge of their physical context. Before the archaeologist can make any statement about the implications of the archaeological finds for the layout of settlements, the nature of activities undertaken at a site, the duration of occupation, the intermittence or permanence of occupation, and other behavioral aspects, she or he must understand the life history of the archaeological site. It cannot be assumed a priori that the archaeological materials have not undergone any changes in their condition, frequency or spatial placement since they were deposited (Schiffer, 1983). To ascertain the mode of site formation by cultural and natural processes and the postdepositional modification of archaeological materials, archaeologists must undertake a number of geological studies. Such studies include careful microstratigraphic investigations, sediment analysis, and the analysis of chemical and physical traces of occupation and transformation by *microarchaeological analysis* (Hassan, 1978; Rosen, 1986). Preparation of lithofacies and isopach maps may also be useful in this respect.

Technology

The determination of the source of archaeological materials is of special importance to archaeologists because it may shed light on the exploitation of local resources or of the existence of trade networks. The composition and texture of archaeological materials may also shed some light on the mode of manufacture and may thus help in identifying cultural groups practicing similar manufacturing techniques. Petrography, mineralogy, ore microscopy, X-ray diffraction, instrumental methods of chemical analysis (neutron activation analysis, X-ray fluorescence, atomic absorption spectroscopy, etc.) have been used in archaeological projects to analyze flint artifacts, obsidian artifacts, ceramics, metals, glass, glaze, faience, and pigments (Tite, 1972; Kempe and Harvey, 1983). The petrography of ceramics (Shepard, 1968) has proven to be specially valuable since ceramics are common in archaeological sites and the method is not expensive.

Conservation

Archaeological sites are resources of tremendous human value, and every effort must be devoted to protect and maintain them. Aside from looting and malicious destruction, archaeological sites are subject to deterioration or total annihilation by natural or artificial geological processes. Changes in atmospheric conditions by pollutant chemicals and particulate matter enhance chemical weathering; changes in groundwater levels as a result of damming can lead to structural instability, salinization, mechanical weathering by alternation of wetting and drying; and increased surface runoff erosion results from destruction of the vegetation cover by overgrazing or motorcycles. Also, offshore currents, waters from hydroelectric dams, incidences of high floods, or earthquakes can hasten the decay or elimination of sites. Tragic examples of these factors are all too common, and the fate of Venice, the Sphinx, and the tomb of Nefertiti in Thebes should be reminders of the ongoing devastation of the human heritage. The role of the geologist lies in determining the specific causes of the phenomena threatening the archaeological sites and in suggesting suitable remedies. The geologist should also anticipate the long-term consequences of geological processes for the important archaeological sites. An international program should be initiated to save those archaeological sites that may in the future be lost to humanity.

FEKRI A. HASSAN

References

Aitken, M. J., 1985, *Thermoluminescence Dating,* New York: Academic Press, 359p.

Antevs, E., 1941, Age of the Cochise culture stages, *Medallion Papers* **29,** 31–56.

Berger, R., and H. E. Suess, eds., 1979, *Proceedings of a Conference, Los Angeles and La Jolla, California, 1976.* Berkeley: University of California Press.

Bordes, F., 1972, *A Tale of Two Caves.* New York: Harper & Row, 169p.

Bryan, K., and L. L. Ray, 1940, Geological antiquity of the Lindenmeier site in Colorado, *Smithsonian Misc. Colln.* **99**(2), 76pp.

Butzer, K. W., 1971, *Environment and Archaeology: An Ecological Approach to Prehistory,* 2nd ed. Chicago: Aldine, 703p.

Butzer, K. W., 1976, *Early Hydraulic Civilization in Egypt: A Study in Cultural Ecology,* Chicago: University of Chicago Press, 134p.

Butzer, K. W., 1977, Geoarchaeology in practice, *Rev. Anthropology* **4,** 125–131.

Butzer, K. W., 1980a, Context in archaeology: An alternative perspective, *Jour. Field Archaelogy* **7,** 417–422.

Butzer, K. W., 1980b, Civilizations: Organisms or system?, *Am. Scientist* **68,** 517–523.

Butzer, K. W., and C. L. Hansen, 1968, *Desert and River in Nubia.* Madison: University of Wisconsin Press, 562p.

Butzer, K. W., and G. LL. Isaac, 1975, *After the Australopithecines: Stratigraphy, Ecology and Culture Change in the Middle Pleistocene.* The Hague: Mouton, 911p.

Coppens, Y., F. C. Howell, G. LL. Isaac, and R. F. Leakey, 1977, *Earliest Man and Environments in Lake Rudolf Basin.* Chicago: University of Chicago Press, 615p.

Daniel, G., 1962, *The Idea of Prehistory.* Harmondsworth: Penguin Books, 186p.

Dolphin, L., 1979, Geophysical prospecting in archaeology, paper presented at the Annual Meeting of the Society for American Archaeology, Vancouver, 17p.

Duplessy, J. C., 1978, Isotopic studies, in J. Gribbin, ed., *Climatic Change.* Cambridge: Cambridge University Press, 46–47.

Eidt, R. C., 1984, *Advances in Abandoned Settlement Analysis: Application to Prehistoric Anthrosols in Colombia, South America.* Milwaukee: The Center of Latin America, University of Wisconsin, 159p.

Farrand, W. R., 1973, New excavations at the Tabun Cave, Mount Carmel, Israel, 1967–1972: A preliminary report, *Paleoorient* **1,** 151–183.

Geer, G. De, 1940, Geochronologia Suecica Principles, *Kungl. Svenska Vetensk, Handl.,* Ser. 3, **18**(6).

Gladfelter, B. G., 1977, Geoarchaeology: The geomorphologist and archaeology, *Am. Antiquity* **42,** 519–538.

Gummerman, G. J., and T. R. Lyons, 1971, Archaeological methodology and remote sensing, *Science* **172,** 126–132.

Harris, E. C., 1979, *Principles of Archaeological Stratigraphy.* London: Academic Press, 136p.

Hassan, F. A., 1974, The Archaeology of the Dishna Plain, Egypt: A study of a Late Palaeolithic settlement, *Geological Survey of Egypt Paper No. 59,* 174p.

Hassan, F. A., 1978, Sediments in archaeology: Methods and implications for palaeoenvironmental and cultural analysis, *Jour. Field Archaeology,* 5, 197–213.

Hassan, F. A., 1979, Geoarchaeology: The geologist and archaeology, *Am. Antiquity,* 44, 267–270.

Hassan, F. A., 1980, Prehistoric settlements along the main Nile, in M. J. Williams and H. Faure, eds., *The Sahara and the Nile.* Rotterdam: A. A. Balkema, 421–450.

Hassan, F. A., 1981, Rapid quantitative determination of phosphate in archaeological sediments, *Jour. Field Archaeology* 8, 384–387.

Hassan, F. A., 1985, Palaeoenvironments and contemporary archaeology: A geoarchaeological approach, in G. Rapp and J. Gifford, eds., *Archaeological Geology.* New Haven, Conn.: Yale University Press, 85–102.

Hopkins, D. A., 1980, Landscape and climate of Beringia during the Late Pleistocene and Holocene time, in W. S. Laughlin and A. B. Harper, eds., *The First Americans: Origins, Affinities, and Adaptations.* New York: Gustav Fleisher, 15–41.

Isaac, G. LL., and E. R. McCown, 1976, *Human Origins.* Menlo Park, Calif.: W. A. Benjamin, Inc., 591p.

Kempe, D. R. C., and P. A. Harvey, eds., 1983, *The Petrology of Archaeological Artifacts.* Oxford: Clarendon Press, 374p.

Kraft, J. C., I. Kayon, and O. Erol, 1980, Geomorphic reconstructions in the environs of Ancient Troy, *Science* 209, 776–782.

Kurtén, B., 1968, *Pleistocene Mammals of Europe.* Chicago: Aldine, 317p.

Lubell, D., F. Hassan, A. Gautier, J.-L. Ballais, 1976, The Capsian escargotières, *Science* 191, 190.

Michel, H. N., and E. K. Ralph, eds., 1971, *Dating Techniques for the Archaeologist.* Cambridge, Mass.: MIT Press.

Michaels, J. W., 1973, *Dating Methods in Archaeology.* New York: Seminar Press, 230p.

Parington, M., 1983, Remote sensing, *Annual Review of Anthropology* 12, 105–124.

Provan, D. M. J., 1971, Soil phosphate analysis as a tool in archaeology, *Norwegian Arch. Rev.* 4, 37–50.

Reed, C. A., 1977, *Origins of Agriculture.* The Hague: Mouton, 1,013p.

Roper, D. C., 1979, The method and theory of site catchment analysis: A review, in M. B. Schiffer, ed., *Advances in Archaeological Method and Theory,* vol. 2, New York: Academic Press, 119–140.

Rosen, A. R., 1986, *Cities of Clay.* Chicago: University of Chicago Press, 167p.

Schiffer, M. B., 1983, Toward the identification of formation processes, *American Antiquity* 48, 675–706.

Shepard, A. O., 1968, Ceramics for the archaeologist, *Carnegie Inst. Washington Publ. 609,* 414p.

Taylor, R. E., 1987, *Radiocarbon Dating.* New York: Academic Press, 214p.

Tite, M. S., 1972, *Methods of Physical Examination in Archaelogy.* London and New York: Seminar Press, 389p.

Vita-Finzi, C., and E. S. Higgs, 1970, Prehistoric economy in the Mount Carmel area of Palestine: Site catchment analysis, *Proc. Prehistoric Soc. (London)* 36, 1–37.

Weymouth, J. W., 1987, Geophysical methods of archaeological site surveying, in M. B. Schiffer, ed., *Advances in Archaeological Method and Theory.* New York: Academic Press.

Wijmstra, T. A., 1978, Palaeobotany and climatic change, in J. Gribbin, ed., *Climatic Change.* Cambridge: Cambridge University Press, 24–45.

Cross-references: *Exploration Geophysics; Geoanthropology; Maps, Logic of; Pedogeochemical Prospecting; Photogeology.* Vol. XIII: *Geochronology; Remote Sensing, Engineering Geology.*

GEOBIOLOGY—See GEOBOTANICAL PROSPECTING; GEOMICROBIOLOGY; LICHENOMETRY.

GEOBOTANICAL PROSPECTING

Botanical methods of prospecting involve the use of vegetation in searching for ore deposits. Although these methods have been used for several centuries, there is much confusion about terminology because there are two distinct methods of botanical prospecting. *Geobotanical* methods are visual and rely mainly on an interpretation of the plant cover to detect morphological changes or plant associations typical of certain types of geologic environments or of ore deposits within these environments. *Biogeochemical* methods (see *Biogeochemistry*), which have been used only since the 1940s, involve chemical analysis of the plant cover to detect mineralization.

Geobotanical methods were first used in Roman times when vegetation was employed in the search for subterranean water. Later the Russian botanist Karpinsky (1841) became the first man to study thoroughly the relationship between plant communities and their geologic substrate. A number of books have appeared on the subject of geobotanical prospecting (Malyuga, 1964; Viktorov et al., 1964; Brooks, 1972, 1983), and the method is now established as a potential tool in mineral prospecting.

The technique of geobotanical prospecting falls into four classifications:

1. Study of plant communities,
2. Indicator plants,
3. Morphological and mutational changes in plants,
4. Aerial geobotanical surveys.

Each of these topics is considered in separate sections.

Study of Plant Communities

As mentioned, the Russian botanist Karpinsky (1841) noticed that specific geologic formations usually carried a characteristic flora that could be used to characterize that substrate. His work has now been highly perfected, particularly in the USSR, where this technique is known as *indicator geobotany.* Karpinsky had noted that in the interpretation of the geology of an area, attention should be paid to the whole community rather than an individual species within it.

Characteristic Floras The ecology of plant communities is greatly influenced by the pH of the soil and by the presence, excess, or deficiency of mineral nutrients. Although almost any type of geologic formation probably has its own *characteristic flora,* only in the more extreme cases is a change of vegetation immediately obvious to a cursory inspection.

Calciphilous (limestone) *floras* comprise calciphilous (lime-loving) as well as *calcicolous* (lime-requiring) species. Limestone soils are usually well drained and well aerated and, hence, because of this beneficial conditioning, can support a rich and varied flora that though often stunted, usually contrasts favorably with poorer floras in surrounding areas. Limestone floras can be best identified by looking for certain genera (e.g., *Dianthus, Fagus, Bromus, Festuca, Linaria,* etc.) that are typical of this type of substrate.

Halophyte floras represent a characteristic plant association found mainly in saline soils containing sodium chloride, sodium carbonate, or sodium sulfate. Halophyte floras have been used extensively in the USSR as a guide to water resources (Chikishev, 1965), and one of the subgroups (*selenium floras*) has been used extensively and successfully in the search for uranium in the western United States.

Selenium floras represent one of the most successful applications of the geobotanical method. Cannon and her co-workers (Cannon, 1957, 1960, 1964; Cannon and Starrett, 1956), by mapping the distribution of typical selenium plants, were able to discover several uranium deposits in the Colorado Plateau area of the United States because of the association of selenium and uranium in the carnotite area of this region.

Serpentine floras represent perhaps one of the most extreme forms of a characteristic flora. Serpentine floras (Brooks, 1987) are typically deficient in numbers and types of species and usually have endemic plants sometimes confined only to a few square kilometers. Plants not confined to serpentine are usually much more stunted on this substrate because of the high magnesium content of the soil, which affects the uptake of calcium by many plants. Other factors that control serpentine floras are the high levels of the toxic elements such as nickel and chromium and the low levels of essential nutrients such as potassium, nitrogen, and phosphorus.

Zinc (galmei) *floras* are found typically in Western Europe and bear some resemblance to serpentine communities. Plant growth is retarded, broadleaf plants are absent, and endemic forms are common. One of the most interesting components of zinc flora is *Viola calaminaria,* which accumulates zinc to a high degree and was used by miners over 100 yr ago in the search for zinc deposits.

Plant Mapping Although some plant communities, such as serpentine floras, are readily distinguishable on a cursory examination of the environment, this is not the case for most communities. In many cases, it will be necessary to carry out some sort of procedure such as plant mapping (see *Vegetation Mapping*) to determine the nature of this community and hence to detect geologic boundaries in the area.

The most accurate way of mapping is to divide the area into sample plots known as *quadrats.* The size of these quadrats is influenced by the size of the so-called *minimal area*—i.e., the smallest area containing a representative selection of most of the plants to be found in the area. Refer to Greig-Smith (1964) for a further discussion of the concept of minimal area. Having selected the quadrats, the number of specimens of each species in each quadrat should then be counted and the data expressed in a suitable form, like that in Fig. 1, which shows how a serpentinite-gabbro contact has been delineated by plant mapping in Western Australia.

If differences in substrate produce only a subtle change in vegetation, even plant mapping will not be sufficient to characterize the substrate. In such cases some form of discriminant analysis may be carried out. Table 1 shows how discriminant analysis of geobotanical data (Nielsen et al., 1973) was used to predict the nature of three substrates on the basis of the distribution.

The area was divided into a belt transect containing 44 quadrats: 26 on amphibolites, 11 on ultrabasics, and 7 on a transitional zone between the two rock types. For each rock type, a computer calculation was made of a regression equation of the form

$$y = l_1x_1 + l_2x_2 + l_3x_3 \ldots + l_{36}x_{36} + C_a.$$

The variables x_1–x_{36} were the abundances of each of the 36 plant species in the quadrats. The constant C_a and coefficients l_1–l_{36} were selected by the computer to maximize the differences in y for all three rock formations. The degree of this differentiation can be conveniently measured by the so-called Mahalanobis D^2 statistic (Mahalanobis, 1936). The magnitude of D^2 is an indication of the

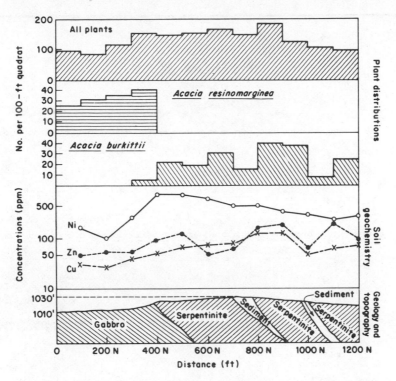

FIGURE 1. Data from belt transect of an area in Western Australia showing the apparent influence of the geology, topography, and soil geochemistry on the distribution of two plant species.

effectiveness of the discrimination. Table 2 shows that with several combinations of species, it was possible to achieve a correct prediction for at least 41 out of the 44 quadrats, whereas plant mapping alone was quite inadequate to discriminate the rock types.

Indicator Plants

Indicator plants indicate by their presence the existence of minerals or of a specific rock type. They are divided into two classes. *Universal indicators* are found over large areas, whereas *local indicators* (often endemic plants) are effective only in a restricted locality.

Primary indicators give a direct response to the mineral that is being sought whereas *secondary indicators* give an indirect indication by responding to another mineral or element that is however associated with the mineral for which search is being made. An example of this is the use of the selenium-indicator *Astragalus* species to discover uranium deposits in areas where uranium and selenium have a geochemical association (Cannon, 1964). Indicator plants have been listed by Brooks (1972, 1979, 1983) and by Cannon (1979) (see Table 2).

Indicators of Copper Among the more successful indicator plants are those that show the presence of copper and selenium. The copper indicator *Becium homblei* (Howard-Williams, 1970) has been successfully used to delineate copper deposits in Zambia and Rhodesia. Its distribution represents a case of interspecific competition with the closely related *B. obovatum*. *B. obovatum* and other competitors cannot grow. Brooks and Malaisse (1985) have described the occurrence and ecology of over 50 copper indicators in Zaïre.

Indicators of Selenium Plant indicators of selenium are numerous and have been used successfully in mineral exploration. Cannon (1957, 1960, 1964) showed that they comprise *Aster venusta* and various species of *Astragalus, Oryzopsis,* and *Stanleya*. The mechanism where these plants act as secondary indicators is that the uranium ore carnotite appears to increase the availability of selenium to plants, and hence, the density of selenium-accumulating plants can be some indication of the presence of uranium mineralization.

Bryophytes as Indicators Bryophytes (mosses) can act as indicators because some species will only grow over certain mineral deposits. An example of this is the well-known copper moss (*Mielichhoferia mielichhoferi*). Persson (1956)

229

TABLE 1. The Use of Discriminant Analysis of Geobotanical Data to Predict the Nature of the Substrate

Variables Used	D^2	Number of Correct Predictions		
		Amphibolites (26 quadrats)	Ultrabasics (11 quadrats)	Transitional Areas (7 quadrats)
22	0.520	11	6	0
25	5.68	2	11	1
22,25	6.20	10	6	1
22,23,25	9.44	15	3	1
21,22,25	7.42	16	7	1
2,22,25	8.30	21	7	1
2,3,22,25	9.13	16	8	1
2,4,22,25	12.14	17	8	1
2,4,15,22,25	21.15	19	9	1
2,4,15,16,22,25	29.10	22	10	1
2,4,15,16,17,22,25	29.14	22	11	2
2,4,15,16,17,22,25,27	29.42	20	10	2
2,4,15,16,17,22,25,30	32.88	21	10	2
2,4,13,15,16,17,22,25,30	41.37	22	10	3
2,4,13,15,16,17,22,25,26,30	42.47	22	10	4
2,4,5,13,15,16,22,25,30,36	45.72	23	11	4
2,4,5,13,15,16,22,25,30,32,36	44.16	22	11	4
1,2,4,13,15,16,17,22,25,26,30	48.24	23	10	3
1,2,4,5,13,15,16,17,22,25,26,30	50.37	24	10	3
1,2,4,5,7,13,15,16,17,22,25,26,30	51.77	24	10	3
1,2,4,5,9,13,15,16,17,22,25,26,30	56.60	24	0	4
1,2,4,5,8,9,13,15,16,17,22,25,26,30	58.71	24	10	4
1,2,4,5,8,9,13,15,16,17,18,22,25,26,30	67.82	24	10	4
1,2,4,5,8,9,13,15,16,17,18,20,22,25,26,30	76.23	24	10	5
1,2,4,5,8,9,13,15,16,17,18,20,22,25,26,30,31	89.76	24	10	4
1,2,4,5,8,9,13,15,16,17,18,20,22,25,26,30,32	81.34	23	9	5
1,2,4,5,8,9,13,15,16,17,18,20,22,25,26,30,34	100.3	24	10	6
1,2,4,5,8,9,13,15,16,17,18,20,22,25,26,30,34,35	119.4	25	9	6
1,2,4,5,8,9,13,15,16,17,18,20,22,25,26,30,34,35,36	137.5	25	10	6
1,2,4,5,8,9,10,13,15,16,17,18,20,22,25,26,30,34,35,36	138.1	25	10	6
1,2,4,5,8,9,11,13,15,16,17,18,20,22,25,26,30,34,35,36	138.2	25	10	6
1,2,4,5,8,9,12,13,15,16,17,18,20,22,25,26,30,34,35,36	140.2	24	10	6
1,2,4,5,8,9,13,14,15,16,17,18,20,22,25,26,30,34,35,36	147.2	25	10	6
1,2,4,5,8,9,13,15,16,17,18,20,21,22,25,26,30,34,35,36	144.3	24	10	6
1,2,4,5,8,9,13,15,16,17,18,20,22,25,26,29,30,34,35,36	142.8	24	10	6
1,2,4,5,8,9,13,15,16,17,18,19,22,25,26,29,33,34,35,36	173.9	26	10	5

Source: Nielsen, 1972
Note: Numbers in first column represent code numbers for 36 plant species.

suggested a novel way of using these mosses in the search for minerals. He examined the collection localities of specimens in the Stockholm Herbarium and arranged for these localities to be examined by other prospecting techniques.

Accumulator Plants Some plants accumulate extraordinarily high amounts of some elements without necessarily being indicators. Examples of this are cobalt accumulators with up to 1.8% in the ash (Duvigneaud, 1959; Brooks and Malaisse, 1985; Brooks et al., 1980), selenium accumulators with up to 4.6% in the ash (Cannon, 1960) and zinc accumulators containing up to 1% in the ash (Reeves and Brooks, 1983). Perhaps the most spectacular examples of elemental uptake are nickel-accumulating species of the genus *Hybanthus*. Severne and Brooks (1972) and Cole (1973) have reported up to 23% nickel in the ash of the Australian species *H. floribundus,* whereas Brooks et al. (1974) have reported 25% of this element in the ash of the New Caledonian species *Hybanthus austro-caledonicus.* Even higher levels (up to 40%) were reported by Jaffré and Schmid (1974) for the New Caledonian plant *Psychotria douarrei.* In Italy, Minguzzi and Vergnano (1948) reported over 10% nickel in the ash of *Alyssum bertolonii.* Some 144 hyperaccumulators of nickel have now been discovered (Brooks, 1987).

Morphological and Mutational Changes in Plants

Changes in the morphology of plants and evidence of disease are useful aids in geobotanical prospecting and have been used as field guides since the eighteenth century. Early workers had to rely on obvious changes such as dwarfism or variation in color. With the increasing sophistication of modern science, however, and with a greater knowledge of plant physiology, many other visual indications of mineralization have been noted and can be used in prospecting. Some of the more important morphological and mutational changes in plants are as follows.

Abnormality of Form Abnormality of form is often a symptom of the presence of boron or radioactive minerals. Buyalov and Shvyryaeva (1961) noted dwarfism and deformation of *Salicorna herbacea* and other plants under the influence of boron. Shacklette (1962) has reported on variations in the bog bilberry (*Vaccinium uliginosum*) under the influence of radioactivity.

Chlorosis (Yellowing) of Leaves Chlorosis of leaves is a common field guide to the presence of excessive amounts of many elements such as chromium, cobalt, copper, manganese, nickel, or zinc. It was noticed as early as in the sixteenth century (Agricola, 1556). Chlorosis is caused by the above elements being antagonistic to iron uptake by plants, and is a symptom of iron deficiency.

Color Changes Color changes in plants are often the result of radioactivity. Shacklette (1964) has observed extensive variations in the color of flowers of *Epilobium angustifolium* under the influence of radioactivity. Color variations in flowers caused by other minerals have been noted by Bazilevskaya and Sibireva (1950) and Malyuga et al. (1959). Gigantism is an unusual phenomenon reported by Shchapova (1938) for *Zostera nana* growing in bituminous areas in the USSR.

Satellite Imagery and Aerial Geobotanical Surveys

A major problem of geobotanical surveys carried out on foot is that it is difficult to survey a large area. This problem is avoided by surveys carried out from the air. Aerovisual surveys, which are popular in the USSR, involve marking on aerial maps the nature of the vegetation cover as determined by a visual inspection from the air (see *Aerial Surveys, General*).

Perhaps the most useful technique in aerial surveys is *infrared photography.* Vegetation has a high spectral reflectance above 800 nm in the near infrared region. Color film sensitive to this part of the spectrum gives an image that shows much more contrast for vegetation and allows for better differentiation and identification of components of the vegetation cover.

Since unhealthy plants give a lower spectral reflectance than healthy plants, it would seem that vegetation affected adversely by mineralization in the substrate should be identifiable in an aerial infrared photograph. In practice, it is seldom possible to detect mineralization by this method because subtle changes in a few individual specimens are apparent only if they can be compared with a large number of surrounding plants of the same species. This situation presupposes a monocrop such as an artificially planted forest, and these conditions are seldom met in nature. Nevertheless, it is usually possible to detect by infrared photography different geologic formations resulting in different plant communities growing on them.

Geobotanical prospecting from the air has now been revolutionized by the development of satellite imagery (Brooks, 1983) where the terrain is scanned at four wavelengths (including infrared) by satellites such as the LANDSAT series. Resolution is about 30 m. The new French SPOT satellites have a much better resolution.

Advantages of Geobotanical Prospecting Methods

1. Once an orientation survey has been carried out, costs are extremely low.
2. Different geologic formations as well as min-

TABLE 2. Plant Indicators of Mineral Deposits

Element	Species	Common name	Family	Locality	Reference
Boron	Eurotia ceratoides (L)	Winterfat	Chenopodiaceae	USSR	Buyalov and Shvyryaeva (1961)
	Salsola nitraria (L)	Saltwort	Chenopodiaceae	USSR	Buyalov and Shvyryaeva (1961)
	Limonium suffruticosum (L)	Statice	Plumbaginaceae	USSR	Buyalov and Shvyryaeva (1961)
Cobalt	Crassula alba (L)	Rattlebox	Crassulaceae	Zaire	Malaisse et al. (1979)
	Crotalaria cobalticola (U)		Leguminosae	Zaire	(Duvigneaud (1959); Brooks et al. (1977)
	Haumaniastrum robertii (U)	Cu flower	Labiatae	Zaire	Brooks (1977)
	Silene cobalticola (U)	Catchfly	Caryophyllaceae	Zaire	Duvigneaud (1959)
Copper	Acalypha dikuluwensis (U)		Euphorbiaceae	Zaire	Duvigneaud and Denaeyer-de-Smet (1963)
	Aeolanthus biformifolius (U)		Labiatae	Zaire	Malaisse et al. (1978)
	Anisopappus hoffmanianus (U)		Compositae	Zaire	Duvigneaud and Denaeyer-de-Smet (1963)
	Armeria maritima (L)	Sea Pink	Plumbaginaceae	Wales	Henwood (1857), Ernst (1969)
	Ascolepis metallorum (U)		Cyperaceae	Zaire	Duvigneaud and Denaeyer-de-Smet (1963)
	Becium homblei (U)	Basil	Labiatae	Zaire/Zambia	Howard-Williams (1970)
	B. Peschianum (U)	Basil	Labiatae	Zaire	Duvigneaud and Denaeyer-de-Smet (1963)
	Bulbostylis barbata (U)	Butterwood	Cyperaceae	Australia	Nicolls et al. (1965)
	B. burchelli (L)	Butterwood	Cyperaceae	Australia	Cole (1971)
	Commelina zigzag (U)		Commelinaceae	Zaire	Duvigneaud and Denaeyer-de-Smet (1963)
	Crotalaria cobalticola (U)	Rattlebox	Leguminosae	Zaire	Duvigneaud (1959)
	C. francoisiana (U)	Rattlebox	Leguminosae	Zaire	Duvigneaud and Denaeyer-de-Smet (1963)
	Cyanotis cupricola (U)		Commelinaceae	Zaire	Duvigneaud and Denaeyer-de-Smet (1963)
	Ecbolium lugardae (L)		Acanthaceae	S.W. Africa	Cole (1971)
	Elsholtzia haichowensis (L)		Labiatae	China	Se Sjue-Tszin and Sjuj Ban-Lian (1953)
	Eschscholzia mexicana (L)	Calif. poppy	Papaveraceae	U.S.A.	Chaffee and Gale (1976)
	Gladiolus actinomorphanthus (U)		Iridaceae	Zaire	Duvigneaud and Denaeyer-de-Smet (1963)
	G. duvigneaudii (U)		Iridaceae	Zaire	Duvigneaud and Denaeyer-de-Smet (1963)
	G. klattianus s.sp. angustifolius (U)		Iridaceae	Zaire	Duvigneaud and Denaeyer-de-Smet (1963)
	G. peschianus (U)		Iridaceae	Zaire	Duvigneaud and Denaeyer-de-Smet (1963)
	G. tshombeanus s.sp. parviflorus (U)		Iridaceae	Zaire	Duvigneaud and Denaeyer-de-Smet (1963)
	Gutenbergia cuprophila (U)		Compositae	Zaire	Duvigneaud and Denaeyer-de-Smet (1963)
	Gypsophila patrinii (L)		Caryophyllaceae	USSR	Nesvetailova (1961)
	Haumaniastrum katangense (U)	Cu flower	Labiatae	Zaire	Duvigneaud and Denaeyer-de-Smet (1963)
	H. robertii (U)	Cu flower	Labiatae	Zaire	Duvigneaud and Denaeyer-de-Smet (1963)
	Helichrysum leptolepis (L)	Everlasting	Compositae	S.W. Africa	Cole (1971)
	Impatiens balsamina (L)		Balsaminaceae	India	Aery (1977)
	Lindernia damblonii (U)		Scrophulariaceae	Zaire	Duvigneaud and Denaeyer-de-Smet (1963)
	L. perennis (U)		Scrophulariaceae	Zaire	Duvigneaud and Denaeyer-de-Smet (1963)
	Lychnis alpina (L)	Campion	Caryophyllaceae	Fennoscandia	Brooks et al. (1979a,b)
	Merceya latifolia (U)		*Bryophyta	Worldwide	Persson (1948)
	Mielichhoferia mielichhoferi (U)	Cu Moss	*Bryophyta	Worldwide	Persson (1948)
	Minuartia verna (L)		Caryophyllaceae	Wales	Ernst (1969)

Element	Species	Common name	Family	Location	Reference
	Oligotrichum hercynicum (U)		*Bryophyta	Alaska	Cannon (1971)
	Pandiaka metallorum (U)		Amaranthaceae	Zaire	Duvigneaud and Denaeyer-de-Smet (1963)
	Polycarpaea corymbosa (L)	Pink	Caryophyllaceae	India	Venkatesh (1964, 1966)
	P. spirostylis (L)	Cu flower	Caryophyllaceae	Australia	Brooks and Radford (1978)
	Rendlia cupricola (U)		Gramineae	Zaire	Duvigneaud and Denaeyer-de-Smet (1963)
	Sopubia metallorum (U)		Scrophulariaceae	Zaire	Duvigneaud and Denaeyer-de-Smet (1963)
	S. neptunii (U)		Scrophulariaceae	Zaire	Duvigneaud and Denaeyer-de-Smet (1963)
	Sporobolus stelliger (U)		Gramineae	Zaire	Duvigneaud and Denaeyer-de-Smet (1963)
	S. deschampsioides (U)		Gramineae	Zaire	Duvigneaud and Denaeyer-de-Smet (1963)
	Tephrosia sp. nov. (L)		Gramineae	Queensland	Nicolls et al. (1965)
	Vernonia cinerea (L)	Ironweed	Compositae	India	Venkatesh (1964, 1966)
	V. ledocteana (U)	Ironweed	Compositae	Zaire	Duvigneaud and Denaeyer-de-Smet (1963)
Iron	*Acacia patens* (L)		Leguminosae	W. Australia	Cole (1965)
	Burtonia polyzyga (L)		Leguminosae	W. Australia	Cole (1965)
	Calythrix longiflora (L)		Myrtaceae	W. Australia	Cole (1965)
	Chenopodium rhadinostachyum (L)		Chen podiaceae	W. Australia	Cole (1965)
	Eriachne dominii (L)		Gramin	W. Australia	Cole (1965)
	Goodenia scaevolina (L)		Goodeniaceae	W. Australia	Cole (1965)
Manganese	*Crotalaria florida* var. *congolensis* (L)		Leguminosae	Zaire	Duvigneaud (1919)
	Maytenus bureauvianus (L)		Celastraceae	New Caledonia	Jaffré (1977)
Nickel	*Alyssum spp.*	Madwort	Cruciferae	S. Europe and Turkey	Brooks et al. (1979a)
	Hybanthus austrocaledonicus (U)		Violaceae	New Caledonia	Brooks et al. (1974)
	H. floribundus (L)		Violaceae	W. Australia	Severne and Brooks (1972), Cole (1973)
Selenium and Uranium	*Lychnis alpina* var. *serpenticola* (L)	Campion	Caryophyllaceae	Fennoscandia	Rune (1953)
	Aster venustus (L)	Woody Aster	Compositae	Western U.S.A.	Cannon (1957)
	Astragalus albulus (L)	Poison vetch	Leguminosae	Western U.S.A.	Cannon (1957)
	A. argillosus (L)	Poison vetch	Leguminosae	Western U.S.A.	Cannon (1957)
	A. confertiflorus (L)	Poison vetch	Leguminosae	Western U.S.A.	Cannon (1957)
	A. pattersoni (U)	Poison vetch	Leguminosae	Western U.S.A.	Cannon (1957)
	A. preussi (U)	Poison vetch	Leguminosae	Western U.S.A.	Cannon (1957)
	A. thompsonae (L)	Poison vetch	Leguminosae	Western U.S.A.	Cannon (1957)
Zinc	*Armeria halleri* (L)	Thrift	Plumbaginaceae	Pyrenees	Palou et al. (1965)
	Hutchinsia alpina (L)		Cruciferae	Pyrenees	Palou et al. (1965)
	Minuartia verna (L)		Caryophyllaceae	W. Europe	Ernst (1968)
	Thlaspi calaminare (U)	Pennycress	Violaceae	W. Europe	Ernst (1968)
	Viola calaminaria (U)	Violet	Caryophyllaceae	W. Europe	Ernst (1968)

Source: Brooks, 1979.

L – local indicator; U – universal indicator; * – Phylum

eralization within them can be detected by geobotanical observations.

3. Satellite imager and aerial methods can be applied to the procedure with consequent saving in time and effort.

4. Indicator plants can sometimes show the presence of mineralization at depth under conditions where other methods would give a negative response.

Disadvantages of Geobotanical Prospecting Methods

1. A high degree of individual skill is needed from workers in this field.

2. Data obtained from orientation surveys are not necessarily of universal application and may have only local significance.

3. In some cases the method can be applied only seasonally, such as when plants are in flower.

4. The method may be applied only where vegetation conditions are favorable.

5. In the West little coordinated research in this field is being undertaken at present.

This listing of advantages and disadvantages is not complete, but it covers the main points. A consideration of paramount importance is the fact that geobotanical prospecting is just one of many techniques and is meant to supplement rather than replace other methods. Until a sufficiently large pool of skilled workers in this field can be trained, it will not be possible to test this procedure adequately or to compare it objectively with other methods of exploration.

ROBERT R. BROOKS

References

Aery, N. C., 1977, Studies on the geobotany of Zawar Mines, *Geobios* **4**, 225–228.

Bazilevskaya, N. A., and Z. A. Sibireva, 1950, Change in the color of the corolla in *Eschscholtzia californica* under the influence of microelements (in Russian), *Bull. Glav. Bot. Sad. Leningrad* **6**, 32–38.

Brooks, R. R., 1972, *Geobotany and Biogeochemistry in Mineral Exploration*. New York: Harper & Row, 292p.

Brooks, R. R., 1977, Copper and cobalt uptake by *Haumaniastrum* species, *Plants and Soil* **48**, 541–545.

Brooks, R. R., 1979, Indicator plants for mineral prospecting—A critique, *J. Geochem. Explor.* **12**, 67–78.

Brooks, R. R., 1983, *Biological Methods of Prospecting for Minerals*. New York: Wiley, 322p.

Brooks, R. R., 1987, *Serpentine and Its Vegetation: A Multidisciplinary Approach*. Portland, Ore.: Dioscorides Press, 454p.

Brooks, R. R., and F. Malaisse, 1985, *The Heavy Metal Flora of Southcentral Africa: A Multidisciplinary Approach*. Rotterdam: Balkema, 199p.

Brooks, R. R., and C. C. Radford, 1978, An evaluation of background and anomalous copper and zinc concentrations in the "Copper Plant" *Polycarpaea spirostylis* and other Australian species of the genus, *Australasian Inst. Mining and Metallurgy Proc.* **268**, 33–37.

Brooks, R. R., J. Lee, and T. Jaffrè, 1974, Some New Zealand and New Caledonian plant accumulators of nickel, *Jour. Ecology* **62**, 493–499.

Brooks, R. R., J. A. McCleave, and E. K. Schofield, 1977, Cobalt and nickel uptake by the Nyssaceae, *Taxon* **26**, 197–201.

Brooks, R. R., R. S. Morrison, R. D. Reeves, T. R. Dudley, and Y. Akman, 1979*a*, Hyperaccumulation of nickel by *Alyssum* Linnaeus (Cruciferae), *Royal Soc. [London] Proc.* **203B**, 287–403.

Brooks, R. R., J. M. Trow, and B. Bølviken, 1979*b*, Biogeochemical anomalies in Fennoscandia: A study of copper, lead and nickel levels in *Melandrium dioicum* and *Viscaria alpina, Jour. Geochem. Explor.* **11**, 73–87.

Brooks, R. R., R. D. Reeves, R. S. Morrison, and F. Malaisse, 1980, *Belgium Royal Soc. Botany Bull.* **113**, 166–172.

Buyalov, N. I. and A. M. Shvyryaeva, 1961, Geobotanical method in prospecting for salts of boron, *Internat. Geology Rev.* **10**, 619–625.

Cannon, H. L., 1957, Description of indicator plants and methods of botanical prospecting for uranium deposits on the Colorado Plateau, *U.S. Geol. Survey Bull. 1030-M*, 399–516.

Cannon, H. L., 1960, Botanical prospecting for ore deposits, *Science* **132**, 591–598.

Cannon, H. L., 1964, Geochemistry of rocks and related soils and vegetation in the Yellow Cat area, Grand County, Utah, *U.S. Geol. Survey Bull. 1176*, 1–127.

Cannon, H. L., 1971, Use of plant indicators in ground water surveys, geologic mapping, and mineral prospecting, *Taxon* **20**, 227–256.

Cannon, H. L., 1979, Advances in botanical methods of prospecting. Part I: Advances in geobotanical methods of prospecting, *Canada Geol. Survey Econ. Geology Rept. 31*, 377–397.

Cannon, H. L. and W. H. Starrett, 1956, Botanical prospecting for uranium on La Ventura Mesa, Sandoval County, New Mexico, *U.S. Geol. Survey Bull. 1009-M*, 391–407.

Chaffee, M. A. and C. W. Gale, III, 1976, The California poppy *(Eschscholtzia mexicana)* as a copper indicator plant—A new example, *Jour. Geochem. Explor.* **5**, 59–63.

Chikishev, A. G., 1965, *Plant Indicators of Soils, Rocks and Subsurface Waters*. New York: Consultants Bureau, 209p.

Cole, M. M., 1965, Biogeography in the service of man, inaugural lecture, Bedford College, London.

Cole, M. M., 1971, The importance of environment in biogeographical/geobotanical and biogeochemical investigations, *Canadian Inst. Mining and Metallurgy Spec. Vol. 11*, 414–425.

Cole, M. M., 1973, Geobotanical and biogeochemical investigations in the sclerophyllous woodland and shrub associations of the Eastern Goldfields area of

Western Australia, with particular reference to the role of *Hybanthus floribundus* (Lindl.) F. Muell, as a nickel indicator and accumulator plant, *Jour. Appl. Ecology* **10**, 269–320.

Duvigneaud, P., 1959, Plant cobaltophytes in Upper Katanga (in Fr.), *Belgium Royal Soc. Botany Bull.* **91**, 111–134.

Duvigneaud, P., and S. Denayer-de-Smet, 1963, Copper and vegetation in Katanga (in French), *Belgium Royal Soc. Botany Bull.* **96**, 93–231.

Ernst, W., 1968, The *Viola calaminaria* association of Westphalia, a heavy metal community at Blankenrode in Westphalia (in German), *Mitt. Flor-Soziol. Arbeitsgemeinschaft* **13**, 263–268.

Ernst, W., 1969, Pollenanalytical evaluation of a heavy metal turf in Wales (in German), *Vegetatio* **18**, 393–400.

Greig-Smith, P., 1964, *Quantitative Plant Ecology.* London: Butterworths, 256p.

Henwood, W. J., 1857, Notice of the copper turf of Merioneth, *Edinburgh New Philos. Jour.* **5**, 61–63.

Howard-Williams, C., 1970, The ecology of *Becium homblei* in Central Africa with special reference to metalliferous soils, *Jour. Ecology* **58**, 745–763.

Jaffré, T., 1977, Accumulation of manganese by species associated with ultrabasic terrain in New Caledonia (in French), *Acad. Sci. Comptes Rendus* **284D**, 1573–1575.

Jaffré, T., and M. Schmid, 1974, Accumulation of nickel in *Psychotria douarrei* in New Caledonia (in French), *Acad. Sci. Comptes Rendus* **278**, 1727.

Karpinsky, A. M., 1841, Can living plants be indicators of rocks and formations on which they grow and does their occurrence merit the particular attention of the specialist in structural geology? (in Russian), *Zhur. Sadovodstva,* Nos. 3 and 4.

Mahalanobis, P. C., 1936, On the generalized distance in statistics, *India Natl. Inst. Sci. Proc.* **12**, 49–55.

Malaisse, F., J. Gregoire, R. R. Brooks, R. S. Morrison, and R. D. Reeves, 1978, *Aeolanthus biformifolius:* A hyperaccumulator of copper from Zaire, *Science* **199**, 887–888.

Malaisse, F., J. Gregoire, R. S. Morrison, R. R. Brooks, and R. D. Reeves, 1979, Copper and cobalt in vegetation of Fungurume, Shaba Province, Zaire, *Oikos* **33**, 472–478.

Malyuga, D. P., 1964, *Biogeochemical Methods of Prospecting.* New York: Consultants Bureau, 205p.

Malyuga, D. P., N. S. Malashkina, and A. I. Makarova, 1959, Biogeochemical investigations at Kadzharan, Armenian SSR (in Russian), *Geokhimiya,* no. 5, 423–430.

Minguzzi, C., and O. Vergnano, 1948, The content of nickel in the ash of Alyssum bertolonii Desv. (in Italian), *Soc. Toscana Sci. Nat. Atti Mem.* **55A**, 49–77.

Nesvetailova, N. G., 1961, Geobotanical investigations for prospecting for ore deposits, *Internat. Geology Rev.* **3**, 609–618.

Nicolls, O. W., D. M. J. Provan, M. M. Cole, and J. S. Tooms, 1965, Geobotany and geochemistry in mineral exploration in the Dugald River area, Cloncurry district, Australia, *Australasian Inst. Mining and Metallurgy Trans.* **74**, 695–799.

Nielsen, J. S., R. R. Brooks, C. R. Boswell, and N. J. Marshall, 1973, Statistical evaluation of geobotanical and biogeochemical data by discriminant analysis, *Jour. Appl. Ecology* **10**, 251–258.

Palou, R., X. de Gramont, J. Magny, and J. Carles, 1965, Two plant indicators of zinc-lead deposits in the Pyrenees, *Soc. Hist. Nat. Toulouse Bull.* **100**, 465–468.

Persson, H., 1948, Studies in "copper mosses," *Jour. Hattori Bot. Lab.* **17**, 1–18.

Reeves, R. D., and R. R. Brooks, 1983, European species of *Thlaspi* L. (Cruciferae) as indicators of nickel and zine, *Jour. Geochem. Exploration* **18**, 275–283.

Rune, O., 1953, Plant life on serpentine and related rocks in the north of Sweden, *Acta Phytogeogr. Suecica* **31**, 1–139.

Se Sjue-Tszin, and Ban-Lian Sjuj, 1953, *Elsholtzia haichowensis* Sun—A plant that can reveal the presence of copper-bearing strata, *Dichzhi Sjuozbao* **32**, 360–368.

Severne, B. C., and R. R. Brooks, 1972, A nickel-accumulating plant from Western Australia, *Planta* **103**, 91–94.

Shacklette, H. T., 1962, Field observations of variations in *Vaccinium uliginosum L., Canadian Field Naturalist* **76**, 162–167.

Shacklette, H. T., 1964, Flower variation of *Epilobium angustifolium* L., *Canadian Field Naturalist* **78**, 32–42.

Shchapova, G. F., 1938, Benthic vegetation of the northeastern bays of the Caspian Sea (in Russian), *Bot. Zhur.* **23**, 122–143.

Venkatesh, V., 1964, Geobotanical methods of mineral prospecting in India, *Indian Mineralogist* **18**, 101.

Venkatesh, V., 1966, Geobotany in mineral exploration, *Steel Mines Rev.* **6**, 3–5.

Viktorov, S. V., Y. A. Vostokova, and D. D. Vyshivkin, 1964, *Short Guide to Geobotanical Surveying.* Oxford: Pergamon, 158p.

Cross-references: *Aerial Surveys, General; Biogeochemistry; Indicator Elements; Vegetation Mapping.*

GEOBOTANY—See GEOBOTANICAL PROSPECTING

GEOCHEMICAL EXPLORATION—See ATMOGEOCHEMICAL PROSPECTING; BIOGEOCHEMISTRY; EXPLORATION GEOCHEMISTRY; HYDROCHEMICAL PROSPECTING; INDICATOR ELEMENTS; LAKE SEDIMENT GEOCHEMISTRY; LITHOGEOCHEMICAL PROSPECTING; MARINE EXPLORATION GEOCHEMISTRY; PEDOGEOCHEMICAL PROSPECTING. Vol. XIII: GEOCHEMISTRY, APPLIED.

GEODATA MANAGEMENT—See COMPUTERS IN GEOLOGY; GEOLOGICAL CATALOGUING; GEOSTATISTICS; MATHEMATICAL GEOLOGY: PUNCH CARDS, GEOLOGICAL REFERENCING; WELL LOGGING.

GEODESY—See SATELLITE GEODESY AND GEODYNAMICS.

GEOELECTRICAL SURVEYS—See EXPLORATION GEOPHYSICS; MARINE MAGNETIC SURVEYS; VLF ELECTROMAGNETIC PROSPECTING.

GEOEXPLORATION—See EXPLORATION GEOCHEMISTRY; EXPLORATION GEOPHYSICS; PROSPECTING.

GEOHISTORY, AMERICAN FOUNDING FATHERS

The study of geology in North America began long after European efforts in earth science were under way. The earliest American geological research was not surprisingly undertaken in large measure by Europeans. Prior to the American Revolution, several European travelers and explorers included observations on U.S. geological curiosities in their accounts (e.g., Wells, 1963), and in 1787 the first major study of U.S. earth science was published in Germany by the Hessian doctor, Johann David Schopf (1787). Such European efforts demonstrated the numerous possibilities for geological investigation afforded by the vast North American continent, and predictably, by the end of the eighteenth century, Americans had also begun to think and write seriously about the geology of their lands.

Samuel Latham Mitchill, a New York doctor, was one of the first Americans to gain recognition for geological studies. Mitchill's (1789) "Geological Remarks on Certain Maritime Parts of New York" includes what is probably the earliest use of the word *geology* published in the United States. His account of the rocks and minerals of the Catskill Mountains of New York (Mitchill, 1795) was reprinted in several periodicals. This New Yorker further stimulated the awakening interest in geology in the United States through his journal, *The Medical Repository*. Between 1798

and 1819 this periodical included almost 200 articles, many by Mitchill, on mineralogy, paleontology, and other aspects of the earth sciences. Thus, Mitchill aided the growth of U.S. geology both by his own research and by encouraging the research of others.

The first decade of the nineteenth century found the United States rapidly developing in the earth sciences. President Thomas Jefferson, who had earlier published his observations on fossil bones (Jefferson, 1799), added his prestige to the study of natural history by supporting the Lewis and Clark expedition of 1804–1806. The French researcher, Constantin-François Chasseboeuf, Comte de Volney, became the first to introduce U.S. geology to a widespread European readership in his *Tableau du Climat et du Sol des Etats-Unis d'Amerique* (Volney, 1803). An 1804 edition of this work was the earliest geological text published in the United States. James Mease's (1807) *A Geological Account of the United States* synthesized the work of Mitchill and Volney and reached a popular audience in the United States. The American public was further exposed to the earth sciences at Charles Wilson Peale's American Museum, where exhibits included rocks, minerals, fossils, and a mounted skeleton of the American mastodon.

Americans' growing awareness of natural science inspired several publications, which in turn advanced the science of geology. The *American Mineralogical Journal,* edited by Archibald Bruce from 1810 to 1814, became the United States' first purely scientific periodical, and though it ceased publication after only four years, it included many valuable articles on the mineral wealth of the United States. In 1818, Benjamin Silliman filled the gap left by Bruce's *Journal* with *The American Journal of Science and the Arts;* this Yale-based publication continued as the showplace for much of the United States' geological development during the next several decades. The first U.S. textbook on the earth sciences was contributed by Parker Cleaveland (1816), a professor of chemistry at Bowdoin College. His *Elementary Textbook of Mineralogy and Geology* altered the course of earth sciences in the United States by rejecting much of the standard Wernerian mineral classification in favor of a division of species based on chemical tests. Another text that influenced U.S. geological development was the New York–published translation of Georges Cuvier's (1818) *Essay on the Theory of the Earth*. Cuvier's *Essay* described the nature and origin of fossil species and presented a classification scheme by which fossil bones and shells might be identified and described. By 1818 American scientists had both the guidelines to classify and identify their geological specimens and the periodical forums in which to publish their observations.

A growing desire and need to understand the distribution of rocks in the United States led to the more frequent investigation of geological materials in their natural setting rather than in the laboratory. In the United States the infant study of *field geology* was championed by William Maclure, a Philadelphia physician and president of the Philadelphia Academy of Natural Sciences. Maclure's (1809) "Observations on the Geology of the United States," which represented the largest geological mapping project attempted anywhere in the world, was remarkable for both its scope and accuracy. The geologist traveled thousands of miles on foot to complete this half-million square mile survey, and his map compares favorably in many respects with modern geological charts of the eastern United States. While his map used the Wernerian (i.e., Neptunian) nomenclature, Maclure did not believe that all rocks had an aqueous origin. His "Essay on the Formation of Rocks" (1818) contains a balanced view of the role of both heat and water in rock formation, and when viewed with Cleaveland's (1816) *Textbook,* it indicates a growing break from the European traditions that dominated earlier geological efforts in the United States.

Progress in earth science research was accompanied by the development of U.S. geological education. Benjamin Silliman founded the first course of earth science study in the United States at Yale, and by 1820 several of his students were active researchers. One of these, Amos Eaton, founded his own school at Troy, New York (now Rensselaer Polytechnic Institute), which had an important scientific curriculum. Other opportunities to study geology and mineralogy were established in most major U.S. cities, and the United States was provided with a growing community of professional geologists.

Opportunities for geological research grew along with the number of trained earth scientists. Governor Stephen Van Rensselaer privately sponsored field investigations of various parts of New York State, including a survey of the route of the Erie Canal (Eaton, 1824). In 1821 Dennison Olmsted undertook a survey of the state of North Carolina, and shortly thereafter, South Carolina financed a cursory geological and agricultural survey, with Lardner Vanuxem as the chief geologist. While some earth scientists worked in the field, others supplied the expanding college geology courses with U.S.-oriented texts. John Lee Comstock presented the first of his several textbooks on mineralogy (1827), while Jeremiah Van Rensselaer (1825), Ebenezer Emmons (1826), and Benjamin Silliman (1829) contributed other basic introductory geology works.

The 1830s was a decade in which U.S.-trained earth scientists were largely occupied with geological surveys sponsored by the several states (see *Geological Surveys, State and Federal*). In 1830 Massachusetts employed Amherst College Professor Edward Hitchcock to conduct a complete geological examination of the Commonwealth and thus became the first state to fund its own scientific survey. Tennessee soon followed by hiring Gerard Troost in 1831, while Maryland sponsored the third such study by James T. Ducatel two years later. The U.S. government held vast territories of unexplored land, and from 1834 to 1836 it commissioned several geological excursions by the controversial G. W. Featherstonaugh. New state surveys were begun in 1835 by J. G. Percival and C. U. Shepard of Connecticut, by Henry D. Rogers of New Jersey, and by his brother William B. Rogers in Virginia. In 1836 Henry Rogers applied his knowledge of New Jersey geology to the Pennsylvania state survey, while Charles T. Jackson commenced the first of his several state surveys in Maine. Delaware followed its Atlantic Coast neighbors by employing James C. Booth in 1837, and in the same year U.S. westward expansion was reflected in geological surveys of Indiana by David Dale Owen and Michigan by Douglass Houghton. Thus, with hopes of increasing both agricultural and mineral production, the states eagerly sought the advice of trained geologists.

As state surveys revealed the nation's mineral wealth, mining companies offered new opportunities for the American earth scientist in the world of business. Rich deposits of gold in Virginia, North Carolina, and Georgia, lead in Missouri, copper in Michigan, and most important, coal and iron in Pennsylvania and adjacent states necessitated numerous local surveys and mine reports. Many of the nation's most prominent geologists, including Benjamin Silliman, Charles Shepard, and Edward Hitchcock, advised mining companies. Others such as Richard C. Taylor and Walter R. Johnson represented a new group of specialists in economic and mining geology.

In spite of the presence of a few specialized geologists in economic geology, the majority of American workers engaged in teaching or state surveys had to be well versed in all aspects of geological inquiry. However, the era of the state surveys saw the introduction of a new concept of specialization in *geological exploration* which revolutionized earth science in the United States. The Natural History Survey of New York, begun in 1836, employed no fewer than eight principal researchers as well as many assistants. Among these were four field geologists, William W. Mather, Ebenezer Emmons, Timothy A. Conrad, and Lardner Vanuxem, as well as a mineralogist, Lewis C. Beck, and a paleontologist, James Hall. Each *field geologist* was assigned a quarter of the state to map, while Beck and Hall studied mineral and fossil specimens from the entire state. This research team approach was new to the earth sciences in the United States and allowed each worker to benefit from the experience and specialties of the others. In particular, James Hall's study of New York State fossils led to a completely new, uniquely American, classification of Paleozoic strata, which had previously been classified largely on the basis of European correlation. This new nomenclature was quickly applied to regions outside New York State as well, and the Hall classification became a standard for much of the subsequent U.S. Paleozoic research.

The desire of geologists to work in groups was further reflected by the organization in 1840 of the Association of American Geologists, which sponsored annual meetings devoted to individual papers followed by group discussions. In this way a tradition of interaction between earth scientists from different regions and specialties was begun. Thus, by the early 1840s geology in the United States had matured into a thriving, self-confident science. Individuals had advanced the discipline through education, publication, mining consultation, and laboratory and field investigation. Specialization within the earth sciences led to increased communication and cooperation between researchers and supported the emergence in succeeding decades of American geologists of international renown.

ROBERT M. HAZEN

References

Cleaveland, P., 1816, *An Elementary Treatise on Mineralogy and Geology.* Boston: Cummings and Hilliard, 668p.

Comstock, J. L., 1827, *Elements of Mineralogy.* Boston: S. G. Goodrich, 76 + 338p.

Cuvier, G., 1818, *Essay on the Theory of the Earth.* New York: Kirk & Mercein, 431p.

Eaton, A., 1824, *A Geological and Agricultural Survey of the District Adjoining the Erie Canal.* Albany, N.Y.: Packard and Van Benthuysen, 163p.

Emmons, E., 1826, *Manual of Mineralogy and Geology.* Albany, N.Y.: Websters & Skinners, 230p.

Jefferson, T., 1799, A memoir on the discovery of certain bones of a quadruped of the clawed kind in the Western parts of Virginia, *Am. Philos. Soc. Trans.* **4,** 345-350.

Maclure, W., 1809, Observations on the geology of the United States, *Am. Philos. Soc. Trans.* **6,** 411-428.

Maclure, W., 1818, Essay on the formation of rocks, *Acad. Nat. Sci. Philadelphia Jour.* **1,** 261-275, 285-309, and 327-344.

Mease, J., 1807, *A Geological Account of the United States.* Philadelphia: Birch & Small, 496p.

Mitchill, S. L., 1789, Geological remarks on certain maritime parts of the state of New York, *Am. Museum* **5,** 123-126.

Mitchill, S. L., 1795, Description of the Blue Mountains in the State of New-York, *New York Mag.* **6,** 465-471.

Schopf, J. D., 1787, *Beytrag zur Mineralogischen Kenntniss des Ostlichen Theils von Nordamerican und seiner Geburge.* Erlangen: J. J. Palm, 194p.

Silliman, B., 1829, Outline of the course of geological lectures given in Yale College, in R. Bakewell, ed., *An Introduction to Geology.* New Haven: H. Howe, 400p.

Van Rensselaer, J., 1825, *Lectures on Geology.* New York: Bliss & White, 358p.

Volney, C.-F. C., Comte de, 1803, *Tableau du climat et du sol des Etats Unis d'Amerique.* Paris: Courcier & Denton.

Wells, J. W., 1963, Early investigation of the Devonian System in New York, 1656-1836. *Geol. Soc. Amer. Special Paper 74,* 74p.

Cross-references: *Geological Surveys, State and Federal; Geology, Philosophy; Geology, Scope and Classification; Popular Geology.* Vol. XIII: *Geology, Applied.*

GEOLOGICAL CATALOGUING

A *catalogue* is a set of records, each comprising all the available information considered relevant to one item or having indicators to that information. *Cataloguing* is the process that produces these records. As Waterson (1972) noted, complete and accurate documentation of all geological collections is essential if the collections are to be of lasting value. This documentation applies equally to personal research collections and museum collections.

In the past, the main emphasis in museum cataloguing has been on developing records of the specimens within a collection. Records concerned with other items of interest to geologists have recently attained increasing importance. These records may be about localities (sites of geological interest), maps, documents, photographs, or thin sections (see *Map and Chart Depositories, Remote Sensing, Societies and Periodicals;* Vol. IVB: *Museums, Mineralogical;* Vol. XIII: *Photogrammetry*).

Each museum has its own particular cataloguing operation, but the fundamental philosophy is much the same (MDA, 1980a). The museum wishes to produce a record that adequately describes each item in its care and to use these descriptions as the basis of a catalogue of the collection. It also wishes to have one or more indexes that act as directories to the catalogue, enabling the museum to answer questions about its collection and environment (Roberts and Light, 1980).

The Record

Data Categories Museum documentation research has concentrated on an analysis of the data categories that comprise a record and about which a curator might wish to record information. There has been a move toward standardization of these categories in the United States, Canada, and the United Kingdom. In the United States, this development is evident in the work of the Museum Data Bank Coordinating Committee. Chenhall (1975) discussed the concepts that this committee felt should be consistently recorded. Parallel work has been undertaken in Canada by the National Inventory Programme and in the United Kingdom by the Museum Documentation Association (MDA, 1980b). A report on the work of these and similar bodies has summarized the data-recording opportunities for a subject area such as geology (Light and Roberts, 1981).

In the case of a geologic specimen, detailed geographic and stratigraphic data defining the collection place and local stratigraphy are of primary importance. Accepted national or international conventions should be followed when using stratigraphic terms. If the geographic and stratigraphic data are not recorded at the time of collection, they may be irretrievably lost and the scientific value of the specimen diminished. The name of the collector and the collection date should also be recorded. The specimen should be numbered in the field at the time of collection. This number may be retained when the specimen is incorporated within a museum collection or superseded by a museum number. The actual form of the number is of secondary importance because its essential attribute is that it acts as a link between the specimen and the data bank. The specimen may be described and identified or named. The physical description is frequently limited to form characteristics. For fossil material, the classified identification is of key importance. In other types of specimens, an established classification system such as the Folk Classification of limestones (Folk, 1959) or the Hey description of mineral species (Hey, 1975) should be followed whenever possible.

Other data of relevance may include historical background information such as previous owners and the person or institution from whom the current owner acquired the specimen. Processes that affect the specimen (such as its conservation and preservation) are also of interest.

These principles are equally applicable to the cataloguing of other types of items. A locality record, e.g., may include data that accurately define the site's geographical boundaries, gives its form, lists any other finds in the same locale, and refers to other germane documents.

Recording Media   Another example of recent moves toward cooperation is found in the recording format used by museums when cataloguing. In the United Kingdom, many museums use standard 5″ x 8″ (A5) and 8″ x 12″ (A4) record cards. The MDA, e.g., produces A5 cards and instructions for documenting/general geological information and detailed mineralogical data (MDA, 1980c, 1980d) and A4 cards for documenting geological localities (GCG, 1980).

Catalogue and Indexes

The use of such recorded data is, alas, often minimal. Few museums have the resources to produce a published catalogue or extensive indexes to their collections. Without such indexes, the accessibility of the data is greatly diminished. Past experience indicates that probably the most valuable indexes are lists compiled according to taxonomic principles, rock and mineral names, place names, stratigraphic terms, etc. Even though such indexes can be produced manually (Smith, 1950), the most economical way of producing a range of indexes is by computer. The Sedgwick Museum of Geology (Cambridge, England), e.g., has experimented with different types of computer applications in this regard. Here, the existing catalogue of 440,000 records has been computerized, and different types of indexes are being produced from the original database (Porter et al., 1977). The records prepared when following this system may be computerized if required. The standard MDA cards have a similar base; they can be used both manually and as a source of data for computer processing. It is likely that geological cataloguing will experience a resurgence as these new techniques are increasingly applied.

D. ANDREW ROBERTS

References

Chenhall, R. G., 1975, *Museum Cataloguing in the Computer Age.* Nashville, Tenn.: American Association for State and Local History, 512p.

Folk, R. L., 1959, Practical petrographic classification of limestones, *Am. Assoc. Petroleum Geologists Bull.* **43**, 1–38.

Geological Curators Group (GCG), 1980, *Geology Locality Card Instructions,* Museum Documentation System, Duxford, Cambridgeshire: Museum Documentation Association, 34p.

Hey, M. H., 1975, *An Index of Mineral Species and Varieties Arranged Chemically,* 2nd ed. (rev.), London: British Museum (Natural History), 728p.

Light, R. and D. A. Roberts, 1981, *International Museum Data Standards and Experiments in Data Transfer.* Duxford, Cambridgeshire: Museum Documentation Association (Occas. Paper 5), 148p.

Museum Documentation Association, (MDA), 1980*a, Practical Museum Documentation,* Museum Documentation System, Duxford, Cambridgeshire, 148p.

Museum Documentation Association, 1980*b, Data Definition Language and Data Standard,* Museum Documentation System, Duxford, Cambridgeshire, 144p.

Museum Documentation Association, 1980*c, Geology Specimen Card Instructions.* Museum Documentation System Duxford, Cambridgeshire, 51p.

Museum Documentation Association, 1980*d, Mineral Specimen Card Instructions.* Museum Documentation System. Duxford, Cambridgeshire, 49p.

Porter, M. F., R. B. Light, and D. A. Robert, 1977, *A Unified Approach to the Computerization of Museum Catalogues.* British Library Research and Development Reports. London: British Library, (Rept. No. 5338 HC), 75p.

Roberts, D., and R. B. Light, 1980, Progress in documentation, Museum documentation, *Jour. of Documentation* **36**(1), 42, 84.

Smith, W. C., 1950, L. J. Spencer's work at the British Museum, *Mineralog. Mag.* **29**, 256–270.

Waterson, C. D., 1972, Geology and the museum, *Scottish Jour. Geology* **8**(2), 129–144.

Cross-references: *Punch Cards, Geologic Referencing.* Vol. XIII: *Conferences, Congresses, and Symposia; Earth Science, Information Sources; Geological Communication; Geological Information, Marketing; Well Data Systems.*

GEOLOGICAL FIELDWORK, CODES

Geological fieldwork is an activity that engages most professional geologists sometime during their career. Even the so-called armchair geologist, or laboratory type has occasion to work in the field. Going into the field to conduct surveys, to collect samples, to carry out mapping programs, to provide ground truth for photointerpretive maps, or for instructional purposes is regarded by many geologists as practically a right. Whether in public domain (or Crown land) or on private land, certain codes of conduct must be observed because access for fieldwork is in fact a privilege, not a right.

Associated with the rapid growth of the earth sciences are ever greater numbers of students and professionals making such field visits. An increase in the numbers of accidents at quarries and mine sites associated with these visiting geologists has prompted legislation in some regions that limits or restricts their activities. Many sites now, e.g., require that all visitors wear some sort of protective gear that, depending on the circumstances, often includes helmets, safety glasses, and perhaps safety boots with steel toes.

The earliest action on this problem appears to have taken place in the United Kingdom in the late 1970s when the Geologists' Association of London prepared a pamphlet, "A Code for Geological Field Work" (1975) in an effort to encourage responsible conduct in the field. The association has distributed over 150,000 copies of the pamphlet throughout the United Kingdom. The code, a modification of the British "Countryside Code," has been well received by industry, government, and landowners. This code, the first of its kind to be implemented on a large scale in the English-speaking world, might well serve as a model for other geological associations. Indeed, this code has already been adapted for use in South Africa, the Netherlands, Norway, and China.

In an effort to ensure that geologists use the countryside responsibly, the Geologists' Association originally provided a code that stipulated ten main points. The following summary is an expansion of the British "Code for Geological Field Work" (Geologists' Association, 1975) that also includes additional remarks on geological field safety and professional outdoor behavior.

Code for Fieldwork

1. Observe and obey local bylaws and ordinances. Remember to leave gates open or shut, as you find them, when you pass through fenced areas. Failure to observe this simple rule can cause additional work and hardship for landowners and could result in death for animals that must get to water or could stray onto highways. If you find a gate open, leave it open, and vice versa. After obtaining permission to enter industrial sites, be sure to follow the recognized procedures, as outlined by the host company.

Be conversant with the particular safety and health requirements if you are entering a new site or environment—e.g., working quarry, open-pit or underground mine, drill site, or offshore platform. Comply with safety rules, blast warning procedures, and any instructions given by officials. Keep a sharp lookout for moving vehicles, cranes, bulldozers, and other mining equipment. On industrial sites, beware of sludge lagoons. Particular care is required on large river and ocean dredges and on offshore drilling platforms.

2. Always seek prior permission before entering private land. When working on farms or ranges in remote regions, it is a good idea to check into the homestead prior to fieldwork. This can be for your protection as well; in case you are injured, your approximate location will be known to someone locally. Those who have been chased around a paddock by a rambunctious stallion or bull can well attest to the advantage of inquiring of the landowner before entry.

3. Do not touch or interfere with machinery or equipment on farms or in quarries, mines, building sites, etc. Never pick up explosives or detonators from rock piles; if found, inform the management immediately.

4. Do not litter fields or roads with rock fragments that might cause injury to livestock or constitute a hazard to pedestrians or vehicles. Because auger holes in paddocks are especially treacherous for livestock, make sure they are backfilled and securely tamped down.

5. Avoid undue disturbance to wildlife and livestock. Plants and animals may inadvertently be displaced or destroyed by careless actions.

6. On coastal sections, consult the local coast

guard service whenever possible. It is imperative to learn of local hazards such as unstable cliffs or tides that might jeopardize excursions out onto coastal flats. Take great care when walking or climbing over slippery rocks below the high-water mark on rocky shores. More accidents to geologists, including fatalities, occur along rocky shorelines than anywhere else. Always wear footgear when wading on the shore. Coral reefs constitute a particular and specialized hazard. Take special precautions when working offshore. Small boats should normally be used only with an experienced sailor. Always wear a lifejacket. Aqualung (Scuba) equipment should be used only by trained divers. Even the use of simple gear such as a snorkel and face mask requires some instruction and practice. When swimming always use the buddy system, and never go into the water alone.

7. When working in mountainous or remote areas, inform someone of your intended route, preferably leaving a map. Be sure that someone is aware of your destination, and check in when you have arrived safely. Such simple procedures help prevent unnecessary worry on the part of others and ensure your safety in the event of inclement weather or other unavoidable delays. Carry at all times a small first-aid kit, some emergency food, survival bag, whistle, flashlight (torch), map, compass, and watch.

8. When exploring underground, especially when exploring karst caverns, be sure you have the proper equipment as well as the necessary experience. Never go alone; use the buddy system and move about in pairs. As in surface work in remote areas, report to someone your departure, location, estimated time away, and when you return. Never enter old mine workings or cave systems unless it has been approved as an essential part of the work. Only do so then by arrangement, with proper lighting and equipment.

9. Do not take risks on insecure cliffs or rock faces. Take care not to dislodge rock or loose soil because other people may be below. When using your geological hammer or pick, try to strike away from yourself and to keep rock chips from flying into your colleagues' eyes.

10. Be considerate of colleagues in the field and of the cultural and natural environments. When collecting, do not render an exposure untidy, defaced, or dangerous for those who follow. Some classic sites have been irrevocably defaced by such inconsiderate carelessness.

Codes for Students and Field Parties

Collecting and field parties, in contrast to lone or paired teams of geologists, should observe in addition the following rules.

1. Students should be encouraged to observe and record but not to hammer indiscriminately.

Innocuous-looking rock outcrops can be dangerous places with large hoards hammering away. Protective eyewear (safety goggles, glasses with plastic lenses) will help avoid injury from flying shards from hammering. Avoid hammering near another person or looking toward another person who is hammering.

2. Keep collecting to an absolute minimum. Avoid removing in situ fossils, rocks, or minerals unless they are genuinely needed for serious study. Never deliberately destroy or deface a unique exposure. Complaints have been made when some rare fossil localities have been totally quarried away by large parties of students. If possible, do your collecting with a camera.

3. For teaching, the use of replicas is recommended. The collecting of specimens should be restricted to those localities where there is a plentiful supply or to talus, scree, fallen blocks, and waste tips.

4. Never collect samples from walls or buildings. Take care not to undermine fences, walls, bridges, or other structures. Beware of traffic when examining road and railway cuttings. Avoid hammering and digging in weathered materials, and do not leave debris on the right-of-way. Railway and main highways are not normally open to large groups unless special permission has been obtained from the appropriate authorities.

5. The leader of a field party is asked to insure that the spirit of a code for geological fieldwork is fulfilled. The field leader should remember that his or her supervisory role is of prime importance. He or she must be supported by adequate assistance in the field. This is particularly important on coastal sections, or over difficult terrain, where parties may tend to become dispersed.

Codes for Professional Conduct

Efforts to strengthen the geological sciences as a profession focus variously on codes of ethics or rules of professional conduct. The guidelines presented in this volume apply to professional practices that involve fieldwork (see *Field Geology*) and reporting (see *Geological Writing; Geologic Reports*), among other activities, including communication among professionals and with the public (see Vol. XIII: *Geological Communication*). Perusal of codes from different professional geoscience associations—e.g., American Institute of Professional Geologists (1984), Australasian Institute of Mining and Metallurgy (Berkman and Ryall, 1976), European Federation of Geologists (Fox, 1984), Geological Association of Canada (1973), Geological Society of Australia (1980), and the American Registry of Certified Professionals in Agronomy, Crops, and Soils (ARCPACS, 1978)—shows a common orientation toward a consensus of relations of professionals to employers and clients, relations of professionals to each other, duties to the profession, and relations to the public. Although each society stresses different approaches to the problem of what constitutes acceptable professional conduct, there is surprising agreement among the associations. Some of the salient points detailed by these societies are summarized here to complement the codes for proper behavior in the field. Professional geoscientists are involved in a range of activities, and these guidelines apply to both office and fieldwork.

Relations with Other Geologists

1. The geologist shall not falsely or maliciously attempt to injure the reputation or business of another.
2. The geologist shall give credit for work done by others to whom credit is due and shall refrain from plagiarism in oral and written communications. He or she shall not knowingly accept credit due another.
3. The geologist should not allow him- or herself any signatures of convenience or any acts of disloyal competition.
4. Rules of loyalty and law must inspire the action of the geologist toward colleagues, employers, and those with whom he or she is working.

Relations with Clients

1. The geologist must avoid all forms of negligence in the exercise of this profession, especially if this behavior involves risks or material or moral damage to his or her client or to the environment.
2. The geologist must always inform the client of the real limits of results likely to be obtained by given professional assistance.
3. The geologist shall protect, to the fullest possible extent, the interest of an employer or client insofar as is consistent with the public safety, geologists' professional obligations, and ethics.
4. The geologist shall not use an employer's or client's information in any way that is competitive, adverse, or detrimental to the interests of the employer or client.
5. The geologist retained by one client shall not accept, without the client's written consent, an engagement by another if the interests of the two are in any manner conflicting.
6. The geologist shall not divulge information given in confidence.
7. The geologist shall engage or cooperate with other experts and specialists whenever the employer's or client's interest would be best served by such service.

8. The geologist who has made an investigation for any employer or client shall not seek to profit economically from the information gained, unless written permission to do so is granted or until it is clear that there can no longer be conflict of interest with the original employer or client. (A publication resulting from a piece of paid research is usually acceptable after a twelve month interval, but this understanding should be made on the original contract or in some written form prior to the job.)
9. The geologist shall not accept a concealed fee for referring a client or employer to a specialist or for recommending geological services other than his or her own.
10. Geologists who find their obligations to employer or client in conflict with professional obligations or ethics should correct such objectionable conditions or resign.

Relations to the Public

1. The geologist's responsibility to the public shall be paramount, and he or she shall avoid and discourage sensational, exaggerated, and unwarranted statements that might induce participation in unwarranted enterprises, or create panic or social unrest.
2. The geologist shall not knowingly permit the publication of his or her reports, maps, or other documents for any unsound or illegitimate undertaking.
3. The geologist shall not give a professional opinion, make a report, or give legal testimony without being as thoroughly informed as might reasonably be expected, considering the purpose for which the opinion, report, or testimony is desired, and the degree of completeness of information on which it is based should be made clear.
4. The geologist may publish dignified business, professional, or announcement cards but shall not advertise his or her work or accomplishments in a self-laudatory, exaggerated, or unduly conspicuous manner.
5. The geologist shall not issue a false statement or false information even if directed to do so by employer or client.

Application of Codes or Rules

Codes of conduct apply to all geologists who belong to professional associations that are affiliated with the major national and international geoscience organizations—e.g., Association of Geoscientists for International Development, American Geological Institute, European Federation of Geologists, or International Union of Geological Sciences. Codes for professional conduct should be respected in all countries where a geologist is working. If an acknowledged professional code exists locally, the geologist must follow it on the condition that its standards are not inferior to those of relevant umbrella organizations.

CHARLES W. FINKL, JNR.

References

American Institute of Professional Geologists, 1984, *Membership Directory.* Arvada, Colo.: 616p.
ARCPACS, 1978, *Registry and Certification Procedures.* Madison, Wis., 14p.
Berkman, D. A., and W. R. Ryall, 1976, *Field Geologists' Manual.* Parkville, Victoria, Australia: Australasian Institute of Mining and Metallurgy, 295p.
Fox, R. A., 1984, European Federation of Geologists: Code of professional conduct, *Episodes* **7**(3), 30–31.
Geological Association of Canada, 1973, *Constitution and By-Laws.* Waterloo, Ontario: Department of Earth Sciences, University of Waterloo, 20p.
Geological Society of Australia, 1980, Rules, code of ethics, list of members, *Geol. Soc. Australia Jour.* **27**(1/2), 15–16.

Cross-references: *Field Geology; Geological Survey and Mapping; Geological Writing; Geologic Reports; Geology, Scope and Classification; Professional Geologists' Associations; Samples, Sampling; Surveying, General.* Vol. XIII: *Engineering Geology Reports; Geological Communication.*

GEOLOGICAL FIELD SURVEYS, MAPPING—See ACOUSTIC SURVEYS; AERIAL SURVEYS; GEOLOGICAL SURVEY AND MAPPING; GROUNDWATER EXPLORATION; HARBOR SURVEYS; LAND CAPABILITY ANALYSIS; MARINE MAGNETIC SURVEYS; MINING PREPLANNING; PROSPECTING; SEA SURVEYS; SURVEYING, GENERAL; SURVEYING, ELECTRONIC; TERRAIN EVALUATION SYSTEMS; VEGETATION MAPPING; VLF ELECTROMAGNETIC PROSPECTING. Vol. XIII: FIELD GEOLOGY; MAPS, ENGINEERING PURPOSES.

GEOLOGICAL HIGHWAY MAPS

Geological highway maps are designed for professional geologists, students, and laypersons—anyone who has an interest in the rocks exposed, and the landforms he or she sees while traveling along the highway. The American Association of Petroleum Geologists (AAPG) has covered all twelve regions of the contiguous United States, as well as Alaska and Hawaii, and one revision has been published and another is in progress. The Canadian Society of Petroleum Geologists has covered the Province of Alberta.

Color printing is used generously, thus permitting greater cartographic detail than otherwise would be possible. Map units are as small as utility and map scale permit, and much generalization is necessary. Correlation conflicts between individuals and between institutions are resolved as far as possible; if they cannot be resolved, the compiler must arbitrate the dispute so there will be no boundary line mismatches. The color scheme and letter symbols are coordinated between the geological maps, the stratigraphic charts, and the cross-sections, as well as between regions. Color coordination, uniform scales, and boundary line matching of map units permit the joining of the maps of the series into a large geological map of the contiguous United States.

The Dallas Geological Society (Texas) originated the program with the publication of the Geological Highway Map of Texas in 1959. Philip F. Oetking was compiler and William C. Kerr, Jr., was cartographer; the work was done under the auspices of a committee chaired by Dan E. Feray. The map was reprinted in 1963 with the addition of a text, "Explanatory Notes," and a columnar section on the front, and physiographic and tectonic maps, W-E and NW-SE structural cross-sections with explanatory text, and a geological history of Texas, on the reverse side.

Subsequently, the project was expanded to cover the entire United States as a program of the AAPG's Geological Highway Map Committee, with Philip F. Oetking as chairman from 1966 to 1967; Dan E. Feray, 1968 to 1969; and H. B. Renfro, 1970 to 1972. Map No. 1 was of the Mid-Continent Region (Kansas–Missouri–Oklahoma–Arkansas), published in 1966. This was followed, in order, by Map No. 2, Southern Rocky Mountain Region (Utah–Colorado–Arizona–New Mexico), in 1967; Map No. 3, Pacific Southwest Region (California–Nevada), in 1968; Map No. 4 Mid-Atlantic Region (Kentucky–Tennessee–West Virginia–Virginia–Maryland–Delaware–North Carolina–South Carolina), in 1970; and Map No. 5, Northern Rocky Mountain Region (Idaho–Montana–Wyoming), in 1972. After the death of Dr. Renfro in 1972, Allan P. Bennison, geological consultant at Tulsa, Oklahoma, was retained as compiler, with Robert H. Dott, Sr., as AAPG coordinator of the Geological Highway Map Program; Cartographer Van K. Higginbotham remained in Dallas. Map No. 6, Pacific Northwest Region (Washington–Oregon), started by Renfro and Feray, was completed and published in 1973, and a revision of the Texas map, nearly completed by Renfro and Feray, with cooperation of Philip B. King (U.S. Geological Survey), was finished and published in 1973 as Map No. 7, the H. B. Renfro Memorial Edition.

Subsequent maps were compiled by Bennison: No. 8, Alaska–Hawaii, 1974; No. 9, Southeastern Region (Alabama–Florida–Georgia–Louisiana–Mississippi), 1975; and No. 10, Northeastern Region (Pennsylvania–Massachusetts–New Hampshire–Rhode Island–Connecticut–New Jersey–Maine–New York–Vermont). The northeastern map was published in 1976 as the National Bicentennial Edition.

Map No. 11 Great Lakes Region (Michigan–Ohio–Indiana–Illinois–Wisconsin) was published in 1982, and the final map of the series, Map No. 12, covering the Northern Great Plains (Minnesota–Iowa–Nebraska–South Dakota–North Dakota) was published in 1984. Map No. 1, Mid-Continent Region (Kansas–Missouri–Oklahoma–Arkansas) was reviesd by A. P. Bennison, assisted by J. M. Webb, and published in 1986. Revision of Map No. 4, Mid-Atlantic Region (Kentucky–Tennessee–West Virginia–Virginia–Maryland–Delaware–North Carolina–South Carolina) is nearly completed.

The maps are printed on sheets of uniform size—70.5 cm × 91.4 cm (27 3/4 in × 36 in)—which are folded into the convenient size of 12 cm × 23.5 cm (4 3/4 in × 9 1/4 in). Rolled copies of most maps are also available. The geological maps, except the Alaska–Hawaii maps, are of uniform scale—1 in. ≈ 30 mi (1 cm ≈ 18.5 km). The front of the sheet contains geological text, correlation charts, and as much other pertinent information as can be accommodated, depending on the area and shape of the region. The back carries items of special interest to travelers in more detail than can be shown on the geological map, (see *Popular Geology*).

Format differs somewhat from region to region, but standard features, in addition to the geological map, include the following:

Chart of Time and Rock Units: The number may range from 1 to 13, depending on the complexity and variety of the geology and availability of space.

Cross-sections: These range from one to four as dictated by the structural complexity. The sections are carried to an arbitrary depth, such as 10,000 ft below sea level, or to a regional basement. In areas that have been extensively prospected for oil and gas with deep borings and seismic exploration, such as the Gulf Coast, cross-sections can be drawn with considerable detail and accuracy; elsewhere, they must depend in part on projection in depth from data observed on the surface. To provide greater utility for motorists, the sections are oriented as closely as possible along major federal and state highways. Sections

are continuous from region to region, and tied together at common points on boundaries of contiguous regions.

Tectonic Map: At a scale of usually 1 in. ≈ 80 mi, this map shows the major structural features of the region and the lines of cross-sections.

Physiographic Map: Scale is the same as for a tectonic map. The map shows the configuration and distribution of landforms.

Land and Sea Changes: A series of small maps summarizes gross changes—uplift and subsidence—through geologic history.

A 1973 innovation provides guides titled *Places of Geological Interest, Gemstone Locations, Where Fossils Can Be Found,* and *National Parks* by means of maps, symbols, lists, and text, all keyed together. Maps, lists, and texts are derived from publications of state geological surveys; offices of parks, recreation and tourism; universities; the National Park Service; the American Automobile Association; and similar agencies and organizations. A list of special references is included in each of the more recent maps. A feature of the map of the Northeastern Region (Map No. 10, Bicentennial Edition) is a discussion of geology as related to the Revolutionary War—mainly battlefields and supplies of mineral raw materials such as iron and copper ores, from which guns, cannons, ammunition, and other, military hardware were manufactured.

Each of the categories of illustrative material is accompanied by an explanatory and interpretive text designed to make the map more interesting and useful. Also included are sources of geological information in the region and agencies where reports and other published material may be purchased. The success of the AAPG's program and the worth of the product have depended in very large measure on the cooperation received from geologists associated with state and federal geological surveys and with educational institutions in the several regions. Their help and material inevitably provided the compiler with much essential information and also with a confidence in the final product that otherwise might not have existed.

ROBERT H. DOTT, SR.

List of Maps

United States

Maps published by the American Association of Petroleum Geologists, Box 979, Tulsa, Oklahoma 74101.

1. Mid-Continent Region (Kansas–Missouri–Oklahoma–Arkansas), 1966, compiled by Philip Oetking, Dan E. Feray, and H. B. Renfro; Van K. Higginbotham, cartographer. Revised edition 1986, compiled by A. P. Bennison, assisted by

J. W. Webb; cartography by Allan Cartography, Medford, Oregon.

2. Southern Rocky Mountain Region (Utah–Colorado–Arizona–New Mexico), 1967, compiled by Philip Oetking, Dan E. Feray, and H. B. Renfro; Van K. Higginbotham, cartographer.
3. Pacific Southwest Region (California–Nevada), 1968, compiled by Dan E. Feray, Philip Oetking, and H. B. Renfro; Van K. Higginbotham, cartographer.
4. Mid-Atlantic Region (Kentucky–Tennessee–West Virginia–Virginia–Maryland–Delaware–North Carolina–South Carolina), 1970, compiled by H. B. Renfro and Dan E. Feray; Van K. Higginbotham, cartographer.)
5. Northern Rocky Mountain Region (Idaho–Montana–Wyoming), 1972, compiled by H. B. Renfro and Dan E. Feray; Van K. Higginbotham, cartographer.
6. Pacific Northwest Region (Washington–Oregon), 1973, compiled by H. B. Renfro and Dan E. Feray, completed by Allan P. Bennison; Van K. Higginbotham, cartographer.
7. Texas, H. B. Renfro Memorial Edition, 1973, compiled by H. B. Renfro, Dan E. Feray, Philip B. King, and Allan P. Bennison; Van K Higginbotham, cartographer.
8. Alaska–Hawaii, 1974, compiled by Allan P. Bennison; Van K. Higginbotham, cartographer.
9. Southeastern Region (Alabama–Florida–Georgia–Louisiana–Mississippi), 1975, compiled by Allan P. Bennison; Van K. Higginbotham, cartographer.
10. Northeastern Region (Pennsylvania–Massachusetts–New Hampshire–Rhode Island–Connecticut–New Jersey–Maine–New York–Vermont), Bicentennial Edition, 1976, compiled by Allan P. Bennison; Van K. Higginbotham, cartographer.

Canada

Maps published by Canadian Society of Petroleum Geologists, 612 Lougheed Building, Calgary, Alberta, Canada.

Alberta, 1975, compiled by Map Committee of the Canadian Society of Petroleum Geologists, Paul C. Jackson, chairman, Rein de Wit, editor; cartography by drafting departments of sponsoring companies and governmental agencies.

Cross-references: *Maps, Logic of; Map Series and Scales; Map Symbols; Popular Geology.* Vol. XIII: *Geological Communication.*

GEOLOGICAL LOGGING—See WELL LOGGING. Vol. XIII: WELL DATA SYSTEMS.

GEOLOGICAL METHODOLOGY

The following survey of basic *geological methodology* is not founded on any comprehensive presentation of the subject because no such works seems to be in existence. The volume edited by

Albritton (1963), with its high-standard contributions by seventeen experts, shows the diversity of topics that would have to be covered by such a review. Besides the relevant articles in this encyclopedia, the volume by Albritton should be consulted throughout, including the annotated bibliography on the philosophy of geology and its unique index to the bibliography. The latter presents the first list of terms inherent in geological and general scientific methodology.

This entry examines the basic patterns of scientific thinking from which the fabric of geological thought evolved and shows some of the logical connections between the threads of that fabric. It also draws attention to apparent gaps in the system of present-day geological thought and to aspects of geological methodology that are in need of research.

About the middle of the nineteenth century, when Charles Lyell wrote and continuously improved his famous *Principles of Geology* (1830), which was based on induction as the important tool of scientific inquiry, the *philosophy of induction* reached its culmination in the works of Mill (1925). At the same time, Whewell, a mineralogist-cum-philosopher, wrote his *History of the Inductive Sciences* (1837) as well as *Philosophy of the Inductive Sciences* (1840). It is significant that in these two important works, *geology* occupies an important position alongside the other major scientific disciplines of the time (see *Geology, Scope and Classification*).

By contrast, a thorough search through the vast modern literature on scientific methodology, written by (professional) philosophers, logicians, or general methodologists, reveals that in most of these works, even the comprehensive ones, geology is either not mentioned or perhaps quoted by name in enumerations of other scientific disciplines. Only in a few of these works, e.g., that by Kraft (1960), are examples of geology and human archaeology discussed, which shows that their methodology or method of inquiry is very similar. This absence of discussion of *geological inquiry* in general works on methodology is the more surprising because Chamberlin (1897, 1933, 1965), by recommending his method of multiple working hypotheses in geological inquiry, chose for geological procedure that basic trend of the *hypothetico-deductive method,* which has since dominated modern science including its most exact disciplines.

In view of this situation, it is a legitimate question whether or not adequate information on geological methodology can be found in the textbooks of geology, which because of the rapid development of modern earth sciences, are published at an ever-increasing rate. It is surprising that in the three-volume textbook *Geology,* written by Salisbury and Chamberlin in 1904–1906, i.e., shortly after the paper on the method of multiple working hypotheses, this procedure is not even mentioned. A survey of the *methodological content* of a considerable number of textbooks on geology, undertaken by the author since 1957, confirms that, with few exceptions, most consist of an encyclopedic presentation of geological knowledge. Few of them convey any coherent idea of the methods by which geologists have arrived at that body of supposed positive knowledge.

The texts analyzed in this way consisted of the following: about 40 geological textbooks, ranging from advanced works to secondary school texts, and about 20 monographs or research papers on methodologically interesting topics. To avoid undue bias in the evaluation of the results that might be affected by predilections or fashions in scientific procedure of different countries, the texts were taken from five different languages, published in ten countries and three continents. The analysis was guided by questionnaires with provision for definitions of *geology,* basic postulates and principles, geological laws, theoretical concepts, methods of procedure, and methodical aids to that aim (cf. Albritton, 1963, 326–363, which gives an idea of relevant topics).

It is significant that in the majority of textbooks, geology is (vaguely) defined as an *activity:* "geology investigates . . . ." Such an activity should be termed *geological inquiry.* Nevertheless, the content of most textbooks consists of (present-day) encyclopedic geological knowledge, without discussing how this knowledge was obtained. It is customary in modern textbooks to complement the verbal text with ample illustrations for conveying information in compact form. Certain geological laws are quoted, and it is left to the reader to apply the law to the interpretation of the pictorial information: "If you *observe* rock units or structures in such and such *spatial* attitude, then you *infer* (i.e., by an act of reasoning guided by the law), the following *sequence in time* of the geological processes that have brought about the above configuration."

The laws in question—e.g., the law of superposition, the law of intrusion, etc.—seemingly and historically have their root in the belief that *inductive reasoning,* i.e., reasoning from the particular to the general, is a valid method of arriving at certain knowledge. When renaming these laws as *rules,* the possibility of exceptions to their validity would readily be admitted. Better still, and in line with the accepted method of inquiry by means of forming and testing hypotheses, the inductive procedure could be used as a preparation for the formulation of a hypothesis adapted to the singular geological object or situation being investigated.

The principle of superposition, utmost importance for stratigraphic inquiry, examines evidence that can be used in deciding which of two adjacent units of layered rocks was, at the time of its emplacement, the upper and which one the lower stratum. Knowledge of this *spatial* relationship is required to draw conclusions as to the *temporal* sequence of deposition or emplacement of these strata. Since geology is in the first place a historical science of nature, and stratigraphic inquiry is the proper procedure for establishing the time sequences of past geological events, the fundamental methodological character of this research is evident.

A similar critical investigation could be carried out on intrusions and dikes. This appears important because of the uncertain origin of dikelike structures that are characteristic of granitized terrains. To this relationship of the *intruding* and the *intruded* rocks the so-called law of intrusion applies, which interprets this spatial relationship in terms of the temporal sequence of the emplacement of the rock units.

Another, spatio-structural, relationship between geological units is included in *rule of inclusions.* This relationship was recognized by Koch (1949, 10) as one of the substantial categories of descriptive natural science, or more appropriately formulated, the *class of states or relations of natural things* that can be recognized and distinguished immediately through sense perception. This relationship *to be included* ranks methodologically alongside the relationships of *superposition* and *intrusion.* It is used, like the latter, to deduce an age relationship between the *included* (the earlier) and the *including* (the later formed).

Geological examples belonging to this class of relations of natural things are known to occur in vastly different dimensions, such as the following: tectonic windows, inliers, roof pendants and xenoliths of all dimensions, phenocrysts in porphyritic rocks, pebbles and fragments of preexisting rocks in sandstones, and inclusions of idiomorphic crystals in larger host crystals. In the case of *liquid inclusions in crystals,* as used for geothermometry, the rule of inference as to the time sequence of the formation is doubtful. Indeed, investigations prompted by appropriate hypotheses have revealed that such liquids, under certain circumstances, can migrate or diffuse into preexisting cavities of crystals (so-called deuteric liquid inclusions), thus making the application of the rule of inclusion doubtful. Similar problems are connected with the occurrence of concretions in sedimentary rocks, which may simulate the relationship of inclusion.

Methodological Aspects of the Principle of Uniformity

The *principle of uniformity,* first published by Hutton (1795), has been widely considered as the cornerstone of geology as a science in the modern sense and is quoted in practically all textbooks of geology. To accept it as a principle of nature means to justify the customary habit of geological reasoning—i.e., to explain processes of the past in terms of observable processes of the present. Hooykaas (1963) scrutinized this principle in its application to various fields of inquiry and concluded that it is not a principle of nature but could be considered as a *procedural principle* for the geological researcher. In the light of the basic presuppositions of the present survey, it should be considered a *postulate,* i.e., an assumption, first unproved, which was confirmed in innumerable instances of post-Proterozoic geology (see *Uniformitarianism*).

The assumption of uniformity is clearly invalidated by an increasing number of geological processes recognized in pre-Paleozoic times. Even during the lifetime of Lyell, vigorous propounder of the principle of uniformity, Sedgwick (cf. Albritton, 1963, 313) had termed the principle an "unwarranted hypothesis." It is certainly objectionable if the principle is presented as a law, which is done in 3 of the 40 textbooks here surveyed.

Method of Multiple Working Hypotheses in Textbooks of Geology

As mentioned, Chamberlin (1897) recommended, as universally applicable, the method of forming and testing *multiple hypotheses.* He thus adapted, at least in principle, geological methodology to the other sciences. In fact, in the first formulation of his method (Chamberlin, 1897, 837), he envisaged its applicability to scientific knowledge in general. Nevertheless, in only one of the geological textbooks here examined is Chamberlin's paper quoted in the reading list for further study—significantly without further comment. His paper, however, is widely quoted in books on field geology, but it is in line with the avoidance of philosophical discussions in such works that no actual applications of the method are discussed. In particular, despite the word-for-word republication of the original paper (Chamberlin, 1931, 1965), there still does not exist a critical reappraisal and full adaptation of that paper to the more advanced methodology of other modern sciences.

Summing up, modern textbooks on geological science cannot be considered as a source of information regarding geological methodology. If it were so, geological mapping, in which the results of geological inquiry are summed up and represented in their most condensed form, should be covered by these texts. Only from the middle 1900s is a trend in this direction noticed, in the form of brief introductions to the elements of geological map making and interpretation. In part, this situation has been overcome by the almost universal habit of personal exposition in the field, organized field excursions, and systematic field training.

It seems improbable that this self-restriction in the task of geological textbooks will be maintained in the future. The work by Fourmarier (1944) contains in its introduction what could be termed *precis of geological methodology*—i.e., a coherent account of the lines of reasoning inherent in a variety of techniques commonly used by the exploring geologist. Regarding textbooks of general science, Anfinson (1963) has given an interesting example of a textbook wholly guided and permeated by methodological reasoning. In the field of earth sciences proper, Strahler (1963) has shown that discussions of methodological procedures (e.g., in a problem of correlation) can be incorporated successfully in a work that represents the current encyclopedic knowledge.

Method of Hypotheses in Research Publications on Pure and Applied Geology

A wide range of research publications on both pure and applied geology shows ample evidence that Chamberlin's method of multiple working hypotheses (1897) has been widely applied to geological inquiry, even if his paper is not expressly quoted and the terms *hypothesis* or *theory* are not expressly used. It is evident, furthermore, that Chamberlin's conception of that method is of an empirical and intuitive origin. In contrast, the "scheme of establishing new corroborated knowledge," proposed by Popper (1963; see *Scientific Method*), is the result of, and logically tied in with, a whole system of philosophical thought that is carried through to its logical consequences.

It is evident that the two or perhaps three alternative hypotheses sometimes described or mentioned in geological research papers represent only a selection of the surviving, best-corroborated hypotheses actually examined during the research work. No mention is made of any model or trial hypotheses considered during the process of inquiry. Practical reasons, particularly the necessary economy of printing space, would in most cases prevent a full account of all the hypotheses used in the inquiry. For the latter reason, the (mostly unprinted) thesis manuscripts in pure and applied geology, for the degree of doctor or master of science, represent a little-explored source of actual methodological procedure in present-day geological inquiry.

The difficulties encountered in research into geological methodology, which involves preparing a coherent survey, are strikingly illustrated by the following example. Fischer (1961) has prepared a comprehensive monograph, "Changes in the Scientific Views on the Formation of Rocks and Mineral Deposits." This work undoubtedly represents a classic in the realm of geological publications—indeed, in the realm of geological thought. Nevertheless, the term *hypothesis* does not appear in the large subject index or in the text, except perhaps by chance. The decisive term in the title of the work, *wissenschaftlichen Anschauung,* or "scientific views," closely corresponds to the etymology of Greek *oewpia,* the theory (or better, hypothesis) of modern scientific procedure.

The extraordinary range and diversity of scientific views, as well as geological objects to which they pertain, all represented, compared, and critically discussed in Fischer's (1961) work, illustrate or corroborate the common basic assumption—namely, that the formation of a hypothesis precedes systematic work, i.e., scientific observation and testing, and that without such a hypothesis as a guide, organized attempts either to corroborate or falsify the hypothesis cannot be carried out. For this reason many teachers (including Popper) recommend hypothesis forming and testing as a routine method for furthering the growth of all scientific knowledge, applicable to every instance of inquiry in progress.

Method of Hypotheses in Present-day Geological Exploration

In view of the difficulty of referring to published examples of the routine application of the method of hypotheses to geological inquiry, it may be permissible to construct such an example from published and personal experience. The example, taken from the recent oil exploration in a geologically little known area of Australia, could equally well be taken from exploration enterprises in other poorly known parts of the world.

At the time oil exploration started in the best potential areas of Australia, existing encyclopedic geological knowledge was scarce, scattered, and poorly coordinated. Exploration was begun by several competing companies at the same time so information coming from neighboring areas was always in need of testing. In cases like this, even the simpler geological situations are complex (Rutten, in Albritton, 1963, 310) and present exceptional difficulties for the explorer. Observable evidence in geological inquiry is typically scarce and only partly preserved or accessible. Experiments, which are of such great importance in the process of testing hypotheses, are hardly possible with geological objects themselves but are performed usually on models only. Geophysical investigations can in a sense be considered as experiments directly applied to geological objects in situ, with the instruments used in the field instead of in the laboratory, but the objects so investi-gated represent *open systems* instead of the *closed systems* used in laboratory experiments.

Once an initial exploratory borehole, brought down to confirm the hypothesis of the presence of a favorable structure at depth, has shown further promise, it is customary to probe the surrounding field with numerous boreholes possibly arranged in the form of a preplanned *grid.* This geometrically regular scheme of procedure facilitates not only the location of boreholes but also the evaluation of further drilling results by means of mathematical statistics; at the same time, it facilitates the application of *interpolation* and *extrapolation* as a means of systematic reasoning about the results obtained from the drilling and inferences drawn therefrom.

The order of sequence of drilling of the various boreholes, however, is individually determined by ad hoc hypotheses as to what further information might be expected from each of them. Even the information on the underground features gained during the progress of drilling any single borehole may affect the continuation of that drilling operation. It may lead to changing the direction or inclination of the borehole, with all these decisions being made on the basis of new or modified trial hypotheses required from day to day and hour to hour.

It is obvious from this discussion that present-day practice in modern applied or exploration geology makes continuous use of a method of inquiry of which Popper (1963) has independently elaborated a theory expressible in strictly logical terms. It is desirable not only that geological procedure in general should take advantage of the discussed methodological foundations but also that the knowledge and discussion of the elements and operations of geological methodology should find their appropriate place in methodological science in general, in geological textbooks, and in particular, in geological education.

In the opinion of the author, students of geology should not have to wait until advanced years of instruction before they learn about the significance of hypotheses in geological inquiry. In an undergraduate course called "Geological Thought," which the author has developed, the significance of *routine hypotheses* in geological inquiry is demonstrated to the freshmen on their first geological excursion. This early field indoctrination is widely employed in basic geological instruction, even in the large universities.

Irrational and Rational Origin of Hypotheses?

Provision must be made for a *rational,* or scientific, procedure of formulating the hypotheses required to further the growth of scientific knowledge. Popper's (1963) scheme (see *Scientific Method*) does not provide for this or extend to the knowledge now established by a well-corroborated hypothesis or large masses of such hypotheses. Such knowledge is steadily being added to the body of established, or encyclopedic, knowledge. Against the background of the latter, new problems or inconsistencies are recognized. Also, the process of testing new hypotheses makes continuous use of established knowledge.

To revert to the problem of inventing new hypotheses, it is widely agreed that such conceptions or new ideas have their origin in irrational, uncontrollable, and even subconscious activities of the brain. This idea is appropriately indicated by the names given to such activities or their outcome (cf. Stachowiak, 1965, 94): brain waves, *Gedankenblitze* (in German); sudden intuitions; sudden, apparently unmotivated changes in the direction of attack on a puzzling problem, etc. As described earlier, field geologists, and in particular, the supervising geologists of an enterprise in exploration geology, are facing situations from day to day that require rapid invention and testing by intensive thought. This requirement apparently severely taxes the powers of imagination of all researchers involved and leaves the younger, less experienced, geologists with the confusing impression that guesswork is involved in geological inquiry.

It is the author's belief, with particular reference to geological trial hypotheses, that what appears to be a sudden brain wave is the outcome of perhaps unconscious associative acts of reasoning. Not only encyclopedic knowledge but also its *degree of logical order* and interrelationships are at the basis of apparently singular conscious acts of thought, such as the aforementioned brain waves. The natural scientist, before setting out on the investigation of a particular problem, has as a basic rule already ordered and conceptually classified the field of possible experience in which the problem is embedded. As mentioned in the entry on scientific method, the author has worked out and illustrated a system of categories or classes of elementary natural phenomena perceivable and discernible through the senses, which can be considered as a natural system of classification or a *natural universe* (termed the *tetraktys*). The system of categories, as a result of its particular method of derivation, is uniquely conformable to the natural organization of the objects of study of the field geologist and mineralogist, as well as to their most common means of observation—namely, sense perception. In addition to its wide applicability to the four basic operations in geological and mineralogical inquiry (see *Scientific Method* and later discussion), the system of categories can also be used, after thorough study of the interrelationships between these categories, as a means of systematically constructing hypotheses in these fields of inquiry.

Deductive Scheme for Hypothesis Construction

A novel scheme for systematically constructing and comparing four contrasting hypotheses pertaining at any time to the interpretation of the same complex object of investigation—in this case, the formation of ore and mineral deposits—was discussed by Amstutz and Bernard (1973). Its range of applicability has since been further illustrated by Amstutz and co-workers. The scheme consists of four *compound theoretical concepts* pertaining to possible (i.e., thinkable) conditions of the spatial and temporal relationships of ore-forming processes with regard to their substratum (host rock). The scheme and its rational use appear related to a *matrix* and to the *operations of reasoning* pertaining thereto, namely

A	Syngenetic	I	Supergene
B	Epigenetic	II	Hypogene

These terms represent four different states or events of *location,* temporal *sequence,* and spatial *directions of movement* of ore-bringing carrier solutions as well as the substratum of ore deposition. The states and events so characterized in abstract, belong to subclasses and combinations (i.e., intersections) of subclasses of the four general categories (or classes) of *spatio-temporal relations* of things and events distinguished by Koch (1949, 14) and symbolized as:

$$(U{:}{:}{:}U')_S; \quad U{:}{:}{:}S; \quad (U{:}{:}{:}U')_T; \quad U{:}{:}{:}T$$

(for all details, see Amstutz, 1959). The significance of this scheme of systematic reasoning lies in the fact that it formalizes and channels systematic reasoning pertaining to each of the four contrasted hypotheses. It thus systematizes deductive reasoning in a typically qualitative, empirical, and historical science such as geology.

Problems of Scientific Ordering Procedures in the Geological and Mineralogical Disciplines

Schemes for ordered or systematic procedure have been developed in the earth sciences since their early beginnings. These schemes apparently were given such significance that A. G. Werner (1750–1817) was considered by many as the founder of the science of mineralogy because of his development of a new rational scheme for the description and classification of minerals.

One may ask, therefore, whether such procedures of ordered manipulation of the content of scientific knowledge have a bearing on the growth of scientific knowledge, as discussed in some detail previously and in the entry on scientific method. Examples given (e.g., those by Amstutz,

1959) suggest that this is the case. There is also the question of whether the development and use of so many schemes for ordered procedure may have something to do with the particular complexity of geological objects of investigation. This complexity is referred to many times by Albritton (1963), but no concise idea of it is given. This may be done by stating in general terms that, as a rule, objects of a comprehensive geological investigation are multicomponent, multiphenomenal, multiconditional, multilevel.

As examples, one may choose a compound igneous complex, as the unit, U, of investigation (Koch, 1949, 3) or, in another case, a *contact metasomatic ore deposit,*—say, U_1. In the latter, the geological subunits, U_1, U_2, etc., distinguishable in that object of investigation, represent as many *components* of the unit U_1. In each one of these— i.e., igneous rocks, sedimentary rocks, contact rocks, varieties of ore—the actual state is investigated by observing and recording the physical and other characters as classified in the categorical scheme. The hypotheses to be formed and tested in that investigation pertain to the conditions of temperature, pressure, presence and absence of water, etc. that may have been prevalent, and often changed, during the geological history of the unit U_1 that is to be unraveled by means of the hypotheses formed for that investigation. The strongest contribution to the complexity of the unit U_1 seems to come from the fact that in a comprehensive geological investigation of it, the research work to be carried out may pertain to a great number of levels of constituents into which each of the listed components may be divided— namely, rock types, constituent minerals of the rocks and ores, chemical elements and compounds constituting these minerals, and in most modern investigations, even the isotropic composition of these chemical elements. Hypotheses proposed to explain the process of formation of a geological unit like U_1 often embrace, and connect, constituents belonging to widely different levels of the hierarchy of constituents, so the hypotheses are of a highly complex nature.

It is obvious that no valid reasoning can be carried out about a complex unit like U_1 without concise identification (and classification) of its many and diverse constituents. It is in keeping with the complex nature of the objects of investigation of a geologist, mineralogist, or petrologist that, despite the marked trend toward specialization, not only is he or she trained in a great variety of geological and auxiliary disciplines but also, in each of them, he or she is acquainted with a large variety of schemes worked out for systematic procedure in one or another of the four basic operations of (descriptive) natural science: (1) scientific observation, (2) identification, (3) classification, and (4) (ordered) description.

Not so long ago, procedures for the identification and classification of fossils, e.g., occupied a particularly large proportion of the training of a geologist. In more recent years, however, there is a strong trend toward reducing the volume of much of the practical work connected with the four basic operations. This drastic curtailment is not without danger to the quality of the training, since the student learns all the procedures singly, in connection with each field of application, and not with reference to *logical principles* and operations inherent in them all. Indeed, to find a common theoretical or methodological basis to all these specialized activities, use should be made of the *logic and relationships of classes*. Quine (1951), in the preface of his work, sees a great future in the application of symbolic logic to the natural and other sciences; Carnap (1960) quotes numerous examples of application of such methods to a diversity of sciences, with the exception of the geological disciplines.

The *categories of descriptive natural science* (Koch, 1949), particularly when considered as classes of elementary phenomena recognizable and discernible by means of sense perception, are readily applicable to all procedures of ordered or systematic investigation of geological and mineralogical objects whether under strictly natural or near-natural conditions. A revised edition of that paper, in preparation, will facilitate these tasks. The brief discussion of the problems presented in this section clearly points to the need for fundamental, methodological research in an important sector of geological inquiry.

Scientific Status of Geology

It is common to speak of the *geological sciences*. The question as to whether geology proper is a science is not discussed in most textbooks. Doubts about the scientific character and status of geology are frequently voiced by distinguished representatives of that discipline (cf. Mackin, in Albritton, 1963, 136; also Read, 1952). Among members of the profession at large, particularly in applied geology, this doubt is widespread and strongly influenced by the supposed guesswork involved in its day-to-day procedures.

The common belief among scientists that only that which is based on exact measurement and mathematical evaluation of exact data has scientific status is detrimental to a discipline such as geology that, despite the ever-increasing use of quantitative means (see *Geostatistics*), expresses so much of its activities, reasonings, and findings in everyday language (Betz, in Albritton, 1963, 193*ff*).

There is really no scientific merit to precisions beyond the requirements of the task in hand. In Popper's (1963) scheme of the growth of knowledge by hypothesis making and hypothesis testing (see *Scientific Method*), the scientific status of the corroborated hypotheses is based on the extent and critical value of all efforts made to corroborate, and particularly to falsify, it. The use or nonuse of quantitative means in these efforts is only incidental to the critical attitude of the researcher. It is this critical attitude and action of testing and retesting of what seems to be scientifically known that decides the scientific character of a unit of inquiry or its outcome. This attitude, then, would also apply to geological inquiry, whether qualitative or quantitative.

Geological Maps—A Symbolism Exclusive to Geology

A survey of exact measurements, experiments, and their mathematical treatment as inherent in the comprehensive investigation of a geological formation would show that the former quantitative procedures pertain essentially to objects of study of disciplines that are auxiliary to geology proper—e.g., minerals, crystals, microscopic mineral composition and textures of rocks, etc. Even the so-called exact results of geophysical measurements must in the end be interpreted and expressed in geological terms proper—i.e., in terms of rock units, their attitudes, structural relationships, etc. in non-mathematical terms.

It is typical of most advanced scientific disciplines that they develop, for their own purposes, symbols and a symbolic language that is essentially restricted to their own use—e.g., quantum mechanics in nuclear physics, chemical structural formulas in chemistry, and new branches of symbolic logic in genetics. It is typical of all these symbolisms, including mathematics and symbolic logic, that once the meaning of the symbols and symbol combinations is explained in terms of ordinary language and the rules of the operations with these symbols are equally explained, then scientific accounts presented in this symbolic language can be read by any competent scientists independent of the language community to which they belong. In addition to being nonverbal, a common characteristic of these symbol languages is that they are capable of conveying information in a highly compact form. The concepts involved in this transposition are intuitive—i.e., of the spatiality and temporality of physical entities or events—and not abstract ones. It seems absurd, therefore, to seek the salvation of geological science in mathematical exactness.

Geological maps (and sections), particularly those containing structural as well as stratigraphic information, represent such a specific nonverbal symbolism (see *Map Symbols*). They show or represent, by means of conventional symbols and their association, the spatial togetherness and in-

terwovenness of geological units of the most diverse kinds and dimensions. Geological maps are read in the form of a discursive verbal description of their very large amount of information. Geological maps are interpreted in terms of an account of temporal sequences of geological events of the past. This interpretation may even include the physical conditions, such as temperature, pressure, or presence or absence of water cover under which these events are inferred to have taken place.

Geological maps are by no means only a condensed symbolic representation of observation statements. The preparation of a geological map involves a consideration, to a greater or lesser extent, of the hypotheses that guided the geologist during the fieldwork (Harrison, in Albritton, 1963, 225). An expert in map interpretation is sometimes capable of interpreting a given map in terms of an alternative hypothesis or hypotheses that were not intentionally embodied in it but that result from the rules of combination and operation of mapping symbols (see *Maps, Logic of*).

As stated, geological map making and its results may embody the bulk of geological information proper as well as pertinent hypotheses available at a certain stage of investigation of given geological formations. Although being based, in a large measure, on qualitative information located (i.e., spatially ordered) by more or less exact surveying data, geological maps represent one of the most ingenious and most compact systems of symbolism developed in the realm of scientific inquiry.

L. E. KOCH

References

Albritton, C. C., Jr., 1963, *The Fabric of Geology.* Reading, Mass.: Addison-Wesley Publishing Co.
Amstutz, G. C., 1959, Syngenese und Epigenese in Petrographie und Lagerstaettenlehre, *Schweizer. Mineralog. u. Petrogr. Mitt.* **39**, 1-84 (English translation in *Internat. Geology Rev.,* Feb.–March, 1961).
Amstutz, C. G., and A. J. Bernard, 1973, *Ores in Sediments.* Berlin: Springer Verlag.
Anfinson, O. P., 1963, *Understanding the Physical Sciences.* Boston: Allyn and Bacon.
Carnap, R., 1960, *Symbolische Logik,* 2nd ed., Vienna: Springer-Verlag.
Chamberlin, T. C., 1897, The method of multiple working hypotheses, *Jour. Geology* **5**, 837–848; reprinted *Jour. Geology* **39**, 155–165 (1931); reprinted *Science* **148**, 754 (1965).
Fischer, W., 1961, *Gesteins-und Lagerstaettenbildung im Wandel der wissenchaftlichen Anschauung.* Stuttgart: E. Schweizerbart'sche Verlagsbuchhandlung.
Fourmarier, P., 1944, *Principes de la Géologie,* 2nd ed., 2 vols. Paris: Masson and Cie.
Hooykaas, R., 1963, *The Principle of Uniformity in Geology, Biology and Theology,* 2nd ed. Leiden: E. J. Brill.

Koch, L. E., 1949, Tetraktys. The system of the categories of natural science and its application to the geological sciences, *Australian Jour. Sci.* **11**(4) (Suppl.), 1–31.

Kraft, V., 1960, *Erkenntnislehre*. Vienna: Springer-Verlag.

Lyell, C., 1830, *Principles of Geology*. London: Murray, 511p.

Mill, J. S., 1925, *A System of Logic*. London: Longmans, Green.

Popper, K. R., 1962, *The Logic of Discovery,* 3rd ed. London: Hutchinson and Co.

Popper, K. R., 1963, *Conjectures and Refutations. The Growth of Scientific Knowledge.* London: Routledge and Kegan Paul.

Quine, W. V. O., 1951, *Mathematical Logic*. Cambridge, Mass.: Harvard University Press.

Read, H. H., 1952, The geologist as a historian, in *Scientific Objective*. London: Butterworths Scientific Publications, 52–67.

Stachowiak, H., 1965, *Denken und Erkennen im kybernetischen Modell*. Vienna and New York: Springer-Verlag.

Strahler, A. N., 1963, *The Earth Sciences*. New York: Harper & Row.

Whewell, W., 1837, *History of the Inductive Sciences,* 3 vols. London.

Whewell, W., 1840, *Philosophy of the Inductive Sciences,* 2 vols. London.

Cross-references: *Cartography, General; Geohistory, American Founding Fathers; Geology, Philosophy; Geology, Scope and Classification; Geostatistics; Map Abbreviations, Ciphers, and Mnemonicons; Map Symbols; Maps, Logic of; Scientific Method.* Vol. XIII: *Geological Communication; Geological Information, Marketing.*

GEOLOGICAL OCEANOGRAPHY—See ACOUSTIC SURVEYS, MARINE; HARBOR SURVEYS; MARINE MAGNETIC SURVEYS; SEA SURVEYS. Vol. XIII: SUBMERSIBLES.

GEOLOGICAL SURVEYS—See GEOLOGICAL SURVEYS, STATE AND FEDERAL. Vol. XIII: INFORMATION CENTERS.

GEOLOGICAL SURVEY AND MAPPING

Field geology (q.v.) is a vast subject not easily described under one heading, and for descriptive purposes it is therefore often split into its different components, such as photogeology (q.v.), field mapping, measurement of sections, specimen collection (see *Field Notes, Notebooks*), etc. In practice there are hazy boundaries between these activities, which all merge into the one process of *geological survey*. In the broadest sense geological survey is basic to all the geological sciences, although for practical purposes one may place geophysical and oceanographic surveys (see *Marine Magnetic Surveys*) in different categories, and topographical survey also is strictly not part of geological survey. However, the latter is a necessary tool, and all geologists should have knowledge of the methods of topographical survey, which are of vital importance in the construction of geological maps. Most important are the plane table survey (see *Plane Table Mapping*), chain (tape and compass) survey (see *Compass Traverse*), and leveling, but one must not forget that modern instrumental developments (see *Geonomy; Surveying, Electronic*), including satellite technology, are rapidly overtaking traditional methods. Details of the methods of field geology and related subjects are clearly too numerous for all to be included in a short entry like this, and for further information refer to the comprehensive accounts by Compton (1962, 1985), Moseley (1981), Barnes (1981), Tucker (1982), and Fry (1984).

With ever-increasing geological knowledge has come increased specialization, and field requirements are different for activities such as mining geology, oil exploration, engineering geology, hydrogeology, and environmental geology (q.v.), to name a few. The surveys will include straightforward geological mapping either on a small or a large scale. The details will involve using aerial photographs (satellite or conventional photographs) and may require compilation of sedimentary logs, collection of fossils, recording of geological structure and complex varieties of metamorphic and igneous rocks, and the plotting of geomorphological information such as glacial or tropical landforms and deposits (till, duricrust, etc.). There is now a vast array of subjects, so much so that extreme specialists in one can hardly communicate with those in others. Different methods are involved, and if one also contemplates the different techniques required for, say, tropical forests and Arctic mountains, one can appreciate that it is no easy matter for a general field geologist to undertake detailed work on any subject in any part of the world. Indeed, geology graduates now require additional specialist training, perhaps a master's degree in a chosen field, but nevertheless there are still basic requirements for most geological surveys, and I describe these in this entry.

In almost all cases geological mapping should start with a photogeological study, and the preparation of a photogeological map. This saves a great deal of valuable field time. The photogeology (q.v.) is followed up in the field, usually using stereo air photographs for plotting data, although exceptions exist to this practice. The survey or mapping can be on a variety of scales depending on its purpose, and for convenience of description I break them down into *reconnaissance* and *detailed surveys.*

Reconnaissance Surveys

This breakdown is subjective: without doubt one person's reconnaissance is another person's detail. At the one extreme, using satellite and other aerial photographs, it may be possible to assess the geology of completely unknown country at a rate of several hundred square kilometers a day, not accurately of course, but sufficiently to determine fundamentals, whereas in areas that have previously been mapped in some detail, but perhaps before modern ideas were developed, a traverse at a scale of 1:10,000 may be considered reconnaissance. In many of the more remote parts of the world, topographic maps are highly inaccurate and small scale. In these regions one has to rely on aerial photographs for topography, including not only the river courses and positions of hills and mountains but also the heights of the latter (using a *stereometer*). In the field exact locations can be plotted using a pocket stereoscope, but the most convenient way to determine height is to use an aneroid. The researcher can obtain good results if he or she makes frequent checks with a base camp of known elevation.

Several examples now follow, starting with rapid reconnaissance of unknown ground and finishing with detailed reconnaissance of an area mapped on a 6 in to the mile scale during the latter part of the nineteenth century.

The Radfan Mountains, Southern Yemen A region 80 km north of Aden in south Arabia, this area is still almost unknown, and at the time of the survey (1965) it was at a medieval stage of existence. It is a semiarid mountainous region with sharp ridges and peaks rising to more than 1,900 m. It is seriously short of water in a number of critical areas, and so the primary object of the survey was not to produce a geological map but to locate sites for water wells. The latter depends upon the former, however, and the following short account attempts to show how both objectives were achieved. It was fortunate that the survey had good logistic support, with aerial photographs at 1:60,000 and 1:12,000 available and a helicopter and a four-wheel-drive vehicle with a military escort also available (there was a certain amount of revolutionary turmoil at the time). The procedure was as follows.

Photogeology In semiarid mountains geology nearly always shows up well on aerial photographs, and this was the case with the Radfan Mountains. It was possible to construct a com-

prehensive photogeological map showing structure and assessing most of the lithologies. This map made it possible to decide a route for a helicopter traverse and to decide which areas to concentrate on for vehicle and foot traverses.

Helicopter Traverse The area flown is shown in Fig. 1, and was completely at low level, following the valleys and skimming the ridges, thus permitting a different perspective to that obtained from the air photos. Photographs were taken to be studied later, the only difficulty being the large amount of information that had to be recorded in a short time. The helicopter was able to land at strategic points, and altimeter heights were noted, specimens collected, and measurements taken on short foot traverses. The photogeology was thus calibrated with the rock types and structure, and since this was primarily a hydrogeological survey, provisional assessment was made of areas where water may be extracted.

Vehicle and Foot Survey The most favorable area for water, and incidentally, the most accessible, was the (dry) Rabwa Valley leading from the Wadi Taym to Thumayr, and the ground survey was concentrated here (there are no flowing rivers or streams in this area). Notice (see Fig. 1) that an important fault follows the Rabwa, bringing basaltic volcanics against sandstones. The fault could be expected to be a water throughway, and the permeable sandstones, dipping obliquely into the fault, with a large catchment, also provided favorable situations for boreholes.

Conclusion Terrains where the geology is little known and that would profit from surveys of the type described here are common over most parts of the Earth's surface. In this case, excluding the time taken to prepare the photogeological map, an area of about 388 km was surveyed in a matter of 2 or 3 days. The survey resulted in the location of productive water wells, particularly along the Rabwa Fault, at low cost, and while a detailed survey would undoubtedly reveal inaccuracies, it would take much longer, perhaps several years, and be much more costly.

Ophiolite Complex of Masirah Island, Oman This is an example of a reconnaissance survey in an arid desert area. The absence of vegetation has resulted in extensive rock outcrop, up to 90% in places, and this is in spite of the low relief; the highest point on the island is less than 330 m. Compared with the last example there was more time for the survey—in fact, about 1,036 km were covered by two geologists in 2 mo—which meant that some of the important areas could be examined in more detail. The methods employed during the survey have been published (Moseley, 1981), and I summarize only the main points here.

Aerial photographs, including satellite photographs, were available on various scales, the most

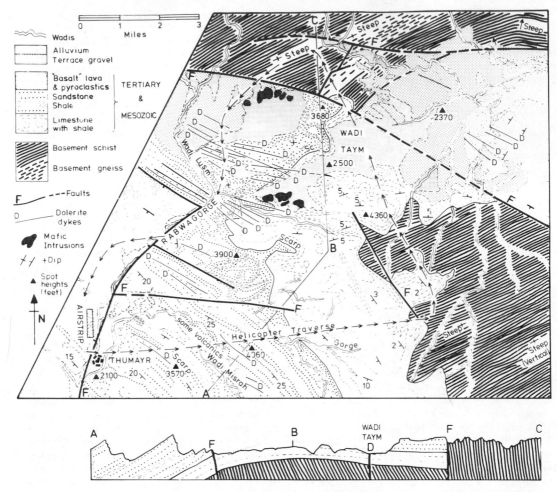

FIGURE 1. A reconnaissance map of part of the Radfan Mountains, South Yemen. The route of a helicopter traverse is shown. Ground traverses (vehicle and foot) were mostly in the southwest between the Wadi Taym and Thumayr. The map of the eastern part of the region is entirely photogeological.

useful being stereo runs at 1:12,000, and it was therefore possible to make a photogeological map before starting the field mapping. It is, of course, important when undertaking projects like this to have a thorough understanding of the problems likely to be encountered in the field, and in this case prior knowledge of ophiolite literature was one of the necessities. The aerial photographs showed that in some areas the geology was simple, and rapid traverses similar to those described in the last section would suffice, but in others there were complexities that would have to be examined in more detail if the true nature of the ophiolite was to be understood. Although still of a reconnaissance nature, more time was to be spent on the last mentioned areas, and we will see how this was achieved by consulting Abbotts (1978, 1979), Moseley and Abbotts (1979), and Moseley (1981), which give relatively detailed ac-

counts of a granite, the sheeted dikes, and the tectonic breccia referred to as the Masirah Mélange. Study of these and other small areas proved critical to the understanding of the ophiolite. This entry gives a further example: that of harzburgite (serpentinized), which is in faulted contact with pillow lava and gabbro. The latter, it was decided later, was part of the sheeted dike complex, but with gabbro dominant and dikes a small proportion of the total. The 1-day survey illustrated by Fig. 2 was accomplished by driving across rough country (possible in this open desert terrain even though there are no roads or tracks) with frequent stops to make foot traverses during which lithologies were noted, specimens collected, and photographs taken. It was necessary to collect many large specimens because of the intended geochemical work; indeed, by the time the survey of the island was complete, 2 tons of rock had been

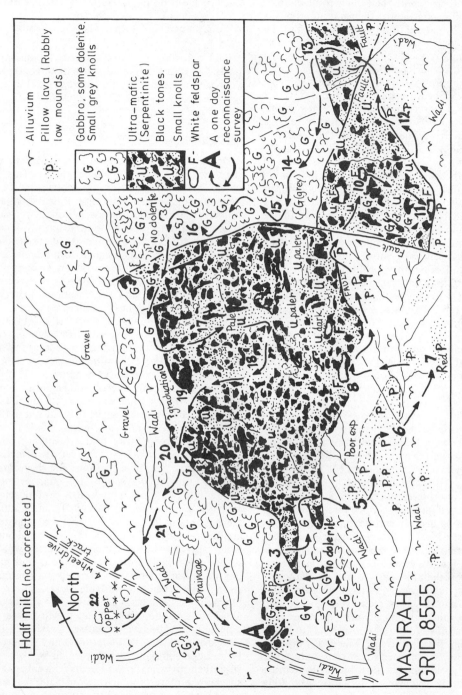

FIGURE 2. A 1-day vehicle and foot reconnaissance survey of part of the Masirah Ophiolite Complex, Oman. The map is a photogeological map, and it determined the route of the traverse. Numbers 1–22 represent stopping points where observations were made and specimens collected.

packed and was awaiting transport. Figure 2 is a simplified copy of the photogeological map of this area of mantle rock with the vehicle route plotted. Only a small proportion of the total outcrop was visited, but by careful route selection, most of the important contacts and rock types were examined, making it possible to interpret the unvisited outcrops with some confidence.

A combination of detailed reconnaissance like that just described, together with rapid traverses to fill in the intervening tracts, made it possible to publish what I believe to be two reliable 1:50,000 maps of Masirah Island (Moseley and Abbotts, 1984). Further to the logistics of the expedition, this expedition was fortunate in Royal Air Force assistance, in that flights to Masirah, a four-wheel-drive vehicle, and camping equipment were provided, and 2 tons of rock was flown to the United Kingdom. Nevertheless, it was still one of those impoverished university expeditions, and had there been, e.g., oil company incentives, there would have been much more logistic support in the field and laboratory backup with specialist geochemists, paleontologists, sedimentologists, and geophysicists involved.

Beacon Tarn, Cumbria, England This example is essentially a student exercise in reconnaissance mapping. It is also an area that was geologically mapped on a 6 in to the mile scale during the original survey of the United Kingdom more than 100 yr ago and shows how a good-quality old map can be quickly updated and new information added. It was already known that the area was made of Silurian slates passing down into graywacke sandstone, and the general strike and dip was also known. There was insufficient information on structure and sedimentology, however, and the object of a reconnaissance like this would be to add sufficient data of this kind so that an up-to-date interpretation could be made.

The Beacon Tarn area is open undulating country covered by heather and bracken, with a relief of about 152 m. The geology shows up well on the aerial photographs, with strike and dip and positions of faults easily seen. Using stereophotographs it is therefore easy to make a photogeological map showing considerably more detail than the original survey maps (Fig. 3; Moseley, 1983). The photographs do not, however, give clear indications of lithology, which has to be obtained from field traverses. The 1-day traverse shown in Fig. 3 is planned to cross as many structures and rock units as possible with frequent stops, when the following observations should be made:

1. Lithological specimens are required and sedimentological observations made.
2. Bedding orientations (compass-clinometer) are taken at frequent intervals.

3. Cleavage orientation is recorded wherever bedding is measured.
4. Cleavage-bedding intersections are measured to give checks on fold plunge.
5. Fold plunge is recorded wherever possible.

It is necessary to obtain sufficient measurements for statistical validity when stereographic plots are made; e.g., is the cleavage axial planar or is it slightly oblique to the axial planes, as in some other parts of Cumbria?

Detailed Survey

Detailed survey is required in areas for which there are existing large-scale geological maps on which much information has previously been recorded. It may be that in some critical areas additional information is necessary for the revision of old hypotheses and the establishment of new ones. This could be considered as a more detailed version of the preceding Beacon Tarn exercise and is rather like building a tower on a plateau of previous knowledge. We know already that the region is interesting, and the additional observations, made in the light of modern concepts, can make new and valuable interpretations possible. Greater and more reliable scientific advances can be made by this method than by surveys of an unknown area where the observations are almost inevitably of a reconnaissance nature.

Areas of economic interest and engineering importance also have to be surveyed in detail so that their financial potential can be realized and so that serious engineering mistakes are not made. There have been many highway, bridge, dam, and other failures in the past. These surveys will include mining areas; site investigations for new dams, harbors, and highways; sites for quarries, sand and gravel workings, open-cast coal developments; and many others. Since the financial stakes are high teams of specialist geological scientists are likely to be employed. Carefully located boreholes may be necessary, many specimens will be required for laboratory investigation, and geophysics will often be used. This listing contrasts strongly with facilities available to, say, university research workers who are more likely to work along the lines described under the first area requiring detailed survey discussed previously but who are typically without adequate resources. Both large companies and research workers restrict their investigations to items relevant to their specialized interests so no one can claim that such areas have been subjected to thorough geological surveys. The nearest approach to thoroughness is usually the government geological survey, but all is not perfect here either since bureaucracy constantly interferes.

The two examples now described refer to work

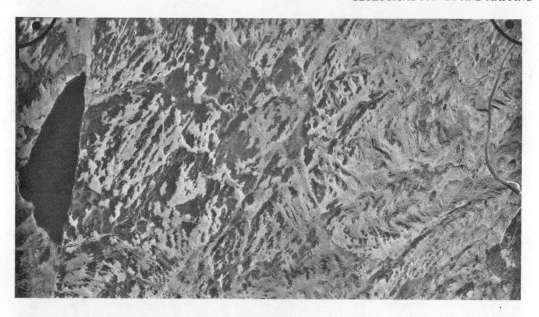

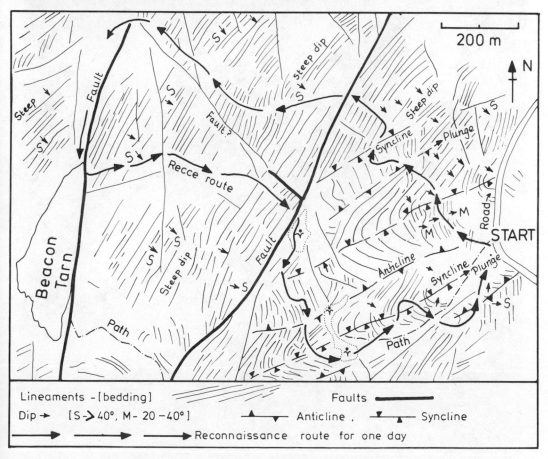

FIGURE 3. An aerial photograph and photogeological map of the Beacon Tarn area, Cumbria, England. The reconnaissance route (on foot) was selected to cross as many lithologies and structures as possible. The rocks are Silurian slates with graywacke sandstone interbeds.

that can be accomplished by single individuals rather than the multidisciplinary but specialist surveys of the giant companies.

High Rigg, Cumbria, England This is part of a region of Ordovician calc-alkali volcanic rocks (the Borrowdale Volcanic Group) for which a detailed map was required as a basis for the study of volcanic processes. Most of the lavas and beds of tuff show up clearly on the stereophotographs, making it possible to construct a reliable photogeological map, which then forms the base map for further geological observations. A final objective is the production of a large scale geological map and an account of the geology (Moseley, 1981). The maps and photographs available were a 1:10,000 topographical map, 1:6,700 stereo air photographs, and 1:1,800 enlargement photographs of the latter. The enlargements were made from negatives purchased from the British Ordnance Survey.

The procedure for this kind of survey is first to prepare photogeological maps on transparent

FIGURE 4. A large-scale aerial photograph (1:1,815) that was used as a base map for plotting geological detail in the Borrowdale Volcanic group, England. Individual rocks as small as 1 m in diameter can be seen.

overlays placed on the enlargement photographs (Fig. 4), and then to add field data as shown in Fig. 5. A constant check has to be kept in the field using the 1:6,700 photographs with a pocket stereoscope to pick out details of small crags and other features, which are not obvious on single prints, no matter how great the enlargement. Even the 1:1,800 scale was not sufficiently large for all the observations to be plotted on the photograph overlays, and the additional field data had to be recorded under locality numbers in field notes. This included more orientations of bedding,

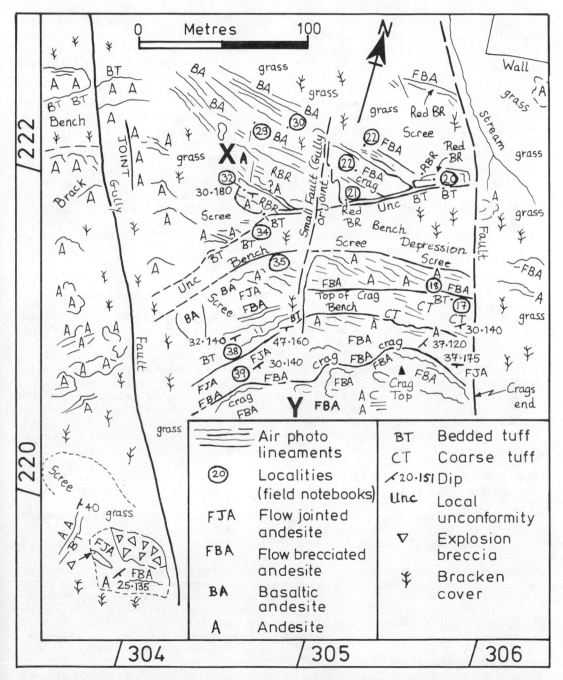

FIGURE 5. Copy of a transparent overlay to Fig. 4 showing geological detail. Further details of items such as measured sections were recorded in field notes.

cleavage, and joints and detailed measured sections showing, centimeter by centimeter, variations in lithology, both vertically and laterally. Specimens had to be collected from precisely located points so that they could be analyzed in the laboratory, petrologically and geochemically. Remember that such fine-grained volcanic rocks cannot be properly identified without geochemical analysis. Other features of importance are visible in Fig. 4 and plotted in Fig. 5, one of which is the intravolcanic unconformity at grid 304222. I can say from personal experience that this unconformity, so obvious on the aerial photographs, is not at all obvious on the ground should one be without the photographs. Finally, a survey of this type would include sections like that shown in Fig. 8. This is essentially a summary of the field notebook observations (see next example).

Galera, Sierra Bernia, in the Pre-Betics of Southeastern Spain It would be wrong to expect that a field method used in one region would be suitable for every other region. Geological fieldwork needs to be flexible and adaptable, and methods have to be devised that are appropriate to local conditions of terrain and climate and to the available topographic maps, air photographs, and other aids. This example has therefore been selected to provide a contrast with High Rigg, just described. The largest scale topographic maps for the Sierra Bernia are 1:25,000, but they show very little detail, and the aerial photographs are even smaller scale, 1:40,000, with no negatives available for good enlargements. Remember that photographic enlargement of prints inevitably results in loss of detail, and if the original prints are of good quality, the best way to see the detail is by optical magnification using a stereoscope. The problem then is that it is difficult to plot this detail on a 1:40,000 scale. Moreover, it is usually the case that the areas where enlargements will be necessary will not be known before the survey starts; they become apparent during the fieldwork. Therefore, there has to be a method of enlarging while working in the field. Structural analysis of the Sierra Bernia fell into this pattern. Several small areas had to be mapped on a large scale, and Fig. 6 is part of one such area (Moseley, 1973). Field time was at a premium and a comprehensive topographic survey was out of the question, so the method outlined in the following, which has also been used in other regions, was adopted.

1. A square grid is drawn on the 1:40,000 photographs using chinagraph (all-purpose) pencil.
2. An enlarged grid is then drawn on paper to form the base of the field map. This can be enlarged to any required scale; e.g., the main map of Fig. 6 represents a ten-time enlarge-

ment from the original 1:40,000 aerial photograph to a scale of 1:4,000 (see Moseley, 1981, Fig. 27, 90, and 91).
3. Within each grid square details are then transferred from the photograph to the new map. They include features seen with a pocket stereoscope and recognizable in the field, such as hilltops, small crags, gullies, prominent trees, areas of cultivation, etc., none of which in this particular example is to be found on the published maps. The method is indicated in outline in Fig. 6. Figure 7 is a section across this area.
4. In the field it is possible to locate oneself precisely, and even at a scale of 1:40,000, individual trees can be seen with a pocket stereoscope.

Geological information can thus be plotted on the new maps as indicated previously. In the case of Galera (see Fig. 6) a new road network postdated both maps and photographs. The method just described made it possible to plot these new roads quickly and with reasonable accuracy. A proper topographic survey would have taken more time than was available, but a number of short tape and compass surveys were necessary for gully and road sections (see next section).

Measurement of Geological Sections

All geological work, whether reconnaissance or detailed, will require measurements of geological sections, and the accuracy will depend on the needs of the survey. In the case of detailed study it may be that centimeter-by-centimeter changes in lithology need to be recorded both vertically (at right angles to bedding, foliation, etc.) and laterally, whereas if the survey is of a reconnaissance nature the measurements will be more approximate. Only by this process can there be final understanding of the origins of the rocks and structures, whether they be igneous, metamorphic or sedimentary. Where exposures are good—e.g., in sea cliffs—very detailed observations of this type will be possible, but it is usually the case that rock exposure is patchy and that small widely separated sections have to be compared with each other, resulting in much inference. The amount of detail and the greater degree of accuracy required for section measurement means that considerable use has to be made of topographic survey methods (see Compton, 1962, or any surveying textbook). The most useful methods are leveling and tape and compass survey, frequently used together since geological sections are rarely on level ground. I have noted that accuracy is important, but the extremely variable nature of geological sequences and structures means that the high degree of accuracy of a topographical survey is not necessary; e.g., a 10 cm bed of sandstone can in-

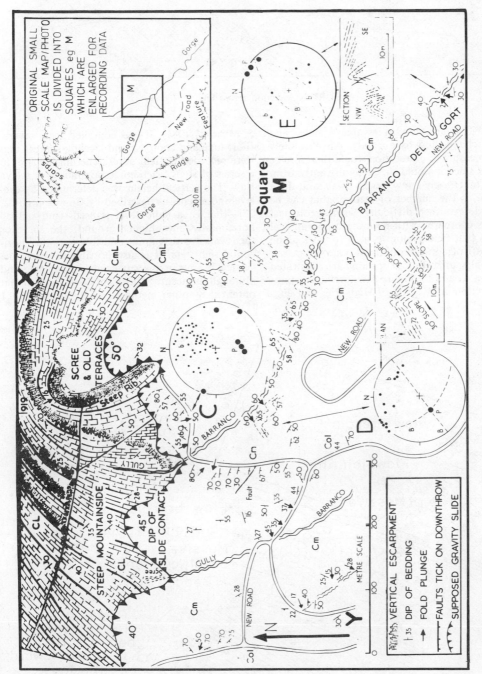

FIGURE 6. A large-scale map of Galera, Sierra Bernia, southeastern Spain, constructed by enlarging a small-scale air photograph. Squares are drawn on the photograph and enlarged to the required M, although the enlargement during the actual fieldwork was much larger than this (times ten, see text).

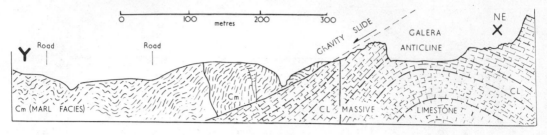

FIGURE 7. A section across Fig. 6.

crease to 15 cm or more in less than a meter, and dip can change just as abruptly. An absolutely precise measure of one section would therefore be pointless since it would differ significantly from another taken only a few meters away, and there is a limit to the number of sections that can be measured. Another difficulty when geological survey is compared with accurate topographical survey is that field geologists have to work alone and a staff person is an unusual luxury.

Examples of measured sections are illustrated by Figs. 8, 9, and 10. Figure 8 is a previously published section across part of High Rigg (see Fig.

5; Moseley, 1981). It was compiled from several short tape and compass and leveling surveys that were recorded in field notes. The hillside rises in a series of steps, each corresponding with a lava flow. The vertical height was determined by eye-level increments as described by Compton (1962, 27)—that is, a hand level is used—and sighted horizontally onto higher ground; the surveyor then walks to that point and takes another horizontal sight; etc. It is also possible to determine vertical heights in this way using a clinometer, and although the accuracy does not compare with that obtained by other methods, where an error of 1

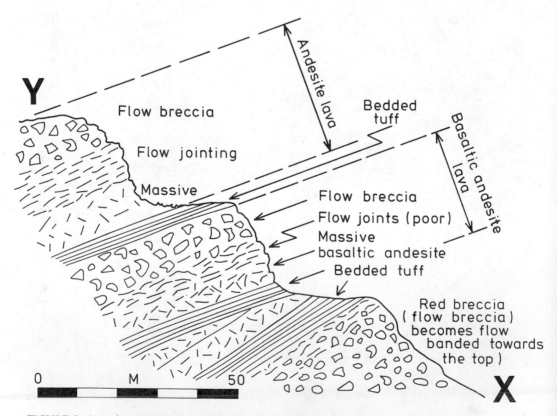

FIGURE 8. A section through lavas and tuffs of the Borrowdale volcanics of High Rigg; see Fig. 5. It was constructed from data recorded in tape and compass and leveling surveys.

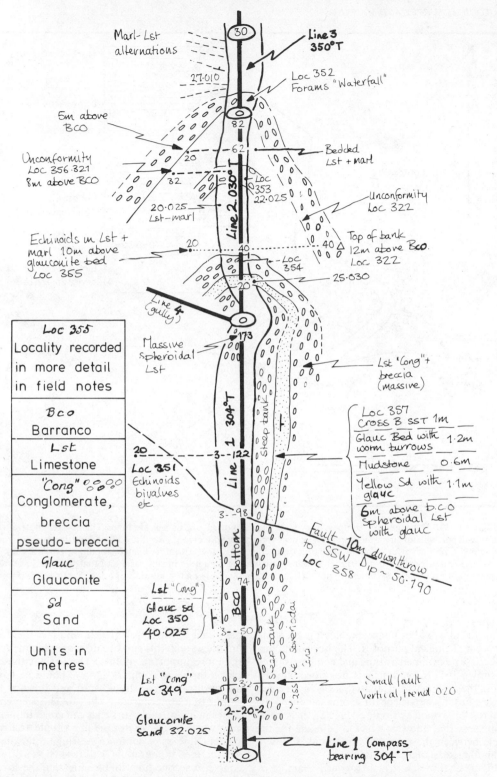

FIGURE 9. An example of a tape and compass survey of the Barranco Salado, southeastern Spain. Three compass lines are shown with measurements in meters. Offsets are taken at right angles to these lines to features of geological interest. A complete survey of this type either would be a closed traverse or would start and finish at known points (see Compton, 1962, 1985).

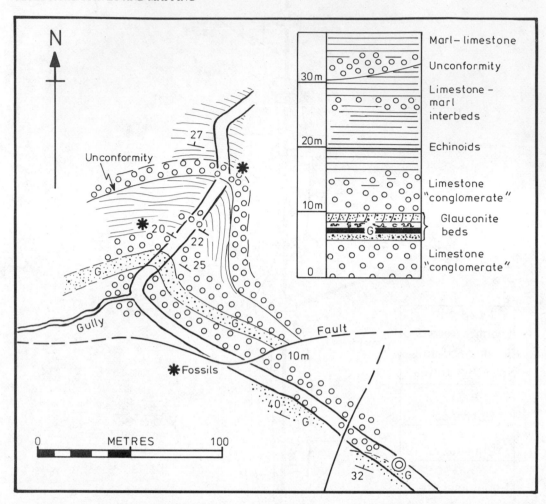

FIGURE 10. The tape and compass survey of Fig. 9 plotted out as a scale map. Thicknesses of strata are determined partly from direct measurement of near vertical rock faces (e.g., locality 357) and partly from horizontal or inclined tape lines with calculations based on average dips of the strata. Readers should appreciate that it is purely accidental for dips to be parallel to water courses (the tape lines in this case) and that tape lines are rarely horizontal, but thickness estimates are not difficult, depending only on the solution of a few right-angled triangles.

cm in 100 m would be unacceptable, it is adequate for most geological purposes. The horizontal distances are determined by tape and compass survey as shown by Fig. 9, which is described shortly. Again the accuracy is less than that of, say, a chain survey, but it is sufficient for geological purposes. Where the ground is sloping (Fig. 8), it is common sense that allowance has to be made for the angle between the tape and horizontal, simply a right–angled triangle calculation.

Figure 9 shows a straightforward example of part of a tape and compass survey where there is no appreciable difference in level along tape lines 1, 2, and 3, although this is not the case with the offsets (e.g., locality 20 in Fig. 9). The method is described by Compton (1962), and it is therefore unnecessary in this entry to do more than mention the more important points. The first of these is that a tape and compass survey should be either a closed traverse or should start and finish at positions that are accurately known. This step eliminates many of the errors that can creep into such a survey. Second, each tape line should follow a straight line as a compass bearing, following a well-defined topographic feature such as a stream course. A second line will be initiated if the stream changes its orientation. Thus in Fig. 9, notice that lines 1, 2, and 3 have different bearings. The third point relates to topographic or geologic features that are some distance from the tape line. These features should be recorded as offsets; the distance is measured by a second tape at right angles to the primary tape. For example, at 122 m on line 1 an offset shows the edge of the barranco at 3 m and locality 351 at 20 m. The fourth point is that it will not be possible to record all the survey details on a survey plan such as Fig. 9, and other details should be recorded under localities in field notes; e.g., under locality 351 there would be observations of lithology, paleoecology of the faunas, and other minor but conceivably important features. It is amazing how much we all miss by slack observation.

Following the field survey the next step is to draw a map and estimate thickness of strata. The method is illustrated by the sedimentary sequence of the Barranco Salado (Figs. 9 and 10). The compass lines and offsets are plotted out to scale against a north point, and all the topographic and geologic features are inserted. Thicknesses of strata are determined either from direct field measurement at right angles to bedding, as is the case with the glauconite beds, or by estimating an average dip between one bedding plane and another—e.g., between localities 354 and 352 (see Fig. 9)—and making a simple calculation based on dip and horizontal distance. One has to bear in mind that the bedding dip is quite likely to make an oblique angle with the tape line and that the tape line may not be horizontal. Solution of such problems, however, can scarcely be regarded as higher mathematics and can be achieved with a few right–angled triangles. A scale section like that shown in Fig. 10 would be a final result.

FRANK MOSELEY

References

Abbotts, I. L., 1978, High potassium granites in the Masirah ophiolite of Oman, *Geol. Mag.* **115**, 415–425.

Abbotts, I. L., 1979, Intrusive processes at ocean ridges; Evidence from the sheeted dike complex of Masirah, Oman, *Tectonophysics* **60**, 217–233.

Barnes, J. W., 1981, *Basic Geological Mapping.* Milton Keynes: Open University Press, 112p.

Compton, R. R., 1962, *Manual of Field Geology.* New York: Wiley, 325p.

Compton, R. R., 1985, *Geology in the Field.* New York: Wiley.

Fry, N., 1984, *Field Description of Metamorphic Rocks.* Milton Keynes: Open University Press, 110p.

Moseley, F., 1966, Exploration for water in the Aden Protectorate, *Royal Engineers Jour.* **80**, 124–142.

Moseley, F., 1973, Diapiric and gravity tectonics in the Pre-Betic (Sierra Bernia) of S.E. Spain, *España Inst. Geol. y Minero Bol.* **84**, 114–126.

Moseley, F., 1981, *Methods in Field Geology.* San Francisco: W. H. Freeman, 211p.

Moseley, F., 1983, Geological field mapping and photogeology, *Geology Teaching* **8**, 82–87.

Moseley, F., and I. L. Abbotts, 1979, The ophiolite mélange of Masirah, Oman, *Geol. Soc. London Jour.* **136**, 713–724.

Moseley, F., and I. L. Abbotts, 1984, A geological map of the Masirah ophiolite complex, Oman, *Overseas Geol. Miner. Resources no. 62*, 1–5, 2 maps.

Moseley, R., J. C. Cuttell, E. W. Lange, D. Stevens, and J. R. Warbrick, 1981, Alpine tectonics and diapiric structures in the Pre-Betic zone of S.E. Spain, *Jour. Struct. Geol.* **3**, 237–251.

Tucker, M. E., 1982, *The Field Description of Sedimentary Rocks.* Milton Keynes: Open University Press, 112p.

Cross-references: *Aerial Surveys; Compass Traverse; Cross-Section; Exposures, Examination of; Field Geology; Field Notes, Notebooks; Fossils; Fractures, Fracture Structures; Geological Fieldwork, Codes; Hard versus Soft Rock Geology; Igneous Rocks, Field Relations; Map Symbols; Maps, Logic of; Mineral Identification, Classical Field Methods; Photogeology; Photointerpretation; Plane Table Mapping; Profile Construction; Samples, Sampling; Sedimentary Rocks, Field Relations; Soil Sampling; Surveying, General; Vegetation Mapping.*

GEOLOGICAL SURVEYS, STATE AND FEDERAL

On March 3, 1964, the U.S. Geological Survey, celebrating its eighty-fifth anniversary, received a congratulatory message from Secretary of the Interior Stewart Udall, who pointed out that "through its geologic, hydrologic, and topographic surveys, and its research on basic principles of earth science . . . [it] has contributed in significant measure to the intelligent development and wise use of this Nation's mineral, water, and land resources" (personal communication to T. B. Nolan, U.S.G.S. Director).

The U.S. Geological Survey originated in 1879 as a result of the consolidation of four scientific surveys made during the period from 1867 to 1879 as part of the exploration of the western United States. A State Geological Survey had already been organized in Massachusetts 49 yr earlier, and between 1830 and 1879, 34 of the 38 states then in existence had established a State Geological Survey. The far-sighted state governing bodies thus had recognized at an early date the great need for an inventory of their mineral wealth.

Geological surveys of a very limited sort had been made in North Carolina in 1824 and 1825 and in South Carolina in 1826, but their only claim to being state geological surveys was that state appropriations were used to carry out the work. A much more significant pre-1830 survey was the detailed systematic study that had been performed by Amos Eaton in Rensselaer County, New York, in 1821. This survey was supported wholly by the private funds of Stephen Van Rensselaer, however, and thus does not qualify as a State Geological Survey.

State Geological Surveys

Massachusetts, in establishing the first public-supported geological survey in the United States, recognized the economic value of geologic knowledge in the future development of the state. Attempting to make internal improvements such as roads and canals and to justify the necessary expenditures of money, the Massachusetts state Legislature recognized that new needs for these expensive transportation facilities must be developed. The governor acknowledged the value of a geological survey by stating:

Much knowledge of the natural history of the country would be thus gained, especially the presence of valuable ores, the extent of quarries, and of coal and limestone, objects of inquiry so essential to internal improvements, and the advancement of domestic prosperity would be discovered, and the possession and advantage of them given to the public. [Hitchcock, 1833, iii]

The first State Geologist, Reverend Edward Hitchcock of Massachusetts, gained the confidence of the Massachusetts state Legislature by his objective reporting, and accordingly the survey was continued, but only until 1842. After that date Massachusetts had no State Geological Survey for more than a century until it was re-established in 1971. Its funding in the 1980s returned to zero—although the office of State Geologist still exists there. Some of the functions of the Massachusetts Geological Survey have been performed by the U.S. Geological Survey since about 1938.

Following the lead of Massachusetts, 20 other states established Surveys in the 1830s and 1840s. Influenced by the *doctrine of mercantilism,* which held that the state has a broad sphere of action in the field of economics, the state legislatures were concerned that geological surveys should produce useful information (Hendrickson, 1961, 361).

Governor Noble of Indiana, positive that the citizens of the state would benefit from a geological survey, told the legislature:

[I]n this State, we have external indications of large beds of coal and other natural deposits; but for want of the proper test of science, their extent and value are unknown. . . . [Further, he was] satisfied that our mineral resources, properly developed, will give employment to thousands, subserve in purposes of commerce, contribute to the support of our public works, and add greatly to the wealth of our citizens and the State. [Indiana House Journal, 1836–37, 26–27]

Another reason for the establishment of state geological surveys was that they would help advance education. By the early part of the nineteenth century there were greater educational opportunities in the United States, with a prevailing view that scientific knowledge should be widely available to the general public. The study of natural science moved from the amateur scholar into the formalized frameworks of colleges and universities, and professors found enthusiastic audiences.

In spite of the liberal attitudes of state legislatures toward science education, and no matter how much they may have wished to promote the economic interests of their citizens, they were always faced with the problem of justifying the expenditure of public funds. In prosperous times when money was easy to appropriate, geological surveys were authorized; during economic hardships, however, expenditures were curtailed and state geological surveys suspended. For example, although 14 state geological surveys were active in 1837, the effect of the financial panic that year was to reduce the total number to only 3 by 1843.

This curtailment of geological services was due, in part, to the near (or complete) bankruptcy of those states that had engaged in extensive programs of internal improvements. Improved economic conditions in the 1850s enabled a number of state geological surveys to be reactivated, reaching a total of 14 in 1855. After a decline in the 1860s the number rose to a high of 18, but the depression of 1873 took its toll, with only 11 surveys surviving in 1885. Thus, it is clear that the final decision as to whether or not a state would support a geological survey was primarily an economic one. Today, the office of State Geologist, or a State Geological Survey exists in all the states plus Puerto Rico (Table 1).

It is difficult to determine objectively, however, what is economically useful information; subjectively, the state geologist has one way of arriving at this opinion, and the members of the state legislature have another. It was important then (as it is now) for the state geological surveys to ensure that state appropriations had been well spent. The state geological survey could accomplish part of this task through its published reports. If a legislator saw, e.g., that the geologists had located coal seams by geography and by position in a sequence of rocks or at a certain depth below the ground surface that the legislator could relate to hills and valleys, if they had determined the thickness and extent of the seam and amount of overburden, if they had given chemical (or proximate) analyses and commented whether the coal could be used for domestic heating or industrial fuel, and if they had said that the seams were workable or not, then the geologists' work was deemed economically useful. Similarly, if a legislator could see in the Survey museum (see Vol. IVB: *Museums, Minerological*) a neat and orderly display of minerals, rocks, and fossils, each well labeled, it answered some of the questions relative to the physical world in which he or she lived; the legislator thus felt that the geologists' work had educational value and was useful.

Conversely, all state geological surveys produce some information that cannot be evaluated in terms of a price tag. Some aspects of basic research conducted by state geological surveys, contrasted with the present emphasis on applied research, were not prominently displayed in most early state geological survey reports. David Dale Owen in Indiana and Edward Hitchcock in Massachusetts both recognized in the 1830s that the primary purpose of a legislative appropriation was to obtain some practical or economic feedback from the survey results. They both also recognized, however, that a state-supported survey could provide additional geological input to the fund of scientific knowledge. Hitchcock, e.g., devoted more than half his final report (1833) to paleontology and other types of pure research and cataloging.

Owen, being sensitive to political opinion of his day, attempted to forestall possible objections by giving lectures on geology and geological surveys to members of the Indiana Legislature, and he prefaced his first report ("Report of a Geological Reconnoisance [sic] of the State of Indiana, Made in the Year 1837, Indianapolis," 1838, 34p.) with a brief account of basic geology so that the untutored reader could understand what was to follow. Both Owen and Hitchcock saw that there was no sharp dividing line between *scientific* and *useful* geology. Owen, e.g., told the legislators how a knowledge of paleontology would enable the miner to determine the lower limits of the potentially coal-bearing strata: no coal would be found below rocks containing the corkscrew fossil *Archimedes*.

In contrast, there were a few geologists who looked on the state-supported survey as a glorious opportunity to make personal collections of minerals, rocks, and fossils and also as a means of enhancing their professional standing by accumulating data that could be published at state expense. Others, not unethically either, paid lip service to the legislative demands for useful results and used the publicly financed survey as a means to the seemingly more important goal of enrichment of knowledge. The classic example of the latter type was James Hall, who served at various times as State Geologist of New York, Iowa, and Wisconsin. He devoted most of his time to paleontology rather than to economic geology; in this pursuit he fared well in New York, but in Iowa

and Wisconsin his surveys were discontinued because of objections to state funds' being spent for expensive books that were of value apparently only to a small group of paleontologists and conchologists (Clarke, 1921, 356-362).

Other State Geologists had similar troubles with their legislatures. John S. Newberry of Ohio, e.g., was unsuccessful in obtaining $60,000 for a third volume of reports on paleontology, and some of his scientists had to work without pay and have their reports published outside, in scientific journals (Orton, 1893, v-vi). The Michigan Geological Survey was curtailed in 1840 "on account of straightened finances and the hostility of the legislature to labors which promised no early practical benefit to the material progress of the State" (Allen and Martin, 1922, 676). In the last 20 or 30 yr of the nineteenth century, conflicts between geologists and legislators subsided, as the idea of a continuing state geological survey with a permanent staff and an alliance with state-supported educational institutions became commonplace.

With the establishment of the U.S. Geological Survey in 1879 and its greater or less cooperation with state geological surveys in subsequent years, much of the pure research was taken over by that federal survey, leaving the state surveys free to tackle various aspects of applied geology. Further, use of the increasing number of professional journals and university publications in the later 1800s enabled the state geologists to become more independent of local politics than in the earlier part of that century.

Federal Geological Surveys

The federal government sponsored a number of exploratory geological surveys in the early part of the nineteenth century. These surveys, which were conducted in the West and mostly by the U.S. Army, were primarily for military purposes and only to a lesser extent for the acquisition of geographic knowledge. An excellent account is given in the multivolume series by Rabbitt (1979, 1980, 1986).

Significant among these early expeditions were those of Lewis and Clark to the Northwest from 1803 to 1807 and of Pike and Long to the Rocky Mountains from 1805 to 1807 and 1819 to 1820. Though the Long expedition supported a geologist, Edwin James, the title "U.S. Geologist" belongs to Featherstonhaugh, who made trips to the Ozark Mountains in 1834 (Rabbitt and Rabbitt, 1954, 741). Of no less significance were the surveys of "mineral lands" in the upper Mississippi Valley from 1839 to 1840 by Owen, who had already made his mark in Indiana.

The explorations of 1867-1879 led to the establishment of the U.S. Geological Survey in

1879. Before 1867, geological investigations were of secondary importance relative to topographical surveying. But in 1867, Congress authorized several expeditions in which geological surveys were to play a prominent role:

The *King Survey* (Geological Exploration of the 40th Parallel) covered a 100 mi wide belt from Cheyenne, Wyoming, to the eastern border of California; the results were published in 1870–1880 in seven volumes and an atlas.

The *Hayden Survey* (Geological and Geographical Survey of the Territories) covered Nebraska, Colorado, New Mexico, Wyoming, Montana, and Idaho, and included work in biological and anthropological sciences as well as geology and topography. The results were published in 1867–1883 in a series of volumes.

The *Powell Survey* (Geographical and Geological Survey of the Rocky Mountain Region) began with a survey of the Colorado River and also included substantial parts of Wyoming, Utah, Nevada, and Arizona. Its subject matter was as broad as that of the Hayden Survey, but only two brief reports were published, in 1877 and 1878.

The *Wheeler Survey* (Geographical Survey West of the 100th Meridian), the only one directed by an Army officer, was similar in scope to the Hayden and Powell surveys, although it was primarily a topographic survey. It included parts of the Dakotas, Nebraska, Kansas, Texas, the Rocky Mountain region, and the Pacific Coast. The results were published in 1875–1889 in several volumes.

The U.S. Geological Survey is now located at the National Center in Reston, Virginia, about 32 km west of the District of Columbia. It has regional offices in Reston; in Menlo Park, California; and in Denver, Colorado, and it also maintains offices in other countries.

Work of State and Federal Geological Surveys Today

Both state and federal surveys have been involved with studies of water resources since 1888 when Powell, then Director of the U.S. Geological Survey, initiated a survey of arid lands, known as the "Irrigation Survey." This interest and activity regarding water resources was added to the extensive investigations of mineral resources that had begun with the State Geological Survey of Massachusetts in 1830. The U.S. Geological Survey has, since 1884, been preparing topographic maps of the United States in cooperation with individual states.

Geological surveys, in all countries of the world, are among the oldest scientific institutions to be financed from public funds . . . [because of] the great advantages which must accrue . . . [from promoting] geological science but also as a work of great practical utility, bearing on agriculture, mining, road-making, the formation of canals and railroads, and other branches of national industry. [Davidson, 1962, 83]

The search for mineral resources has always been a primary concern of geological surveys. Geological considerations have long been an important adjunct to civil engineering projects since such information is relevant to the planning, design, and construction of dams and other earthworks, roadbeds, airstrips, canals and waterways, and foundations for large structures. With the development of greater population densities in urban and suburban areas, geological surveys have had to deal more and more with problems associated with environmental pollution, mineral and water supply (see *Groundwater Exploration*), and environmental hazards. Current environmental concern has caused increasing involvement by state geological surveys and to a lesser extent, the U.S. Geological Survey, in subsurface disposal of liquid wastes, toxic waste disposal, and nuclear waste disposal.

Geological surveys are also now emphasizing the cultural aspects of geology—i.e., understanding the physical environment as the home of people. Both state and federal geological surveys, through their published and unpublished reports and collections and catalogues of minerals and fossils, contribute to the edification of professionals who will continue to provide a wide range of geological services in the future.

State and federal geological surveys have become more active with nongeologic audiences in recent years (see *Popular Geology*), making maps (q.v.) and interpretive reports that aid in decision making by government entities such as those involved in land-use planning and management, for both surface and also subsurface environmental protection, in coastal and inland areas. Other services such as topographic maps and aerial photos (together with other kinds of imagery from high-altitude airplanes and spacecraft) (see *Photogeology*) are provided by the U.S. Geological Survey, whose total program was described for the Congress by Agnew (1975).

Environmental impact assessments of the effects of federal, state, and local projects have consumed an increasing amount of time and attention by governmental geological surveys in the 1970s and 1980s. Many state geological surveys have been given additional responsibilities in regard to reclamation of surface-mined lands, both public and private, within their borders.

Several state geological surveys now have regulatory mandates over one or more geologic resources, such as oil and gas development and production (7 states), surface-mine reclamation (6), water (5), dam safety (2), and other purposes (5) such as sedimentation pollution, and drilling for geothermal energy. Such programs can be a mixed blessing to a scientific research organization, because regulatory activities tend to focus a spotlight (which can be largely negative) of public opinion on the objective studies by the survey's professional staff—and with the consequent hazard of political interference in its work. The U.S. Geological Survey, in the 1970s, has encountered such hostility by those regulated as well as by "public interest" groups, that the Survey lost its Conservation Division completely, through the formation of a new agency, the Minerals Management Service, in the Department of the Interior, in 1982.

In the 1970s funding of state geological surveys increased, and their staffing levels were augmented by approximately 800 people (to a total of 2,800), much of it in response to needs for groundwater protection and other environmental issues. The U.S. Geological Survey's work force of about 9,000 people, however, was not enlarged commensurate with its increased appropriations, which caused the survey to embark on a small amount of contract research.

In the 1970s and early 1980s both state and federal geological surveys have become more attuned to societal concerns while at the same time they have attempted to carry on substantial programs of scientific/engineering research and analysis—the purpose for which they were initially established. Unfortunately, in the mid-1980s shrinking budgets have eroded personnel and programs of both the U.S. Geological Survey and many of the state geological surveys. Therefore, the states have been hard pressed to continue responding to the growing public needs for geologic answers, and the U.S. Geological Survey has de-emphasized its 1879 charge to perform geologic mapping.

ALLEN F. AGNEW

TABLE 1. Addresses of State Geologists, May 1987

Alabama
205–349–2852
Ernest A. Mancini
State Geologist and Oil and Gas Supervisor
Geological Survey of Alabama
P.O. Box O
Tuscaloosa, AL 35486

Alaska
907–494–7625
Robert B. Forbes
State Geologist and Director
Division of Geological and Geophysical Surveys
794 University Avenue, Basement
Fairbanks, AK 99709

Arizona
602–621–7906
Larry D. Fellows
State Geologist and Assistant Director
Bureau of Geology and Mineral Technology
Geological Survey Branch
845 N Park Avenue
Tucson, AZ 85719

Arkansas
501–371–1448
Norman F. Williams
State Geologist and Director
Arkansas Geological Commission
Vardelle Parham Geology Center
3815 W Roosevelt Road
Little Rock, AR 72204

California
916–445–1923
James F. Davis
State Geologist
Department of Conservation
Division of Mines and Geology
1416 Ninth Street, Room 1341
Sacramento, CA 95814

Colorado
303–866–2611
John W. Rold
Director and State Geologist
Colorado Geological Survey
1313 Sherman Street, Room 715
Denver, CO 80203

Connecticut
203–566–3540
Hugo F. Thomas
Director and State Geologist
Natural Resources Center
Department of Environmental Protection
State Office Bldg., Room 553
Hartford, CT 06106

Delaware
302–451–2833
Robert R. Jordan
State Geologist and Director
Delaware Geological Survey
University of Delaware
Newark, DE 19716

(continued)

TABLE 1. (*continued*)

Florida
904-488-4191
Charles W. Hendry, Jr.
State Geologist
Bureau of Geology and
Department of Natural Resources
903 W Tennessee Street
Tallahassee, FL 32304

Georgia
404-656-3214
William H. McLemore
State Geologist
Georgia Geologic Survey
Department of Natural Resources
19 Martin Luther King, Jr., Drive, SW
Atlanta, GA 30334

Hawaii
808-548-7533
Manabu Tagomori
Manager and Chief Engineer
Division of Water and Land Development
Department of Land and Natural Resources
P.O. Box 373
Honolulu, HI 96809

Idaho
208-885-6195
Maynard M. Miller
Director and State Geologist
Idaho Geological Survey
University of Idaho
Moscow, ID 83843

Illinois
217-333-4747
Morris W. Leighton
Chief
Illinois State Geological Survey
Natural Resources Building
615 E Peabody Drive
Champaign, IL 61820

Indiana
812-335-2862
Norman Hester
State Geologist and Director
Indiana Geological Survey
611 N Walnut Grove
Bloomington, IN 47405

Iowa
319-335-1574
Donald L. Koch
Director and State Geologist
Department of Natural Resources
Geological Survey Bureau
123 N Capitol Street
Iowa City, IA 52242

Kansas
913-864-3965
Lee C. Gerhard

Director and State Geologist
Kansas Geological Survey
1930 Constant Avenue A, West Campus
The University of Kansas
Lawrence, KS 66046

Kentucky
606-257-5863
Donald C. Haney
State Geologist and Director
Kentucky Geological Survey
311 Breckinridge Hall
Lexington, KY 40506-0056

Louisiana
504-342-6754
Charles G. Groat
Director and State Geologist
Louisiana Geological Survey
P.O. Box G
Baton Rouge, LA 70893

Maine
207-289-2801
Walter A. Anderson
Director and State Geologist
Maine Geological Survey
Department of Conservation
State House Station 22
Augusta, ME 04333

Maryland
301-554-5503
Kenneth N. Weaver
Director
Maryland Geological Survey
2300 St. Paul Street
Baltimore, MD 21218

Massachusetts
617-727-9800
Joseph A. Sinnott
State Geologist
Executive Office of Environmental Affairs
20th Floor
100 Cambridge Street
Boston, MA 02202

Michigan
517-334-6923
R. Thomas Segall
State Geologist
Geological Survey Division
Department of Natural Resources
Box 30028
Lansing, MI 48909

Minnesota
612-627-4780
Priscilla C. Grew
Director
Minnesota Geological Survey
2642 University Avenue
St. Paul, MN 55114-1057

TABLE 1. (*continued*)

Mississippi
601–354–6228
Edwin E. Luper
Director and State Geologist
Bureau of Geology
Department of Natural Resources
P.O. Box 5348
Jackson, MS 39216

Missouri
314–364–1752
James Hadley Williams
Director and State Geologist
Department of Natural Resources
Division of Geology and Land Survey
P.O. Box 250
Rolla, MO 65401

Montana
406–496–4180
Edward T. Ruppel
Director and State Geologist
Bureau of Mines and Geology
Montana College of Mineral
Science and Technology
Butte, MT 59701

Nebraska
402–472–3471
Perry B. Wigley
Director
Conservation and Survey Division
Institute of Agriculture & Natural Resources
The University of Nebraska
Lincoln, NE 68588–0517

Nevada
702–784–6691
John H. Schilling
Director and State Geologist
Bureau of Mines and Geology
University of Nevada
Reno, NV 89557–0088

New Hampshire
603–862–3160
Eugene Boudette
State Geologist
Office of State Geologist
117 James Hall
University of New Hampshire
Durham, NH 03824

New Jersey
609–292–1185
Haig F. Kasabach
State Geologist
New Jersey Geological Survey
Division of Water Resources
CN–029
Trenton, NJ 08625

New Mexico
505–835–5420
Frank E. Kottlowski
Director
Bureau of Mines and Mineral Resources
Campus Station
Socorro, NM 87801

New York
518–474–5816
Robert H. Fakundiny
State Geologist
New York State Geological Survey
State Museum, Empire State Plaza
3136 Cultural Education Center
Albany, NY 12230

North Carolina
919–733–3833
Stephen G. Conrad
Director and State Geologist
Division of Land Resources
Dept. Nat. Resources & Community Dev.
P.O. Box 27687
Raleigh, NC 27611

North Dakota
701–777–2231
Sidney B. Anderson
State Geologist
North Dakota Geological Survey
University Station
Grand Forks, ND 58202–8156

Ohio
614–265–6605
Horace R. Collins
Chief and State Geologist
Department of Natural Resources
Division of Geological Survey
Fountain Square, Bldg. B
Columbus, OH 43224

Oklahoma
405–325–3031
Charles J. Mankin
Director
Oklahoma Geological Survey
University of Oklahoma
830 Van Vleet Oval, Room 163
Norman, OK 73019

Oregon
503–229–5580
Donald A. Hull
Director and State Geologist
Dept. of Geology and Mineral Industries
910 State Office Building
1400 SW Fifth Avenue
Portland, OR 97201–5528

(*continued*)

271

TABLE 1. (*continued*)

Pennsylvania
717-787-2169
Donald M. Hoskins
State Geologist
Bureau of Topographic and Geologic Survey
Department of Environment Resources
P.O. Box 2357
Harrisburg, PA 17120

Puerto Rico
809-724-8774
Ramon M. Alonso
Director
Servicio Geologico de. P.R.
Department de Recursos Naturales
Apartado 5887, Puerta de Tierra
San Juan, PR 00906

Rhode Island
401-792-2265
J. Allan Cain
State Geologist and Chairman
Department of Geology
The University of Rhode Island
Kingston, RI 02881

South Carolina
803-737-9440
Norman K. Olson
State Geologist and Chief
South Carolina Geological Survey
Harbison Forest Road
Columbia, SC 29210

South Dakota
605-677-5227
Merlin J. Tipton
State Geologist
South Dakota Geological Survey
Department of Water and Natural Resources
Science Center, USD
Vermillion, SD 57069-2390

Tennessee
615-742-6691
William T. Hill
State Geologist
Department of Conservation
Division of Geology
Customs House, 701 Broadway
Nashville, TN 37219-5237

Texas
512-471-1534
William L. Fisher
Director and State Geologist
Bureau of Economic Geology

The University of Texas at Austin
University Station, Box X
Austin, TX 78712-7508

Utah
801-581-6831
Genevieve Atwood
State Geologist and Director
Utah Geological and Mineral Survey
606 Black Hawk Way
Salt Lake City, UT 84108-1280

Vermont
802-244-5164
Charles A. Ratte
State Geologist
Office of State Geologist
Agency of Environmental Conservation
103 South Main, Center Building
Waterbury, VT 05676

Virginia
804-293-5121
Robert C. Milici
State Geologist and Commissioner
Division of Mineral Resources
P.O. Box 3667
Charlottesville, VA 22903

Washington
206-459-6372
Raymond Lasmanis
State Geologist and Oil and Gas Supervisor
Department of Natural Resources
Geology and Earth Resources Division
Olympia, WA 98504

West Virginia
304-594-2331
Robert B. Erwin
Director and State Geologist
Geological and Economic Survey
Mont Chateau Research Center
P.O. Box 879
Morgantown, WV 26507-0879

Wisconsin
608-262-1705
Meredith E. Ostrom
State Geologist and Director
Geological and Natural History Survey
3817 Mineral Point Road
Madison, WI 53705

Wyoming
307-742-2054
Gary B. Glass
State Geologist
Geological Survey of Wyoming
P.O. Box 3008, University Station
Laramie, WY 82071

References

Agnew, A. F., 1975, The U.S. Geological Survey: prepared by Congressional Research Service for Senate Interior and Insular Affairs Committee. *U.S. Congress, 94th, 1st Session, Committee Print*, December, 139p.

Allen, R. C., and H. M. Martin, 1922, A brief history of the geological and biological survey of Michigan, 1837–1920, *Michigan Historical Magazine* **6**, 675–750.

Back, W., 1959, Emergence of geology as a public function, 1800–1879, *Washington Acad. Sci. Jour.* **49**, 205–209.

Clarke, J. A., 1921, *James Hall of Albany*. Albany, N.Y. 565p.

Davidson, C. F., 1962, Geology in the service of mankind, *Impact of Science on Society* **12**, 83–101.

Fisher, J. L., 1974, New directions for energy policy and analysis, in *Earth Science in the Public Service*. Washington, D.C.: U.S. Geological Survey (Prof. Paper 921), 7–11.

Hendrickson, W. B., 1961, Nineteenth-century State geological surveys: early government support of science, *Isis* **52**, 357–371.

Hitchcock, E., 1833, *Report on the Geology, Mineralogy, Botany, and Zoology of Massachusetts*. Amherst, 700p.

Leighton, M. M., 1951, Natural resources and geological surveys. *Econ. Geology* **46**, 563–577.

Merrill, G. P., 1924, *The First Hundred Years of American Geology*. New Haven, Conn.: Yale University Press, 734p. (esp. 127–552).

Nolan, T. B., 1959, The United States Geological Survey and the advancement of geology in the public service, *Washington Acad. Sci. Jour.* **49**, 209–214.

Orton, E., 1893, *Report of the Geological Survey of Ohio*, vol. 7, Ohio Geological Survey, v–vi.

Owen, D. D., 1838, *Report of a Geological Reconnoissance [sic] of the State of Indiana Made in the Year 1837. . . .* Indianapolis, 34p.

Rabbitt, J. C., and Rabbitt, M. C., 1954, The U.S. Geological Survey: 75 years of service to the nation, *Science* **119**, 741–758.

Cross-references: *Maps, Environmental Geology; Maps, Physical Properties.* Vol. XIII: *Earth Science, Information and Sources; Economic Geology; Geological Communication; Geology, Applied; Medical Geology; Urban Geology.*

GEOLOGICAL WRITING

The art and techniques of *geological writing* are conveniently set out by the editorial staff of *Geotimes* (Cochran et al., 1979). Also of great value is a classic book, *Suggestions to Authors of Reports of the United States Geological Survey* (Bishop et al., 1978). Some writers prefer the third edition. No attempt is made to duplicate these remarkable volumes here. This entry selects some key points from these books for writers to consider.

Audience

For the beginner, practice is everything. A patient, wise teacher is one who will guide and correct while taking care not to stultify or discourage. Young (and some old) writers are very sensitive to criticism, and it should be provided with care and discretion. How many potentially fine scientists have been put off or demoralized by some enthusiastic but insensitive teaching assistant? Combined field experience and writing is an ideal way to practice (Norris, 1983).

Then you may have a long report, perhaps even a university thesis, running up to five hundred typewritten pages. These long reports are normally published only by state or federal surveys. As commercial books they would not appeal to audiences large enough to justify the publishing cost. Another solution is to cut up the report into shorter segments and rewrite them into separate papers, stressing one or another aspect of the survey. The bulletin or journal of your national or regional geological society might be a suitable place to publish, but if the main thrust is toward a specific subdiscipline such as structure, petrology, sedimentology, geomorphology, etc., there are excellent international journals dealing with each subject.

Other types of papers are the ideas, or trial-balloon, essay; or a general report, by literature surveys or extended travels and fieldwork; or long-term laboratory or computer investigations. Such a report presents general conclusions of widespread interest. Good places for both idea papers and the generalizing reviews are the *Journal of Geology* (Chicago) and the *American Journal of Science* (New Haven) or one of the European equivalents; excellent examples that reach different geographic audiences are the *Geologische Rundschau* (Germany), *Geologiska Föreningens Förhandlingar* (Stockholm), *Geological Magazine* (Cambridge, England) and several others from the United Kingdom.

Perhaps you are aiming at a popular or semipopular audience. This is a tough hurdle and not for the beginner, but *Scientific American, Natural History,* and the *American Scientist* are avenues for reaching a wide public. Or perhaps you plan a full-length book. Cloth-bound volumes typically fall into one of three categories: (1) the semipopular, travelogue-adventure-with-science style of work, for which, if they are well done, sales are pretty good today; (2) the advanced-level treatise on some topical subject; and (3) the instructional textbook. Take care: before spending too much time, talk to the publishers. They will want to see a table of contents and a trial chapter.

Style

In Latin a *stilus* was a pointed instrument for writing. From this came our word *style*. It has gradually come to mean the manner of writing or, for that matter, any kind of manners. We could say that a poor writing style is a form of bad manners because it irritates the readers and wastes their time; it is quite commonly difficult to comprehend and may then be misleading. If the pursuit of science is the search for truth, and your report leads to false inferences, you are set on a course of self-destruction. An elegant style need not be flowery. The crystallographic simplicity and purity of the diamond are the secrets of its luster and ultimately of its value.

Style can refer to anything; e.g., my library has a book entitled *Styles of Folding* (Johnson, 1977), but here we shall discuss writing. It is easier to cite examples of poor style than it is to lay down criteria for good style. As basic reading, one might start with a few volumes of *Geotimes,* reading the last page in each monthly issue, the "Geologic Column" by Robert L. Bates.

In the United States, common standard reference is the University of Chicago's *Manual of Style* (13th ed., 1982), but its recommendations are not universally endorsed. A careful writer might benefit from having one or two special volumes giving recommended word usages and nuances; excellent examples are Evans and Evans (1957), Bernstein (1965), and Fowler (1965).

The writing of a technical or scientific paper requires not only that the research involved is accurate but also that the reporting of this research be accurate as well as perhaps interesting. If the presentation puts the desired audience to sleep, something is seriously wrong.

There are standard works on how to write a scientific paper (e.g., Trelease, 1951; Souther, 1957; O'Connor, 1975). The last, published in the Netherlands, reflects the gradual worldwide takeover of scientific communication by the English language, which reflects the large numbers of English-speaking scientists in North America, the United Kingdom, Australia, New Zealand, India, South Africa, and former British Colonial areas. At major international congresses, in the 1950s and 1960s, papers would be presented in French, German, Russian, or whatever. The anglophone audiences would take the opportunity to enjoy a surreptitious nap or slip out briefly for an extra coffee break. Scientists from the non-English-speaking countries have finally gotten the message. They had to learn the language or go unheard. For those scientists, it has often meant going back to school and attempting to discover the best English style. This is where problems arise. Many native English speakers have learned farmyard or marketplace English as children, but never style. Scientific style is in fact both a discipline and an art with rather clear and firm rules, which are not very daunting if tackled as a technical skill to be studied and mastered.

For detailed suggestions and guidance, the reader is referred to the entry *Style, Writing.* A highly specialized professional copy-editing style is required to get a manuscript ready for the press. For this the reader may be referred to a book entitled *Words into Type* (Skillin et al., 1974). Style is not everything. If you have not really discovered anything and if you have no new ideas, the most elegant style will hardly save the day, unless you are producing strictly an inferior piece of work chiefly for profit.

The same is true for speaking before an audience. The finest piece of research can be a flop when presented in a high-speed reading or in a squeaky voice or mumbled with the head pointed down. If you are not a good speaker, it is perfectly feasible to take voice lessons. It is a compliment to your audience to take the trouble to speak clearly and slowly and preferably without notes, but with illustrative slides. It is said that no slide should have more than eight words on it. Ideas are best promulgated in small bite-sized capsules. Geological audiences are almost invariably good-natured, but take care lest your presentation puts them to sleep.

Title

The label on the package is often what sells the product in the commercial world. The label "Rolls Royce" sells cars. Your book or article title should be not only eye-catching and short and sweet but also tell what the writing is about. In titles one should avoid nineteenth century modesty that sounds pedantic in the twentieth century: e.g., "A tentative approach to an examination of the XYZ problem." I noticed in the geology library at Columbia University about three inches of cards giving titles of works that began with the words *studies of.* This wording makes it difficult and time consuming for readers to find studies on a particular topic.

Inasmuch as most geological discussions involve rocks, age, and place, it is a good idea to get this information into your title, briefly; thus, "Lower Cambrian tillites from eastern Senegal, West Africa." That would be fine for an article, but you can shorten the title as you expand the topic and the area. For example, you might do a global review, emerging from the same starting point—e.g., "Early Paleozoic Glaciation"—by implication, you are going to deal with the whole story, all over the world.

Examples of horrible titles abound in the literature. A particularly common type nowadays is perpetrated in profusion by marine geologists—

e.g. (an imaginary case), "A preliminary report on the distribution of *Globotruncana oblonga* in magnetic zones 11 and 12 between lat. 11°47′ and 17°01′S, and long. 145°09′W and 147°28′W, taken on legs 2015 and 2016 of expedition NEMO." This all may be true, but it is stylistically boring and uninformative except to a small coterie of cognocenti. A more appropriate title could be "Early Miocene pelagic foraminifera of the mid-Pacific, SW of Hawaii." One should avoid cookbook and housekeeping details.

A book can have a subtitle. The main thing is to get something brief for the title that can be printed on the front cover and spine. The printer cannot squeeze in a long story. But inside, on the title page, to explain the headline-type title, you can put in a colon and subtitle, thus, "Megacycles: Long-Term Episodicity in Earth and Planetary History" (Williams, 1981). Subtitles, however, should be avoided in journal papers. As a test for your title, ask a librarian how it will be indexed.

Authorship

Any book or paper needs an author, not merely to write it but because all computer, bibliographic, and indexing agencies use author names and dates as a form of identification. It should also carry the author's postal address. If Jones writes more than one paper in one year they should be identified as "Jones (1984a, 1984b, 1984c . . .)."

If Jones was on an expedition and wants to mention all his friends, including the cook and the mess-steward, he may (although it is unnecessary) list all their names. The twenty-five authors can cover half the first page with their names and affiliations. Pity the poor indexer or librarian who has to copy them all down. But what does the busy scientist do who cites this reference? He or she simply uses Jones et al. (*et alia,* Latin for "and others"). The best way to handle this sort of complex situation is to use discretion and judgment. Never have more than three authors. Divide the topics among those best qualified, and elect one or two of the group to write the general summary.

Style requirements apply also to the in-text references to the writer. I can use the first person singular when it is appropriate and not egotistical: "I was invited to join the *Nemo* expedition," but not, "I suddenly had a brilliant flash of insight." If you do not want to use the first person, you can write, "In the writer's opinion . . . ." Remember, when someone else is looking at this paper, that is what that reader will say in discussing the question. For multiple authorships by all means say "we think," "we observed," but never use the editorial or royal *we*.

Abstract

After the title, author's name, and affiliation, usually comes an abstract. Almost universal in journals, they should be encouraged also in professional books. The purpose is to help the busy reader. It is just another example of good manners, which is really the secret of style.

Many instruction sheets have been written about how to prepare an abstract. It should be informative and contain in capsule form the essential data in the article. It should be short: about 250 words. It should not be narrational; e.g., "The writer describes his laboratory investigations into certain minerals discovered on his 1910 field season. He will outline some alternative hypotheses of their genesis." The reader and the indexer want to know which minerals, what ages, what processes, what place? An alternative, factual version could be "Calcite and aragonite minerals from the Late Cretaceous of Wyoming, studied by electron probe, suggest epidiagenetic emplacement during the early Holocene or last Interglacial." The number of words is identical.

Many international journals require abstracts in one or two other languages. Try to get a neighbor or colleague to help out. Write a letter to one of the French or German scientists cited in the references. He or she would be interested in your work and probably glad to help.

Citations

Citations are in-text references to a collection at the end of the article, chapter, or book. The approved method is to name author and date (in a large book, the specific page number is optional but essential in the case of verbatim quotation). Examples are: "Evans and Evans (1957) object to the use of *presently* to mean *now*. Others agree that this usage is obsolete (Bernstein, 1967)." Note that no period should appear before or inside the parenthesis; this is possibly the most universal mistake in punctuation. A large number of names, such as in a historical review, can be gathered together in a catchall citation: "The historical foundations of this study were laid by Steno, Hall, Darwin, and Lyell (see references in Green, 1982)."

Certain journals, notably in physics and chemistry (and including *Nature* and *Science*) utilize the numerical reference (1), which saves them space but wastes the readers' time because they have to flip the pages constantly to and fro to find out who did what. Further, those references are listed in sequence of citation, so that if you are looking, say, for that reference to Jones (1921), you may have to search through 137 numerical references because they are not alphabetical. For the serious researcher the numerical reference is poor style.

The same taboo is true for footnotes. If you are looking for a given reference, you may have to search through a whole book, page by page. Furthermore, footnotes raise the cost of printing about ten times and are likely to get misplaced. Notes or commentaries that are interruptive or not germane to the general theme can be handled in two economical ways. A short comment can be introduced in a separate paragraph by "Incidentally, with respect to . . . ." If a large number of notes are needed, they can be indicated by numeral citations and gathered together before the references. The careful writer should check what is the journal's or publisher's policy in such matters.

Unpublished reports and personal communications should both be treated with great care. Careless use of them can make lifelong enemies. Never cite these bits of information without writing to the persons concerned for their approval. Quite likely it will turn out that the material is in fact available in some other place and formally published. You can find addresses in all the major geological societies—American, British, French, German, Australian. If you are not a member, their secretaries can sell you a copy.

References

References are ideally arranged alphabetically at the end of the text. The arrangement of references is covered by an International Standards recommendation (with the aid of UNESCO), but few journal or book editors have heard of it and each follows his or her company's preferred style. You should check with your publisher about what style to use.

Numerical citations are normally listed in the sequence in which they appear, with the attendant difficulties outlined in the preceding section. Another way is simply to arrange the references in alphabetical order and number them consecutively for citation in the text. When preparing a text, it is best to use the author and date system.

How does one keep track of references? In these days of library and home computer services, it is convenient to call in your subject item, range of dates, or even author names and get a quick printout. The books or papers still need to be perused, selected, and relisted. This can be done also on your own equipment, but it is often convenient to document the information on index cards that you can carry with you and that require no expensive hardware. Then the selected cards can be very simply alphabetized and typed into the word processor.

Some journals and publishers are happy with an extremely brief form of referencing. In the earth sciences where descriptive methodology is, or has been, the standard technique for more than a century, the full title of the article is highly desirable. Its use can save the researcher hours of wasted time that might be spent in the library hunting up false leads and dead ends.

The high-speed copier (xerography) has revolutionized much referencing. Instead of cards, one can simply copy the front page (with reference and abstract), with as much of the article or its illustrations as seems immediately useful. In this way one can collect large amounts of reference material that (properly filed) can be returned to over the years to jog the memory.

Illustrations and Tables

This is not the place for a detailed discussion of illustrations and tables, important supplements to the written material. If you are not good at drafting, the solution is easy: take a course, and practice. It requires lots of hard work, but once mastered, the drawing of figures is enormously pleasing. With computers they can be developed and tested in many ways, often leading to unexpected new discoveries.

Indexing

Authors often think that indexing is simply a chore. If you do your index, here are some important tips. Compare key words with definitions in the American Geological Institute (AGI) *Glossary.* Often they may have been used in ambiguous or even erroneous ways. Sometimes a key synonym has been completely omitted.

Some journals require that you list key words to assist their computer indexing. For books I like to see a complete alphabetic author index and separate subject index. They do not need to occupy much space because they can be printed in very small type, but they save hours of work for the researcher. To every careful writer and indexer, I personally say "thank you."

RHODES W. FAIRBRIDGE

References

Bernstein, T. M., 1967, *The Careful Writer: A Modern Guide to English Usage.* New York: Atheneum, 487p.

Bishop, E. E., E. B. Eckel, et al., 1978, *Suggestions to Authors of Reports of the United States Geological Survey,* 6th ed. Washington, D.C.: U.S. Govt. Printing Office, 273p.

Cochran, W., P. Fenner, and M. Hill, eds., 1979, *Geowriting: A Guide to Writing, Editing and Printing in Earth Science.* Falls Church, Va.: American Geological Institute, 80p.

Evans, B., and C. Evans, 1957, *A Dictionary of Contemporary American Usage.* New York: Random House, 567p.

Fowler, H. W., 1965, *A Dictionary of Modern English Usage,* 2nd ed. London: Oxford University Press, 725p.

Johnson, A. M., 1977, *Styles of Folding: Mechanics and Mechanisms of Folding Natural Materials.* Amsterdam and New York: Elsevier, 406p.

Norris, R. M., 1983, Field geology and the written word. *Jour. Geol. Education* **31,** 184.

O'Connor, M., 1975, *Writing Scientific Papers in English.* Amsterdam: North Holland, 108p.

Skillin, M. E., R. M. Gay, and other authorities, 1974, *Words into Type,* 3rd ed. Englewood Cliffs, N.J.: Prentice-Hall, 583p.

Souther, J. W., 1957, *Technical Report Writing.* New York: Wiley, 70p.

Trelease, S. F., 1951, *The Scientific Paper: How to Prepare It, How to Write It,* 2nd ed. Baltimore, Md.: Williams & Wilkins, 163p.

Williams, G. E., 1981, *Megacycles: Episodicity in Earth and Planetary History.* Stroudsburg, Pa.: Hutchinson Ross Publishing, 434p.

Cross-references: *Field Notes, Notebooks; Geological Cataloguing; Geologic Reports.* Vol. XIII: *Geological Communication.*

GEOLOGIC REPORTS

Nature and Purpose

Scientific and technical information is transmitted among professional geologists in the form of published articles appearing in the formal literature of the science (see Vol. XIII: *Geological Communication*) and as unpublished geologic reports. Both the literature and reports contain written compilations, geologic maps, geologic sections, and charts, tables, photographs, and drawings. The formal literature is concerned with documentating various scientific aspects of the geology of an area or region. In contrast, geologic reports are prepared to provide specific information needed to plan and conduct an organized activity, usually engineered construction or mineral resource development.

Geologic reports are compiled by individual geological consultants or by consulting firms. The report is compiled at the request of a client who requires specific geologic information to be able to plan or design a development. Client and consultant meet to discuss the development plans and to agree on a scope of work that defines the level of effort, scheduling, and cost of the geologic report. The report will be made up of findings taken from the literature and a variety of original field geologic mapping and explorations, all designed to provide the specific information and recommendations needed by the client.

Projects require geologic information at a number of stages or phases of concept development. In the beginning, there will be a site selection, followed by a feasibility study, a preliminary design, a final design, formulation of construction contract documents, and occasionally, assistance in construction monitoring or operational aspects of the project.

Key elements of the report describe the geologic controls over the components and functionality of the project. In the beginning, the report defines the geologic setting, made up of engineering geologic units (similar to rock-stratigraphic units, but primarily recognizing soil and rock units of similar physical characteristics and engineering properties). Each of the units is characterized in terms of its impact on the project and is then characterized by its areal extent, depth of burial, and thickness. Groundwater must also be characterized by vadose and saturated zones, piezometric surfaces, and seasonal fluctuations.

Illustrations are used in the report to clarify the writer's message and to insure that the writer's interpretation is integrated into the project plans. Prime among the illustrations is the geologic map; an engineering geologic, hydrogeologic, or economic geologic map, in most cases. Sometimes the maps are combined, such as when hydrogeologic conditions are shown on an engineering geologic map. Mineral resource development reports normally include engineering geologic and hydrogeologic maps as part of the resource development geologic report (see *Maps, Environmental Geology; Maps, Physical Properties*).

Geologic sections and geotechnical profiles are used in the report to portray concepts of subsurface interpretations (see *Cross-Sections*). Sections represent smaller-scale and deeper interpretations of subsurface geology. Horizontal scales smaller than about 1:5,000 are common. Profiles are used to portray those geologic elements and geotechnical characteristics [such as standard penetration test blow counts (*N* numbers), Rock Quality Designation (RQD) numbers, and core recovery percentages] that are critical to the placement of excavations, cuts, foundations, or underground openings along the line of profile or in its immediate vicinity. Scales larger than 1:5,000 are common; usually in the 1:500 to 1:2,500 scale range (see *Map Series and Scales*).

An important facet of the report is the characterization and presentation of the physical characteristics and engineering properties of the engineering geologic units. These measurements are tabulated and presented as averages and ranges and are often supported by charts relating variable properties, as determined through laboratory or field testing.

The report should also identify, cite, and summarize all important references in the geologic literature of the site and its site area. The *site area* usually comprises an 8 km radius of the actual construction or resource development site. The value of a good geologic report will be to provide all the necessary geologic data and reasonable interpretations and conclusions that are needed in the planning, design, and construction of the project.

Planning the Report

Good planning leads to the development of a program of field investigations and report content that fills the client's data needs. Planning begins with discussions between the consultant and the client, in which the client, the client's engineer, and the geologic consultant discuss the objectives of the proposed project. Having posed necessary questions, the consultant formulates a proposal for conduct of the geologic investigation. The scope of work includes an estimate of the professional labor and funding required to complete the fieldwork, office evaluations, and report writing. Once approved, the proposal becomes the basis for payment of the consultant.

To complete the investigation efficiently, the consultant often receives assurances of *deliverables* by the client. Typical items are site access, survey control over explorations, topographic coverage at a construction-related scale (usually larger than 1:2,500), and some forms of specialized laboratory testing.

After formulating the scope, planning continues through review of the references in all available geologic literature relating to the site area. The investigation is conducted according to a definite plan, however, all incoming observations are examined and evaluated as soon as possible. Limited changes to the investigation plan are made from time to time to take advantage of new findings or conditions not previously known.

Organizing the Report

Geologic reports are tailored to the client's information needs. However, the organization of reports usually follows a well-defined pattern, in which the purpose of the report is made clear, the manner of the investigation is explained, and the data findings are developed and explained. The report concludes with a series of recommendations that pertain to the use of the geologic data in the client's project. The following elements are usually used in the reports:

Letter of transmittal, in which the consultant identifies the client's agent and terms under which the report is tendered;
Executive summary, in which the salient discoveries and recommendations of the investigation are made known to the client;
Contents, in which the main elements of the report are listed in the sequence in which they are presented;
Introduction/background, in which the conditions of conduct of the investigation are disclosed;
Project objectives, in which the consultant's understanding of the client's objectives is reiterated, as the basis of the investigation;
Field investigation, in which the methods of geologic mapping and field explorations and testing are described and referenced to recognized standards;
Field testing and sampling, in which the methods and standards for conduct of field and laboratory testing are described and referenced to recognized standards;
Evaluation process, in which the methods of assessment of geologic data are identified;
Conclusions and recommendations, in which the client's need for geologic advice is satisfied by specific findings and interpretations;
Appendixes, containing support data that are not efficiently translated into words in the main body of the report.

The geotechnical report is a more specialized product, compiled in collaboration between an engineering geologist and a geotechnical engineer. The elements of a typical geotechnical report, written for a client who is planning a conventional high-rise building in an urban area, are given in Table 1.

In cases in which the project encounters complications, the consultant is sometimes engaged to monitor construction activities. A report is compiled in which the geologic conditions discovered in the course of construction or mineral recovery are recorded, along with their effect on the activity. Recommendations and suggestions are provided to the client during the particular construction activity. The construction-phase report also serves as a basis for negotiation of additional payments to the contractor, should unexpected geologic conditions, in terms of the contract documents provided at the time of competitive bidding for construction appear. An as-constructed geologic report also proves invaluable in developing remedial actions in the event the project fails to function according to its design objectives.

Data Types

Data contained in geologic reports are of two basic types: factual and interpretive. Most observations are factual, as in the cases of field and laboratory test data. Geologic maps, sections, or profiles are mainly interpretations. The map, i.e., portrays a two-dimensional continuum in which the land surface in the area of interest is blanketed by geologic map units delimited by geologic contacts. It is unusual for the geologist-author to know the absolute location of all of the contacts. The map is therefore an attempt, by interpretation, to estimate the planar positioning of all the engineering geologic units.

TABLE 1. Elements of Geotechnical Reports

Letter of Transmittal
  Submittal phrase
  Title and location of project
  Manner in which the geotechnical investigation was planned and scoped; with whom and when
  Summary of site geological/geotechnical conditions
  Generalized site-grading recommendations
  Generalized foundation recommendations
  Signature of responsible member of management, plus project geologist or geological/geotechnical engineer

Executive Summary

Contents

Body of Report
  Introduction
    Background of project
    Objectives of project
    Objectives of investigation
    Location of structural components and explorations (reference a site plan and index map)
    Deliverables from the consultant
    Items to be provided by the client and/or owner
    Participation by named subcontractor consultants
    Reference to items contained in appendixes of report

  Structural Considerations
    Reference to owner site layout as a figure
    Elements of owner layout plan; separate structures, heights, and structural loads
    Column/wall loads
    Levels of foundation bearing loads
    Column spacing
    Depths/levels of site excavations
    Nature of cuts/fills on site

  Site Conditions
    Existing structures/facilities
    Adjacent structures/facilities
    Suspected conditions: those not encountered in exploration but reasonably expected

  Field Investigation
    Methods
    Equipment
    Quantities

  Site Geologic Conditions
    Classification of engineering geologic units: type, depth, thickness, areal extent, implications, impact on exploration effort, etc.
    Top-of-rock; method of identification
    Geologic structure/discontinuities

  Field Testing and Sampling (brief, see appendixes)
    Methods
    Equipment
    Quantities

  Groundwater Conditions
    Nature and occurrence of groundwater: vadose and saturated zones
    Expected depth to piezometric surface at time of exploration
    Expected seasonal variations
    Perched groundwater/multiple confined aquifers
    As-encountered conditions in site explorations

  Laboratory Testing (brief, see appendixes)
    Methods
    Equipment
    Quantities

(*continued*)

TABLE 1. (*continued*)

Seismic Risk Assessment (or other geologic constraint)
  Historic nature of risk
  Parameters defining the risk
  Quantification of the risk for designer

Recommendations (mainly geotechnical in nature)
  Bearing capacity suggested for foundation load-carrying elements for each affected engineering geologic unit
  Methods of identifying recommended foundation conditions, as observed during foundation excavation
  Optional foundation recommendations and depths of load-bearing strata
  Allowable bearing values for each foundation option
  Anticipated settlement (maximum, ultimate); variances at locations on site
  Approximate time of settlement; necessity for staged construction; estimates of differential settlement between structural elements
  Lateral loads on retention wall systems: active/passive pressures
  Structural and nonstructural backfill requirements
  Excavation parameters and anticipated difficulties
  Seepage: expected presence and methods of accommodation; ranges of flow anticipated
  Need for shoring or underpinning of adjacent structures
  Wall and floor subdrain provisions
  Bedding support for floor slabs
  Mitigation of geologic constraints

Construction Monitoring
  Necessity
  Types recommended

Conclusions
  Availability of consultant for continued involvement
  Need for consultant to review final plans prior to construction or to review changes in plans as known at time of geologic investigation

Figures
  Index map
  Site layout plan (plot plan)
  Others, as indicated

Tables

Appendixes
  Site exploration
  Laboratory testing
  Field testing

In compiling bid documents for major construction contracts, many engineering firms are now separating factual from interpretive data. This separation sometimes takes the form of two different reports, or may be accomplished by chapter separation. Separation further assists the client in assessing the higher reliability/lower risk factual data from the somewhat less reliable/higher risk (in terms of unknown factors) interpretations.

As the geologist attempts to satisfy the client's need for subsurface data, the degree of interpretation becomes even more pronounced. Sections and profiles are estimates compiled from surface outcrop mapping, exploratory trenches and test pits, boreholes, and geophysical surveys. None of the exploratory techniques provides a clear picture of the absolute geologic nature of the section; its geologic picture is clearly an interpretation based on scientific knowledge and professional experience.

Additional examples of interpretive data are determinations of the extent of control that the various geologic units may have on the construction process. Groundwater conditions, in particular, are presented as interpretations, as is the upper surface of hard rock, known as top-of-rock, excavation below which represents an additional effort and expense in reaching the desired construction grade. Although the ranges and averages of observed laboratory tests on rock properties are factual, recommendations making use of such data to the extent that the properties will affect or control construction represent interpretations. Disclosures of geologic problems, also known as *geologic hazards* or *geologic constraints,* are interpretations made to assist the client in developing appropriate design and construction plans.

When separated, factual and interpretive data give report users a clear indication of the degree of assuredness felt by the geologist-author. The separation is most important in geologic reports that will become part of contract documents used in competitive construction bidding.

Legal Considerations

Professional geologists, in the act of compiling reports, are subject to a variety of legal constraints (Patton, 1981). The main source of liability comes from statements made in geologic reports. Should the project encounter problems, the owner or contractor may choose to claim *negligence* on the part of the geologist. The courts recognize that the work of any professional should be judged on the basis of exercise of *ordinary care* in the course of performing duties associated with the contractual charge from client to consultant (see *Legal Affairs*).

The geologic report becomes the most important form of evidence as to the worth and accuracy of the geologist's work. The report must contain honest representations of factual data and be without attempts to conceal or misrepresent conditions that exist and that may be discovered by others, either during construction or in later investigations. Good geologic training and a thorough appreciation of the ethics of professional practice are usually sufficient to guide the professional geologist to proper conduct of the investigation. The greatest exposure to liability, however, comes in the wording and context of the geologic report.

Inappropriate wording becomes the snare of liability. At best, the use of words that imply unwarranted promises and lead to misinterpretation by the client or contractor may result in termination of work or loss of future assignments to the geologist (see *Geological Writing*). Most geologists have been trained to provide the fullest extent of service possible. This desire to please can often lead to two types of inappropriate wording: (1) unwarranted promises of work to be delivered in the course of compiling the geologic report (work that cannot be accomplished within the time or funding to be provided by the client) and (2) inaccurate descriptions of conditions observed or anticipated at the site.

In general, inappropriate words are of two types: (1) common words that connote a higher degree of certainty than is possible in geologic work and (2) words that are superlative or absolute in their commonly understood meaning.

Wording that connotes a notion of certainty is represented by terms such as *approve* and *inspection.* Use of *approve* often imparts a feeling that the geologist accepts the condition as being wholly appropriate to the client's need. A better term would be *review,* connoting that the geologist considered the condition or observation. *Inspection* may imply that the geologist has carried out an absolutely detailed examination of a condition or process, say, the logging of rock core during exploratory boring carried out by a contract drilling firm. In reality, the geologist has *observed* the process of the drilling firm and does not warrant the safety or accuracy of the process but merely provides a record of some factual aspects of the exploration (measurements and descriptions of core recovered). A representative listing of words to be avoided in report writing, together with preferable alternatives, has been prepared by the Association of Soil and Foundation Engineers (ASFE, 1980).

Summary

Geologic reports represent the lasting effort of geological work on particular projects. The reports should carefully outline scope of work and should meet the client's expressed need for geologic information in the siting, planning, design, and construction of engineering or mineral development projects. The report will be consulted in the event of problems in construction and should not contain wording, conclusions, or recommendations that are unwarranted or outside of the scope of work for which the geologist has contracted with the client. The wording should be direct and straightforward and evidence a clear understanding of the client's objectives in construction. The report should be carefully reviewed for wording and accuracy of statements before release to the client. A statement should be made in concluding the report that any changes in the client's concept of development or in design for the project should be reviewed, by the geologist–author, for possible impacts on the recommendations provided in the present report.

ALLEN W. HATHEWAY

References

Association of Soil and Foundation Engineers, 1980, *A Guide to Establishing Quality Control Policies and Procedures in Geotechnical Engineering Practice.* Silver Spring, Md.

Brown, G. A., and R. J. Proctor, 1981, *Professional Practice Guidelines.* Short Hills, N.J.: Association of Engineering Geologists.

Cochran, W., P. Fenner, and M. Hill, 1979, *Geowriting: A Guide to Writing, Editing and Printing in Earth Science.* Falls Church, Va.: American Geological Institute, 80p.

Hatheway, A. W., 1981, Contracts, proposals and negotiations, in G. A. Brown and R. J. Proctor, eds. *Professional Practice Guidelines.* Short Hills, N.J.: Association of Engineering Geologists, 3.1–3.45.

Henderson, G. V., 1981, Report writing, in G. A. Brown and R. J. Proctor, eds., *Professional Practice Guidelines.* Short Hills, N.J.: Association of Engineering Geologists, 6.1–6.14.

Kiersch, G. A., and A. B. Cleaves, eds., 1969, *Legal Aspects of Geology in Engineering Practice,* vol. 7, Case Histories in Engineering Geology. Boulder, Colo.: Geological Society of America, 112p.

Patton, J. H., Jr., 1981, Professional liability, in G. A. Brown and R. J. Proctor, eds., *Professional Practice Guidelines.* Short Hills, N.J.: Association of Engineering Geologists, 2.1–2.9.

Rose, D., 1965, A civil engineer reads a geology report, *Geotimes* **10,** 9–12.

Cross-references: *Field Notes, Notebooks; Geological Writing.* Vol. XIII: *Geological Communication; Geological Information, Marketing.*

GEOLOGIC SYSTEMS, ENERGY FACTORS

Geology is concerned with understanding the origin and operation of individual Earth events, such as the eruption of a volcano, erosion of a valley, building of a coral reef, and uplift of a mountain range. Certain of these events or processes taken singly may be elucidated most satisfactorily by direct application of physics, chemistry, or biology to the Earth. The central core of geology, however, comprises more than the single events; it includes also the interpretation of the sequence and interrelationship of Earth events since the time of origin of the Earth or, perhaps better, our galaxy (see *Geology, Scope and Classification*). In mathematical symbolism, the summation and integration of Earth events may be expressed as follows:

$$\text{Earth history} = \int_{\text{birth of galaxy}}^{\text{present}} E_s dt$$

where E_s refers to geologic events arranged in the succession in which they occurred and dt is the conventional expression from integral calculus representing a function of time.

The full significance of the unique sequence of Earth events, as they actually occurred, may be illustrated by the following analogy. As a house, built of 10,000 bricks laid with mortar in a particular ordered succession, is functionally much more than the sum of 10,000 individual bricks plus mortar, so is the actual sequence of events more than the events taken individually. In terms of communication theory, the ordering of geologic events represents a lower state of *entropy* than if they were chaotically jumbled.

Energies Driving Geologic Processes

Curiously, geologic (Earth) history is fundamentally not Earth-centered (except for the materials of the Earth) as might be expected, but instead is basically energy-centered. Geologic events, such as manifestations of the dominant processes of geology (volcanism, gradation, diastrophism, and metamorphism) and biospheric activity, are found intrinsically to be responses of Earth materials to energy, which is nongeologically controlled, imposed on them. Table 1 classifies Earth-driving energies and geologic effects produced by them.

Table 1 is essentially self-explanatory. Solar energy is obviously external to the Earth, but rotational, orbital, and gravitational energies, although classified as external, are partly internal as well.

Energy-Centered View of the Earth

To view the Earth from an energy-centered perspective, we detach ourselves from our planet and look at it from a remote distance. It then appears as a not very imposing little ball spinning about the sun at a distance of 150 million km continuously and inescapably exposed to the sun's radia-

TABLE 1. Energies and the Geologic Effects Produced by Them

External energy
  Solar (nuclear)
    Gradational processes of geology
      Aqueous, aeolian, glacial
    Living plants and animals
    Combustible fuel deposits and water power
    Wind, stomrs, and weather
  Rotational and orbital
    Coriolis effect
    Tides
  Gravitational (jointly external and internal)
    Gradational processes
    Mineral phase changes
Internal energy
  Thermal (primitive, nuclear, frictional, chemical reaction)
    Volcanism
    Diastrophism
    Anamorphism
    Continental migration—plate tectonics
    Polar wandering(?)
  Chemical—mineral
    Metamorphism
    Nuclear power
  Magnetic
  Electrical

tion. The wide space beyond the Earth's atmosphere is a very cold temperature sink, and since the Earth's level of radiated and absorbed energy is far above that of its surroundings, the Earth becomes a type of heat engine that must respond inexorably to every calorie of energy to which it is exposed. From this viewpoint, the surface of the Earth is indeed no master of what it "wants" to do (apology for the unscientific phraseology).

The area of the Earth's disc facing the sun is approximately 12×10^7 km^2, whereas a sphere (ellipse) outlined by the Earth's orbit at 150 million km has an area of about 2.5×10^{17} km^2. Thus, the Earth intercepts less than one-billionth of the radiated energy (cosmic, alpha, beta, gamma, magnetic, infrared, ultraviolet, visual, etc. spectrum) of the sun. This intercepted energy averages about 2 cal cm^{-2} min^{-1}, or about 7.5×10^{15} kwh/hr. About half the light energy is absorbed (the Earth's albedo is 0.44); of this, about 15% is absorbed by the bare rock Earth, and about 85% is used in evaporative and transpirative processes. Despite the insignificant part of the sun's energy that is converted to geologic work, almost all the work of the physical agents and the gradational processes (Keller, 1954) owe their activity to this source of energy. These are elaborated in some detail in Table 2.

The *biosphere* in relation to energy deserves special comment. Energy (probably solar), plus a favorable ratio of the elements C, H, O, N, P, K, S, and Mg (in chlorophyll) and others, especially the minor nutrient or trace elements (q.v. in Vol. XII, Pt. 1), gave origin to living, primitive plants. These new products of energy—i.e., living plants—opened the way for an additional type of energy capture by the product—namely, photosynthesis, which stores energy in an unstable form on the Earth. Living organisms are capable of propagating and expanding their kind, thus compounding and increasing the storage of energy and "negative entropy" (Schrodinger, 1945).

The absorption or capture of solar energy by organisms is astoundingly inefficient, however. Calculations and estimates indicate that of 1,000 light quanta impinging on plants, only about one is trapped and its energy stored. Herbivores consuming plants capture about 10% of the energy in the plants, and carnivores convert about 10% of the energy captured by the herbivores. Thus, if humans extract 10% of the energy in beefsteak, they recovers only 1 light quantum of 100,000 that fell on the field of corn or alfalfa. Although there is no indication that humans can increase this efficiency, they can take steps to increase food production by using energy to modify climate, another manifestion of energy. Proposals to this end include the blasting of rock barriers to divert ocean currents and thereby change the temperature and rainfall of certain parts of the globe, desalinizing of water to convert arid wastes to humid landscapes and thus change the pattern of atmospheric moisture, and modifying the distribution of rainfall by cloud seeding.

Apart from these relative details, we may speculate further on broader biologic issues, such as how energy may drive the course of organic evolution (Blum, 1951) in the future. Will *Homo sapiens* continue to flourish, or will their pool of energy be depleted by normal evolutionary overspecialization; unfavorable mutation, either normal or induced by artificially concentrated radioactivity; self-destruction by automated machines that become uncontrollable (mechanical energy); self-destruction by artificially released radiation (nuclear energy); or other means? If humans destroy themselves, will a new type of organism arise by the combination of energy—say, not with tetrahedrally coordinated carbon, the core element now in body tissue, but with silicon or boron? However catastrophic this destruction may be to self-centered *Homo sapiens,* it can be viewed as merely one phase of energy transfer across the Earth during the passage of time.

TABLE 2. Geologic and Major Cultural Products of Solar Energy (against Cold External Space)

Physical agents and processes
  Wind
    Hurricanes
    Dust storms
    Loess deposits
    Sand dunes
    Volcanic dust (ash) deposits
    Natural sand blasting
    Water waves
      Shoreline erosion and deposition
    "Weather," cyclonic and jetstream storms
    Windmill power
  Rain, snow, sleet, hail
    Stream and sheet erosion
      Slopes, valleys, canyons
    Hydroelectric power and water power
    Groundwater solution and precipitation
    Glacial erosion and deposition
  Thermally induced volume changes
Biosphere
  Living organisms and their mechanical and chemical activity
  Petrologic destruction and construction
  Degradation and aggradation
  Negative entropy of life
    Propagation of the species
    Mutants
    Self-destruction
Fossil fuel

Rotational and Gravitational Energies

Without the Earth's *rotational energy,* the source of which is obscure, geology would not ex-

283

TABLE 3. Geoglogic Products of Rotational
and Gravitational Energies

Coriolis movement
  Air movements
  Oceanic movements
  Stream meanders (?)
Tides
  Erosion and deposition along shores
  Electric power (Brittany; Passamaquoddy?)
  Braking effect on earth's rotation
Downslope movement (lower potential energy)
  Mass wastage
  Erosion and deposition of rock detritus
Retention of atmosphere and hydrosphere
Compression of rocks and minerals
  Changes to heavy-type phases
    Gabbro to eclogite
    Quartz to coesite
    Carbon to diamond

ist as it now does. *Gravity,* whose origin and possible change are still a challenge to physicists (Gamow, 1961), nevertheless is one of the most enduring and permanent forces operating on geologic work. Table 3 summarizes some of the effects of rotation and gravity.

Although opinion differs as to the efficacy of the *Coriolis action* in development of meander in streams, there is no question about its very important role in the movements of the air and oceans. By the action of gravity, the potential energy of Earth materials due to location is transferred in part to kinetic energy, and a vast amount of gradational geologic work is done. Intense forces of gravity shrink the Earth's interior and produce heavy phases in minerals.

Internal Thermal Energy

At the beginning of the 1900s the thermal energy of the subcrustal Earth was attributed almost solely to primitive heat, which had been declining for about 20 m.y. to 40 m.y., according to Lord Kelvin's calculations (Burchfield, 1975). The future outlook was for weakening internal vigor and consequent relative passiveness. Geology was apparently confronted with the problem of how enough heat could have been conserved against loss by conduction to have energized the geologic activity found recorded in the rocks.

In 1906, however, Lord Rayleigh discovered that nuclear disintegration of certain isotopes of U, Th, and K evolved heat, and thereupon the problem of geologic heat was reversed (Burchfield, 1975). Because heat is evolved in the Earth's interior in excess of that lost by conduction, the problem became to find how the Earth, in order to remain near equilibrium, can dissipate thermally or geologically this excess heat. It does so via volcanism and diastrophism. The Earth in ef-

fect thus becomes unstably hot as a nuclear energy internal combustion heat engine.

The nuclear fuel for this engine occurs most abundantly in granite, from which about 2.2×10^{-13} cal g^{-1} sec^{-1} are evolved, but it also occurs in basalt, which evolves heat at about 1/7 the rate of granite, and in dunites and stony meteorites at about 1/100 this rate (Verhoogen, 1960). Thus, the mean heat loss by conduction to the Earth's surface is about 1.2×10^{-6} cal cm^{-2} sec^{-1}, from which it follows that a column of granite 20 km thick, or basalt 140 km thick, would liberate nuclear heat equivalent to its loss by conduction. The granitic crust above the Mohorovičić discontinuity (Moho) ranges in thickness from about 30 km to 40 km on continents, whereas in oceanic areas 5–8 km of rock overlie the Moho; the peridotitic (?) mantle extends downward about 2,900 km. Thus, apparently fuel-containing rock occurs clearly in excess of the amount capable of furnishing heat equivalent to loss by conduction, and this excess is expended eventually in volcanism and diastrophic uplift.

If excess heat were to be retained and temperature built up indefinitely, a widespread subcrustal liquid zone might develop, but geologic experience indicates that before such a condition is widely generated, thinning and weakening of the crust tends to allow a leaking out of the excess thermal energy as a volcanic eruption, fissure flow, or intrusion. A volcano is in effect, therefore, a safety valve and quick cooler for the overheated internal engine of the Earth, as listed in Table 4. For example, Mt. St. Helens, in a timely reminder, confirmed that the Earth continued to be remarkably energetic in 1980. One of the relatively small tectonic plates (Juan de Fuca) has been moving eastward under the Oregon–Washington region, generating earthquakes and transferring thermal energy upward with astounding

TABLE 4. Geologic Effects of Internal Thermal Energy

Primitive and nuclear
  Volcanism
    Volcanoes and flows (safety valves)
    Intrusions (slow coolers)
    Magmatic ores
  Diastrophism
    Expansion
    Convectional overturn (?)
  Anamorphism
    Sillimanite, garnet, amphibole, etc.
  Continental drift, plate tectonics
  Polar wandering (?)
Chemical-mineral energy
  Metamorphism
  Nuclear power
Magnetic energy

force when overheated. Any region on the Earth's surface where tectonic plates are in discordance cannot escape earthquakes, earth movements, and possibly eruptive activity. By the same token, an intrusion, either magmatic or migmatitic (Walton, 1960), is a slow cooler.

Verhoogen (1960) estimated that about 3×10^{15} g of volcanic material has been erupted annually in recent times and that the annual heat loss from it, plus that of hot springs, natural steam, and fumaroles, totals about 10^{18}–10^{19} calories. This is perhaps 1/20–1/50 the heat lost by conduction, but the total heat dissipated at the Earth's surface is estimated to be 1/10–1/100 of the radiogenic heat evolved.

Additional energy expended by *diastrophic uplift* may be estimated by assuming it is equivalent to that of degradation and by calculating the loss in potential energy of the material carried to the sea from the mean elevation of the continents. This amount is estimated by Verhoogen (1960) to be about 10^{17}–10^{18} cal/yr. The source of this energy is manifestly internal.

Compelling evidence points toward a nuclear-active source for much internally evolved heat. Cautious observers note, however, that liquid lava, presumably liquefied by nuclear heat effects, is not significantly richer in nuclides than solid rock.

Cyclic heating and cooling of subcrustal zones

of the Earth may be accompanied by expansion and contraction. Arguments for a shrinking Earth have been well presented by Landes (1952). Conversely, evidence for a differentiating Earth and expanding crust has been presented by Menard (1960) and Heezen (1960) who describe a rising mid-ocean ridge and high heat flow beneath it.

Seafloor spreading and the movement of continents and large crustal plates (plate tectonics) demonstrate tremendous amounts of internal energy doing geologic work. For example, the subduction of oceanic crust beneath a continent, like in eastern Asia or western South America, or of one continental crust shoved beneath another, like India wedging beneath southern Asia and producing the colossal Himalaya Range, are illustrations of surprisingly immense global effects of internal energy. This newest, revolutionary branch of geology, *plate tectonics,* is in a first, descriptive stage; quantification of energies that drive the process is an inviting field of research to follow.

Geologic processes move in a direction that tends to yield equilibrium or, more commonly, a steady state. The processes may be reversible in direction or cyclic, depending on the nature of energy that is dominant. A score or more of cyclic (waveform, or closed-circuit) geologic reactions are easily tabulated, of which two of the larger are illustrated in Figs. 1 and 2—the rock cycle and the geomorphic cycle. These show that the Earth

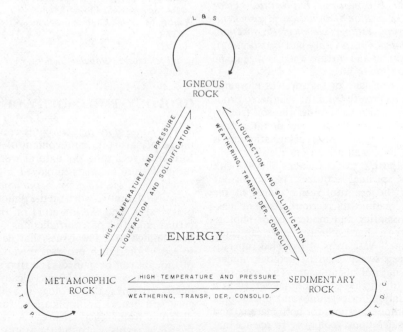

FIGURE 1. The rock cycle. Any rock type may be converted to any other rock type or be reconstituted within its own type by the action of appropriate energy.

FIGURE 2. The geomorphic cycle. Internal energy tends to unlevel the Earth's surface by processes of volcanism and diastrophism, whereas solar energy (and gravity) tends to level the Earth's surface by gradation.

is indeed a slave to energy and that geologic history is a record of the Earth's response to non-geologically controlled energy imposed on it.

Economic and Cultural Contributions of Geology

Earth materials provide the raw physical materials of our modern culture. The concepts of geologic discipline are essential parts of modern civilization.

Early cave dwellers lived in caves naturally eroded in limestone, sandstone, loess, and other rocks (see *Geoanthropology*). Today's *Homo sapiens* carve these same rocks into desired shapes and build their modern caves to suit their fancy. Metals are refined from geologic ores to fabricate teaspoons, battleships, electronic vacuum tubes, sometime gold-standard coins, and artificial satellites. Geologists find petroleum, coal, natural steam, and uranium minerals for fuel and other purposes (see *Atmogeochemical Prospecting, Geobotanical Prospecting, Exploration Geochemistry, Exploration Geophysics, Hydrochemical Prospecting, Lithogeochemical Prospecting, Pedogeochemical Prospecting,* and *Prospecting*). Pure groundwater and surface water is a geologic fluid indispensible to life.

All food, from land or sea, derives inorganic nutrients for its growth from the weathered products of rocks, which yield either the soil (solids) or dissolved ions in soil solutions or the ocean. Therefore, a geologic process, rock weathering, is a most basic one to all life on our planet.

The good Earth provides humans likewise with aesthetic and spiritual resources. We go on our vacations to the beautiful scenic places of the Earth for recreation (re-creation). Ancient masterpieces of painting and modern color photography capture the Earth's beauty so we can enjoy it in our homes. Vast areas of troubled superstition have been swept clean by geologic understanding of natural phenomena. The paleontologic record of organic evolution has served as a sound foundation for religious and philosophic thinking. Without question, for both material and spiritual considerations, geology is the science that is closest to the full human life.

W. D. KELLER

References

Blum, H. F., 1961, *Time's Arrow and Evolution.* Princeton, N.J.: Princeton University Press, 222p.

Burchfield, J. D., 1975, *Lord Kelvin and the Age of the Earth.* New York: Science History Publications, 260p.

Fowler, W. A., 1957, Formation of the elements, *Sci. Monthly* **85,** 84–100.

Gamow, G., 1961, Gravity, *Sci. Am.* **204**(3), 94–111.

Heezen, B. C., 1960, The rift in the ocean floor, *Sci. Am.* **203**(4), 98–115.

Keller, W. D., 1954, The energy factor in sedimentation, *Jour. Sed. Petrology* **24,** 62–68.

Landes, K. K., 1952, Our shrinking globe, *Geol. Soc. America Bull.* **63,** 226–239.

Menard, H. W., 1960, East Pacific Rise, *Science* **132,**(3441), 1737–1748.

Powell, C. McA., and P. J. Conaghan, 1973, Plate tectonics and the Himalayas, *Earth and Planetary Sci. Letters* **20,** 1–12.

Schrodinger, E., 1945, *What Is Life?* New York: Macmillan.

Sullivan, W., 1974, *Continents in Motion.* New York: McGraw-Hill, 399p.

Verhoogen, J., 1960, Temperatures within the earth, *Am. Scientist* **48,** 134–160.

Walton, M., 1960, Granite problems, *Science* **131**(3401), 635–645.

Cross-reference: *Popular Geology.*

GEOLOGY, PHILOSOPHY OF

The *philosophy of geology* is concerned with the scope, structure, and content of scientific information regarding the natural world and with the analysis of methods employed in gaining such information (see *Geology, Scope and Classification*). No definitive work on the philosophy of geology exists, and the literature on the subject is diffuse and full of contradictions. Attention is here limited to a few issues that need to be resolved before much progress toward unification of thought can be gained. The first such issue is raised by the proposition, uncritically repeated in many textbooks, that geology is a historical science and hence, by implication, fundamentally different from other sciences. A second set of questions revolves around the status of the principle of the uniformity of nature, long hailed as a basic geological tenet. Finally, there is the question regarding the contribution of geology to the mainstream of modern scientific and social thought. The first consideration relates to scope, the second to method, and the third to judgments of value.

Scope of the Geological Sciences

Geology is now subdivided into more than a hundred fields, and the ordering of these fields in a way that will disclose how the parts relate to the whole, and the whole in turn to other sciences, is in itself a problem for speculative philosophy. Bliss (1952) has suggested a possible solution in his classification of disciplines according to four categories: sciences, histories, arts or technologies, and philosophies. Within the natural sciences, information on particular subjects is gathered and organized according to logical systems of classification and ordering. The histories treat of the evolution of their subject matter, and the arts and technologies are concerned with the application of scientific and historical information to practical problems. The philosophies reflect on the methods and systems of the sciences, histories and arts.

In terms of Bliss's (1952) classification, geology is science, history, *and* art. *Physical geology* is a collective name for a set of sciences treating of the composition, structure, shape, and dynamics of the Earth. *Historical geology* is another name for the evolution of the Earth. *Economic geology* applies the principles of physical and historical geology to the search for ores and fuels and to other practical ends. And as the Earth is the subject of geology, so geology is the subject of the philosophy of geology, which is a critique of the methods and findings of geologists. How geology, thus subdivided according to mode of thought, is joined with other closely related sciences is suggested by Table 1.

Thus the proposition that geology is a historical science is misleading. Physical geology treats of Earth things—minerals, rocks, structures, landforms, etc.—as they are, and of Earth processes—weathering, erosion, etc.—as they may be observed to bring about changes in the present scene. Conversely, it would be fair to say that historical geology is a model historical science, the first such science to be established, and the aspect of geology that has most profoundly affected the modern view of the natural world.

It is a common observation that the present surface of the Earth is constantly changing in response to the work of streams and other agencies of erosion and deposition. "You can't step in the same stream twice" is an ancient maxim that reminds us that the *configurational* aspects of nature (to use Simpson's, 1963, term) are ephemeral and transitory. The immediate aim of historical geology is to reconstruct in temporal sequence the past configurations of the Earth from which the present one has emerged.

To reconstruct the past configurations of the Earth and its communities of plants and animals, historical geology works backward. The present is taken as the key to the past, in both the theoretical and the operational senses. First, it is assumed that the laws of physics and chemistry are not bound by time but rather are descriptive of the *imminent* properties of nature, again to use Simpson's (1963) terminology. Second, it is agreed that present models of change in the configurational aspects of nature shall be used to explain the solid geometry of rocks to the limits of their powers of resolution. In either sense, historical geology relies on a principle of simplicity, as elucidated by Goodman (1967).

Principle of the Uniformity of Nature

One of the most ambiguous of principles known to science, uniformity has been called a scientific law, a postulate, a dogma, a principle, or an operational prescription for applying the logical principle of simplicity to geologic problems. Uniformity has often been interpreted to mean that the configurations of nature succeed each other at a slow and uniform pace. This view is contrary to ordinary experience, which shows,

TABLE 1. Classification of Geological and Related Disciplines according to Bliss' System of Analysis

Philosophies	Sciences		Arts
Philosophy of physics	Physics	—	Physical engineering (e.g., mechanical, civil)
Philosophy of chemistry	Chemistry	—	Chemical engineering
Philosophy of astronomy	Astronomy	Cosmogeny	Applied astronomy (e.g., celestial navigation)
Philosophy of geology	Physical geology	Historical geology	Economic geology
Philosophy of biology	Biology (in sense of neontology)	Paleontology	Applied biology (e.g., medicine)
Philosophy of anthropology	Anthropology	Archeology	Applied anthropology

Source: Modified from Bliss, 1952.

e.g., that under many climatic regimes erosion is a series of minor catastrophic episodes. The principle of uniformity (see *Uniformitarianism*) is viable on either of two grounds. It may be regarded as an assumption to the effect that the laws of chemistry and physics are timeless, in the sense that they must have applied to changes in state during the Precambrian as well as the present. Or this principle may be interpreted as meaning that one must not multiply present dynamic models for reconstructing past geometric configurations in nature *without necessity*. In either case, uniformity may be regarded as a corollary of the logical-methodological principle of simplicity: it is vain to do with more (laws or models) whatever can be done with fewer. Note here that recent recognition of past configural changes due to the impact and explosion of extraterrestrial bodies was a necessary addition to dynamical models.

If the principle of uniformity is to be identified with the principle of simplicity, which operates in all disciplines and not only in geology, the question arises as to whether geology has any principles uniquely its own. The answer must be in the affirmative, unless or until it can be shown that the principle of faunal succession, e.g., is reducible to physical or chemical laws. While the final answer to this question is held in abeyance, yet another philosophical problem may be posed. What has geology contributed to that body of knowledge that is worth everybody's knowing?

Geological Ideas

The geological ideas that have been most effective in liberating social opinion from prejudice and superstition (see *Geomythology*) are related to the concepts of time and change in the natural world (see *Geological Methodology*). These ideas may be summarized in the following propositions. Compared with the span of human history, the Earth is almost incomprehensibly old. Successive episodes in the long history of the Earth are revealed in the spatial relationships of rocks and fossils. This history is a story of change in the configurational aspects of nature: biological and geographical. According to the simplest interpretation of the fossil record, all forms of life are kin. And, in the course of organic evolution, extinction of species has been the usual outcome. These propositions are neither obvious nor simple, and in their combined impact on the scientific thought and social opinion of the past two centuries, they rank with the Copernican revolution.

Writings

Nineteenth-century monographs on the philosophy of geology by Jobert (1847), Page (1863), and Vogelsang (1867) are now of more interest to the historian than to the philosopher. Whewell,

(1858, 1872), was a responsible critic of uniformitarianism, and he clearly comprehended the difference between the scientific and historical aspects of geology. In our century, Bubnoff's *Grundprobleme der Geologie* (1954) probes for the roots of geological thought in an unpretentious manner that lends charm to the erudition of this small volume.

With regard to logical methodology, American geologists are likely to refer to the works of G. K. Gilbert, T. C. Chamberlin, Douglas Johnson, J. Hoover Mackin, and George Gaylord Simpson. Gilbert (1896) pointed to analogies as the sources of scientific hypotheses; Chamberlin (1897) formalized Gilbert's approach in his method of multiple working hypotheses. Johnson (1933) disclosed the complicated interplay of induction and deduction involved in the analytic method, and Mackin (1963) drew sharp contrasts between analytical and empirical approaches to geological problems. Simpson (1963, 1970) has effectively analyzed the connections between uniformitarianism, in its various guises, and other "-isms," including gradualism, historicism, and evolutionism.

Few modern philosophers of science have taken geology as a principal source of models and examples. Hugh Miller, T. A. Goudge, S. E. Toulmin, David Kitts, and I. Bernard Cohen are exceptions. The first two authors (Miller, 1939; Goudge, 1961) have made a strong case for the cultivation of historical science as a meeting ground for the synthesis of historical and scientific principles. Toulmin (1962) argues that since geology was the first science to treat of the evolution of its subject matter, it was the first science to grow up. Kitts (1963, 1966, 1974) has explored the meaning of "geologic time" and has analyzed the workings of historical explanation in geology. Cohen (1985) has identified the emergence of the theory of continental drift and plate tectonics as a genuine scientific revolution of twentieth-century vintage.

CLAUDE C. ALBRITTON, JR.

References

Albritton, C. C., Jr., ed., 1975, *Philosophy of Geohistory: 1785–1970*. Stroudsburg, Pa.; Dowden, Hutchinson and Ross, 386p.
Bliss, H. E., 1952, *A Bibliographic Classification*, 2 vols. New York: H. W. Wilson.
Bubnoff, S. von, 1954, *Grundprobleme der Geologie*, 3rd ed. Berlin: Akademie-Verlag, 234p. (Translation: *Fundamentals of Geology*, Edinburgh: Oliver and Boyd, 1963, 287p.)
Chamberlin, T. C., 1897, The method of multiple working hypotheses, *Jour. Geology* **5**, 837–848.
Cohen, I. B., 1985, *Revolution in Science*. Cambridge, Mass. and London: Harvard University Press, 791p.
Gilbert, G. K., 1896, The origin of hypotheses, illus-

trated by the discussion of a topographic problem, *Science* **3**, 1–13.

Goodman, N., 1967, Uniformity and simplicity, in C. C. Albritton, Jr., ed., *Uniformity and Simplicity.* Geol. Soc. America, Special Paper 89, 93–99.

Goudge, T. A., 1961, *The Ascent of Life: A Philosophical Study of the Theory of Evolution.* Toronto: University of Toronto Press, 263p.

Gould, S. J., 1965, Is uniformitarianism necessary? *Am. Jour. Sci.* **263**, 223–228.

Jobert, A. C. G., 1847, *The Philosophy of Geology,* 2nd ed. London: Simpkin, Marshall and Co., 184p.

Johnson, D. W., 1933, The role of analysis in scientific investigation, *Geol. Soc. America Bull.* **44**(3), 461–494.

Kitts, D. B., 1963, Historical explanation in geology, *Jour. Geol.* **71**, 297–313.

Kitts, D. B., 1966, Geologic time, *Jour. Geology* **74**, 127–146.

Kitts, D. B., 1974, Physical theory and geological knowledge, *Jour. Geology* **82**, 1–23.

Mackin, J. H., 1963, Rational and empirical methods of investigations in geology, in C. C. Albritton, Jr., ed., *The Fabric of Geology.* Stanford, Calif.: Freeman, Cooper and Co., 135–163.

Miller, H., 1939, *History and Science: A Study of the Relation of Historical and Theoretical Knowledge.* Berkeley: University of California Press, 201p.

Page, D., 1863, *The Philosophy of Geology: A Brief Review of the Aim, Scope and Character of Geological Inquiry.* Edinburgh: William Blackwood and Sons, 160p.

Simpson, G. G., 1963, Historical science, in C. C. Albritton, Jr., ed., *The Fabric of Geology.* Stanford, Calif.: Freeman, Cooper and Co., 24–48.

Simpson, G. G., 1970, Uniformitarianism. An enquiry into principle theory, and method in geohistory and biohistory, in M. K. Hecht and W. C. Steere, eds., *Essays in Evolution and Genetics in Honor of Theodosius Dobzhansky.* New York: Appleton-Century-Crofts, 43–96.

Toulmin, S. E., 1962, Historical inference in science: Geology as a model for cosmology, *Monist* **47**(1), 142–158.

Vogelsang, H. P. J., 1867, *Philosophie der Geologie und mikroskopische Gesteinsstudien.* Bonn: Max Cohen, 229p.

Whewell, W., 1858, *History of Scientific Ideas, being the First Part of the Philosophy of the Inductive Sciences,* 3rd ed., vol. 2. London: John S. Parker, 324p.

Whewell, W., 1872, *History of the Inductive Sciences from the Earliest to the Present Time,* 3rd ed., 2 vols. New York: D. Appleton.

Cross-references: *Geohistory, American Founding Fathers; Geological Methodology; Geology, Scope and Classification; Serendipity.*

GEOLOGY, SCOPE AND CLASSIFICATION

Geology is the science of the Earth, the organized study of its dynamics, composition, and related systems. The subject is broadly concerned with the origin and operation of earth features and events, and the integrated sequence of events since the first record of them in the rocks.

In the Aristotelian classification of terrestrial materials, fire, earth, water, and air encompassed all energy and matter. *Gaea* (earth) referred to all natural solid matter, so *geology* has become logically the study of natural materials as well as of the Earth. Such a materials science is also appropriate to solid matter such as extratelluric dust, meteorites, and to the nature and composition of the moon and the planets; it is logical that geologists, trained in the analysis of hard earthy materials should now have a section, Astrogeology, within the U.S. Geological Survey (oddly enough, in the Department of the Interior).

The first use of the term *geologia* was by Richard de Bury (1473, according to Adams, 1938). The science of earthly things, it was specifically the antithesis of *theology,* the study of spiritual things. At that time it meant the study of law, which was the only nontheological discipline then recognized, but "geologia" and its students happily have since become channeled into the pure science of the Earth. According to Adams, *geology* in its modern sense was well established by the end of the seventeenth century.

Individual subdivisions of geology treat of many subdisciplines that are covered by other volumes in this Encyclopedia of Earth Science series. Basic is the science of the Earth's composition or materials (mineralogy, q.v., in Vol. IVB) and geochemistry. Others treat of mineral organization into rocks, igneous and metamorphic petrology, sedimentology (q.v., Vol. VI), and sedimentary petrology; of the Earth's surface structure (tectonics and geotectonics); of its history (stratigraphy); of its organic evolution (paleontology, q.v., Vol. VII); of its oceans and seas (oceanography); of its scenery and topographic evolution (geomorphology, q.v., Vol. III); and of its internal character and its machinery or dynamic processes. Numerous secondary and related disciplines may be further distinguished: economic (or applied) geology, geohydrology and hydrogeology, engineering geology, and military geoscience (Fig. 1).

All these topics are treated separately in the Encyclopedia of Earth Sciences series. In the general area, cross-reference may be made to the broader synthesis of earth science (of which geology is a central part), to discussions of earth science teaching such as the Principle of Parsimony, or Simplicity Principle (the philosophic base), and to geological methodology. Information is also to be found on geologic source materials, geologic communication, and similar specialized professional aspects. The fundamental question of causes, or what makes geological processes work, is treated in the entry on *Geologic Systems, Energy Factors.*

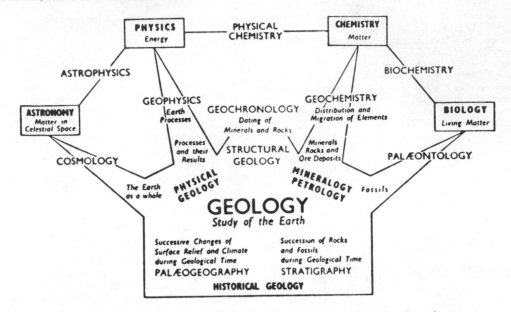

FIGURE 1. Subdivisions of the science of geology and their relations to other sciences.

From this analysis it might appear that geology is merely a casual grouping of numerous disciplines. However, they are tightly held together by common parameters. Thus, mineralogy is fundamental to petrology and economic geology. Paleontology is a sine qua non for stratigraphy. Sedimentational features such as cross-bedding and mud cracks are essential for the tectonicist to distinguish top and bottom. If structural geologists are not well informed in paleontology, they may confuse the stratigraphic sequence and their elegant structures may be upside down. Igneous petrologists seem to live in a world of high temperatures, but with the discovery of granitization, it is now evident that sediments and connate seawater are often their raw materials; Barth (1962) has written: "Igneous processes on the continents cannot exist in pure form; they are interrupted by sedimentary processes, and it is sedimentation that causes differentiation. . . . Variety among igneous rocks is a result of sedimentary processes." Yet, in universities, sedimentology is almost invariably taught by someone other than an igneous petrologist.

In economic geology (q.v.), while the battle of the neptunists and plutonists has waxed and waned, in the first half of the twentieth century, the plutonists seemed to have gained unchallenged ascendency. But in the mid-century decade, Baas-Becking has shown that bacteria can synthesize chalcopyrite, and sedimentologists have discovered slump structures (penecontemporaneous seafloor phenomena) in sulfide ore deposits, formerly taken to be drag folds. Thus, the various subdisciplines become inextricably interwoven.

Fundamental Divisions

The fundamental divisions of geology fall into the three following categories.

Physical Geology This division is largely the application of the fundamental laws of physics and chemistry to the processes and materials of the Earth. These processes and materials are closely united by being subject to a common external heat source (the sun), to a common dynamic control (celestial mechanics), and to a common inheritance—spatial and material within the Earth. Here we have what Holmes (1965) calls "*machinery* of the earth."

The dynamic, physical processes include magmatism, volcanism, tectonism, weathering, erosion and transportation, sedimentation, diagenesis, metamorphism—the so-called geologic cycle; all are interrelated. The external processes are called *epigene* and internal ones *endogene*. They are studied under various subdisciplines, in specialist areas, dictated largely by the techniques appropriate to them: mineralogy largely by the methods of chemistry and, more recently, of solid-state physics; petrology largely by optical microscopy by the scanning electron microscope and by trace-element geochemistry; weathering by the methods of soil science and geomorphology; sedimentation through oceanography; and tectonics by regional surveys and in the laboratory by the technology of engineering testing and mechanics.

Historical Geology This involves the interpretation of the materials and processes of physical geology in terms of relative time and the integration of the stratigraphic record with the paleontologic record—i.e., the evolution of the Earth and its biota. Thus the time dimension is added to the physics and chemistry of the Earth. This historical dimension is the almost unique contribution of geology (along with astronomy) to the knowledge of material things. Natural laws exist equally well here (e.g., "the geosynclinal trough is the father of the mountain," Bucher's Law 20), and the initiation and evolution of reproducible life is one of the great miracles of geologic history.

Regional and Applied Geology This category treats the mapping of the earth's surface and its accessible bedrock and the application of geological knowledge for human use. In the 1970s and 1980s it has come to include the mapping of ocean bottoms and of other planetary surfaces. Many kinds of maps can be constructed: the general survey map (the standard base of reference for all branches of geology in spatial terms), the structural contour maps, isopach and facies maps (largely required for stratigraphic purposes, much used commercially by the exploration divisions in the oil and gas industry), the mineral resources maps (required by the mining industries), construction materials maps (required for civil engineering purposes), hydrogeologic maps (employed by water supply organizations), among others (see *Maps, Environmental Geology; Maps, Physical Properties*).

Applied or economic geology is what the name implies: the application of theory and experience to the practical use and economy of humankind. In the history of geology it is not surprising that the study of mineralogy grew largely out of the search for metals or for precious gems. Some of the earliest centers of learning were in the old *Bergakademien* or *Montanschule* of central Europe and mining schools of the United Kingdom and North America. Universities in some parts of the world (e.g., in Latin America) often still train their geologists within their schools of engineering, while the paleontologists (naively regarded as being without much economic significance) may be found attached to the schools of arts.

Geology as a Science

The material scope and classification of the geological sciences (or geosciences) have been discussed. But what makes geology quite distinctive? It is sometimes said that a well-trained geologist can study chemistry to become a geochemist, but a chemist turning to geochemistry—while assuredly making useful contributions to geology—often makes serious errors in not thinking geologically. The well-rounded geologist is perhaps comparable to the good physician in medicine who utilizes the discoveries of the pharmaceutical chemist. Each, indeed, is complementary.

Geology in general differs from other sciences in that the opportunities for reproducible experimentation are distinctly limited. Rarely can the system be regarded as closed. Most often our models contain artificially separated parameters within immense open systems. Scale models are often useful for demonstration purposes only, the mathematical problems of scaling being formidable (Hubbert, 1937). Simple and even complex procedures in mineral analysis are little more than an extension of chemistry; high-pressure testing of rocks for structural deformation is an extension of mechanics. But the grand, wonderous processes—the evolution of a new phylum in paleontology, the evolution of a geosyncline to a mountain system, or the drifting apart of two continents—are in no way reproducible at will.

Much geological thinking is by the method of *reasoning by analogy* (see *Geological Methodology*). We wrinkle the tablecloth to simulate folding of mountains and often thereby introduce false ideas. Yet even Hubbert (1973) was not above demonstrating the mechanics of low-angle faults with the aid of a cold beer can on a wet table. The grand notions of geology need to be visualized by means of familiar small-scale examples. We think of beeswax or wet mud for rock flowage concepts (following the lead of Cloos, 1953). We think of convection cells in a beaker of water over a Bunsen burner. We think of fluvial erosion by watching trickles of water on a laboratory stream table. Rigorous analysis calls for close scrutiny of the scales and materials, but the pattern of the working hypothesis is established by the analogy modes.

In historical geology, paleogeography, and paleontology, we constantly make deductions and reach decisions "on the basis of inadequate data in an atmosphere of ambiguity. It is this ability, frequently mentioned, which makes it difficult for the student whose undergraduate education was mainly in mathematics, physics, or chemistry to adapt readily to graduate study in geology" (Hambleton, 1963). This sober judgment of scattered and sometimes grossly imperfect data, so often called for in professional economic geology, e.g., in the siting of a million dollar oilwell, is akin to the choice of military commanders who must make rapid tactical decisions in the face of enemy pressure, when their intelligence data are woefully imperfect yet delay may mean annihilation. Life is too short to allow for the collection of all the data, and in any case, in geology much of them are often lost to erosion or are deeply hidden. We must reach the decision, however; we must risk something to get that oilwell soon because we need the oil this year, not in a few decades.

It is also true that geologists use the *evidence of repetition* as the proof of a model (see *Scientific Method*). A physicist might find this difficult to follow on a geologist's scale, but a geologist, on approaching a strange mountain system for the first time, feels immediately at home when he or she recognizes successive belts of molasse, flysch, nappes and metamorphic core zones. Slowly, geologists have discovered that there are certain standard patterns for mountain ranges, just as for fossil assemblages of certain age, and because of the countless repetition of these observations in varied mountains and in varied formations, they feel almost as confident in prediction as their colleagues in physics in their experiments to demonstrate the fundamental laws. We say almost as confident because of the immense complexity of the natural setting; just as the physician learns to be cautious because no two human patients are ever identical, the geologist tends to hedge his or her statements with probabilities and possibilities because no two mountains are identical. The visual similarities of stratigraphic microfacies are widely used as an aid to correlation although the detailed analysis of the rock might be far too time consuming to be practical (Fairbridge, 1954).

The accumulation of a vast store of observational experience provides what Glangeaud (1949) has called the *collective dialectic* (see discussion by Van Bemmelen, 1961). A dialectic in philosophy is a method of logical procedure, based on the question, answer, and resolution system. The collective data of geology are stored in libraries and on maps, exchanged at congresses and symposia, and developed under team projects. Even working alone, the geologist may see her- or himself interrogating the Earth. Cloos wrote a charming and illuminating book called *Conversation with the Earth* (1953), and in the Bible (Job) one may read: "Speak to the earth and it shall answer thee."

Normal scientific research calls for the process of observation (diagnosis)–empirical rule (tentative)–deduction–working hypothesis–deduction (prognosis). These steps are commonly repeated to provide multiple data. If appropriate, the working hypothesis is tested by experiment. The repeatability of the experiment is accepted as proof of that particular deduction. Multiple observations are thus integrated into reliable *empirical rules,* which after multiple testing may become codified as *natural laws.* Otherwise, they remain what Bucher (1933) has dignified as "opinions"—i.e., rules that seem to work but that require further testing. A certain incompleteness is inevitable with any historic record; in geology every basin subsidence leads to burial and every uplift leads to erosion and total effacement.

Chamberlin (1897) developed the concept of the multiple working hypothesis, which is a necessary antidote to the one-shot deduction, the so-called intuitive flash that may be brilliantly correct but more often not. There is a certain danger, however, in assuming that the selection of multiple hypotheses, rejected one by one, will eventually leave one gem that perfectly fits the data. Much more likely is the situation where one may select a possibly minuscule part of each hypothesis and reconstruct these data onto an integrated, complex hypothesis, a critical and *selective eclectic methodology* (see *Geology, Philosophy*). Nature is not normally as aesthetically pure and simple as some physicists seem to wish. The building of a mountain belt calls for a complex analysis.

Van Bemmelen (1961) reminds us that there are classicists and romanticists in geology as in other fields:

[T]he former being inclined to design schemes and to use consistently the deductions from working hypotheses; the latter being more fit for intuitive discoveries of functional relations between phenomena and therefore more able to open up new fields of study.

[Van Bemmelen's footnote]: It may be that many "romanticists" possess the so-called "harmonic" temperaments often found in blood group A people, while the "classicists" would have the "rhythmic" temperaments of those belonging to blood group B. This most interesting statistic correlation between character and blood group has been derived from the investigations by L. Bourdel and J. Genevay, recently published in their book *Groupes sanguins et temperaments* (Paris: Ed. Maloine, 1960).

Examples of both character types are Werner and Hutton (Wegmann, 1958). Werner was a classicist. At the end of the eighteenth century he postulated the theory of *neptunism,* according to which all rocks including granites, were deposited in primeval seas. It was an artificial scheme, but as a classification system, it worked quite satisfactorily at the time. Hutton, Werner's contemporary and opponent, was more a romanticist. His concept of *plutonism* supposed continually recurrent circuits of matter, which like gigantic paddle wheels raise material from various depths of the Earth and carry it off again. This is a very flexible system that opens the mind to accept the possible occurrence in the course of time of a great variety of interrelated plutonic and tectonic processes.

There is a trend, headed by the active and systematic, the classicist type of scientists, to consider physics and chemistry as the only true natural sciences, for example, the only sciences in which most relations can be expressed mathematically and in which the magnitude of the variable and constant factors can be measured. Many in this group may compare geologists and biologists to barbaric stone-age scientists, who, more or less hopefully, struggle along to reach the remote but ultimate goal of total "quantification" and "mathematization." By such classicists the latter are accepted only out of courtesy in the exclusive circle of the exact natural sciences. [Wright, 1958]

Geology offers a unique something to the net total of science that no other discipline can approach, however: the *time* parameter. The complete history of an entire planet is something not even the astronomers can aspire to. The nature of the Earth's interior, now under active study by the geologists, may well disclose some new principles of physics and physical chemistry of which our learned colleagues in those disciplines have not dreamed. The manner of its evolution is the tantalizing goal. With the capacity at our disposal to measure time gauged in years times 10^6 or 10^9, we have an extraordinary instrument.

Absolute dating has now made possible a totally new time factor: cycles. These range from the annual (varve layers) and the sunspot cycle (beach ridges), up to the 10^8 periods of our galaxy, the megacycles (Williams, 1981). Geology is now developing as a *predictive science,* with the potential for anticipating the earthquake or the volcanic eruption (Drake and Maxwell, 1981).

As the philosopher Bronowski has remarked: "The eye is not a neutral instrument but modulates the input." The geologist sees the mountain as a four-dimensional dynamic entity. The engineer or the poet sees it quite differently.

Many geologists rebel against what Link (1954) has called "robot-geology" (i.e., strictly quantitative geology), since the human mind still can handle vastly complex appraisals and make decisions. At a conference on piloted satellites versus unpiloted satellites, the opinion of the experts was that the person in control was a more economic proposition (from the weight/efficiency viewpoint) than a vast mobile computing laboratory. Many routine observations certainly can be machine sensed and recorded, but humans can apply selectivity and observe the unexpected.

Correlations that are achieved by humans so rapidly that they appear intuitive may be crosschecked by statistical methods and computers (see *Geostatistics*). According to Melton (1958), three types of correlation structures may be distinguished:

1. A type with causal relations in which certain forces have a subduing effect on others (cycles with negative feedback);
2. A type in which the causal relations have an amplifying effect (cycles with positive feedback);
3. A type in which no functional relations between the coinciding phenomena can be discerned. Even if no numerical data are fed into these structures, their very recognition can greatly aid the so-called geological art.

There is clearly a vital need for both the romanticists and the classicists in geology, a necessary symbiosis of inspired stimulation and disciplined control.

In a report on a high school educational project [AGI (American Geological Institute) Earth Science Curriculum Project], a text and teachers' manual entitled *Investigating the Earth* was prepared. While manuals of physics and chemistry can often be prepared satisfactorily by single scientists, it was indicated that *Investigating the Earth* was essentially a committee accomplishment. While it is well known that committees are useful instruments in democratic social procedures, they are not celebrated for their efficiency. Duplication of effort, argumentation, discussion, etc. are accepted as inefficient, but since the birth, education, lifetime, and death of the human organism are inefficient, it is a method not inappropriate to its inventor. Ultimately it works; the vast accumulation of geological data is committee handled—sieved, classified, and stored. From it, the great natural laws of geology may be discerned.

Democracy is not a basis for the achievement of scientific judgment. The opinion of the majority may well be false. But love of the Earth, of nature in its totality from macrocosm to microcosm, is a good prerequisite to that achievement, a respect for integrity and honesty is vital, and the human value of generosity is somewhere essential in the real scientist's makeup. Knowledge of a person, of oneself, cannot be formalized, but collected data of science can be organized and, furthermore, transmitted. In this way, we contribute to posterity.

Geological Education

While there are almost as many definitions of geology as there are geologists, there is an equally wide diversity of opinion as to what makes an educated geologist. Stimulated by a number of factors, which combined in the mid-twentieth century to present a serious crisis in the establishment of conservative geology, numbers of committees and dedicated persons have seriously studied the question of geological education.

Multiple factors precipitated the crisis. First, there is the very evolution of the methodology. There has been a slow but definite shift from the boots-and-hammer approach to the white lab coat and the shiny boxes of electronics; as a result, overcautious or static-minded practitioners have been passed by, they have become obsolescent.

Second, there is the changed scope of geology, the emergence of the marine geologist, the astrogeologist, etc. New professors are needed and new curricula have to be devised.

Third, there is the wholesale abandonment by

the geological establishment of disciplines that were formerly part of their professional expertise. Most of engineering geology (foundation testing; construction material study; test boring; dam-site, tunnel, and transportation surveys) has been simply left to the engineers, and only a few universities still give courses in engineering geology, so the demand for a routine service geologist (with a master's degree) in this vast field has dropped almost to zero. Water-supply geologists are still occasionally called for, but the bulk of hydrologists are trained engineers, not geologists; another broad field of employment thus fades.

The fourth factor is a chronic overproduction of fuels—petroleum, natural gas, coal, uranium—and in almost all ore deposits. The truth is, geologists have been just too clever in providing the world with all the fruits of the solid Earth. In the 1950s, thousands of oil geologists were laid off. In the 1980s political barriers created new demands, but exploration tends to be cyclical. Once discovered, an oil field has no further need of the geologist—the engineers continue drilling and study the hydraulic pressures and best economic development of the field, but the geologist may well be fired. It is often the same in coal or metal mining: once the geologists have proven the reserves, the mining engineers or foremen, take over to exploit the seams or the lode. It is easy and understandable for some members of the profession to feel bitter about a situation that is not of their making. But there is an inevitability about the evolution of a science just as there is about the cold facts of economics.

On March 26, 1961, the oil correspondent to the *The Lubbock Avalanche Journal* wrote: "Does geology have the will to survive?" The vigorous approach of the various educational committees, the key role geologists have won in oceanography and in space science, and the enormously growing demand for earth science teachers are answers to the saddened oil field owner. Indeed, in petroleum geology the new approach is beginning to pay off, and the relatively well-paid geology Ph.D. takes a respected place in that rigorous profession.

What is the new approach in geological education? A very helpful survey was published by Hambleton (1963) on behalf of an AGI conference study. Two distinct aspects of education are involved: (1) Geological education for what? What kind of professional niche is the geologist going to fill? Apart from essential introductory work, how and when should he or she specialize? (2) Geological education with what? With what kind of preliminary prerequisite training must the budding geologist be equipped before even embarking on advanced training?

Some purists have suggested that so much foundation work in mathematics and the pure sciences is needed to make an expert today that the would-be geologist should not be permitted to spend the undergraduate years doing any geological courses at all; such luxuries can wait until the graduate years.

The economics of both small colleges and large universities often requires that a geology department devote a considerable effort to so-called service courses in basic physical and historical geology or possibly in the new earth science. Two contrasting schools of thought exist—the one that presents courses such as adventures in the composition and philosophy of a science, which is designed to stimulate and contribute to the liberal education of the nondedicated general student, and another that presents a rigorous basic technical geology course, which is planned for the needs of the dedicated would-be professional geologists, but regardless of the fact that perhaps 95% of the class has no intention of taking a second year of geology and is profoundly bored by the technology. It would seem a wise procedure to leave the rigorous technology only to those who return in that second year, hungry for advanced learning. It is a fact that many famous geologists became interested in the subject only after they became acquainted with the field thanks to a stimulating exponent, presenting material to a beginning group.

Hambleton (1963) summarized this complex problem:

It is obvious that there can be many kinds of educational philosophies for geology. Nevertheless, recurrent themes appeared in reports from all teams. Interestingly, the size of a school had very little relation to the philosophy expressed at the school. There was nearly unanimous agreement that our educational goals should be taught within the framework of courses built around principles. The production of an educated geological scientist should be the aim of our departments of geology. The training should be broad enough so the student should be able to attack most problems with some degree of success. There should be emphasis on problem-solving and we need to encourage students not to be afraid to make decisions.

Training in the humanities, the social sciences, and languages is needed to help develop critical thinking as well as educated geologists. Courses that require the student to communicate in writing and orally are especially valuable. Liberal arts training often produces the mental attitudes that result in the best critical thinkers. Because geology depends to a large extent on the application of the principles of mathematics, biology, chemistry and physics to the problems of the earth, training in these sciences is essential and these fields must be incorporated in and become part of courses in geology. In general, the cognate sciences are not a part of geological training at present.

Undergraduate training should include the subject matter of the classic geology courses. Considerable emphasis should be placed on field work, field trips, and field relationships. This is especially true for the future laboratory specialist who will need this background for geologic interpretation of laboratory data.

There was much feeling that courses must be designed to equip the geologist to deal with the "untidy" problems that characterize his field, that is, he must learn to use approaches where multiple working hypotheses are necessary, reasoning by analogy is important, and decisions are reached on the basis of inadequate data in an atmosphere of ambiguity. It is this ability, frequently mentioned, which makes it difficult for the student whose undergraduate education was mainly in mathematics, physics, or chemistry to adapt readily to graduate study in geology.

Because of the necessity to prepare for geology properly the undergraduate career should be considered as preprofessional education; specialization should come at the graduate level. Some schools have suggested the desirability of a five-year program. Undergraduate curricula should be very broad. However, geology must be offered to students at all levels if we wish to attract students and make them aware of geological problems.

In summary, we need people who can solve problems with the tools of modern science and who have a broad base in related science, who know how to think as well as how to do, who have been trained in fundamental principles and ideas, who have broad acquaintance with field work.

The new approach to *learning methods* also emerges from the AGI surveys reported on by Hambleton (1963):

(a) Interest the student in research and attach him to a faculty project as early as possible. Let him learn by doing and emphasize independent study.

(b) Stretch the student's mind as far as possible.

(c) Get students into field observation as early as possible and make field trips a learning as well as a teaching experience.

(d) The best learning takes place when the students begin to teach each other. Hence, one should provide the opportunity for general student discussions. The case history approach can be used to advantage.

(e) The teacher should be a learner along with the students.

(f) It is time to move such material previously held at the graduate level into the undergraduate program.

(g) Emphasis on memory alone is generally undesirable.

(h) Team teaching, courses taught by several different faculty members who are experts in their specialties, can be used to great advantage if properly organized and handled.

(i) Faculty debates on unsettled or controversial questions may be a valuable teaching tool for students who have enough background to understand why there can be difference of opinion.

(j) Perhaps programmed or machine teaching can be used to advantage in getting students familiar with facts, leaving class time for the instructor to discuss theory and interpretation. The use of closed-circuit television should be explored more thoroughly.

(k) Departments should be concerned with the welfare of unusually good students.

(l) Students should feel free to talk to staff outside of class hours, and there should be close relationship between students and staff.

Regarding the means of communicating the role of geologist, responses can be summarized by the statement, "Be a geologist and let them observe one in action." In other words, be something that the student can emulate. Students should read geologic literature, both old and new, and should take part in geologic research projects at an early stage in their training. There should be opportunity for students to observe and work with practicing geologists who are engaged in research problems. Many departments require a senior thesis and some departments have lower classmen serve as assistants to upper classmen.

In the years following the AGI surveys, an added factor has been felt: women. Only in the late twentieth century has the job market at last made a place for them. A special study of West German geological education prepared by Pilger (1964) also serves to illustrate that the problems of American geologists are not unique.

Finally, a note is appropriate on the role of the geologist and the geological group in the community. A fine textbook emphasizing this role in society was written by Menard (1974), who later exemplified the new integrated approach when he temporarily relinquished marine geology to become director of the U.S. Geological Survey. This action-reaction relationship can work in two ways. First, geologists can play a vital role in civic affairs. They can apply their professional judgment in conservation, in the location and maintenance of parks, in ensuring good water supply and waste disposal, in the sound location of highways, and in applying some common sense to public transport problems. They can assist the geological activities of the Boy Scouts. Geologists can give lectures and open talks on radio and television and can encourage mineral and so-called rock-hound clubs (see *Popular Geology,* with references).

Second, the community can help the geology group, be it in high school, college, or university. An advisory group or geology circle can be organized among the businesspersons, farmers, engineers, or scientists of a community to stimulate interest, to raise money for equipment, research projects, regional mapping, etc., and to aid conservation.

It would be interesting to know how many millions of dollars are spent yearly by uninformed yet well-heeled individuals simply drilling water wells through hundreds of feet in solid granite for want of a friendly geologist who could easily show them a likely belt of alluvial gravels. Or the townships that authorize highways to be blasted through massive rock barriers when a small detour would call for nothing more than a bulldozer operation. Or the coastal communities that spend their fortunes nourishing their beaches with thousands of tons of sand, only to see them washed away in winter storms. Or the fishing center that skims the cream off its profits by constantly dredging a channel that a good coastal geologist could easily relocate for them.

Conclusion

Geology is a science that has relatively few practitioners. Members of the component societies in North America number somewhat over 40,000. About an equal number are operating in the USSR, and perhaps a similar figure would cover the rest of the world. Such a world group is never going to be a great political force. They have one peculiarity in common, however, that can hardly be said for any other group, except perhaps the astronomers. They are all devoted to their calling. There is something approaching fanaticism in their affection and respect for the Earth. They enjoy a spiritual oneness with nature that places them in a privileged, happy, yet responsible position among their fellow creatures.

RHODES W. FAIRBRIDGE

References

Adams, F. D., 1938, *The Birth and Development of the Geological Sciences.* New York, 506p. (Dover reprint, 1954.)
Albritton, C. C., Jr., ed., 1963, Philosophy of geology: A selected bibliography and index, in *The Fabric of Geology.* Reading, Mass.: Addison-Wesley, 262-363.
Anderson, C. A., 1963, Simplicity in structural geology, in C. C. Albritton, Jr., ed., *The Fabric of Geology.* Reading, Mass.: Addison-Wesley, 175-183.
Barth, T. F. W., 1962, Ideas on the relationship between sedimentary and igneous rocks, *Geochemistry,* no. 4, 337-341.
Bucher, W. H., 1933, *The Deformation of the Earth's Crust.* Princeton, N.J.: Princeton University Press, 518p. (reprinted 1957, New York: Hafner.)
Chamberlin, T. C., 1897, The method of multiple working hypotheses, *Jour. Geology* 5, 837-848. [See also *Science* 148, 754-759 (1965).]
Cloos, H., 1953, *Conversation with the Earth* (trans. from German). New York: Knopf, 413p.
Drake, C. L., and J. C. Maxwell, 1981, Geodynamics—Where are we and what lies ahead? *Science* 213, 15-22.
Fairbridge, R. W., 1954, Stratigraphic correlation by micro-facies, *Am. Jour. Sci.* 252, 683-694.
George, T. N., 1965, *University Instruction in Geology: A Comparative Survey of Curricula and Methods in the Universities of Czechoslovakia, Federal Republic of Germany, France, United Kingdom, U.S.A., U.S.S.R.* Paris: UNESCO, 171p.
Glangeaud, L., 1949, Epistémologie des sciences naturelles et structure du réel, *Congr. Internat. Philosophie des Sciences (a5),* 1, 127-138.
Hambleton, W. W., 1963, The status of undergraduate geological education, *Geotimes* 8(4), Part 2.
Holmes, A., 1965, *Principles of Physical Geology,* 2nd ed. London: Nelson, 1288p.
Hubbert, M. K., 1937, Theory of scale models as applied to the study of geologic structures, *Geol. Soc. America Bull.* 48, 1459-1520.
Link, W. K., 1954, Robot geology, *Am. Assoc. Petroleum Geologists Bull.* 38, 2411.
Melton, M. A., 1958, Correlation structures of morphometric properties of drainage basins and their controlling agents, *Jour. Geology* 66, 442-460.
Menard, H. W., 1974, *Geology, Resources, and Society.* San Francisco: W. H. Freeman, 621p.
Pilger, A., 1964, Das Studium der Geologie in der Bundesrepublik Deutschland und in Westberlin mit Vergleichen aus anderen Landern, *Deutsch. Geol. Gesell. Zeitschr.* 115, 1-32.
Van Bemmelen, R. W., 1961, The scientific character of geology, *Jour. Geology* 69, 453-463.
Wegmann, E., 1958, Das Erbe Werner's und Hutton's, *Geologie* 1(3/6), 531-559.
Williams, G. E., ed., 1981, *Megacycles.* Benchmark Papers in Geology, vol. 57. Stroudsburg, Pa.: Hutchinson Ross Publishing Co., 434p.
Wright, C. W., 1958, Order and disorder in nature, *Geologists' Assoc. Proc.* 69, 77-82.

Cross-references: *Cartography, General; Geohistory, American Founding Fathers; Geological Methodology; Geology, Philosophy; Geostatistics; Maps, Environmental Geology; Petroleum Geology; Photogeology; Popular Geology; Prospecting; Remote Sensing, General; Scientific Method; Serendipity; Surveying, General; Terrain Evaluation Systems; Uniformitarianism; Well Logging. Vol. XIII: Earth Science, Information and Sources; Economic Geology; Geological Communication; Geology, Applied; Geotechnical Engineering; Maps, Engineering Purposes; Medical Geology; Military Geoscience; Mineral Economics; Oceanography, Applied; Urban Geology.*

GEOMAGNETICS, GEOMAGNETIC SURVEYS—See EXPLORATION GEOPHYSICS; MARINE MAGNETIC SURVEYS; VLF ELECTROMAGNETIC PROSPECTING; WELL LOGGING. Vol. XIII: MAGNETIC SUSCEPTIBILITY, EARTH MATERIALS.

GEOMAGNETISM AND PALEOMAGNETISM

Geomagnetism is the study of the Earth's magnetic field, while *paleomagnetism* is defined as the study of the history of the geomagnetic field. Because a record of the geomagnetic field is often recorded by rocks during their formation, magnetism is the only geophysical property of the Earth that can be measured and studied over significant periods of Earth history.

Historical Perspective

Magnetism as a physical property was probably known to ancient humans, although perhaps the earliest record of magnetic repulsion came from ancient Egypt. Much of this knowledge was gained through observation of the properties of lodestone, a rock composed almost entirely of the mineral magnetite. The property of magnetized needles to point to the north was probably known by the Chinese hundreds of years before Christ, however, this knowledge did not reach the West until the twelfth century A.D. (Tarling, 1983). Peter Peregrinus, a Frenchman, studied the magnetic properties of lodestone spheres in the thirteenth century, introducing the concepts of *poles* in which *lines of equal force* converge. William Gilbert (1600) expanded on these studies and was the first to compare the geomagnetic field to the field associated with uniformly magnetized spheres of lodestone. Around the same time, it was discovered that compass needles did not point to true north but a few degrees one way or the other and that this magnetic *declination* also varied in time. Although it is often thought that modern science began with chemistry or physics, Gilbert's pioneering work shows that the first experimental science was actually geomagnetism.

The deflection of compass needles by magnetized rocks no doubt led to interest in the magnetization of rocks, although little studying was done before the nineteenth century. Late in the century Folgerhaiter concluded that volcanic rocks attain their magnetization on cooling and that the direction of magnetization parallels that of the Earth's field. This conclusion provided the fundamental observation of paleomagnetism: Rocks retain a record of the ancient geomagnetic field. Further research on volcanic rocks, sediments baked by lavas, fired clays, and bricks confirmed that all these materials acquired magnetizations parallel to the field on cooling and that they maintained these directions of magnetizations for extended periods of time.

As studies proceeded researchers noted that the magnetization of some rocks was in the opposite direction of the present field—i.e., their magnetic vectors pointed to the south instead of to the north—leading to the proposal that the Earth's field had reversed in the past. Matuyama (1929), in careful studies of Japanese lavas, concluded that rocks formed during the Early Quaternary were all reversely magnetized while those formed during the Late Quaternary were all magnetized parallel to the present field (normally magnetized). This finding provided perhaps the first information regarding the ancient history of the geomagnetic field.

Further developments in paleomagnetism ensued following the development of magnetometers sensitive enough to permit measurement of the magnetizations of many different rock types, statistical methods for data analysis, and methods of magnetic "cleaning" of rocks, mostly during the 1950s and 1960s. The growing quantity of data during this period led to the confirmation of Wegener's (1924) hypothesis of continental drift and the establishment of a geomagnetic time scale (e.g., Cox et al., 1963). Recognition of marine magnetic anomaly patterns in conjunction with this time scale led the way to confirmation of the seafloor spreading hypothesis (Vine and Matthews, 1963).

Advances in geomagnetism have similarly been driven by improvements in data collection techniques and a growing data set. Establishment of magnetic observatories on several continents in the 1800s provided Gauss with sufficient data to perform a spherical harmonic analysis of the field, confirming the internal origin of the main dipole. In the 1950s and 1960s routine collection of marine magnetic data led to delineation of the crustal contribution to the geomagnetic field. Most recently, acquisition of magnetic data by satellites have greatly increased our understanding of both internal and external fields. As more years pass since the development of modern equipment and its installation in observatories, time-related changes in the geomagnetic field will be better understood.

Geomagnetism

The geomagnetic field can be thought of, to a first approximation, as a magnetic dipole at the center of the Earth that possesses a dipole moment of 7.94×10^{22} Am^2 (SI units), with the dipole axis inclined at 11.5° to the Earth's geographic axis. This dipole accounts for 90% of the geomagnetic field observable at the surface of the Earth, the rest being attributable to nondipole components of the internal field or generated by conduction in the upper mantle caused by electric currents originating in the upper atmosphere. Spherical harmonic analysis shows that this field must be generated near the center of the Earth, and it is currently believed that its probable cause is fluid motions of conductive nickel-iron melt in the fluid outer core.

Neither the dipole nor the nondipole components of the field are constant in time; this time variation of the field is termed the *secular variation*. Studies of the present-day main field imply several features of the secular variation (Stacey, 1977):

A decrease in the dipole moment of 0.05% per year.
Westward precession of the dipole at a rate of 0.05° longitude per year.
Rotation of the dipole toward the spin axis at 0.02° per year.
Westward drift of the nondipole field at 0.2° per year.
Growth and decay of features of the nondipole field.

The study of the motions of conducting fluids and their generation of magnetic fields is called *magnetohydrodynamics*. Models of the generation of the geomagnetic field are computationally complex but ultimately must account for what we know of the main field behavior, i.e., the character of the secular variation, the quasi-random periodicity of field reversals, and the general coincidence of the magnetic and geographic poles averaged over time.

At a given point on the surface of the Earth, it is most convenient to consider the geomagnetic field in terms of its vector components. The three quantities commonly used to describe the field uniquely are *declination, inclination,* and *intensity*. Intensity refers to the magnitude (or length) of the magnetic vector, declination the angle of the horizontal projection of the vector with respect to the horizontal, the convention being positive downward. The *north magnetic pole* is thus defined as the point on the surface of the Earth where the inclination is $+90°$ and conversely, the *south magnetic pole* is the point where the inclination is $-90°$. Note the difference between these and the *geomagnetic poles,* the surface points where the best fit dipole axis intersects. Charts of the geomagnetic field therefore consist of contour lines of equal values of the three quantities, *isogonic* lines for declination, *isoclinic* lines for inclination, and *isodynamic* contours for total field intensity.

Although of less direct interest to geologists, much recent effort has gone into study of the external magnetic field—the magnetic field in the atmosphere and the region of outer space surrounding the Earth. The configuration of the external field is a result of interaction between the main dipole field and the solar wind. One result of this interaction is that the field intensity around the Earth is higher on the side facing the sun than on the night side. The *magnetosphere,* as the affected region is called, consists of charged particles that are held in space by the Earth's field but that respond to the solar wind. Since the solar wind varies in time, so does the external field. These variations can be observed on the Earth's surface as short period fluctuations in the total field vector, magnetic storms, and the *aurora borealis*. Although normally a minor component of the total field, these external influences may periodically make magnetic surveys impossible.

Paleomagnetism

Rock Magnetism There are many applications of paleomagnetism in the earth sciences today, but most depend on the ability of rocks and sediments to acquire a magnetization parallel to the magnetic field in which they formed or were altered, this field presumably being the ancient geomagnetic field. It has been demonstrated that in many cases this magnetization is very stable—i.e., it can remain frozen in the rock for billions of years. In other cases, rocks can become partly or totally remagnetized during geologic events in their history long after their formation. Much of the effort of paleomagnetic laboratory work is directed toward establishing the stability of the magnetization in a given rock unit and determining the fidelity of the record of the ancient field within the rock.

Rocks owe their remanent magnetic properties to the presence of small amounts of magnetic minerals within them, the most common being the iron oxides, magnetite and hematite. A magnetized rock is said to have a *natural remanent magnetization* (NRM). In an igneous rock, these minerals spontaneously acquire a magnetization parallel to the Earth's field on cooling below a point known as the Curie temperature ($578°C$ for pure magnetite). This as known as a *thermal remanent magnetization* (TRM). A TRM may also be acquired by a sedimentary rock during a later reheating episode or by archeological materials such as bricks or clay pots that are baked in kilns. Sediments can acquire an NRM as the magnetic field orients detrital grains of magnetite or hematite. This is referred to as a *detrital remanent magnetization* (DRM). Chemical changes occurring during metamorphism, diagenesis, or some other process may form magnetic minerals. These minerals will also be magnetized parallel to the ambient field. Predictably, this process is called *chemical remanent magnetization* (CRM).

Two or more of these events often may have occurred during a rock's history, resulting in an NRM that is a superimposition of several components. The paleomagnetist must sort out such complexities where they occur or, alternatively, establish that the observed NRM was formed at the same time that the rock was formed. These separate components of NRM typically reside in different minerals within the same rock sample or within different grain size ranges of the same mineral. Several laboratory techniques are employed to separate magnetization components, all of which employ progressive demagnetization of the rock to remove initially the less stable components while retaining the stablest until the rock is completely demagnetized. Methods include subjecting the rock to successively higher alternating fields (AF demagnetization) or heating in steps to progressively higher temperatures (*thermal demagnetization*) (McElhinny, 1973). Another technique involves dissolution of the oxide minerals in acid solutions (*chemical demagnetization*) (Roy and Park, 1974). When two or more components remain superimposed, mathematical techniques are often employed to separate them. Two examples are principal component analysis (Kirschvink, 1980) and remagnetization circles (Halls, 1976). Additional magnetic techniques and petrographic studies are also useful in NRM analysis.

Plate Tectonics Paleomagnetism is perhaps best known for its contributions to the study of plate tectonics and continental drift (q.v.). The fundamental assumption involved in these applications is that the time-averaged geomagnetic field is produced by a geocentric axial dipole so that the calculated paleomagnetic pole coincides with the paleogeographic axis (Hospers, 1954). Studies have shown that over the past 20 m.y. this is indeed the case when poles are averaged over a few thousand years (e.g., McElhinny, 1973). Extension of this hypothesis farther back in time is based on the principle of uniformitarianism (q.v.). The geometric configuration of such a dipole allows us to determine paleolatitude from magnetic inclination by the simple formula:

$$\tan(\text{inclination}) = 2\tan(\text{latitude})$$

Analysis of *apparent polar wander* (APW) paths can also be used to trace paleo plate motions. An APW path for a plate is constructed by averaging paleomagnetic poles for each time period going back in time. It is called *apparent* polar wander because if the geocentric axial dipole hypothesis holds, the polar wander reflects motion of the plate with respect to the geographic pole rather than shifting of the geographic pole. Since there can be only one geographic pole for a given time, differences in paleomagnetic poles for the same time from different continents mean that they have moved relative to one another since that time. Since magnetic vectors will always point to the north or south, however, paleolongitude differences cannot be determined by paleomagnetism. Thus, although two plates may have moved with respect to one another in the past, unless the motions have components of relative rotation or paleolatitude change, they may not be reflected in the paleomagnetic data.

Magnetostratigraphy and Magnetic Dating The history of geomagnetic field reversals has been fairly well established back into the Late Mesozoic. Periods of normal and reversed polarity *chrons* and the dates of field reversals were initially determined by comparing paleomagnetic polarity of lava sequences with potassium-argon ages in the same sequences (e.g. Cox et al., 1963; McDougall and Tarling, 1963). Because of the errors associated with isotopic dating, however, these methods are only useful for the past 5 m.y. Extension of the time scale involved calculating seafloor spreading rates for the past 5 m.y. and then dating the remaining record of seafloor reversals by assuming constant spreading rates and extrapolation (Heirtzler et al., 1968). An important discovery during these procedures was that reversals occur at nearly random intervals. This meant that any piece of seafloor with magnetic anomalies covering several chrons will show a unique reversal pattern that can fit into only one place in the magnetic polarity time scale. Thus, it is unnecessary to have a record of each reversal up to the present day to date pieces of oceanic basement.

Similar techniques have been applied to stratigraphic sections exposed on land and to deepsea sediment cores. These studies determine magnetic polarity for the section by sampling and measuring in the laboratory. Once the pattern of reversals for the section is delineated, it can be fit into the polarity time scale in a similar manner to the marine magnetic anomaly pattern. Since these sections may contain subtle unconformities and variations in sedimentation rate, approximate paleontological dates are often useful adjuncts. The major contribution of paleomagnetism is that a reversal boundary represents an event that occurred simultaneously throughout the world. Hence, the time-transgressive nature of many lithologic and paleontologic boundaries is not a problem and can be detected. Magnetostratigraphy also permits correlation of sections containing different biota and correlation of nonfossiliferous units. Over the past 150 m.y. reversals have occurred, on average, at a rate of 3 per million years, which defines the time resolution to be expected in magnetostratigraphy.

Because the last reversal occurred approximately 700,000 yr ago, reversal stratigraphy is not applicable to younger rocks and sediments. Studies of lake sediments (e.g., Creer, 1977) have revealed correlatable features in the paleomagnetic record of the secular variation of the geomagnetic field. These can be used for correlations back to 25,000 yr. Similarly, fired bricks and baked soils from dated archeological sites provide a high-resolution record of the secular variation. These records are not only useful for correlation purposes but also are essential for delineating long-term behavior of the geomagnetic field.

Other Applications Many other geological problems can be addressed by taking advantage of what we know of rock magnetic characteristics and our ability to measure them in the laboratory. In many instances paleomagnetic studies provide a method of measuring a larger suite of samples more quickly and less expensively than other techniques such as petrographic study and isotope measurement.

Local structural problems can be studied by observing relative deviations of remanent magnetization vectors from each other or from directions inferred from APW paths. This method is useful particularly for discovery of rotations about vertical axes that cannot be resolved by structural techniques (e.g., Beck, 1980; Hrouda, 1982; Sparisou et al., 1984).

Another method employed in structural analysis makes use of the shape anisotropy of magnetite with respect to induced magnetization. This is known as *anisotropy of magnetic susceptibility* or AMS. When rocks deform, grains within them will either rotate or grow into preferred orientations. Preferential directionality of this sort can be easily mesured by the AMS method, delineating different strain modes (e.g., Hrouda, 1982). The AMS can be expressed as an ellipsoid that often mimics the strain ellipsoid, at least in terms of shape. AMS measurement is considerably quicker than optical techniques of measuring preferred crystallographic orientations.

Thermomagnetic studies have been used to calculate temperatures and depths of intrusions and their contact aureoles. When a dike or sill intrudes country rock of sufficiently older age, it will acquire a different NRM direction than the unheated country rock. In the contact zone, however, the country rock will either be totally or partly remagnetized depending on its distance from the intrusive. Where it is partly remagnetized the maximum attained temperature can be determined using thermal demagnetization analysis and, thus, a temperature-distance profile constructed. Assuming a cooling model, the width of the heated zone will be a function of the burial depth at the time of intrusion (McClelland-Brown, 1981; Schwarz, 1977).

AMS methods have been used to study sedimentary and igneous fabrics as well as structural fabrics. Ellwood and Ledbetter (1977) have determined paleocurrent direction in pelagic sediments with the technique. It can also be used to detect bioturbation or other perturbations of primary depositional fabrics (Ellwood, 1983). In igneous rocks AMS has been used to determine flow direction and late stage tilting (Ellwood, 1978; Ellwood and Whitney, 1980).

Different rock magnetic characteristics have been employed as correlative tools in distinguishing between subtle differences in rock type, degree of alteration, degree of serpentinization, and different sources of granitic melts (Park, 1983). They have even been used to correlate obsidian artifacts with their source volcanoes and thus to trace ancient trade routes. Paleomagnetism has grown from a once esoteric study of the magnetism of rocks to a powerful tool in broadly ranging geophysical and geological studies.

DANN J. SPARIOSU

References

Beck, M. E., Jr., 1980, Paleomagnetic record of plate-margin tectonic processes along the western edge of North America, *Jour. Geophys. Research* 85, 7115-7131.

Cox, A., R. R. Doell, and G. B. Dalrymple, 1963, Geo-

magnetic polarity epochs and Pleistocene geochronology, *Nature* 198, 1049-1051.

Creer, K. M., 1977, Geomagnetic secular variations during the last 25,000 years: an interpretation of data obtained from rapidly deposited sediments, *Royal Astron. Soc. Geophys. Jour.* 48, 91-110.

Ellwood, B. B., 1978, Flow and emplacement direction determined for selected basaltic bodies using magnetic anisotropy measurements, *Earth and Planetary Sci. Letters* 41, 254-264.

Ellwood, B. B., 1983, The effect of bioturbation on the magnetic fabric of some natural and experimental sediments, *EOS* 64, 219.

Ellwood, B. B., and M. T. Ledbetter, 1977, Antarctic bottom water fluctuations in the Vema Channel: Effects of velocity changes on particle alignment and size, *Earth and Planetary Sci. Letters* 35, 189-198.

Ellwood, B. B., and J. A. Whitney, 1980, Magnetic fabric of the Elberton Granite, northeast Georgia, *Jour. Geophys. Research* 85, 1481-1486.

Gilbert, W., 1600, *De Magnete.*

Halls, H. C., 1976, A least-squares method to find a remanence direction from converging remagnetization circles, *Royal Astron. Soc. Geophys. Jour.* 45, 298-304.

Heirtzler, J. R., G. O. Dickson, E. M. Herron, W. C. Pitman, and X. LePichon, 1968, Marine magnetic anomalies, geomagnetic field reversals and motions of the ocean floor and continents, *Jour. Geophys. Research* 73, 2119-2136.

Hospers, J., 1954, Rock magnetism and polar wandering, *Nature* 173, 1183.

Hrouda, F., 1982. Magnetic anisotropy of rocks and its application in geology and geophysics, *Geophys. Surveys* 5, 37-82.

Irving, R., 1964, *Paleomagnetism and Its Application to Geological and Geophysical Problems.* New York: Wiley, 339p.

Kirschvink, J. L., 1980, The least squares line and plane and the analysis of paleomagnetic data, *Royal Astron. Soc. Geophys. Jour.* 62, 699-718.

Matuyama, M., 1929, On the direction of magnetization of basalt in Japan, Tyozen, and Manchuria, *Japan Acad. Proc.* 5, 203-205.

McClelland-Brown, E. A., 1981, Paleomagnetic estimates of temperatures reached in contact metamorphism, *Geology* 9, 112-116.

McDougall, I., and D. H. Tarling, 1963, Dating of polarity zones in the Hawaiian Islands, *Nature* 200, 54-56.

McDougall, J., and D. H. Tarling, The magnetic provenancing of some Mediterranean obsidians, *Jour. Arch. Sci.* in press.

McElhinny, M. W., 1973, *Paleomagnetism and Plate Tectonics.* London: Cambridge University Press, 358p.

Nagata, T., 1965, Main characteristics of the geomagnetic secular variation, *Jour. Geomagnetism and Geoelectricity* 17, 263-276.

Park, J. K., 1983, Paleomagnetism for geologists, *Geoscience Canada* 10, 180-188.

Roy, J. L., and J. K. Park, 1974, The magnetization process of certain red beds: Vector analysis of chemical and thermal results, *Canadian Jour. Earth Sci.* 11, 437-471.

Schwarz, E. J., 1977, Depth of burial from remanent

magnetization: The Sudbury irruptive at the time of diabase intrusion (1250 ma), *Canadian Jour. Earth Sci.* **14**, 82–88.

Spariosu, D. J., D. V. Kent, and J. D. Keppie, 1984, Late Paleozoic motions of the Meguma terrane, Nova Scotia; New paleomagnetic evidence, in R. Van der Voo, N. Bonhommet, and C. Scotese, eds., *Plate Reconstruction from Paleozoic Paleomagnetism.* AGU Geodynamics Series, vol. 12. Washington, D.C.: American Geophysical Union, 82–98.

Stacey, F. D., 1977, *Physics of the Earth* New York: John Wiley & Sons, 414p.

Vine, F. J., and D. H. Matthews, 1963, Magnetic anomalies over oceanic ridges, *Nature* **199**, 947–949.

Wegener, A., 1924, *The Origin of Continents and Oceans.* Translated by J. G. A. Skerl. London: Methuen.

Cross-references: *Compass Traverse; Marine Magnetic Surveys; Plate Tectonics and Continental Drift; Sea Surveys; Surveying, General; VLF Electromagnetic Prospecting.* Vol. IVB: *Magnetic Minerals.* Vol. VI: *Anisotropy in Sediments.* Vol. XIII: *Magnetic Susceptibility, Earth Materials.*

GEOMATHEMATICS—See GEOSTATISTICS; MATHEMATICAL GEOLOGY.

GEOMECHANICS—See ARTIFICIAL DEPOSITS AND MODIFIED LAND; BLASTING AND RELATED TECHNOLOGY; CITIES, GEOLOGIC EFFECTS; CRATERING, ARTIFICIAL; LANDSLIDE CONTROL; MINE SUBSIDENCE CONTROL; PLATE TECTONICS AND CONTINENTAL DRIFT; SATELLITE GEODESY AND GEODYNAMICS. Vol. XIII: EARTHQUAKE ENGINEERING; RESIDUAL STRESS, ROCKS; RHEOLOGY, SOIL AND ROCK; ROCK MECHANICS.

GEOMETRONICS—See ACOUSTIC SURVEYS, MARINE; AERIAL SURVEYS; EXPLORATION GEOPHYSICS; HARBOR SURVEYS; MARINE MAGNETIC SURVEYS; PHOTOGEOLOGY; PHOTO-INTERPRETATION; REMOTE SENSING, GENERAL; SEA SURVEYS; SURVEYING, ELECTRONIC; VLF ELECTROMAGNETIC PROSPECTING; WELL LOGGING. Vol. XIII: REMOTE SENSING, ENGINEERING GEOLOGY

GEOMICROBIOLOGY

In the broadest terms, *geomicrobiology* is that branch of earth science that deals with the effects of microorganisms on geological processes. It is thus a subunit of *geobiology (biogeology)* that considers the role of all organisms in geological systems. Geomicrobiology is closely related to biogeochemical prospecting and overlaps considerably with biogeochemistry (q.v.). The terms *biogeonomy, lithobiology, lithobiontics,* and *geological microbiology* (Kuznetsov et al., 1963) are also related to the subject. The term *microgeology* (Ehrenberg, 1854) is sometimes applied to *micropaleontology* (q.v. in Vol. VII). The occurrence and description of fossil bacteria is also regarded as a part of paleontology, whereas geomicrobiology involves the description and analysis of dynamic processes—e.g., the cycling of material as influenced by microbial activities. Textbooks and reviews related to geomicrobiology were compiled by Davis (1967), Krumbein (1972, 1978, 1983), Kuznetsov et al. (1963), Silverman and Ehrlich (1964), Trudinger and Swaine (1979), Trudinger and Walter (1980), and Zajic (1969).

The main fields of study in geomicrobiology are extremely varied and wide ranging. The salient lines of experimental work include, e.g., the following topics: microbial weathering of rocks (including building stones) and soil formation processes; microbial influence on the genesis of rocks and ore deposits; microbial processes related to mineral formation and breakdown; microbiology of processes resulting in metal enrichment, described as microbial leaching; microbiology of the important mineral cycles (e.g., sulfur, nitrogen, phosphorus, silica, iron, manganese, and parts of the carbon cycle, q.v. in Vol. IVA and Vol. XII, Pt. 1); microbiology of carbonate equilibria (solution and precipitation); carbohydrate microbiology and its application to geology and environmental geology; groundwater microbiology; and microbial processes associated with chemical and atmospheric evolution.

The process of *mineralization* of an organic form of an element into the inorganic form is often governed by microbial processes. Immobilization of inorganic nutrients, oxidation and reduction of elements, volatilization or fixation of elements that can be transferred into a gaseous phase, and formation of geological deposits—i.e., rocks or ore deposits when influenced by microbial processes—are the contents of geomicrobiology. The production of organic chelating compounds (see Vol. XII, Pt. 1: *Metal-Complexing Micromolecules*), the absorption of microorganisms or their products to mineral particles, and changes of the inorganic environment produced by this activity as well as microbial isotope fractionation are also relevant topics. The energy transfer processes of all *chemolithotrophs*—i.e., microorganisms taking their energy from inorganic compounds—are also considered as part of the field of geomicrobiology. Thus, geomicrobiological studies are related to most of the important processes that take place within the upper part of the lithosphere where it is in close contact with the biosphere. Because microorganisms are the most resistant and widely distributed organisms in the biosphere, the scope of geomicrobiology is much wider than related disciplines such as geobiology, soil microbiology (q.v. in Vol. XII, Pt. 1), aquatic microbiology, atmospheric microbiology, and others, though these fields of study are related to the general subject.

Microorganisms have established ecological niches in geological environments where no other organism can survive. Bacteria survive at high hydrostatic pressures (more than 1,000 atm) as well as at the low pressures of the stratosphere. Growth occurs at less than 0.1 atm, which corresponds to Martian atmospheric pressure. They can survive, and some of them even metabolize, in the boiling waters of hydrothermal environments (Brock, 1978). They thrive in the hypersaline conditions of the Dead Sea and solar ponds (Krumbein and Cohen, 1974, 1977). They also live in the salty and dry environments of salt mines on traces of organic matter preserved in the salts.

Limiting Conditions in Geomicrobiology

Because the lithosphere, atmosphere, and hydrosphere overlap with the biosphere, forming a borderline life zone, it is important to indicate the limits of microbial growth. The following physicochemical parameters are the limits of metabolic activity (q.v. in Vol. XII, Pt. 1); many microorganisms can survive under even more extreme conditions in the dormant stage.

Bacteria metabolize at temperatures as low as -20 to $-40°C$ and they have been found to metabolize actively at temperatures close to the boiling point, and under high hydrostatic pressure, they can even metabolize at temperatures up to 105°C. The pH tolerance of microorganisms is very broad; fungi and some chemolithotrophic bacteria, e.g., metabolize at extremely low pH values. The pH range in which microbial metabolism is known to continue is from 0.5 to 13. The low values are favorable for fungi, while many cyanobacteria can survive at extremely high pH values. The oxidation-reduction potential ranges extend from approximately -450 mV to $+850$ mV. Experiments with marine bacteria have shown that they can metabolize still at pressures of 1,000 atm or more. The concentration of salts in the culture medium and in the environment is ruled by the solubilities of elements in water and the dielectric constant of water. It is well estab-

lished that bacteria survive and metabolize in brines that are close to complete dryness with extremely low water activity values (Shilo, 1979). The Dead Sea is the best studied habitat of *halophilic microorganisms* (30% salinity). Studies of "Permian" bacteria (Dombrowski, 1963; Bien and Schwartz, 1965) have, e.g., shown that water supply is the main factor that limits survival and metabolic activity in high salt concentrations. Water is also the most important limiting factor for microbial activity in the lithosphere. Wherever groundwater and water vapor of subcritical pressure occur, microorganisms are able to thrive. Microorganisms also live in environments almost devoid of water such as oil and jet fuels where the activity of water is extremely low. Bacteria multiply in dry rocks at depths where the geothermal gradient still allows water vapor to exist.

Bacteria are able to withstand extremely high doses of radiation—i.e., up to 10^6 r. Ultraviolet radiation is not as harmful to many bacteria as to macroorganisms, so they can survive in the outermost reaches of the Earth's atmosphere. Particular mineral environments also create an ecological niche for specialists that are highly adapted to special conditions.

Organisms Important in Geomicrobiology

The taxonomy of microorganisms is complicated because no natural (evolutionary) taxonomic system has been developed. Bergey's *Manual of Determinative Bacteriology* is the best source of information for bacterial names (Buchanan and Gibbons, 1974). Many microorganisms can be regarded as geological agents, and the following selection is only an indication of some of the more important ones.

Phototrophic Bacteria Bergey's *Manual of Determinative Bacteriology* 1974 is already outdated in respect to the *phototrophic bacteria*. All phototrophic bacteria are very important agents in geomicrobiology. They may be grouped today into the following groups:

1. Rhodospirillales including all obligately anoxygenic and obligately anaerobic phototrophs. Rhodospirillales have bacteriochlorophylls as light harvesting pigments.
2. Cyanobacteria including facultative anoxygenic and obligately oxygenic and facultative anaerobic as well as obligately aerobic genera. Cyanobacteria have chlorophyll a and phycobiliproteins as light harvesting pigments.
3. The obligately oxygenic and obligately aerobic photosynthetic bacteria with chlorophyll a and chlorophyll b but entirely prokaryotic structure and function. These bacteria are tentatively call Prochloron and have only been found as symbionts of didemnid ascidians.

Rhodospirillales The most important organisms in these groups of phototrophic bacteria are the Rhodospirillaceae, the Chromatiaceae, and the Chlorobiaceae, with *Rhodospirillum, Chromatium,* and *Chlorobium* as the most prominent genera. Most of the photosynthetic sulfur bacteria are capable of using H_2S as an electron donor for photosynthesis. Many are capable of assimilating CO_2 alone or other sources of organic carbon. According to ferredoxine analyses and other criteria, they belong to the first organisms developed on earth. They produce sulfur or sulfates from H_2S, and most of them live anaerobically.

Filamentous Sulfur Bacteria These bacteria are grouped into different orders. The sulfur-depositing *Beggiatoa* belongs to the order of gliding bacteria, together with other filamentous sulfur bacteria such as *Thiothrix* and *Thioploce. Thiospira* is considered to be filamentous by some Russian authors but, according to Bergey's *Manual,* belongs to the *Spirilla* with flagellae (Buchanan and Gibbons, 1974).

Cyanobacteria Cyanobacteria are also known as cyanophytes or blue-green algae. Cyanobacteria belong to the oldest taxa on Earth. They were the most important organisms in Precambrian rocks and formed some of the most interesting fossils of the Paleozoic, stromatolites. Recently, it has been shown that some cyanobacteria are capable of exerting both oxygenic and anoxygenic photosynthesis by using photosystems I and II in the case of oxygenic and photosystem I exclusively in the case of anoxygenic photosynthesis (Cohen, 1975). This was an important discovery because it is related to the development of an oxygenated atmosphere and to the evolution of organisms. An organism capable of producing organic matter with H_2S and H_2O as electron donors probably had many advantages in the early history of the Earth. Cyanobacteria are major factors in carbonate genesis, in iron deposition, as well as in the accumulation of organic matter in Recent sediments as examples from many intertidal environments have shown. They are excellent candidates for hydrocarbon sources and petroleum source rocks. They survive at high salinities, in low and high oxygen environments, and at very low light intensities. They can live at the bottom of lakes below two distinct plates of photosynthetic bacteria at light intensities of less than 0.2% of the surface light penetrating the water column (Cohen et al., 1977, Krumbein and Cohen, 1974, 1977).

Chemolithotrophic Bacteria In contrast to *chemoorganotrophic* (heterotrophic) bacteria, the chemolithotrophic bacteria derive their energy from the oxidation of reduced inorganic compounds such as sulfur, nitrogen, iron, manganese, and possibly antimony compounds. In many cases also hydrogen is an energy and electron donor.

Filamentous Chemolithotrophs Filamentous chemolithotrophs are the ensheathed bacteria of the *Sphaerotilus–Leptothrix* group (Dondero, 1975) that oxidize iron and manganese. *Crenothrix* and *Streptothrix* belong to this group as well. For all the aforementioned bacteria, real chemolithotrophy has not been established. Researchers assume in many cases, however, that they derive energy from oxidation of iron and manganese but require organic substances as a carbon source for growth. The same is suggested for members of the group of prosthecate bacteria and bacteria with appendages and holdfasts, commonly grouped as *budding bacteria.*

Budding Bacteria Prominent examples of this group are *Metallogenium,* the symbiotic iron and manganese oxidizer, and *Gallionella,* the iron oxidizer. Several of the *Hyphomicrobia* and *Pedomicrobia* also oxidize manganese or iron.

Nitrobacteriaceae These bacteria oxidize nitrogen compounds, produce NO_2 and NO_3 from ammonia and nitrite, and assimilate CO_2. They require no organic carbon. *Nitrobacter, Nitrococcus,* and *Nitrosomonas* are the most important genera.

Sulfur Oxidizers (Thiobacteria) These bacteria oxidize various reduced sulfur compounds and produce sulfur or sulfuric acid. Many of them need traces of organic matter or live a *mixotrophic existence;* i.e., they use sulfur compounds and reduced carbon compounds as energy source, while CO_2 is the main carbon source.

Most prominent is *Thiobacillus thioxidans. Thiobacillus denitrificans* is facultative anaerobic and under these conditions produces gaseous N_2 and NO. *Thiobacillus ferrooxidans,* formerly *Ferrobacillus ferrooxidans*, oxidizes reduced sulfur compounds and is involved in iron oxidation. Other sulfur oxidizers are the recently discovered thermophilic *Sulfolobus, Thiovolum, Thiospira,* and *Thiomicrospira.*

Siderocapsaceae These bacteria are able to deposit iron from ferrous iron solutions, and all are *organotrophs;* i.e., they use organic carbon compounds. Their only common feature is their ability to deposit iron and manganese; otherwise the four taxa differ largely. The validity of the taxa has been questioned because many heterotrophic bacteria are able to deposit iron and manganese hydroxides. The most common genera are *Siderocapsa, Naumanniella, Ochrobium,* and *Siderococcus.*

Methane-Producing Bacteria These bacteria are strict anaerobes that produce methane from CO_2 by using electrons generated from H_2 (chemolithotrophic) or formate and by fermenting acetate and methanol (chemoorganotrophic). A very peculiar physiologic group of geological importance includes *Methanobacterium, Methanosarcina,* and *Methanococcus* as dominant genera.

They have been grouped together with *Halobacterium* into the archaebacteria, which are especially offsprings of the evolution of the eukaryota.

Sulfate-Reducing Bacteria All sulfate reducers are very important since chemical sulfate reduction does not occur in natural habitats. The main genera, *Desulfovibrio* and *Desulfotomaculum* (formerly *Sporovibrio desulfuricans*), belong to different families; *Desulfovibrio* is strictly anaerobic, *Desulfotomaculum* may be facultative aerobic.

Large amounts of reducible sulfur compounds are transformed to H_2S, thus creating the sulfide environment in many sediments, soils, and rocks. Recently several other groups of sulfate- and sulfur-reducing bacteria have been found and described, e.g., *Desulfuromonas* (Widdel and Pfenning, 1984).

Antimony Oxidizing Bacteria Chemolithotrophic oxidation of antimony has been described by Lyalikova (1974), who named the organism *Stibiobacter*. The affirmation of the new genus is still lacking and may be debatable since, like the case of iron sulfides, sulfide ores are involved, and the presence of *Thiobacillus Y* is needed.

All the organisms mentioned previously belong almost exclusively to the geologic environment, being either chemolithotrophs or photolithotrophs or involved in geologically important electron-transferring actions. The following organisms are grouped here because many genera of them are of some geological importance.

Organotrophic Bacteria Many of the organotrophic soil bacteria living on a wide variety of substrates are largely changing the physical-chemical environment and thus producing geologically important reactions; e.g., many heterotrophic soil bacteria produce keto-gluconic acid, which attacks silicate minerals.

Actinomycetes These filamentous or rock-shaped bacteria live in geological environments. They influence solubilities of many minerals to a large degree, especially in the soil.

Fungi The imperfect fungi, and especially the Penicilliaceae, are important factors in the cycling of geological materials; e.g., a large variety of fungi are almost as effective in iron and manganese oxidation as *Thiobacillus ferrooxidans* and the *Sphaerotilus–Leptothrix* group. Many fungi oxidize NH_4^+ and H_2S. Green algae, diatoms, eukaryotic algae, dinoflagellates, coccolithophorids, and many other microalgae are effective in cycling geological material and forming geologically important deposits.

Lichens Many of the lichens are rock dwelling organisms. These complex thallophytic plants are made up of an alga and a fungus growing in symbiotic association on a solid surface such as a rock. Some are endolithic and only their fruiting bodies are visible. They are of great importance to geomicrobiological processes (Krumbein, 1969, 1972; Jackson and Keller, 1970).

Protozoa Radiolaria, flagellates, ciliates, rotatorians, and many other protozoans are involved in cycling silica, carbonate, and many other compounds. Because they catalyze and influence several geomicrobiological processes, they are included in this list of microorganisms that are of geological importance.

Microorganisms in Geological Processes

Because such a large variety of different organisms and communities influence geochemical cycles, this section only indicates some of the more important and peculiar factors pertinent to geomicrobiology.

Rock Weathering From polar to tropical regions, in deserts, on high mountains, and on many buildings or free standing sculptures of historical importance, a variated microflora is developing that invades more than 2–3 cm of the uppermost layer of the rocks. Some microorganisms, such as epilithic lichens, are not detrimental because they protect rock surfaces from physicochemical weathering. Chemolithotrophic, heterotrophic, and many photolithotrophic microorganisms are, however, active in rock decay, where they dwell in rock crevices, bore, and dissolve rock material. They form new minerals and even destroy objects of art by creating detrimental coating, crusts, and films. Krumbein (1969, 1972) has indicated some of the detrimental influences on building stones and other natural rocks.

Silverman and Ehrlich (1964) have compiled a list of minerals that are dissolved and precipitated by microorganisms. Quantitative values of manganese iron transfer in the weathering of rocks, including the mobilization of silica, have been summarized by Krumbein (1969) and Berthelin (1971). Many fungi in soil and on rocks mobilize large amounts of sodium from feldspars, wherever organic acids and chelating compounds of organotrophic bacteria and lichens are able to mobilize and immobilize alumina, iron, alkali, and earth-alkali elements. Silica is also mobilized and transported by bacteria. Weathering of cherty rocks in desert regions is frequently associated with endolithic lichen. The formation of a rock varnish often is related to epilithic and endolithic microbial "solution fronts" (Krumbein, 1969). The destruction of the famous ancient paintings in the cave of Lascaux was caused by the growth of algae and fungi that resulted in the deterioration of the colors, the pigments of which were combined with organic compounds. Organotrophic and chemolithotrophic bacteria were also softening and liquefying the hard carbonate sinters supporting the paintings. Accounts in the Bible also refer to rock maladies and report deterioration of building stones (Genesis III). Many of the processes of microbial decomposition are still unknown, and the quantitative effect on important rocks is still debatable (Krumbein, 1973). Modern methods such as scanning electron microscopy combined with electron dispersive X-ray analysis may help to clarify the role of microorganisms in the destruction of rocks and in soil formation.

The application of protective coatings containing antibiotics, formaldehyde, or hydrophobic substances might inhibit the microbial attack and reduce weathering of rock artifacts because these organisms need at least small amounts of water. Such protective measures are probably too expensive for large-scale application and, in many cases, require additional research.

Rock Formation No sedimentary rock has developed since Precambrian time without the involvement of hundreds of thousands microorganisms per gram of sediment, at least in the early stages of sediment formation. Two important biological factors associated with sedimentation include sulfate reduction and carbonate deposition (Berner, 1971; Goldhaber and Kaplan, 1974, Krumbein, 1979). Almost all sedimentary rocks that pass through a phase of organic enrichment undergo complicated steps of microbial mineral solution and formation. Several of the processes involved have been described for stromatolitic rocks (Krumbein and Cohen, 1974), anoxic sediments (Kaplan and Rittenberg, 1964), and sulfur-bearing sediments (Kuznetsov et al., 1963; Jorgensen, 1978). Concretions, fossils, birdseye structures, and pseudostylolites are examples of structural features caused by bacterial activity in sediments. The most prominent effects of microbial activity tend to be best preserved in siliceous deposits, carbonates, siltstones, claystones, and many sandstones. Figures 1 and 2 illustrate the population of sediments by different microorganisms and the minerals deposited in these microhabitats.

Mineral Deposits

Enrichment of economic minerals is frequently attributed, at least in part, to metabolic activity in the surface zone. The enrichment factor, in relation to average concentrations, often reaches 50,000 times or more in some cases. The enrichment of gold by organotrophic bacteria, e.g., has been reported to occur in the *rhizosphere* (Pares, 1964). Many of the factors leading to enrichment by geomicrobiological processes have been described by Silverman and Ehrlich (1964).

The genesis of sulfide ores is regarded as being almost exclusively caused by microorganisms.

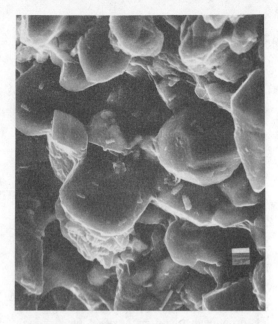

FIGURE 1. Calcium carbonate sediment in an algal mat environment. Bacteria are growing partly in the form of microcolonies on the minerals. Some of the bacteria exhibit initial steps of crystallization into carbonate (?) particles (bar = 5μm).

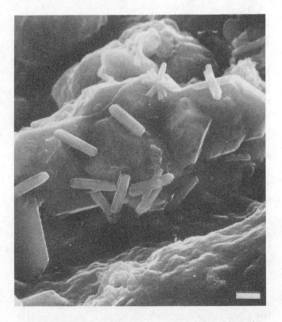

FIGURE 2. In comparison to Fig. 1, it is evident that some of the structures shown here are lithified bacteria. Noncalcified bacteria are visible as well. Three different species of bacteria are present. A star-shaped colony (middle center) looks like a bundle of aragonite. Electron dispersive X-ray analysis indicates that these stars are actually bacterial cells in different stages of calcification (bar = 2μm).

Iron hydroxide ores, oolitic iron ores, lead deposits, Kupferschiefer deposits, and many other sulfide ores are believed to be the result of *Desulfovibrio* activity in anoxic sediments in basins similar to the Black Sea. Itabirite formations and banded iron formations of Precambrian shield areas do appear to be bacteriogenic because much evidence suggests that these banded iron ores were formed in areas that are comparable to modern stromatolitic environments (Hartman, 1984), or under the influence of primitive plankton blooms. That many manganese ore deposits are bacteriogenic is based on evidence of microbial influences on manganese nodule formation. LaRock and Ehrlich (1975), Nealson (1981) and Schneider (1975), e.g., suggest that microorganisms are involved in the oxidation and precipitation of manganese ores in the deep sea. The ability of microorganisms to form acids, chelating compounds, hydrogen, sulfide, and oxides is an important evolutionary development that must have played a major role in former periods of Earth history. It thus seems likely that forming processes played a less prominent role after the development of the eukaryotic cell, which drastically altered the nature of the biosphere. In fact, Lovelock (1979) suggests, with good evidence, that life, as a regularity of nature, is controlling completely the geochemical cycles of the crust and in the atmosphere.

Microbial Leaching

The important industrial process of microbial leaching is applied on a large scale in the United States. Research groups in other countries are also investigating possible methods for commercial leaching of low-grade copper, zinc, and uranium ores. Most research is concerned with acidophilic *Thiobacilli,* but the use of heterotrophic bacteria, especially *Metallogenium* and fungi, is also being investigated (Schwartz, 1977; Murr et al., 1978; Trudinger and Walter, 1980).

Mineral Cycles

As indicated in most tests that deal with soil microbiology (q.v. in Vol. XII, Pt. 1), general microbiology, and biogeochemistry (Trudinger and Swaine, 1979; Fenchel and Blackburn, 1979), the mineral cycles (q.v. in vol. IVA and XII, Pt. 1) of carbon, nitrogen, sulfur, iron, and manganese at the Earth's surface, in both aquatic and terrestrial environments, are governed mainly by microbial activities. Sulfur, nitrogen, and carbon are reduced and transformed to organic compounds, recombined, and mineralized, oxidized, and again reduced in complicated biogeochemical cycles that are ruled by anaerobic and aerobic, organotrophic and chemolithotrophic, and phototrophic

microorganisms. The main mineral cycles are well documented, but the cycles of silica and phosphorus are less well established. Silica, e.g., is also incorporated, dissolved, precipitated, and reduced into organic compounds by microorganisms (Lauwers and Heinen, 1974). Microbial mobilization and transport appear to play at least a role in the development of many siliceous sedimentary cherts and deep-sea deposits or in desert environments.

Microbiology of Carbonates and Carbonate Deposits

It has become evident that many microbial processes are involved in the formation of structureless carbonate rocks. Beachrock (Krumbein, 1979), stromatolites, fine-grained carbonate sediments, and possibly many of the structureless fossil limestones are produced by the decay of organic matter rather than by photosynthetic processes. The Black Sea carbonates are an excellent example of the changing concepts of carbonate formation. Researchers originally thought that these carbonates are produced by the activity of sulfate-reducing bacteria (*Desulfovibrio*). Müller and Blaschke (1969), however, later showed that coccolithophorids contributed to the formation. In addition Krumbein (1974) and Krumbein and Cohen (1974) have found that many of the carbonate granules that are more frequent than the coccolithophorids may be of bacterial origin, forming in a decay environment of sedimenting organic matter.

The destruction and dissolution of carbonate rocks often is associated with coastal deterioration by microorganisms (Schneider, 1967). In karst areas and desert environments, the solution of carbonates by endolithic lichens and bacteria is frequently more rapid than the initiation of karst formation.

W. E. KRUMBEIN

References

Berner, R. A., 1971, *Principles of Chemical Sedimentology.* New York: McGraw-Hill, 240p.
Berthelin, J., 1971, Alteration microbiologique d'une arène grantique. Science du Sol. *Assoc. Français Étude Sol Bull.* **1**(suppl.), 11–29.
Bien, E., and W. Schwartz, 1965, Geomikrobiologische Untersuchungen, VI. Über das Vorkommen lebender und toter Bakterienzellen in Salzgesteinen, *Zeitschr. Allg. Mikrobiologie* **5,** 185–205.
Brock, T. D., 1978, *Thermophilic Microorganisms and Life at High Temperatures.* New York: Springer, 465p.
Buchanan, R. E., and N. E. Gibbons, 1974, *Bergey's Manual of Determinative Bacteriology.* Baltimore: Williams & Wilkins, 410p.
Cohen, Y., 1975, Dynamics of Procaryotic Photosynthetic Communities of the Solar Lake, Ph.D. thesis, Hebrew University, Jerusalem, 125p.

Cohen, Y., W. E. Krumbein, M. Goldberg, M. Shilo, 1977a, Solar Lake (Sinai) 1. Physical and chemical limnology, *Limnology and Oceanography* **22**, 597–608.

Cohen, Y., W. E. Krumbein, M. Shilo, 1977b, Solar Lake (Sinai) 2. Distribution of photosynthetic microorganisms and primary production, *Limnology and Oceanography* **22**, 609–620.

Cohen, Y., W. E. Krumbein, M. Shilo, 1977c, Solar Lake (Sinai) 3. Bacterial distribution and production, *Limnology and Oceanography* **22**, 621–634.

Cohen, Y., W. E. Krumbein, M. Shilo, 1977d, Solar Lake (Sinai) 4. Stromatolitic cyanobacterial mats, *Limnology and Oceanography* **22**, 635–656.

Davis, J. B., 1967, *Petroleum Microbiology.* Amsterdam: Elsevier, 605p.

Dombrowski, H. J., 1963, Lebende Bakterien aus dem Paläozoikum, *Biol. Zentral-blatt* **82**, 477–484.

Dondero, N. C., 1975, The Sphaerotilus–Leptothrix group, *Ann. Rev. Microbiology* **29**, 407–428.

Ehrenberg, C. G., 1854, *Zur Mikrogeologie—Das Erden und Felsen schaffende Wirken des unsichtbar kleinen selbstandigen Lebens auf der Erde.* Leipzig: Voss, 374p.

Fenchel, T., and T. H. Blackburn, 1979, *Bacteria and Mineral Cycling.* London: Academic Press, 225p.

Goldhaber, M. B., and I. R. Kaplan, 1974, The sulfur cycle, in E. D. Goldbert, ed., *The Sea*, vol. 5, *Marine Chemistry.* New York: John Wiley & Sons, 569–655.

Hartman, H., 1984, The evolution of photosynthesis and microbial mats: A speculation on the banded iron formations, in Y. Cohen, R. H. Castenholtz, and H. O. Halvorson, eds., *Microbial Mats: Stromatolites.* New York: Liss, 449–453.

Jackson, A. T., and D. W. Keller, 1970, A comparative study of the role of lichens and "inorganic" processes in the chemical weathering of recent Hawaiian lava flows, *Am. Jour. Sci.* **269**, 446–466.

Jorgensen, B. B., 1978, A comparison of methods for the quantification of bacterial sulfate reduction in coastal marine sediments, *Geomicrobiology Jour.* **1**, 29–47.

Kaplan, I. R., and S. C. Rittenberg, 1964, Basin sedimentation and diagenesis, in E. D. Goldbert, ed., *The Sea*, vol. 3. New York: Wiley Interscience, 963p.

Krumbein, W. E., 1969, Über den Einflub der Mikroflora auf die exogene Dynamik (Verwitterung und Krustenbildung), *Geol. Rundschau.* **58**, 333–363.

Krumbein, W. E., 1972, Role des microorganismes dans la génèse, la diagénèse et la dégradation des roches en place, *Rev. Ecol. Biol. Sol* **9**, 283–319.

Krumbein, W. E., 1973, Über den Einflub von Mikroorganismen auf die Bausteinverwitterung—eine ökologische Studie, *Deutsch. Kunst- und Denk- malspflege* **31**, 54–71.

Krumbein, W. E., 1974, On the precipitation of aragonite on the surface of marine bacteria, *Naturwissenschaften* **61**, 167.

Krumbein, W. E., ed., 1978, *Environmental Biogeochemistry and Geomicrobiology.* Ann Arbor, Mich.: Ann Arbor Science, 1055p.

Krumbein, W. E., 1979, Photolithotrophic and chemoorganotrophic activity of bacteria and algae as related to beachrock formation and degradation (Gulf of Aqaba, Sinai), *Geomicrobiology Jour.* **1**, 139–203.

Krumbein, W. E., ed., 1983, *Microbial Geochemistry.* Oxford: Blackwell.

Krumbein, W. E., and Y. Cohen, 1974, Biogene, klastische und evaporitische Sedimentation in einem mesothermen monomiktischen ufernahen See (Golf von Aqaba), *Geol. Rundschau* **63**, 1035–1064.

Krumbein, W. E., and Y. Cohen, 1977, Primary production, mat formation and lithification: Contribution of oxygenic and facultative anoxygenic cyanobacteria, in E. Flugel, ed., *Fossil Algae.* Berlin: Springer, 37–56.

Kuznetsov, S. I., M. V. Ivanov, and N. N. Lyalikova, 1963, *Introduction to Geological Microbiology* (transl. from Russian). New York: McGraw-Hill, 252p.

LaRock, P. A., and H. L. Ehrlich, 1975, Observations of bacterial microcolonies on the surface of ferromanganese nodules from Blake Plateau by scanning electron microscopy, *Microbial Ecology* **2**, 84–96.

Lauwers, A. M., and W. Heinen, 1974, Bio-Degradation and Utilization of Silica and Quartz, *Arch. Microbiol.* **95**, 67–78.

Lovelock, J. A., 1979, *Gaia. A New Look at Life on Earth.* Oxford: Oxford University Press, 157p.

Lyalikova, N. N., 1974, Stibiobacter senarmontii—a new antimony-oxidizing microorganism, *Mikrobiologiya* **43**, 941–948.

Müller, C., and R. Blaschke, 1969, Zur Entstehung des Tiefsee-Kalkschlammes im Schwarzen Meer, *Naturwissenschaften* **56**, 561–562.

Murr, L., A. E. Torma, and J. A. Brierley, eds., 1978, *Metallurgical Applications of Bacterial Leaching and Related Microbiological Phenomena.* New York: Academic Press, 526p.

Nealson, K. H., 1981, The microbial manganese cycle, in W. E. Krumbein, ed., *Microbial Geochemistry.* Oxford: Blackwell, in press.

Parès, Y., 1964, Action de Serratia marcescens dans le cycle biologique des métaux, *Annales Inst. Pasteur* **107**, 132–143.

Schneider, J., 1967, Biological and inorganic factors in the destruction of a lime stone coast, *Contr. Sedimentology* **6**, Schweizerbarth'sche Verlagsbuchhandlung, Stuttgart, 112p.

Schneider, J., 1975, Manganknollen—Rohstoffquelle und Umweltproblem für die Zukunft, *Umschau* **75**, 724–726.

Schwartz, W., ed., 1977, *Conference on Bacterial Leaching 1977.* Weinheim: Verlag Chemie, 270p.

Shilo, M., 1979, *Strategies of Microbial Life in Extreme Environments,* Dahlem Konferenzen, Weinheim: Verlag Chemie.

Silverman, M. P., and H. L. Ehrlich, 1964, Microbial formation and degradation of minerals, *Adv. Appl. Microbiology* **6**, 153–206.

Trudinger, P. A., and D. J. Swaine, ed., 1979, *Biogeochemical Cycling of Mineral-Forming Elements.* Amsterdam: Elsevier, 612p.

Trudinger, P. A., and M. R. Walter, 1980, *Biogeochemistry of Ancient and Modern Environments.* Berlin: Springer, 723p.

Widdel, F., and N. Pfennig, 1984, Dissimilarity sulfate- or sulfur reducing bacteria, in N. R. Krieg and J. G. Holt, eds., *Bergey's Manual of Systematic Bacteriology,* vol. 1. Baltimore: Williams & Wilkins, 663–679.

Zajic, J. E., 1969, *Microbial Biogeochemistry.* New York: Academic Press.

Cross-references: *Biogeochemistry; Geobotanical Prospecting; Lichenometry.* Vol. XII, Pt. I: *Carbon Cycle; Enzyme Activity; Nitrogen Cycle; Phophorus Cycle; Soil Microbiology; Sulfur Transformations.*

GEOMYTHOLOGY

Geomythology explores the relationship between geology and various kinds of folklore. There are at least five ways in which these apparently disparate fields are interrelated. First are the imaginative attempts by early or primitive peoples to explain various features of their natural environment such as landforms of striking aspect or recurrent phenomena like earthquakes and volcanic eruptions. These are *geomyths* proper and are what folklorists call *etiological,* or explanatory, myths. They are the most common type of geologically inspired folklore.

The second category includes traditions that embody the memory of real events. These are *legends* proper. Often the germ of fact that underlies a legend has become so embellished, usually with supernatural elements, that it is difficult, if not impossible, to discern the historical content and therefore difficult, if not impossible, to determine whether a given tale is a myth or legend in the sense those terms are used here. The most interesting, and at the same time the most speculative, aspect of geomythology is the attempt to trace legends to their possible factual source.

The third category is not folklore in the strict sense, because it is a purely literary invention, but it is somewhat analogous to the real folk tales, or *märchen* of oral tradition, in that it is intended to amuse and has never been seriously believed, as myths and legends were and often still are believed. This type has been termed *fakelore.*

A fourth category might be called *factlore.* It includes ideas long considered to be pure folklore that turn out to have some scientific basis.

Finally, there is a reciprocal relationship whereby geology has benefited from folklore. This relationship is largely manifested in connection with nomenclature, but occasionally a legend has been of more direct help to the geologist. Examples of all these types of interrelationships are given below, together with a discussion of the place of Atlantis in geomythology.

Geomyths

A typical example of a true geomyth is provided by Devils Tower in northeastern Wyoming, a volcanic neck or plug of phonolite that stands out sharply as an erosion remnant above its less resistant surroundings. Columnar jointing is so perfectly developed that it looks artificial to one not familiar with this rather common phenomenon, developed as lava cools and contracts. There are two versions of the legend of the origin of Devils Tower, but in both of them the butte was formed when a party fleeing from a giant bear appealed to their gods for aid. The rock on which they stood suddenly shot up to a great height, carrying them out of the bear's reach, and the fluting of its sides represents the claw marks made by the frustrated animal as it tried to reach the people on top (Fig. 1).

Folklore concerning the causes of earthquakes almost universally attributes the shaking of the ground to the movement of some being or creature who supports the Earth or lives within it, or in a few cases treads heavily on it. Myths accounting for volcanic eruptions in general also usually involve some creature imprisoned in the Earth whose exhalations and activity are manifested as volcanism, including volcanic earthquakes.

Only rarely do geomyths of this type bear any resemblance to the geologic facts concerning the actual cause of the feature they purport to explain and then essentially by coincidence. But shrewd observation on the part of the Polynesians who settled the Hawaiian Islands is evident in their tale of the coming of the volcano goddess Pele to their land. They say that she first sought a fiery home on Kauai, the northwesternmost of the major is-

FIGURE 1. The legend of Devils Tower, Wyoming. (Courtesy of U.S. National Park Service)

lands, then moved southeastward down the chain, successively digging craters in search of fire on Oahu, Molokai, Maui, and finally Hawaii, where she resides to this day in Halemaumau, the fire pit of Kilauea volcano. In fact, the volcanism of the Hawaiian Islands is progressively younger from northwest to southeast, with recent activity concentrated at Mauna Loa and Kilauea on Hawaii. That this is the case could easily have been comprehended by the natives of the islands, from the relative degree of weathering and erosion of the older and younger volcanoes and the difference in thickness of the vegetation cover developed on their slopes.

Legends

There cannot be any doubt that some traditions from the distant past might be based in part on fact, for legends complete with supernatural elements have been engendered by well-documented events that occurred quite recently. For instance, the Tarawera eruption in New Zealand in 1886 destroyed a Maori village, killing more than 100 people. The disaster was immediately interpreted as punishment for some transgression by the villagers, and two legends soon arose. One blames the eruption on the fact that the people of the village that was destroyed broke a taboo of the sacred mountain by eating forbidden wild honey. The other legend attributes it to a demon named Tamaohoi, long imprisoned in the mountain, who was summoned forth by one of his descendants, the *tohunga* (priest or medicine man) Tuhoto, to punish the villagers for their decline in morals. The *tohunga* of the village at the time of the eruption was named Tuhoto, and he did have an ancestor named Tamaohoi.

The prevalence of flood traditions around the world is often cited as evidence that a worldwide catastrophe must have occurred during the memory of humankind. However, if one remembers that floods in the plural are a practically worldwide phenomenon and that flood traditions are conspicuously absent in the very parts of the world least likely ever to have suffered a catastrophic flood (such as Egypt and central Asia), it seems far more likely that many flood traditions are derived from independent events. Others are geomyths rather than legends, inspired, i.e., by the finding of fossil shells in rocks high above present sea level. Noah's deluge, demonstrably derived from a much earlier Sumerian legend, is believed by many biblical scholars to be the memory of a disastrous inundation in Mesopotamia in about 3000 B.C. The fact that since the beginning of Christianity missionaries have been zealously carrying the story of Noah to all corners of the Earth—so that in many cases it merged with preexisting flood traditions or in others was trans-

planted virtually intact to a new setting—has also played a large part in creating the impression that there must have been a very widespread inundation a long time ago. However, volcano legends from widely separated parts of the world show similarities just as striking, which suggests that those who witness eruptions experience much in common, just as do the survivors of floods; perhaps if volcanoes were not restricted to certain parts of the Earth and eruptions were as common as floods, there would also be a tradition of a worldwide volcanic cataclysm.

Fakelore

The Paul Bunyan stories are an example of fakelore. For instance, before Paul and his companion Babe the Blue Ox came along, the Colorado River was known as Old Contrary because in some stretches it was a mile wide and a foot deep and others a foot wide and a mile deep. Hitching Babe to a bulltongue plow, Paul widened it where it was deep and deepened it where it was wide just to even things up.

Factlore

The association of the catfish with earthquakes in Japan might prove to be an instance of factlore. Folklore had it that the wriggling of a great catfish inside the Earth was the cause of earthquakes. In line with this belief were numerous reports of unusual behavior of real catfish shortly before major earthquakes. Such reports were so persistent that scientists began to wonder if in fact the fish were somehow sensitive to minute stress-induced changes in the Earth. Experiments seemed to indicate that catfish in tanks in the laboratory were more sensitive to certain stimuli about 6–8 hr before a shock, providing the water circulating through the tanks passed through the ground (Hatai and Abe, 1932). The sensitivity was thought to have something to do with natural electrical currents in the Earth. While the results of these experiments cannot by any means be considered conclusive, seismic stresses do have measurable effects on the electrical and magnetic fields of the Earth, and those effects are currently being investigated in connection with the search for means of predicting earthquakes. Thus, it may turn out that certain creatures, among them perhaps the catfish, are sensitive enough to those effects that they can sense impending shocks. In that case, the association of the catfish with earthquakes would not be entirely fortuitous.

Misconceptions

Chief among the widely prevalent misconceptions is the notion that in an earthquake, yawning chasms can open up and swallow anyone or anything that happens to be in the wrong spot, then close up again, squashing the unfortunate creature or object. Of all the dangers to be feared in connection with earthquakes, that is the very least. Cracks commonly do open up in the ground, but they are always superficial and cases where they have closed up again are so rare that they are almost nonexistent. Furthermore, there is not a single absolutely unquestionable case of any person or animal being swallowed up in such a crack; the two best authenticated instances—a cow in the San Francisco earthquake of 1906 and a woman in the Fukui earthquake in Japan in 1948—have other plausible explanations.

Terminology

Geology is indebted to folklore for the word *volcano,* from the god Vulcan of Roman mythology, and for *Pele's hair* and *Pele's tears,* terms from Hawaiian folklore referring to fine-spun volcanic glass and droplets of lava whipped from the surface of moving flows or from firefountains. Other instances of nomenclature derived from folklore can be cited.

Geologists have also considered legends in assessing volcanic hazards in the Pacific northwest and New Zealand because they frequently constitute a record of activity within the past few hundred years, before the coming of the white race. A Maori legend telling of a red glow in the sky over Rangitoto Island in Auckland Harbor, the youngest cone in the vicinity of that important city, suggested that the cone had been active since the Maoris arrived in New Zealand about 800 yr ago; this information was subsequently confirmed by radiocarbon dating of its youngest eruption as 225 ± 110 years ago—too young for comfort. In another case, a local legend helped to pin down the date of the last eruption on Lipari, in the Eolian Islands in the Tyrrhenian Sea, to within about 50 yr.

Atlantis

It is not certain just where Atlantis belongs in geomythology. It is not part of the traditions of any group of people; the first mention of it anywhere is in two of Plato's *Dialogs,* the *Timaeus* and the *Critias.* If it is a legend embodying a fragment or fragments of historical fact, we have to assume that it reached Plato in a much distorted form and then decide which elements of his description should be taken literally and which as distortions. On that basis a case can be made for almost any part of the Earth. The best candidate yet proposed as the prototype of Atlantis (in that it combined both the essential elements, a civilization advanced for its time and a catastrophic destruction involving submergence) is the Minoan civilization on Crete and other Aegean islands. It

collapsed with unusual rapidity at approximately the same time that there was a Krakatoa-like eruption of the nearby Santorini (Thera) volcano that ended with a caldera collapse in the center of the island. But it has by no means been proved that the general destruction on Crete was a direct result of the eruption, and in fact some evidence suggests that the eruption occurred about 50 yr sooner (Doumas and Papazoglou, 1980; Vitaliano and Vitaliano, 1974, 1978). Conversely, if Atlantis is a myth, in the sense that Plato invented it to prove a philosophical point, it is possible that he incorporated some bits of fact from various sources, in which case the (to him mythical) Minoans and the geography of Crete could have been among those sources. But that alone would not remove Atlantis from the realm of myth.

Further examples of the interrelationship between geology and folklore and fuller details concerning most of the examples given here can be found in Vitaliano (1973).

DOROTHY B. VITALIANO

References

Doumas, C., and L. Papazoglou, 1980, Santorini tephra from Rhodes, *Nature* **287,** 322–324.
Hatai, S., and N. Abe, 1932, The responses of the catfish, *Parasilurus asotus,* to earthquakes, *Imperial Academy (Tokyo) Proc.* **8,** 375–378.
Vitaliano, D. B., 1973, *Legends of the Earth: Their Geologic Origins.* Bloomington: Indiana University Press, 305p.
Vitaliano, C. J., and D. B. Vitaliano, 1974, Volcanic tephra on Crete, *Am. Jour. Archaeology* **78,** 19–24.
Vitaliano, D. B., and C. J. Vitaliano, 1978, Tephrochronological evidence for the time of the Bronze Age eruption of Thera, in *Thera and the Aegean World, Second International Scientific Congress, Santorini, Greece, August 1978, Acta,* 217–219.

Cross-references: *Popular Geology; Serendipity.* Vol. XIII: *Geological Communication.*

GEONOMY

Wiechert used the term *geonomy* in the early 1900s, although it is not clear whether he coined the word or whether it is of still earlier origin. Field was an advocate of the term, using it frequently from about 1938 onward in various publications (e.g., in the *Transactions of the American Geophysical Union*). According to his definition, geonomy embraced the broad field of earth science, including even certain aspects of biology. It has been reintroduced a number of times, especially since the 1950s, each time with a somewhat different meaning. The term is included in some dictionaries and encyclopedias and omitted in others, suggesting that it is not universally accepted or precisely defined.

In 1964, Beloussov suggested that geonomy refers to the various disciplines involved in the study of the upper mantle. Van Bemmelen (1964), a long-term supporter of the word, applied it to the study of geodynamic phenomena on a global scale. He later, however, used it to cover the entire range of earth science, including geology, geophysics, geochemistry, and many other minor disciplines.

There has been considerable discussion on the desirability of replacing *geosciences* or *earth sciences* by *geonomy,* but there has been no consensus of opinion. The result of a questionnaire inserted in twelve journals in the geosciences indicated a small majority in favor of introducing the term to denote collectively all the sciences that deal with the solid Earth, the atmosphere, and the hydrosphere. The various arguments for and against the adoption have been summarized by Manten (1969).

J. A. JACOBS

References

Beloussov, V. V., 1964, The upper mantle project, *Research in Geophysics* **2**, 560.

Manten, A. A., 1969, Geonomy, geology or geo-sciences? *Earth-Sci. Rev./Atlas* **5A**, 88.

Van Bemmelen, R. W., 1964, Phénomènes géodynamiques à l'échelle du globe (géonomie), à l'échelle de l'écorse (géotectonique) et à l'échelle de l'orogenèse (tectonique), *Soc. Belge Géologie, Paléontologie et Hydrologie Mém.* **8**, 126p.

Cross-references: *Geohistory, American Founding Fathers; Geology, Philosophy; Geology, Scope and Classification.*

GEOPHILATELY

Philately, or stamp collecting, is an old hobby. Stamps illustrating geologic features are common; mountains, volcanoes, glaciers, lakes, waterfalls, rivers, minerals, and fossils are frequently illustrated. The number of stamps of geologic interest runs into the hundreds. *Scott's Standard Postage Stamp Catalogue* (1970), one of the important reference books of philately, is published yearly. *Lithos* and *Minereaux et Fossiles,* two French journals, regularly report new stamp issues that highlight geological features. The articles typically contain full-color illustrations and provide descriptive narratives for each stamp.

Devils Tower, Wyoming

The violet three-cent Devils Tower National Monument stamp was issued on September 24,

For this entry, the figures have been reproduced and the text adapted from R. L. Nichols, 1971, *Jour. Geol. Educ.* **19**, 176–181.

FIGURE 1. A violet stamp showing Devils Tower, Wyoming.

1956, to commemorate the fiftieth anniversary of the federal law providing for the protection of American natural antiquities (Fig. 1). Devils Tower (see *Geomythology*), a Wyoming landmark, is a pillar of phonolite porphyry that rises more than 300 m above the surrounding plain. It intrudes Triassic and Jurassic rocks and, on geologic evidence, is thought to be Tertiary in age (Darton and O'Harra, 1907). A potassium-argon age determination indicates that it is approximately 40 m.y. old (Bassett, 1961), which is consistent with its geologically determined age. Darton and O'Harra (1907) and Jaggar (1900) believed it to be an exhumed eroded laccolith. Dutton and Schwartz (1936) and Hunt (1967) consider it to be a volcanic neck. The tower is composed of huge columns, the size of which makes them almost unique because unbroken columns are hundreds of feet long and average 6 feet in diameter.

Crater Lake, Oregon

The national parks were publicized by a series of stamps issued in 1934. Crater Lake is shown on one of these stamps (Fig. 2).

Crater Lake is nearly circular, approximately 10 km in diameter, 600 m deep, and unbelievably blue. It occupies a caldera that is nearly 1,200 m deep, since the lake is surrounded by steep cliffs in places 620 m high. The caldera was formed by

FIGURE 2. A blue stamp showing Crater Lake, Oregon.

the collapse of a composite cone after its magma chamber had been emptied by pumice explosions and pyroclastic flows (Williams, 1942). The C^{14} determinations indicate that the collapse occurred about 6,000 yr ago when Indians were living in that part of Oregon. Following the collapse, Wizard Island, a cinder cone 240 m high, was formed. Tree-ring studies indicate that it is more than 800 yr old. The walls of the caldera are notched by U-shaped valleys cut by glaciers that flowed down the slopes of the volcano before its collapse. Llao Rock, which is part of the rim, is a transverse cross-section of a lava flow that ran down the outer slope of a volcano before the collapse and buried a U-shaped valley. The stamp, appropriately blue in color, shows the lake, Wizard Island, the steep cliffs, and Llao Rock.

The story revealed by the rocks that outcrop on the walls of the caldera is just as interesting as the history of the caldera. A part of the geologic history of the volcano that occurred before the collapse is shown by the stratigraphic sequence seen on the walls of the caldera at Pumice Point (Fig. 3). Intermittent pyroclastic activity, two glacial episodes, the extrusion of two lava flows, and the formation of a soil zone are indicated (Atwood, 1935). Elsewhere, upright carbonized tree trunks suggest nuée ardente (glowing cloud) deposits.

Old Faithful, Yellowstone National Park

Old Faithful is one of the series of stamps issued in 1934 to commemorate National Parks Year. It is an attractive blue, vertical, five-cent stamp (Fig. 4).

The geysers of the world are found mainly in Yellowstone National Park, Iceland, and New Zealand. There are approximately 200 active geysers in Yellowstone—more than in all the rest of the world. The most famous is Old Faithful, which was discovered geologically in 1870, where observers noticed that its interval of eruption was more regular than that of the other geysers. A study of many thousands of eruptions shows that the average interval is about 65 min. It throws a column of water between 32 m and 56 m in the air, depending on the violence of the eruption and the wind velocity (Marler, 1969). Its duration of eruption varies from 1.5 min to 5 min. It discharges approximately 39 m of water at a temperature above 92°C. Most of this water does not drain back into the crater but starts a long journey toward the Atlantic Ocean.

Many partly petrified stumps, logs, pine needles, and cones are found embedded in the geyserite of which the cone of Old Faithful is

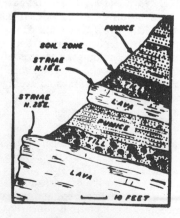

FIGURE 3. Stratigraphic section seen on the walls of the caldera of Crater Lake. (Atwood, 1935).

FIGURE 4. A blue stamp showing Old Faithful, Yellowstone National Park.

composed. They prove that after the cone had been built to approximately its present size, a period of hundreds of years of thermal inactivity ensued during which trees grew on the cone. This period was followed by renewed thermal activity that killed the trees.

Geyserite has been found in the glacial deposits of Yellowstone. Thermal activity occurred, therefore, as far back as interglacial time and has had a duration of many tens of thousands of years. The size of the cone of Old Faithful, coupled with knowledge of the rate of geyserite deposition, suggests that the cone is thousands of years old.

Ship Rock, New Mexico

A four-cent maroon U.S. stamp showing Ship Rock was issued on January 6, 1962, to commemorate the fiftieth anniversary of the admission of New Mexico to statehood (Fig. 5).

Ship Rock is a landmark approximately 450 m high. It is composed of volcanic breccia and tuff, which are cut by ramifying dikes. It is a volcanic neck—the tombstone of a volcano that has been almost completely destroyed. Only a part of its more resistant conduit filling remains. The dikes that radiate from Ship Rock are more resistant than the country rock. They project tens of meters above the surrounding plain and are among the most spectacular dike ridges in the United States (Fig. 6).

The area around Ship Rock is of great geologic interest. Uranium, helium, gas, oil, and vast resources of coal are found nearby. Laccolithic mountains, volcanic necks, mesas, buttes, hogbacks, natural bridges, and stream terraces are also present in the area.

Grand Canyon, Arizona

An attractive brown six-cent U.S. stamp was issued in 1969 to commemorate the one-hundredth anniversary of John Wesley Powell's ex-

FIGURE 6. A photograph showing one of the dikes that radiate from Ship Rock, New Mexico. (Professor Marshall Schalk)

pedition down the Colorado River (Fig. 7). The Colorado is one of the five great rivers of the United States. It rises in the northern Rockies and empties into the Gulf of California. It is the author of the Grand Canyon: by cutting slowly through long periods of time it has formed this great canyon. The Grand Canyon is approximately 320 km long, 6–30 km wide, and in places more than 1,525 m deep. It is, however, only one of many canyons through which the Colorado flows. The Black, Boulder, Marble, Glen, and Cataract canyons are others, but none equal the Grand Canyon in size, majesty, beauty, or the geologic story revealed. A view of the Grand Canyon from Yavapai Point is one of the grandest. It is perhaps equalled in the United States only by Half Dome in Yosemite Valley, Death Valley from Dante's View, Mt. Rainier from a plane flying into Seattle, the sea in a northeast storm off the coast of Acadia National Park in Maine, or the glacial or mountain scenery in Alaska and the Rockies.

The canyon is a chasm not only in space but also in time. On hiking down into the canyon, the

FIGURE 5. A maroon stamp showing Ship Rock, New Mexico.

FIGURE 7. A stamp showing Major Powell and his men on the Colorado River in 1869.

start is made on the Kaibab Limestone, approximately 200 m.y. old. In perhaps 4 hr, the bottom of the Tapeats Sandstone, immediately above the Inner Gorge, which is nearly 600 m.y. old, can be reached. For every minute of travel down the trail, the hiker has walked backward in time about 2 y.m.

The Colorado River is characterized by rapids and quiet stretches. The tributary streams, when in flood, deposit bouldery debris in the main valley. The rapids are found on the downstream side of these deposits and the quiet stretches on the upstream side. Before the Glen Canyon Dam was built, these deposits were periodically removed by the Colorado River during its big floods. In an interesting speculation, Professor Troy Péwé of Arizona State University has suggested that even the largest discharges through the Glen Canyon Dam will not be able to remove these deposits. They will therefore grow large. The rapids will develop into cascades and perhaps even into small waterfalls, and quiet stretches into small ponds.

The age of the Grand Canyon is something that intrigues the tourist. Field studies and potassium-argon dating indicate that the downcutting started sometime after 10.6 ± 1.1 m.y. ago and that by 1.16 ± 0.18 m.y. ago the Grand Canyon was approximately as large as it is now (Péwé, 1968; McKee et al., 1968).

Matterhorn, Switzerland

A Swiss thirty-centime stamp, issued on June 1, 1965, showing the Matterhorn, is shown in Fig. 8. It has a dark red background, and the rest of the stamp is white, green, and slate. It was issued to commemorate the centenary of the first ascent of the Matterhorn. The pleasing colors and its stark simplicity are striking.

The Matterhorn is perhaps the most famous and majestic mountain in the Alps. It stands

FIGURE 9. A Swiss stamp showing quartz crystals.

alone, is 4,481 m high, and reaches upward like the spire of a church. It was the last of the great mountains of Switzerland to be conquered. The name of Edward Whymper will be forever associated with the Matterhorn. He reached the summit on July 14, 1865, with six companions after seven unsuccessful attempts. The great tragedy of the first ascent was that four climbers died on the descent when the climbing rope parted and they fell 1,210 m.

The rocks in the Matterhorn are not in their normal stratigraphic sequence. Those in the upper part of the mountain have been thrust over those in the lower part. It is called a mountain without roots. The Matterhorn has given a part of its name to the geologic feature of which it is such a fine example; it is a *horn*. It has been sculptured out of a larger mass by the quarrying and plucking of the glaciers that have moved down its steep slopes. Switzerland has also issued an attractive series of stamps illustrating minerals and fossils (Fig. 9).

North Borneo

A rose-red and light-green ten-cent North Borneo stamp issued on February 1, 1961, features a map showing Borneo, Sumatra, and Java. The Sunda Strait between Java and Sumatra, in which the volcanic island of Krakatoa is located, is clearly shown in Fig. 10.

This volcano, after being dormant for 200 yr, erupted in August 1883. The whole configuration of the island was changed by the explosion (Fig. 11). Where the Earth's surface was 609 m above sea level before the eruption, it was more than 304 m below sea level after it. Sea waves 20–30 m high rolled up on the coasts of Java, Sumatra, and the nearby island and killed 40,000 people (Fig. 12). The waves crossed the Indian Ocean, rolled around the Cape of Good Hope, moved north-

FIGURE 8. A stamp showing the Matterhorn, Switzerland.

FIGURE 10. A rose-red and light-green North Borneo stamp showing Borneo, Java, Sumatra, and adjacent areas.

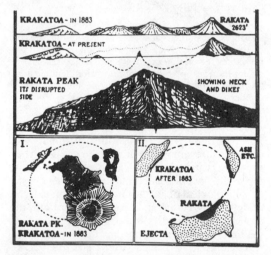

FIGURE 11. Cross-sections and maps showing Krakatoa before and after the eruption of 1883. (Royal Society Report)

FIGURE 12. A map showing in black the areas that were inundated by the sea waves from the 1883 eruption of Krakatoa. (Royal Society Report)

FIGURE 13. A brown-olive stamp featuring Mt. Ararat, Armenia.

ward in the Atlantic, entered the English Channel, and affected the tidal gauge record at Plymouth, England.

The sounds from the eruption were heard 4,800 m away and over more than one-twentieth of the Earth's surface. Had Krakatoa been located at San Francisco, the sounds would have been heard at Anchorage, Honolulu, Mexico City, and Chicago. Dust thrown into the upper atmosphere circled the planet and was responsible for the beautiful sunrises and sunsets of 1883 and 1884. An atmospheric shock wave circled the planet three times and was picked up by barometers all over the world.

Mt. Ararat, Turkey

A brown-olive 25,000-ruble stamp featuring Mt. Ararat with a winter snowline was issued by the Soviet Socialist Republic (Armenia) (Fig. 13). Ararat is an extinct volcano in Turkey. It has an altitude of 5,151 m and dominates the Armenian plateau, above which it rises approximately 4,000 m. It is composed of andesite, obsidian, and scoria. It has no crater, but well-preserved parasitic cones are found on its flanks. A permanent snowfield and one glacier are present.

To raise sea level so that Ararat was submerged, as told in the Bible, was truly a miracle because if all the moisture in the atmosphere fell as rain, it would cover the Earth with about 10 cm of water.

R. L. NICHOLS

References

Atwood, W. W., Jr., 1935, The glacial history of an extinct volcano, Crater Lake National Park, *Jour. Geology* **43**, 142–168.

Bassett, W. A., 1961, Potassium-argon age of Devils Tower, Wyoming, *Science* **134**(3487), 1373.

Darton, N. H., and C. C. O'Harra, 1907, Devils Tower folio, *U.S. Geol. Survey Folio 150.*

Dutton, C. E., and G. M. Schwartz, 1936, Notes on the jointing of the Devils Tower, Wyoming, *Jour. Geology* **44**, 717–728.

Hunt, C. B., 1967, *Physiography of the United States.* San Francisco: W. H. Freeman and Company, 224.

Jaggar, T. A., 1900, The Laccoliths of the Black Hills, *U.S. Geol. Survey 21st Ann. Rept.*, Part III, 163–303.

Marler, G. D., 1969, *The Story of Old Faithful.* Washington, D.C.: Yellowstone Library and Museum Association, in cooperation with National Park Service, U.S. Department of the Interior, 49p.

McKee, E. D., W. K. Hamblin, and P. E. Damon, 1968, K-Ar age of Lava Dam in Grand Canyon, *Geol. Soc. America Bull.* **79**, 133–136.

Péwé, T. L., 1968, *Colorado River Guidebook, Lees Ferry to Phantom Ranch.* Phoenix, Ariz.: Lebeau Printing Company, 78p.

Scott's Standard Postage Stamp Catalogue—the Encyclopedia of Philately, 1970, New York: Scott Publications.

Williams, H., 1942, The geology of Crater Lake National Park, Oregon, with a reconnaissance of the Cascade Range southwest to Mount Shasta, *Carnegie Inst. Washington Pub. 540.*

Cross-reference: *Popular Geology.*

GEOSCIENCE INFORMATION—See GEOLOGICAL CATALOGUING; MAP AND CHART DEPOSITORIES; PUNCH CARDS, GEOLOGIC REFERENCING; REMOTE SENSING, SOCIETIES AND PERIODICALS. Vol. XIII: CONFERENCES, CONGRESSES, AND SYMPOSIA; EARTH SCIENCE INFORMATION SOURCES; INFORMATION CENTERS.

GEOSTATISTICS

Literally, *geostatistics* refers to the statistics of the Earth. In general usage, however, it refers to the applications to, and use of, statistics in the solution of geological problems (Gary et al, 1972). Associated terms include *mathematical geology, geomathematics,* and *geometrics. Mathematical geology* and *geomathematics* are essentially equivalent and include the applications of mathematics to geological problems (Gary et al., 1972). The International Association for Mathematical Geology was formed in 1968 and publishes a journal (Merriam, 1969). According to the first president of the association (Vistelius, 1969, 1), mathematical geology "is a new science . . . brought

about by a happy marriage between mathematics and geology." Agterburg, in his comprehensive textbook, states that "geomathematics, in its broadest sense, includes all applications of mathematics to the earth's crust" (1974, p. 1). Several symposia and colloquia have been concerned with special areas of geostatistics (e.g., Merriam, 1970, McCammon, 1975), and an extensive listing of articles, textbooks, and symposia dealing with geomathematics (or mathematical geology) and geostatistics is included in Griffiths (1970). *Geometrics* is a term proposed as a geological analogue of biometrics, econometrics, psychometrics, and sociometrics (Griffiths, 1958); these terms imply that quantification (e.g., measurement) is a basic requirement and leads to greater rigor in definition of concepts (Griffiths, 1970, 1974).

A special usage of the term *geostatistics* has been coined by Matheron (1962, 1963, 1971), who defines it as follows:

Geostatistics, in their most general acceptation, are concerned with the study of the distribution in space of useful values for mining engineers and geologists, such as grade, thickness or accumulation, including a most important practical application to the problems arising in ore-deposit evaluation. [Matheron, 1963]

The procedures devised by Matheron are not, however, limited to applications in mining and ore-deposit evaluation but may be applied wherever observations, geological or otherwise, do not represent independent random samples. Since most geological observations possess some degree of dependence either in space or time or both, this aspect of geostatistics, the study of regionalized variables, is a very important development (see Agterburg, 1974; Davis, 1973; Clark, 1979).

Some form of geostatistics has been practiced since the beginnings of geology, and its origin may be traced from the early associations of Lyell, Galton, and Darwin (see Fisher, 1953, on Lyell's use of counting fossils as the basis of stratigraphic ordering of the Cenozoic, e.g., Griffiths in Milner, 1962, 565), through the emphasis on quantitative analysis by Sorby (1880) and Mackie (1897) and then via the study of accessory minerals (e.g., Milner, 1962) and grain-size analysis (Udden, 1898; Wentworth, 1929). It was, however, in the mid-1930s that present practice had its roots and it was W. C. Krumbein who initiated this phase and catalyzed its modern development (Krumbein and Pettijohn, 1938). Since that time, geostatistics has grown and flourished to become an accepted area of specialization in earth science.

The content of geostatistics includes specification of appropriate geological random variables and their frequency distributions (Krumbein and Graybill, 1965; Griffiths, 1967); the problems of sampling (Griffiths, 1971) and estimation (see Vol. VI: *Sediment Parameters,*); univariate analysis, which applies various tests for nonrandomness to one or more sets of single variables such as in the use of analysis of variance, chi square, and Student's *t* test (Griffiths, 1967; Krumbein and Graybill, 1965); bivariate analysis, wherein variables are treated in pairs and the form (regression) and degree (correlation) of their linear association are specified (Griffiths, 1968); and multivariate analysis, which commences with a rectangular data matrix ($n \times p$) of p variables measured on n samples and proceeds through the square ($p \times p$) covariance matrix to the square ($p \times p$) matrix of zero-order (linear) correlations. The first question is then to determine the number of linearly independent vectors in this matrix—i.e., the rank of the matrix. The procedure is some form of factor analysis, usually principle components analysis (e.g., Agterburg, 1974; Davis 1973).

Subsequent questions may concern classification and nomenclature using one of the many types of cluster analyses (Davis, 1973; Siegal and Griffiths, 1975), or the question may concern the relationships between a set of predictor variables (X_{ij}) and a dependent variable to be predicted as in multiple regression (Davis, 1973). This procedure may be extended to include a specification of the relationships between two or more sets of variables as in canonical correlation (Lee, 1969). A common characteristic of the fields of geology and geography is their interest in spatial variation so that the preceding techniques may be extended to variation in two dimensions, and multiple regression then becomes trend-surface analysis (Krumbein and Graybill, 1965; and Davis, 1973).

The time dimension leads to another set of procedures called time series analysis in which the lack of independence between successive observations is explicitly recognized as the salient characteristic of the set of observations. This is a relatively new area of development in geostatistics, and the procedures (Kendall, 1973) include univariate, bivariate, and multivariate analyses analogous to that for independent random samples. Matheron's *geostatistique* is one series of developments properly in this domain (Matheron, 1962). Appropriate use of *regionalized variables* extends *time series analysis* to the spatial domain.

Finally, the evolution and growth of geostatistics and geomathematics have been accelerated by the parallel growth of electronic computers since the mid-1950s. Ready accessibility to computer facilities is now essential to the extensive use of statistics in any field, and geostatistics is no exception (see Merriam, 1975). Geostatistics also formed the basis for wider applications of mathematics to problem solving in the geosciences, encompassing the use of operations research, cybernetics, and systems analysis (e.g., Griffiths, 1968, 1972, 1974).

JOHN C. GRIFFITHS

References

Agterburg, F. P., 1974, *Geomathematics: Mathematical Background and Geoscience Applications,* Developments in Geomathematics, vol. 1. New York: Elsevier, 595p.

Clark, I., 1979, *Practical Geostatistics.* London: Applied Science Publisher, 129p.

Davis, J. C., 1973, *Statistics and Data Analysis in Geology.* New York: John Wiley & Sons, 550p.

Fisher, R. A., 1953, The expansion of statistics, *Royal Soc. Statistics Jour.* **A116.** (Reprinted in *Am. Sci.* **42,** 275–282, 1954.)

Gary, M., R. McAfee, Jr., and C. O. Wolf, 1972, *Glossary of Geology.* Washington, D.C.: American Geological Institute, 805p.

Griffiths, J. C., 1958, Geometrics in petroleum petrography, *Producers Monthly* **22,** 40.

Griffiths, J. C., 1962, Statistical methods in sedimentary petrography, in H. B. Milner, ed., *Sedimentary Petrography,* vol. 1. London: Allen & Unwin, 565–617.

Griffiths, J. C., 1967, *Scientific Method in the Analysis of Sediments.* New York: McGraw-Hill, 508p.

Griffiths, J. C., 1968, Operations research in the mineral industries, in *Proceedings of a Symposium on Decision-making in Mineral Exploration.* British Columbia Research Council, 5–9. (Also in *Western Miner* **41**(2), 22–26, 1968.)

Griffiths, J. C., 1970, Current trends in geomathematics, *Earth Sci. Rev.* **61,** 121–140.

Griffiths, J. C., 1971, Problems of sampling in geoscience, *Inst. Mining and Metallurgy Trans.* **B80,** 345–356.

Griffiths, J. C., 1972, Cybernetics–geomathematics interaction, in P. Fenner, ed., *Quantitative Geology,* Special Paper 146. Boulder, Colo.: Geological Society of America, 87–94.

Griffiths, J. C., 1974, Quantification and the future of geoscience, in D. F. Merriam, ed., *The Impact of Quantification on Geology,* Geology Contribution No. 2. Syracuse, N.Y.: Syracuse University Press, 83–101.

Kendall, M. G., 1973, *Time Series.* London: Chas. Griffin & Co., 197p.

Krumbein, W. C., and F. A. Graybill, 1965, *An Introduction to Statistical Models in Geology.* New York: McGraw-Hill, 475p.

Krumbein, W. C., and F. J. Pettijohn, 1938, *Manual of Sedimentary Petrography,* New York: Appleton-Century-Crofts, 549p.

Lee, P. J., 1969, *Fortran IV Programs for Canonical Correlation and Canonical Trend–Surface Analysis,* Computer Contribution No. 32. Lawrence, Kansas: State Geological Survey, The University of Kansas.

Mackie, W., 1897, On the laws that govern the rounding of particles of sand, *Edinburgh Geol. Soc. Trans.* **7,** 298–311.

Matheron, G., 1962, Traite de geostatistique applique *[France] Bur. Recherches Géol. et Minières Mém. 14,* 333p.

Matheron, G., 1963, Principles of geostatistics, *Econ. Geology* **58**, 1246–1266.

Matheron, G., 1971, *The Theory of Regionalized Variables and Its Applications,* Cahiers de Centre de Morphologie Mathematique No. 5. Paris: Ecole des Mines, 212p.

McCammon, R. B., ed., 1975, *Concepts in Geostatistics.* New York: Springer-Verlag, 168p.

Merriam, D. F., ed.-in-chief, 1969, Editor's remarks, *Internat. Assoc. Math. Geology Jour.* **1**(1), 5.

Merriam, D. F., ed., 1970, *Geostatistics: A Colloquium.* New York: Plenum Press, 177p.

Merriam, D. F., ed.-in-chief, 1975, *Computers and Geosciences* **1**(1/2, July), 117p.

Milner, H. B., 1962, *Sedimentary Petrography,* vol. 1, *Methods in Sedimentary Petrography.* London: Allen & Unwin, 603p.; vol. 2, *Principles and Applications.* London: Allen & Unwin, 715p.

Siegal, B., and J. C. Griffiths, 1975, Classification of glacial tills by computer using the CLUS program, *Computers and Geosciences* **1**(1/2), 65–74.

Sorby, H. C., 1880, The structure and origin of non-calcareous stratified rocks, *Geol. Soc. London Quart. Jour.* **36**, 46.

Udden, J. A., 1898, *The Mechanical Composition of Wind Deposits,* Publication No. 1. Rock Island, Ill.: Augustana Library, 69p.

Vistelius, A. B., 1969, Preface, *Internat. Assoc. Math. Geology Jour.* **1**(1), 1.

Wentworth, C. K., 1929, Method of computing mechanical composition types of sediments, *Geol. Soc. America Bull.* **40**, 771–790.

Cross-references: *Geological Methodology.* Vol. XIII: *Computerized Resources Information Bank; Well Data Systems.*

GEOTECHNOLOGY—See BLASTING AND RELATED TECHNOLOGY; BOREHOLE DRILLING; CRATERING, ARTIFICIAL. Vol. XIII: RAPID EXCAVATION AND TUNNELING; TUNNELS, TUNNELING.

GOSSAN, BOXWORK—See PROSPECTING, Vol. XIII: MINERAGRAPHY.

GRAVITY SURVEYS—See EXPLORATION GEOPHYSICS.

GROUNDWATER EXPLORATION

Most small-yield wells are located for homeowners by *well drillers* (see *Borehole Drilling*) without the involvement of professional hydro-geologists. Drillers commonly have knowledge of local groundwater conditions, and a homeowner typically has only a limited area to site the well. When large well yields are required for irrigation, industry, or municipalities, however, it is usually necessary to mount a geotechnical search for the location and development of the best aquifer. To be successful the hydrogeologist must have knowledge of both the geologic conditions of the area and their effect on the storage and movement of groundwater systems. The total character of the environmental setting determines which exploration techniques are necessary to site and develop the well. For a general overview of these techniques, see Davis and de Wiest (1966, 260–317), Freeze and Cherry (1979, 306–312), and Walton (1970, 55–81).

Aquifer Conditions

Groundwater can occur in all sediment and rock types—unconsolidated materials, as well as sedimentary, igneous, and metamorphic rocks. It is stored and can move through openings between grains or crystals, fractures and fissures, and cavities. The search for a sufficiently large *aquifer* (a body of earth material capable of yielding groundwater in sufficient quantity for a production well), therefore, must be directed to locating those geologic features that will maximize storage and flow of groundwater (Huntoon and Lundy, 1979). The volume of groundwater that an aquifer can supply for a sustained period of time is a function of its porosity, permeability, transmissivity, and storativity or storage capacity (Dagan, 1967). Such properties are dependent on the type and amount of material, size and arrangement of particles or crystals, sorting, compaction, cementation, and amount and connection of open spaces.

In some groundwater investigations it may also be necessary to determine the distribution of aquicludes and aquitards. An *aquiclude* consists of earth material that is highly impervious to groundwater flow, whereas an *aquitard* has material that permits slow leakage of groundwater but is insufficient to produce a well. These conditions especially must be known when dealing with semiconfined or confined (*artesian*) systems.

So many variables determine groundwater conditions in the geologic environment that aquifer properties at one site may differ significantly from aquifer conditions at another locale. A relatively impermeable rock can become permeable through deformation. An initially permeable rock may later become impermeable by cementation. Irregularities in rock or sediment continuity may enhance or restrict groundwater movement. Faults may inhibit or enhance flow, and joints and fissures may divert flow trajectories. Indeed the character, thickness, and lateral extent of rock and sediment bodies that contain aquifers can be highly diverse, but these are the conditions the hydrogeologist must assess in his or her exploration for groundwater.

Lithology and Groundwater

Sand and gravel deposits are the major source for high-yielding wells. These materials are predominantly formed by fluvial processes so are geologically restricted to areas of the riverine environment (see Vol. XIII: *Alluvial Plains, Engineering Geology*) (except near coastal plains). They occur in current watercourses and valleys, in abandoned or buried valleys, or in plains and intermontane valleys. Abandoned or buried valleys may initially produce abundant groundwater supplies, but sustained yield may be less than in present stream valleys due to an absence of extensive recharge.

Soluble rocks such as limestone and dolomite have highly contrasting permeability and transmissivity depending on the character of the secondary solution openings and the jointing fabric. Some are capable of producing large amounts of groundwater supplies.

Groundwater supplies from crystalline rocks (igneous and metamorphic) are also highly variable. Volcanic terrane with lava flows can occasionally produce high-yield wells because of lava tubes, fractures, and permeable zones at the tops of some flows. Wells in rocks such as granite, gneiss, and gabbro usually discharge less than 3×10^{-4} m^3/s, and these yields are mostly dependent on effective joint sets.

Exploration Methods

Many factors must be considered in the design of an exploration program for groundwater supplies. The degree of complexity of the geologic setting may determine which methods should be used. The depth to the water is also important because if it lies 100 m or more deep, the aquifer conditions become more difficult to interpret. Whether the well is to be a wildcat (located at a site where there are no other nearby wells), the amount of water needed and the amount of money available for the search also dictate the degree of sophistication of the exploration program. The types of exploration methods may be classified on the basis of surface and subsurface exploration and whether the techniques used are primarily geologic, hydrologic, or geophysic.

Surface Methods

Geology In using geology methods the *hydrogeologist* is primarily interested in mapping the area. To do this he or she may employ a wide range of equipment and techniques. Remote sensing (q.v.), including aerial photographs (see *Photogeology; Photointerpretation*) may reveal important general relationships (Mollard, 1973). This mapping is then supplemented by detailed ground mapping of the types of rocks, their structural and stratigraphic relationships, and the geomorphology of the area. The permeability of the different rock units, the nature of fracture systems, and the distribution at the surface and at depth must be determined. The mapping of *lineaments* (extended linear features) has been shown to be very important in locating groundwater supplies in Pennsylvania and New England. Folds, faults, unconformities, facies changes, and other geologic conditions may prove vital in the assessment of groundwater properties.

Geophysics Although many different geophysical-type instruments are available for surface study of subsurface phenomena (Wachs et al., 1979), the three principal techniques used for groundwater exploration are electrical resistivity, seismic refraction, and gravity (see *Well Logging*).

The *electrical resistivity method* involves the measurement of the voltage drop between lines of current electrodes and potential electrodes that have been emplaced into the soil (Kelly, 1962; Kosinski and Kelly, 1981). The *apparent resistivity* (the ratio of measured voltage to the applied current) is measured. This value depends on the positioning and spacing of the electrodes and the character of the earth materials. This method is especially useful when the thickness-to-depth ratio of sand and gravel deposits is high and their resistivity is distinctly different from overlying and underlying strata. Resistivity methods are also useful in detecting the presence of shallow groundwater and delineation of buried valleys. Such methods are not applicable for areas with complex geology, numerous inhomogeneities, and deep water tables.

The *Seismic refraction* method involves the detonation of an array of small dynamite charges and the measurement of the time lapse between the blasts and the arrival time of the shock waves at the seismograph (Bonini, 1959). A shock wave may also be artificially produced by blows from a hammer or a machine that pounds on the ground. The properties of subsurface material determine the path and velocity of the shock wave, thus providing a signature for their characteristics. *Reverse shooting* is the most common seismic techniques (see Vol. XIII: *Seismological Methods*). Geophones are equally spaced along a line, and explosive charges are detonated in succession starting at the ends. The seismic wave arrivals are plotted in a time versus distance graph that then allows calculation of velocity, apparent slope of lithologic units, and depth to bedrock. This seismic method has been most successfully used in two-layer problems, such as determination of the salt- and fresh-water interface, location of sand and gravel under clay, and position of buried valleys. It is not applicable in complex geology settings, in areas where materials all give low velocities, and where bedrock is deep.

The *gravity method* does not have wide application because it cannot be used for detailed studies. It is usually restricted to the determination of large-scale features where only two contrasting lithologies are present. For example, its most extensive use has been in location of buried valleys (Hall and Hajual, 1962; Stewart, 1980). By use of the *gravimeter,* density changes in near-surface materials are evaluated by determining changes in gravity. These changes are rarely greater than one-ten thousandth of the normal gravity field. A traverse is made across the study, and calculations made for a variety of corrections, such as topographic changes. The resultant numbers reflect differences in the gravitational attraction of the underlying materials.

Water dowsing also called *witching* or *divining,* is another approach to water discovery that is used by some people. Although the method is not scientific, it is still in use in many parts of the United States and throughout the world. The Dowser uses a nearly infinite variety of materials, but the most common "equipment" is a forked twig. The two parts of the fork are held in the hands and when moved over a water vein the stem reportedly violently signals the presence of water. Testing and experiments performed by dowsers indicate a low level of success, which also confirms statistics that compare dowsers with other methods of groundwater search (Vogt and Hyman, 1959).

Geobiological and geochemical techniques may also prove helpful in exploration under certain conditions. For example, phreatophytes can aid in determining the range of depths to the water table. Some plants may also provide an indication of the quality of underground water (see Vol. XIII: *Geomorphology, Applied*).

Subsurface Methods

Surface exploration methods may provide an intuitive understanding of subsurface conditions, but all the data need to be interpreted and extrapolated to a particular model developed by the hydrogeologist. Only by a *test drilling program* can there be assurance of the real status of subsurface sediments, rocks, and their structures. Logging of the materials can take a variety of forms. Samples of the materials are studied so that the log will show their character, as well as thickness and depths encountered. Depending on the sampling method porosity can be determined, and under certain conditions with further laboratory testing, an idea about their permeability can also be obtained.

A range of measurements may be obtained by use of a borehole geophysics program (see *Well Logging*) (Dyck et al., 1972). The hole may be electrically logged by either a *spontaneous potential log,* with movement of a single electrode, or by a *resistivity log,* using of one or more electrodes. Care must be taken so that the use of drilling mud or water reaching the borehole does not obscure the true electrical properties of the rock formation. Investigators hope that the readings will reflect the normal relationships so that rock resistivity is inversely proportional to its porosity and to the ionic concentration of the interstitial fluid. Secondary porosity caused by fractures and solution opening in crystalline rocks and limestone produce low electrical resistivity. However, secondary porosity in fine-grained sedimentary rocks is more difficult to detect because clay minerals decrease resistivity, thus masking porosity changes.

Radioactive logs provide another technique that can be useful under certain conditions for determining rock properties (Patten and Bennett, 1963). Two different logging types can be obtained by use of radioactive instrumentation—*gamma-ray logs,* and *neutron logs*. Radioactivity logs can be useful in identifying some rock types but provide little information on water content. An important advantage of such logging is that it is the only method that can obtain data of rock characteristics for wells already cased. Porosity is a property that unfortunately, can be determined only by the neutron logging method. Since this type of logging determines the abundance of hydrogen atoms, it can provide a reflection of the amount of water contained in the rock unit.

Acoustic logs offer another possibility for estimation of rock porosity. The speed of sound in rocks is directly proportional to rock porosity. The method has been applied to moderately porous, well-compacted sandstones, limestones, and dolomites (Walton, 1970).

Temperature logging is dependent on the thermal conductivity of the rocks. Measurements are made using an electrical thermometer, and variations in temperature are then attributed to rock properties or differences in the recharge area to the aquifer. For example, the temperature logs of wells penetrating several different aquifers may show large differences. This is the case in the sand and gravel aquifers near Binghamton, New York, where in some highly pumped wells the top aquifer has wide seasonal variations because of river recharge; the deeper aquifers show constant temperatures as water was withdrawn from storage.

Additional data for rock properties and groundwater flow characteristics can be obtained in several other ways. *Borehole cameras* with television attachments are now available to take photographs of the rock. Changes in borehole diameter, caused by differences in rock swell properties, can be measured by calipers moved down the well. *Flow velocity logs* can be obtained under some circumstances by flow meters.

Hydrology Methods

Observation, determination, and measurement of the flow properties of groundwater constitute the realm of the *hydrologic studies*. Tracer experiments can be conducted in attempts to calculate the flow direction and velocity of a water plume (Davis et al., 1980). Some type of dye, such as fluorescein or other types of slugs, are introduced into the probable recharge area. Observation boreholes are installed downgradient from the site and periodically sampled to detect the arrival time and dilution amount of the artificially induced slug.

Pumping tests provide the best single method for calculation of well performance and characteristics of the groundwater system (Walton, 1978). Such tests should be conducted when there is the necessity for a high-yielding well and development costs are not a consideration. A single well pumping test involves pumping the well and measuring the decline of the water table as a function of time. It is important to keep discharge as nearly constant as possible. The longer the pumping test lasts, the greater the precision in determining *transmissivity* ($T = Kb$, where T is transmissivity, K is the hydraulic conductivity as determined by character of the materials, and b is aquifer thickness). Such tests often last several days and even more than a week (Fetter, 1980).

Increased information about aquifer properties can be obtained by pumping tests that employ both a discharge well and a series of observation wells. The observation wells are equipped with *piezometers,* and it is important that they extend the full depth of the aquifer. Water levels are measured in the observation wells on the same schedule as in the discharge well. As the cone of depression expands, it affects water levels in the observation wells. With cessation of pumping, water levels continue to be measured during the recovery period, and such information can provide valuable data in compilation of the final properties of the system. These tests not only determine transmissivity but also storativity or storage capability, or *specific storage,* which equals the volume of water a unit volume of aquifer releases from storage with a unit decline in hydraulic head. In addition other characteristics of the aquifer can be determined—e.g., whether the groundwater system is unconfined, semiconfined, or confined.

DONALD R. COATES

References

Bonini, W. E., 1959, Seismic-refraction method in ground-water exploration, *A.I.M.E. Trans.* **211,** 485–488.

Dagan, G., 1967, A method of determining the permeability and effective porosity of unconfined anisotropic aquifers, *Water Resources Research* **3,** 1059–1071.

Davis, S. N. and R. J. M. de Wiest, 1966, *Hydrogeology.* New York: John Wiley & Sons, 463p.

Davis, S. N., G. M. Thompson, H. W. Bently, and G. Stiles, 1980, Ground-water tracers—a short review, *Ground Water* **18**(1), 14–23.

Dyck, J. H., W. S. Keys, and W. A. Meneley, 1972, Application of geophysical logging to ground-water studies in southern Saskatchewan, *Canadian Jour. Earth Sci.,* **9,** 8–94.

Fetter, C. W., Jr., 1980, *Applied Hydrogeology.* Columbus, Ohio: Charles E. Merrill, 488p.

Freeze, A. R., and J. A. Cherry, 1979, *Groundwater.* Englewood Cliffs, N.J.: Prentice-Hall, 604p.

Hall, D. H., and Z. Hajual, 1962, The gravimeter in studies of buried valleys, *Geophysics* **27**(6), Part 2, 939–951.

Huntoon, P. W., and D. A. Lundy, 1979, Fracture controlled ground-water circulation and well siting in the vicinity of Laramie, Wyoming, *Ground Water* **17**(5), 463–469.

Kelly, S. F., 1962, Geophysical exploration for water by electrical resistivity, *New England Water Works Assoc. Jour.* **76,** 118–189.

Kosinski, W., and W. E. Kelly, 1981, Geoelectric soundings for predicting aquifer properties *Ground Water* **19,**(2), 163–171.

Mollard, J. D., 1973, *Landforms and Surface Materials of Canada: A Stereoscopic Airphoto Atlas and Glossary.* Regina, Canada: J. D. Mollard and Associates Ltd.

Patten, E. P., and G. D. Bennett, 1963, Application of electrical and radioactive well logging to ground-water hydrology, *U.S. Geol. Survey Water-Supply Paper 1544-D,* 60p.

Stewart, M. T., 1980, Gravity survey of a deep buried valley, *Ground Water* **18**(1), 24–30.

Todd, D. K., 1980, *Groundwater Hydrology.* New York: John Wiley & Sons, 535p.

Vogt, E. Z., and R. Hyman, 1959, *Water Witching U.S.A.* Chicago: University of Chicago Press, 248p.

Wachs, D., A. Arad, and A. Olshina, 1979, Locating ground water in the Santa Catherina area using geophysical methods, *Ground Water* **17**(3), 258–263.

Walton, W. C., 1970, *Groundwater Resource Evaluation.* New York: McGraw-Hill, 664p.

Walton, W. C., 1978, Comprehensive analyses of water-table aquifer test data, *Ground Water* **16,** 311–317.

Cross-references: *Acoustic Surveys, Marine: Borehole Drilling; Exploration Geophysics; Hydrochemical Prospecting; Well Logging.* Vol. XIII: *Electrokinetics; Geomorphology, Applied; Hydrodynamics, Porous Media; Hydrogeology and Geohydrology; Hydrology; Seismological Methods; Wells, Water.*

H

HARBOR SURVEYS

A *harbor* can be defined as a coastal body of water deep enough for ships and sheltered from winds, waves, and currents. Depending on the vagaries of nature, economics, and politics, coastal populations have usually clustered around harbors, and industrial growth has concentrated about these population centers. Consequently, the natural attributes of some modern *seaports* do not match the developments that have overtaken them; e.g., *channels* that comfortably passed the largest sailing vessels in the late 1800s are too shallow for a modern supertanker, and existing wharves are often not equipped to handle container cargo. Morever, many inhabitants of a coastal city today regard the harbor area primarily as an obstacle to be crossed by bridges or tunnels, as a recreational facility, or as a receptacle for municipal and industrial effluent. Such modern developments and attitudes create the need for harbor surveys.

In general, *harbor surveys* are undertaken to evaluate the nature of the harbor bottom as it may affect shipping or engineering projects or to assess factors usually classed as environmental, such as water quality and movement. Surveys for either purpose have common problems and each can benefit from an organized approach in planning and execution and in the interpretation and presentation of survey data. The following observations touch on standard survey methods for requirements mainly within the province of the earth scientist.

Preparing for the Survey

The first step in any survey is to collect all available background information, be it only an ancient *hydrographic chart* or a map of the coastline. For an established harbor, considerable data of a specific nature may be located in various offices throughout the port town or city or may be on file with one or more government agencies. Such data may include sounding charts, reports from geophysical surveys, logs for boreholes (q.v.), and collections of air photos. The air photos can yield clues about long-term, progressive changes in the shape and area of the harbor and aid in determining whether these changes relate to the seasons, the tides, or the weather. Published information of a general nature, such as reports and maps dealing with geology, geography, geophysics, groundwater, agriculture, etc., can reveal important aspects of the local setting in terms of regional trends and processes. For example, the type of bedrock underlying the harbor often can be inferred from published geological maps.

Once collected, background data should be synthesized and compiled at a map scale appropriate to the survey area. Then, the objectives of the survey should be reviewed in terms of the relevant tools and techniques available. A question that needs to be answered at this stage is, Is the survey really necessary? The survey data will possibly be superseded by a drilling program to be conducted later in any case, or perhaps the data can be obtained by probing with a long pole rather than with a more sophisticated and costly geophysical tool.

Survey Tools

Surveys connected with engineering projects usually require evaluation of the nature and thickness of sediments over bedrock, the configuration of the bedrock surface, and often, the composition and structure of the bedrock. The most commonly applied geophysical technique for assessing these properties is the seismic reflection method (McQuillan and Ardus, 1977; Trabant, 1984) (see *Acoustic Surveys, Marine*); the simplest seismic reflection device is the conventional *echo sounder*. The morphologic characteristics of the harbor bottom as revealed by echo sounder profiles often define variations in composition of bottom sediment and probable locations of bedrock outcrops (King, 1967). Depending on the power and frequency characteristics of the transducer, some echo sounders can be used to map layering in soft sediment. In general, however, more powerful, lower frequency, energy sources (e.g., sparkers and air guns) are required to ensure consistently deep subbottom penetration (Fig. 1).

Water depth imposes a partly limiting condition on the reflection seismic profiling method as applied in harbor surveys. The strong multiple reflections generated by reverberation of seismic pulse energy between the water surface and water bottom appear on the seismic record at a time equivalent to twice the water depth (see Fig. 1).

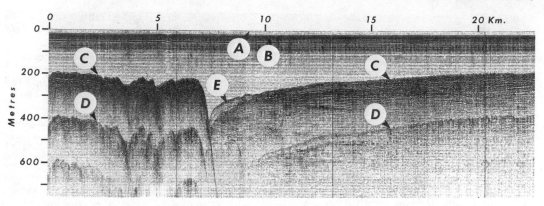

FIGURE 1. A seismic profile (air gun) showing Precambrian crystalline bedrock (0–7.5 km) in fault contact with Cenozoic shales and sandstones (7.5-23 km). (After Grant, 1970, fig. 3). A, sea surface; B, direct wave (air gun to hydrophone); C, sea bottom; D, first multiple reflection; E, unconsolidated sediments.

Unless the subbottom reflections returning at later times are very strong, they will be obscured by the multiple reflections. Thus, it is not practical to anticipate definitive seismic results from depths below this limit.

There is an increasingly wide variety of subbottom profiling systems to choose from, and manufacturers' brochures invariably contain impressive samples of their records. Beware the locale of such records relative to the intended survey area, however, as a sediment section composed of silts and muds tends to yield much better records than those obtained from coarser sediments. Sands and gravels, especially glacial deposits, can be particularly difficult to penetrate seismically. Severe problems with seismic penetration may be encountered even in very soft sediments if they are charged with gas (methane). Methane generation is common in organic-rich muds and silts, and such gas often produces a very effective barrier to seismic energy (Keen and Piper, 1976).

Deep towing of seismic energy sources and receivers is a direct and efficient means of gaining better resolution of the bottom profile and subbottom reflectors (Bidgood, 1974). This approach is constrained to some extent by water depth and bottom relief, and for optimum effect in areas of irregular bottom topography, it requires a fairly elaborate—and probably bulky—winch system (see Vol. XIII: *Oceanography, Applied*).

Among the newest seismic tools are side-scan sonar systems, which utilize a line array of transducer elements to generate a narrow beam of high-frequency sonic pulses (McQuillan and Ardus, 1977; Trabant, 1984). The *sonar beam* is directed obliquely at the bottom normal to the direction of the survey line; usually a beam is emitted from both sides of the fish so that a double swath of the bottom is surveyed. Side-scan

sonar is particularly useful for detecting bottom irregularities such as rock outcrops, boulders, and wrecks, and it is a very effective tool for mapping variations in bottom sediment. Problems with these systems can arise if the water column is sharply layered (changes in temperature or salinity), because the obliquely fired sonar beam may be refracted or channelled by such boundaries. The tow depth of the side-scan fish can sometimes be adjusted to escape this effect.

The more direct approach for evaluating bottom sediments is by sampling, which may be done with a *gravity corer, grab sampler,* or *drilling advice* (Sly, 1969). It is usually more efficient to design a sampling program on the basis of prior seismic surveying. In some surveys it may be practical to employ divers to collect samples; often it is useful to have a diver check the action of the sampling device on the bottom, to observe whether samples recovered are indeed representative. Engineering requirements may specify measurement of in situ properties of the sediments such as sheer strength or bearing capacity (see Vol. XIII: *Marine Sediments, Geotechnical Properties*). The engineers usually supply or specify the tools necessary for such measurements, and they may be responsible for processing sample material collected. Universities or government agencies in the area may have facilities for performing such analyses.

A positioning system is an integral part of any standard survey (see *Satellite Geodesy and Geodynamics; Surveying, Electronics*), since sediment samples and seismic records are of little value if their location is unknown (Ingham, 1974). Various types of electronic positioning systems are available, all of which can give very accurate results (Munson, 1977). Firms specializing in providing survey positioning are fairly common, and subcontracting this aspect of the survey to a reputable firm may be a sensible step. In general, the small portable electronic systems require line-of-sight operation. It is possible to monitor the position of a survey vessel from shore using conventional survey instruments, if accurate synchronization of fix times is arranged between ship and shore. The survey vessel may also be located with respect to fixed points on land by horizontal sextant or to accurately located marker buoys. If positioning measures fail or are absent for some reason, profile locations can sometimes be established by reference to bathymetry, provided that a good bathymetric chart is available.

The platform for the survey—a barge or a boat—can usually be found locally. The main requirements for a survey vessel are that it be large enough to carry equipment and personnel and small enough to maneuver in the survey area. Available vessels often lack a proper electrical supply for operating survey equipment, and so surveyors may have to be prepared to mount a portable generator. If samples are to be collected in the course of the survey, a winch or A-frame may also have to be installed. If the survey vessel is an open boat, a temporary hut to shelter electronic equipment may be required because rain and spray will be a problem.

The three basic elements of survey hardware—geophysical and geological equipment, a survey positioning system, and a survey vessel—may be rented separately or as a package through a consultant firm. Each element can be rented with an operator; there are rental purchase schemes if purchase of equipment is contemplated, although electronic systems tend to become obsolete fairly quickly and also require skilled maintenance personnel. Some items of sampling equipment—e.g., grab samplers and corers—may be available by purchase, or they can be built locally. If not required following the survey, such items might be donated to the geoscience department of the nearest university.

The Survey

If the nature of the bottom sediments is a critical parameter to be surveyed, and if side-scan sonar is used for this purpose, the scanning range of the side-scan sonar will probably determine the spacing of survey tracks. Assuming that a seismic profiling system is operated concurrently, the special requirements of the survey will dictate whether interlining is required for subbottom control. If the basic survey pattern is a series of parallel tracks, at least a few cross-lines should be run.

Ideally, a combination of seismic profiles and side-scan sonar records will provide a map of bedrock outcrops and bottom sediment units (see Vol. XIII: *Remote Sensing, Engineering Geology*). In subsequent sampling the first objective will be to verify these unit boundaries and test for variations within units that may not be revealed by the seismic data. If the variations are not tolerable within survey requirements, it may be necessary

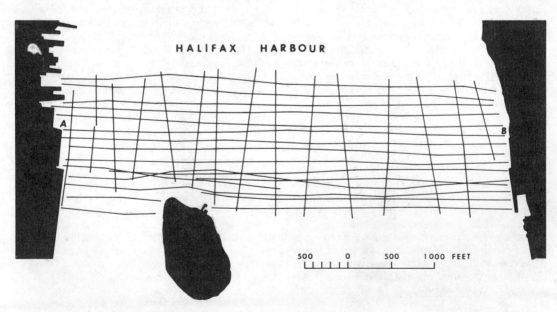

FIGURE 2. Track plot of a seismic (sparker) survey over a proposed tunnel location (Halifax Harbour, Nova Scotia, Canada).

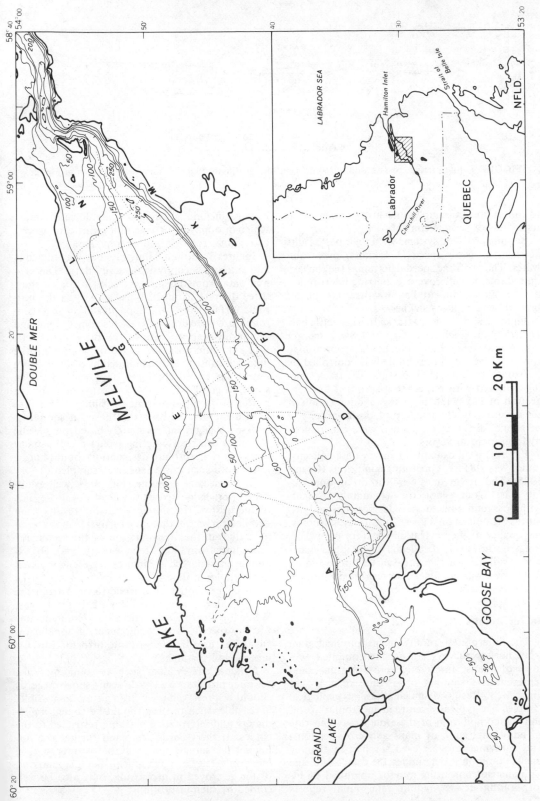

FIGURE 3. Reconnaissance survey tracks (seismic and magnetic profiling) plotted on a bathymetric diagram (Lake Melville, Labrador, Canada).

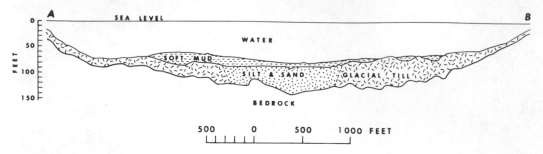

FIGURE 4. Interpretive cross-section based on a seismic profile through the survey area shown in Fig. 2.

to collect bottom samples on a grid or line pattern regardless of seismic control.

Seismic profiling systems yield their best results at relatively slow speed (3–4 knots) over calm water. The profiling speed determines the time required to complete coverage, barring interruption of the survey due to bad weather, equipment breakdowns, and sea sickness.

Figure 2 shows survey tracks for a specific harbor study in which the survey tool was a *sparker profiling system* (Grant, 1967). Track positioning for the survey in this example was controlled by horizontal sextant measurements. The track spacing achieved with respect to a specified line separation of 125 ft (38 m) gives some indication of the accuracy of this method of positioning. The personnel directing the launch were experienced hydrographic surveyors.

Figure 3 is an example of survey data collected under very different operating conditions (Grant, 1975). The survey lines were run on an opportunity basis from a large trawler, mainly at night, with positioning based on radar bearings and distances to points on shore. For survey purposes, the quality of the navigation was very poor. Fortunately, however, detailed hydrographic charts were available for the entire survey area, and on this basis it was possible to correct the positioning in a very satisfactory manner.

Survey Results

The interpreted results of geophysical data from harbor surveys are usually compiled as a suite of maps, interpretive cross-sections, and representative samples of original records (see Figs. 3–8). Results from analyses of geological samples are typically returned in tabular form, and the compilation of these data may generate an additional suite of maps and cross-sections. The bathymetric chart used for plotting survey tracks commonly determines the map scale; the bathymetric contouring may be upgraded by incorporating new sounding data. Final reports

usually include a smaller scale location map, which also may be used to plot significant aspects of regional geology, topography, etc.

Figures 3 through 8 illustrate one way to depict and interpret geophysical survey data. The seismic data from the survey tracks in Fig. 2 show three discrete sediment types as well as the bedrock surface (Fig. 4). *Isopach* and *surface contour maps* were drawn for each sediment unit, and a contour map was drawn for the bedrock surface. The interpretation based on seismic profile data was subsequently calibrated by drilling. More recent surveys in this area have used side-scan sonar to good effect, and the superior resolving power of deep-tow seismic profiling systems has been demonstrated on a trial basis. It should be stressed that an interpretation based on seismic character can clearly define separate sedimentary units but that absolute designation should not be made until such units have been sampled.

In the second example (Fig. 3), a well established port is located at the head of Goose Bay, to the west of the survey area, and results from detailed surveys in that area furnished many valuable clues in the interpretation of the reconnaissance lines from Lake Melville (Figs. 4–8). For example, gas-charged sediments were encountered in boreholes at the head of the lake; some of the disturbances noted on seismic records from the survey (Figs. 5 and 6) can be attributed to gas. While it is clear that this survey is of limited value for any specific local requirement, it nonetheless furnishes useful background information for more detailed studies.

Original survey data such as sounding rolls, seismic profiles, and side-scan sonar records should be preserved in as clean a state as possible; if feasible, such records should be copied immediately and interpretive marking restricted to the copies. If the records are of no further use following the survey, they may be of interest to a local university or government. Sample material, if not destroyed in processing, may also be welcomed by such institutions.

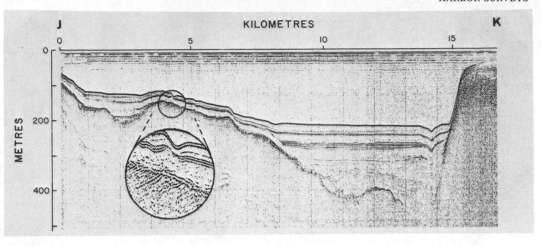

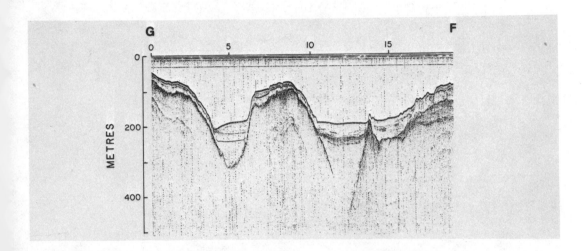

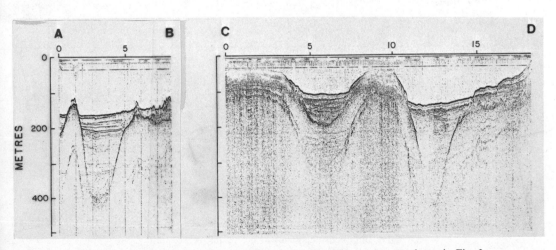

FIGURE 5. Representative seismic profiler records from the survey area shown in Fig. 3.

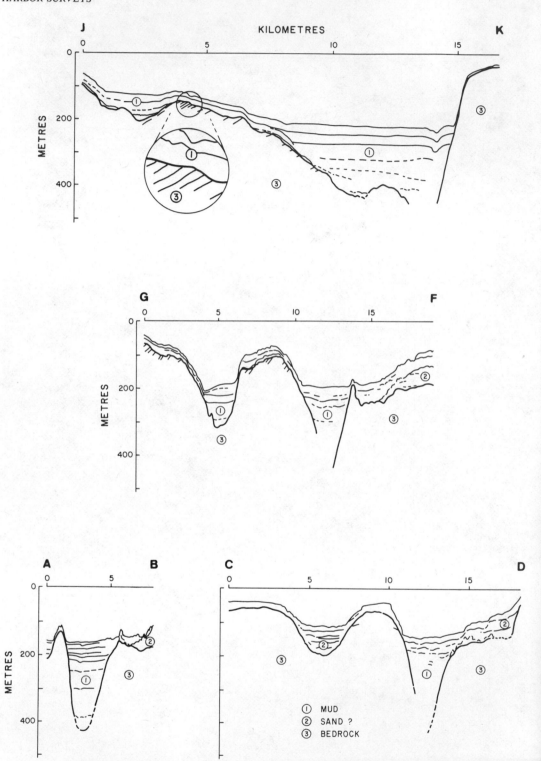

FIGURE 6. Interpretive diagrams of the seismic records in Fig. 5.

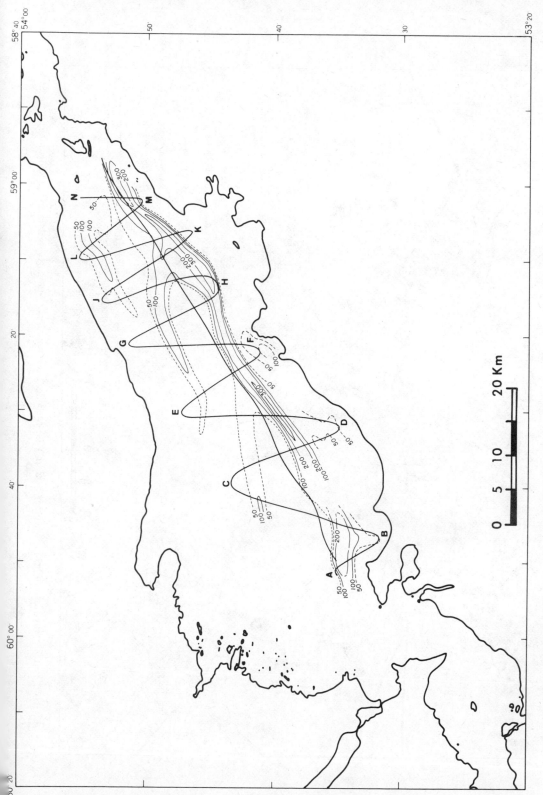

FIGURE 7. Contours of apparent thickness of unconsolidated sediments as interpreted from seismic profiler coverage shown in Fig. 5.

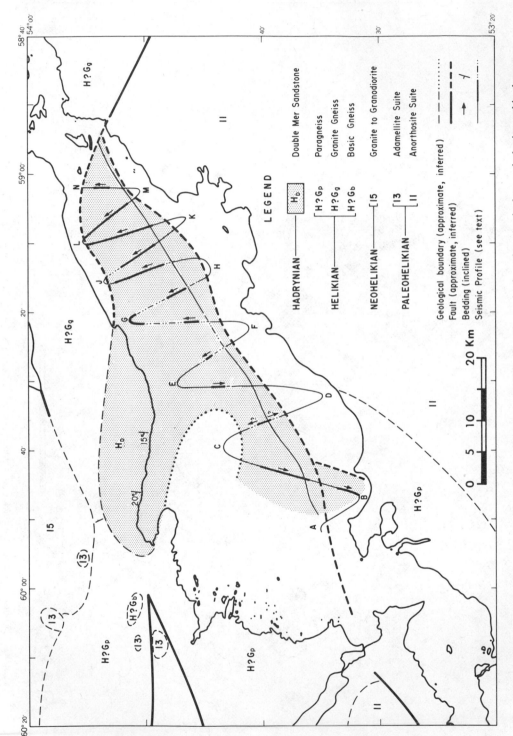

FIGURE 8. Interpretive map of bedrock geology beneath Lake Melville, based on seismic and magnetic data and physiographic character.

All aspects of any harbor survey can be contracted to a consultant firm—at a price—and a very good survey may result. The extra charge for the experience of the consultant may offset the cost of mistakes, providing the consultant has this experience. Whatever the financial arrangements, presurvey research is never wasted, and a sound knowledge of the problems to be solved, the appropriate tools to be used, and their limitations will contribute measurably to a successful survey.

ALAN C. GRANT

References

Bigood, D. E. T., 1974, A deep towed sea bottom profiling system, in *Engineering in the Ocean Environment,* vol. 2. Halifax, N.S.: IEEE International Conference, 96-l07.

Grant, A. C., 1967, A continuous seismic profile from Halifax Harbour, Nova Scotia, *Maritime Sediments* **3,** 64.

Grant, A. C., 1970, Recent crustal movements on the Labrador shelf, *Canadian Jour. Earth Sci.* **7,** 571-575.

Grant, A. C., 1975, Seismic reconnaissance of Lake Melville, Labrador, *Canadian Jour. Earth Sci.* **12**(12), 2103-2110.

Ingham, A. E., 1974, *Hydrography for the Surveyor and Engineer.* London: Crosby Lockwood Staples, 139p.

Keen, M. J., and D. J. W. Piper, 1976, Kelp, methane, and an impenetrable reflector in a temperate bay, *Canadian Jour. Earth Sci.* **13**(2), 312-318.

King, L. H., 1967, Use of a conventional echo-sounder and textural analyses in delineating sedimentary facies—Scotian Shelf, *Canadian Jour. Earth Sci.* **4,** 691-708.

McQuillan, R., and D. A. Ardus, 1977, *Exploring the Geology of Shelf Seas.* London: Graham and Trotman Ltd., 234p.

Munson, R. C., 1977, Positioning systems, *Report on the work of WG414B* (U.S. Dept of Commerce). Presented at the Fifteenth International Congress of Surveyors, Stockholm, Sweden, 33p.

Sly, P. G., 1969. Bottom sediment sampling, *Proc. l2th Conf. Great Lakes Res.* International Association of Great Lakes, 883-989.

Trabant, P. K., 1984, *Applied High Resolution Geophysical Methods.* Boston: International Human Development Corp., 265p.

Cross-references: *Acoustic Surveys, Marine; Satellite Geodesy and Geodynamics; Sea Surveys; Surveying, Electronic.* Vol. XIII: *Coastal Engineering; Marine Sediments, Geotechnical Properties; Oceanography, Applied; Seismological Methods.*

HARDROCK VERSUS SOFTROCK GEOLOGY

The terms *hardrock* and *softrock* are a part of the jargon used in referring to the particular geological setting in which a geologist works. Specialists in metalliferous mineral deposits or in igneous and metamorphic petrology (q.v. in Vol. V) may be called *hardrock geologists,* while those engaged in exploration for oil, natural gas, or coal and others who work mostly in sedimentary rock terranes are likely to be called *softrock geologists.* A miner taking ore from a quartz vein may be called a *hardrock miner,* and one who mines bituminous coal or gypsum a *softrock miner.* These arbitrary labels are based on the premise that igneous and metamorphic (crystalline) rocks—granite and gneiss, e.g.—are harder than sedimentary rocks, such as limestone and shale. This is true in general, but the distinction is specious when applied to some rocks. Calcite in limestone is quite as hard as that in marble, and the grains of quartz in sandstone are as hard as those in quartzite. Moreover, the quartz of sandstone in the softrock group is harder than calcite in marble or the feldspar and mica in granite of the hardrock group. *Tenacity,* the capacity to resist blows from a hammer or pick without breaking, and *friability,* the tendency to crumble into small particles, are properties quite distinct from *hardness,* although they are often confused with hardness, especially by nongeologists. Hardness is the capacity to scratch; the harder of two substances can scratch the softer. Although not precise, *hardrock* and *softrock* are convenient and frequently used terms in the informal geological lexicon.

ROBERT B. HALL

Cross-references: Vol. XIII: *Rock Mechanics; Rocks, Engineering Properties.*

HYDRAULIC MODELS, SURVEYS—See ALLUVIAL SYSTEMS MODELING; CANALS AND WATERWAYS, SEDIMENT CONTROL; PLACER MINING. Vol. XIII: CHANNELIZATION; HYDROMECHANICS; RIVER ENGINEERING; URBAN HYDROLOGY.

HYDROCHEMICAL PROSPECTING

Even ancient history books refer to mineral prospecting by qualitative testing of water from seeps, springs, streams, wells, and bores. Two basic problems, however, have hindered quantitative analysis in water samples in modern geochemical prospecting. The first problem is that the concentration levels of indicator elements (q.v.) in natural water samples are very low (usually a few parts per billion), thereby posing both analytical difficulties and a serious risk of contamination during sample processing (e.g., glass or plastic containers can either contribute elements to or subtract elements from the sample unless suitable precautions are taken). The second problem is that the chemistry of natural water is very sensitive to weather conditions and to its local environment. Consequently, geochemists have shown an understandable reluctance to use water samples in exploration. However, recent analytical advances have made it feasible to detect suitable indicators by using practical sample sizes of 100–500 ml of water (samples are typically concentrated by evaporation or solvent extraction prior to analysis), and refined interpretation techniques have reduced the difficulties of interpreting water chemistry. This development has stimulated interest in stream- and lake-water sampling in glaciated terrains and groundwater sampling in arid regions.

Surface Waters in Glaciated Terrain

Brundin and Nairis (1972), e.g., investigated the relative effectiveness of sediment sampling, organic matter sampling, and water sampling in the glaciated terrain of northern Sweden. Their conclusion was that organic matter sampling gave the best information but that water sampling was quite suitable for outlining anomalies of Mo, Zn, and U; water proved to be a poor medium for detection of Pb and Cu anomalies. The Swedish investigation was prompted by disadvantages of more conventional stream sediment sampling in glaciated terrain (mainly, lack of suitably fine sediment over long distances and widely ranging contents of scavenger substances).

Canadian terrain offers similar difficulties, and Boyle et al. (1971), after reviewing current theory and practice of hydrogeochemical prospecting, predicted an increasing role for samples of water from streams, lakes, and other sources. Sampling of lakes, both for water and sediment (see *Lake Sediment Geochemistry*), is receiving attention for several reasons. First, lakes cover relatively large areas of Canada and could conceal mineralization. Second, the lakes are receptacles for water and sediment drawn from substantial catchment areas, thereby constituting composite sample sites for low-density reconnaissance. Finally, the lakes provide economical access for sampling by floatplane or helicopter.

Surface Waters in Nonglaciated Terrain

Experiments have been conducted on the use of stream-water samples in nonglaciated terrain, but in general there is less justification, and stream sediment sampling remains more economical than other geochemical methods and results can be interpreted with greater confidence. Two notable exceptions to this generalization concern F and U

or its daughter products. Field determinations of F present at low concentrations in stream water can be made in the field using a fluoride-ion-sensitive electrode, which is similar in size and operation to a pH meter. Despite the fragile nature of the electrode, very useful results have been obtained under fully operational conditions in rugged high-rainfall terrain in Australia. Rn, the highly mobile daughter product often used to detect concealed U mineralization by soil gas sampling (see *Atmogeochemical Prospecting*), is readily trapped and dissolved by water so that analysis of Rn in stream and groundwaters is also an appropriate prospecting method in nonarid areas (Andrews and Wood, 1972; Michie et al., 1973).

Groundwater

In arid regions where there is inadequate drainage for sediment surveys and unsuitable overburden for soil sampling (eolian detritus, silcrete, calcrete, laterite) (see *Pedogeochemical Prospecting*), the choice for geochemical sampling is reduced to bedrock or groundwater. Bedrock dispersion patterns are usually small, and sample collection may necessitate expensive drilling. In contrast, water samples collected from bores sunk for domestic or stock use are available cheaply and may delineate large hydromorphic anomalies that subsequently can be examined in detail by bedrock sampling or by water sampling of prospecting boreholes sunk on a grid pattern. The paucity of literature on groundwater sampling conveys a false impression of inactivity. Numerous surveys have been conducted for U, Cu, Pb, Zn, and other elements but without conspicuous success. In many cases it is not clear whether failure can be attributed to lack of mineralization or failure to detect mineralization. More care is needed to interpret anomalies in water than in rock, soil, or sediment because anomalies are generally of low contrast and need to be interpreted by refined procedures, such as regression analyses, that take into account important factors such as pH, Eh, temperature, associated soluble species, etc., which compete to control the abundance of any given indicator element.

One documented example of water sampling that led to the discovery of additional mineralization is a springwater sampling program in the southwest Wisconsin area of the Mississippi Valley Pb-Zn district (DeGeoffroy et al., 1968; DeGeoffroy and Wu, 1970). An area of about 2,250 km^2 was investigated using 7,210 water samples in which the Zn content ranged between 0.03 ppm and 3.50 ppm. Interpretation based on regional trends and local deviations outlined many known Zn deposits and some new mineralizations.

Two recent contributions from the USSR also indicate activity in groundwater sampling (Kraynov, 1971; Naumov et al., 1972). Both papers highlight the need for careful interpretation of anomalies. The paper by Naumov et al. in particular is important in providing a critical discussion of the behavior of Cu, Bi, Ag, Pb, Zn, and SO_4^{2-} in groundwater surrounding deposits in a desert area. Bi and Zn proved to be far superior to the other elements in producing large and discernible dispersion patterns.

Sulfate abundance in either surface or groundwaters has been proposed frequently as a potential indication of oxidizing sulfide deposits. A method proposed by Dall'Aglio and Tonani (1973) for discriminating between SO_4^{2-} produced by oxidizing sulfides and that attributable to marine sources (e.g., gypsum) may prove quite important. Their method consists of plotting or regressing SO_4^{2-} versus Ca or Cl to distinguish streamwater samples containing more SO_4^{2-} than can be accounted for by derivation from sedimentary sulfate sources. Similar methods of interpretation of SO_4^{2-} content in groundwater collected from sulfide mineralized, but gypseous, arid terrain in Australia are claimed to show promising results.

Indicator Elements

In summary, the following indicators have shown the most promise: U, Rn, He, Mo, Zn, Bi, F, SO_4^{2-}. The elements Cu and Pb, frequently used in rock, soil, or sediment surveys, seem unsuitable for most water surveys. Refined interpretation must be used since true anomalies do not necessarily contain the greatest abundance of the geochemical indicator.

A. S. JOYCE

References

Andrews, J. N., and D. F. Wood, 1972, Mechanism of radon release in rock matrices and entry into groundwaters, *Inst. Mining and Metallurgy Trans.* **81**, B206–B209.

Boyle, R. W., E. H. W. Hornbrook, R. J. Allan, W. Dyck, and A. Y. Smith, 1971, Hydrogeochemical methods—application in the Canadian Shield, *Canadian Inst. Mining and Metallurgy Bull.* **64**, 60–71.

Brundin, N. H., and B. Nairis, 1972, Alternative sample types in regional geochemical prospecting, *Jour. Geochem. Exploration* **1**, 7–46.

Dall'Aglio, M., and F. Tonani, 1973, Hydrogeochemical exploration for sulphide deposits: Correlation between sulphate and other constituents, in M. J. Jones, ed., *Geochemical Exploration 1972.* London: Institute of Mining and Metallurgy, 305–314.

DeGeoffroy, J., and S. M. Wu, 1970, Design of a sampling plan for regional geochemical surveys, *Econ. Geology* **65**, 340–347.

DeGeoffroy, J., S. M. Wu, and R. W. Heins, 1968, Selection of drilling targets from geochemical data in the southwest Wisconsin Zinc area, *Econ. Geology* **63**, 787–795.

Kraynov, S. T., 1971, The effect of the acidity-alkalinity of groundwaters on the concentration and migration of rare elements, *Geochemistry International* **8**, 828–836.

Michie, U. McL., M. J. Gallagher, and A. Simpson, 1973, Detection of concealed mineralization in northern Scotland, in M. J. Jones, ed., *Geochemical Exploration 1972.* London: Institute of Mining and Metallurgy, 117–130.

Naumov, V. N., D. N. Pachadzhanov, and T. I. Burichenko, 1972, Copper bismuth, silver, lead and zinc in waters of the supergene zone, *Geochemistry International* **9**, 129–134.

Cross-references: *Exploration Geochemistry; Lake Sediment Geochemistry.* Vol. XIII: *Geochemistry, Applied.*

HYDROGRAPHY—See CANALS AND WATERWAYS, SEDIMENT CONTROL; HARBOR SURVEYS.

I

IGNEOUS ROCKS, FIELD RELATIONS

Field Characteristics of Lavas

Lava (from the Italian *lavare* 'to wash') was originally applied to streams of water, and in the eighteenth century in Neopolitan dialect to streams of molten rock from Vesuvius. The term now is used both for the molten material that erupts from volcanoes and to the rock that forms on solidification of this material. Most lavas are silicate liquids and range in composition from komatiite, an ultramafic lava rich in Mg, Fe, and Ca, to rhyolite, a felsic lava rich in Si, Al, and alkali elements. Basalt, a mafic lava, is the most common type.

The main types of lava and their chemical and physical characteristics are given in Table l. There is a close relationship between the physical properties of lavas and their chemical compositions. Ultramafic and mafic lavas erupt at high temperatures and have relatively high densities and low viscosities (i.e., they are highly fluid). They tend to have low concentrations of dissolved water and other volatile phases. Felsic lavas have lower temperatures, lower densities, higher viscosities, and usually higher water contents when they erupt.

Important field characteristics include color, hardness, texture and volcanic structures. These features often allow the identification of the composition of the lava and the type of environment in which it erupted.

Excellent descriptions of the types of structures that develop in modern volcanic rocks are given in many standard texts (e.g., MacDonald, 1972; Williams and McBirney, 1979; Fisher and Schminke, 1984). Discussion of other field characteristics, those in older deformed and metamorphosed lavas, in particular, are less common, being found only in specialized publications.

Color The color of a lava directly reflects its composition. Mafic and ultramafic lavas are composed mainly of dark-colored ferromagnesian (mafic) minerals such as olivine and pyroxene, and these lavas are dark gray to black on fresh unweathered surfaces. Intermediate to felsic lavas, if largely crystalline, contain a high proportion of light-colored minerals like feldspar and quartz and usually have pale gray, tan, or buff colors. Obsidian and other glassy felsic lavas are exceptions in that they are often dark brown or green or almost black.

Metamorphism and weathering change the color. Subaerial mafic lavas oxidize and turn reddish-brown. During metamorphism, the mafic minerals alter to chlorite, actinolite, and epidote, which are green minerals that impart their color to the rock. Metamorphosed mafic lavas typically become dark green, intermediate lavas paler green, and felsic lavas pale yellow to brown or white. Carbonate or other types of metasomatism however, may produce misleading colors that complicate identification.

Hardness and Magnetic Character Mafic lavas, especially when they are altered, are composed of relatively soft minerals and can often be scratched with a geologic hammer. Felsic lavas are usually harder. Ultramafic lavas, and some basalts, contain abundant magnetite (iron oxide) either as a primary phase or as an alteration product of olivine (see *Minerals and Mineralogy*). This mineral gives the rock a magnetic character that can be measured with hand magnet or a compass.

Texture By *texture* is meant the general physical appearance of a lava, especially the size, shape, and arrangement of constituent mineral grains. The texture of a lava is related to its composition and cooling conditions. Most lavas cool rapidly during eruption, with the result that all but the central parts of thick flows are fine grained

TABLE 1. Compositions and Physical Properties of Lavas

Name	Composition		SiO$_2$	Al$_2$O$_3$	MgO	Temperature (°C)	Viscosity (Poises)
			(Weight%)				
Rhyolite	Felsic	Acid	70	14	0	800?	10^{12}
Andesite	Intermediate	Alkaline	60	17	3	1,000	10^4
Basalt	Mafic	Basic	50	15	7	1,200	10^3
Komatiite	Ultramafic		45	8	25	1,500	10

or glassy. In many volcanic rocks, individual mineral grains are no larger than 0.5 mm and cannot be distinguished in the field.

In low-viscosity (relatively fluid) mafic and ultramafic lavas, mineral grains nucleate easily and grow quickly: most basalts and komatiites are largely crystalline, being glassy only at rapidly chilled margins of flows or fragments. In the more viscous felsic lavas, nucleation and crystal growth are sluggish, and many rhyolite bodies are glassy throughout.

Many lavas are *porphyritic*. This term refers to a texture in which a portion of the mineral grains is significantly larger than the rest of the rock. The larger grains, called *phenocrysts,* are thought to crystallize during slow cooling before eruption, and the finer grains, or *groundmass,* during rapid cooling following eruption. Andesites are commonly porphyritic, with phenocrysts of plagioclase, pyroxene, or hornblende, and basalts may contain olivine, pyroxene, or plagioclase. Phenocrysts are often large enough to be recognized in the field and give a good indication of the composition of the host lava.

Vesicles are rounded, often spherical cavities originally filled with gas exsolved from the lava (Fig. 1). Their diameter varies from microscopic to several centimeters but is typically around 1–5 mm. At great water depths, as at mid-oceanic ridges, high water pressures inhibit exsolution of dissolved gases. Deep-water lavas are essentially nonvesicular or contain very small and infrequent vesicles. In this way they can be distinguished in the field from shallow-water and subaerial lavas (those that erupt on land rather than under water), which commonly contain abundant, relatively large vesicles. During alteration and metamorphism, vesicles fill with secondary minerals such as quartz, carbonate, chlorite, or zeolite, in which case they are called *amygdules.* Vesicles may be distributed throughout lava bodies, but they are often concentrated in zones close to the tops of flows. At flow bases, elongate *pipe vesicles* are sometimes found.

Certain textures are restricted to a specific type of lava. *Flow banding,* alternating layers with different color or phenocryst content produced during flowage of lava, is common only in glassy felsic rocks. *Spinifex* texture, characterized by elongate skeletal plates or needles of olivine or pyroxene, is restricted to komatiites and related basalts.

Morphology and Structure in Volcanic Rocks Like textures, the morphology of volcanic flows and the structures in them provide information about both composition and eruption environment. The types of structures depend strongly on the viscosity of the lava, which as described earlier, correlates closely with composition. Fluid, low-viscosity basalts tend to erupt

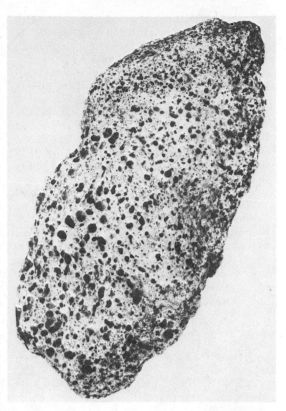

FIGURE 1. A highly vesicular lump of basalt. The vesicles range from less than 1 mm to more than 10 mm in diameter. (From Decker and Decker, 1982; photo by James Griggs, U.S. Geological Survey)

rapidly from narrow fissures and, except when large volumes erupt or the lava becomes ponded, form flows that are thin and long. For example, Hawaiian basalt flows are only 2–15 m thick but travel for tens of kilometers. Komatiites and Si-poor mafic lavas have lower viscosities and form still thinner flows. In contrast, andesite flows tend to be short and thick. Dacites and rhyolites rarely erupt as lavas, and when they do, they form domes and spines.

More commonly, felsic magmas erupt explosively as fragmental or *pyroclastic* (fire-broken) rocks. There are two reasons for this: (1) intermediate to felsic lavas frequently contain relatively large amounts of dissolved water and other volatile components, and (2) they have high viscosities that inhibit the escape of the volatile phases. In felsic magmas ascending toward the surface, gases exsolve but cannot escape; pressure builds up; and explosive eruptions result.

Structures in Lavas Flows

By far the most useful indicator of subaqueous volcanic eruption is *pillow* lava. Pillows are elongate, bulbose, tube- or sacklike bodies of lava that form by repeated budding and extrusion of lava at the fronts of underwater flows (Fig. 2). Diameters typically are around 1 m, although much larger and much smaller examples are known. In many exposures, especially the flat, glaciated outcrops common in Precambrian terrains, pillows appear to be isolated sacks. This is largely due to the nature of field exposures. Pillow lavas usually consist of piles of interconnecting, budding and branching, twisting tubelike bodies, and in the two-dimensional sections represented by field outcrops, they appear as separate, isolated bodies (Fig. 3).

Pillow rims are glassy where they were chilled by seawater. In the interiors concentric layering develops, which is often emphasized by bands rich in vesicles. Radial jointing is a feature of most pillows but curiously is rarely observed in Pre-

FIGURE 2. Pillow lava from 2,700 m below the surface of the ocean at the Mid-Atlantic Ridge. (From Decker and Decker, 1982; photo by W. B. Bryan, Woods Hole Oceanographic Institution)

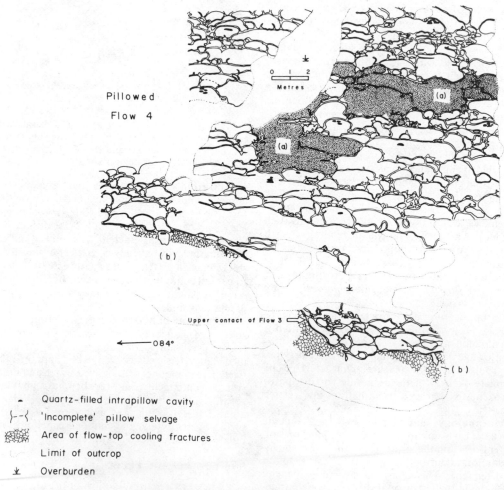

Pillowed Flow 4

(a)

(a)

(b)

Upper contact of Flow 3

←——— 084°

(b)

- Quartz-filled intrapillow cavity
}--{ 'Incomplete' pillow selvage
Area of flow-top cooling fractures
Limit of outcrop
↓ Overburden

FIGURE 3. Sketch of an outcrop of Archean pillow lava from Manitoba, Canada. (From Hargreaves and Ayres, 1979; copyright © 1979 by the National Research Council of Canada)

cambrian examples. Pillows are common only in basalts and occur far less frequently in komatiites, andesites, and other types of lavas.

Not all subaqueous flows are pillowed. In most submarine sequences, some massive lava is present. The term *massive* means uniformly textured lava without obvious volcanic structures. Massive lava forms simple sheet flows or occurs as parts of composite flows. In some of these, the upper part of the flow is pillowed and the lower part is massive; in others, the reverse is the case, and in others, massive lava grades laterally into pillows. Another indicator of subaqueous eruption is *hyaloclastite,* a fragmental volcanic rock made up of broken pieces of volcanic glass and less common blocks and irregular blobs of lava. Hyaloclastites form when hot lavas are suddenly chilled and thermally shattered as they erupt into water. They usually are associated with lava flows, forming irregular layers or filling the spaces between pillows. Hyaloclastites can sometimes be distinguished from pyroclastic rocks by the uniformly glassy or fine-grained nature of their clasts and the uniform composition of these fragments. Pyroclastic rocks commonly contain fragments of exotic rock types ripped from the vent or crater walls during eruption.

In subaerial lavas, pillowlike structures do occur (they are called *lava toes*) but they are not common. Most subaerial flows are massive or composed partly or entirely of blocks and fragments. Two main types are distinguished on the basis of surface forms. *Pahoehoe* is highly fluid lava with a thin, smooth elastic skin that is commonly dragged by internal movement of lava into billowing flows or ropy structures (Fig. 4). *Aa* are

more viscous flows with thick crusts that are broken up during movement of the lava into extremely jagged, clinkery, spinose fragments. Some Hawaiian flows grade from pahoehoe near the vent to aa farther down slope. In these cases, the change in character reflects increasing lava viscosity probably caused by cooling and loss of volatile phases, or an increasing crystallinity. In Iceland and elsewhere relationships between the two lava types are more complicated.

As lava cools and contracts, shrinkage cracks or joints form. The joints commonly produce five- or six-sided columns oriented perpendicular to the surface of the flow. *Columnar joints* (Fig. 5) range in diameter from a few centimeters to more than 3 m and are up to 30 m long. They form in both subaerial and subaqueous lavas, usually within tabular sheet flows.

Some thicker mafic and ultramafic flows differentiate by gravitative settling of crystals to form *layered* flows. These typically have lower olivine or pyroxene cumulate layers and an upper gabbroic layer beneath a thin chilled crust.

Pyroclastic Rocks

Space allows only an abbreviated account of these complicated rocks. For more information see Fisher and Schminke (1984) or Williams and McBirney (1981).

FIGURE 4. Lava flows of aa (left) and pahoehoe, both issued from Kilauea volcano in Hawaii. (From Decker and Decker, 1982; photo by Barbara Decker)

FIGURE 5. Columnar jointed basalt from Giants Causeway, Ireland. (From Press and Siever, 1974; copyright © 1974 by W.H. Freeman)

TABLE 2. Types of Tephra and Pyroclastic Rocks

	Size of Frag- ments (mm)	Name of Pyro- clastic Rock
Ash	2	Tuff
Lapilli	2–64	Lapilli tuff, tuff breccia, spatter
Blocks and bombs	64	Agglomerate (when rich in bombs); otherwise, volcanic breccia

The material blown through the air from an erupting volcano is called *tephra*. Tephra comes in a variety of sizes (see Table 2 for the names that are used) and shapes (Fig. 6). Pieces of lava that were molten when erupted are rounded; the larger pieces, called bombs, often have flow lines and spindle shapes acquired during their flight through the air. Blocks of older lava plucked from crater walls during eruption are typically angular.

Pyroclastic rocks accumulate from airfall deposits, or they may be erupted as pyroclastic flows. Tephra in airfall deposits usually becomes sorted according to fragment size during eruption and accumulation. Airfall tuffs and lapilli tuffs

FIGURE 7. Volcanic ash layers in a road cut on Oshima volcano, Japan. (From Decker and Decker, 1982; copyright © 1982 by W.H. Freeman)

(see Table 2) are made up of layers or beds (Fig. 7) with uniform fragment size or with a fragment size that varies systematically from top to bottom of a bed. *Graded bedding,* as this is called, and other sedimentary structures such as cross-bedding and erosion structures (see *Geologial Structures;* Vol. VI: *Bedding Genesis*) are common in airfall deposits—in particular in those that erupt subaerially but accumulate in shallow water.

Pyroclastic flows are very hot dense clouds of volcanic ash and gases that are ejected from craters and flow extremely rapidly down the sides of volcanoes. They produce poorly stratified deposits. The fragments are commonly so hot when they accumulate that they weld together to form a hard, compact rock called *welded tuff* or *ignimbrite*. When welding is extreme, all fragmentary character is lost and the rock is easily mistaken for massive lava.

Loosely consolidated tephra is often transported and redeposited by the action of water.

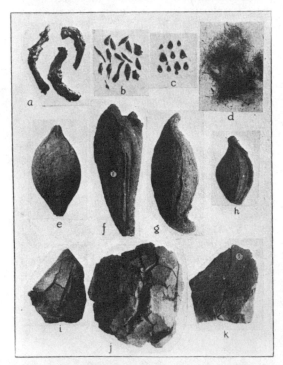

FIGURE 6. Some common forms of pyroclastic ejecta (tephra). (*a*) Ribbon bombs, (*b*) small ribbons, (*c*) Pele's tears, (*d*) Pele's hair, (*e–h*) almond- and spindle-shaped bombs, (*i–j*) breadcrust bombs, (*k*) block. (From Williams and McBirney, 1981; copyright © 1981 by W.H. Freeman)

TABLE 3. Summary of Way-Up Indicators

Tops of pillows are normally rounded and convex, and bases have a toelike form or are concave where they are molded to conform to the shape of underlying pillows (Fig. 3).

Glassy rims may be thicker at the tops of pillows than at the bottoms.

Fine-grained chill zones and breccias are better developed and thicker upper parts of flows.

Vesicles are usually more abundant at the tops of pillows and lava flows. Those at flow tops are rounded or have irregular shapes; at bases, pipe vesicles may be found.

Spinifex texture develops at the top, and olivine or pyroxene cumulates occur at the bases of some ultramafic or mafic flows.

Sedimentary structures such as graded bedding and cross-bedding show the way up in airfall pyroclastic deposits.

TABLE 4. Distinctions between Lavas and Intrusions

	Volcanic Rocks	Intrusions
Grain size	Fine (except in thick flows)	Medium to coarse (except at chilled margins)
Grain size variations	Asymmetric: zones with fine grain size thicker at top than at bottom of flow	Symmetric: fine-grained zones at top and bottom
Glassy rocks	Common	None
Vesicles	Common	Very rare
Breccias	Hyaloclastites and pyroclastics contain glassy or fine-grained, commonly vesicular fragments; pyroclastic rocks have sedimentary structures	Fragments in magmatic or fault breccias have the grain size of intrusive rocks, are rarely vesicular, do not have sedimentary structures
Pillows	Common in subaqueous lavas	Occur very rarely in shallow subsurface dikes
Columnar joints	Common	Only in shallow sills
Contacts	Conformable with surrounding rocks	Conformable or nonconformable

Volcanic mud flows (*lahars*), streams, and currents in lakes and oceans redistribute and sort the volcanic material to produce sedimentary rocks.

Way-Up Indicators in Volcanic Rocks

Volcanic rocks contain many features that allow a field geologist to distinguish the tops from the bottoms of lava flows—i.e., to decide which way was up. This information is very useful when mapping folded or faulted volcanic sequences because, by monitoring changes in directions of the tops, complexities in the structure can be sorted out. Table 3 lists a number of these features.

Distinction between Lavas and Instrusions

Listed in Table 4 are features that allow lavas to be distinguished from intrusions.

N. T. ARNDT

References

Decker, R., and B. Decker, 1982, *Volcanoes*. San Francisco: W. H. Freeman, 244p.
Fisher, R. V., and H. U. Schmincke, 1984, *Pyroclastic Rocks*. Berlin: Springer-Verlag, 472p.
Hargreaves, R., and L. D. Ayres, 1979, Methodology of Archean metabasalt flows, Utik Lake, Manitoba, *Canadian Jour. Earth Sciences* **16**, 1452–1466.
MacDonald, G. A., 1972, *Volcanoes*. Englewood Cliffs, N.J.: Prentice-Hall, 510p.
Press, F., and R. Siever, 1974, *Earth*. San Francisco: W. H. Freeman, 945p.
Williams, H., and A. R. McBirney, 1981, *Volcanology*. San Francisco: W. H. Freeman, 397p.

Cross-references: *Lithogeochemical Prospecting; Photointerpretation; Sedimentary Rocks, Field Relations; Tephrochronology.*

INDICATOR ELEMENTS

Indicator elements have been used for centuries as a guide in geochemical prospecting for various types of mineral deposits. Witness, for instance, the fact that the ancient prospectors associated quartz (silicon) with gold and galena (lead) with silver. Many other instances of these associations could be quoted from the old literature, but only since the 1950s have indicator, or *pathfinder, elements* received detailed attention. One of the earliest papers on the subject is that of Warren and Delavault (1959).

Concentrations of a single element rarely occur in the Earth. More generally a suite of elements is concentrated in a particular deposit because of certain intrinsic chemical properties that depend essentially on their electronic constitution and hence their position in the periodic table. More specifically the migration and concentration of elements are controlled by the Eh-pH conditions, complexation phenomena, hydrolytic reactions, colloidal phenomena, biological reactions, adsorption and base exchange reactions, diffusion, solubility in molten silicates, and various other parameters. In addition to these purely chemical features are others of a physical nature that concentrate the resistate minerals during weathering and the formation of placers. Because of all these interacting processes, it is not usually possible to predict a priori what elements will be concentrated together. Fortunately, most elemental associations and indicators are known from purely empirical data that have been gathered during the long history of geochemistry. Elements with the most diverse chemical properties are frequently concentrated together; why, we do not know. In geochemical prospecting, therefore, the empirical data are of fundamental importance since they can be used in a practical way to predict the probable occurrence of elements that may not be suspected in deposits. By the same token certain elements may serve as indicators of others that have a particularly low abundance in deposits. Thus, arsenic in rocks, soils, and vegetation may indicate the presence of gold deposits, and nickel may be indicative of platinum metals in certain terranes.

The efficiency of an indicator element in *geochemical prospecting* depends on a number of factors, chief of which is its intimate association with the elements in the deposits that are sought. The indicator element should also have a low abundance (clarke) compared with the element sought, thus ensuring a low background for the indicator element and conspicuous contrasting anomalies in geochemical surveys. In addition, certain indicator elements should have favorable chemical features, ensuring relatively wide dispersion due to volatility, solubility, or other characteristics if hydrochemical (see *Hydrochemical Prospecting*), stream sediment, or atmogeochemical surveys (see *Atmogeochemical Prospecting*) are utilized; when other surveys are employed such as those based on soils or heavy minerals, features such as low dispersion, formation of resistate minerals, etc. are desirable. Finally, indicator elements should admit of easy determination analytically. A good example of many of these features is arsenic, which has a low abundance (3 ppm) yet which may be concentrated in gold and other deposits up to low percentages (Boyle and Jonasson, 1973). Arsenic has high dispersive power under hydrothermal conditions, and its hypogene aureoles and trains are commonly broad and extended; under supergene conditions the element has a relatively low dispersive power, in particular where soils and weathered residuum are rich in iron. Arsenic is also relatively easily determined analytically compared with gold and a number of other elements with which it is commonly associated.

This entry discusses the use of indicator elements in the various types of geochemical surveys only in a general way, placing much more emphasis on the various elemental associations in mineral deposits. The data presented are, however, not comprehensive and do not take into account all the known elemental associations in nature such as those that exist in soils, natural waters, plants, etc. The information presented, however, is deemed adequate for use in most types of geochemical prospecting surveys. Those seeking more details should consult the publication by the writer on indicator elements (Boyle, 1974a) and the more recent textbooks on geochemical prospecting noted in the references and selected bibliography. In the following summary the arrangement of the elements is in accordance with the grouping in the periodic table (Table 1).

Group IA: Hydrogen, Lithium, Sodium, Potassium, Rubidium, and Cesium

Hydrogen (q.v. in Vol. IVA) is one of the most common elements in the Earth, being bound mainly in the various states of water and other complexes in the hydrosphere and biosphere, in water and hydroxyl groups in minerals, and in the various solid and liquid hydrocarbon substances. Small amounts of hydrogen also occur in the free state in various rocks, in fumaroles and volcanoes, and in natural gases; in the atmosphere the element is rare, averaging about 0.1 ppm. Hydrogen can be obtained for commercial purposes from water or acids by a variety of reduction reactions: *viz.* reaction of an active metal such as sodium with water, reaction of a metal and an acid, reaction of hydrides with water, electrolysis, and reduction of water by carbon (water gas or producer gas reaction). From hydrocarbons, hydrogen can be obtained by various thermal and thermochemical cracking processes. There are no specific indicators of hydrogen in nature.

Hydrogen is produced as a result of the radiolysis of water. Natural ground- and stratal waters pervading or migrating through radioactive deposits containing uranium and thorium may become enriched in elemental hydrogen. In this respect hydrogen can be considered as an indicator of uranium and thorium deposits and also of hydrocarbon deposits enriched in these two radioactive elements.

Lithium (q.v. in Vol. IVA) exhibits its greatest enrichment in pegmatites, granitic dikes, and certain granites in which it is present in spodumene, lepidolite, petalite, eucryptite, triphylite-lithiophilite, amblygonite, and various other rarer lithium-bearing minerals. It is also found in certain greisen zones and dikes in which it occurs mainly in lepidolite and zinnwaldite (q.v. in Vol. IVB). The greisens are frequently associated with tin, tungsten, and molybdenum veins in granitic rocks, gneisses, and metasediments. The element also occurs in commercial quantities in certain brines and other natural waters. It may also be present in certain clay-rich saline evaporites in considerable quantities and is concentrated in wad and other manganese oxides and in some Mg-rich clays and shales. In clays it is usually present as a lithium montmorillonite (hectorite) (q.v. in Vol. XII, Pt. 1). The element is relatively rare in carbonatites, but in some of these deposits it may occur in amounts up to 100 ppm. Certain members of peralkaline (agpaitic) massifs are also enriched in lithium.

TABLE 1. Periodic Table of the Elements

IA	IIA	IIIB	IVB	VB	VIB	VIIB	VIII	VIII	VIII	IB	IIB	IIIA	IVA	VA	VIA	VIIA	O
1 H 1.008																	2 He 4.003
3 Li 6.939	4 Be 9.012											5 B 10.81	6 C 12.011	7 N 14.007	8 O 16.00	9 F 19.00	10 Ne 20.183
11 Na 22.990	12 Mg 24.31											13 Al 26.98	14 Si 28.09	15 P 30.974	16 S 32.064	17 Cl 35.453	18 A 39.948
19 K 39.102	20 Ca 40.08	21 Sc 44.96	22 Ti 47.90	23 V 50.94	24 Cr 52.00	25 Mn 54.94	26 Fe 55.85	27 Co 58.93	28 Ni 58.71	29 Cu 63.54	30 Zn 65.37	31 Ga 69.72	32 Ge 72.59	33 As 74.92	34 Se 78.96	35 Br 79.909	36 Kr 83.80
37 Rb 85.47	38 Sr 87.62	39 Y 88.91	40 Zr 91.22	41 Nb 92.91	42 Mo 95.94	43 Tc (99)	44 Ru 101.1	45 Rh 102.91	46 Pd 106.4	47 Ag 107.870	48 Cd 112.40	49 In 114.82	50 Sn 118.69	51 Sb 121.75	52 Te 127.60	53 I 126.90	54 Xe 131.30
55 Cs 132.91	56 Ba 137.34	57 La 138.91	72 Hf 178.49	73 Ta 180.95	74 W 183.85	75 Re 186.2	76 Os 190.2	77 Ir 192.2	78 Pt 195.09	79 Au 196.97	80 Hg 200.59	81 Tl 204.37	82 Pb 207.19	83 Bi 208.98	84 Po (210)	85 At (210)	86 Rn (222)
87 Fr (223)	88 Ra (226)	89 Ac (227)															

			58 Ce 140.12	59 Pr 140.91	60 Nd 144.24	61 Pm (145)	62 Sm 150.35	63 Eu 151.96	64 Gd 157.25	65 Tb 158.92	66 Dy 162.50	67 Ho 164.93	68 Er 167.26	69 Tm 168.93	70 Yb 173.04	71 Lu 174.97
			90 Th 232.04	91 Pa (231)	92 U 238.03	93 Np (237)	94 Pu (242)	95 Am (243)	96 Cm (245)	97 Bk (249)	98 Cf (249)	99 E (254)	100 Fm (252)	101 Mv (256)	102 No (253)	103 Lw

Note: Atomic weights are based on ^{12}C.

341

In pegmatites, granitic dikes, and certain granites, lithium may be accompanied, in addition to the more common elements such as Na, K, Al, and Si, by Rb, Cs, Be, Sc, rare earths, Y, Nb, Ta, Ti, Zr, B, Sn, W, Ga, Tl, Mn, P, F, and more rarely by U, Th, Bi, (As), (Sb), and Mo.* The most common associates and the best indicators are Rb, Cs, Be, Nb, Ta, B, P, F, and Sn. In greisens the most distinctive associated elements are Rb, Cs, Be, W, Sn, Mo, Bi, Ti, Ga, B, and F.

In brines and saline evaporites, such as those in Searles Lake, California, lithium is present as the phosphate or chloride and is accompanied by Na, K, B, W, P, F, Br, Cl, I, SO_4^{2-}, CO_3^{2-}, and S. There are no particular indicators except perhaps B and W. All natural spring, underground, and saline waters should be tested directly for lithium if a source is sought in these materials. One of the richest sources of lithium in the world is the underground brines of Silver Peak, Nevada, an old silver mining area. These brines are some seven times as rich in lithium as the brines of Great Salt Lake, another large source of lithium. Lithium wad and other manganese dioxides are comparatively rare but may indicate the presence of lithium deposits in the geological terrane, especially where they occur at the orifices of springs. The wads contain a veritable host of elements, among which Ba, Sr, Ti, rare earths, As, Sb, W, Mo, Co, and all the other heavy metals may be represented.†

Lithium-bearing clays and shales are not common but should be carefully considered as a future source of the element. The lithium mineral in certain clays is hectorite, a lithium clay mineral belonging to the montmorillonite group. In lithium-bearing clays and shales the element is usually accompanied by higher than average amounts of magnesium, fluorine, and boron, all of which could serve as indicators of the element. Hydrothermally altered tuffs in some districts may carry above normal contents of lithium, accompanied by elements such as Cs, Rb, Be, and F. For the elemental associates of lithium in carbonatites, see the section on niobium.

*In these and other lists of elemental associates the common gangue elements such as Si, Al, Fe, Ca, etc. are omitted unless they are of importance in geochemical prospecting methods. It should also be understood that not all the elements are concentrated with the particular element in question. Omissions of particular elements are frequently the rule rather than the exception. Elements listed in parentheses are only rare associates or indicators.

†All wads, manganese oxide deposits, and limonitic deposits should be analyzed during geochemical surveys as a matter of routine. They are great adsorbers of many elements and may, therefore, indicate the presence of these elements in the geological terrane in which they occur.

Sodium and *potassium* (q.v. in Vol. IVA) are closely allied in nature and rarely occur to the exclusion of one another. Both elements are strongly enriched in granitic rocks and pegmatites, in which they occur mainly in the feldspars and micas. Nepheline syenites and carbonatites are enriched in both sodium and potassium, but the K to Na ratio varies widely in the rocks that make up these complexes, commonly from 0.3 to 8. The *fenitization process* associated with many carbonatites involves significant introductions of both sodium and potassium, in particular the former. Sodium and potassium are obtained commercially from saline evaporites, saline lakes, alkaline lakes and marshes (sodium), and underground brines, in which they are usually present as chlorides, nitrates, sulfates, carbonates, or borates. All saline evaporites, saline lakes, salinas, salt flats, alkaline lakes, caliche, salty earths, and brines should be analyzed directly for the presence of sodium and potassium salts followed by mineral identifications if a source is sought in these materials.

Salt precipitates at the orifices of springs carrying sodium and potassium may be indicative of deposits of halite, sylvite, and other K or Na salts in the underlying bedrocks. Similarly, springs and precipitates and associated pans, enriched in sodium and potassium salts, particularly sodium carbonate (nahcolite, trona, gaylussite, etc.), may indicate the presence of young carbonatites as in the carbonatite regions of Tanzania in Africa and elsewhere in the world. Such precipitates commonly carry unusual elements such as Ba, Sr, Ce, Nb, etc., all indicators of carbonatites. See also the section on niobium. Many of the springs associated with carbonatites are highly charged with CO_2.

Potassium is strongly adsorbed by clay minerals, and hence, certain clays and shales may be greatly enriched in the element. Such materials may be utilized as fertilizer components in the future. Direct analysis of potassium is suggested for clays and shales suspected to contain higher than average amounts of the element.

Soda metasomatism is a feature of many types of carbonatites as noted earlier. Higher than average contents of sodium in rocks, soils, tills, waters, and vegetation may, therefore, indicate the presence of these bodies. Sodium is also enriched in the wall-rock alteration zones of many types of gold and other metallic deposits (albitization), and hence, the element may be a useful indicator of these deposits. Potassium, likewise, is frequently enriched in the wall-rock alteration zones of certain gold deposits, and both sodium and potassium may be enriched in the alteration zones associated with pegmatites, base metal, and many other types of deposits. Use of the K_2O/Na_2O ratio may be useful in assessing proximity to gold, silver, uranium, and base metal mineralization in many types of epigenetic deposits since this ratio increases significantly and consistently as mineralization is approached where potash metasomatism is manifest. Where albitization or soda metasomatism is prevalent, the ratio remains constant or decreases (Boyle, 1974b).

Potassium has a natural radioactive isotope, ^{40}K, whose abundance is approximately 0.0ll9% of the total amount of the element present. ^{40}K decays in a complex manner by negative beta emission, K-capture, and gamma emission. Because of these phenomena, potassium-enriched rocks and zones are slightly radioactive and can be detected by radiometric methods. *Rubidium* (q.v. in Vol. IVA) follows potassium closely in its geochemistry and is concentrated in most potassium minerals. No rubidium minerals are known. Most of the rubidium of commerce is obtained from lepidolite and carnallite. Other possible sources are pollucite, leucite, zinnwaldite, various potassium micas and feldspars, adularia, and certain varieties of beryl.

Rubidium is concentrated in certain pegmatites, granitic dikes, and granites; in greisens that may be associated with tin, tungsten, and beryllium deposits; and in potash-rich alteration zones associated with gold, silver, and base metal mineralization. The element is also frequently enriched in potassium-bearing brines (up to 600 ppm) and mineral waters and in potassium-rich saline evaporites. Certain potassium-rich clays and shales may contain above average contents of rubidium.

The most common elemental associate of rubidium in all geological materials is potassium, and it is the best indicator of the element. In pegmatites, greisens, and granites the common associated elements are Li, Cs, Be, Sc, Y, rare earths, Nb, Ta, Ni, Zr, Mo, W, Sn, Bi, Tl, Ga, B, P, and F and, more rarely, U, Th, (As), (Sb), and Mo. Of these, the best indicators are Cs, Li, Ta, Nb, Be, Sn, P, and F. In natural waters and saline evaporites rubidium may be accompanied by K, Li, Cs, (W), B, Cl, and the various other readily soluble elements such as Na, Ca, Mg, etc. In some clays, shales, and tuffs (the tuffs are often hydrothermally altered), rubidium enrichments are commonly accompanied by higher than normal amounts of Li, Cs, Be, rare earths, and F.

Cesium (q.v. in Vol. IVA) is the least abundant of the common alkalis. It forms the mineral pollucite, $(Cs,Na)_2Al_2Si_4O_{12} \cdot H_2O$, which commonly resembles quartz and may be easily mistaken for it. Other possible sources are lepidolite, microcline feldspar, adularia, certain varieties of beryl (vorobievite), rhodizite (a complex Be borate), cesium-kupletskite (a complex Nb silicate), potassium micas, carnallite, and other potassium minerals. The present commercial source of cesium is pollucite.

Cesium follows potassium and rubidium closely in its geochemistry and is concentrated in certain pegmatites, granitic dikes, and granites where it is present in pollucite, the potassium and lithium minerals, beryl, and certain greisens associated with tin, tungsten, and beryllium deposits. The element may also be enriched in certain brines and other natural waters and in potassium-rich saline evaporites. Certain potassium-rich clays, shales, and altered tuffs may contain above average amounts of cesium.

The most common elemental associates of cesium in practically all geologic materials are potassium and rubidium, and both are good indicators of the element. In pegmatites, granitic dikes, greisens, and granites, the common associated elements are Li, Rb, Be, Sc, Y, rare earths, Nb, Ta, Ti, Zr, W, Ga, Tl, Sn, β, P, and F and, more rarely, U, Th, Bi, (As), (Sb), and Mo. Of these, the best indicators are Li, Rb, Be, Ta, Nb, Sn, P, and F. In natural waters and brines and in saline evaporites, the commonly associated elements are the same as those for lithium and rubidium. All brines, mineral waters, and saline evaporites should be analyzed for all the alkalis on a routine basis since some of these elements may be present in these materials in commercial quantities.

Group IB: Copper, Silver, and Gold

Copper (q.v. in Vol. IVA) occurs in nine principal types of deposits as follows:

1. Disseminated copper sulfides and native copper in shales and their metamorphic equivalents (Kupferschiefer and Zambia types): In certain of the copper shale deposits (e.g., Kupferschiefer), copper is associated with a veritable host of chalcophile elements, among which the most abundant are Ag, Zn, Cd, Pb, Mo, Re, Co, V, Mn, Se, As, and Sb. Of these, the best indicators appear to be Ag, Zn, Pb, Mo, and Co. In some copper shale deposits (e.g., White Pine, Michigan), the only associated elements exhibiting significant enrichments are Ag, Co, and Ni.

2. Disseminated copper sulfides and native copper in sandstones, sandy shales, and conglomerates (red bed type): The common associated elements in these deposits are Ag, Pb, Zn, Cd, Hg, V, U, Ni, Co, P, Cr, Mo, Re, Se, As, Sb, Mn, and Ba. Among these, Ag, Pb, and Ba are nearly universal and provide the best indicators. Co, Ni, Mo, As, and Sb occur in many deposits and are useful indicators in some districts. U, V, P, Cr, and Se, while present in trace amounts in most deposits, reach high concentrations only within certain metallogenic provinces.

3. Native copper deposits in amygdaloidal basalts and associated quartzites and conglomerates: These deposits have few elemental associates. Among the most common are Ag, B, As, and S; less common and local are Pb and Zn. The minerals calcite, epidote, prehnite, datolite, and zeolites are characteristic.

4. Disseminated copper sulfides, arsenides, etc. in or associated with monzonite, various acid porphyries, and other granitic intrusives (porphyry copper type): The most common elemental associates in these deposits are Mo, Re, and Fe. Other associates, frequently of local occurrence, are B, Zn, Pb, Ag, Hg, Au, As, and Sb; W is rare. Te is enriched in some regions.

5. Copper sulfides in skarn-type deposits: In these, copper is accompanied by a number of elements among which the most frequent are Fe, Mn, Zn, Pb, Ag, Cd, Mo, W, Au, Sn, Bi, Te, As, and (Sb); Co and Ni are rare. B and F may occur in some deposits. The calcium-magnesium-manganese-iron silicates are characteristic.

6. Massive and disseminated copper sulfides associated with nickel, cobalt, and iron sulfides in or near basic igneous rocks such as gabbros and norites (Sudbury type): These deposits have a characteristic suite of elements that accompany the copper. Most frequent are Ni, Co, Fe, and As. Elements in trace amounts include Pt metals, Ag, Au, Bi, Se, and Te. Pb, Zn, Cd, Sn, Be, and Sb are rare. The best indicators of such deposits are Ni, Cu, Co, As, and the Pt metals.

7. Massive and disseminated copper sulfides associated with iron, lead, and zinc sulfides in volcanic and/or sedimentary terranes: In these, copper is associated with and indicated by the following elements: Zn, Pb, Cd, Ag, Fe, As, and Sb. Other elements present in the deposits, often in traces, are Au, Mo, W, Re, Co, Ni, B, Ga, In, Tl, Ge, Sn, Bi, Se, and Te. Hg is a good indicator of these deposits and has been used effectively in geochemical surveys in some districts. The Archean massive sulfide bodies, usually rich in copper and zinc, are commonly low in lead. Barium is enriched in some of the massive and vein-type copper deposits.

8. Copper sulfides, arsenides, etc. in veins, lenses, pods, disseminations, etc. in faults, fractures, drag folds, crushed zones, and shear zones: The associated elements and indicators are essentially the same as those noted in items 5, 6, and 7.

9. In carbonatites such as at Palabora, South Africa, when the orebody averages about 0.7% Cu: In carbonatites copper is normally present in chalcopyrite, bornite, cubanite, and chalcocite. Some of these minerals are late in the various stages of formation of the carbonatites. Elements associated with the copper stage are Fe, S, Pb, Ni, Co, Zn, Cd, Au, and Ag. Some bodies exhibit enrichments of W and Bi with Au. For the other elemental associates in carbonatites, see the discussion under the element niobium.

Silver (q.v. in Vol. IVA) is won from various types of copper, lead, zinc, gold, and nickel deposits. For the indicator and associated elements in these deposits, see the discussion accompanying each of these elements. Particularly good indicators of the presence of silver in these deposits are Pb, Zn, Cd, Hg, Tl, Cu, Au, Ba, Mn, Bi, Se, Te, As, and Sb. Manganiferous siderite or calcite is a good indicator mineral in many silver deposits.

Much silver is also obtained from native silver deposits, those containing the nickel-cobalt arsenides (Cobalt type), in particular. The characteristic elemental associates of silver in these deposits are Ni, Co, Fe, S, As, Sb, Bi, and U. The last element is restricted to deposits in specific mineral belts. The deposits also contain some Ba, Cu, Zn, Cd, Pb, and frequently a little Hg, the last mainly in allargentum, dyscrasite, and native silver. Among these various elements, the best indicators in geochemical surveys utilizing rocks, soils, vegetation, stream sediments, and waters are Ni, Co, As, Sb, and Bi. Where uranium occurs in the deposits it may serve as an indicator element in geochemical surveys, and the veins can also be located by their high radioactivity. The usefulness of Ba, Cu, Zn, Cd, and Pb as indicators of native silver deposits depends on their content in the veins. Mercury may be useful in rock and soil surveys where it occurs in the deposits.

Silver is a constituent of several copper, uranium, and vanadium deposits in sandstones. These deposits are generally referred to as the red bed type. The characteristic elements concentrated in these deposits include U, V, Sr, Ba, Cr, Mo, Re, Fe, Co, Ni, Cu, Ag, Au, Zn, Cd, Pb, P, As, Sb, S, and Se. Some of these deposits are similar to the red bed sandstone deposits discussed under copper, uranium, and vanadium. For further discussion of silver deposits and geochemical methods of their discovery, see Boyle (1968).

Gold (q.v. in Vol. IVA) is won both from deposits mined essentially for the element and as a by-product of the mining and treatment of nickel, copper, zinc, lead, and silver ores. The following types of primary (endogene) and secondary (exogene) deposits, exploited mainly for gold, can be distinguished (Boyle, 1979):

1. Auriferous skarn-type deposits.

2. Auriferous pegmatites, coarse-grained granitic bodies, and porphyry dikes and sills: These are comparatively rare.

3. Gold-silver and silver-gold veins, stockworks, lodes, mineralized pipes, and irregular silicified bodies in fractures, faults, shear zones, sheeted zones, and breccia zones essentially in volcanic terranes.

4. Auriferous veins, lodes, sheeted zones, and saddle reefs in faults, fractures, bedding-plane

discontinuities and shears, drag folds, crushed zones, and openings on anticlines essentially in sedimentary terranes; also replacement tabular and irregular bodies developed near faults and fractures in chemically favorable beds such as limestones, dolomites, and calcareous shales and schists.

5. Gold-silver and silver-gold veins, lodes, stockworks, silicified zones, etc. in a complex geological environment, comprising sediments, volcanics, and igneous or granitized rocks.

6. Disseminated and stockwork gold-silver deposits in igneous, volcanic, and sedimentary rocks:

Disseminated and stockwork gold-silver deposits in igneous bodies.

Disseminated gold-silver deposits in volcanic flows and associated volcaniclastic rocks.

Disseminated gold-silver deposits in volcaniclastic and sedimentary beds:

Deposits in tuffaceous rocks and iron formations,

Deposits in chemically favorable sedimentary beds.

7. Gold deposits in quartz-pebble conglomerates and quartzites.

8. Eluvial and alluvial placers.

The quartz-pebble conglomerate deposits provide the bulk of the world's production of gold, almost 55%. The other deposits, mainly the various vein and disseminated types, and eluvial and alluvial placers now provide the remaining 45% of the production.

For the elemental associates of gold in deposits from which the element is won as a by-product, see the discussions under nickel, copper, zinc, lead, silver, platinum, and niobium.

The elemental associates in the skarn-type gold deposits are the same as those discussed for similar deposits of copper, lead, zinc, and tungsten. There is frequently abundant arsenic and minor to trace amounts of antimony and bismuth in the gold deposits, and these elements, especially arsenic, provide good indicators of gold and silver in most types of geochemical prospecting surveys.

The deposits in categories 3–6 are invariably quartz veins, irregular bodies, saddles, and stockworks; quartz-sulfide veins and impregnation zones; silicified and pyritized zones, and disseminated gold-sulfide zones. The principal associated elements and indicators in these deposits are SiO_2, Ag, As, Sb, S, and Fe; in addition are Cu, Ba, Zn, Cd, Hg, B, Tl, U, Sn, Pb, Bi, Se, Te, Mo, W, F, Co, Ni, and (Pt metals), depending on the metallogenic belt in which the deposits occur. Some deposits, especially those of Tertiary age, carry considerable amounts of Se, Sb, Pb, Zn, Cu, Hg, and Cd. Precambrian deposits tend to have a high Au/Ag ratio, much greater than 1, whereas those of Tertiary age have a low Au/Ag ratio, usually less than 1. There are numerous exceptions to this generalization.

Many deposits in sedimentary rocks and some in volcanic terranes contain B, and a few in all rock types have W, Mo, Te, and Bi. A few deposits in basic and intermediate igneous and volcanic rocks are marked by enrichments of Cr in their wall-rock alteration zones. Gold-quartz deposits in basic and intermediate igneous intrusives and in basic and intermediate lavas are usually enveloped by characteristic potassium-rich, desilicated, and carbonated zones. Other deposits of this type are marked by a development of albite in their wall rocks and hence exhibit a typical soda metasomatism. Adularia is developed in the wall rocks of some deposits, alunite in others. Silicification, pyritization, and arsenopyritization are widespread in numerous gold-quartz deposits in all types of rocks.

Two well-defined types of auriferous veins are recognized in certain (young) metallogenic belts: gold-silver telluride veins and gold-silver selenide veins. The gold-silver telluride veins are marked by enrichments of Ag, Fe, Te, S, and F. In some there are also Zn, Cd, Cu, Sb, As, Hg, Mo, and W. The Au/Ag ratio is normally much greater than 1, although in certain veins it is much less than 1. The common associated elements in the gold-silver selenide veins are Se, Ag, Fe, Cu, Pb, Zn, Hg, Sb, and As. The Au/Ag ratios are low, usually less than 1 in these deposits.

The principal elemental associates of gold in the quartz-pebble conglomerate deposits are Fe, S, Ag, U, Th, and rare earths. In some deposits U, Th, and rare earths occur only in traces (e.g., Tarkwa, Ghana). The iron and sulfur are combined mainly in pyrite, which is ubiquitous in some deposits; in the Tarkwa deposits the iron occurs mainly in the oxide form. The Au/Ag ratio is always greater than 1 and averages about 9 in the Rand. Other elements associated with gold in the quartz-pebble-conglomerate type of deposit are As, Cu, Pb, Zn, Hg, Co, and Ni. Of these, probably only arsenic is a suitable indicator, although the other elements may serve the same purpose if they are present in some quantity.

The heavy minerals associated with gold in eluvial and alluvial placers are familiar and need not be considered in any detail here. The most frequent associated heavy minerals are magnetite, hematite, ilmenite, pyrite, arsenopyrite, garnet, zircon, and monazite; in addition, cassiterite, platinoids, native bismuth, amalgam, native copper, cinnabar, scheelite, wolframite, barite, galena, stibnite, and sulfosalts are present in some areas. None of these minerals are specific indicators of gold, but some such as arsenopyrite, native bismuth, scheelite, stibnite, and sulfosalts suggest its presence. Constant scrutiny and analysis of quartzites, conglomerates, and other resistate sediments should be maintained during prospecting so that investigators do not miss fossil gold placers.

Group IIA: Beryllium, Magnesium, Calcium, Strontium, Barium, and Radium

Beryllium (q.v. in Vol. IVA) is a typical element of the pegmatites and certain granites but is also found in a diverse number of other occurrences. It is commonly concentrated in certain types of albitized, fluoriferous, sericitized, and greisenized granitic rocks, also in greisen zones in metasediments and in certain schists and gneisses often with fluorine minerals such as topaz and fluorite. Beryllium is enriched in certain members of peralkaline (agpaitic) complexes; in certain skarn-type (taconite) deposits; in some tin, tungsten, and molybdenum quartz veins; in small amounts in certain gold-quartz and manganese-lead-zinc veins; in veins and segregations of barylite and eudidymite associated with syenitic intrusions; and in veins, segregations, and disseminations containing minerals such as beryl, bertrandite, phenakite, euclase, berylliferous saponite, etc. in certain altered tuffs, limestones, and other rocks. Certain carbonatites also carry enriched amounts of beryllium in both the fenitized zones (up to 1,000 ppm at Alno in Sweden) and in separate masses or disseminations of barylite. Beryllium may also be enriched in minerals such as cordierite (up to 2% BeO) in some metasedimentary rocks and veins, in certain types of zircon, in sodalite and nepheline, and in willemite in zinc deposits of the type of Franklin Furnace, New Jersey (United States). The element is also found in higher than normal amounts in some of the zeolites such as apophyllite in veins and rocks, but such occurrences are rare. During the processes of weathering and sedimentation, beryllium tends to follow aluminum and may be slightly enriched in clays, bauxites, and shales; it is also found in residual manganese deposits. Some coal ashes have relatively high enrichments of beryllium.

Beryllium forms numerous minerals, the most important commercially being beryl. The minerals chrysoberyl, gadolinite, bavenite, bertrandite, phenacite, euclase, helvite-danalite, beryllonite, bromellite, hambergite, rhodizite, genthelvite, barylite, herderite, and swedenborgite may also provide a source of beryllium if present in sufficient quantities. Minerals such as allanite, alvite, aminoffiite, bavenite, fergusonite, idocrase, milarite, and tengerite may carry 1% or more Be.

The elemental associates of beryllium in pegmatites and certain granites are Li, Rb, Cs, Sc, Y, rare earths, Nb, Ta, Mo, W, B, Tl, Sn, P, Bi, F, U, and Th. Of these, Li, B, Nb, Ta, and F are probably the best indicators in geochemical prospecting surveys, although the others may prove useful in certain areas. The beryllium minerals found characteristically in granite pegmatites are beryl, gadolinite, chrysoberyl, euclase, phenacite, and bertrandite. In nepheline syenite pegmatites, the suite comprises meliphanite, leucophanite, eudidymite, epididymite, hambergite, and helvite.

In albitized, fluoriferous, sericitized, silicified, and greisenized granites, the principal associates of beryllium are F, Li, Rb, Cs, and B. Sn, W, Mo, and P may also be present, and there may be traces of all these elements that accompany beryllium in pegmatites. In greisen zones in metasediments beryllium is accompanied by the same suite of elements as in greisenized granites.

Certain genthelvite deposits with associated phenacite occur in veins and stockworks in alkaline muscovite-biotite granites in the USSR and elsewhere. These deposits are marked by local silicification and biotitization. Associated minerals are sphalerite, pyrite, galena, and molybdenite. The best indicator elements of these deposits appear to be Na, K, Si, S, Zn, Fe, Mn, Pb, and Mo.

Beryllium is a frequent constituent of skarn deposits (tactites) in which it occurs principally in barylite, helvite-danalite, phenakite, bromellite, euclase, and vesuvianite (idocrase). In addition, minerals such as grossularite, epidote, cordierite, allanite, and axinite may be beryllium bearing. Some varieties of allanite, for instance, may contain up to 4% BeO. The principal associates of beryllium in skarn-type deposits are Fe, Mn, Sc, rare earths, P, F, B, Mo, W, Sn, Bi, Ba, Sr, As, Cu, Pb, and Zn. There is often a coherence between zinc, manganese, and beryllium in skarn-type deposits, a feature that is emphasized by the relatively high contents of beryllium in vesuvianite, willemite, and other minerals at Franklin Furnace, New Jersey. To avoid missing beryllium in skarn deposits, bulk samples and mineral concentrates should be analyzed directly for the element.

Certain tin-, tungsten-, and molybdenum-bearing quartz veins and segregations frequently carry some beryl or helvite, phenakite, bertrandite, and other beryllium minerals in some areas. Zinnwaldite, the lithium mica, tends to be enriched in beryllium in these deposits. Topaz is a common mineral in some of these deposits. For the associated and indicator elements in these deposits, see the discussions under lithium, tin, tungsten, and molybdenum.

Beryllium is not common in gold-quartz veins, but those containing fluorine (fluorite) may contain small quantities of the element. Likewise, certain manganese-lead-zinc veins containing rhodonite, rhodochrosite, other manganese minerals, sulfides, and helvite may contain small amounts of beryllium, mainly in the helvite.

Certain types of fluorite-bearing veins, stockworks, and replacement deposits developed in fractures and brecciated zones commonly in limestones or calcareous shales carry beryllium, generally in bertrandite, beryl, phenakite, or chrysoberyl. Associated minerals are fluorite, hematite, calcite, quartz, chert, often opal, and topaz. Scheelite occurs in some of these deposits. The fluorite is commonly purple, and some of the deposits are weakly radioactive. Most deposits occur near quartz-feldspar porphyry and granitic stocks. The principal associated and indicator elements are F, W, Mn, and (U). Ancillary indicators are Rb, Cs, Li, Zn, Pb, Sn, Sc, and rare earths in certain deposits of this type.

Beryl and most of the other beryllium minerals are commonly closely associated with quartz in one form or another. In some areas, however, beryl may occur in carbonate veins often with barite and fluorite. The emeralds of Colombia occur mainly in bituminous limestones commonly in carbonate veins. Helvite is sometimes found in a variety of manganiferous veins bearing rhodonite, rhodochrosite, pyrite, galena, and sphalerite.

There is frequently a coherence between barium and beryllium in nature, a feature that is emphasized by the relatively large deposit of barylite in veinlets of quartz and albite at Seal Lake, Labrador. Such occurrences are invariably related to carbonatites or other types of alkali-rich intrusives that are highly fluxed with fluorine and chlorine. The fenites and various rocks associated with carbonatites may contain a variety of beryllium minerals in small amounts. For the various elements associated with beryllium in carbonatites, see the discussion under niobium.

Emeralds, beryl, and a number of other beryllium minerals are relatively resistant to weathering and tend to collect in eluvial and alluvial placers, particularly where the weathering has been deep and prolonged. Some of the placers represent more than one cycle of erosion and deposition. Beryllium mineral placers are rarely commercial except for the gem varieties of beryl.

Beryllium is a good indicator of its deposits in most types of geochemical surveys. Other specific and useful indicators of the element are F, Li, Rb, Cs, Sn, W, and a variety of other elements commonly accompanying beryllium such as Ba, Sr, B, Sc, Y, rare earths, U, Th, Nb, Ta, P, Ti, Mo, and Mn. Fluorine is almost universal in beryllium deposits and is the best specific indicator.

Magnesium (q.v. in Vol. IVA) is a characteristic element of intermediate, basic, and ultrabasic igneous and metamorphic rocks and of sedimentary carbonate rocks, especially those containing dolomite. Certain carbonatites contain within their complexes masses of medium-to-coarse-grained dolomite or dolomitic limestone (dolomitic sövite). The principal magnesium minerals used commercially are olivine, obtained from ultrabasic igneous rocks and used for refractories; asbestos and talc, also obtained from ultrabasic rocks (serpentinites) and altered metadolomites and used for a great variety of purposes; brucite, the hydroxide, obtained from crystalline limestones, dolomites, and serpentinites and used for refractories and the production of the metal; dolomite, obtained mainly from sedimentary deposits; magnesite, used in refractories and for the extraction of the metal and won mainly from epigenetic veins and lenses and also from sedimentary deposits; and various salts, particularly the sulfates and chlorides, used for the extraction of the metal and various other purposes and obtained from ocean bitterns, various underground brines and mineral waters, salt lakes, and saline evaporites. Other magnesium minerals of commercial importance include serpentine (ornamental stone); biotite and phlogopite (for ceramic and electrical purposes); glaucophane, amphibole, and pyroxene (for roofing granules, aggregate, etc.); and spinel and garnet (gems and abrasives). There are no particularly diagnostic indicator elements for magnesium deposits except the element.

Calcium (q.v. in Vol. IVA) is a characteristic element of intermediate and basic igneous rocks and of the sedimentary carbonate rocks and evaporites. Certain carbonatites contain within their complexes masses of nearly pure medium-to-coarse-grained calcium carbonate (sövite). The calcium minerals and rocks of commerce require little discussion. They include calcium carbonate (calcite) in its various forms obtained from sedimentary carbonate deposits; various salts, principally the chlorides obtained from natural brines and saline evaporites; and the sulfates that occur as gypsum and anhydrite in sedimentary evaporite beds. Other calcium minerals of commercial importance include the various feldspars, pyroxenes, amphiboles, epidote, wollastonite, garnets, and zeolites. These have various uses. Calcium minerals such as fluorite, apatite (phosphate rock), and scheelite are discussed under the sections on fluorine, phosphorus, tungsten.

Calcium-bearing rocks in a terrane are often indicated by higher than normal contents of calcium in the soils, tills, stream sediments, and natural waters and by the relatively high pH (~ 8.5) of these materials. A high sulfate content that is enriched in the isotope ^{34}S in soils, tills, stream sediments, and natural waters may indicate the presence of gypsum and anhydrite in the terrane. Natural waters and their precipitates, stratal brines, and evaporites should be analyzed directly for calcium, if a source of the element is sought in these materials.

Strontium (q.v. in Vol. IVA) is widely distributed in rocks—in particular granites, carbonate rocks, and gypsum and anhydrite. Some carbonatite complexes are enriched in strontium. The only minerals of commercial importance are the sulfate, celestite, and the carbonate, strontianite. The former is the more abundant.

Both celestite and strontianite occur in epigenetic veins and lenses in faults and fractures and in stratiform deposits consisting of irregular lenses, pods, and strings of concretions. Both types of deposits are generally found in sedimentary terranes, particularly in red bed sequences; in evaporite basins; and in volcanic sedimentary basins, often with abundant tuffs. Examples of the first two environments are numerous, including the Enon deposit in Nova Scotia (celestite, Lower Windsor Formation); Port au Port Peninsula, Newfoundland (celestite, Upper Mississippian Codroy Formation); Yate District, Gloucestershire, England (celestite, Keuper Marl); Germany (celestite, Zechstein); and the USSR (celestite, Permian of the Russian Platform). The stratiform bodies appear to be of sedimentary or diagenetic origin in some places; in others they are obviously replacement bodies formed after lithification of gypsum, anhydrite, marl, and other carbonate rocks. The vein-type bodies appear to have been precipitated from stratal waters, the strontium and other accompanying constituents being derived from the piles of evaporites, carbonate rocks, and shales in which the deposits usually occur.

Celestite is a common constituent of the cap rocks of salt domes and is also found in certain sulfur deposits that seem to have been derived by the reduction of gypsum and anhydrite. So far as is now known to this writer these types of deposits are not commercial.

The elements accompanying strontium in the vein-type deposits are Ba, Pb, Zn, Ca, Fe, S, and F. These elements are combined mainly in barite, galena, sphalerite, anhydrite, gypsum, pyrite, hematite, and fluorite. In the bedded deposits the associated elements are much the same except that in some of the deposits there are small amounts of boron in borates and often some native sulfur.

Barium (q.v. in Vol. IVA) is widely distributed in igneous and igneous-type rocks, particularly in the more acid varieties such as syenites and granites in which the element occurs mainly in the potash feldspars. Barium is also commonly enriched in the various syenitic rocks of alkaline complexes and carbonatites) and in the carbonate masses of the latter. Certain shales and sandstones may be enriched in barium, and the element is often concentrated in small amounts in manganiferous sediments and precipitates. Some coals are relatively rich in barium. The only minerals of commercial importance are the sulfate, barite, and the carbonate, witherite; the former is the more abundant. Barite is a common gangue mineral in many types of mineral deposits; witherite occurs only rarely as a gangue mineral. Barytocalcite, nitro-

barite, and the various silicates, phosphates, and vanadates of barium are rare minerals and cannot be considered commercial sources of the element.

Barite and witherite deposits are found in the following forms:

1. As veins and replacements, containing essentially barite with minor amounts of fluorite, calcite, siderite, ankerite, dolomite, quartz, and sulfides, and as veins and lenses of witherite with the aforementioned minerals.
2. As bedded deposits containing essentially barite with minor amounts of chert, quartz, dolomite, siderite, strontianite, witherite, pyrite, and iron oxides.
3. As disseminated and massive barite forming the gangue of various types of vein and massive sulfide deposits, particularly those containing lead, zinc, copper, and silver. The barite may be won as a by-product in some cases.
4. Residual deposits of barite in unconsolidated materials such as soil, clay, decomposed rocks, and other residuum. The barite in these deposits is derived from deposits of the first three types.

The vein and replacement types of barite deposits occur in igneous, sedimentary, or metamorphic rocks of all types. There is, however, a tendency for these deposits to occur in limestones, calcareous shales, and sandstones. Numerous deposits of this type are in red bed sequences or in series of rocks containing evaporites—e.g., Walton, Nova Scotia.

The bedded deposits are invariably in sequences of limestones, dolomites, and shales. Some of these deposits may be of sedimentary origin, although most are probably of replacement origin. The large bedded deposits at Magnet Cove in Arkansas are apparently of replacement origin related in some manner to the Magnet Cove alkali syenite-carbonatite complex.

The disseminated and massive barite bodies accompanying or intimately mixed with various types of vein and massive sulfide bodies are widespread, examples being found at the Buchans Mine in Newfoundland, the Meggen deposit in Germany, and other deposits in Eastern Europe and elsewhere. The vein-type and discordant massive deposits are epigenetic; those in stratiform bodies have been the subject of prolonged debate. Some investigators hold to a sedimentary or volcanic exhalative origin, others to a replacement origin. The latter is more probable.

Witherite deposits occur as veins and irregular lenses in sedimentary rocks, particularly limestone, calcareous shales, and sandstone. Some witherite deposits are bedded or elongated lenses in sequences of limestone and shale. These types of deposits are rare.

Barite is a frequent associated mineral in certain carbonatites and is present in some rare-earth carbonate veins. In the latter it is associated with bastnaesite, the cerium fluocarbonate, as at Mountain Pass, San Bernardino County, California. Barite may be commercially recoverable from some of these deposits. Some of the bedded deposits associated with alkali syenite-carbonatite complexes such as at Magnet Cove, Arkansas, provide large commercial sources of barite. Witherite is rare in most carbonatites, and no commercial deposits of the mineral are known in these complexes.

The residual deposits of barite are widespread in many parts of the world. They derive from all the aforementioned deposits. Most are lumps, balls, and irregular masses in the highly weathered residuum forming the regolith on limestones, sandstones, evaporites, carbonatites, etc.; also in karsts, sinkholes, and solution cavities on these rocks.

In most types of barium deposits the element is associated with Ca, Sr, Mg, Fe, Mn, and Pb. In some deposits there may also be enrichments of Zn, Cd, Ag, Au, Cu, V, As, and Sb. In others, Hg and Sb accompany the barium. F is a characteristic element in many deposits. Rare earths, Th, and U accompany barium in the rare-earth carbonate deposits, and Nb and the other elements commonly concentrated in carbonatites accompany the element in these deposits (see also the discussion under niobium).

Radium (q.v. in Vol. IVA) is a highly radioactive daughter element of uranium and thorium and can be used as an indicator element of deposits of these two elements in geochemical prospecting surveys (Boyle, 1982*a*). The content of radium in uranium ores is variable and depends on the uranium content and the age of the deposit. The ratio Ra/U is approximately 3.5×10^{-7} by weight in ores older than 500,000 yr.

A word of caution about the use of radium and its daughter elements (e.g., radon) as indicators of uranium (and thorium) deposits is necessary. High radium (and radon) contents in various earth materials are not necessarily indicative of uranium (or thorium) deposits. On the contrary, many are often the result of chemical enrichments associated with granites, pegmatites, black shales, gneisses, and many other rocks containing only low contents of uranium (and thorium). Furthermore, it should be recognized that thorium and uranium are members of the actinide series (Groups IVB and VIB respectively), whereas radium is a member of the alkali earth group (Group IIA) and radon is an inert gas. All have different geochemistries, from which it follows that the daughter elements (radium, radon, etc.) may become markedly separated from their parents in groundwater systems and during weathering and

other exogene processes. This feature should always be kept in mind during the interpretive phase of geochemical prospecting, particularly in relating radium (and radon) anomalies to possible uranium (and thorium) deposits. This is the reason that certain radium (or radon) anomalies are found without corresponding uranium (or thorium) anomalies. Similarly, uranium anomalies may exist without noticeable radioactivity. In this case the uranium mineralization or enrichment is young and is said to be out of radioactive equilibrium with its daughter products. Radioactive equilibrium is reached at about 500,000 yr at which time the Ra/U constant is about 3.5×10^{-7}.

Group IIB: Zinc, Cadmium, and Mercury

Zinc (q.v. in Vol. IVA) occurs in eight principal types of deposits. *Cadmium* (q.v. in Vol. IVA) normally accompanies zinc and is won from its ores as a by-product. Most zinc-cadmium deposits also contain lead. The types of zinc deposits are as follows:

1. Disseminated sphalerite in shales: These deposits accompany the copper shales (Kupferschiefer type) and are not usually exploited.

2. Disseminations, knots, and concretions of sphalerite in sandstones, quartzites, and shales: These deposits contain galena and copper and iron sulfides. Examples are known at Maubach, Germany, and elsewhere.

3. Zones of disseminated sphalerite, veins of sphalerite, replacement deposits of sphalerite, pods of sphalerite, etc. in carbonate rocks that often show the effects of dolomitization and silicification: The deposits are commonly known as the Mississippi Valley type. They contain considerable quantities of lead; copper and silver are minor.

4. Skarn-type zinc deposits: These deposits frequently have considerable quantities of lead, copper, silver, and some gold.

5. Franklinite-willemite-zincite deposits (of the Franklin Furnace, New Jersey, type): These deposits probably belong in category 4 as they have frequently been referred to as pyrometasomatic deposits in the older literature. The mineral association is most complex and contains, in addition to the three principal zinc minerals noted, a large assortment of zinc silicates, manganese silicates, axinite, scapolite, calcite, garnet, rhodochrosite, fluorite, sphalerite, galena, arsenopyrite, chalcopyrite, and loellingite. The deposits at Franklin Furnace and Sterling Hill, New Jersey, occur in crystalline limestone and coarse gneisses of Precambrian (Grenville) age.

6. Veins and replacement deposits of sphalerite in various types of rocks: Accompanying minerals are usually galena and iron, copper, and silver

sulfides and sulfosalts. Barite may be present as a gangue mineral.

7. Massive sulfide deposits containing essentially sphalerite and iron, copper, and silver sulfides and sulfosalts: Barite may be present as an important mineral in some deposits. Some of these deposits occur in volcanic terranes, others in sedimentary terranes, and still others in mixed volcanic and sedimentary terranes. Deposits of Precambrian Archean age commonly lack lead; younger deposits are typically enriched in lead.

8. Irregular masses and replacement bodies of smithsonite, calamine, hydrozincite, or willemite: These commonly occur in carbonate rocks and represent reprecipitation of zinc and other constituents derived from the oxidation of primary disseminations, of sphalerite in carbonate and other rocks. The bodies may occupy the same locale as the primary disseminations or they may be slightly removed laterally or vertically (downward) from the primary ores. Examples of types of orebodies are widespread in the Tintic district, Utah (smithsonite); in the Beltana district of South Australia (willemite); and in the Moresnet district of Belgium (calamine).

The most frequent elemental associate of zinc and cadmium is lead. In some deposits there may also be enrichments of Fe, Mn, Cu, Ag, Au, Ba, Sr, B, F, As, Sb, Bi, Mo, Ga, In, Tl, Ge, Hg, Co, Ni, and Sn. The Mississippi Valley–type deposits have few elemental associates accompanying zinc, lead, and cadmium. Only Mg, Si, and Fe are enriched in most orebodies in the dolomitized, silicified, and pyritized (marcasite) parts. In some deposits there may be slight to moderate enrichments of Cu, Ag, Ni, Co, Ba, and occasionally Sb. F is enriched in some deposits; Hg is low in most deposits. The bodies of secondary (supergene) zinc minerals such as smithsonite and willemite are characterized by few elemental associates of zinc. There is commonly some Cd and a little Pb; rarely Sb, As, Co, Ni, Cu, and Ba in any quantity other than trace or minor amounts.

Mercury (q.v. in Vol. IVA) is obtained almost entirely from the sulfide, cinnabar, although the native metal occurs in some ores, and the element has been obtained from livingstonite, metacinnabarite, mercurian tetrahedrite and tennantite, and a few of the other rarer mercury minerals such as calomel.

Cinnabar and the other types of mercury deposits occur in fissure veins, and stockworks and as disseminations, impregnations, and replacements along faults or in brecciated zones. The common host rocks are sandstone, limestone, dolomite, calcareous shale, chert, serpentinite, andesite, basalt, trachyte, and rhyolite. Commonly associated metallic minerals are stibnite, realgar, orpiment, arsenopyrite, chalcopyrite, pyrite, and marcasite. Some veins have a ferberite and/or

scheelite-stibnite association; others contain quartz, tourmaline, and cinnabar. Some lead-zinc and copper ores may contain cinnabar, and mercury minerals, particularly cinnabar, coloradorite (HgTe), and tiemannite (HgSe), may occur in certain gold-quartz and silver veins. The native silver, allargentum, and dyscrasite in the Cobalt-type Ni-Co arsenide deposits are often enriched in mercury (up to 4.7%). The gangue in most mercury deposits is invariably opal, chalcedony, or quartz and carbonates. Barite and fluorite are usually restricted to individual deposits. Native sulfur is abundant in some deposits, and gypsum occurs in others. Argillic alteration, silicification (development of opal), pyritization, and sericitization are common. Carbonatization of serpentinite is marked in some deposits. Introduction of bituminous substances into the fractures, faults, and breccia zones and into the adjacent wall rocks is common in some deposits.

Most of the mercury deposits of the world occur in mercuriferous belts that correspond to zones of instability or dislocation of the earth and are often marked by the presence of hot springs and other volcanic or thermal activity. One of these belts (Pacific Belt) follows the Cordilleras of the Americas and the eastern flank of the Australasian and Asian continents; another belt (Mediterranean Belt) follows the line of Alpine (Tertiary) folding from Spain and North Africa through the Mediterranean to Indonesia and eastward where it apparently coalesces with the first belt. A number of mercury occurrences and deposits are located outside these belts. Examples are a number of deposits that lie in the Tien Shan and Sayan mountains and northeastward through the Baikalides to the Pacific Belt; the Gortdrum deposit in Eire where mercurian tetrahedrite is abundant; Clyde Forks, Ontario, where mercurian tetrahedrite and cinnabar occur in Precambrian rocks; the Archean cinnabar deposits associated with the so-called Antimony Line in the Murchison Greenstone Belt, South Africa; and a number of deposits in Arkansas where cinnabar is disseminated in Pennsylvanian sandstones and shales. All these deposits have one point in common: they are located in zones marked by deep faulting and shearing. Probably deep faulting is common to all mercury occurrences whether in the mercuriferous belts or not.

Most mercury deposits are of Cenozoic age, although there are exceptions to this generalization as witnessed by the South African deposits mentioned earlier. The grade of the deposits ranges from 0.5% to 6% Hg, the last figure being that for the famous Almaden mine in Spain.

A survey of the elemental associates of mercury in its deposits indicates that As and Sb are particularly characteristic. Others that may be enriched in individual deposits include Cu, Ag, Au,

Sr, Ba, Zn, Cd, B, Tl, Ge, Pb, Bi, Se, Te, Mo, W, and F. Some deposits in serpentinites have relatively high concentrations of Ni in minerals such as nickelian pyrite and millerite.

Group IIIA: Boron, Aluminum, Gallium, Indium, and Thallium

Boron (q.v. in Vol. IVA) is widely distributed in igneous, metamorphic, and sedimentary rocks but usually only in small amounts. The element is characteristic in pegmatite dikes and in certain mineral veins and other types of deposits. The common boron minerals in rocks, pegmatites, and mineral deposits include tourmaline, axinite, datolite, dumortierite, and danburite, none of which provides a commercial source of the element.

The commercial boron minerals are the hydrated sodium, magnesium, and calcium borates and boric acid. Boric acid occurs as the mineral sassolite; some of the important borates are kernite, tincalconite, borax, ulexite, priceite, boracite, and colemanite. Dumortierite is used in ceramics.

Borates and similar boron compounds are obtained for commercial purposes from the following sources: (1) brines of saline lakes and marshes; (2) encrustations around playas, salt lakes, and salt marshes; (3) bedded deposits formed in ancient salt lakes, salt marshes, and playas; (4) hot springs and fumaroles; and (5) disseminations, beds, and lenses of borate minerals in potassium salts in marine evaporites and irregular beds and disseminations of borate minerals associated with bedded evaporites, salt domes, and diapirs.

Boron is a good indicator of all types of borate deposits and can be used effectively in most types of geochemical prospecting surveys for deposits in which the element occurs.

Mineral deposits characterized by the development of boron minerals or boron metasomatism are listed in Table 2.

Aluminum (q.v. in Vol. IVA) metal is obtained commercially principally from bauxite. Other sources are aluminous laterite, clay and shale, nepheline syenite, and other rocks relatively enriched in the element.

Other commercial minerals of aluminum include corundum, emery, cryolite, dawsonite, topaz, and in particular, alunite. The natural alums are mainly hydrated double sulfates of potassium, sodium, and ammonium. Some of these, such as potash alum, and soda alum, are highly soluble and occur as surface evaporation products in arid regions or in sheltered places. Certain springs may be charged with aluminum sulfates, and their precipitates may contain the various alums. In dry areas irregular bedded deposits of alum minerals may originate from springs.

Alunite, $KAl_3(SO_4)_2(OH)_6$, is most commonly

TABLE 2. Characteristic Boron Minerals in Deposits

Deposit	Characteristic Boron Minerals
Pegmatites (simple and complex types containing various rare elements such as Li, Nb, Ta, Sn, etc.); high-temperature quartz veins	Tourmaline, dumortierite; cappelenite, tritomite, nordenskiöldine, hambergite, homilite, and datolite in certain nepheline-syenite pegmatites.
Skarns (including those containing Be, Fe, Cu, Zn, Pb, Sn, Bi, rare earths, Mo, and W)	Tourmaline, axinite, danburite, dumortierite, kotoite, ludwigite, nordenskiöldine
Greisenized, sericitized, and albitized granites and similar rocks containing Be, Sn, and W	Tourmaline, dumortierite
Tin-tungsten veins and lodes	Tourmaline, axinite, dumortierite
Gold-quartz veins and lodes	Tourmaline
Polymetallic (Cu, Pb, Zn, Ag, etc.) veins and lodes	Tourmaline, axinite, dumortierite
Uranium-thorium vein, stockwork, and disseminated deposits	Tourmaline, axinite
Porphyry Cu and Mo deposits	Tourmaline, dumortierite (only certain deposits)
Native copper deposits in basalts, conglomerates and sandstones	Datolite, axinite
Native silver deposits	Datolite (rare), axinite
Mercury deposits	Tourmaline (rare)
Carbonatites	Ludwigite
Bedded borates	Borates
Hot springs and steam jets	Boric acid (sassolite), borates

associated with acid volcanic or intrusive rocks and is formed where these rocks have been highly altered by sulfuric acid vapors or waters. Some alunite deposits may also owe their origin to the alteration of potassic aluminous rocks by sulfuric acid solutions derived from the oxidation of pyrite and other sulfides. Alunite generally occurs disseminated in the rocks or in well-defined veins. Some deposits contain essentially only alunite, others are pervasive alterations associated with gold-silver and other metallic deposits such as at

Goldfield, Nevada, and elsewhere. Large deposits
of alunite are known in Italy, Mexico, the USSR,
and Cedar City, Utah, in the United States.

Cryolite (Na_3AlF_6) occurs in commercial quantities at only a few places in the world, particularly at Ivigtut, West Greenland, and at Miask,
Ilmen Mountains, USSR. At Ivigtut the cryolite
is found in an irregular massive body in porphyritic granite. It is accompanied by microcline,
quartz, fluorite, siderite, topaz, pyrite, arsenopyrite, galena, sphalerite, molybdenite, and a
number of rare aluminum fluorite minerals. Cryolite is used as an electrolyte in the Hall process
for the reduction of aluminum oxide to the metal.
See also the section on fluorine.

Dawsonite, $NaAl(CO_3)(OH)_2$, is a basic carbonate of sodium and aluminum. It occurs in a
variety of settings including veins and impregnations and extensive beds. In the latter form it occurs in the Green River Formation of the western
United States. Topaz, $Al_2SiO_4(OH,F)_2$, is a possible source of alumina and fluorine (see section
on fluorine).

Aluminum is accompanied in its various deposits by a veritable host of elements, none of
which is particularly indicative of its presence.
Cryolite deposits may be indicated by F in geochemical surveys. Aluminum is indicative of deposits in which it is enriched.

Gallium, indium, and thallium (q.v. in Vol.
IVA) are rare elements. Gallium is enriched in certain sphalerite deposits, some copper shales,
bauxite, some tin ores, and certain coals. Indium
is often enriched in sphalerite ores, certain tin and
tungsten ores, and a variety of sulfides. Thallium
is enriched in potassic feldspars, galena, and a variety of sulfides and sulfosalts in polymetallic deposits. The elements that accompany gallium and
indium in deposits are the same as those listed in
the sections on aluminum, zinc, lead, copper, tin,
and tungsten. The elemental associates of thallium in deposits are mainly K, Fe, Mn, Pb, Zn,
Cd, Cu, As, Sb, Bi, Ag, Au, In, Ga, Ge, Sn, S,
and Se.

Groups IIIB, IVB: Scandium, Yttrium, Rare Earths (Lanthanides), Thorium, and Uranium (Actinides)

Scandium (q.v. in Vol. IVA) is a rare metal
closely associated in nature with yttrium and the
rare-earth metals. It is widely diffused in rocks of
all types, particularly in those containing an
abundance of ferromagnesian minerals in higher
than average amounts. The element also occurs in
a great variety of minerals in higher than average
amounts, including cassiterite, wolframite, tantalite-columbite, euxenite, monazite, eschynite,
keilhauite, zinnwaldite, beryl, lepidolite, biotite,

wiikite, zircon, garnet, and allanite. Some varieties of beryl (bazzite) are relatively rich in scandium, containing up to 10%. Some of the minerals of carbonatites including baddeleyite,
pyrochlore, alkalic pyroxenes and amphiboles,
and kimzeyite are rich in scandium with contents
ranging from 700 ppm to 4,000 ppm. The most
common mineral of scandium is thortveitite,
$(Sc,Y)_2Si_2O_7$.

Scandium oxide and metal are obtained from
thortveitite ore, yttrium and rare-earth metal ores,
and the residues of wolframite and cassiterite
concentrates. The resource of scandium in various uranium ores and phosphate deposits is very
large.

The common elemental associates of scandium
in its deposits are Li, Rb, Cs, Be, Y, La, rare
earths, Ti, Zr, Hf, Nb, Ta, Mo, W, Fe, B, Sn, Bi,
U, Th, P, and F. Of these the best indicators are
Y, La, rare earths, U, Th, W, Sn, Ta, Nb, P, and
F. Scandium should prove useful as an indicator
of its deposits, and those of the rare earths, in
soil, till, stream sediment, and vegetation surveys.

Yttrium, lanthanum, cerium (q.v. in Vol. IVA),
and the other *rare-earth (lanthanide) metals* are
widely diffused in all types of igneous, sedimentary, and metamorphic rocks. Concentrations occur, however, only in certain peralkaline massifs
and complexes, certain pegmatites, certain veinlike deposits, some skarns, certain quartz-pebble
conglomerates, some carbonatites, phosphorites,
and placers.

The principal types of deposits of these metals
are as follows:

1. Peralkaline massifs and complexes: Examples of deposits in this category include the Lo-
vozero Massif, Kola Peninsula, USSR, and the
Ilimaussaq Complex, Greenland. All are
characterized by a veritable host of Y, La, and
rare-earth-bearing minerals. The light lanthanides
are commonly concentrated in the niobo-tita-
nates, titanosilicates, and phosphates; the heavy
lanthanides and Y favor the various zirconium-
bearing minerals (e.g., zircon, eudialyte, etc.).
Thorium and uranium are commonly associated
with Y, La, and the rare earths and are concentrated in minerals such as loparite, rinkolite, eudialyte, pyrochlore, sphene, apatite, zircon, thorite, and steenstrupine.

2. Pegmatites, migmatites, and coarse-grained
granitic rocks: The common minerals in these deposits are monazite, thorite, uraninite, cerite, allanite, euxenite, pyrochlore, samarskite, fergusonite, gadolinite, yttrofluorite, xenotime,
thortveitite, gagarinite, wiikite, thalenite, and
thucholite. The elements commonly enriched with
Y, La, Ce, and the other rare earths in pegmatites
and similar deposits are Li, Rb, Cs, Be, Sr, Ba,
Sc, Ti, Zr, Hf, Nb, Ta, Mo, W, B, Sn, Bi, U, Th,
P, and F. There is a marked and nearly constant

affinity with thorium, uranium, phosphorus, and fluorine.

3. Skarn deposits: In these the minerals apatite, flourite, garnet, monazite, allanite, uraninite, stillwellite, caryocerite, mosandrite, thucholite, epidote, vesuvianite, scheelite, and sphene are the principal carriers of Y, La, the rare-earth elements, thorium, and uranium. A commercial example of this type of deposit is the Mary Kathleen in Queensland, Australia, mined mainly for uranium. The principal elements associated with Y, La, and the rare earths in skarn deposits are F, P, Ti, U, Th, Sc, and the numerous other elements concentrated in skarn.

4. Monazite veins or lodes: The principal example of this unusual type of deposit is in South Africa near Nieuwe Rust, Van Rhynsdorp Division. The orebody is an irregular veinlike body in a shear zone in altered granite of Precambrian age. The main minerals are apatite, monazite, zircon, and iron and copper sulfides. Its origin is obscure. The elements concentrated are rare earths, Th, Zr, P, Fe, and Cu.

5. Rare-earth carbonate veins and masses: Some of these are closely associated with carbonatites. The principal example of this type of deposit is near Mountain Pass, San Bernardino County, California. The orebody consists of bastnaesite (cerium fluocarbonate), barite, apatite, pyrite, galena, chalcopyrite, fluorite, carbonates, parisite, monazite, allanite, cerite, and sahamalite. The elements concentrated include rare earths, Th, Ba, Sr, Pb, Cu, P, and F. The deposit exhibits certain affinities with carbonatites.

6. Calcite-fluorite-apatite, calcite-fluorite-apatite-biotite-pyroxene, and calcite-biotite-apatite veins and segregations: These deposits are common in the Precambrian Grenville Province of eastern Canada. For details of these deposits see the sections on uranium and fluorine.

7. Fluorite and apatite veins and lodes: Some varieties of fluorite and apatite are enriched in the rare earths. Their potential as a commercial source is unknown (see also the discussions under fluorine and phosphorus).

8. Carbonatites: These deposits frequently contain higher than average amounts of rare earths, Th, and U in a variety of titanium, niobium, and other minerals—in particular, calcite, dolomite, bastnaesite, pyrochlore, perovskite, apatite, monazite, sphene, zircon, allanite, zirconium-garnet, and fluorite (see further the discussion under niobium). Carbonatites are generally enriched in the cerium subgroup of rare earths as compared with the yttrium subgroup. Carbonatites are not usually considered a primary commercial source of rare-earth elements, but the elements can be recovered as by-products in some cases.

9. Pyritic quartz-pebble conglomerates: The principal examples of these deposits are the Elliot Lake pyritic quartz-pebble conglomerates of Ontario where the rare earths, U, and Th occur in a brannerite-uraninite-monazite mineral assemblage. The principal elements concentrated are U, Th, Sc, Y, rare earths, Fe, As, and S. The yttrium group of rare earths predominates in the deposits. Similar deposits, although not as rich in thorium and rare earths, occur in the Witwatersrand of South Africa and at Jacobina in Bahia, Brazil. These deposits are greatly enriched in Au and Ag and contain uranium in amounts ranging from 70 ppm to 960 ppm.

10. Phosphorites: Most of these deposits contain traces of rare earths, as well as scandium, uranium, and thorium. The potential by-product reserve of these elements in the phosphorites is enormous (see also the section on phosphorus).

11. Placers: These have long been a source of rare earths and thorium minerals, of which the most important are monazite and xenotime. Associated minerals, often of commercial importance in beach, stream, and other types of placers, include ilmenite, leucoxene, magnetite, cassiterite, rutile, zircon, staurolite, garnet, and gold.

The best indicators of Y, La, and rare-earth deposits in geochemical surveys utilizing soils, tills, stream sediments, and vegetation are the elements. Phosphorus and fluorine are also excellent indicators, and many of the associated elements in the deposits noted previously can probably be used under certain conditions. The use of these elements, however, such as Ba, Sr, Ti, Nb, U, and Th, should be preceded by adequate testing by pilot surveys for their suitability in any given area.

Thorium (q.v. in Vol. IVA) is widely distributed in most rocks but only in trace amounts. The deposits of thorium are essentially the same as those mentioned earlier for yttrium, lanthanum, and the rare earths and in the next discussion for uranium. The elements accompanying thorium are, likewise, the same as those for yttrium, lanthanum, the rare earths, and uranium in most deposits. Thorium may be indicated by its radioactivity, and the element can be used as an indicator of its deposits in most types of geochemical surveys except those based on water analyses because of its general low mobility in natural waters (Boyle, 1982*a*).

Uranium (q.v. in Vol. IVA) is widely diffused in igneous, sedimentary, and metamorphic rocks, being enriched in the more acidic igneous and metamorphic rocks, in the black carbonaceous shales, and in phosphorites (Boyle, 1982*a*).

The principal types of uranium deposits are as follows:

1. Certain granitic rocks: These may contain higher than normal amounts of uranium and thorium (5 ppm U, 15 ppm Th). Most are not of commercial importance at present but provide a large reserve of uranium and thorium ore for the future. Such bodies are indicated by their higher than normal radioactivity and by enrichments of U, Th, Y, La, rare earths, and frequently P, F, and Zr.

2. Pegmatites and other coarse-grained feldspar-quartz-mica-pyroxene segregations and migmatites: The principal uranium minerals are uraninite, uranothorite, euxenite, pyrochlore, and fergusonite. Such bodies are marked by their higher than normal radioactivity and by enrichments of U, Th, Sc, Y, rare earths, and frequently Fe (hematite), Nb, Ta, Mo, Ti, Zr, Hf, P, and F.

3. Skarn and hornfels deposits: These are relatively rare, a commercial example being the Mary Kathleen deposit in Queensland, Australia. The uranium-bearing minerals are mainly uraninite, allanite, and monazite; occasionally, thucholite. There are commonly much magnetite, garnet, and various other typical skarn minerals. The characteristic elements associated with uranium are mainly Th, Y, La, and rare earths, as well as the usual elements such as Mo, Bi, W, Cu, Zn, etc. commonly found in skarn and hornfels deposits.

4. Calcite-fluorite-apatite, calcite-fluorite-apatite-biotite-pyroxene, and calcite-biotite-apatite veins and segregations: These deposits are common in the Grenville Province of the Precambrian of eastern Canada. The principal radioactive minerals are uraninite, uranothorite, allanite, betafite, zircon, and fergusonite. These bodies are marked by their higher than average radioactivity and by enrichments of U, Th, Y, La, rare earths, P, F, Nb, Ta, Ti, Zr, and Hf.

5. Veins, lodes, pipes, and disseminations: These deposits may occur in practically any rock type, although most are present in sedimentary rocks, gneisses and schists, and granitic bodies. The principal primary radioactive mineral is uraninite (variety pitchblende); davidite is an important ore mineral in some deposits; brannerite is rare in commercial quantities in these deposits. In some deposits the accompanying minerals are calcite, hematite, quartz, pyrite, chalcopyrite, and a few selenides. The principal enriched elements in these deposits are U, Ca, Mg, Fe, S, Cu, Mo, and Se; thorium and rare earths are only slightly enriched in the ores in contrast to the uranium deposits in granitic rocks, pegmatites, etc. In other deposits the accompanying minerals comprise calcite, quartz, hematite, native silver, chalcopyrite, galena, sphalerite, arsenopyrite, abundant nickel-cobalt arsenides, native bismuth, and various antimonides and sulfosalts. The characteristic elements concentrated in these deposits include U, Ca, Mg, Fe (hematite and pyrite), S, As, Cu, Mo, Ag, Pb, Zn, Ni, Co, Bi, Sb, and Hg.

6. Sandstone deposits: These occur in porous beds, rolls and pinch-outs, fractures and brecciated zones, zones containing abundant coalified wood, and various other sites amenable to precipitation of minerals from circulating waters. The principal primary uranium mineral is uraninite, which is associated with a variety of copper and vanadium minerals. Near the surface a host of supergene uranium, vanadium, and copper minerals is usually developed. The characteristic elements concentrated in these deposits are numerous and diverse and include U, V, Sc, rare earths, Be, Ga, Ge, Li, Sr, Ba, Cr, Mo, Fe, Co, Ni, Cu, Ag, Au, Zn, Cd, Pb, P, As, Sb, S, and Se. Some of these deposits are similar to the red bed sandstone deposits discussed under vanadium, copper, and silver.

7. Pyritic quartz-pebble conglomerates: These deposits are the same as those discussed under item 9 in the section on rare earths. Typical examples are the deposits in the Elliot Lake area of Ontario, Canada. The main primary uranium minerals are brannerite and uraninite. The principal elements concentrated are U, Th, Sc, Y, La, rare earths, Fe, As, P, and S; in some deposits (e.g., Witwatersrand, South Africa), Au and Ag are enriched and are won as major products from the deposits.

8. Carbonatites: Some of these deposits contain small amounts of uranium, mainly in apatite, zircon, pyrochlore, monazite, perovskite, sphene, and baddeleyite; also less frequently in thorianite, thorite, cerite, allanite, and eschynite (see the further discussion under niobium). Carbonatites are not considered to be a commercial source for uranium, although the element could probably be obtained from certain deposits as a by-product.

9. Black shales: Certain black carbonaceous shales are considerably enriched in uranium and a host of other elements, particularly those of a chalcophile character, including Ag, Zn, Cu, Pb, Ni, Co, As, S, and other elements.

10. Phosphorites: In some areas these rocks are enriched in uranium, rare earths, and a great variety of other elements, particularly those of a chalcophile character such as Ag, Pb, Zn, and Cu (see also the discussion in the section on phosphorus).

11. Coal and lignite: Some varieties of coal and lignite contain higher than average amounts of uranium.

12. Calcretes, dolocretes, gypcretes, phoscretes, and various other types of canga, kunkar, and caliche deposits: A typical example of this type of deposit is at Yeelirrie, Western Australia. The matrix of the deposits is usually carbonates, gypsum, limonite, clay, sand, etc.; the principal uranium minerals are disseminated carnotite, tyuyamunite, soddyite, or uranocircite. Elements concentrated in calcrete and similar types of deposits include K, Ca, U, V, P, and Mo, the last only in certain deposits.

13. Placers: Monazite placers and others con-

taining minerals such as xenotime and fergusonite may contain small amounts of uranium (see also the description under rare earths). Few placers are of economic importance for uranium.

Uranium is an excellent indicator of its deposits in practically all types of geochemical surveys, including those based on radioactivity (Boyle, 1982a). Depending on the types of deposits and the elements concentrated in them as outlined here, a number of other indicator elements can also be used effectively in geochemical prospecting for uranium. These include Th, rare earths, P, F, Co, Ni, As, Sb, V, Ag, Cu, Mo, and a number of others noted previously. Some of these are useful in soil and stream sediment surveys. Others may be useful in water surveys. Selenium indicator plants, and the selenium content of certain plants, may be useful in locating deposits of the sandstone type.

Group IVA: Carbon, Silicon, Germanium, Tin, and Lead

Carbon (q.v. in Vol. IVA) is concentrated in living matter and its fossil equivalents including coal, petroleum, natural gas, various solid hydrocarbons, and several types of carbonaceous and graphitic shales and schists. Graphite veins and lodes are common in certain high grade metasediments. The gem and abrasive varieties of carbon—i.e., diamond and bort—are residents of kimberlite pipes and also frequently occur in placers, both modern and fossil. The carbonates and carbonate rocks are so common they require no discussion.

Hydrocarbon surveys of ground- and stratal waters, springs, soils and glacial materials, and ocean and lake sediments have been used increasingly in prospecting for petroleum and natural gas with moderate success.

Graphite veins and lodes are sometimes indicated by higher than normal contents of graphite in soils and glacial materials, also in stream sediments. The work involves examination of these materials by microscopic methods since there is no chemical way of differentiating graphite as such.

Diamonds are particularly resistant to weathering and go into the eluvium and alluvium where they can be recovered by washing and collection on a greased mat. Common associates in placers are quartz, gold, platinum, zircon, magnetite, rutile, brookite, beryl, topaz, corundum, tourmaline, and pyrope. None of these are specific mineral indicators except pyrope, chrome diopside, and picoilmenite (Mg ilmenite), which are particularly characteristic of most kimberlites.

Kimberlites and carbonatites appear to be genetically connected, the kimberlites being an intermediate phase between ultrabasic rocks and carbonatites as indicated by their trace and minor element compositions. Thus, kimberlites tend to be enriched in Mg, Fe, Mn, Ni, Co, Cr, and Ti as well as in Ba, Sr, Rb, P, F,* Nb, Zr, rare earths, and some of the other elements commonly found in carbonatite complexes. From this it follows that elements such as Ni, Co, Cr, Ba, Nb, and F, may be useful indicators of diamondiferous kimberlites utilizing soils, glacial materials, stream sediments, vegetation, and probably water as sampling media.

Silicon (q.v. in Vol. IVA) is the most abundant electropositive element in the rocks of the Earth's crust. The deposits of silicon are so well known that they need little discussion here. Briefly, they comprise the following:

1. Silicates of all types used for a great variety of purposes too numerous to mention here: The silicates of commercial importance, not discussed under the other elements in this entry, include feldspar, mica, nepheline, pyrophyllite, kyanite, andalusite, sillimanite, dumortierite, talc, wollastonite, and asbestos.

2. Quartzite, sandstone, and silica sand used for flux, production of silica brick, glass making, and a great variety of other purposes.

3. Bentonite and bleaching clay.

4. Clay for brick and ceramics.

5. Sand and gravel.

6. Diatomite and radiolarite.

7. Chert and flint.

8. Quartz and quartz crystal.

9. Geysersite and siliceous sinter associated with hot springs.

There are no specific indicators of silicon deposits except the element.

Germanium (q.v. in Vol. IVA) is a relatively rare semimetallic element that forms several rather uncommon minerals, of which the best known are argyrodite, Ag_8GeS_6, germanite, $Cu_{26}Fe_4Ge_4S_{32}$; and renierite, $(Cu,Zn)_{11}(Ge,As)_2Fe_4S_{16}$. None of the germanium minerals form deposits of commercial importance, although in some deposits—e.g., Tsumeb, Namibia, and Mansfeld, Germany (Kupferschiefer)—they may be present in sufficient amounts to make recovery of germanium as a by-product feasible. Germanium is also a trace to minor constituent in various sulfides and sulfosalts, particularly sphalerite, cinnabar, enargite, tetrahedrite-tennantite, pyrargyrite, and some tin-bearing sulfides. Magnetite may be enriched in germanium in some deposits; siliceous sinters and opal are concentrators of the element in places.

Germanium is obtained industrially from sphalerite concentrates and germanium-bearing sulfide ores during smelting and electrolytic reduction. A potential source of commercial germanium is certain coals and lignites that contain relatively large amounts of the element in some

*Personal communication, D. R. Boyle

beds. Certain magnetite ores have low concentrations of germanium in places.

The elements accompanying germanium in deposits are the same as those mentioned for zinc, mercury, silver, tin, copper, arsenic, antimony, and the other elements commonly found in polymetallic deposits.

Tin (q.v. in Vol. IVA) is a common metal that forms several minerals of commercial importance, including cassiterite, stannite, teallite, cylindrite, and franckeite. Cassiterite is the most important industrial source of tin.

Deposits of tin are varied and include the following:

1. Pegmatites and coarse-grained granitic bodies: The principal mineral is cassiterite. The characteristic elements concentrated are Sn, W, Ta, Nb, Bi, As, Be, Sc, rare earths, B, F, Li, Rb, Cs, and Mo. There is a nearly constant association of tin and tungsten in pegmatites. The association with lithium is also marked.

2. Skarn deposits: Few of these are commercial sources of tin. The principal mineral is cassiterite. The elements concentrated include Sn, W, B, F, Be, Cu, Pb, Zn, As, Mo, and Fe. There is usually a close association between Sn, W, B, and F in most skarn deposits.

3. Veins, lode deposits, and disseminations in greisenized, sericitized, and chloritized zones: The principal tin mineral is cassiterite, with stannite as an important constituent of certain veins. The characteristic elements concentrated are Sn, W, Mo, Li, Rb, Cs, Be, Sc, Fe, Cu, Zn, Cd, Pb, B, As, Bi, S, P, and F. Veins in which stannite is abundant contain these same elements with the addition of Ag, Au, Ga, In, Tl, Ge, and Sb.

4. Disseminations and quartz (opal) stockworks in rhyolite flows, quartz-feldspar porphyries, ignimbrites, etc. that are commonly sericitized and opalized: The principal tin minerals are wood-tin and concretionary cassiterite. The elements concentrated are mainly Sn, SiO_2, Bi, Fe, (W), and F. Few of these are economic sources of tin. Examples are widespread in the Tertiary rhyolites and ignimbrites of Mexico. The Mount Pleasant deposit in southern New Brunswick, Canada, also falls into this class. This deposit occurs in intensely fractured and altered rhyolitic volcanic rocks, tuffs, pyroclastics, and porphyries where zones of silicification, chloritization, epidotization, greisenization, and fluoritization contain a veritable host of minerals including cassiterite, stannite, molybdenite, wolframite, scheelite, fluorite, topaz, kaolinite, sphalerite, pyrite, chalcopyrite, galena, roquesite, arsenopyrite, bismuthinite, and bismuth. The elements concentrated in this deposit include Li, Rb, Cs, Cu, Ag, Zn, Cd, Sc, Y, rare earths, B, In, Sn, Pb, Ti, P, As, Bi, Nb, Ta, S, Mo, W, F, and Fe.

5. Cassiterite pipes: These are unusual types of deposits that take the form of chimneys and pipes frequently in granitic rocks, gneisses, and sedimentary rocks. Some are developed in limestones and bear a certain relationship to the skarn deposits. The principal mineral in the deposits is cassiterite. There is a close association of Sn, B, F, and sometimes As in the cassiterite pipes. W is enriched in some of the pipes.

6. Massive sulfide deposits: Some of these contain traces of tin in cassiterite and as a constituent of sphalerite, galena, and other sulfides. None of these deposits can be considered as large commercial sources of tin, but some supply important amounts of the metal as a by-product. The best indicators of tin in massive sulfide deposits appear to be B, Bi, In, Tl, Cd, Ge, Sb, and As.

7. Placers: Both eluvial and alluvial placers provide most of the world's supply of tin. The principal mineral is cassiterite with smaller amounts of wolframite and the numerous other minerals found in placers. Deposits of this type are developed extensively in Malaysia, Indonesia, Thailand, Zaire, and Nigeria.

Tin is a good indicator of its deposits in most types of soil, glacial overburden, stream and lake sediment, and vegetation surveys. The natural compounds of the element are relatively insoluble, and hence water surveys are not suitable for geochemical prospecting for tin deposits. Depending on the types of deposits and the concentration of elements in them, other indicators of tin are effective in most types of geochemical surveys. These include, in particular, W, Li, B, Be, Nb, Ta, P, and F. For the polymetallic types of tin deposits, Cu, Pb, Zn, Ag, Cd, As, Sb, and Bi may be useful indicators.

Lead (q.v. in Vol. IVA) is obtained from a variety of deposits, most of which also contain Ag, Zn, Cd, and Cu. These deposits have been discussed under these elements and need not be repeated here. The most common elemental associates of lead are Zn, Cd, Ag, Cu, Ba, Sr, V, Cr, Mn, Fe, Ga, In, Tl, Ge, Sn, As, Sb, Bi, Se, Hg, and Te and, more rarely, B and F. All these elements can be used as indicators of lead deposits in geochemical surveys, but the best are Zn, Cd, Ag, Cu, Ba, As, and Sb. The natural compounds of lead are relatively insoluble, and hence water surveys utilizing lead as an indicator are rarely effective.

Group IVB: Titanium, Zirconium, and Hafnium

Titanium (q.v. in Vol. IVA) is widely diffused in most rock types, being present in the largest amounts in basic rocks. Numerous titanium minerals are known, but those of commercial importance include only rutile, anatase, leucoxene, brookite, and ilmenite. Titaniferous magnetite and sphene are common, but they are not normally used as an ore of titanium. Their deposits provide a titanium reserve for the future. In addition, a large number of relatively rare minerals contain titanium, such as mosandrite, perovskite, brannerite, pyrochlore, and euxenite. Some of these are ores of Nb, rare earths, Th, and U.

Titanium is notable for the large number of unusual and complex types of primary deposits. Some of these (e.g., ilmenite deposits) are associated with gabbroic bodies, especially anorthosites; others are affiliated with alkali syenites and/or carbonatites. In these bodies the commercial minerals are rutile, brookite, and leucoxene. In some deposits ilmenite and perovskite can probably be recovered as a by-product. Several complex deposits contain an unusual assortment of minerals of which apatite is most common. An example is the remarkable rocks, referred to as nelsonites, that occur in Nelson County, Virginia; these are rich in rutile, ilmenite, and apatite. The gangue is quartz, plagioclase, pyroxene, and hornblende. Titanium minerals are also concentrated in residual deposits and placers, and the element is a constituent of certain bauxites and laterites.

The elements most commonly concentrated in primary titanium deposits are Ti, Fe, Ca, F, and P. Some contain minor amounts of Fe, Cu, and other sulfides. The residual and placer deposits contain mainly rutile, ilmenite, magnetite, and other common resistate minerals concentrated during weathering processes.

Titanium is a good indicator of its deposits in soil, till, and stream sediment surveys. Its use in vegetation is uncertain, although it is known that some plants tend to concentrate higher than average amounts of titanium. The natural compounds of titanium are all relatively insoluble, and hence the use of the element in water surveys is not effective. The use of ancillary indicators in geochemical prospecting for titanium deposits has not received much attention. It would seem that phosphorus may be a good indicator for certain deposits and iron for others.

Zirconium and *hafnium* (q.v. in Vol. IVA) are two elements that invariably occur together in nature and are obtained for commercial purposes from the same sources. Zirconium forms several minerals, but only zircon, the silicate, and baddeleyite, the oxide, provide an industrial source for the element. Hafnium forms no known minerals but is a variable constituent of both zircon and baddeleyite.

The principal commercial sources of zirconium and hafnium are zircon and baddeleyite in beach sands, dune sands, stream placers, and eluvial deposits. The principal minerals accompanying the zirconium minerals are ilmenite, leucoxene, rutile, staurolite, tourmaline, sillimanite, kyanite, and quartz. Most of these minerals, including zircon, derive from the deep weathering of granite, granitic pegmatites, peralkaline massifs, and other rocks in which zircon is an accessory mineral. The baddeleyite deposits of the Pocos de Caldas area, Brazil, occur mainly in eluvial and alluvial deposits and consist of fibrous baddeleyite, amorphous ZrO_2, zircon, and altered zircon. These minerals are derived mainly from the deep pervasive weathering of an alkalic complex. Some baddeleyite and amorphous ZrO_2 also apparently occur in veinlike masses in these deeply weathered rocks.

Disseminations and concentrations of zircon, eudialyte, catapleiite, lavenite, baddeleyite, and other zirconium minerals frequently occur in peralkaline massifs and complexes and in their pegmatites; they also occur in certain masses of apatite and in pyroxene-mica-apatite veins, or pegmatitelike bodies. Some of these minerals could perhaps be obtained as by-products of the mining of these bodies. The Ilimaussaq Peralkaline Complex in Greenland contains zones with up to 25% eudialyte.

Carbonatite complexes are enriched in zirconium and hafnium, the principal minerals of these elements being zircon, zirconian garnet (kimzeyite), and baddeleyite. At Palabora, South Africa, baddeleyite is a by-product of copper mining.

Zirconium is a good indicator of its deposits in soil, till, and stream sediment surveys. Its use in vegetation surveys has not been investigated extensively, although it is known that some plants near zirconium deposits contain higher than normal amounts of the element. Zirconium is a nearly constant constituent of natural water, and some spring waters are relatively enriched in the element. There are no data, however, on the effectiveness of hydrogeochemical surveys for locating zirconium deposits.

Zirconium and hafnium are accompanied by Th, U, Ti, Nb, Ta, Be, P, F, and rare earths in peralkaline massifs and complexes and their pegmatites and by Ti, Th, rare earths, P, and locally, Sn and W in placer deposits. Each of these diagnostic elements may serve as indicators of zirconium under favorable conditions. The elements characteristically concentrated in carbonatites (see the section on niobium) may also be useful in indicating zirconium deposits. Some zirconium and hafnium deposits are radioactive due to the presence of associated Th and U.

Group VA: Nitrogen, Phosphorus, Arsenic, Antimony, and Bismuth

Nitrogen (q.v. in Vol. IVA) is concentrated in the atmosphere, in coal, and in certain nitrate deposits. It is obtained for commercial purposes from all three. Most of the commercial nitrogen compounds are obtained by a process of fixing atmospheric nitrogen. A small amount of nitrogen compounds are recovered as by-products of the coke industry in the form of aqua ammonia or ammonium sulfate. Considerable quantities of nitrates are still obtained from nitrate deposits, particularly those in Chile. A little nitrogen gas is obtained from certain nitrogen-bearing natural gas wells. Some nitrogenous fertilizer materials are obtained from guano, by-products of meat and fish plants, cottonseed and other seed mills, and processed sewage.

The principal types of nitrate deposits are cave, playa, and caliche deposits. Only the last is of commercial importance. They occur in very arid regions, usually bordering salars and playa mud flats, and consist of beds and irregular lenses a few inches to several feet thick that are often overlain by loose sand and gravel. The nitrates are associated with a large number of other salts that frequently show a gradation in chemical composition downward in the regolith. The salts concentrated in the nitrate deposits are mainly those of the elements Na, K, Li, Mg, Ca, Sr, Ba, Cr, Mn, B, N, Cl, S, Br, and I.

Searches for nitrate deposits should be guided mainly by geological principles, and all materials suspected of containing commercial nitrates should be analyzed. Groundwaters and springs high in soluble nitrates may be useful guides under certain conditions. Bromine, iodine, and boron analyses may be useful in testing caliche, waters, and other materials for concentrations of these elements associated with nitrate deposits.

Phosphorus (q.v. in Vol. IVA) is widely distributed in all rocks but usually only in small amounts. The basic rocks tend to have the highest contents of the element. The most common and commercial phosphate minerals are those of the apatite series, amblygonite, and monazite.

Phosphorus is concentrated in a great variety of deposits and is also won as a by-product of various industrial processes. The principal commercial sources of phosphates are the massive apatite deposits associated with peralkaline massifs and complexes (e.g., Khibiny massif, USSR) and the sedimentary phosphorites.

Briefly, the types of phosphate deposits are as follows:

1. Apatite veins, pockets, and segregations in or associated with metamorphic pyroxenites and other granitized sediments (Grenville type): These deposits are irregular veins, chimneys, and segregations of apatite frequently in a gangue of calcite, pyroxene, and phlogopite. The country rocks are metamorphic pyroxenite, crystalline limestone, or biotite gneiss. The principal elements concentrated in these deposits are Ca, Mg, Fe, Si, S, rare earths, Ti, Zr, P, F, Cl, and C (graphite).

2. Apatite lenses, veins, and disseminations in or associated with gabbros and anorthosite (nelsonites): These deposits commonly have much rutile and ilmenite (see also the section on titanium).

3. Apatite lenses, disseminations, veins, etc. in or associated with alkaline and peralkaline massifs and complexes, carbonatites, and similar bodies: Apatite is the principal mineral, occurring frequently as an admixture with nepheline, various other alkali silicates, titaniferous magnetite, magnetite, sphene, and pyrochlore. The characteristic elements concentrated in these deposits are mainly Na, Ca, Sr, Ba, Fe, Ti, V, Nb, Ta, rare earths, Zr, Th, U, P, and F (see also the section on niobium).

4. Marine phosphorites: These are categorized as shell phosphates and chemically or organically precipitated phosphorites. Only the latter are exploited commercially. The principal mineral of importance in them is collophane, a microcrystalline variety of apatite. Where the marine phosphate beds are weathered and leached, they commonly give rise to pebble and other types of residual rock phosphate deposits. Phosphorites contain a large number of diverse elements, among which U, V, F, Se, As, rare earths, Cr, Ni, Zn, Mo, and Ag are most commonly enriched.

5. Bog phosphate deposits: Certain bog iron ores and peats may be greatly enriched in phosphorus, mainly in vivianite, the hydrous iron phosphate, and wavellite, the hydrous aluminum phosphate. Most of these deposits are small and often low grade.

6. Guano deposits: The two types of guano deposits are insular (and coastal) guano and cave guano. Insular and coastal guano have been mined on many islands and coasts in the Pacific, Indian, and Atlantic oceans. Cave guano is of restricted occurrence, and few commercial deposits are known.

7. Residual and reworked (placer) deposits: These deposits are extremely varied and result from a combination of both mechanical and chemical reworking processes involving all the foregoing types of deposits. In some places lean rock phosphates have been enriched solely by the weathering out and reconcentration of phosphatic nodules. In other places there has been much solution and reprecipitation of phosphatic material on weathered and liberated nodules and irregular masses. The phosphates concentrated in this manner include the aluminum, iron, and calcium phosphates.

8. Miscellaneous sources of phosphates: These include phosphatic chalks, phosphatic nodules on the ocean floors, and by-product phosphates from steel plants, amblygonite and monazite reduction, meat and fish packing plants, and sewage.

Arsenic (q.v. in Vol. IVA) is widely diffused in most rocks but only in amounts of a few parts per million. Most of the arsenic in rocks is present in sulfides, particularly pyrite, or in accessory minerals such as apatite. The common minerals of arsenic are native arsenic, realgar, orpiment, niccolite, loellingite, cobaltite, arsenopyrite, and a great variety of sulfosalts of which the most common are tennantite and enargite.

Most of the arsenic of commerce is obtained as a by-product of the smelting and reduction of lead, copper, silver, and gold ores. Arsenic is a constituent of a great variety of deposits as detailed in the paper by Boyle and Jonasson (1973). It is particularly concentrated in certain skarn, gold-quartz, polymetallic, disseminated porphyry copper (enargite), and massive sulfide bodies of varied genesis; also in certain native silver deposits (Cobalt type), some containing uranium (Great Bear Lake–Jachymov type). The element is a particularly useful indicator in geochemical surveys for locating a variety of deposit—in particular, those of gold, silver, copper, lead, and zinc. Under certain conditions it may be useful as an indicator for nickel and cobalt deposits and for those of tin, tungsten, uranium, mercury, and platinum metals.

Antimony (q.v. in Vol. IVA) is widely diffused in many types of mineral deposits in trace, minor, and major amounts. It tends to be concentrated in those deposits containing sulfides, either as separate antimony minerals or as a minor or trace constituent of a great number of sulfides and sulfosalts. The common antimony minerals in endogene (primary) mineral deposits are native antimony, stibnite, and a great variety of sulfosalts, of which the most common are tetrahedrite, jamesonite, boulangerite, bournonite, polybasite, and pyrargyrite. The most common supergene antimony minerals in deposits are kermesite, senarmontite, stibiconite, nadorite, and bindheimite. In some deposits pyrargyrite may be of supergene origin.

Much of the antimony of commerce is obtained as a by-product of the smelting and reduction of lead, copper, silver, and gold ores. Some stibnite veins and their supergene minerals are mined essentially for antimony.

Antimony is a good indicator of deposits in which it is highly concentrated and, in addition, is a useful indicator of various other types of deposits in which it is a minor constituent, particularly those of silver, copper, gold, lead, zinc, and mercury (Boyle and Jonasson, 1984).

Bismuth (q.v. in Vol. IVA) occurs only in very small amounts in rocks, typically less than 0.1 ppm. Most of the bismuth of commerce is obtained from a variety of polymetallic deposits, especially those containing molybdenum, tin, tungsten, copper, lead, silver, and gold. The most common primary ore minerals of bismuth are native bismuth and bismuthinite; of less importance are cosalite, galenobismutite, and tetradymite.

The element is also a trace constituent of a great variety of minerals from some of which—e.g., galena, tetrahedrite-tennantite, enargite, jamesonite—it is won as a by-product of metallurgical processes. The principal secondary or supergene minerals of bismuth are bismutite and bismite.

The principal types of deposits in which bismuth is a constituent are as follows:

1. Pegmatites, aplites, mineralized quartz pipes, and stockworks: These generally occur in granitic bodies, acidic porphyries, or high-grade metamorphic schists and gneisses. The characteristic elements concentrated with bismuth are Li, Rb, Cs, Be, Sc, rare earths, Cu, Ti, Nb, Ta, Mn, Mo, W, B, Sn, U, Th, As, P, and F.

2. Greisen zones, dikes, and irregular bodies developed mainly in granitic rocks, quartz porphyries, or metasediments; also certain feldspathized zones in granitic rocks and metasediments: The characteristic elements in these deposits are the same as those mentioned in item 1.

3. Disseminations of bismuth, bismuthinite, and other bismuth minerals in highly altered rhyolitic flows, quartz-feldspar porphyries, tuffs, ignimbrites, etc.: The characteristic deposit of this type is at Mount Pleasant, southern New Brunswick. For a description of this deposit, and the elements concentrated in it, see the section on tin.

4. Skarn deposits: In these the bismuth is associated with Cu, Pb, Zn, Sn, W, Mo, Ni, Au, Ag, and the various other elements concentrated in skarn.

5. Bismuth in tin, molybdenum, and tungsten deposits: A large part of the world's production comes from this source. The elements with which bismuth is closely associated in these deposits include Sn, W, Sb, Ag, Fe, As, Pb, Mo, Cu, Zn, Ba, P, B, and F (see also the sections on tin, molybdenum, and tungsten).

6. Bismuth concentrations in complex nickel-cobalt-arsenide deposits: The elements concentrated in these deposits are Bi, U, Fe, Co, Ni, Mo, Cu, Ag, Zn, Cd, Pb, Hg, As, and Sb (see also the sections on silver, cobalt, and uranium).

7. Bismuth concentrations in lead-zinc-silver and copper deposits: The bismuth is obtained as a by-product of these ores, the element being present mainly as a trace or minor constituent of galena, tetrahedrite-tennantite, enargite, jamesonite, and various other sulfosalts and sulfides.

8. Bismuth concentrations in gold deposits: There is usually insufficient bismuth in these deposits to warrant recovery of the metal.

9. Cosalite deposits: Cosalite is found in a variety of deposits, particularly in veins, but also in skarn deposits and certain pegmatitelike bodies. The usual associated elements are Pb, Zn, Cu, As, Sb, and Co.

10. Secondary (supergene) bismuth deposits: These are derived from the preceding types of deposits and contain bismuthite, bismite, arseno-bismite, and corroded native bismuth and bismuthinite.

Geochemical surveys are not generally performed to discover bismuth deposits, but the element is often a good indicator of various types of deposits, particularly those containing Sn, W, Mo, U, Cu, Pb, Zn, Au, and Ag.

Group VB: Vanadium, Niobium, and Tantalum

Vanadium (q.v. in Vol. IVA) is widely dispersed in most rock types but only in small amounts. The basic rocks usually contain the highest amount of the element. Vanadium forms a large number of minerals of which the most common are patronite, rauvite, carnotite, tyuyamunite, roscoelite (vanadium mica), descloizite, pucherite, and vanadinite. Titaniferous magnetite also contains some vanadium, the amount ranging from 0.1% to 1% V_2O_5.

Vanadium can be obtained for commercial purposes from the following types of deposits:

1. Titaniferous magnetite deposits associated with gabbros and anorthosites and in places with carbonatites: The principal ore mineral is magnetite. The characteristic elements concentrated are V, Ti, Fe, and P (see also the section on niobium).

2. Vanadium-bearing uranium deposits in sandstone terranes: The principal primary minerals are vanoxite, patronite, rauvite, roscoelite, corvusite, and montroseite, and the secondary (supergene) minerals are carnotite, tyuyamunite, and various other vanadium oxides and vanadates. The elements concentrated in these deposits are numerous and diverse and include U, V, Sr, Ba, Cr, Mo, Fe, Co, Ni, Cu, Ag, Au, Zn, Cd, Pb, P, As, Sb, S, and Se. Some of these deposits are similar to the red bed sandstone deposits discussed under uranium, copper, and silver.

3. Deposits associated with asphalt and other solid hydrocarbons: The best known deposit of this type occurs at Minas Ragra, Peru, where calcium vanadate and patronite (VS_4) fill cracks and fissures in shale and are associated with masses of asphalt material. The characteristic elements concentrated in these deposits are V, S, C, Ni, Fe, and Ca.

4. Vanadium in various polymetallic deposits: These are commonly copper, copper-lead-zinc, or lead-zinc deposits, the oxidized parts of which yield descloizite, vanadinite, mottramite, and other lead, zinc, and copper vanadates. These deposits provide rich vanadium ores in places. These oxidized ores are enriched in Cu, Pb, Zn, V, Mo, Ag, Au, and As.

5. Vanadium in phosphorites and vanadiferous shales: These deposits provide a large reserve

of vanadium for the future (see the discussion on phosphorites under phosphorus).

6. Vanadium in sedimentary iron ores: Certain sedimentary iron ores are enriched in vanadium. The characteristic accompanying elements are Mn and P.

7. Vanadium in petroleum, asphalt, and tar sands.

8. Vanadium in coal ashes.

9. Placers containing vanadiferous-titaniferous magnetite.

Vanadium is a good indicator of its deposits and can perhaps be used in some districts to indicate certain types of Cu, U, Pb, Zn, and Ag deposits as detailed here.

Niobium (q.v. in Vol. IVA) is a rare metal that is concentrated in certain pegmatites, albitized granites, carbonatites, and the eluvial and alluvial placers derived from these bodies. The principal minerals of niobium are pyrochlore, fergusonite, the columbite-tantalite series, the tapiolite-mossite series, betafite, the euxenite-polycrase series, niobian perovskite, samarskite, the eschynite-priorite series, and niocalite. Most of the niobium of commerce is obtained from columbite-tantalite and pyrochlore. Perovskite-bearing complexes provide a reserve for the future.

The principal deposits of niobium are as follows:

1. Granitic and syenitic pegmatites and certain coarse-grained and fine-grained muscovite granites: These usually contain columbite-tantalite, fergusonite, euxenite-polycrase, samarskite or eschynite-priorite. The characteristic elements concentrated in these deposits include Nb, Ta, Sn, W, Sb, Bi, Li, Be, Ti, Rb, Cs, rare earths, U, Th, B, Zr, Hf, P, and F.

2. Albite-biotite granites and albite-riebeckite granites: The principal minerals in these bodies are tantalite-columbite, pyrochlore, cassiterite, zircon, thorite-orangite, xenotime, monazite, cryolite, and topaz. The elements concentrated include Nb, Ta, Sn, W, Zr, Th, U, rare-earths, P, Al, and F.

3. Carbonatites: These alkaline rock-carbonate complexes and their fenitized zones form large ringlike bodies, elongated oval bodies, sills, and dikes and contain a most diverse assemblage of minerals. The principal niobian minerals are pyrochlore and perovskite. Some contain economic concentrations of copper minerals. The elements commonly concentrated in carbonatites are Na, K, Fe, Ba, Sr, rare earths, Ti, Zr, Hf, Nb, Ta, U, Th, Cu, Zn, P, S, and F. Some complexes are marked by higher than average contents of Ag and Pb and others by low concentrations of Be and Li; Mo and W occur in a number of complexes in anomalous amounts.

4. Bauxites and laterites: Certain deposits of bauxite, bauxitic clays, and laterite derived from the weathering of alkalic rocks and carbonatites are slightly enriched in niobium.

5. Eluvial and alluvial deposits derived from the preceding types: The principal minerals obtained are tantalite-columbite, pyrochlore, cassiterite, zircon, baddeleyite, xenotime, and monazite.

Niobium is relatively immobile in most natural settings, and hence the element is an excellent indicator of its deposits in soil, till, and stream sediment surveys. By contrast, water analyses using niobium as indicator are not effective. The various elements accompanying niobium in the deposits mentioned can perhaps be used as indicators for niobium deposits, although there is little detailed information on their effectiveness. Some investigators have found that Cu, Zn, Pb, Mo, V, U, Th, Sn, Li, Rb, Ba, Sr, Be, rare earths, and P in soils and stream sediments are reliable indicators of niobium deposits. Some carbonatites and other types of niobium deposits are radioactive due to the presence of small amounts of thorium and uranium. Many carbonatites have a high magnetic profile due to the presence of magnetite. Many of the carbonatites of Recent and Tertiary age, as well as some of those of older vintage, are indicated by alkaline cold, warm, and hot springs charged with CO_2 and enriched in Na, K, Ca, Mg, Cl, HCO_3^-, SO_4^{2-}, F (fluoride), and H_2S. Some of these waters are radioactive due to the presence of radium and radon.

Tantalum (q.v. in Vol. IVA) is a rare metal that commonly accompanies niobium in granitic and syenitic pegmatites, coarse- and fine-grained granites, albite-biotite and albite-riebeckite granites, and greisen zones in granitic rocks and metasediments. Some carbonatites are relatively enriched in tantalum, the element accompanying niobium in its various minerals. The tantalum of commerce is obtained mainly from columbite-tantalite, wodginite, and other tantalum-bearing minerals in pegmatites and eluvial and alluvial deposits derived from them.

The minerals of tantalum are essentially the same as those of niobium, and the geochemistry of the two elements is remarkably similar because of the effects of the lanthanide contraction in the periodic table. The foregoing discussion on niobium applies, therefore, to most geochemical surveys organized to search for deposits of tantalum.

Group VIA: Oxygen, Sulfur, Selenium, and Tellurium

Oxygen (q.v. in Vol. IVA) is the most abundant element in the Earth's crust and occurs combined in many forms. Oxygen gas for industrial purposes is obtained from the atmosphere by liquefaction processes.

Sulphur (q.v. in Vol. IVA) is widely diffused in the rocks of the Earth's crust, being concentrated mainly in the sulfides. It is also a major constituent of many sulfates, sulfosalts, and a host of other complex minerals. Industrial sulfur is obtained principally from native sulfur deposits and from natural gases containing hydrogen sulfide.

The principal deposits of sulfur and its compounds are:

1. Gypsum and anhydrite deposits.

2. Massive sulfides and other deposits containing pyrite, pyrrhotite, and other sulfides: The element is obtained from the smelter gases.

3. Native sulfur deposits associated with the cap rocks of salt domes and diapirs.

4. Native sulfur deposits associated with gypsum deposits and probably of biogenic sedimentary origin.

5. Native sulfur deposits associated with solfataric activity and probably partly of volcanic origin.

6. Sulfur in sour natural gas and petroleum: Most of the sulfur is present as H_2S, but some is contained in the organic petroleum base. The sulfur is obtained by oxidation of H_2S and during petroleum refining processes.

Those searching for sulfur deposits should be guided by geological principles, and all materials considered to contain sulfur should be analyzed directly for the element. Natural gases should, likewise, be tested directly for their sulfur content. Sulfur springs and solfataras may be good indicators of native sulfur deposits in volcanic areas.

The sulfate content of underground and surface waters, stream sediments, soils, and glacial materials may be used as an indicator of both sulfide- and sulfate-bearing deposits of many kinds. Thus, high sulfate contents in these various materials may be indicative of the presence of oxidizing sulfides and sulfosalts or of deposits such as evaporites, polymetallic sulfide deposits, porphyry-type copper-molybdenum deposits, etc. that are subjected to leaching by ground- or stratal waters.

Sulfur has two common isotopes, ^{32}S and ^{34}S, whose ratio has been much used in problems dealing with the origin of deposits containing sulfur in its many compounds. In general, the primary (hypogene) sulfates in most evaporites, veins, and other sulfate-bearing metallic deposits are enriched in ^{34}S; in contrast, sulfates of supergene origin derived from the oxidation of primary sulfides generally reflect the sulfur isotopic ratios of the primary sulfides that are commonly much less enriched in ^{34}S than are their accompanying hypogene sulfates. It is thus possible by using $^{34}S/^{32}S$ ratios to differentiate hypogene from supergene sulfates. Furthermore, sulfides (mainly pyrite and pyrrhotite) in the country rocks commonly have different sulfur isotopic ratios than those of their contained epigenetic mineral deposits, and the derived sulfates are, likewise, different isotopically. From this it follows that waters leaching mineralized zones may be differentiated from those leaching barren rocks by sulfur isotopic analyses of their contained sulfates. The same holds true for the sulfur isotopic composition of spring precipitates and in some places for the sulfate component of stream sediments. Finally, sulfur isotopic ratios of primary sulfides may be a clue to proximity to orebodies during drilling. In some sulfide mineralized belts researchers have noticed that the ^{34}S content of the sulfides in the wall rocks and alteration zones increases as the sulfide-bearing orebodies are approached.

Selenium (q.v. in Vol. IVA) is widely diffused in rocks but in only small amounts. The element is typically concentrated in the sulfides. Numerous selenium minerals are known, but none forms commercial deposits. Most of the selenium of commerce is obtained from refining processes of polymetallic ores, especially those of copper and nickel.

Geochemical surveys are not usually mounted to discover selenium deposits, but the element is often a useful indicator for several types of deposits. These include uranium-vanadium-copper-silver-molybdenum deposits of the red bed sandstone type, certain pitchblende-selenide deposits, gold-silver selenide deposits, and various types of polymetallic ores, especially those containing copper, mercury, bismuth, and silver.

Tellurium (q.v. in Vol. IVA) occurs in rocks in amounts measured in tenths or hundredths of parts per million, the average being about 0.08 ppm. The shales, particularly the carbonate-bearing and pyritic varieties, carry the largest amounts of tellurium (1–2 ppm). The element forms a number of tellurides, especially with gold, silver, lead, bismuth, mercury, copper, nickel, and platinum metals. These occur in a variety of deposits, of which the most common are quartz-wolframite-bismuth veins and stockworks; quartz-cassiterite-sulfide veins and stockworks; massive nickel-copper-cobalt sulfides associated with basic igneous rocks; various polymetallic massive and vein-type sulfide deposits in volcanic and sedimentary terranes; most types of gold-silver deposits, particularly the skarn, quartz vein, lode, and stockwork types; and volcanic deposits of native sulfur. Tellurates and tellurites are also known in the oxidized zones of gold and various polymetallic ores, especially those containing copper, nickel, and lead.

Tellurium is obtained commercially from refining processes of polymetallic ores especially those containing copper, nickel, and lead; also from certain telluriferous gold ores, for example, from the Emperor Mine, Fiji.

Geochemical surveys are not often carried out to discover tellurium deposits since the element rarely occurs in deposits in amounts greater than a few tens or hundredths of parts per million. Tellurium may, however, be a relatively good indicator of certain types of deposits containing silver, gold, bismuth, nickel, cobalt, tungsten, tin, molybdenum, copper, and platinum metals when rocks, soils, glacial materials, and stream sediments are used as the sampling media. The mineralization in certain zones containing disseminated (porphyry) copper deposits is frequently enriched in tellurium.

Group VIB: Chromium, Molybdenum, and Tungsten

Chromium (q.v. in Vol. IVA) is widely diffused in most rocks, being present in the highest amounts in the basic and ultrabasic igneous varieties. The only ore mineral of chromium is chromite, and practically all commercial deposits of the mineral are associated with ultrabasic rocks and occur as ribbons, layers, irregular masses, pipes, streaks, lenses, and disseminations in these rocks. Many of these orebodies are closely connected with serpentinization and talcification of the ultrabasic bodies. Weathering of the primary deposits commonly gives rise to residual concentrations of chromite as in New Caledonia. There is a slight concentration of chromium in some of the uranium-vanadium sandstone-type deposits, the element being present largely in chrome micas (mariposite). Slight concentrations of chromium are also noticeable in certain gold-quartz and polymetallic sulfide deposits in basic and ultrabasic rocks. In these deposits the chromium is present mainly in the chrome micas (fuchsite and mariposite) in the wall-rock alteration zones. Chrome micas may also occur in various greenschists of sedimentary origin. The content of chromium in fuchsite ranges up to 3.5% and in mariposite up to 0.7%. The chrome garnet, uvarovite, commonly found in serpentinites and in contact rocks rich in chromium, may contain up to 20% Cr. Lead chromate (crocoite) and other chromates occur in some of the oxidized zones of lead and other sulfide deposits traversing basic and ultrabasic rocks.

Chromium is a good indicator of its deposits in practically all types of geochemical surveys with the exception of those based on natural waters. Nickel and cobalt are suitable indicators of chromite deposits in soil and stream sediment surveys in some areas; perhaps also in water and vegetation surveys, although there are no suitable data on their effectiveness. Some chromium deposits are enriched in arsenic (mainly as nickel arsenides) (e.g., Malaga, Spain), and others contain platinum metals. These elements may, therefore, be suitable indicators in some districts. Heavy mineral surveys of soils and stream networks followed by mineralogical and spectrographic analyses are particularly useful in prospecting for chromite deposits where there has been some weathering. Even in glaciated regions this method offers considerable success if used judiciously. Wad, limonite, and other precipitates at the orifices of springs are enriched in chromium in the vicinity of chromite deposits. Bodies of ultrabasic rocks containing chromite are commonly denuded of vegetation, their soils supporting only grasses and scattered trees. Whether these deleterious effects on vegetation are due to available chromium or to nickel or magnesium is uncertain.

Molybdenum (q.v. in Vol. IVA) is a trace constituent of most rocks at the parts per million range. It is slightly enriched in acid igneous rocks and in black shales. Several minerals of molybdenum are known, but only molybdenite is of commercial importance at present. Deposits of powellite, $Ca(Mo)O_4$, and wulfenite, $PbMoO_4$, have been worked in the past.

Molybdenum is concentrated in the following types of deposits:

1. Pegmatites and aplites: In these deposits molybdenum occurs as molybdenite. The characteristic elements concentrated are Mo, W, Re, Cu, Sn, Be, B, P, F, Zn, Bi, and Fe. Few of these deposits are of commercial interest.

2. Greisen zones, greisen dikes, greisenized stockworks, and sheared, feldspathized (albitized, microlinized) zones mainly in granitic rocks but also in metasediments and acid porphyries: The economic mineral is molybdenite. The characteristic elements concentrated are the same as those in item 3 following.

3. Quartz veins, quartz pegmatites, quartz segregations, and stockworks: The economic mineral is molybdenite. The characteristic elements concentrated are Mo, W, Re, Sn, Li, Be, Bi, Fe, Cu, Zn, Pb, B, P, and F.

4. Skarn deposits: Molybdenite is the principal molybdenum mineral. The characteristic elements concentrated in skarn deposits include Mo, W, Re, Bi, Fe, Cu, Au, Ag, Co, Ni, Be, Ti, Zn, Cd, B, As, and S.

5. Disseminated deposits principally in monzonite but also in various acid porphyries, granite, diorite, schists, etc.: Some of these constitute the so-called porphyry copper deposits from which molybdenite is won as a by-product. Others contain only molybdenite as the economic mineral, with wolframite as a by-product in some deposits. The characteristic elements concentrated in these deposits include Mo, Re, Cu, Ag, Be, Fe, W, (Sn), (Th), (U), (Nb), Zn, (Pb), As, B, F, and (P).

6. Carbonatites: Molybdenum is enriched in some carbonatites, principally in the fenite zones,

in amounts up to 500 ppm, the element being present mainly in molybdenite, precipitated in some complexes during late-stage mineralization processes. For the elements accompanying molybdenum in carbonatites, see the section on niobium.

7. Disseminations and stockworks in rhyolite flows, quartz-feldspar porphyries, tuffs, ignimbrites, etc.: The principal molybdenum mineral is molybdenite. For further descriptions of these deposits and their elemental associates, see the section on tin.

8. Copper shales and red bed sandstone deposits: These carry some molybdenite in places, but the amounts are low. For the accompanying elements in these deposits see the discussion of these deposits under the sections on copper, silver, and uranium.

9. Gold-quartz, pitchblende, and various polymetallic deposits: Many of these deposits carry small amounts of molybdenite especially if they contain copper minerals. None has been found that contains sufficient molybdenite to constitute a molybdenum orebody.

10. Oxidized molybdenum deposits: These contain mainly wulfenite, $PbMoO_4$; molybdite, MoO_3; powellite, $Ca(Mo)O_4$; and/or ferrimolybdite (hydrated molybdate of Fe). They are formed by the oxidation of polymetallic deposits bearing molybdenite. Few are now of commercial importance, but in the past some yielded small tonnages of molybdenum ore.

Molybdenum is an excellent indicator of its deposits and can be used in all types of geochemical surveys utilizing soils, till, stream sediments, waters and their precipitates, and vegetation. The best ancillary indicator elements of molybdenum appear to be Cu, W, and Bi. The other elements noted earlier as being concentrated in molybdenum deposits can probably be used in certain areas.

Tungsten (q.v. in Vol. IVA) is widely diffused in most rocks but only in the tenths of parts per million range. Acid rocks such as granites, pegmatites, and quartz-feldspar porphyries have the highest contents of tungsten. Many minerals of tungsten are known, but the most important commercial primary ones are scheelite, wolframite, huebnerite, and ferberite. There are in addition a number of secondary (supergene) tungsten minerals, among which the most common are tungstite, cuprotungstite, powellite, raspite, stolzite, ferritungstite, and anthoinite.

Tungsten deposits can be classified as follows:

1. Pegmatites and aplites: The principal minerals are wolframite, huebnerite, and ferberite, with subsidiary scheelite. The principal elements concentrated are W, Mo, Re, Sn, Cu, As, Nb, Ta, Bi, Li, Be, Rb, Cs, B, Sc, rare earths, F, and Mn.

2. Greisen zones, greisen dikes, greisenized quartz stockworks, and sheared, feldspathized (albitized, microclinized) zones mainly in granitic rocks but also in metasediments and acid porphyries: The economic mineral is wolframite; scheelite is relatively rare but occurs in quantity in some deposits. The characteristic elements concentrated are W, Mo, Re, Be, Sc, rare earths, As, Sn, Li, Bi, Fe, Cu, Zn, Pb, B, P, and F.

3. Disseminations and quartz (opal) stockworks in rhyolite flows, quartz-feldspar porphyries, tuffs, ignimbrites, etc. that are commonly sericitized and opalized: The principal tungsten mineral is wolframite, rarely scheelite. For descriptions of these deposits see the section on tin.

4. Skarn deposits: The principal tungsten mineral is scheelite; wolframite, huebnerite, and ferberite are often present but less abundant. The main elements concentrated are W, Mo, Re, Sc, rare earths, Bi, Cu, Pb, Zn, Fe, S, As, Au, Ag, B, and F.

5. Veins mainly with a quartz gangue: There is great variation in these deposits. Some contain essentially cassiterite and wolframite with small amounts of copper, lead, zinc, and iron sulfides. Others are wolframite-molybdenite veins, scheelite-gold-quartz veins, scheelite-wolframite gold-silver veins, scheelite (wolframite) copper veins, and scheelite-wolframite silver-copper-lead-zinc veins. The characteristic elements of these deposits are likewise varied and include Li, Rb, Cs, Be, B, Sc, rare earths, U, Mo, Re, W, Mn, Fe, Cu, Ag, Au, Zn, Cd, Ga, In, Tl, Ge, Sn, Pb, As, Sb, Bi, P, S, and F.

6. Disseminations and veinlets of scheelite in talc (soapstone) bodies: The origin of these occurrences of scheelite is uncertain. They occur in parts of North Carolina and are described by Bentzen and Wiener (1973).

7. Placers derived from the preceding types of deposits: The principal minerals won from these deposits are cassiterite, wolframite, tantalite-columbite, and scheelite.

8. Certain alkaline brines or evaporite deposits in present-day playas, salt pans, salt lakes, or their ancient equivalents in arid regions: The brine of Searles Lake in California contains about 70 ppm $WO_3$3. Similar brines in arid regions of the USSR are enriched in tungsten. Few deposits of this type are exploited at the present time because of difficulties in extracting the tungsten. They provide, however, a large reserve for the future. In the brines and evaporites the tungsten is closely associated with As and Mo and also with Li in places (see also the section on lithium).

Tungsten is an excellent indicator of its deposits and can be used in all types of geochemical surveys based on rocks, soils, tills, and stream sediments. Good ancillary indicators of tungsten include Sn, Mo, and Bi. The effectiveness of the other elements enriched with tungsten in its deposits is relatively unknown. Those that are most common in tungsten deposits, including B, F, As, Li, and Cu, are useful indicators in some districts.

Group VIIA: Fluorine, Chlorine, Bromine, and Iodine

Fluorine (q.v. in Vol. IVA) is a trace constituent of most rocks, occurring mainly in apatite and in the micas, amphiboles, and tourmaline where it substitutes for the OH^- groups and O^{2-}. The element also occurs in fluorite in some rocks. The principal minerals of fluorine are fluorite, topaz, and cryolite, Na_3AlF_6. Fluorite is the principal commercial mineral containing fluorine.

Fluoriferous deposits are widely distributed in various types of rocks, including the following:

1. Fluorite-bearing granitic bodies, rhyolites, and quartz-feldspar porphyries: The fluorite occurs as disseminations and patches in the matrix of the rocks. Few, if any, of these deposits can be considered to be commercial sources of fluorite at the present time.

2. Greisen zones, greisen dikes, feldspathized zones, sericitized zones, and silicified zones mainly in granitic rocks but also in various schists and gneisses: Fluorite and topaz are common accessory gangue minerals in these deposits, which are mined mainly for tin, molybdenum, and tungsten. The fluoriferous minerals are not usually recovered, although in some deposits they could be won as by-products (see also the sections on lithium, tin, molybdenum, and tungsten).

3. Carbonatites: Many of these complexes are rich in fluorite, but the mineral is not generally recovered; it could constitute a by-product in some deposits. The mineral is commonly a late-stage mineral in these complexes. Associated fluorides in some carbonatites are cryolite, weberite, and other aluminofluorides. The fluocarbonates—in particular bastnaesite and parisite—that occur in some carbonatite complexes and in associated carbonate masses comprise a large reserve of fluorine. For the elements associated with fluorine in carbonatites, see the section on niobium.

4. Calcite-fluorite-apatite veins and masses: These unusual deposits are common in the high-grade metamorphic terranes of the Precambrian Grenville Province of Canada. None has been exploited for fluorite, but some contain up to 20% CaF_2. The elements characteristically concentrated in these deposits are Ca, F, P, U, Th, Zr, Ti, Mo, and rare earths (see also the sections on rare earths and uranium).

5. Cryolite deposits: The mineral cryolite is relatively rare; it occurs in some carbonatite complexes and in a large lenselike (pegmatitic) mass at Ivigtut, Greenland. There, the mass occurs in a granitic rock in gneiss and appears to be associated in some manner or other with the many alkaline rocks and carbonatite complexes in south-western Greenland. Those searching for cryolite deposits should concentrate on districts of alkali-rich rocks and carbonatite complexes.

6. Veins, lodes, pockets, pipes, mantos, disseminations, and stockworks carrying mainly fluorite: These deposits have a widespread distribution and occur in many rock types, being particularly common in carbonate rocks, shales, and sandstones and in granites, rhyolites, quartz porphyries, and other acidic igneous rocks. The elements accompanying fluorine in these deposits are Ca, Fe, S, Si, Ba, Sr, Pb, and Zn; some deposits have Sn, W, Mo, and Cu, and others may contain Th and U. A few carry Au, Ag, and Te (e.g., Cripple Creek, Colorado).

7. Bedded, layered, and irregular stratiform deposits mainly in limestones and sandstones: In these deposits the fluorite appears to replace the carbonate rocks or the cement in the sandstones. The characteristic elements accompanying fluorine are Fe, Zn, Pb, Sr, and Ba.

8. Phosphorites and other types of phosphate deposits: Large reserves of fluorine are present in these deposits mainly in the form of fluorapatite (see the section on phosphorus).

9. Miscellaneous sources of fluorine: Fluorite, a gangue mineral in many types of vein deposits including gold deposits (e.g., Cripple Creek, Colorado), and various polymetallic deposits, especially lead-zinc deposits, tin deposits, tungsten deposits, and molybdenum-copper deposits. The mineral can probably be recovered as a by-product from some of these deposits. Topaz is a gangue mineral in a large number of pegmatites, veins, and greisens containing tin, tungsten, molybdenum, and beryl minerals.

10. Spring deposits: The travertines of both hot and cold springs sometimes contain fluorite, although rarely in commercial quantities. Associated minerals are carbonates, barite, pyrite, opal, limonite, and wad.

11. Residual deposits: Under certain weathering conditions fluorite disintegrates and dissolves only slowly and tends to collect in the weathered residuum and eluvium near its primary deposits. Some of these residual deposits have been exploited in the past.

Good indicators of most low-temperature fluorite deposits are Ba, Sr, Pb, Zn, and Cu. High-temperature fluoriferous deposits may be indicated by Sn, W, Mo, Nb, Ta, rare earths, Th, and U.

Fluorine is a persistent constituent of almost all types of mineral deposits, occurring in quantities from traces to minor and major amounts. The element is, therefore, almost the universal indicator of mineralization. The other elements of almost universal occurrence in mineral deposits are boron and sulfur. These three elements, used in conjunction with one another, provide a formidable array of indicators in geochemical prospecting for practically all types of deposits. Where one element fails to serve as an indicator, another will serve the purpose in most cases.

Chlorine (q.v. in Vol. IVA) is obtained mainly from sodium chloride or halite deposits and from brines of underground and surface origin. The sea is an enormous reservoir of chlorine.

Some nepheline syenites are enriched in chlorine, the element being present mainly in sodalite. Certain serpentinized ultrabasic bodies are rich in chlorine (up to 0.8% in the serpentine) as are also numerous granitic intrusives (up to 0.5% in the biotite). In the serpentinized bodies, the chlorine is largely present in serpentine and asbestos, probably replacing the OH groups; its source is most probably the marine sediments that enclose most of these rocks. In the granitic rocks the bulk of the chlorine is present in micas, commonly biotite, where it substitutes for the OH groups. In all rocks there is always some chlorine in NaCl in liquid inclusions. Certain coals are rich in chlorine, with amounts up to 1% Cl.

Geochemical surveys are not usually mounted to discover halite deposits. Those searching for such deposits should be guided by geological principles. The presence of brine springs or of brines in drillholes may be a useful guide to halite deposits.

Chloride brines, both cold and hot, issuing from springs and present in faults are common in regions of epigenetic mineral deposits containing gold, silver, and the various base metals. They are particularly abundant in Tertiary and Quaternary volcanic regions but also occur in older rocks, some even of Precambrian age. The relationship of these brines to the epigenetic deposits is uncertain, but some investigators think that the metals are transported as chlorides. Whatever the relationship, the occurrence of high amounts of chlorides in ground- and spring waters should be taken as an indication of mineralization in volcanic and associated sedimentary terranes. Along the same line it has been suggested that the chlorine content of intrusives can be used as an indicator of mineralization (Stollery et al., 1971; Parry, 1972). Thus, fecund intrusives rich in chlorine in a district may be a general guide to the presence of associated magmatic-hydrothermal deposits. The water-extractable chloride in granitic rocks appears, however, to be of little use as an indicator of plutons that have associated mineralization (Kesler et al., 1973). High chloride contents in ultrabasic and basic intrusives and in waters leaching zones of these bodies may suggest the presence of asbestos and nickel-cobalt-copper and sulfide deposits, the latter due to the redistribution of nickel, cobalt, and copper, during serpentinization and other hydration processes.

Springs and groundwaters enriched in chloride commonly mark the presence of carbonatite complexes, especially those of Recent age but also those of older vintage in some terranes (see also the section on niobium).

In hot arid regions chlorine tends to concentrate in the oxidized zones of silver, copper, mercury, and lead deposits where the element is bound as the relatively insoluble calomel, $HgCl$; chlorargyrite, $AgCl$; and Hg chlorides, double chlorides, bromides, iodides, and oxychlorides. Some of these minerals collect in the soils, eluvium, and nearby stream sediments where leaching has not been too extensive.

Bromine (q.v. in Vol. IVA) is obtained mainly from the sea, stratal brines, bitterns, surface saline deposits, and evaporite salt deposits. There is an inexhaustible source in the oceans. Those searching for bromine sources should analyze brines, surface saline deposits, and evaporite salt deposits directly for the element.

The geochemistry of bromine in epigenetic deposits and their associated primary halos and host rocks is not well known. The element occurs in igneous-type rocks in amounts ranging from 0.5 ppm to 2 ppm, in sediments in amounts up to 50 ppm and increasing with the amount of organic matter, and there is some evidence that slight concentrations take place in a variety of epigenetic sulfide and other mineral deposits. In hot arid regions bromine tends to concentrate in oxidized zones of silver, copper, mercury, and lead deposits where the element is bound as the insoluble bromargyrite, $AgBr$, and as a minor constituent in a great variety of Ag, Cu, Pb, and Hg chlorides, double chlorides, iodides, and oxychlorides.

Iodine (q.v. in Vol. IVA)) is obtained as a by-product of Chilean nitrates, from natural brines and mineralized waters, from oil field waters, and from kelp ash. Some coals also have small quantities of iodine, and some saline soils are enriched in the element. The Chilean nitrates contain iodine in the form of the minerals dietzeite, $Ca_2(IO_3)_2(CrO_4)$, and lautarite, $Ca(IO_3)_2$ (see also the section on nitrogen).

Those searching for sources of iodine should analyze brines, mineralized water, oil field waters (some of which contain up to 150 ppm iodine as in Java), natural salts, coals, caliche, and saline soils directly for the elements.

The geochemistry of iodine in epigenetic deposits and their associated primary halos and host rocks is not well known. The element occurs in igneous-type rocks in amounts ranging from 0.05 ppm to 0.5 ppm, in sediments up to 1 ppm and increasing with the amount of organic matter, and there is some evidence that slight concentrations take place in a variety of epigenetic sulfide and other mineral deposits. Certain porphyry copper

and massive sulfide deposits are said to be relatively enriched in iodine. Some sulfide deposits, however, appear to be lower in iodine content than their host rocks. In hot arid regions iodine tends to concentrate in the oxidized zones of silver, copper, mercury, and lead deposits where the element is bound as insoluble iodargyrite, AgI, and as a minor or trace constituent in a great variety of Ag, Cu, Pb, and Hg chlorides, double chlorides, bromides, and oxychlorides.

The use of iodine as an indicator of epigenetic mineral deposits requires detailed study since some of the results quoted in the literature are contradictory as to the effectiveness of the element.

Group VIIB: Manganese and Rhenium

Manganese (q.v. in Vol. IVA) is widely distributed in rocks in minor amounts. The basic igneous rocks contain the most manganese, and the element is highly concentrated in certain types of manganiferous sediments.

Manganese ores are of several types including veins and lenses of various manganese oxides and manganiferous siderite in rocks of diverse types, nodules and masses of manganese oxides in porous limestone beds and brecciated zones, sedimentary deposits of various manganese oxides in sedimentary terranes, sedimentary deposits of manganese oxides associated with iron formations, manganiferous shales, metamorphosed sedimentary deposits consisting of a variety of manganese oxides, and residual deposits derived from all the foregoing types of deposits. Parts of the floors of the oceans are covered with manganese nodules and irregular masses that not only are a vast reserve of the element but also contain variable amounts of metals such as Cu, Co, Ni, Mo, and W that can probably be removed economically with further research.

Manganiferous deposits and areas are typically reflected in the soils, tills, and stream sediments by the occurrence of black manganese oxides and wad. Practically all types of geochemical surveys can be used to outline these areas.

In passing, note that manganese is a general indicator of certain types of lead-zinc-silver deposits and gold veins. In the former, manganiferous siderite is one of the principal gangue minerals. In the gold deposits the gangue minerals are rhodochrosite, rhodonite, and other manganese minerals such as johannsenite and alabandite; these minerals are particularly characteristic of Tertiary deposits.

Hydrous manganese oxides in soils, spring precipitates, and drainage sediments tend to concentrate many elements, mainly those that form cations or cationic complexes. This feature is partly due to adsorption processes promoted by a negative charge that the colloidal hydrous manganese oxides carry and partly due to coprecipitation processes. The list of elements concentrated by hydrous manganese oxides is long and includes practically all the metals and some of the non-metals as well. The following elements exhibit a marked concentration by hydrous manganese oxide (wad): Li, Sr, Ba, Ra, Mo, W, Co, Ni, Cu, and Ag.

Rhenium (q.v. in Vol. IVA) is a trace constituent of certain niobium, tantalum, tungsten, rare earth, and platinum group minerals. It is frequently enriched in molybdenite in amounts up to 3,000 ppm or more. Three poorly differentiated minerals of rhenium are known: dzhezkazganite (lead rhenium sulfide), ReS_2 (?), and rhenium oxide, Re_2O_7 (?). None of these minerals is of commercial importance. Rhenium is enriched in molybdenite deposits, some porphyry copper–molybdenum deposits, copper-bearing shales and sandstones, and certain types of Cu, Pb, Zn, and U deposits. The element is produced commercially as a by-product of certain copper shale ores and from the treatment of flue dust and residues obtained from the roasting and smelting of molybdenite ores. The elements accompanying rhenium are the same as those discussed under molybdenum, copper, niobium, tantalum, tungsten, rare earths, and platinum group metals. Selenium and tellurium often show a correlation with rhenium in some deposits.

Rhenium has not been used extensively as an indicator of deposits in which it occurs mainly because the element is difficult to determine accurately, except by neutron activation methods, at its normal concentration in natural materials (0.005 ppm). Some cupriferous and uraniferous sandstones and shales, however, may average about 25 ppm Re. Where enrichments of rhenium are encountered in weathered residuum, soils, glacial material, and drainage sediments, molybdenum, tungsten, copper, and a variety of other polymetallic sulfide deposits should be suspected in the terrane. Enriched amounts of rhenium in plant ash may also indicate the presence of rhenium-bearing rocks and deposits. Meyers and Hamilton (1961) found from 50 to 500 ppm Re in the ash of a number of plants growing in the uraniferous sandstone regions of the western United States.

Group VIII: Iron, Nickel, Cobalt, and Platinum Metals

Iron (q.v. in Vol. IVA) is widely diffused in most rocks, being concentrated mainly in the basic varieties and in certain ferriferous sedimentary rocks.

Iron ores are of several types, including veins and massive lenses of various iron sulfides, particularly pyrite and pyrrhotite, iron oxides, and siderite; skarn deposits of magnetite and hematite; sedimentary deposits of various iron oxides in sedimentary terranes; deposits associated with the great iron formations of Precambrian and later ages; iron shales and sandstones; metamorphosed sedimentary deposits consisting of a variety of iron oxides and silicates; and residual deposits of limonite and other iron oxides derived from weathering of the foregoing types of deposits or from the weathering of basic and intermediate rocks and iron-bearing sediments. The last type of deposit constitutes the lateritic ores found in tropical regions. In addition to these types we may add the hematite and magnetite masses and disseminated bodies commonly found in carbonatite complexes. Some of these are now worked for iron ores; magnetite is recovered from others as a by-product of copper and niobium mining.

Iron deposits and iron-bearing rocks are reflected in a terrane by enriched amounts of limonite and other iron oxides in the soil, till, and drainage sediments, also by the presence of limonitic precipitates, transported gossans, and bog iron deposits in or near water courses.

Nickel (q.v. in Vol. IVA) is widely dispersed in rocks of all types, being concentrated particularly in those of basic and ultrabasic igneous origin. The element forms a number of minerals of which the most important industrially are pentlandite, nickeliferous pyrrhotite, millerite, niccolite, chloanthite, gersdorffite, garnierite, nepouite, and nickeliferous serpentine (serpophyte).

The following types of nickel deposits can be distinguished:

1. Massive sulfide lenses and disseminated deposits containing essentially pentlandite, pyrrhotite, pyrite, and chalcopyrite, commonly with small amounts of platinoid minerals: Most of these deposits are associated with basic or ultrabasic rocks. The principal elements concentrated are Ni, Co, Fe, Cu, Au, Ag, Pt metals, Se, Te, As, and S. Some deposits in this classification contain the nickel in a very dispersed state in three principal ways: (1) as a substitution of magnesium in various minerals—e.g., olivine and serpentine, (2) in finely divided nickel sulfides, (3) and in Ni-Fe alloys. These bodies are not usually economic.

2. Veins and lenses of sulfides containing millerite, gersdorffite, chalcopyrite, pyrrhotite, and pyrite: The characteristic elements concentrated are Ni, Co, Fe, Cu, and S. Some of these deposits are intimately associated with basic and ultrabasic rocks; others exhibit only a general relationship.

3. Veins containing niccolite, gersdorffite, chloanthite, and various other nickel-cobalt arsenides and sulfides: Associated minerals are native silver, arsenic, calcite, bismuth, galena, sphalerite, chalcopyrite, pitchblende, and various other minerals. The characteristic elements concentrated are Ni, Co, Ag, Fe, Cu, Pb, Zn, As, Sb,

S, Bi, and U, (see also the sections on silver and uranium).

4. Residual (supergene) nickel-cobalt deposits: These are of two types not readily differentiated from each other. One type contains the nickel essentially in garnierite, nepouite, and nickeliferous serpentine (serpophyte), and the other (laterite) contains nickel mainly in a complex of hydrous iron, manganese, and nickel oxides and indefinite supergene silicates. Both types are often found together and are gradational or intermixed. Some of these deposits occur at the present erosion surface; others are fossil and represent ancient buried erosion surfaces. The characteristic elements concentrated are Ni, Co, Fe, Mn, and Cr.

Nickel is a good indicator of its deposits in practically all types of geochemical surveys. The ancillary indicators of nickel deposits are Cu, Co, As, Cr, and platinum metals. Chromium can be particularly useful in deeply weathered terranes. Nickel can be used as an indicator of a number of other types of deposits in which, with cobalt, it occurs in minor quantities. These include various types of uranium deposits, native silver deposits, certain nickeliferous gold deposits, and platinum metal deposits (see the following sections on cobalt and platinum metals).

Cobalt (q.v. in Vol. IVA) forms a number of primary minerals, the most important of which are cobaltite, smaltite, linnaeite, carrollite, skutterudite, safflorite, and glaucodot. Secondary or supergene ore minerals include asbolan, heterogenite, sphaerocobaltite, and erythrite.

Cobalt follows nickel closely, and considerable amounts of the element are obtained for commercial uses from the nickel deposits described previously. In addition to these, large amounts of cobalt are also obtained from certain bedded copper shales (Zambia), from native silver-nickel-cobalt arsenide deposits, from certain primary and oxidized copper sulfide veins, from certain veins containing lead-zinc ores, and from veins mined essentially for gold (gold-cobalt type).

The elements accompanying cobalt are diverse and depend essentially on the types of deposits. Grouped, these include (massive nickel-copper deposits) Ni, Co, platinum metals, Fe, Cu, Ag, Au, Se, Te, and S; (native silver-nickel-cobalt arsenide deposits) Ni, Co, Ag, Fe, Cu, Pb, Zn, As, Sb, S, Bi, and U; (copper-cobalt sulfide ores) Cu, Co; (lead-zinc cobalt ores) Pb, Zn, Cd, Ag, Co; (gold-cobalt deposits) Co, Au, Ag; and (laterites) Ni, Co, Fe, Mn, and Cr.

Cobalt is a good indicator of its deposits and can be used in practically all types of geochemical surveys. The use of the various ancillary indicator elements depends on the type of deposit sought. Those most indicative of cobalt in a general way are Ni and Cu.

Cobalt is a particularly good indicator of a number of types of deposits in which it occurs as a minor constituent, mainly because it has a relatively high mobility under normal weathering conditions. The deposits indicated by cobalt include those of copper (shales and some vein-type deposits), uranium (vein types and disseminated types in shales and some sandstone terranes), nickel-copper (massive sulfide deposits), native silver-Ni-Co arsenide vein-type deposits, certain Pb-Zn veins and lodes, certain types of cobaltiferous gold deposits, and platinum metal deposits of various types (see the following section).

The platinum metals (q.v. in Vol. IVA) include ruthenium, rhodium, palladium, osmium, iridium, and platinum. All are found in close association with one another. The principal minerals carrying these metals are native platinum; various platinum metal alloys and intermetallic compounds; sperrylite, $PtAs_2$; cooperite, $(Pt,Pd,Ni)S$; braggite, $(Pt,Pd,Ni)S$; laurite, RuS_2; and stibiopalladinite, Pd_5Sb_2. A number of other rarer platinoid minerals containing Te, As, Sb, Sn, Hg, and Bi are known. The only rocks slightly enriched in platinum metals are ultrabasic rocks (dunites, peridotites, and serpentinites). These rocks average about 65 ppb Pt and 20 ppb Pd with corresponding lower values for Ru, Rh, Os, and Ir in the average range of 0.7, 5, 5, and 6 ppb respectively. In sulfidic and chromiferous layers and segregations in ultrabasic-basic rocks, these values are usually augmented by factors of 10–100 and in places by a factor of 1,000. For comparison, ultrabasic rocks average about 4 ppb Au and their sulfidic segregations from 40 ppb to 400 ppb Au.

The principal types of deposits in which the platinum group metals are concentrated include the following:

1. Disseminated platinoid minerals mainly in layered and differentiated lopoliths, laccoliths, sills and dikes of peridotite, pyroxenite, dunite, and other ultrabasic rocks and their metamorphic equivalents and only rarely in gabbro, norite, and other basic rocks and their metamorphic equivalents: Most of the platinoid minerals in these occurrences comprise the natural alloys with smaller amounts of the various arsenides, sulfides, antimonides, and so on. These minerals occur principally in small segregations, lenses, and irregular masses of coarse-grained iron-rich dunite and wehrlite; dark schlieren and iron-rich veinlike masses; chromitite segregations; and irregular masses and pipes of coarse-grained (pegmatitic) hortonolite-dunite as in item 2 following. The platinum metal grade of these types of occurrences is low, typically only background or slightly enriched, usually less than ten-fold. The pipes and iron-rich coarse-grained segregations assay in the range of 0.7 ppm to 34 ppm or more total platinoids.

2. Hortonolite-dunite pipe deposits: These rocks are unique and occur mainly in South Africa where they are supposedly a differentiate of the Bushveld Igneous Complex, although in places they cross the stratification of the complex. Some investigators have suggested that the hortonolites are pegmatite analogues of the complex. The principal platinoid minerals are the native alloys associated with sperrylite, cooperite, and other platinoid minerals; the characteristic elements concentrated are Pt metals, Cr, As, Sb, and (Bi). Enrichments of Pt metals may exceed 35 ppm in parts of the pipes.

3. Disseminated platinoid minerals in coarse-grained (pegmatitic) pyroxenite and other ultrabasic and basic phases with sulfur and chromitite segregations in lopoliths, laccoliths, sills, and dikes: These deposits are common in South Africa, an example being the Merensky Reef, a supposedly differentiated lower part of the Bushveld Complex. In the oxidized zones of most of these deposits, the main economic mineral is native platinum with much palladium and minor quantities of rhodium, osmium, iridium, gold, and silver. In the primary zones the minerals are chromite, magnetite, pyrrhotite, pentlandite, chalcopyrite, cubanite, millerite, and nickeliferous pyrite. The platinoid minerals in the primary zone include the various native alloys, sperrylite, braggite, stibiopallandinite, laurite, and others; some of the platinoids are probably also finely disseminated or in solid solution in the pyrite, pyrrhotite, and pentlandite. The principal elements concentrated are Pt metals, Ag, Au, Cr, Fe, Cu, Ni, Co, and S. The grade of the Merensky Reef is roughly about 0.25 oz/ton platinoids (8 ppm), 0.20% Ni, and 0.10% Cu. The Stillwater Complex in Montana also falls in this category. It contains a number of members that are enriched in the platinum-group metals and gold. Values up to 8 ppm Pt, 11 ppm Pd, 1.7 ppm Rh, 0.5 ppm Ir, 1 ppm Ru, and 0.1 ppm Au have been reported (Page et al., 1976).

4. Minor concentrations of platinoids in nickel-copper ores of the Sudbury type: These deposits are associated with basic rocks and contain the platinum mainly as sperrylite. A number of other platinum metal minerals are usually present, and there may also be some solid solution of platinum metals in the various sulfides. The principal associated elements are Ni, Co, Cu, Fe, As, Ag, and Au. There are also minor enrichments of Te, Bi, Sn, and Sb commonly correlative with the platinum metal contents. The grade of these deposits is roughly 0.005 oz/ton Pt (0.17 ppm) and 0.008 oz/ton Pd (1.27 ppm). Some of the Noril'sk ores in the USSR are exceptionally rich in platinum metals (up to 0.3 oz/ton, or 10 ppm).

5. Skarn deposits associated with basic and ultrabasic rocks: These deposits are developed where basic and, in particular, ultrabasic rocks intrude carbonate rocks. Some are in shear zones and fractures in other types of host rocks near ultrabasic and basic intrusives. Examples are known in South Africa and elsewhere. One deposit in Sumatra occurs in a skarn body in limestone near a granodiorite contract (Hundeshagen, 1904). It was characterized by the presence of much garnet (grossularite), wollastonite, and enrichments of Pt metals, Au, Ag, and Cu. Serpentinites occur nearby. The principal platinum minerals in skarn deposits are sperrylite, cooperite, and stibiopalladinite; associated minerals are copper and nickel sulfides. The principal elements concentrated are Pt metals, Ni, Co, Fe, and Cu, sometimes Au and Ag.

6. Concentrations of platinum metals in various types of copper, lead, silver, and gold veins and lodes: The platinum metals in these deposits are won as a by-product. There is commonly a close association between copper and platinum metals in these deposits. If there is no copper in the deposits, the platinoids are either rare or absent. Some of the cupriferous platinoid-rich ores are arseniferous or antimoniferous, and some are characterized by moderate enrichments of Te and Bi. Most of these deposits occur in shear zones, faults, or crushed zones in or near peridotites, serpentinites, and other ultrabasic and basic rocks or their metamorphosed equivalents. There is ample evidence to support the thesis that the platinoids (and the Cu, Ag, Au, etc.) were extracted from the basic and ultrabasic intrusives and their host rocks by some process of metamorphic secretion. An excellent example of this possibility is provided by the Cuniptau Mine near Goward, Ontario. Mertie (1969) describes the copper-platinum deposit of the Salt Chuck Mine north of Ketchikan, Alaska. At this mine the chalcopyrite-chalcocite-gold ore, with a significant platinum metal content, occurs in fractures and disseminations in pyroxenite, gabbro, and gabbro pegmatite. The oxidized zones of some of these deposits are enriched in platinum metals, commonly in plumbojarosite, scorodite, and similar minerals. One good example of this enrichment occurred in the Yellow Pine district, southern Nevada (Knopf, 1915). Another example is the New Rambler Mine, Medicine Bow Mountains, Wyoming, where Pt and Rh are enriched and Pd depleted in the weathered portion of the deposit (McCallum et al., 1976).

7. Unusual platinum metal concentrations in pegmatites and quartz veins: These are rare, although authenticated examples are known in Colombia, the USSR, South Africa, and Ontario. Their origin is an enigma. Some of those in South Africa occur in brecciated quartz lodes occupying faults of post-Karroo age in felsite and felsite tuff comprising part of the Bushveld Igneous Complex. Various platinoid minerals, including native platinum, are associated with specularite, chrome chlorite, and sericite. None of the occurrences in this category are of commercial value.

8. Minor concentrations of platinum metals in uraniferous veins and other deposits: The Nicholson Mine and some other prospects in the Uranium City uraniferous belt of Saskatchewan contain small concentrations of platinoids. There, the platinoids appear to be associated with selenides. This association—platinoids-selenium—is a common one that is noticeable at Sudury, Ontario; Tilkerode, Harz, Germany; Artonvilla Mine, Messia, South Africa (Mihalik et al., 1974); and elsewhere.

9. Porphyry (disseminated) copper ores: Some of these contain small amounts of platinoids, mainly palladium (see the section on copper).

10. Copper shales and schists (Kupferschiefer–White Pine–Zambia type): Some of these deposits contain trace amounts of platinoids associated with the copper and Ni-Co minerals. Rich copper sections of the Kupferschiefer run about 1 ppm Pt and 1 ppm Pd respectively (see the section on copper).

11. Auriferous and uraniferous quartz-pebble conglomerates: The gold ores of the Witwatersrand contain trace amounts of platinum metals. Some 80% of the platinoid minerals present in the conglomerates are (Ir, Os, Ru) alloys; the remainder comprise sperrylite, hollingworthite, michenerite, and others (Feather, 1976) (see the section on gold).

12. Conglomerates resulting from the erosion and denudation of terranes characterized by the occurrence of lopoliths, laccoliths, sills, and dikes of ultrabasic and basic rocks: Some of the gold and platinum-metal stream and river placer deposits of the Choco in Colombia were winnowed from Tertiary conglomerates and associated sediments slightly enriched in platinoids and gold derived from ultrabasic terranes.

13. Gossans and residual deposits: Most gossans and residual accumulations of massive and disseminated sulfide bodies in or associated with basic and ultrabasic igneous rocks commonly contain enriched amounts of platinoids. Some laterites on these rocks are likewise enriched in these metals in Ethiopia and elsewhere. Chromite commonly accompanies the platinoid mineral nuggets and residual particles in these deposits. Platinum is also reported in the bauxite of Tungar Hill and Dhangawan in India; its nature is not known.

14. Placers: Platinoid alloys pass directly from their deposits into eluvial and alluvial placers. The various platinoid minerals, including sperrylite and the sulfides, sulfide-arsenides, antimonides, tellurides, and selenides, may do likewise, although under conditions of intense weathering they tend to liberate the metals that accrete in the oxidized zones and in the eluvium to give dust, small nodules, and nuggets of platinoid alloys that ultimately collect in eluvial and alluvial placers. These types of placers are invariably found in terranes of ultrabasic and basic rocks, especially serpentinites. The native platinoid alloys and various platinoid minerals in placers are accompanied by chromite, chrome magnetite, ilmenite, native gold, native copper, and the various other minerals often found in placers. The famous platinum sands of Choco, Colombia, carry about 0.5 oz platinoids per ton and about the same amount of gold.

Little published information is available on geochemical prospecting for the platinum metals (Boyle, 1982b). Those searching for such deposits should concentrate on areas containing layered ultrabasic rocks, differentiated ultrabasic dikes and lenses, and nickel-cobalt deposits, especially of the Sudbury type. All types of copper deposits should be carefully scrutinized for platinum metals, especially those in or near ultrabasic rocks. Probably the best strategy is to prepare heavy concentrates of the gossans, soils, glacial materials, and stream sediments of an area and to follow this by detailed mineralogical, spectrographic, and chemical analyses. Chromite is a good indicator for platinum metals in some areas. Other good indicators would appear to be Ni, Co, Cu, and V. There is also commonly a coherence of the platinum metals with one or more of the following elements: Ag, Au, Sn, Pb, As, Sb, Bi, Te, Se, Hg, and P. Some of these elements may be useful in lithochemical, pedochemical, hydrochemical (water and drainage sediment), heavy mineral, gossan, and biogeochemical surveys carried out to discover platinum metal deposits. The various platinoids may be useful in differentiating gossans that indicate economic Ni-Cu and platinum metal-bearing deposits from those that are developed on barren sulfides or rocks. In Australia the economic Ni-Cu deposits are marked by gossans enriched in the platinoids.

The search for platinoid-bearing conglomerates and associated sediments should be focused on terranes of known or suspected conglomerates, particularly those of a fluviatile or deltaic origin containing pebbles and cobbles derived from ultrabasic and basic rocks. Most of these conglomerates will be of Tertiary age, but those of older vintage should not be neglected. Heavy mineral separates should be obtained from such conglomerates, and these should be studied mineralogically and analyzed for platinum metals and gold.

Group O: Helium, Neon, Argon, Krypton, Xenon, and Radon

This group contains the inert gases, all of which are found in the atmosphere and in trace amounts in rocks. Some, especially helium, are obtained from natural gases held in the rocks. The others are obtained from liquefaction of the gases in the atmosphere. Those searching for these inert gases in various natural gases should analyze them directly for the various members of the group.

Radon, the radioactive member of the series, is composed of the three isotopes ^{219}Rn, ^{220}Rn, and ^{222}Rn derived respectively from decay processes in the actinium, thorium, and uranium families. The longest lived of these isotopes is ^{222}Rn (3.8 days). The radon content of water, till, and soil has been used as an indicator of uranium and thorium deposits in a number of uraniferous belts (Boyle, 1982a).

Helium consists of two isotopes, ^{3}He and ^{4}He. The latter is constantly produced, as alpha particles, in the decay processes associated with uranium, thorium, and samarium. ^{3}He derives from the decay of tritium, and its abundance in natural helium is exceedingly low (1.38×10^{-4}%). Concentrations of helium in natural waters, rocks, soils, etc. may signal the presence of uranium and thorium deposits (Boyle, 1982a).

R. W. BOYLE

References and Selected Bibliography

Bentzen, E. H., and L. S. Weiner, 1973, Scheelite discovered in certain soapstone deposits in the Blue Ridge of Madison County, North Carolina, *Econ. Geol.* **68**, 703-707.

Beus, A. A., and S. V. Grigorian, 1977, *Geochemical Exploration Methods for Mineral Deposits,* A. A. Levinson, ed., R. Teteruk-Schneider, trans., Wilmett, Ill.: Applied Publishing Ltd., 287p.

Boyle, R. W., 1968, The geochemistry of silver and its deposits, *Canada Geol. Survey Bull. 160,* 264p.

Boyle, R. W., 1974a, Elemental associations in mineral deposits and indicator elements of interest in geochemical prospecting (rev.), *Canada Geol. Survey Paper 74-45,* 40p.

Boyle, R. W., 1974b, The use of major elemental ratios in detailed geochemical prospecting utilizing primary halos, *Jour. Geochem. Exploration* 3(4), 345-369.

Boyle, R. W., 1979, The geochemistry of gold and its deposits (together with a chapter on geochemical prospecting for the elements), *Canada Geol. Survey Bull. 280,* 584p.

Boyle, R. W., 1982a, *Geochemical Prospecting for Thorium and Uranium Deposits.* Amsterdam: Elsevier, 498p.

Boyle, R. W., 1982b, Gold, silver, and platinum mineral deposits in the Canadian Cordillera—Their geological and geochemical setting, in A. A. Levinson, ed., *Precious Metals in the Northern Cordillera.* Association of Exploration Geochemists, 1-19.

Boyle, R. W., 1982c, Geochemical methods for the discovery of blind mineral deposits, Part 1, *Canadian Inst. Mining and Metallurgy Bull.* **75**(844), 123-142; Part 2, **75**(845), 113-132.

Boyle, R. W., and I. R. Jonasson, 1973, The geochemistry of arsenic and its use as an indicator element in geochemical prospecting, *Jour. Geochem. Exploration* 2(3), 251-296.

Boyle, R. W., and I. R. Jonasson, 1984, The geochemistry of antimony and its use as an indicator element in geochemical prospecting, *Jour. Geochem. Exploration* 20(3), 223-302.

Brooks, R. R., 1972, *Geobotany and Biochemistry in Mineral Explorations.* New York: Harper & Row, 290p.

Brooks, R. R., 1983, *Biological Methods of Prospecting for Minerals.* New York: Wiley, 322p.

Desborough, G. A., and B. F. Leonard, eds., 1976, An issue devoted to platinum-group elements, *Econ. Geology* 71(7), 1129-1480.

Feather, C. E., 1976, Mineralogy of platinum-group minerals in the Witwatersrand, South Africa, *Econ. Geology* 71(7), 1399-1428.

Fletcher, W. K., 1981, *Analytical Methods in Geochemical Prospecting,* Handbook of Exploration Geochemistry, vol. 1, G. J. S. Govett, ed. Amsterdam: Elsevier, 255p.

Ginzberg, I. I., 1960, *Principles of Geochemical Prospecting: Techniques of Prospecting for Non-ferrous Ores and Rare Metals,* V. P. Sokoloff, trans. New York: Pergamon Press, 311p.

Govett, G. J. S., 1983, *Rock Geochemistry in Mineral Exploration,* Handbook of Exploration Geochemistry, vol. 3, G. J. S. Govett, ed. Amsterdam: Elsevier, 461p.

Granier, C. L., 1973, *Introduction à la prospection géochimique des gîtes métallifères.* Paris: Masson et Cie, 143p.

Howarth, R. J., 1983, *Statistics and Data Analysis in Geochemical Prospecting,* Handbook of Exploration Geochemistry, vol. 2, G. J. S. Govett, ed. Amsterdam: Elsevier, 437p.

Hundeshagen, L., 1904, The occurrence of platinum in wollastonite on the Island of Sumatra, Netherlands, East Indies, *Inst. Mining and Metallurgy Trans.* **13**, 550-552.

Joyce, A. S., 1976, *Exploration Geochemistry,* 2nd ed. Adelaide, S.A.: Techsearch Inc., and Australian Mineral Foundation, Glenside, S.A., 220p.

Kemp, J. F., 1902, Platinum and associated metals, *U.S. Geol. Survey Bull. 193,* 91p.

Knopf, A., 1915, A gold-platinum lode in southern Nevada, *U.S. Geol. Survey Bull.* **620-A**, 1-18.

Kesler, S. E., J. C. Van Loon, and C. M. Moore, 1973, Evaluation of ore potential of granodiorite rocks using water-extractable chloride and fluoride, *Canadian Inst. Mining and Metallurgy Bull.* **66(730)**, 56-60.

Levinson, A. A., 1974, *Introduction to Exploration Geochemistry.* Calgary, Alberta: Applied Publishing Ltd., 612p.

Levinson, A. A., 1980, *Introduction to Exploration Geochemistry,* 2nd ed. Wilmette, Ill.: Applied Publishing Ltd., 615-924 (the 1980 supplement).

McCallum, M. E., R. R. Loucks, R. R. Carlson, E. F. Cooley, and T. A. Doerge, 1976, Platinum metals associated with hydrothermal copper ores of the New Rambler Mine, Medicine Bow Mountains, Wyoming, *Econ. Geology* 71(7), 1429-1450.

Marchant, J. W., 1978, A review of the history and literature of groundwater hydrogeochemical exploration for ores, *University of Cape Town Geochemical Exploration Unit Report No. 5,* 322p.

Mertie, J. B., 1969, Economic geology of the platinum metals, *U.S. Geol. Survey Prof. Paper 630,* 120p.

Meyers, A. T., and J. C. Hamilton, 1961, Rhenium in plant samples from the Colorado Plateau, *U.S. Geol. Survey Prof. Paper 424-B,* B-286-B-288.

Mihalik, P., J. B. E. Jacobsen, and S. A. Hiemstra, 1974, Platinum-group minerals from a hydrothermal environment, *Econ. Geology* **69**(2), 257-262.

Naldrett, A. J., ed., 1979, Nickel-sulphide and platinum-group element deposits, *Canadian Mineralogist* **17**(2), 141-154.

O'Neill, J. J., and H. C. Gunning, 1934, *Platinum and Allied Metal Deposits of Canada,* Economic Geology Series No. 113. Ottawa, Ontario: Geological Survey of Canada, 165p.

Page, N. J., J. J. Rowe, and J. Haffty, 1976, Platinum metals in the Stillwater Complex, Montana, *Econ. Geology* **71**(7), 1352-1363.

Parry, W. T., 1972, Chlorine in biotite from Basin and Range plutons, *Econ. Geology* **67**(7), 972-975.

Rouse, A. W., H. E. Hawkes, and J. S. Webb, 1979, *Geochemistry in Mineral Exploration,* 2nd ed. New York: Academic Press, 657p.

Shima, M., 1970, *Chikyu Kagaku Tanko-ho [Geochemical Prospecting Methods],* 2nd ed. Tokyo: Kyoritsu Publishing Co. Ltd., 273p.

Siegel, F. R., 1974, *Applied Geochemistry.* New York: Wiley-Interscience, 353p.

Stollery, G., M. Borcsik, and H. D. Holland, Chlorine in intrusives: A possible prospecting tool, *Econ. Geology* **66**(3), 361-367.

Warren, H. V., and R. E. Delauvault, 1959, Pathfinding elements in geochemical prospecting, in *Symposium de Exploracion Geoquimica,* vol. 2, International Geological Congress, Twentieth Session, Mexico, 1956, 255-260.

Cross-references: *Exploration Geochemistry; Lake Sediment Geochemistry; Lithogeochemical Prospecting; Marine Exploration Geochemistry; Pedogeochemical Prospecting; Prospecting.*

INFORMATION AND DOCUMENTATION—See FIELD NOTES, NOTEBOOKS; GEOLOGICAL CATALOGUING; GEOLOGICAL SURVEYS, STATE AND FEDERAL; MAP AND CHART DEPOSITORIES; PROFESSIONAL GEOLOGISTS' ASSOCIATIONS; REMOTE SENSING, SOCIETIES AND PERIODICALS. Vol. XIII: CRIB; EARTH SCIENCE INFORMATION SOURCES; GEOLOGICAL INFORMATION, MARKETING; INFORMATION CENTERS.

INFORMATION RETRIEVAL—See COMPUTERS IN GEOLOGY; FIELD NOTES, NOTEBOOKS; GEOLOGICAL CATALOGUING; GEOLOGICAL HIGHWAY MAPS; GEOLOGIC REPORTS; MAP AND CHART DEPOSITORIES; MAPS, ENVIRONMENTAL GEOLOGY; MAPS, PHYSICAL PROPERTIES; PUNCH CARDS, GEOLOGIC REFERENCING; WELL LOGGING.

L

LAKE SEDIMENT GEOCHEMISTRY

Stream sediment samples, considered representative of the average trace metal content of a drainage basin, form the basis of many reconnaissance geochemical exploration surveys. In some areas, however, stream sampling (see *Hydrochemical Prospecting*) may not be practical if drainage is poorly developed or accessibility to streams is difficult, particularly in forested terrain. In contrast, lakes are easily reached by float-equipped aircraft. Moreover, by virtue of their position in depressions on the landscape, lakes act as traps for mechanical and chemical loads carried by streams and groundwater. Consequently, with the possible exception of interwatershed groundwater movement, lake trace element levels should also reflect material within the catchment.

Reconnaissance lake sediment surveys offer the prospect of rapidly and inexpensively assessing the mineral potential of a region; however, their application to exploration has been investigated vigorously only since 1970. Prior to pioneering papers in North America by Schmidt (1956) and Arnold (1970), studies were undertaken by limnologists interested primarily in physical (Mortimer, 1949) or biological (Gorham et al., 1974) properties of lakes. Chemical studies were commonly restricted to the most abundant anions and cations (Brunskill and Ludlam, 1969), and with a few exceptions, systematic studies of minor elements were not undertaken. In more recent years, however, both environmental monitoring studies and investigations of the application of lake sediment sampling to mineral exploration have expanded our knowledge of trace element behavior in lakes.

Factors Affecting the Metal Content of Lake Sediments

Timperley and Allan (1974) have proposed a model relating the trace metal content of lakes with a chemical balance between input and outflow of suspended or dissolved material (Fig. 1). Metals released from bedrock or overburden enter the drainage network attached to solid particles or as ionic or complexed soluble species. The relative proportion of each inflowing to a particular lake is dependent on geological, geomorphological, pedological, topographical, climatic,

and erosional conditions of the surrounding drainage basin. Despite the fact that two lakes may be similar in size, shape, depth, and relative position on a landscape, however, limnological factors modifying primary distributions of trace elements deposited by streams or groundwater must also be considered. Lake sediment anomalies consequently do not necessarily form in response to mineralized or metal-rich bedrock in the watershed, despite the fact that such units are observed to be actively discharging abnormal amounts of trace elements. Conversely, anomalous levels of an element may accumulate even though unusually high concentrations of metal are not present within the catchment. In view of the strong possibility of equivocal interpretation of the genesis of metal-rich zones, factors such as sediment texture and composition or lake environment that may lead to the development of false anomalies must be carefully evaluated.

If a lake is to reflect a metal-rich source in the catchment, trace elements (see *Indicator Elements*) must migrate toward and accumulate within the lake. The most obvious input is suspended or dissolved loads carried by streams. Suspended particles are deposited following a regular gradation of coarse through fine sizes. Heavy grains settle near sites of inflow to form deltas that may extend tens or hundreds of meters into the lake where slumping of forest beds has occurred. Elsewhere around the lake margin, bank erosion and winnowing of fines by waves leaves nearshore sediment relatively sand- and gravel-rich (Thomas et al., 1973) and, in this respect, similar to deltaic deposits (see Vol. VI: *Deltaic Sediments*). Conversely, silt, clay, and finely divided organic matter are deposited more uniformly near the center of the lake (Mortimer, 1949; Frink, 1969; Schoettle and Friedman, 1973).

Normal textural gradations may be modified by hydraulic activity of lake water. For example, turbidity currents, initiated by a sudden increase in bank or shore erosion (Ludlam, 1974), transport sand and pebbles toward the center of the lake. Similarly, rapid introduction of sand and woody fragments to the center of the lake during flood periods is probably a major short term factor affecting sediment texture. Turbulent currents acting below wave base are a less impressive but still important factor (Clark and Bryson, 1959) that may either arrest the settling of particulates

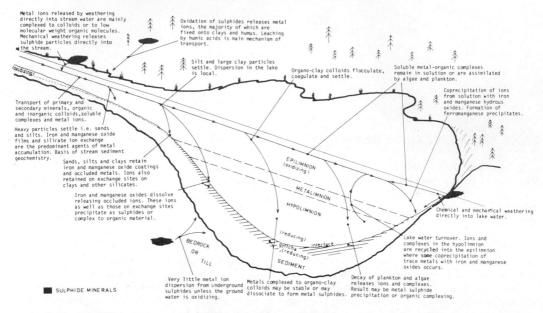

FIGURE 1. The drainage system involved in mineral weathering, metal transport, and metal accumulation in lake sediments. (After Timperley and Allan, 1974)

(Schoettle and Friedman, 1973) or mix recent muds with lake bottom parent material (Thomas et al., 1973). An absence of recent sedimentation may favor hydrous Fe and Mn oxide enrichment (Schoettle and Friedman, 1973) and, in view of the potential scavenging ability of this component, may preferentially extract trace elements relative to other nearby sediments.

Expected sediment textural relations also may be altered by disproportionately large streams entering small lakes. Where lake flushing is rapid (Rawson, 1960; Gorham et al., 1974), the metal content of sediment is likely to be directly related to inflowing stream sediment levels (Hoffman and Fletcher, 1981a). In contrast, groundwater influx and limnological factors probably control trace element patterns in lakes having no significant stream inflow.

Constituents of lake sediment, by virtue of diverse origins, concentrations, and physical and chemical properties, commonly have trace element contents peculiar to that component. For example, the negative correlation of Cu and Zn with sand found by Spilsbury and Fletcher (1974) in nearshore sediment illustrates how a compositional parameter, sand content, can strongly influence trace metal levels. This is because the sand fraction is comprised predominantly of quartz and feldspar grains, which are usually impoverished in these elements relative to clay minerals or sesquioxides.

Organic matter, because of its importance to biologists and limnologists and its association as the predominant factor controlling sediment texture, has been studied most intensively. The organic component of lake sediment is derived from terrestrial, degraded terrestrial, and planktonic sources, each of which has widely differing characteristics and properties. Trace element content of undecomposed vegetation mechanically introduced into the lake, therefore, would be expected to differ from the metal content held in soluble *humic acid complexes.* The latter, released in the watershed by decay of vegetation and pedogenesis, appear particularly effective in leaching soils (Rashid, 1972) and decaying leaf litter (Kimball, 1974) of their trace metals prior to entering the lake. Similarly, algae and plankton produced within the lake probably contain a trace element suite that is distinguishable from land-derived organic matter.

Classification of organic material into recognizable classes on the basis of an *allochthonous* or *autochthonous* source (Sanger and Gorham, 1970) may reduce variability attributable to provenance factors and aid in differentiating metal enrichment, reflecting a peculiar variety of organic constituent rich in one or more elements. A similar discussion has been presented by Mackereth (1966), which describes the origin, mode of introduction and accumulation, and properties of the sesquioxide fraction.

Though mechanical inputs to lakes and physical size-sorting processes are most apparent, metals introduced as soluble ions and complexes in runoff (Timperley and Allan, 1974) or groundwater (Brunskill and Harriss, 1969; Mehrtens et al., 1973) also may exert a significant influence on trace element patterns. Cu, Zn, and other metals are removed from water following flocculation of humic and fulvic acids, algal and planktonic assimilation, coprecipitation with, or adsorption onto, hydrous Fe and Mn oxides, and by adsorption onto clay mineral surfaces.

Removal of metal from water by organic chelating agents or algae is probably important in some lakes (Sanger and Gorham, 1970). Formation of metal-organic complex, however, is in part dependent on the character of the organic molecules (Schindler et al., 1972). Similarly, the genera of the plankton also affects metal uptake (Parker and Hassler, 1969; Bachmann and Odum, 1960; Knauer and Martin, 1973). Within oligotrophic lakes (low N, P, and nutrient sources and high dissolved oxygen content), biological productivity is low, and consequently, significant quantities of organic matter are unlikely to be deposited compared with eutrophic lakes (high N, P, and nutrient sources and low dissolved oxygen content). Despite possible enhanced extraction efficiency in eutrophic lakes, great productivity may also result in organic matter's acting as a trace metal dilutant, particularly where sediment is comprised essentially of organic constituents (Hoffman and Fletcher, 1981*a*).

An alternative mechanism to biological action for extraction of trace elements from lake water involves precipitation of a sulfide phase (Timperley and Allan, 1974). Sulfide precipitation, however, is probably not independent of biological activity, as plankton growth depletes the water of its oxygen content that, in turn, facilitates bacterial reduction of sulfate ions to sulfide species (Gorham, 1958). Measurements reported by Matson (1968) indicate that a substantial lowering of ionic levels of trace elements occurs several days prior to an algal bloom; however, the cause of this reduction remains unexplained. Trace elements may presumably become either bound to organic matter by partaking in complexation reactions and remain in solution or associated with plankton, or they may precipitate with sulfide ions and settle to the lake floor.

Many chemical reactions within the lake appear to depend on the oxygen content of water. Mortimer (1949) outlines a model relating changes in water temperature and oxygen levels with the oxidation potential of lake water and sediment. Briefly, following convective homogenization of lake water in spring or fall, surface water (*epilimnion*) either heats or cools and becomes less dense than bottom water (*hypolimnion*). Oxygen, slowly consumed by organisms in the decay of organic matter, cannot be replenished from the atmosphere because of density stratification. In some lakes, oxygen is completely exhausted. Reducing conditions consequently initiate conversions such as ferric to ferrous and manganic to manganous forms, increasing hydrous oxide solubility and concomitantly decreasing the scavenging capability associated with remaining sesquioxide solid phases. *Solubilization,* however, is rarely complete, and a fraction of the Fe and Mn remains in its higher oxidation state (Coey et al., 1974) to retain its adsorbing properties and the associated trace metals.

Metal accumulation zones within lakes may reflect either metal-rich lithologies (or mineralized bedrock) within the catchment or limnological processes (Garrett and Hornbrook, 1976). Of particular concern is the recognition of those trace-element enrichment patterns related to mineralized bedrock. In conflict with this objective are limnological factors that may enhance background levels or diminish anomalously high values. Examples illustrating the former include metal accumulation via scavenging by hydrous Fe and Mn oxides, organic matter, and clay mineral sediment constituents (Hoffman and Fletcher, 1979); whereas for the latter, dilution by quartz and feldspar sand is particularly important. In view of the fact that sediment reducing conditions typical of organic-rich lakes, ferrous iron, sulfide, and other reduced ions migrate upward or downward within diagenetic solutions until a counter ion is found with which to precipitate. Mackinawite (FeS) or greigite (Fe_3S_4) (Doyle, 1968) is most commonly produced and gives the sediment a greenish-gray color. Although not considered by Berner (1969), trace elements are probably also immobilized as sulfides (Timperley and Allan, 1974).

Because of physical, chemical, and biological parameter variability of lakes within the same geological, geomorphological, topographical, and climatic environment, trace metal levels and distributions must be evaluated with respect to conditions peculiar to each basin (Table 1). The fact that two lakes lie in close proximity does not necessarily mean that they behave identically to similar inputs of trace elements.

Metal accumulation zones within lakes may reflect either metal-rich lithologies (or mineralized bedrock) within the catchment or limnological processes (Garrett and Hornbrook, 1976). Of particular concern is the recognition of those trace element enrichment patterns related to mineralized bedrock. In conflict with this objective are limnological factors that may enhance background levels or diminish anomalously high values. Examples illustrating the former include metal accumulation via scavenging by hydrous Fe and Mn organic matter and clay mineral sediment constituents (Hoffman and Fletcher, 1979); whereas for the latter, dilution by quartz and feldspar sand is particularly important. In view of the

TABLE 1. Summary of Watershed, Lake, and Sediment Parameters Likely
to Have a Strong Influence on Trace Metal Levels of Lake Sediments

Watershed	Lake	Lake Sediment
Climate	Lake area and shape	Sample homogeneity
Topography	Maximum lake length	Consistency: sample water content
Lake elevation	Maximum lake width	Oxidized or reduced mud
Geology	Lakeshore character: boggy versus sandy	Sesquioxide content and concretions
Surficial deposits	Lakeshore length	Organic matter decomposition
	Distribution of islands	Ancient bottom sediment
Types and extent of vegetative cover	Number of major and minor inflowing streams and proximity to sample site	Deltaic sands or gravels
Soils	Proximity of sample site to outflow	Nearshore sands or gravels
Extent of flushing	Lake water color	Turbidites
	Watershed area	Algal mat
		Calcium carbonate
		Shell fragments
		Hydrogen sulfide odor

fact that sediment properties and limnological environments vary considerably across the floor of a single lake, the influence of metal-controlling parameters over a restricted portion of the lake floor must be recognized at an early stage of the survey to ensure that spurious values do not override regional trends. In summary, for successful application of lake sediment sampling to exploration, factors such as scavenging, dilution, sedimentation rates, emergence of groundwater, and oxidation-reduction reactions must not be permitted to obscure those metal levels reflecting mineralized bedrock.

History of Lake Sediment Geochemistry

Although anomalous stream sediment dispersion trains are often observed to be truncated on entering a lake and both Schmidt (1956) and Arnold (1970) demonstrated that the trace element content of lake sediments reflects proximity to mineralization, Allan (1971) was the first to report results of a regional geochemical lake-sediment-sampling program. His study, in the permafrost environment of the Coppermine River area, Northwest Territories, Canada, involved collection of inorganic, nearshore sediment at one sample per 25 km². The Coppermine River basalt, containing numerous copper showings and one major deposit, was successfully outlined. Regional surveys by Allan and his co-workers of the

Canadian Geological Survey restricted sample collection to nearshore inorganic sediment in inflow and outflow bays on the premise that these areas contain similar trace metal concentrations as adjacent stream sediments and hence are directly related to catchment material (Dyck, 1971). Sampling in the shallow water of the nearshore environment is also amenable to rapid collection procedures involving impressing a plastic or metal tube into the lake floor. However, in climatic zones where trees commonly line the shore and inorganic sediment is rare, this sampling technique may not be practical.

Results of subsequent regional surveys by Allan et al. (1973) in the Bear and Slave provinces, Northwest Territories, and Dyck (1974) in the Beaverlodge District, Saskatchewan, confirmed that lake sediment geochemistry effectively indicated areas of trace metal enrichment related to mineralization or bedrock geology. The survey by Allan was particularly outstanding because of the resulting discovery of a Pb-Zn-Cu-Ag prospect associated with a gossan zone in volcanic terrain (Nichol, 1975). Similarly, Hoffman and Fletcher (1972) found that the Rayfield River syenite stock (south-central British Columbia), containing bornite disseminated along fractures, was associated with regionally enhanced Cu levels in lake sediments. Further, a correspondence was found between the highest Cu grades in bedrock and the maximum Cu content in lake sediment.

In temperate climates, workers have assumed that a gray-green, water-saturated, finely divided ooze, referred to as *gyttja* (Timperley and Allan, 1974), is representative of the trace metal content of the drainage basin (Davenport et al., 1975). Physically, gyttja appears homogeneous. Samples retrieved from the center of the lake would seem most suitable to avoid the strong shoreline influences. However, sampling the center of the lake requires devices such as a mud snapper, Ekman dredge, or phleger corer capable of reaching to several tens of meters. Despite increased difficulty of sample collection over nearshore programs, lake-center surveys have attained a fair degree of success and proven economically viable (Nichol, 1975). For example, Davenport et al. (1975) were able to outline several Zn-rich zones overlying favorable limestone formations in western Newfoundland. Similarly, Mehrtens et al. (1973) found enhanced Mo levels in lake sediments downslope from molybdenite mineralization in the central interior of British Columbia.

Interpretation of lake sediment data, however, can present problems. Trace metal distributions obtained following analysis of organic-rich sediment may fail to recognize previously defined mineralization. Coker and Nichol (1975), extending the study of lake sediment geochemistry to the southern Canadian Shield of Ontario, were unable to confirm mineralized areas as anomalous unless Zn/Mn and Ni/Mn ratios were considered. This situation was attributed to the scavenging effect of hydrous Fe-Mn oxides for trace elements. Gleeson and Hornbrock (1975) encountered a different type of problem relating the metal content of lake sediment with mineralized ultramafic bodies. In their case, the absence of a lake sediment anomaly appeared to result from deep glacial overburden acting as a barrier to secondary dispersion. Basal till sampling, however, was effective.

Numerous workers suggest statistical techniques be used to minimize the influence of sediment composition on trace metal levels. For example, the metal content of lake sediment may be associated with organic matter (Shimp et al., 1971; Ruch et al., 1970; Schleicher and Kuhn, 1970; Collinson and Shimp, 1972; Thomas, 1973; Kennedy et al., 1971). Allan et al. (1973) and Dyck (1974) suggest regression analysis be employed to minimize variability attributable to organic matter scavenging. However, strongly positive correlations between organic matter and trace elements may reflect other factors. Water content associated with 1 gm of dry weight gyttja substantially increases the volume of the sample compared with inorganic sediment. This larger volume contains sediment particulates in a finely dispersed state and is more effective in adsorbing metals than the more compact sands. Further-

more, the greater volume of the sample also gives the appearance of holding a higher concentration of a sedimented metal precipitate. In either case, regression analysis erroneously attributes higher metal concentrations to some property of the organic component. Nevertheless, Davenport et al., (1975) and others have found regression analysis to be useful in enhancing contrast by reducing variability caused by limnological factors.

Studies of lake sediment geochemistry are still in their preliminary stages. Detailed sampling programs of individual lakes are necessary to establish the degree of trace element homogeneity within the lake basin, which in turn, will allow for definition of an optimum sampling density for reliable regional evaluation of drainage basin trace metal levels. Sediment compositional parameters must be examined so that conditions where metal accumulation reflects mineralized bedrock can be clearly distinguished from limnological enhancements.

Summary

Many reconnaissance lake sediment programs indicate that sediment metal levels identify the anomalous character of a catchment. Nevertheless, other cases exist where abnormal concentrations of metal in lakes cannot be attributed to metal-rich or mineralized bedrock. Conversely, examples may be cited where recognized mineralization was not reflected in nearby lakes. Equivocal interpretation of lake sediment data, therefore, represents a formidable problem. For some surveys, definition of anomalies may be resolved by employing statistical techniques such as regression analysis. Nevertheless, despite possible interpretive difficulties, lake sediment sampling in remote or treed areas provides a most attractive survey procedure for rapid regional mineral potential evaluation. Additional and detailed studies, however, are necessary to establish, first, if metal levels in lakes indeed reflect mineralized bedrock of the catchment and, second, assuming such a connection exists, the probability that the lake sediment geochemical survey will delimit the drainage basin as regionally anomalous.

Acknowledgments

The author is grateful to Rio Tinto Canadian Exploration Limited for field assistance that helped to confirm the importance of some of the lake parameters and to W. K. Fletcher, for his critical review of this paper. The research work was also supported by a National Research Council of Canada Centennial scholarship during 1972 and 1973.

STAN HOFFMAN

References

Allan, R. J., 1971, Lake sediment: A medium for regional exploration of the Canadian shield, *Canadian Inst. Mining and Metallurgy Bull.* **64**(715), 43–59.

Allan, R. J., E. M. Cameron, and C. C. Durham, 1973, Lake geochemistry—A low sample density technique for reconnaissance geochemical exploration and mapping of the Canadian Shield, in M. L. Jones, ed., *Geochemical Exploration 1972* (Proceedings of the Fourth International Geochemical Symposium). London: Institute of Mining and Metallurgy, 131–160.

Arnold, R. G., 1970, The concentrations of metals in lake waters and sediments of some Precambrian lakes in the Flin Flon and La Ronge areas, *Saskatchewan Research Council Geology Div. Circ.,* 430p.

Bachmann, R. W., and E. P. Odum, 1960, Uptake of Zn and primary productivity in marine benthic algae, *Limnology and Oceanography* **5**, 349–355.

Berner, R. A., 1969, Migration of iron and sulfur within anaerobic sediments during early diagenesis, *Am. Jour. Sci.* **267**, 19–42.

Brunskill, G. J., and S. D. Ludlam, 1969, Fayetteville Green Lake, New York, I. Physical and chemical limnology, *Limnol. Oceanogr.* **16**, 817–829.

Brunskill, G. J., and R. C. Harriss, 1969, Fayetteville Green Lake, New York. IV. Interstitial water chemistry of the sediments, *Limnology and Oceanography* **16**, 858–861.

Clark, D. B., and R. A. Bryson, 1959, An investigation of the circulation over second point bar, Lake Mendota, *Limnology and Oceanography* **4**, 140–144.

Coey, J. M. D., D. W. Schindler, and P. Weber, 1974, Iron compounds in lake sediments, *Canadian Jour. Earth Sci.* **11**, 1489–1493.

Coker, W. B., and I. Nichol, 1975, The relation of lake sediment geochemistry to mineralization in the northwest Ontario region of the Canadian Shield, *Econ. Geology* **70**, 202–218.

Collinson, C., and N. F. Shimp, 1972, Trace elements in bottom sediments from upper Peoria Lake, middle Illinois River, a pilot project, *Illinois Geol. Survey Environmental Geology Notes 56,* 21p.

Davenport, P. H., E. H. W. Hornbrook, and J. Butler, 1975, Regional lake sediment geochemical survey for zinc mineralization in western Newfoundland, in I. L. Elliott and W. K. Fletcher, eds., *Geochemical Exploration 1974,* Developments in Economic Geology, vol. 1. Amsterdam: Elsevier, 555–578.

Doyle, R. W., 1968, Identification and solubility of iron sulfide in anaerobic lake sediment, *Am. Jour. Sci.* **266**, 980–994.

Dyck, W. 1971, Lake sampling re stream sampling for regional surveys, Report of activities, November 1970 to March 1971, *Canada Geol. Survey Paper 71-1,* Pt. B, 70–71.

Dyck, W., 1974, Geochemical studies in the surficial environment of the Beaverlodge area, Saskatchewan, Part E. A area, Saskatchewan: Lake versus stream sampling, *Canada Geol. Survey Paper 74-32,* 21–30.

Frink, C. R., 1969, Chemical and mineralogical characteristics of eutrophic lake sediment, *SSSA Proc.* **33**, 369–372.

Garrett, R. G., and E. H. W. Hornbrook, 1976, The relationship between zinc and organic content in centre-lake bottom sediments, *Jour. Geochem. Exploration* **5**, 31–38.

Gleeson, C. F., and E. H. W. Hornbrook, 1975, Semi-regional geochemical studies demonstrating the effectiveness of till sampling at depth, in I. L. Elliott and W. K. Fletcher, eds., *Geochemical Exploration 1974,* Developments in Economic Geology, vol. 1. Amsterdam: Elsevier, 611–630.

Gorham, E., 1958, Observations on the formation and breakdown of the oxidized microzone at the mud surface of lakes, *Limnology and Oceanography* **3**, 291–298.

Gorham, E., J. W. G. Lund, J. E. Sanger, and W. E. Dean, Jr., 1974, Some relationships between algal standing crop, water chemistry, and sediment chemistry in the English Lakes, *Limnology and Oceanography* **19**, 601–617.

Gorham, E., and D. J. Swaine, 1965, The influence of oxidizing and reducing conditions upon the distribution of some elements in lake sediments, *Limnology and Oceanography* **100**, 268–279.

Hoffman, S. J., and W. K. Fletcher, 1972, Distribution of copper at the Dansey-Rayfield River property, south-central British Columbia, *Canadian Jour. Geochem. Exploration* **1**, 163–180.

Hoffman, S. J., and Fletcher, W. K., 1979, Extraction of Cu, Zn, Mo, Fe and Mn from soils and sediments using a sequential procedure, in J. R. Watterson and P. K. Theobald, eds., *Geochemical Exploration 1978.* Toronto: Association of Exploration Geochemists, 289–299.

Hoffman, S. J., and W. K. Fletcher, 1981a, Detailed lake sediment geochemistry of anomalous lakes on the Nechako Plateau, central British Columbia. A comparison of trace metal distributions in Capoose and Fish Lakes, *Jour. Geochem. Exploration* **14**, 221–244.

Hoffman, S. J., and W. K. Fletcher, 1981b, Organic matter scavenging of copper, zinc, molybdenum, iron and manganese, estimated by a sodium hypochlorite extraction (ph 9.5.), *Jour. Geochem. Exploration* **15**, 549–562.

Horne, R. A., and C. H. Woernle, 1972, Iron and manganese profiles in a coastal pond with an anexio zone, *Chem. Geology* **9**, 299–304.

Kennedy, E. J., R. R. Ruch, and N. F. Shimp, 1971, Distribution of mercury in unconsolidated sediments from southern Lake Michigan, *Illinois Geol. Survey Environmental Geology Notes 44,* 18p.

Kimball, K. D., 1974, Seasonal fluctuations of ionic copper in Knights Pond, Massachusetts, *Limnology and Oceanography* **18**, 169–172.

Knauer, G. A., and J. H. Martin, 1973, Seasonal variations of cadmium, copper, manganese, lead and zinc in water and phytoplankton in Monterey Bay, California, *Limnology and Oceanography* **18**, 597–604.

Ludlam, S. D., 1974, Fayetteville Green Lake, New York. The role of turbidity currents in lake sedimentation, *Limnology and Oceanography* **19**, 656–664.

Mackereth, F. J. H., 1966, Some chemical observations on postglacial lake sediments, *Royal Soc. London Philos. Trans.* (Series B) **250**, 165–213.

Matson, W. R., 1968, Trace metals, equilibrium and kinetics of trace metal complexes in natural media, Ph.D. dissertation, MIT, 258p.

Mehrtens, M. B., J. S. Tooms, and A. G. Troup, 1973, Some aspects of geochemical dispersion from base metal mineralization within glacial terrain in Norway, North Wales and British Columbia, Canada, in M. L.

Jones, ed., *Geochemical Exploration 1972* (Proceedings of the Fourth International Geochemical Symposium). London: Institute of Mining and Metallurgy, 105–116.

Mortimer, C. H., 1949, Underwater soils: A review of lake sediments, *Jour. Soil Sci.* 1, 63–73.

Mothersill, J. S., and P. C. Fung, 1972, The stratigraphy, mineralogy, and trace element concentrations of the Quaternary sediments of the northern Lake Superior basin, *Canadian Jour. Earth Sci.* 9, 1735–1755.

Mothersill, J. S. and R. J. Shegelski, 1973, The formation of iron and manganese-rich layers in the Holocene sediments of Thunder Bay, Lake Superior, *Canadian Jour. Earth Sci.* 10, 571–575.

Nichol, I., 1975, Promising future in store for lake sediment reconnaissance, *The Northern Miner,* March 6, 44–46.

Parker, M., and A. D. Hassler, 1969, Studies on the distribution of cobalt in lakes, *Limnology and Oceanography* 114, 229–242.

Rashid, M. A., 1972, Role of quinone groups in solubility and complexing of metals in sediments and soils, *Chem. Geology* 9, 241–248.

Rawson, D. S., 1960, A limnological comparison of twelve large lakes in northern Saskatchewan, *Limnology and Oceanography* 5, 195–211.

Ruch, R. R., E. J. Kennedy, and N. F. Shimp, 1970, Distribution of arsenic in unconsolidated sediments from southern Lake Michigan, *Illinois Geol. Survey Environmental Geology Notes 37,* 16p.

Sanger, J. E., and E. Gorham, 1970, The diversity of pigments in lake sediments and its ecological significance, *Limnology and Oceanography* 15, 59–69.

Schindler, J. E., J. J. Alberts, and K. R. Honick, 1972, A preliminary investigation of organic-inorganic associations in a stagnating system, *Limnology and Oceanography* 17, 952–957.

Schleicher, J. A., and J. K. Kuhn, 1970, Phosphorus content in unconsolidated sediments from southern Lake Michigan, *Illinois Geol. Survey Environmental Geology Notes 39,* 15p.

Schmidt, R. C., 1956, Adsorption of Cu, Pb and Zn on some common rock forming minerals and its effect on lake sediments, Ph.D. dissertation, McGill University, 181p.

Schoettle, M., and G. M. Friedman, 1973, Organic carbon in sediments of Lake George, New York: Relation to morphology of lake bottom, grain size of sediments, and man's activities, *Geol. Soc. America Bull.* 84, 191–198.

Shimp, N. F., J. A. Schleicher, R. R. Ruch, D. B. Heck, and H. V. Leland, 1971, Trace element and organic carbon accumulation in the most recent sediments of southern lake Michigan, *Illinois Geol. Survey Environmental Geology Notes 41,* 25p.

Spilsbury, W., and K. Fletcher, 1974, Application of regression analysis to interpretation of geochemical data from lake sediments in central British Columbia, *Canadian Jour. Earth Sci.* 11, 345–348.

Thomas, R. L., 1973, The distribution of mercury in the surficial sediments of Lake Huron, *Canadian Jour. Earth Sci.* 10, 194–204.

Thomas, R. L., A. L. Kemp, and C. F. M. Lewis, 1973, The surficial sediments of Lake Huron, *Canadian Jour. Earth Sci.* 10, 226–271.

Timperley, M. H., and R. J. Allan, 1974, The formation and detection of metal dispersion halos in organic lake sediments, *Jour. Geochem. Exploration* 3, 167–190.

Cross-references: *Hydrochemical Prospecting; Indicator Elements; Marine Exploration Geochemistry; Prospecting.* Vol. XIII: *Geochemistry, Applied.*

LAND CAPABILITY ANALYSIS

"To define and achieve good use of land may well be the most fundamental of all environmental objectives" (Council on Environmental Quality, 1974). Indeed, the appropriate use of land and resources is not only fundamental but also a complex and difficult problem. A multitude of factors interact, many of which are not scientific or technical but rather social, cultural, or legal concerns. Planning commissions, government agencies, and sometimes individuals are given the responsibility of compiling and evaluating data that will influence future *land use.* Because land use is most often a choice from a number of alternatives, it is imperative that recommendations and decisions are based on data that are properly evaluated and integrated. Of particular concern is the consideration of whether a certain area is capable of sustaining the use to which it is to be committed without incurring serious adverse effects.

Land capability can be ascertained through an evaluation of the physical aspects of the area. The first step in this process is to *inventory* and map elements such as topography, soils, geology, vegetation, and water resources. This step, although basic, is critical, and the data collected must adequately meet the intended purpose. Elaborate methods of analysis, computer or otherwise, and plans based on them provide little value if the data base is not adequate. Of course, the thoroughness of the inventory will depend on the goals of the project as well as on availability and limitations of time, personnel, equipment, and existing data. In conjunction with the inventory, it is necessary to determine potentially hazardous areas that may be sensitive to environmental disruption, as well as places of special scenic or historic significance.

Following collection and organization, the basic data are evaluated and compared prior to determining the capability of an area for various uses. The results of the capability analysis can then be combined with the suitability analysis of social, cultural, and legal factors to form recommendations for a comprehensive land-use plan.

Elements of Computer Capability Analysis

In the past, evaluation of basic physical data consisted of manually preparing overlays and composite maps or in some cases merely making a best judgment estimate of the relationships among variables. Because manual comparison and drafting is slow and tedious, more efficient and faster methods of manipulation have been sought. The computer provides such a method, being used to store, retrieve, and integrate the data and to produce the final products (see *Geostatistics*).

Computer approaches to capability analysis have been developed and refined. Although the specifics of programs may differ, the basic elements and procedures are similar (Fig. 1). One such system is the Ohio Capability Analysis Program (OCAP), developed by the Ohio Department of Natural Resources in cooperation with the Ohio State University Systems Research Group. Reference here is made to OCAP as a general example. Several other major systems are listed in the bibliography (Allen, 1973; Colorado Environmental Data System, 1973; Fabos and Caswell, 1977; IRIS, 1972; Schlesinger et al., 1979; Tomlinson et al., 1976).

OCAP consists basically of a computer storage file of physical data, a program to analyze the data, and a program to map the stored data and generate information from the analysis program.

TABLE 1. Basic Data Variables

Land use	Drift thickness
Political boundaries	Glacial geology
Sanitary sewer service areas	Surficial materials
Water service areas	Sand and gravel deposits
First-order watersheds	Depth to groundwater
Slope	Groundwater resources
Soil	Groundwater recharge areas
Land cover	Geologic conditions affecting solid-waste disposal
Surface water	
Elevation	
	Bedrock resources

Source: Modified from Dunn and Marshall (1974).

After inventorying the physical data, such as soils, geology, water, and vegetation, the information is placed into the computer storage bank. Because the basic physical data elements collected are usually presented on maps, it is necessary to convert the mapped data into computer format (Table 1).

Converting base maps into computer data requires a reference system to locate map positions. OCAP uses a series of numbered lines that are ov-

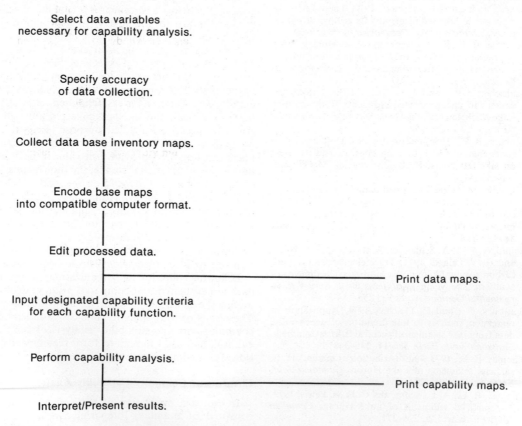

FIGURE 1. Steps in Ohio Capability Analysis Program. (Modified from Planning Services Section, 1974)

erlaid on the base maps. Other systems may use a grid for location or other advanced techniques. The conversion of the resource data is then carried out by use of a digitizing coordinatograph interfaced with a keypunch, which punches the data on computer cards. Each separate variable or unit shown on the map is assigned a unique computer symbol and combined with a distance code to indicate the length of line along which the area is found. When all the lines are digitized, the area covered by each variable is defined (Fig. 2).

The encoded data are edited and corrected, and at this point maps of the basic physical elements can be produced by the computer. The data are stored on tape or disk for later retrieval. The digitized data from each map are called a data set, and all data sets together make up the complete study area data file.

To employ the system for a capability analysis, it is first necessary to specify the area of interest, input and output map scales, legend information, and map symbolism. The basic input data sets,

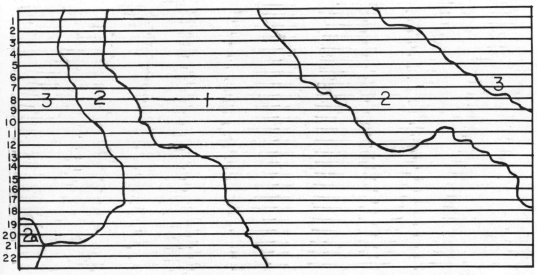

BASE SOILS MAP OVERLAID WITH LINE GRID

(1653 3)   (2603 1)   (4550 2)   (11000 1)

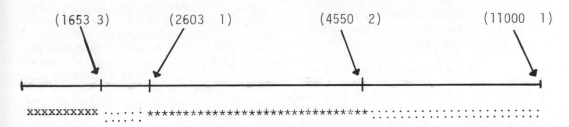

LINE 10 REDUCED AND PRINTED *

UNIQUE SOILS NUMBER	UNIQUE SOILS NUMBER
1	:
2	*
3	x

* Line 10 was reduced to numbered format, then printed in computer format.

FIGURE 2. Comparison of base map and digitized format. (Modified from Planning Services Section, 1974)

such as slope, depth to bedrock, permeability, etc., are chosen and combined to determine capability by using what are called functions. A *function* has been defined as "a measure of the capacity of a unit of land for a particular use" (Planning Services Section, 1974). Each of the basic variables used is assigned a weighting factor based on its considered importance to the particular analysis. Each category on each of the basic data maps is given a utility rating that indicates the degree of its effect in the analysis. For example, a 2–6% slope may be more restrictive than a 0–2% slope for many uses. The utility rating of each of the categories is multiplied by the proper weighting factor, and the resultant score of the sum of the products is assigned to each computer cell. Scores for each of the cells on the map are grouped into capability categories and the map printed by the computer. Figure 3 illustrates the development of a function and how the data from the basic maps result in *capability maps*.

Several approaches are generated to produce different types of maps. For example, the map may show those areas with severe, moderate, or slight limitations to residential development. Such a map shows that an area may have limitations but does not indicate which factors are most limiting. Another approach is to analyze the data from a problem point of view. An example would be organization of the input data to indicate where installation problems, such as excavation difficulties, or maintenance problems, such as wastewater disposal, are most limiting (see Fig. 4).

Problems of Computer Capability Analysis

The application of computers to capability analysis is not troublefree. Some problems are highly technical, involving the operation and programming of the equipment, while others concern the collection, evaluation, and use of the data. Three general problem areas exist. The first involves data collection and availability, the second data input to the computer system, and the third the integration and evaluation of the data variables.

In the first instance, the need for large amounts of specific and accurate data is not unique to computer-based systems. However, much of the basic data collected and mapped by soil scientists, geologists, biologists, and other specialists do not readily lend themselves to direct computer evaluation. A geologic map, e.g., commonly shows the distribution of rock types, their ages, and something about their physical characteristics. Very often those responsible for compiling and analyzing data from many sources are unable to assimilate the importance of such information. Even if relevance and importance are realized, data in this form are difficult to evaluate and enter into computer systems. It is often necessary, e.g., to know how the information and units outlined on a geologic map are related to bearing strengths, permeability, or ease of excavation. One way to overcome such problems is for the specialists conducting the inventory to supply maps of specific variables such as *rippability*. Another method is to develop a matrix of properties important to the particular type of analysis and to assign either absolute or relative values to each mapped unit. For example, depending on the detail of the basic information, it may be possible in one case to state that a unit has a specific bearing strength, while in another case permeability may be ranked as good, fair, or poor. Scientists collecting data for computer capability studies realize that traditional forms of presentation may have to be modified, and they must therefore continually look for ways to maximize use of their material by making it more understandable and applicable.

The second problem area concerns data capture, or data input to computer storage, usually the slowest part of the process after field data collection. In addition to the time problem the minimum size of the grid cell may in some systems be so large that significant errors are introduced to the maps (Ferris and Fabos, 1974). Automated line-following scanners (Schlesinger et al., 1979) and image digitizers are employed to reduce data capture time and improve accuracy. Considerable progress has been made in all aspects of computer mapping and personal or mini computers are now also used (Carter, 1984).

Finally, the data must be properly integrated and evaluated. The computer specialists responsible for integrating the data and conducting the analysis must fully understand which variables should be combined and how they should be weighted. This procedure requires close cooperation with the collecting scientists and field personnel. A false sense of accuracy may be imparted to a study through the use of a computer, however, and plans based on poorly designed programs or on the analysis of generalized data can lead to misuse of land and resources.

Benefits of Computer Capability Analysis

The great worth of computer capability analysis is that a large number of variables can be handled and compared rapidly and efficiently. In addition, maps can be produced mechanically without time-consuming hand drafting.

At the present time most analysis programs are not intended to be site specific; i.e., they should not be used as a substitute for on-site tests and evaluations. They indicate to users areas that are most capable or most limited for particular uses, areas that should be preserved, or areas where

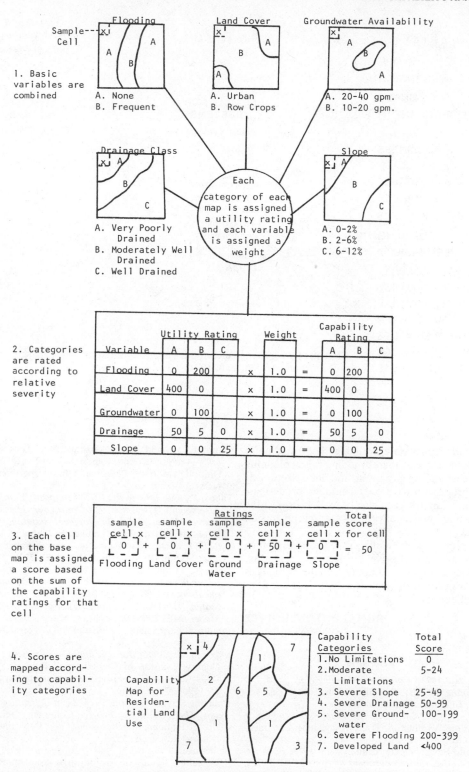

FIGURE 3. Development of a function for residential land use. (Modified from Maxson, 1975)

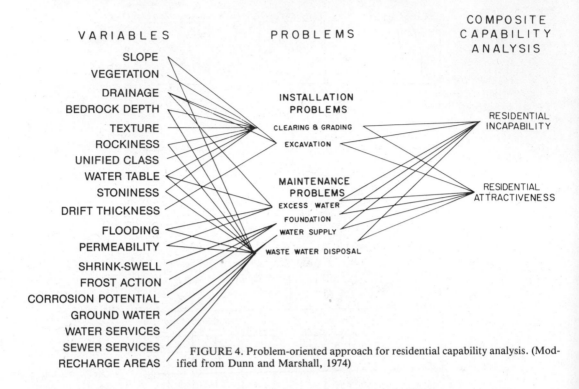

VARIABLES

SLOPE
VEGETATION
DRAINAGE
BEDROCK DEPTH
TEXTURE
ROCKINESS
UNIFIED CLASS
WATER TABLE
STONINESS
DRIFT THICKNESS
FLOODING
PERMEABILITY
SHRINK-SWELL
FROST ACTION
CORROSION POTENTIAL
GROUND WATER
WATER SERVICES
SEWER SERVICES
RECHARGE AREAS

PROBLEMS

INSTALLATION PROBLEMS
CLEARING & GRADING
EXCAVATION

MAINTENANCE PROBLEMS
EXCESS WATER
FOUNDATION
WATER SUPPLY
WASTE WATER DISPOSAL

COMPOSITE CAPABILITY ANALYSIS

RESIDENTIAL INCAPABILITY

RESIDENTIAL ATTRACTIVENESS

FIGURE 4. Problem-oriented approach for residential capability analysis. (Modified from Dunn and Marshall, 1974)

economic deposits are available. They also point out problem areas, thereby aiding in the planning of further testing programs and the selection of appropriate construction techniques or designs. Capability analysis can also be used to predict the effects and impacts of proposed developments on an area.

In spite of the difficulties involved in establishing systems and effectively collecting, presenting, and evaluating data, computer capability analysis holds promise and should find continued and increased application.

RONALD D. STIEGLITZ

References

Allen, N., 1973, *Computer Mapping for Land Use Planning: COMLUP.* Ogden, Utah: U.S. Forest Service, Intermountain Region, 140p.

Carter, J. R., 1984, *Computer Mapping: Progress in the'80's.* Washington, D.C.: Association of American Geographers, 86p.

Colorado Environmental Data System, 1973, *Final Report to Colorado Department of Natural Resources,* Fort Collins, Colo.: Colorado State University, 442p.

Council on Environmental Quality, 1974, *Fifth Annual Report.* Washington, D.C.: U.S. Government Printing Office.

Dunn, T., Jr., and D. C. Marshall, 1974, *Land Capability Analysis* (County Report No. 1). Lake County, Ohio: Division of Planning, Ohio Department of Natural Resources.

Fabos, J., and S. Caswell, 1977, *Composite Landscape Assessment: Assessment Procedures for Special Resources, Hazards and Development Suitability:* Part II of the Metropolitan Landscape Planning Model (METLAND). Massachusetts Agricultural Experiment Station (Research Bull. No. 637), 323p.

Ferris, K. H., and J. Fabos, 1974, *The Utility of Computers in Landscape Planning: The Selection and Application of a Computer Mapping and Assessment System for the Metropolitan Landscape Planning Model (METLAND).* Amherst: Massachusetts Agricultural Experiment Station and USDA Forest Service (Publ. No. 617), 116p.

Maxson, G., 1975, *Land Capability Analysis,* Stark County: Division of Planning and Research, Ohio Department of Natural Resources (County Rept. No. 3), 119p.

Planning Services Section, 1974, *Land Capability Analysis, The Wolfe Creek Project.* Columbus, Ohio: Office of Planning and Research, Ohio Department of Natural Resources, 55p.

Schlesinger, J., B. Ripple, and T. Loveland, 1979, Land capability studies of the South Dakota automated geographic information system (AGIS), in *Computer Mapping in Natural Resources and Environment.* Cambridge: Harvard Library of Computer Graphics, 1979 Mapping Collection, vol. 4, 105–114.

Tomlinson, R. F., H. W. Calkins, and D. F. Marble, 1976, *Computer Handling of Geographical Data.* Paris: The Unesco Press, 214p.

Cross-references: *Maps, Environmental Geology; Terrain Evaluation Systems.*

LAND CLASSIFICATION—See TERRAIN EVALUATION SYSTEMS. Vol. XIII: TERRAIN EVALUATION, MILITARY PURPOSES.

LANDSLIDE CONTROL

Undesirable movements of soil or rock materials are by and large a reaction of these materials to gravity, sometimes supplemented by seismic activity. The study of the form of actual slides is the first step in their control and prevention. Slide geometries and mechanisms are controlled by many factors, of which geology, hydrology, and topography are the more important. Given a sufficiently low failure strength, a slide may occur on almost any slope, however shallow. The only requirement is that the failure strength be reduced to a value that can be overcome by gravitational force. The form of the slide can give an indication of how weakening takes place.

Most slides, either in rock or soil, can be identified as one or a combination of three basic types. These are falls, rigid body movements, and flows. Each has a characteristic mechanism of movement and final shape.

Falls take place where there is little or no lateral support on a very steep or overhanging slope. *Ravelling* is probably the most common form, consisting of intermittent rock falls on a small scale. This type is most troublesome along road cuts. Less commonly, an entire section of a cliff will collapse with little or no frictional sliding. Rates of movement are usually high as free fall takes place for some portion of the total displacement. The result of such a slide is a loose pile of rubble at the base of the slope.

Rigid body movements and flow are actually the end members of a spectrum of slide behavior. On the one extreme, *rigid body movements* involve sliding of a coherent rock or soil body along some sort of failure surface or zone, either curved or planar. Depending on the shape of the failure surface, the motion is rotational or translational. *Rotational slides* are most common in soils where the failure surface is free to form in such a manner as to allow the least amount of work to be done to accomplish the motion. *Translational slides* are most common in rock where some

planar zone of weakness restricts movement of the whole slide mass to one direction.

On the other extreme, *flows* involve a considerable decrease in cohesion throughout the slide mass, allowing large amounts of internal deformation. This type of behavior is usually associated with a high water content. A relatively distinct failure surface or boundary between flowing and nonflowing material usually exists, but the slide mass behaves much as a liquid. A rock mass may also flow but not due to a high water content. At the instant sliding begins, a rock mass may move as a single body. It will, however, very quickly break up along other planes of weakness within the mass. Thus, many parts of the slide are moving independently, and from a distance it may appear to behave as a fluid.

Most slides in soil involve some flowage. The motion of the upper part as it moves steeply downward may bear a fair approximation to rigid body movement, but the lower part will usually move by flowage after it is forced beyond the initial position of the base of the slide. Water held between soil particles of clay and silt size has a profound influence on soil deformation because of the small size of particles compared to the spaces between them. In a moist soil, failure often involves a gradual loss of cohesion because of the viscous nature of the soil particle–water interaction. In rock, failure involves a large reduction in resistance to movement and is usually sudden. Soil may behave this way if it is relatively dry or composed of sand and coarser particles.

In a soil, other variables such as chemical composition, particle shape, and particle fabric also have a considerable influence on how a soil will deform under a given stress. It would be most desirable to be able to predict the position of failure surfaces based on the stress field existing in a soil mass. To determine a stress field, it is necessary to know the stress-strain relation for a given material. Results from the laboratory testing of soils, however, do not closely fit any of the idealized deformation models that relate stress to strain, probably because of the complex physical nature of soil materials. In fact, the estimation of slope stability is made possible by avoiding the problem of soil deformation and considering instead only the state of a soil mass at failure.

Calculating slope stability without knowledge of the stress conditions throughout the soil requires that a failure surface be assumed. Once a potential failure surface has been selected, the actual stresses operating along this hypothetical surface can be resolved from the soil overburden load. The shear stress can then be compared to the shear strength of the soil (see Vol. XIII: *Soil Mechanics*). The stability then becomes a measure of the amount by which the shear stress that tends to cause failure is exceeded by the shear stress required just to maintain the slope in equilibrium—i.e., the failure strength. When the two are equal, a state of limiting equilibrium or incipient failure exists. The ratio between the two is called the *factor of safety*. In the design of slopes, a factor of safety well above a value of one is desirable to remove all doubt concerning margins of error. Many methods of calculating factors of safety exist, most of them based on the limiting equilibrium state (Janbu, 1973; Lambe and Whitman, 1969).

A stability analysis (see *Slope Stability*) may be made of a certain slope and the factor of safety found to be tolerable above one. With time, however, the following factors may operate to increase the stresses conducive to failure and/or decrease the shearing resistance of the soil. Landslide *prevention* may be possible by predicting the occurrence of these factors and taking steps to eliminate them. Landslide *control* may be possible by identifying the causative factors and similarly eliminating them.

Factors Leading to an Increase in Shear Stress

Removal of Lateral or Underlying Support In nature, underlying support is most commonly removed by running water or by waves undercutting slopes. Depending on the shape and steepness of a slope, undercutting will result in the removal of material that is acting to oppose failure. The washing out of granular material may take place if a high groundwater table is present adjacent to a slope. Seepage erosion at the site of groundwater discharge will have the same effect as undercutting. Artificial cuts are formed very rapidly and require special design measures. Similarly, the lowering of the water level in a lake or reservoir requires careful control. In this instance a rapid water level decrease results in a removal of support.

Increased Loading An increase in the weight on the upper portion of a potential failure surface will result if a load or surcharge is placed at the top of a slope. The consequent decrease in stability will depend on what proportion of the added weight is felt as shear along the potential failure surface. Such loading can result from natural processes such as accumulations of water or snow. Material transported downhill from a failure on the upper part of a slope can also accomplish loading. Artificial pressures consist of ore stock piles, mine tailings, waste dumps and buildings. Internal loading takes place when irrigation, rain, or runoff water is absorbed into the slope material.

Transitory Earth Stress These stresses are vibratory in character. Earthquakes are the most effective type. They cause lateral accelerations that can act to increase the shear stress along a poten-

tial failure surface. Vibrations from passing traffic or construction operations may also be a consideration.

Factors Leading to a Decrease in Shearing Resistance

Pore-Water Pressure Increase This is one of the most common causes of instability. The presence of a hydrostatic head produces a reduction in the effective normal stress acting across a given potential failure surface. Rainfall is the most common source of water, and permeable or fissured soils are most conducive to its collection. Rapid drawdown of reservoirs will result in an excess of hydrostatic head if the water table does not lower at the same rate as the reservoir water level.

Loosely packed, saturated soils may densify if subjected to cyclic loading, with a consequent increase in pore-water pressure. If the pore pressure increases to equal the overburden pressure, the shear strength of the soil decreases to that of water, and liquefaction (see Vol. XII, Pt. 1: *Thixotropy, Thiotropism*) takes place. Much laboratory and field evidence exists for the liquefaction of sands and silts (Seed, 1973), but there is little about the behavior of clays.

Material Changes Changes in the water content may cause a decrease in shear strength due to chemical and/or physical processes. The presence of water may lead to the weathering of clays as bonds between particles are weakened. Changes in the fabric of sands and silts to produce a denser packing may cause liquefaction if enclosed pore water cannot immediately escape (see *Soil Fabric*). Contraction accompanying the drying of soil causes fissuring, which in turn allows water greater freedom of access. Of course fissuring also weakens a soil.

Soil Movement Creep, a very slow movement in soil throughout a given horizon, may take place continuously. The rate is controlled by rainfall and freeze-thaw cycles. While creep represents a steady-state response of soil to gravitational loading, it may develop into a landslide. Changes in clay fabric during creep may cause a decrease in shear strength great enough to allow failure on a discreet surface. In actuality, failure surfaces are usually thin zones of disturbed material. Movement along the zone produces the disturbance that will progressively weaken the zone material. This weakening will probably reduce the factor of safety for recurrent movement (Skempton, 1964). It is important to find this zone if a slope has already failed so that the zone's strength can be taken into account.

The same principles of shear stress increase and shear strength decrease apply to rock slopes. Establishing a potential failure surface is, however, much more difficult. Rock is far too strong to fail internally from gravity-produced stresses of the magnitude available near the surface. Failure will probably take place along a preexisting surface or zone of weakness. These zones include fractures, faults, bedding planes, foliation in metamorphic rocks, and brecciated zones. To determine the stability of a rock mass, these features must be located. Unfortunately many usually exist and determining their strength is difficult.

The design of slopes is usually aided by stability calculations. The conditions of soil strength and pore pressure, however, may vary once the slope is constructed, thereby changing the stability. Because of these changes it may be necessary to consider both long-term and short-term stability. Immediately on completion of construction, the normally consolidated soil (i.e., soil that has not been previously buried to a greater depth than at present) of an excavated slope will be adapting to a new stress condition. Lateral pressure will have been removed, allowing the slope to compact. As this takes place, cracks and pores tend to close, producing an increase in pore pressure if the soil is partly or fully saturated and of low permeability. In fact, as the soil creeps outward, pore pressure may rise to equal the overburden pressure. As this condition is approached, the strength of the soil decreases until ultimately it is equal only to the soil cohesion. Such a condition is very conducive to failure and is usually quickly reached—i.e., short term. The undrained direct shear and triaxial tests simulate this condition, known as the $0 = 0$ condition (Lambe and Whitman, 1969).

If the soil is *overconsolidated*, (i.e., has been buried to a greater depth than at present) excavation may cause an expansion. This results in the opening of fissures and pores with a consequent increase in permeability. Pore pressures will increase if these openings fill with water but will probably not become abnormal (see Vol. XIII: *Soil Mechanics*). The swelling of the soil will likely be a long-term process, so the admittance of water will be gradual. Shear strength will also probably decrease as swelling and water content increase. If sliding occurs during the swelling process, it is classified as an intermediate failure; if it occurs at or near the end of the swelling period, it is a long-term failure. Stability analysis for these types of failures should utilize drained tests (i.e., at a displacement rate slow enough to allow the constant dissipation of pore-water pressures).

Landslide Prevention and Control

Stability calculations are a form of landslide control in that they give some indication of potential hazard. Anticipated strength or load changes can be taken into account and slopes designed accordingly. Unforeseen changes resulting in landsliding will require corrective measures, however. Although the determination of slope stability requires thorough laboratory testing of soil and rock materials, some idea of whether a slope may fail can be gained simply by visual observation. On soil slopes, bulging lobes near the base of soft or very wet ground indicate instability. Soil creep can be detected by looking for inclined tree trunks, sloping fence posts, or displaced retaining walls. A rapidly moving slide mass will probably re-establish stability through its own momentum, although the removal of support for material farther up the hill must be considered. Slowly moving masses are probably controlled by water content and will continue moving as long as water is available.

Slope instability in rock is much more difficult to detect because few features indicate incipient failure. Checking the tops of rock slopes may prove worthwhile because cracks often open up, although weeks or months may pass before failure occurs, if ever. Installation of movement-detecting devices will signal increases in the rate of movement that may indicate impending failure. Calculation of a factor of safety for a rock slope is very difficult because failure usually takes place along fractures or other surfaces of weakness that are difficult to locate and test for strength.

With an understanding of why landslides occur, measures can be decided on to enable the safe utilization of slopes. Such measures, however, must be applied according to the type of material present. Two general approaches are available: (1) either the strength of a slope can be increased or (2) the gravitational potential reduced. Brute force methods, consisting simply of removing a slope or reducing its angle, are simplest; however, these methods may be unnecessarily expensive. In the case of rock, it can be very expensive because blasting is usually required (see *Blasting and Related Technology*). Special precautions may be necessary in restricting the area to which blasted rock is thrown. If it is decided that removal of a rock slope is impractical, methods are available for strengthening a rock mass in place. Rock bolts or cables depend on the strength of steel in tension to bind the outer portion of a rock mass to firm rock in the interior (see Vol. XIII: *Rock Structure Monitoring*). In excavations, rock bolts may be used for artificially restoring some of the overburden pressure, thereby decreasing strain release and crack opening. The bolts are inserted through drilled holes and cemented in place. Their life depends on what precautions have been taken to protect them from the corrosive effects of water. Spraying of cement or grouting on rock slope surfaces is a more economic alternative but tends to be most effective for preventing intermittent rock fall (ravelling). Wire nets can be spread over slopes to contain falling rocks, or fences may be erected at the base to catch fallen rocks.

Soil slopes are usually dealt with differently because of the relative ease with which their profile can be modified. The stability of a soil slope may be increased substantially by reducing the soil load at the head or increasing it at the toe. Either action opposes the tendency of a soil mass to move along a curved potential failure surface. It is usually least expensive to deposit soil removed from the head of a slope at the base. In general, because of a soil's lower maximum strength, slope angles in soil must be lower than those for rock.

The utilization of a proper slope angle in soils may not be compatible with certain construction requirements. Where a steep slope is necessary, such as for a railway or road cut, a retaining wall will be required to maintain it. This increases the stability of a steep soil slope by opposing downward and outward movement. *Retaining walls* are usually made of concrete or corrugated steel sheets but can be very costly because of design and maintenance requirements. *Piles* are used as a less expensive alternative but are commonly found to be ineffective. Movement of soil masses tends to overturn piles, or soil may simply flow between them.

Besides simply resisting slope movement, other methods may be applied that counteract indirect causes of slides. The most important of these is *slope drainage*. Water contained in a slope can, as described previously, contribute to instability. Drainage is accomplished either by the installation of underground drains to remove water from the subsurface or ditches dug to carry the water away before it has a chance to seep in or both. Seepage erosion can be prevented by constructing sand or gravel filters at the site of the seepage. These filters tend to relieve the concentration of seepage as it leaves the slope. Incipient slides commonly develop cracks at their heads, which provide easy entry for water. Such features should be covered or filled in. Vegetation may also assist in drying a slope.

An improvement in the reliability of slope stability calculation techniques may be possible if the significant mechanical properties of slope materials can be determined. In most engineering analyses the most significant inputs are the strength properties determined from very small samples of the slope mass. In extrapolating these measurements to the field, it is assumed that a complete potential slide will behave essentially as one of the laboratory specimens; i.e., movement must take place simultaneously at all points on the predicted slide surface. Rigidity need not necessarily be assumed as vertical shearing tractions in the slide mass can be taken into consideration with the Morgenstern–Price (1965) method. The method

by which these tractions are considered, however, is rather arbitrary in that a function describing the lateral variation of this shearing is randomly picked. The function that yields the lowest factor of safety is used.

These procedures would probably give the most reliable results in a slope of homogeneous isotropic material. There, the same mechanical properties throughout a slide mass that results in a smooth stress distribution would be conducive to a smooth failure surface. However, slide masses are rarely so ideal, and the morphology of the failure surfaces may be irregular if variations in material properties exist laterally or horizontally throughout a slope. Detecting changes in mechanical properties will be aided by considering the geology of the slope in question. It has been demonstrated that geology can control slide susceptibility on a regional scale—i.e., slides confined to given geologic units. Conversely, it remains to be found how the geology on a relatively small scale may control failure. By taking local geologic conditions into account, departures from idealized conditions assumed in stability analyses may be detectable. Furthermore, the appropriateness of existing analysis techniques to geologic conditions could be determined.

Geologic conditions in slopes composed of unindurated materials may determine failure mechanisms in several ways:

1. Variation may be present in failure strength characteristics of different strata. This variation is taken into account by conducting standardized strength tests on the different materials. However, the angle of the failure surface to the maximum stress direction is not incorporated in estimating the shape of the assumed failure plane.

2. Stability calculations on previously failed slopes take into account the strength of material incorporated in a failure zone but do not account for changes in the shape of the original failure surface. Large movements along the failure zone may result in its cross-sectional length being reduced as irregularities are smoothed. Decreasing the irregularity of the failure zone would decrease the amount of work needed for movement to occur along that zone. An understanding of mechanisms by which a slide might smooth its failure zone and the reaction of various materials to these mechanisms would be necessary.

3. The total material involved in a slide may not fail at once. Multiple failure may occur where strata of differing mechanical properties are present. The failure of one portion of a slope may render another portion unstable depending on whether or not strata of differing mechanical properties are present. The failure of one portion of a slope may render another portion unstable depending on whether another portion becomes unsupported or loaded. Strengths of individual

strata relative to the weakest strata might provide an estimate of susceptibility to multiple sliding. Pore-water conditions would have to be taken into account because they would probably be altered when an initial failure occurred.

These possible departures from idealized slide mechanisms may require that stability analyses be modified to take them into account. The geology of any potential slope failure site would influence the determination of where these possible modifications to stability analyses would be needed. To determine the validity of these modifications, a three-stage approach would be required:

1. Case histories would be examined to determine if geology were considered in the analysis as input to the stability calculation technique. Geologic information should be used to judge what raw data are needed to make the calculation as reliable as possible.
2. Case histories with enough implicit geologic information available in the raw data collection would be modified to incorporate the previously mentioned mechanisms. This modification would allow an estimate of the importance of these mechanisms by the way their incorporation changed the factors of safety.
3. A field study of failed slopes would be necessary in the modification of stability calculations. The study of case histories may reveal where data relating to geology are lacking. The geology of slide localities would be determined and the stability determinations made by gathering data that seem pertinent from the geologic standpoint.

LARRY D. DYKE
CHRISTOPHER C. MATHEWSON

References

Brawner, C. O., and V. Milligan, 1971, *Stability in Open Pit Mining,* Proceedings of the First International Conference of Stability in Open Pit Mining, American Institute of Mining, Metallurgy and Petroleum Engineering, 242p.

Coates, D. R., 1970, *Rock Mechanics Principles.* Ottawa, Ont.: Department of Energy, Mines, and Resources (Mines Branch Monograph 874), 6.1–6.27.

Coates, D. R., 1977, *Landslides.* Boulder, Colo.: Geological Society of America, 278p.

Cooke, R. U., and J. C. Doornkamp, 1974, *Geomorphology in Environmental Management.* Clarendon Press, 413p.

Eckel, E. G., ed., 1958, *Landslides and Engineering Practice.* Washington, D.C.: Committee on Landslide Investigations (Highway Research Board Spec. Rept. 29, NAS-NRC Pub. 544), 232p.

Hoek, E., and J. W. Bray, 1977, *Rock Slope Engineering,* 2nd ed. Hertford, England: Stephen Austin and Sons Limited, 402p.

Janbu, N., 1973, Slope stability computations, in R. C. Hirschfeld and S. J. Poulos, eds., *Embankment-Dam Engineering.* New York: John Wiley & Sons, 47–86.

Lambe, I. W., and R. V. Whitman, 1969, *Soil Mechanics.* New York: John Wiley & Sons, 352–373.

Lo, K. Y., 1965, Stability of slopes in anisotropic soils, *Am. Soc. Civil Engineers Jour. Soil Mechanics and Found. Div.* **91,** 85–106.

Morgenstern, N. R., and V. E. Price, 1965, The analysis of the stability of general slip surfaces, *Géotechnique* **15,** 79–93.

Peck, R. B., 1967, Stability of natural slopes, *Am. Soc. Civil Engineers Jour. Soil Mechanics and Found. Div.* **93,** 403–417.

Schuster, R. L., and R. J. Krizek, eds., 1978, *Landslides—Analysis and Control.* Washington, D.C.: National Academy of Sciences (Transportation Research Board Special Report 176), 234p.

Seed, H. B., 1973, Landslides caused by soil liquefaction, in National Academy of Sciences, *The Great Alaska Earthquake of 1964* (Engineering Vol.). Washington, D.C.: Committee of the Alaska Earthquake of the Division of Earth Sciences, 73–119.

Skempton, A. W., 1964, Long-term stability of slopes, *Géotechnique* **14,** 77–102.

Skempton, A. W., and J. N. Hutchinson, 1969, Stability of natural slopes and embankment foundations, *7th Internat. Conf. on Soil Mechanics and Found. Eng.,* 261–340.

Terzaghi, K., 1950, Mechanisms of landslides, in *Application of Geology to Engineering Practice.* Boulder, Colo.: Geological Society of America (Berkey Vol.), 83–125.

Zaruba, Q., and V. Mencl, 1969, *Landslides and Their Control.* Amsterdam: Elsevier, 205p.

Cross-references: *Blasting and Related Technology; Soil Fabric.* Vol. XIII: *Consolidation, Soil; Geomorphology, Applied; Rheology, Soil and Rock; Rock Mechanics; Rock Slope Engineering; Rocks, Engineering Properties; Rock Structure Monitoring; Soil Mechanics.*

LAND SUBSIDENCE—See MINE SUBSIDENCE CONTROL; MINING PREPLANNING.

LEGAL AFFAIRS

The role of geology in legal affairs has grown since the 1960s in particular. This rapid expansion has been coincident with the new wave of environmentalism that started at that time. During this period environmental law has flourished, and numerous statutes have been passed in the United States that give legitimacy to the movement. These new legal mandates greatly expand the former power base of earlier governmental acts, especially in matters relating to pollution and waste materials. The involvement of geologists with environmental law runs the entire gamut of land, water, and air. In some instances geologists help write and develop the law. In other cases they provide studies that are mandated by law, assist in monitoring processes, and provide reports and courtroom testimony in environmental lawsuits.

Historical Roots

Water was the mainspring for many early civilizations, and rules governing it were common. The *Code of Hammurabi,* ca. 2300 B.C., provided strict regulations in its use. In A.D. 534 Justinian laid the foundation of Roman common law and set the stage for the principle of riparian rights (that those who own property along the banks of water bodies have certain vested rights in the water). A royal edict was decreed in A.D. 1306 for England when air pollution became a serious problem. It curtailed the use of coal in London, and violation was punishable by death. In 1536 England passed the Public Nuisance Law, which set a precedent for this type of environmental damage. In France the Napoleonic Code of 1804 provided property owners with rights to the minerals and waters on or under their land. Thus, it became the basis for mining law and also those groundwaters that are not considered riparian type.

In the United States environmental law stems from a blend of English common law, the Constitution with the Bill of Rights, and detailed statutes that regulate, set standards, or provide management policy about specific parts of the environment. To correct a perceived injustice in environmental matters, the usual lawsuits prior to the 1960s were based on principles that involved nuisance, negligence, or trespass. Since the 1960s, however, large numbers of cases stem from owner's rights under the theory of eminent domain or the trust doctrine.

Much of the new land-use legislation stems from the governmental ethic of public trust: that certain lands and waters are retained in the public domain for the use and enjoyment of everyone. Such principles were first established with the beginning of the national park system with Yellowstone Park in 1872. However, a somewhat paradoxical twist was also promulgated as the principle of sovereign immunity, which pervaded the courtroom atmosphere into the mid 1900s. This concept held that governmental agencies were largely immune from lawsuits by citizens. An additional impediment to bringing a suit against the government was the issue of *standing.* Thus, a person or group that wants to sue must be specially damaged by some action related to government jurisdiction. Even in a class action suit the plaintiffs must prove their members will personally suffer damages. Michigan, in 1970, was the first state to recognize that individuals could have standing to sue the government on environmental grounds. Also, prior to 1970 the U.S. Supreme Court invariably heard environmental lawsuits wherein damages were in terms of only money or health impairment. Since 1970, however, they have extended the scope of possible lawsuit hearings to include aesthetic, conservation, and recreational values.

Environmental statutes are intended to inform and reform the way people treat the environment. In most cases when rules of conduct need to be altered relevant to environmental matters, some type of governmental action is mandated. This may involve the issuing of licenses or permits, regulating new construction, adjudicating disputes, and enforcing mandated decisions.

Laws for Geologists

Geologists become involved in a wide range of environmental law, and even some statutes are written that specifically require geologists be placed into the decision-making process (Kiersch and Cleaves, 1969). In *Pasadena* v. *Alhambra* (1949) the Supreme Court of California stipulated that *hydrologists* were to set policy to determine water allocation for the region. The southern California grading ordinances, such as the *Los Angeles Grading Ordinances* of 1952, 1963, and 1969, require that *engineering geologists* be actively engaged in decisions relating to hillside development. The *Pennsylvania Solid Waste Management Act* of 1969 mandates geological inspection, analysis, and approval of sites that are proposed for burial of solid refuse. Nuclear Regulatory Commission rules concerning siting of nuclear power plants requires the involvement of geologists throughout the entire period of preliminary to final investigations of the site (see Vol. XIII: *Nuclear Plant Siting, Offshore*). The 1972 guidelines issued by the New York State Department of Transportation, titled "Requirements for Geologic Source Reports," is another example whereby governmental agencies extend their powers as administrative overseers of environmental affairs with codified regulations for adherence by users. Thus, in New York, before a mining company can qualify to use mineral aggregates on state highways, it must file and receive approval of a special report that must be written by a certified geologist.

Mining Law The *Napoleonic Code* of 1804 has dominated rock, mineral, and metal rights in the Western world. Nineteenth-century monarchies, however, excluded gold and silver, which was reserved for the Crown as a regalian right. In the United States the California gold rush of 1849 helped establish new local mining laws that later formed the basis for common law. These practices became incorporated into federal statutes in 1866 and then into the *Mining Act* of 1872. Principles were adopted that allowed that discovery and working gave the right to a mineral and that the discoverer had the right to continue mining a vein even if it continued under the surface claim of another person.

The *Federal Mineral Leasing acts* of 1917 and 1920 removed coal, petroleum, oil shale, potash, phosphate, and sodium salts from mining controls. The *Multiple Mineral Development Law* of 1954 attempted to resolve differences between the mining laws and the leasing laws. This law led to the *Multiple Use Act* of 1955 that stipulated that underground mining was subordinate to surface mining of lands that contained proved and marketable sand, stone, gravel, pumice, pumicite, cinders, and clay.

Because of oil and natural gas mobility, laws dealing with petroleum differ from laws dealing with fixed minerals. Courts have held that fluid hydrocarbons are minerals, but their mobility prevents them from being the absolute property of the landowner until they are extracted and put in his or her possession (*Funk* v. *Holdeman,* 1866).

Surface-Water Law   In the United States two different theories govern ownership rights on rivers: the *riparian doctrine* and the *prior appropriation doctrine.* Under riparian law, water-use privilege accompanies ownership of land adjacent to the watercourse. Such a right exists even though the water may not be physically used by the property owner. When water is used, however, it must be reasonable and beneficial to the owner. Typical lawsuits establishing these principles were *Colburn* v. *Richards* (1816) and *Merritt* v. *Brinkerhoff* (1820). Another important precedent was set by *Strobel* v. *Kerr Salt Co.* (1900) whereby the relative economic merits and investment value of property owners should not be a consideration in determining rights. Thus, large industries must consider the rights of smaller parties. The riparian system of water law is especially used in climates producing abundant water, as in eastern United States. It is not practical, however, in dry climates where water is scarce and rivers do not flow throughout the year, as in much of western United States.

In the more arid parts of the West, prior appropriation establishes first rights to water. Here, land ownership is viewed as irrelevant, and instead, water rights are acquired by those who first use the water for some beneficial purpose. The classic case of *Tartar* v. *Spring Creek Water* (1855) helped legitimize the principle that priority in time determined the order for water use when it is insufficient in amount to satisfy all who wish to use it. The *Desert Land Act* of 1877 decreed such

rights to those states for nonnavigable waters in the public domain.

Groundwater Law   Less is known about the movement and occurrence of subsurface water than of surface water, so laws are in greater conflict when controversies arise over groundwater. The usual decision is to regard percolating waters as part of the property, thus subject to ownership in accordance with Napoleonic Code. *Acton* v. *Blundell* (1843) provided the court precedent in the United States. If water is judged to flow in underground streams, however, the courts have usually ruled that riparian doctrine holds, so that the landowner above such waters must share its benefits with others. With the ever-increasing demands for groundwater in arid regions, other approaches for water use have been made. In the premier case of *Pasadena* v. *Alhambra* (1949) the court formulated a new doctrine of mutual prescription whereby original water users, along with later users, were provided correlative rights on the basis of an apportionment formula devised by the state.

Land-Use Law

Many laws that govern land utilization have been passed on all government levels. Such legislation is aimed at assigning land-use priorities or for providing health, safety, welfare, and sustenance of citizens.

Water and Soils   Prior to 1849 flood control had been considered only a local problem, but the *Swamp Land acts* of 1849 and 1850 recognized the severity of flooding. These acts set the stage for statewide planning of flood-prone areas and led to the establishment of the *Mississippi River Commission* in 1879 whose mandated duty was to improve navigation and "prevent destructive floods." The *Flood Control Act* of 1936 and the *Soil Conservation Act* of 1935 put the federal government in the business of large-scale land managers. The Department of Agriculture (through the Soil Conservation Service) was to assume responsibility for upstream flood control measures, and the Army Corps of Engineers was authorized to undertake projects on navigable rivers throughout the United States. The scope of the Department of Agriculture was enlarged in 1954 with passage of the *Watershed Protection and Flood Prevention Act,* which allowed this agency to construct flood prevention structures that would impound up to 4.9×10^6 m^3 of water if costs did not exceed \$250,000.

Another type of breakthrough in environmental land law occurred in 1902 with the *Reclamation Act* which established irrigation in the West as a national policy for implementation. Although lawsuits challenging such powers were brought against the government, the Supreme Court affirmed the principle of governmental promotion for the general welfare through "large-scale projects of reclamation, irrigation, or other internal improvement" (*United States* v. *Gerlach Livestock Co.,* 1950). An unusual expansion of governmental prerogatives for the general welfare occurred in 1933 with enactment of the *Tennessee Valley Authority Act,* which charged the new agency with a land development and rehabilitation program that would involve land use, water control, and soil conservation throughout the region.

Range and Forest Lands The first federal environmental legislation of forested lands was passed in 1891 with the *Forest Reserve Act* and was followed in 1897 with the *Forest Management Act.* These acts were aimed at the West, and their purpose was to ensure a continuous timber supply and to provide favorable conditions for stream flow. The *Weeks Act* of 1911 laid the groundwork for federal programs on a countrywide basis, and the *Clarke–McNary Act* of 1924, the *McSweeney–McNary Act* of 1928, and the *Sustained Yield Forest Management Act* of 1944 combined to provide a comprehensive environmental package of the legal stewardship of the government of forest lands. The 1944 act summarized this posture by stating their purpose was "in the maintenance of water supply, regulation of stream flow, prevention of soil erosion, amelioration of climate, and preservation of wildlife." Again in 1960 the *Multiple Use–Sustained Yield Act* emphasized the necessity of maintaining forests in perpetuity for the purpose of assuring (1) timber from trees, (2) forage from grazing lands, (3) recreation, (4) wildlife management, and (5) protection of rivers and consideration for their constant flow.

The *Taylor Grazing Act* of 1934 was the first federal legislation that directed attention to a public policy for the establishment of grazing districts and the raising of forage crops. It provided the secretary of the interior with powers to undertake erosion and flood control measures that would protect and rehabilitate grazing lands.

Parks and Wilderness Governmental attitudes concerning the establishment of parklike areas were greatly influenced by Roman law and the English commons, which held that certain lands be set aside for the people as a public trust. Passage of the *Yellowstone Act* of 1872 by Congress made Yellowstone National Park the first of its kind in the world, and put the government into the business of preserving terrain for recreational purposes. The New York legislature enacted a bill in 1885 to establish a forest area that was to remain wild forest land permanently, and this led to the Adirondack State Park. This forever wild concept culminated in the *Wilderness Act* of 1964, which set aside millions of acres throughout the

United States for the purpose of benefitting "the American people . . . [with] an enduring resource of Wilderness."

Additional environmental legislation was implemented by federal laws such as the *Land and Water Conservation Act* of 1965 and the *Wild and Scenic Rivers Act* of 1968. These bills amplified the government's resolve to plan and preserve lands for outdoor recreation and the "enjoyment of present and future generations."

Environmental Hazards and Law

There is a strong relationship of some legislation with disaster events. Such statutes are aimed at the prevention or mitigation of possible disasters in hazard-prone areas. Thus, a body of law has developed around environmental hazards such as floods, earthquakes, landslides, and even volcanic activity. Geologists have now become fully involved in all phases of the legislation, from helping to write the scientific aspects of laws to the mapping and prediction of geologic materials and processes that may pose threats to humans and their property.

Three federal laws will become increasingly important in terms of hazard understanding and prevention of disasters. The *National Dam Inspection Act* of 1972 authorized the corps of engineers to provide inspection of and to inventory dams. Funds unfortunately were not allocated to the program until after the disaster of the Georgia Toccoa dam failure in 1977. The *Federal Disaster Relief Act* of 1974 requires that before any state or local government can receive a loan it must agree to study the hazard and act to mitigate it. Perhaps the strictest federal legislation is formulated in the *Earthquake Hazards Reduction Act* of 1977. It authorizes the president to establish and maintain a coordinated program that aims to predict seismic hazard, design structures that are earthquake-resistant, develop models for land-use decisions, and undertake research on hazard mitigation.

Floods Flooding can occur from both natural and human-related causes. Although Congress passed a *Flood Control Act* in 1917 for remedial work along the Mississippi and Sacramento rivers, it took major disasters to prompt action on comprehensive flood legislation. The disastrous Mississippi Basin flood in 1927 prompted passage of the Flood Control Act of 1928, which authorized the federal government to undertake massive protective work along the Mississippi River. The major floods in 1935 and 1936, however, provided a quantum jump in the power provided under the Flood Control Act of 1936 to permit the corps of engineers to construct flood-constraining structures on any navigable river throughout the country.

Failure of the St. Francis dam, California, in 1928 caused flooding that killed more than five hundred persons and did extensive property damage. This event led to an amendment to the State Water Resources Code in 1929 that required that all private dams be inspected by geologists. Dams under federal jurisdiction and those not along a natural watercourse were exempt from such regulations until the 1960s. The bursting of the Baldwin Hills reservoir in 1963 drowned five people and caused property losses of $12 million. Soon after, the *California State Water Resources Code* was again amended, and this time it included a requirement for geologic inspection of all reservoirs, storage areas, and dams under state jurisdiction, regardless of location.

In 1968 Congress passed the *National Flood Insurance Act,* which was part of the *Housing and Urban Development Act* of 1968, that provided low-cost insurance for homeowners in the 100- and 500-year floodplains. Such insurance could be obtained only after state and local governments had adopted and enforced floodplain land use measures. Although amendments to the act were included in 1969, the catastrophic damages from tropical storm Agnes in 1972 resulted in the much more comprehensive Flood Disaster Protection Act of 1973 (also inserted as an amendment to the Housing and Urban Development Act of 1973). By this time mudslides were also being included in the legislation, and it stipulated that to receive insurance local ordinances must prohibit any construction of land modification within the 100-year floodplain that would impair stream flow unless such benefits clearly outweighed the potential harm.

Floods have resulted in a wide range of lawsuits. Coates (1976) described the case of *Demoski* v. *State of New York* (1957) in which property damages were suffered by the plaintiff because channelization of the Susquehanna River had increased water velocity and during flood times ruined part of his homesite. The court awarded money for the replacement of property lost. In *Drewett* v. *Aetna Casualty and Surety Co.* (1976) a homeowner applied for flood insurance while a flood was in progress and sued the company for damages from the flood. The court, however, denied the claim on the principle of "loss in progress" and held the company was not liable.

An unusual lawsuit developed in the case of *Connelly* v. *State of California* (1970). The plaintiff owned several marinas along the Sacramento River and, with the threat of heavy rains in the region, telephoned the California Department of Water Resources to find the predicted river height so he could adjust his docks. The river rose higher than predicted, causing damage to the docks. Plaintiff claimed the state prediction was flawed due to negligence in its preparation. The state argued that weather predictions were not statements of fact and thereby should be released from liability based on discretionary immunity. The court upheld the plaintiff and ruled the state's arguments invalid because of negligent preparation of their data.

Earthquakes In 1933 Long Beach, California, suffered a 5.5 Richter scale earthquake that caused damage to many brick school buildings and other structures. The earthquake fortunately occurred late in the day, after classes had been dismissed, or there would have been great loss of life. This reality led to passage a few months later of the *Field Act* of 1933. It required all new public school buildings to be constructed to resist earthquakes. The Kearn County, California, 7.7 Richter scale earthquake in 1952 confirmed the importance of earthquake-resistant construction because the new buildings suffered relatively little damage compared with those built prior to the 1933 regulations.

The 6.5 Richter scale earthquake that struck the San Fernando, California, area on January 9, 1971, killed 65 people and caused property damage of more than $500 million. This resulted in the *California Hospital Safety Act* of 1972, which stated that all hospital construction design must be approved by an engineering geologist and then reviewed by the California Division of Mines and Geology. In addition the *Alquist Priolo Geologic Hazard Zones Act* of 1973 (later renamed the *Alquist Priolo Special Studies Zone Act*) required delineation of all potential hazardous fault zone areas in California. Such zones were defined as being along those faults with surface displacement within the past 11,000 years. New structures and real estate development for human occupancy are prohibited from a 50 ft (15.2 m) zone on either side of the fault. In addition, cities and counties are now required to obtain approval by filing a geological report prepared by a registered geologist before any development can occur in a surface fault rupture zone.

Landslides The 1951–1952 winter rainstorms that deluged Los Angeles County, California, caused extensive property damage from the mobilization of mud, rock, and debris. These events triggered the *Grading Ordinance* of 1952, the first to require the involvement of geologists in the planning and design of developments on hillsides. Site evaluation was mandated near vertical cuts; unusually high fills were to be eliminated; and slope cut and fill could be no greater than 1:1 and 1.5:1 respectively. By January 1956, the ordinances had been modified to require geology reports prior to issuance of building permits on hillsides.

The Palos Verdes peninsula in southern California is the site of the ancient Portuguese Bend landslide. In the 1956–1957 period the landslide became reactivated and destroyed 150 houses. The homeowners brought suit against Los Angeles County, claiming construction of Crenshaw Boulevard overloaded the slide-prone terrain and caused movement of 134,000 m³ of soil and rock. In the case of *Albers* v. *County of Los Angeles* (1965) the judge found an absence of county negligence or contributory negligence but awarded claimants $5,360,000 for damages on the basis of inverse condemnation, because the California Constitution states "private property shall not be taken or damaged for public use without just compensation." Because of the Los Angeles ordinances requiring geology reports at hazardous sites, many poorly trained people were employed by various groups and developers to write reports. This situation finally led to the development in California of uniform standards and a licensing program in the certification of qualified geologists.

Heavy rains again struck southern California in February 1962, causing millions of dollars in damages. Stricter grading ordinances were passed in 1963 that required even greater involvement by geologists, such as during the construction period of new developments. In the final certification the geologist must declare that he or she has inspected the grading and that in his or her opinion all excavations are made in accordance with the design and are stable as determined by the science of engineering geology (Jahns, 1969).

In the State of Washington a motorist was injured when landslide materials struck his vehicle. In *Boskovich* v. *King County* (1936) a lower court ruled the plaintiff was entitled to recover damages because of governmental failure to post warning signs. However, on appeal another court found the county not liable by stating that a warning sign would not have prevented the accident. The court also stipulated that the county is not the insurer of road safety and was not guilty of negligence. Although the doctrine of sovereign immunity was not specifically mentioned, its overtones influenced the ruling. The results of this case should be compared with those of *Brown* v. *McPhersons Inc.* (1975). In this lawsuit plaintiffs were local residents who had suffered losses because of a landslide. A geologist had warned the state of the hazardous conditions, but this notification went unheeded. The state was found to be negligent in not warning the citizens, the realtors, or the developers of the probable dangers.

Eminent Domain

Governments have always used their power to acquire private lands in the interest of the public good. In the United States, land acquisition by governmental condemnation cannot be for whimsical or nonbeneficial uses. The rights of property owners to obtain *just compensation* are assured under the Fifth Amendment of the Constitution. However, numerous environmental lawsuits have resulted from condemnation practices when:

1. Plaintiffs believe abnormal environmental damage will occur. Some of the classic cases include disputes arising out of lawsuits involved with the Hetch Hetchy in California, Boundary Waters Canoe Area in Minnesota, Mineral King in California, and the Florida jet port (Coates, 1980, 681–682).
2. Plaintiffs feel that the monetary value of their property was not adequately assessed (Coates, 1971, 1976, 1980, 687–688).

Environmental Trends

In addition to the statutes and case histories already discussed, there have been new advances in statutes since the 1960s that seek to ensure a less polluted environment or that address the problems of land use in different ways. With the exception of the *Rivers and Harbors Act* of 1899 (*Zabel and Russell* v. *U.S. Army Corps of Engineers,* 1971), the great majority of U.S. legislation dealing with pollution and waste management has been enacted since 1960. The following acts typify these statutes: *Water Pollution Control acts* of 1961 and 1972; *Clean Air acts* of 1963, 1966, 1967, and 1977; *Solid Waste Disposal Act* of 1965; *Clean Water Restoration Act* of 1966; *Air Quality Act* of 1967; *Water Quality Improvement Act* of 1970; *Clean Water Act* of 1972; *Environmental Pesticide Control Act* of 1972; *Resources Conservation Recovery Act* of 1976; *Toxic Substances Control Act* of 1976; *Surface Mining Control and Reclamation Act* of 1977.

Since the 1960s various strategems have been used by local, state, and federal governments to apply constraints to poorly planned developments or those land-use practices that are not harmonious with the environment. States such as Hawaii, Maine, and Vermont were among the first to apply statewide restrictions on land use. The first land-use legislation passed by Congress was the *Coastal Management Act* of 1972, which provided the basis for all coastal states to plan and manage the coastal environment.

Of all legislation passed by Congress, none has had the scope or the detail of the *National Environmental Policy Act* of 1970 (NEPA). For the first time *environmental impact statements* (EIS) became an integral part of the planning procedure for all state and local agencies for any project using federal funds that might significantly affect the environment (Anderson, 1973). Thus, prior to construction a proper EIS that provides details on the character of possible impacts of the project must be submitted. Because most projects involve earth materials and processes, geologists become involved in the investigations. The NEPA has not had a smooth course, however, and has given no absolute guarantee that the environment will always be safeguarded. Prodding government bureaucracy to act may be difficult and often needs some nudging by requiring mandatory deadlines authorized through federal court action. For example, in *Illinois* v. *Costle* (1979) a Washington, D.C., district judge voiced dissatisfaction with having to issue an injunction commanding the Environmental Protection Agency to meet deadlines under the Resources Conservation Recovery Act of 1976. Well-meaning statutes are not self-implementing. We need a national will to protect the environment from the threatening health and pollution hazards which this Act addresses.

The following two case histories reveal other aspects of the EIS. In a residential section of Asheville, North Carolina, a group of homeowners were affected by construction of an interstate highway that had received approval under the NEPA. They filed a lawsuit (*Mountainbrook Homeowners* v. *Adams,* 1979) that the construction and disposal of rock was not consistent with plans as specified under the EIS and thereby diminished the value of their property. Whereas the EIS stipulated materials removed by the construction would be placed on the land to follow the natural contours and blend with the local topography, instead the 30 m piles of soil and rock were placed in unsightly mounds that diverted the flow of water in adjacent streams. The federal district court dismissed the complaint, finding that the federal court had no jurisdiction to order a remedy on behalf of citizens. The court also ruled that the NEPA was primarily a planning and decision-making document and was not designed to create benefits for a special group of plaintiffs. Furthermore, restriction on the use of real estate should be governed by state and local law, and there was no evidence that Congress intended such matters to be resolved by federal action.

The case of *Red Line Alert* v. *Adams* (1980) shows the importance of geologists working closely with planners because in this case there was a communications gap. This lawsuit involved a change in the way a Boston subway extension was to be constructed. In the original EIS, a 10 m deep tunnel was to have been made by a boring machine. However, the successful low bidder proposed a 24 m deep tunnel that would require blasting, a technique not mentioned in the EIS. The court held that failure to discuss such a method required the filing of a supplemental EIS. It further ordered a hearing to determine if an injunction was warranted pending the successful development of the supplemental EIS. Coates

(1976) also described the *Capital Region Citizens Committee and Others* v. *John A. Volpe and Others* (1973) case in which a group sought an injunction against the building of Interstate 88 because of failure to have on file a proper EIS.

DONALD R. COATES

References

Acton v. *Blundell,* 12 M. & W. 324 (1843).
Albers v. *County of Los Angeles,* 62. Cal. 2d 250, 42 Cal. Rpt. 89 (1965).
Anderson, F. R., 1973, *NEPA in the Courts: A Legal Analysis of the National Environmental Policy Act.* Washington, D.C.: Resources for the Future, 324p.
Boskovich v. *King County,* 188 Wash. 63 (1936).
Brown v. *McPhersons, Inc.,* 86 Wash. 2d 293, 545 R.2d 13 (1975).
Coates, D. R., 1971, Legal and environmental case studies in applied geomorphology: Environmental geomorphology, in Coates, ed., *Publications in Geomorphology.* Binghamton: State University of New York, 223–242.
Coates, D. R., 1976, Geomorphology in legal affairs of the Binghamton, New York, metropolitan area, in Coates, ed., *Urban Geomorphology.* Boulder, Colo.: Geological Society of America (Spec. Paper 174), 111–148.
Coates, D. R., 1980, *Environmental Geology.* New York: John Wiley & Sons, 701p.
Colburn v. *Richards,* 13 Mass. 419 (1816).
Connelly v. *State of California,* 3 Cal. 3d 744, 84 Cal. Rpt. 257 (1970).
Demoski v. *State of New York* (N.Y. Sup. Ct.), 12 Misc. 2d 416 (1957).
Drewett v. *Aetna Casualty and Surety Co.,* 539 F.2d 496 (4th Cir. 1976).
Funk v. *Holdeman,* 53 Pa. St., 229 (1866).
Hurley v. *Kinkaid,* 285 U.S. 95 (1932).
Illinois v. *Costle,* 9 ELR 20243 (D.D.C. 1979).
Jahns, R. H., 1969, Seventeen years of response by the City of Los Angeles to geologic hazards, *Proc. Geology Hazards and Public Problems Conf.,* 283–295.
Kiersch, G. A., and A. B. Cleaves, eds., 1969, Legal aspects of geology in engineering practice, *Geol. Soc. America Eng. Geology Case Histories No. 7,* 112p.
Merritt v. *Brinkerhoff,* 17 Johns. 306 (1820).
Mountainbrook Homeowners Association v. *Adams,* 9 ELR 20686, (W.D.N.C. 1979).
Pasadena v. *Alhambra,* 33 Cal. 2d 908, 3, 207 R.2d 17 (1949).
Red Line Alert v. *Adams,* 14 ERC 1417 (D. Mass. 1980).
Strobel v. *Kerr Salt Co.,* 164 N.Y. 303, 58 N.E. 142 (1900).
Tartar v. *Spring Creek Water and Mining Co.,* 5 Cal. 395 (1855).
United States v. *Gerlach Livestock Co.,* 339 U.S., 723, 738 (1950).
Zabel and Russell v. *U.S. Army Corps of Engineers,* 276 F.2d 764, (5th Cir., U.S. Sup. Ct. 1971)

Cross-references: *Environmental Engineering; Geology, Scope and Classification.* Vol. XIII: *Geology, Applied.*

LICHENOMETRY

Lichenometry is the technique of dating deposits, rock surfaces, or other substrates by measuring sizes and/or abundance of lichens. The technique, used mainly on recent glacial deposits, can be useful over a range of a few hundred to several thousand years, depending on the lichen.

Although aspects of lichen growth have long been studied by botanists, geomorphologists first established long-term growth rates and considered practical applications. Principles of lichenometry were originally set forth by Roland Beschel (1950, 1961). The technique was gradually adopted by other geomorphologists in Europe and North America, and has become a widely used dating tool, especially where radiocarbon dating is not feasible.

Principles

The fundamental premise is that certain crustose lichens grow radially at slow rates over long periods of time, so the largest lichens indicate the oldest substrates. Individual lichens, to be useful, must occur as discrete, roughly equidimensional thalli (bodies), because growth rates are difficult or impossible to measure when lichens are fibrous or irregularly shaped or when closely spaced thalli inhibit each other's radial growth. Thus, the most useful lichens are those that occur as round individuals (Fig. 1) and that grow for the longest period of time before stagnating or being weathered off the substrate. Radial growth of certain crustose lichens has been demonstrated over short terms by photographic time studies and over longer terms by correlation of maximum lichen sizes with known ages of substrate.

Another premise used in lichenometry is that the area of substrate covered by lichen thalli increases with time. This is a result of the continual

FIGURE 1. Large, round thallus of *Rhizocarpon macrosporum,* one of several long-lived, yellow-green varieties of *Rhizocarpon*. Scale is 15 cm long.

establishment of new thalli, as well as continued growth of already established thalli. The premise seems to be valid in some cases, although on rocks that weather rapidly, such as some limestones, lichens have difficulty in becoming established and there may not be much relationship between lichen cover and age of the rock surface.

Techniques

If absolute growth rates are not known, lichen sizes or degree of lichen cover can be used for relative dating of substrates. For example, Benedict (1968) found that lichens on young moraines in the Indian Peaks area of Colorado fell into three broad size categories. From this distribution he concluded that there were three ages of moraines, the oldest being the one with the largest lichens.

For absolute dating, growth rates must be determined. Direct measurement of growth necessitates periodic measurements of specimen diameters, preferable through photographic means (Hooker and Brown, 1977). Indirect measurement involves measuring lichen diameters of particular species on surfaces of known age and constructing a growth curve from the data. Points on the curve representing a species' early growth may be had from historically dated surfaces, such as tombstones, rock walls, mine tailings, and road cuts. Older points on the curve must be based on substrates that have been dated by radiocarbon (e.g., Denton and Karlen, 1973), dendrochronology (e.g., Luckman, 1977), or other means (see Vol. IVA: *Geochronometry*). All the points on a particular growth curve should be from sites that are environmentally similar (e.g., rock type, climate) because a species can have different growth rates in different environments. Similarly, once the growth curve has been constructed, any substrates to be absolutely dated must be in environmentally similar sites to those used in the curve.

There is some debate surrounding the best procedure for size sampling on deposits bearing multitudes of lichens. On any particular deposit there will exist a wide range of sizes of lichens of a particular species because new individuals continue to colonize the substrate through time. The single largest lichen of the species should be the oldest and should represent the best approximation of the age of the deposit. For these reasons Beschel (1950, 1961) advocated using only the largest thallus of a species for dating purposes. Some other lichenometrists (e.g., Innes, 1984) have advocated averaging the largest sizes on a deposit. Averaging allows for the possibility that a few old lichens were transported to the deposit without being destroyed and are actually older than the deposit and for the possibility that the very largest lichen(s) on a deposit may be growing abnormally fast because of an abnormally favorable microenviron-

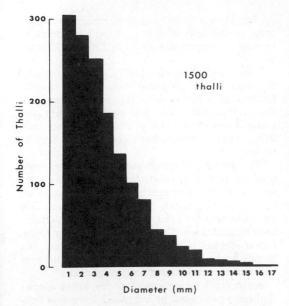

1500 thalli

FIGURE 2. Size-frequency histogram of *Rhizocarpon geographicum* thalli growing on mudflow levees in the Front Range of Colorado. The distribution implies continuous reproduction and suggests that none of the thalli were survivors of the mudflow. (From Benedict, 1967, courtesy of the International Glaciological Society)

ment. Various other techniques have been tried to get around the possibility of anomalous lichen sizes. For example, Benedict (1967) used a size-frequency distribution to show that transported lichens probably are not present on the deposits he studied (Fig. 2), while Hedgson (1978) rejected the largest thallus on a deposit if it were less than 10% larger than the second largest thallus.

A deposit or other substrate should ideally be examined in its entirety and the largest lichen size(s) recorded. In cases where complete examination of a sequence of deposits is not practical, some workers have set up sample grids or sample areas on each deposit and measured only the lichens in the sample areas. Innes (1985) recommends recording the largest lichen diameters in each of at least 10 sample areas totalling at least 400 m². He also recommends measuring the long diameters of elliptical thalli, although some other workers have used short diameters or diameters of largest inscribed circles. Detailed discussions of sampling techniques can be found in Lock et al. (1979) and Innes (1984, 1985).

Problems

The value of lichenometric dating is to a certain extent a matter of opinion. Jochimsen (1966) feels that ecological variations and sampling problems are so serious that lichenometry based on thalli

sizes is unworkable. Among the more serious problems she identifies are (1) in very few cases are lichen thalli perfectly circular; more commonly they are ellipsoid, sinuate, or netlike. Such nonuniform thalli development suggests that growth usually proceeds in an irregular manner. (2) Most lichens used in lichenometry are very difficult to identify precisely as to species in the field, making identification mistakes common in some cases. (3) Lichen growth rates can vary significantly due to environmental factors, which can vary from site to site within a deposit, as well as from deposit to deposit. Innes (1985) lists several factors that can influence growth rate: substrate lithology and stability, moisture availability, temperature, light intensity, altitude, aspects, wind exposure, snow cover, pollution, and interference from other vegetation. Worsley (1981) reiterates some of the concerns of Jochimsen (1983) and also points out an absence of reproducibility in many published sample designs.

These problems are widely recognized and necessitate cautious application of the method in any particular area. Most investigators who have worked with lichenometry, however, do not subscribe to Jochimsen's (1973) extreme view that the problems render the method essentially useless. The fact that many lichen growth curves have been successfully constructed indicates that there is a good correlation between elapsed time and thallus size in many cases. Lichenometry is never extremely accurate and may not be useful at all in some places even where lichens are common. Yet it has been established as a generally workable technique for relative dating and rough absolute dating.

The technique is most useful when deposits to be dated are a few hundred years old or less because beyond that range independently dated substrates that can be used for a growth curve are hard to find. Most published growth curves that extend back a thousand years or more consist in their older ranges of a straight line drawn through and extrapolated past very few (or one) points. The limitations of such curves must be kept in mind. It is not known with any certainty whether or not most lichen growth rates are constant during old age. Neither is there much information on the length of time individuals of particular species can grow before reaching senescence.

Usable Species

The lichens used most commonly for dating are the yellow-green lichens of the subgenus *Rhizocarpon*. The species that appears most commonly in the literature is *R. geographicum* (L.) DC.; however, identification to species level is quite difficult in the field and it is probable that many other species have been called *R. geographicum*.

Confusion of species leads to a subsidiary problem: Comparison of different published *Rhizocarpon* growth curves is difficult because a few workers have used only *R. geographicum,* others have used any yellow-green *Rhizocarpon* species (and said so), while others have used any yellow-green lichen and called it *R. geographicum.* A recent discussion of taxonomic problems and nomenclature is that of Innes (1983). A key to *Rhizocarpon* species is provided in the classic taxonomic work of Runemark (1956); other references are provided by Innes (1985). Some of the other *Rhizocarpon* species that have been used for dating purposes are *R. alpicola* Körb, *R. superficiale* (Schaer.) Vain., and *R. tinei* (Tornab.) Run.

Confusion aside, the *Rhizocarpon* lichens have the advantages of being common on siliceous substrates throughout the world, living for relatively long times, and commonly occurring in the discrete, roughly round thalli.

Growth curves of lichens referred to as *Rh. geographicum* were originally thought to be linear, but several curves have been published that show an initial phase of relatively rapid growth (the so-called great period) followed by a long period of slower growth (e.g., Fig. 3). Similar curves have been constructed for some other lichens (e.g., Fig. 4).

Growth rates for *Rh. geographicum* vary considerably from region to region, apparently because of environmental differences. A table of growth rates published through 1973 was compiled by Webber and Andrews (1973). Rates denoted as great period rates (either by the original author or through interpretation by Webber and Andrews) vary from 2 mm to 90 mm/yr, while the great period lasts 40–300 yr. Rates considered to be post–great period vary from 2 mm to 30 mm/yr.

Miller and Andrews (1972) present evidence that suggests *Rh. geographicum* continues to grow for at least 9,500 yr in eastern Baffin Island, although their oldest conclusively dated lichen substrate is less than 3,000 yr old. Most other published growth curves have considerably smaller ranges.

The various *Rhizocarpon* species have been used on quartzite, chert, shale, and a variety of igneous rocks. The species are not common on pure carbonate rocks, however, probably due to an intolerance for calcium in high concentrations. Osborn and Taylor (1975) concluded that *Xanthoria elegans* is the only useful lichen on carbonate rocks in the Canadian Rockies; this species unfortunately has a limited lifespan.

Apart from *Rhizocarpon* species, lichen species that have been used or suggested for use in lichenometry are, according to Lock et al. (1979), *Aspicilia caesiocinerea* Nyl.; *A. cinerea* (L.) Kbr.; *Caloplaca cinericola* (Hoe) Darb.; *Diploschistes anactinus* (Nyl.) Zahlbr.; *D. scruposus* (L.) Norm.; *Lecanora atrobrunnea* (Ram.) Schaer.; *L. badia* (Hoffm.) Ach.; *L. polytropa* (Hoffm.) Rabenh.; *L. thomsonii* H. Magn.; *Lecidea lapicida* Ach.; *L. paschalis* Zahl.; *L. promiscens* Nyl.; *Sporostasia testudinea* (Ach.) Mass.; and *Xanthoria elegans* (Link) Th. Fr. In addition, foliose lichens belonging to the genera *Alectoria, Physcia, Umbilicaria,* and *Usnea* have been used in some cases.

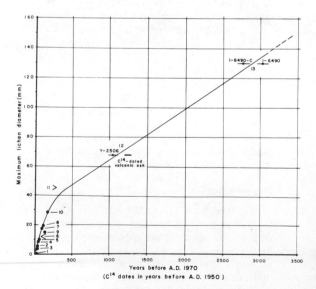

FIGURE 3. Growth curve for *Rhizocarpon geographicum* in the White River Valley and Skolai Pass areas, Alaska. Points on the curve were dated using radiocarbon, dendrochronology, and historical records. (After Denton and Karlen, 1975, reproduced by permission of the Regents of the University of Colorado)

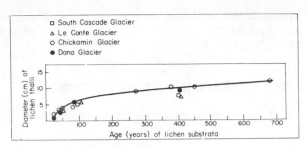

FIGURE 4. Growth curve for *Lecanora* cf. *cinerea* in the Dome Peak area, North Cascade Range, Washington. Points were dated using dendrochonology and historical records. (From Miller, 1969, reproduced by permission of the Regents of the University of Colorado)

Applications

Lichenometry has been used mainly on Holocene glacial deposits, and the technique has made significant contributions to the understanding of glacial advances, and thus climate, in late Holocene time. Beschel (1961) outlines other applications, including timing of glacier recession, timing of sea level and lake changes, dating of rockfall and solifluction events, and comparison of weathering rates of different substrates. Beschel also considers that lichen sizes can be used to determine variations in climate in space. To do this lichen sizes must be measured on several independently dated surfaces in a region; variations in the calculated growth rate are assumed to reflect variations in macro- or microclimate.

More recent applications of lichenometry have been dating of river floods and volcanic events. Discussion of these and other appications, with references, can be found in Innes (1985).

GERALD OSBORN

References

Benedict, J. B., 1967, Recent glacial history of an alpine area in the Colorado Front Range, U.S.A.: I. Establishing a lichen-growth curve, *Jour. Glaciology* 6, 817–832.

Benedict, J. B., 1968, Recent glacial history of an alpine area in the Colorado Front Range, U.S.A.: II. Dating the glacial deposits, *Jour. Glaciology* 7, 77–87.

Beschel, R. E., 1950, Flechten als Altersmasstab rezenter Moranen, *Zeitschr. Gletscherkunde. u. Glazialgeologie* 1, 152–161.

Beschel, R. E., 1961, Dating rock surfaces by lichen growth and its application to glaciology and physiography, in G. O. Raasch, ed., *Geology of the Arctic,* vol. 2. Toronto: University of Toronto Press, 1044–1062.

Denton, G. H., and W. Karlen, 1973, Lichenometry: Its application to Holocene moraine studies in southern Alaska and Swedish Lapland, *Arctic and Alpine Research* 5, 347–372.

Hodgson, D. M., 1978, Lichenometric Studies of Three Glaciers in the Yukon Territory of Canada. University of Edinburgh, B.Sc. dissertation.

Hooker, T. N., and D. H. Brown, 1977, A photographic method for accurately measuring the growth of crustose and foliose saxicolous lichens, *Lichenologist* 9, 65–75.

Innes, J. L., 1983, Use of an aggregated *Rhizocarpon* "species" in lichenometry: An assessment, *Boreas* 12, 183–190.

Innes, J. L., 1984, The optimal sample size in lichenometric studies, *Arctic and Alpine Research* 16, 233–244.

Innes, J. L., 1985, Lichenometry, *Progress in Physical Geography* 9, 187–254.

Jochimsen, M., 1973, Does the size of the lichen thalli really constitute a valid measure for dating deposits?, *Arctic and Alpine Research* 5, 417–424.

Luckman, B. H., 1977, Lichenometric dating of Holocene moraines at Mount Edith Cavell, Jasper National Park, Alberta, *Canadian Jour. Earth Sci.* 14, 1809–1822.

Miller, C. D., 1969, Chronology of Neoglacial moraines in the Dome Peak area, North Cascade Range, Washington, *Arctic and Alpine Research* 1, 49–66.

Miller, G. H., and J. T. Andrews, 1972, Quaternary history of northern Cumberland Peninsula, east Baffin Island, N.W.T., Canada. Part VI: Preliminary lichen growth curve for *Rhizocarpon geographicum, Geol. Soc. America Bull.* 83, 1133–1138.

Osborn, G., and J. Taylor, 1975, Lichenometry on calcareous substrates in the Canadian Rockies, *Quat. Res.* 5, 111–120.

Runemark, H., 1956, Studies in *Rhizocarpon, Opera Botanica* (Lund) 2, 1.

Webber, P. J., and J. T. Andrews, 1973, Lichenometry: A commentary, *Arctic and Alpine Research* 5, 295–302.

Worsley, P., 1981, Lichenometry, in A. Goudie, ed., *Geomorphological Techniques.* London: George Allen and Unwin, 302–305.

Cross-references: *Tephrochronology.* Vol. IVA: *Geochronometry.* Vol. XII, Pt. 1: *Radiocarbon Dating; Radioisotopes.* Vol. XIII: *Geochronology.*

LITHOBIOLOGY—See GEOMICROBIOLOGY; LICHENOMETRY.

LITHOGEOCHEMICAL PROSPECTING

Lithogeochemical prospecting for minerals includes any method of *mineral exploration* that employs one or more chemical properties of rocks. The method is based on *primary geochemical dispersion patterns* of igneous, metamorphic, and sedimentary rocks that are now exposed at the ground surface, (e.g., Dunlop et al., 1979).

Geochemical patterns are classified as *syngenetic* if they formed at the same time as the enclosing rock or *epigenetic* if material was introduced later (i.e., hydrothermally) into a preexisting matrix, (Krauskopf, 1967). Syngenetic patterns formed in sedimentary rocks (except in drainage sediments, overburden, and soils) are sometimes included in the syngenetic type even though they are actually secondary dispersion patterns preserved in the sedimentary rock. Epigenetic geochemical patterns may be further subdivided to the origin, type of emplacement, matrix, and form of dispersion (Mason, 1958).

Exploration Sequence

Orientation Survey Fundamental to the interpretation of geochemical data is the establishment of normal *background* distributions for the elements under consideration. Background variations as well as causes and distinctions between such variations due to mineralization must be appreciated. A most important prerequisite for a geochemical exploration program is an *orientation survey* (Hawkes and Webb, 1962). Orientations are normally divided into two phases. Phase 1 determines background distributions whereas Phase 2 focuses on elemental distributions in the vicinity of known mineralization.

Phase 1 In Phase 1, an extensive sampling program must be organized to establish trace element variations attributed to (1) stratigraphic position of the rock or of the group of rocks that hosts mineralization, (2) geographic location, (3) compositional differences in rock type, and (4) local differences within a single outcrop.

To provide a statistically significant unbiased estimate of trace element variation due to these factors, the sampling program must be organized as follows: (1) the *rock chip* survey lines must be arranged to cross the main rock groups more or less at right angles to the strike on the basis of equal geographic distribution; (2) the lines must avoid proximity with known mineralization; (3) all obviously different rock types (e.g., pillows, dikes, etc.) must be sampled; (4) the samples must

be collected at regular intervals considering the extent of the outcrop and accessibility. In the example of geochemical prospecting in Cyprus (Fig. 1) for sulfides, twenty lines were located over the northeast and southwest outcrops of the volcanic rocks that included sulfide mineralization at a sampling interval of 200 m; and (5) at many localities samples must be collected from various parts of the same rock type that show obvious differences in texture or composition—e.g., of pillows and dikes from the center to the chilled edge and interstitial material from between the pillows. Samples must also be collected from oxidized regions. Increasing oxidation may show significant anomalies in mineralized areas, especially where sulfide mineralization occurs.

Phase 2 Orientation surveys in the vicinity of active and disused mines assist in geochemical rock-sampling surveys. Orientation helps determine the nature of primary dispersion patterns and associations of elements. *Pathfinder elements* (see *Indicator Elements*) guide searches for buried ore deposits. Orientation surveys over known ore deposits should include sections showing specific ore–wall rock relations and also test the consistency of apparent relationships of pathfinder elements. Thus, orientation surveys follow these general procedures: (1) the number of radial sampling traverses, for a distance of at least 1 km from mineralization, should consider the geology and the structure of the area. Specific detailed sections should study contacts and mineralized zones. (2) Similar traverses should be organized in representative ore deposits, incorporating the main varieties of ores—e.g, Cu-rich and Cu-poor pyrites. (3) If the results from orientation surveys are not conclusive, it is advisable to locate representative orebodies and arrange for additional orientations to help evaluate the results of a geochemical survey because concentrations of elements are normally directly related to lithological composition. Such a correlation was found in the Cyprus igneous ophiolitic complex between Ni- and olivine-bearing rocks (Pantazis and Govett, 1973).

Exploration Survey The initial selection of areas should be based on a thorough review of known geology and records of past mining and prospecting activities. Particular attention should be given to possible types of ore deposits, the distribution of favorable rocks, geologic structures, and other conditions that limit the application of a rock chip survey. Thus, the most promising regions selected should constitute core study areas. Depending on the size of the area, the first stage may be broad-scale reconnaissance exploration to help decide which segments have the best mineral potential and which parts can be eliminated as relatively unfavorable. The broad-scale reconnaissance survey, based on a low sample density,

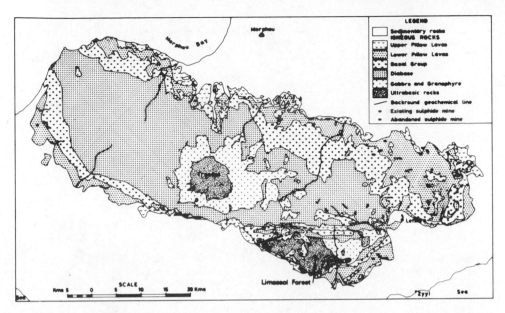

FIGURE 1. General geology of the Troodos Ophiolite Complex, showing the background geochemical lines.

should appraise the mineral potential of the area by determining regional elemental distributions (Garrett et al., 1980). The most promising regions are then followed up by detailed surveys that locate primary or secondary geochemical anomalies.

Analytical Procedures For wet analysis the rock samples are crushed in a grinder, passed through an 80-mesh screen, and brought into solution as follows:

1. Routine procedure: A 1 g portion of sample is digested in a Pyrex tube in 5 ml of concentrated nitric acid. After boiling for 15 min on a magnetic stirring hot plate, distilled water is added to bring the volume up to 23 ml. Boiling continues for a further 15 min. After cooling, distilled water is added, bringing the total volume to 25 ml. The tube, stoppered with a screw-cap, is then centrifuged.

2. Total element extraction: A 1 g portion of sample is weighed in a porcelain crucible, heated in a muffle furnace at 750°C for 1 hr, cooled, and transferred to a Teflon evaporating dish with as little water as possible. After evaporation to near dryness, 10 ml of hydrofluoric acid are added, and the dish is left covered overnight. After evaporation to about 5 ml, 10 ml of perchloric acid are added and evaporated to near dryness; then 25 ml of 5% hydrochloric acid are added, and the sample is covered and heated. After cooling the sample is transferred to a 50 ml volume flask and made up to volume with 5% hydrochloric acid.

The performance of this method for the total element must include U.S. II Geological Survey

standard samples (1 in each batch of 12 samples), and the overall precision of the routine acid extraction technique must be controlled by the inclusion of standard samples (2 or 3 in each batch of 50 samples) from the country rocks.

Evaluation of Results in Terms of Prospecting Techniques

Successful geochemical prospecting depends on the recognition of elemental distribution patterns that are different in mineralized areas (i.e., anomalous) from patterns in nonmineralized areas (i.e., background). To recognize subtle differences, it is frequently advantageous to treat the data statistically (Pantazis, 1981). The smaller the differences sought between background and anomalous situations, the more vigorous the mathematical treatment must be. In general, this demands that great care be taken to ensure that mixed populations are not used to characterize background conditions and that the nature of the background distribution is understood.

Threshold Values, Frequency Distributions Conventional treatments of geochemical prospecting data demand that a *threshold value* be determined that defines the lower limit of anomalous values. A procedure often used is to define threshold as the mean of background observations plus two times their standard deviation. Histogram plots are used to study population distributions of each base metal. Figure 2 shows a computerized histogram plot.

Comparison of *modal values* is a particularly

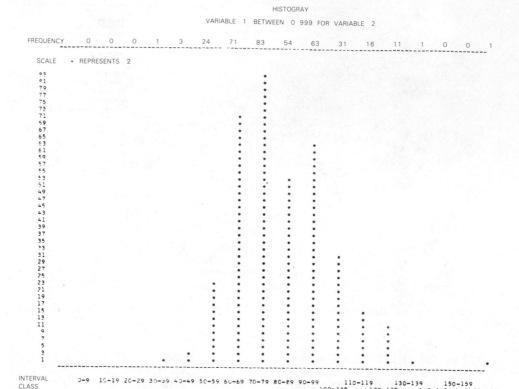

FIGURE 2. Frequency distribution histogram for zinc in pillow lavas of the Mathiati Mine area (inner and outer zones) prepared by computerized methods.

useful approach when population distributions are skewed. Samples from mineralized areas may be classified as greater than or less than background modal values. Based on elemental frequency distributions, it is sometimes possible to group background and mineralized samples into empirical classes. Figure 3 shows three different classes for cobalt and nickel background curves in Cyprus. Similarly, Fig. 4 shows background distributions for copper and zinc.

Consideration of individual elements, either as absolute values or in terms of their distribution, is of little value as a general exploration technique. Exceptions include studies of the geographical distribution of elements (Fig. 5).

Percentile Mapping Contour mapping of elements is greatly improved if the classes are based on major changes of slope of the cumulative curve. Such groupings can be achieved by a computerized percentile program. A major advantage of this method is that geochemical class limits are related to lithology and are not arbitrary. When combined with a mapping program, elemental threshold values can be directly plotted to pinpoint anomalous zones.

Ratios of Elements Variation of element ratios often assists in interpreting the data associated with mineralization. A determinative function derived from means of specific groups also helps to classify rocks into mineralized or background groups.

A visual portrayal of elemental ratios can be performed using color separation techniques, which are usually employed in combination with percentile mapping to enhance the relations between classes derived from the latter. In this way, two or three elements can be dealt with simultaneously using an equal number of basic colors. Color overlaps and varying concentrations of elements produce a number of different colors.

Nonlinear methods of mapping (such as the computer program NLMA1) often prove successful in separating mineralized from unmineralized samples. Such programs study the relation between more than three elements and can plot them on two-dimensional X-Y plots (Fig. 6).

Discriminant Functions The discriminant function is a multivariate classifying function used to assign an unknown individual with several measured variable attributes $X_1, X_2, \ldots X_k$ to

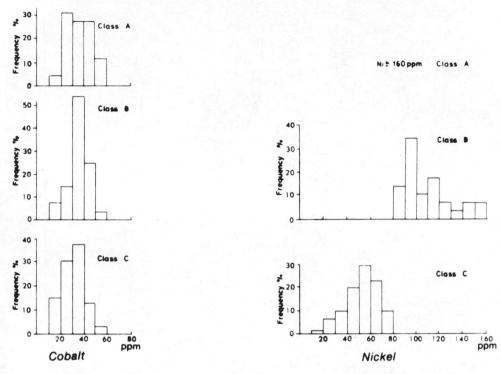

FIGURE 3. Frequency distribution histograms for cobalt and nickel in Troodos pillow lavas, Cyprus.

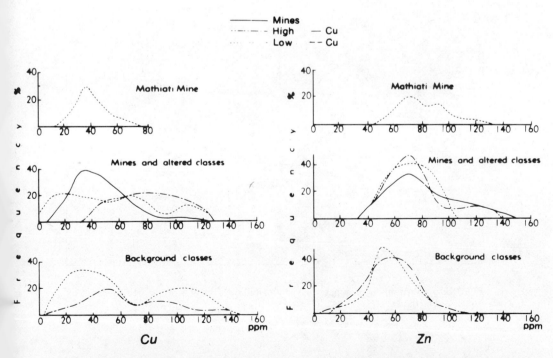

FIGURE 4. Frequency distribution curves for copper and zinc in Troodos pillow lavas, Cyprus.

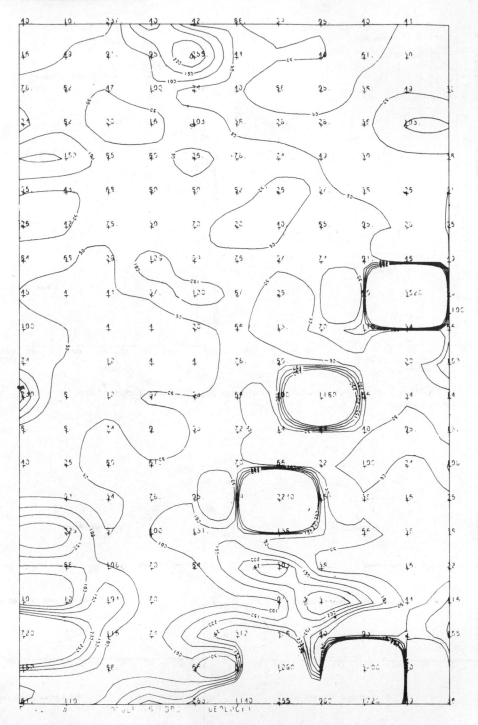

FIGURE 5. Distribution of copper in Troodos pillow lavas of Kornos area prepared by computerized methods.

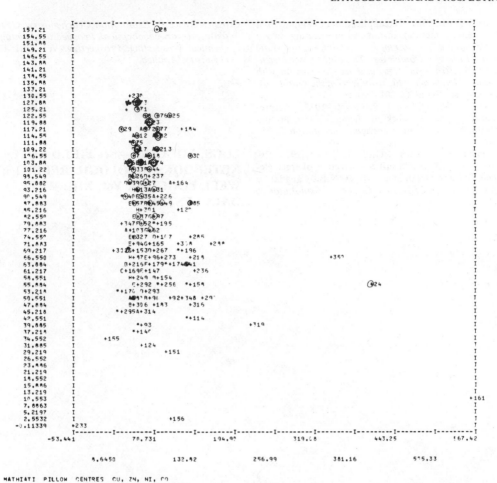

MATHIATI PILLOW CENTRES CU, ZN, NI, CO

FIGURE 6. Nonlinear mapping of the pillow lavas of the Mathiati Mine of the inner (circled) and outer zones produced by the nonlinear mapping algorithm (NLMAl).

one of several groups on the basis of a linear function representing all *k* variables and weighing the variables according to their contribution to the discrimination. Moreover, the discriminant function takes into consideration the scatter, or variance of each of the *k* variables in each group, and is chosen to maximize the ratio of the difference between sample means and standard deviation within the groups (Whitehead and Govett, 1974).

Given adequate data to calculate the original defining functions, the discriminant function is a much more sensitive and reliable tool for classifying individual samples than the determinative function. The efficiency of the discriminant function depends, however, on the assumption that the groups selected for calculating the function are different.

THEODOULOS M. PANTAZIS

References

Dunlop, A. C., E. P. Ambler, and E. T. Avila, 1979, Surface lithogeochemical studies about a distal volcanogenic massive sulphide occurrence at Limerick, New South Wales, *Jour. Geochem. Exploration.* **11**, 285-297.

Garrett, R. G., V. E. Kane, and R. K. Zeigler, 1980, The management and analysis of regional geochemical data, *Jour. Geochem. Exploration* **13**, 115-152.

Govett, G. J. S., and T. M. Pantazis, 1971, Distribution of Cu, Zn, and Co in The Troodos Pillow Lavas, Cyprus, *Inst. Mining and Metallurgy Trans.* **80**, B27-B46.

Hawkes, H. E., and J. S. Webb, 1962, *Geochemistry in Mineral Exploration.* New York: Harper & Row, 415p.

Krauskopf, K. B., 1967, *Introduction to Geochemistry.* New York: McGraw-Hill, 721p.

Mason, B., 1958, *Principles of Geochemistry.* New York: Wiley, 3l0p.

Pantazis, T. M., 1981, The application of computerised geochemical methods in mineral prospecting, in *An International Symposium on Metallogeny of Mafic and Ultramafic Complexes: The Eastern Mediterranean-Western Asia Area, and Its Comparison with Similar Metallogenic Environments in the World,* vol. 2, Athens: UNESCO, 281–290.

Pantazis, T. M., and G. J. S. Govett, 1973, Interpretation of a detailed rock geochemical survey around Mathiati Mine, *Jour. Geochem. Exploration* **2,** 25–36.

Whitehead, R. E. S., and G. J. S. Govett, 1974, Exploration rock geochemistry—detection of trace element halos at Heath Steele Mines (N.B., Canada) by discriminant analysis, *Jour. Geochem. Exploration* **3,** 371–386.

Cross-references: *Exploration Geochemistry; Geostatistics; Hydrogeochemical Prospecting; Pedogeochemical Prospecting; Prospecting.* Vol. XIII: *Geochemistry, Applied.*

LOGS, LOGGING—See FIELD NOTES, NOTEBOOKS; GEOLOGIC REPORTS; WELL LOGGING. Vol. XIII: WELL DATA SYSTEMS.

M

MAGNETIC EXPLORATION—See
EXPLORATION GEOPHYSICS;
MARINE MAGNETIC SURVEYS; VLF
ELECTROMAGNETIC PROSPECTING.
Vol. XIII: OCEANOGRAPHY, APPIED.

MAP ABBREVIATIONS, CIPHERS, AND MNEMONICONS

Definitions

Abbreviations are some sort of contracted or shortened word or phrase that is formed by leaving out letters or words. Distinctions are sometimes made between abbreviations and suspensions. In a true *abbreviation,* the end of a word is lopped off (e.g., *lat.,* for latitude; *cem.,* for cemetery; *diam.,* for diameter; *est.,* for estuary). In cases where the interior of the word is removed, the shortened form is properly referred to as a *suspension,* as, e.g., in *LH,* for lighthouse; *Mts,* for mountains; or *WW,* for waterworks. *Ciphers* are often simply regarded as keys to a code, whereas *mnemonicons* are words, phrases, formulas, or other aids that are designed to help in remembering. Soil maps, as prepared by the U.S. Soil Conservation Service, are perhaps among the best examples of mnemonic usage where complicated scientific names of soils are keyed to alphanumeric map symbols. The names of the specific soils are comprised by parts that designate specific attributes of the soil; e.g., the formative element Alb derived from the Latin *albus* (meaning 'white') is indicated by the mnemonicon *albino* to connote the presence of an albic horizon. Whatever the form of shortening and symbolization, familiarity with or at least access to lists of abbreviated forms that are likely to be encountered should prove useful to field geologists.

Abbreviations on Maps and in Field Notes

Maps contain a wide variety of abbreviations and symbols (see *Map Symbols*). Although most symbols are identified in the map legend or key, many abbreviations are assumed to be common knowledge. Forms of shorthand notations or specialized applications can, however, be especially obscure. Geotechnical engineering classifications, e.g., may be rather cryptic when notations are freely used without keys. For those not fully conversant with engineering classifications dealing, e.g., with shales (q.v. in Vol. XIII) and weathered materials or units specifically related to Atterberg limits (see Vol. XIII: *Atterberg Limits and Indices*) and the Unified Soil Classification System (q.v. in Vol. XIII) or well logging (q.v.) will probably find some of these technical notations a bit troublesome to comprehend.

The use of abbreviations and symbols in formal writing and reporting has become less frequent than in the past. Since World War II there has been a prolific expansion of government agencies and geoscience research organizations, particularly in the United States, Australia, and Western Europe. The increase in technical writing (see *Geologic Reports;* Vol. XIII: *Geological Communication*) has produced a plethora of abbreviations that are found especially on maps and other sorts of pictorial documents, including remotely sensed data.

With the advent of new knowledge and development or modification of technical material, users are often hard pressed to keep up with the fashions of their particular fast-changing fields. Although abbreviations are most frequently used in tabular matter, they find wide application on maps and charts of all sorts. Abbreviations are used on maps where space is insufficient for words to be spelled out or where the spelled-out words would congest the map or otherwise detract from its appearance. Preferably, only approved abbreviations should be used, but it is often necessary to improvise some abbreviations.

Map Abbreviations The general abbreviations listed in Table 1, e.g., were prescribed for use on all maps of the United States in 1940. Items contained in the list are relevant to surveys that employ old maps in investigations of changes in the morphology of portions of the Earth's surface or rates of surficial processes.

Mineral and Rock Localities Standard abbreviations (letter symbols) used to designate mineral and rock localities, as employed within the Australian Bureau of Mineral Resources, are summarized in Table 2. Included in this table are some symbols that have been adopted by the U.S. Geological Survey, but there is no complete conformity among different surveys.

Topographic Maps In the U.S. Geological Survey, the abbreviations used on topographic maps are fairly well standardized. The list in Table 3 is derived from many different sources and includes universally recognized abbreviations as well as some that have been devised specifically for use on topographic and related types of maps. As a general rule, the names of states, territories, and possessions of the United States are spelled in full when standing alone. When they follow the name of a city or any other geographical term, it is preferable to spell them out except in lists. Two forms of abbreviations for names of states are shown in Table 4. The two-letter form is authorized by the U.S. government for use with ZIP code addresses and is increasingly finding other applications as a convenient notation.

Compound Words Table 5 lists, but is by no means complete, compound words for use on maps. This list is based on U.S. Government Printing Office (GPO) rules but contains some exceptions that have been established through long usage and that are more appropriate for map use than the forms prescribed by the GPO.

Government Agencies Also useful when dealing with government agencies is knowledge of the accepted abbreviations for agencies and organizations, including international groups. The increasing tendency to use abbreviations for governmental organizations often contributes to an already confusing array of alphabet-soup abbreviations. Although few of the abbreviations listed in Table 6 actually appear on maps, the list will be helpful in map editing and related duties such as researching and evaluating sources of technical and background information.

Field Maps and Notes Some abbreviations that are useful on field maps and in field notebooks are listed in Table 7. Although this list is not official in the sense that it is adopted by any specific geological survey, it does provide some general guidelines as to how a field geologist might compile abbreviations and map notes. Because field maps and notebooks are often proprietary and may be referred to again and again long after the project has been completed, they should be comprehensible to other geologists. It is thus usually expedient to avoid making abbreviations that are too obscure. Keys to the abbreviations should always be provided (with the map or in field notes), no matter how clear or simple the contraction may seem at the time of invention.

Geographical Coordinates When referring to points of the compass on maps or in field notes, the following system may be used:

Cardinal: N, E, S, W;
Intercardinal: NE, SE, SW, NW;
Others: NNE, ENE, ESE, SSE, etc.; N by E, NE by N, NE by E, E by N, etc.

In notes and reports tabulations of coordinates may follow one of the following systems:

lat. 45°15′30″ N   long. 99°17′45″ W
lat. 45°15-30 N   long. 99°17-45 W
lat. 45°15.5′ N   long. 99°17.75′ W
lat. 45°15; N   long. 99°17;75 W

In any of these systems, periods may be omitted after *lat.* and *long.* Maps require a grid system of one sort or another, and most maps have more than one system that is indicated in the individual sheet margins. Some grids may have been devised by a particular organization for internal use, while others are part of state, national, or global schemes. Some military grids, such as those of the Royal Australian Survey Corps, and others are based on the UTM (Universal Transverse Mercator) map projection. Such systems provide numeric values for northings and eastings in addition to the usual lines of latitude and longitude.

Concluding Remarks

These tables of abbreviations give but an indication of the range of possibilities that exists for shorthand notations for map use and in field notes. Specialists in all branches of the geosciences maintain eclectic lists of novel contractions and abbreviations. Technical handbooks and style guides (e.g., U.S. Geological Survey Staff, 1958; Blackadar et al., 1980; Dietrich et al., 1982) should be consulted for state-of-the-art information relating to scientific notation (see also Vol. II: *Units, Numbers, Constants, and Symbols*). For the geologist in the field, the following manuals will provide additional access to subjects that incorporated abbreviations, ciphers, and mnemonicons: Lahee (1961), Berkman and Ryall (1976), Bates and Kirkaldy (1977), Ahmed and Almond (1983).

CHARLES W. FINKL, JNR.

(The reference list appears on page 422.)

TABLE 1. General Abbreviations Prescribed for Use on U.S. Military Maps

Abbreviation	Term or Feature	Abbreviation	Term or Feature
A	Arch or arroyo	Use with Hydrographic Symbols	
abut	Abutment	*General*	
b	Brick	Bn	Beacon
BS	Blacksmith shop	RK	Rock
bot	Bottom	Wk	Wreck
Br	Branch	LMP	Levee mile post
br	Bridge	LS	Levee station
C	Cape	TH	Towhead
cem	Cemetery	NRS	Naval radio station
con	Concrete	NRC	Naval radio direction-finder (radio compass station)
cov	Covered		
Cr	Creek	PD	Position doubtful
cul	Culvert	PA	Position approximate
DS	Drug store	ED	Existence doubtful
E	East		
Est	Estuary	*Relating to lights*	
f	Fordable	F	Fixed
Ft	Fort	FL	Flashing
GS	General store	Occ	Occulting
gir	Girder	Alt	Alternating
GM	Grist mill	Gp	Group
i	Iron	R	Red
I	Island	W	White
Jc	Junction	G	Green
kp	King-post	B	Blue
L	Lake	Sec	Sector
Lat	Latitude	(U)	Unwatched
Ldg	Landing	ev	Every
LSS	Life-saving station	m	Miles
LH	Lighthouse	min	Minutes
Long	Longitude	sec	Seconds
Mt	Mountain	vis	Visible
Mts	Mountains		
N	North	*Relating to buoys*	
nf	Not fordable	*C*	Can
p	Pier	*N*	Nun
pk	Plank	*S*	Spar
PO	Post office	*HS*	Horizontal stripes
Pt	Point	*B*	Black
qp	Queen-post	*R*	Red
R	River	*W*	White
RH	Roundhouse	*VS*	Vertical stripes
RR	Railroad	*G*	Green
2d	Second	*Y*	Yellow
S	South	*Ch*	Checkered
s	Steel		
SH	Schoolhouse	*Relating to fog signals*	
SM	Sawmill	*FB*	Fog bell
Sta	Station	*FD*	Fog diaphone
St	Stone	*FG*	Fog gun
str	Stream	*FH*	Fog horn
TG	Tollgate	*FS*	Fog siren
3d	Third	*FT*	Fog trumpet
Tres	Trestle	*FW*	Fog whistle
tr	Truss	*SB*	Submarine fog bell
WT	Water tank		
WW	Waterworks	*Relating to bottoms*	
W	West	*Cl*	Clay
w	Wood	*Co*	Coral
		G	Gravel

(*continued*)

TABLE 1. (*continued*)

Abbreviation	Term or Feature	Abbreviation	Term or Feature
Relating to bottoms			
M	Mud	lt	Light
Oz	Ooze	rky	Rocky
P	Pebbles	rot	Rotten
S	Sand	sft	Soft
Sh	Shells	sml	Small
SP	Specks	spk	Speckled
St	Stones	stf	Stiff
brk	Broken	str	Streaky
Cal	Calcareous	vol	Volcanic
crs	Coarse	bk	Black
dec	Decayed	br	Brown
dk	Dark	bu	Blue
fly	Flinty	gn	Green
fne	Fine	gy	Gray
grd	Ground	rd	Red
gty	Gritty	wh	White
hrd	Hard	yl	Yellow
lrg	Large	stk	Sticky

Source: Basic Field Manual: Conventional Signs, Military Symbols, and Abbreviations (FM21-30). Washington, D.C.: U.S. Government Printing Office, 44–47.

TABLE 2. Letter Abbreviations for Mineral and Rock Localities

Abbreviation	Mineral or Rock Locality	Abbreviation	Mineral or Rock Locality
Ae	Agate	Cf	Clay—Fuller's earth
Rc	Crushed rock aggregate	Ck	Clay—kaolin
Gr	Aggregate other than crushed rock. Uses (e.g., concrete aggregate, road metal) and rock type may be indicated in reference.	Cg	Clay—pigment
		Cp	Clay—pottery
		Cw	Clay—white, other than pottery clay
		Cy	Clay—use not specified
Ag	Silver	C	Coal
At	Alunite	Co	Cobalt
An	Anhydrite	Nb	Columbium (niobium)
Sb	Antimony	Cu	Copper
Ap	Apatite	Cn	Corundum
As	Arsenic	Cv	Covellite
Aa	Asbestos—amphibole	Di	Diamond
Ac	Asbestos—chrysotile	Dt	Diatomite
Ad	Asbestos—crocidolite	Dl	Dolerite
Az	Azurite	Do	Dolomite (suitable for industrial purposes)
Ba	Barite		
Bx	Bauxite	E	Emery
Be	Beryllium	Ep	Epsomite
Bi	Bismuth	Fe	Iron
Bo	Borax	Fs	Feldspar
Bn	Bornite	Fl	Fluorite
Bs	Building stone (rock type(s) may be indicated in reference)	G	Galena
		Gs	Gems (sapphire, topaz, zircon—to be specified in reference)
Cd	Cadmium		
Ct	Natural cement	Gl	Glauconite
Cc	Chalcocite	Au	Gold
Ch	Chalcopyrite	Gt	Graphite
Cr	Chromium	Gr	Gravel—aggregate devoid of sand
Cb	Clay—bentonite	Gp	Gypsum
Cx	Clay—fillers	He	Hematite
Cl	Clay—heavy products (brocks, tiles, pipes)	Hg	Mercury
		Im	Ilmenite

TABLE 2. (*continued*)

Abbreviation	Mineral or Rock Locality	Abbreviation	Mineral or Rock Locality
J	Jarosite	RA	Radioactive minerals, unspecified
Ck	Kaolin	RE	Rare-earth minerals
K	Potassium	Rc	Crushed rock aggregate
Ky	Kyanite	Rr	Rip-rap
Pb	Lead	Rm	Road materials, other than aggregate
Ls	Limestone (suitable for industrial purposes)	R	Rutile
		Na	Salt (sodium chloride)
Li	Lithium	Sb	Antimony
Ms	Magnesite	Sf	Sand, foundry
Mt	Magnetite	Sg	Sand, glass
Ml	Malachite	Sd	Sand, building
Mn	Manganese	Ss	Sandstone (suitable for industrial purposes)
Ma	Marble		
Hg	Mercury	Se	Selenium
Mi	Mica	Sh	Shale (suitable for industrial purposes)
Mo	Molybdenum	St	Siderite
Mz	Monazite	Si	Silica
Na	Salt (sodium chloride)	Ag	Silver
Ne	Nepheline	Sl	Slate (suitable for industrial purposes)
Ni	Nickel	Sn	Tin
Nb	Niobium (columbium)	Sp	Sphalerite
Oc	Ochre	Sr	Strontium
Ol	Oil shale or torbanite	S	Sulfur
Op	Opal	T	Talc, steatite
Os	Osmiridium	Ta	Tantalum
Pb	Lead	Th	Thorium, other than monazite
Pe	Perlite	Sn	Tin
Ph	Phosphate rock	W	Tungsten (wolfram, scheelite)
Pt	Platinum	U	Uranium
K	Potassium	V	Vanadium
Pz	Pozzolan	Ve	Vermiculite
Pu	Pumice	Wh	Whiting
Py	Pyrite	Zn	Zinc
Pp	Pyrophyllite	Zr	Zirconium
Qc	Quartz crystal		
Qt	Quartzite (suitable for industrial purposes)		

Source: From Berkman and Ryall (1976, 160–161).

TABLE 3. Abbreviations for Use on Topographic Maps of the United States

Name or Term	Abbreviation	Name or Term	Abbreviation
Abandoned	Aband	Altitude	Alt
Above	Abv	Anchorage	Anch
Abutment	Abut	Ancient	Anc
Academy	Acad	And	& (proper names)
Administration	Adm		
Aeronautical	Aero	Antenna, Antennae	Ant
Agency	Agcy	Aquarium	Aquar
Agriculture	Agr	Approximate(ly)	Approx
Agricultural	Agr	Aqueduct	Aque
Airfield	Afld	Archipelage	Arch
Airport	Aprt	Arroyo	A
Airway	Awy	Artesian Well	Art Well
Air Force Base	AFB	Asphalt	Asph
Alkali	Alk	Association	Assn
Alluvial	Alluv	Astronomical	Astr
Alternate	Alt	Astronomy	Astr

(*continued*)

TABLE 3. (*continued*)

Name or Term	Abbreviation	Name or Term	Abbreviation
Atlantic	Atl	Construction	Constr
Atoll	At	Continent	Cont
Auxiliary	Aux	Cooperative	Coop
Auxiliary Meridian	Aux Mer	Coral	Crl
Avenue	Ave	Corner	Cor
Average	Av	Corporation	Corp
Aviation	Avn	Correction	Cor
Awash (Rock)	Awsh	County	Co
Bank	Bk	Court	Ct
Bar	Bar	Courthouse	CH
Bay	B	Cove	C
Bayou	B	Covered	Cov
Beacon	Bn	Crater	Ctr
Bearing	Brg	Creek	Cr
Bench Mark	BM	Crevasse	Crev
Bluff	Blf	Crevice	Crev
Boat Harbor	B Hbr	Crossing	Xing
Bog	Bog	Culvert	Cul
Bottom	Bot	Customhouse	Cus Ho
Boulevard	Blvd	Dam	Dam
Boundary	Bdy	Degree(s)	Deg
Braided (Stream)	Brd	Department	Dept
Branch	Br	Depression	Depr
Breakwater	Bkwr	Desert	Des
Bridge	Br	Depot	Dep
Brook	Bk	Destination	Dest
Building	Bldg	Destroyed	Dest
Bureau	Bu	Diameter	Diam
Business Route	B-R	Director	Dir
Bypass	Byp	Discontinued	Discon
Cableway	Cblwy	Distance, Distant	Dist
Campground	Cpgrd	District	Dist
Canal	Can	Ditch	D
Canal Zone	C.Z.	Divide	Div
Canyon	Can	Division	Div
Cape	C	Dock	Dk
Capitol	Cap	Dockyard	Dkyd
Carline	CL	Dormitory	Dorm
Cartography	Cartog	Drive	Dr
Castle	Cas	Dunes	Dns
Cathedral	Cath	East	E
Causeway	Cswy	Electric(al)	Elec
Cave	Cv	Electricity	Elec
Cemetery	Cem	Elevated	Elev
Center	Ctr	Elevation	Elev
Central	Cent	Elevator	Elev
Channel	Chan	Emergency	Emer
Chapel	Ch	Engineer	Engr
Chemical(ly)	Chem	Engineering	Engr
Chemistry	Chem	Entrance	Entr
Chimney	Chy	Equator	Eq
Church	Ch	Equipment	Eqpt
Civil, Civilian	Civ	Equivalent	Equiv
Clearance	CL	Escarpment	Escrp
Cliff	Clf	Esker	Esk
Coast Guard Station	CGS	Established	Estab
College	Col	Estimated	Est
Colony	Col	Estuary	Est
Commercial	Cml	Executive	Exec
Company	Co	Experimental	Exper
Concrete	Conc	Exposed Wreck	Exp Wk
Conservation	Consv	Extension	Ext

TABLE 3. (*continued*)

Name or Term	Abbreviation	Name or Term	Abbreviation
Extinct	Ext	Hollow	Hol
Factory	Fcty	Hook	Hk
Falls	Fls	Horizontal	Hor
Fathom	Fm	Hospital	Hosp
Fault	Flt	Hour	Hr
Federal	Fed	House	Ho
Feet, Foot	Ft	Hummock	Hum
Ferry	Fy	Hydraulic(s)	Hyd
Field	Fld	Hydrography	Hyd
Fill	Fill	Inch	In
Filtration	Fltr	Incorporated	Inc
Firebreak	FB	Indefinite	Indef
Fishery	Fish	Indian	Ind
Fjord	Fjd	Industrial	Ind
Flat	Fl	Industry	Ind
Flood	Fl	Inspection	Insp
Flume	Flm	Institute	Inst
Ford	Fd	Institution	Inst
Forest	For	Interior	Int
Fork	Fk	Intermittent	Int
Fort	Ft	Intracoastal	Intracstl
Fortifications	Fts	Irrigation	Irr
Foundry	Fdry	Island	I
Gage	Ga	Islands	Is
Gaging Station	Gaging Sta	Islet	It
Gardens	Gdns	Isthumus	Isth
Gatehouse	GH	Jetty	Jty
General	Gen	Junction	Junc
Geodesy, Geodetic	Geod	Junior	Jr
Geographic(al)	Geog	Kilometer(s)	Km
Geography	Geog	Laboratory	Lab
Geologic(al)	Geol	Lagoon	Lag
Geology	Geol	Lake	L
Geometric(al)	Geom	Land Grant	Ld Gt
Geometry	Geom	Landing	Ldg
Geophysical	Geoph	Landing Field	Ldg Fld
Geophysics	Geoph	Landing Strip	Ldg Str
Geothermal	Geothrm	Lane	La
Glacier	Gl	Landmark	Ldmk
Government	Govt	Large	Lge
Grade	Gr	Latitude	Lat
Grant	Gt	Left	L
Gravel Pit	Gr Pit	Levee	Lv
Great	Gt	Library	Libr
Ground	Grd	Lifesaving Station	LSS
Guide Meridian	G Mer	Light	Lt
Gulch	Gl	Lighthouse	LH
Gulf	Gf	Lithograph	Litho
Gymnasium	Gym	Lithography	Litho
Hammock	Hmk	Little	L
Hachures	Hach	Location	Loc
Harbor	Hbr	Location Monument	LM
Head	Hd	Lock	Lk
Headquarters	Hdqrs	Logarithm	Log
Height	Ht	Longitude	Long
Heights	Hts	Lookout	LO
High School	HS	Lookout Tower	LT
High Water	HW	Low Water	LW
Higher High Water	HHW	Lower Low Water	LLW
Highway	Hy	Magazine	Mag
Historic(al)	Hist	Magnetic	Mag

(*continued*)

TABLE 3. (*continued*)

Name or Term	Abbreviation	Name or Term	Abbreviation
Mainland	Mnlnd	Peninsula	Pen
Maintenance Station	Maint Sta	Penstock	Pnstk
Mangrove	Mngrv	Perennial	Per
Manufacturing	Mfg	Pilot Station	Pil Sta
Marine	Mar	Pipeline	Pl
Marsh	Msh	Plateau	Plat
Maximum	Max	Pocosin	Poc
Meadow	Mdw	Point	Pt
Meander Corner	MC	Pond	Pd
Mean Higher High Water	MHHW	Population	Pop
Mean Lower Low Water	MLLW	Position	Pos
Mean Low Water	MLW	Possession	Poss
Mean Sea Level	MSL	Post Office	PO
Mean Tide Level	MTL	Power	Pwr
Memorandum	Memo	Powerhouse	PH
Memorial	Mem	Preliminary	Prelim
Meridian	Mer	Preserve	Presv
Meter	M	Principal Meridian	Prin Mer
Middle	Mid	Private	Priv
Mile, Miles	Mi	Project	Proj
Milepost	MP	Projection	Proj
Military	Mil	Prominent	Prom
Mineral	Min	Promontory	Prom
Mineral Monument	MM	Property	Prop
Mine(s)	Mns	Proposed	Prop
Minimum	Min	Province	Prov
Minute(s)	Min	Provisional	Prov
Monument	Mon	Public	Pub
Moraine	Mor	Public School	PS
Mount	Mt	Pumping Station	Pump Sta
Mountain	Mtn	Quadrangle	Quad
Mountains	Mts	Quadrant	Quad
Municipal	Muni	Quarantine	Quar
Museum	Mus	Quary	Qry
National	Nat	Race (water)	Race
National Guard	NG	Radio	Rad
Nautical	Naut	Radio Beacon	R Bn
Navigable, Navigation	Navig	Radio Detection and Ranging	Radar
Neck	Nk	Radio Tower	R Tr
North	N	Railroad	RR
Northeast	NE	Railway	Ry
Northwest	NW	Ramp	Rmp
Number	No	Range(s) (Land)	R (Rs)
Numbers	Nos	Range(s) (Hypsographic)	Rng (Rngs)
Object	Obj	Rapids	Rap
Observatory	Obsv	Ravine	Rav
Obstruction	Obstr	Reef	Rf
Ocean	O	Reference Mark	RM
Orchard	OR	Reference Mount	RM
Original	Orig	Reflector	Ref
Outcrop	OC	Refuge	Rfg
Overhead	Ovhd	Reproduction	Repro
Overpass	OP	Reservation	Res
Pacific	Pac	Reservoir	Res
Parallel	Par	Restricted	Restr
Park	Pk	Ridge	Rdg
Parking	Pkg	Right	R
Parkway	Pkwy	Rimrock	Rmrk
Pass	Pass	River	R
Passage	Pass	Road	Rd
Pavilion	Pav	Rock	Rk
Peak	Pk	Rocky	Rky

TABLE 3. (*continued*)

Name or Term	Abbreviation	Name or Term	Abbreviation
Roundhouse	RH	Territory	Terr
Route	Rte	Theater	Thtr
Ruins	Rns	Thoroughfare	Thoro
Run	R	Tidal Flat	Tid Fl
Saddle	Sad	Timber	T
Saint	St	Tollgate	TG
Sainte	Ste	Topographic(al)	Topog
Saints	SS (SS Peter & Paul Ch)	Topography	Topog
		Tower	Tr
Sanitarium	San	Township, Land	T
Sanitorium	San	Township, Civil	Twp
Sand	Sand	Townships	TS
School	Sch	Track	Tk
Science	Sci	Training	Tng
Seawall	SW	Tramway	Tram
Second(s)	Sec	Transit	Trans
Section	Sec	Transmission	Trans
Sections	Secs	Transmitter	Trans
Seminary	Sem	Transportation	Transp
Sewage	Sew	Tributary	Trib
Shoal	Shl	Tundra	Tund
Siding	Sdg	Tunnel	Tun
Signal Station	Sig Sta	Truck Route	T-R
Slip	Slp	Turnpike	Tnpk
Slough	Slu	Under construction	UC
Small	Sml	United States	U.S.
Society	Soc	University	Univ
Sound	Sd	Valley	Val
Sounding	Sdg	Vertical	Vert
South	S	Vertical Angle	VA
Southeast	SE	Vertical Angle Bench Mark	VABM
Southwest	SW	Viaduct	Viad
Spring	Spr	Village	Vil
Springs	Sprs	Vineyard	Vnyd
Spur	Sp	Volcano, Volcanic	Volc
Square	Sq	Warehouse	Whs
Standard	Std	Wash	Wsh
Standard Parallel	Std Par	Water	W
Station	Sta	Waterhole	WH
Strait	Str	Water Level	WL
Strata	Strata	Water Line	WL
Stream	Str	Water Surface	WS
Street	St	Water Tank	WT
Stripmine	SM	Water Tanks	WTs
Subdivision	Sub	Water Tower	WT
Submerged	Subm	Water Towers	WTs
Substation	Substa	Waterway	Wwy
Supplement	Supp	Waterworks	WW
Surface	Surf	Weather Service Signal Station	WS Sig Sta
Survey	Surv	Wells	Wells
Syphon	Syph	West	W
Tailings	Tail	Wharf	Whf
Technical, Technology	Tech	Wharves	Whvs
Telegraph	Teleg	Windmill	WM
Telephone	Tel	Witness Corner	WC
Television	TV	Wreck	Wk
Temperature	Temp	Wreckage	Wks
Temporary	Temp	Yard	Yd
Terminal	Term	Yards	Yds
Terrace	Ter	Year	Yr

Source: U.S. Geological Survey Staff, 1979, *Technical Instructions of the National Mapping Division.* Reston, Va.: U.S. Geological Survey, 6–13.

TABLE 4. Abbreviations for States, Protectorates, and Commonwealths

State	Abbreviation	ZIP CODE	State	Abbreviation	ZIP CODE
Alabama	Ala.	AL	Nebraska	Nebr.	NB
Alaska	Alaska	AK	Nevada	Nev.	NV
Arizona	Ariz.	AZ	New Hampshire	N.H.	NH
Arkansas	Ark.	AR	New Jersey	N.J.	NJ
California	Calif.	CA	New Mexico	N. Mex	NM
Colorado	Colo.	CO	New York	N.Y.	NY
Connecticut	Conn.	CT	North Carolina	N.C.	NC
Delaware	Del.	DE	North Dakota	N. Dak.	ND
District of Columbia	D.C.	DC	Ohio	Ohio	OH
Florida	Fla.	FL	Oklahoma	Okla.	OK
Georgia	Ga.	GA	Oregon	Oreg.	OR
Guam	Guam	GU	Pennsylvania	Penn.	PA
Hawaii	H.I.	HI	Puerto Rico	P.R.	PR
Idaho	Idaho	ID	Rhode Island	R.I.	RI
Illinois	Ill.	IL	Samoa	Samoa	—
Indiana	Ind	IN	South Carolina	S.C.	SC
Iowa	Iowa	IA	South Dakota	S. Dak.	SD
Kansas	Kans.	KS	Tennessee	Tenn.	TN
Kentucky	Ky.	KY	Texas	Tex.	TX
Louisiana	La.	LA	Utah	Utah	UT
Maine	Maine	ME	Vermont	Vt.	VT
Maryland	Md.	MD	Virginia	Va.	VA
Massachusetts	Mass.	MA	Virgin Islands	V.I.	VI
Michigan	Mich.	MI	Washington	Wash.	WA
Minnesota	Minn.	MN	West Virginia	W. Va.	WV
Mississippi	Miss.	MS	Wisconsin	Wisc.	WI
Missouri	Mo.	MO	Wyoming	Wyo.	WY
Montana	Mont.	MT			

TABLE 5. Word Compounds for Use on Maps of the U.S. Geological Survey

aboveground	airstrip	badlands	bellboy
above-water	airway	bagpipe	bell buoy
acidworks	alleyway	baggage room	bench mark
acre-foot	angle iron	bakeshop	bighorn
adderstongue	anthill	ballroom	billboard
aide-de-camp	anybody	bandstand	birdhouse
airbase (photog.)	anyone	barroom	birdseye
air base (military)	anything	barbershop	birthplace
airborne	anywhere	bargehouse	bitterroot
aircraft	aquaplane	barnyard	bittersweet
airdrome	arborvitae	barrelhead	blackberry
airduct	arborway	baseball	blackfoot
airfield	areaway	base line	blackthorn
airfreight	arrowhead	basketball	blockhouse
airgap	ashheap	basswood	blowhole
airhole	ashpit	batwing	bluebell
airlane	axhead	bathhouse	blueberry
airlift		battlefield	bluebird
airline (aviation)	backbone	battleground	bluefish
airmail	backbreaker	battleship	bluegill
airman	backdrop	bayberry	bluegrass
airpark	backfill	beachhead	blue line
airphoto	backfire	bearwallow	blue-line
airplane	backpack	bedrock	blueprint
airport	backstop	beechnut	boardinghouse
airpower	backwater	beefsteak	boarding school
airship	backyard	beehive	boardwalk

TABLE 5. (*continued*)

boathouse	cherrystone	duckpond	gasworks
boatyard	chickenhouse	dugout	gatehouse
bobcat	chicken yard	dugway	gateway
bobsled	chokecherry	dyeworks	glassworks
bobwhite	churchyard		goldenrod
bogland	clamshell	earthbound	goldfield
boilerhouse	claybank	earthfill	gold mine
bonefish	claypit	earthquake	gooseberry
borderland	clayworks	earthslide	grandstand
border line	clearinghouse	earthwall	grapevine
boxelder	clockwise	earthwork	gravel pit
boxwood	cloverleaf	east-northeast	graveyard
brassworks	clubhouse	east-southeast	greenbrier
breakwater	coalfield	eastward	greenhorn
breastworks	coal mine	elderberry	greenhouse
breezeway	coalpit	elkhorn	gridline
brickkiln	coastline	enginehouse	gristmill
brickwork	codfish	engine shop	groundhog
brickyard	commonwealth	engine yard	guardhouse
bridgehouse	copperhead	entryway	guesthouse
broadcast	copper mine	everglade	guideline
broadside	copperplate	evergreen	
broadway	cornerstone	expressway	hackberry
brotherhood	cottonfield		hailstone
brushwood	cotton mill	fairground	half moon
buckbrush	cottonseed	farmhouse	halftone
buckeye	cottonwood	farmland	hayfield
buckhorn	courthouse	faultline	haystack
buckskin	courtyard	fence line	hazelnut
bucktail	crabapple	fenland	headgate
buckthorn	cranberry	ferryboat	headland
buckwheat	creekbed	ferry slip	headquarters
bulkhead	crossarm	ferryway	headwall
bulldog	crossbar	fieldhouse	headwater
bullfrog	crossframe	firebreak	heathland
bullhead	crosshatch	firehouse	hedgerow
bullpen	crosslines	firewall	hencoop
bumblebee	crossroad	fisherman	henhouse
bunkhouse	crosstie	fishpond	herringbone
burdock	crosswalk	fishtrap	hickorynut
busline	crowfoot	fishweir	highland
buttercup	customhouse	flagpole	high tide
butterfly	cutoff	flatland	high water
buttermilk		floodgate	highway
butternut	damsite	floodmark	hillside
bypass	dancehall	flood plain	hilltop
	daybeacon	floodwall	hogback
cableway	deadline	flyway	hog's-back (geol.)
campfire	diamondback	foghorn	homestead
campground	dockyard	football	honeysuckle
campsite	dogwood	footbridge	horseshoe
canebrake	downslope	foothill	hothouse
canvasback	downstream	footnote	houseboat
carbarn	drag strip	footpath	huckleberry
carline	drainpipe	forebay	
carriageway	drawbridge	foreshore	iceberg
catbird	drawspan	freeway	icefield
catbrier	drive shaft	frogpond	intake
catfish	drive-in theater	frostline	interstate
cattail	driveway		iron mine
catwalk	drugstore	gasfield	ironwood
causeway	drydock	gashouse	ironworks
centerline	drywash	gasline	
checkpoint	duckblind	gas main	jackrabbit
		gas well	

(*continued*)

TABLE 5. (*continued*)

keystone	parkway	sagebrush	speedway
	pass point	saltmarsh	spillway
lakebed	peelcoat	saltpeter	stagecoach
landforms	penstock	saltpond	stageline
land line	permafrost	salt water	standpipe
landmark	pierhead	saltwater	steamboat
land office	pigweed	saltworks	steamplant
landscape	pipeline	sandbar	steamship
landslide	planetable (surv.)	sand dune	steelworks
layout	plateholder	sandhill	stereomodel
lifeboat	playfield	sandpit	stereopair
light buoy	playground	sandspit	stockyard
lighthouse	plumbline	sawmill	storehouse
lightship	plumb point	sawtooth	straightedge
light-table (carto.)	poison-ivy	schoolhouse	streambed
limekiln	poison-oak	scribe coat	streamflow
limestone	pondlily	seabed	streetcar
lookout	poorhouse	seaboard	substation
lowland	potash	seacoast	sugar beet
low tide	pothole	sea gate	sugarcane
low water	powerhouse	seagoing	sugarloaf
lumberyard	powerline	seagull	sugar maple
	powerplant	sea level	superhighway
mainland	pressplate	sea lion	suffline
mangrove	pricklypear	seaplane	swampland
manmade	pronghorn	seaport	sweetbrier
mapmaking	proofread	seashore	sweet corn
marshland	proofsheet	seawall	sweetfern
meetinghouse	pulpmill	sea water	sweetgum
metalworks	pumphouse	seaway	switchback
milepost	pussywillow	sewerline	switch tower
milldam		sewer pipe	switchyard
millpond	quicksand	shadberry	
millstone	quicksilver	shellfish	tableland
moonlight		shipway	tabletop
moonshine	racecourse	shipwreck	tanbark
mudflat	racetrack	shipyard	tarpit
	raceway	shoreline	tarworks
neatline (map)	ragweed	shotgun	taxiway
northbound	railhead	sidelap	tenderfoot
northeast	railroad	sidewalk	thronapple
north-northeast	railway	signpost	thoroughfare
north-northwest	ranchhouse	sinkhole	through road
northwest	range line	skidway	throughway
	raspberry	skyline	thumb screw
offset	rattlesnake	skyway	thunderbolt
offshore	redfern	slateworks	tidal flat
oilfield	redfish	slaughterhouse	tideflat
oil shale	redwood	sluicebox	tide gauge
oil well	ricefield	sluice gate	tideland
open-air theater	right-of-way	sluiceway	tiderace
opencut (mining)	rimrock	smokestack	tidewater
open pit	riverbank	snowfield	timberline
outline	riverbed	snowline	toll bridge
overall	river bottom	snowshed	tollgate
overlay	riverside	snowslide	tollhouse
overlook	roadbed	soapstone	toll road
overpass	roadhouse	sourdough	tombstone
overprint	roadway	southbound	tomcat
oxbow	rockfill	southeast	townhall
oysterbed	rockslide	south-southeast	township
oystershell	roominghouse	south-southwest	townsite
	roundhouse	southwest	towpath
packinghouse	runway	spearhead	trafficway
papermill			

TABLE 5. (*continued*)

trainshed	walkway	waterhole	west-southwest
tramway	warehouse	water level	whirlpool
trolley line	wasteland	waterlily	wildcat
truckline	waste pipe	waterline (map)	wildfowl
truckway	wastewater	water line (text)	wildlife
turnpike	wasteway	watershed	windbreak
turntable	watercourse	watertank	windmill
	watercress	water tower	wingwall
underground	waterfall	waterway	woodbine
underpass	waterfowl	waterworks	woodland
upstream	waterfront	wellhouse	workhouse
	water gage	westbound	
vineyard	watergate	west-northwest	

Source: U.S. Geological Survey Staff, 1979, *Technical Instructions of the National Mapping Division.* Reston, Va.: U.S. Geological Survey, 19–23.

TABLE 6. Abbreviations for Government Agencies and Organizations

Agency	Abbreviation
Administrative Division (USGS)	AD
Advance Research Projects Agency (DOD)	ARPA
Agency for International Development	AID
Agricultural Stabilization and Conservation Service	ASCS
American Congress on Surveying and Mapping	ACSM
American Petroleum Institute	API
American Red Cross	ARC
American Society of Cartographers	ASC
American Society of Civil Engineers	ASCE
American Society of Photogrammetry	ASP
American Society for Testing Materials	ASTM
Appalachian Regional Commission	ARC
Board on Geographic Names	BGN
Bonneville Power Administration	BPA
Bureau of Indian Affairs	BIA
Bureau of Labor Statistics	BLS
Bureau of Land Management	BLM
Bureau of Mines	BM
Bureau of Reclamation	BR
Bureau of the Census	BC
Central Intelligence Agency	CIA
Chief, Topographic Division	CTD
Civil Aeronautics Board	CAB
Commission on Civil Rights	CCR
Commodity Credit Corporation	CCC
Commodity Futures Trading Commission	CFTC
Computer Center Division (USGS)	CCD
Conservation Division (USGS)	CD
Council of Economic Advisers	CEA
Defense Mapping Agency	DMA
Aerospace Center	DMAAC
Hydrographic/Topographic Center	DMAH/TC
Delaware River Basin Commission	DRBC
Department of Defense	DOD
Earth Resources Observation System	EROS
Employee's Compensation Appeals Board	ECAB
Engineer Research and Development Laboratories	ERDL
Environmental Data Service (NOAA)	EDS
Environmental Protection Agency	EPA
Environmental Research Laboratories (NOAA)	ERL
EROS Data Center (LIA)	EDC

(*continued*)

TABLE 6. (*continued*)

Agency	Abbreviation
Farm Credit Administration	FCA
Farmers Home Administration	FHA
Federal Aviation Administration	FAA
Federal Communications Commission	FCC
Federal Deposit Insurance Corporation	FDIC
Federal Emergency Management Agency	FEMA
Federal Energy Regulatory Commission	FERC
Federal Highway Administration	FHWA
Federal Home Loan Bank Board	FHLBB
Federal Housing Administration	FHA
Federal Insurance Contributions Act (Social Security)	FICA
Federal Maritime Commission	FMC
Federal Mediation and Conciliation Service	FMCS
Federal Reserve System	FRS
Federal Trade Commission	FTC
Fish and Wildlife Service	FWS
Food and Agriculture Organization (United Nations)	FAO
Food and Drug Administration	FDA
Foreign Agriculture Service	FAS
General Accounting Office	GAO
Geography Program (USGS LIA)	GP
Geologic Division (USGS)	GD
Department of Health, Education, and Welfare	HEW
Heritage Conservation and Recreation Service	HCRS
Department of Housing and Urban Development	HUD
Inter-American Defense Board	IADB
Inter-American Geodetic Survey	IAGS
Interdepartment Radio Advisory Committee	IRAC
Internal Revenue Service	IRS
International Atomic Energy Agency	IAEA
International Boundary Commission (U.S.–Canada)	IBC
International Boundary and Water Commission (U.S.–Mexico)	IBWC
International Civil Aviation Organization	ICAO
International Cartographic Association	ICA
International Finance Corporation	IFC
International Labor Organization	ILO
International Telecommunications Union	ITU
International Trade Commission	ITC
International Typographic Composition Association	ITCA
Institute of Radio Engineers	IRE
Joint Committee on Printing	JCP
Joint Mapping and Photography Committee	JMPC
Land Information and Analysis Office (USGS)	LIA
Library of Congress	LC
Maritime Administration	MA
Military Airlift Command	MAC
Mississippi River Commission	MRC
Missouri River Basin Commission	MRBC
National Academy of Sciences	NAS
National Aeronautics and Space Administration	NASA
National Archives and Records Service	NARS
National Bureau of Standards	NBS
National Cartographic Information Center	NCIC
National Geodetic Survey (NOAA, NOS)	NGS
National Labor Relations Board	NLRB
National Oceanic and Atmospheric Administration	NOAA
National Ocean Survey (NOAA)	NOS
National Park Service	NPS
National Science Foundation	NSF
National Security Council	NSC
National Society of Cartographers	NSC
National Weather Service (NOAA)	NWS
Naval Oceanographic Office	NAVOCEAO

TABLE 6. (*continued*)

Agency	Abbreviation
North Atlantic Treaty Organization	NATO
Office of National Petroleum Reserve (USGS)	ONPR
Office of Personnel Management	OPM
Office of the Chief of Engineers	OCE
Pan American Institute of Geography and History	PAIGH
Public Buildings Service	PBS
Public Health Service	PHS
Publications Division (USGS)	PD
Research and Development Board	RDB
Resource and Land Investigations Program (USGS LIA)	RALI
Rural Electrification Administration	REA
Securities and Exchange Commission	SEC
Selective Service System	SSS
Small Business Administration	SBA
Social Security Administration	SSA
Soil Conservation Service	SCS
Southeast Asia Treaty Organization	SEATO
Strategic Air Command	SAC
Surveys, Investigations, and Research	SIR
Tennessee Valley Authority	TVA
Topographic Division (USGS)	TD
United Nations	U.N.
United Nations Educational, Scientific and Cultural Organization	UNESCO
United States Department of Agriculture	USDA
United States Air Force	USAF
United States Army	USA
United States Coast Guard	USCG
United States Corps of Engineers (Army)	USCE
United States Employment Service	USES
United States Forest Service	USFS
United States Geological Survey	USGS
United States Information Agency	USIA
United States Marine Corps	USMC
United States Navy	USN
United States Postal Service	USPS
Veterans Administration	VA
Water Resources Division (USGS)	WRD
World Bank	WB
World Health Organization	WHO

Source: U.S. Geological Survey, *Technical Instructions of the National Mapping Division.* Reston, Va.: U.S Geological Survey, Technical Information Office, 13–16.

TABLE 7. Abbreviations that Can Be Used
on Field Maps and in Field Notebooks

Term	Abbreviation
Rocks and Minerals	
agglomerate	aggl.
alkali feldspar	alk. feld.
andalusite	andal.
andesite	and.
basalt	bas.
biotite	bi.
breccia	brc.
conglomerate	congl.
cordierite	cord.
dolerite	dol.
gabbro	gab.
garnet	gar.
gneiss	gn.
granite	gr.
hornblende	hb.
hornfels	hf.
ignimbrite	ig.
kyanite	ky.
limestone	lst.
marble	mbl.
microgranite	mgr.
migmatite	migt.
mudstone	mdst.
muscovite	musc.
mylonite	myl.
pegmatite	peg.
phyllite	phyll.
plagioclase	plag.
pyroxene	px.
quartz	qtz
quartzite	qtzite
rhyolite	rhy.
sandstone	sst.
schist	sch.
sillimanite	sill.
siltstone	stst.
slate	slt.
syenite	sy.
Descriptive Terms	
amygdaloidal	amgy.
arenaceous	aren.
argillaceous	argil.
bedded	bdd
cross bedded	xbdd
interbedded	ibdd
calcareous	calc.
cleaved	clvd
crystalline	xtline
folded	fldd
foliated	flotd
jointed	jntd
linear	linr
lineated	lintd
planar	plnr
porphyritic	porph.
veined	vnd.
vesicular	vesic.
volcanic	volc.

Source: From Ahmed and Almond (1983, 70–71).

References

Ahmed, F., and D. C. Almond, 1983, *Field Mapping for Geology Students*. London: Allen and Unwin, 71p.

Bates, D. E. B., and J. F. Kirkaldy, 1977, *Field Geology*. New York: Arco, 215p.

Berkman, D. A., and W. R. Ryall, 1976, *Field Geologists' Manual*. Parkville, Victoria: Australasian Institute of Mining and Metallurgy, 295p.

Blackadar, R. G., H. Dumych, and P. J. Griffin, 1980, Guide to authors—A guide to the preparation of geological maps and reports. *Geol. Survey Canada Misc. Rept. 29,* 66p.

Dietrich, R. V., J. T. Dutro, and R. M. Foose, 1982, *AGI Data Sheets for Geology in the Field, Laboratory, and Office*. Falls Church, Va.: American Geological Institute.

Lahee, F. H., 1961, *Field Geology*. New York: McGraw-Hill, 926p.

U.S. Geological Survey Staff, 1958, *Suggestions to Authors of the Reports of the United States Geological Survey,* 5th ed. Washington, D.C.: U.S. Government Printing Office, 255p.

Cross-references: *Cartography, General; Field Notes, Notebooks; Geological Highway Maps; Geological Survey and Mapping; Geological Writing; Geologic Reports; Map Colors, Coloring; Map Series and Scales; Map Symbols; Maps, Logic of; Prospecting; Rock Color Chart; Samples, Sampling; Well Logging.*

MAP AND CHART DEPOSITORIES

Most drafted maps are prepared for reproduction by some kind of mechanical process to obtain multiple copies. This entry is concerned not with the making of maps in the field (see *Geological Survey and Mapping*) or with the means of mass reproduction but with the types of maps and charts that are already available; it also mentions some of the mapping agencies that mass produce maps and reviews the status of world map collections and depositories.

One of the primary purposes of field geology is the preparation of field maps, although there are differences in practice depending on whether the map being produced is basic, derived, or for a special purpose (Keates, 1973). A basic geologic map is constructed from an original field survey (see *Cartography, General*) and is likely to be at large or medium scale. Many maps, in contrast, are derived mainly from existing maps. This is true of many medium- and small-scale maps. Special-purpose maps typically combine features of basic and derived maps. With large-scale geological maps, e.g., the topographic base will often be taken from existing topographic maps. The geological information will then be constructed on this base either in the field (in rough form) or from notebooks (see *Field Notes, Notebooks*) soon after its collection. Field geologists thus often use other maps in the preparation of new basic maps or in the revision of existing maps.

The availability and sources of maps is of crucial interest to geologists in general but is particularly relevant to the field geologist and prospector. National, state, or provincial surveys should be consulted to determine the availability of geological maps (see *Field Geology*). Many surveys maintain open-file reports and provide access to works in progress and preliminary maps. Such advance information is often extremely useful, especially in remote areas that lack the kind of documentation so typical of developed regions.

Types of Classification

The purpose of a map, as well as the nature of its content, provides a primary basis of classification. Topographic maps usually form the background for thematic maps on geologic topics, surficial materials, soil, and similar subjects. Maps may be classified according to scale (see *Map Series and Scales*), content (see *Maps, Logic of*), or derivation (publishing agency).

World Status of Maps and Charts

Before World War I, only a few countries such as the United Kingdom, France, and Germany had detailed maps covering their whole national areas. Today, many countries have complete coverage of their territories. About a third of the world's land area has been mapped at scales of 1:75,000 and greater. Another third is covered by medium-scale (up to 1:25,000) topographic maps.

World series maps are commonly on the 1:1,000,000 scale; they include the aeronautical charts and International Map of the World. Poorly mapped areas of the world are indicated on such maps by broken or dashed lines. Nautical chart coverage of the world was largely complete for areas bordering the continents and most islands by the end of the nineteenth century, but some gaps remain. With satellite navigation, some deep-sea areas have been accurately mapped in recent years. Arctic and Antarctic regions and parts of the South Pacific and South Atlantic are the most deficient in detailed coverage. The Defense Mapping Agency, in association with the British Admiralty and other chart-producing countries, maintains worldwide bathymetric coverage that is constantly updated. The National Ocean Service maintains charts of U.S. coastal waters. The International Hydrographic Organization, based in Monaco, produces revisions to the General Bath-

ymetric Chart of the Oceans as new data are accumulated.

Aerial Photographs and Remotely Sensed Data

Aerial photographs provide a useful supplement to topographic maps. Where maps are not available, aerial photographs may serve as substitutes (see Vol. XIII: *Remote Sensing, Engineering Geology*). Most of the world is covered by aerial photography (Dickinson, 1979), but such photography is classified for security reasons in many countries.

In the United States the National High Altitude Photography Program (NHAP) makes contemporary aerial photographs, both black and white and color infrared, and these are easily accessible. The NHAP Program, a federal multiagency activity coordinated by the U.S. Geological Survey, was initiated in 1978. Fourteen federal agencies currently participate in the NHAP by providing funds for the development of the photographic database. The contributing agencies include the Agricultural Stabilization and Conservation Service, Soil Conservation Service, Statistical Reporting Service, Forest Service, National Oceanic and Atmospheric Administration, Defense Mapping Agency, Bureau of Indian Affairs, Bureau of Land Management, Bureau of Mines, Fish and Wildlife Service, National Park Service, Office of Surface Mining, U.S. Geological Survey, and the Tennessee Valley Authority. Satellite imagery covers the world and is generally available for many regions (but not in Eastern Bloc countries) in color, false color, or black and white. The EROS Data Center in Sioux Falls, South Dakota, maintains extensive collections of LANDSAT imagery as well as photography from Skylab, Apollo, and Gemini spacecraft.

Types of Maps, Charts, and Atlases Available

Although field geologists are primarily interested in geological maps, as defined in the broadest sense, there are additional kinds of maps that provide useful information. The principal map types of use to geologists include, e.g., aeronautical, political districts, highways (national and secondary), historical, hydrographic (coastal areas, inland waters, foreign waters), national forests, vegetation, public land survey plats, soil, and topographic. These kinds of maps, and others, are often assembled in atlases (see later discussion). The *National Atlas of the United States of America* (1970, published by the U.S. Geological Survey), e.g., contains contributions from all government mapping agencies. Most countries have centers where detailed information on exist-

TABLE 1. Some Primary Mapping Agencies for Selected Countries

Country	Mapping Agency
Australia	Bureau of Mineral Resources, Geology and Geophysics
	Division of National Mapping, Department of National Development
Brazil	Servico Geografico do Exercito
Canada	Geological Survey of Canada
	Surveys and Mapping Branch, Department of Energy, Mines and Resources
Chile	Instituto Geografico Militar
Federal Republic of Germany	Institut für Landeskunde Geodasie
	Bundesanstalt für Geowissenschaften und Rohstoffe
	Niedersächsischen Landesamtes für Bodenforschung
France	Bureau de Récherches Géologiques et Minères
	Institut Geographique National
United States	Agricultural Stabilization and Conservation Service
	Bureau of Land Management
	Bureau of Mines
	Corps of Engineers
	Defense Mapping Agency
	Environmental Protection Agency
	Fish and Wildlife Service
	Forest Service
	Geological Survey
	National Ocean Survey
	Soil Conservation Service
United Kingdom	Ordnance Survey and Directorate of Overseas Surveys
USSR	Glavnoye Upravlenie Geodezii i Kartografi

ing map series and related data may be obtained (Table 1). This service is performed in the United States by the Map Information Office of the U.S. Geological Survey. This survey publishes and distributes state indexes that show map coverage and give ordering information.

Military Maps Military agencies are important mapping centers in many countries and regional territories. For countries such as the United States and Australia, both civilian and military organizations collaborate in developing their respective programs and in the actual mapping. In parts of the Middle East, South America, Africa, Eastern Europe, and Southeast Asia, it is sometimes difficult to obtain geological maps because they may be regarded as classified information. It is almost impossible to procure many different kinds of maps on the USSR, especially topographic and geological maps. The Army Map Service in Washington, D.C., however, will supply topographic maps of the USSR.

Professional Societies Large societies such as the American Geographical Society, the National Geographic Society, and the Royal Geographical

Society (London) are important map-making and collection centers. Technical societies such as the American Congress on Surveying and Mapping, American Society of Photogrammetry, American Society of Civil Engineers, and others support certain mapping programs.

Atlases, Folios, Photo Albums Atlases often provide another source of archival information that is useful to field geologists. Although an atlas is technically a book of maps, compilations may take many forms. Some atlases are bound in book form while others are issued piecemeal, one map at a time, or serially at irregular intervals. Serial issues have the advantage of providing information and updating it as it becomes available. Individual maps, usually produced at the same scale and in a predetermined format, can be added to a file or binder as new acquisitions, revisions, or printings. Bound volumes such as the *The Times Atlas of World History* (Maplewood, N.J.: Hammond, 1979) and the *Atlas of Man* (New York: St. Martin's, 1978), serve as a backdrop to modern exploitation of mineral resources in remote areas. World history has been influenced markedly by humankind's quest for mineral wealth (Poss, 1975), and it seems reasonable to assume that future adventures of the miner and prospector as well as sophisticated explorationist will be affected by the lay of the land and extent of socioeconomic systems. Other important sources of initial interest include, e.g., the *Bertelsmann Hausatlas* (Gutersloh, West Germany: Bertlesmann, 1969), the *Life Pictorial Atlas of the World* (New York: Time Inc., 1961), the *Times Atlas of the World* (Boston: Houghton Mifflin, 1967), and the *National Geographic Atlas of the World* (Washington, D.C: National Geographic Society, 1975). Excellent regional atlases have been issued by Readers Digest. Atlases focusing on specific world regions or topics are too numerous to document here. Further information on sources are given in Drecka and Tuszynska-Rekawek (1964) and more recently by Ristow (1976). Most atlases of general interest to the field geologist belong to national, regional, or special-purpose collections.

National and Regional Atlases

National atlases portray the physical, economic, and social conditions of the country. With few exceptions, most atlases represent diligent efforts to produce a work worthy of national acclaim. Such works of national pride are printed on an excellent grade of paper and employ harmonious blends of colors that reflect geographic and cartographic scholarship (Yonge, 1957). National atlases have been produced by official agencies in more than forty countries. Federal agencies, professional organizations, and commercial firms ordinarily participate in the production of a national atlas. In the United States, e.g., an interagency National Atlas Committee (renamed the Committee on the National Atlas of the United States) coordinated efforts by eighty-four divisions and bureaus of the Departments of Agriculture, Commerce, Defense, Interior, and Transportation. The American Association of Museums, American Geographical Society, Hammond Incorporated, and Rand McNally & Company also played key roles in the seven-year development scheme.

A recent trend among atlases published in languages less familiar than English is the provision for translated map titles, legends, and text material. Map scales and grid systems differ from one country to another. The Universal Transverse Mercator (UTM) grid is commonly used in many regions as are various military grids. Marginal information is now more complete than it was in the past. Maps lacking complete identification data may still be troublesome, especially when issued singly over a period of years as part of a folio.

European national atlases tend to be high-quality productions with regard to toponomy and statistics. Covering virtually every aspect of physical, political, historical, and social geography, the following editions are of particular note: *Atlas von Niederösterreich* (Vienna: Freytag-Berndt u. Artaria, 1951), *Atlas de Belgique* (Brussels: Comité National de Géographie, 1951), *Atlas de France* (Paris: Comité National de Géographie, 1953), *Atlas over Sverige* (Stockholm: AB Kartografiska Institutet, 1953), *Atlas Polski* (Warsaw: Centralny Urzad Geodezji i Kartografii, 1953), and *Atlas de Portugal* (Coimbra: Aristides de Amroim Girao, 1941). Comprehensive atlases have also been prepared for Czechoslovakia (*Atlas Republiky Ceskoslovenske*), Denmark (*Atlas of Denmark*), Finland (*Atlas of Finland*), Germany *(Deutscher Planungsatlas)*, Italy (*Atlante fisico-economico d'Italia*), and Norway (*Okonomisk-geografisk atlas over Norge*). In most of these kinds of works, useful information pertaining to general geology, geomorphology, gravity, soil cover, relief, and hydrography, among many other geologically related topics is found.

Much of Asia, Africa, and Central and South America is sadly deficient in national atlases. Many new nations, now free of European hegemony, have not attempted to develop a national atlas. Notable early exceptions include the *Atlas of Israel* (Jerusalem: Department of Surveys, 1956), *Atlas du Maroc* (Rabat: Comite de Geographie du Maroc, 1954), *Atlas Estadistico de Costa Rica* (San Jose: Ministerio de Economia y Hacienda, 1953), *Republica Dominacana: Album Estadistico Grafico* (Cuidad Truhillo: Direccion

General de Estadistica, 1944), and *Atlas Estadistico do Brasil* (Rio de Janeiro: Departamento Nacional do Cafe, 1941).

The first comprehensive atlas of Australia, the *Atlas of Australian Resources* (Sydney: Angus and Robertson, 1953), was prepared as part of a national effort to assess natural resources and determine the potential for development. Produced at a scale of 1:6,000,000, these looseleaf color maps were accompanied by additional diagrams and bibliographical notes. The atlas proved most successful and evolved into a series covering topics such as dominant land use, croplands, vegetation regions, soils, geology, climatic regions, temperatures, and rainfall.

Examples of North American atlases focus on the *National Atlas of the United States* (Washington, D.C.: National Research Council and National Academy of Sciences, 1956) and, more recently, the *National Atlas of the United States* (Washington, D.C.: U.S. Geological Survey, 1970). The standard scale of the 1956 atlas was 1:10,000,000, although other scales were used for special needs. Maps showing geology, soils, and major land uses of the nation were particularly effective and informative, possibly because the cartographers chose to deploy subtle harmonious colors and dot patterns to advantage in many ten-color maps. The newer 1970 national atlas contains more than eighty thematic maps, many accompanied by textual material. These kinds of thematic maps help realize the eventual achievement of worldwide coverage as envisaged by the National Atlas Commission of the International Geographical Union. Of particular interest to geologists are the special subject maps (scales of 1:7,500,000, 1:17,000,000, and 1:34,000,000 presented to facilitate comparison of data) that deal with landforms (relief, physiography, physiographic divisions, land-surface form), geophysical forces (gravity, earthquakes, magnetism), geology (tectonic features, glacial geology, karstlands and caverns), marine features (coastal landforms, ocean sediments and currents, tides, sea temperature and salinity, wave heights), soils, vegetation, climate, and water (surface water, floods and drought, groundwater, water resources).

Compendiums of special-purpose maps serve geoscientists with restricted interests and needs. Special atlases dealing with almost every aspect of earth science are legion. The scope for increased specialization is almost unlimited as scientists probe deeper into the workings of natural systems, especially as they are affected by misadventures or modern civilization. Atlases geared to applied (field) studies are becoming increasingly popular and necessary. The following examples illustrate a range of applied (field) topics.

The *Concise Atlas of World Geology and Min-eral Deposits* (London: Mining Journal Books, 1980) gives a broad view of world geology and mineral deposits. The color maps show geological structures, stages and processes of Earth history, and important mineral areas and deposits. The maps are accompanied by tables outlining recent production and reserves of principal metals, nonmetals, and energy minerals.

Snead's (1982) photographic atlas of *Coastal Landforms and Surface Features* treats most types of coasts found around the world. The photographic record of nearly all types of features provides a reference base for comparative and illustrative purposes (see also the *Encyclopedia of Beaches and Coastal Environments,* Schwartz, 1982). The atlas and encyclopedia support an expanding interest in coastal environments that has been stimulated by storm damage and pollution. Increased tanker traffic in the Straits of Florida, e.g., created concern of collision or groundings that would cause a major oil spill to reach coastal habitats in southern Florida. The *South Florida Oil Spill Sensitivity Atlas* (Miami: South Florida Regional Planning Council, 1981) contains a series of twenty-three coastal maps that rank the sensitivity of coastal segments to oil pollution. The Environmental Sensitivity Index (ESI) rates shorelines and wildlife sensitivity to spilled oil on a scale of 1 to 10 for eleven shoreline types. The *Serial Atlas of the Marine Environment* (Spinner, 1969) is another example of an effort to assimilate marine resource data in cartographic form for planning purposes. Much broader-based special purpose atlases include the *Rand McNally Atlas of the Oceans* (New York: Wiley, 1977), the *Times Atlas of the Oceans* (New York: Van Nostrand Reinhold, 1984), and the *Ecological Atlas of Foraminifera of the Gulf of Mexico* (Woods Hole, Mass.: Marine Science International, 1981).

Mission to Earth: Landsat Views the World (Short et al., 1976) contains an atlas that reviews the practical applications of LANDSAT to resource evaluation and decision-making activities. The role of space platform data-gathering capabilities is brought to the fore in examples of geological mapping, mineral resource evaluation, water inventory, coastal and wetlands management, and land-use assessment. The main picture gallery, an atlas of color photographs of the Earth taken from space, covers a wide range of surface features. The standard mode for each scene is a 1:1,000,000 scale format that covers about 34,000 km^2 of territory. This type of photographic atlas demonstrates a wide range of uses for space imagery.

International Organizations

Many societies and other types of organizations are engaged in activities associated with maps and mapping. In general, they encourage cooperation through meetings and publication of professional papers in the journals. Standardizations of map treatments and conventional signs (see *Map Symbols*), as well as promotion of progress in technical processes, are additional objectives of such groups.

The United Nations Office of Cartography plays various roles in the activities noted previously. The Inter-American Geodetic Survey is a special unit of the U.S. Army Corps of Engineers. It was organized to assist with the completion of mapping in the Americas. The Pan American Institute of Geography and History has sponsored regular meetings on cartography. The International Hydrographic Bureau serves as a clearinghouse for information related to hydrography and charting. Other groups that promote progress in the various aspects of mapping and charting are the International Association of Geodesy, the International Cartographic Association, the International Civil Aviation Organization, the International Geographical Union, the International Federation of Surveyors, the International Society for Photogrammetry and Remote Sensing, and the International Union of Geodesy and Geophysics.

Map Collections

The worldwide list given in Table 2 is limited to primary map and chart depositories. Geographical libraries consisting primarily of monographic collections are excluded. Table 2 also includes locations of national collections, regional directories, and military holdings (within the parameters of security restrictions). Although map collections in Canada, the Federal Republic of Germany, the United Kingdom, and the United States account for nearly half the total listings, only selected collections are included in the list. More complete listings for these countries are given in Kramm (1959), Lewanski (1965), Busse and Ernestus (1972), Bergen (1973), and Ristow (1976).

CHARLES W. FINKL, JNR.

(The reference list appears on page 435.)

TABLE 2. A Selected List of World Map and Chart Collections

ARGENTINA
Mapoteca
Instituto Geografico Militar
Cabildo 381
Buenos Aires

AUSTRALIA
A.C.T.
Map Collection
Dept. of Geography
Australian National University
Canberra 2601

Map Library
Div. of National Mapping
Dept. of National Development
AMP Building, Hobart Place
Canberra City 2601

New South Wales
General Reference Library
Library of New South Wales
Macquarie Street
Sydney

Map Library
Department of Geography
University of Sydney
Sydney

Department of Geography Map
Library
University of New England
Armidale 2351

Map Library
Geography Department
University of Newcastle
Shortland 2308

South Australia
Map Collection
State Library of South Australia
Box 419, G.P.O.
Adelaide 5001

Map Library
Department of Geography
University of Adelaide
North Terrace
Adelaide 5000

Geography Map Library
School of Social Sciences
Flinders University of South
Australia
Stuart Rd.
Bedford Park 5042

Queensland
Map Collection
Department of Geography
University of Queensland
St. Lucia, Brisbane 4067

Map Collection
State Library of Queensland
William Street
Brisbane 4000

Tasmania
Map Collection
Department of Geography
University of Tasmania
Box 252C, G.P.O.
Hobart

Part of Reference Library
State Library of Tasmania
91 Murray Street
Hobart

Victoria
Map Collection
University of Melbourne Library
Parkville 3052

Map Section
Reader Services Department
State Library of Victoria
304 Swanston Street
Melbourne 3000

Map Collection
Department of Geography
Monash University
Wellington Road, Clayton 3168

Western Australia
Map Collection of State Reference
Library Western Australia
3 Francis Street
Perth

Map Library
Reid Library
University of Western Australia
Nedlands 6009

AUSTRIA
Geographische Abteilung
Österreichisches
Ost-und Südosteuropa-Institut
Josefsplatz 6
A-1010 Wien 1

(continued)

TABLE 2. (*continued*)

Kartensammlung
Geographisches Institut der
   Universität
Universitätsstrasse 7/5
A-1010 Wien

Kartensammlung
Österreichische Nationalbibliothek
Josefsplatz
A-1014 Wien

Kartensammlung des Kriegsarchivs
Stiftgasse 2
A-1070 Wien

Bibliothek
Österreichische Geographische
   Gesellschaft
Karl Schweighofergasse 3
A-1070 Wien

BELGIUM
Section des Cartes
Bibliotheque Royale Albert Ier
Boulevard de l'Empereur, 4
B 1000 Bruxelles

Afdeling van de Kaarten en Plans
Centrale Bibliotheek van de
   Rijksuniversiteit
Rozier 9, B-9000 Gent

BOLIVIA
Carta Geológica de Bolivia
Centro de Documentación del Serv.
   Geol. de Bolivia
Casilla 2729
La Paz

BRAZIL
Seção de Mapas
Arquivo Nacional, Praca da
   Republica
Rio de Janeiro, ZC 20,000

Seção de Iconografia
Biblioteca Nacional
Avenida Rio Branco, 219/39
   Centro ZC-21
Rio de Janeiro, 20,000

São Paula
Mapoteca, Biblioteca
Serviço de Comunicações Tecnico-
   Cientificas
Instituto Geográfico e Geologico
Rua Antonio de Godoi
122 - 9° andar, São Paulo

BULGARIA
Cartographic Collection
Ivan Vazov Public Library
Plostad Săedinenie 1
Plovdiv

Special Collections
Sofie University Library
Boul. Rusky 15
Sofia

Cartographic Collection
Science & Technical Information
   Service
Cartographic Institute
Boul. 9 Septemvri 219
Sofia

Map and Graphic Arts Department
Cyril and Methodius National
   Library
Coul. Tolbuhin 11
Sofia

CAMEROON
Bibliothèque Nationale
B.P. 1053
Yaoundé

CANADA
Alberta
Map Collection
University of Calgary Library
2920 24th Avenue, N.W.
Calgary T2N 1N4

University Map Collection
Department of Geography
University of Alberta
Edmonton 7

British Columbia
Map Division
University of British Columbia
   Library
Vancouver 8

Manitoba
Map Division
Provincial Archives of Manitoba
Room 247, Legislative Building
Winnipeg R3C 0V8

Nova Scotia
Public Archives of Nova Scotia
Coburg Road
Halifax

Map Collection, Library
Dalhousie University
Halifax

Ontario
Map Library
McMaster University
Hamilton

Map Library, Dept. of Geography
University of Western Ontario
London

Departmental Map Library
Department of Energy, Mines, and
   Resources
Ottawa K1A 0E9

Map Library
Geological Survey of Canada
601 Booth Street
Ottawa K1A 0E8

Map Library, University of Ottawa
   Library
65 Hastey
Ottawa K1N 6N5

National Map Collection
Public Archives of Canada
395 Wellington Street
Ottawa K1A 0N3

Map Collection, Archives of
   Ontario
Parliament Buildings
Toronto 182

Map Library, Dept. of Geography
University of Toronto
100 St. George Street
Toronto 181

Map Room, History Section
Metropolitan Toronto Central
   Library
214 College Street
Toronto M5T 1R3

Québec
Cartothèque
Bibliothèque de l'Université Laval
Québec 10e

Section Cartes et Gravures
Archives Nationales du Quebec
Parc des Champs de Bataille

Départment des Cartes et Plans
Bibliothèque Nationale du Québec
1700, Rue Saint-Denis
Montreal 129

Map Collection
McGill University
P.O. Box 6070
Montreal 101

CHILE
Mapoteca National
Instituto Geografico Militar
Castro No. 354
Santiago

REPUBLIC OF CHINA
Taiwan
Maps Collection Section
China Research Institute of Land
   Economics
1, South Tung Hwa Road
Taipei, Taiwan

CZECHOSLOVAK REPUBLIC
Map Collection
Geographical Institute
Slovak Academy of Sciences
Stefánikova 41
Bratislava

Map Collection
National Technical Museum
Kostelní ul. 42
170 78 Praha 7, Letná

TABLE 2. (*continued*)

Mapová sbírka B.P. Molla v
Universitní knihovně v Brně
(Map Collection, University
Library, Brno)
Universitní knihovna
Leninova ul. 5-7, 601 87-Brno

Matica Slovenska (Slovak National
Library)
036 52 Martin

Národní Knihovna (National
Library)
Klementinum 190
110 01 Praha 1

Oddělení pro hospodářské dějiny a
hisotrickou geografii UCSD
Nové Město, Panská 7
110 00 Praha 1

DENMARK
Kort-og Billedsamling
Statsbiblioteket
Universitetsparken
8000 Arhus C.

Koertsamling
Det Kongelige Bibliotek
Christians Brugge 8
Kobenhavn

Kortsamling
Geografisk Instituts
Karaldsgade 68
2100 Kobenhavn

FINLAND
Maanmittaushallituksen
Kartografinen Arkisto
(Cartographic Archive of the
National Board of Survey)
Kirkkokatu 3
00170 Helsinki 17

National Department, Helsinki
University Library
00170 Helsinki 17

The Nordenskiöld Collection
Helsinki University Library
00170 Helsinki 17

FRANCE
Bibliothèque Ministère d'Etat
chargé de la Défense Nationale
231, Boulevard Saint-Germain
Paris, 7ᵉ

Bibliothèque du Génie
39 Rue de Bellechasse
75007 Paris

Bibliothèque
Ecole de Mines
60 Boulevard St. Michel
75272 Paris Cedex 06

Bibliothèque Nationale et
Universitaire
6 Place de la Republique
67000 Strasbourg

Bibliothèque
Observatoire de Paris
61 Avenue de l'Observatoire
75014 Paris

Cartothèque
Centre de Géographie
Universités de Paris, I, IV, VII
191 Rue Saint-Jacques
75005 Paris

Cartothèque de l'Institut
Géographique National
2 Avenue Pasteur
94160 St. Mandé

Cartothèque et Photothèque du
Service de Documentation et de
Cartographie Géographiques
191, Rue Saint Jacques
75005 Paris

Collections des Cartes et Plans
Archives Nationales
60, Rue des Francs-Bourgeois
75003 Paris

Départment des Cartes et Plans
Bibliothèque Nationale 58 Rue de
Richelieu
75084 Paris Cedex 02

GERMANY
Abteilung 3011, Hessisches
Hauptstaatsarchiv
Mainzer Strasse 80
62 Wiesbaden (BRD)

Badisches Generallandesarchiv
Nördl. Hildapromenade 2
75 Karlsruhe (BRD)

Bibliothek, Bundesanstalt u.d.
Niedersächsischen Landesamtes
fur Bodenforschung
Postfach 23 01 53
D-3 Hannover 23 (BRD)

Bibliothek des Instituts für
Auslandsbeziehungen
Charlottenplatz 17
7000 Stuttgart 1 (BRD)

Deutsches Hydrographisches
Institut
Bernhard-Nocht-Strasse 78
2 Hamburg 4 (BRD)

Geheimes Staatsarchiv
Preussischer Kulturbesitz
Archivstrasse 12-14
D-1 Berlin 33 (BRD)

Geographische Kartensammlung
Freie Universität Berlin
Grunewaldstrasse 35
D-1 Berlin 41 (BRD)

Kartenabteilung, Bundesarchiv-
Militärarchiv
Wiesentalstrasse 10
D7803 Freiburg i. Briesgau (BRD)

Kartenabteilung
Deutsche Staatsbibliothek
Unter den Linden 8
108 Berlin (DDR)

Kartenabteilung
Hauptstaatsarchiv Stuttgart
Konrad-Adenauer-Strasse 4
7000 Stuttgart 1 (BRD)

Kartenabteilung, Hessisches
Staatsarchiv Marburg
Friedrichsplatz 15
355 Marburg/Lahn (BRD)

Kartenabteilung
Ibero-Amerikanisches Institut
Preussischer Kulturbesitz
Gärtnerstrasse 25-32
1 Berlin 45 (BRD)

Kartenabteilung
Staatsbibliothek (Preussischer
Kulturbesitz)
Reichspietschufer 72-76
Postfach 1407
D-1 Berlin 30 (BRD)

Kartenabteilung
Staatsarchiv Koblenz
Karmeliterstrasse 1/3
5400 Koblenz (BRD)

Kartenarchiv
Parlamentsarchiv des Deutschen
Bundestages
Kurt-Schumacher-Strasse 8
5300 Bonn (BRD)

Kartensammlung
Bayerische Staatsbibliothek
Abholfach, bzw. Ludwigstrasse 16
D-8000 Munchen 34 (BRD)

Kartensammlung
Sektion Geographie der Martin-
Luther-Universität Halle-
Wittenberg
Domstrasse 5
402 Halle (DDR)

Kartensammlung
VEB Hermann Haack
Geographisch-Kartographische
Anstalt Gotha/Leipzig
Justus-Perthes-Strasse 3/9
58 Gotha (DDR)

Kartensammlung, Bayerisches
Hauptstaatsarchiv, Div. I
Postfach 20 05 07
D-8 Munchen 2 (BRD)

Kartensammlung der Deutschen
Bücherei
Deutscher Platz
701 Leipzig (DDR)

(*continued*)

TABLE 2. (*continued*)

Kartensammlung der Staats- und
  Universitäts-bibliothek Hamburg
Moorweidenstrasse 40
2 Hamburg 13 (BRD)

Kartensammlung der
  Württembergische
  Landesbibliothek
Konrad-Adenauer-Strasse 8
Postfach 769
7000 Stuttgart 1 (BRD)

Kartensammlung des
  Niedersächsischen
  Hauptstaatsarchivs
Am Archiv 1
3 Hannover (BRD)

Kartensammlung, Die Deutsche
  Bibliothek
Zeppelinallee 6
D-6 Frankfurt am Main (BRD)

Kartensammlung
Geographisches Institut der
  Universität Köln
Albertus-Magnus Platz
5 Köln 41 (BRD)

Kartensammlung
Geographisches Institut
Universität Göttingen
Herzbergerstrasse 2
34 Göttingen (BRD)

Kartensammlung
Geographisches Institut der
  Universität Tübingen
Schloss
D-7400 Tubingen (BRD)

Kartensammlung
Geographische Zentralbibliothek
Geographisches Institut der
  Akademie der Wissenschaften
  der DDR
Georgi-Dimitroff-Platz 1
701 Leipzig (DDR)

Kartensammlung
Institut für Angewandte Geodäsie
Kennedyallee 151
D06 Frankfurt am Main (BRD)

Kartensammlung
Institut für Landeskunde
Michaelsstrasse 8
Postfach 130
D-53 Bonn-Bad Godesberg (BRD)

Kartensammlung
Kriegsarchiv
Bayerisches Hauptstaatsarchiv, Div.
  IV
Postface 20 05 07
8 Munchen 2 (BRD)

Kartensammlung
Sektion Geographie
Humboldt-Universitat
Universitätsstrasse 3b
108 Berlin (DDR)

Zentralbibliothek des Zentralen
  Geologischen Institutes
Invalidenstrasse 44
104 Berlin (DDR)

HUNGARY
A Kartográfiai Vállalat Térkép- és
  Adattára
(Map and Data Collection of the
  Cartographia)
XIV. Bosnyák tér 5
H 1443 Budapest Pf. 132

Békés megyei Levéltár, Békés
  Megyei Térképgyüjtemény
(Archives of Bekes Territory, Bekes
  Map Collection)
Petöfi Sándor Tér 3
5601 Gyula, Pf. 17

Bp. Föváros Levéltára
Térképtár
Városház u. 9-11
1052 Budapest, V

Library
Geographical Research Institute
Hungaria Academy of Sciences
Népkostársaság u. 62
Budapest VI

Magyar Állami Foldtani Intézet
  Térképtára
(Geological Map Collection,
  Hungarian Geological Institute,
  Geological Survey of Hungary)
Népstadion-út 14
1143 Budapest, XIV

Magyar Földrajzi Társaság
Könyvtára és Térképtára
(Library and Map Collection,
  Hungarian Geographical
  Association)
Népköztársaság utja 62
1062 Budapest VI

Map Collection
Baranya County Archives
Kossuth L. u. 11
7621 Pécs

Map Collection
National Archives of Hungary
Bécsikapu tér 4
1014 Budapest

Map Collection
War Institute and Museum
Kapisztrán tér 2
1250 Budapest

Országos Széchényi Könyvtár
  Térképtára
(Map Department, Szechenyi
  National Library)
Régi Fóti ut 77
H-1152 Budapest XV

Pest Megyei Levéltár
  Térképgyüjteménye
(Map Collection of Pest Territorial
  Archives)
Városház utca 7
1364 Budapest, V

Somogy Megyei Levéltár Kéziratos
  térképgyüjteménye
(Manuscript Map Collection of the
  Archives of Somogy Territory)
Rippl-Ronai tér 1
7401, Kaposvar

Térképgyüjteménye, Hajdu-Bihar
  Megyei Levéltár
(Map Collection, Archives of
  Hajdu-Bihar Territory)
Vörös Hadsereg utja 20
4001 Debrecen, Pf. 39

Vas megyei Levéltár
  térképgyüjteménye
(Map Collection of the Vas
  Territory Archives)
9701 Szombathely, Pf. 78

Zala megyei Levéltár térkép
  részlege
(Map Section of the Archives of
  Zala Territory)
Széchényi tér 3
Postafiók 1110
8901, Zalaegerszeg

INDIA
Map Section, Records Branch
National Archives of India
New Delhi

National Library, Belvedere
Calcutta-27

ISRAEL
Dept. of Surveys, Ministry of
  Labour
P.O. Box 14171
Tel-Aviv 61140

ITALY
Biblioteca
Istituto Geografico Militare
Via Cesare Battisti 10
50100 Firenze

Biblioteca Nazionale Marciana
S. Marco 7
30124 Venezia

Cartoteca
Touring Club Italiano
Via Adamello 10
20139 Milano

REPUBLIC OF IVORY COAST
Departement des Documents
  Speciaux
Bibliotheque Nationale
Abidjan

TABLE 2. (*continued*)

JAPAN
Map Room
Humanities Section
Reference and Bibliography
 Division
National Diet Library
10-1, 1-chome
Magata-Cho, Chiyoda-Ku

Geographical Division
University Museum
The University of Tokyo
7-3-1, Hongo
Bunkyo-Ku, Tokyo

Map Source Section
Map Management Division
Geographical Survey Institute
24-13 3-chome
Higashiyama, Meguro-Ku, Tokyo

KENYA
Department of Geography
University of Nairobi
P.O. Box 30197
Nairobi

LUXEMBOURG
Departement de la Reserve
 Precieuse
Section: Cartes et Plans
Bibliotheque Nationale de
 Luxembourg
37, Boulevard F. D. Roosevelt
Luxembourg

MALAWI
Geological Survey Department
P.O. Box 27
Zomba

MALAYSIA
Map Library
Geography Department
University of Malaya
Pantai Valley
Kuala Lumpur

MEXICO
Mapoteca
Direccion Gral. de Geografia y
 Meteorologia
Av. Observatorio No. 192
Mexico, D.F., Z.P. 18

NETHERLANDS
Kaartenzaal
Koninklijk Instituut voor den
 Tropen
Mauritskade 63
Amsterdam-0

Bibliotheek
Topografische Dienst
Westvest 9
Delft

Bodel Nijenhuis Collection,
 Universiteitsbibliothek
Rapenburg 60-74
Leiden

Map Collection
Geographical Institute
State University
Transitorium II, Heidelberglaan 2
Utrecht

PAPUA NEW GUINEA
Dept. of Geography, Map
 Collection
University of Papua and New
 Guinea
Box 4820, University

NEW ZEALAND
Geography Library
University of Auckland
Private Bag
Auckland 1

Map Library
University of Waikato
Hamilton

Alexander Turnbull Library
The National Library of New
 Zealand Map Collection
P.O. Box 8016
Wellington

NIGERIA
Maps and Manuscripts Section
University of Lagos Library
Akoka Yaba, Lagos

Map Collection
Ibadan University Library
Ibadan

NORWAY
Geografisk Instituut
University of Oslo
P.O. Boks 1042
Blindern, Oslo 3

Kartsamlingen
Universitetsbiblioteket i Bergen
Möhlenprisbakken 1
5000 Bergen

POLAND
Cartographic Collection
Czartoryski Library
National Museum in Kraków
 (Muzeum Narodowe w
 Krakowie)
ul. sw. Marka 17
31-018 Kraków

Sekcja Kartografii
Biblioteka Uniwersytecka w Łodzi
(Cartographic Section, Library of
 the University of Lodz)
ul. Matejki 34/38
Łódź, Poland

Dział Zbiorow Specjalnych.
 Oddział Kartografii
Ksiaznica Miejska im. M.
 Kopernika
(Cartography Subsection, Kopernik
 City Library)
ul. Wysoka 16
Toruń

Sekcjz Zbiorów Kartograficznych,
 Biblioteka Główna
Uniwersytetu Mikołaja Kopernika
(Division of Cartographical
 Collections, Library of the
 Nicholas Copernicus University)
ul. Chopina 12/18
Torun

Zbiory Kartograficzne
Archiwum Glownego Akt Dawnych
(Map Collection, Central Archives
 of History)
ul. Długa 7
Warszawa

Abiory Kartograpficzne Biblioteki
 Narodowej
(Map Department of the National
 Library)
ul. Hankiewicza 1
Warszawa 22

Zbiory Kartograficzne, Centralna
 Biblioteka Wojskowa
(Map Section, Central Army
 Library)
Al. I Armii WP 12a
Warszawa

Uniwersytet Wroclawski, Oddział
 Zbiorów Kartograficznych
(Map Section, Library, University
 of Wrocław)
ul. sw. Jadwigi 3/4
Wrocław

Zbiory Kartograficzne Archiwum
 Państwowego Miast Wroclawia i
 Wojewodztwa Wroclawskiego
(Cartographic Collection of the
 State Archives in Wroclaw)
ul. Pomorska 2
Wrocław

PHILIPPINES
Bureau of Coast and Geodetic
 Survey
421 Barraca Street, San Nicolas
Manila

RHODESIA
Map Collection
National Archives of Rhodesia
P/B 7729, Causeway
Salisbury

Map Library
University of Rhodesia
P.O. Box MP 167, Mount Pleasant
Salisbury

(*continued*)

431

TABLE 2. (*continued*)

SINGAPORE
Library
Institute of Southeast Asian Studies
Cluny Road

Map Library
Department of Geography
University of Singapore
Bukit Timah Road

SOUTH AFRICA
South African Library
Queen Victoria Street
Cape Town

Geography Library
University of Witwatersrand
Johannesburg

Johannesburg Public Library and
Africana Museum
Market Square
Johannesburg

Government Archives, Natal
Private Bag 9012
Pietermaritzburg

The Natal Society Library
P.O. Box 415
Pietermaritzburg

Map Collection
Geological Survey of South Africa
Private Bag 112
Pretoria

Map Collection
State Library
P.O. Box 397
Pretoria

Map Library
Trigonometrical Survey
P.O. Box 624
Pretoria

SPAIN
Cartoteca Histórico y Moderna
Sección de Documentacion
Servicio Geográfico del Ejército
Prim. No. 8, Madrid (4)

Sección de Geografía y Mapas
Biblioteca Nacional
Avenida de Calvon Sotelo 20
Madrid 1

SWEDEN
Antikvarisk-Topografiska Arkivet
Box 5405
S-114 84 Stockholm

Kart-och planschavdelningen
Uppsala Universitetsbibliotek
Box 510
751 20 Uppsala 1

Kart- & Planschavdelningen
Kungliga Biblioteket
(Map and Print Division, Royal
Library)
Box 5039
S-102 41 Stockholm 5

Kungl Krigsarkivet (Royal Military
Record Office)
Fack, S-104 50 Stockholm 80

Map Collection
University Library of Göteborg
Box 5096
S-402 22 Göteborg 5

SWITZERLAND
Department des Estampes et Cartes
de la Bibliotheque
Publique et Universitaire
1211 Geneve 4

Kartensammlung
Staatsarchiv Zürich
Predigerplatz 33
CH-8001 Zürich

Kartensammlung
Zentralbibliothek Zürich
8025 Zürich

Kartographisches Institut
Eidg. Techn. Hochschule
Leonhardstr. 33
CH-8806 Zürich

Schweizerische Landesbibliothek
Hallwylstrasse 15
CH-3003 Bern

Stadt- und Universitätsbibliothek
Bern
Münstergasse 61
3000 Bern

Zieglersche Kartensammlung
Öffentliche Bibliothek der
Universität
Schönbeistr. 18/20, Ch-4056 Basel

UNITED KINGDOM
England
Chart and Globe Collection
The National Maritime Museum
Greenwich, London SE10 9NF

Department of Manuscripts, British
Library
Great Russell Street
London WC1B 3DG

Essex Record Office
County Hall
Chelmsford, Essex CM1 1LX

Guildhall Library
Basinghall Street
London EC2P 2EJ

Hydrographic Department
M.O.D. Taunton
Somerset

Map Collection, Department of
Geography
The University
Hull HU6 7RX

Map Collection
India Office Library and Records
Foreign and Commonwealth Office
197 Blackfriars Road
London SE1 8NG

Map Collection, Royal
Geographical Society
1 Kensington Gore
London SW7 2AR

Map Collection
University of London Library
Senate House, Malet Street
London WC1

Map Library
Department of the Environment
Government Offices, Great George
Street
Whitehall, London SW1P 3AH

Map Research and Library Group
MCE(RE) Ministry of Defence
Block 'A' Government Buildings
Hook Rise South, Surbiton KT6
7NB

Map Room
British Library
Great Russell Street
London WC1B 3DG

Map Room
University Library
Cambridge CB3 9DR

Map Section
Bodleian Library
Oxford University
Oxford OX1 3BG

National Maritime Museum
Library
National Maritime Museum
Greenwich, London SE10

Public Record Office
Chancery Lane
London WC2A 1LR

School of Geography
University of Oxford
Mansfield Road
Oxford OX1 3TB

Technical Services
Directorate of Overseas Surveys
Kingston Road
Tolworth, Surbiton, Surrey KT5
9NS

West Sussex Record Office
County Hall
Chichester, Sussex PO19 1RN

TABLE 2. (*continued*)

Northern Ireland
Map Collection
Department of Geography
The Queen's University of Belfast
Belfast BT7 1NN

Public Record Office
66 Balmoral Avenue
Belfast BT9 6NY

Scotland
Map Room
National Library of Scotland
George IV Bridge
Edinburgh EH1 1EW

Register House Plans
Scottish Record Office
H.M. General Register House
P.O. Box 36
Edinburgh EH1 3YY

UGANDA
Department of Geography Map
   Collection
Makerere University
Box 7062
Kampala

UNITED STATES OF AMERICA
Alaska
Map Collection
Elmer E. Rasmuson Library
University of Alaska
Fairbanks

California
Map and Atlas Collection
Bancroft Library
University of California
Berkeley 94720

Map Room
General Library
University of California
Berkeley 94720

UCLA Map Library
University of California
Los Angeles 90024

Central Map Collection
Stanford University Libraries
Stanford 94305

Colorado
Map Library
University of Colorado
Guggenheim Room 8
Boulder 80302

Connecticut
Map Collection
Yale University Library
Box 1603 A Yale Station
New Haven 06520

Florida
Map Library
Library East
University of Florida
Gainesville 32611

Maps Division
R.M. Strozier Library
Florida State University
Tallahassee 32306

Georgia
Map Room
Science Library
University of Georgia Libraries
Athens 30602

Illinois
Chicago Historical Society Library
North Avenue and Clark Street
Chicago 60614

Map Collection
The Newberry Library
60 West Walton Street
Chicago 60610

Map Collection
Northwestern University Library
Evanston 60201

Map Library
Illinois State University
Normal 61761

Map and Geography Library
418B Library
University of Illinois
Urbana 61801

Indiana
Map Library
Department of Geography
Kirkwood Hall
Indiana University
Bloomington 47401

Kansas
Map Library
Spencer Research Library
University of Kansas
Lawrence 66044

Kentucky
Map Collection
Geology Library
100 Bowman Hall
University of Kentucky
Lexington 40506

Louisiana
Map Library
School of Geoscience
Louisiana State University
Baton Rouge 70803

Maryland
Map Library
National Ocean Survey (U.S. Dept.
   of Commerce)
6001 Executive Boulevard
Rockville 20852

Massachusetts
Winsor Memorial Map Room
Harvard College Library
Cambridge 02138

Michigan
Map Room
Harlan Hatcher Graduate Library
University of Michigan
Ann Arbor 48104

William L. Clements Library
University of Michigan
Ann Arbor 48104

Map Room
Detroit Public Library
5201 Woodward Avenue
Detroit 48202

Map Department
Western Michigan University
   Library
Kalamazoo 49001

Minnesota
Map Division
Wilson Library
University of Minnesota
Minneapolis 55455

New Hampshire
Map Section
Baker Library
Dartmouth College
Hanover 03755

New Jersey
Map Division
Princeton University Library
Princeton 08540

New York
Manuscripts and History Library
New York State Library
State Education Department
Washington Avenue
Albany 12234

Department of Maps, Microtexts,
   and Newspapers
John M. Olin Library
Cornell University
Ithaca 14853

Map Collection
History Division
Brooklyn Public Library
Grand Army Plaza
Brooklyn 11238

(*continued*)

TABLE 2. (*continued*)

Map Collection
Dag Hammarskjold Library
United Nations
P.O. Box 20, Grand Central
   Station
New York 10017

Map Department
American Geographical Society
156th Street and Broadway
New York 10032

Map Division
New York Public Library
5th Avenue at 42nd Street
New York 10018

Map and Print Room
New York Historical Society
170 Central Park West
New York 10024

The University Map Room
School of International Affairs
Columbia University
420 W. 118th Street
New York 10017

Ohio
General Reference Department
Cleveland Public Library
325 Superior Avenue
Cleveland 44114

Map Collection
Kent State University Libraries
Kent 44242

Map Division
History and Literature Department
Public Library of Cincinnati and
   Hamilton County
800 Vine Street
Cincinnati 45202

Oregon
Map Library
University or Oregon Libraries
Eugene 97403

Pennsylvania
Map Collection
Pattee Library
The Pennsylvania State University
University Park 16802

Map Collection
Social Science & History
   Department
Free Library of Philadelphia
Logan Square
Philadelphia 19103

Tennessee
Map Information and Records Unit
Maps and Surveys Branch
Tennessee Valley Authority
200 Haney Building
Chattanooga 37401

Texas
Edwin J. Foscue Map Library
Science/Engineering Library
Southern Methodist University
Dallas 75275

Vermont
Map Room
Guy W. Bailey Library
University of Vermont
Burlington 05401

Virginia
Archives Map Collection
Virginia State Library
12th and Capitol Streets
Richmond 23219

Library
U.S. Geological Survey
National Center
Reston 22092

Washington
Map Center, Suzzalo Library
University of Washington
Seattle 98195

Washington D.C.
Cartographic Archives Division
The National Archives of the
   United States
8th and Pennsylvania Avenue,
   N.W.
Washington 20402

Department of Defense Map
   Library
Topographic Center, Defense
   Mapping Agency
6500 Brooks Lane
Washington 20315

Geography and Map Division
Library of Congress
Washington 20540

Map Library
National Geographic Society
17th and M Streets, N.W.
Washington 20036

Nautical Chart Library
Defense Mapping Agency
   Hydrographic Center
Washington 20390

Wisconsin
Division of Archives and
   Manuscripts
State Historical Society of
   Wisconsin
816 State Street
Madison 53706

Map and Air Photo Library
Science Hall
University of Wisconsin
Madison 53706

USSR
Cartography Department
M.E. Saltykov Shchedrin State
   Public Library
D-69 Sadovaya ul. 18
Leningrad

Cartography Department
V.I. Lenin State Library of the
   USSR
Prospekt Kalinina 3
Moscow

Cartographic Section
Library
Academy of Sciences of the USSR
Birzhevaya Liniya, 1
199164 Leningrad, V-164

Library
Academy of Sciences of the USSR
Geographical Society of the USSR
Grivtsova, 10
Leningrad, 190 000

WALES
Department of Geography
University College
Swansea, Glamorganshire

Department of Prints, Drawings,
   and Maps
The National Library of Wales
Aberystwyth, Cards

YUGOSLAVIA
Map Collection
National and University Library
Marulicev trg 21
41000 Zagreb

Narodna i univerzitetska biblioteka
   "Kliment Ohridski"
Ul. "Goce Delcev"
Skopje

Sources: Compiled from Ristow (1976) and irregular publications of *GeoCenter,* Internationales Landkartenhaus GmbH, D-7000 Stuttgart 80, West Germany.

References

Bergen, J. V., 1973, *Map Collections in the Midwestern Universities and Colleges.* Dekalb: Western Illinois University, Geography Department, University Map Library, 55p.

Bergquist, W. E., E. J. Tinsley, L. Yordy, and R. L. Miller, 1981, Worldwide directory of national earth-science agencies and related international organizations, *U.S. Geol. Survey Circ. 834,* 87p.

Busse, G. von, and H. Ernestus, 1972, *Libraries in the Federal Republic of Germany.* Wiesbaden: Harrassowitz, 308p.

Dickinson, G. C., 1979, *Maps and Aerial Photographs.* London: Arnold, 348p.

Drecka, J., and H. Tuszynska-Rekawek, 1964, *National and Regional Atlases.* Warsaw: Polish Academy of Sciences, Institute of Geography, l55p.

Keates, J. S., 1973, *Cartographic Design and Production.* London: Longman, 240p.

Kramm, H., 1959, *Verzeichnis Deutscherr Kartensammlungen.* Wiesbaden: Harrassowitz, 84p.

Lewanski, R. C., 1965, *Subject Collections in European Libraries, A Directory and Bibliographic Code.* New York: Bowker, 789p.

Poss, J. R., 1975, *Stones of Destiny.* Houghton: Michigan Technical University, 253p.

Ristow, W. W., 1976. *World Directory of Map Collection.* Munich: Verlag Dokumentation, 326p.

Schwartz, M. L., ed., 1982, *The Encyclopedia of Beaches and Coastal Environments.* Stroudsburg, Pa.: Hutchinson Ross, 940p.

Short, N. M., P. D. Lowman, Jr., S. C. Freden, and W. A. Finch, Jr., 1976, *Mission to Earth: Landsat Views the World.* Washington, D.C.: National Aeronautics and Space Administration, 459p.

Snead, R. E., 1982, *Coastal Landforms and Surface Features.* Stroudsburg, Pa.: Hutchinson Ross, 247p.

Spinner, G. P., 1969, *A Plan for the Marine Resources of the Atlantic Coastal Zone.* New York: American Geographical Society, 80p (including *Serial Atlas of the Marine Environment*).

Yonge, E. I., 1957, National atlases: a summary, *Geographical Jour.* **67**(4), 570–578.

Cross-references: *Geological Cataloguing; Geological Surveys; State and Federal; Map Series and Sales; Maps, Logic of.* Vol. XIII: *Information Centers.*

MAP COLORS, COLORING

Color is used on geologic maps to delineate the distribution of various rocks (stratigraphy, lithology) and other features. It is also employed to advantage on other specialized maps that deal with aspects of geomorphology (physiography), structure, mineral deposits, soils, hydrology, and bathymetry. The requirements of a good color map include legibility, economy of production, and good taste. Tradition often prescribes certain associations of color with particular implications, which being thus established, control other associations. Dark colors are often used for igneous rocks, light shades for sedimentary.

Understanding the rationale of map coloring requires some general knowledge of colors. There are three *primary colors*—red, blue, and yellow. They may be more accurately referred to in terms of pigments or, in printing, as process colors: magenta, cyan, and yellow. If pairs of these colors are combined in equal volumes, they yield the standard *secondary colors:*purple, orange, and green. If the standard secondary colors are again mixed in equal volumes in pairs, standard *tertiary colors* result: russet from purple and orange, olive from purple and green, and citrine from orange and yellow. And finally, each primary forms one-half of the respective tertiary color, but the presence of the other two primaries modifies its brilliance.

Legibility

Distinctions of color should be pronounced so that they are clearly recognized. Color contrast depends on hue and tint (or shade), on flat colors, and also on pattern. The normal *hue* has the full available strength of its constituent colors; a *tint* is lighter. A *shade,* conversely, is darker because it reflects less light than the normal hue. *Shades* may be produced by adding black but, more artistically, by mixing a complementary tertiary color with the normal hue. In color printing, tints result where the white of the paper shines through the hue, which is thereby diluted. Shades are produced by overprinting grays, which are usually tints of olives or russets strong in blue.

Because contrast depends on the juxtaposition of unlike hues, it is strongest between any two pure primaries. Contrast is equally striking, though less likely to be offensive, between a primary and the secondary that is composed of the other two primaries. Contrast is less pronounced between secondary hues and is markedly less between tertiaries. Black and white provide the strongest contrasts; in relation to colors, these subdue the distinction between tints and shades. Thus, where strong contrast is desired, juxtapose primary hues, or a primary with its complementary secondary, or a light tint of a hue against a dark shade of the same or another hue. Distinctness may be achieved by lesser contrasts. Greater contrasts are usually reserved for small areas for emphasis.

Economy of Printing

Economical production is based on the least possible number of impressions or printings while maintaining the required distinctions of color. Because it is possible to print one color over another on the lithographic press, a relatively small number of printings may yield many distinctions. In most colored illustrations, three colors (red, yellow, and blue or green) usually suffice, reinforced sometimes by black. Only numerous and very re-fined differences require many printings. The Geologic Map of North America (Goddard, 1965), e.g., was produced by 12 printings, which yielded 42 distinct effects. Two more printings were required for the black and blue of the base map.

Patterns

Conspicuous patterns are used on geologic maps because they greatly increase the range of recognizable distinctions. Such patterns are a concession to technical requirements and may offend good taste unless they are skillfully designed. The types of patterns used should suggest the igneous, sedimentary, metamorphic, or superficial character of the rocks (see *Map Symbols*). The U.S. Geological Survey recommends patterns of circular figures (dots and circles) for alluvial, glacial, and eolian formations; parallel straight lines for sedimentary formations deposited in seas, lakes, or other bodies of water; hachure patterns (short dashes irregularly spaced) for metamorphic rocks; and patterns of angular figures (triangles, rhombs) for igneous rocks, as employed, e.g., on the *Geologic Map of the United States* (King and Beikman, 1974). Patterns are most easily achieved by overprints on flat colors, which has been done on recent maps of the U.S. Geological Survey and modern maps published in the USSR and elsewhere. Color patterns employed by the French BRGM (Bureau de Récherches Géologique et Minières) on geologic maps to show bedrocks, lava types, unconsolidated surficial rocks, and weathering crusts (Table 1) provide an effective means of imparting a great deal of information to the map user in a succinct form. These patterns for lithological composition are combined with dark-gray signs and dark-brown hatching overprinted on background colors. Examples of color signs and symbols used on geomorphological maps are summarized in Table 2. For these thematic maps, specific colors are assigned to various kinds of morphostructures, form complexes generated by exogenic processes, and microforms (see Vol. III: *Geomorphic Maps*). Black color signs are also used over colored backgrounds, increasing the range of possibilities for cartographic display.

Symbolization

Colors on a map should be identified by letter-number symbols. This is especially important for the many people who are color blind. Such alphanumeric identification assists the user in comparing the map with its legend. Colored maps without symbols are handicapped because the viewer must match bands of color with one of an assortment of similar colors in the legend. The simplest form of symbolization is by numbers.

Numbers are appropriate where there are only a few units but become confusing where there are 50 or more. More common are single or multiple letters or letters combined with numbers. Specifications for the *Geologic Map of Europe* adopted by the International Geologic Congress call for Roman lower case letters to designate ages of strata, modified by suffix numbers. Different kinds of eruptive rocks are shown by Greek letters. Many geologic maps, such as those of the U.S. Geological Survey, express general age by Roman letters that represent the geologic systems, series, etc. Long strings of modifying suffixed letters should be avoided because they become confusing.

Map Appearance

In addition to being informative, a map must be pleasing to look at. Color patterns must not clash or draw undue attention to certain parts of the map. Neither should they be boring or uninteresting. In general, maps that are pleasing to view result from an association of light tints, particularly those of tertiary hues in large areas, with appropriate contrasts of bright primary or secondary hues that may be proportionately more brilliant because the areas covered by them are smaller. Thus, the essential of good taste is adaption to purpose. Good examples of handsomely colored special-purpose maps are legion. The *Geological Map of the Grand Canyon National Park* (Grand Canyon Natural History Association, 1976), e.g., affords a pleasing appearance through the use of effective color display that is distinct but not gaudy. The international *Soil Map of the World* (FAO/UNESCO, 1972) is another example of fine cartographic display in color.

Principles for Coloring Geologic Maps

Few hard-and-fast rules apply universally to map coloring. Many countries, particularly those in Western Europe, have a long tradition of cartographic expertise. Whatever guidelines or rules are followed, the final result must be evaluated from the map appearance and ease of use. Aside from their scientific value, good maps demand a high degree of artistic appreciation. Map coloring, then, is as much an art as a science or technical skill. Responding to tradition, techniques of reproduction, and aesthetic attributes, different cultures have developed color combinations that satisfy their particular needs and requirements. Although there is a modicum of international cooperation in attempts to standardize colors for geologic maps, no single system has proved satisfactory for all needs. The U.S. Geological Survey, providing an American point of view, follows a basic set of principles that were described by Bailey Willis in 1912. These guidelines for use of

TABLE 1. Selected Color Patterns and Signs for Lithological Composition,
as Used on European Geomorphological Maps

I. Bedrocks /deep-grey signs/

1. acid
2. basic
3. volcanogene-sedimentary
4. metamorphic

intrusive and
effusive
rocks

5. flish
6. carbonate
7. molassa

II. Lava types /dark-grey signs/

1. blister lava
2. ropy-lava
3. block-lava

4. fan-columnar lava
5. "giants cause way"

III. Loose rocks[a] /dark-grey signs/

1. conglomerates
2. blocks
3. boulders
4. rock debris
5. pebbles beds
6. single pebbles
7. marine pebble beds
8. gravel
9. coarse sand
10. fine sand
11. sandy loam
12. loam

13. loess-type loam
14. clay
15. varved clay
16. loess
17. silt
18. coquina and other
    marine biogene rocks
19. oolites and other
    chemical sediments
20. solar salt
21. peat and sapropels
22. coal
23. volcanic ash
24. travertine

IV. Crusts of weathering /dark-brown hatching/

1. kaolin
2. lateritic

3. redbed
4. mottled

Source: Demek and Embleton (1976), 29.

[a] The loose sediment lithology is to be shown only on accumulative relief features/accumulative slopes, sheets, fans, etc.

color on geologic maps (for maps of states at 1:500,000 scale and of quadrangles at scales between 1:24,000 and 1:250,000) are briefly discussed in relation to other approaches employed elsewhere.

A first principle is that the relationships of color and relative age should be invariable; i.e., when applied to distinguish subdivisions of a sequence of strata, colors should always be used in a definite succession that expresses relative age.

For example, the sequence red, purple, violet, blue, green, and yellow could be adopted to represent the succession of formations, groups, or series of sedimentary rocks from older to younger. Let the order of the colors be invariable according to this principle, no matter what part or how much of the geologic column is represented. Then red will always represent something older than something shown in purple, violet, blue, green, or yellow. Blue will always stand for something older

TABLE 2. Background Colors and Tints, Hatches, Signs, and Symbols for Use on Geomorphological Maps

black colour lines

Outlines of non-structural "living" forms expressed in the relief	bright red colour signs and symbols	Outlines of prepared inherited structural forms	black colour signs and symbols

SECTION I

I.1. DEEP FRACTURES AND FAULTS   red colour lines
I.2. MORPHOSTRUCTURES AND THEIR ELEMENTS [1],[2]

	Background /colours, tints/	Hatchings and signs		Background /colours, tints/	Hatchings and signs
1.2.1. TECTONICAL					
1.2.1.1. CRESTS AND SUMMIT SURFACES OF RIDGES, MASSIFS AND RANGES; MASSIFS AND RANGES	terracotta-red	deep red	1.2.3.2. BASIN BOTTOMS AND BASINS	dark-crimson	dark-brown
1.2.1.2. BASIN BOTTOMS AND BASINS [3]	pale-pink	-"-	1.2.3.3. SLOPES	crimson, bright pink	-"-
1.2.1.3. SLOPES OF MORPHOSTRUC- TURES [2]	brown-reddish	-"-	1.2.4. MUD VOLCANIC		
1.2.2. VOLCANIC			1.2.4.1. SUMMIT SURFACES OF HILLS AND RANGES, RANGES	dingy-crimson	black
1.2.2.1. CRESTS AND SUMMIT SURFACES OF RIDGES, MASSIFS AND RANGES	crimson	dark red	1.2.4.2. BASIN BOTTOMS [3]	light-crimson dingy	-"-
1.2.2.2. BASIN BOTTOMS AND BASINS [3]	dark-crimson	-"-	1.2.4.3. SLOPES	pale-pink, dingy	-"-
1.2.2.3. SLOPES AND THEIR MICRO- FORMES RELIEF	crimson and pink	-"-	1.2.5. MORPHOSTRUCTURES PREPARED BY DENUDATION		
1.2.3. TECTONICAL-VOLCANIC /un- touched by the post-volcanic tectonics/			1.2.5.1. CRESTS AND SUMMIT SURFACES OF MASSIFS, RANGES AND RIDGES; RANGES AND RIDGES	hatching on the coloured back- ground reflecting denudation type	-"-
1.2.3.1. CRESTS AND SUMMIT SURFACES OF RIDGES, MASSIFS AND RANGES	light-crimson	dark-brown	1.2.5.2. BASIN BOTTOMS AND BASINS [3]	-"-	-"-
			1.2.5.3. SLOPES AND THEIR MICRO- FORMES RELIEF	-"-	-"-

SECTION II

II.   FORM COMPLEXES, FORMS AND THEIR ELEMENTS GENERATED BY EXOGENE PROCESSES

	Background /colours and tints/	Hatchings and signs		Background /colours, tints/	Hatchings and signs
II.1. Fluvial, azonal	green and yel- low-green	dark-green	II.7. Nival-congelifluction frost action and permafrost	violet	dark lilac
II.2. Fluvial, arid and semi-arid	yellow	dark-brown	II.8. Karst and suffosional forms	dark-orange and orange	dark-brown
II.3. Fluvial-solifluctional /in humid tropics/	brown-orange	-"-	II.9. Aeolian		bright-orange dotted pattern black
II.4. Marine and lacustrine	blue	dark-blue	II.10. Biogene		black
II.5. Nival-glacial	lilac	dark-lilac	II.11. Relief features indirectly caused by the activity of man /lithodiageneous forms/		black
II.6. Glacial and fluvioglacial	light-lilac	-"-	II.12. Anthropogene		black

SECTION III

III.   SLOPES [4] OF MODELLED MORPHOSTRUCTURES, VALLEYS AND MICROFORMS FOR SLOPES

	Background /colours, tints/	Hatchings and signs		Background /colours, tints/	Hatchings and signs
III.1. Gravity, azonal	brown	black	III.5. Formed by sea activity	blue	deep-blue
III.2. Fluvial, azonal	green-brown and olive	dark-green	III.6. Nival-glacial and fluvioglacial	pink-lilac	dark-lilac
III.3. Fluvial, arid and semi-arid	yellow-brown	dark-brown	III.7. Nival-congelifluction, frost, action and permafrost	blue-violet pink-violet	-"-
III.4. Fluvial-solifluctional /in humid tropics/	yellow and yellow-brown	-"-	III.8. Karst	bright-orange	dark-brown
			III.9. Biogene		black

RANGES FOR SLOPE STEEPNESS [5]
grey colour of variable deepness depending upon the slope angle of dip
2 - 5°   6 - 15°   16-25°   26 - 35°   36 - 55°   more than 55°

IV. Other symbols various colour

1/ When the geomorphological boundaries coincide with outlines of active morphostructures, they are shown with red lines of respective drawing /see Table II/

2/ Background colour is used to show major elements of morphostructures and relief features. Elements of meso- and micro-relief which can not be shown on the 1:200,000 - 1:1,000,000 maps by the background colour are suitably shown with signs and symbols of respective colour: darker tints of the same colour are used in case their genesis is same as that of the major elements: different colour is used for meso- and micro-features if their genesis is different. The colours are chosen after the Table by D. IVANIA /1962/.

3/ Accumulative bottoms /floors/ of such basins are mostly shown by background colouring showing the accumulation type /see below/. The same concerns denudational basins.

4/ Accumulative slopes are painted in colours reflecting the main accumulational factor, the color being of the lightest tint. The signs of related sediments are also shown on these slopes. Tectonical accumulation forms are shown with dark-red symbols /seismical landslides, tectonical slips and collapses/.

5/ Grey colour of various intensity does not interfere with the background colour showing the slope genesis on printed maps. On raw maps the slope steepness is better be shown by variations of main colour deepness.

6/ This pattern is imposed on the background colour reflecting the primary genesis of the relief features, on which the Aeolian forms are superimposed /alluvial, plains, marine plains, etc./. The pattern reflects the genesis and type of the Aeolian features /see Table II./.

Source: Demek and Embleton (1976).

438

than something shown in green or yellow. In principle, the essential features of sequence and structures would be immediately known.

A second basic principle is that color and geologic time divisions are measurably independent; i.e., no color should be regarded as being exclusively indicative of any particular time division. This principle is a radical departure from general usage in countries that have well-established color schemes (King and Beikman, 1974). In Europe, e.g., blue means Jurassic, but in the United States it stands for Carboniferous. Thus, blue in each country represents something different, and to a viewer in one country examining the maps of another, it suggests something other than it should. What is needed is flexibility of usage, which is essential to a general color scheme designed for application in different provinces. It is hoped that a certain orderliness will eventually take the place of the disorder that is forced by arbitrary usages on any student of the maps of several countries. It is emphasized that this principle is not intended to give license to use any color whatever for any period or to contravene the principle of utility, which requires that a color scheme for a given province shall be as nearly uniform as practicable.

The broad features of the so-called American color system were established by J. W. Powell, the second director of the U.S. Geological Survey. In his first official report (1882), Powell announced a scheme of stratigraphic nomenclature, map coloring, and patterns to be used in survey publications. Detailed specifications for usage were issued later after areas of diverse geology in many parts of the country had been sampled for mapping (Powell, 1890). The principal differences in the American and international map color systems are evident in the stratified sedimentary rocks. The intrusive and volcanic rocks in both systems are shown in more brilliant tints, with a preference for the reds and oranges. The two map-coloring systems are compared in Table 3. The original proposals for each are followed by samples of subsequent usage.

The European international geologic color scheme resulted from prolonged consideration by the international committee that was charged by the Geological Congress with the duty of preparing the *Geological Map of Europe* (initiated in 1964). This map uses the order of prismatic colors from purple through blue and green to yellow for that part of the scheme that relates to Triassic and post-Triassic terranes. It also employs an arbitrary principle that Mesozoic terranes should be distinguished from Paleozoic terranes by a contrast of light and shade, the Paleozoic terranes being indicated by dark colors. This color scheme delineates the geology of Europe in a handsome style and is well accepted also to portions of western North America where Mesozoic and Tertiary formations occupy large areas in contrast to the Paleozoic terranes.

This European color scheme is, however, not particularly well suited to lands in which Paleozoic terranes are predominant and minutely subdivided. The density of colors selected for the Paleozoic tends to produce maps that give a gloomy impression and that sometimes are illegible because of the juxtaposition of so many dark colors. Inasmuch as the range of prismatic colors from purple, blue, and green to yellow is preempted in the European color scheme for Mesozoic and Tertiary terranes, and the reds assigned to the ancient crystalline and eruptive rocks, the remaining choice of colors available for the Paleozoic is too limited for satisfactory discriminations (King and Beikman, 1974). Also, in the Precambrian the number of formations recognized in the cratonic regions of North America and Australia, e.g., greatly exceeds those distinguished in Europe. The sheets for Australia, part of the *Geological Map of the World* (Bureau of Mineral Resources, Geology, and Geophysics, 1965), lamentably display some inappropriate aspects of the European system for Australian conditions. In the more recent *Tectonic Map of Australia* (Tectonic Map Committee, Geological Society of Australia, 1971) an approximation of the American system is employed to produce a much clearer and more pleasing picture of the regional geology. The *Geological Map of Canada* (Geological Survey of Canada, 1969) follows a prismatic color scale, but the blue colors are extended downward to the base of the Paleozoic, reserving the red, orange, and brown colors for the Precambrian in which rocks of many kinds must be differentiated. These considerations illustrate a need for flexibility in the design of color schemes for maps of different continental regions. In sum, the adoption of a universal color scheme for geologic maps can be achieved only to a qualified degree.

Color Schemes for Large-Scale Maps

The preceding considerations apply to general maps on a small scale, maps on the scale of 1:1,000,000, the standard international base (see *Map Scales and Series*). A cartographer who has to construct such a map should consider the geologic development of the province the map is to represent, in whole or in part, and should devise the application of the color scheme in a way that gives the greatest range of colors where there is the greatest development of the geologic column.

Detailed maps present special conditions that may be simple or comprehensive. If they are simple, it is desirable that the colors correspond with those that represent the same general time divisions on the general map. If the detailed map shows terranes that are greatly subdivided, the

TABLE 3. Comparison between the American and International Systems of Coloring Stratified Rocks on Maps

System	American color system			International color system		
	U.S. Geol. Survey 2d Ann. Rept. 1881[1]	U.S. Geol. Survey Folio 227 1945[2]	Geologic Map of United States 1974[3]	2d and 3d Internat. Geol. Cong. Bologna and Berlin 1881, 1885[4]	Eastern Siberia 1964[5]	Geologic Map of France 1968[6]
Quaternary	Gray	Brownish yellow	Gray / Pale yellow	Undecided	Gray	Gray
Tertiary	Yellow	Yellow ocher	Light yellow / Pale brown / Pale flesh / Dark yellow / Greenish yellow	Yellow	Light yellow / Yellow / Greenish yellow	Light yellow / Dark yellow / Orange yellow
Cretaceous	Green	Olive green	Olive green / Yellow green / Cool green	Green	Green	Green
Jurassic		Blue green	Blue green	Blue	Blue	Blue
Triassic		Peacock blue	Peacock blue	Violet	Violet	Violet and purple
Permian		Light blue	Cool blue	Gray	Warm brown	Gray
Pennsylvanian	Blue	Blue	Gray		Dark gray	Dark gray
Mississippian			Warm blue			
Devonian		Blue gray	Blue	Brown	Cool brown	Brown
Silurian	Purple	Blue purple	Purple	Greenish gray	Olive green	Olive gray
Ordovician		Red purple	Rose and pink		Pale green	Olive green
Cambrian		Brick red	Red and coral		Rose	Warm brown
Precambrian	Brown	Brownish red / Gray brown	Yellow brown / Brown / Bluish gray / Brick red	Rose	Brown	Pale brown

Source: King and Beikman (1974), 27.
[1]Powell, 1882, p. xi–lv.
[2]Butts, 1945, specifications of folio series on inside covers (*Geologic Atlas of the United States*).
[3]King and Beikman, 1974, this report and map.
[4]For summary, see Frazer, 1888.
[5]Kransky, 1964.
[6]Service de la Carte Géologique du France, 1968.

range of colors must be increased. Colors should be selected to correspond with colors that indicate the general time divisions on the general maps. Pure brilliant hues, as a rule, should be reserved for igneous rock; shades and tints belonging to a single minor division of the color sequence should preferably be used so that darker effects represent the lower and lighter effects the upper terranes, in such manner that gradations from dark to light correspond with conformable sequences and contrasts indicate systematic distinctions.

CHARLES W. FINKL, JNR.

References

Bureau of Mineral Resources, Geology, and Geophysics, 1965,*Geological Map of the World; Australia and Oceana* (scale 1:5,000,000). Canberra, Australia, sheets 6, 7, 11, 12.

Butts, C., 1945, Description of the Hollidaysburg and Huntingdon quadrangles, Pennsylvania, *U.S. Geol. Survey Geol. Atlas Folio 227.*

Demek, J., and C. Embleton, 1976, *Guide to Medium-Scale Geomorphological Mapping.* Brno: International Geographical Union, Commission on Geomorphological Survey and Mapping, 339p.

FAO/UNESCO, 1972, *Soil Map of the World* (scale 1:5,000,000) Rome: UNESCO.

Frazer, P., 1888, A short history of the origin and acts of the International Congress of Geologists, and of the American Committee delegates to it, *American Geologist* 1, 3-11, 86-100.

Geological Survey of Canada, 1969, *Geological Map of Canada* (scale 1:5,000,000), R. J. W. Douglas, comp. Ottawa, Ont.: Canada Geological Survey (Map 1250A).

Goddard, E. N., 1965, *Geologic Map of North America* (scale 1:5,000,000). Washington, D.C.: North American Geologic Map Committee, U.S. Geological Survey.

Grand Canyon Natural History Association, 1976, *Geological Map of the Grand Canyon National Park, Arizona* (scale 1:62,500) Washington, D.C.: Williams and Heintz Map Corp.

King, P. B., and H. M. Beikman, 1974, Explanatory text to accompany the geologic map of the United States, *U.S. Geol. Survey Prof. Paper 901,* 40p.

Kransky, L. I., 1964, *Geological Map of the Northwestern Part of the Pacific Mobile Belt within the U.S.S.R.* (scale 1:5,000,000) Moscow: Ministry of Geology of the R.S.F.S.R.

Powell, J. W., 1882, *Report of the Director: U.S. Geological Survey 2nd Annual Report for 1880-1881.* Washington, D.C., xi-lv.

Powell, J. W., 1890, *Report of the Director: U.S. Geological Survey 10th Annual Report for 1888-89.* Washington, D.C., 1-8.

Service de la Carte Géologique du France, Bureau de Récherches Géologique et Minières (BRGM), 1968, *Carte Géologique de la France* (scale 1:1,000,000). Paris.

Tectonic Map Committee, Geological Society of Australia, 1971, *Tectonic Map of Australia and New Guinea* (scale 1:5,000,000). Sydney.

Willis, B., 1912, Index to the stratigraphy of North America (map by B. Willis and G. W. Stose; scale 1:500,000), *U.S. Geol. Survey Prof. Paper 71,* 894p.

Cross-references: *Cartography, General; Geological Survey and Mapping; Map Abbreviations, Ciphers, and Mnemonicons; Map Symbols; Maps, Logic of.*

MAPS, ENVIRONMENTAL GEOLOGY

The development of the field of environmental geology (q.v.) can perhaps be related to the growth of concern for the natural environment and the inclusion of factors related to the natural environment in the urban planning process. This development was accompanied by an increase in interdisciplinary studies that brought geologists into working contact with planners, landscape architects, engineers, soil scientists, and public officials.

A growing concern for the natural environment in the United States brought about environmental studies of all kinds in the 1960s. The term *environmental geology* may have been first used by the Illinois Geological Survey in the series of maps and reports titled Environmental Geology Notes, first published in 1965 (Hackett and Hughes, 1965). Environmental geology has a well-developed meaning; a succinct portion of the definition is "the collection, analysis and application of geologic data and principles to problems created by human occupancy and use of the physical environment" (Gary et al., 1974, p. 231). Environmental geology deals with the entire spectrum of people's use of the Earth, both in cities and in rural and primitive regions.

Environmental geology includes the traditional fields of engineering geology and the part of economic geology that pertains to mineral resources. Much of environmental geology is an outgrowth of engineering geological mapping that was developed in Europe over the past several decades (Legget, 1973). It plays an important role in nonurban projects such as siting for power plants, dams, highways, and mines and educates lay people about the location and value of mineral resources. Environmental geology encompasses urban geology, and because populated areas have a greater interaction with the physical environment, urban geology has been the prime focus of environmental geology (Flawn, 1970).

The key word in environmental geology is *application.* Environmental geology involves the application of earth science information to human needs and problems. It involves the study of landforms, Earth materials, and Earth processes. The purpose of environmental geology is overall long-range physical planning and development of the most efficient and beneficial use of the land.

Several publications brought out during the 1970s provide valuable reading about *environmental geology maps*. An excellent coverage of the subject can be found in *Cities and Geology* (Legget, 1973). Chapter 3, "Planning and Geology," familiarizes the reader with the history and development of geology and engineering geology as applied to land planning and resource management in various countries. One of the books that contributed to the development of environmental geology in the United States is *Environmental Geology, Conservation, Land-use Planning, and Resource Management* (Flawn, 1970). Another important book, which reviews case studies as well as the role of geology in society, is *Environmental Planning and Geology* (Nichols and Campbell, 1971). Worthwhile reading that provides the perspective of a landscape architect and planner is *Design with Nature* (McHarg, 1969). Probably the best collection of examples of environmental geology maps by the U.S. Geological Survey is *Nature to Be Commanded...* (Robinson and Spieker, 1978).

Environmental geology, by definition, deals with many disciplines to apply earth science information to resources, engineering, water usage, safety, health, etc. Environmental geology maps, therefore, are varietal. Maps may be of basic data for a specific environmental concern, such as bedrock, soil, landforms, or groundwater. Environmental geology maps may interpret basic data such as mineral quality and quantity, permeability, landslide types, or groundwater recharge areas. Last, maps may show specific planning decisions such as cost per acre for reclamation, size of area required for septic drain fields, slope stability risk zones, or water yield. There is no sharp break separating *basic* and *interpretive maps* (see *Maps, Logic of*). There is instead a continuum from the very basic, such as bedrock maps, to the highly interpretive, such as relative slope stability (Mathewson and Font, 1974; Spangle et al., 1976).

Environmental geology mapping has dealt mainly with the youngest of Earth materials and the most active of earth processes. It is on the youngest Earth materials, the surficial geology, that people live, build most of their structures, and deposit their wastes. The active processes that work to modify the surface of the Earth naturally tend to become geologic hazards when people, their structures, and activities interact with Earth processes. People interact with Earth processes either passively, such as living close to an epicenter of an earthquake, or actively, such as decreasing the stability of slopes through road construction. Most environmental geology maps have sought to help mitigate problems people have or could have with the geologic environment. To do this, the traditional geologic map format has been modified to facilitate the interpretive needs of planners, developers, local governmental officials, and the general public.

In many cases existing geologic maps have been reinterpreted to provide more specific information for planning purposes. Most often, however, it is necessary to collect basic data to make bedrock and surficial geologic maps that are specifically designed for planning interpretations. A vital part of mapping for planning purposes is the collection of data pertinent to the applications. These usually include physical properties of the rocks and surface deposits, such as texture, mineralogy, lithology, fractures, permeability, consolidation, shear strength, and others; categorization and description of representative landforms, especially as they relate to the distribution of rocks and surface deposits; documentation of downslope movements on the different Earth materials under different natural and artificial conditions; mapping of areas of relatively active erosion and deposition; documentation of both surface- and groundwaters; and other data according to local needs.

The combination of types of data needed brings together studies of bedrock, soils, water, and engineering. The environmental geology mapper needs to be broadly based in studies of an area, and ideally the work is accomplished by an interdisciplinary team. Many geotechnical firms now employ such teams for environmental assessment and planning for engineering design.

Much of the interpretation of the location and extent of Earth materials and processes is made through geomorphological study and mapping. Geomorphology, the study of landforms, relates Earth processes, land surface morphology, and types of Earth materials to one another. The importance of geomorphology to environmental geology maps cannot be understated. It provides much of the scientific basis and interpretive tools for making a surficial geologic map from isolated points of data. Analysis of the landforms and development of a hypothesis for the interrelationships of landforms, Earth materials, and Earth processes is what allows the geologist to predict the occurrence of a mappable surficial unit and delineate it with some degree of confidence on a map.

For reviews and examples of applications geomorphology, the following are but a few sources: *Applied Geomorphology* (Hails, 1977), *Geomorphology in Environmental Management* (Cooke and Doornkamp, 1974), *Landslides* (Coates, 1977, Parts 2 and 5), and the series of proceedings volumes of the Annual Geomorphology Symposium, Binghamton, New York, the first volume of which is *Environmental Geomorphology* (Coates, 1971).

Examples

Legget (1973) describes geological maps applied to regional and local planning, including examples from Canada, Europe, Australia, the Far East, and the United States. Environmental geology maps have been published by numerous agencies within the United States, many times in cooperation with one another. The U.S. Geological Survey, the U.S. Department of Housing and Urban Development, state geological surveys, counties, and cities have produced studies of geology as a factor in planning. Examples of these maps include the San Francisco Bay Region studies (e.g., Nilsen et al., 1979; Atwater, 1978); the Environmental Geologic Atlas of the Texas Coastal Zone (e.g., Fisher et al., 1973); the Illinois State Geological Survey Environmental Geology Notes (e.g., Hunt and Kempton, 1977); the Vermont Geological Survey, Geology for Environmental Planning Series (e.g., Stewart, 1974); the Nevada Bureau of Mines and Geology Environmental Series (e.g., Luza, 1974); the Idaho Geological Survey (Othberg, 1987); and U.S. Geological Survey geologic hazards maps (e.g., Witkind, 1972). There are other excellent maps, many of which are published or open-filed by federal, state, and local agencies. Also, an increasing number of graduate thesis projects map geology for environmental purposes (e.g., Kehew, 1977; Fiksdal, 1979). Many environmental geology maps are prepared by private firms for local areas and site investigations (e.g., Nicoll, 1971; Smith, 1980).

Two selected examples of environmental geology maps are shown in Figs. 1 and 2. Figure 1 is part of Map 4 of five maps in a geologic and environmental atlas by the Maryland Geological Survey (Cleaves et al., 1974). In addition to this map, the atlas presents a geologic map (Map 1), a landform map (Map 2), an estimated thickness of overburden map (Map 3), and a mineral resources and mined land inventory map (Map 5). The atlas is fairly representative of the types of map information included in environmental geology investigations. The map legend (Fig. 1) is a good example of interpretations of the capabilities and limiting properties of geology that affect land planning and development. It does not, however, highlight specific geologic factors that are hazardous for certain other areas, such as swelling clay (Crosby et al., 1978), earthquakes (Atwater, 1978), or landslides (Nilsen et al., 1979).

Figure 2, an example from a coastal zone atlas, demonstrates the flow of interpretations from surficial geologic mapping to a geologic hazard map, "slope stability," and to a geologic resource map, "sand and gravel resources." The application in this case was directed toward the coastal environment, and in addition to the geologic fac-

tors, the atlas presents maps of coastal flooding, critical biological areas, coastal drift, and land cover/land use (Washington State Department of Ecology, 1979). The surficial geologic map is accompanied in the atlas by a discussion of mapping methods and land-use applications. The atlas also presents technical descriptions of the geologic units plus a table of engineering properties of geologic units.

The slope stability map shows an interpretation of the relative potential for landslides. As discussed in the text, the geologic factors that were considered in making the map are the physical properties of the geologic units and their relationships to one another, the natural and cut slopes found on these geologic units, the presence or absence of water available to percolate into and through Earth materials, the wave erosion and deposition category of the coastline, and the interpretation of former landslide movements. Stable slopes (S) are usually gentle, dry, or underlain by highly competent Earth materials. Intermediate slopes (I) have less than critical factors but could become critical and subject to landsliding if one or more factors change, either naturally or artificially. Unstable slopes (U) are steep, and the geology, groundwater, and wave erosion factors are interpreted to be critical, or the slopes show evidence of landsliding. Modified slopes (M) have undergone artificial cutting and filling and probably have unpredictable slope stability. Onsite geological and engineering investigations are recommended for development in intermediate areas and should be standard practice in unstable and modified areas.

The sand and gravel resources map is a direct derivative of the physical properties of the geologic units. The map does not address the problems of access and transportation of aggregates, which greatly affect the mineability of a resource; rather, the map shows the distribution and relative quality of known and inferred sands and gravels. In this case, the presence of sand is deleterious to the quality of gravel because sand is so plentiful that it is, in effect, a waste product of aggregate operations in this coastal region. The effects of oxidation due to groundwater and weathering degrades the quality of both sand and gravel. The highest quality sand is found in geologic units relatively high in sand but low in silt, clay, and effects of oxidation. The highest quality gravel is found in geologic units relatively low in sand, silt, and clay and in which oxidation effects are normally absent.

Summary

Environmental geology maps may vary greatly in their topical and areal scope, but they have a common purpose: that relevant earth science in-

EXPLANATION OF MAP UNITS

Terrain underlain by crystalline rock

Maximal constraint conditions

1 High water table conditions. Part or all of area may be subject to flooding. Shown as Unit f on Map of Estimated Thickness of Overburden, Unit 1 on Landform Map, and Qal on Geologic Map.

2 Overburden estimated at 0-5 feet in thickness. Includes areas with slopes greater than 12°, shown as Units S2 and S3 on Landform Map. Mantle instability may occur if slope is disturbed.

3 Overburden estimated at 0-5 feet in thickness. Slope of the land generally less than 12°, and may be less than 6°.

Intermediate constraint conditions

4 Variable thickness of overburden. Terrain is underlain by Cockeysville Marble; abrupt local changes in overburden thickness; rock pinnacles and residual boulders commonly occur; overburden commonly is a sandy carbonate residuum. Corresponds to Unit d on Map of Estimated Thickness of Overburden.

5 Overburden estimated at 5 to 20 feet in thickness. Saprolite comprised of quartz-clay minerals-ferric oxides and hydroxides. Rock fragments rare to common, depending upon parent material.

Minimal constraint conditions

6 Depth of overburden estimated to exceed 30 feet. Slopes less than 12° and commonly less than 6°. Saprolite composed of quartz-clay minerals-ferric oxides and hydroxides, with the exception of Long Green Valley, where dolomite sand and other residuum from weathering of marble occurs.

Terrain underlain by sedimentary deposits

7 Areas in which sand and gravel predominate. Excavation characteristics and stability of cut slopes are variable due to abrupt horizontal and vertical changes in distribution of sand-gravel and clay. Ground-water seepage in predominantly sand layers results in severe slope erosion. After periods of prolonged precipitation localized perched water table conditions are common, particularly if clay seams and lenses are present. On the other hand excavations in, or cut slopes through, dry sand may result in bank failures triggered by vibration.

8 Like 7, except underlain by marble which will constrain construction activities if excavation penetrates through the sedimentary cover.

9 Areas in which clay predominates. Excavation characteristics and stability of cut slopes are variable due to abrupt horizontal and vertical changes of clay and sand present at cut slopes and vertical banks in clay may be stable over short periods of time. However, jointing in the clay commonly results in bank failures if cut is left open for an extended period.

10 Like 9, except underlain by marble which will constrain construction activities if excavation penetrates the overlying clay.

11 Iron cemented layers exceeding 0.25 meters in thickness. These layers may require blasting. Indicated by dashed line. Known occurrences limited to the Hillendale and Putty Hills areas.

12 Areas in which gravel, sand, and clay of fluvial origin are mixed together with poorly sorted, low density materials of colluvial origin. Bank failures may occur if land is modified by construction.

FIGURE 1. *Left:* Portion of map showing geologic factors affecting land modification. *Right:* Explanation for the map, which describes constraint conditions associated with the units, such as high water table, overburden thickness, slope instability, and texture of surficial materials. (From Cleaves et al., 1974)

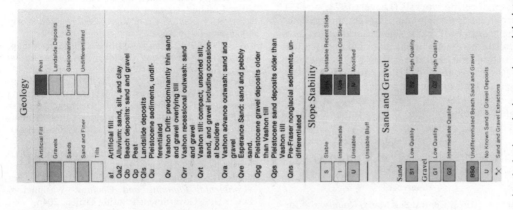

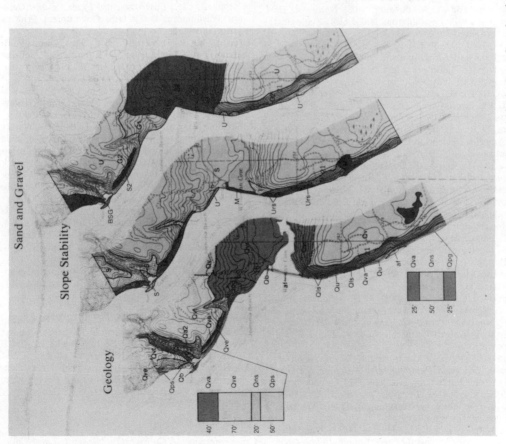

FIGURE 2. *Left*: Series of three maps of the same coastal area within the Puget Sound region showing the basic geologic mapping and the interpretations of slope stability and sand and gravel resources. *Right*: Explanation for the three maps. (Adapted from Washington State Department of Ecology, 1979)

formation be made available and be applied to land-planning and -management activities. "To conserve the soundness and productive power of a region for the use of future generations is a basic requirement of overall planning and one that should be observed as a guide in the prosecuting of all engineering works" (Legget, 1973, p. 173). Environmental geology maps are an integral part of the transfer of basic scientific knowledge to wise resource development and civil engineering. It is fitting that the title of the U.S. Geological Survey's professional paper on earth science maps applied to land and water management is *Nature to Be Commanded . . .*, an excerpt from Francis Bacon's quote, "Nature to be commanded must be obeyed" (*Novum Organum*, 1620).

KURT L. OTHBERG
PAMELA PALMER

References

Atwater, B., 1978, Central San Mateo County, California: Land-use controls arising from erosion of seacliffs, landsliding, and fault movement, in G. D. Robinson and A. M. Spieker, eds., *Nature to Be Commanded....* Washington, D.C.: U.S. Geological Survey (Prof. Paper 950), 11–20.

Cleaves, E. T., W. P. Crowley, and K. R. Kuff, 1974, *Geologic and Environmental Atlas of the Towson Quadrangle.* Baltimore: Maryland Geological Survey, 5 maps, scale 1:24,000.

Coates, D. R., ed., 1971, *Environmental Geomorphology,* Publications in Geomorphology. Binghamton: State University of New York, 262p.

Coates, D. R., ed., 1977, *Landslides,* Geological Society of America Reviews in Engineering Geology, vol. III. Boulder, Colo., 278p.

Cooke, R. U., and J. C. Doornkamp, 1974, *Geomorphology in Environmental Management: An Introduction.* Oxford: Clarendon Press, 413p.

Crosby, E. J., W. R. Hansen, and J. A. Pendleton, 1978, Guiding development of gravel deposits and of unstable ground, in G. D. Robinson and A. M. Spieker, eds., *Nature to be Commanded....* Washington, D.C.: U.S. Geological Survey (Prof. Paper 950), 29–41.

Fiksdal, A. J., 1979, Geology for land-use planning in part of Pierce County, Washington, M.S. thesis, Portland State University.

Fisher, W. L., L. F. Brown, Jr., J. H. McGowen, and C. G. Groat, 1973, *Environmental geologic atlas of the Texas coastal zone—Beaumont–Port Arthur area.* Austin: Bureau of Economic Geology, University of Texas, 93p.

Flawn, P. T., 1970, *Environmental Geology: Conservation, Land-use Planning, and Resource Management.* New York: Harper & Row, 313p.

Gary, M., R. McAfee, Jr., and C. L. Wolf, eds., 1974, *Glossary of Geology.* Washington, D.C.: American Geological Institute, 805p.

Hackett, J. E., and G. M. Hughes, 1965, Controlled drilling program in northeastern Illinois, *Illinois Geol. Survey Environmental Geology Notes,* no. 83, 42p.

Hails, J. R., ed., 1977, *Applied Geomorphology: A Perspective of the Contribution of Geomorphology to Interdisciplinary Studies and Environmental Management.* New York: Elsevier, 418p.

Hunt, C. S., and J. P. Kempton, 1977, Geology for planning in DeWitt County, Illinois, *Illinois State Geol. Survey Environ. Geol. Notes No. 83,* 42p.

Kehew, A. E., 1977, Environmental geology of Lewiston, Idaho, and vicinity, Ph.D. dissertation, University of Idaho.

Legget, R. F., 1973, *Cities and Geology.* New York: McGraw-Hill, 624p.

Luza, K. V., 1974, Reno folio physical properties map, *Nevada Bur. Mines and Geology Environmental Series,* Reno Area, scale 1:24,000.

Mathewson, C. C. and R. G. Font, 1974, Geologic environment: Forgotten aspect of land use planning process, in H. F. Ferguson, ed., *Geologic Mapping for Environmental Purposes.* Boulder, Colo.: Geological Society of America (Eng. Geology Case Histories, No. 10), 23–28.

McHarg, I., 1969, *Design with Nature.* New York: Doubleday/Natural History Press, 197p.

Nichols, D. R., and C. C. Campbell, eds., 1971, *Environmental Planning and Geology.* Washington, D.C.: U.S. Government Printing Office, 204p.

Nicoll, G. A., 1971, Reconnaissance engineering geology for master planning, in D. R. Nichols and C. C. Campbell, eds., *Environmental Planning and Geology.* Washington, D.C.: U.S. Government Printing Office, 170–175.

Nilsen, T. H., R. H. Wright, T. C. Vlasic, and W. E. Spangle, 1979, Relative slope stability and land-use planning in the San Francisco Bay region, California, *U.S. Geol. Survey Prof. Paper 944,* 96p.

Othberg, K. L., 1987, Landforms and surface deposits of Long Valley, Valley County, Idaho, *Idaho Geological Survey Map 5,* scale approximately 1:62,500.

Robinson, G. D., and A. M. Spieker, eds., 1978, Nature to be Commanded . . . , Earth-science maps applied to land and water management, *U.S. Geol. Survey Prof. Paper 950,* 95p.

Smith, M., 1980, *Geotechnical Map of the Gig Harbor Peninsula, Pierce County, Washington.* Tacoma: Robinson, Noble and Carr, Inc., scale 1:24,000.

Spangle, W., and Associates; F. B. Leighton and Associates; and Baxter, McDonald and Company, 1976, Earth-science information in land-use planning—Guidelines for earth scientists and planners, *U.S. Geol. Survey Circ. 721,* 28p.

Stewart, D. P., 1974, Geology for environmental planning in the Milton–St. Albans region, Vermont, *Vermont Geol. Survey Environmental Geology,* no. 5, 48p.

Washington State Department of Ecology, 1979, *Coastal Zone Atlas of Washington,* vol. 7. Pierce County: Washington Department of Ecology (Publ. No. DOE 77-21-7), plates PI 1–PI 21C, 5p.

Witkind, I. J., 1972, Map showing seiche, rockslide, rockfall, and earthflow hazards in the Henrys Lake quadrangle, Idaho and Montana, *U.S. Geol. Survey Map I-781-C,* scale 1:62,500.

Cross-references: *Cartography, General; Land Capability Analysis; Maps, Logic of; Maps, Physical Properties; Terrain Evaluation Systems.* Vol. XIII: *Geomorphology, Applied; Urban Geomorphology.*

MAPS, LOGIC OF

Any map consists of two inseparable parts: (1) a graphic portrayal of areal, linear, or point information and (2) an explanation in words concerning the essential attributes of the items portrayed. The term *attributes* refers to the names, definitions, properties, characteristics, and mutual relations of areal units. A map is thus a classification, a method of presenting areal distribution patterns as a scale model of the real world.

Maps as a Kind of Classification

The process of classification proceeds by grouping individuals into larger units and by dividing larger already recognized units or a heterogeneous population into smaller classes. Both processes of grouping and division are subject to logical rules that must govern the classification of natural objects. According to Grigg (1967), classifications should be designed for a specific purpose because they rarely serve two purposes equally well.

The purposes of a map are related to the two processes of classification. According to Beckett (1968, p. 53), a map is made "in order to be able to make more precise statements about the mapped subdivisions of the region than we can about the region as a whole." This statement is true, but mapping also includes the grouping of small areas into larger units so that we can make statements about the group that are more general than those we can make about its components. In these two intents, and in their combination, lie all the reasons for mapping. Every map, however, occupies some part of a field of contest that has at one end the goal of attainment of perfectly detailed information about the attributes possessed by specified areas and at the other end the goal of complete knowledge of location of all areas that have one or more attributes of interest (Fig. 1).

In Fig. 1, operations leading to grouping and synthesis presuppose the existence of discrete individuals that can be welded into new, more inclusive individuals. Operations leading to analysis and logical division largely consist of a search for, and precise definition of, manageable, useful individuals; this search presupposes the existence of or creates concepts by which individuals can be defined or recognized.

The information that goes into the construction of a map may be analyzed by arranging it in the form of an array or matrix that shows locations or places in rows and the attributes of those places in columns (Fig. 2A). There are essentially two different ways of making meaningful groupings of such an array of data. Regrouping of the rows (places) in Fig. 2A into those of Figure 2B

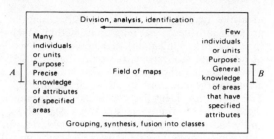

FIGURE 1. Field of purposes of maps. The two goals, (A) attainment of precise knowledge of attributes of specified areas and (B) general knowledge of the areas having specified attributes, are approached by opposing methods of classification: division and grouping. (From Varnes, 1974)

ATTRIBUTES

	1	2	3	4	5	6	7	8	9	10
A		X	X	X		X	X	X		X
B	X	X	X			X	X	X	X	X
C	X			X		X		X	X	
D		X	X			X	X	X		X
E	X	X	X			X	X	X		X
F		X		X				X		X
G	X	X	X		X		X		X	X
H		X	X				X			
I		X	X	X		X	X	X	X	

A

ATTRIBUTES

	1	2	3	4	5	6	7	8	9	10
A		y	X	X		X	X	X		X
D			X	X		X	X	X		X
I		X	X	X		X	X	X	X	
B	X	X	X			X	X	X	X	X
E	X	X	X			X	X	X		X
G	X	X	X		X		X		X	X
C	X			X		X		X	X	
F		X		X				X		X
H		X	X				X			

B

ATTRIBUTES

	3	7	1	9	2	4	5	6	8	10
A	X	X			X	X		X	X	X
B	X	X	X	X	X		X	X		X
C			X	X		X		X	X	
D	X	X					X	X	X	X
E	X	X	X	X	X		X	X		
F					X	X			X	X
G	X	X	X	X	X		X			X
H	X	X			X					
I	X	X			X	X	X		X	X

C

FIGURE 2. Matrices showing (A) attributes of places, (B) places grouped by similar attributes, (C) attributes grouped by similar places. (From Varnes, 1974)

identifies two aggregates of places that may be grouped into new areal map units, within which the places have similar, although not identical, sets of attributes. Conversely, the array may be used for a topical rather than an areal study by re-grouping the columns, as shown in Figure 2*C*. This operation identifies two pairs of attributes that covary at certain places: 3 and 7 perfectly, 1 and 9 almost perfectly. The reason for covariance can then become the subject of a geologic investigation.

In the construction of a geologic map of a previously unmapped area, the land is usually classified by a grouping procedure; i.e., individual places are grouped according to selected attributes until areal units are outlined that are satisfactory for the intended purpose of the map. Conversely, after some classification by grouping has established attributes that define the map units, a boundary between two units may be extended into unknown territory by the process of logical division—i.e., by applying differentiating criteria to distinguish and thus to separate newly observed places.

Geologic Maps

There are many different kinds of geologic maps, and their content and mode of graphic portrayal depend not only on the purpose for which the map is intended but also on the interest and ability of the mapper and the time and funds at his or her command. Because the materials that compose the solid Earth are not homogeneous, they can be subdivided or grouped in manifold ways. Moreover, no two geologists will independently produce identical maps of the same area, even if they have similar abilities and support. As Harrison (1963, p. 226) stated:

For most people, the geological map, with its scheme of contrasting colours, apparently unequivocal structural symbols, and sharply drawn contacts between rock units, creates the impression that it is, like most other types of maps, a factual and objective record of data derived from observations made on different classes of rocks clearly distinguishable from each other by well-defined physical characteristics. For most geological maps this impression is fallacious. A good geological map is much more than an objective presentation of the distribution of rock units, their structure and their relations; it is also a subjective presentation of interpretations based on a multitude of observations and, to a greater or less degree, based on theories and prejudices held at the time the map was made.

The making of a geological map is thus a complex procedure that involves continued interplay of logical deduction and induction, observation and inference, division and grouping—all carried on within limits imposed by purpose and scope, requisite scale and accuracy, cartographic require-ments, and the time and funds available. There must inevitably be compromises between the logical ideal and expedience.

Guiding Principles

When constructing a geologic map, the geologist is guided by certain logical principles. Among these are the postulates summarized by Gilluly et al. (1952) that apply to sequences of sedimentary rocks:

The Law of Superposition: In any pile of sedimentary strata that has not been disturbed by folding or overturning since accumulation, the youngest stratum is at the top and the oldest at the base.

The Law of Original Horizontality: Water-laid sediments are deposited in strata that are not far from horizontal and parallel or nearly parallel to the surface on which they are accumulating.

The Law of Original Continuity: A water-laid stratum, at the time it was formed, must continue laterally in all directions until it thins out as a result of nondeposition or until it abuts the edge of the original basin of deposition.

The Law of Truncation by Erosion or Dislocation: A corollary of the previous principle, that a stratum that ends abruptly at some point other than the edge of the basin in which it was deposited must have had its original continuation removed by erosion or displaced by a fracture in the Earth's crust.

These are general rather than absolute rules; exceptions are not common and can usually be easily recognized.

With the development of more than 150 yr of geologic mapping, a large body of knowledge has accumulated that permits a vast array of logical deductions and inductive inferences to be drawn with regard to geologic processes, not only in sedimentary terrains but also in more complex igneous and metamorphic regions. Today some are the subject of intense research and controversy while others have become well established. The methodology for constructing geologic maps is presented in textbooks by Lahee (1961) and Low (1957). Logical procedures in the design and use of digital geographic information systems has been described by Robinove (1986).

Types of Geologic Maps

The most common type of geologic map shows the areal distribution of rock units as outlined areas. Structural features such as faults, folds, bedding, joints, or foliation are depicted by special symbols (see *Map Symbols*). Bedrock and surficial deposits may be shown together or on separate maps. Ideally, all information should be at a scale appropriate to the purpose of the map and superposed on topographic or planimetric base maps. To display better the inferred three-dimensional relations of bedrock or surficial units, the plan map is often accompanied by one or more cross-sections that show the mapped units as they would appear if exposed in a vertical cut along the line of section. The geologic map must have an explanation that shows the colors or patterns used for each rock unit (arranged in order by age), together with brief summaries of their lithology, and descriptions of all linework and symbols. A separate text usually gives more complete descriptions and interprets the geologic development of the area. Thus, a prime purpose of a general geologic map is to contribute to the formation of broad concepts and to stimulate inferences or predictions that may guide future investigations.

Geologic maps constructed for special purposes emphasize specific features or show additional information that may not appear on a general geologic map. Among these are *resource maps* that show mines and metalliferous or mineral fuel deposits, wells, oil and gas fields, water supplies, or construction materials. *Structure contour* maps are made to show the configuration of a subsurface stratum of particular interest. *Isopach maps* show the thickness of a selected bed or series of beds by contours. *Geophysical maps* show properties such as the intensity of magnetic or gravity fields. *Paleogeologic maps* show the distribution of rock units at some selected time in the ancient past. Maps prepared for *land-use planning* may emphasize some selected properties of near-surface materials or areas subject to geologic hazards.

Interpretive Maps

Geologic maps, by virtue of their complexity and specialized nature, are prepared mainly for use in the geologic sciences rather than by the general public. Because general geologic maps are, however, basic documents that give an inventory of "what is where" at or near the Earth's surface, they are used to help solve problems associated with resource development, alternative uses of the land, engineering construction, or avoidance of hazards. These needs often require that a geologic map be interpreted for purposes different from those for which the map was constructed, requiring a reclassification of the geologic map units according to attributes that may or may not have been essential to their original delineation. If these newly required attributes were not essential, then their presence or absence must be inferred without gathering new information by a process of logical induction that requires special geological experience.

The derivation from a conventional geologic map of information or other maps applicable to the needs of civil engineering dominantly involves a transformation of meaning of the geologic boundaries. The previously delineated units are assigned new essential attributes of engineering performance, behavior, or use. The logic and success of this transformation depend on the degree of accuracy and reliability required, on how closely the properties of interest covary with the originally mapped boundaries, on how much additional information is available, and on how heterogeneous the originally mapped units are with respect to these properties.

Refer to Fig. 2*B* and suppose that a simple geologic map was constructed showing areal units composed of places A-D-I (combined into one areal unit), unit B-E-G, and separate units composed of places C, F, and H. Suppose also that the necessary and sufficient attributes to define unit A-D-I are numbers 3, 4, and 6. Assume that the map is to be used for determining areas suitable for some use, such as locating a foundation site for heavy structures, which requires attributes 7 and 8. The unit A-D-I could be judged suitable either if the attributes 7 and 8, which were nonessential to its delineation, were recorded and mentioned in the map explanation or text or if the presence of attributes 7 and 8 could be reasonably inferred from the presence of 3 or 4 or 6. If field examination did not record attributes 7 and 8, one must rely on past experience or logical induction; i.e., where attributes 3 and 4 are both present, then both 7 and 8 are likely to be present also.

Problems in Cartography

In addition to presenting problems of classification and interpretation, geologic maps also involve some cartographic problems such as the selection of colors or patterns to represent rocks of various ages and lithologies and the letter or number systems used to identify formations. Some systems are arbitrary—e.g., the range of colors for sedimentary rock map units of certain ages adopted by the U.S. Geological Survey. Sometimes the map patterns or overprints used to indicate lithology have a certain logical analogy to the texture of the geologic material—e.g., a dot pattern for sand or closely spaced circles for gravel.

A difficult cartographic problem arises when it is desirable to depict graphically the succession of deposits at a certain place or area. Three methods have been used: (1) an overprinted pattern to show the distribution of, e.g., a thin cover of surficial material such as blown sand or colluvium; (2) a stripe system in which the presence of an underlying material is shown by narrow stripes of its color or pattern within the color or pattern of the overlying material; and (3) a unitized system in which a single color and/or pattern is used to identify an entire unique succession of near-surface deposits. One of the difficulties encountered with special cartographic methods, particularly the last mentioned, is that of achieving proper visual emphasis while assuring that the most important map features are perceived by the map reader in their proper emphasis. Careful attention must be given, therefore, to the purpose of the map, not only in arriving at the classification of map units but also in determining the manner in which they are depicted.

DAVID J. VARNES

References

Beckett, P. H. T., 1968, Method and scale of land resource surveys in relation to precision and cost, in G. A. Stewart, ed., *Land Evaluation—CSIRO Symposium.* South Melbourne, V.C., Australia: Macmillan, 53–63.

Gilluly, J., A. C. Waters, and A. O. Woodford, 1952, *Principles of Geology.* San Francisco: W. H. Freeman, Co., 631p.

Grigg, D., 1967, Regions, models and classes, in R. J. Chorley and P. Hagget, eds., *Models in Geography.* London: Methuen and Co., 461–509.

Harrison, J. M., 1963, Nature and significance of geological maps, in C. C. Albritton, Jr., ed., *The Fabric of Geology.* Boulder, Colo.: Geological Society of America, Addison-Wesley Publ. Co., 225–232.

Lahee, F. H., 1961, *Field Geology,* 6th ed. New York: McGraw-Hill, 926p.

Low, J. W., 1957, *Geologic Field Methods.* New York: Harper & Row, 489p.

Robinove, C. J., 1986, Principles of logic and use of digital geographic information systems, *U.S. Geological Survey Circular 977,* 19p.

Varnes, D. J., 1974, The logic of geological maps, with reference to their interpretation and use for engineering purposes, *U.S. Geol. Survey Prof. Paper 837,* 48p.

Cross-references: *Cartography, General; Map Series and Scales; Map Symbols; Maps, Engineering Purposes; Maps, Environmental Geology; Maps, Physical Properties; Surveying, General.* Vol. XIII: *Geological Communication.*

MAPS, PHYSICAL PROPERTIES

Regional physical properties maps that display average or relative physical and engineering properties of geologic materials have been prepared by a number of state and federal governmental agencies. The Nevada Bureau of Mines and Geology (1976) and the Texas Bureau of Economic Geology (Brown et al., 1976, 1977, 1980; Fisher et al., 1972, 1973; McGowen et al., 1976a, 1976b) have included physical properties maps as part of their environmental geology programs (see *Maps, Environmental Geology*) for their respective states. Their physical properties maps are designed to give planners, architects, engineers, and other users insight into the physical and engineering properties of geologic units. These maps are particularly useful in urban areas.

A *physical properties map* can be designed to provide regional data for a variety of physical uses such as domestic-sewage-waste disposal and road construction. The data are applicable not only to the land's surface but also to a maximum depth of 15 m. Many specific types of information may be obtained from a physical properties map. For example, a unit's relative permeability and drainage characteristics can be incorporated within the map's explanation, which in turn may be used to determine a unit's capability for domestic-sewage-waste disposal.

Preparation

A regional physical properties map at a scale in the range of 1:24,000 to 1:250,000 is initially prepared from detailed geologic maps. A physical properties map for the Reno 7.5 min quadrangle was prepared by grouping the geologic units into two broad categories, consolidated and unconsolidated materials (Fig. 1). The consolidated ma-

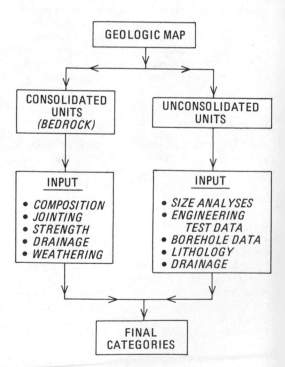

FIGURE 1. Flow chart illustrating the development of the final categories on a physical properties map from a geologic map.

terials, comprising chiefly bedrock units, were subdivided into eight categories based on lithology, weathering and jointing characteristics, and strength properties. The unconsolidated materials, chiefly stream and lacustrine units, were subdivided into nine categories based on variations in lithology, permeability, and bearing value. The final categories were established by combining with the geologic map data borehole data, engineering test data, and size analyses. The principal contributors of data were state and federal agencies and soils engineering consulting firms. Where engineering data were not available or did not exist, size analyses, Atterberg limits, and in some cases, unconfined compression tests were conducted by the Nevada Bureau of Mines and Geology's personnel.

Six parameters were used in formulating the map categories: (1) permeability, (2) drainage, (3) relief, (4) shrink-swell potential, (5) compressibility, and (6) bearing strength. The parameters were determined from quantitative testing as well as from estimates derived from sieve analyses (Fig. 2). The quantitative data were determined from standard penetration tests, unconfined compression tests, permeability, and consolidation tests. A complete sieve analysis accompanied with Atterberg limits, when appropriate, was used to categorize materials according to the Unified Soil

Classification. Once a material's category was established, permeability, shrink-swell potential, compressibility, and bearing strength were estimated.

These six parameters used in describing the engineering and physical properties for each map group were selected because they could be readily determined from size analyses, on-site inspection, and soils engineering project reports. These parameters provided basic information that can be used to ascertain potential physical-use problems indigenous to the Reno area. The availability of test data and regional geologic conditions determines what parameters are worth presenting for a particular region.

Development

A portion of the geologic map of the Reno quadrangle (Bonham and Bingler, 1973) is shown in Fig. 3. Seven units are displayed on the geologic map, including andesite, pyroxene diorite, alluvial fan deposits, two outwash deposits, lacustrine deposits, and alluvium. Tertiary pyroxene andesite flows and hypabyssal diorite stocks are locally overlain by Quaternary alluvial fan deposits of Peavine Mountain. The alluvial fan deposits of Peavine Mountain are composed of poorly sorted gravelly to clayey silt. The central

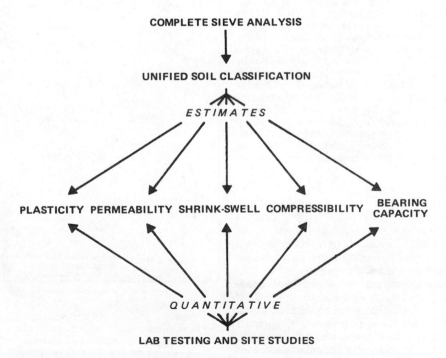

FIGURE 2. Parameters used in the formulation of a physical properties map category. Parameters are derived from both quantitative and qualitative analyses.

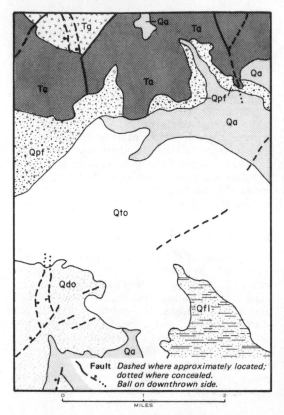

FIGURE 3. Geologic map of a portion of the Reno quadrangle. Qa, alluvium: poorly sorted clayey to silty sand; Qfl, floodplain and lake deposits: medium- to thin-bedded clayey silt and sand; Qto, Tahoe Outwash: large boulder to cobble gravel, sandy gravel, and gravelly sand; Qdo, Donner Lake Outwash: deposits similar to Tahoe Outwash except weathered to depths of 4 ft (120 cm) or more; Qpf, alluvial fan deposits of Peavine Mountain: poorly sorted, gravelly to sandy and clayey silt; Tg, granitic stock: hypabyssal stock composed of several intrusive phases ranging in composition from pyroxene diorite to granodiorite porphyry; Ta, Alta Formation: pyroxene andesite flows, flow breccia, and laharic breccia. (After Bonham and Bingler, 1973)

and southwestern parts of the map display Pleistocene outwash deposits. The outwash deposits consist of boulder- to cobble-sized gravel, sandy gravel, and gravelly sand that are locally overlain by thin Quaternary floodplain and lacustrine deposits. The floodplain and lacustrine deposits crop out in the southeastern part of the Reno quadrangle and consist of medium- to thin-bedded clayey silt and sand interbedded with discontinuous layers of silt and peat. Young Quaternary alluvium composed of poorly sorted clayey to silty gravelly sand locally overlies the outwash deposits in the northern and southern parts of the mapped area.

The geologic units were regrouped into six categories that formed the basic map units for the physical properties map (Fig. 4). Group 1 materials include pyroxene andesite flows, flow breccia, laharic breccia, and hornblende andesite. These materials have moderate to closely spaced joints, fracture permeability, low compressibility and shrink-swell potential, impervious drainage, good bearing value, and gentle to very steep relief.

Group 2 materials consist of hydrothermally altered andesite and diorite. The dominant clay minerals are montmorillonite, kaolinite, and sericite; pyrite and other sulfide minerals may be present in unweathered parts of the altered rocks. These materials have moderately to very closely spaced joints, low to moderate compressibility, low to moderate shrink-swell potential, poor drainage, moderate to high water-holding capacity, medium to high plasticity, low to moderate bearing value, and gentle to very steep relief.

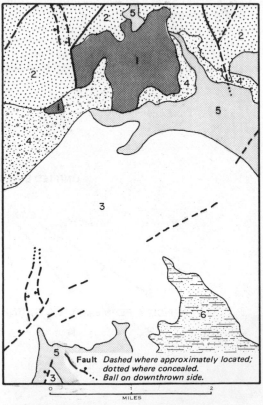

FIGURE 4. Physical properties map of a portion of the Reno quadrangle. Group 1, dominantly andesite; Group 2, hydrothermally altered andesite and diorite, fan deposits of Peavine Mountain; Group 5, Quaternary alluvium; Group 6, Quaternary floodplain and lake deposits. (From Luza, 1974)

Group 3 materials are composed of Donner Lake and Tahoe Outwash deposits. The deposits consist of large boulder- to cobble-sized gravel, sandy gravel, and gravelly sand. They have medium to high permeability, low compressibility, low shrink-swell potential, excellent drainage, good bearing value, low plasticity, and low relief.

The materials in Group 4 consist of alluvial fan deposits that are composed of poorly sorted sandy gravel to clayey silt. The dominant clay mineral is montmorillonite. Group 4 lands have impervious to low permeability, low to moderate compressibility, medium to high shrink-swell potential, fair to impervious drainage, good to poor bearing value, low to moderate plasticity, and gentle relief.

Group 5 is composed of poorly sorted clayey to gravelly sand. These materials are poorly bedded to unbedded and have very low to medium permeability, low to moderate compressibility, low to medium shrink-swell potential, excellent to poor drainage, good to poor bearing value, low to medium plasticity, and low relief.

Quaternary floodplain and lake deposits make up Group 6. They consist of thin sheetlike deposits medium- to thin-bedded, clayey silt and sand, and discontinuous layers of silt and peat. This group has impervious to fair drainage, fair to very poor bearing value, low to high plasticity, susceptibility to liquefaction, and low relief.

In the discussion of each group, terms such as *low to medium permeability* to *high shrink-swell potential* are used. Luza (1974) provided an interpretive guide in the explanation of the Reno quadrangle physical properties map. The guide was intended to give semiquantitative information on several categories so that the user could make relative comparisons between each group.

Six major categories are given in the guide (Fig. 5): (1) shrink-swell potential, (2) compressibility, (3) plasticity, (4) relief, (5) joints and fractures, and (6) relative values of permeability. The terms for the first three parameters are based on Atterberg limits. The ranges are empirically derived and thus provide only a guide for relative comparison. The terms indicating relief were chosen to follow the format of an existing slope map for the Reno quadrangle. The joint and fracture terminology and spacing were those McGlade et al. (1972) used for rocks in Pennsylvania. The spacing may be adjusted according to the fracture-frequency conditions at a particular locality. The last category, permeability, uses terms and permeability coefficients that were empirically derived by Terzahi and Peck (1948) for gravel, sand, silt, and clay. The values for permeability should be used only for relative comparisons; verification of absolute values must be conducted either in the laboratory or in the field. The guide to various parameters discussed in the explanation text of Fig. 5 may also

be incorporated in an accompanied text as long as the reader understands what the terms *high* and/or *low* mean.

Special-Use Applications

Each group category shown on the physical properties map may be evaluated in terms of se-

Shrink-swell Potential[1]

Term	Shrinkage limit	Plasticity index
Probably low	$>$12	0—15
Probably moderate	10—12	15—30
Probably high	0—10	$>$30

[1] Holtz, W.G., and Gibbs, H.J., 1956.

Compressibility[2]

Term	Compression index	Liquid limit
Slight to low	0—0.19	0—30
Moderate	0.20—0.39	31—50
High	$>$0.39	$>$50

[2] Sowers, G.B., and Sowers, G.F., 1961.

Plasticity[3]

Term	Plasticity index	Dry strength
Non-plastic	0—3	very low
Slightly plastic	4—8	slight
Medium plastic	9—30	medium
Highly plastic	$>$30	high

[3] Sowers, G.B., and Sowers, G.F., 1961.

Relief

Term	Percent slope
Low	0—5
Gentle	5—15
Moderate	15—30
Steep	30—50
Very steep	$>$50

Joints and Fractures[4]

Term	Spacing
Very widely spaced	10 feet
Widely spaced	3—10 feet
Moderately spaced	1—3 feet
Closely spaced	2 inches—1 foot
Very closely spaced	$<$2 inches

[4] McGlade, W.G., Geyer, A.R., and Wilshusen, J.P., 1972

Relative Values of Permeability[5]

Term	Coefficient of permeability (cm/sec)	Typical material
Very permeable	k$>$$10^{-1}$	coarse gravel
Medium	k = 10^{-1}—10^{-3}	sand, fine sand
Low	k = 10^{-3}—10^{-5}	silty sand
Very low	k = 10^{-5}—10^{-7}	silt
Impervious	k$<$$10^{-7}$	clay

[5] Terzahi, K., and Peck, R.B., 1948.

FIGURE 5. Guide used to make relative comparisons of various parameters between each group.

lected activities and land uses. Activities and land uses may vary from one locality to another, depending on terrain, geologic conditions, and local building codes and ordinances. Table 1 illustrates in chart form some selected land uses for the Reno quadrangle. The symbols +, O, and X indicate whether each category is satisfactory, has possible problems, or is undesirable for a particular use. Similar notations were used by Fisher et al. (1972) in their discussion of special-use applications for the Environmental Geologic Atlas of the Texas Coastal Zone–Galveston–Houston area.

TABLE 1. Evaluation of the Natural Suitability of Physical Properties Groups
for Selected Activities and Land Uses

GENERAL PHYSICAL PROPERTIES	GEOLOGIC MAP UNIT(S)	(1) Road Construction—base material	(2) Fill Material—backfill	(3) Septic Systems	(4) Foundations—heavy	(5) Foundations—light	(6) Buried cables and pipes	(7) Excavatability
Group 1 Dominantly andesite: moderate to very closely spaced joints, fracture permeability, low compressibility and shrink-swell potential, impervious drainage, good bearing value, gentle to very steep relief	Includes pyroxene andesite flows, flow breccia, laharic breccia, and hornblende-biotite andesite, Tertiary Alta and Kate Peak Formations	O	O	X	+	+	+	X
Group 2 Hydrothermally altered rocks, dominant clay minerals are montmorillonite, kaolinite, and sericite; pyrite and other sulfide minerals may be present in unweathered parts of the altered rock, moderately to very closely spaced joints, low to moderate compressibility, low to moderate shrink-swell potential, poor drainage, moderate to high water-holding capacity, medium high plasticity, low to moderate bearing value, gentle to very steep relief	Altered andesite and diorite	X	O	O	O	O	X	+
Group 3 Large boulder to cobble gravel, and gravelly sand, medium to high permeability, low compressibility, low shrink-swell potential, excellent drainage, good bearing value, low plasticity, low relief	Includes Pleistocene Donner Lake and Tahoe outwash deposits	O	O	O	+	+	O	O
Group 4 Poorly sorted sandy gravel to clayey silt, dominant clay mineral is montmorillonite, impervious to low permeability, low to moderate compressibility, medium to high shrink-swell potential, fair to impervious drainage, good to poor bearing value, low to moderate plasticity, gentle relief	Quaternary alluvial fan deposits of Peavine Mountain	O	O	O	O	O	O	+
Group 5 Poorly sorted clayey to silty gravelly sand, poorly bedded to unbedded, very low to medium permeability, low to moderate compressibility, low to medium shrink-swell potential, excellent to poor drainage, good to poor bearing value, low to medium plasticity, low relief	Quaternary alluvium	O	O	O	X	O	O	+
Group 6 Thin, sheet-like deposits of medium- to thin-bedded, clayey silt and sand; contains discontinuous layers of silt and peat, impervious to very low permeability, moderate to high compressibility, low to medium shrink-swell potential, impervious to fair drainage, fair to very poor bearing value, low to high plasticity, susceptible to liquefaction, low relief	Quaternary floodplain and lake deposits	X	O	X	X	O	O	O

(1) Road construction: base material—low compressibility, low shrink-swell potential, high shear strength; meets type 1, 2, and 3 specifications for Nevada road and bridge construction.
(2) Fill material: backfill—below topsoil, low to moderate shrink-swell potential, 100% passing 3-inch sieve (U.S. Sieve Series).
(3) Septic systems—moderate permeability, low to moderate shrink-swell potential, and good subsurface drainage.
(4) Foundations: heavy—high load-bearing strength, low shrink-swell potential, and good drainage.
(5) Foundations: light—low shrink-swell potential.
(6) Buried cables and pipes—low shrink-swell potential and low corrosivity.
(7) Excavatability—ease of digging with conventional machinery.
+ Satisfactory
O Possible problems
X Undesirable (will require special planning and engineering).

The major categories under the land-use heading in Table 1 are keyed to specifications and criteria used in the evaluation process. Such a table should be augmented with a written discussion, particularly when a group has a number of undesirable characteristics. For example, in Table 1, Group 6 materials may have possible excavation problems. Although all Group 6 materials are easily excavated with conventional machinery, in some areas groundwater seasonally fluctuates close to the surface. Therefore, either those areas ought to be avoided during unseasonally high water levels or the water table should be artificially lowered before excavation proceeds. Only after a careful on-site inspection will one be able to delineate these possible problem areas.

In summary, regional physical properties maps can be derived from geologic maps, provided the geologic information is augmented with some laboratory test data. A physical properties map illustrates only average physical properties and should not serve as a substitute for on-site investigations.

KENNETH V. LUZA

References

Bonham, H. F., Jr., and E. C. Bingler, 1973, Geologic map of the Reno quadrangle, *Environmental Folio,* Reno area, No. 4A, map g, Nevada Bur. Mines and Geology, scale 1:24,000.

Brown, L. F., Jr., J. L. Brewton, T. J. Evans, J. H. McGowen, C. G. Groat, and W. L. Fisher, 1980, *Environmental Geologic Atlas of the Texas Coastal Zone—Brownsville-Harlingen Area.* Texas Bureau of Economic Geology, 140p.

Brown, L. F., Jr., J. L. Brewton, J. H. McGowen, T. J. Evans, W. L. Fisher, and C. G. Groat, C. G., 1976, *Environmental Geologic Atlas of the Texas Coastal Zone—Corpus Christi Area.* Texas Bureau of Economic Geology, 123p.

Brown, L. F., Jr., J. H. McGowen, T. J. Evans, C. G. Groat, and W. L. Fisher, 1977, *Environmental Geologic Atlas of the Texas Coastal Zone—Kingsville Area.* Texas Bureau of Economic Geology, 131p.

Fisher, W. L., L. F. Brown, Jr., J. H. McGowen, and C. G. Groat, 1973, *Environmental Geologic Atlas of the Texas Coastal Zone—Beaumont-Port Arthur Area.* Texas Bureau of Economic Geology, 93p.

Fisher, W. L., J. H. McGowen, L. F. Brown, Jr., and C. G. Groat, 1972, *Environmental Geologic Atlas of the Texas Coastal Zone—Galveston-Houston Area.* Texas Bureau of Economic Geology, 91p.

Holtz, W. G., and H. J. Gibbs, 1956, Engineering properties of expansive clays, *Am. Soc. Civil Engineers Trans.* **121,** 641-663.

Luza, K. V., 1974, Physical properties map of the Reno quadrangle, *Nevada Bur. Mines and Environmental Folio,* Reno area, No. 4A, Map j, Nevada Bur. Mines and Geology, scale 1:24,000.

McGlade, W. G., A. R. Geyer, and J. P. Wilshusen, 1972, Engineering characteristics of the rocks of Pennsylvania, *Pennsylvania Geol. Survey Environmental Geology Rep. 1,* 200p.

McGowen, J. H., L. F. Brown, Jr., T. J. Evans, W. L. Fisher, and C. G. Groat, 1976a, *Environmental Geologic Atlas of the Texas Coastal Zone—Bay City-Freeport Area.* Texas Bureau of Economic Geology, 98p.

McGowen, J. H., C. V. Procter, Jr., L. F. Brown, Jr., T. J. Evans, W. L. Fisher, and C. G. Groat, 1976b, *Environmental Geologic Atlas of the Texas Coastal Zone—Port Laraca Area.* Texas Bureau of Economic Geology, 107p.

Nevada Bureau of Mines and Geology, 1976, Reno quadrangle, *Environmental Folio,* Reno area, No. 4A, 52p.

Sowers, G. B., and G. F. Sowers, 1961, *Introductory Soil Mechanics and Foundations.* New York: Macmillan, 386p.

Terzahi, K., and R. B. Peck, 1948, *Soil Mechanics in Engineering Practice.* New York: John Wiley & Sons, 566p.

Cross-references: *Cartography, General; Environmental Engineering; Maps, Environmental Geology, Maps, Logic of.* Vol. XIII: *Maps, Engineering Purposes.*

MAP SERIES AND SCALES

Earth scientists have long recognized the value of maps as a major communications medium for conveying scientific data and details to potential users. Through the use of graphic symbols, positioned in their relative locations, geologic maps are valuable in conveying qualitative and/or quantitative data in an amount deemed necessary by the geologist to represent his or her findings. A map allows the presentation of information that is impossible to convey accurately by the printed word and facilitates the presentation of three-dimensional information on a two-dimensional plane (see *Map Symbols*).

Geologic maps have been published in a variety of formats and/or series in the U.S. Geological Survey beginning in 1882. The need for specialized studies has resulted in an expansion of the map series. The amount of detail to be shown on a map is determined by the scale, and consequently, the scales of maps vary. As the scale of the map is reduced, the geologist must generalize information. This factor, the audience to be reached, and the time and cost to process the report, the adequacy of a particular format to present the data properly, and the format or publication series used for reports in adjacent areas all influence the selection of the particular map series and scale.

Early Geologic Map Publications

The U.S. Congress, in June 1878, called on the National Academy of Sciences to take into consideration the methods and expenses of conducting all surveys of a scientific nature, to report to Congress a plan for surveying and mapping the territories of the United States, and to recommend a suitable plan for publication and distribution of reports, maps, documents, and other results of surveys. In response to their recommendations, Congress created the U.S. Geological Survey in the Act of March 3, 1879, and included the following authorization: "The publications of the Geological Survey shall consist of the annual report of operations, geological and economic maps illustrating the resources and classification of the lands, and reports upon general and economic geology and paleontology."

In 1882 the first monographs were published as part of the *Second Annual Report* of the U.S. Geological Survey. These monographs, or geological studies, were also issued in separate form and became part of the geologic folios in 1894. The initial folio of Livingston, Montana, consisted of four maps and, like each of the folios to follow, was named for the quadrangle it covered. The folios varied from four to as many as nine maps and portrayed topography, geology, underground structure, and mineral deposits of the area. If conditions in an area warranted, maps showing the economic geology, oil and gas, and artesian water were also included. The folios were replaced by the Geologic Quadrangle Map Series in 1949.

Map Scales

The recommended publications scales for geologic maps correspond with the topographic maps published by the U.S. Geological Survey. Large-scale maps (1:24,000 and 1:20,000) provide detailed map information for densely settled areas. Medium-scale maps (1:62,500 and 1:63,360) are adequate for general use. Small-scale maps (1:125,000, 1:250,000, and 1:500,000) cover large areas on a single sheet for most of the United States and are used for regional and statewide maps.

Geological maps prepared outside the United States employ a wide variety of map scales. These appear to be determined by one or a combination of the following factors: the level of economic development of a nation, total land mass to be mapped, use of the metric versus the English system of measurement, and mapping programs previously instituted by members of colonial or commonwealth systems of government.

Mapping programs of highly developed nations reveal an emphasis on large-scale maps. For example, West Germany and Switzerland are extensively mapped at a 1:25,000 scale. Other European nations rely more on 1:50,000 and 1:100,000

scales. Geologic maps in South America, Africa, and Australia are predominantly at 1:100,000, 1:200,000, and 1:500,000 scales. These smaller-scale maps fulfill the needs for nations in an early stage of development; however, their needs will change in accordance with their pace of economic development.

The English unit of measurement, the inch, provides the oddly numbered representative fractions on maps of those nations once part of the English Commonwealth. Scales such as 1:63,360 and 1:253,440 are listed among those of some of the following: Canada, Ceylon, India, Ireland, Malaysia, New Zealand, and others. Present programs in these nations are concerned with the metric units, but past political associations are reflected in these map scales. Metric scales are utilized throughout most of the world and provide convenient units of measurement. Scales of 1:25,000, 1:50,000, 1:100,000 etc. are commonly used. Special-purpose geologic maps illustrating tectonic, bedrock, and like studies are often prepared at 1:1 million, 1:2.5 million, and 1:5 million scales.

The USSR and mainland China have extensively mapped their vast areas at smaller scales and concentrated areas at larger scales. While many of these nations possess topographic maps, which are the bases for geologic mapping at larger scales, information available on current geologic maps reflects the impressions outlined here.

Current Map Series

Geologic Quadrangle Map Series This series provided a continuation of the geologic quadrangle maps begun in the early folios. Early publications were unnumbered, but within 5 yr the series was identified by a distinguishing prefix, GQ, and given numbers commencing with the twenty-sixth map. The Geologic Quadrangle Maps provide the basis for a geologic atlas of the United States and are multicolor maps of high-quality accuracy and detail of geologic information. This series represents a principal publication commitment of the national geologic mapping program and meets the public's increasing needs for geologic maps.

The scale of the Geologic Quadrangle Map is governed by the standard topographic quadrangles issued by the U.S. Geological Survey. The 7 1/2 min quadrangle at 1:24,000 scale and the 15 min quadrangle at 1:62,500 scale are basic scales for the coterminous United States; Alaska's 15 min quadrangles are at 1:63,360 scale. The topographic quadrangles that serve as the base maps for the geologic mapping are printed in three or more colors, the basic data including contours in brown, drainage in blue, and features of human origin in black. Recently published GQ maps have combined some of these features to print in one color.

General geologic data covering the entire area of the quadrangle are presented in this series. The basic format provides uniformity among these maps and includes an explanation, a brief text, selected references, and supporting illustrations that may include columnar sections, structure sections, and other graphics. Each map is considered a single sheet and must fit a specific paper size of 34 in × 50 in; consequently, the number of supporting illustrations is determined by their importance and the availability of space. Bedrock and surficial geology of an area may be published separately.

Miscellaneous Investigations Series This map series, designated by the prefix I, was first published in 1955. The series was established to provide an outlet for geologic studies that do not fall logically into any other established series. Its wider range of subject matter includes engineering geology subjects, paleotectonic maps, moon and planetary maps, photogeologic maps, and geologic studies of foreign areas.

Maps in this series may range from simple black-and-white maps to full-color maps. The base maps may be in any of the scales used commonly in the Geological Survey and may include mosaicked portions of standard quadrangles. Unlike the Geologic Quadrangle Series, the contours, streams, and roads are combined and printed in a single screened color, usually black.

Mineral Investigations Series During World War II and the years immediately preceding it, information concerning the extent and mode of occurrences of domestic deposits of strategic minerals was needed quickly. To avoid delays attendant on formal publications, maps were issued in preliminary form, and accompanying texts were mimeographed. As of February 20, 1950, the series of preliminary maps was superseded by two series of numbered maps—Mineral Investigations Field Studies, carrying the prefix MF, and the Mineral Investigations Resources, prefixed by MR. Both these series originally concerned mineralized areas, where the MF maps showed mines, prospects, and mineralized areas and the MR maps were commodity oriented. Derivative maps from Geologic Quadrangle Maps most recently have been published in the MF series. The MF map is limited to a maximum of six to eight geologic units and is printed in small quantities as a black-and-white map.

To speed up publications, the geologist-drafted copy and a typewritten explanation are used as copy for the printed map. These maps are published on a complete and combined base unless a simplified base is necessary. The scales of these maps vary according to the detail required. Maps in the MR series are typically small scale, and U.S. or regional maps are at 1:500,000 to 1:5 million. The MF series maps vary from large to medium scales.

Oil and Gas Investigations Map Series This series began in 1943 and concerns areas with oil and gas potential. The first 109 maps of this series are preliminary maps, and in 1951 the series began to carry the prefix OM. The maps are prepared at various scales and portray data that may include oil and gas fields, test wells, salt domes, pipelines, stone deposits, stratigraphy, photogeology, and tectonic structures.

Oil and Gas Investigations Charts, the OC series, record stratigraphic sections of potential oil- and gas-bearing beds. These charts are normally at large map scales with vertical scales variable from 1 in to 20 ft to 1 in to 500 ft.

Coal Investigations Map Series The Coal Investigations Map Series, the C series, was initiated in 1950 and includes maps and sections of coal-bearing rocks. These coal resources maps and preliminary geologic studies are prepared mostly at the standard large and medium scales.

Geophysical Investigations Map Series Using the prefix GP since 1950, maps in this series show the results of field studies by geophysical methods. Gravity maps, aeromagnetic maps, and aero-radioactivity maps are published as black-and-white or multicolor maps. A number of the black-and-white maps have been printed on translucent paper to enable the user to overlay the geophysical data on the corresponding geologic map. The majority of published GPs are at standard quadrangle scales.

Conclusion

As research priorities change and emphasis is placed on the timeliness of reports, the current list of map series may undergo further revision to meet future needs. Of the geologic map series listed in this entry, four are infrequently used due to limited subject material and/or changes in the emphasis of current mapping programs.

It is important to note the existence of geologic maps that are not classified in any of the listed series—i.e., state geologic maps (Brewer and Gelpke, 1980), maps from the *National Atlas of the United States,* geologic maps included in the Hydrologic Investigations Atlases, and the small-scale Geologic, Oil and Gas, Tectonic, Basement Rock Structure, and Bouguer Anomaly maps of the United States at 1:2.5 million scale and of North America at 1:5 million scale.

The major needs of the geologic mapping program of the U.S. Department of the Interior's Geological Survey continue, however, to be served by the Geologic Quadrangle Series, Miscellaneous Investigations Series, and Geophysical Investigations Series.

ROBERT J. FERENS

References

Brewer, T., and R. Gelpke, 1980, State-wide maps listed, *Geotimes* **25**, 19–25.

Gerlach, A. C., ed., 1970, *The National Atlas of the United States of America.* Washington, D.C.: U.S. Government Printing Office, 417p.

U.S. Geological Survey, 1962, *Publications of the Geological Survey 1879–1961,* 1962–1970 ed. Washington, D.C.: U.S. Government Printing Office.

U.S. Geological Survey, 1958, *Suggestions to Authors of the Reports of the United States Geological Survey,* 5th ed. Washington, D.C.: U.S. Government Printing Office, 255p.

Cross-references: *Cartography, General; Geological Highway Maps; Map Abbreviations, Ciphers, and Mnemonicons; Map Symbols; Maps, Environmental Geology; Maps, Physical Properties.* Vol. XIII: *Geological Communication; Maps, Engineering Purposes.*

MAP SYMBOLS

A symbol is an abstraction or pictorial representation of something else. Symbols on a map consist of discrete points, lines, or shaded areas; they have size, form, and (usually) color. Map symbols present information collectively, leading to appreciation of form, relative position, distribution, and structure. Locations of symbols on a map are controlled by positions on the ground, an element of cartography that cannot be changed (Keates, 1980). By digitization, symbols are conveniently stored and reproduced by computer.

Although the origin of symbols used in communication is lost in antiquity, they receive ever-increasing use in technical applications. Letters of the alphabet are essentially symbols of voice sounds, and numerals convey the concept of precise quantities. Because words alone cannot identify or describe areal details, abstract and pictorial symbols have become essential ingredients of maps. The symbols used by ancient cartographers were pictorial and highly fanciful; dangerous reefs, e.g., were often symbolized by pictures of broken and sinking boats. Pictures of sea monsters warned mariners of regions that were believed to be dangerous for one reason or another. Picture maps were a first step in the development of the modern symbol maps seen today.

The oldest known topographical map is a clay tablet from Nuzi in northeast Iraq that dates back to the third millennium B.C. (Harvey, 1980). Similar kinds of maps, and associated developmental stages, can be traced throughout history, viz. from Egypt in the second millennium, from the heartlands of the Assyrian Empire in the seventh and sixth centuries, from outposts of Greek civilization in the fifth and fourth centuries; from the Roman Empire from the second to sixth century A.D., and from Italy from the thirteenth to fifteenth century. These early picture maps continued to evolve through local field plans, town plans, and bird's-eye views. By the late third millennium, the Babylonians had mastered the idea of the scale map. These plans were used for drawing building sites on small clay tablets that included bar scales. By A.D. 267 the Chin emperors in China had constructed true-scale maps of the whole empire based, in part, on the so-called *tao li* principle. Using tao li, the lengths of derived distances are fixed by pacing out the side of right-angled triangles (the length of the third side of the triangle, which cannot be walked over, is thus determined).

Real advances in the transition from picture maps to symbol maps did not occur until the 1700s when it became a cartographic trend to use first hachures and later contour lines to mark relief, abandoning altogether the pictorial representation of hills. Harvey (1980) reports that this benchmark effort involved the application to dry land of a system developed by marine cartographers from the 1730s to show depth of water. Hachuring was, however, already deployed before contour lines were used on land maps. The map of Breisgau published by J. B. Homann of Nuremberg in 1718 was among the first to use systematic hachuring. Deeper shading, produced by drawing the lines broader and closer together, could be used to show either height or degree of slope. A turning point in cartographic expression came in 1799 in a book by J. G. Lehmann, *Darstellung einer neuer Theorie de Bezeichnung der schiefen Flachen (Exposition of a New Theory of Making Relief),* published in Leipzig. Lehmann elevated hachuring to the status of a precise system. Assuming vertical illumination, the deepest shading was used to mark the steepest slopes. Since Lehmann's time, horizontal instead of vertical hachure lines have appeared in combination with contour lines—e.g., in the official 1:50,000 map of Switzerland (1944).

Symbol maps, although having a basis in picture maps, became conventionalized in the topographic maps of medieval Europe. Early symbols included dots or tiny circles to mark towns, crossed hammers for mines or quarries, a crozier for abbeys, a chalice for Hussite towns, crossed keys for Catholic towns, crowns for royal possessions, and horse's heads for feudal lands. Once symbols stopped being pictorial, however, some explanation of their meaning was required. One of the earliest maps with a key or legend was one of Franconia (1533) by Peter Apianus (Harvey, 1980).

As the techniques and instruments of map making evolved (see *Cartography, General*), the symbolization of Earth features became more sophisticated. A simple geometric figure might convey a concept that would require many words to elucidate. Although a majority of symbols used on contemporary maps are abstract, many are pictorial. The symbols shown on U.S. Geological Survey topographic maps for natural features such as marshes, woods or brushwood, rivers, and glaciers suggest a plan view of these features. Cultural features such as schools, cemeteries, churches, and the like are represented by stylized pictures.

Of all map symbols, perhaps the closest link between modern topographic maps and their ancestors is the use of color. The British Ordnance Survey maps, e.g., still follow the almost universal color traditions in topographic maps for the past 500 yr: arable land is distinguished from pasture by brown and green; water is blue; and buildings and towns are gray, pink, or red (see *Map Colors, Coloring*).

Map Scale and Symbolization

The scale of a map determines the size and shape of the symbols that can be used (see *Map Series and Scales*). On a large-scale map, e.g., the actual ground plan and scale dimensions of buildings may be shown. For a small-scale map, conversely, all buildings are shown by the same symbol regardless of their shape and size. Because all maps are constructed to a scale, it is essential that the format be decided in advance of mapping. The format may be a whole page in a book, part of a page, a separate map requiring folding, a wall map, or a sheet of almost any conceivable size. Shape of the area to be mapped may be an important consideration. Robinson et al. (1978) also advise that an important concern of map scale may be the projection on which the map will be made. When a projection choice has been narrowed to suitable possibilities, the variations in the shapes of the mapped area on different projections can be matched against the format to see which will provide the best fit and maximum scale.

Field geologists must bear in mind relations between compilation scale and finished map scale. If the map is to be reproduced from an original drawing that does not allow changing scale, compilation will have to be done at the finished scale. As a general rule, the complexity of the map makes compilation at a larger scale desirable or even imperative. Lettering and inked drawings can always be reduced to advantage. At the larger scale, drafting is easier and reduction sharpens the lines. The mapmaker must also decide whether to compile on an already drawn projection or base map or to construct a projection and compile the entire map. For a simple field map, the worksheet constitutes a rough draft manuscript. The worksheets should provide all information that is necessary for the final drafting by scribing or tracing (see *Field Notes, Notebooks*).

Symbol Types *Symbolization* is the graphic coding of essential map characteristics. The coding process is critical to the success of cartographic communication. On the one hand, judicious simplification and classification can be negated by poor symbolization. On the other hand, good symbolization can impart an incorrect visual impression of precision and accuracy to poorly simplified or classified data. Thus, the mapmaker must attempt to apply map symbols that are recognized by convention for the preparation of geological maps (see examples given in Tables 1–38).

Symbols have two basic limitations: physical and psychological. The limitations are imposed on the visual variables by the equipment, materials, and skill of the mapmaker. The physical limitations include factors such as the format size, line widths, and availability of lettering styles and sizes, color screens, preprinted symbols, dimensionally stable film or plastics, specialized symbol templates, and the draftsperson's and machines' ability to use these cartographic tools. The psychological aspects of graphic limits are a function of the abilities of the map-reading audience. For maximum effectiveness in communicating the map message, pictorial symbols should tell a good deal without the necessity of a map legend, but a key is usually provided. Where the data are of equal rank or significance, the symbols should be roughly equivalent in size. Associative symbols, because they may be quite diagrammatic, almost always require a key (Keates, 1982). This category employs a combination of geometric and pictorial characteristics to produce easily identifiable symbols, as seen on many geological maps. Geologists also make use of geometric symbols (circles, triangles, squares, diamonds, stars, etc) where the dimension of the shape is the most important factor. Color can be used as a secondary dimension to differentiate the data, as occurs when some of the data classes are related (Keates, 1982).

Mapping qualitative areal data is relatively straightforward. When employing patterns, the cartographer may differentiate by varying both orientation and arrangement. Care must be exercised not to create a visually complex display that causes excessive eye movement or confusion. In geology, it is sometimes necessary to map nominally a distribution through two levels of its classification. Referred to as subdivisional organization, hues can be used to distinguish the major categories and to differentiate subclasses within categories by the use of patterns. Hue often shows the major breakdown of the distribution into its first-order classes (see *Map Colors, Coloring*). Pattern is subtler and lends itself more readily to differentiation at the second order of classification.

Lettering *Lettering* a map means the preparation of all names, numbers, and other topographical material. It requires careful consideration of many factors. Robinson et al., (1978) identify seven major headings in their planning checklist:

1. Style of type
2. Form of type
3. Size of type
4. Contrast between the lettering and its ground— i.e., the color of type
5. Method of lettering
6. Positioning of the lettering
7. Relation of the lettering to reproduction

Style refers to the design character of the type, including line thickness and serifs. The bewildering range of type styles available to the cartographer may be summarized in terms of three main groups of lettering characters: (1) classic or oldstyle typefaces, (2) modern styles, and (3) sans serif, the latter sometimes called Gothic as opposed to Roman (Nesbitt, 1957). Fancy or ornate letter forms are difficult to read and should be avoided. Although bold lettering may be more visible, it might overshadow other equally important data. Because type may not always be the most important element on a geologic map, lightline type may be an effective choice in some circumstances. As a rule, the fewer the styles, the better the harmony and the more handsome the map appearance.

Form refers to whether letters are composed of capitals, lowercase, slant, upright, or combinations. Although there is no hard-and-fast rule, past practice has established that more important names and titles should appear in capitals and less important names and places in lowercase letters. Because capitals are more difficult to read than lowercase letters, greater use of well-formed lowercase letters improves the legibility of a map. On topographic maps, e.g., hydrographic and topographic features are normally labeled in slant or italic, whereas cultural features are identified in upright forms.

Type size refers to the letter height, designated in points, with one point being nearly equal to 0.35 mm. It is important to remember that 4- or 5-point type probably comes close to the lower limit of visibility for the average viewer. Such small type sizes should be avoided in favor of larger sizes that enhance the findability of names on a map. The mapmaker's choice of type size is, however, limited by use of the largest sizes possible that are still consistent with good design (Brockemuel and Wilson, 1976).

The methods of lettering include hand lettering and mechanical means whereby the type of lettering is affixed to the map. Geologists have at their disposal a wide range of methods for lettering maps. Stick-up lettering is ordinarily printed on a thin plastic sheet, the back of which is coated by a colorless wax. Words, numbers, or letters are burnished onto the artwork by rubbing with a smooth object. Mechanical lettering devices enable those unskilled in freehand lettering to produce acceptable neat lettering. The best devices require a special pen, with ink feeding through a small tube while the pen is guided with the aid of a template.

The positioning of the lettering involves the placement of the type. Map reading is greatly affected by the placement of type. When properly placed, the lettering clearly identifies the phenomenon to which it refers, without ambiguity. Positioning of the type also affects the graphic quality of the map as to selections of type styles and forms. The object to which a name applies should be easily recognizable; the type should not conflict with other map material; and the overall appearance should not be stiff and mechanical.

The elements of visibility and recognition are among the major guides against which the choices and possibilities are to be measured. They are the final determination of what constitutes a handsome map.

Standardization of Cartographic Representations

Although map symbols have been standardized voluntarily and through application of the cartographic presentations of national geological surveys, the symbols are continually being revised. In the United States, e.g., many federal agencies have established stringent rules within their organizations, while others have adopted conventional usages. Even though there are fairly uniform practices among state and federal agencies, colleges and universities, and private industries, no agency has standardization as its primary function. Thus, one finds frequent departures from the general practice, especially in the case of special-purpose maps (see, e.g., *Maps, Environmental Geology;* Vol. XIII: *Maps, Engineering Purposes*). Many countries have adopted the map symbolization employed by the U.S. Geological Survey, while others have devised systems that better suit their particular needs. Conversely, some of the symbols deployed in the United States were used in Europe long before the establishment of U.S. federal agencies. Through the continuing efforts of international scientific societies, the standardization of major categories of geologic and lithologic symbolization is gradually being affected. There are, however, still numerous conflicts of usage, especially among the hundreds of symbols used to represent lithologies.

Topographic Map Symbols

Symbols used on topographic maps span a wide range of abstraction and pictorial representation. Those shown in Table 1 are a summary of basic symbols used by the U.S. Geological Survey for topographic maps at the following scales: 1:24,000, 1:25,000, 1:20,000, 1:62,500, and 1:63,360. Topographic symbols used for map scales in the ranges of 1:100,000, 1:250,000, 1:500,000 and 1:1,000,000, are modified for small-scale areal representation. Although the range of topographic map symbolization given in Table 1 is complete, the stylized pictorial representation of natural biophysical and cultural features is representative of most modern topographic maps. Notations on this type sheet do warn, however, that variations will be found on older maps.

Surficial Materials

In most field surveys, subdivision or classification of surficial materials is ordinarily subordinated to simple designations such as soil, drift, sand, caprock, etc. In areas of significant surficial cover that obscures the underlying bedrock (see *Saprolite, Regolith, and Soil*), generalizations of subsurface geology are often obtained from costeans, trenches, and excavations. For the purposes of geophysical or geochemical surveys, it is often advantageous to map different kinds of surficial materials. In cases where geologists wish to use surface geochemistry as a critical exploration tool—e.g., soil surveys (see *Pedogeochemical Prospecting*)—it is helpful to differentiate different kinds of surficial materials. One modest example, summarized in Table 2, lists the main kinds of surface mapping units once used by International Nickel Australia Limited (Perth, Western Australia) for parts of the West Australian craton. Here, letter symbols were used to designate Quaternary soil materials and crusts and remains of Tertiary deep weathering profiles.

A more complete symbolization of surficial materials is provided by the Bureau de Récherches Géologiques et Minières (BRGM), Service Géologique National, in their guide for preparation of geologic maps at a scale of 1:50,000 (BRGM, 1975). The symbols shown in Table 3 provide mapping units for detrital materials, silicious detritus, carbonates, hydrocarbons, evaporites, residuum, and fossils, among others.

The Geological Survey of Canada, in another example, maintains a separate legend for surficial geology maps (Blackadar et al., 1980). This leg-

end (Table 4) separates surficial materials on the basis of environments: nonglacial, proglacial, and periglacial. The main symbols for surficial geological maps and diagrams are summarized in Table 5 with brief descriptions.

Still a different treatment of surficial materials is offered by the German Bundesanstalt für Geowissenschaften und Rohstoffe, und Niedersächsisches Landesamt für Bodenforschung. This part of their cartographic legend provides symbolization for different grain sizes of unconsolidated materials as well as symbols for organic accumulations and lake deposits (Table 6). These kinds of signs, symbols, and patterns are used by state geological surveys in the Federal Republic of Germany, but all map elements are not finalized.

Geological Map Symbols

The following extended tables present the standard basic geological symbols and the accessory symbols that make up a geological map. Those given in the following tables, along with more specialized symbols, provide a selected summary of the standard geological symbols used by the Australian Bureau of Mineral Resources (BMR, 1978). These symbols are intended for use by geologists and draftspersons at all stages of map preparation and to appear on published maps. The following guidelines are followed in the selection of these symbols:

1. Symbols should be simple, easy to draw, clear, and reproducible on printed maps.
2. The symbol that creates a mental image of the object or concept represented is preferred to alternate symbols.
3. As far as possible, symbols portraying related objects or concepts should have common characteristics—e.g., the group of symbols for oil and gas wells.
4. Where it is compatible with the points outlined already, an established symbol is retained.
5. There must be a general symbol that can be used where knowledge is incomplete—e.g., a fault symbol that can be used where only the strike of the fault is known.

Note that the BMR warns that symbols designed for large-scale or special-purpose maps are, in general, unsuitable for use on general-purpose maps. Their indiscriminate use on maps such as the 1:250,000 series should be avoided. Geologists compiling special-purpose maps should try to conform to the five points set out previously or relate symbols to those adopted for general-purpose maps.

Symbols used on geological maps prepared by the Australian Bureau of Mineral Resources are provided in the summary tables showing geological boundaries (Table 7), bedding (Table 8), folds

461

(Table 9), faults (Table 10), joints (Table 11), metamorphic foliation (Table 12), cleavage (Table 13), lineation (Table 14), banding in igneous rocks (Table 15), photogeology (Table 16), geophysics (Table 17), fossil and other sites (Table 18), mineral deposits and workings (Table 19), petroleum occurrences and development (Table 20), hydrology (Table 21), volcanoes (Table 22), and topography and landforms (Table 23).

Details of a European approach are summarized in a selection of geologic map symbols used by the Geological Survey of the Federal Republic of Germany. Symbolization is normally divided into two basic groups, depending on use and final presentation of the geological maps—viz., symbols used on colored maps and lithologic symbols used for black-and-white geological maps and columnar sections. Some symbols are associated with alphanumeric codes while others are free of specific designation. These tables have been abstracted from Bender's (1981) handbook on applied geology. The symbols are grouped in the following categories: boundaries and borders, including symbols, signs, and patterns free of alphanumeric identification (Table 24); main groups of hard rocks (Table 25); and igneous and metamorphosed rocks (Table 26).

Symbols recommended for publication by the U.S. Geological Survey are summarized in a series of selected tables. The symbols represented include contacts (Table 27); contours and isopleths (Table 28); planar features (Table 29); linear features and joints (Table 30); faults (Table 31); folds (Table 32); fossils (Table 33); monoclines and minor fold axes (Table 34); sections (sedimentary lithologic patterns for columnar sections) (Table 35); synclines (Table 36); underground workings and exploration (including oil and gas wells) (Table 37); veins, ore, wall-rock alteration, dikes, and surface opening and exploration for large- and small-scale maps (Table 38).

Conclusion

Maps are pictorial representations that enable users to contemplate and analyze areas of the Earth's surface in the convenient form of a two-dimensional sheet. Maps permit study of the distribution of features and materials compiled from extended and specialized observations and precise measurements that are set down in cartographic form to provide a standard reference for many decades. These miniaturized versions in two dimensions of the real three-dimensional world are highly artificial representations that are subject to various qualifications and limitations. Because a map makes locational statements about phenomena, features, and materials, that are designated by the symbols, its usefulness—i.e., whether a map fulfills its stated purpose—can be judged only by the map user. The adequacy of geological maps largely depends on the skilled use of signals and signs. A signal, requiring a single predetermined response, is not open to various interpretations. Symbols, regarded as characterizing signs, are subdivided according to some aspect of the designatum they represent. The characterizing signs or symbols used on a map to designate specific features or phenomena through the use of concepts also provide a basis for distinguishing different kinds of maps. Geological maps are special-purpose maps that employ special symbols that are scarcely comprehensible to the map user who has little knowledge of geological concepts. This specialization of subject matter must be learned before the map can be fully understood and has nothing to do with the cartographic theory. The specialized structure of geological map symbols makes these thematic maps highly interesting and useful to the professional geoscientist. In the final analysis, geological maps as a means of communication depend on the degree of expertise used in the deployment of geological map symbols.

CHARLES W. FINKL, JNR.

References

Bender, F., 1981, *Geologische Geländeaufnahme, Strukturgeologie, Gefügekunde, Bodenkunde, Mineralogie, Petrographie, Geo-Chemie, Paläontologie, Meeresgeologie, Fernerkundung, Wirtschaftsgeologie.* Stuttgart: Enke, 147p.

Blackadar, R. G., H. Dumych, and P. J. Griffin, 1980, *Guide to Authors—A Guide for the Preparation of Geological Maps and Reports.* Ottawa: Geological Survey of Canada, 66p.

BMR (Bureau of Mineral Resources, Geology and Geophysics), 1978, *Symbols Used on Geological Maps.* Canberra, unpaginated.

BRGM (Bureau de Récherches Géologiques et Minières), 1975, *Botes d'Orientation pour l'Establissement de la Carte Géologique a 1:50,000.* Orleans, France: Service Géologique National, 240p.

Brockemuel, H. W., and B. Wilson, 1976, Minimum letter size for visual aids, *Prof. Geographer* **28**(2), 185–189.

Harvey, P. D. A., 1980, *The History of Topographical Maps, Symbols, Pictures and Surveys.* New York: Thames and Hudson, 199p.

Keates, J. S., 1980, *Cartographic Design and Production.* London: Longman, 240p.

Keates, J. S., 1982, *Understanding Maps.* London: Longman, 139p.

Nesbitt, A., 1957, *The History and Technique of Lettering.* New York: Dover, 173p.

Robinson, A., R. Sale, and J. Morrison, 1978, *Elements of Cartography.* New York: Wiley, 448p.

Cross-references: *Cartography, General; Field Notes, Notebooks; Geological Highway Maps; Geological Survey and Mapping; Geological Reports; Map Abbreviations, Ciphers, and Mnemonicons; Map Colors, Coloring; Maps, Physical Properties; Vegetation Mapping; Well Logging.*

TABLE 1. Summary of Topographic Map Symbols Used by the U.S. Geological Survey, as Revised March 1978

Primary highway, hard surface	
Secondary highway, hard surface	
Light-duty road, hard or improved surface	
Unimproved road	
Road under construction, alinement known	
Proposed road	
Dual highway, dividing strip 25 feet or less	
Dual highway, dividing strip exceeding 25 feet	
Trail	

Railroad: single track and multiple track	
Railroads in juxtaposition	
Narrow gage: single track and multiple track	
Railroad in street and carline	
Bridge: road and railroad	
Drawbridge: road and railroad	
Footbridge	
Tunnel: road and railroad	
Overpass and underpass	
Small masonry or concrete dam	
Dam with lock	
Dam with road	
Canal with lock	

Buildings (dwelling, place of employment, etc.)	
School, church, and cemetery	Cem
Buildings (barn, warehouse, etc.)	
Power transmission line with located metal tower	
Telephone line, pipeline, etc. (labeled as to type)	
Wells other than water (labeled as to type)	Oil Gas
Tanks: oil, water, etc. (labeled only if water)	Water
Located or landmark object; windmill	
Open pit, mine, or quarry; prospect	x
Shaft and tunnel entrance	

Horizontal and vertical control station:

Tablet, spirit level elevation	BM △5653
Other recoverable mark, spirit level elevation	△5455
Horizontal control station: tablet, vertical angle elevation	VABM △95/9
Any recoverable mark, vertical angle or checked elevation	△3775
Vertical control station: tablet, spirit level elevation	BM ×957
Other recoverable mark, spirit level elevation	×954
Spot elevation	×7369 ×7369
Water elevation	670 670

Boundaries:

Boundaries: National	
State	
County, parish, municipio	
Civil township, precinct, town, barrio	
Incorporated city, village, town, hamlet	
Reservation, National or State	
Small park, cemetery, airport, etc.	
Land grant	
Township or range line, United States land survey	
Township or range line, approximate location	
Section line, United States land survey	
Section line, approximate location	
Township line, not United States land survey	
Section line, not United States land survey	
Found corner: section and closing	
Boundary monument: land grant and other	
Fence or field line	

Index contour		Intermediate contour
Supplementary contour		Depression contours
Fill		Cut
Levee		Levee with road
Mine dump		Wash
Tailings		Tailings pond
Shifting sand or dunes		Intricate surface
Sand area		Gravel beach

Perennial streams		Intermittent streams
Elevated aqueduct		Aqueduct tunnel
Water well and spring		Glacier
Small rapids		Small falls
Large rapids		Large falls
Intermittent lake		Dry lake bed
Foreshore flat		Rock or coral reef
Sounding, depth curve		Piling or dolphin
Exposed wreck		Sunken wreck
Rock, bare or awash; dangerous to navigation		

Marsh (swamp)		Submerged marsh
Wooded marsh		Mangrove
Woods or brushwood		Orchard
Vineyard		Scrub
Land subject to controlled inundation		Urban area

Available from: U.S. Geological Survey, 1200 South Eads Street, Arlington, Virginia 22202.

Notes: Various colors, not reproduceable here, are used to distinguish similar symbols. Variations will be found on older maps.

TABLE 2. Surface Mapping Units for the West Australian Craton

Symbol	Reference	Description
Crusts, Caps, and Pans		
QDf	Ferricrete	⎫
QDs	Silcrete	⎬ Duricrusts
QDc	Calcrete	⎭
QJc	Jasper, chrysoprase, chert	Caps
QPp	Siliceous red-brown hardpan	Soil pans
Transported (Soil) Materials		
QAs	Alluvial spread (sheet or blanket deposit with/-out surface gravel veneer)	⎫ Alluvial deposits
QAc	Alluvial channel	⎭
QGi	Ironstone gravels (pisolites, concretions)	⎫
QEs	Earthy soil materials (> 30% gravels)	⎬ Colluvial deposits
QTs	Coarse scree or talus with/-out manganese-limonite stains	⎭
QOd	Dunes (lunettes)	⎫
QOdg	Kopi mounds, platforms	⎪
QOds	Sand mounds, platforms	⎬ Eolian deposits
QOs	Wind-blown sheets	⎪
QOsg	Kopi sheets	⎪
QOss	Sand sheets	⎭
In Situ Weathered (Soil) Materials		
TLc	Intact laterite profile	⎫
TLm	Mottled zone	⎬ Laterite deep weathering profile
TLp	Pallid zone	⎭
TDw	Deep weathering profile or portion thereof	⎫ Non-indurated weathered materials ⎭

Note: For maps showing surface geology.
Data courtesy of International Nickel Australia Limited, Perth, Western Australia.

TABLE 3. Symbolization for Maps Showing Surficial Geology

1 - Roches détritiques

1.1   argile

1.2   argile[1]

1.3   argile à anhydrite

1.4   argile gypsifère

1.5   argile sableuse

1.6   arkose

1.7   brèche monogénique

1.8   brèche polygénique

1.9   conglomérat monogénique

1.10   conglomérat polygénique

1.11   sable (points non alignés)[1]

1.12   sable fin (id.)

1.13   sable moyen (points non alignés)

(1) Sur la carte il est préférable de choisir les figurés les moins denses pour éviter d'altérer la teinte du fond.

TABLE 3. (*continued*)

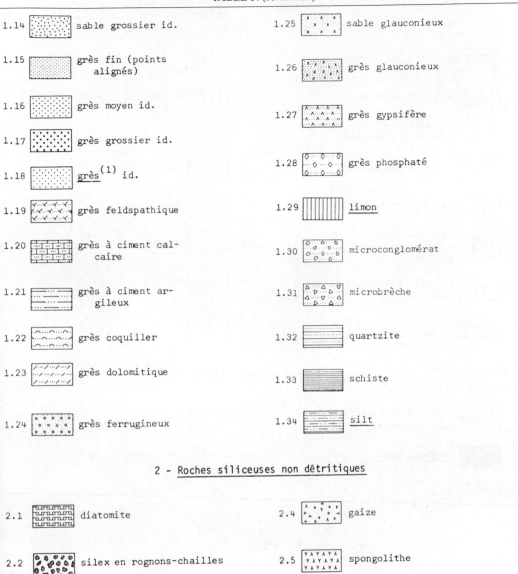

1.14   sable grossier id.

1.15   grès fin (points alignés)

1.16   grès moyen id.

1.17   grès grossier id.

1.18   grès[1] id.

1.19   grès feldspathique

1.20   grès à ciment calcaire

1.21   grès à ciment argileux

1.22   grès coquiller

1.23   grès dolomitique

1.24   grès ferrugineux

1.25   sable glauconieux

1.26   grès glauconieux

1.27   grès gypsifère

1.28   grès phosphaté

1.29   limon

1.30   microconglomérat

1.31   microbrèche

1.32   quartzite

1.33   schiste

1.34   silt

2 - Roches siliceuses non détritiques

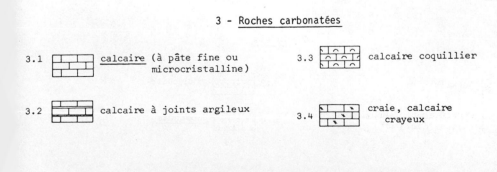

2.1   diatomite

2.2   silex en rognons-chailles

2.3   silex en lits

2.4   gaize

2.5   spongolithe

2.6   meulière

3 - Roches carbonatées

3.1   calcaire (à pâte fine ou microcristalline)

3.2   calcaire à joints argileux

3.3   calcaire coquillier

3.4   craie, calcaire crayeux

(*continued*)

TABLE 3. (*continued*)

3.5   calcaire argileux

3.6   calcaire bioclastique

3.7   calcaire construit

3.8   calcaire à pellets (calcaire grumeleux)

3.9   calcaire noduleux

3.10 calcaire oolithique

3.11 calcaire pisolithique, à oncholites ou à pelotes algaires

3.12 calcaire sableux (calcaire "gréseux")

3.13 calcaire à silex, à chailles calcaire siliceux

3.14 calcaire dolomitique (ou dolomie calcaire)

3.15 calcaire à entroques

3.16 calcaire glauconieux

3.17 dolomie

3.18 falun

3.19 lumachelle

3.20 marne

3.21 marne

3.22 marnes dolomitiques

4 - Hydrocarbures solides et charbons

4.1   calcaire asphaltique ou bitumineux

4.2   charbon, lignite

4.3   tourbe

4.4   grès asphaltique ou bitumineux

4.5   schistes bitumineux

5 - Evaporites

5.1   anhydrite

5.2   gypse

5.3   sels de potasse (sylvinite, carnallite)

5.4   sel gemme (halite)

6 - Divers

6.1   bauxite

6.2   encroûtement ferrugineux (Alios, etc.)

6.3   glauconite

6.4   phosphate

TABLE 3. (*continued*)

7 - <u>Matériaux épars</u>

(Voir également Note d'orientation n° 7 : "Notation et représentation des formations superficielles")

7.1 <u>blocs</u>

7.5 <u>concrétions et débris ferrugineux</u>

7.2 <u>silex intacts</u>

7.6 <u>silex brisés</u>

7.3 <u>débris de roches carbonatées</u>

7.7 <u>chailles</u>

7.4 <u>débris de roches siliceuses</u>

7.8 <u>sables</u>

8 - <u>Fossiles</u>

8.1 algues

8.2 ammonites, goniatites, nautiles

8.3 archaeocyathus

8.4 bélemnites

8.5 brachiopodes

8.6 bryozoaires

8.7 characées

8.8 lamellibranches

8.9 orthocères

8.10 poissons

8.11 conodontes

8.12 crinoïdes

8.13 échinides

8.14 foraminifères pélagiques

8.15 foraminifères benthiques

8.16 gastéropodes

8.17 graptolites

8.18 stromatolites

8.19 ostracodes

8.20 polypiers (formes globuleuses)

8.21 polypiers (formes rameuses)

8.22 polypiers (formes aplaties)

8.23 rudistes

8.24 serpules

8.25 spongiaires

(*continued*)

TABLE 3. (continued)

8.26	Y	spícules de spongiaires		8.28	⌀	végétaux
8.27	◠	trilobites		8.29	⬭	vertébrés

9 - Stratifications, figures sédimentaires et structures diverses

9.1	≡	stratification régulière
9.2	〰	stratification irrégulière
9.3	⫽⫽⫽⫽	stratification oblique
9.4	∿	rides (ripple-marks) symétriques
9.5	∿	rides asymétriques
9.6	⫻	striation
9.7	⬭	septaria
9.8	⋀⋀	stylolites
9.9	⁒	figures de dessication (mud-cracks)
9.10	⬳	stratifications entrecroisées
9.11	▬	hard-ground, paléosol

9.12	⬯	lentilles
9.13	⬰	flute-cast, load-cast, groove-cast et autres figures directionnelles de courant
9.14	🝔	glissements (slumpings)
9.15	○	nodules
9.16	⋀⋀⋀	"cone in cone"
9.17	⟳⟲	karst
9.18	🝔	remplissage de poche ou de fissure
9.19	⊢∿	niveau de transgression, lacune de sédimentation, sans discordance appréciable
9.20	⊢∿⊣	discordance de ravinement
9.21	⊢∿⊣	discordance angulaire
9.22	⊢─⊣	contact tectonique

Source: Bureau de Récherches Géologiques et Minières (BRGM), Service Géologique National, Orleans, France (2nd ed. rev. July 1975).

TABLE 4. Legend for Canadian Surficial Geology Maps Showing the Main Geological Environments, Materials, and Their Alphanumeric Codes

SURFICIAL DEPOSITS

Nonglacial Environment

Colluvial sediments: Nonsorted debris, 1–10 m thick, mantling lower slopes and valley floors, soliflucted and washed from upslope weathered rock areas; deposition active since deglaciation but basal sediments probably date from earlier nonglacial intervals

7b   Calcareous colluvium: Slightly stony sandy muds derived from weathered carbonates (unit 1b)

7a   Granitic colluvium: Muddy sands derived from weathered gneiss (unit 1a)

Fluvial sediments: Gravel and sand, 2–20 m thick, deposited on floodplains and fans

6b   Seasonally flooded sediments

6a   Terraced sediments above present flood zone

Marine sediments: Gravel, sand, silt, and clay, 1–100 m thick, deposited in deltaic, beach, and nearshore environments during regression of postglacial sea

5c   Beach sediments: Gravel and sand, 1–5 m thick, forming ridges and swales. (5c1: Site of beach sediment containing shells older than last glaciation)

5b   Nearshore sediments: Silt and find sand, 1–5 m thick, forming plains

5a   Deltaic sediments: Coarsening upward sequences of clay, silt, sand, and gravel, 10–100 m thick, with flat, terraced, dissected, or gullied surfaces. Some deltas at marine limit are glaciomarine features (5a1: Delta thought to predate last glaciation on basis of elevation and radiocarbon-dated driftwood)

Proglacial and Glacial Environment

4   *Glaciolacustrine sediments:* Clay, silt, and sand, less than 1 m thick, deposited in glacier-dammed lakes; surface mimics form of underlying weathered rock or colluvium

Glaciofluvial sediments: Gravel and sand, 1–100 m thick, deposited beneath and in front of the marginal zone of a glacier

3b   Proglacial outwash: Gravel and sand, 1–10 m thick, terraced, deposited on floodplains and fans

3a   Kames: Gravel and sand, 5–100 m thick, forming conical hills and ridges

Glacial Environment

Till: Nonsorted debris, 0.5–20 m thick, with contrasting vegetation covers reflecting undetermined compositional differences; dominantly lodgment till

2b   Till veneer: 0.5–2 m thick, surface mimics form of underlying rock surface (2b1: 30–60% vegetation cover; 2b2: 1–5% vegetation cover)

2a   Till blanket: 2–20 m thick, gently rolling surface, fluted in places, 30–60% vegetation cover

Nonglacial (Periglacial) Environment

Residuum: Residual soils of various textures, about 1 m thick, overlying metamorphic and sedimentary rocks on smooth, gentle slopes interrupted by tors, cryoplanation terraces, and nivation hollows

1b   Platy felsenmeer and silty, sandy rubble mantling limestone, dolostone, and sandstone; minor gravel mantling conglomerate

1a   Felsenmeer, composed of blocks 1–2 m across, with interstitial grus, mantling gneiss

ROCK

Rock: Rock of various lithologies and ages; hilly and hummocky with basins, steep slopes, and cliffs produced by glacial scouring

Rc   Unconsolidated sedimentary rock: Quartz sandstone of late Cretaceous to early Tertiary age, unweathered

Rb   Consolidated sedimentary rocks: Limestone, dolostone, and sandstone of late Precambrian to early Devonian age with discontinuous veneer of rubble and till, about 0.5 m thick

Ra   Igneous and metamorphic rocks: Gneiss, granite, and minor quartzite of Precambrian age, relatively unweathered

Left margin labels: QUATERNARY — POST LAST GLACIATION, LAST GLACIATION, PRE LAST GLACIATION; PRE-QUATERNARY

Source: From Blackadar, Dumych, and Griffin (1980).

TABLE 5. Principal Symbols for Maps and Figures, Showing Surficial and Bedrock Geology,
Used by the Geological Survey of Canada

Description	Symbol	Description	Symbol
Depressional lineament following a structural feature		Intermediate shoreline features or trimlines; escarpment, cut bench etc.	
Small bedrock outcrop		Escarpment	
Castellated outcrop, tor		Buried valley	
Cirque		Delta	
Drumlins, drumlinoid ridges, crag and tail, flutings, etc. parallel to ice flow; undifferentiated		Ice-wedge polygons (use where visible on medium scale air photos)	
Drumlins (on large scale maps, on small scale maps)		Pingo	
Crag and tail		Active thermokarst backwasting slope	
Striae (ice flow direction known, unknown)		Active layer failures, retrogressive thaw flow slide, etc. (large feature, small)	
Roche moutonnée or rock drumlin		Palsen, palsa	
Lineation caused by floating ice		Cryoplanation terrace	
Moraine ridge (major, minor)		Thermokarst depression (large, small)	
Ice thrust bedrock ridge		Landslide scar (large, small)	
Ice contact face		Active coastal aggradation (off a poorly consolidated coast, off a consolidated coast)	
DeGeer moraines (on large scale map scallops follow ridge, on small scale they are schematic)		Active coastal escarpment (in poorly consolidated material, in consolidated material)	
Lateral moraines		Rock glacier	
Esker (direction of flow known or assumed, unknown)		Piping depression	
Crevasse filling		Failing slope	
Conical gravel hill		Avalanched slopes, avalanche track	
Ice-contact delta		Steep gully wall	
Kettle holes (small, large)		Solifluction lobes	
Abandoned channel (large, small, sidehill)		Dunes (inactive, active)	
Beach		Concentration of boulders	
Limit of submergence		Sink hole (small, large)	

TABLE 5. (*continued*)

Description	Symbol
Fossil (animal) locality (marine, terrestrial)	Ⓕ Ⓘ
Gravel occurrence	Ⓖ
Gravel pit (active, abandoned)	✕ ✕
Ground observation point	◉
Shallow drill site	•
Buried organic (plant) locality	•
Location of stratigraphic section of special interest	✦
Observation made from the air	o
Geological boundary, (defined, gradational, assumed)	
Composition and/or genesis of material largely uncertain	?
Radiocarbon date	Date Material Lab no Elevation

Description	Symbol
Drift-covered area	
Rock outcrop, area of outcrop, probable outcrop, float, frost heaved rock	
Geological boundary (defined, approximate, assumed)	
Geological boundary (gradational, inferred or metamorphic)	
Limit of geological mapping	
Limit of area surveyed with aircraft	
Flow contact	
Bedding, tops known (horizontal, inclined, vertical, overturned, dip unknown)	
Bedding, tops unknown (inclined, vertical, dip unknown)	
Bedding, general trend (dip unknown, top unknown; dip and top known; dip known, top unknown)	
Bedding, estimated dip (gentle, moderate, steep)	g, m, s
Primary flow structures in igneous rock (horizontal, inclined, vertical, dip unknown) If a supplementary symbol is needed use	
Schistosity, gneissosity, cleavage, foliation (horizontal, inclined, vertical, dip unknown) Second generation (horizontal, inclined, vert)	
Schistosity, gneissosity, cleavage, foliation, general trend	
Gneissosity, cleavage, foliation (horizontal, inclined, vertical, dip unknown)	
Foliation (horizontal, inclined, vertical, dip unknown)	
Banding (inclined, vertical, dip unknown)	
Axial plane of minor fold (horizontal, inclined, vertical, dip unknown)	
Lineation (horizontal, inclined, inclined but plunge unknown, vertical)	
Layering (in intrusive rocks)	
Lineation, axes of minor folds (horizontal, inclined, vertical)	

(*continued*)

TABLE 5. (continued)

Symbol description	Symbol	Symbol description	Symbol
Drag-fold (arrow indicates plunge) Drag-fold in gneissosity		Rock dump or tailings	
Minor fold (arrow indicates plunge)		Quarry or mine; rock trench and stripped area Quarry or mine (abandoned)	
Multiple fold (arrow indicates plunge, inclination of axial plane known, unknown) Multiple fold (plunge unkown)	25 15	Mine or mineral prospect (lead, zinc)	Pb Zn
Structural trend (from air photographs)		Mineral prospect, mineral occurrence (manganese)	✕ 3 ✕ Mn
Lineament (from air photographs)		Mineral isograd Other alternatives when more than one	
Fault (defined, approximate, assumed)		Trench Open cut, axial	
Fault (inclined, vertical)		Borehole	●BH ●BH2
Fault (solid circle indicates downthrow side, arrows indicate relative movement)		Diamond-drill hole Surface projection of geology inferred	●DDH
Thrust fault (teeth in direction of dip;defined) (teeth indicate upthrust side)		Gossan	G G G
Thrust fault (approximate, assumed)		Show of oil and gas (abandoned)	
Fault zone, shear zone; schist zone (width indicated)		Show of gas (abandoned)	
Shearing and dip		Show of oil (abandoned)	
Vein fault (defined, assumed)		Gas producer	
Mineralized bed or seam (hematite)	hem hem	Oil producer	●
Dyke, vein, or stockwork (defined, approximate, assumed)		Location of drilling	○
Joint (horizontal, inclined, vertical, dip unknown)		Dry (abandoned)	
Anticline (defined, approximate) Antiform		Shaft, raise, winze Shaft (abandoned)	
Syncline (defined, approximate) Synform			
Anticline and syncline (overturned)			
Anticline or syncline (arrow indicates plunge)			
Antiform or synform			
Locality where age has been determined, in millions of years	Ⓐ 1400		
Location of measured section			

Source: From Blackadar, Dumych, and Griffin (1980).

TABLE 6. Examples of Symbols and Signs for Unconsolidated (Loose, Friable) Rocks, Soil Materials, and Organics, Used in the Federal Republic of Germany

	/X	Steine		//L	Lehm im allgemeinen, Geschiebelehm
	/X/hg	Schutt, z.B. Hangschutt		//Lo	Löß
	/G	Kies im allgemeinen		//Lol	Lößlehm
	/gG	Grobkies		/T	Ton im allgemeinen bzw. tonig
	/mG	Mittelkies		/K	Kalk im allgemeinen bzw. kalkig
	/fG	Feinkies		/H	Torf im allgemeinen
	/S	Sand im allgemeinen		/Hn	Niedermoortorf
	/gS	Grobsand		/Hh	Hochmoortorf
	/mS	Mittelsand		/F	Mudden (Unterwasser-bildungen) im limnischen Bereich i.a.
	/fS	Feinsand		//y	künstliche Auffüllung, Aufschüttung
	/U	Schluff		A	Abraum

Source: From Bundesanstalt für Geowissenschaften und Rohstoffe, und Niederschsisches Landesamt für Bodenforschung, FRG.

TABLE 7. Geological Boundaries Used on Maps of the Australian Bureau of Mineral Resources, Geology and Geophysics (BMR), Showing Standard Basic Symbols, Special Symbols, and Symbols Used Mainly on Tectonic and Structural Maps

Symbol No.	Symbol	Description	Notes	Drafting Specifications (mm)
2.1 STANDARD BASIC SYMBOLS				
2.1.0 General symbols				
	Bibl. 12, 23, 38, 44, 82, 131, 139, 151, 152			
2.1.1	————	Geological boundary, position accurate	Add dip if known, e.g.	0.15
.2	– – – –	Geological boundary, position approximate	Symbols for approximate, inferred, and concealed boundaries may be omitted from the map reference and this note inserted: 'Where location of boundaries, folds, and faults is approximate, line is broken; where inferred, queried; where concealed boundaries and folds are dotted, faults are shown by short dashes'	
.3	?– – – ?	Geological boundary, inferred		
.4		Geological boundary, concealed		
.5	? · · · · ?	Geological boundary; inferred, concealed		
.6	Y / O	Boundary between two intrusive rocks; Y is younger, O is older	Ordinary geological boundary; Y and O placed at point where relationship observed. See also igneous intrusive boundary symbols used mainly on tectonic and structural maps. Bibl. 18	
.7	ᴡᴡᴡᴡᴡ	Sedimentary facies boundary or Sedimentary facies change	Generally schematic only See also symbol 2.2.3 Bibl. 10, 103, 139, 151, 166, 174	

.8	**Unconformities**		
	See also symbols 2.3.4 to 2.3.7		
.9	Unconformity		Observed. Top of U is towards younger rocks. Bibl. 12, 14, 23, 103, 145
.10	Unconformity		Use only in sections, mainly rock relationship diagrams, where the short length of unconformity often makes symbol 2.1.9 impracticable. Bibl. 12, 14, 39, 55, 56, 86, 139, 153
.11	**Miscellaneous (mainly small) rock units**		
	For metamorphic aureole, see Introductory Notes, Metamorphic Rocks		
.12	Dyke or vein		Alternative symbols. If vein widens to mappable width, at scale of mapping, may show as in third symbol. Colour may be used, and if so, dots may not be needed. Type of rock or filling by initial, e.g. q — quartz; d — dolerite. Where different veins or dyke bodies or systems need to be distinguished, use either a dual letter system, e.g. dl, do, dt, or if age relationship is known, d1, d2, d3. If appropriate the age and name of the body may be indicated as for other rock bodies, e.g. Pdw—Wurugoi Dolerite, of Proterozoic age. Other abbreviations as in BMR List of Standard Abbreviations. Bibl. 36, 37, 57, 88, 107
.13	Dyke or vein		
.14	Dyke or vein		
.15	Sill		Where sill is not of mappable width. Type of rock to be shown in lower case, e.g. —— d —— dolerite sill. Bibl. 17
.16	Outcrop		To be used where lithology is not to be indicated. May be used for special types of outcrop, e.g. gossan. Bibl. 18
.17	Outcrop	X Dlc	Outcrop too small to be shown at map scale. Give formal name of rock unit. If this cannot be done, show lithology by suitable letter symbol (explained in map reference). Use lower case to distinguish from the symbols for rocks and minerals of economic interest shown in Section 14

(continued)

475

TABLE 7. (continued)

Symbol No.	Symbol	Description	Notes	Drafting Specifications (mm)
.18	— · — · —	Miscellaneous boundary	For example, limit of dyke swarms. See also symbols 12.2.17 and 18.1.27 See also Bibl. 124 (oil field, gas field)	
.19	— · — · —	Geological boundary, transitional	Dividing, for example, granitic and metamorphic types. Bibl. 10	

2.2 SPECIAL SYMBOLS
2.2.0 General symbols

.1	············· a	Alluvial boundary	A miscellaneous boundary symbol for superficial deposits on large-scale map. Use a letter symbol appropriate for type of deposit	

.2 Metamorphic boundaries

.3	ʌʌʌʌʌ	Metamorphic facies boundary	Generally schematic only, on rock relationship diagram. Bibl. 10, 151 See also Bibl. 100	
.4	"biotite" (line)	Metamorphic isograd, with index mineral(s)	Red line .5mm / Single mineral denotes higher-grade side of isograd. Mineral pair e.g. biotite may be shown if desired; garnet	
.5	"biotite" ο ο ο ο ο ο ο	Metamorphic isograd, with index mineral	Black or coloured. Bibl. 119, 151 / in this case, isograd marks incoming of higher-grade mineral	

476

.6 Metamorphic aureole, contact aureole

See Introductory Notes, Metamorphic Rocks

.7 Marker bed, marker band

See Bibl. 60

2.3 SYMBOLS USED MAINLY ON TECTONIC AND STRUCTURAL MAPS

2.3.0 Igneous intrusive boundaries

See also symbol 2.1.6

.1	Intrusive boundary	Longer barbs indicate the older rock which was invaded. Bibl. 18
.2	Intrusive boundary with contact metamorphic zone indicated	Spacing of fine red lines indicates relative intensity of contact metamorphism. See also Bibl. 36, 104 See also Introductory Notes, Metamorphic Rocks.

2.3.3 Unconformities, disconformities

See also Bibl. 34, 151

.4	Angular unconformity	Top of U or V is towards younger rock. See also Bibl. 118
.5	Disconformity	

(continued)

TABLE 7. (continued)

Symbol No.	Symbol	Description	Notes	Drafting Specifications (mm)
.6		Angular unconformity	Use only in sections, mainly rock relationship diagrams, where the short length of unconformity often makes symbols 2.3.4 and .5 impractical	
.7		Disconformity		
.8	Basins, domes			
.9		Boundary of depositional basin	Teeth facing basin. Bibl. 35, 168, 171, 180	
.10		Basin, depositional	Compare with tectonic (deformational) structures, Symbols 4.3.8 and .9 Bibl. 25, 72, 135, 136	
.11		Dome, depositional		
.12	Transgressive basin margin			
	Use appropriate unconformity symbol Bibl. 23			

Bibl. 81, 101, 166, 170. See also Bibl. 112

Bibl. 165, 166, 171, 174. See also Bibl. 23 See also 2.3.0, .3, .12, .13

.13	Structure contours, form lines		
.14	Structural contour, position accurate	Distinguished from bedding trend-line and basement contours (symbol 2.3.16) if necessary by varying thickness of line or by colour	400 — 0.15
.15	Structural contour, position approximate		300 — 2.0, 0.5
.16	Structural form line or basement contour	Constructed to the basement or a discrete horizon in structural basins compiled on the basis of drilling. In metres from sea level. Information based on methods other than drilling	250 — 0.15
.17	Structural form line or basement contour		800 — 0.5, 1.0
.18	Boundaries of major structures		
.19	Boundaries of major structural units	Anticlinorial and synclinorial zones, blocks, etc. Inferred boundaries shown by broken lines	0.1
.20	Concealed boundaries of major structural units		1.25, 0.5
.21	Boundaries of secondary category structural units	Sub-basins, segments of major structural units, etc.	0.5, 0.2, 6.0

Source: From Bureau of Mineral Resources (BMR) (1978).

TABLE 8. Bedding Symbols Used on Geological Maps and for Airphoto Interpretations by the Australian Bureau of Mineral Resources, Geology and Geophysics (BMR), Showing Strike and Dip, Undulating and Folded Strata, Tops of Beds, and Direction of Movement (Water, Ice)

Symbol No.	Symbol	Description	Notes	Drafting Specifications (mm)
3.1 STANDARD BASIC SYMBOLS				
3.1.0 General symbols				
		Bibl. 12, 15, 25, 33, 34, 39, 40, 61, 73, 76, 92, 139		
		For airphoto interpretation of bedding see 11.1.1 to 11.1.11		
3.1.1		Strike and dip of strata	Normally used to represent observed bedding data; 3.1.4 only used where a strong doubt about facing direction is held by observer	
.2		Prevailing strike and dip of strata, or Strike and dip of strata; dip not measured	Bibl. 19	
.3		Strike of strata, dip not determined	Bibl. 98	
.4		Strike and dip of strata, facing not known	See footnote at end of this sub-section. Bibl. 14, 18, 39	
.5		Vertical strata		
.6		Vertical strata showing facing		
.7		Vertical strata, facing not known	Bibl. 9	
.8		Horizontal strata	See also 3.2.1 Symbol should be orientated north/south	

480

	Symbol	Description	Notes	
.9		Strike and dip of overturned strata	See footnote at end of this sub-section	
.10		Strike and dip of strata / Showing proved direction of facing based on the sedimentary structures	Optional. May be used to distinguish new mapping in area where facing was not proved in earlier mapping } Bibl. 20 See footnote at end of this sub-section	
.11		Strike and dip of overturned strata		
.12		Curving dip	Angles indicate maximum and minimum dip Bibl. 25	
.13		Dip-slope	Observed. Arrow as long as exposed slope. Bibl. 29, 34. To show outline, may combine with trend-line symbol 3.1.14	
.14		Trend-line	May use 'trend of bedding' (or strata). See also 11.1 For plunge added to trend-line see 4.1.14	
.15		Trend-line showing dip		
.16		Curved outcrop of uniformly dipping strata	See also 3.1.13 and Notes	
.17 Undulating, folded strata				
.18		Generalised strike and overall dip of crumpled, undulating strata	The overall dip is in a direction generally parallel to the fold axes. Bibl. 14, 58, 102	
.19		Overall dip of gently folded strata	Bibl. 67	
.20		Overall dip of strongly deformed strata	Bibl. 25, 50, 88	

(continued)

481

TABLE 8. (continued)

Symbol No.	Symbol	Description	Notes	Drafting Specifications (mm)
.21		Symbols to indicate top of beds (facing)		
		Bibl. 14, 34, 40, 84, 115, 135		
.22		Top of bed (or stratum)	General symbol. Arrow shows facing; base at point of observation	
.23		Top of bed indicated by cleavage/ bedding relationship		
.24		Top of bed indicated by cross-bedding	Not generally used on standard maps	
.25		Top of bed indicated by graded bedding		
.26		Facing of lava-flow top or Facing of pillow lava top	Base of arrow at point of observation	
.27		Direction of movement (water, ice)		
		Bibl. 16, 25, 130		
.28		Direction of movement of sediment-bearing currents	Symbols to indicate how determined: f = fluting, r = asymmetrical ripple marks or other "structures", x = cross-stratification. Bibl. 10, 50, 130	
.29		Direction of movement of sediment-bearing currents, sense unknown	Bibl. 113	
.30		Glacial striae showing direction of movement	Centre of circle is point of observation (generally not on bedding plane) Bibl. 88	
.31		Glacial striae, direction of movement not known	Bibl. 29, 51, 108	

Footnote
GSWA use the symbol ⊢ 'dip and strike of strata' as an observation giving orientation of bedding surface in the Archaean of WA, without any implication regarding facing. Other symbols (○→ , ↻→) indicate facing. Usually the facing of only 1% or 2% of the total dip measurements can be established. The symbol ↦ is not used, as it would imply that when ⊢ is used, the direction

482

of facing is known. The symbols ⊤ ⊤ ⊤ ⊤ are regarded as unnecessary and are not endorsed. To clarify continued use of ⊤ on their Archaean maps GSWA suggest that a note 'facing not implied' could be added in the reference. Where basement Archaean is overlain by Proterozoic or Phanerozoic rocks which appear on the same map sheet, the note could be changed to 'facing not implied in Archaean rocks'

3.2 SPECIAL SYMBOLS

3.2.0 General symbols

.1 Horizontal inverted strata

.2 Strike and dip of foresets

.3 Range of strike and dip of irregular foresets

.4 Dip on exhumed erosion surface

Not to be combined with dip and strike of strata. Large-scale or special maps only *(applies to .2 and .3)*

For specialist, generally geomorphological maps only *(applies to .4)*

.5 Combined symbols

Bibl. 15, 25, 88, 137. See also cleavage 8.2.1 to .3

.6 Strike and dip of strata with plunge of mineral elongation

.7 Strike and dip of strata parallel to foliation; with plunge of mineral elongation

.8 Strike and dip of strata parallel to foliation; with horizontal mineral elongation

.9 Strike and dip of overturned strata with plunge of mineral elongation

Proved direction of facing based on sedimentary structures may be indicated by dot on dip symbol

For arrowheads applicable to the other types of lineation see under Lineation 9.0

For 'unmeasured' and 'prevailing' see Introductory Notes

Source: From Bureau of Mineral Resources (BMR) (1978).

483

TABLE 9. General and Special Map Symbols for Folds and Related Structures Used by the Australian Bureau of Mineral Resources, Showing Signs for Anticlines, Synclines, Monoclines, Flexures, and Basin and Dome Structures

Symbol No.	Symbol	Description	Notes	Drafting Specifications (mm)
		For generalised strike and dip of folded strata, average dip of folded strata see 3.1.18 to .20		
		4.1 STANDARD BASIC SYMBOLS		
		4.1.0 General symbols		
		Bibl. 12, 38, 44, 82, 126, 131, 139, 151, 153, 174		
.1		Anticline	Position accurate	
.2		Anticline, position approximate	For usage of 'accurate', 'approximate', 'inferred', see Introductory Notes under 'Classification of Geological Boundaries, Fold Axes, and Faults'. If trace of axial surface rather than crest or trough or hinge-line mapped indicate by small f, e.g.	
.3		Anticline, inferred		
.4		Anticline, concealed		
.5		Anticline, fold axis horizontal		
.6		Anticline, showing trend and plunge of axis	Upright fold only (anticline has near-vertical axial surface); for inclined fold see symbol 8. For 'plunge not measured' see Introductory Notes. Bibl.19,40,104	
.7		Anticline, showing dip of axial surface	Instead of 'axial surface' the less general term 'axial plane' is commonly used	
.8		Anticline, showing dip of axial surface, with trend and plunge of fold axis	For plunge not measured: see Introductory Notes	

484

No.	Name	Description
.9	Syncline	The symbol for syncline may be modified in the same way as the symbol for anticline to indicate position approximate, inferred, and concealed, and to show trend and plunge of axis and dip of axial surface
.10	Monocline	Line marks position of upper line of dip change. Lower dip reading is that of plane of flexure. On large-scale maps a second line may be added to indicate lower line of dip change. See also 4.2.2, fault-induced monocline Bibl. 28, 49, 66, 85, 88, 110
.11	Overturned anticline	Overturned anticline showing dips of limbs
.12	Overturned anticline showing dip of axial surface and trend and plunge of axis	For plunge not measured: See 1.4.3
.13	Overturned syncline	Dip of axial surface, and trend and plunge of axis may be shown in the same way as for overturned anticline
.14	Plunge symbol on trend-line	In small-scale maps where folding indicated by observed bedding trend-lines. Bibl. 70, 89, 158. Compare with 3.1.16. No value means unmeasured, or photo-interpreted trend of plunge
.15	Locality of superposed folds	i.e. Intersecting folds of different ages. Bibl. 17
.16 Facing of strata not known		
.17	Antiform	Modified as in symbols 4.1.1. to 4.1.4 to show position approximate, inferred, and concealed
.18	Synform	
.19	Overturned antiform	Dip of axial plane, and trend and plunge of axis may be shown in the same way as for anticlines and synclines. Bibl. 24

(continued)

485

TABLE 9. (continued)

Symbol No.	Symbol	Description	Notes	Drafting Specifications (mm)
.20		Overturned synform		
.21		Monoform		
.22		Minor folds; nose of fold		
		Bibl. 12, 14, 25, 41, 61, 94, 104, 133, 139		
.23		Minor anticline with plunge	Minor fold; nose of fold. Symbols 4.1.24 and 4.1.26 for upright folds only. Where trend and plunge of axis not known, see 4.2.4 and .5. For minor folds with lineation see 4.2.22 to .24. For plunge not measured see 1.4.3	
.24				
.25		Minor syncline with plunge		
.26				
.27		Drag-fold with plunge	Bibl. 61, 115	
.28		Minor antiform with plunge	Suitable on large-scale maps; otherwise, symbols 4.1.17 and	
.29			For plunge not measured see 1.4.3	
.30		Minor synform with plunge	.18, with plunge. Symbols .28	
.31			and .30 for upright folds only	

486

.32		Plunge of fold axes or Trend of plunge fold axis if value is not shown	Bibl. 14, 25, 32, 39, 62, 81 See also Bibl. 34. To be used where strata too tightly folded to show individual folds. See also 4.2.3. For vergences and lineation see 4.2.22 to .24. See also Bibl. 12 for axial plane trace on tectonic sketch; also 4.3.1 to .4

4.2 SPECIAL SYMBOLS
4.2.0 General symbols

.1		Axis of asymmetrical anticline	Steeper flank shown by double arrow. See also 4.1.7
.2		Fault induced monocline	Bibl. 90. See also Bibl. 95
.3		Closely spaced folds	Rarely used. See Tectonic Sketch, Bibl. 162
.4		Anticline; proved direction of facing indicated by dot	Trend and plunge of axis is not known. Where plunge of fold axis is known, and defines attitude of fold, use 4.1.23 to .26
.5		Syncline; proved direction of facing indicated by dot	
.6		Reclined anticline, showing trend and dip of axial surface	The attitude of the fold is not shown by the plunge of the fold axis (which is in a vertical plane); hence direction of facing is shown
.7		Reclined syncline, showing trend and dip of axial surface	
.8		Reclined fold	Facing of strata not known

(continued)

TABLE 9. (*continued*)

Symbol No.	Symbol	Description	Notes	Drafting Specifications (mm)
.9		Axial plane trace of fold showing direction of dip where known	A fold interpreted on the basis of reversed repetition of strata. The facing of the strata, and the form and attitude of the fold are not known	
.10		Inverted anticline	Or synformal anticline	
.11		Overturned inverted anticline	Or synformal overturned anticline	
.12		Inverted syncline	Or antiformal syncline	
.13		Overturned inverted syncline	Or antiformal overturned syncline	
.14	Minor folds			
.15		Kink fold with plunge		
.16		Minor fold with dip of axial plane		
.17		Minor fold overturned with dip of axial plane and plunge of fold axis		
.18		Minor anticline with plunge parallel to crenulation	Arrowhead indicates lineation type paralleling fold axis. See also 4.1.23 to .26	
.19		Minor syncline with plunge parallel to crenulation		
.20		Plunge of fold axis with parallel crenulation, and strike and dip of axial plane		

488

.21 **Vergence**

Bibl. 15

.22 Showing 'S' vergences

.23 Showing 'Z' vergences

.24 Plunge of Folds — Showing 'M' vergences

.25 Showing overturned 'S' vergence

.26 Showing overturned 'Z' vergence

View down plunge for limbs 'Right way up'
Style of arrow-head indicates type of lineation
parallel to fold axis (See Section 9)

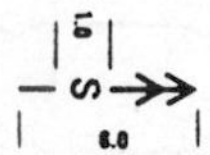

View down plunge for overturned limbs
Arrow-head indicates plunge

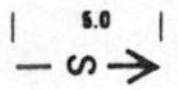

4.3 SYMBOLS USED MAINLY ON TECTONIC AND STRUCTURAL MAPS

4.3.0 General symbols

.1 Anticline

.2 Anticline, concealed

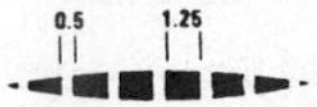

.3 Syncline

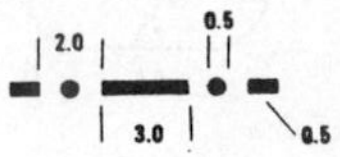

.4 Syncline, concealed

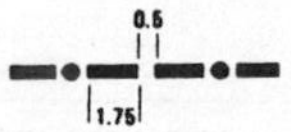

.5 Flexure

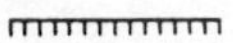

Bibl. 172. Line thickness indicates magnitude of
structure schematically only. See also Bibl. 12 for
axial plane trace on tectonic sketch

Bibl. 176

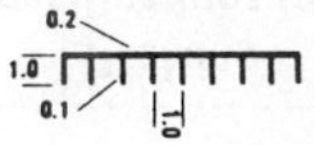

.6 **Basin and dome structures, depositional**

See 2.3.8 to .11

(continued)

489

TABLE 9. *(continued)*

Symbol No.	Symbol	Description	Notes	Drafting Specifications (mm)
.7 Basin and dome structures, tectonic				
.8		Basin, tectonic; centrocline	Bibl. 166, 176 Compare with depositional basins and domes. See also symbols 4.3.12 and .13	
.9		Dome, tectonic; pericline		
.10		Brachyanticline	Bibl. 156	
.11 Basin and dome structures, character unknown				
.12		Structural depression } Character unknown		
.13		Structural elevation	The double tick is meant to imply that facing is not known; see 3.1.4	

Source: From Bureau of Mineral Resources (BMR) (1978).

490

TABLE 10. General and Special Map Symbols for Faults Used by the Australian Bureau of Mineral Resources, Showing Faults, Shear Zones, Thrusts, and Plunge of Lineations on Fault Planes

Symbol No.	Symbol	Description	Notes	Drafting Specifications (mm)
5.1 STANDARD BASIC SYMBOLS				
		See note, page 1, on the style of arrows and arrow-heads for structural elements		
5.1.0 General symbols				
		Bibl. 12, 38, 44, 82, 126, 131, 139, 151		
.1		Fault		
.2		Vertical fault	Bibl. 29, 31 See also Bibl. 86	
.3		Inclined fault, sense of displacement not known; direction and amount of dip shown	Position accurate	
			Double tick distinguishes this symbol from 5.1.8 (normal fault)	
.4		Fault, position approximate		
.5		Fault, inferred	Established fault, movement unknown	
.6		Fault, concealed	Show dip if known. For usage of 'accurate', 'approximate', 'inferred', and 'concealed' see Introductory Notes, Classification of geological boundaries, fold axes and faults (1.4.5)	
.7		Fault, inferred and concealed		
.8		Normal fault	Tick, on downthrow side, indicates direction of dip Bibl. 10, 11, 172	
.9		High-angle reverse fault	Triangle, on upthrow side, indicates direction of dip. Bibl. 39, 68, 144, 172 — Dip > 45°	
.10		Thrust-fault	Dip < 45°	

(continued)

491

TABLE 10. (*continued*)

Symbol No.	Symbol	Description	Notes	Drafting Specifications (mm)
.11		Fault showing relative horizontal displacement or, Strike-slip fault showing relative horizontal displacement	Bibl. 29, 34, 52, 73, 144	
.12		Fault (U, D indicate relative movement; up, down)	True character unknown; only relative displacement indicated Bibl. 5, 29, 34, 52, 87, 124	To show approximate, inferred, and concealed, modify the symbol as for symbols .1 to .7
.13		Normal fault — Showing relative movement of downthrown block		
.14		Reverse fault — (Showing relative movement of downthrown block)		
.15		Fault showing width in metres		
.16		Fault zone with crushing	Bibl. 8 Showing dip where required	
.17		Fault with slickensides	Bibl. 25	
.18 Shear zone, shearing, schistosity, mylonite, breccia				
.19		Sheared zone	Bibl. 12, 14, 34, 82, 93, 166	
.20		Inclined sheared zone		
.21		Shearing or Schistosity	Broad area of shearing or schistosity. Bibl. 48, 93, 102 117, 164. See also Bibl. 19. See also Metamorphic Foliation, 7.2.2 to .5	

492

.22	Shear zone		Wide shear zone
.23	Mylonite zone		
.24	Fault with breccia		Bibl. 73 See also Bibl. 75, 105
.25	Fault containing; m-mylonite, br- breccia, q- quartz		Bibl. 11, 17 See also Notes 2.1.12 to .14

5.2 SPECIAL SYMBOLS
5.2.0 General symbols

.1	Line of faulted outcrop too small to be shown		Used on Dummer and Crossland 1:250 000 sheets to show narrow ferruginised and silicified outcrops, along pre-Cainozoic faults, that protrude through Cainozoic cover
.2	Low-angle thrust, Triangles are on upper plate and indicate direction of dip		
.3	Low-angle thrust fault, ⊢—indicates upper plate		An alternative symbol for large-scale maps Bibl. 12, 85, 104, 146, 153, 163, 167
.4	'Trailing edge' trace of low-angle thrust fault		Where landsurface slopes are steeper than that of a thrust-fault plane, trailing edge traces of the fault-plane (assuming that the movement is of the upper plate) may be exposed. Dip symbols should be on the upper plate and should indicate general fault plane dip direction. Where the direction of the fault trace is sharply at variance with the strike of the fault plane, the triangle may be set in a break of the fault trace, provided there are sufficient other symbols to indicate the upper plate
.5	Klippe		
.6	Window in fault plane exposing rocks of lower plate		

(continued)

493

TABLE 10. (*continued*)

Symbol No.	Symbol	Description	Notes	Drafting Specifications (mm)
.7		Post-intrusive fault along intrusive boundary	Substitute for 'i' the letter symbol appropriate for the intrusive, e.g. 'q' for quartz	
.8		Intrusive boundary along pre-existing fault	The symbol used by Tas GS on Hobart 1:50 000 sheet. Compare with symbol for intrusive igneous boundary, symbol 2.3.1	
.9 Combined symbols				
.10		Plunge of lineation on fault plane	Detailed maps	
5.3 SYMBOLS USED MAINLY ON TECTONIC AND STRUCTURAL MAPS				
.1		Normal fault. Hachures show downthrow and indicate direction of dip	Bibl. 51, 176	
.2		High-angle reverse fault	Dip more than 45°. Bibl. 95, tectonic sketch	Triangles commonly spaced two to four times width of base
.3		Thrust-fault	Dip 45° or less	

Source: From Bureau of Mineral Resources (BMR) (1978).

TABLE 11. Standard and Special Map Symbols for Joints Used by the Australian Bureau of Mineral Resources, Showing Strike and Dip of Joints, Joint Patterns, and Traces of Joints on Inclined Surfaces

Symbol No.	Symbol	Description	Notes	Drafting Specifications (mm)
6.1 STANDARD BASIC SYMBOLS				
Bibl. 14, 25, 60, 106, 137				
.1		Strike and dip of joint		
.2		Prevailing strike and dip of joint or Strike and dip of joint; dip not measured	Indicate which in reference	
.3		Vertical joint		
.4		Horizontal joint		
.5		Joint pattern	Generally airphoto interpretation. See also 6.2.2 Bibl. 51	
6.2 SPECIAL SYMBOLS				
6.2.0 General symbols				
.1		Trace of joint on inclined surface, showing direction of dip	Applicable particularly to mining and engineering geology plans. True strike and dip of joint defined by that of inclined surface	
.2		Joint pattern	Suitable for large-scale maps for special purposes, e.g. engineering geology	
.3 Columnar jointing				
.4		Trend and plunge of columnar jointing		

Source: From Bureau of Mineral Resources (BMR) (1978).

TABLE 12. Map Symbols for Metamorphic Foliation (Other than Cleavage) Used by the Australian Bureau of Mineral Resources, Showing Strike and Dip of Foliation, Multiple Folding Episodes, and the General Trends of Foliation

Symbol No.	Symbol	Description	Notes	Drafting Specifications (mm)
7.1 STANDARD BASIC SYMBOLS				
		See note, 1.4.1, on the style of arrows and arrowheads for structural elements		
.1		Strike and dip of foliation		
.2		Prevailing strike and dip of foliation, or Strike and dip of foliation; dip not measured		
.3		Vertical foliation	Bibl. 12, 14, 19, 31, 34, 93, 107, 135	
.4		Horizontal foliation		
.5		Strike of foliation, dip indeterminate or not determined		
7.2 SPECIAL SYMBOLS				
7.2.0 Multiple folding episodes				
		Black plate, but colour may be used for different generations		
.1		Strike and dip of foliation, first folding episode	The number of ticks indicates whether 1st, 2nd or 3rd generation in a sequence of foliation episodes	

Specialised terminology for multiple deformation applicable only to large-scale maps. Ticks can indicate generation of foliation as shown. Other features of first generation have solid black heads; second generation lineations can have open heads

496

		Symbol	Reference
7.2.2 General trend of foliation			
.3	General trend of foliation in low-grade metamorphics		
.4	General trend of foliation in medium to high-grade metamorphics		Bibl. 117, 154, 172, 176 (western & central Australia)
.5	General trend of foliation in migmatite and gneiss		
7.2.6 Combined symbols			
.7	Strike and dip of foliation, with trend and plunge of crenulation		Bibl. 12, 60
.8	Strike and dip of foliation, with trend and plunge of crenulation, and of mineral elongation or alignment		Bibl. 12, 53, 135
.9	Strike and dip of foliation, with horizontal mineral elongation or alignment		In the same way, symbols for other types of lineation may be combined with the foliation symbols. For 'not measured' and 'prevailing' see Introductory Notes 1.4.3 See also symbols 3.2.6 to .9
.10	Vertical foliation with plunge of lineation		
.11	Horizontal foliation and trend of horizontal lineation		

Source: From Bureau of Mineral Resources (BMR) (1978).

TABLE 13. Map Symbols for Cleavage Used by the Bureau of Mineral Resources, Showing Strike and Dip of Cleavage, Prevailing Strike and Dip of Cleavage, and Vertical and Horizontal Cleavage

Symbol No.	Symbol	Description	Notes	Drafting Specifications (mm)
8.1 STANDARD BASIC SYMBOLS				
	Bibl. 14, 29, 77, 110, 137, 148			
.1		Strike and dip of cleavage	Type of cleavage should be specified in reference	
.2		Prevailing strike and dip of cleavage or Strike and dip of cleavage, dip not measured	Indicate which in reference	
.3		Vertical cleavage		
.4		Horizontal cleavage		
.5		Strike of cleavage, dip not determined		
8.2 SPECIAL SYMBOLS				
8.2.0 Combined symbols				
	Bibl. 14, 15, 21, 22, 88, 102			
.1		Strike and dip of cleavage, with apparent dip of bedding (or strata) on cleavage plane		
.2		Strike and dip of strata, with apparent dip of cleavage on bedding plane	Also, under lineation: Strike and dip of bedding with trend and plunge of bedding-cleavage intersection (c.f. symbol 9.2.3)	
.3		Strike and dip of bedding and of cleavage	Strike of bedding and cleavage coincident	

Source: From Bureau of Mineral Resources (BMR) (1978).

498

TABLE 14. Standard and Special Map Symbols for Lineation Used by the Australian Bureau of Mineral Resources, Showing Lineation, Mineral Elongation, Bedding, Crenulation, Flow Lineation, and Platy Flow Structures

Symbol No.	Symbol	Description	Notes	Drafting Specifications (mm)
9.1 STANDARD BASIC SYMBOLS Bibl. 16, 17, 137				
.1		Lineation	Type not specified. Bibl. 12, 39, 93	
.2		Mineral elongation (or alignment)	Lineation may be expressed in orientation of minerals, pebbles (GSWA). See also Bibl. 32, 64	
.3		Bedding — cleavage intersection		
.4		Crenulation	Bibl. 12	
.5		Trend and plunge of lineation	Plunge to be measured in the vertical plane; point of observation at base of arrow. Symbol .6 used in combination with other symbols (see 9.2.0). A specific type of lineation may be shown (Symbols .1 to .4) For 'unmeasured' and 'prevailing' see Introductory Notes, 1.4.3	
.6				
.7		Vertical lineation		
.8		Horizontal lineation		

Note:
For single symbols 9.1.1 to 9.1.4, 9.2.8, the point of observation is at the base of arrow which indicates trend. A full circle at base of arrow for single symbol is used when plunge as well as trend is measured. (i.e. 9.1.5)

(continued)

499

TABLE 14. (continued)

Symbol No.	Symbol	Description			Notes	Drafting Specifications (mm)
9.2 SPECIAL SYMBOLS						
9.2.0 Combined symbols						
	Bibl. 15, 25, 60, 88					
.1		Strike and dip of	foliation	with trend and plunge of lineation	A specific type of lineation may be shown (Symbols 9.1.2 to .4) Bibl. 43, 137	
.2			cleavage			
.3			bedding			
.4		Strike and dip of bedding, with horizontal lineation			For 'not measured' and 'prevailing' see Introductory Notes, 1.4.3	
.5		Vertical bedding, with plunge of lineation			See also symbols 3.2.6 to .9 and 7.2.7 to .11	
.6		Strike and dip of foliation, with lineation measured in plane of foliation			A specific type of lineation may be shown, measured in plane of foliation, cleavage, bedding. However, as standard practice, it is urged that plunge be shown, not pitch	
.7	Flow lineation					
.8		Trend and plunge of flow lineation			See Note above / This arrow should remain as a general purpose symbol, with particular usage shown in the map reference. See also Bibl. 32	
.9		Flow lineation, horizontal			For 'not measured' and 'prevailing' see Introductory Notes	
.10	◇	Flow lineation, vertical				◇

Source: From Bureau of Mineral Resources (BMR) (1978).

TABLE 15. Standard Basic Map Symbols for Banding in Igneous Rocks Used by the Australian Bureau of Mineral Resources

Symbol No.	Symbol	Description	Notes	Drafting Specifications (mm)
10.1 STANDARD BASIC SYMBOLS				
Bibl. 16, 17, 20, 25, 39, 74, 91, 94, 99, 135, 137				
.1		Strike and dip of platy flow structure	Extrusive or intrusive igneous rocks. Use sparingly on standard maps. 'Banding' or 'flow banding' may be used instead of 'platy flow structure'. Suitable for eutaxitic foliation in ignimbrite	
.2		Strike and dip of platy flow structure, dip not measured or Prevailing strike and dip of platy flow structure	For 'unmeasured' and 'prevailing' see Introductory Notes	
.3		Vertical platy flow structure		
.4		Horizontal platy flow structure	Note by GSWA: Classification here should be descriptive: not genetic	
.5		Platy flow structure, dip indeterminate or not determined		

Additional symbols may be used for distinguishing various types of planar structures: Note by GSWA: These symbols should be restricted to layering in intrusive rocks, e.g. gabbroic complexes. Bibl. 13, 60, 61, 139

Source: From Bureau of Mineral Resources (BMR) (1978).

501

TABLE 16. Map Symbols for Photointerpreted Structural Data Used by the Australian Bureau of Mineral Resources, Showing Bedding, Lineaments, and Joints

Symbol No.	Symbol	Description	Notes	Drafting Specifications (mm)
11.1 PHOTO-INTERPRETED STRUCTURAL DATA				
		Black plate. Bibl. 25, 41		
11.1.0		Bedding		
		See also Bedding symbol 3.1.14; photo-interpreted symbol 11.2.1		
.1		Strike and dip of strata, dip not estimated		
.2		Strike and dip of strata, dip less than 5°		
.3		Strike and dip of strata, dip 5°-15°	Bibl 9, 34, 56, 60, 76, 89, 93, 139	
.4		Strike and dip of strata, dip 15°-45°	*Numbers, e.g. indicate dip measured from airphotos and/or radar imagery. Dips (photo-) estimated at intervals along the strike may be supported by one or more (photo-) measured dips	
.5		Strike and dip of strata, dip greater than 45°		
.6		Vertical strata	*Add when applicable	
.7		Horizontal strata		

502

.8	Trend-line		
.9	Bedding, combined symbols		
.10	Trend-line showing dip of 5° - 15°	Use appropriate photointerp. dip symbol or 3.1.15	
.11	Lineament		
.12	Lineament	Photo-interpreted; very commonly fault or joint	
.13	Joints		
	See joint symbol 6.1.5		

11.2 PHOTOGRAMMETRY

.1	Photo-centre point	Photo-centre points on maps (when required) quoting run and full number of first and last photograph of each run on each map sheet; only last two figure quoted for intermediate photographs
.2		

Source: From Bureau of Mineral Resources (BMR) (1978).

TABLE 17. Map Symbols for Geophysics Used by the Australian Bureau of Mineral Resources,
Showing Seismic Traverse Lines, Reflecting and Refracting Horizons, Interpreted Basement, Radiometric
Contours, Gravity Features, and Magnetic Contours

Symbol No.	Symbol	Description	Notes	Drafting Specifications (mm)
12.1 SEISMIC	Bibl. 26, 38			
.1		Seismic traverse line		
.2		Seismic reflecting horizon, showing velocity in metres per second		
.3		Seismic refracting horizon	Bibl. 38, 134, 150 Section only Black plate	
12.2 INTERPRETED BASEMENT				
.1		Magnetic basement		
.2		Gravity basement		
12.3 RADIOMETRIC CONTOURS		See Bibl. 193 (plate 42)		

12.4 GRAVIMETRY

	Symbol		Notes
.1	○	Gravity station	combined with elevation in metres (grey plate) and/or bouguer anomaly in milligals (12.4.3)
.2	●	Gravity station – other than BMR	Bibl. 38, 57, 83, 131
.3	-32.4	Bouguer anomaly in milligals	Plus or minus value
.4	+20	Isogal – Index	
.5		Isogal – intermediate	Purple plate
.6		Gravity anomaly – relative high	
		Gravity anomaly – relative low	See also Bibl. 77, 193 (plate 43)

12.5 MAGNETIC CONTOURS, ANOMALIES, BASEMENT CONTOURS
See Bibl.18, 23, 78, 154, 165 (plate 3) Black plate

	Symbol		Notes
.1	+20	Magnetic contours (nT)	Magnetic contours may be printed in reverse on the back of the map and be seen in relation to the geology by placing the map over a light source. The following information should appear on the map sheet: Magnetic contours are based .... the survey was made at an altitude of .... above sl. contour interval .... nt
.2		Magnetic 'low'	

(continued)

TABLE 17. (continued)

Symbol No.	Symbol	Description	Notes	Drafting Specifications (mm)
12.6 GEOLOGY INTERPRETED FROM GEOPHYSICAL DATA				
.1	g ——	Fault interpreted from geophysical data	gv, gravity data; m, aeromagnetics. Suitable for single colour map. See also Bibl. 54, 101, 111, 112, 118, 125, 170	
.2		Geological boundary	See 2.1.4	
.3	←⋯⋯⋯→	Anticline	Purple plate (bibl. 151); See 4.1.4	
.4	←⋯⋯⋯→	Syncline	Interpreted from geophysical data — alternatively, black plate with symbol suitably annotated (bibl. 124, 127)	See 4.1.4 See 5.1.6 Bibl. 81 See 3.1.14
.5	▪ ▪ ▪ ▪ ▪	Fault		
.6	— —	Trend-line		
.7	——	Lineation or Linear feature	See 11.1.13	
.8	g ——	Linear feature	Suitable for single-colour map	

Source: From Bureau of Mineral Resources (BMR) (1978).

506

TABLE 18. Map Symbols for Fossil and Specimen Localities Used by the Australian Bureau of Mineral Resources, Showing Localities, Type Sections and Locations of Geological Sections, and Drill Holes

Symbol No.	Symbol	Description	Notes	Drafting Specifications (mm)
		13.1 FOSSIL AND SPECIMEN LOCALITIES		
		Bibl. 33, 34, 37, 45, 51, 52, 56, 113, 139.		
.1	♢	Fossil locality	If desired may use following additional notations on special or large scale maps, but not on standard series maps:	
.2	●	Macrofossil locality	m – Marine e.g. ♢m } Bibl. 73, 83, 112 b – Brackish water f – Fresh water	
.3	⊛	Microfossil locality	(φ) sparse, e.g. — abundant, e.g. ● To refer to report or explanatory notes, add number to locality, as for symbol .11 below	
.4	φ	Plant fossil locality		
.5	▯	Fossil wood locality		
.6	⊗	Palynomorph locality	Includes pollen spores, acritarchs, chitinozoans, dinoflagellato thecae and cysts, certain colonial algae and other acid-insoluble microfossils	
.7	⬥	Vertebrate fossil locality		
.8	※	Trace fossil locality		
.9	⊓	Stromatolite locality		
.10	◎	Oncolite locality		

(continued)

507

TABLE 18. (continued)

Symbol No.	Symbol	Description	Notes	Drafting Specifications (mm)
.11	x⁹	Locality visited. Number is authors reference	Alternatively give a different description if appropriate number is .... See also 2.1.17 (outcrop)	
.12	×C134	Specimen locality, with reference number		×C134
.13	(symbol)	Sample locality for isotopic age determination or Radiometric age in m.y.	Place sample number against symbol, and key in reference or on back of sheet giving age and method of determination; or, place age in m.y. in sloping figures against symbol. Bibl. 9,18,20,33,49,114	
.14	$X \frac{266}{Rb-Sr}$	Isotopic age in m.y. (Symbols 13.1.14 – .17 not usually shown in general geological maps)	Cross indicates locality; figures and letters give age in m.y. and method used; for carbon dating age is in thousands of years. To show age by alternative methods at the same site the symbol may be amended e.g. x 265±10 (309±10) — Radiometric age in m.y. by potassium — argon (and rubidium — strontium) methods	$\frac{266}{Rb-Sr}$
.15	$X \frac{320}{K-Ar}$			
.16	$X \frac{560}{Pb}$			
.17	$X \frac{300}{fis.fr}$			

13.2 TYPE SECTIONS AND LOCATIONS OF GEOLOGICAL SECTIONS

Symbol No.	Symbol	Description	Notes	Drafting Specifications (mm)
.1	(symbol)	Type section	If section short, show by joined triangles; if long, use two triangles joined by a thin continuous line. Arrow points mark ends of section Bibl. 13, 17, 22, 34, 58, 78, 136	
.2	(symbol)	Type locality		

.3	B5 (⊢———⊣)	Measured section, with reference number	Bibl. 24, 36, 52	B5 (⊢———⊣) 0.2 / 0.15
.4	C ⊢———⊣ D	Location of geological section	Geological section may appear in surround of map, or elsewhere	C ⊢———⊣ D 0.15 / 0.15

13.3 DRILLHOLES

Bibl. 8, 23, 94, 139, 140, 169

.1	⊗	Drillhole	General symbol. Use letter symbol to distinguish purpose:	⊗ 0.2 / 1.5

CH   Corehole

DD   Diamond-drillhole   For use on small scale maps e.g. 1:250,000, 1:100000 scale

HD   Hand-drillhole   See Petroleum Exploration and Development Wells, Water Bores,

PD   Percussion-drillhole   Mining (and other) Drillholes

RD   Rotary-drillhole

S   Structure hole   Inclined holes may be shown as for inclined drillhole for mining

SH   Scout hole

SSH   Seismic-shot hole

St   Stratigraphic hole

Letter symbols may also show important minerals and rocks encountered in drillhole e.g. Ph-Phosphate rock ⊗St Op-Opal ⊗Op

TABLE 19. Map Symbols for Mineral Deposits and Workings Used by the Australian Bureau of Mineral Resources, Showing Mineral Deposits and Developments and Mine Plans and Sections

Symbol No.	Symbol	Description	Notes	Drafting Specifications (mm)
		14.1 MINERAL DEPOSITS AND MINERAL DEVELOPMENTS Bibl. 7, 9, 13, 40, 61, 73, 93, 94, 120, 128, 153, 155		
.1		Minor mineral occurrence	May be of mineralogical interest only Letter symbols indicate mineral	0.5 —●
.2		} Unworked deposit	May have been tested by drill or costean. Deposit located at centre. Symbols .3 is an alternative; deposit located by black dot	2.5 ○
.3				0.5 —●— 0.15 ○
.4		Prospect or mine with little production	Letter symbols may be used to indicate minerals, e.g.	
.5		Abandoned prospect, no production or Abandoned mine with little production	Letter symbols may be used to indicate minerals and what work was done, e.g. a, g, r Examples: a-auger drilling; c-costeaning; d-rotary, percussion, diamond drilling; g-geochemical survey; r-ground radiometric survey	
.6		Mine, minor		
.7		Mine, major		
.8		Open cut, quarry; minor		
.9		Open cut, quarry; major		
10		Alluvial workings, minor		

.11		Alluvial workings, major	
.12		Mine, not being worked, or Abandoned mine	
.13		Open cut, quarry; not being worked	
.14		Alluvial workings, not being worked or Abandoned alluvial workings	
.15		Battery, smelter	Several types of treatment plant can be differentiated by number and key in margin
.16		Battery, smelter; not operating or Site of former battery	
.17		Limit of exploration activity or Extent of prospected area	Symbols for prospects, mines, and minerals may be shown within the area

14.2 MINE PLANS AND SECTION

Not Shown on General Purpose Maps

.1		Main shaft (showing number of compartments and depth in metres)
.2		Shaft — extending above and below level
.3		Shaft — accessible extending below plan level

(continued)

TABLE 19. *(continued)*

Symbol No.	Symbol	Description	Notes	Drafting Specifications (mm)
.4	◩	Shaft — accessible, extending above plan level		◨
.5	▭	Shaft — inaccessible		▭
.6	◲	Head of rise or winze		◲
.7	◩	Foot of rise or winze	Add 'N' if inaccessible	◩
.8	⊠	Rise or winze, extending through level		⊠
.9	◩ ²⁵ ₃₅	Inclined shaft — small-scale maps	Showing inclination and length in metres	◧
.10	◩ ²⁵ ₃₅	Inclined shaft — large-scale maps		◨
.11	◖	Cross-section of cross-cut or drive; same side of plane of section as observer		◖
.12	◹	Cross-section of cross-cut or drive; opposite side of plane of section from observer		◹
.13	⊠	Cross-section of cross-cut or drive extending across plane of section		⊠

.14		Ore chute	
.15	stoped above	Plan ⎫ Stopes	
.16, .17	Stoped	Section ⎭	
.18		Lagging or cribbing along drive, etc	
.19		Workings caved or otherwise inaccessible	
.20		Filled workings	
.21		Portal and approach of tunnel or adit	
.22			
.23	2	Costean or trench, showing depth in metres	
.24	2		

On large-scale maps may be possible to indicate within stoped area year(s) worked, tonnage removed and grade of ore recovered

(continued)

TABLE 19. (continued)

Symbol No.	Symbol	Description	Notes	Drafting Specifications (mm)
.25		Open cut or quarry		
.26				
.27	□ 2	Prospecting pit, showing depth	Large-scale maps only	
.28		Sample line	Type of sample may be indicated by letters, e.g. c – channel, h – chip, b –bulk,	
.29	✕	Grab-sample locality		
.30	145	Roof (back)	Elevation of underground workings	
.31	145	Floor		
.32		Natural surface (in sections)	Where lithology and dip of strata not shown	
.33	DD O	Drill hole, showing projection in horizontal plane and inclination	Arrow points down, i.e. for hole drilled upwards symbol is DD – diamond drill, HD – hand drill, PD – percussion drill	
.34	DD O 187	Vertical drill hole showing depth in metres	See Drill Holes, under sub-section 13.3	
.35		Dump		
.36		Information projected onto a section: from near side		
.37		from far side		

From Bureau of Mineral Resources (BMR) (1978).

TABLE 20. Map Symbols for Petroleum Occurrences and Development Used by the Australian Bureau of Mineral Resources, Showing Natural Occurrences of Oil and Gas, Exploration and Development Wells, and Names of Petroleum Exploration and Development Wells

Symbol No.	Symbol	Description	Notes	Drafting Specifications (mm)
15.1 NATURAL OCCURRENCES OF OIL AND GAS				
Bibl. 85, 141				
.1		Gas seep		
.2		Oil seep		
.3		Oil and gas seep or show	Black plate. Colour may be used in special purpose maps	
.4		Oil seep reported (by geologist) but not relocated		
.5		Gas seep reported (by geologist) but not relocated		
.6		Oil and gas seep reported (by geologist) but not relocated		
.7		Mud volcano without hydrocarbons		
.8		Oil field	Bibl. 183. No standard symbol is proposed. A line in a suitable colour either broken or unbroken may be used to show limit of field. The area may be hachured at compiler's discretion. Different colours may be used for oil fields and gas fields	
.9		Gas field		

(continued)

515

TABLE 20. (*continued*)

15.2 EXPLORATION AND DEVELOPMENT WELLS

Bibl. 58, 80, 92, 124, 151, 153

Description to be prefaced by the relevant term, i.e. Petroleum exploration or Petroleum development. The same symbol is used for exploration and development wells except when both types of well appear on a map, in which case the letter d should be added to the symbols portraying development wells. The Qld Dept of Mines has three divisions: exploratory, appraisal, and development

Symbol No.	Symbol	Description	Notes	Drafting Specifications (mm)
.1	⊙	Petroleum exploration well, proposed site	Where any of these wells has been completed as: 1) Water well; add W (on blue plate) to appropriate symbol 2) Water Injection well; add WI (on black plate) to appropriate symbol 3) Control well (water injection); add WIC (on black plate) to appropriate symbol The same north, south, east and west points are added to symbols 15.2.5 to .7 to show that the well is abandoned without any production taking place	2.0 / 0.25
.2	○	Petroleum exploration well, drilling		○ 0.2
.3	◇	Petroleum exploration well, dry, abandoned	SSH — Seismic shot-hole Bibl. 82 W — Completed as a water bore (on blue plate)	3.0
.4	○St	Stratigraphic hole (For petroleum exploration)		St / 3.75
.5	◑	Petroleum exploration well with show of oil	Proposed by SA Mines Dept: 1) For well abandoned without production, see Notes to symbol 15.2.3 2) Recommended that, on general purpose maps, a show of oil be defined as a measurable quantity, at least 1 litre recovered in pipe; a show of gas at least gas at least gas to surface	◑
.6	☼	Petroleum exploration well with show of gas		☼
.7	◒	Petroleum exploration well with show of oil and gas		◒
.8	●	Oil well		●

.9	●	Oil well, abandoned
.10	☼	Gas well
.11	✳	Oil and gas well
.12	☼	Gas and condensate well

The same oblique line is added to symbols .10, .11 and .12 to show that the well is abandoned after production

Where wells have been specifically drilled for the purpose of:
1) Water Injection
2) Control (water injection) use the symbol for Well dry, abandoned and add WI or WIC (on black plate)

For suspended oil well, oil and gas well, gas well, gas and condensate well, GSWA shows the appropriate symbol with a horizontal line through it. South Aust. Dept of Mines suggests that 'drilling suspended' be placed beside the appropriate symbol

15.3 NAMES OF PETROLEUM EXPLORATION AND DEVELOPMENT WELLS

Petroleum well names should be set out in a fashion similar to the following examples:

Amoseas Balfour 1

BMR Longreach 2

Farmout Drillers Alice River 1

LOL 6 (Balmoral)

Source: From Bureau of Mineral Resources (BMR) (1978).

TABLE 21. Map Symbols for Hydrology Used by the Australian Bureau of Mineral Resources, Showing Natural Features, Water Bores and Wells, Names of Water Bores and Wells, and Water Tanks and Dams

Symbol No.	Symbol	Description	Notes	Drafting Specifications (mm)
		To be used for general geological maps. For additional symbols to those listed below, use standard symbols compiled by Division of National Mapping for 1:100 000 and 1:250 000 map series. Hydrogeological maps require many additional symbols and may require the use of more than one colour. Bibl. 12, 16, 51, 56, 66, 69, 76, 93, 100, 127, 132		
16.1 NATURAL FEATURES				
.1		Lake, lagoon, or waterhole		
.2		Intermittent lake		
.3		Spring	General symbol	
.4		Spring, intermittent	Main mineral constituent may be shown by appropriate mineral or chemical symbol	
.5		Hot spring, M-mud spring	i.e. hotter than blood heat (37°C)	
.6		Salt spring, salinity not measured	Other details, e.g. salinity in ppm, chemical content and yield, not normally shown on general geological map but may be shown if desired. Alternatively if a large number of bores, wells and springs are shown on the sheet, they may be numbered and a key with specifications and data placed in margin or on back of sheet	
.7		Spring, salinity less than 1500 ppm	i.e. potable	
.8		Spring, salinity 1500-10 000 ppm	i.e. suitable for stock only	
.9		Spring, salinity greater than 10 000 ppm	i.e. unsuitable for stock	

.10	Sinkhole	Add SH if possibility of confusion with quarry, etc. To appear in same colour as watercourses on coloured maps. Bibl. 29. See also Bibl. 147		
.11	Waterhole; soak	If on watercourse, may be shown thus		
.12	Waterhole, soak; persistent	Bibl. 16, 91, 106, 121, 148, 153		
.13	Rockhole			
.14	Ephemeral water-table pool	See also symbol 16.1.1		
.15	Salinity less than 1500 ppm			
.16	Salinity 1500-10 000 ppm	See note on potability opposite. To be used with symbols .1, .2, .14		
.17	Salinity greater than 10 000 ppm	symbols 16.1.7 to .9		

16.2 WATER BORES AND WELLS

.1	Bore	General symbol. Water quality not specified	
.2	Abandoned bore	May use letter code to show reason why abandoned (see also symbol for 'abandoned saline bore').	

(continued)

TABLE 21. (continued)

Symbol No.	Symbol	Description	Notes	Drafting Specifications (mm)
.3		Bore, salinity less than 1500 ppm	i.e. potable for humans	
.4		Bore, salinity 1500-10 000 ppm	i.e. usable for stock	
.5		Saline bore, salinity greater than 10 000 ppm	i.e. unusable for stock	
.6		Abandoned saline bore	Use appropriate salinity symbol	
.7				
.8		Bore with windpump		
.9		Bore with pump engine		
.10		Artesian bore, flowing	Salinity, abandoned artesian bore, hot water bore, etc. may be shown as for non-artesian bore.	
.11		Artesian bore, ceased to flow	Two aquifer systems in the same area may be distinguished by using modified symbols, e.g. with symbols 16.1.15 to .17	
.12		Sub-artesian bore	Non-flowing. Confined water has risen above the watertable, but not to the surface	
.13		Well	General symbol. Salinity, abandoned well etc. may be shown as for bore	

Notes (for .3):

Bibl. 69

Additional data may be shown on special maps, e.g. Depth of bore - (213). Descriptive data may be shown by use of letter symbols, e.g.:

H – hot water, i.e. hotter than blood heat 37°C

Td – abandoned for technical reasons (drilling or equipment problems)

Nf – abandoned because of diminished flow

Su – sulphur-bearing

Hy – hydrogen-sulphide-bearing

16.3 NAMES OF WATER BORES AND WELLS

Where a bore or well has no identification other than a 'number name' leave the abbreviation 'No.' in title e.g. No. 2 Bore (or if preferred for a particular sheet, Bore No. 2). N.B. Be consistent on each sheet

Where a 'number name' is preceded by a descriptive or station name the abbreviation 'No.' is deleted, e.g. Canobie No. 3 Bore becomes Canobie 3 Bore

Where a bore or well is known synonymously by both a 'number name' and a descriptive name, retain the abbreviation 'No.' and bracket the descriptive name, e.g. No. 3 (Bloodwood) Bore

Some organisations prefer to use the registered number assigned to a bore by the relevant State or Territory water authority

16.4 WATER TANKS AND DAMS

.1	□ T	Water tank	(e.g. steel, concrete, masonry, galvanised iron)
.2	□ E	Earth tank or 'dam'	Bibl. 71, 88. See also Bibl. 65
.3	□ S	Water storage	To be used where water storage facilities are known to exist; but type of storage is uncertain
.4	⊲	Dam on stream	

Source: From Bureau of Mineral Resources (BMR) (1978).

521

TABLE 22. Map Symbols for Volcanoes Used by the Australian Bureau of Mineral Resources, Showing Eruptive Centers, Crater Walls, Thermal Areas, Lava Flows, and Pyroclastic Flows

Symbol No.	Symbol	Description	Notes	Drafting Specifications (mm)
		Letters may be added to indicate chemical or petrological characteristics, e.g. 't' — tholeiitic suite; 'ca' — calc alkali suite. Volcanic symbols should be shown in red where practicable. The symbols are applicable particularly to Cainozoic volcanic phenomena with topographical expression Bibl. 63, 123, 129, 158		
.1		Volcanic neck, volcanic pipe, volcanic vent, extinct.	General symbol. May add 'v' to avoid confusion with bore or oil well	
.2		Major eruptive centre with recorded eruption	e.g. Central crater or vent of volcano	
.3		Major eruptive centre with no recorded eruption	The rays for eruptive centres should be arranged symmetrically to give a stylised representation of a cone	
.4		Minor eruptive centre with recorded eruption	e.g. Satellite cone, cumulodome	
.5		Minor eruptive centre with no recorded eruption		
.6		Crater wall, caldera wall or escarpment related to volcanism	On 1st Ed. maps shown on special red plate, on background colour of the under-layer	
.7		Thermal area	'H' indicates thermal activity	

522

| .8 | | Lava flow |
| .9 | | Pyroclastic flow |

True outline shown
May be annotated for lava type if known and date of eruption

Source: From Bureau of Mineral Resources (BMR) (1978).

TABLE 23. Selected Map Symbols for Topography Used by the Australian Bureau of Mineral Resources, Showing Natural Features, Topographic Survey Stations, Elevations, and Contours

Symbol No.	Symbol	Description	Notes	Drafting Specifications (mm)
		For additional symbols to those listed below use standard symbols compiled by Division of National Mapping for 1:100 000 and 1:250 000 map series		
		18.1 NATURAL FEATURES		
		See also Sections 16 and 17. Bibl. 13, 47, 58, 59, 60, 93, 116, 142, 151. Restraint should be exercised in use on general geological maps		
.1		Swamp, marsh		
.2				

(continued)

TABLE 23. (continued)

Symbol No.	Symbol	Description	Notes	Drafting Specifications (mm)
.3	y	Mangroves		
.4		Rock ledge, coral reef	Letter symbol for rock unit used where required	
.5				
.6	*	Rocks awash		
.7	+	Rock submerged		
.8		Edge of raised reef terrace	To be included on Brown plate to differentiate from all other similar line weights on Black plate. Bibl. 79	
.9		Coastline	Blue	
.10	Qm	Strandline	As a geological boundary	
.11			Long lines of trend, line weight on the topo plate	

		Brown	
.12	Sand ridge or dune		0.15
.13	Claypan, saltpan		0.1
.14	Escarpment	If produced by faulting, may letter 'Fault scarp' along top. Alternatively letter 'Scarp' along appropriate fault symbol. Other descriptive words may be added to symbol e.g. raised reef, cliff. Compare with fault symbol 5.3.1 Bibl. 22	0.25 0.1 1.0
.15	Alluvial terrace	On Brown plate. See also topographic symbol 18.1.8	0.15 0.5 0.75 0.1
.16	Alluvial fan	Bibl. 94. Hand drawn to suit the individual land feature See also 18.1.24	0.15
.17	Landslips showing heel of slip and direction of movement	Tas. GS mentions problem in showing landslides true to scale on detailed specialist maps Bibl. 1 See also 18.1.18-.20	0.2 1.0
.18	Landslips	Numerous slips. May be used to show extensive area of slips on general purpose maps or distribution of small slips on large scale maps	0.15

(continued)

525

TABLE 23. (continued)

Symbol No.	Symbol	Description	Notes		Drafting Specifications (mm)
.19		Landslip	Large slip, showing scar and tongue of colluvium, and small slips		
.20		Landslip	Brown Plate. Large area — escarpment symbol and boundary on topoplate show limit of landslip scar; boundary and letter symbol (Qs) on geol. plate show area of landslip debris. Dot intensity decreasing towards base of landslip scar	For large-scale maps only. Because the features portrayed	
.21		Cliff	See also Bibl. 99	may have great	
.22		Cirque	Bibl. 94	diversity in size and	
.23		Moraine		shape, which can be shown, the symbols	
.24		Alluvial fan	Bibl. 94	are hand drawn	

526

.25	Sinkhole		See 16.1.12 and 13.1.14 to .17 Choice of symbol depends on size of feature	
.26	Astrobleme, impact structure, cryptoexplosion		If possible should be to scale Bibl. 172	
.27	Miscellaneous boundary		A boundary to be used where necessary for topographic units such as limit of Pleistocene glaciation, basalt wall, lava tunnel	

18.2 TOPOGRAPHIC SURVEY STATIONS, ELEVATIONS AND CONTOURS
Bibl. 13, 33, 47, 59, 60, 93, 97, 116, 131, 138, 151

.1	Trigonometrical station			
.2	Astronomical station			
.3	Elevation in metres, accurate		Datum to be specified: e.g. mean sea level, Derby	
.4	Elevation in metres, approximate			
.5	Contour in metres		Brown	
.6	Formline in metres		Same as contour, but broken line	
.7	Bathymetric contour in metres		Blue	

Source: From Bureau of Mineral Resources (BMR) (1978).

527

TABLE 24. Examples of Symbol-Free Signs (and Colors) Used by the Niedersächisches Landesamt für Bodenforschung, Showing Contours, Contacts, Thickness Contours, Profile Lines, Faults, and Dip and Strike

Signaturfarbe:

	schwarz	Schichtgrenze nachgewiesen/vermutet/im Untergrund
	rot	Schichtgrenze, geophysikalisch nachgewiesen
	schwarz	Faziesgrenze
	frei	Mächtigkeitslinien
	blau	Profillinien, Verlauf des Pofilschnitts
	schwarz (frei)	tektonische Störung, allgemein nachgewiesen/unter Bedeckung/vermutet
	schwarz (frei)	Flexur nachgewiesen/vermutet
	schwarz (frei)	Abschiebung (Zähnchen zeigen in die Einfallsrichtung der Störungsfläche), nachgewiesen/unter Bedeckung/vermutet
	schwarz (frei)	Aufschiebung (Zacken zeigen in die Einfallsrichtung der Störungsfläche), nachgewiesen/unter Bedeckung/vermutet
	schwarz (frei)	Blattverschiebung (Zähnchen zeigen in die Einfallsrichtung der Störungsfläche), nachgewiesen/unter Bedeckung/vermutet
	schwarz (frei)	Streichen und Einfallen der Schichtung (mit Angabe der Streichrichtung und des Einfallwinkels)
	schwarz (frei)	Streichen und Einfallen der Schichtung bei überkippter Lagerung (mit Angabe des Einfallwinkels)
	schwarz (frei)	Lagerung der Schichten horizontal/senkrecht (**saiger**)
	schwarz (frei)	Streichen und Einfallen der Schieferung (mit Angabe der Streichrichtung und des Einfallwinkels)

TABLE 24. (*continued*)

	schwarz (frei)	Streichen und Einfallen der Knickschieferung, Schubklüftung
	schwarz (frei)	Muldenachse nachgewiesen/vermutet
	schwarz (frei)	Sattelachse nachgewiesen/vermutet
30 · 30	schwarz (frei)	Einfallen der Sattel-/Muldenachse mit Angabe des Einfallwinkels
	schwarz (frei)	Rutschstreifen auf Harnischflächen
	schwarz	Pinge
	schwarz	Bruchfeld
	schwarz oder in Abhängigkeit von abgebauten Rohstoffen	Bergbau außer Betrieb; Grube, aufgelassen
	schwarz oder in Abhängigkeit von abgebauten Rohstoffen	Bergbau in Betrieb
	schwarz	geologisch wichtige Stelle (Naturdenkmal = ND)
26 ○	frei	Bohrung mit Erläuterungsnummer oder Symbol oder Farbe (Innenfeld) der tiefsten erreichten Schicht, mit Angabe der Endteufe
17 ⊗	frei	Entnahme für Proben (mit Nummer aus den Erläuterungen)
18	frei	Handschacht (mit Nummer aus den Erläuterungen
	frei	Stollenmundloch (evtl. mit Verlauf der Strecken)
19 *f*	frei	Fossilfundpunkt, allgemein (mit Nummer aus den Erläuterungen)

(*continued*)

TABLE 24. (*continued*)

	frei	limnische Fossilien
	frei	marine Fossilien
	frei	Mikrofossilien
	frei	Pflanzenreste i. a.
	frei	Knochen
	frei	Artefakt-Fundpunkt
	rot	Endmoränen-Hauptstreichrichtungen nachgewiesen/vermutet
	rot	Kames-Streichrichtung nachgewiesen/vermutet
	rot	Åsar nachgewiesen/vermutet
	rot	Stauchungsrichtung, glaziär
	rot	Gletscherschrammen
	rot	große Glaziärgeschiebe (Findlinge), ferntransportiert
	rot	Geschiebebestreuung
	braun	große Geschiebe aus dem Nahbereich
	rot	Drumlins

TABLE 24. (*continued*)

	grün	Doline
	braun	Pingo
	gelb	Schlatt (Windausblasungswanne)
	rot	Söll
	blau	Subrosionssenke
	blau	Umgrenzung von Salzstöcken in der Tiefe
	nach Alter	Abrißkante (von abgerutschten Schollen, Bergstürzen usw.)
	nach Alter	Abrutschmasse, Bergsturz (Innenfläche: Farbe des abgerutschten Gesteins und petrographische Signaturen)
	freie Wahl, nicht rot	Junge Erdrutschungen über bekanntem Untergrund
	rot	Scholle, glaziär verfrachtet (Innenfläche: Farbe und Signatur des verfrachteten Gesteins)
	grün	Geländestufe, durch fluviatile Erosion bedingt (z.B. Terrassenkante)
	rot	Geländestufe, durch Glazialerosion bedingt
	blau	Geländestufe, durch Abrasion bedingt (z.B. Kliff)
	grau	Rinne, fluviatil
	blau	Rinne, marin

(*continued*)

TABLE 24. (*continued*)

	rot	Rinne, glazial
	grün	Uferwall, fluviatil
	blau	Uferwall, marin (Strandwall)
	frei außer rot	Transportrichtung bei der Sedimentation i. a.; Transportrichtung bei Detraktionsbildungen, z.B. Muren
	rot	Transportrichtung bei der Sedimentation im glaziären Bereich
	rot	Sander-Schüttungs-Richtung
	frei	Richtung des äolischen Transports
	frei	Schuttkegel-Transportrichtung (dicke Seite der Fächerteile distal)
	frei	Schwemmkegel-Transportrichtung
	schwarz	Höhle
	blau	Quelle allgemein
	blau (oder in Abhängigkeit des Salzes	Mineralquelle
	blau (oder in Abhängigkeit der Temperaturen)	Quelle mit erhöhten Temperaturen (Therme)

Source: From Bender (1981)

TABLE 25. Examples of Map Symbols for the Main Groups of Hard Rocks Used by the Niedersächsisches Landesamt Für Bodenforschung

◇◇◇	/−b	Breccie	▽▽▽▽▽	/−ti	Kieselschiefer und kieselige Gesteine, allgemein
○○○○○	/−c	Konglomerat	▭	/−k	Kalkstein
· · ·	/−s	Sandstein, allgemein	▭	/−koo	Kalkoolith, Oolithkalkstein
• • •	/−gs	Grobsandstein	▭		Travertin
· · ·	/−ms	Mittelsandstein	▭	/−d	Dolomitstein
· · ·	/−fs	Feinsandstein	▭	/−m	Mergelstein
·· ··	/−u	Schluffstein, Siltstein	△ △	/−ah	Anhydritstein
≡≡	/−t	Tonstein	▽ ▽	/−y	Gipsstein
≡≡≡	/−ts	Tonschiefer	▭ ▭ ▭	/−na	Steinsalz
· ·	/−g	Grauwacke, Arkose, Feldspatsandstein (/−sf)	• • •	/−la	Laterit
· · ·	/−q	Quarzit	■	/−ko	Kohlegestein

■□ /−ez   Eisenerze i.a.

Source: From Bender (1981).

TABLE 26. Examples of Map Symbols for Igneous and Metamorphic Rocks Used by the Niedersächsisches Landesamt für Bodenforschung

Magmatite

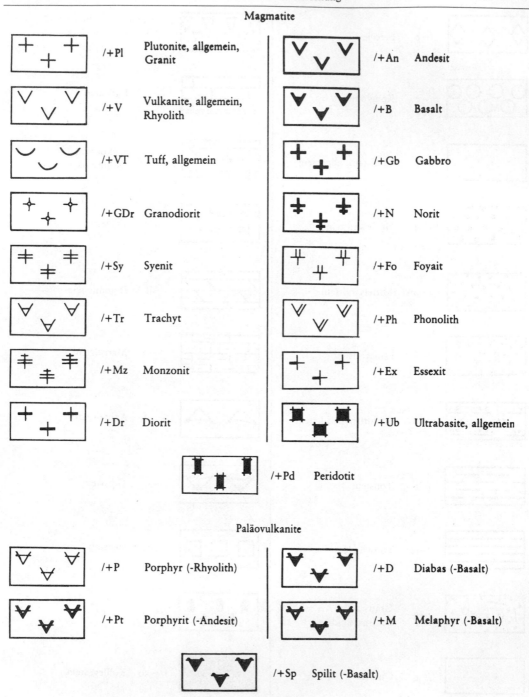

/+Pl	Plutonite, allgemein, Granit
/+V	Vulkanite, allgemein, Rhyolith
/+VT	Tuff, allgemein
/+GDr	Granodiorit
/+Sy	Syenit
/+Tr	Trachyt
/+Mz	Monzonit
/+Dr	Diorit
/+An	Andesit
/+B	Basalt
/+Gb	Gabbro
/+N	Norit
/+Fo	Foyait
/+Ph	Phonolith
/+Ex	Essexit
/+Ub	Ultrabasite, allgemein
/+Pd	Peridotit

Paläovulkanite

/+P	Porphyr (-Rhyolith)
/+Pt	Porphyrit (-Andesit)
/+D	Diabas (-Basalt)
/+M	Melaphyr (-Basalt)
/+Sp	Spilit (-Basalt)

TABLE 26. (*continued*)

Metamorphite
Orthogesteine

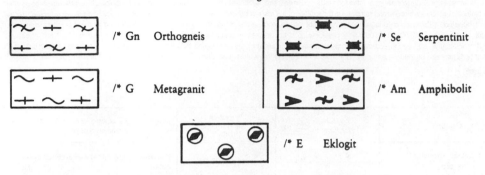

/* Gn   Orthogneis

/* Se   Serpentinit

/* G   Metagranit

/* Am   Amphibolit

/* E   Eklogit

Paragesteine
Regionalmetamorphose

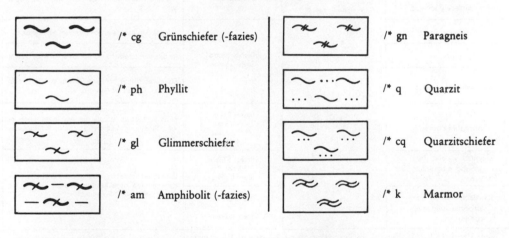

/* cg   Grünschiefer (-fazies)

/* gn   Paragneis

/* ph   Phyllit

/* q   Quarzit

/* gl   Glimmerschiefer

/* cq   Quarzitschiefer

/* am   Amphibolit (-fazies)

/* k   Marmor

Ultrametamorphose (auch Anatexis)

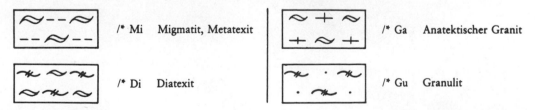

/* Mi   Migmatit, Metatexit

/* Ga   Anatektischer Granit

/* Di   Diatexit

/* Gu   Granulit

Kontaktmetamorphose

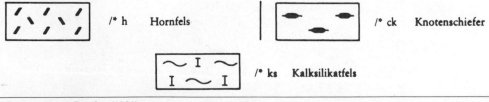

/* h   Hornfels

/* ck   Knotenschiefer

/* ks   Kalksilikatfels

Source: From Bender (1981).

TABLE 27. Map Symbols for Contacts Used by the U.S. Geological Survey, Showing Dips, Locations, and Inferred Positions Determined by Magnetic Survey

CONTACTS

Boundaries between geologic formations or other rock units.   Symbols should be combined to fit available space where practical.   Preferred phrasing when several types of contacts are mapped and combined in the explanation: *Dashed where approximately located; short dashed where inferred; dotted where concealed.*   Contact line symbols signify accuracy of location or character of exposure; only solid line contacts used for maps at scales smaller than 1:125 000 (1:250 000; 1:500 000; 1:1 000 000).   Generally solid line implies accuracy of placement within distance represented by 1 mm at scale of map.   If symbols give engineering accuracy of location of contact, standard used in mapping should be given in italics.   Coal and other economically important beds may also be used as contacts.   Line widths: All contact line widths generally 0.12 mm; 0.10 mm in congested areas.   Dip values: Show in ULCI–6 type

Contact	KJs	Template 8 Letter symbols U–8	Triangles indicate selected localities where contact was well exposed at time of mapping
Contact, showing dip	90 45	Template 37, 8	If known, show top side of vertical contact by single line and 90
Overturned contact, showing dip	65	Template 1(½), 37, 8	
Contact, approximately located		Dashes 3.5 mm Space 0.5 mm	Not surely located within distance represented by 1 mm at scale of map
Indefinite contact			Insufficient data to establish contact with certainity
Inferred contact		Dashes 1.5 mm Space 0.5 mm	No data to establish contact but contact must be present
Gradational contact			Continuous change from one lithology or rock type to another. Contact arbitrary
Concealed contact	(KJs)	Dashes 0.5 mm Space 0.5 mm Letter symbols U–8	Must be beneath mapped geologic units, water or ice. Symbols in parentheses indicate the concealed bedrock
Contact, located by ground magnetic survey		Dashes 5.5 and 1.5 mm Space 0.5 mm	Contacts determined by instrumentation or by other than conventional surface geologic methods may require special symbols for differentiation
Contact, located by airborne magnetic survey		Dashes 5.5 and 1.5 mm Space 0.5 mm	

Source: From U.S. Geological Survey, Branch of Technical Illustrations, Technical Standards Section, internal files.

TABLE 28. Map Symbols for Contours and Isopleths Used by the U.S. Geological Survey, Showing Structure Contours, Outcrop Points, Magnetic Contours, Flight Paths, Isoradioactivity Contours (Isograds), Gravity Contours (Isogals), Gravity Stations, Isopachs, and Mineral Isograds

CONTOURS AND ISOPLETHS

Generally printed in red or other contrasting color but may be shown in black where the basic geology and base map are simple.   Label and make every 5th contour heavier.   May be used for many kinds of geologic data.   In geophysical maps, give nature of contoured data in map title.   Line widths: Use 0.38 mm line width for heavy (index) contours and 0.20 mm for light (intermediate) contours.

DESCRIPTION	SYMBOL	NOTES	
Structure contours *Drawn on top (or base) of geologic horizon. Dashed where control is poor. Contour interval 20 feet. Arrow indicates direction of dip*	——2000—— —	Dashes 5 mm Space 0.5 mm Value Ul–8 or Ul–10 Template 37, 8	Structure contours not shown as concealed; may be omitted in areas of no information. Arrows used only where index contours fail to show dip. Contour interval may be in meters, at authors discretion
Outcrop point *Used for structural control*	x	U–10–l/c	
Magnetic contours *Showing total intensity magnetic field of the earth in gammas relative to arbitrary datum. Hachured to indicate closed areas of lower magnetic intensity, dashed where data are incomplete. Contour interval 20 gammas*	——2000—— — 	Dashes 5 mm Space 0.5 mm Value Ul–8 or Ul–10 Hachures: line width same as light contours Length of tick: 1.2 mm	Show at least two hachures on small closed contours; otherwise use 20 mm space between hachures. Pattern with 2=12% screen, instead of hachures, when numerous closed contours are shown; check with Division
Maximum or minimum intensity *Location measured within closed high or closed low*	X 2864	X ULC–10 cap Value ULCl–8 at 45°	
Flight Path *Showing location and spacing of data*	— — — —	Line width 0.12 mm Length 6 mm	Space as shown by author
Isoradioactivity contours (or isograds) *Interval 50 counts per second (airborne survey). Interval 50 microroentgens per hour (ground surveys)*	————		
Gravity contours (or Isogals) *Dashed where control is poor. Contour interval 1 milligal*	—— — — —		
Gravity station and number	. 635	Template 4 Number U–8	
Isopachs *Dashed where control is poor. Interval 3 meters*	—— — — —		
Mineral isograds *Metamorphic zones indicated by mineral names*	SILLIMANITE STAUROLITE	Names U–8 caps	

Source: From U.S. Geological Survey, Branch of Technical Illustrations, Technical Standards Section, internal files.

TABLE 29. Examples of Map Symbols for Planar Features Used by the U.S. Geological Survey, Showing Strike and Dip of Beds, Foliation or Schistosity, and Cleavage

PLANAR FEATURES

Planar symbols (strike and dip of beds, foliation or shistosity, and cleavage) can be combined with linear symbols to record data observed at same locality by superimposing symbols. Coexisting planar symbols are shown intersecting at point of observation. All combinations of planar and linear symbols used on map need not be shown in explanation. A statement "Planar and linear symbols may be combined" beneath PLANAR FEATURES AND LINEAR FEATURES in explanation is adequate. Examples of combined planar and linear features and coexisting planar features may be shown in explanation. Line width: Use 0.12 mm line width for all symbols. Dip values: Set in ULCI–6 type.

ATTITUDE OF BEDS

DESCRIPTION	SYMBOL	NOTES	
Strike and dip of beds	45	Template 11L, 11W	
Strike and direction of dip of beds			
Approximate strike and direction of dip of beds		Template 11L, 11W, (opaque center)	
Strike and dip of beds *Top of beds known from sedimentary features*	30	Template 11L, 11W, 4	Used only on maps where the top of beds is not always known
Strike and dip of overturned beds	65	Template 11L, 1(½), 37	
Strike and dip of overturned beds *Top of beds known*	65	Template 11L, (1½), 37, 4	
Strike of vertical beds		Template 11L, 11W	
Strike of vertical beds *Dot indicates top of beds*		Template 11L, 11W, 4	
Component of dip *Dot marks point of observation*		Template 11L, 4, 27	Do not use if symbols for lineation in metamorphic rocks are used on map
Horizontal beds	⊕	Template 35, 35, 14, (opaque ends)	
Strike and dip of beds and plunge of slick-ensides	40 60	Template 11L, 11W, 29L, 5	
Crumpled, plicated, crenulated, or undulatory beds and average dip	55	Template 15, 15, 15, 15(side)	

FOLIATION OR SHISTOSITY

Strike and dip of foliation	20	Template 11L, 5(½)	
Strike and direction of dip of foliation			
Strike of vertical foliation *Relationship of foliation (or shistosity) to bedding not shown in outcrop*		Template 11L, 5	
Horizontal foliation		Template 15, 17 (4 times)	
Strike and dip of foliation and parallel bedding	10	Template 11L, 22W, 25(½)	

TABLE 29. (*continued*)

FOLIATION OR SHISTOSITY (CON'T)		
DESCRIPTION	SYMBOL	NOTES
Strike of vertical foliation and parallel bedding	✦	Template 11L, 29L, 15
Strike and dip of foliation and parallel overturned bedding	⟱ 40	Template 11L, 1(½), 35, 22L, 8
Horizontal foliation and bedding	⊕	Template 29L, 29L, 33, 25
CLEAVAGE		
Strike and dip of cleavage	15	Template 11L, 11W, 11W
Strike of vertical cleavage	⊢—⊣	Template 11L, 11W, 11W
Horizontal cleavage	⊣⊢	Template 29L, 29L, 22W (4 times)
Inclined	—— —▲— ═▲═	11, 25(½) 11, 5(½) 19(⅛), 5(½)
Vertical	—⊦⊦— —⬦— ═⬦═	11, 25 11, 5 19(⅛), 5(½)
Horizontal	╫	19(⅛) twice

Contrasting symbols can be used to distinguish between different kinds of planar structures (slip cleavage, compositional layering, flow structure). Type of planar structure should be specified in explanation

Source: From U.S. Geological Survey, Branch of Technical Illustrations, Technical Standards Section, internal files.

TABLE 30. Examples of Map Symbols for Linear Features and Joints Used by the U.S. Geological Survey, Showing Lineations, Foliation, Slip Cleavage, and Strike and Dip of Joints

LINEAR FEATURES			
May be combined with the above planar symbols as shown. Symbols are joined at point of observation.			
Bearing of plunge of lineation		Template 11L, 8	
Vertical lineation		Template 32, 32	Use open symbol in combination with line symbols
Horizontal lineation		Template 11L, 8, 8	
Strike and dip of foliation and plunge of lineation		Template 11L, 10, 29L, 23	
Vertical foliation showing horizontal lineation		Template 11L, 8, 8, 5	
Strike and dip of foliation showing horizontal lineation		Template 11L, 8, 8, 5(½)	
Strike and dip of beds and plunge of lineation		Template 11L, 11W, 29L, 23	
Vertical foliation and vertical lineation		Template 5, 11L	
Strike of vertical foliation showing plunge of lineation		Template 11L, 8, 5	
Vertical beds showing horizontal lineation		Template 11L, 22L, 8, 8	
Horizontal beds, showing trend of horizontal lineation		Template 11L, 22L, 14, 8, 8	
Vertical beds showing plunge of lineation		Template 11L, 22L, 8	
Approximate strike of folded beds showing plunge of fold axes		Template 15, 15, 15, 22W	
Attitude of foliation and overturned beds, strikes parallel but dips differ		Template 11L, 1(½), 29L, 22W, 8	
Double lineation		Template 11L, 10, 29L, 29L, 23, 23	
Strike and dip of beds intersecting slip cleavage		Template 11L, 11W, 11W, 11L, 11W	
Strike and dip of beds intersecting slip cleavage		Template 11L, 11W (3 times)	
JOINTS			
Open symbols may be contrasted with closed symbols to separate unmineralized and mineralized joints.			
Strike and dip of joints		Template 11L, 15(½)	
Strike and direction of dip of joints			
Strike of vertical joints		Template 11L, 15	
Horizontal joints		Template 15, 11L, 11L	
Strikes and dips of multiple joints		Template 11L, (3 times), 15(½) (3 times)	

Source: From U.S. Geological Survey, Branch of Technical Illustrations, Technical Standards Section, internal files.

TABLE 31. Examples of Map Symbols for Faults Used by the U.S. Geological Survey, Showing Faults, Lineaments, Shear Zones, Fault Breccia, and Subsurface Faults

FAULTS

Same line conventions used for faults as for contacts. Preferred phrasing when several line conventions are used for faults and combined in the explanation: *Dashed where approximately located; short dashed where inferred; dotted where concealed; queried where doubtful.* U, *upthrown side;* D, *downthrown side.* Dips shown where observed or known. Line widths: Fault line widths generally 0.38 mm; 0.3 mm on complex maps. Relative importance of faults may be shown by different widths of lines and suitable explanations. Dip values: Show in ULCI–6 type.

DESCRIPTION	SYMBOL	NOTES	
Fault	OWL CREEK FAULT	Named faults U–6 caps; use ULC if congested	
Fault, showing dip	90 65	Template 37, 8	
Fault, approximately located	— — — — —	Dashes 3.5 mm Space 0.5 mm	Not surely located within distance represented by 1 mm at scale of map
Inferred fault	– – – – – – –	Dashes 1.5 mm Space 0.5 mm	Evidence for fault only indirect
Probable fault	– – –?– – – –?– – – –?–	Dashes 1.5 mm Space 0.5 mm Space for ? 2.5 mm ? U–9	Queries, spaced three or more dashes apart, indicate uncertainty of existence, not location. Probable is more definite than doubtful
Doubtful fault			
Concealed fault	·················	Dots 0.5 mm Space 0.5 mm	Must be concealed by overlying mapped geologic unit, water or ice
Hypothetical fault	·····?······?······?····	Dots 0.5 mm Space 0.5 mm ? U–9	Existence from indirect geologic evidence; could be explained by causes other than faulting
Fault, located by ground magnetic survey	— – — – — –	Dashes 5.5 mm and 1.5 mm Space 0.5 mm	
Fault, located by airborne magnetic survey	— - — - — -	Dashes 5.5 mm and 1.5 mm Space 0.5 mm	
Fault, or lineament from aerial photographs *Not checked or identified on ground*	— · — · — ·	Dashes 5.5 mm and 0.5 mm Space 0.5 mm	
Lineament	▬▬▬▬▬	Line width 0.3 mm Names U–8	Used on small-scale tectonic maps. Add lineament name where possible
Fault *Showing bearing and plunge of grooves, striations, or slickensides*	60	Template 29L, 5	Plunge measured in vertical plane. Identify type of evidence observed in italic statement
Fault *Showing dip and amount of displacement in meters.* U, *upthrown side,* D, *downthrown side*	80 D U	Template 37, 8 U & D U–6	High-angle, used in combination with dip arrow to indicate apparent normal or reverse movement. Dip is in italic. Displacement is in upright.

(continued)

TABLE 31. (*continued*)

DESCRIPTION	SYMBOL	NOTES
Fault *Bar and ball on down-* *thrown side*		Template 22W, 4	Generally used where space does not allow U and D symbols without confusion; do not use bar and ball and U/D on same map
Fault *Showing relative hori-* *zontal movement*		Template 11L, 39, 11L, 39	
Fault *Showing bearing and* *plunge of slickensides* *on fault plane. D,* *downthrown side*	D 60 Normal D 60 Reverse	Template 29L, 5 D U–6	
Normal fault *Hachures on apparently* *downthrown side*		Template 11 (L, spacing; W, tick)	Used on tectonic maps or where space does not permit use of U and D
Reverse fault *R, upthrown side*	œ	R U–6	Angle of dip originally greater than 45° but precise value indeterminate. Hanging wall believed to have moved upward with respect to footwall
Thrust fault *T, upper plate*	45	Template 37, 8 T U–6	Angle of dip originally less than 45°. Dip of fault, where known, shown by barbed arrow
Thrust fault *Sawteeth on upper* *plate*		Template 32	Symbol emphasizes fault; spacing of teeth may separate thrust faulting of different ages. May be limited to major thrust faults. Sawteeth may be spaced up to 13 mm apart on long thrust faults
Overturned thrust fault *Sawteeth in direction* *of dip; bar on side* *of tectonically* *higher plate*		Template 31, extend ends of triangle to fault	
Fault (shear or mylonite) zone, showing dip	55	Line width 0.12 mm Template 37, 8	Show relative movement by U and D or arrows
Fault breccia		Line width 0.12 mm Template 10 or Pattern 401	Extent may be outlined by faults or shown only where observed. Used as overprint for broad areas of fault breccia
Fault, intruded by dike		Template 35, 35	Use on small-scale black-and-white map or for narrow dike. On colored maps show dike in color and fault movement by U and D
Fault, intruded by dike		Template 35, 35 Pattern 4 = 29%	Use on large-scale black-and-white map for dike of sufficient width to be mapped. Former location of fault shown. Dikes usually shown in color
Subsurface fault		Dashing same as regular faults	Show in same color as structure contours where contours are offset along a dipping fault

Source: From U.S. Geological Survey, Branch of Technical Illustrations, Technical Standards Section, internal files.

TABLE 32. Examples of Map Symbols for Folds Used by the U.S. Geological Survey,
Showing Anticlines, Domes, and Antiforms

FOLDS

Same line conventions used for folds as for contacts and faults.   Preferred phrasing when more than one line convention used for folds: *Dashed where approximately located; short dashed where inferred; dotted where concealed; queried where doubtful.*   Line widths: Fold line widths 0.26 mm; 0.20 mm may be used if folds are congested.   Dip values: Show in ULCI–6 type.

ANTICLINES

DESCRIPTION	SYMBOL		NOTES
Anticline *Showing trace of crestal plane. Dashed where approximately located*		Template 11L, 8, 8	On detailed geologic maps of asymmetric folding and high relief, trace of axial surface may be shown
Anticline *Showing trace of crestal plane and direction of plunge*		Template 31, 11L, 8, 8	
Anticline *Showing trace of crestal plane and plunge*			
Asymmetric anticline *Showing trace of crestal plane and plunge. Short arrow indicates steeper limb*		Template 11L, 8, 8, 31	
Asymmetric anticline *Showing dip of limbs and plunge*			
Overturned anticline *Showing direction of dip of limbs and plunge*		Template 14 (½), 37, 37, 8, 8, 31	
Inferred anticline or **Probable anticline**		Dashes 1.5 mm Space 0.5 mm	Use inferred or probable, not both. Based on indirect geologic evidence; location probably not within distance represented by 1 mm at scale of map
Doubtful anticline		Dashes 1.5 mm Space 0.5 mm ? U–9	Queries indicate doubt of existence of anticline from available data; location may also be in doubt
Concealed anticline		Dot 0.5 mm Space 0.5 mm	Must be beneath a mapped geologic unit or covered by water or ice. Not shown where extension of known anticline is obvious
Dome		Template 35, 35, 23 (4 times) Line width 0.12 mm	Generally used on small-scale tectonic maps
Inverted anticline		Template 1 (½), 1 (½), 37, 37, 8, 8	Beds inverted near trough
Antiform *Drawn on foliation, cleavage or bedding*		Template 29L, 27, 27	Convex upward; structure in metamorphic rocks or in bedded rocks where tops are not known

Source: From U.S. Geological Survey, Branch of Technical Illustrations, Technical Standards Section, internal files.

TABLE 33. Examples of Map Symbols for Fossils Used by the U.S. Geological Survey

CARTOGRAPHIC TECHNICAL STANDARDS

Replaces T. S. P.		Subject: GEOLOGIC SYMBOLS	T. S. Paper	4.03.1
Dated		Fossil Symbols	Effective	4/28/80

The fossil symbols listed below are available in two sizes on wax-backed material for geologic reports. The two sizes allow flexibility in illustrating both thick- and thin-bedded sections. These symbols should only be used with the consent of the report and/or Geologic Map Editor.

Symbols adapted from the Standard Legend, Bataafse Internationale Petroleum Maatschappij, N.V.			
FOSSIL NAME	SYMBOLS	FOSSIL NAME	SYMBOLS
Algae		Ostracods	
Ammonites		Plant remains	
Belemnites		Spicules	
Brachiopods		Sporomorphs	
Bryozoa		Stromatoporoids	
Charophytes		Trilobites	
Conodonts		Vertebrates	
Corals		Wood, silicified	
Crinoids		Marine fossils	
Echinoids		Brackish-water fossils	
Fish remains		Fresh-water fossils	
Fish scales		Fossils sparse	
Foraminifera in general		Fossils abundant	
Foraminifera, larger		Symbols other than those adapted from the standard legend but in common use by USGS	
Foraminifera,smaller,benthonic		General microfossils	
Foraminifera,smaller,pelagic		Calcareous microfossils	
Fossils in general		Silicoflagellates	
Gastropods		Diatoms	
Graptolites		Dinoflagellates	
Lamellibranchs (Pelecypods)			

Source: From U.S. Geological Survey, Branch of Technical Illustrations, Technical Standards Section, internal files.

TABLE 34. Examples of Map Symbols for Monoclines and Minor Fold Axes Used by the U.S. Geological Survey, Showing Monoclines, Anticlinal and Synclinal Bends, and Fold (Major and Minor) Axes

MONOCLINES		
May be classified as inferred, probable, doubtful, or concealed by same line conventions used for anticlines and synclines. Line widths: Make all line widths 0.26 mm ‹		
DESCRIPTION	SYMBOL	NOTES
Monocline *Showing trace and direction of plunge. Dashed where approximately located*		Template 11L, 8, 8, 31
Anticlinal bend *Showing trace and direction of plunge. Dashed where approximately located*	A	Template 11L, 8, 31 A U–6
Synclinal bend *Showing trace and direction of plunge. Dashed where approximately located*	S	Template 29L, 8, 31 S U–6

(Notes spanning anticlinal and synclinal bends: Use on large-scale, detailed maps where anticlinal and synclinal bends diverge sufficiently to be mapped)

MINOR FOLD AXES		
Line widths: Make all line widths 0.12 mm		
Minor anticline *Showing plunge*	20	Template 11L, 1(½), 32
Minor syncline *Showing plunge*	50	Template 11L, 1(½), 32
Minor fold axis *Showing plunge*	FA 15	Template 11L, 32 FA U–6
Minor fold axis, horizontal	FA	Template 11L, 32, 32
Minor folds *Showing plunge of axes*	25	Template 35, 8

(Notes for minor anticline, syncline, fold axis: Plunge measured in vertical plane)

(Notes for minor folds: Used where beds are too tightly folded to show axes of individual folds separately. Used to indicate sense of observed folds)

Source: From U.S. Geological Survey, Branch of Technical Illustrations, Technical Standards Section, internal files.

TABLE 35. Examples of Map Patterns for Columnar Sections, Lithologic Cross-Sections, and Other Illustrations Used by the U.S. Geological Survey

BRANCH OF TECHNICAL ILLUSTRATIONS
TECHNICAL STANDARDS SECTION

Replaces T.S. Paper		Effective Date	11/1/71	T.S. Paper No.	12.02.3
Subject	SECTIONS - Sedimentary lithologic patterns for columnar sections				

These patterns are generally accepted for columnar sections. Use the definitions as guidelines for selecting patterns for lithologic cross sections and other illustrations.

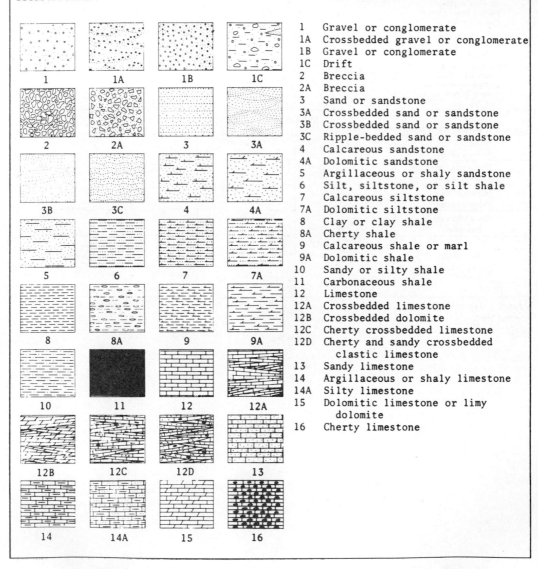

1   Gravel or conglomerate
1A Crossbedded gravel or conglomerate
1B Gravel or conglomerate
1C Drift
2   Breccia
2A Breccia
3   Sand or sandstone
3A Crossbedded sand or sandstone
3B Crossbedded sand or sandstone
3C Ripple-bedded sand or sandstone
4   Calcareous sandstone
4A Dolomitic sandstone
5   Argillaceous or shaly sandstone
6   Silt, siltstone, or silt shale
7   Calcareous siltstone
7A Dolomitic siltstone
8   Clay or clay shale
8A Cherty shale
9   Calcareous shale or marl
9A Dolomitic shale
10 Sandy or silty shale
11 Carbonaceous shale
12 Limestone
12A Crossbedded limestone
12B Crossbedded dolomite
12C Cherty crossbedded limestone
12D Cherty and sandy crossbedded
       clastic limestone
13 Sandy limestone
14 Argillaceous or shaly limestone
14A Silty limestone
15 Dolomitic limestone or limy
       dolomite
16 Cherty limestone

TABLE 35. (*continued*)

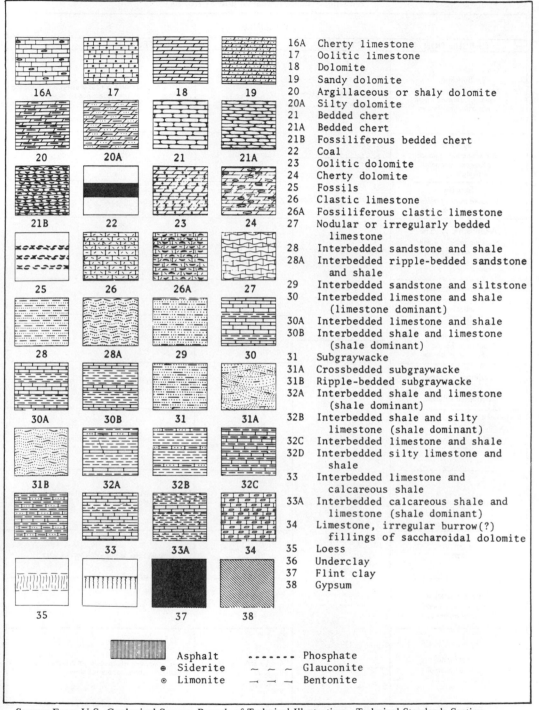

16A	Cherty limestone	
17	Oolitic limestone	
18	Dolomite	
19	Sandy dolomite	
20	Argillaceous or shaly dolomite	
20A	Silty dolomite	
21	Bedded chert	
21A	Bedded chert	
21B	Fossiliferous bedded chert	
22	Coal	
23	Oolitic dolomite	
24	Cherty dolomite	
25	Fossils	
26	Clastic limestone	
26A	Fossiliferous clastic limestone	
27	Nodular or irregularly bedded limestone	
28	Interbedded sandstone and shale	
28A	Interbedded ripple-bedded sandstone and shale	
29	Interbedded sandstone and siltstone	
30	Interbedded limestone and shale (limestone dominant)	
30A	Interbedded limestone and shale	
30B	Interbedded shale and limestone (shale dominant)	
31	Subgraywacke	
31A	Crossbedded subgraywacke	
31B	Ripple-bedded subgraywacke	
32A	Interbedded shale and limestone (shale dominant)	
32B	Interbedded shale and silty limestone (shale dominant)	
32C	Interbedded limestone and shale	
32D	Interbedded silty limestone and shale	
33	Interbedded limestone and calcareous shale	
33A	Interbedded calcareous shale and limestone (shale dominant)	
34	Limestone, irregular burrow(?) fillings of saccharoidal dolomite	
35	Loess	
36	Underclay	
37	Flint clay	
38	Gypsum	

Asphalt	⸻ Phosphate
⊕ Siderite	∼ ∼ ∼ Glauconite
⊙ Limonite	⊣ ⊣ ⊣ Bentonite

Source: From U.S. Geological Survey, Branch of Technical Illustrations, Technical Standards Section, internal files.

TABLE 36. Examples of Map Symbols for Synclines Used by the U.S. Geological Survey, Showing Synclines, Basins, and Synforms

SYNCLINES		
DESCRIPTION	**SYMBOL**	**NOTES**
Syncline *Showing trace of trough plane. Dashed where approximately located*		Template 11L, 8, 8 Dashes 3.5 mm Space 0.5 mm	On detailed geologic maps in areas of asymmetric folding and high relief, trace of axial surface may be shown
Syncline *Showing trace of trough plane and direction of plunge*		Template 31L, 11L, 8, 8	
Syncline *Showing trace of trough plane and plunge*			
Asymmetric syncline *Showing trace of trough plane and plunge. Short arrow indicates steeper limb*		Template 11L, 8, 8, 31	
Asymmetric syncline *Showing dip of limbs and plunge*			
Overturned syncline *Showing direction of dip of limbs and direction of plunge*		Template 14(½), 37, 37, 8, 8, 31	
Inferred syncline or Probable syncline		Dashes 1.5 mm Space 0.5 mm	Based on indirect geologic evidence. Location probably not within distance represented by 1 mm at scale of map
Doubtful syncline		Dashes 1.5 mm Space 0.5 mm Space for ? 2.5 mm ? U–9	Queries indicate doubt of existence
Concealed syncline		Dot 0.5 mm Space 0.5 mm	Must be beneath mapped geologic unit or covered with water or ice. Not shown where extension of known syncline is obvious
Basin		Line width 0.12 mm Template 35, 35, 23 (4 times)	Generally used on small-scale tectonic maps
Inverted syncline *Arrows show direction of dip of limbs*		Template 1(½), 1(½), 37, 37, 8, 8	Beds inverted near crest
Synform *Drawn on foliation, cleavage or bedding*		Template 11L, 27, 27	Convex downward: structure in metamorphic rocks or in bedded rocks where tops are not known

Source: From U.S. Geological Survey, Branch of Technical Illustrations, Technical Standards Section, internal files.

TABLE 37. Examples of Map Symbols for Underground Workings and Exploration Used by the U.S. Geological Survey, Showing Shafts, Raises, Ore Shuts, Stopes, Drift Cribbing, Drill Holes, and Oil and Gas Well Locations

UNDERGROUND WORKINGS AND EXPLORATION		
Symbols drawn to scale on large-scale maps.		
DESCRIPTION	SYMBOL	NOTES
Shaft at surface	◪	
Shaft, above and below surface	⊠	
Bottom of shaft	⊠	Show bottom of sump by note on map of lower level
Inclined workings, above and below level *Chevrons point down*	≫≫ ▶ ≫≫	Spacing of chevrons may indicate steepness; place at regular intervals: 1, 2, 3, or 2, 4, 6, etc. meters
Winze or head of raise	◨	
Raise or winze extending through level	⊠	
Raise or foot of winze	⊠	
Ore chute	―□―	
Stope		Can be explained by note, "Stoped above" or "Stoped below"
Elevation of roof or back	―▽―	
Elevation of floor or sill	―△―	
Lagging or cribbing along drift		
Saved or otherwise inaccessible workings		
Drill hole	∘	Give inclination of hole + or − in degrees in note and show vertical projection of bottom of hole to map
OIL AND GAS WELLS		
Symbols for wells drilled for oil and gas are made up of seven compatible basic symbols which may be superimposed as necessary to show reported conditions		
Drilling well or Well location	∘	
Dry hole or Abandoned well	◇	
Gas well	☆	
Oil well	●	
Show of gas	✧	
Show of oil	◈	

(continued)

TABLE 37. (*continued*)

DESCRIPTION	SYMBOL	NOTES	
Shut in well			
Well *Showing vertical projection of bottom of hole, total depth, and surface altitude in meters* 5000 ○ - - - TD 2000		Type U	
Dry hole *Showing formation and altitude at surface, formation at bottom of hole, and total depth in meters* ◇ Km 2809 Kd 4996		Type U	

Source: From U.S. Geological Survey, Branch of Technical Illustrations, Technical Standards Section, internal files.

TABLE 38. Examples of Map Symbols for Veins, Ore, Wall-rock Alteration, and Dikes
(Including Surface Opening and Exploration for Large- and Small-Scale Maps)
Used by the U.S. Geological Survey

VEINS, ORE, WALLROCK ALTERATION, AND DIKES		
Shown in color, generally red, only where necessary to differentiate types and grade.		
DESCRIPTION	SYMBOL	NOTES
Vein, showing dip	 50	Line width 0.38 mm Template 3 (dots) Pattern 406 Template 37, 8 Value ULCI–6	Give mineralogy and grade of mineralization in percent metal or oxide, or weight per ton by notes. Can also be shown in solid
Ore body		Line width 0.12 mm Pattern 406	
Mineralized stringers or veinlets		Line width 0.26 mm Template 4	
Altered wallrock *Showing intensity of alteration by con- centration of dots*		Pattern by hand	
Dike		Line width 0.38 mm X's UBC–9–I/c	May be shown in color without X's where essential to distinguish different rock types
ORE IN SEDIMENTARY ROCKS AND SEDIMENTARY FEATURES CONTROLLING ORE DEPOSITION		
Strike of roll *Showing geometric configuration in cross section*		Template 11L, 14(½), 30, 30, 23, 23	Explain configuration by note
Direction of plunge of cross-stratification in sandstone *Showing direction of flow of depositing stream*		Template 32 (fill in half)	
Fossil Log		Template 11	
Lineation trend		Template 11L, 8 (fill in half)	
Festoon trend		Template 11L, 1(¼), 14(½)	
SURFACE OPENINGS AND EXPLORATION—LARGE-SCALE MAPS		
Symbols drawn to scale on large-scale maps. Line width 0.12 m		
Vertical shaft			
Inclined shaft			
Portal or adit			
Portal and opencut			
Trench			
Prospect pit or open cut			
Mine dump			

(continued)

TABLE 38. (*continued*)

DESCRIPTION	SYMBOL	NOTES	
Drill hole *Showing name and number*	○ PAHUTE 2	Type U caps	
Drill hole *No geologic data available*	○ ND	Type U caps	
Diamond-drill hole	○ DDH	Type UBCI caps	
Drill hole, low-grade ore	φ	Give definition of low and high grade in explanation	
Drill hole, high-grade ore	♦		
Drill hole, inclined *Showing bearing and inclination; surface position and altitude; vertical projection of bedrock surface, bottom of hole, and thickness of overburden; and length of hole, in meters*	45° ○ 3620 45	Type Numbers U	Combine drill-hole collar symbols as required with vertical projection to map

SURFACE OPENINGS AND EXPLORATION—SMALL-SCALE MAPS

Symbols not drawn to scale on map. Vary size of symbols with density of data.

Shaft	◪		
Abandoned shaft	◪ A	Type U	
Inclined shaft	◪		
Tunnel, adit, or slope	—<		
Inaccessible tunnel, adit, or slope	—<<		
Strip mine		Pattern 226 at 45°	Pattern shows stripped area
Trench	>—<		
Prospect pit or outcrop	x		
Sand, gravel, clay, or placer pit	✕		
Abandoned sand, gravel, clay, or placer pit	✕		
Mine, quarry, glory hole, or open pit	⋇		
Abandoned mine, quarry, glory hole, or open pit	⋇		

Source: From U.S. Geological Survey, Branch of Technical Illustrations, Technical Standards Section, internal files.

MARINE EXPLORATION GEOCHEMISTRY

Geochemistry has been used successfully on land for many years in the exploration for deposits of metals, nonmetals, and hydrocarbons. More recently it has been used for metal, non-metal, and hydrocarbon exploration in the marine environment. *Marine exploration geochemistry* includes exploration techniques based on the systematic measurement and evaluation of chemical properties of natural components in the ocean environment: rock, sediment, brine, water, gas, and life forms. Major ($>1\%$), minor ($<1\%$ to $>0.1\%$), and trace ($<0.1\%$) amounts of an element or assemblage of elements and/or molecules and compounds are the quantities most often measured and assessed with the purpose of highlighting materials and areas or zones with greater than normal contents of the factor or factors being measured—i.e., the so-called *geochemical anomalies*. These anomalies may be related to metal or nonmetal mineralization or to hydrocarbon (petroleum and/or natural gas) accumulation at or beneath (shallow or deep) the hydrosphere-lithosphere interface. Whether the mineralization or accumulation targeted by geochemical and other techniques can be worked economically depends on many factors, among which the most important may be the extent and grade of the deposit (for the major component and important secondary recoverable products), the technology available or to be developed for the actual mining operation, the process available for extracting the commodities sought and its potential for coping with a nonuniform ore supply if such conditions should arise, the depth of water and bottom conditions, the size and uniformity of size of materials to be mined, the distance to processing and extraction facility, and the weather conditions in the mining area. The ultimate decision as to whether a deposit will be exploited is strictly economic: will the deposit yield a profit at an acceptable rate for a long enough period of time for the investors?

During the 1970s, the research in marine mineral prospecting for hard minerals has emphasized the development of in situ real-time analytical techniques, the evaluation of nontraditional sample types (seaweed, Sharp and Bolviken, 1979; mineral suspensates, Siegel and Pierce, 1978, 1982) in the offshore exploration for onshore mineralization, paleoreconstructions of Pleistocene drainage systems, and the extraction of dissolved species in sea water by physical adsorption and selective membrane filtration. Much of the impetus for such research has resulted from increased prices on world markets, the stringency and costs of environmental safeguards on land-based mining, processing and transportation requirements, and the possible interruption of supply because of political decisions. Any venture in the marine environment or on land requires that the commodity being mined be competitive on the world market (existent and projected) unless a government decides that it is in the national interest to provide a strategic subsidy in all phases of a research and development prospecting program, especially when a high-risk venture is proposed.

Marine mining (q.v.), founded on sound geological work, and thereafter abetted by geophysical (see *Exploration Geophysics*) and geochemical (see *Exploration Geochemistry*) techniques, has been a healthy industry for many years. More than 100 companies worldwide are currently engaged in various aspects of exploiting mineral resources from the seafloor. Great stimulus was given to marine exploration and exploitation with the advent of offshore petroleum production (see *Petroleum Geology*) which in 1984 supplied more than 38% of the western nations' oil needs (p.c. American Petroleum Institute) and may supply as much as 50% of such needs by the turn of the century or shortly thereafter. Efforts in the area of marine exploration have increased greatly during recent years as the specter of population growth, mineral depletion on land, and the increasing dependence of a country on extranational sources of commodities have become evident to even the most casual observers and planners for the continued effort to maintain and extend the quality of human life and a comfortable existence on our planet.

McKelvey and Wang (1970) presented a series of maps showing the locations of existing marine mining projects and areas for future exploration, and the role of geochemistry in the expansion of marine exploration has become more important. The reasons for the expansion are several:

1. Private companies or consortia view mining in the marine environment as a profit-making venture. Indeed, at the beginning of the 1970s, successful operations were being run (and had been run for many years) for the products listed in Table 1 (at the commodity prices in effect at that time).
2. Individual nation users, and importers, wish to lessen their dependence on a supplier or a limited group of suppliers.
3. Individual nations or groups of nations wish to develop an export commodity capability. The existing marine geographic zones being mined are in the relatively shallow marine environment, and these environments and those with rather immediate or long-range potential for a mining operation are given in Table 2. The commodities being mined or that could be mined when technological advances are made

TABLE 1. Selected Commodities Extracted from Seawater, Beaches and Nearshore Placers,
Seafloor and Subfloor

Subfloor	Seafloor	Beaches and Nearshore Placers	Seawater
Oil and gas	Sand and gravel	Tin (Cassiterite)	Magnesium
Sulfur	Shell and aragonite mud	Diamonds	Magnesium compounds
Salt	Precious coral	Gold	Salt
Barite	Phosphorite (U)[a]	Platinum	Water
Iron ore		Ilmenite	Bromine
Coal		Rutile	Iodine
Potash		Zircon	Heavy Water
		Monazite	
		Magnetite	
		Chromite	
		Columbite-Tantalite	
		Scheelite	
		Wolframite	

Note: The aggregate value in 1979 U.S. dollars of these commodities exceeded $100 billion.
[a]Phosphorite contains uranium that may be extractable.

and when the economic climate is appropriate are also give in Table 2.

Geochemistry has become increasingly important in marine exploration expansion by mining interests, for several reasons. One is the experience of geochemists in recognizing the physical-chemical-biological conditions necessary for the formation of a given commodity or group of commodities and hence the ability to indicate from existing knowledge and/or geochemical analyses potentially propitious areas to be studied to find a given deposit. The geochemist has experience in working with a great variety of natural materials and is thus able to match analytical method with sample to obtain the accurate and precise results on chemical composition that are important to decision making in the mining industry. In addition, geochemists who work on marine environment projects may be able to enhance the success of an exploration program by recognizing the representativeness of a sample with respect to a relatively large area. The effectiveness in mapping the extent of deep marine environment surficial deposits has increased tremendously during the 1970s with notable advances in precision depth recording and in deep-sea bottom photography and television capabilities. Thus, with bottom samples taken by any one of a number of free-fall sampling devices, coring devices, dredges, etc., the geochemist can obtain samples of the seafloor to analyze their potential yields of commodities and then, with deep-sea observation capabilities, determine if the materials may be representative of much greater areas than that represented by the sample alone.

Indeed, consideration of the economic potential of the marine environment from a mining point of view indicates adequate future supplies in terms of commodities required for human survival; this, because only about 3% of the ocean floor has been extensively and satisfactorily surveyed from a geological-geochemical aspect, and this 3% has yielded significant mineral wealth.

Shallow Marine Environment

Smith (1986) gives a review of geochemical exploration in the shallow marine environment. Although the exploration for placer and other nearshore deposits depends in great part on paleoreconstructions of Pleistocene environments when sea level was more than 100 meters lower than it is now, and on mineralogy and sedimentological processes (Kogan, 1977; Owen, 1980), geochemistry may have an important future role in the exploration for them. For example, Owen (1978) used the concept of exploration window in the exploration for platinum-bearing placers along the coast of the Bering Sea. This exploration window is a combination of geochemical and size fraction parameters established by regression analysis that targets depositional environments for noble metals (e.g., platinum and gold). For platinum, for example, significant statistical correlations between manganese, zinc, copper, nickel, vanadium, and the >1.0 millimeter size fraction indicates placers deposited in high energy zones. For placers deposited in low energy zones, there is a good correlation between platinum, cobalt, copper, zinc, and the silt and clay size fractions. Certainly, as Smith (1986) points out, geochemistry is useful in identifying the influence of biogenic materials and secondary processes such as adsorption in these nearshore environments so that they can be separated from the geochemical

TABLE 2. Mineral Deposits and Their Location(s) in the Marine Environment

Zone	Commodities of Interest
Beaches	Placer deposits of Au, Pt, diamonds, magnetite, ilmenite, zircon, rutile, columbite, chromite, cassiterite, scheelite, wolframite, monazite, quartz, calcium carbonate, sand, and gravel
Continental shelves	Calcareous shell deposits, precious coral, phosphorite, glauconite, barium sulfate nodules, sand, and gravel; placer deposits in drowned river valleys or cassiterite, Pt, Au, and other minerals
Subseafloor rocks	Oil, gas, sulfur, salt, coal, iron ore, and possibly other mineral deposits in veins and other forms as in the rocks on land
Seawater	Common salt, magnesium metal, magnesium compounds, bromine, potash, soda, gypsum, desalinized water, heavy water and, potentially, sulfur, strontium, and borax; most other elements are found in seawater, and given recently developed extraction techniques, seawater is a potential source of uranium, molybdenum, and other commodities
Deep-sea floor	Manganese nodules—as a source of manganese, iron, nickel, cobalt, copper, molybdenum, vanadium, and possibly several other metals
	Hot brines and associated sediments—as a source of gold, silver, copper, zinc, lead, and possibly several other metals
	Animal remains—as a possible source of phosphates and metals such as tin, lead, silver, and nickel
	Clays—for structural uses and possibly also for alumina, copper, cobalt, nickel, and other metals
	Calcareous oozes—as cement rock and other calcium carbonate applications
	Siliceous oozes—as silica and in diatomaceous earth applications
	Zeolites—as a source of potash

Source: Siegel (1974), slightly modified from Mero (1967).

signal originating from economic mineral accumulations.

Deep-Sea Mineral Deposits

A great impetus to deep ocean exploration for hard minerals has come from the recognition of the extent and amount of manganese nodule material on the deep seafloor and the potential value of deposits of these nodules as estimated from geochemical analyses. Also, the oil embargo fiasco of 1973–1974 made many of the industrialized countries aware of the fact that petroleum and other commodities essential to steady development could be cut off for any political or capricious reason that could easily arise in the present geopolitical environment.

Manganese nodules (also called ferromanganese nodules) are spheroid-shaped, baseball-sized concretions present on the ocean floor. Horn et al. (1972) present a series of maps showing the nodule distribution. Nodule contents of manganese, copper, nickel, and cobalt, plus a suite of other elements (Table 3) make them attractive for mining ventures in areas where they abound. The Pacific Ocean may contain as much as 1.5 trillion tons of these nodules, although only 1% may be commercially mineable, and it is estimated that they are forming at a rate of about 10 million tons per year. The area of greatest economic promise is an east-west belt 200 km wide south of Hawaii in water deeper than 4,000 m. According to Mielke (1975) there are more than 40 sites, each with a reserve of 75 million tons, that are amenable to mining. At a 3 million ton annual extraction, each site would have a 25 yr operation. The rate of formation could be greater than the rate of mining, so the nodules could present an essentially inexhaustible supply of several elements (Table 4). The geochemist must establish the grade of the nodules, which can vary greatly between oceanic provinces and within the same province (Table 5), and define areas where the nodules meet minimum abundance (5 kg/m²) and

TABLE 3. Pacific Manganese Nodules: Weight Percentages (Dry-Weight Basis)
from 54 Samples

Element	Average	Maximum	Minimum
Manganese	24.2	50.1	8.2
Iron	14.0	26.6	2.4
Silicon	9.4	20.1	1.3
Aluminum	2.9	6.9	0.8
Sodium	2.6	4.7	1.5
Calcium	1.9	4.4	0.8
Magnesium	1.7	2.4	1.0
Nickel	0.99	2.0	0.16
Potassium	0.8	3.1	0.3
Titanium	0.67	1.7	0.11
Copper	0.53	1.6	0.028
Cobalt	0.35	2.3	0.014
Barium	0.18	0.64	0.08
Lead	0.09	0.36	0.02
Strontium	0.081	0.16	0.024
Zirconium	0.063	0.12	0.009
Vanadium	0.054	0.11	0.021
Molybdenum	0.052	0.15	0.01
Zinc	0.047	0.08	0.04
Boron	0.029	0.06	0.007
Yttrium	0.016	0.045	0.033
Lanthanum	0.016	0.024	0.009
Ytterbium	0.0031	0.0066	0.0013
Chromium	0.001	0.007	0.001
Gallium	0.001	0.003	0.0002
Scandium	0.001	0.003	0.001
Silver	0.0003	0.0006	

Source: Cardwell, P. H. Extractive metallurgy of ocean nodules. *Mining Congress Journal,* November 1973, p. 38.

Note: In addition to the elements given, cadmium, tin, arsenic, and bismuth are also found in manganese nodules.

tenor requirements (average of 2.25% Ni plus Cu with a 1.8% minimum) for a given operation (Frazer, 1980). In one operation manganese is the commodity sought; in others, copper and/or nickel and/or cobalt are the principal metals being derived from the nodules. Cobalt-bearing manganese crusts are found at shallower depths than manganese nodules, on sea mounts for example, and are of great economic interest. They form in waters with 0.9–2.2 ml/l oxygen that are enriched in manganese (Halback et al., 1982; Aplin and Cronan, 1985). Cronan (1986) reports that waters with <2.2 ml/l oxygen are found south of 20°S latitude and projects that the crusts with high contents of cobalt will be found in the Pacific Ocean north of 15°S latitude at depths of between 300 m and 2500 m. In the case of the United States and projections for 1985 made by the U.S. Bureau of Mines, nodule mining could eliminate the U.S. dependence on foreign manganese sources completely and reduce imports for nickel, copper, and cobalt by 10%, 6.5%, and 70%, respectively (Table 6). The same general relations might be true for other industrialized nations that suffer a national scarcity of one or more of these metals.

In addition to nodules, deep-sea deposits of metals associated with hot brines and sediments are also being evaluated by geochemists and geologists for possible mining exploration, especially in the Red Sea. It is estimated that 30 million metric tons of Fe, 2.5–3.2 million metric tons of Zn, 0.5–1 million metric tons of Cu, 0.8–1 million metric tons of Pb, 4,500–9,000 tons of Ag, and 45 tons of Au, worth more than $4 billion, is present in the sediments on the 130 km² floor of the Atlantis II deep (Hahlbrock, 1979; McKelvey, 1980). The Preussag AG company directed the exploration program and developed equipment to extract the metalliferous mud from the sea bottom. Pilot study results have been satisfactory and it is expected that exploitation will begin before the end of the 1980s. Other hot spots of brines associated with metal-rich sediments in closed, deep-sea basin environments will undoubtedly be found during marine exploration programs in which geochemistry will be a major contributor.

The finding, observing, and sampling of hydrothermal vents and associated massive polymetallic sulfide deposits along the spreading cen-

TABLE 4. Estimated Reserves of Metals in Manganese Nodules of the Pacific Ocean (tonnages in metric units)

Element	Amount in Nodules (billions of tons)	Reserves in Nodules at 1964 Rate of Consumption (years)	Approximate World Land Reserves (years)	Reserves in Nodules / Reserves on Land	Rate of U.S. Consumption in 1974 (millions of tons/year)	Rate of Accumulation in Nodules (millions of tons/year)	Rate of Accumulation / Rate of U.S. Consumption	World Consumption / U.S. Consumption
Mg	25	600,000	—[a]	—	0.04	0.18	4.5	2.5
Al	43	20,000	100	200	2.0	0.30	0.15	2.0
Ti	9.9	2,000,000	—[a]	—	0.30	0.069	0.23	4.0
V	0.8	400,000	—[a]	—	0.002	0.0056	2.8	4.0
Mn	358	400,000	100	4,000	0.8	2.5	3.0	8.0
Fe	207	2,000	500	4	100	1.4	0.01	2.5
Co	5.2	200,000	40	5,000	0.008	0.036	4.5	2.0
Ni	14.7	150,000	100	1,500	0.11	0.102	1.0	3.0
Cu	7.9	6,000	40	150	1.2	0.055	0.05	4.0
Zn	0.7	1,000	100	10	0.9	0.0048	0.005	3.5
Ga	0.015	150,000	—	—	0.001	0.001	1.0	—
Zr	0.93	+100,000	+100	1,000	0.0013	0.0065	5.0	—
Mo	0.77	30,000	500	60	0.025	0.0054	0.2	2.0
Ag	0.001	100	100	1	0.006	0.00003	0.005	—
Pb	1.3	1,000	40	50	1.0	0.009	0.009	2.5

Source: Mero (1972).
[a]Present reserves are so large as to be essentially unlimited at present rate of consumption.

557

TABLE 5. Average Analyses of Manganese Nodules (in percent of dry weight)

Region	Nickel	Copper	Manganese	Cobalt
North Pacific siliceous ooze	1.28	1.16	24.6	0.23
North Pacific red clay	0.76	0.49	18.2	0.25
South Pacific elevations	0.41	0.13	14.6	0.78
South Pacific abyssal plain	0.51	0.23	15.1	0.34
North Atlantic	0.38	0.15	14.2	0.34
South Atlantic	0.48	0.15	18.0	0.31
Indian Ocean	0.50	0.19	14.7	0.28

Source: United Nations. Economic significance, in terms of seabed mineral resources, of the various limits proposed for national jurisdiction. Committee on the peaceful uses of the seabed and the ocean floor beyond the limits of national jurisdiction, A/AC.138/87, June 1973; 39 pages plus appendixes; from Mielke, 1975.

TABLE 6. Projected U.S. Consumption and Percent of Imports Satisfied by Nodule Mining Operations in 1985 (short tons metal unless noted)

Metal	Estimated Annual Increase in U.S. Consumption (percent)	Projected U.S. Consumption	Estimated Recovery from Nodules by U.S. Firms and Subsidiaries	Percent of U.S. Consumption from Nodules	Percent of U.S. Imports from Nodules
Manganese	2.0	2,200,000[a]			
		1,075,000[b]			
		55,000[c]	518,000[c]	100.0	Exports[c]
Nickel	3.0	337,000	49,700	15.0	19.0
Copper	3.5	3,360,000	41,400	1.2	6.5
Cobalt	2.6	12,200	8,280	68.0	70.0

Source: U.S. Bureau of Mines, 1975, *Commodity Data Summaries.* Washington, D.C.: U.S. Government Printing Office, 41, 45, 97, 111.

[a]Ore.
[b]Ferromanganese.
[c]Pure metal.

ters of mid-oceanic ridges was the great geologic-geochemical discovery of the 1970s. Thermal waters and associated sulfide mineralization at the vents were first observed and sampled from the submergible ALVIN in 1977 at depths up to 3000 m, approximately 320 km northeast of the Galapagos Islands. A second, more spectacular, active hydrothermal system was discovered in 1978 along the oceanic ridge crest off Baja, California, Mexico, at 21°N latitude at the East Pacific Rise, the most active spreading center in the oceans at 7.5 cm per year. The active vents at the Galapagos ridge axis emitted shimmering waters with temperatures at 17°C versus 2°C for the surrounding sea water. In comparison, the hydrothermal vents at 21°N latitude emitted waters jetting out at temperatures of 380°C to 420°C and have been called black smokers because of the clouds of iron sulfide precipitating from them (Waldrop, 1980; Speiss, 1980; Cronan, 1986). The deposits from the black smokers may form large chimneylike structures 3 m high and 1 m or more across at the base.

At the spreading centers where basalt rises and discharges to form new crust, cold oceanic bottom waters percolate down through the fractured and porous rocks on the ridge margins until they near the molten basalt system very deep beneath the ridge crest. The sea water superheats and reacts chemically with the basalt leaching out from it iron, manganese, copper, nickel, and other elements while reducing sulfate to sulfide. When the metal-charged heated solutions rise, there is a gradual mixing of hot, acidic, reducing hydrothermal waters with cold, alkaline, oxidizing sea water and this results in the precipitation of two types of metal-rich deposits. One is rich in polymetallic sulfides dominantly of pyrite and marcasite that are associated with phases rich in zinc (sphalerite) and copper (chalcopyrite). The other is dominated by iron oxides and hydroxide, mainly as goethite and limonite (Francheteau et al., 1979; Hekinian et al., 1980). Important economic metals associated with these deposits include cobalt, lead, silver, and cadmium. Although many polymetallic deposits associated with marine hydrothermal vent systems have been discovered, only the deposits found along the Juan de Fuca ridge system off the northwest coast of the United States fall within the EEZ of a coastal state and may be an area for hard mineral mining in the future.

Malahoff et al. (1983) point out the similarity between a hydrothermal system such as that at 21°N latitude and the Troodos deposit, Cyprus, which has been mined for copper and associated metals for centuries. He suggests that prospecting on land would benefit from a careful investigation of the hydrothermal vent system deposits and related rocks. Cronan (1983) proposes that multielement analysis of marine sediments (e.g., for Mn, Fe, Ni, Cu, Zn, Ca, and Al) can be useful in geochemical exploration for hydrothermal deposits in the southwest Pacific Ocean and that results may be extended to exploration projects on land. For example, Bignell et al. (1976) and Cronan (1986) observed halos of diminishing concentrations of Fe and Mn or of the Fe/Mn ratio in sediments around potential economic sea floor sulfide deposits to a distance of several kilometers from the hydrothermal source.

Geochemical Analysis Equipment

The cost-efficiency of marine exploration geochemistry programs has been hampered by the need to send collected samples to shore laboratories for analysis. This difficulty was alleviated somewhat by establishing shipboard laboratories for on-site analysis, but such analysis still required sample preparation and treatment previous to analysis. Thus, there was still a delay in getting analytical results and in determining the following step in the exploration project. Much research, then, has been directed into in situ deep-sea analysis equipment or rapid shipboard analysis that does not require that a sample be weighed or specially prepared.

Instrumental techniques have been adapted or are now being researched for the specific purpose of geochemical prospecting on site in the marine environment or in situ in that environment. Friedrich et al. (1974) described a radioisotope energy-dispersive X-ray fluoroscope technique (EDX) that is in use on land and adapted to shipboard for the analysis of Mn, Fe, Co, Ni, Cu, and Zn in manganese nodules. The X-ray spectra from the nodules were excited by a 30 mCi ^{238}Pu source and read by an X-ray detection system composed of a 30 mm^2 effective-area Si (Li) detector with a measured energy resolution of 195 eV for Mn K X-rays, standard nuclear electronics, a 1,024-channel analyzer, and a data readout unit (IBM typewriter). The system was cooled in a 30 l liquid nitrogen Dewar Flask. Precision values were less than 3% for Mn, Fe, Ni, Cu, and Zn and about 5% for Co. The technique is operational, excellent analytically, and can be carried out without weighing and specially preparing a sample. However, the sample must be crushed to chips a few

millimeters in diameter, dried at 110° C for about an hour, and pulverized in an agate mortar. Therefore, although the method provides data within a short time after collection so that evaluations can be made on board to guide further exploration, it does not provide immediate in situ analysis.

A system for in situ analysis has been developed and is being researched. The analytical technique is based on thermal neutron activation of material with ^{252}Cf and interpretation of the resulting capture gamma spectrum. The analysis of the capture gamma spectrum after sample activation with ^{252}Cf has been successful in tests with the transition elements, chlorine, gold, manganese, and mercury, and Senftle (personal communication) noted that nickel was readily detectable under about 140 cm of overburden. He proposed the use of an in situ marine system for activating large-volume-area samples (versus a single nodule, e.g.) reading the capture gamma emissions with a Ge-Li detector, and using a computer for rapid data reduction (5 min lag) and subsequent mapping, which would facilitate immediate program operation planning and changes. In 1973, Battelle-Northwest Laboratory reported successful testing of the first self-contained unit for on-site in situ analysis of seabed minerals by such a technique. The researchers on the project believe that the instrument will detect and quantitatively analyze 30 elements on the seafloor, delivering the concentrations as a computer printout. The Battelle-Northwest instrument package is lowered from a surface ship for operation, with depth capability limited only by winch capacity and cable length. Senftle (1970) suggested that the system be included on an ocean-bottom submersible unit that could be directed by onboard scientists.

More recently, in a feasibility study of the use of in situ capture gamma-ray analyses for seabed exploration, Senftle et al. (1976) made neutron-capture gamma-ray spectra of bottom sediments at four sites (representing five samples) in Long Island Sound using a ^{252}Cf neutron source and a Ge(Li) detector. The probe they used is much more compact than that used by Battelle-Northwest. Penetration of the probe into the bottom sediment was about 35 cm. The probe was connected to a barge-mounted truck that contained a 4,096-channel analyzer, a minicomputer, a magnetic tape recorder, and other electronic instruments. The water depth ranged from 15 m to 30 m. When neutron-capture gamma-ray data were compared with X-ray fluorescence analyses of the bottom sediments, several problems were evident. First, chlorine is a major source of interference and did not give the same spectrum in seawater and in the bottom sediments. Second, anomalous results between spectra of the bottom sediments from the various stations were attributed to the fact that the measured capture gamma-ray spectra are the result of both thermal and epithermal resonance capture; substantial epithermal capture causes identification and calibration problems for quantitative analysis. Senftle et al. (1976) suggest that the use of several small sources arranged around a central detector could improve the counting statistics and reduce the seriousness of the problems.

The use of *pathfinders* (see *Indicator Elements*), or guides to economic deposits of selected metals or nonmetals, rather than using a major component of the commodity is limited in the marine environment. For the most part, samples must be taken and analyzed, and only then do we have a good indication of the presence of any economic commodity from which we must set up costly programs to determine the quantity present. We should emphasize here that great costs are involved in taking samples from the marine environment, especially in the bathyl and abyssal depths. However, this problem does not always exist to the same degree.

Summerhayes et al. (1970) realized that uranium is often included in phosphate deposits in trace quantities and proposed that radioactivity of samples could be used as guides to submarine phosphate deposits. Tests were made with a submergible scintillometer in water up to 1,000 m deep from a stopped ship. Similarly, compositional mapping of bottom sediments has been described by Bastin (1973) and Miller and Symous (1973). Their data were sufficiently positive to give impetus to the development of equipment with continuous recording capability that could be towed by a boat. Analysis of total radioactive content of marine sediments was made with a towed bottom probe by Beckman and Abdullah (1974). In these studies a scintillation counter coupled with a count-rate meter or three-channel spectrometer was used. With the Ge(Li) detectors now in use, it is possible to resolve gamma-ray lines from both natural and induced radioactivity that were previously irresolvable with Na(Tl) detectors. These Ge(Li) detectors were used for offshore in situ bottom sediment analyses by neutron activation and by X-ray fluorescence (Noakes et al., 1974; Wogman et al., 1975; Senftle et al., 1976).

More recently, Miller et al. (1977) described in detail a seabed gamma-ray spectrometer that was towed on the continental shelf of the United Kingdom for more than 10,000 km at speeds up to 7 knots. The radiometric data were used to produce contour maps of the total gamma-ray activity and the potassium, uranium, and thorium activities of the seabed. The systematic traversing showed that different rock and sediment types tend to have characteristic levels of radioactivity and radio-element ratios; i.e., the equipment is sensitive to change in superficial sediment composition. Thus, geological and geochemical mapping by the towed seabed gamma-ray spectrometer on the continental shelf is a reality. This method may be useful in marine mineral exploration for anomalous seabed radioactivity due to outcropping radioactive mineralization, uranium-bearing phosphorites, and radioactive detrital minerals commonly associated with placer deposits (monazite, zircon).

Research on the recovery of uranium from seawater is going on in Japan, the United States, England, the Federal Republic of Germany, India, and the USSR. If uranium can be won from seawater in a cost-effective process (in the future), the problem of running out of uranium as a fuel for the fission nuclear reactors is overcome and the pressing need to develop breeder reactors or fusion reactors is alleviated. The recovery process most often studied involves uranium extraction by physical adsorption onto a crypto-crystalline inorganic adsorber composed of titanium hydroxide, $Ti(OH)_4$. Great volumes of seawater are put into contact with the adsorbent after which the uranium-loaded $Ti(OH)_4$ is returned to the seawater for another uranium loading and the uranium is recovered from the effluent. Laboratory experiments are being made to establish the most efficient (cost-effective) thickness and form (flat-bedded or stacked in tubes) of the adsorbent, and a shellfish product, chitosan, is being evaluated as an adsorber or an adsorber constituent (Driscoll, 1978).

Still another remote-sensing device is being researched for deep-sea mineral exploration, although it does not involve chemical determinations. Magnuson et al. (1981) are investigating the feasibility of using acoustic signals from a surface survey ship to locate deposits of manganese nodules on the ocean floor and to infer module size and weight density. As has been noted, current prospecting techniques involve the lowering of sensors, dredging equipment, television cameras, and other equipment to the ocean floor. These techniques are time consuming and costly. The new method being researched involves transmitting acoustic pulses from a surface ship to the ocean floor where they are reflected back to a receiver on the ship. The pulses received when nodules are on the ocean bottom are different from those when nodules do not exist, so their presence or absence is easily discerned. The potential new prospecting method is both faster than the methods being used and allows the scientists to determine to some extent the quality of the deposits.

Hydrocarbons in the Marine Environment

Marine exploration geochemistry for hydrocarbons (petroleum and natural gas) has been going on for some time, especially in areas deemed propitious by geological and geophysical work for spudding in a well group. Natural gas seeps at sea have been detected by so-called hydrocarbon sniffers, which are essentially sophisticated gas chromatographic units that determine the presence, amount, and type(s) of natural gas in waters overlying the ocean floor (Sackett, 1977). Such a unit may operate on shipboard by analyzing waters brought to the ship by underwater pumps and hose units, or it may be towed by the ship at selected depths down to the sediment/water interface and continually record gasometric data. In 1970, Ingerson reported that major companies exploring for hydrocarbons in the Gulf of Mexico were using geochemical techniques in their operations and that successes using the geochemical techniques (with complimentary geological and geophysical data) had been noted in the North Sea and off Gabon. The *Glomar Challenger* program has activated intense interest in deep-sea exploration for hydrocarbons because of its finds of indications of oil associated with deep-sea sediments and salt domes in the Gulf of Mexico and other oceanic areas.

The detection of natural gas seeps at sea has been carried out by a combination of high-resolution seismic systems (see *Acoustic Surveys, Marine*) that produce three-dimensional pictures that show shallow seafloor structures and gas seeps (Tinkle et al., 1973). Two types of seismic systems were used for the investigations that resulted in the recognition of natural gas seeps in the ocean: a 3.5 kHz vertical scanning subbottom profiler, and a side-scanning bottom profiler. The subbottom profiler was used to obtain diagrammatic representations of the bottom topography and of the upper section to about 300 m. Tinkle et al. explain that the high-resolution capability of the seismic records is the result of using an extremely short, high power sound pulse that emanates from a towed transducer package as a cone-shaped beam directed vertically downward. The reflected signal is subjected to a time-varied gain, which enhances the recording of strong reflections from greater depth. Individual sedimentary layers, faults, domal structures, mud bumps, and gas seeps are thus readily detectable. The gas seeps are ideal for detection by this seismic system because of the response of sound waves to the density contrast between seawater and rising bubbles of gas, which expand with the decreasing pressure as they rise in the water column.

The side-scanning system records bottom topography and aids in identifying features directly on the bottom or immediately above it. Tinkle et al. (1973) described it as a towed transducer unit connected by cable to a shipboard recorder; by using a very short pulse-length signal of about 100 kHz frequency, generated laterally from the transducer in a fan-shaped beam perpendicular to the gravel direction, small details and minor bottom features can be detected. This system is obviously useful for the lateral detection of gas seeps.

More powerful seismic units with pulse lengths and other acoustic features designed for deeper penetration are not suited for investigating relatively subtle features such as gas bubbles in the water column but are used to check on the intermediate structure underlying seeps located by the higher-frequency seismic sources cited earlier. There may be a correlation between seep areas and known production zones. Analyses of the gases in the seeps may allow the identification of gases escaping from deeper parts of the sedimentary section where economic accumulations occur (Reed and Kaplan, 1977).

Geochemical seep detection for offshore oil and gas exploration has also been discussed by Sigalove and Pearlman (1975), Bernard et al. (1976), and Cline and Holmes (1977). Most recently, Kvenvolden et al. (1981) have reported on geochemical prospecting for hydrocarbons in the outer continental shelf, southern Bering Sea, Alaska, analyzing hydrocarbon gases from the top 0.5 m of sediment using shipboard gas chromatography. They distinguished the possible presence of thermogenic hydrocarbons with the ratios of methane to ethane plus propane, and ethane to ethene. In the first case, when $C_1/(C_2 + C_3)$ has a value of less than 50, a thermogenic source is indicated; in the second case, when $C_2/C_2:1$ is greater than 1, a thermogenic source is indicated.

Research being carried out on gases dissolved in subsurface waters associated with formations on the continent known to contain hydrocarbon accumulations may be applicable to the marine environment. In general, dissolved gases decrease in concentration in formational waters with increasing distance from the deposit (Zorkin, 1969; Zarella, 1969).

Although gasometrically determined anomalies may be able to predict the presence of hydrocarbons in the subsurface, and these may be related to appropriate geological traps, they are unable to predict whether there is an exploitable quantity in the subsurface. This can be demonstrated only by drilling test wells and determining flow characteristics.

Conclusion

Finally, it should be mentioned that, at present, marine exploration, academic or applied, is not generally restricted by national, artificial geographical limits of control, whereas exploitation of natural resources (inorganic and organic) from the marine environment is restricted irregularly by the nations of the Sea Conference. The Exclusive

Economic Zones (EEZs) that were created by the 1982 Law of the Sea Convention may encourage exploration of the continental shelfs and the deep areas that fall within the EEZs. The EEZ is a region that extends 200 nautical miles from the coast. It contains vital natural resources, organic and inorganic, on the sea floor, within the subbottom, and in the overlying waters. Most of the areas falling in the EEZ have not been explored so that the resources there and the resources potential have not yet been defined. According to the Convention, each coastal state is responsible for the development and judicious management of its marine resources and for the protection of the environment. The Convention also gives coastal states jurisdiction over that part of the continental margin that extends into the sea beyond the EEZ. Where exploration has demonstrated areas on the floor beneath the high seas that have mineral exploration potential, as in the case of the manganese nodules, no sovereignty reigns. Law of the Sea Conferences have resulted in slow progress insofar as establishing norms for harvesting the vastness of oceanic wealth in a well-planned, economically correct manner that would benefit all nations of the world. Indeed some nations knowingly are overfishing both areas and species to the detriment of the entire international community. This behavior is certainly not expected of developed nations and nations that purport to put environmental conservation and protection at the forefront of national and international priorities.

The simple question that exists, then, is who or what body has a right to restrict ocean mining beyond nationally established limits and grant concession and collect royalties? If, indeed, an answer to this question is agreed on, then it would have to extend to ocean fishing as well. What is needed then, ideally, is a new, total world awareness and reorganization of priorities directed toward the conservation of renewable and depletable international resources and toward a more equitable distribution of a portion of the financial and technological gains developed from the mining and/or harvesting of such resources between the peoples of the world. Whether this aim is attainable before irreparable damage to the world is done is a purely political question, a question of survival.

FREDERIC R. SIEGEL

References

Aplin, A., and D. S. Cronan, 1985, Ferromanganese oxide deposits from the Central Pacific Ocean, *Geochim. et Cosmochim. Acta* **49**, 427–451.

Bastin, A., 1973, Natural radioactive tracers and their use in Belgium, in *Tracer Techniques in Sediment Transport*. Vienna: International Atomic Energy Agency.

Beckman, H., and H. Abdullah, 1974, Nuclear logging methods in marine geology, *Industries Atomiques et Spatiales* **2**, 35–48.

Bernard, B. B., J. M. Brooks, and W. M. Sackett, 1976, Natural gas seepage in the Gulf of Mexico, *Earth and Planetary Sci. Letters* **31**, 48–54.

Bignell, R., D. S. Cronan, and J. S. Tooms, 1976, Metal dispersion in the Red Sea as an aid to marine geochemical exploration, *Inst. Mining and Metallurgy Trans.* **85**, B273–B278.

Cline, J. D. and M. L. Holmes, 1977, Submarine seepage of natural gas in Norton Sound, Alaska, *Science* **198**, 1149–1153.

Cronan, D. S., 1983, Metalliferous sediments in the CCOP/SOPAC region of the Southwest Pacific with particular reference to geochemical exploration for the deposits, *CCOP/SOPAC, Tech. Bull.* **4**.

Cronan, D. S., 1984, Criteria for the recognition of areas of potentially economic manganese nodules and encrustations in the CCOP/SOPAC area of the Central and Southwestern Pacific, *South Pacific Marine Geology Notes* **3**, 1–17.

Cronan, D. S., 1986, Geochemical exploration for deep sea mineral deposits, in I. Thornton and R. J. Haworth, eds., *Applied Geochemistry in the 1980s*. New York: Halsted Press, 241–259.

Driscoll, M. J., 1978, Recovery of uranium from seawater, *MIT Progress Rept. No. 2*.

Francheteau, J., and 14 coauthors, 1979, Massive deep-sea sulfide ore deposits discovered on the East Pacific Rise, *Nature* **277**, 523–528.

Frazer, J. Z., 1980, Resources in seafloor manganese nodules, in J. Kildow, ed., *Deepsea Mining*. Cambridge, Mass.: MIT Press, 41–83.

Friedrich, G. H. W., H. Kunzendorf, and W. L. Pluger, 1974, Ship-borne geochemical investigations of deep-sea manganese nodule deposits in the Pacific using a radioisotope energy-dispersive system, *Jour. Geochem. Exploration* **3**, 303–317.

Hahlbrock, U., 1979, Mining metalliferous mud in the Red Sea, *Ocean Industry*, May, 45–48.

Halbach, P., F. T. Manheim, and P. Otten, 1982, Co-rich ferromanganese deposits in the marginal seamount regions of the Central Pacific Basin—Results of Midpac '81, *Erzmetall.* **35**, 447–453.

Horn, D. R., B. M. Horn, and M. N. Delach, 1972, Distribution of ferromanganese deposits in the world ocean, in D. R. Horn, ed., *Ferromanganese Deposits on the Ocean Floor*. Washington, D.C.: National Science Foundation, 9–17.

Ingerson, E., 1970, Geochemistry, *Geotimes* **14**, 13–14.

Kogan, B. S., 1977, Methods for geochemical prospecting in coastal marine placers (in Russian), *Izvestiya Vysshikh Uchebnykh Zavedeniy, Geologiya I Razvedka* **7**, 52–64.

Kvenvolden, K. A., T. M. Vogel, and J. V. Gardner, 1981, Geochemical prospecting for hydrocarbons in the outer continental shelf, southern Bering Sea, Alaska, *Jour. Geochem. Exploration* **14**, 209–219.

McKelvey, V. E., 1980, Seabed minerals and the Law of the Sea, *Science* **209**, 464–472.

McKelvey, V. E., 1986, Subsea Mineral Resources, *U.S. Geol. Survey Bull. 1689 (Mineral and Petroleum Resources of the Ocean)*, 106p.

McKelvey, V. E. and F. F. H. Wang, 1970, World subsea mineral resources (preliminary maps), *U.S. Geol. Survey Misc. Geol. Inv., Map I-632,* plus discussion to accompany Map I-632.

Magnuson, A. H., K. Sundkvist, Y. Ma, and V. Smith, 1981, Acoustic sounding for manganese nodules, *Offshore Tech. Conf.* OTC 4133, 147–161.

Malahoff, A., R. W. Embley, D. S. Cronan, and R. Skirrow, 1983, The geological setting and chemistry of hydrothermal sulphides and associated deposits from the Galapagos Rift at 86°W, *Marine Mining* 4, 123–137.

Mero, J. L., 1967, *Marine Sciences and Industrial Potential Symposium,* Transference of Technology Series No. 2. Austin: College of Business Administration, University of Texas, 263p.

Mero, J. L., 1972, Potential economic value of ocean floor manganese nodule deposits, in D. R. Horn, ed., *Ferromanganese Deposits of the Ocean Floor.* Washington, D.C.: National Science Foundation, 191–203.

Mielke, J. E., 1975, *Ocean Manganese Nodules,* Prepared for Committee on Interior and Insular Affairs, U.S. Senate, by the Congressional Research Service. Washington, D.C.: U.S. Government Printing Office, 203p.

Miller, J. M., and G. D. Symons, 1973, Radiometric traverse of the seabed off the Yorkshire coast, *Nature* 242, 184–186.

Miller, J. M., P. D. Roberts, G. C. Symons, N. H. Merrill, and M. R. Wormald, 1977, A towed sea-bed gamma-ray spectrometer for continental shelf surveys, in *International Symposium on Nuclear Techniques in Exploration, Extraction, and Processing of Mineral Resources.*, Vienna: International Atomic Energy Agency, 447–463.

Noakes, J. E., J. L. Harding and J. D. Spaulding, 1974, Locating offshore mineral deposits by natural radioactive measurements, *Marine Technology Soc. Jour.* 8, 36–39.

Owen, R. M., 1978, Geochemistry of platinum-enriched sediments: Application to mineral exploration, *Marine Mining* 1, 259–282.

Owen, R. M., 1980, Quantitative models of sediment dispersal patterns in mineralised nearshore environments, *Marine Mining* 2, 231–249.

Reed, W. E., and I. R. Kaplan, 1977, The chemistry of marine petroleum seeps, *Jour. Geochem. Explor.* 7, 255–293.

Sackett, W. M., 1977, Use of hydrocarbon sniffing in offshore exploration, *J. Geochem. Explor.* 7, 243–254.

Senftle, F. E., 1970, Mineral exploration by nuclear techniques, *Mining Cong. Jour.* 6p.

Senftle, F. E., A. B. Tanner, P. W. Philbin, J. E. Noakes, J. D. Spaulding, and J. L. Harding, 1976, In-situ capture gamma-ray analyses for sea-bed exploration, in *Nuclear Techniques in Geochemistry and Geophysics.* Vienna: International Atomic Energy Agency, 75–91.

Sharp, W. E. and B. Bolviken, 1979, Brown algae: A sampling medium for prospecting fjords, in J. R. Watterson and P. K. Theobald, eds., *Geochemical Exploration 1978.* Association of Exploration Geochemists, 347–356.

Siegel, F. R., 1974, *Applied Geochemistry.* New York: Wiley-Interscience, 353p.

Siegel F. R., and J. W. Pierce, 1978, Geochemical exploration using marine mineral suspensates, *Modern Geology* 6, 221–227.

Siegel, F. R., and J. W. Pierce, 1982, Marine samples in the search for onshore mineralization: bottom sediment, suspended mineral matter, Chilean Archipelago, *Modern Geol.* 8, 87–93.

Sigalove, J. J., and M. D. Pearlman, 1975, Geochemical seep detection for offshore oil and gas exploration, *7th Ann. Offshore Tech. Conf.,* OTC 2344, 95–102.

Summerhayes, C. P., B. H. Hazelhoff-Roelfzema, J. S. Tooms, and D. B. Smith, 1970, Phosphorite prospecting using a submersible scintillation counter, *Econ. Geology* 65, 718–723.

Smith, P. A., 1986, Exploration geochemistry in the shallow marine environment, in I. Thornton and R. J. Haworth, eds., *Applied Geochemistry in the 1980s.* New York: Halsted Press, 212–240.

Speiss, F. N., and 21 coauthors, 1980, East Pacific Rise: hot springs and geophysical measurements, *Science* 207, 1421–1432.

Tinkle, A. R., J. W. Antoine, and R. Kuzela, 1973, Detecting natural gas seeps at sea, *Ocean Industry,* April, 139–142.

Tooms, J. S., 1972, Potentially exploitable marined minerals, *Endeavor* 31, 113–117.

U.S. Bureau of Mines, 1975, *Commodity Data Summaries 1975.* Washington, D.C.: U.S. Government Printing Office, 193p.

Waldrop, M. M., 1980, Hot springs and marine chemistry, *Mosaic* 11, 8–14.

Wogman, N. A., H. G. Rieck, and J. R. Kosorok, 1975, In situ analysis of sedimentary pollutants by X-ray fluorescence, *Nuclear Instruction and Methods* 128, 561–568.

Zarella, W. M., 1969, Application of geochemistry to petroleum exploration, in W. B. Heroy, ed., *Unconventional Methods in Exploration for Petroleum and Natural Gas.* Dallas: Southern Methodist University Press, 29–41.

Zorkin, L. M., 1969, Regional regularities of underground water gas contents in petroleum-gas basins, *Geol. Zhur.* no. 2.

Cross-references: *Acoustic Surveys, Marine; Harbor Surveys; Lake Sediment Geochemistry; Sea Survey.* Vol. XIII: *Geochemistry, Applied; Oceanography, Applied; Submersibles.*

MARINE GEOCHEMISTRY—See MARINE EXPLORATION GEOCHEMISTRY.

MARINE MAGNETIC SURVEYS

Modern *marine magnetic surveys* involve measuring the Earth's total magnetic field while towing a *magnetometer* behind a moving ship. The magnetometer is now standard equipment on most marine surveys (see *Sea Surveys*) and is complementary to other geophysical systems such as seismic profilers, echo sounders (see *Acoustic Surveys*), and gravimeters. Modern marine magnetometers provide a continuous record of the Earth's total magnetic field along the ship's track. Magnetic values can be plotted on a map and then contoured for interpretation. With contoured magnetic maps, it is possible to identify and delineate *magnetic anomalies,* which in conjunction with other geophysical data, can be used to interpret the structural and lithologic character of the ocean bottom.

During the early 1900s it was necessary to use wooden ships to make accurate magnetic measurements at sea. The best known ship of this type was the *Carnegie,* which was used for magnetic surveys for 20 years before it was destroyed by fire in 1929. Prior to 1940, magnetic measurements at sea were made on board ship with land-type instruments modified for sea use. Gimbals were used to stabilize the magnetic instruments, with only partial success (Bullard and Mason, 1963).

Towed Magnetometers

The fluxgate, proton precession, and rubidium-vapor magnetometers are all capable of measuring the Earth's total magnetic field to a high degree of accuracy at sea. During the early 1950s, the *fluxgate magnetometer* was first used to measure the total field while being towed behind a ship. This magnetometer has a rod of ferromagnetic material that acts as the core of one or more windings, with the windings connected to alternating-current circuits. As the magnetic field changes, so does the magnetism of the core, which is measured by the change in flux produced by the alternating current. Although the fluxgate magnetometer may be towed behind a ship, its disadvantages are that it requires frequent calibration and lacks stability.

The *nuclear precession magnetometer* (Fig. 1), which was first adapted for sea use several years after the fluxgate magnetometer, is now considered to be more sensitive and more easily handled than the fluxgate magnetometer. In a nuclear precession magnetometer, a coil is wound around a bottle containing a liquid such as water. A direct current is passed through the coil for several seconds to orient the axes of the protons in the liquid parallel to the axis in the coil. This field is then eliminated by switching off the current. A signal is produced by the precession of aligned protons about the Earth's total magnetic field. The field can usually be measured to an accuracy of 1 part in 50,000; however, the accuracy of the typical survey is somewhat less because of the difficulty of making the proper corrections.

The *rubidium-vapor magnetometer* has only

FIGURE 1. Nuclear precession magnetometer manufactured by Varian Associates.

recently been available for marine magnetic use. It is very highly sensitive to the Earth's magnetic field.

Navigation

On a marine magnetic survey, no system is more essential than the ship's positioning system. Because the measurements supplied by the magnetometer will be of questionable value if the ship's position cannot be plotted accurately, most survey ships are now equipped with one or more of the numerous electronic systems now available (see *Surveying, Electronic*).

There are three basic types of positioning systems. In the *circular,* or *ranging, system,* a master station is placed aboard the ship, and two slave stations are placed on shore. With this system, the land width is constant and the position of the ship is determined by the intersection of two circular arcs measuring the distance from each slave station to the ship. This system typically has a short range and high accuracy and is best suited for nearshore or coastal surveys.

Another type of electronic positioning system is called a *hyperbolic system.* In this system a master station and two slave stations are on shore. For a fix, it is necessary to get an intersection of two hyperbolic lines on the plotting sheet. This system typically has a long-range capability of hundreds of nautical miles and is accurate to within several hundred meters. The type of electronic positioning system used will depend on factors such as distance from shore, accuracy required, mobility of operation, and costs. With the advent of satellite navigation in 1969, most well-equipped survey ships carry this system to augment the hyperbolic system.

The ship's position is plotted every 3–5 min on the plotting sheet that displays the hyperbolic or circular navigational lines. A log is maintained of the ship's speed, course changes, sea state, wind speed, and wind direction. This log is used to smooth the ship's track to adjust for erratic fluctuations in the track that result from instrument or plotting error.

Survey Design

Prior to commencing a marine magnetic survey, the ship's proposed track line should be constructed on the navigational plotting sheet. Existing magnetic contour sheets should be consulted so that the track lines are laid out at right angles to the magnetic trend. Track line spacing depends on time available, quality of navigation, scale of the plotting sheet, and type of information needed. To correlate magnetic anomalies accurately, surveys should not be run with a line spacing greater than 10 nautical mi. A detailed survey is generally conducted on a line spacing of less than 1 mi. Cross check lines should be run periodically during the survey to determine if the magnetic values correspond where the cross check lines intersect the regular track lines. If the magnetic values at the intersection of the two track lines do not agree within 50 gamma, the navigational system is probably not operating properly.

Surveying

The magnetometer *fish* is normally towed two to three ship lengths astern to avoid the magnetic disturbance caused by the ship (Fig. 2). During the survey, the fish is positioned just below the sea surface, with the depth depending somewhat on the speed of the ship. When the ship stops, the fish must be hauled aboard the ship so that the cable does not get caught in the screws or tangled with the other equipment. The cable is hauled in by an electrically powered winch and is wound on a large drum.

When possible, survey operations are maintained around the clock. The magnetic equipment must be monitored constantly. The moving paper tape of the magnetometer on which a pen makes a continuous profile (Fig. 3) of the total magnetic

FIGURE 2. Magnetometer fish with cable wound on drum.

FIGURE 3. Continuous magnetic profile automatically drawn on moving paper tape.

intensity should be examined and annotated at least every 30 min with information such as time, date, time zone, ship's name, and phase range in gammas. Every 24 hr this profile is removed from the instrument and is scaled for magnetic values, typically at 50 gamma intervals or less and including highs and lows. The measured field values are recorded in a log book with their respective time values.

Data Processing

After the magnetic values are plotted on the smoothed track line, they are contoured at an interval of 50–100 gamma. Before the magnetic anomaly map is considered completed, the regional field must be subtracted from the observed field.

During processing of the raw magnetic data, it may be important to make a variety of corrections. There are three types of time variations: (1) normal daily variation, (2) magnetic storms, and (3) long-term secular variations. Although the average daily variation is 30–50 gamma, this is not significant enough to warrant correction except in certain areas where the variation is extreme.

Computers are now available for linking navigation systems, magnetometers, and other geophysical instrumentation so that magnetic data may be displayed with digital plotting equipment. For example, the computer and plotter can prepare a plotting sheet at any desired scale and then make a continuous plot of the ship's track. As the plotter marks the ship's position, the magnetic data are simultaneously plotted on the sheet.

Oceanic Magnetism

Magnetometers are capable of sensing rock magnetism several kilometers below the ocean floor. Most of the magnetism comes from the basaltic oceanic basement because the overlying sediments are too weakly magnetized to affect the magnetometer significantly. The magnetic properties of the oceanic basement depend on the amount of magnetite, thickness of the magnetized layer, and its depth below the surface.

The oceanic basement is magnetized to the depth of the *Curie point isotherm*. This isotherm is 575°C and is situated about 10 km depth at the oceanic ridges where heat flow is high and about 20 km depth throughout the rest of the ocean basins.

Marine magnetic surveys are concerned only with measurements of the total field, not with magnitude or direction of horizontal and vertical components. The magnitude of the total field ranges from about 30,000 gamma at the magnetic equator to about 60,000 gamma at the magnetic poles.

Magnetic Anomalies

In 1961, Raff and Mason published the results of a magnetic survey in the northeast Pacific Ocean. This detailed survey was one of the first to delineate accurately linear magnetic anomalies. The publication of these magnetic anomaly maps immediately led to widespread speculation into the cause of the anomalies. Some of the important features common to these anomalies are as follows:

1. They are associated with and parallel to the mid-oceanic ridges.
2. The highest anomaly magnitudes occur along the axis of the ridge.
3. The magnitude of the anomalies tends to decrease away from the ridges
4. Anomalies are symmetrical about the axis of the ridge
5. The magnitude of the anomalies generally ranges from several hundred to several thousand gammas.

6. The width of the anomalies ranges from 10 km to 50 km.

Where magnetometers are dragged close to the seafloor, the wavelengths of the anomalies are much shorter but still linear and parallel to the ridge. Thus, the greater the distance between the oceanic basement and the magnetometer, the greater will be the tendency for local anomalies to be smoothed or attenuated (Heirtzler, 1970).

Vine and Matthews (1963) proposed that these linear magnetic anomalies were caused by alternating strips of seafloor being magnetized in opposite directions. After new fluid basalt moves up along dikes parallel to the axis of the ridge and the anomalies, the basalt is magnetized in the direction of the Earth's magnetic field when it cools below the Curie temperature. Since the Earth's magnetic field reverses periodically, each successive anomaly strip represents alternately normal and reversed magnetism.

Linear magnetic anomalies represent some of the most convincing evidence in support of the theory of global plate tectonics. Linear anomalies are now used to date oceanic basement and determine spreading rates.

TERRY S. MALEY

References

Bullard, E. C., and R. G. Mason, 1963, The magnetic field over the oceans, in M. N. Hill, ed., *The Sea,* vol. 3. New York: Wiley-Interscience, 175–217.

Heirtzler, J. R., 1970, Magnetic anomalies measured at sea, in A. E. Maxwell, ed., *The Sea,* vol. 4. New York: Wiley-Interscience, 85–128.

Raff, A. D., and R. G. Mason, 1961, Magnetic survey off the West Coast of North America, 40 N Latitude to 52 N Latitude, *Geol. Soc. America Bull.* **72,** 1267–1270.

Vine, F. J., and D. H. Matthews, 1963, Magnetic anomalies over oceanic ridges, *Nature* **199,** 947–949.

Cross-references: *Acoustic Surveys, Marine; Exploration Geophysics; Marine Mining; Satellite Geodesy and Geodynamics; Sea Surveys; Surveying, Electronic; VLF Electromagnetic Prospecting. Vol. XIII: Magnetic Susceptibility, Earth Materials; Oceanography, Applied; Seismological Methods.*

MARINE MINING

Almost all the mineral commodities mined on land are found also within the marine environment (Table 1), although few of them are economically recoverable. The reason for this has been mostly a lack of knowledge of the geological marine environment and a lack of incentive to develop the resources in that environment. This situation is changing, however, largely due to dis-

TABLE 1. Summary Classification of Marine Mineral Deposits

Dissolved Deposits	Unconsolidated Deposits			Consolidated Deposits
	Continental Shelf, 0–200 m (Littoral)	Continental Slope, 200–3,500 m (Bathyal)	Deep Sea, 3,500–6,000 m (Abyssal)	
Seawater: Fresh water Metals and salts of: Magnesium Sodium Calcium Bromine Potassium Sulphur Strontium Boron Uranium Other elements *Metalliferous Brines:* Concentrations of: Zinc Copper Lead Silver	*Nonmetallics:* Sand and gravel Lime sands and shells Silica sand Semiprecious stones Industrial sands Phosphorite Aragonite Glauconite *Heavy Minerals:* Magnetite Ilmenite Rutile Monazite Chromite Zircon Cassiterite *Rare and Precious Minerals:* Diamonds Platinum Gold Native copper	*Authigenics:* Phosphorite Ferromanganese oxides and associated minerals Metalliferous mud with: Zinc Copper Lead Silver	*Authigenics:* Ferromanganese nodules and associates Cobalt Nickel Copper *Sediments:* Red clays Calcareous ooze Siliceous ooze	*Disseminated, massive, vein, tabular, or stratified deposits of:* Coal Ironstone Limestone Sulfur Tin Gold Metallic sulfides Metallic salts Hydrocarbons

Source: After Cruickshank, in Cruickshank and Marsden, 1973.

coveries of new minerals in the deep sea bed and to international interest or controversy over their ownership (United Nations, 1982). Two basic geologic divisions of the sea bed affect the development of mining systems: (1) the continental margins from the coast to depths of several hundred meters and (2) the deep sea bed at depths of several thousands of meters. On the political side, these approximate areas are becoming known and accepted as the Exclusive Economic Zone (EEZ) within approximately 200 mi of land and the Area of the Seabed Beyond the Limits of National Jurisdiction (the Area). The minerals and the methods of mining differ in each (Cruickshank, 1982).

Mining operations consist of certain common basic functions, however, whether on land or under the sea, and involve surficial deposits of unconsolidated material or hard-rock orebodies deep within the Earth's crust. These six basic functions or requirements are the *infrastructure,* or permanent bases of operations; *operational prerequisites,* or basic requirements for carrying out the field operations; specialized systems required for *exploration and characterization; exploitation*

functions including the needs for *environmental protection and restoration;* and the problems of *operational safety* common to each (Cruickshank and Marsden, 1973).

Infrastructure

In marine mining, *infrastructure* refers in general to the construction of capital works in advance of mining operations, including docks, platforms, or ships; mining equipment; mill or metallurgical plant; and preparation of dump sites and storage facilities. Transportation facilities may involve the laying of pipelines, the construction of roadways or tracked systems, and boosters, terminal, or transfer stations. The supply of utilities to permanent stations either on- or offshore requires the construction of power plants and transmission lines, and similarly, personnel and their families will require living quarters and sanitation and recreational facilities. Many of these aspects are discussed in greater detail under other headings, but some have no basis for detailed study. For example, there is an advanced requirement and technology for platforms at sea but no current need for sea bed vehicle transportation as a counterpart to terrestrial rail or roadway systems. These needs will come, however, and their inclusion in this general statement is implied. All operations at sea originate and terminate onshore. Until permanent habitations are established on the sea bed or on massive floating platforms, this condition will remain valid. However, the consideration of marine mining operations brings no new concepts or needs to the common requirements of a shore base.

Operational Prerequisites

For all mining activities at sea, whether exploration or exploitation, there are certain basic operational prerequisites. These include a knowledge of the weather, a platform from which to carry out the mission, a source of power, and a way to determine precise location. The ability to maintain position at any desired location is mandatory for the accomplishment of most activities requiring contact with the seafloor (see Vol. XIII: *Oceanography, Applied*).

Platforms A platform is the fundamental requirement of any operation at sea. The three basic functions of platforms used in marine mineral exploration and exploitation are (1) to support equipment for observation and measurement, (2) to support tools and equipment for working with seafloor materials, and (3) to transport people and materials (Table 2). Platforms may be airborne, floating, or in contact with the bottom; they may be self-propelled or stationary; or they may be manned or unmanned. Floating platforms range in size from massive bulk carrier units weighing in excess of 300,000 T to an individual diver. Fixed platforms range from small drilling structures built out from shore to massive constructed islands covering many hectares.

Power Systems Power sources for use in marine mining operations are subject to many of the same limitations that apply to remote sites on land. Selection of the optimum power source must be weighed against factors such as amount of power required, including peak power and sustained load; availability of suitable generating equipment; and cost.

Power sources for marine use are many and varied and are constantly being improved. Most existing offshore mining operations use conventional power sources such as portable diesel electric generators, and there is a trend to the use of gas turbine generators in the offshore oil industry. These generators may be considered in any similar mining ventures. A serious limitation associated with the transmission of electric power at great ocean depths is the difficulty of providing watertight cable connectors. The trade-offs between high-voltage AC and high-voltage DC must be considered also with regard to transmission loss. The use of nuclear energy has been considered for future marine operations, and the utilization of energy from the ocean may be considered. For submerged instrumentation and measurement, many small portable power sources are available and reliable, and a continuous supply of power aboard floating platforms is available using conventional sources.

Anchoring The basic methods for maintaining the position of a floating platform are static anchoring and dynamic anchoring. The use of static anchors, which may be set out in multiple arrays, placed or drilled into the bottom, or thrust down as a rigid spud from the platform, involves contact with the sea bed. Dynamic systems utilize electronic positioning with multiple thrusters to maintain the vessel's position without bottom contact and are particularly useful in water depths greater than 500 m.

Exploration and Characterization

Since the 1960s, marine mineral exploration programs have increased, and much information is available on the tools and techniques used. Over

TABLE 2. Classification of Platforms in Use for Marine Mining Operations According to Function

Platform Function	Airborne	Sea Surface	Submerged	Bottom Contact
Measuring				
Exploration	X	X	X	X
Survey	X	X	X	X
Environmental monitoring	X	X	X	X
Handling				
Drilling	—	X	—	X
Dredging	—	X	—	X
Construction	X	X	X	X
Transporting				
Personnel	X	X	X	—
Supplies	X	X	X	—
Products	—	X	—	—

70 new or ongoing exploration programs were reported in 1972 and included operations in three major oceans and off the shores of 20 countries (Cruickshank, 1973). Major emphasis has been on coal, tin, titanium minerals, and gold, all in nearshore, relatively shallow, areas. There have been selective projects in deep water to sample manganese nodules in the Atlantic and Pacific oceans and metalliferous muds in the Red Sea. An exploration effort has four major categories: (1) geophysical sensing, (2) sampling, (3) characterization, and (4) deposit evaluation, which are usually carried out in this sequence. The last two are office procedures utilizing the data obtained in the initial stages at sea.

Geophysical Sensing Prospecting for mineral deposits, investigation of geologic structures, and measurement of environmental characteristics may be assisted by geophysical surveys (see *Sea Surveys*). Most subsurface structures and mineral deposits can be located if detectable differences in their physical properties exist, and many characteristics of the environment can be specified by such measurements. The major classical methods of geophysical sensing involve the measurement of density, magnetism, electrical conductivity (self-potential, resistivity, and induced potential), and elasticity (proportional to the velocity of wave propagation including seismic refraction and reflection, acoustic subbottom profiling, and sidescan sonar (see *Acoustic Surveys, Marine; Harbor Surveys; Marine Magnetic Surveys*). Other methods involve the measurement of both natural and induced radioactivity, thermal conductivity, and chemical activity. Although the latter have not been widely used in mining exploration, their use is becoming more widespread with refined technology. Optical methods of sensing include visual observations, using television or cameras in submersibles, aircraft, or spacecraft. Only those methods dependent on elastic, magnetic, and optical properties are freely applicable with existing equipment.

Sampling Evaluation of marine mineral deposits requires complete environmental characterization, which involves the sampling of the superjacent waters, the seafloor, and the subbottom. Different tools and techniques are required for each. Water samples may be required for trace element analysis, pollution control monitoring, or characterization of water masses. Particular layers in the water column may be sampled in bulk with suction apparatus or by the use of sample bottles. Surficial sampling of the seafloor is usually carried out using drag dredges or some form of mechanical grab. Dredges vary in recovery volume from a fraction of a yard to several cubic meters and are designed to be dragged along the seafloor. Large-diameter pipe dredges may be used to collect hard rock from the walls of steep submarine slopes or bottom. The grab will take gross samples of the seafloor surface, the size of which is limited by the capacity of the hoisting equipment. Tube-type corers are used for seafloor substrate sampling and range from simple open tubes (free-fall) to piston corers with liners. They are only usable in very soft sediments, and penetration is limited to a few meters.

In sandy or gravelly material interlocking grains make it necessary to apply mechanical power for drilling or vibrating. Power may be applied as impulsive, percussive, vibratory, rotary, or oscillatory jet motion, or it may involve high-pressure water or air jets or a combination of these. Depths of penetration are limited to tens of meters, and water depth capabilities are usually about the same. Methods for quantitative sampling of consolidated deposits by drilling to depths of thousands of meters have been perfected for minerals exploration but are not in general use. Deposits of high-unit-value materials may be sampled by coring, whereas low-unit-value materials are usually sampled by coring and/or sludge sampling.

Characterization Deposit characterization sampling is also carried out to determine the engineering properties of the seafloor materials and to relate these to geophysical survey data and subsequent mineral evaluation and even to the design of the mining system. Fewer samples will be needed for characterization than for evaluation, but undisturbed samples are required. Box corers are frequently used for this task in soft sediments.

Exploitation Systems

The systems required for minerals exploitation have five components, regardless of the mineral or method. These are the *mining or gathering* of ore, the *lifting* of the ore to surface, its *transportation,* the *treatment* to a usable commodity, and the *cycle of materials.*

Mining Methods The four basic methods of mining hard minerals include scraping the surface, excavating a pit, tunneling into the deposit, and fluidizing through boreholes. All deposits on land are mined by one or more adaptations of these, and sea bed minerals are amendable to the same basic methods (Table 3).

Penetration or disturbance of the seafloor will vary in degree according to the method chosen and the particular deposit, but for the same volume of material, scraping involves the greatest surface area, and fluidizing the least. Subsequent transport of the mined material to the next stage of treatment will involve requirements such as vertical hydraulic lifts, mechanical lift, seafloor pipelines, etc.

Scraping Deposits that are on or near the surface of the sea bed will be amendable to a gathering process that involves some form of mechanical

TABLE 3. Basic Methods for Mining Sea Bed Hard Minerals

	Method			
	Scraping	Excavating	Tunneling	Fluidizing
Typical example	Earth moving	Open-pit mining	Underground mining	Solution mining
Marine mining applications	Bulldozing Scraping Dragline dredging Trailer suction dredging	Bench and blast Bucket ladder dredging Cutter suction dredging	Adit entry Shaft entry	Frasch process Gasification Solution
Amenable seabed deposits	Sand and gravel Phosphorite	Oozes Crusts	Hardrock deposits	Metalliferous muds
	Shells Mn nodules	Outcropping Hardrock deposits		Potash Coal Sulfur
		Heavy mineral placers Sand and gravel		Metal sulfides

device scraping the sea bed to gather the ore into a position for lifting to the surface or other form of treatment. In general, penetration of the device is very much less than its horizontal movement. In its simplest form the action may be likened to raking or shoveling or, in some cases, vacuum cleaning. Where hard material is present, the scraping action may be preceded by ripping or blasting.

One example of the scraping method is the use of drag suction dredges for the exploitation of sand and gravel. Another is the systems developed for recovering metalliferous oxide nodules from the deep sea bed at depths of 5,000 m or more. Three major tests have been made at these depths, two of them utilizing hydraulic or air lift with a towed, passive collector, or self-propelled collector driven by Archimedian screws and containing a crushing device to slurry the nodules (Welling, 1981; U.S. Dept. of Commerce, 1981). The third used buckets. A continuous bucket line system employs a loop of flexible cable about 15 km long with collecting buckets attached at intervals. The loop can be suspended from the bow and stern of the mining ship or from two vessels on a close parallel course (Masuda et al., 1971; Gauthier and Marvaldi, 1975).

Excavating This word implies the removal of a large quanity of material from a limited surface area, leaving a hole of substantial dimension. Conventional dredges are limited by design to about 35 m water depth but tests in the Red Sea used a suspended vibratory suction dredge at a depth of 2,200 m to collect a bulk sample of

16,000 metric tons of metalliferous sulfide muds from the seafloor (Mustafa and Amann, 1980). Harder materials would require the application of other techniques such as ripping with large toothed machines similar to bulldozers or blasting with explosives or other agents. A normal technique for open-pit mining of unconsolidated materials is by dredging either by mechanical or hydraulic methods. Alluvial deposits of heavy minerals such as tin or gold are commonly dredged using bucket ladders. For deposits on nonmetallics such as shells or sand and gravels, hydraulic cutterhead dredges are typically used.

If hard rock is being mined, blasting and removal by dredge is an applicable method. Deposits amenable to this type of mining would be thick surficial deposits of indurated sediments or outcropping hard-rock deposits such as the metalliferous sulfide deposits at active spreading centers in mid-ocean ridges. One such undersea mine operating at the present time, for barite in Alaska, is in very shallow water close to land (Thompson and Smith, 1970). However, the basic techniques are applicable at any depths. More sophisticated methods of hard-rock surficial mining may be developed in the future where all handling and processing would take place on the seafloor, and the resulting mine layout might be similar to present-day terrestrial operations.

Tunneling Undersea mines worked from tunnels (see Vol. XIII: *Tunnels, Tunneling*) originating on dry land or artificial islands are quite common close to shore, particularly for undersea coal mines. In very deep water or at great distance from shore, no such operations have been seriously planned. Concepts have been described and preliminary designs prepared by the U.S. Navy in their Rocksite Project (Austin, 1966), in which it was proposed to develop a one atmosphere human-operated installation in the subsea bed, using lock tubes to provide access from the seafloor. This method would be very costly in terms of maintaining life support and transference of mined and waste materials across the seafloor interface but is technically feasible at the present time in shallow water.

Layout on the sea bed might be similar to some underground mines on land with regard to mined materials disposal, but all other activities such as shops and milling would most likely be underground.

Fluidizing Hard-rock deposits that are amenable to hydrometallurgical treatment of their ores are potentially extractible by fluidizing methods. The valuable constituent is transformed to a fluid phase in place and removed from the ground through a borehole (Marine Board, 1975) (see *Borehole Mining*). The prime example, the Frasch process for the recovery of sulfur, in which superheated water is pumped into the deposit to melt the sulfur so that it may be pumped out of the ground, is similar to that used for the recovery of oil and gas. There are major problems in dealing with toxic or corrosive solvents and in fracturing or otherwise providing a flow path for the solvent through the deposit. These problems are being overcome on land, and they should be applicable in more sophisticated form to deep-sea-bed deposits.

Lifting The process of lifting the mined material from the sea bed to its next place of treatment, which will be most commonly at sea level on board a platform or on land, may be accomplished by two basic methods. *Mechanical lifts* include devices such as drag buckets or grabs raised and lowered in repetitive fashion or constructed to carry material continuously to the surface with bucket lines or conveyors. The other basic method is *hydraulic*, where the mined material is carried in a fluid flow that is contained within a restrictive barrier such as a pipeline. The fluid may be water, gas, drilling mud, petroleum derivative, or any other suitable material or combination. Flow may be induced by pumping or injection of materials of light density such as air (airlift) or other gas or by the injection of liquids (jet lift) or light solids at high velocity into the stream. Other methods such as the application of electrically induced flow may be developed.

Transportation Analogous to the lifting phase of operation, transportation of mined materials can be accomplished in bulk containers, in continuous flow through pipes, or on conveyors.

Many choices are available for surface, submerged, or seafloor transportation or any combination. The final choice depends on the overall system of which it is a part.

Ore Treatment Extraction of metal from ores is basically a two-stage operation. First, a process is required to separate the valuable material from the waste. This is accomplished by utilizing differences in physical properties, such as particle size and specific gravity, magnetic properties, electrical properties, elastic properties, or radioactive properties.

The second stage involves the extraction of the metal or usable product from the ore. These processes are based on physicochemical properties and typically require the addition of chemical or other external agents. Extractive processes are more complex than the separation processes. Development of the ore treatment requirements is tailored specifically for each orebody (Taggert, 1945; Dames and Moore, 1977). While the initial ore separation processes are frequently carried out on site, all the extractive plants for marine mining operations are presently located in shoreside facilities.

Cycle of Materials All industrial operations can be assumed to have an input of raw materials, manufactured materials (or supplies), and energy. They can similarly be assumed to have an output of product and waste materials in gaseous, liquid, or solid form. The disposal of these materials presents some unique problems for marine mining operations, and the specifics are different for each operation. Most waste materials in mining operations are solid materials from the orebody or from its development.

Environmental Protection and Restoration

Up to this writing there are few precedents for the restoration of mining areas on the sea bed. Environmental protection of the mining area is of major concern, however, and much work has been carried out in the United States and elsewhere to ensure that disturbances from marine mining are minimized (U.S. Dept. of the Interior, 1974; U.S. Dept. of Commerce, 1981).

The major concern would be that the operation would not leave any conditions that might be detrimental to life or property over the long term. Planning for operations requires the prediction of the postoperational state, the avoidance of negative impacts, and the implementation of postoperational monitoring. Negative impacts include the creation of hazards to benthic organisms and other forms of marine life, disruption of navigation or commercial fishing, or alteration of coastal forms by changes in erosional or depositional patterns.

The wide variety of potential mining methods again makes generalization difficult, and the amount of restoration required on a mine site would depend largely on the geographic location, the regional environment, and the mining method. These factors would be examined in detail in a preoperational analysis for each proposed mining operation.

Operational Safety

The possibility of accident in any industrial operation must always be anticipated. Accidents may occur from natural causes at sea such as storms, tsunamis, earthquakes, slides, or even fish attacks, or they may be caused by human error, equipment failure due to any of the preceding, or fire, explosion, spillage, inundation, or collapse. The major concern in anticipating these calamities is the safety of human life, followed by concern for the natural environment and property, including waste of the resource. Design features that may be built in to avoid accidental loss are part of the initial development of any mining project.

The Future

Although marine mining has been selectively practiced since the early part of the century, it is still in a formative period. New discoveries of marine minerals, particularly in the deep sea bed, new technologies developed by the aerospace and offshore oil industries, depletion of accessible mineral deposits on land, and concern for the continental environment tend to enhance the prospects for commercial development of marine minerals in the future. The apparent marine resources of most mineral commodities exceed those for the same commodities on land, and the apparent resources of dissolved minerals in seawater exceed those for all others combined, by a factor of three (Cruickshank, 1978). By the middle of the next century, marine mining should be a substantial industry, supplying much of the world's fast-growing mineral requirements.

MICHAEL J. CRUICKSHANK

References

Austin, C. R., 1966, Manned undersea structures, the rock site concept, Naval Ordnance Test Station, (Tp 4162), 44p.

Cruickshank, M. J., 1962, Exploration and Exploitation of Offshore Mineral Deposits, M.Sc. thesis, Colorado School of Mines, 185p.

Cruickshank, M. J., 1973, Mining and Mineral Recovery: Undersea Technology Handbook Directory. Houston: Compass Publications, A15–28.

Cruickshank, M. J., 1978, Technological and Environmental Considerations in the Exploration and Exploitation of Marine Minerals, Ph.D. dissertation, University of Wisconsin, Madison, 217p.

Cruickshank, M. J., 1982, Recent studies on marine mineral resources, paper presented at the Oceanology International Conference, Brighton, England, March 2–9.

Cruickshank, M. J., and R. W. Marsden, 1973, Marine mining, in A. B. Cummins and I. A. Innes, eds., Mining Engineering Handbook. New York: Society of Mining Engineers, Chapter 20.

Cruickshank, M. J., C. M. Romanowitz, and M. P. Overall, 1968, Offshore mining—Present and future, Eng. and Min. Jour. 60(1), 84–91.

Dames & Moore and E.I.C. Corp., 1977, Description of Manganese Nodule Processing Activities for Environmental Studies, vol. 1. Processing Systems. Rockville, Md.: U.S. National Oceanic and Atmospheric Administration, 132p.

Gauthier, M. A., and J. H. Marvaldi, 1975, The two-ship CLB system for mining polymetallic nodules, paper presented at the Oceanology International Conference, Brighton, England.

Lee, C. O., et al., 1960, The Grand Isle Mine, Mining Eng. 12(6), 578–590.

Marine Board, 1975, Mining in the Outer Continental Shelf and in the Deep Ocean. Washington, D.C.: U.S. National Academy of Sciences.

Masuda, Y., M. J. Cruickshank, and J. L. Mero, 1971, Continuous bucket line dredging at 12,000 feet, Offshore Technology Conf. Proc. (OTC 1410), 837–858.

Mustafa, Z., and H. M. Amann, 1980, The Red Sea pre-pilot mining test 1979, Offshore Technology Conf. Proc. (OTC 3874), 197–210.

Taggert, A., 1945, Handbook of Mineral Dressing. New York: John Wiley & Sons, 1,915p.

Thompson, R. M., and K. G. Smith, 1970, Undersea lode mining in Alaska, paper presented at the Second Annual Offshore Technology Conference, Houston, Texas, April 22–24 (Paper No. OTC 1312).

United Nations, 1980, Draft Convention on the Law of the Sea (informal text). New York: U.S. Department of State, Office of the Law of the Sea Negotiations, 180p.

U.S. Department of Commerce, 1981, Deep Seabed Mining. Washington, D.C.: NOAA (Final Programmatic Environmental Impact Statement), vol. I, 221–260.

U.S. Department of the Interior, 1974, Proposed outer continental shelf hard mineral mining operating and leasing operations, draft environmental statement, Washington, D.C., 362p.

Welling, C. G., 1981, An advanced design deep sea mining system, Offshore Technology Conf. Proc. No. 4094, 247–255.

Cross-references: *Acoustic Surveys, Marine; Borehole Mining; Floating Structures in Waves; Harbor Surveys; Marine Magnetic Surveys; Plate Tectonics, Mineral Exploration; Sea Surveys.* Vol. XIII: *Coastal Engineering; Marine Sediments, Geotechnical Properties; Oceanography, Appied; Pipeline Corridor Evaluation.*

MATHEMATICAL GEOLOGY

Mathematical geology, or geomathematics, in its broadest sense, deals with all applications of mathematics to the Earth's crust (Agterberg, 1974, 1979b, 1982). In mathematical geology emphasis is given to geological process-models that are modeled mathematically. Geomathematics deals primarily with methods of calculation and computer algorithms for problem solving in geology. Because of the irregular spatial distribution of the observations (including boreholes), the diversity of geoscientific information, and the imperfect record of events that took place in the course of geologic time, geologists have to cope with significant amounts of uncertainty. Their two main sources of concepts and methods for modeling by means of random variables are the scientific disciplines of (1) mathematical statistics and (2) geostatistics (theory of regionalized random variables) as developed by Matheron (1965, 1976). Progress in geomathematics is closely linked to the usage of computers for automated data handling, and a separate subsection is devoted to this topic.

Applications of mathematics and statistics to solve geological problems were initially made by a small number of mathematically inclined geologists (cf. Merriam, 1981b). Playfair (1802) calculated the depth at which beds would occur downdip from surface measurements. He also computed subareas occupied by certain rock types in a larger area. Probably the best known example of an early application of statistical reasoning in geology is Lyell's (1833) definition of the epochs of the Tertiary from relative abundances of taxa that are still in existence today compared to all taxa found in the series studied.

The first mathematical geologists extensively using statistics for solving a variety of problems include Krumbein (1969), Vistelius (1980), Chayes (1956), and Griffiths (1970). These authors made applications to many different attributes of both sedimentary and igneous rocks. One of their objectives was to develop methods for assessing the probability of occurrence of undiscovered mineral deposits (hydrocarbon and metal resources evaluation). This has remained a primary concern of geomathematicians.

Important contributions to geology made by mathematical statisticians include the following. Kolmogorov (1951) developed a stochastic process-model to explain the logarithmic normal distribution of the thickness of beds in certain sequences of strata. Fisher (1953) devised a statistical method for calculating vector means with measures of their precision. The topic of statistical treatment of directional features was further pursued by Watson (1981), who also considered statistical problems with homogeneous strains. Aitchison (1981) has proposed a new approach to correlation of proportions such as the major chemical constituents of rocks.

The theory of regionalized random variables has been developed by Matheron (1965) since the 1950s, building on work by de Wijs (1951) and Krige (1951). The latter two authors pioneered the development of methods for ore reserve valuation. Because of this the expressions *de Wijsian model* and *kriging* have been coined. In a de Wijsian model it is assumed that the spatial variability of metal concentration values is independent of block size (similarity assumption). Kriging is a method to predict the metal concentration of a block from observed values in its vicinity.

Recent Developments

A comprehensive review of the role of data processing in the geological sciences has been provided by Howarth and Leake (1980). These authors concluded that the geological sciences are in a state of rapid transition from the classical qualitative approach, involving largely the use of map and hammer, to a quantitative evaluation of measurements of great variety that overlap substantially into the physical, chemical, and biological spheres.

A collection of review articles by different authors on computer applications in the earth sciences for the period 1971–1980 is contained in Merriam (1981a). Subdisciplines treated separately in this book are paleontology, paleoecology, biostratigraphy, stratigraphic analysis, seismic reflection methods, petroleum exploration, oceanography, petrology, geochemistry, field geology, map analysis, mineral resources evaluation, and mining geology. There have been significant recent developments in each of these fields.

Dobrin (1981) reviewed the use of computers in seismic reflection prospecting. According to Dobrin, new recording and processing technology has brought us much closer than anyone would have predicted in 1970 to the ultimate goal of extracting the same geological information that would be retrievable if there were a borehole to the maximum depth of interest at every shotpoint. Under proper conditions it is now feasible to detect gas deposits on seismic record sections by making use of the fact that reflections from the top of gas-filled sands have a higher amplitude than those from water- or oil-filled sands. These relative reflection amplitudes on seismic records can be recognized only through digital recording and processing (also see Wren, 1980).

Several books have been published on the use of geomathematics in the study of sediments (Schwarzacher, 1975; Merriam, 1976; Gill and Merriam, 1979; Cubitt and Reyment, 1983). Statistical and mathematical techniques are important in the search for oil (Harbaugh et al., 1977; Miall, 1980) and minerals (Weiss, 1979), as well as for the precise calculation of ore reserves (Guarascio et al., 1976; David, 1977; Journel and Huijbregts, 1978; Verly et al., 1983; Matheron and Armstrong, 1987) and broken ore sampling (Gy, 1979). Twomey (1977) has dealt with the mathematics of inversion in remote sensing. Mardia (1972) has treated directional statistics. Singer and Mosier (1981) classified over 100 papers on regional mineral resource assessments of nonfuels according to method used and form of product. A variety of geomathematical techniques are presented in Prissang and Skala (1979). Future trends of geomathematics are outlined in Craig and Labovitz (1981).

Several research projects on quantitative techniques in the earth sciences that are producing new results are being conducted under the auspices of the International Geological Correlation Programme (IGCP), which commenced in 1972. IGCP focuses on the worldwide organization and distribution of knowledge about geological resources and environment. Its research projects on quantitative methods and data processing in geological correlation include "Standards for computer applications in resource studies" (IGCP project 98, 1975–1980; see Cargill and Clark, 1978), "Quantitative stratigraphic correlation techniques" (IGCP project 148, 1976–1983; see Agterberg, 1982), and "Design and generation of world data base for igneous petrology" (IGCP project 163, 1977–1984; see Chayes, 1981). The purpose of these projects is not only to develop new geoprocessing methods but also to apply them toward the solution of large-scale problems.

Geologists have constructed qualitative models for many of the processes that have taken place. An example is provided by the model of a Kuroko-type deposit developed since the 1960s in economic geology (see Franklin et al., 1981). Observations on the relatively young (Miocene) Kuroko orebodies in Japan (see, e.g., Lambert and Sato, 1974) have resulted in the model that a deposit of this type was formed syngenetically on the bottom of the sea in close association with felsic volcanics. This process resulted in a lenticular body of massive sulfides with a recognizable feeder pipe below it and clay alteration zones around it. Gannicott et al. (1979) used this model to detect chemical alteration patterns around the older (Jurassic) Seneca deposit near Harrison Lake, British Columbia, by statistical analysis of noisy data.

Most of the publications referenced in the preceding paragraphs are in the field of statistical treatment of geoscience data. Other mathematical modeling consists of the application of differential equations from the theory of continuum mechanics and other theories of physics and chemistry by means of which it is attempted to describe geological processes in a deterministic manner. Whitten (1981) has pointed out that many geomathematical applications in structural geology are now in the field of finite-element fold modeling. He discussed many papers in this field published during the 1970s. Geophysicists continue to make advances in modeling gravimetric and other responses from orebodies of known size and shape using potential theory. Geophysical applications of finite-element modeling for electrical and electromagnetic data in three dimensions have been reviewed by Pridmore et al. (1981).

Theory for the flow of fluids is used in the modeling of water movement. For example, Simons (1972), as well as Bonham-Carter and Thomas (1973), predicted the three-dimensional pattern of steady wind-driven currents in Lake Ontario. The topic of northern lake modeling has been reviewed by Fox et al. (1979), and a book on fluid transport in porous media has been published by Dullien (1979). Freeze and Cherry (1979) discussed mathematical modeling of groundwater resources. The mathematical treatment of geomorphological processes and sedimentation is receiving attention (see Komar, 1976, 1980).

Some types of petrological processes are amenable to both empirical statistical treatment and a theoretical approach based on physical chemistry. The calculation of mass transfer among minerals and aqueous solutions in geochemistry provides an example. LeMaitre (1982) discussed petrological mixing models in which the coefficients of metamorphic reactions are estimated by means of least squares fitting. In contrast, Helgeson and Murphy (1983) use requisite mass transfer equations for isothermal-isobaric processes involving stoichiometric models. These authors have expressed the reversible dissolution of a single mineral as a function of time and effective surface area. Five examples of new developments (quantitative stratigraphy, spatial analysis, applications of image analysis, use of multivariate statistical techniques in mineral resources evaluation, and geostatistical crustal abundance models) are discussed in more detail in the next section.

Selected Applications

Quantitative Stratigraphy Stratigraphy is concerned with the original succession and age relations of rock strata. It covers the form, distribution, lithologic composition, fossil content, and other properties of layered rocks (cf. Hedberg, 1976). The three best known categories of stratigraphic classification are (1) lithostratigraphy, in which strata are organized into units based on their lithologic character; (2) biostratigraphy, with units based on fossil content; and (3) chronostratigraphy, with units based on the age relations of the strata.

As in other fields, the rapid growth of information in biostratigraphy and lithostratigraphy (e.g., well logs) has led to an increased demand for quantification of information for machine handling or graphic display. Quantitative stratigraphy is useful because it helps to organize the information in novel ways.

Special properties of the paleontological record form the basis of biostratigraphy. The properties include first appearance (entry), range, peak occurrence, and last appearance (exit) of fossil taxa. These events differ from physical or chemical events in that they are unique and nonrecurrent and their order is irreversible. Paleontological correlation for geological studies depends on comparing similar fossil occurrences in or between regimes and is commonly referred to as a paleontological zonation. Recent reviews of quantitative biostratigraphic correlation techniques have been published by Hay and Southam (1975), Worsley and Jorgens (1977), Van Hinte (1978), Brower (1981), Hudson and Agterberg (1982), Agterberg (1984), and Goldstein et al., (1985). Successful large-scale applications of these methods have been made by Blank and Ellis (1982), Gradstein and Agterberg (1983), and Doeven et al. (1982), D'Iorio (1986), and Williamson (1987).

The observed order of biostratigraphic events is different from place to place. In correlating wells drilled for oil, occurrences of the same event in different wells normally are connected by straight lines in stratigraphic profiles or fence diagrams. If there is a reversal in order for two events in two wells, these lines will cross. The cross-over frequency for pairs of events therefore provides a measure of inconsistency.

Gradstein and Agterberg's (1983) data base consisted of 206 Cenozoic foraminiferal data events (highest occurrences only) in 22 wells along the Canadian Atlantic margin. On the basis of this information, Agterberg and Nel (1982a and b) have developed a computer program for the ranking and scaling of events, the RASC program, which produces two types of biostratigraphical answers: (1) the optimum sequence of events along a relative time scale and (2) the optimum clustering, based on the cross-over frequencies of events, weighted for the number of occurrences, using the optimum sequence.

The scale along which the clustering is performed can be related to the numerical time scale. This permits study of the depositional history of the Canadian Atlantic margin (Agterberg and Gradstein, 1983, and Gradstein et al., 1985).

Spatial Analysis A comprehensive review of methods of spatial statistics has been provided by Ripley (1981), including statistical theory for constructing contour maps. Other books specifically dealing with spatial analysis are by Davis and McCullagh (1975) and Gaile and Willmot (1984). For example, Agterberg and Chung (1973) used kriging for the extrapolation of observed patterns of spatial variation of average sulfur content in existing collieries along the coast toward submarine coal reserves of the Lingan Mine on Cape Breton Island, Nova Scotia. Results were obtained by using two models, kriging (Model 1) and universal kriging (Model 2), for sulfur content of the harbor seam, which is about 2 m thick. The difference between these two models is that in kriging (Model 1) it is assumed that the same covariance function is valid in the entire study area, whereas in Model 2, a regional trend is subtracted from the data before this assumption is made. The uncertainty of the predicted values is relatively large, as indicated by standard deviations for the predicted values.

The coal on Cape Breton Island is used for thermal power generation as well as for the manufacturing of metallurgical coke. A metallurgical coal should have no more than 1% sulfur after washing. The new wash plant at Sydney can economically reduce coal with 3% sulfur to this level (Hacquebard, 1979). For planning purposes it is desirable to predict as precisely as possible the reserves of offshore coal (e.g., significant stone partings could prevent mining) and the average sulfur content of the reserves to regulate the supply of metallurgical coal.

In 1972, when the spatial analysis was performed, the reserves had not been sampled because offshore drilling for this purpose was too expensive. However, in 1977 six offshore boreholes were drilled. Five of these are located in areas with relatively low predicted average sulfur content according to the contours constructed in 1972 or when these contours are projected toward the east. However, as pointed out by Hacquebard (1979, 54) to have positive proof of the existence of extensive low-sulfur areas offshore requires more samples than have been available to date. Other applications of this type of spatial analysis have been described by Agterberg (1984).

Applications of Image Analysis Methods of image analysis have been reviewed by Watson (1975), Underwood et al. (1976), and Ripley (1981). Matheron (1975) has developed a random set theory to formalize the measurements that can be made by image analyzers. Useful techniques have been developed by Serra (1982). Some of these methods were applied to a contour map of the thickness of the Sparky Sandstone in the Lloydminster area, Alberta (Agterberg, 1980). Maps representing geological features commonly have a small scale like topographical maps. Hence, a considerable amount of smoothing and generalization is normally required. The theory of fractal analysis by Mandelbrot (1977, 1982) can be usefully employed for this purpose. Mandelbrot's approach has resulted in a wealth of computer algorithms for simulating random spatial phenomena (e.g., Carpenter, 1980; Fournier and Fussell, 1980).

Geologists are concerned with the study of morphology for a variety of reasons. Preferred orientations of map patterns may reflect the directions of specific geological processes that were operative in the past (e.g., sedimentation on a delta, structural deformations of rock formations, or change of shape of microfossils in the course of evolution). Rock textures with different types of spatial relationships between pores and mineral grains also provide a fertile ground for the application of automated methods of image analysis, especially when porosity and permeability of rocks are to be determined.

Geological applications of image analysis have been described in Agterberg and Fabbri (1978), Agterberg (1979b, 1981b), Fabbri (1981), and Fabbri and Kasvand (1981). These applications are for images of geoscience maps as well as thin sections of rocks. An alternative approach for the study of rock textures is to model sequences of mineral grains as Markov chains (Vistelius, 1980; Vistelius et al., 1983). Ehrlich and Weinberg (1970) have developed a method of characterizing grain shapes by Fourier series expansion of their boundaries in a thin section.

Use of Multivariate Statistical Techniques in Mineral Resources Evaluation This topic has been reviewed in Agterberg (1981a, 1981b), Agterberg and David (1979), Chung and Agterberg (1980), and Harris and Agterberg (1981), where further references are given. The objective of these methods is to evaluate various geological environments for their probability to contain different types of mineral deposits and to define targets for mineral exploration. Botbol et al. (1978) and McCammon et al. (1983) used characteristic analysis, which is a version of principal components analysis. These authors preprocessed geochemical variables by constructing second-derivative maps similar to those used in exploration geophysics (Dobrin, 1976). Bonham-Carter and Chung (1983) have found that considering the spatial character of geochemical variables may result in significant improvements of results obtained by multivariate statistical analysis. These authors removed trends from the data and applied kriging to the residuals.

The kriged values were then used in the multivariate analysis.

During the early 1980s several interactive graphic computer systems have been developed for probabilistic prognosis (Chung, 1983; Duda et al., 1981; Fabbri, 1981; Harris and Carrigan, 1980; McCammon et al., 1983). These systems have in common that various types of geoscience data are converted into patterns that are useful to predict occurrence of undiscovered mineral deposits. There are considerable differences in the relative amounts of factual and subjective judgments required as input. Duda et al. (1981) and Harris and Carrigan (1980) use networks of decision rules and allow for contradictory subjective opinions of individual geologists to cope with the uncertainties related to geological projections and processes. The method of Duda et al. (1981) is called an *expert system* in computer science. It is an application of artificial intelligence, which is a subfield of computer science. In contrast, Chung (1983), Fabbri (1981), and McCammon et al. (1983) operate on map data systematically quantified for regions and use statistical methods for obtaining the derivative patterns.

Geostatistical Crustal Abundance Models Various types of geochemical abundance models have been reviewed in Agterberg (1981c), Harris and Agterberg (1981), Schuenemeyer and Drew (1983), and Harris (1984). The primary objective of these models is to aid in mineral resource evaluations for large regions and on a worldwide basis. For example, a lognormal model originally developed by Agterberg and Divi (1978) for the Canadian Appalachian Region is based on the following three hypotheses, which can be used separately in practical applications:

1. The concentration values of some scarce metals (e.g., Cu, Pb, Zn, U, and Au) in blocks of the same weight sampled from the upper part of the Earth's crust satisfy a lognormal distribution with logarithmic variance that depends only on the homogeneous spatial autocorrelation function of the metal considered.
2. The sizes (ore tonnages) and mean grades (concentration values) of the mineral deposits for the scarce metal in a large region satisfy a bivariate lognormal (size-grade) distribution. The logarithmic variance of the metal's concentration values in small blocks is the same for all mineral deposits considered.
3. The observed bivariate lognormal size-grade distribution for the scarce metal in known mineral deposits is the result of economic filtering of another lognormal distribution, which includes the subeconomic mineral deposits in the region as well as the known mineral deposits.

These three hypotheses can be formulated in mathematical terms, and a larger set of relationships between metal concentration values and sizes of blocks of ore or rock can be derived. The geochemical and mineral deposit data commonly available for larger regions can be described by using these relationships, and speculative projections can be made regarding the subeconomic ore in the region. These projections are speculative because, even when the geostatistical model fits the parameters of the known ore deposits, this does not necessarily imply that it would also fit those of the unknown mineral resources.

Computers in Geology

Today one can buy a microcomputer with as much speed and precision as the digital computers commonly used in the early 1960s. Large computers are millions of times more powerful than the first experimental computers of the late 1940s and are still becoming faster, more accurate, more reliable, and about 25% less expensive per year. The development of our capabilities to make intelligent use of these machines has been slow, however (see *Computers in Geology*).

The hardware of a computer consists of memory, a central processing unit (CPU), and peripheral equipment. The term *software* refers to the instructions and computer programs. The memory can hold a large quantity, frequently millions, of numbers and words that can be manipulated in the CPU. Magnetic tapes, disks, and drums in the peripheral equipment are used to extend the memory of a computer to billions of facts, to which a user has almost instantaneous access.

Special peripheral equipment can increase the input capabilities by giving the digital computer a numeric representation of the physical phenomenon it is experiencing. For example, a geologic map could be digitized automatically as about a million gray-level values measured for separate picture points whose locations are memorized. Other peripheral devices such as plotters and television-like cathode-ray tubes (CRTs) facilitate graphical display, which can be full color or monochromatic. These technological advances have been accompanied by advances in computer-based statistical analysis, improved data management techniques, and the development of new methods of pattern analysis. The following two illustrative examples deal with the probability of occurrence of mineral deposits.

Example of Image Analysis Figure 1*A* shows the occurrence of acidic volcanic rocks as represented on geological maps with a scale of about 1:125,000 of the Bathurst–Newcastle area in the Canadian Appalachian region. The C-shaped area of rhyolitic rock represented in black is surrounded by metabasalt and sedimentary rocks

with which it belongs to the Tetagouche Group of Middle–Late Ordovician age. The rhyolitic core has been referred to by some geologists as a basin structure and by others as a dome. Skinner (1974) has argued convincingly that the C shape is the result of two periods of folding. According to Skinner, the Tetagouche Group was folded into northwesterly trending recumbent folds overturned toward the southwest during the late Ordovician Taconic Orogeny, then refolded with an approximately northeasterly trending axis during the Devonian Acadian Orogeny. The rhyolitic core would be the youngest part of the Tetagouche Group and stratigraphically underlain by the metabasalt and sedimentary rock. Skinner suggested an ignimbritic (pyroclastic flow) origin for most of the acidic volcanics. Figure 1*F,* which is for the same area as Fig. 1*A* shows the occurrences of 40 volcanogenic massive sulfide deposits located at the centers of the black squares. Most of these mineral deposits occur within or adjacent to rhyolite crystal tuff, although the host rock is generally a chlorite schist associated with a magnetic iron formation.

Agterberg and Fabbri (1978) have studied the relationship between the patterns of Figs. 1*A* and 1*F* after digitizing them on a flying spot scanner and by using Minkowski operations (cf. Serra, 1982) such as dilatation (Figs. 1*B* and 1*C*) and erosion (Figs. 1*D* and 1*E*). Two dilatations of the 40 mineral deposits are shown in Figs. 1*F* and 1*G*. Figure 1*H* is the intersection between the patterns of Figs. 1*A* and 1*G*. The area of each black pattern can be measured precisely by counting the black picture points. These measurements could be used to answer, e.g., the following five questions: (1) What is the probability that a very small area underlain by acidic volcanic rocks (as shown on a 1:125,000 geological map) contains a sulfide deposit? (2) What is this probability if the very small area is not more than a given distance removed from at least one contact between acidic volcanics and other rock units? (3) What is the probability that a square cell of a given size placed at random on the study area contains at least one sulfide deposit? (4) What is the latter probability given the proportion of the cell underlain by acidic volcanics? (5) What is the proportion of acidic volcanics in a cell of given size for which the probability that this cell contains at least one mineral deposit is a maximum? For answers of these questions, the reader is referred to Agterberg and Fabbri (1978).

Example of Multivariate Analysis The second example (originally from Agterberg et al., 1972) is illustrated in Fig. 2. Cells measuring 10 km on a side were used to quantify geological and geophysical data for an area underlain by Archean rocks within the Abitibi Volcanic Belt on the Canadian Shield. The percentage of rock types

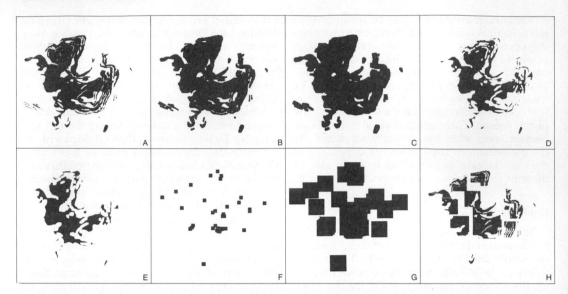

FIGURE 1. Image analysis of acidic volcanic rocks (A) and volcanogenic massive sulfide deposits (F) in the Bathurst–Newcastle area, New Brunswick. B and C represent the first two dilatations of the pattern of acidic volcanics (A); D and E are the first two erosions of A; F and G are 9-point and 19-point dilatations of the original pattern of 40 points for mineral deposits; H represents the area of coincidence of A and G. A single dilatation (or erosion) consisted of adding (or subtracting) adjoining picture points along all boundaries. The distance between original picture points was 259 m in the north-south and east-west directions. North points upward. The frame measures approximately 80 km on a side. (From Agterberg and Fabbri, 1978)

shown on geological maps with a scale of about 1:250,000 was determined for each of the cells (by point counting). The objective of this project was to determine for each cell the conditional probability that it contains massive sulfide deposits with copper, this probability being conditional on the rock-type percentage values and some geo-physical parameters. Conditional probabilities of this type can be estimated by a variety of multi-variate statistical methods. The model used for Fig. 2 is stepwise multiple regression.

Each probability estimated for a single cell can be interpreted as the expected number of events for that cell. An event consists of the occurrence

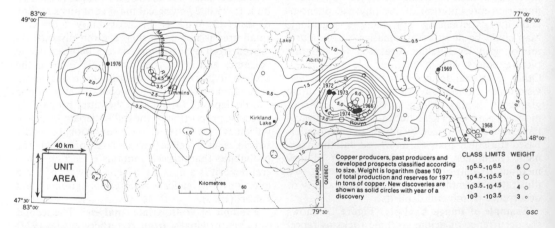

FIGURE 2. Contoured probability index map for expected number of events per 40 km unit area in the Abitibi area, with new discoveries shown for comparison. An event represents one or more copper deposits per 10 km cell. The contoured value defines a random variable with an approximately binomial distribution for the probabilities of having 0, 1, 2, . . . events per surrounding 40 km cell. Stepwise regression technique was applied to 55 lithological and geophysical variables defined for the 10 km cells belonging to the Universal Transverse Mercator grid. (After Agterberg et al., 1972)

of one or more mineral deposits in that cell. The advantage of this interpretation is that expected values can be added for groups of adjacent cells, and this gives expected numbers of events in larger unit areas that can be contoured. Figure 2 shows contoured expected values for copper deposits with the unit area measuring 40 km on a side. The contours were scaled with respect to a relatively well-explored region for which it was assumed that all 10 km cells with and without deposits had been correctly identified. The values in Fig. 2 are relatively high when there are many contacts between acidic volcanics and other rocks in the surrounding 40 km unit area and when mafic volcanics and acidic intrusives are also present in this vicinity. The Bouguer anomaly is above its regional average in these places.

The contours in Fig. 2 were constructed in 1971 on the basis of cumulative production and reserves of copper until 1968. The event was defined as the occurrence of one or more copper deposits, each containing at least 1,000 short tons (907 metric tons) of copper per 10 km cell. Later discoveries of the same type are shown for comparison in Fig 2. Their locations with respect to the contours illustrate the prognostic value of probability index maps. A more detailed report on this hindsight study was given in Agterberg and David (1979).

Conclusion

The application of statistics to facts in geoscience is subject to many difficulties. The great fluctuations in the weights to be assigned to separate observations at the surface of the Earth present significant problems that cannot be solved without the incorporation of concepts involving geological statistical manipulations of field data where statistically inclined geologists are not satisfied with the lack of formal proof associated with most geoscience concepts.

Statistical models in geoscience should be essentially spatial in that specific features are to be correlated with features both in their vicinity and occurring at greater distances. For example, on a local scale the mineral deposits of Fig. 1F are most strongly correlated with sedimentary rocks, whereas on a 1:125,000 geological map, most of them fall on acidic volcanic rocks. On an even more regional basis, their location is correlated with the contacts between acidic volcanics and other rocks. This is illustrated by the fact that the probability that a cell measuring 10 km on a side contains one or more mineral deposits is almost zero if it is fully underlain by acidic volcanics.

The spatial characteristics of geoscience facts and the importance of the nonobservable and frequently unknown processes that led to their origin dictate that interpretations of statistics in geoscience differ significantly from those of other types of statistics. The new problems that arise offer a wide ranging and challenging field of research.

FREDERIK P. AGTERBERG

References

Agterberg, F. P., 1974, *Geomathematics*. Developments in Geomathematics, vol 1. Amsterdam: Elsevier, 596p.

Agterberg, F. P., 1979a, Algorithm to estimate the frequency values of rose diagrams for boundaries of map features, *Computers and Geosci.* 5, 215–230.

Agterberg, F. P., 1979b, Statistics applied to facts and concepts in geoscience, *Geologie en Mijnbouw* **58**, 201–208.

Agterberg, F. P., 1980, Mineral resource estimation and statistical exploration, *Canadian Soc. Petroleum Geologists Mem. 6*, 301–318.

Agterberg, F. P., 1981a, Application of image analysis and multivariate analysis to mineral resource appraisal, *Economic Geology* 76, 1016–1031.

Agterberg, F. P., 1981b, Cell-value distribution models in spatial pattern analysis, in R. G. Craig and M. L. Labovitz, eds., *Future Trends in Geomathematics*. London: Pion, 4–28.

Agterberg, F. P., 1981c, Geochemical crustal abundance models, *Soc. Mining Engineers Trans.* **268**, 1823–1830.

Agterberg, F. P., 1981d, IGCP Project 148—Evaluation and development of quantitative stratigraphic correlation techniques, *Geol. Correlation* 9, 57–58.

Agterberg, F. P., 1982. Recent developments in geomathematics, *Geo-Processing* 2, 1–32.

Agterberg, F. P., 1984, Trend-surface analysis, in G. L. Gaile and C. J. Wilmott, eds., *Spatial Statistics and Models*. Dordrecht: Reidel, 147–171.

Agterberg, F. P., ed., 1984, Theory, Application and comparison of Stratigraphic Correlation Methods, *Computers and Geosciences 10*, 1–183.

Agterberg, F. P., and C. F. Chung, 1973, Geomathematical prediction of sulphur in coal, New Lingan Mine area, Sydney Coalfield, *Canadian Mining and Metall. Bull.*, October, 85–96.

Agterberg, F. P., C. F. Chung, A. G. Fabbri, A. M. Kelly, and J. S. Springer, 1972, Geomathematical evaluation of copper and zinc potential of the Abitibi area, Ontario and Quebec, *Canada Geol. Survey Paper 71-91*, 55p.

Agterberg, F. P., and M. David, 1979, Statistical exploration, in A. Weiss, ed., *Computer Methods for the 80's*. New York: American Institute of Mining, Metallurgy and Petroleum Engineers, 90–115.

Agterberg, F. P., and S. R. Divi, 1978, A statistical model for the distribution of copper, lead and zinc in the Canadian Appalachian Region, *Econ. Geology* 73, 230–245.

Agterberg, F. P., and A. G. Fabbri, 1978, Spatial correlation of stratigraphic units quantified from geological maps, *Computers and Geosci.* 4, 285–294.

Agterberg, F. P., and F. M. Gradstein, 1983, System of interactive computer programs for quantitative stratigraphic correlation, *Geol. Survey Canada, Current Research, Paper 83-1A*, 83–87.

Agterberg, F. P., and L. D. Nel, 1982a, Algorithms for the ranking of stratigraphic events, *Computers and Geoscience* **8,** 69–90.

Agterberg, F. P., and L. D. Nel, 1982b, Algorithms for the scaling of stratigraphic events, *Computers and Geoscience* **8,** 163–189.

Aitchison, J., 1981, A new approach to null correlations of proportions, *Jour. Mathematical Geology* **14,** 175–190.

Blank, R. G., and C. H. Ellis, 1982, The probable range concept applied to the biostratigraphy of marine microfossils, *Jour. Geol.* **90,** 415–433.

Bonham-Carter, G. F., and C. F. Chung, 1983, Integration of mineral resource data for Kasmere Lake area, Northwest Manitoba, with emphasis on uranium, *Jour. Mathematical Geology* **15,** 25–45.

Bonham-Carter, G. F., and J. H. Thomas, 1973, Numerical calculation of steady wind-driven currents in Lake Ontario and the Rochester Embayment, *Proc. 16th Conf. Great Lakes Research,* 640–662.

Botbol, J. M., R. Sinding-Larsen, R. B. McCammon, and G. B. Gott, 1978, A regionalized multivariate approach to target selection in geochemical exploration, *Econ. Geology* **73,** 534–546.

Brower, J. C., 1981, Quantitative biostratigraphy, 1830–1980, in D. F. Merriam, ed., *Computer Applications in the Earth Sciences.* New York: Plenum, 63–103.

Cargill, S. M., and A. L. Clark, eds., 1978, Standards for computer applications in resource studies, *Jour. Mathematical Geology* **10,** 405–642.

Carpenter, L. C., 1980, Computer rendering of fractal curves and surfaces, Computer Graphics, *Quart. Assoc. Computing Mach.* (Spec. SIGGRAPH'80 issue), 9–15.

Chayes, F., 1956, *Petrographic Model Analysis.* New York: Wiley, 113p.

Chayes, F., 1981, IGCP Project 163: Design and generation of a world data base for igneous petrology, *Geol. Correlation* **91,** 198–200.

Chung, C. F., 1983, SIMSAG: Integrated computer system for use in evaluation of mineral and energy resources, *Jour. Mathematical Geology* **15,** 47–58.

Chung, C. F., and F. P. Agterberg, 1980, Regression models for estimating mineral resources from geological map data, *Jour. Mathematical Geology* **12,** 473–488.

Craig, R. G., and M. L. Labovitz, eds., 1981, *Future Trends in Geomathematics.* London: Pion, 318p.

Cubitt, J. M., and R. A. Reyment, eds., 1983, *Quantitative Stratigraphic Correlation.* Chichester, England: Wiley, 320p.

David, M., 1977, *Geostatistical Ore Reserve Estimation,* Developments in Geomathematics 2, Amsterdam: Elsevier, 364p.

Davis, J. C., and M. J. McCullagh, eds., 1975, *Display and Analysis of Spatial Data.* London, Wiley, 378p.

De Wijs, H. J., 1951, Statistics of ore distribution, *Geol. en Mijnbouw* **30,** 365–375.

D'Iorio, M. A., 1986, Integration of foraminiferal and dinoflagellate data sets in quantitative stratigraphy of the Grand Banks and Labrador Shelf, *Canadian Petroleum Geology Bull.* **34,** 277–282.

Dobrin, M. B., 1976, *Introduction to Geophysical Prospecting,* 3rd ed. New York: McGraw-Hill, 630p.

Dobrin, M. B., 1981, Use of computers in seismic reflection, in D. F. Merriam, ed., *Computer Applica-*

tions in the Earth Sciences. New York: Plenum, 145–168.

Doeven, P. H., F. M. Gradstein, A. Jackson, F. P. Agterberg, and L. D. Nel, 1982, A quantitative nannofossil range chart, *Micropaleontology* **28,** 85–92.

Duda, R. O., T. Gasching, R. Reboh, and B. M. Wilber, 1981, Operations manual for the Prospector consultant system, Final Report. Stanford: Stanford Research Institute International, 75p.

Dullien, F. A. L., 1979, *Porous Media. Fluid Transport and Pore Structure.* New York: Academic Press, 396p.

Ehrlich, R., and B. Weinberg, 1970, An exact method for characterization of grain shape, *Jour. Sed. Petrology* **40,** 205–212.

Fabbri, A. G., 1981, Image processing of geological data, Ph.D. dissertation, University of Ottawa, 385p.

Fabbri, A. G., and T. Kasvand, 1981, Applications at the interface between pattern recognition and geology, *Sciences de la Terre, Série Informatique Géologie, No. 15,* 87–111.

Fisher, R. A., 1953, Dispersion on a sphere, *Royal Soc. [London] Proc.* **217A,** 295–305.

Fournier, A., and D. Fussell, 1980, Stochastic modeling in computer graphics, Computer Graphics, *Quart. Assoc. Computing Mach.* (Spec. SIGGRAPH'80 issue), 1–8.

Fox, P. M., J. D. LaPerriere, and R. F. Carlson, 1979, Northern Lake modeling: A literature review, *Water Resources Research* **15,** 1065–1072.

Franklin, J. M., D. F. Sangster, and J. W. Lydon, 1981, Volcanic-associated massive sulfide deposits, *Econ. Geology,* 75th Ann. Vol., 485–627.

Freeze, R. A., and J. A. Cherry, 1979, *Groundwater.* Englewood Cliffs, NJ: Prentice-Hall, 604p.

Gaile, G. L. and C. J. Wilmott, eds., 1984, *Spatial Statistics and Models.* Dordrecht: Reidel, 482p.

Gannicott, R. A., G. A. Armbrust, and F. P. Agterberg, 1979, Use of trend surface analysis to delimit hydrothermal alteration patterns, *Canadian Mining and Metall. Bull.,* June, 82–89.

Gill, D., and D. F. Merriam, 1979, *Geomathematical and Petrophysical Studies in Sedimentology.* Oxford: Pergamon, 267p.

Gradstein, F. M., and F. P. Agterberg, 1983, Models of Cenozoic foraminiferal stratigraphy—northwestern Atlantic Margin, in J. M. Cubitt and R. A. Reyment eds., *Quantitative Stratigraphic Correlation.* London: Wiley, 119–182.

Gradstein, F. M., F. P. Agterberg, J. C. Brower, and W. S. Schwarzacher, 1985, *Quantitative Stratigraphy.* Dordrecht: Reidel, 632p.

Griffiths, J. C., 1970, Current trends in geomathematics, *Earth-Sci. Rev.* **6,** 121–140.

Guarascio, M., M. David, and C. J. Huijbregts, eds., 1976, *Advanced Geostatistics in the Mining Industry.* Dordrecht: Reidel, 461p.

Gy, P. M., 1979, *The Sampling of Particulate Materials, Theory and Practice,* Developments in Geomathematics 4. Amsterdam: Elsevier, 431p.

Hacquebard, P. A., 1979, Geological Appraisal of the Coal Resources of Nova Scotia, *Canadian Inst. Mining and Metallurgy Trans.* **82,** 48–59.

Harbaugh, J. W., J. H. Doveton, and J. C. Davis, 1977, *Probability Methods in Oil Exploration.* New York: Wiley, 269p.

Harris, D. P., 1984, *Mineral Resources Appraisal—*

Mineral Endowment, Resources, and Potential Supply. Oxford: Clarendon Press, 445p.

Harris, D. P., and F. P. Agterberg, 1981, The appraisal of mineral resources, *Econ. Geology,* 75th Ann. Vol., 897-938.

Harris, D. P., and F. J. Carrigan, 1980, A probabilistic endowment appraisal system based upon the formalization of geologic decisions—A general description, *U.S. Dept. Energy Open File Report GJBX-112(80),* 114p.

Hay, W. W., and J. R. Southam, 1978, Quantifying biostratigraphic correlation. *Ann. Rev. Earth Sci.* **6,** 353-375.

Hedberg, H. D. ed., 1976, *International Stratigraphic Guide.* New York: Wiley, 200p.

Helgeson, H. C., and W. M. Murphy, 1983, Calculation of mass transfer among minerals and aqueous solutions as a function of time and surface area in geochemical processes. Computational approach. *Mathematical Geol.* **15,** 109-130.

Howarth, R. J., and B. E. Leake, 1980, The role of data processing in the geological sciences. *Sci. Progress,* **66,** 295-329.

Hudson, C. B., and F. P. Agterberg, 1982. Paired comparison models in biostratigraphy, *Mathematical Geology* **14,** 141-159.

Journel, A. G., and C. J. Huijbregts, 1978, *Mining Geostatistics.* London: Academic Press, 600p.

Kolmogorov, A. N., 1951, Solution of a problem in probability theory connected with the problem of the mechanism of stratification, *Am. Math. Soc. Trans.* **53,** 8p.

Komar, P. D., 1976, *Beach Processes and Sedimentation.* Englewood Cliffs, N.J.: Prentice-Hall, 429p.

Komar, P. D., 1980, Settling velocities of circular cylinders at low Reynolds numbers, *Jour. Geology* **88,** 327-336.

Krige, D. G., 1951, A statistical approach to some basic valuation problems on the Witwatersrand, *South African Inst. Mining and Metall. Jour.* **52,** 119-139.

Krumbein, W. C., 1969, The computer in geological perspective, in D. F. Merriam, ed., *Computer Applications in the Earth Sciences.* New York: Plenum, 251-275.

Lambert, J. B., and T. Sato, 1974, The Kuroko and associated ore deposits of Japan—A review of their features and metallogenesis, *Econ. Geology* **69,** 1215-1236.

LeMaitre, R. W., 1982, *Numerical Petrology.* Amsterdam: Elsevier, 281p.

Lyell, C., 1833, *Principles of Geology,* vol. 3. London: Murray, 398p.

McCammon, R. B., J. M. Botbol, R. Sinding-Larsen, and R. Bowen 1983, Characteristic analysis—1981— Final program and a possible discovery, *Mathematical Geology* **15,** 85-108.

Mandelbrot, B. B., 1977, *Fractals: Form, Chance and Dimension.* San Francisco: W. H. Freeman, 365p.

Mandelbrot, B. B., 1982, *The Fractal Geometry of Nature.* San Francisco: W. H. Freeman, 461p.

Mann, C. J., 1981, Stratigraphic analysis: Decades of revolution (1970-1979) and refinement (1980-1989), in D. F. Merriam, ed., *Computer Applications in the Earth Sciences.* New York: Plenum, 211-242.

Mardia, J. V., 1972, *Statistics of Directional Data.* London: Academic Press, 357p.

Matheron, G., 1965, *Les Variables Régionalisées et Leur Estimation.* Paris: Masson, 306p.

Matheron G., 1975, *Random Sets and Integral Geometry.* New York: Wiley, 261p.

Matheron, G., 1976, Forecasting block grade distributions: The transfer functions, in M. Guarascio, M. David, and C. J. Huijbregts, eds., *Advanced Geostatistics in the Mining Industry.* Dordrecht: Reidel, 237-251.

Matheron, G., and M. Armstrong, eds., 1987, *Geostatistical Case Studies.* Dordrecht: Reidel, 264p.

Merriam, D. F., ed., 1976, *Quantitative Techniques for the Analysis of Sediments.* Oxford: Pergamon, 250p.

Merriam, D. F., ed., 1981a, *Computer Applications in the Earth Sciences. An Update of the 70's.* New York: Plenum, 385p.

Merriam, D. F., 1981b, Roots of quantitative geology, *Syracuse Univ. Geol. Contr.* **8,** 1-15.

Miall, A. D., ed., 1980, Facts and principles of world petroleum occurrence, *Canadian Soc. Petroleum Geologists, Mem. 6,* 1,003p.

Playfair, J., 1802, *Illustrations of the Huttonian Theory of the Earth.* London: Cadell and Davies, 528p. (Reprinted in 1956 by Dover Publications, New York.)

Pridmore, D. F., G. W. Holmann, S. H. Ward, and W. R. Sill, 1981, An investigation of finite-element modeling for electrical and electromagnetic data in three dimensions, *Geophysics* **46,** 1009-1024.

Prissang, R., and W. Skala, eds., 1979, *Beitrage zur Geomathematik. Berliner Geowissenschaftliche Abhandlungen,* vol. 15, Berlin: Reimer, 126p.

Ripley, B. D., 1981, *Spatial Statistics.* New York: Wiley, 252p.

Schuenemeyer, J. H., and L. J. Drew, 1983, A procedure to estimate the parent population of the size of oil and gas fields as revealed by study of economic truncation, *Mathematical Geology* **15,** 145-161.

Schwarzacher, W., 1975, *Sedimentation Models and Quantitative Stratigraphy,* Developments in Sedimentology 19. Amsterdam: Elsevier, 382p.

Serra, J., 1982, *Image Analysis and Mathematical Morphology.* New York: Academic Press, 610p.

Simons, T. J., 1972, Multi-layered models of currents, temperature, and water quality parameters in the Great Lakes, in A. K. Biswas, ed., *International Symposium on Modelling Techniques in Water Resources Systems,* Vol. 1. Ottawa: Environment Canada, 150-159.

Singer, D. A., and D. L. Mosier, 1981, A review of regional mineral resource assessment methods, *Econ. Geology* **76,** 1006-1015.

Skinner, R., 1974, Geology of Tetagouche Lakes, Bathurst and Nepisiguit Falls Map-areas, New Brunswick, *Geol. Survey Canada Mem. 371,* 133p.

Twomey, S., 1977, *Introduction to the Mathematics of Inversion in Remote Sensing and Indirect Measurements,* Developments in Geomathematics 3. Amsterdam: Elsevier, 243p

Underwood, E. E., R. de Wit, and G. A. Moore, eds., 1976, Fourth International Congress for Stereology, *U.S. Natl. Bur. Standards Spec. Pub. 431,* 540p.

Van Hinte, J. E., 1978, Geohistory analysis—Applications of micropaleontology in exploration geology, *Am. Assoc. Petroleum Geologists Bull.* **62,** 201-222.

Verly, G., M. David, A. G. Journel, and A. Marechal,

eds., 1983, *Geostatistics for Natural Resources Characterization.* Dordrecht: Reidel, 1092p.

Vistelius, A. B., 1980, *Fundamental Mathematical Geology. Definition of Terms and Description of Methods,* Akad. Nauk. SSSR, Leningrad, 389p. (in Russian).

Vistelius, A. B., F. P. Agterberg, S. R. Divi, and D. D. Hogarth, 1983, A stochastic model for the crystallization and textural analysis of a fine-grained granitic stock near Meach Lake, Gatineau Park, Quebec, *Geol. Survey Canada, Paper 81-21,* 62p.

Watson, G. S., 1975, Texture analysis, *Geol. Soc. America Mem. 142,* 367–391.

Watson, G. S., 1981, The interaction of statistics and geology, *Syracuse Univ. Geol. Contr.* **8,** 17–27.

Weiss, A., ed., 1979, *Computer Methods for the 80's in the Mineral Industry.* New York: American Institute of Mining Metallurgy and Petroleum Engineers, 975p.

Whitten, E. H. T., 1981, Trends in computer application in structural geology: 1969-1979, in D. F. Merriam, ed., *Computer Applications in the Earth Sciences.* New York: Plenum, 323–380.

Williamson, M. A., 1987, A quantitative foraminiferal biozonation of the Late Jurassic and Early Cretaceous of the East Newfoundland Basin, *Micropaleontology 33,* 37–65.

Worsley, T. R., and M. L. Jorgens, 1977, Automated biostratigraphy, in A. T. S. Ramsay, ed., *Oceanic Micropaleontology,* Vol. 2. London: Academic Press, 1201–1229.

Wren, A. E., 1980, Geophysics: Target shooting for the drill, *Canadian Soc. Petroleum Geologists Mem. 6,* 219–242.

Cross-references: *Computers in Geology; Geostatistics.* Vol. XIII: *Computerized Resources Information Bank.*

METHODOLOGY—See
ENVIRONMENTAL MANAGEMENT; FIELD GEOLOGY; GEOLOGICAL METHODOLOGY; MAPS, LOGIC OF; MINING PREPLANNING; SCIENTIFIC METHOD.

MINERAL EXPLORATION—See
ATMOGEOCHEMICAL PROSPECTING; BIOGEOCHEMISTRY; BLOWPIPE ANALYSIS; EXPLORATION GEOCHEMISTRY; EXPLORATION GEOPHYSICS; GEOBOTANICAL PROSPECTING; GROUNDWATER EXPLORATION; HYDROCHEMICAL PROSPECTING; INDICATOR ELEMENTS; LAKE SEDIMENT GEOCHEMISTRY; LITHOGEOCHEMICAL PROSPECTING; MARINE EXPLORATION GEOCHEMISTRY; MARINE MAGNETIC SURVEYS; MINERAL IDENTIFICATION, CLASSICAL FIELD METHODS; PLATE TECTONICS, MINERAL EXPLORATION; WELL LOGGING.

MINERAL IDENTIFICATION, CLASSICAL FIELD METHODS

Field identification of minerals is limited by the portability and maintenance of equipment. Accuracy of identification increases with facilities and thus increases in a hierarchy of outcrop to camp to analytical laboratory. This entry is confined to outcrop and camp methods. A field worker commonly has on the outcrop a pick, a pocket knife with a magnetized blade, a 10 power hand lens, and a plastic dropping vial of hydrochloric acid. Mineral identification in such circumstances is an educated guess based on physical properties, occurrence, and association. The physical properties that can be checked are habit, color, luster, tarnish, cleavage, and hardness relative to a knife. *Habit* refers to the typical physical appearance of a mineral, e.g., millerite is often in hairlike fibers, leucite in trapezohedral crystals, and erthyrite in a powder or scales.

Mineral occurrence is the natural environment in which minerals are found. Some minerals and elements have very limited occurrence, e.g., lithium minerals are largely confined to granitic pegmatites and to some salt lake deposits. Other minerals and elements are ubiquitous. It is easier to state the occurrences from which quartz and pyrite are absent than those in which they may be found. The term *association* is used to indicate the minerals one might expect to find with a given mineral. This is another way of looking at mineral environment. The sum of the chemical compositions of a group of minerals found together is the chemical composition of the rock; its chemical environment and associated minerals must conform to such chemistry. For example, assume a rock contains 25% quartz, 68% alkali feldspar, and 6% mica. Green glassy mineral garins found in this rock are not olivine, as olivine and quartz would have reacted to form other minerals. This is negative evidence, but allows one to reject olivine, the most common green glassy mineral. A hard black granular mineral occurring in a vein with quartz, calcite, and pyrite is not chromite. A mineral with a similar appearance occurring in a dark colored rock mass associated with olivine, serpentine, talc, and a carbonate might well be chromite. There are more than 2,000 minerals, and based on physical properties alone, each of them may be confused with other minerals. If knowledge of occurrence and association is added to physical properties, the identification can often be narrowed to a few minerals.

A camp, even in a wilderness, can have equipment for blowpiping (see *Blowpipe Analysis*) and simple wet chemical tests. Approximate identifications will have been made on the outcrop, and the first camp effort is to confirm that identification. The tenor of an ore can also be evaluated if one is prospecting. Most confirmatory tests (flame tests, bead tests, drop tests, and colorimetric tests) test for specific elements. There are also special property tests such as fluorescence and magnetism. *Flame* and *bead tests* are discussed in the entry *Blowpipe Analysis*.

Drop tests (Feigl, 1958) are a series of tests in which a small amount of the unknown mineral is dissolved in acid and a drop of the acid is placed on a glass slide. Drops of reagents are placed adjacent to the mineral solution, mixed, and any reaction noted. The reaction may be a precipitate, color change, or crystal growth. A series of tests are run and the results of these plus occurrence, association, and physical properties should produce an accurate identification of the mineral. For example, the lithium iron manganese phosphate, triphylite, is a mineral of granitic pegmatites found associated with quartz, alkali feldspar, and usually other pegmatite minerals such as tourmaline, micas, and apatite. It is gray to pinkish brown and has good cleavage and a hardness on Moh's scale of 4 1/2–5. Triphylite is soluble in hydrochloric acid, and the solution gives a yellow precipitate with ammonium molybdate (phosphate), a brown precipitate with ammonium hydroxide (iron), and no reaction with either ammonium oxalate or barium chloride. It becomes magnetic on heating in the reducing flame of a blowpipe (iron), gives a purple color to a borax bead (manganese), and has a red lithium flame. Not all minerals have so many positive tests, but essentially all minerals may be identified, although varieties may not be recognized.

Colorimetric tests are those in which a colored solution indicates the presence of the element in question and the intensity of color indicates the amount. Diphenylthiocarbazone, abbreviated to dithizone, is an example of a colorimetric agent used in prospecting for heavy metals (Fischer, 1930). A carbon tetrachloride solution of dithizone is an intense green color. Mixed with a neutralized nitric acid solution of an ore or any neutral aqueous solution, the color will determine the presence and concentration of copper, lead, zinc, silver, mercury, cobalt, or nickel. This reagent is most sensitive to zinc, which can be detected in concentrations of 50 ppm (0.005%). Concentrations are determined by the amount of reagents

used and by the color, which with increasing concentration, progresses from green to green-blue to blue to purple to red-purple to purplish red to red. In actual prospecting a series of color standards is made up daily from standard solutions of known concentration, and concentration in field samples is determined by matching colors. A number of new tests for elements appear annually in the various chemical periodicals, and although most require well-equipped, clean laboratories, some, e.g., tests for gold by Shahine and Mahmoud (1976) and by Christora and Ivanova (1976), could be developed as field techniques. Most of the field-adaptable tests, however, are one-element tests and are more useful to prospectors for one element than for general mineral identification.

Although it is commonly considered to be a laboratory technique, the examination of crushed mineral grains in oil immersion with a petrographic microscope is readily accomplished in a field camp. This technique is very effective with the silicates, which are the most difficult minerals to identify by classical chemical means.

A kit containing the equipment and reagents necessary for most chemical tests need not be too bulky as one such kit measures about 30 x 40 x 20 cm. Although no single book covers all up-to-date methods, Smith's (1953) *Identification and Qualitative Chemical Analysis of Minerals* is quite good; this volume covers analytical techniques and mineral descriptions but lacks data on drop tests and most organic reagents such as dithizone. Plans for the field testing kit mentioned are covered in this book. Occurrence, association, physical properties, and some chemical tests are covered in all standard mineralogy textbooks.

IRVING S. FISHER

References

Christora, R., and M. Ivanova, 1976, A highly selective and sensitive method for identification of gold, *Mikrochimica Acta* **2**, 349–355.
Feigl, F., 1958, *Spot Tests in Inorganic Analysis.* Amsterdam: Elsevier, 600p.
Fischer, H., 1930, Mikrochemischer nachweis einiger schwermetalle durch tropfenreaktion mit dithizon (Diphenyl–Thiocarbazon), *Mikrochemie* **8**, 319–329.
Shahine, S. A., and R. M. Mahmoud, 1976, A selective and sensitive spot test for gold, *Mikrochimica Acta* **1**, 89–92.
Smith, O. S., 1953, *Identification and Qualitative Chemical Analysis of Minerals,* 2nd ed. New York: Van Nostrand, 385p.

Cross-references: *Blowpipe Analysis; Field Geology; Geology, Scope and Classification; Indicator Elements; Lithogeochemical Prospecting; Soil Sampling.* Vol. XIII: *Mineragraphy.*

MINERALOIDS

Minerals are naturally occurring crystalline compounds. Other naturally occurring substances lack long-range atomic order; they are the *mineraloids*. They are optically isotropic and they do not diffract X-rays. They range from gem material such as amber and opal to fossil fuels such as coal and petroleum. They may be of biotic—e.g., chitinous shells—or abiotic—e.g., obsidian—origin.

For convenience they may be divided into two groups—an inorganic group lacking carbon (Table 1) and an organic group containing essential carbon (Table 2). Water, a naturally occurring inorganic compound, may be considered the melt of the mineral ice and is not treated here. Native mercury, although a liquid at room temperature, is considered to be a mineral because it crystallizes at $-38.9°C$, a temperature attained in terrestrial environments.

Inorganic Mineraloids

Precipitates Allophane (q.v.in Vol. IVB) is a hydrous aluminum silicate with a variable silica:alumina ratio and water content that may contain up to 10% phosphate by weight. Amorphous both to X-rays and to light, it has a refractive index between 1.40 and 1.49. The color is highly variable from sample to sample—white, blue, green, pink, or colorless. Its density ranges from 1.65 gm/cm^3 to 2.15 gm/cm^3 and its hardness from 1 to 3 on the Mohs scale. It may be massive to granular or in submicroscopic pellets.

Allophane may be derived from weathering of volcanic glass, from hydrothermal alteration of other minerals, or from alteration of aluminosilicates by descending sulfate-bearing meteoritic waters. It is commonly associated with gibbsite and halloysite.

Collophane (q.v. in Vol. IVB) is a calcium phosphate with variable amounts of fluoride, hydroxide, and carbonate. It is mostly cryptocrystalline with the structure of apatite. The color is commonly off-white to gray, yellow, or brown with a dull luster. Although apatite defines hardness 5 on the Mohs scale, collophane has a hardness of 3-4. It has a density of 2.5-3 gm/cm^3.

Collophane is the major constituent of phosphorite, chief source for phosphate for chemical industries and agriculture. Its layered, commonly colloform, structure suggests that it was precipitated from upwelling ocean currents at a pH below that which would cause calcite to precipitate. The rarity of conditions under which collophane can precipitate alone is reflected in the relative rarity of economically valuable phosphate deposits (Jensen and Bateman, 1979, 159).

Limonite (q.v.in Vol. IVB) is a mixture of iron hydroxides and oxyhydroxides with variable amounts of water and noncrystalline material. Its typical color is yellow ochre, although it may be dark brown to black. Although it may attain a hardness of 5 1/2, many samples are quite soft and scratchable. Depending on its compactness, its density ranges from 3.3 gm/cm^3 to 4.3 gm/cm^3.

TABLE 2. Organic Mineraloids

Abiotic
Chondritic carbon: Alkanes, aromatics, amino acids
Pegmatitic organic matter
Nonfuel gases: Carbon dioxide, hydrogen sulfide, helium

Biotic
Fossil fuels: Coal, petroleum, methane
Resins: Amber, copalite, resinite
Waxes
Asphalts: Bitumen
Kerogen: Oil shales
Chitin: Shells

TABLE 1. Inorganic Mineraloids

Precipitates
Allophane: Aluminous silicious gel
Collophane: Massive, fine-grained or amorphous phosphate
Limonite: Massive, oolitic, or pisolitic oxides or oxyhydroxides that may be amorphous or cryptocrystalline mixtures [alpha FeO(OH)] or lepidocrocite [gamma FeO(OH)]
Opal: Hydrous disordered and nearly amorphous to well-ordered crystalline microspheres
Psilomelane (wad): Earthy mixture of amorphous or finely crystalline manganese oxides and hydroxides
Silica gels with variable amounts of water in lakes and kaolinite deposits

Glasses
Volcanic: Felsic to mafic lava quenched at or near the Earth's surface
Meteorologic: Lightning-fused rock or soil (fulgurites) or sand (lechatelierite)
Meteoritic: Impact-fused rock or soil and tectites

Metamict minerals
Pitchblende and other uranium or thorium bearing mineraloids

Limonite is the chief constituent of gossans (see Vol. XIII: *Mineragraphy*) marking ore deposits, and is brought about by weathering of iron-bearing sulfide minerals (Jensen and Bateman, 1979, 244). It may precipitate from lagoon or bog waters; such oolitic limonite provides the bulk of the iron ore for Western Europe (Jensen and Bateman, 1979, 402–404).

Opal (q.v. in Vol. IVB) is predominantly silica with less than 9% water by weight. Under an electron microscope, opal reveals a structure of aggregates of minute crystallites randomly oriented. These crystallites may be well-ordered low crystobalite (opal-C), disordered low crystobalite and low tridymite (opal-CT), or nearly amorphous to X-rays (opal-A). The brilliant play of colors of gem opal results from regular packing in layers of spherules 150–300 micrometer in diameter. Refractive indices range from 1.11 to 1.46 and densities from 2 gm/cm^3 to 2.25 gm/cm^3.

Opal may result from hydrothermal activity or as dehydration of colloidal silica precipitate. Replacing wood at a cellular level, it becomes petrified wood. Diatom shells are opaline, and diatomite is used as a filler in paper, paint, fertilizer, pesticides, and plastics; as an abrasive; and in the manufacture of refractory materials.

Psilomelane (wad) covers otherwise unidentified black manganese oxide and hydroxide minerals and amorphous materials commonly with variable amounts of the oxides admixed with it. It is dark brown to black and soft enough to rub off on the fingers when handled. Determination of the exact mineralogy of psilomelane requires detailed chemical and X-ray analysis.

Psilomelane occurs as a chemical precipitate in shallow marine sediments and as submarine manganese nodules (Jensen and Bateman, 1979, 154, 179). Black stain on many limestone cliffs results from manganese weathering out of the carbonate and being oxidized.

Silica gel is a clear to turbid jellylike substance that occurs in some kaolinite deposits and seasonally in some African lakes. The latter occurrence is attributed to seasonal rainfall in areas underlain by strongly alkaline rocks such as carbonatites. During the dry season these waters become extremely alkaline and can dissolve abundant silica. Heavy rains dilute the water and lower its pH, causing the gel to precipitate. These gels may be parent materials for some opal and chert.

Glasses Natural glasses have chemical compositions equivalent to rocks from which they were formed but lack minerals, except for some crystallites in some volcanic glasses. Their lack of crystallinity is attributed to cooling rates so rapid that crystal nucleation and growth cannot occur. Glasses are unstable compared to minerals and weather quickly; few are older than Cretaceous in age.

Volcanic glasses range in chemical composition from that of granite (obsidian) to that of basalt (basaltic glass). They result from quenching of lava on extrusion. Vesiculated glasses are called pumice if cell walls are thin with vesicles not interconnected and scoria if the walls are thick with vesicles interconnected. Pitchstone is a hydrated glass with a dull resinous luster, and perlite is a partially hydrated glass with numerous concentric cracks in it. Volcanic glasses react with water to form clays or zeolites.

Lightning striking bare rock, soil, or sand generates enough heat to cause local melting and subsequent quench to glass. Two centimeter glass puddles (Fig. 1) are much in evidence on some mountaintops. Glass tubes up to 5 cm in diameter and a half meter long are formed when lightning strikes wet soil (fulgurites, Fig. 2) or quartz sand (lechatelierite).

Meteorite impact may release sufficient energy to destroy crystal structure either by shock waves or heat. Tektites are aerodynamically shaped glasses up to a few centimeters in size that are thought to result from impact-generated melt splashed off the Earth's surface.

Metamict Minerals Alpha-particle decay of uranium and thorium series elements has sufficient energy to knock individual ions out of their structurally stable position, locally destroying the crystallinity of the host mineral. Given enough time, metamictization renders a crystal amor-

FIGURE 1. Grenville age (1000 million years old) granite from the Pedlar Massif with a puddle of glass formed by lightning strike in May 1980. Location is Tyro, Nelson County, Virginia. Scale is 15 cm. (Photo, K. Frye)

585

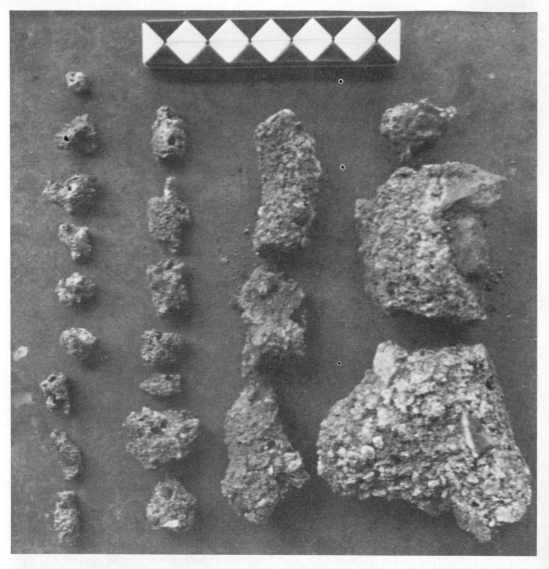

FIGURE 2. Fulgurite melted in June 1981 from the same spot as in Fig. 1 after collection of the cobble. The tube was fused to a depth of nearly .5 m before the steam explosion extracted it from the wet soil. (Photo, K. Frye)

phous to light and to X-rays and reduces its density measurably. Annealing between 500° and 900°C commonly restores crystallinity, in many cases accompanied by a visible flash of light. Most minerals containing essential uranium or thorium in their composition are metamict (q.v. in Vol. IVB) as are some zircons containing traces of these elements. Pitchblende is metamict uraninite (uranium oxide).

Organic Mineraloids

Abiotic Carbonaceous chondrites, although not a major class of meteorites, may be a key to theories of planetary origins in the solar system because of their unusual mineralogy (see Vol. IVB: *Meteoritic Minerals*). They differ from other meteorites by the presence of hydrated minerals, carbonate minerals, and noncrystalline carbon. Their carbon mineraloids include alkanes, aromatic hydrocarbons, and racemic amino acids.

Some granite pegmatites from widespread localities contain a brown organic substance much resembling coal, whereas others contain a bituminous substance containing carbonyl complexes of uranium and thorium (see Vol. IVA: *Organic Mineraloids*). Sources of these igneous organic complexes remain obscure.

Gases other than methane may be trapped in the same sorts of formations as house petroleum and natural gas. Carbon dioxide, presumably released by decarbonation reactions between limestones or dolomites and igneous intrusions, is tapped and used as a stripping agent for enhanced oil recovery. Helium, either primordial or a result of alpha decay, is associated with methane in many gas fields. Both hydrogen sulfide and sulfur dioxide are associated with fumaroles and hydrothermal activity.

Biotic Fossil fuels—coal, oil, and natural gas—result from diagenetic alteration of plant material that failed to decay on death. An abiotic mantle source for some methane has been proposed but is considered unlikely or irrelevant by most geologists.

Resins (q.v. in Vol. IVB) are secreted by trees and shrubs on injury and may be geologically preserved as amber, some with insects, seeds, and pollen preserved within. Resins preserved in coal seams are known as resinite; a wide variety of local names are used for other fossil resins. Valued as a gem, an extensive prehistoric trade developed between North Sea amber occurrences and Mediterranean peoples. Use of fossil resins in varnishes has now been largely replaced by synthetic materials.

Waxes are chemically resistant natural substances associated with some brown coals and with some vertebrate fossils, but sea wax may be an algal product (see Vol. IVA: *Organic Mineraloids*).

Natural distillation of petroleum can leave behind residues with high molecular weight to form asphalts and other bitumens in sedimentary rocks or as surface seeps. Asphaltites are more polymerized than asphalt, but this transition is gradational. Metabitumite is a hard, insoluble mineraloid associated with contact metamorphism.

Kerogens are solid bituminous mineraloids dispersed in some shales that yield oil on destructive distillation. They appear to derive from swampy environments where putrefaction was incomplete.

Chitin is an amorphous nitrogenous material that forms carapaces on insects and trilobites.

KEITH FRYE

References

Hem, J. D., 1972, Colloids, in R. W. Fairbridge, ed., *The Encyclopedia of Geochemistry and Environmental Sciences,* Encyclopedia of Earth Sciences, Vol. IVA. New York: Van Nostrand Reinhold, 180–182.

Jensen, M. L., and A. M. Bateman, 1979, *Economic Mineral Deposits,* 3rd ed. New York: Wiley, 593p.

Jones, J. B., and E. R. Segnit, 1981, Opal, in K. Frye, ed., *The Encyclopedia of Mineralogy,* Encyclopedia of Earth Sciences, Vol. IVB. Stroudsburg, PA.: Hutchinson Ross, 330–333.

Langenheim, R. L., 1981, Resin and amber, in K. Frye, ed., *The Encyclopedia of Mineralogy,* Encyclopedia of Earth Sciences, Vol. IVB. Stroudsburg, Pa.: Hutchinson Ross, 445–447.

Mueller, G., 1972, Organic mineraloids, in R. W. Fairbridge, ed., *The Encyclopedia of Geochemistry and Environmental Sciences,* Encyclopedia of Earth Sciences, Vol. IVA, New York: Van Nostrand Reinhold, 823–830.

Pabst, A., 1981, Metamict state, in K. Frye, ed., *The Encyclopedia of Mineralogy,* Encyclopedia of Earth Sciences, Vol. IVB. Stroudsburg, Pa.: Hutchinson Ross, 239–240.

Turner, S. and P. R. Buseck, 1981, Todorokites: A new family of naturally occurring manganese oxides, *Science* **212,** 1024–1027.

White, W. A., 1981, Allophane, in K. Frye, ed., *The Encyclopedia of Mineralogy,* Encyclopedia of Earth Sciences, Vol. IVB. Stroudsburg, Pa.: Hutchinson Ross, 539.

Cross-references: *Mineral Identification, Classical Field Methods; Minerals and Mineralogy; Petroleum Exploration Geochemistry.*

MINERALS AND MINERALOGY

Minerals are naturally occurring crystalline compounds; *mineralogy* is the science of these materials. As such, it shares much of its subject matter with metallurgy, crystal chemistry, physical chemistry, and other sciences that deal with solid materials.

Crystal Architecture

Compounds are composed of elements of which 85 may react to form minerals. Each element is composed of atoms that have identical electronic configurations about their nuclei. Each atom may gain or lose valence electrons, thereby becoming ionized. Ions of opposite charges form periodic arrays such that electrostatic charges are neutralized over short atomic distances. Each ion thus surrounds itself with as many ions of opposite charge as space permits.

Although many bond models are used to describe the variety of behavior of crystalline materials, the ionic bond is the most widely applicable in minerals. Obvious exceptions exist such as metallic bonding of native gold and hybrid orbital bonding of diamond. Less obvious is the fact that the all-important silicon-oxygen bond is only about 40% ionic in silicate minerals.

The two most important properties of ions are their size and their electrostatic charge. Although the electronic configuration of each ionic species is unique, sizes and charges are not. Hence, one ion may substitute for another without disturbing the structural periodicity of a mineral. As a result of similarity in size and charge, a complete crystal solution may exist between a pure ferrous iron end member and a pure magnesium end member as in the biotite series between annite and phlogopite.

Charge on a substituting ion need not be the same as on the one being replaced if any difference can be made up locally by a coupled substitution. Thus, as calcium replaces sodium in the plagioclase series, aluminum replaces silicon on a one to one basis. A difference of two electrostatic charges during substitution is far less common.

Although long-range periodicity is a defining characteristic of minerals, structurally perfect crystals cannot exist. All minerals contain structural as well as chemical defects. These defects may occur at points, along rows of ions or across planes, or they may permeate a volume. The major difference between microcline and sanidine is in a high degree of ordering of aluminum and silicon in the former and their disorder in the latter.

Since may different ions have similar sizes, many different mineral compositions may have the same or very similar structures. Similarity of mineral structure constitutes the basis for the classification of mineral species into mineral groups. Indeed, differences among species properties within a group may be so subtle that only the mineral group can be identified in the field. Members of the garnet and amphibole groups fall into this category.

Depending on temperature and pressure of formation, a compound may adopt more than one structure (see Vol. IVB: *Barometry, Thermometry*). In metamorphic terranes the presence of one or more of the isochemical minerals kyanite, andalusite, and sillimanite gives information about the temperature and pressure at which a mineral assemblage crystallized.

Once a mineral has crystallized into a particular crystal shape, chemical changes can occur without altering the form of the crystal. Oxidation and leaching can give hematite pseudomorphous after pyrite. The presence of beta quartz papamorphs in a lava flow indicates that this silica crystallized at temperatures above 573°C even though it inverted to alpha quartz on cooling through its inversion temperature. Not all natural materials are crystalline; these are *mineraloids*.

Mineral Structures

Nonsilicate minerals other than native elements may be classified on a basis first of cation to anion ratio and second of size ratios. This classification is extended to cover cases where more than one kind of cation site is occupied. Many two-cation minerals, however, are structural derivatives of one-cation minerals; e.g., ilmenite is a half-breed derivative of corundum in that layers of aluminum ions are alternatively replaced by titanium and ferrous iron.

One result of structural classification of non-silicate minerals is a grouping of apparently dissimilar minerals. Halite, galena, and periclase belong to the same isostructural group even though one is a halide, one a sulfide, and one an oxide. The apatite group contains silicates and vanadates in addition to phosphates.

Silicate minerals are classified on a basis of the number of shared oxygen ions between silicon ions. Each silicon ion is surrounded by four oxygen ions in the shape of a tetrahedron. When one oxygen ion is bonded to two silicon ions, these two tetrahedra are described as sharing corners. Since each tetrahedron has four corner oxygen ions, zero to four corners may be shared.

When no corners are shared, electrostatic neutrality is balanced by other cations between neighboring tetrahedra. Such is the case in structures of the garnet and olivine groups of minerals. Sharing of one tetrahedral corner creates a dimer; minerals of the epidote group are characterized by isolated tetrahedra and dimers in their structure.

Sharing of two tetrahedral corners gives either infinite chains of linked tetrahedra, as in pyroxene and pyroxenoid groups, or closed rings of six tetrahedra each, as in the tourmaline group. Ring structures of other sizes exist, but they are rare.

If two chains are laid side by side such that alternate tetrahedra in each chain share corners between chains, half these tetrahedra share two corner oxygen ions and half share three (Fig. 1). This double chain arrangement characterizes the structure of the amphibole group.

Additional single chains of tetrahedra can be added to form structures with multiple chains until, eventually, there is a two-dimensional sheet of tetrahedra, each sharing a corner oxygen with three other tetrahedra. This structure characterizes the clay, the serpentine, the mica, and the

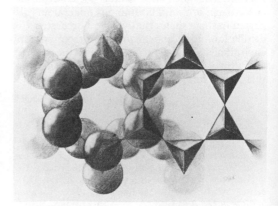

FIGURE 1. Double chain of silica tetrahedra representing the basic building unit of the amphibole group. Representation of spherical ions on the left melds into tetrahedral coordination polyhedron on the right. (Dan Bollman, artist)

chlorite groups. An additional complexity arises from the fact that aluminum ions can replace some silicon ions in tetrahedral coordination, causing a different electrostatic charge balance.

Finally, tetrahedra may share all four corner oxygen ions with neighboring tetrahedra in a variety of three-dimensional patterns. Quartz and its polymorphs have only silicon ions in their tetrahedra, but in the feldspar, feldspathoid, and zeolite group structures, aluminum replaces some tetrahedral silicon. Since these structures do not represent optimal ionic packing, they do no persist to any great depth in the Earth's mantle.

Crystal Symmetry

Just as wallpaper shows symmetrical patterns in its periodic repetition, three-dimensional periodicity inherently generates symmetry in three dimensions. Crystal symmetry was one of the first aspects of mineralogy to be studied in a systematic and essentially modern fashion. Symmetrical forms of crystals are but an outward manifestation of the symmetrical patterns of crystal structures.

Different elements of symmetry (Fig. 2) and combinations thereof permit the different forms—i.e., symmetrically related crystal faces—of the 32 crystal classes. Not all forms are equally probable, however, and each mineral species has certain forms that it commonly displays—i.e., its habit. Although belonging to the same crystal class, minerals of the garnet group typically display trapezohedra and rhombic dodecahedra, whereas spinel mineral crystals typically show cubes or octahedra.

Habit also includes elements of symmetry that almost, but not quite, exist in a structure. This

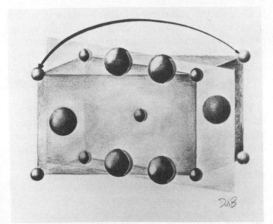

FIGURE 2. Mirror element of symmetry in rutile structure. The large spheres represent oxygen and the small ones titanium. The shaded vertical plane represents the mirror, and the arrow points to two symmetrically related spheres. (Dan Bollman, artist)

situation permits some crystals to have two or more components that are related by an element of symmetry that is not part of their group of symmetry elements. The result is crystal twinning, which is commonly recognized by the presence of re-entrant angles in the sample. Swallowtail twins of gypsum constitute an easily recognizable example.

Although most specimens of minerals lack well-developed forms, their universal internal periodicity permits them to act as diffraction gratings for X-rays. Since each mineral species has a unique periodicity, each gives a unique set of angles at which diffraction occurs, and minerals of each group give similar patterns of diffraction angles. Minerals that cannot be uniquely identified in the field can be identified in the laboratory by their diffraction patterns.

Mineral Properties

Although the strongest chemical bonds determine the type of mineral structure present, many characteristic mineral properties used for mineral identification are commonly determined by the weakest bonds present.

Density Density is a measure of the amount of mass in a given volume, commonly given in grams per cubic centimeter. Specific gravity is a measure of the same property but is the ratio of the weight of a sample to the weight of an equal volume of water at 4°C where water attains its maximum density. Methods of measuring mineral density are given in Vol. IVB. With practice, fairly good estimates of density can be made by hefting a sample.

Mineral density varies only by approximately one order of magnitude for most minerals. Ice at less than 1 g/cm³ and native platinum at approximately 20 g/cm³ mark the extremes, with most rock-forming minerals between 2.6 g/cm³ and 3 g/cm³. Variation in density is remarkably small when compared with atomic masses, which range from 1 for hydrogen to 238 for uranium. Part of the reason for this is that most of the volume of rock-forming minerals is lightweight oxygen and structures are not efficiently packed. Sulfides are systematically denser than oxygen-bearing minerals because sulfur has a higher atomic mass. Native elements and alloys are the densest minerals because of efficient packing and lack of dilution by low-weight anions.

Variations in density of minerals making up crystal rocks cause variations in local gravitational attractions. These variations can be measured with a gravimeter and contoured as gravitational highs and lows on a map. Features exposed at the surface, such as basalt dikes, can be interpreted directly, but buried gravity anomalies, such as fault zones beneath a sedimentary cover, may fit more than one model and require other data for unique interpretation.

Elasticity *Strain* is produced in a mineral when force—i.e., stress—is applied to it. On the atomic scale, this means that forces other than those of the chemical bonds are present. *Elastic deformation* occurs when strain is proportional to stress, the proportionality being an elastic stiffness coefficient. In general, the shorter and stronger chemical bonds are in a mineral, the greater their stiffness. Since chemical bonding may vary greatly with differences in crystallographic direction in some minerals, such as chain silicates, elastic stiffness is also anisotropic in these minerals.

Elastic stiffness coefficients increase with an increase in pressure and decrease with an increase in temperature. Since both temperature and pressure increase with an increase in depth within the Earth, negative temperature change partially offsets positive pressure change in these coefficients of deeply buried minerals.

Seismic wave velocities increase with an increase in elastic stiffness, making it possible to interpret seismic traveltimes in terms of mineralogic changes in buried rock structures. Although analysis of seismic data utilizes large computers, final interpretation of these data requires human geological judgment (see Vol. XIII: *Seismological Methods*).

Plastic Deformation Although mineral stiffness with respect to short-term stresses such as seismic waves increases with depth, long-term stresses such as those associated with plate tectonics produce anelastic deformation, which is irreversible. Plastic deformation of minerals occurs along well-defined crystallographic planes in specific directions. Such deformation may be preserved as thin deformation lamellae or as deformation twinning that can be observed with the polarized-light microscope. More commonly, however, it is annealed out by recrystallization.

Anelastic deformation occurs in brittle minerals subject to stress under high confining pressure. Native metals and alloys, in contrast may be deformed without loss of cohesion at surface pressure and temperature. These minerals are malleable in that they may be flattened with a hammer. Many of them are also ductile in that they may be extruded as a wire. Yellow sulfide minerals are truly "fool's gold" in that none of them is malleable.

Thermal Conductivity How warm or cold a mineral feels at a particular temperature depends on how effectively it conducts heat. Minerals conduct heat principally by two mechanisms—electron transport and phonon waves. Although all solids conduct heat by phonons, only metals and semimetals have free electrons. Their conductivities are one to two orders of magnitude greater than those of insulators.

A glass bauble at room temperature feels warmer to the touch of a trained tongue than does the crystalline gem it tries to imitate because the disorder of glass gives it measurably lower thermal conductivity than an ordered crystal. Diamond is an extreme example of this phenomenon. Graphite, conversely, has thermal conductivity parallel to the layers that is four times its conductivity across the layers. This strongly anisotropic thermal character makes graphite useful in manufacture of heat shields.

Thermal Expansion Although most minerals expand on heating, some do so in a dramatic and irreversible fashion that can be used as an aid in their identification. Minerals of the zeolite group bubble and intumesce, and vermiculite exfoliates on application of heat. Thus, a 100 gm miniature propane torch may be a valuable addition to a field kit (see *Blowpipe Analysis*).

Hardness Mineralogic hardness is a measure of resistance to scratching as related to the Mohs scale (q.v., Vol. IVB) or the Povarennykh scale. Hardnesses of 5 and under may be estimated by the ease with which the minerals are scratched with a steel knife blade. For field measurements of hardnesses greater than 5, small fragments of quartz (7), topaz (8), corundum (9), and diamond (10) may be cemented to opposite ends of two small rods.

Toughness Just as hardness defines a mineral's resistance to scratching, toughness defines its resistance to breakage (see Vol. IVB: *Jade*). Some tough minerals, in decreasing order, are carbonado diamond, nephrite jade, jadeite jade, chrysoberyl, corundum, quartz, diamond, forsterite, beryl, garnet, zircon, and topaz. No toughness scale has been defined.

One factor reducing mineral toughness is the presence of *cleavage,* a tendency to break along preferred crystallographic directions. Numbers of cleavage directions may be one in the case of micas and clays; two in pyroxenes, amphiboles, and feldspars; three in halite and calcite; four in fluorite; or six in sphalerite.

Parting is similar to cleavage in that planar surfaces are produced. It differs in that some specimens of a mineral species show it whereas others do not. Some samples of corundum display basal parting, some rhombohedral, and others none at all. Parting, in some cases, may be related to twinning.

Electrical Properties Crystals with a unique polar axis contain a built-in spontaneous electric polarization that charges with changes with temperature so that they develop electrostatic charges on opposite ends of their polar axis. Minerals of the tourmaline group are the most prominent exhibitors of this *pyroelectricity.*

Pyroelectric crystals are also piezoelectric in that electrostatic polarity may be induced by stress. Conversely, electric charge applied to a polar axis of a piezoelectric crystal causes dimensional change at 90° to this polar axis. Thus, electric current oscillation can be converted into acoustic vibrations and vice versa, leading to a variety of uses of quartz oscillators in the electronics industry.

Although most common rock-forming minerals are insulators, rocks made up of these minerals have measurable electrical resistivities (see Clark, 1966, 553). Resistivity varies as a function of mineralogy, water content, salinity, and frequency of the external electrical field. Measurement of variation of resistivity down a borehole allows a geologist to make stratigraphic interpretations without necessarily being able to see those rocks being measured (see Vol. XIII: *Electrokinetics*).

Magnetic Properties Magnetic characteristics of minerals arise from interactions of electron spins between ions within a crystal structure and between these spins and any external magnetic field. A mineral may thus create its own magnetic field or be attracted or repelled by an external field gradient (see Vol. IVB: *Magnetic Minerals*).

Colorless rock-forming minerals are slightly repelled by an external field; they are *diamagnetic*. Ferromagnesian minerals and semiconductors such as pyrite are slightly attracted by an external field; they are *paramagnetic*. Some transition metal oxides and sulfides are slightly attracted to an external field because of magnetic ordering within their structures; they are *antiferromagnetic*.

Magnetic ordering that results in unpaired aligned electronic spins imparts a net magnetic polarity to a crystal structure. Magnetite is the most important of these natural ferrimagnets. *Aeromagnetic maps* may be used to delineate different rock types on the basis of the magnetic characteristics of their contained minerals (see *Aerial Surveys*). In addition, magnetic separators can be used to separate minerals with different magnetic characteristics.

When the geomagnetic field is impressed on magnetic minerals at the time of rock formation, the direction of that field may be recorded and preserved in a rock. In the cases of some basalts and diabases this remnant magnetism is strong enough to deflect a compass laid on the outcrop so that two otherwise indistinguishable (in the field) basalts with opposite polarities may be reliably recognized.

Radioactivity The geologically most important radioactive elements for the generation of crustal heat are uranium, thorium, and potassium. *Aeroradiation maps* may be used to delimit rock bodies with differing radioactivity. Radiation counters of various types are used in mineral exploration.

Each radioactive decay has a unique decay constant that allows calculation of the time at which a system, individual mineral or whole rock, became closed to both parent and daughter isotopes. This calculation permits assignment of radiometric ages to rocks. Since different isotope radiometric "clocks" are reset at different temperatures, more than one radiometric event may be recorded by different isotope analyses.

Surfaces In terms of defects, a surface of a mineral grain is the most drastic defect of all; its structure terminates there. In general, order at a surface differs significantly from that in the interior of a mineral grain. A surface phase is responsible for some colors and lusters as well as some chemical properties of a mineral.

A major characteristic of a surface phase is its *wettability*. Water is a polar molecule, and polar surfaces, such as most silicates and carbonates, are easily wetted; most sulfides are nonpolar and poorly wetted.

Sulfide ore minerals may be separated from carbonate and silicate gangue minerals in a process called *flotation*. Crushed ore is mixed with water with small amounts of oil and a foaming agent. Rising air bubbles lift the poorly wetted sulfide grains into the froth while the well-wetted gangue minerals sink.

A grease table utilizes the same principle to trap poorly wetted diamonds while the well-wetted gangue is carried away. Other poorly wetted minerals include zircon, beryl, topaz, spinel, corundum, and rutile.

Color The causes of color are beyond the scope of this entry (see Nassau, 1978). In most pale pastel nonmetallic minerals, color derives from chromophores that are not essential to a mineral and not diagnostic of a species. Colors of metallic and semimetallic minerals tend to be rather specific. Spectra from high-altitude remote sensing tend to be mostly of moss and lichen and not of rock outcrops.

Luminescence Luminescent colors are those emitted by a mineral after activation energy has been supplied in the form of heat (thermoluminescence), rubbing (triboluminescence), or radiant energy (fluorescence or phosphorescence). As with color, luminescent chromophores may or may not be essential to a mineral species (see Nassau, 1978).

Luster The quality of reflectance from a mineral surface determines its luster. Total reflection of the entire visible spectrum gives metallic luster, and total reflection of part of the spectrum imparts colored semimetallic luster. Pearly luster results from a multiplicity of layers within a specimen whereas greasy luster indicates a distinct layer. The brightness of vitreous to adamantine lusters is a function of the refractive index of a mineral.

KEITH FRYE

References

Bragg, L., G. F. Claringbull, and W. H. Taylor, 1965, *Crystal Structures of Minerals.* Ithaca, N.Y: Cornell University Press, 409p.

Clark, S. P., Jr., 1966, Handbook of physical constant, *Geol. Soc. America Mem. 97,* 587p.

Deer, W. A., R. A. Howie, and J. Zussman, 1962–1963, *Rock-Forming Minerals,* 5 vols. London: Longmans.

Fleischer, M., 1983, *Glossary of Mineral Species.* Carson City, Nev.: Mineralogical Record, 202p.

Frye, K., 1974, *Modern Mineralogy.* Englewood Cliffs, N.J.: Prentice-Hall, 325p.

Frye, K., ed., 1981, *Encyclopedia of Mineralogy,* Encyclopedia of Earth Sciences, Vol. IVB. Stroudsburg, Pa.: Hutchinson Ross, 794p.

Megaw, H. D., 1973, *Crystal Structures: A Working Approach.* Philadelphia: Saunders, 563p.

Nassau, K., 1978, The origins of color in minerals, *Am. Mineralogist* **63**, 219–229.

Newnham, R., 1975, *Structure-Property Relations.* New York: Springer-Verlag, 234p.

Cross-references: *Aerial Surveys; Blowpipe Analysis; Exploration Geophysics; Geomagnetism; Lithogeochemical Prospecting; Mineral Identification, Classical Field Methods; Mineraloids.* Vol. XIII: *Mineragraphy; Mineral Economics.*

MINES, MINING—See BOREHOLE MINING; MARINE MINING; MINE SUBSIDENCE CONTROL; MINING PREPLANNING; PLACER MINING.

MINE SUBSIDENCE CONTROL

Mine subsidence is the settlement of a part of the Earth's crust due to removal of subsurface solids or underground excavation. Settlements may vary in magnitude from complete collapse or substantial lowering to small distortions of the ground surface.

Mine subsidence control is the use of techniques to prevent or reduce subsidence movements to avoid or minimize damage to surface structures, to limit fracturing of strata above the mine, or to control subsurface stresses to assist mining operations while maximizing extraction. These techniques are also referred to as *ground control, surface stabilization, or subsurface stabilization.* Subsidence control techniques for active mining have been developed most extensively

in European coal fields by empirical and theoretical studies.

Mine Subsidence Mechanisms

The extraction of subsurface ores, minerals, or rock materials by mining results in a series of openings beneath the ground surface. The existence of these voids in a previously continuous rock mass creates a potentially unstable condition, with a tendency to fill the voids by collapse of the mine roof, heave of the mine floor, and/or crushing of the pillars of material remaining on the sides of or between the mine openings (Fig. 1). The collapse into the mine openings will progress upward to the surface unless bulking of broken strata, arching, or bridging by a strong rock stratum occurs at some point between the mine level and the ground surface. In some cases, surface movements will occur concurrently with the mining process, while in other cases, appreciable surface movements will not occur until perhaps tens of years later, after progressive weathering, crushing, spalling, or softening of the pillars, mine roof, or mine floor. Figures 2, 3, and 4 illustrate examples of subsidence damage over abandoned bituminous coal mines in Pennsylvania and West Virginia. Subsidence may result from mining operations such as solution mining of salt or sulfur; block caving; sublevel caving; or stope mining of metallic ores, coal, and other rock materials; and room-and-pillar or longwall mining of coal, trona, iron ore, or the other bedded deposits.

When the ore or other rock is removed in the mining operation, the surrounding rock or soil tends to move into the opening, particularly downward from the force of gravity. The amount and nature of the movements depend on stresses (overburden, tectonic, etc.), their magnitude and distribution around the opening; on the strength and discontinuities (joints, faults, bedding planes, cavities) of the rock or soil around the mine and above it; and on support placed in the mine to hold up the roof or restrain the sides or floor from moving. When a sufficiently large area is mined, downward movements of the ground surface will occur. There are also associated horizontal components of movement, and tensional and compressional strains develop, as indicated in Fig. 5.

The tensional and compressional strains and differential movements are often severest near the edge of the mined area and cause damage as shown in Fig. 2 and 3. Given a large enough extraction area, there exists a critical width of extraction at which subsidence at the center of the mined area will reach the maximum possible. The tension-compression strain in this central area will be null. The affected area at the surface usually extends beyond the extraction area, and at an in-

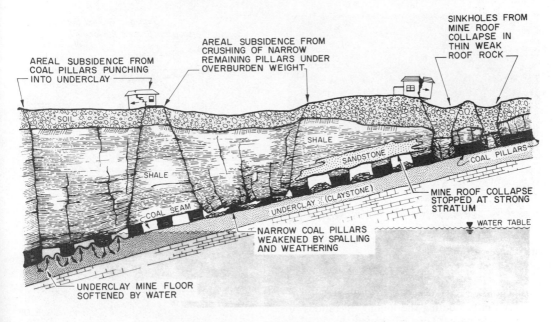

FIGURE 1. Examples of mine subsidence mechanisms.

clination that is termed the *angle of draw* (Fig. 5).

Ground

The design and application of mine subsidence control measures involve the disciplines of geology, mining engineering, rock mechanics, and civil engineering. The specific measures selected for subsidence control will depend on the type of mining conditions, whether the mine is active or abandoned, and the degree of control or protection required for mine workings or surface structures. Measures for controlling or preventing mine subsidence may be divided into four general categories:

1. Leaving blocks or pillars of rock or mineral in the mine for support;

FIGURE 2. Building damaged by subsidence above abandoned coal mine.

FIGURE 3. Interior of damaged building shown in Fig. 2.

FIGURE 4. Sinkhole developed in alluvium above abandoned coal mine.

2. Placing supports (such as piers, cribbing, or posts) either during the mining operations or after mining has been completed;
3. Partial or complete backfilling of mine voids during or after mining;
4. Arranging mining geometry and sequence of operations (sometimes by mining more than one seam or level at a time) to cancel out or selectively position stresses, strains, and surface movements, thus keeping the strains and movements within acceptable limits.

For purposes of discussion, subsidence control measures or methods for active mines are described separately from those for abandoned mines, although some of the methods are similar. While the described methods are directed primarily at protecting surface structures and facilities, some of the methods are applicable for controlling stresses and movements in active mining and permitting maximum extraction.

Subsidence Control Methods for Active Mines Surface area may be protected by *mineral abandonment*—i.e., by leaving an unmined block of mineral beneath the area requiring protection. *Partial mining or extraction,* by leaving a series of pillars, may be utilized to limit surface movements to very small amounts. A pattern of pillars may be left, as shown in Fig. 6, when a room-and-pillar type of mining is being used. The area of pillars (or unmined block) left must be larger than the area to be protected at the surface to account for the angle of draw for the particular situation. The selection of the pillar and room dimensions depends on overburden pressure, strength of the pillars, strength of mine roof and floor, and susceptibility of pillars, floor, and roof to deterioration over time.

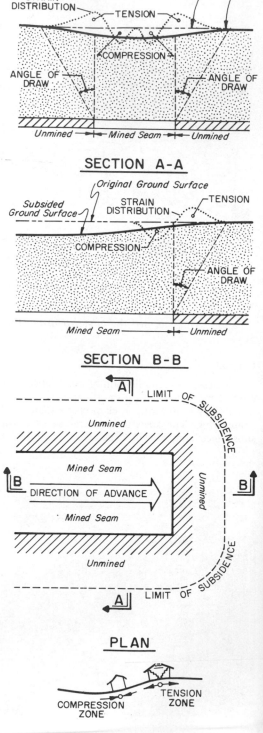

FIGURE 5. Subsidence and strain profiles from mining large width in seam.

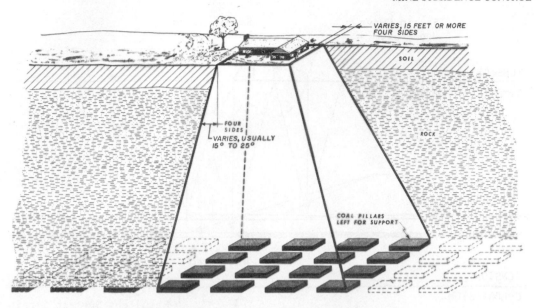

FIGURE 6. Partial mining with pillars left for surface support.

Where longwall, shortwall, or other complete extraction mining methods for wide areas are being used, alternating mined panels and long pillars may be employed, with dimensioning and spacing of the panels and pillars to cause overlapping and canceling strain profiles from adjacent mined panels and to reduce the magnitude of subsidence. Small pillars or cribbing that will crush partially or yield gradually can be used along the margins of extraction areas to provide a transitional subsidence zone that will smooth out the subsidence profile.

Backfilling of mine voids by hand packing, mechanical, hydraulic, or pneumatic methods is used to reduce movements in the mine, as well as to distribute stresses to prevent crushing of pillars or rock bursts from stress concentrations at the working mine face. The effectiveness of backfilling will depend on the strength and compressibility of the backfill material as well as the stresses on it and other mine conditions.

Many different materials have been used for backfilling—waste mine rock, process tailing, sand, etc. Sometimes cement is added to increase the strength of the fill material.

When complete extraction mining is conducted with relatively wide working faces, the following methods may be applicable for subsidence control: directional control, selective location, and multiple-seam and stepped-face means of harmonious extraction. *Directional control* involves orienting the direction of mining advance for a face wider than the critical width (area of null strain effects from side boundaries) parallel to the short dimension of the surface structure under which the mining is being conducted. This will minimize the intensity of damage effects that depend on the lengthening or shortening of the ground surface from the advancing face and the length of the surface structure. Extraction areas can also be selectively located so that surface structures are not positioned in maximum strain zones.

Where two or more seams are being mined at the same time, the mining of each seam can be located and sequenced so that the resultant tensile and compressive strains from each seam cancel each other, as illustrated in Fig. 7. A similar canceling of strains can be achieved when mining one seam by mining a staggered series of faces proceeding in the same direction. The widths of faces and distances between faces are engineered so that strains from the leading faces are canceled or reduced by strains produced by the following faces. Backfilling or other methods may be used to reduce subsidence along boundaries where the strains do not cancel.

Subsidence Control or Surface Stabilization Techniques for Abandoned Mines Subsidence control can be achieved by filling void spaces in and above the mine or by selective support of structures or areas. *Selective support* methods are utilized to supplement existing mine pillars. Included are piers constructed within the mine, deep foundations, and grout columns.

Piers of concrete, concrete block, timber crib-

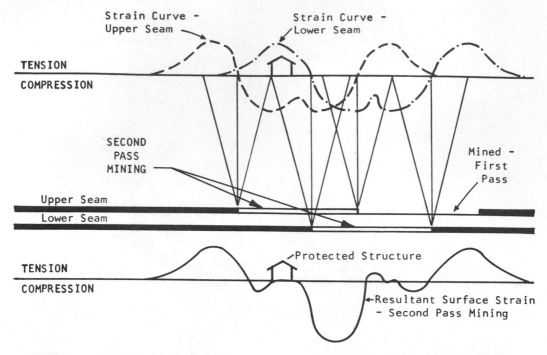

FIGURE 7. Multiple-seam harmonious mining to cancel strains for protection of surface structure.

bing, or other materials may be constructed within the mine to support the roof where the mine is accessible and conditions permit safe work. Access may be gained through mine entries or through large-diameter shafts drilled from the surface.

Drilled piers extending from the ground surface to the base of the mine can be used to support structures. Drilled piers are generally economically feasible to depths up of about 20 m. *Piling* installed in predrilled holes can sometimes be used for support. Both drilled piers and piling must be designed to take account of the corrosive acid water environment present in some mines. A corrosion-resistant steel alloy may be used for piling. One drawback is the possibility of buckling or shearing of deep foundations from lateral loads induced by subsidence. Another drawback of deep foundations is that they support only the structure and not the adjacent driveways, sidewalks, sewers, and utilities. For safety, flexible connections for gas lines and other utilities should be employed.

Grout columns may be constructed, where the mine is not accessible, to supplement existing mine pillars and to fill fractures above the mine (Fig. 8). A grout column is constructed by drilling a hole (typically about 15 cm in diameter) to the base of the mine. Gravel or crushed stone is placed down the hole to form a truncated conical pile in

the void at the mine level. A grout mixture of cement, fly ash, or sand and water is then injected into the gravel and broken mine rock (gob) through a pipe (perforated at the lower end) lowered to the bottom of the hole (into the gravel) and then gradually withdrawn. Additional gravel is added as needed during grout injection to develop sufficient mine roof contact for the grout column.

Filling Methods

Where caving of the mine roof is likely, the mine voids are filled to leave little or no room for caving. Filling also maintains the existing pillar support by confining the pillar sides and by protecting them from spalling and weathering. Filling methods have some varying effectiveness depending on completeness of filling and compressibility of the fill material.

Hydraulic flushing can be done by *controlled* or *blind* flushing or placing methods. Materials such as mine waste, sand, or fly ash (produced by coal-fired power plants) are sluiced or pumped into the mine voids in a water-solids mixture. Where the mine is accessible to personnel, workers can place the material in a controlled manner using bulkheads to contain the material (Fig. 9).

Blind flushing involves placing a slurry of solids and water through holes drilled to the mine,

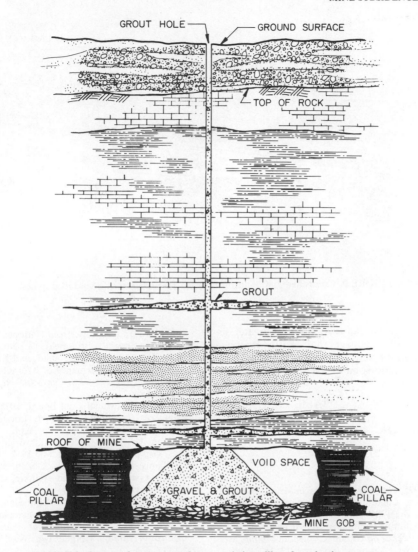

FIGURE 8. Grout column to supplement existing pillars for selective support.

without control of where the flushing material is deposited (Fig. 9). Once a hole is filled, the operation moves to the next hole. *Grouting* is a special flushing application where cement is added with the solids to give additional compressive strength to the fill material. *Pneumatic injection* of materials with compressed air can be used for filling mine voids. Fly ash is commonly used in blind injection through a borehole. The fly ash is usually transported and pumped by a bulk tanker similar to that used for transporting cement.

A technique for eliminating the subsidence problem is *overexcavation and backfilling*. The mine overburden is stripped to the base of the mined seam, and engineered backfill is compacted in layers to the desired grade. This technique is usually limited to shallow depths unless there is enough mineral remaining in a large area to make strip mining an economic advantage.

Architectural-Construction Measures

Architectural design and construction measures can sometimes be used in association with or as an alternative to subsidence control to protect surface structures. These include rigid foundation slabs or mats to bridge areas of differential movement, construction joints to separate large structures into smaller sections that can move independently, use of building materials such as

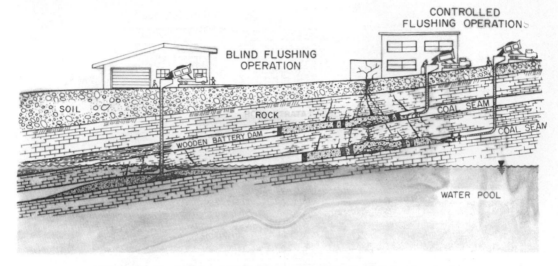

FIGURE 9. Controlled and blind flushing for stabilization of abandoned mines.

wood and steel that can accommodate greater distortions without damage than brittle materials such as masonry and plaster, and provisions for flexible connections and lines for gas and other utilities.

JAMES C. GAMBLE
RICHARD E. GRAY

References

Bell, F., 1975, *Site Investigations in Areas of Mining Subsidence.* London: Newnes-Butterworths, 168p.

Bieniawski, Z., 1981, Improved design of coal pillars for U. S. mining conditions, in *Proceedings, 1st Annual Conference on Ground Control in Mining.* Morgantown: West Virginia University, 13–34.

Boscardin, M., 1980, Building Response to Excavation-Induced Ground Movements, Ph.D. thesis, University of Illinois at Urbana, 279p.

Brauner, G., 1973, Subsidence due to underground mining, *U.S. Bur. Mines Inf. Circ. 8571,* 56p; and *U.S. Bur. Mines Inf. Circ. 8572,* 53p.

Gray, R., and R. Bruhn, 1984, Coal mine subsidence—Eastern United States, in T. Holzer, ed., *Man-Induced Land Subsidence.* Boulder, Colo.: Geological Society of America, 123–149.

Gray, R., and R. Bruhn, 1982, Subsidence above abandoned coal mines, in Y. Chugh and M. Karmis, eds., *Proceedings of Symposium on State-of-the-Art of Ground Control in Longwall Mining and Mining Subsidence, Hawaii.* New York: AIME, SME, 271p.

Gray, R. E., J. C. Gamble, R. J. McLaren, and D. J. Rogers, 1974, State of the art of subsidence control, Report ARC-73-111-2550, prepared by General Analytics, Inc., for Appalachian Regional Commission and Pennsylvania Department of Environmental Re-

sources, 265p (available from National Technical Information Service, Springfield, Va., No. PB-242465).

Hill, J. R. M., M. M. McDonald, and L. M. McNay, 1974, Support performance of hydraulic backfill, A preliminary analysis, *U.S. Bur. Mines Rept. Inv. 7850,* 12p.

Great Britain, National Coal Board, 1975, *Subsidence Engineers' Handbook,* 2nd ed. London: National Coal Board, Production Department.

Littlejohn, G., 1979, Consolidation of old coal workings, *Ground Engineering* 12(4), 15–21.

McWilliams, P. C., and A. E. Gooch, 1975, Ground support systems in block-cave mining, A survey, *U.S. Bur. Mines Inf. Circ. 8679,* 46p.

Mahar, J., and G. Marino, 1981, Building response and mitigation measures for building damages in Illinois, in *Proceedings, 1st Annual Conference on Ground Control in Mining.* Morgantown: West Virginia University, 238–252.

O'Rourke, T., and S. Turner, 1979, A critical evaluation of coal mining subsidence patterns, in *Proceedings, 10th Ohio River Valley Soils Seminar on the Geotechnics of Mining.* Lexington, Kentucky, 1–8.

Price, D. G., A. B. Malkin, and J. L. Knill, 1969, Foundations of multi-story blocks on the coal measures with special reference to old mine working, *Quart. Jour. Eng. Geology* 1, 271–322.

Sinclair, J., 1963, *Ground Movement and Control at Collieries.* London: Pitman, 349p.

Tomlinson, M. J., 1963, *Foundation Design and Construction.* New York: Wiley, 165–183.

Wardell, K., 1953, Some observations on the relationship between time and mining subsidence, *Inst. Mining Engineers Trans.* 113(1953–1954), 471–483.

Cross-references: *Mining Preplanning.* Vol. XIII: *Foundation Engineering; Grout, Grouting; Rocks, Engineering Properties; Rock Structure Monitoring.*

MINING ENGINEERING GEOLOGY—
See BOREHOLE MINING; COAL
MINING; MARINE MINING; MINE
SUBSIDENCE CONTROL; MINING
PREPLANNING; PLACER MINING.

MINING GEOPHYSICS—See
EXPLORATION GEOPHYSICS.

MINING METHODS—See BOREHOLE
MINING; COAL MINING; MARINE
MINING; PLACER MINING. Vol. XIII:
CAVITY UTILIZATION; SHAFT
SINKING; TUNNELS, TUNNELING;
URBAN TUNNELS AND SUBWAYS.

MINING PREPLANNING

Restoration of surface-mined lands is at the forefront of environmental concerns. The value of turning mined land into productive land has aesthetic, social, and economic implications to society. Along with a growing worldwide trend toward the reclamation of surface-mined land has evolved the relatively new concept of *reclamation preplanning.* Mathewson (1980) summarized the applications of engineering geology to coal mine planning. Planning as a component of mine design and operation, however, has been practiced since the advent of the mining engineer. Extensive research, textbooks, and symposium proceedings exist on this aspect of mining.

Preplanning for reclamation is consideration of the post mining land use prior to mining and implementation of reclamation processes into the mining program to accomplish the desired goal. As the need for mined energy, mineral, and construction resources increases, so must the amount of land subject to surface mining increase to meet the growing demand. With this growth comes the responsibility to protect our greatest natural resource, the land.

Preplanning may take many forms, depending on a variety of parameters. Among these are included the type of mining, the unique physical and geological characteristics specific to a geographical area, the location of the mine relative to land use needs, and state and federal laws pertaining to each area.

Types of Surface Mining

With respect to reclamation, the principle types of surface-mining operations may be divided into

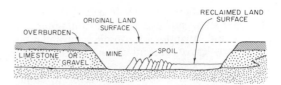

FIGURE 1. Schematic cross-section of net volume loss mining. Reclaimed surface is lower than surrounding topography.

two general categories: (1) net volume loss and (2) no net volume loss mining.

In *net volume loss mining,* the volume of spoil produced is insufficient to refill completely the excavation created by the permanent removal of the desired resource. After mining has ceased, a topographic depression, or hole, remains (Fig. 1). Principal net volume loss operations include quarries, sand and gravel operations, and large-scale open-pit mines.

No net volume loss mining occurs when the overburden is much thicker than the mined material. The total volume of overburden increases during removal due to an increase in the void ratio of the overburden. When this material is backfilled and graded, expansion of the overburden compensates for the volume of material extracted and results in little or no net volume loss. The land may be regraded to roughly its original contours. In some instances, the expansion may be great enough to create a low hill where none existed prior to mining (Fig. 2). Principal no net volume loss operations include some coal and lignite strip mines.

Unique Physical Characteristics

Preplanning for reclamation requires consideration of the unique physical and geological characteristics of the area. The physical and geological characteristics of an area such as Wyoming, and consequently, its associated reclamation problems, obviously differs from those of Alabama. Characteristics that should be considered in preplanning for reclamation include premine topography, annual precipitation and evap-

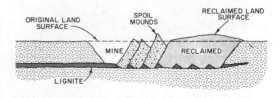

FIGURE 2. Volume of overburden is sufficient to replace material taken out of the pit or trench.

oration, surface and subsurface drainage characteristics, soil and resource contaminants, quality and quantity of topsoil, consolidation characteristics of backfill material, and native vegetation.

Topography is a controlling factor in reclamation. Many state laws now require shaping of the mined area to premine contours or shaping of the land for higher and better uses. Differences in topography may result in significant differences in reclamation costs when regrading to premine contours.

A high amount of precipitation is advantageous where revegetation of the reclaimed land is desired, but it may cause erosion and sedimentation problems. Regions of low rainfall may have difficulty revegetating without a substantial amount of irrigation.

Surface and subsurface drainage characteristics are important in determining the impact of pollutants on overall water quality and the determination of the groundwater system that will form at a reclaimed mine site. The premined groundwater system, the stratigraphy of the overburden, and the mechanical properties of the overburden material all interact to produce four basic hydrogeologic types (Fig. 3) in the reclaimed mine (Kennedy, 1981; Mathewson et al., 1982). Surface drainage may be altered, excessive amounts of sedimentation may enter the surface drainage, or

groundwater recharge may be changed as a result of improper mining and reclamation techniques.

Knowledge of potentially toxic materials in the overburden and in the resource being mined is important to the success of any reclamation procedure. Control of these contaminants must be considered so that water quality is not degraded and successful revegetation of the land is accomplished. Iron sulfides, a common toxic-forming mineral, can oxidize and react with water to produce sulfuric acid. If acid water is allowed to contaminate the reclaimed land or to run off into nearby areas, the potential for revegetation is drastically reduced.

It may be necessary to segregate and save the topsoil for redistribution over the reshaped land to enhance revegetation during reclamation. If topsoils are very thin, or much like the underlying material, there may be no advantage in separating them. The segregation process adds a substantial cost to the reclamation.

Consolidated characteristics of the backfill material determine when construction on the reclaimed land is safe. Studies by Schneider (1977) and Rangel (1979) indicate that subsidence of reclaimed overburden is related to the characteristics of the overburden and water balance in the area (Armstrong, 1987). Sands and blasted rock experience very little settlement, while claystones may settle by as much as 20% of the original

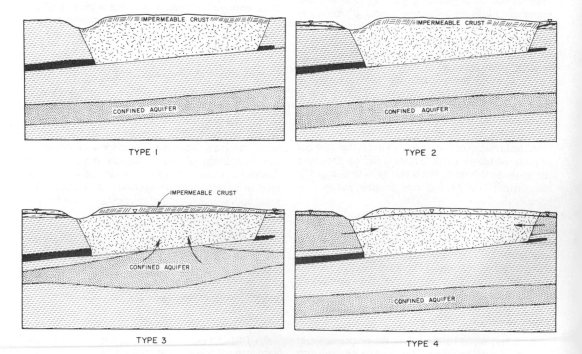

FIGURE 3. Classification of hydrogeologic types in reclaimed surface mines.

height. Measurable settlement in claystones appears to cease after 5 yr, indicating that building construction on this reclaimed land might be feasible after a short waiting period.

In many cases, the native vegetation of an area determines what plants will be used in revegetation. The most often attempted method used to reclaim an area is to revegetate with fast growing grasses. The difficulty involved in re-establishing a more natural flora is dependent on many of the characteristics listed here. The usual procedure may involve experimenting with numerous species of local plants until a species or combination of species proves successful.

Location of Mine

The location of the mine determines, for the most part, the potential land use after mining. Location may be divided into three major categories: rural, urban, and nonpopulated.

Rural The highest and best land uses in rural areas are typically agricultural. Consequently, when reclaiming land in rural areas, operators must consider what agricultural use is best suited for the particular area they are mining. The lignite mining operation at the Big Brown site in Fairfield, Texas, is a good example of what can be done to bring surface-mined land back to productive agricultural use. Planting a mixture of coastal Bermuda grass for hay and clover as a nitrogenous supplement has been very successful within 2 yr after reclamation.

After the land has been backfilled and graded, the soil usually has a low organic matter content. Other land, such as feedlots, are high in organic content and usually present a problem in waste disposal. Disposal of feed-lot waste at a nearby mine, for example, could solve both problems. Numerous other options are also possible.

Quarrying and open-pit operations in rural areas present the problem of finding a use for the resultant hole. Small pits might be graded and replanted for grazing purposes. If the water table in the area is sufficiently high, small fishing and recreation lakes or stock watering tanks warrant consideration. Ohio has had success reclaiming sand and gravel pits as water recreational areas for a number of years (Klosterman, 1974). Larger pits are more difficult to reclaim because of the large amount of fill material that would be required. The U.S. Army Corps of Engineers is, however, investigating the possibility of using fine-grained, but nutrient-rich, dredged material from coastal dredging operations as potential fill for abandoned pits and quarries (Saucier, 1975). The success of this program will depend on the distance of transport and geochemical considerations of leachate and groundwater contamination.

Urban Reclaimed land in urban areas has a different potential. Because of low profit margins and high transportation costs, sand and gravel operations and, to a large extent, limestone quarries are located near urban areas. In Denver, Colorado, and Dallas, Texas, some worked-out pits have been used as sanitary landfills prior to industrial or commercial development. Before sanitary landfills are initiated, it is necessary to investigate the possibility of groundwater pollution.

Pits occupying large areas may be suitable as construction sites. The floor of a limestone quarry in Dallas, Texas, e.g., was leveled near the end of the quarry operation to provide an excellent site for a regional postal center. Other possible uses might include industrial complexes, shopping centers, and apartment developments. The prime consideration in reclamation for any project involves grading and shaping of the land to fit the future use.

Nonpopulated Regions with a low population density and sparse vegetation cover may not be amenable to the type of reclamation projects described in the previous categories. Backfilling and regrading the land to produce a surface morphology similar to the one that existed prior to mining appears to be the only feasible practice for such areas at this time.

Sparsely populated regions with a dense vegetation cover requires more careful consideration in preplanning. Much work relevant to reforestation of surface-mined lands has been conducted by the U.S. Forest Service, which initiated a research program in 1937 that was designed to investigate possible means of reducing damages to forests during mining and to restore forest resources on strip-mined land as soon as possible after mining (Sawyer, 1962). Research with various varieties of trees in different geographical regions continues today.

Laws and Regulations

Numerous state and federal laws have been enacted since the 1970s regarding the reclamation of surface mined land. World and national interests concerned with the development of coal resources, in particular, have focused a great deal of attention on reclamation because a large amount of land will be subjected to surface mining.

In addition to the reclamation requirements placed on coal producers, many states are requiring that other mineral and resource industries restore the land that they mine. North Dakota, South Dakota, Washington, and Wyoming have passed laws that require the rehabilitation of the land for any surface-mined material. Many states also require a mining plan, a reclamation plan, which is an indication of the intended land use prior to the development of resources.

Examples of Progress in Preplanning

Worldwide On a worldwide basis, reclamation has flourished for a number of years in smaller countries where the need for conserving land has been greater than in the United States. Mining laws in Germany have been in existence since the early 1920s (Knabe, 1964). German experience with soil contamination, since a high percentage of toxic soils are associated with coal, led to the implementation of soil segregation practices, which have been required since 1940 to assure that no toxic spoil is placed on the ground surface. The usual practice is to segregate the topsoil from the poor subsurface soils, which must be covered with fertile material. To accomplish the desired soil separation, a seven-step mining technique has been adopted (Knabe, 1964). In this technique, the initial lowering of the groundwater table tends to reduce the possibility of spoil sliding. The topsoil is then separated and stored for later use. After excavation of the deposit, the spoil is transported and dumped at a predetermined location. With the cessation of mining operations, the stored topsoil is placed on the spoil heap and the area graded and revegetated.

Underground mining operations also require preplanning for reclamation. In some instances these mines require partial or total filling to protect the land surface from subsidence (see *Mine Subsidence Control*). The fill material, normally sand, is usually surface mined at nearby locations, and the reclamation of these pits requires additional preplanning. Preplanning measures and reclamation procedures in Poland for this indirect result of deep-coal mining are discussed by Hutnik et al. (1974). The principal method of reclamation of these worked-out sand pits is reforestation, accomplished by re-creating, as far as possible, the natural conditions for forest development by grading the pit slopes, applying fertilizers, and redistributing topsoils stored prior to mining. Alteratively, the sand pits may be used as water supply facilities or multipurpose reservoirs for water sports centers and water catchments for industrial, domestic, and agricultural uses.

In Jamaica, interest in reclamation has centered on problems associated with the large-scale mining of bauxite since 1952 (Davis and Hill, 1972). Due to the small size of the island, 11,525 km², and a dominant agricultural society, the need for bringing land back to productive agricultural use is great. Where it is not possible to reclaim mined-out land, the mining companies must reclaim an equivalent area elsewhere that has not previously been used for agricultural purposes. An alternative is to pay the government a set sum for every acre mined that is not reclaimed. Mining procedures in Jamaica require removal and stockpiling of the topsoil. After mining, the land is shaped to a desired form and the stored topsoil is redistributed over the area. After shaping, a crop or forest is established on the reclaimed land. Reclamation is not complete until the Department of Mines examines the area and issues a certificate of approval.

United States Reclamation of strip-mined lands in the United States began on a small scale in Indiana in 1918 when individual mine operators experimented with different methods of replanting spoil areas (Sawyer, 1962). The first state law requiring reclamation was passed by West Virginia in 1937 (Knabe, 1964). Today most states have laws requiring reclamation of certain types of strip mining.

The Surface Mining Control and Reclamation Act of 1977 (Public Law 95-87) established the Office of Surface Mining within the Department of the Interior and authorized it to establish rules, regulations, and procedures to control surface-mining impacts from coal mining operations in the United States. Each state that has a coal mining industry is required to establish state rules that comply with federal. No federal rules have been established for other mining operations in the United States, although numerous states have established state rules for uranium, sand and gravel, quarries, and other mining.

Reclamation in the United States has resulted in extensive research into the productive capability and the geotechnical aspects of reclaimed land. These factors are extremely important in determining a postmine land use. Successful examples of preplanning in the United States are numerous. These successes include the development of water recreational areas in the Midwest; reforestation of some land in the East, Midwest, and South; and agricultural productivity in all these areas.

We by no means intend to imply that all surface-mined lands are successfully reclaimed or that attempts have been made to reclaim all mined land. Large areas of surface-mined land are, in fact, unreclaimed. With the expected addition of vast amounts of land subject to surface coal mining in the western United States and the federal and state requirements for reclamation comes the realization of the need for preplanning.

A Preplanning Concept

Preplanning is a long-term project. The legal requirements of returning the land to a suitable condition and assuring that detrimental environmental effects of mining are minimized present a challenge to the mining industry. To comply with federal and state regulations, significant forethought is required.

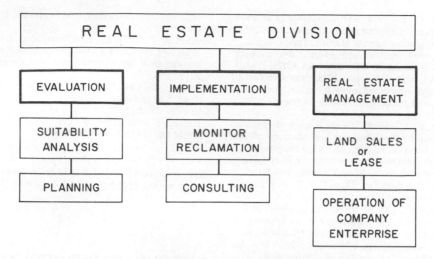

FIGURE 4. Phases and duties of the real estate division.

In dealing with large areas of land, mining companies may often overlook possibilities in the real estate market. The potential monetary gain available through proper preplanning and development in reclamation may be significant.

To make the reclamation process as beneficial as possible, a branch of the company should be organized whose corporate charge is to make reclaimed land profitable. Preplanning in the early stages of a mining operation should be used as a guide for acquisition, development, reclamation, and management of mined land. A real estate division, organized within the mining company, should consider the specific phases of evaluation, implementation, and real estate management.

Evaluation Figure 4 shows the breakdown of the various phases within a real estate division. The evaluation phase covers land suitability analysis and preplanning to determine the optimum postmine land uses. This may require no more than planning from environmental or urban planning maps, or it may require extensive engineering property testing to establish a data base.

Contact with development concerns and local and state authorities may lead to a viable preplanning program. If, e.g., a developer considers a location suitable for possible use in future projects, then preplanning and reclamation could be oriented along these lines. Likewise, it is often possible to develop city or state recreational areas on mined-out lands by coordinating reclamation procedures with proprietary and municipal planning schedules. The possibilities are limited only by imagination and geology. The final plan for reclaimed land use is, of course, flexible.

Implementation This phase basically involves monitoring the reclamation process to assure its proper development. Private consultants, local and state officials, management personnel of development concerns, and others may be involved with the postmine land use. Reclamation procedures must operate within the limits of the law, and all aspects should be continually monitored during this phase.

Real Estate Management The operational phase of real estate management will most likely occur near the end of the reclamation process. Negotiation of the sale or lease of the land to the prospective client is the major concern.

If the company decides to use the postmined land in a private enterprise, the real estate division might operate the enterprise. The possibility of a private enterprise should normally be explored by mining operators as a potential investment venture.

LARRY L. MINTER
CHRISTOPHER C. MATHEWSON

References

Armstrong, S. C., 1987, Engineering Geologic Analysis of Reclaimed Spoil at a Southeast Texas Gulf Coast Surface Lignite Mine, M.Sc. thesis, Texas A&M University, College Station, Texas.

Davis, C. E., and V. G. Hill, 1972, Reclamation of mined out bauxite lands in Jamaica, 1952–1970, paper presented at the American Institute of Mining Engineers Annual Meeting, San Francisco, California.

Hutnick, R. J., Z. Harabin, and Z. Strzyszcz, 1974, Restoration of coal-mined land in Poland, *Papers, Second Research and Applied Technology Symposium on Mined Land Reclamation*, 242–252.

Kennedy, J. L., III, 1981, Hydrogeology of Reclaimed Gulf Coast Lignite Mines, Ph.D. thesis, Texas A&M University, College Station, Texas.

Klosterman, G. E., 1974, Preplanning—Key to reclamation, paper presented at the American Institute of Mining Engineers Annual Meeting, Dallas, Texas.

Knabe, W., 1964, Methods and results of strip mine reclamation in Germany, *Ohio Jour. Sci.* **62**(2), 75–105.

Kosovac, M., and S. Kundu, 1978, Iron Ore Co. of Corroda's computerized analysis method speeds mine planning and pit designs, *Mining Engineering* **30**(7), 767–770.

Mathewson, C. C., 1980, The application of engineering geology to the evaluation and planning of coal mines, in G. O. Argall, Jr., ed., *Mine Planning & Development*. Technical Papers from the First International Symposium on Mine Planning and Development, Beijing/Beidaihe, China Coal Society, and the Chinese Society of Metals, San Francisco, California.

Mathewson, C. C., K. W. Brown, L. E. Deuel, and D. G. Kersey, 1982, *The Impact of Surface Lignite Mining on Surface- and Ground-water Quality in Texas* (Final Report, August). Washington, D.C.: U.S. Bureau of Mines (Contract No. J0295016), 82p.

Rangel, J. E., 1979, The effect of stratigraphy and soil plasticity on the settlement characteristics of reclaimed surface mined land, M.Sc. thesis, Texas A&M University.

Saucier, R. T., 1975, Dredged material as a natural resource—Concepts for land improvement and reclamation, *Geol. Soc. America Abs. with Programs* **7**(7), 1259.

Sawyer, L. E., 1962, Mined area restoration in Indiana, *Jour. Soil and Water Conservation* **17**(2), 65–67.

Schneider, W. J., 1977, Analysis of the densification of reclaimed surface mined land, M.Sc. thesis, Texas A&M University.

Cross-references: Vol. XIII: *Rock Mechanics; Rocks, Engineering Properties; Rock Structure Monitoring.*

MUSEUMS AND DEPOSITORIES—See MAP AND CHART DEPOSITORIES.

O

OFFSHORE NUCLEAR PLANT PROTECTION

The design of structures to protect offshore nuclear power plants requires an understanding of the nature and magnitude of environmental forces that are markedly more intricate than for conventional onshore construction and considerably more sophisticated than that for previously constructed onshore structures. For example, the static design alone of any bottom-seated structure, such as a protective *breakwater,* must be evaluated for the effects of appropriate load combinations of gravity with wind and wave (steady state and storm), earthquake, and tornado (Fischer et al., 1975). In addition, all less severe environmentally generated loads with combinations of probability greater than 10^{-7} must also be taken into account. Once these forces of nature are understood, however, bottom-seated facilities such as protective breakwaters are amenable to design by accepted engineering analyses (advanced though these may be).

A further point of interest is that while, in theory, floating nuclear power plants (FNPs) may be protected by a number of different types of breakwaters (e.g., cellular caisson-type systems), at the present time both economics and engineering state-of-the-art dictate that the major portion of these breakwaters be constructed of natural materials. Some of the advantages of *rubble-mound structures* relate to their relatively uncomplicated design; their mass and shape, which offer the maximum resistance to wave action; and their bulk, which results in relatively low foundation loading.

A closer look at the interaction between the structure and the foundation reveals still further advantages to be derived from the flexibility of such structures. Flexibility, combined with a responsible design, immunizes *rockfill structures* from catastrophic failure. Flexibility is the key to ensuring structural integrity in the event of the protective structure's having to tolerate potentially damaging stresses such as those associated with differential foundation settlement, toe scour, or earthquake-induced liquefaction.

Inevitably, large rubble-mound structures require enormous quantities of high-quality rock. Because of the substantial economic, logistic, and environmental considerations that such quantities of rock present, extensive investigations are necessary. Preliminary studies must be undertaken to isolate the geological availability of sources of suitable rock in terms of minimizing the impact of its workability and transportation on the environment. Existing quarries, in suitable areas, must be identified and evaluated as potential suppliers by comparing rock quality, in situ characteristics of the rock mass as these relate to breakage in optimum size ranges, production capability, haulage, and anticipated environmental and other difficulties. Further investigations must establish, for a short list of quarries, representative petrographic descriptions and engineering index properties (e.g., specific gravity, compressive strength, and durability). A study should be made of the service records of rock that has been used in heavy ocean construction. In the final phase of investigation, large-scale blasting and sizing tests are essential to predict how the rock will break and to establish realistic material modeling criteria for use in field density determinations and laboratory triaxial testing (Fig. 1).

The timing of the overall investigation is most important: first, to ensure that rock of suitable quality is used in the breakwater and that the sources of borrow selected are favorably located in terms of economic haulage; second, to verify that realistic values of the density and the angle of internal friction of the rockfill are incorporated into the design; and third, to establish a responsible basis for construction bids and thereby permit the appointed contractor the maximum lead time to mobilize personnel and equipment, stockpile rock, and plan the complicated logistics of the construction operation (Fischer et al., 1975).

This entry outlines, for FNP breakwaters, the suitability of rock types, the economic and environmental considerations associated with quarrying and transportation, the availability of materials, and aspects of all follow-up investigations. The earth scientist charged with undertaking similar investigations, but for a smaller breakwater or one not required to meet the demands of the U.S. Nuclear Regulatory Commission (formerly the Atomic Energy Commission) may interpret the following discussion accordingly.

FIGURE 1. Some aspects of investigations for FNP breakwaters. (*a*) Trial blasts to predict workability, (*b*) mechanical sizing to determine gradation, (*c*) triaxial testing to determine the angle of internal friction of the rockfill, (*d*) typical breakwater to protect a two-unit (2,300 MW) plant.

b

c

d

Suitable Materials

In the simplest terms, the natural materials suitable for constructing all breakwaters are those that have a source favorably located for economically producing fragments of rubble of the required sizes and those types of rock sufficiently tough to withstand both the short-term processes associated with quarrying, transportation, and placing and the longer-term processes associated with its weathering during the life cycle of the breakwater.

The following detailed discussion on the suitability of rock materials has general application.

Geological Suitability Geologically, the most suitable materials are derived from massive, fine-grained igneous rock such as diabase and basalt, the massive metamorphic rocks such as greenstone and quartzite, and the massive sedimentary rocks such as hard sandstone and limestone.

Coarser-grained igneous and metamorphic rocks such as granite and granitic gneiss are also generally suitable. Highly foliated rocks such as schist, micaceous gneisses, and some amphibolites, as well as thinly bedded, sheared, and excessively jointed rocks, are not suitable because these are likely to shatter on blasting, producing a large number of flat fragments and a high percentage of fines (Treasher, 1964).

The importance of the characteristics of natural discontinuities in the rock mass cannot be overstressed since these control not only the limiting upper size of fragments and the ability of the rock to break into angular roughly equidimensional fragments showing suitable surface roughness but also the distribution of fragment sizes (gradation) in the production blast. Such properties, if favorable, contribute significantly to a relatively dense rockfill with a relatively high angle of internal friction.

Suitable breakwater stone is further geologically characterized by being resistant to physical and chemical weathering. However, except under extreme climatic conditions (polar or tropical), weathering is a very long-term process, and the possibility of its influencing any sound rock placed in a breakwater during its 40 or 50 yr life cycle is extremely remote (Watson et al., 1974).

In addition to rock rubble, suitable aggregate stone is necessary for the concrete to manufacture *caissons* and *primary armor dolosse* (Fig. 2). Such stone should be hard, tough, durable, strong, of the proper gradation, and free from excess silt, soft or coated grains, mica, harmful alkali, and organic matter. A wide variety of igneous, metamorphic, and sedimentary rocks meet these qualifications. The properties of the rock mass do not affect the engineering suitability of aggregate stone since blasted material is further processed by crushing and screening (Marachi et al., 1972).

American Society for Testing and Materials (ASTM) and state standards are well established for quantitatively evaluating concrete aggregate and are essential.

Geotechnical Suitability Favorable geological rock types relate to materials with favorable *engineering index properties,* properties such as mass per unit volume, compressive strength and durability in terms of mechanical and chemical stability, and a function of the mineral composition, texture, and the macro- and microstructure of the rock element. Satisfactory laboratory-determined parameters are associated with petrographically determined features such as fine grain size; a lack of microfractures, or well-developed cleavage; hard stable minerals; an isotropic mineral arrangement; and good crystal interlock. Conversely, less suitable rocks are associated with coarser grain size, the presence of microfractures, unstable soft minerals, and an abundance of minerals with well-developed cleavage; a well-defined anisotropic mineral arrangement; a clastic texture; and other deleterious properties. Engineering index parameters are determined on the basis of laboratory tests standardized by the ASTM.

It may be emphasized at this point, however, that the results of these tests are no more than index parameters and that like petrographic characteristics should be used only to provide guidelines in selecting suitable rock and not to establish rigid diagnostic criteria for condemning, needlessly, potential useful sources. Many rocks having one or more of the features cited as being characteristic of poorer material have been satisfactorily used in breakwater systems (Marsal, 1973). Again, a study of the service records of rock previously furnished by potential suppliers for offshore construction is essential for establishing a realistic evaluation of a particular rock source.

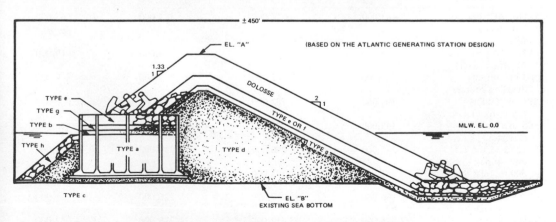

FIGURE 2. Typical cross-section of a main breakwater.

Design Suitability Design suitability of rock is evaluated in terms of the efficiency with which individual rock fragments recombine and act as a single rockfill mass. The critical design parameter in this respect is the angle of internal friction, which defines the strength of the rockfill structure. The highest angle of internal friction attainable is naturally desirable. It cannot be overemphasized that the tests used to determine this parameter should reflect the gradation of the core material to be used and the confining stresses, relative density, and loading conditions anticipated in the breakwater.

In summary, favorable lithologic conditions contribute to favorable engineering index properties, which in turn contribute to a favorable design value. Provided that rock meets the general criteria outlined in the opening paragraph of this section, it would be suitable for use in breakwaters. However, it follows from the discussion that the better the quality of rock, the more desirable would be the design parameters for the breakwater. An optimum breakwater design incorporates the steepest slopes and, hence, the smallest size. A reduction in size is desirable in that it reflects a corresponding reduction in cost and environmental impact associated with obtaining and moving borrow.

Environmental and Economic Aspects

The materials required for breakwater construction merit environmental consideration in that, first, within the framework of the offshore siting concept, the activity associated with quar-

rying and the partial transportation of rock borrow, although of a transitory nature, represents one of the few major areas of land-based environmental impact, and, second, the range of potential impact is very broad and is one over which the engineer has, in most instances, considerable control.

In general, the key to environmental and economic success relates to planning the maximum use of water haulage to convey rock borrow from selected quarries to proposed site locations. The careful selection of existing quarries, suitably located for the water transportation of rock or the provision of sufficient lead time to find, license, and develop new quarries on the coast or on a navigable waterway will invariably result in the greatest economic saving and minimum negative impact on the environment.

Quantity of Materials To consider the economic and environmental aspects of this topic in more detail, it is necessary to know the quantity and sizes of materials that are required for typical FNP breakwaters. Consequently, the approximate quantities of suitable materials, computed for four rubble breakwaters (each enclosing two floating nuclear units, 2,300 MW) at typical locations on the East and Gulf coasts, are shown in Table 1 for a typical breakwater design (see Fig. 2). In the case of Site A, calculations for probable crest and basin elevations are based on location in the minimum water depth that would satisfy lower-bound siting criteria (without excessive dredging). Site A, therefore, reflects the minimum size of breakwater and hence, the least quantities of construction materials. In contrast,

TABLE 1. Approximate Volumes in Cubic Meters and Classification of Rock Required for Typical Rubble-Mound Breakwaters

Types of Material	Typical Sites			
	A—New Jersey	B—N. Carolina	C—Florida	D—Texas
(a) No. 200 sieve to 3 in. (caisson fill)	403,000	512,000	587,000	566,000
(b) No. 10 sieve to 6 in. (filter layer)	19,000	19,000	19,000	19,000
(c) 3 in. to 6 in. (leveling course)	9,000	9,000	9,000	9,000
(d) 0 to 400 lbs. (core stone)	1,305,000	1,782,000	2,221,000	2,087,000
(e) 6 to 12 ton (armor stone)	445,000	479,000	502,000	496,000
(f) 10 to 15 ton (armor stone)	226,000	290,000	334,000	321,000
(g) 600 to 1600 lbs. (stone)	251,000	304,000	330,000	323,000
(h) 1.5 to 4 ton (armor stone)	166,000	217,000	226,000	221,000
Totals	2,825,000	3,612,000	4,228,000	4,042,000

calculations for Sites B, C, and D are based on location in the maximum practical depth of water and reflect the maximum amount of materials required.

Quarrying For the largest breakwater considered, the land committed to supply 5.7 million m³ in an average 4 bench quarrying operation would be about 6 ha. Because the area to be quarried is relatively small, the environmental effects of working it depend on the location and status of the source. Obviously, developing and opening a new quarry would initially have a far greater negative impact on the environment than reopening an idle quarry or expanding an existing operation. In the densely populated northeastern states, it is considered impractical to open a new quarry.

In the less densely populated southern states, the opening of a new quarry presents a feasible concept. On the positive side, the influx in jobs and money created by rock-quarrying operations would certainly be welcomed in many areas. Provision may have to be made, however, for a lead time in excess of 2 yr to find, investigate, license, and develop such a source.

Transportation Typical transportation cycles from quarries to construction sites relate to the following conditions: (1) quarries with direct access to water haulage, (2) quarries located within a short truck or rail haul of water transportation, and (3) quarries relying to a significant extent on rail haulage. Because of anticipated environmental problems and the high costs associated with trucking, extended truck hauls are not considered feasible.

Negative impact resulting from transportation of stone from the quarry to a typical site includes, at worst, emissions of gaseous pollutants from internal combustion engines, vehicular and engine noise, increased traffic congestion, and oil spills in rivers, harbors, and along coastal areas. The greater the transportation distance, the more pronounced these effects become. Adverse impact may be minimized by making the maximum use of barge haulage. The Federal Water Pollution Control Act, substantially amended in 1970 and again in 1972, imposes on the vessels traversing U.S. navigable waters water pollution control laws, which are among the most stringent and most rigidly enforced laws of any nation in the world. The increase in waterway traffic as a result of transporting borrow to a proposed FNP construction site would not be significant since the waterways carry more than 16% of the nation's freight (American Waterways Operators, 1973).

Parameterized Costs The parameterized costs of rock delivered to the construction site for a typical two unit station are most difficult to outline in a generic sense because of the great number of variables influencing the exact design of the breakwater, the basis of selection of a supplier, the basis on which the contract bid is let, and the contractor's criteria for distributing contingency allowances. The cost of rock in the breakwater also varies according to the size of fragments, quality of rock, location of quarry, and method of transportation. It is estimated that to make consideration of a rubble-mound breakwater a practical economic consideration, the bulk of the material (i.e., the corestone) should not exceed $15 a ton in place. With careful planning it may be possible to achieve even lower costs. Larger fragments require individual handling and would consequently be significantly more expensive. Fragments in the 10–15 ton range, in place, may be expected to cost $25–35 a ton. When it is considered that the smaller size fragments cost between $2 and $6 a ton at the quarry and the larger sizes about $8 a ton, the relatively large cost of handling and transportation is apparent. Transportation charges are greatest for truck hauling, which averaged 8¢ per ton-mile in 1971. In comparison, rail loading rates are lower, averaging 1.6¢ per ton-mile in 1971, and river barge hauling rates are even lower, averaging 0.3¢ per ton-mile in 1972 (American Waterway Operators, 1973). Note that if the design of the breakwater can tolerate a shot-loaded core as opposed to rubble requiring mechanical scalping, a considerable saving may be achieved for the volumes required.

As far as supplies to offshore breakwaters affecting the regional quarry output and influencing the price of rock, this was found to be insignificant. The 2 million tons per year required during construction represents, in the case of Site A (see Table 1), only a 2% increase in regional quarry output.

Availability of Materials

Detailed preliminary studies indicate that there are adequate sources of suitable natural materials available, both in the United States and in neighboring countries, to confirm the economic and environmental practicality of rubble-mound protective breakwaters for FNPs on the East and Gulf coasts. In addition, it is considered economically feasible to supply borrow to most East and Gulf coast points by a predominantly water haul. It is obvious that sources suitably located for barge haulage would not necessarily be the nearest sources to projected sites but would inevitably be competitively located because of the favorable costs of barging.

The following brief and generalized discussion on the availability of materials is abstracted from a more detailed presentation by Watson et al. (1974), delineating sources of suitable rock both in the United States and neighboring countries and identifying major suppliers.

Northeastern States A broad sweep of the geological brush indicates that the northeastern

United States are adequately endowed with sources of highly suitable basalt, diabase, dolomitic limestone, and granite. Some 137 major suppliers are known in the coastal and adjacent states between Maine and Maryland inclusive.

Central Eastern States Farther to the south, the more suitable rock (mainly granite, granite gneiss, and limestone) is separated from the sea by the semiconsolidated and unconsolidated Tertiary sediments of the Coastal Plain. Access to the sea is provided via rail and navigable rivers.

Southeastern States The obvious first choice for rock supply for a projected site off the Florida coast would be the Caribbean Islands. Not all Caribbean Islands are feasible, however, for political reasons, and back-up sources on the continent would have to be ensured.

Numerous suppliers with rail access are found in the Georgia and South Carolina Piedmont. Such quarries could furnish granite and gneiss.

Central and Western Gulf Here, sources, although distant, are again numerous. Possible sources include the igneous (mainly basalt, some granite) and metamorphic outcrops in Mexico, the limestones of the Edwards–Comanche Plateau, and the igneous and metamorphic rocks of the Llano area, which would be railed to Corpus Christi.

Other sources of rock may be the Paleozoic sedimentary rocks of Arkansas, which would be barged via the Arkansas and Mississippi rivers to the Gulf, or the Paleozoic carbonates of the Alabama Valley and Ridge, which could be barged down the Alabama or Tombigbee rivers to Mobile.

Field and Laboratory Studies

Field Assuming that a comprehensive literature study has been completed by engineering geologists with an understanding of the requirements of stone for an FNP breakwater, a three-phase field program is recommended:

Phase 1 Quarries shortlisted in a report based on literature studies because of suitable rock types, scale of operation, and the proposed transportation of materials imposing the minimum impact on the environment are visited during this phase. A visual geotechnical evaluation is made of the rock. Quarry records are examined, available test data selected, and anticipated technical and environmental problems noted. Comprehensive records are compiled in this way for all quarries investigated. These studies are concluded in a report evaluating quarries as potential suppliers, by comparing rock quality, the in situ characteristics of the rock mass as they relate to breakage in optimum size ranges; production capability; and haulage availability, especially with respect to water haulage and anticipated environmental and other difficulties. If it is apparent in the light of this investigation, that a significant economic and/or environmental advantage is to be gained by opening a new quarry, then an appropriate program of exploration should be undertaken.

Phase 2 Once the most suitable existing quarries have been isolated or a new source explored, a detailed phase of more quantitative data collection commences. Selective *geotechnical mapping* is undertaken to identify any anomalous zones within the quarry and to identify areas and techniques most suitable for obtaining the various sizes required. Mapping further establishes the basis for representative sampling of the quarry. Samples are collected for petrographic analysis and laboratory index testing, and this work undertaken. It is strongly recommended that Schmidt Hammer rebound readings be taken in the field (Schmidt, 1968). This test is quick and easy to perform and is a useful index in that the rebound number R may be correlated with compressive strength. The service records of rock that have been used in heavy ocean construction are evaluated during this phase of investigation.

Phase 3 Representative trial blasts and large-scale sizing tests are undertaken at all supply and back-up sources. The data obtained furnish essential input for laboratory triaxial tests to determine the angle of internal friction of the fill. Because of the large size of the fill materials required, it is not practical to perform tests on the actual material and so tests rely on smaller-scale model samples that must have a gradation curve parallel to, or modeled on, that of the representative field sample.

Naturally, trial shots are carefully planned so that they are representative of a production blast, and blasting details are recorded. Volumes in situ and in the pile are surveyed, and the shot pile is representatively sampled and samples accurately sized. Finally, supplementary data on the shape and roughness of particles are attached.

Laboratory

Index Tests It is important to determine engineering index parameters to ensure, quantitatively, that the rock satisfies certain minimum requirements and to provide a positive basis for evaluating whether rock of superior quality may provide a more economical design.

Specific gravity (ASTM C 97) is, e.g., an important index property since the specific gravity of a rock is directly related to its resistance to movement. An illustrative example is discussed by Treasher (1964) who compares two similar size stones, one having 10% greater density in air than the other. The heavier stone, when immersed in water, has 50% more resistance against movement by wave action than the lighter stone. A specific gravity of greater than 2.6 is desirable. Granite gneiss commonly has a specific gravity in excess of 2.7, good quality dolomitic limestone in excess

TABLE 2. Approximate Maximum in Situ Volumes (Cubic Meters) of Material
That May Be Committed for Excavation

Description of Material	Site A	Site B	Site C	Site D
Caisson sand and filter layer[a]	363,000	456,000	521,000	503,000
Breakwater core stone and armor[b]	1,780,000	2,280,000	2,680,000	22,558,000
Contingency stone[c]	890,000	1,140,000	1,340,000	1,279,000
Waste and overburden[d]	890,000	1,140,000	1,340,000	1,279,000
Concrete aggregate[a] (rough estimate)	290,000	325,000	345,000	340,000
Maximum amount of materials to be quarried (core stone, armor aggregate, waste, and overburden—does not include caisson sand or filter layer material)	3,850,000	4,885,000	5,705,000	5,456,000

[a]Volumes determined corrected for a swell factor of 16% (Killebrew, 1968).
[b]Volumes determined corrected for a swell factor of 35% (Killebrew, 1968).
[c]It is recommended that arrangements should be made for securing an additional 50% more core rock and armor.
[d]Waste and overburden are estimated to be one-third of the volume of core and armor stone quarried (Treasher, 1964). Much of the waste may be crushed and sold to other parties.

of 2.8, and fine-grained igneous rocks such as diabase and basalt often show specific gravities greater than 2.9. An index of the strength of the rock element is provided by its compressive strength (ASTM C 170), which should exceed 25,000 psi.

An estimation of the susceptibility of rock to chemical weathering may be obtained by observing its ability to remain intact when subjected to alternate cycles of oven drying and immersion in sodium sulfate and/or magnesium sulfate solutions (ASTM C 88). It is recommended that five cycles of the test be run and that both quantitative and qualitative examinations be undertaken. It is desirable that the quantitatively determined losses do not exceed 2.0%. A test procedure, similar to that outlined in ASTM C 88 but replacing the magnesium sulfate and/or sodium sulfate solution with seawater (for 20 cycles), is considered appropriate for rock to be used in sophisticated offshore structures.

The durability of the rock, in terms of its resistance to abrasion, is estimated by observing the percentage wear of particles in a test sample subjected to extensive abrasion when rotated with steel spheres in a drum—the Los Angeles machine (ASTM C 535). Percentage wear should not exceed 25%. Again, test results should be interpreted with care. Coarse-grained rocks often show relatively high losses as a result of the partial impact inflicted by the spheres and should not necessarily be condemned. To evaluate durability further, the Aggregate Impact and Wet Attrition tests described in British Standards (BS 812-1951) are recommended.

Absorption (ASTM C 97), which provides a measure of a rock's porosity and in turn its potential resistance to the action of freezing and

thawing, ideally should not exceed 1%. If this value exceeds 1% or it is intended to use the breakwater as a protective FNP structure over the period of 100 yr (i.e., replace initial units after 40 yr) or in very cold or extreme climates, then freeze and thaw tests of the type described in the German Standard, DIN 4226, should be run. In considering the 100 yr life, which incidentally, is highly feasible, other additional index tests would also be recommended.

Design Tests To obtain a reliable estimate of the strength of the breakwater by testing materials in a laboratory situation, all conditions in the field should be simulated with a high degree of precision. This relates especially to tangible parameters such as gradation, density, loading, and confining stresses.

Even if these conditions are correctly simulated, uncertainties, typical of those inherent in laboratory situations, remain. For example, the size of the specimen influences results. Marachi et al. (1972) found that, keeping all other variables constant, the friction angle for larger specimens was consistently smaller. Extrapolating their findings to incorporate the size of material used in the field tends to suggest that a reduction in the laboratory-determined friction angle of between 2° and 3° may have to be considered.

The ratio of particle size to cell size is also an important consideration. Present state-of-the-art tends to point to a maximum particle size not exceeding one-sixth of the specimen diameter. In performing tests, however, it is interesting to note that considerable degradation of particles occurs early in the loading cycle.

Simulating density and gradation in the laboratory is also critical, since these parameters greatly influence the angle of internal friction. It may be stressed that materials in a typical breakwater may have a relatively low density and that such fills, with high void ratios, attain their maximum principal stress ratio at appreciable strains (Marsal, 1973). Stability analysis should therefore ensure that the large strains required to develop shearing resistance are taken into account. Confining stress is also an important consideration. The maximum anticipated effective stresses in vertical and horizontal directions are on the order of 70 psi and 35 psi, respectively.

In summary, it appears that the angle of shearing resistance tends to:

Increase with increasing relative density,
Decrease with decreasing confining stress,
Increase with individual particle crushing strength,
Increase with better gradation,
Decrease slightly as the maximum particle size increases,
Be higher for materials composed of angular particles rather than rounded particles.

Prior to the availability of laboratory-determined values, it is suggested that, for good quality rock, preliminary design envelopes be developed for angles of shearing resistance of between 37° and 42°.

This discussion was in no way intended to be comprehensive but to suggest that considerable care should be exercised in planning, undertaking, and interpreting the results of laboratory procedures.

Conclusion

The selection of suitable sources of rock borrow for the construction of breakwaters to protect FNPs or other important offshore structures must be based on intensive investigations that are commenced in parallel with site selection studies (Fischer et al., 1975) well in advance of finalizing the breakwater design.

Although the quantities of rock for FNP breakwaters are not large in comparison with certain earth/rockfill dams (the Tarbela Dam, presently under construction in Pakistan, requires 186 million m³), the requirements on quality in terms of the safety criteria specified by the Nuclear Regulatory Commission are stringent.

In conclusion, it may be stated that the maximum safety, the lowest overall cost, and the least impact on the environment may be achieved only by careful planning, a good understanding of the environment, a sound knowledge of the engineering and geological requirements of borrow for FNP breakwaters, the provision of sufficient lead time to apply a logical sequence of sound engineering investigation procedures and alternative cost benefit studies, and the generation of suffi-

cient geotechnical data to implement selection of the most suitable borrow source(s) and most optimum breakwater design.

Acknowledgments

The technical content in this entry was developed in investigations on the Atlantic Generating Station for the Public Service Electric and Gas Co. of New Jersey and studies for Offshore Power Systems, who presently have the world's first FNPs under construction in Jacksonville, Florida.

IAN WATSON
JOSEPH A. FISCHER

References

American Waterways Operators, 1973, *Big Load Afloat: U.S. Domestic Water Transportation Resources.* Washington, D.C., 158p.

Fischer, J. A., et al., 1975, Macro-engineering aspects of floating nuclear power plant siting, Proc. MTS/IEEE Conf. and Exposition, Ocean 75, 634–642.

Fischer, J. A., I. Watson, H. Singh, et al., 1975, Design forces for offshore nuclear power plant construction, *Am. Soc. Civil. Engineers Proc. Specialty Conf., Ocean Engineering III.*

Killebrew, C. E., 1968, Tractor shovels, tractor dozers, tractor scrapers, in E. P. Pfleider, ed., *Surface Mining.* New York: American Institute of Mining, Metallurgical and Petroleum Engineers, 463–477.

Marachi, D. N., C. K. Chan, and H. B. Seed, 1972, Evaluation of properties of rockfill materials, *Am. Soc. Civil Engineers Proc., Jour. Soil Mechanics and Found. Div.* **98,** 95–114.

Marsal, R. J., 1973, Mechanical properties of rockfill, in *Embankment Dam Engineering.* New York: Wiley, 100–200.

Schmidt, E., 1968, *Design of Concrete Test Hammer.* Zurich, Switzerland: Proceq Sa, 14p.

Treasher, R. C., 1964, Geologic investigations for sources of large rubble, in P. D. Trask and G. A. Kiersch, eds., *Engineering Geology Case Histories,* Vol. 4. 273–286.

Watson, I., W. H. McLemore, and T. E. Mills, 1974, Sources and requirements of rock rubble-breakwaters for offshore nuclear power plants, *Assoc. Eng. Geologists: Programs and Abs.* 36–37.

Cross-references: Vol. XIII: *Coastal Engineering; Marine Sediments, Geotechnical Properties; Nuclear Plant Siting, Offshore; Oceanography, Applied; Rock Mechanics; Rocks, Engineering Properties.*

OPEN SPACE

A good working definition for *open space* is provided in the 1959 statutes of California (Sec. 1, Chap. 12, Div. 7, Title 1):

[A]ny space or area characterized by (1) great natural scenic beauty or (2) whose existing openness, natural condition, or present state of use, if retained, would enhance the present or potential value of abutting or surrounding urban development, or would maintain or enhance the conservation of natural or scenic resources.

Although the *Federal Housing Act* of 1961 defines open spaces in somewhat similar terms as undeveloped urban lands that have value for parks and recreation, conservation of land, and other natural resources of scenic and historic purposes, many dissimilar attitudes and perceptions exist about such land. Environmentalists view it as sacred, whereas business interests consider it land that remains after all other properties have attained their highest and best commercial use. When unused, governments have traditionally viewed open space as a waste of potential tax revenue. Open space that has been retained for some societal purpose is termed deliberate while land waiting for the developmental process is called *vacant.*

Open space takes many forms and related terms include *green belt, new town,* and *garden city.* The scale of open space ranges from small areas within a planned community to large tracts beyond city and suburban limits. Since this entry treats open space as a land resource that needs careful management for its preservation, it automatically becomes the domain of the geologist.

Historical Background

The principle of open space dates back at least to the thirteenth century B.C. when the Levitical cities were surrounded by pasture lands for use by city dwellers. The city of Gezer contained 9 ha with an open land belt 15 times larger. The 1515–1516 writings of Sir Thomas More provide green belt ideas where the imagined cities of Utopia were surrounded by agricultural lands; when cities were filled, new cities were built beyond the agricultural belts. In England the open space concept was rooted in the royal forest and hunting preserves that evolved into public places, the commons, for gathering firewood, grazing animals, and recreation. When faced with increasing pressures from landlords and business to partition the urban commons, Parliament passed the *Enclosure Act* of 1845, which placed restrictions on commons-type land. With the help of the *Green Belt Act* of 1938 and the *Town and Country Act* of 1947, the original 74 London commons, each averaging 65 ha have expanded to 2,170 km² in 1959 with concentric rings of agricultural and recreation land mixtures. The Moscow green belt, e.g., constitutes 30% of the city's total area.

The Garden City concept introduced in 1898 by Ebenezer Howard represents another open space approach. The term *new town* has become associated with those new cities where environmental considerations for open space are planned

(Coates, 1974, 378–382). Howard's ideas were embodied in the 1903 construction of Letchworth, England, and new towns now exist in many countries including Sweden, the USSR, France, Finland, and West Germany. Columbia, Maryland, and Reston, Virginia, are the best examples in the United States.

The smallest scale for open space retention occurs on the community level and is currently known as PUD (Planned Unit Development) (Huntoon, 1971), or open space subdivision (Foster and Kuplesky, 1972). Such developments should not be confused with the subdivision clusters of row house construction, or Levittowns, that mushroomed after World War II. Instead, lot sizes under PUD are smaller than the conventional zoning code allowance, but the developer agrees to use the residual open space for parks, recreation, and scenic sites. Radburn, New Jersey, built in 1929, served as the prototype for such developments, and Heritage Village, Connecticut, built in 1964, has become the community most mimicked since then. Ramapo, New York, provides a larger scale for this design whereby, in the process of so-called density averaging, it has obtained 53 ha of public parks.

The Problem

During urbanization local governments have typically translated adjacent lands into taxation domains in terms of industry, commerce, residences, and public facilities. The basic assumptions were that (1) economic and population growth are necessary, (2) unlimited amounts of land are available for development, and (3) any negative expansion effects are easily solved by human ingenuity. By the 1950s such postulates were finally undergoing critical examination and the country was awakening to a new conservation mood. Land had been disappearing fast and was becoming increasingly costly (Whyte, 1959), pollution was being recognized as an unwanted product of unrestrained growth (Baxter, 1974); and the visual and natural amenities of the landscape were being destroyed by the newly named process of *urban sprawl* (Whyte, 1958).

Since 1970 urban sprawl has been annually consuming more than 1 million acre (405,000 ha) of land and causing massive desecration and degradation of contiguous lands. In the 1960s Chicago expanded at a rate of 40 km²/yr and paved over important mineral resource areas (Risser and Major, 1967). Alfors et al. (1973) calculated that by the year 2000 California will have lost $17 billion in mineral resources because of constraints placed on production by the urbanizing process's using three times as much land as cities of the 1920s; one-half of Los Angeles is devoted to automobile selling, servicing, and transportation

613

systems. Rantz (1970) showed that most of the $62 million flood damages in 1969 from torrential rains in southern California were localized in areas where urban sprawl was not matched with zoning codes for hazard areas. In a similar California study, Slosson (1969) determined that of landslide losses in 1969 totaling $6.5 million, only $182,400 in damages occurred in areas where geologists had been thoroughly involved in the developments.

Geology Inventories and Priorities

Geologists are becoming increasingly involved in showing that the quality of the open space environment can no longer be taken for granted, that earth materials and processes need careful management, and that reassessment of priorities is necessary to meet the challenge of any new type of growth pattern in the urban region. This new environmental attitude is aimed at showing that urbanization disturbs the land, interferes with natural processes; produces excessive long-range damages to renewable, nonrenewable, and landscape amenity resources; and consumes excessive amounts of the landscape.

Before open space decisions can be properly made, inventories must be prepared and priorities established. Open space serves many functions, which can be classified into the following units (Williams, 1969): (1) resource production in forestry, agriculture, rocks and minerals, water supply and use; (2) preservation of natural and human resources, fish and wildlife habitats, geologic features, historic and cultural sites; (3) health, welfare, and well-being for water and air quality, waste disposal, recreation, visual amenities; (4) public safety regarding floods, unstable hill slopes, airplane flight paths, fire zones; (5) corridors for power transmission and transportation systems; and (6) urban expansion for commerce, industry, housing, and public service. As shown in numerous studies, many of these functions are interrelated, and input for their allocation and understanding rests on geologic and earth-science-related data.

Prior to the 1950s most geologic studies and maps were prepared for use by other geologists. However, starting in the 1950s the U.S. Geological Survey started producing maps and tables of the San Francisco Bay area, understandable to laypeople, that described the economic geology and engineering properties of materials. On January 1, 1970, the survey joined with the U.S. Department of Housing and Urban Development in additional inventorying of all geologic features needed in urban planning for the San Francisco Bay area. Since 1975 the survey has completed environmental planning maps of many communities that show features such as flood-prone areas, soils, landforms, and depth to water character-

istics. Other early reports that stressed geologic considerations in land uses and open space planning include those by McComas (1968), Mozola (1970), and Christiansen (1970).

Inventories that stress the importance of geologic parameters when land is viewed as a resource in open space planning take two different forms. The first type is characterized by the Metropolitan Washington Council of Government's (1968) inventory of the region for the purpose of determining *environmental constraints to development*. For example, shallow bedrock, position of water table, ecologically fragile areas, etc. were considerations that determined the type of construction that a location could sustain. The second type of inventory has become known as the McHarg approach (or ecologic approach, geologic approach) whereby planning priority is placed on *preservation and protection of natural processes*. Since the early 1950s McHarg and his followers (McHarg, 1969: Wallace, 1970) have placed highest priority on open space, which includes all surface water and riparian lands as far back from water bodies as possible. This includes wetlands, floodplains to the 50 yr recurrence level, and all aquifer recharge areas. Slopes steeper than 12° should remain undeveloped.

The scale of urban inventories ranges from such small towns as Vestal and Owego, New York (Town of Vestal, 1974; Gentili, 1974), to counties and megalopolis-type regions (Syracuse–Onondaga County, 1973; Nassau County Environmental Management Council, 1974). All of New York State has been inventoried by the New York State Office of Planning Coordination (Foster and Kuplesky, 1972). Land use was divided into 10 main categories, and each contains 23 subheadings and 134 categories.

Reports that deal with the establishment of land-use priorities for open space range from simple listings of recommendations (Kent, 1970; Gratzer, 1972; Town of Vestal, 1974) to long textual discussions (Platt, 1972) and computerized modeling (Fabos, 1973; Tilmann et al., 1975). McHarg uses a combination of techniques that include a series of map overlays, and they can be keyed to a matrix system in which the various environmental components are ranked into numbered classes. Little (1968) shows that when a municipal official is faced with a choice of whether to preserve a natural process, community amenity, or recreation, the decision is easy because they are all interrelated.

By making natural processes the cornerstone of planning, other landscape elements are kept in greater balance. Landslide-prone topography (see *Landslide Control*) is best retained as open space for water recharge areas. This will also cause less pollution and reduction in sediment yield (Guy, 1970). When aquifer systems receive maximum recharge and streams remain unchannelized, river regimen is stabilized and flooding reduced to a minimum (Leopold, 1968). Proper management practices in agricultural land conservation, restriction in floodplain occupancy, and retention of wetlands (U.S. Corps of Engineers, 1974) are additional methods for the maintenance of the integrity of open space and thus provide the necessary ingredients for working with natural systems.

Mineral resource development has many attendant problems when associated with mining activities near communities (Risser and Major, 1967; Flawn, 1971) For example, aggregate, so necessary in construction industries, such as sand, gravel, and rock, are not recognized in federal legislation as mineral substance and so are not applicable under the mineral location laws governing the public domain. Although simultaneous multiple land use is possible for subsurface mining, surface materials, when mined, often create hardships because of nuisances such as noise, dust, vibrations, traffic, and ugliness. Political pressures for their shutdown can become intense. Public education and understanding can aid in developing plans for sequential multiple use and flexible zoning programs initiated for such resource areas. Thus, such lands can be restored to other beneficial uses after the extraction period has ended. Denver provides a worthy example where a sand and gravel pit was mined out, then used as a landfill site until the late 1950s, after which it became the site for the Denver Coliseum.

Governmental Action and Laws

Starting in the 1960s, people, the courts, and legislative bodies increasingly recognized the importance of maintaining balance in natural systems. Since ecosystems cannot speak or act for themselves in legal proceedings, various surrogate organizations became their advocates and championed the view that natural features and objects have certain legal rights. The large number of lawsuits, as well as their significance, that challenged unnecessary and environmentally harmful developments was instrumental in the passage of the *National Environmental Policy Act* of 1970 and the establishment of the Environmental Protection Agency. Furthermore, states such as Hawaii, Vermont, and Maine passed land-use laws whereby the land was to be treated as a natural resource.

The dual rationale for these trends was that humankind no longer was to be considered as omnipotent and that destruction of nature ultimately created feedback mechanisms that proved harmful to humans as well. In Tiburon, California, a property owner brought suit against the city to allow him to develop an open space land parcel

(*Agins* v. *City of Tiburon,* 1979). In 1968 Agins had purchased 2 ha (5 acre) along a ridge overlooking San Francisco. After the purchase the city prepared a comprehensive land-use plan and sought to condemn the property for the management of open space. During the same period the city adopted a zoning ordinance that allowed no more than five houses for such a land parcel. Agins sued under the theory of inverse condemnation, claiming such action deprived him of his rightful income. The case first went to the California Supreme Court, which ruled that under California law inverse condemnation is not an available remedy. In a unanimous decision, when appealed to the U.S. Supreme Court, the Court concurred that a zoning ordinance that limits development in the name of conservation does not necessarily violate the constitutional rights of the affected property owners.

Once priorities have been established, the protection and preservation of open space is accomplished by one or a combination of purchase zoning restrictions, and prohibitions, taxation relief, and other techniques. The *California Open Space Act* of 1959 provided pathfinding legislation for cities and counties to acquire open space lands, and the *California Land Conservation Act* of 1965 allowed cities to require that subdividers donate park land or make payments in lieu of them. In 1961 Congress passed the *Housing Act* and started providing financial aid for open space projects, and in 1965 the Bureau of Outdoor Recreation through the Land and Water Conservation Fund initiated a program for purchase of land if it was devoted to parks and recreation facilities. Such programs operate on a 50–50 matching basis, and an example of such usage is the Susquehanna Riverside project in the Binghamton Metropolitan area where $200,000 was spent to preserve riverbanks and public use in the floodplain corridor. During the 1960s Westchester County, New York, acquired nearly 1,100 ha of open space lands at a cost of $5 million, and the *New York Environmental Quality Bond Act* of 1972 authorized $175 million, out of the $1.15 billion total funding, for purchase of lands in the public interest.

A combination of zoning laws and ordinances has proved to be an effective legal weapon on the local scene to prohibit or restrict development in environmentally sensitive and hazardous areas. Zoning that leads to protection of natural processes is called "conservation zoning" (Heller, 1971, 70–71; Kaiser et al., 1974, 155–157). Examples of such zoning include floodplain zoning, coastal zoning, wetland zoning, stream bank zoning, and steep-slope zoning. In the latter case management of hillsides in southern California comes under the southern California system of grading ordinances (Jahns, 1969), whereby geologists are deeply involved in the planning and

615

monitoring of developments. Other types of zoning that affect open space include large lot zoning, agricultural zoning, impact zoning, and multiple-use zoning.

Tax relief is becoming an increasingly popular method for the preservation of open space. New York's General Municipal Law (1963, Section 247) states that a town can accept less than fee right in open land and in return can provide property tax relief. The usual practice is to reduce property valuation if the owner agrees to leave his or her lands undeveloped for a stated period of time. The amount of tax is then based on the market value of the products from the land instead of as real estate operating under the highest and best use principle. California and Pennsylvania passed similar legislation in 1965, and Minnesota followed in 1967.

A variety of other laws have also been passed that affect the management of open space. The Hawaii Land Use Law of 1961 and the Maine Site and Development Law of 1970 are typical statutes that provide constraints to construction. In 1972 Boca Raton, Florida, passed a referendum that established a ceiling of growth for the construction of new dwellings. Other attempts to inhibit growth included action taken by Governor Tom McCall of Oregon to reduce the 1972 travel advertising budget to urge visitors not to come to Oregon. In New Mexico a citizens group called the Undevelopment Commission propagandized against immigration to that state.

The role of the geologist is now firmly established in the analysis of open space and those planning strategies that can most effectively serve to manage this land resource (Alfors et al., 1973; Legget, 1973; Coates, 1974; Cooke and Doornkamp, 1974; LaFleur, 1974). The special expertise a geologist brings to such evaluations is the ability to predict which practices will cause the greatest disruption of the land-water ecosystem and which lands are most hazardous and to determine the location of natural resources. Thus, knowledge of geomorphology, economic geology, engineering geology, and hydrology needs to be blended in all appraisals for potential land-use changes of the open space environment.

DONALD R. COATES

References

Agins v. City of Tiburon, 24 Cal. 3rd 266 (1974).

Alfors, J. T., J. L. Burnett, and I. E. Gay, Jr., 1973, Urban geology master plan for Califronia, California Div. Mines and Geology Bull. 198, 112p.

Baxter, W. F., 1974, People or Penguins—The Case for Optimal Pollution. Columbia University Press, 110p.

Christiansen, E. A., ed., 1970, Physical environment of Saskatoon, Canada, Natl. Research Council Canada Pub. No. 11378, 68p.

Coates, D. R., ed., 1974, Environmental Geomorphology and Landscape Conservation: Vol. II, Urban Areas. Stroudsburg, Pa.: Dowden, Hutchinson & Ross, 454p.

Cooke, R. U. and J. C. Doornkamp, 1974, Geomorphology in Environmental Management. London: Oxford University Press, 413p.

Fabos, J. G., 1973, Model for landscape resource assessment: Part I of the Metropolitan Landscape Planning Model, Univ. of Mass. Agri. Exp. Sta. Res. Bull. No. 602, 141p.

Flawn, P. T., 1971, Mineral resources and multiple use land, in D. R. Nichols and C. C. Campbell, eds., Environmental Planning and Geology. Washington, D.C.: U.S. Govt. Printing Office, 22–27.

Foster, J. and H. Kuplesky, 1972, The Open Space Subdivision. Albany: New York State Office of Planning Services, 28p.

Gentili, J., comp., 1974, Owego Environmental Study. Ithaca, N.Y.: College of Architecture, Art and Planning, Cornell University, 103p.

Gratzer, M. A., 1972, Land use inventory for open space planning in eastern Connecticut, Univ. Conn. Agri. Exp. Sta. Res. Rept. No. 38.

Guy, H. P., 1970, Sediment problems in urban areas, U.S. Geol. Survey Circ. 601-E, 8p.

Heller, A., ed., 1971, The California Tomorrow Plan. Los Altos, Calif.: William Kaufmann, Inc., 120p.

Huntoon, M. C., Jr., 1971, PUD: A Better Way for the Suburbs. Washington, D.C.: The Urban Land Institute, 71p.

Jahns, R. H., 1969, Seventeen years of response by the City of Los Angeles to geologic hazards, Proceedings of the Geological Hazards and Public Problems Conference. Washington, D.C.: U.S. Govt. Printing Office, 283–295.

Kaiser, E. J., and numerous others, 1974, Promoting Environmental Quality through Urban Planning and Controls. Washington, D.C.: U.S. Environmental Protection Agency, 441p.

Kent, T. J., Jr., 1970, Open Space for the San Francisco Bay Area. Berkeley: Institute of Governmental Studies, University of California, 85p.

LaFleur, R. G., 1974, Glacial geology in rural land use planning and zoning, in D. R. Coates, Ed., Glacial Geomorphology. Publications in Geomorphology. Binghamton: State University of New York, 374–388.

Legget, R. F., 1973, Cities and Geology. New York: McGraw-Hill, 624p.

Leopold, L. B., 1968, Hydrology for urban land planning—A guidebook on the hydrologic effects of urban land use, U.S. Geol. Survey Circ. 554, 18p.

Little, C.E., 1968, Challenge of the Land. New York: Pergamon Press, 151p.

McComas, M. R., 1968, Geology related to land use in the Hennepin Region: III. State Geol. Survey Circ. 422, 24p.

McHarg, I. L., 1969, Design with Nature. Garden City, N.Y.: Natural History Press, 197p.

Metropolitan Washington Council of Government, 1968, Natural Features of the Washington Metropolitan Area. Washington, D.C.: 49p.

Mozola, A. J., 1970, Geology for environmental planning in Monroe County, Michigan, Mich. Geol. Survey Rept. Inv. 13, 34p.

Nassau County Environmental Management Council, 1974, Nassau County Environmental Plan Report. Mineola, N.Y., various pages.

Platt, R. H., 1972, The open space decision process, Chicago University Dept. Geography Research Paper 142, 189p.

Rantz, S. E., 1970, Urban sprawl and flooding in southern California, *U.S. Geol. Survey Circ. 601-B*, 11p.

Risser, H. E., and R. I. Major, 1967, Urban expansion—An opportunity and a challenge to industrial mineral producers, *Illinois State Geol. Survey Environ. Notes 16*, 19p.

Syracuse-Onondaga County Planning Agency, 1973, *Onondaga County Environmental Plan:* Vol. I. *Environmental Policy Plan,* and Vol. II. *Environmental Inventories and Analysis.* Syracuse, 173p.

Slosson, J. E., 1969, The role of engineering geology in urban planning, *Colorado Geol. Survey Spec. Publ. No. 1,* 8–15.

Tilmann, S. E., S. B. Upchurch, and G. Ryder, 1975, Lane use site reconnaissance by computer-assisted derivative mapping, *Geol. Soc. Amer. Bull.* **86,** 23–34.

Town of Vestal, 1974, *The Town of Vestal Open Space Inventory.* Vestal, N.Y., 85p.

U.S. Corps of Engineers, 1974, *A Community Decision.* Baltimore, Md.: U.S. Army Engineer District, 30p.

Wallace, D. A., 1970, *Metropolitan Open Space and Natural Process.* Philadelphia: University of Pennsylvania Press, 199p.

Whyte, W. H., Jr., 1958, Urban sprawl, *Fortune* **57**(January), various pages.

Whyte, W. H., Jr., 1959, A vanishing resource: Open space, now or never, *Landscape Architecture* **50**(1), 8–13.

Williams, E. A. (directed), 1969, *Open Space: The Choices before California.* San Francisco: Diablo Press, 187p.

Cross-references: *Environmental Engineering; Legal Affairs.* Vol. XIII: *Economic Geology; Urban Geology; Urban Geomorphology; Urban Hydrology.*

ORBITAL, AERIAL PHOTOGRAPHY— See AERIAL SURVEYS; PHOTOGEOLOGY; PHOTOINTERPRETATION; REMOTE SENSING, GENERAL; SATELLITE GEODESY AND GEODYNAMICS. Vol. XIII: PHOTOGRAMMETRY; REMOTE SENSING, ENGINEERING GEOLOGY.

P

PEDOGEOCHEMICAL PROSPECTING

In mineral exploration geochemical methods are of great importance (Rose et al., 1979; Levinson, 1974, 1980). Several types of surveys are used, applying chemical analysis of soils, rocks (see *Lithogeochemical Prospecting*), stream sediments (see *Hydrochemical Prospecting*), vegetation (see *Biogeochemistry; Geobotanical Prospecting*), and air (see *Atmogeochemical Prospecting*). The choice of which type of survey to apply is influenced by the particular project and the characteristics of the area. *Pedogeochemical prospecting* is normally used in follow-up surveys where other methods have indicated the possibility of mineralization (Matheis, 1981; Viaene et al., 1981).

Pedogeochemical prospecting is based on the fact that buried mineral deposits in certain areas are marked by geochemical features in soils overlying the mineralized area. The essential condition for the formation of *geochemical anomalies* in soils is the migration of elements in secondary dispersion pathways.

During the formation of mineral deposits, primary dispersion processes result in primary distribution patterns of elements in the orebody and its surrounding environment (Shaver, 1986). In secondary dispersion, elements are redistributed due to oxidation, weathering, solution, transport, precipitation, and redeposition. Thus, a *secondary dispersion halo* is developed around the orebody. If this halo extends upward to the soil surface, the orebody can be detected by soil geochemical surveys.

Secondary Dispersion Patterns in Soils

The behavior of various elements during secondary dispersion processes is controlled by relative mobilities. Mobile elements can be transferred in soluble form by ground- or surface water. During migration, frequent exchange processes occur between the solutions and the solid phases they contact. Immobile elements remain in place as insoluble clastic materials and can be transported only by mechanical processes (see *Indicator Elements*).

The dominant factors that affect the solubility of ore minerals and the mobility of elements are the presence of free oxygen and CO_2, the pH of the solution, and the presence of other substances in solution. If translocated elements are locally enriched—e.g., by formation of secondary minerals or by adsorption on clay minerals and organic matter—they become part of a secondary dispersion halo.

The occurrence of indicator elements (q.v.) in the insoluble phase of soils is usually restricted to orebodies lying near the surface. The relationship between parts of different dispersed phases of elements occurring in soils and the depth of a mineral deposit is illustrated in Fig. 1. If the deposit is deep-seated, concentrations of elements in other phases also become lower. Sometimes only mobile gaseous phases contain indicators of hidden ore (Friedrich, 1975).

The intensity of the geochemical anomaly depends on numerous factors such as type of mineralization, type of wall rock, climate, and soil condition. Soil-forming processes can profoundly influence secondary dispersion in soils. In im-

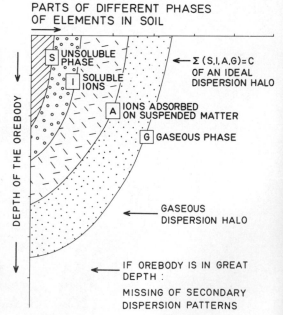

FIGURE 1. Schematic diagram showing parts of different dispersed phases of elements occuring in soil in relation to the depth of the orebody. (From Friedrich, 1975)

mature soils, profile variations are not well defined. With progressive differentiation of soil horizons, the pattern of metal redistribution becomes more pronounced. In freely drained soils, many mobile metals tend to become impoverished in the A horizon and enriched in the B horizon. This kind of distribution is usually observed in podzols and podzolic soils, where the dominant leaching agent is downward-percolating rain water. In latosols, metals are partly coprecipitated with Fe and Mn in ferroginous B horizons. Where drainage is impeded, metal distributions show a tendency toward accumulation in organic-rich A1 horizons. In arid environments, metals can be enriched in horizons characterized by evaporation. For most metals, however, redistribution in desert environments is relatively restricted in view of the high pH. If such particularities are taken into consideration, soil geochemical survey techniques can be applied almost worldwide.

Geochemical Anomalies in Soils

The normal abundance of elements in the unmineralized material of a given area is referred to as *background*. A *geochemical anomaly* is a deviation from this background. Anomalies that are related to orebodies are termed *indicator elements*. Figures 2 and 3 show examples where the

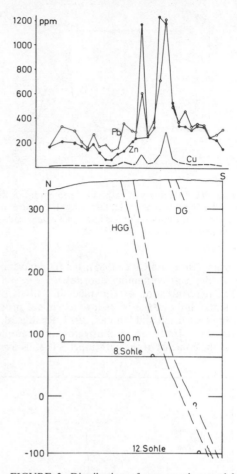

FIGURE 3. Distribution of copper, zinc, and lead above hidden vein-type deposit, Bad Grund, Harz, Germany. (From Steinkamp, 1976)

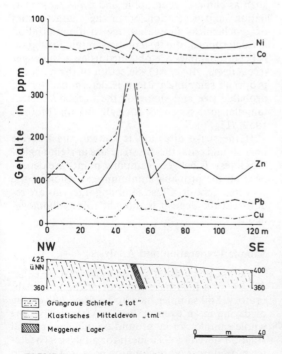

FIGURE 2. Distribution of lead, zinc, copper, cobalt, and nickel in soils above the Meggen stratiform massive sulfide orebody, Germany. (From Hilmer, 1972)

primary contrast between ore and unmineralized host rock is reflected in an anomaly contrast in the overlying soils (Hilmer, 1972; Steinkamp, 1976). In many cases the significant anomalies are set in a broad area of higher-than-background values. This range of elevated values, which must be distinguished from the real anomaly, is called the *threshold*.

Where the principal ore metals cannot be easily traced, *pathfinder* elements or compounds (see *Indicator Elements*) can be used to guide the search for ore deposits, if they have a characteristic distribution with respect to mineralization (Bradshaw and Lett, 1980). Pathfinders indicating some types of mineral deposits are Hg (heavy metal deposits with Hg-bearing sulfides, e.g., sphalerite and tetrahedrite; Hg deposits; and barite and fluorite deposits); As (Au deposits); SO_2, S, H_2S (massive sulfide ore deposits); He (Mo/Cu

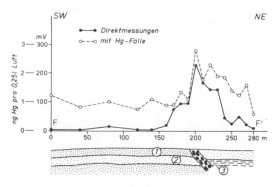

FIGURE 4. Mercury as pathfinder element in soil air above the Dreislar barite vein, Germany. 1, 2, and 3 are different sedimentary rock units (graywackes and slates). (From Friedrich and Wallner, 1975)

deposits) (Butt and Gole, 1985); Rd (U deposits); and F (deposits containing fluorite).

The determination of Hg traces in soils is of particular importance in pedogeochemical prospecting (Friedrich and Wallner, 1975; Ryall et al., 1981) (Fig. 4). Figure 5 illustrates an example where the contrast of the pathfinder S in soils over

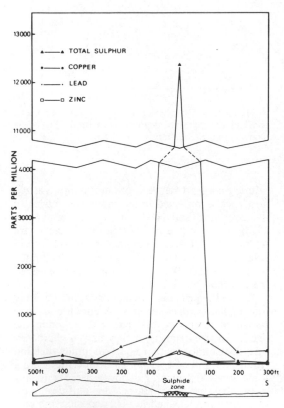

FIGURE 5. Sulfur as pathfinder element in soil above a massive sulfide deposit, Newfoundland. (From Meyer and Peters, 1973)

an ore deposit is much more distinctive than the contrast of the ore metals (Meyer and Peters, 1973).

Contamination

Nonsignificant anomalies are sometimes produced by contamination, of which there are many sources. The most common are mining activities, industries of all types, metal-rich agricultural chemicals, construction activities, etc. Therefore, the extent of possible contamination must be determined when planning geochemical exploration in old mining districts and in areas of industrial activity or intense farming.

Fieldwork

In the course of an *orientation survey* the pedogeochemical characteristics of the area are studied first to obtain information concerning the local conditions. Based on this information, suitable operational techniques are selected. Of particular importance are those factors that must be taken into consideration when interpreting the analytical results. The nature of the soils and their relation to bedrock are of particular importance. Samples for preliminary analyses should be obtained from a number of soil profiles (test pits) to interpret the relative impact of pedogenic factors such as climate, geographic and topographic situation, and vegetation. When the characteristics of geochemical anomalies due to mineralization are determined, the systematic sampling program is begun. Samples are obtained using a soil auger (see *Augers, Augering*). Selection of the most appropriate sampling pattern is determined by the probable size and shape of the orebody. A rectangular grid system is normally set up (Mather, 1959) (Fig. 6).

If the strike direction is known, the traverse lines should cross the ore structure at right angles. The intervals between sampling points are dependent on the probable minimum extension of the expected deposit. Close sampling intervals are required if the deposit is thought to be inhomogeneous.

Sample Preparation and Analysis

Sample preparation and analysis are done either directly in a field base camp or at the home laboratory. Soil samples have to be dried (air drying or drying in an oven at 70°C). After drying, the whole sample can be ground or the clay mineral fraction sieved to −80 mesh for analysis. The latter procedure is more common because there is no need for machines and the sieving procedure results in an enrichment of the trace elements bound to the clay minerals.

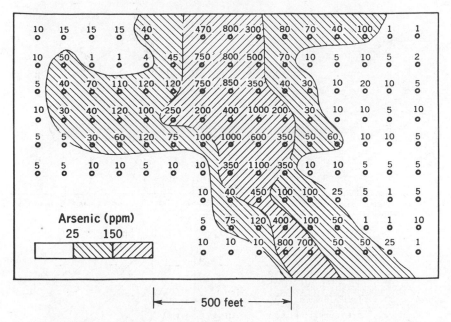

FIGURE 6. Square grid used to detect irregular patterns of arsenic in residual soil overlying arsenical gold mineralization, Sierra Leone. Data on −80 mesh fraction. (From Mather, 1959, in Hawkes and Webb, 1962)

For exploration purposes, several analytical methods to determine trace elements are available. An easy and rapid method is determining the *cold-extractable heavy metal content.* Specific elements can be determined by use of special reagents. In mobile or permanent laboratories, atomic absorption spectrophotometers are commonly used to determine of the trace elements. Depending on the chemical treatment of the samples, cold-extractable, acid-soluble, or total-metal contents can be determined. Choice of methodology depends on the prospecting plan and calculations of the geochemical behavior of the collected samples. Spectrographic analyses were common in the past, but atomic absorption spectrophotometry is now more widely used.

Statistical Treatment and Interpretation of Data

Interpretation of geochemical data requires that background and threshold values be determined in each survey area because such values covary with lithology. The data that are treated statistically must be representative of comparable parent material. Modern statistical treatments employ computer analyses (Chork and Govett, 1985). Trend surface analysis, e.g., is essential for ascertaining the geochemical landscape—i.e., the interpretation of the distribution of single data in the two dimensions of a map. Figure 7 shows the computer-drawn trend surfaces of copper values in the Maubach district, Germany (Kulms, 1972).

Figure 8 presents an example from the Harz district, Germany, where the geochemical landscape is plotted relative to a three dimensional pattern (Steinkamp, 1976).

All anomalies that indicate the possibility of a deposit of economic interest should be investigated in detail. If deposits are confirmed, and if they are verified by geophysical methods, exploration is continued by a drilling program.

G. H. W. FRIEDRICH
R. HERMANN

References

Bradshaw, P. M. D., and R. E. W. Lett, 1980, Geochemical exploration for uranium using soils, *Jour. Geochem. Exploration* 13, 305–319.
Butt, C. R. M., and M. J. Gole, 1985, Helium in soil and overburden gas as an exploration pathfinder—an assessment, *Jour. Geochem. Exploration* 24, 141–174.
Chork, C. Y., and G. J. S. Govett, 1985, Comparison of interpretations of geochemical soil data by some multivariate statistical methods, Key Anacon, N.B., Canada, *Jour. Geochem. Exploration* 23,213–242.
Friedrich, G., 1975, Moglichkeiten und Grenzsen geochemischer Exploration, in *Schriften der GDMB* 28, 233–256.
Friedrich, G., and P. Wallner, 1975, Quecksiilber-Gashofe im Bereich der Erzvorkommen Moschellandsberg, Bensberg und Dreislar, *Erzmetall* 28(1), 13–16.
Hawkes, H. E., and J. S. Webb, 1962, *Geochemistry in Mineral Exploration.* New York: Harper & Row, 415p.

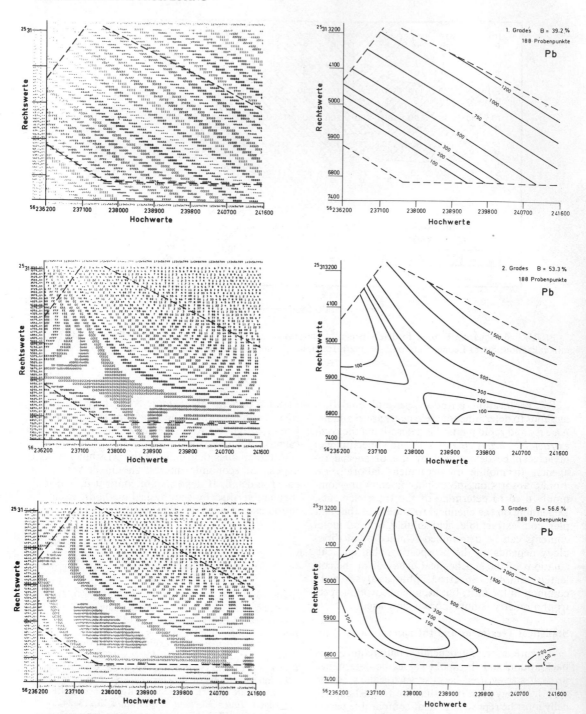

FIGURE 7. Computer-drawn trend surface analysis of lead in soil, Maubach district (Pb-Zn mineralization in Triassic sandstone), Germany. (From Kulms, 1972)

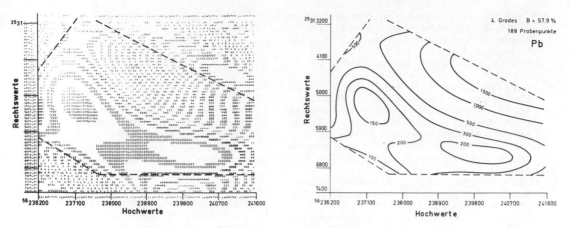

FIGURE 7. (*continued*)

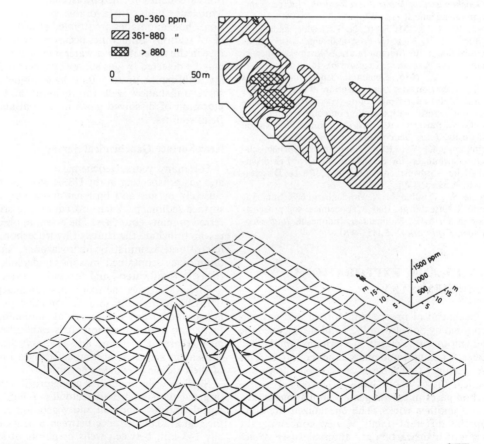

FIGURE 8. (*A*) Map showing distribution of zinc in soils in area with anomalous lead-zinc-copper values, Bad Grund mining district, Germany. (*B*) Computer-drawn geochemical landscape showing distribution of zinc in soils (same area as mentioned in Fig. 8*A*). (From Steinkamp, 1976)

Hilmer, E., 1972, Geochemische Untersuchungen im Bereich der Lagerstatte Meggen, Rheinisches Schiefergebirge, Ph.D. Dissertation, Aachen, University of Technology, 162p.

Kulms, M., 1980, Die Verteilung der Elemente Pb, Zn, Cd, Hg, Cu, Co, Ni, Mn und Fe in den Boden der Lagerstattengebiete Maubach und Bleialf sowie in den Flusswassern und Flusssedimenten des Triasdreiecks von Maubach-Mechernich-Kall, Eifel. Ein Beitrag zur geochemischen Lagerstattenprospektion, Ph.D. Dissertation, Aachen, University of Technology, 195p.

Levinson, A. A., 1974, *Introduction to Exploration Geochemistry*. Calgary: Applied Publishing Ltd., 612p.

Matheis, G., 1981, Trace-element patterns in lateritic soils applied to geochemical exploration, *Journ. Geochem. Exploration* 15, 471–480.

Mather, A. L., 1959, Geochemical prospecting studies in Sierra Leone, D.I.C. thesis, Imperial College, London.

Meyer, W. T., and R. G. Peters, 1973, Evaluation of sulphur as a guide to buried sulphide deposits in the Notre Dame Bay area, Newfoundland, in M. J. Jones, ed., *Prospecting in Areas of Glacial Terrain*. London: Institute of Mining and Metallurgy, 55–66.

Rose, A. W., H. E. Hawkes, and J. S. Webb, 1979, *Geochemistry in Mineral Exploration*, 2nd ed. London: Academic Press, 657p.

Ryall, W. R., K. M. Scott, G. F. Taylor, and G. P. Moore, 1981, Mercury in stratabound copper mineralization in the Mammoth area, Northwest Queensland, *Jour. Geochem. Exploration* 16, 1–11.

Shaver, S. A., 1986, Elemental dispersion associated with alteration and mineralization at the Hall (Nevada Moly) quartz monzonite-type porphyry molybdenum deposit, with a section on comparison of dispersion patterns with those from Climax-type deposits, *Jour. Geochem. Exploration* 25, 81–99.

Steinkamp, K. W., 1976, Explorationsgeochemische Untersuchungen im Bereich sulfidischer Erzvorommen im nordwestlichen Oberharz, Ph.D. Dissertation, Aachen, 130p.

Viaene, W., T. Suhanda, N. Vandenberghe, Y. Sunarya, and R. Ottenburgs, 1981, Geochemical soil prospecting in northwest Kalimantan, Indonesia, *Jour. Geochem. Exploration* 15, 453–470.

PETROLEUM EXPLORATION GEOCHEMISTRY

Geochemical prospecting for oil and gas was proposed by Sokolov in 1929 and soon gained the recognition of scientists in Germany, the United States, Rumania, and many other countries. This prospecting (q.v.) is based on the well-known theoretical theses about halos of dispersed hydrocarbon gases under oil-gas accumulations in layers of mother rocks. The specialized literature contains different points of view concerning the origin of hydrocarbon gas anomalies (see *Atmogeochemistry*) in subsurface horizons of a section. Calculations of diffusion interchange of gases based on the measurement of the physical param-

eters of sediments are the most well founded. Theoretical questions of anomalous gas formation by filtration processes beginning as a result of disturbance of gas-waterproof layer solidity (break disturbances, microfissuring and macrofissuring, hydrodynamical windows) are less well known.

My calculations show that with an overlying sediment thickness of about 2,000 m and a diffusion coefficient equal to 10^{-6} cm^2/sec, the time of establishment of the stationary gas flow in subsurface layers reaches 1 by an age considerably beyond most known gas-oil-bearing rocks. This author proposes to use, in practice of gas-survey works, so-called frontal gas concentrations caused by unstationary diffusion gas flow—i.e., such anomalous concentrations that can be discovered in the upper part of continuous gas halos by means of modern highly sensitive instruments.

Geochemical investigations of deep and shallow drilling wells at known oil and gas deposits in different areas of the USSR established that epigenetic gas anomalies in the upper parts of sections are mainly formed by ascending filtration of hypogene fluids. In favorable geological conditions these anomalies are situated at different levels of sections. Sometimes occurring near the surface in the zone of active water interchange, gases may be detected by gas survey on structural drilling with depths up to 600 m, by gas-biochemical survey on shallow wells (up to 30 m), and by exploration of dissolved gases in natural and artificial sources.

Near-Surface Geochemical Surveys

For many years, geochemical methods of oil and gas prospecting in the USSR were guided exclusively by gas and bitumen anomalies in subsurface sediments. Such procedures are still preferred in many countries. The essential objects of research include the study of hydrocarbon gases, microflora, assimilating hydrocarbons, and also bitumoides, contained in soils, undersoil, basic rocks, groundwaters, and formation waters.

In certain cases the oxidizing-reducing potential (Eh) of environmental surroundings is determined for special conditions of anomalous gas formation and destruction. It is established that an increase of methane and its heavy homolog content in a rock is accompanied by a lowering of the oxidizing-reducing potential.

Shallow wells are drilled along profiles that are targeted with regard to tectonics, relief, lithological features, and soil and vegetation cover of the region. The distance between profiles is usually 1–3 km; between wells or points of surface sample collection, it is 100–300 m. Locations contaminated by oil by-products and wastes of enterprises must not be included in the survey areas.

In the case of peat bogs and marshy regions, drill holes must disclose the presence of basement rocks. The most reliable results of *gas surveys* are obtained from lithologically homogeneous rocks. A survey of samples from bored wells shows that the reliability of geochemical data is reduced because there are losses of gas that cannot be accounted for.

Sometimes *soil luminescent-bitumen surveys* are used in conjunction with other geochemical methods. Successful use of this method in the USSR led to the discovery of deposits near Bogorodskoyed (Saratov region) and Kamyshk–Burun (Predkavkazye). In a number of cases, however, anomalous bitumen content of soil was conditioned by increased concentrations of humus (organic matter) or sometimes by pollutants. Today, surveys of water sources in conjunction with subsoil surveys are widely used. In certain cases anomalous accumulation of gas hydrocarbons is indicated by a lowered sulfate content in groundwaters. The Bogdanovskoye and Monastyryschchenskoye oil and gas deposits in the Ukraine may be cited as examples of economic deposits that were discovered by water and ground surveys in subsurface layers of the hydrocarbon anomalies determined previously.

In favorable geological conditions gas-biochemical exploration methods may be based on the combined determination of the content of dissipated hydrocarbon gases and bacteria oxidized by them in subsurface sediments and water. Besides the microorganisms that are determined from certain hydrocarbon components (methane, ethane, and others) from water and rock biogenesis, the ratio of certain species of bacteria that assimilate gaseous and vaporous hydrocarbons is also taken into consideration.

Figure 1 shows gas-biochemical anomalies that were discovered during a survey of subsurface water in the Ukraine (Mikhailovskoye and Novogrygoryevskoye oil deposit areas). Both deposits

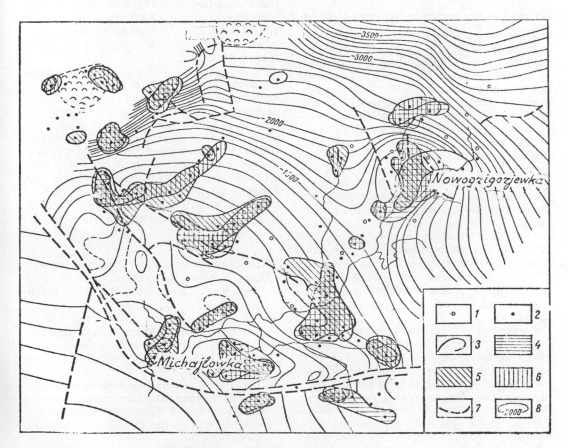

FIGURE 1. Map of gas-biochemical anomalies in the Mikhailovskoye-Novogrygoryevskoye region. (Data from I. P. Krivitzky, T. M. Seletzky and I. V. Krachkovsky). (1) Geothermic wells; (2) water-sample selection sites, (3) anomaly contours by oxidizing bacteria; (4) methane; (5) vaporus hydrocarbons; (6) mixture of ethane, propane, and butane; (7) break disturbances; (8) stratoisotypes along the foot of the Bashkir stage.

are characterized by bacterial anomalies, but several of the discovered anomalies were caused by unloading of deep formation waters along faults.

For relatively inexpensive surveying operations, subsurface geochemical surveys as a whole are characterized by low-density sampling. In the zone of active gas and water interchange, hydrocarbon anomalies are influenced by certain natural factors, mainly atmospheric agents and groundwater. The occurrence of residual gas anomalies depends on climatic conditions, depth of upper waterproof bedding relief of the locality, landscape, etc. Because of these conditions, subsurface gas-biochemical surveys have been deployed in the USSR in the 1970s in reconnaissance. There is, however, detailed studying of the most promising regions where detailed geochemical surveys of reference horizons would be expedient. Areas characterized by the most pronounced hydrocarbon gas anomalies and microbiological indexes are regarded as the most prospective.

Oil and Gas Prospecting by Reference Horizons

The geochemical exploration of gas horizons is included in the array of geochemical surveys on the basis of which final evaluation of prospects for studying oil- and gas-bearing areas is made and also the placement of wells. Such investigations are made by special structural wells with depths of 200–300 m. Gas saturation in rocks is studied with gas samples extracted by a thermovacuum method from score and mud-laden fluids that are hermetically sealed immediately after their recovery from the well. Such work focuses on finding the observations and estimations of epigenetic hydrocarbon anomalies located in the zone of restricted gas and water interchange (catagenesis).

The possibility of gas survey by reference horizons was forecast by Sokolov in 1934 (Sokolov et al., 1971) while analyzing carrotage by *schlam* of deep wells. On this basis Sokolov proposed constructing gas profiles in areas of gas anomalies using lithologically homogeneous layers and horizons.

Application of the method to many oil and gas deposits in the USSR led to the next principal scheme of such anomalous formation (Fig. 2). In tectonically active areas, fissures provided pathways for periodically renewed deep fluid migration (see *Plate Tectonics, Mineral Exploration*). Tectonic block movement took place subsequently. At that time fluids migrated into upper horizons. Formation water filling fissures and carrying dissolved gases selectively saturate only such collector layers that have high permeability, porosity, and minimal formation pressure. Formation energy of an upper water-saturated hori-

zon is more intensively exhausted while formation waters become partially unloaded. During tectonic movements, fissure systems create conditions for interchange of water-bearing layers with the Earth's surface or overlying collector horizons. Unloading of upper waters substitutes for deep gas-saturated water layers that follow deep fissures and, having surplus pressure, laterally penetrate formations that are being drained. Due to sharp reduction in pressure, part of the dissolved gases are isolated into free phases that saturate overlying gas-waterproof horizons that occur along microfissures. Subsequently, slow gas interchange between collector formations and adjoining poorly permeable horizons takes place by diffusion. These processes continue for a long time after the conducting system of fissures is found to be cured.

Thus, in tectonically active areas, gas hydrocarbon anomalies located in upper horizons of a section form in two stages: (1) a short-term pulsation stage of ascending filtration of deep layer fluids along fissures and their lateral penetration into collector layers and (2) long-term filtration and diffusion of gas interchange between gas-saturated collector layers and overlying (or underlying) horizons.

In platform areas, where microfissure and macrofissure systems form in gas-waterproof formations during the process of anticlinal structure development, they function as pathways for ascending migration of hydrocarbon gases from lower deposits to upper horizons. The formation spaces in which epigenetic hydrocarbon anomalies are fixed at shallow depths are used as reference to search for gas-metrical horizons. Base horizons are determined primarily by the total number of geological factors directly influencing the processes of gas anomaly formation and destruction:

1. The reference horizon represented by separate thick carbonate or terrigene layers or lithological bundles must be continuous over the entire area.
2. The reference horizon must be located in the watertower—i.e., be overlain by a gas-watertight layer that ensures isolation from the destructive influence of atmospheric agents for the gas anomaly.
3. The content of authigene organic matter (especially bitumenoids) in reference horizons should not exceed background values because that can lead to the formation of false gas anomalies.
4. A stratigraphic section characterized by wide concentration of hydrocarbon gases is assumed to be a reference horizon in the study area.

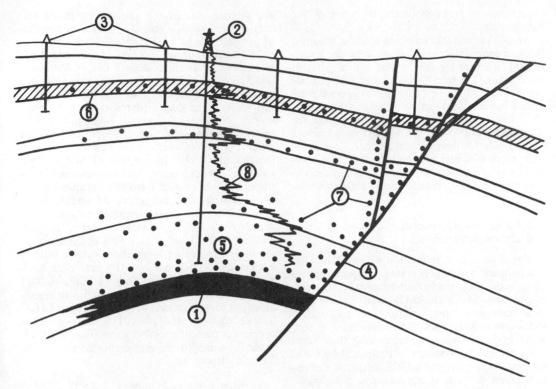

FIGURE 2. Principal scheme of hydrocarbon anomaly formation in reference to gas-meter horizons. (1) Oil-gas accumulation, (2) deep well, (3) structural geochemical wells, (4) break disturbance, (5) uninterrupted gas halo under deposit, (6) reference gas-meter horizon, (7) dispersed hydrocarbon gases, (8) curve or rock gas saturation by deep well.

Every oil- or gas-bearing area or region has distinctive geological-geochemical characteristics in the thick sediments that determine conditions of existing and migratory flow of hydrocarbon gases ascending from deposits. Therefore, concrete oil- or gas-bearing geochemical criteria can be used in areas of known deposits and also to search effectively unproductive areas.

Rock Gas Saturation

Analyses of gas extracted from score and washing fluid of structural geochemical wells is a main source of information about rock gas saturation. Experience has shown that geochemical studies of both score and washing fluid have their virtues and shortcomings; neither gives values of rock gas saturation in layer conditions. The following factors misrepresent the influence of score on gas saturation: outstripping of infiltration of solution and pressing back of gas into layers, rising temperature from mechanical loosening and drilling in rocks, gas losses while score is lifted to the surface until it is hermetically sealed, and incomplete gas extraction from rock in the laboratory.

Gas saturation of washing fluid occurs during the drilling of gas-saturated rocks. It varies with the degree of fluid penetration into the gas-saturated rock and the amount of gas dissolved in the rock. The level of drilling fluid gas saturation does not remain constant due to the following factors: mass interchange with well wall rocks, gas apportionment into free phase during the solution lifting process, gas losses during solution selection in the well mouth, and influence of chemical reagents and lubricants.

Numerous false factors cannot practically be taken into account during the drilling process. Therefore, when wells are studied by gas-metrical methods using score and washing fluid, particular attention is given to the maintenance of common conditions of selection, storage, sample decontamination, and analysis of extractive gases. In the processes of geological interpretation of obtained data, preference is given to geochemical values expressed by relative quantities—contrast of gas anomalies, relative content of certain hydrocarbon components, and the ratio of their concentrations. Through error averaging, it is possible to calculate the average gas saturation concentrations throughout the thickness of the base horizon.

Areas that are characterized by data obtained through geophysical exploration as having an anticlinal structure are then drilled with a series of structural geochemical wells located along crossing profiles. At the first turn one to two wells are drilled in the arched section of the anticline. Samples of score and washing fluid are then taken for the whole section. Gas extracted from cores and drilling solution immediately goes to analysis so that layers with high concentrations of hydrocarbon gases can be selected. Later wells are concentrated around the revealed hydrocarbon anomalies.

Estimation of Gas Anomalies and Their Interpretation

The quantitative estimation of gas anomalies can be used to ascertain average values of anomalous and background gas concentrations in a reference horizon for the study area. The ratio of these quantities determines the main quantitative index-anomaly contrast that is calculated by separate hydrocarbon components and their summary content. Independent of geological data, siz of area, and region studied, the calculation of hydrocarbon anomaly contrast is done as follows.

Without Regard to Structural Factors This method of anomaly estimation is used when there is a lack of geochemical data for contouring the anomaly field or when investigations have been on one cross-cut or lengthwise profile in three to four wells. In this case all gas analytic data obtained on selected base horizons, irrespective of the well from which they were obtained, are systematized in the common massif. Following standard methods of statistical analysis of primary data, the drawings of gas concentration distribution and cumulative curves can be constructed. The gas concentrations are calculated in cm^3/l of washing fluid or in cm^3/kg of selected rock. The value of average-weighted background gas concentration is determined by cumulative drawing; it should be approximately the same for 60–80% of the samples. By averaging other samples with higher gas saturation, the average anomalous concentration of hydrocarbon component in a study reference horizon is calculated.

In practice, the common qualitative estimation of anomalous gas field data on methane and H.H. [heavy hydrocarbons, i.e., sum of heavy gases (C_2-C_4)] is used; sometimes penthane and gexane vapors are included but not always in measurable quantities. The ratio of averaged values of anomalous and relative background gas concentrations calculated for a specific area determines the contrast of gas anomaly in a selected reference horizon. The ratio of averaged values of anomalous

and relative background gas concentrations calculated for a specific area determines the contrast of gas anomaly in a selected reference horizon.

The calculation of contrast of anomalous gas effects on the Tasbulat deposit (South Mangyshlak) is an example. Geochemical investigations were conducted there in four structural geochemical wells 500 m deep situated along a cross-cut profile. Neogene, Paleogene, and Upper Cretaceous sediments were found. Senonian–Turonian sediments were chosen as the reference horizon for this area. The distribution of hydrocarbon gases in the reference horizon was not even throughout the profile; fixed concentrations in the working well were not great. Maximum background concentrations determined by cumulative drawing were CH_4—0.0040 cm^3/kg, H.H.—0.0006 cm^3/kg, in more than 80% of the analyzed samples. Proceeding from these values, relative background concentrations, by reference horizons, were calculated as follows: CH_4—0.0015 cm^3/kg, H.H.—0.0004 cm^3/kg, and also anomalous gas concentrations were calculated as CH_4—0.0192 cm^3/kg, H.H.—0.0170 cm^3/kg. Thus proper contrast of revealed anomalous gas fields in the Senonian Turonian reference horizon on the Tasbulat deposit is:

Methane: 0.0192/0.0015 = 12.8
H.H.: 0.0170/0.0004 = 42.5

With Regard to Structural Factors This method is used in cases when the number of drilled wells is sufficient for a complete or partial contouring of gas anomaly on the study structure but where no information exists about regional gas background. Here, also, relative background and anomalous gas concentration is calculated for the chosen reference horizon, and the coefficient of gas anomaly contrast is computed. However, as distinct from the first method, two groups of wells—background and anomalous—are picked out by average values of reference horizon gas saturation. On the gas distribution map, constructed from the sum of heavy gases, background wells are situated outside the anomalous gas field and contoured by the line of maximum background concentration. If the anomaly situation is controlled by a structural arch, then most of background wells will be situated on the far wings and periclines of the rise. In that way the influence of structural factors on gas distribution in a reference horizon is automatically taken into account while calculating anomaly contrast.

Maximum background gas concentration is determined by cumulative curves constructed on the whole massif of gas-analytical data relevant to the chosen reference horizon. By calculating an average background concentration, all data on the content of uninvestigated components of background wells are taken into account.

The second way to estimate proper gas anomaly contrast is illustrated by an example from the Sakar–Changa gas deposit in central Kara-Kum (Fig. 3). This deposit is arch situated on a gentle anticline. Its north wing is complicated by a fault of sublatitudinal extent. The gas deposits are located in Neokomian–Aptian terrigene sediments found in 1,200–1,300 m depths. The gases extracted from a score of deep structural wells (up to 600 m) were investigated. Organic matter and its bituminous part were studied. The gas anomaly showed up in Senonian carbonate sediments situated in the arch part of the rise and screened by a break disturbance.

The contrast of methane for the methane anomaly was calculated as follows. The anomalous methane field in the Senonian reference horizon was contoured by isogas 0.1 cm³/kg on the map to show the distribution of methane and heavy gases. This limit of concentrations was met by nearly 50% of the samples taken in cumulative drawings. Therefore, it is statistically correct to use the value 0.1 cm³/kg as a maximum background concentration of methane. The average relative background concentration of methane was calculated by averaging the values from wells outside the anomalous field, in the wings and pericline of the structure. The background value consisted of 0.006 cm³/kg. The average anomalous content of methane calculated by the group of anomalous wells situated in the arch of the rise is equal to 0.33 cm³/kg. Thus, proper contrast of gas anomaly by methane is 0.330/0.006 = 55. In this way the contrast of a gas anomaly by sum of heavy hydrocarbons is calculated.

Calculation of Absolute Anomaly Contrast
The third method of hydrocarbon anomaly evaluation, based on the calculation of average

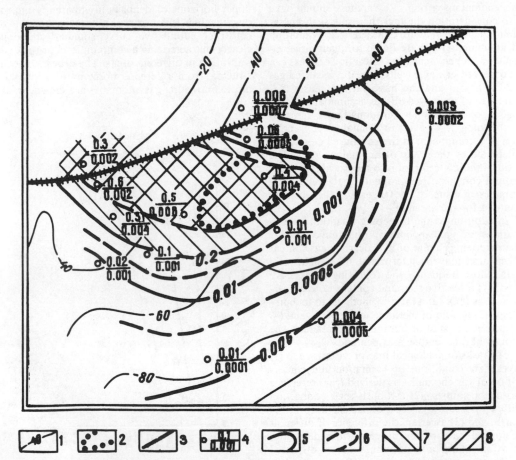

FIGURE 3. Sakar–Changa deposit. Methane and heavy hydrocarbons distribution in the Senonian reference horizon. (1) Isohypses of the Karashor stage four; (2) contour of gas deposits; (3) break disturbance; (4) structural geochemical well, numerator content of methane, denominator content of heavy hydrocarbon gases (cm³/kg); (5) isolines of methane content; (6) isolines of sum of heavy hydrocarbon gases content; (7) methane anomaly; (8) anomaly by sum of heavy hydrocarbon gases.

values for the study region of background gas concentration in a chosen reference horizon, is the most informative and preferred method. For such calculations, it is necessary to have geochemical material from a minimum of two investigated areas in the region with a sufficient number of background wells to contour the anomalous gas fields. In this case, as in the second method, extrapolation of gas-analytical data is used. On the basis of the extrapolation, a common cumulative curve is constructed. While calculating average background concentration, all contents of the component are taken into account from traces to maximum background and usually consists of 60–80% of the whole sample number.

On the map of gas distribution of study areas, wells are background in which the average content of hydrocarbon components does not exceed the average background concentration calculated by common drawing. The other wells belong to the anomalous group. Researchers should take into consideration the fact that only such anomalous gas fields that are characterized by a group of anomalous wells are liable to quantitative estimation. A heightened gas saturation of rocks can be observed in single wells even if a localized gas field is absent and can have various causes (i.e. influence of destroyed oil-gas accumulations, unloading of formation water along faults, errors in methodology). An average anomalous concentration of a component in the contoured gas field is calculated by the group of anomalous wells. The ratio of this concentration to the calculated background content of components in the region determines a value of coefficient of the absolute contrast for the anomaly.

In addition to the foregoing quantitative estimations of gas anomalies, qualitative indexes characterizing change of hydrocarbon gas content in an area may be interpreted for geochemical data. Most frequently the following indexes are used: (1) a relative methane content in hydrocarbon gases (%); (2) a ratio of methane to its homologs, (3) a ratio of methane to propane, and (4) a ratio of the sum of butane and pentane to the sum of normal butane and pentane.

The tasks of geological interpretation are to (1) investigate conditions of formation and conservation of gas anomalies as derived from reference gas-meter studies, (2) establish their connection with supposed oil-gas accumulations deep in the Earth, and (3) ascertain the influence of different geological factors on processes of migration, transformation, and accumulation of dispersed hydrocarbon gases. The degree of anomalous gas field displacement contours and their shape are stipulated by many causes. The most important of them are the contrast between structural forms in the conditions of arch traps, tectonic and lithologic restrictions or screens in sections of the

study area, the position of points or zones of deep hydrocarbon gas flowing under relatively high arches, and the transmission of structural forms to a section. It should be remembered that the incidence of anomalous gas-field contours with deposit contours is practically unknown. Thus, it is necessary to take into consideration a sum of geological-geophysical data about the deep structure of the study area while choosing exploration well locations.

The possible location of deep-source hydrocarbon deposits and migrational direction of gas along collector layers or of anomalous field formation may be found through quantitative analyses of gas distribution in a reference horizon. In most cases displacement direction of lighter components in the gas field is lateral along the reference horizon.

As geochemical data on study regions are collected and information about gas saturation of the upper horizons of deposits in apparently unproductive areas is compared, it becomes possible to construct cumulative curve standard drawings clearly characterizing a common level of rock gas saturation in different areas. The drawing is constructed from reference horizons in which the most contrasting gas anomalies are shown. In ex-

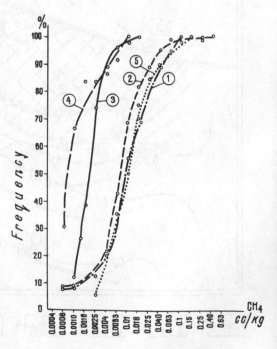

FIGURE 4. Standard drawing of methane saturation or reference horizons, South Mangyshlak area. (1) Gas-oil deposit, Taschbulat; (2) gas-oil deposit, Zhetybay; (3) apparently unproductive structure, Senek; (4) unproductive Manata exploration area; (5) productive Kansu exploration area.

ploration areas where anomalous gas fields are not found, gas-analytic data of simultaneous sediments are used to heighten their similarity.

The standard cumulative curve drawing for areas of Mangyshlak is represented in Fig. 4. This area is characterized by methane saturation of reference horizons in the Zhetybay and Taschbulat deposits and also in the apparently unproductive Senek structure. Substantially lower levels of gas content in Senek were reflected by the considerable displacement to the left of the cumulative curve, along the concentration scale of the drawing.

The cumulative curve drawing for the Kansu area characterized the area as highly productive. Based on geochemical investigations, this area was recommended for deep drilling. This activity produced industrial-quality wells. The curve for the Manata exploration area is also characteristic; it shows gas evenly displaced to lower levels. The structural drilling in the Manata area did not confirm the presence of closed anticlinal folds as was proposed by geophysical data.

Such standard drawings may be used for preliminary evaluation of oil-gas-bearing formations in the exploration areas of the study region. But their presence does not exclude necessary and thorough analyses of all geological-geochemical materials; not only discovering, contouring, and quantitatively estimating anomalous gas fields in reference horizons but also detailed qualitative analyses of gas distribution throughout the area.

A. A. GEODEKYAN
V. A. STROGANOV

References

Geodekyan, A. A., and V. A. Stroganov, 1973, Geochemical prospecting for oil and gas using reference horizons, *Jour. Geochem. Exploration* **2**, 1–9.

Sokolov, V. A., A. A. Geodekyan, G. G. Grigoryev, A. Ya. Krems, V. A. Stroganov, L. M. Zorkin, M. I. Zeidelson, and S. Ya. Vainbaum, 1971, New methods of gas surveys, gas investigations of wells and some practical results, in *Geochemical Exploration* (Spec. Vol. I). Montreal: Canadian Institute of Mining and Metallurgy, 538–543.

Cross-references: *Atmogeochemical Prospecting; Exploration Geochemistry; Geological Survey and Mapping; Indicator Elements; Petroleum Geology; Prospecting; Remote Sensing, General; Sedimentary Rocks, Field Relations; Terrain Evaluation Systems; Well Logging.*

PETROLEUM GEOLOGY

Petroleum geology is general geology applied to the search for and development of petroleum resources (*petroleum* is the collective noun for oil and gas). It centers around the stratigraphy and structure of sedimentary basins because virtually all the known petroleum reserves are in sedimentary basins. The methods of petroleum geology involve the more specialized topics of paleontology (q.v. in Vol. VII), sedimentology (q.v. Vol. VI) and geochemistry (see *Exploration Geochemistry;* Vol. XIII: *Geochemistry, Applied*)—all aspects being integrated. Geophysics has increased in importance since about 1960 at the expense of field geology as land prospects became scarcer and exploration moved offshore to the continental shelves. Since about 1970, technological advances in geophysical exploration (in which computers played an important part) have greatly improved the quality of the results. Seismic reflection surveys can provide an accurate picture of the subsurface structure and stratigraphy and, under favorable circumstances, may even reveal the presence and position of gas accumlations.

The basic data of petroleum geology come from the study of the rocks exposed at the surface and of the data provided by geophysical surveys and boreholes. Borehole data are mainly in the form of logs (see *Well Logging*) showing changes with depth of physical properties, such as electrical resistivity, of the rocks penetrated by the borehole; but these are supplemented by samples of the rocks obtained during drilling, which may also contain fossils.

Petroleum geology has regional character. Some areas of the world are very similar, such as the Niger Delta and the U.S. Gulf Coast; western Canada, Mexico, and parts of Libya. But northwestern Europe is quite different from southeast Asia, and the Middle East is quite different from the rest of the world. Petroleum geologists from one part of the world would not necessarily agree with those from another on what the cardinal features of petroleum geology are. The insights revealed by petroleum geology in both regional and local character are important to the broader understanding of geology and the geology of the world because the field deals with a dimension rarely available in surface geology on a scale never available outside the great mountain chains.

Geological Context of Petroleum

A sedimentary basin, in its simplest terms, is an area that accumulated sediment at a relatively greater rate than the surrounding areas (accumulating it to a significantly greater thickness in a given period of time). The accumulation of sediment is largely by virtue of subsidence. The nature of the sediment that accumulates depends on the position of the subsiding area relative to the land from which the sediment is derived. If the sedimentary basin lies under the land, terrestrial sediments accumulate; if under the sea, marine.

And if, as is commonly the case, the subsiding area is near the coastline, terrestrial sediments accumulate in part of the sedimentary basin and marine sediments in other parts. These physiographic environments have characteristic sediments: the higher the energy of the environment (in general), the coarser the grain of the sediment and the better its porosity and permeability. [A sedimentary rock with spaces between the grains is said to be porous: the measure of porosity is the proportion of pore space in unit bulk volume of the sedimentary rock. If fluid can be made to pass through the pore spaces, the rock is said to be permeable; a measure of permeability is the rate at which a specified fluid can be passed through a sample of standard dimensions under standard conditions (see Levorsen, 1967, Chapter 4).] Petroleum reservoirs are commonly sedimentary rocks associated with nearshore environments.

Sedimentary basins develop over several tens of millions of years, and during this time the relative distribution of land and sea over the sedimentary basin may change, with the result that the sediments of one environment become superimposed on those of another. The normal history of a sedimentary basin begins with a period of marine transgression, during which the sea becomes more extensive at the expense of the land, and the finer-grained sediments tend to accumulate on top of the coarser (Fig. 1). It ends with a period of regression, during which the shoreline moves back in a seaward direction and coarser sediments tend to accumulate on top of the finer. There are many fluctuations within the larger cycle, and these contribute to the alternation of rock types in the vertical sequence.

This distribution of rock types in time and space is of fundamental importance for petroleum, and the reservoirs (the coarser grained) may be sealed by the finer-grained sediment. The second reason for the importance of these broad patterns is that the petroleum geology of transgressive sequences tends to be different in character from that of regressive sequences—a matter to which we shall return when the principles have been introduced.

Petroleum geology is concerned largely with the geology of the subsurface, and the approach to *subsurface geology* differs from that to surface geology in some important respects. Perhaps most important, it is concerned with the pore spaces of the sedimentary rocks rather than with the nature of the mineral grains because it is concerned with the contained fluids rather than the solids. The pore spaces contain water, usually saline water, which can move through the pore spaces if there is permeability and a driving force. If there is gas with the water, it too will move, but because it is

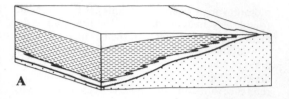

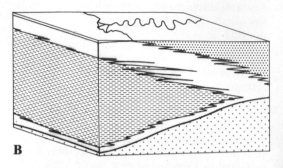

FIGURE 1. Block diagram showing (*A*) accumulation of a transgressive sequence and (*B*) a transgressive sequence followed by a regressive sequence. Neritic facies, dashes; neritic-paralic, fine dots; paralic-terrestrial, coarse dots. (From Chapman, 1983)

less dense than the water, it tends to rise through the water until it is restrained by a fine-grained bed, such as a bed of mudstone or shale (in applied geology, *shale* is often used to refer to any argillaceous sediment). It will then follow this layer until it either reaches the surface to form a seepage and is dissipated or arrives in a position in which the driving force is insufficient to move it farther. There it accumulates, forming a gas reservoir underlain by water. The characteristics of the accumulation are related to the quantity of gas, the porosity of the reservoir, and its geometry. Oil likewise accumulates above water but below gas because the density of oil is less than that of water but more than that of gas. We are thus concerned with three major topics: the origin and generation of petroleum, its migration, and its accumulation.

Origin of Petroleum

The origin and migration of petroleum are still understood in only rather general terms (although this statement would be disputed by many geochemists) and there is wide disagreement on many matters (see Hedberg, 1964; Hunt, 1979). There would, however, be general agreement on three points: (1) most petroleum originates from organic matter that accumulated with fine-grained sediment under quiet conditions deficient in oxygen; (2) it is generated during burial under the influences of time, heat, pressure, and probably catalysts; and (3) it is expelled from the source rock during compaction. (The history of science, however, teaches us that majorities are not always right!)

Petroleum geochemists regard the conditions obtaining above depths of 1,500 m (5,000 ft) as inadequate for the generation of significant quantities of petroleum. Some petroleum geologists, on geological evidence, favor a shallower ceiling of about 600 m (2,000 ft) (see, e.g., Weeks, 1958, 31). It matters for exploration economy that this problem be resolved. This resolution requires the identification of the source rock for many shallow accumulations of oil and gas (for there are many large accumulations at depths of about 1,000 m).

Migration of Petroleum

The migration of petroleum is better understood than its origin, but still only in general terms. Migration through the fine-grained source rock is referred to as *primary migration:* its passage through the coarser, more permeable rocks is referred to as *secondary migration.* Primary migration takes place as a result of mechanical driving forces induced by compaction of the source rock under gravity. It is thus directed vertically (upward *or downward*) in the direction of decreasing potential energy. Primary migration probably takes place as a separate phase because the petroleum appears to have acquired its essential character (oil or gas, perhaps also composition, density, etc.) before primary migration is complete and secondary migration begins. The main reasons for this view are that the fine-grained sediments are chemically more active than the permeable rocks of secondary migration and entrapment, the physical environment is more rigorous, and secondary migration paths can be very short. There is also evidence in some oil fields that changes of oil quality have taken place in the reservoir, presumably during further burial to higher temperatures and pressures. Secondary migration takes place laterally within the more permeable rocks and continues (as mentioned previously) until the force is insufficient to drive the petroleum farther or until it dissipates at the surface.

The main reason why petroleum generation and primary migration are poorly understood is that they are processes that take place so slowly under constantly changing chemical and physical conditions that they cannot realistically be modeled in the laboratory. Secondary migration does not suffer from this problem because when we produce an oil or gas well, the oil and gas demonstrably move through the reservoir rock to the well. There are problems, but there do not seem to be great difficulties in moving petroleum through permeable rocks.

Accumulation of Petroleum

Petroleum accumulation is achieved when the configuration of the permeable and impermeable sedimentary rocks along the migration path is such that further migration is reduced or stopped. More technically, such traps are spaces of minimum potential energy of the petroleum with respect to its physical environment. They are of infinite variety but nevertheless fall into three main groups: (1) anticlinal traps, (2) fault traps, and (3) stratigraphic traps. That is also the order of quantitative importance, but most accumulations are in faulted anticlines, which are referred to as *structural traps.*

Anticlinal traps (Fig. 2) have the following essential features: the contours on the top of the reservoir are closed, so that the reservoir forms a space of low potential energy for petroleum in the gravitational and hydrodynamic fields; the reservoir rock is both porous and permeable; and the reservoir is capped, or sealed, by a fine-grained rock that is usually mudstone or shale but may be marl or evaporite. Such a trap may have a single reservoir (referred to as a *pool* sometimes) or many reservoirs. Each will have an oil/water contact, or gas/water contact, that is a horizontal or slightly inclined interface. If there are both oil and gas, there will also be a gas/oil contact. This contact is originally horizontal, but production patterns affect it on account of the reduction of reservoir pressure and the consequent expansion of the gas.

The oil/water and gas/water contacts of different reservoirs are usually at different levels. When they are found to coincide, it is assumed that there is some communication between the reservoirs with common oil/water or gas/water contacts.

The essential features of a *fault trap* (Fig. 3) are an inclined reservoir rock that is sealed above by a cap rock and is truncated by a fault. The fault itself may seal the reservoir, or it may bring fine-grained sediments to this position. A fault trap also requires some lateral barrier along the reservoir to prevent lateral migration, such as another fault, or commonly, the reservoir is folded along the fault.

Fault traps, like *anticlinal traps,* may have one or more reservoirs, and the oil/water or gas/water contacts are usually different but may be the same. Indeed, if two wells encounter the same petroleum reservoir rock with the water contacts at different levels, it is usually argued that a fault separates them.

In many petroleum accumulations it is difficult

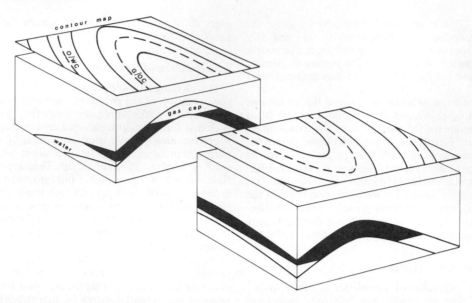

FIGURE 2. Block diagram of petroleum in an anticlinal trap. The upper surface shows its representation on a map. (From Chapman, 1983)

to establish whether it is primarily an anticlinal or a fault trap because so many are faulted anticlines. They are then classified as structural traps. Most of the world's known petroleum accumulations are in *structural traps*, but that may be because we tend to look for structural traps.

The other important type of accumulation is the *stratigraphic trap*. Stratigraphic traps, as the name suggests, are due to a restraint on petroleum migration that arises from stratigraphic causes, such as permeability or porosity changes in the reservoir rock, formation of reservoirs around a buried hill, convergence of rock units or lenticular reservoirs, truncation of a reservoir by an unconformity with overlying fine-grained sediment, and the important class of organic reefs. The com-

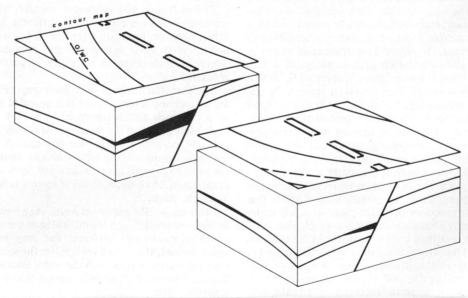

FIGURE 3. A typical fault trap showing the oil/water contact and displacement of strata trapping oil along the fault between confining layers. (From Chapman, 1983)

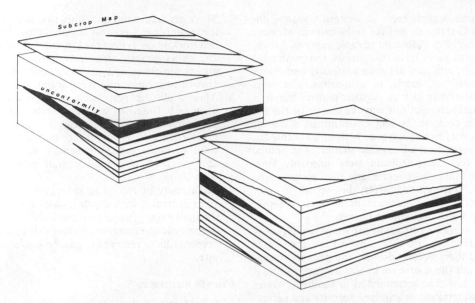

FIGURE 4. Block diagram of petroleum in a stratigraphic (unconformity) trap and the representation of the subcrop on a map. (From Chapman, 1983)

monest stratigraphic trap is probably the wedging or lenticular sand reservoir, but the most important are the *unconformity traps* (Fig. 4) and the *organic reefs* (Fig. 5). Some stratigraphic traps are also folded, some also faulted; but organic reefs are rarely folded, and we discuss this important exception later.

Once the geometry of a reservoir is known, the volume of oil in place can be estimated. But within the pore spaces, the water is never entirely displaced by oil or gas, so the volume of oil or gas in place is the volume of the pore space less the volume of water. Unfortunately, the oil and gas in place are not fully recoverable at present, and more than half is left in the ground. *Secondary recovery methods,* such as water flooding of the

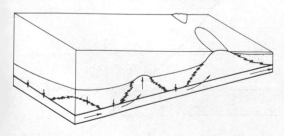

FIGURE 5. Block diagram of the accumulation of a transgressive sequence with reefs (arrows indicate movement of pore fluids). When reef growth in the area is terminated and a fine-grained rock accumulates over the reefs, they could become petroleum traps.

reservoir from wells around the accumulation, have been important in increasing the proportion recovered, and considerable research effort is devoted to the economic recovery of the remainder because this would more than double the usable reserves.

The sizes of petroleum accumulations vary greatly. Many factors determine the minimum economic size, the most important being the amount of recoverable oil or gas; the cost of developing, producing, and transporting it; and the price of the products. The role of the petroleum geologist is to find and appraise the accumulations with the fewest possible wells.

Stratigraphic Nature of Petroleum Provinces

Petroleum is distributed evenly neither over the world nor through the rocks of different ages. But statistics are distorted by the occurrence of about half the world's known reserves in rocks of Mesozoic and Tertiary ages in the Middle East. This imbalance also serves to illustrate the regional nature of petroleum geology, the character of which (leaving the Middle East aside) is determined largely by whether the petroleum-bearing sequence is transgressive or regressive (see Fig. 1).

Transgressive Sequences Transgressive sequences may be of sands and mudstones or carbonates and mudstones, but almost all important carbonate sequences are transgressive, and they are typically unfolded. We find important petroleum accumulations in unfolded Silurian organic reefs of the north-central United States, unfolded

Devonian organic reefs in western Canada, unfolded Cretaceous reeflike carbonates in Mexico, and unfolded Paleocene organic reefs in Libya. The fossil reefs form the porous and permeable reservoir, and they are sealed laterally and above by mudstones, marls, or evaporites. The reefs were originally rich in organic matter, but it is most unlikely that this was the source of the petroleum because the reef environment is one of high energy rich in oxygen, both unfavorable for the preservation of organic matter. The source rocks for the petroleum were probably fine-grained mudstones and marls that accumulated before, during, and after the period of reef growth, at some distance from the reefs in deeper, quieter water deficient in oxygen.

The reefs of petroleum provinces are typically transgressive, growing upward as the water around them became deeper, and primary migration from the source rocks was downward into a carrier bed that accumulated in shallower water and had retained or acquired porosity and permeability. The secondary migration in the carrier bed followed the driving force and the permeability paths, both normally toward the land of the time. If these paths led to a *fossil reef* (see Fig. 5) that had been drowned and covered by mudstone, then the petroleum accumulated because the pore-space of the reef material formed a space of minimum potential energy. Such a history is also indicated by the fact that while fossil reefs tend to be in groups, not all the reefs in a group necessarily contain petroleum. And reefs that are entirely surrounded by mudstone (i.e., those that did not grow up from a carbonate platform) contain little or no petroleum.

Regressive Sequences The petroleum geology of regressive sequences has many points of interest for both the petroleum geologist and the geologist working on sedimentary basins in other contexts (see Chapman, 1983). It helps our understanding to look at young regressive sequences because these are relatively simple and because the principles we draw from them may be extended to the older, more complicated sequences.

Regressive sequences consist dominantly of sand and mudstone, and young regressive sequences, such as those in the U.S. Gulf Coast, the Niger Delta, and many areas of southeast Asia, have several features in common:

1. The marine mudstones beneath the sandy overburden tend to have pore-water pressures much greater than the pressure that would be exerted by a column of salt water extending to that depth from sealevel or from the water table below the land surface (sometimes they are found to be about twice as great as the normal hydrostatic pressure). Such mudstones are usually less dense than normal for their depth.

2. There are faults and anticlines that grew while the sediment in them was accumulating, so that bed thicknesses reflect the structure. These are known as growth structures (growth faults, growth anticlines; sedimentary basins are really growth structures on a large scale).

3. Almost all the petroleum in regressive sequences is found in structural traps and usually in the deeper sands close to the abnormally pressured mudstones (but they may extend upward to much shallower sands). Fields with multiple reservoirs are common (some with over 100).

4. The density of the oil in superimposed reservoirs decreases with depth in such a way that the shallower, younger, tend to have heavy oil; deeper, older, reservoirs, lighter oil; and even deeper, older, reservoirs, gas or gas condensate.

Growth Structures

To understand these features and their interrelationships, we look first at growth structures, which are best illustrated by growth faults. A *growth fault* is a normal fault (usually) across which the thickness of correlative rock units changes abruptly, with the thicker sequence on the downthrown block (Fig. 6). Movement of the fault during sediment accumulation is considered to be the cause of the abrupt changes in thickness. But the data allow us to make two important inferences: (1) that the sediment accumulates from the traction load on the seafloor in the neritic environment and (2) that the sedimentary column was subsiding relative to some baselevel of energy (it is probably sufficiently accurate to regard this as subsidence relative to sea level).

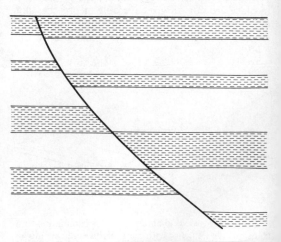

FIGURE 6. Idealized section through a growth fault. Each rock unit is thicker on the downthrown side. (From Chapman, 1983)

The rate of sediment accumulation by settling from suspension is governed by the physical properties of the sediment and the water. No plausible reason has been found why these should change abruptly across a fault in the accumulated sediments. This difficulty is removed if sediment accumulates from the traction load on the seafloor (muds as well as sands). The downthrowing block accumulates more of this sediment than the upthrowing block, implying that not all the sediment supplied to an area accumulates in it. The clue here is Barrell's concept of baselevel (see Chapman, 1983). Each sedimentary particle is moved by the energy of the water until it arrives in a position from which it cannot be moved farther. There it accumulates. Some particles reach that position near a growth fault, but more reach it on the downthrowing block than on the upthrowing block.

The concept of *subsidence* of the accumulating sediment relative to baselevel explains this situation. But since sediment also accumulates on the upthrowing block, we infer that the upthrowing block is also subsiding relative to base level—that *both* blocks are subsiding but one faster than the other and so accumulating sediment at a greater rate. This view is also supported by the common observation that the upthrown block of one growth fault is the downthrown block of another.

The association of growth faults with young regressive sequences is therefore important evidence that the regressive sequence accumulates during subsidence by virtue of more sediment's being supplied to the area than can be removed by current and wave action (present-day deltas are examples of this). *Growth anticlines* are anticlines in which rock units are thicker on the flanks than on the crest. Such anticlines grew while subsiding and accumulating sediment, the flanks subsiding faster than the crest. Like growth faults, they are important in that they form traps before most of the petroleum has been generated.

Compaction of Sediments

We now move to the matter of compaction of the sediments during subsidence and burial and its effect on the pore fluids. We begin with the observation that *compaction* is a diagenetic process in sediments that tends to increase their bulk density and mechanical strength and to reduce their porosity. It is a process that can proceed only if the pore fluids can be compressed or expelled (mechanical aspects only are considered here because these seem sufficient for understanding the principles).

Both sands and muds compact initially at shallow depth by the rearrangement of the grains, but once a substantial degree of grain contact has been achieved, sand compaction is very different from mud compaction. Further loading of a clean quartz sand leads to some elastic deformation of the grains and a small consequent loss of porosity and permeability. If the load is removed within a relatively short time, the strain in the sand grains is removed elastically and the sediment recovers its original properties. Over longer periods of time, part of the strain becomes permanent, but the sand retains good porosity and permeability.

Muds, conversely, compact plastically under gravity load of the overlying sediments once a substantial degree of grain contact has been achieved. This process, which is largely irreversible, also tends to reduce the mud's porosity, but more significant, the constriction of the pore spaces as the mud becomes mudstone seriously and permanently reduces its permeability. This contrast in the changes of permeability has an important influence on the expulsion of pore fluid (without which compaction cannot take place). In sands, the *pore-fluid expulsion* required for normal compaction is easily achieved, but in mudstones, the loss of permeability may lead to retarded fluid expulsion and, hence, retarded compaction. If the pore fluid cannot escape sufficiently rapidly for normal compaction to take place, the load of the overlying sediments is taken partly by the pore fluids and their pressure is correspondingly greater than normal hydrostatic.

Undercompaction means that the mudstone has greater porosity, smaller bulk density, and less mechanical strength than a normally compacted mudstone at the same depth. Since the compaction of a mudstone is essentially irreversible, these properties are retained from a shallower depth of burial rather than acquired at the depths found at present. The presence of abnormally pressured mudstones at depths as shallow as 600 m, in places even shallower, supports this inference.

The consequences of undercompaction in a young regressive sequence are that the sequence becomes mechanically unstable soon after the permeable, sandy overburden begins to accumulate and may remain mechanically unstable over a few tens of millions of years while the sequence is buried. This early deformation is seen in growth structures, which are formed before the bulk of the pore fluid can escape from the mudstones.

Regressive Associations

There is, therefore, a logical link in the association between young regressive sequences, abnormally high pore-fluid pressures, and growth structures. In the context of petroleum geology, this association is extended to include petroleum in structural traps rather than stratigraphic. The migration of petroleum is evidently a consequence of the compaction of the source rock. The migration paths are determined by the fluid poten-

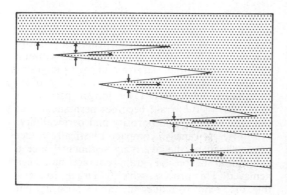

FIGURE 7. Schematic section through a regressive sequence showing the directions of pore-fluid migration. Vertical scale is grossly exaggerated. (From Chapman, 1983)

tial gradients (the driving force) and the overall permeability of the system. Gravitational compaction leads to vertical primary migration from the mudstones upward and downward into the more permeable sands. Secondary migration is lateral within the sands and, on account of the sands' geometry, toward the land (Fig. 7). The *migration paths* therefore pass through the growth structures that exist long before petroleum migration is complete and may grow during the migration.

Some Current Problems

The actual processes of petroleum migration and accumulation are certainly more complex than those summarized here. In particular, the generation and migration of petroleum in carbonate sequences, in which solution processes rather than a mechanical compaction are probably dominant, require much further study. Research around the world into the various components will no doubt lead to fuller understanding. Interesting current problems can be illustrated by the final association mentioned with regressive sequences—that oil in successive reservoirs tends to be 'lighter' (less dense) with depth.

Many geochemists regard this situation as a combination of two factors: *biodegradation* of the crude oil in the reservoirs at shallow depth due to bacterial action (the bacteria converting low-molecular-weight hydrocarbons to higher molecular weight) and *thermal alteration* of the crude oil as it is buried to higher temperatures. But there is ambiguity in the data, because the direction of increasing temperature, pressure, and age is, in a regressive sequence, also the direction of sedimentary rocks of deeper marine environments. Thus, the association may have a stratigraphic

cause or contribution. This problem will be largely resolved if some oil fields with multiple reservoirs in a transgressive sequence can be found and the role of biodegradation properly assessed.

Petroleum geology is an essential part of geology because it involves the acquisition and interpretation of data that are not available in surface outcrop. It contributes not only a mass of data from many parts of the world (including many, such as the continental shelves and deeper plateaux, that would otherwise remain unknown) but also general concepts. For example, we have seen that the evidence of subsurface geology is that anticlines of young regressive sequences are formed by vertical deforming forces that result from mechanical instability in the stratigraphic sequence, not the horizontal forces supposed from surface geology. And the regressive sequences are not the result of orogenic uplift of the sedimentary basin but of orogeny outside the subsiding sedimentary basin. The rising mountains generate sediment that more than fills the space created by subsidence.

Petroleum is a finite resource, and about half the original reserves have probably been found. Increasing scarcity will make petroleum geology more important because it will remain a valuable natural material long after the motor car has another fuel.

RICHARD E. CHAPMAN

References

Chapman, R. E., 1983, *Petroleum Geology.* Amsterdam: Elsevier, 415p.
Hedberg, H. D., 1964, Geologic aspects of origin of petroleum, *Am. Assoc. Petroleum Geologists Bull.* **48,** 1755-1803.
Hunt, J. M., 1979, *Petroleum Geochemistry and Geology.* San Francisco: W. H. Freeman & Co., 617p.
Levorsen, A. I., 1967, *Geology of Petroleum,* 2nd ed. San Francisco: W. H. Freeman & Co., 724p.
Weeks, L. G., ed., 1958, *Habitat of Oil.* Tulsa, Okla.: American Association of Petroleum Geologists, 1384p.

Cross-references: *Marine Exploration Geochemistry; Well Logging.* Vol. XIII: *Economic Geology; Formation Pressures, Abnormal; Geochemistry, Applied; Hydrodynamics, Porous Media; Rock Mechanics.*

PHOTOGEOLOGY

The task of *air photo interpretation* is the identification of features or objects exhibited or reflected by the Earth's surface and faithfully reproduced in the photo image. Some features or objects are so clearly defined that they can be identified directly; the nature of others can be determined only after a careful analysis of their photo expression. Thus, photogeology enables earth scientists to obtain geological information through the study and analysis of *aerial photographs*. A stereomodel is formed by viewing overlapping photos through a *stereoscope*.

In areas where the geological conditions are closely reflected in the geomorphology, it may be possible to recognize geological features quite unequivocally from the direct evidence presented by the photo image. In such areas, formational contacts and structural relationships are easily established, and much routine mapping can be done directly on the aerial photographs. In poorly exposed areas with little relief contrast and an extensive soil mantle or overburden, the topographic surface may have a less direct relationship to geology. Where direct evidence is absent, or has to be supplemented, physiographic analysis may provide clues to structure or even lithology.

Interpretation Elements

Successful photogeological interpretation requires special attention to interrelated elements such as outcrops, landforms, drainage, vegetation, and cultivation. Images on aerial photographs of these elements are expressed in terms of tone, texture, pattern, slope, and shape (Miller, 1961; Ray, 1960). *Tone* refers to each distinguishable shade variation from black to white. *Texture* describes the composite appearance presented by an aggregate of unit features too small to be individually distinct. It is a product of their individual color, size, spacing, arrangement, and shadow effects. *Pattern* refers to a more or less orderly spatial arrangement of particular elements shown on the photo and implies the characteristic repetition of certain general forms or relationships. It applies to the disposition of features such as rock outcrops, streams, and fields. *Slopes* and slope changes are elements to which all landforms can be reduced.

Photointerpretation and Fieldwork

Photogeology is not an alternative to fieldwork; a close integration of field survey and photointerpretation is required to obtain reliable results (Allum, 1966). Photointerpretation (q.v.) and preparation of preliminary maps, which include all information available from previous investigations, should normally precede field mapping. Preliminary photogeological studies can improve plans for *field surveys* and reduce considerably the amount of fieldwork (Ray, 1960). An important advantage of *structural mapping* from aerial photographs is the overall view of *regional tectonics* that can be obtained at an early stage (Ray, 1960).

Lithologic Interpretation

Sedimentary areas often provide more lithological and structural information than regions underlain by igneous and metamorphic rocks. The inherent inhomogeneity of sedimentary terrain causes differential erosion between hard and soft beds. Salient photo characteristics of sedimentary rock types are shown in Fig. 1. As a result resistant beds show a greater relief than less or non-resistant strata, and stratification will be clearly expressed in the topography (Fig. 2). Table 1, which groups photo characteristics for igneous rocks, is also a guide to a uniform and systematic approach to photogeological interpretation. *Intrusive igneous rocks,* which lack stratification, tend to be homogeneous over large areas. They do, however, frequently exhibit pronounced joint and fracture patterns (Fig. 3). *Extrusive igneous rocks* are usually characterized by distinct landforms such as volcanic cones, craters, lava flows, and tuff plains (Fig. 4). *Metamorphic rocks* may retain characteristics of their parent rocks, and they also often exhibit properties related to metamorphic processes—e.g., cleavage, schistosity or foliation, banding, and fracturing (Fig. 5).

Structural Interpretation

In structural interpretation the aim is to mark attitudes, distribution, and continuity and discontinuity of key horizons and to deduce structural relations from this information. Resistant, relief-forming beds consisting of sandstone or quartzite often make good marker beds. Under favorable climatic conditions limestones may be useful as well in this regard. In practice, however, any horizon that can be traced over a large area and that has prominent photo characteristics—e.g., contrasting color and phototone, distinctive vegetation, specific land use, or drainage pattern—can become useful to structural interpretation (Allum, 1966). The often close relations between structure and relief—i.e., slopes and slope changes—indicate that structural interpretation based on aerial photographs depends to a large extent on understanding and analysis of the morphology. Drainage analysis, lineament patterns, and tonal and vegetational arrangements should corroborate the morphological interpretation. Relationships between relief forms and geological structure will be especially clear in regions of relatively young tectonics. Under such conditions even dense tropical vegetation does not obscure landform-structure relations.

Fold structures, often complicated by faulting or thrusting, can be detected from repetition of beds with opposed dips, except in isoclinal folds. If the fold axis is horizontal, both flanks will be approximately parallel. The flanks either curve away from a plunging synclinal axis or curve and

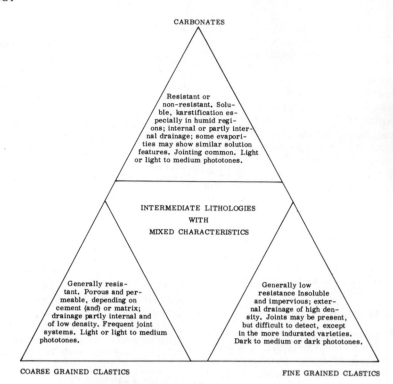

FIGURE 1. The salient photocharacteristics of the principal sedimentary rock types.

TABLE 1. The Main Photocharacteristics of the Principal Igneous Rock Types

	ACIDIC		INTERMEDIATE		BASIC AND ULTRABASIC	
	Felsic minerals dominant			Ferromagnesian minerals abundant		
	GRANITE	GRANODIORITE	SYENITE	DIORITE	GABBRO	PERIDOTITE
INTRUSIVES	72% Silica Quartz Orthoclase	66% Silica NA-CA Plagioclase	59% Silica Alkali-Feldspar	57% Silica NA-CA Plagioclase Hornblende	48% Silica CA-NA Plagioclase Pyroxene	41% Silica Feldspar absent Olivine
are	Main photo-characteristics					
Holocrystalline	Massive and uniform, frequently strongly jointed. Topography in humid regions and at low altitudes subdued or hummocky, in dry regions or at high altitudes sharp and jagged. Drainage joint-controlled or dendritic. Light and light to medium phototones.		Intermediate		Massive and uniform, in many respects similar to granite, resistance probably lower and jointing less intensive. Dark phototones.	
EXTRUSIVES may be:	RHYOLITE	DACITE	TRACHITE	ANDESITE	BASALT	OLIVINE BASALT
Porphyritic Microcrystalline Hypocrystalline or Holohyaline	PUMICE OBSIDIAN		PITCHSTONE		TACHYLITE	
	PYROCLASTS		PYROCLASTS		PYROCLASTS	
	Main Photo-characteristics					
	Strato volcanos and cinder cones, blocky flows predominate, spatter cones, blisters and lava tunnels. Light and light to medium phototones.		Intermediate		Wide-spread fissure eruptions and shield volcanos, ropey flows predominate, frequent columnar jointing. Dark and dark to medium phototones.	

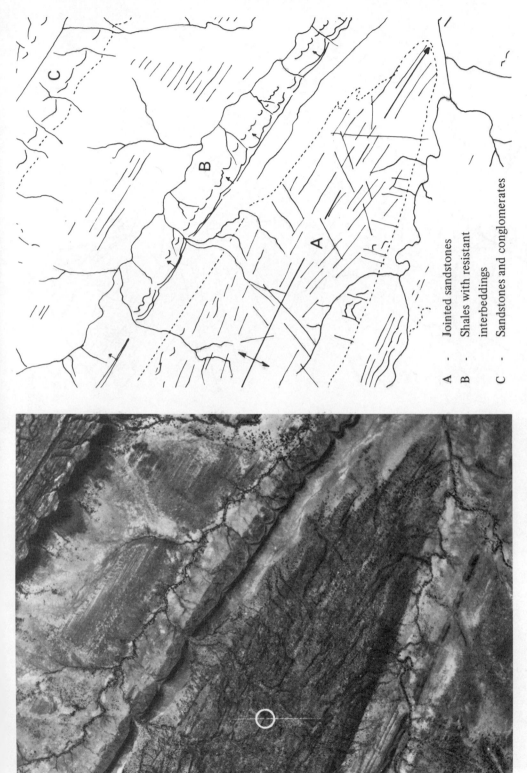

A - Jointed sandstones
B - Shales with resistant interbeddings
C - Sandstones and conglomerates

FIGURE 2. Folded sediments, Amadeus Basin, Australia, 1:49,000.

641

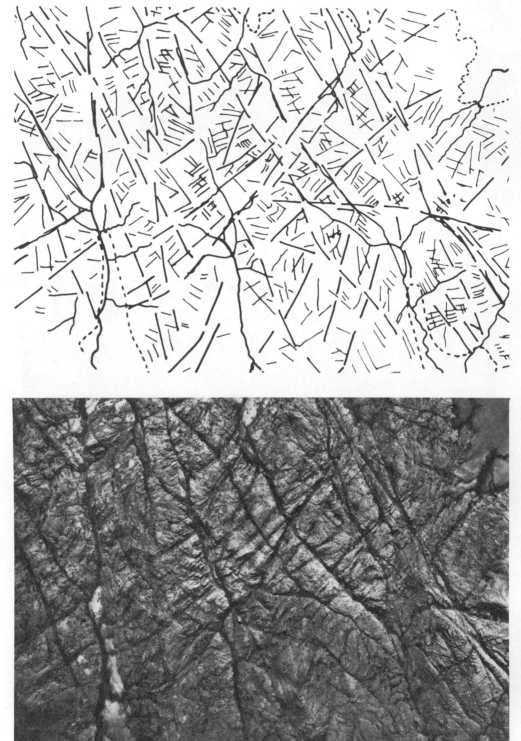

FIGURE 3. Granite, with strongly developed joint and fracture patterns, Wyoming, United States, 1:28,000.

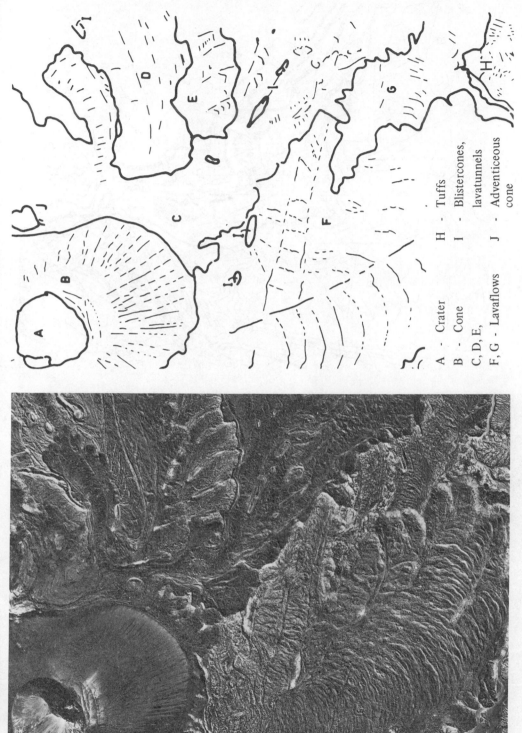

A - Crater   H - Tuffs
B - Cone   I - Blistercones,
C, D, E,   lavatunnels
F, G - Lavaflows   J - Adventiceous
                       cone

FIGURE 4. Paricutin volcano, with lava flows and tuffaceous deposits, Mexico, 1:8,000.

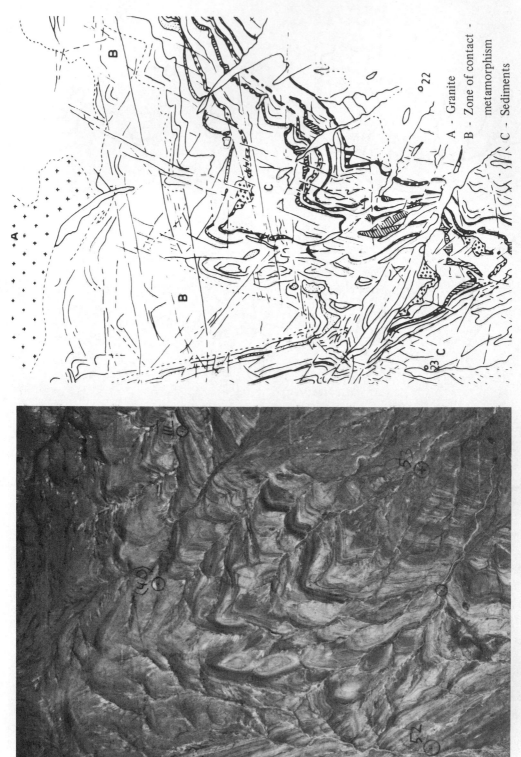

A - Granite
B - Zone of contact -
    metamorphism
C - Sediments

FIGURE 5. Contact metamorphism in sediments, Montana, United States, 1:20,000.

close around a plunging anticlinal axis, forming V- or U-shaped noses. Thus, both flanks of a fold will form linear or curved outcrop belts that are opposed in attitude. If such belts are followed throughout an area and their strikes and dips ascertained, three-dimensional relationships can be determined and a structural framework established (see Fig. 2). In poorly exposed areas drainage patterns or anomalies may give clues to the structure. Streams on fold flanks may, e.g., flow parallel to the strike, following depressions between more resistant beds, while smaller tributaries run down dip and face slopes, giving rise to trellis-type patterns. Beds curving around a fold nose sometimes control the curvature of main streams. Radial and annular drainage patterns or combinations are often associated with fold structures.

Faults, especially *high-angle faults,* frequently appear as straight or curved lines on aerial photographs. Such lineaments often escape notice by the observer on the ground but may stand out clearly on the photograph. Sharp changes of the phototone are frequently related to linear features suggestive of faulting but caused by alignments or abrupt changes in morphology, drainage patterns, or vegetational boundaries. Typical examples include the displacement or termination of resistant key beds, sudden changes in their strike and dip, fault-line scarps, and rectilinear depressions. Fault- or fracture-controlled drainage has straight and parallel or, where two different directions prevail, angular forms. *Low-angle faults* often show only vague surface traces, and continuous lineations may be absent. Low-angle reverse faults and thrusts, which tend to have irregular and strongly curving traces, can be especially difficult to interpret from aerial photographs alone, even in well-exposed areas.

Nonangular unconformities or disconformities, where the beds below and above the unconformable contact are parallel or nearly so, may be very difficult to map from aerial photographs. In contrast, *angular unconformities,* where the beds below the unconformable contact have undergone tilting or folding and erosion prior to the deposition of younger beds, will usually stand out clearly. Especially difficult to recognize are unconformable surfaces that, together with the pre- and postunconformity beds, have been affected by a younger folding phase.

Mineral Exploration

Aerial photographs, when combined with field observation, provide valuable information in the search for environments where accumulations of mineral substances, either fuel or ore, might be expected (Ray, 1960). Petroleum is associated with marine and paralic deposits where the sequence contains potential source and reservoir strata. Reservoir rocks must provide favorable conditions for accumulation. The convergence of beds, facies changes, and unconformities indicates the possible presence of stratigraphically controlled traps (see *Petroleum Geology*). The location of structural traps—i.e., folds and faults—depends on detailed tectonic mapping. The majority of clearly exposed structures was discovered long ago. Attention then focused on physiographic mapping for the detection of subsurface structures hidden below unconsolidated cover. However, much of the current exploration on continental shelves or in deep basins with discordant sediments concealing potential reservoirs admittedly presents less opportunity for direct interpretation of terrestrial photogeological features.

When exploring for *metalliferous deposits,* photogeologists need to consider the nature of different types of deposits because it influences methods of mapping. Careful study of known ore deposits in the general area and searches for ancient mine workings are essential to successful photogeological investigations. Special attention should be given to the melding of geophysical survey data (see *Exploration Geophysics*) with photo interpretation. *Ultramafic intrusives* offer a favorable environment for metalliferous ores, while others are associated with contacts, dikes, and fault or fracture zones, along peripheries of *granitic intrusives.* The search for *stratabound deposits,* such as sedimentary copper, lead, and zinc ores; deposits in volcanic and volcano-sedimentary copper, lead, and zinc ores; and deposits in volcanic and volcano-sedimentary settings demands an unravelling of the stratigraphic relations.

Geologic information derived from aerial photos has made useful contributions to exploration geology. *Placer deposits* (see *Placer Mining*) are associated with both contemporary and paleo river systems, seashores, and playas. Physiographic events play an important role in the accumulation of detrital metals and heavy minerals. Highly significant geomorphological and sedimentological information, which is used to reconstruct the erosional history and conditions of deposition of the transported material, may be obtained from aerial photographs. *Residual ores* result from lateritic weathering in tropical climates and are usually associated with peneplains and plateaus that have passed through a cycle of geomorphic maturity to old age. Clues to epidiagenetic mineralization can often be found on aerial photographs in the form of relict land-surface morphology, lateritic residuals (breakaways), or paleodrainage patterns on replaned surfaces.

645

Recent Developments

Black-and-white photographs are proven exploration tools. They are difficult to surpass in availability and spatial resolution. Contrary to some opinions, there can be little doubt that for quite some time to come they will play an important part in geological mapping. Furthermore, color, infrared, false color, and multiband photographs; improvements in resolution capabilities; application of special enhancement techniques; and increase in scale ranges associated with hyperaltitude photography all present many new possibilities for photographic systems (Fig. 6).

Since about 1960 several new types of sensors have been introduced (see *Photointerpretation*). The entire electromagnetic spectrum is utilized from short wavelengths at which gamma rays are emitted to long wavelengths at which radar operates. Some of these sensors are termed *active,* as in radar, where a signal of known characteristics is transmitted, reflected, and received. Others, such as thermal infrared scanners or aerial cameras, are regarded as passive because electromagnetic energy is only received by the sensors (Swain and Davis, 1978).

Color photographs cover the same range of wavelengths (400–750 nm; nanometer = 10^{-9} m) as normal black-and-white photographs. Because the human eye can detect more separate colors than shades of black and white, such photographs may contain additional information and can be interpreted with greater accuracy.

Black-and-white infrared photographs cover a range of wavelengths between 400 nm and 1,100 nm, permitting them to also record radiant energy beyond the visual spectrum. Black-and-white infrared photographs can be employed in areas covered by haze and are especially suited for recording land/water boundaries. In regions covered by dense tropical forest they are useful when preparing detailed drainage maps. Hydrologists also find infrared techniques useful in soil-moisture investigations.

The spectral range of *false color photography* falls between 500 nm and 1,000 nm. Such film is sensitive to green, red, and infrared bands, which on the photographs are rendered as blue, green, and red respectively. Due to the relationship between type and condition of vegetation and its infrared reflectivity, the greatest use of false color photographs is in vegetation surveys. In geology their advantages are less apparent, unless there is

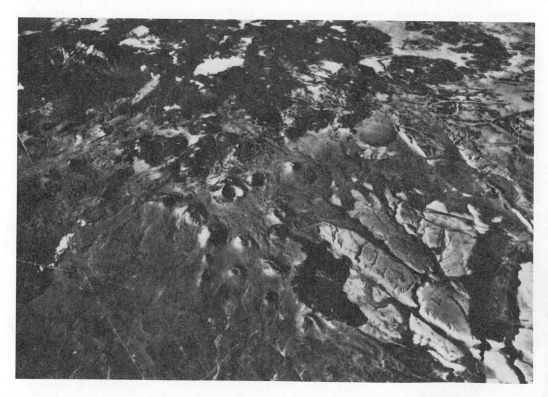

FIGURE 6. High-altitude photograph of San Francisco Volcanic Field. Approximate scale in foreground, 1:240,000.

a direct relationship between vegetation and rock or soil.

Multiband photography offers a possibility for increased definition of spectral signatures for various earth materials. In this technique, a number of photographs are taken simultaneously, each with a different film and filter combination, so that different parts of the spectrum are recorded separately. Precise quantitative information on the spectral signature of a feature in the various bands can be obtained by means of densitometry. Various enhancement techniques are used to assist visual interpretation.

Sidelooking radar (SLAR) has been a routinely operational system since about 1970. Radar, having its own source of energy, is not limited by adverse light or weather conditions and operates even at night. Pulses of electromagnetic energy are transmitted and reflected back as signals from terrain targets. These are recorded on photographic film. Notwithstanding the use of relatively long antennas, the swath width of SLAR systems is limited because of the increasing loss of resolution from near to far range. In *synthetic aperture radar* (SAR) it is possible to obtain a good resolution in all ranges and use a short antenna. SLAR and SAR images contain information on the morphology and roughness of a given terrain and on the relative reflectivity of its con-

stituent materials (Fig. 7). Overlapping images can be viewed stereoscopically. All-weather capability is important in those parts of the world where persistent cloud cover prevents or hampers the taking of aerial photographs. With the help of SLAR, large parts of South America and the Far East, e.g., could be topographically and geologically mapped (Mekel, 1978).

Thermal scanners operate in the 3,000–13,000 nm range. These devices observe the terrain in successive swath lines that are perpendicular to the flight infrared photography. Strong atmospheric absorption limits band use at 3,000–5,000 nm and 8,000–13,000 nm. Heat differences on the order of 0.1°C can be detected. Thermal information is valuable to hydrological studies and related fields, in regions of volcanic activity, and possibly also in the detection of faulting associated with groundwater-controlled temperature effects.

Multispectral scanners (MSS) provide data in a multiband mode similar to those obtained from multiband camera systems. The received energy is optically separated into discrete wavelength bands, and the signals from each band are recorded on magnetic tape. This information can be used to produce separate images for each band or can be computer processed to determine the spectral signature of the various terrain targets. Apart from cameras and other equipment, both

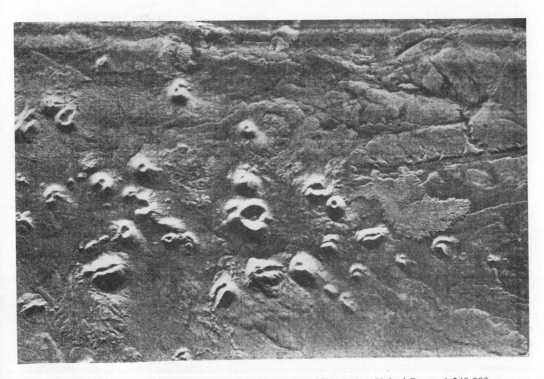

FIGURE 7. Radar image of San Francisco Volcanic Field, Arizona, United States, 1:240,000.

TABLE 2. Photogeological Interpretation Chart

Charac-teristics	Photocharacteristics						Morphologica			
	Tone			Texture			Drainage			
Units	Rock	Vegetation	Cultivation or Other Human Influence	Rock	Vegetation	Cultivation or Other Human Influence	External or Internal	Pattern	Density	Valley c Gully Cross-Sectior
A										
B										

The following descriptive terms may be used.

Tone	Texture	Drainage
Dark	Coarse, fine	*External or Internal*
Dark gray	Smooth, rough	Persistent drainage lines
Medium gray	Even, uneven	Interrupted drainage lines, karst
Light gray	Banded	phenomena
Light	Speckled	No surface drainage
	Granular	
	Linear	*Pattern*
	Blocky	Dendritic
	Matted	Parallel
	Woolly	Trellis
		Radial
		Annular
		Meandering
		Anastomotic
		Contorted
		Anomalcus
		Palimpsest

Density
  Very low
  Low
  Medium
  High
  Very high

Valley or Gully cross-section
  Shallow   U-form
  Deep   Gentle V-form
               Sharp V-form

ERTS (now LANDSAT) and Skylab satellites have been equipped with multiband scanner systems: four bands (green, red, and two in infrared) and thirteen bands respectively (Swain and Davis, 1978); due to a failure, however, the Skylab MSS never functioned. LANDSAT-2 and -3 were equipped with scanner systems similar to those of LANDSAT-1. SEASAT-1 was especially designed to observe ocean currents. Apart from other equipment, it carried SAR and thermal scanners. After functioning for a short period, SEASAT-1 became silent due to a power failure. LANDSAT-D, scheduled for 1980, but not yet launched, has been planned to carry a thematic mapper. Some observers argue, however, that it will be more useful to develop a satellite system based on multispectral linear arrays. Such a system allows for an arrangement of scanners that will ensure a considerable overlap of successive images so that they can be viewed stereoscopically. LANDSAT and Skylab data can contribute to an understanding of megatectonics and regional relationships. In

Expression					Cover			Conclusions	
Rock Properties									
Resis-tance	Bedding	Attitude	Jointing	Bound-aries	Surficial Material	Vegeta-tion	Cultivation or Other Human Influence	Probable Lithology and Structure	Field-Check

Rock Properties

Resistance
 Very low
 Low
 Moderate
 High
 Very high

Bedding
 None
 Very massive
 Massive
 Well bedded
 Very well bedded

Attitude
 Horizontal
 Gentle   5–29°
 Moderate   0–59°
 Steep   60–85°
 Vertical

Jointing
 None   Not persistent
 One direction   Low density
 Several directions Medium density
 Persistent   High density

Boundaries
 Sharp   Persistent
 Vague   Not persistent

Note: Resistance and attitude combined with climate and duration largely determine such configurations that may be described as plane, rolling, hilly, mountainous or table-land, cuesta, hogback ridges.

Cover

Surficial material
 None
 Very thin
 Thin   Fine grained
 Moderate   Coarse grained
 Thick

Vegetation
 None
 Sparse   Scattered
 Moderate   Patchy
 Dense   Aligned
 Very dense

Cultivation
 None
 Rare
 Common
 Frequent
 Intense

Conclusions

Probably:

Sediments
 Fine-grained clastics
 Coarse-grained
  clastics
 Carbonates

Intrusives
 Batholiths
 Laccoliths
 Sills
 Dikes

Extrusives
 Lava flows
 Tuffaceous deposits

Metamorphics

Structure
 Gently folded or
  faulted
 Moderately folded or
  faulted
 Intensively folded or
  faulted

areas where geological maps are of poor quality or lacking, the satellites can be profitably used in the reconnaissance stage, provided such areas are well exposed and favorable atmospheric conditions prevail.

Automatic thematic mapping from multiband information may be possible for the relatively simple conditions found in land use and crop types. It seems less feasible in the more complicated situations geology is apt to present because the morphological expression, so important in visual stereoscopic interpretation, cannot be considered by the computer.

J. F. M. MEKEL

References

Allum, J. A. E., 1966, *Photo-Geology and Regional Mapping.* Oxford: Pergamon Press, 107p.

Bandat, H. F. von, 1962, *Aero-Geology.* Houston: Gulf Publ. Co., 350p.

Bodechtel, J., and H. G. Gierloff-Emden, 1969, *Weltraumbilder der Erde.* Munchen: List Verlag, 176p.

Colwell, R. H., 1960, Photo-interpretation in geology, in B. E. Tator, ed., *Manual of Photographic Interpretation*. Washington, D.C.: American Society of Photogrammetry, 169–342.

Mekel, J. F. M., 1978, The use of aerial photographs and other images in geological mapping, *ITC Textbook of Photo-interpretation,* v. 8(1), 2nd ed., Vol. I 206p., Vol. II 190p.

Miller, V. C., 1961, *Photo-Geology.* New York: McGraw-Hill, 248p.

Parker, R. B., ed., 1973, ERTS issue, *Wyoming Univ. Contr. Geology* 12(2), 110.

Ray, R. G., 1960, Aerial photographs in geologic interpretation and mapping, *U.S. Geol. Survey Prof. Paper 373,* 230p.

Reves, R. G., ed., 1975, *Manual of Remote Sensing, 2* vols. Falls Church, Va.: American Society of Photogrammetry, 2,144p.

Sabins, F. F., 1978, *Remote Sensing, Principles and Interpretation.* San Francisco: W. A. Freeman and Co., 426p.

Swain, P. H., and S. M. Davis, eds., 1978, *Remote Sensing, the Quantitative Approach.* New York: McGraw-Hill, 396p.

Cross-references: *Aerial Surveys; Hard- versus Soft-Rock Geology; Photointerpretation; Remote Sensing, General; Remote Sensing, Societies and Periodicals; Satellite Geodesy and Geodynamics; Terrain Evaluation Systems.* Vol. XIII: *Hydrogeology and Geohydrology; Remote Sensing, Engineering Geology.*

PHOTOGRAPHIC INTERPRETATION— See PHOTOGEOLOGY; PHOTOINTERPRETATION. Vol. XIII: PHOTOGRAMMETRY.

PHOTOINTERPRETATION

Definition

Photointerpretation is the study of ground objects and patterns with the aid of aerial photographs. Air photography was first attempted in France by A. Laussedat in the early 1850s shortly after the invention of the photographic process by Daguerre (1839). In early days, kites and captive balloons were used to obtain pictures from the air, and although photointerpretation predates aviation, its further history is closely associated with the development of aircraft. Photos taken from rockets and orbiting satellites have become available.

Basic Principles

The photointerpreter, while looking at a photograph, passes almost unconsciously through a number of activities in which eye and brain participate. The detection of objects is followed by their identification (recognition). These two steps together can be brought under the heading of "photoreading." Systematic analysis of the photographic image, together with extrapolation and deduction, subsequently is required for a full understanding of the facts observed. When all these activities are combined, they ultimately lead to *photointerpretation* in the true sense of the word.

A thorough knowledge of the types of features and patterns observed is an essential prerequisite for the interpretation of aerial photographs. This technique is therefore most useful in the hands of a specialist. Photointerpretation takes a somewhat similar position in research as, e.g., microscopy. It is certainly not an independent natural science as is sometimes claimed (Lueder, 1959). A knowledge of the region photographed may also substantially add to the reliability of the interpretation. Even then, however, field checking is indispensable.

Types of Air Photos

Most air photos are taken with the axis of the camera pointing vertically downward. These *vertical air photos* are normally flown in a *strip,* or *run,* and each photo overlaps the preceding one by about 50%. This is conditio sine qua non for the stereoscopic study of air photos. The three-dimensional image so obtained is of great importance for proper interpretation. Oblique photos are occasionally made, either high obliques depicting the horizon or low obliques on which the horizon does not appear.

Although black-and-white panchromatic film emulsions are most common, other film types may also be used to advantage. Black-and-white infrared film has found applications, e.g., in hydrological surveys and in coastal mapping. A fixture of this film type is that pictures of good quality can be obtained where panchromatic photography fails due to atmospheric conditions. Three-emulsion-layer true color film and color infrared or false color film are also used. The latter film type has been successfully applied to the detection of diseased crops and trees as well as giving a clue to soil-moisture conditions, pollution, etc.

Narrow-band photography, sensitive only for a certain part of the visible and photographic near-infrared spectrum, is also successfully applied. The sensitivity range is selected in concordance with reflection characteristics of the objects of interest to obtain optimum results. Multispectral photography is based on the same philosophy but has a wider scope. Several narrow-band photographs in various parts of the spectrum are produced for the area. The identification of objects is facilitated by *tonal signatures,* which result from reflection characteristics in different wavelengths.

Equipment

Given the importance of stereoscopy it is logical that stereoscopes of all sorts are essential instruments for the photointerpreter. Lens stereoscopes are often used in the field, whereas mirror stereoscopes find their main application in the laboratory. Binoculars with 2.5–4.5X and, in exceptional cases, 8X or more magnification, are used in conjunction with the mirror stereoscope for detailed study of the photographic image. In modern *zoom stereoscopes,* independent zooming of the left and the right oculars solves the problem of obtaining stereovision using photos with different scales. Stereoscopes for the simultaneous study of a photo pair by student and instructor (double scanning) also exist.

Parallax bars devised for measuring height differences in the stereo model and templates or other devices for measuring terrain slopes and/or geological dips are other important tools. Graphs facilitate the calculations required. Slope measurement can also be achieved by putting a bar or a platen of adjustable inclination in the stereoscopic image and comparing its inclination with the terrain slope.

Gray tone (density) differences can be determined by using a gray scale or, more precisely, by a *densitometer* or a microdensitometer. Patterns also can be analyzed by using graphs, e.g., for texture types, crown density, etc. For more advanced studies, computer analysis or laser analysis of the density patterns is preferred.

For the transfer of information from aerial photographs to base maps, so-called sketch masters with semisilvered mirrors permitting simultaneous viewing of air photo and map and projectors have also been developed. The base map may be assembled using slotted template or slotted arm equipment or by more precise photogrammetric means.

Applicability and Use

The interpretability of aerial photographs is influenced by distortion of the photographic image as seen under the stereoscope. The stereoscopic relief is not identical to the true relief but is usually exaggerated. Another source of distortion stems from the fact that air photos are central and not orthographic projections of the Earth's surface. Displacements, radiating from the central point of the photos, occur in the photographic image due to relief.

The interpretability of aerial photographs is also strongly affected by terrain conditions. A feature can be rendered visible by its contrasting gray tone, photographic density, or relief char-

acteristics that are translated into stereoscopic relief. If a feature is not sufficiently contrasting in density and/or relief with its surroundings, it can be traced only from the air on the basis of associated phenomena, vegetation, and land use that are characteristic for the feature concerned.

Photointerpretation, for obvious reasons, has always been of interest to the military. It received great impetus particularly during World War II. Apart from the recognition of objects of military importance from the air, full emphasis was given to terrain evaluation (q.v.) for *trafficability studies.* Applications in military engineering are numerous. Military interpreters were the first to enter the field of remote sensing in the 1940s.

Earth Sciences Geologists were the first earth scientists to discover the value of photointerpretation for their work (see *Photogeology*). In the late 1920s and early 1930s, aerial photographs became a recognized tool in geology, particularly oil exploration. The largest photogeological survey before World War II was conducted in New Guinea from 1935–1938 by Royal Dutch Shell. The use of air photos in soil science dates from the 1940s (Goosen, 1967); hydrologists, among others, followed suit.

The amount of detail obtainable varies from one area to another and depends on soil conditions, vegetative cover, relief, and lithology. Because intense land utilization tends to obscure natural features, a good understanding of the possibilities and limitations inherent to the method is essential. The characteristics of the photographic material used play an important role in this relation.

It should be noted that the study of landforms, geomorphology, is fundamental to photointerpretation in all the earth sciences. The objects of these studies (rocks, soils, water, vegetation) are often not visible from the air, and interpretations therefore are normally based on associations with specific landforms.

Geological structures, e.g., are rendered visible mostly through their expression in surface relief. Both gray tone (density) differences depicting the horizontal extent of landforms and relief features are used (Miller, 1961). If the surface area covered by a certain rock type or structural element, however, is too small to develop a characteristic landform, it cannot be correctly interpreted regardless of the photoscale employed. The situation is somewhat enhanced in sedimentary terrains where interpretation is still possible on the basis of position within geological structures (see *Photogeology*).

The fact that soils are almost completely invisible on aerial photographs poses special problems in soil science. The soil surface is usually covered by vegetation, and the soil profile cannot be seen from the air at all. Soil scientists thus have to arrive at conclusions through a systematic analysis of the aerial photographs and by deductions on the basis of associated phenomena such as drainage pattern, landform, and vegetation. Because landforms are among the most conspicuous phenomena seen on aerial photographs, geomorphology also plays an important role in pedological photointerpretation.

Surface-water features, in contrast, are directly visible on aerial photographs, and the drainage net, swamps, springs, and zones with impeded drainage can be mapped in detail. Groundwater conditions, not directly visible from the air, are evaluated by studying lithology, slope forms, and vegetative cover. Estimates of permeability on the basis of drainage density are of particular interest. Geomorphological approaches and especially morphometric parameters are frequently used in photohydrological work to determine discharge estimates and sediment yield.

Other Applications A long tradition of photointerpretation exists in forestry. Apart from forest inventory, much stress is placed on forest mensuration. The great importance of measurements of tree height and crown diameter has also given photointerpretation in forestry a strong photogrammetric accent. In agriculture, the accent is on crop survey from the air and on the detection of crop failure due to pests. The use of aerial photographs in the social sciences is in its infancy—in particular, in urban studies. It is remarkable that considerably more progress has been made by archaeologists in the study of ancient settlements (see *Geoarchaeology*).

New Developments in Photointerpretation

Spacecraft Photography Since about 1975, numerous and often spectacular developments were related to pictures taken from ultrahigh altitude using rockets and satellites. Such pictures are sometimes telemetered to Earth—e.g., by Tiros and other weather satellites. These television pictures have introduced photointerpretation techniques to meteorological studies. The pictures taken from Mercury and Gemini spacecraft were recovered after the flight and processed in photographic laboratories. The more recent ERTS and Skylab programs produced repetitive photographic coverage of large parts of the globe and have opened new vistas for sequential photointerpretation of dynamic phenomena. The very small-scale pictures obtained from these altitudes reveal major structures and phenomena of the Earth's crust that cannot be studied adequately by other means.

Remote Sensing Another interesting new development involves expansion of the field of interpretation beyond the visible spectrum. The visible spectrum is only a fraction of the total electromagnetic spectrum. Sensors have been developed to record the electromagnetic energy of several other wavelengths and frequencies. Because no photographic processes are involved, the term *remote sensing* has come into use for many kinds of aerospace techniques. Thermal infrared imagery, radar sensing systems, and multispectral scanning are among the most promising remote-sensing techniques. Airborne geophysics also deserves mention in this context as several of these nonimaging remote sensors are valuable to mineral exploration (Reeves et al., 1975).

Automation Automated handling of data from aerospace has become an invaluable aid to photointerpretation. Most information recorded by satellites is stored on tape because the image, which may subsequently be produced, cannot depict all recorded information simultaneously. Also, the rapid production of deformed images may interfere with visual interpretation. The quantitative results of computer analysis can, however, be conveniently incorporated in geographical data banks.

Conclusion

It should be evident that subtle interactions between the human eyes and brain, in all complexity, cannot be replaced by a machine. Certain well-defined tasks can, however, be approximated by automation. The identification of objects and their classification into groups is particularly promising in this regard. Systematic analysis of images will be more difficult, and the deductions required are now beyond the potential of automation. Gray tone (density) is the easiest to automate of the objects concerned. The texture of objects offers more complex problems. Results have also been obtained with size and shape characteristics. Stereoscopic height and slope, so essential in conventional photointerpretation, are difficult to automate.

H. TH. VERSTAPPEN

References

Avery, T. E., 1967, *Interpretation of Aerial Photographs*. Minneapolis, Minn.: Burgess Publishing Company, 392p.

Goosen, D., 1967, Aerial photointerpretation in soil survey, *FAO Soils Bull. No. 6*. Rome: FAO, 55p.

Lueder, D. R., 1959, *Aerial Photographic Interpretation*. New York: McGraw-Hill, 462p.

Miller, V. C., 1961, *Photogeology*. New York: McGraw-Hill, 248p.

Reeves, R. G., A. Anson, and D. Landen, 1975, *Manual of Remote Sensing*, 2 vols. Falls Church, Va.: American Society of Photogrammetry, 2,144p.

Smith, J. T., Jr., 1968, *Manual of Color Aerial Photography*. Falls Church, Va.: American Society of Photogrammetry, 550p.

Strandberg, C. H., 1967, *Aerial Discovery Manual*. New York: Wiley, 121p.

Verstappen, H. Th., ed., 1970, *Aerospace Observation Techniques.* New York: Pergamon Press, 100p.

Cross-references: *Aerial Surveys; Photogeology; Remote Sensing, General.* Vol. XIII: *Gemorphology, Applied; Remote Sensing, Engineering Geology.*

PLACER MINING

Placer deposits are masses of gravel or other residual or detrital material that contain one or more valuable minerals that have been concentrated by weathering and mechanical processes. The concentrated minerals are, therefore, resistant to weathering and abrasion and are typically of greater specific gravity, 3.3–22, than the most common minerals, which have specific gravities of about 2.6–2.7. The minerals with high specific gravity are colloquially called *heavy minerals.* Examples of metals and minerals recovered from placer deposits are given in Table 1.

Placer gold deposits played an important role in early human history and in the pioneering stages of settlement of new regions because they could readily be mined on a small to moderate scale using simple handmade equipment. Thus, they provided quick assets for the early settlers and thereby paid for much of the early road building in new areas and for the migration of people who subsequently turned to lode mining, agriculture, or other long-term means of livelihood.

Geology of Placers

The geologic requirements for the formation of a placer deposit are a source of mixed minerals, one or more of which is valuable; a means of

TABLE 1. Examples of Minerals and Metals Recovered from Placer Deposits

High Unit Value	Low Unit Value
Gold	Iron ore
Platinum	Titanium minerals
Gems	
Diamonds	Phosphates
Rubies	Glass sand
Sapphires	Feldspar
Zircon	Gravel and sand
Tin ore	
Tungsten ore	
Tantalum ore	
Niobium ore	
Monazite and other rare earth and thorium minerals	
Nongem zircon	
Kyanite	
Beryl	

transportation and concentration; and a moderately small area of deposition. To have the necessary enhanced value, the volume of the source material must be reduced to a much smaller volume of placer material by the removal of a substantial proportion of the minerals that are of little or no value in the original source. Placers have been classified by environment and processes of formation: residual, eluvial, stream, flood-gold, beach, marine desert, fossil, and miscellaneous. These types can be further divided by physical form; thus, stream placers may be subdivided into gulch, valley, terrace or bench, deltaic-fan, and buried-channel placers. The buried-channel placers, in turn, may be subaqueous or dry, deep or shallow; they are called *fossil placers* if they are of early geologic time, such as the Tertiary placers of California.

Residual deposits result from the crumbling and erosion of the source material with little migration of the valuable minerals. Valuable minerals become more concentrated as the enclosing valueless material is removed by moving wind or water. These residual deposits are characterized by a marked decrease downward in values, which may be very high at the surface, a fixture known as *grading* in Guyana.

Eluvial deposits represent concentrations of heavy minerals that have moved downhill from the original source of minerals or from a residual deposit. This movement affords further opportunity for washing out barren silt and clay and for settling the heavy minerals toward the base of the deposit. Hence, the values increase downward, in contrast with residual deposits.

Stream deposits are those that have been formed by the mechanical action of streams in winnowing and concentrating the heavy minerals. The densest minerals, especially those in very small grains, settle downward, coming to rest upon bedrock or a compact layer of clay. Stream placers are by far the most numerous kind of placer. *Flood-gold deposits* are thin stream deposits along the convex sides and upper ends of gravel bars in streams. Although small, they may be renewed by floods, and so they can be mined repeatedly, yielding far more metal than is in them at any one time. This characteristic of renewal makes them similar to some *beach placer deposits,* whose heavy mineral content is renewed by wave action during storms or periods of high water.

Marine placers are of two types: those formed on land and subsequently submerged by a rise in sea level, as with some offshore tin placers near Malaya, and deposits formed on the sea floor by the winnowing action of waves, storms, and deep currents, as with the gravel and titanium mineral deposits on the Atlantic continental shelf of North America.

Desert placers are formed by torrential stream flow and wind action in alluvial fans and basins alongside mountain ranges, reflecting the present environment, as compared with *fossil placers* that may now be in deserts but that resulted from other climates.

Any of the main types of placers may be found in unexpected places as products of earlier climates or processes. Thus, they may be found along ancient beach lines or along stream deposits on terraces, and they may be overlain by younger sedimentary or volcanic rocks, as in the famous Tertiary placer deposits of California.

Evaluation of Placers

The value of a placer deposit is based on the recoverable mineral or metal in a unit volume of material. This value is determined for the material from the surface of the ground to bedrock and includes both the total amount of mineral in the section and the distribution through this vertical distance. The weighted average of the contained values is calculated for the total depth or for specific vertical intervals to give the values of the deposit.

The values for the separate increments of depth are determined by samples obtained from pits, shafts, or drill holes, as dictated by local conditions. All increments of depth from top to bottom of a deposit must be evaluated both for volume of material removed from the sample and for the contained values. No seemingly barren layers can be left unevaluated because they nearly always must be mined to recover underlying material, which increases costs and lowers the average grade of the ore that is processed. All the material removed from each depth interval in a pit or section of a channel should be processed to recover the valuable mineral to avoid errors that commonly result from splitting (Wells, 1973). The *grade* must be determined by physical methods so that only the recoverable mineral in the sample is measured. Assays that report the total metal content, both recoverable and nonrecoverable, usually yield metal values that are too high (Wells, 1973, 91). Truly quantitative determinations of the value of precious metal placer deposits are difficult or impossible to make. The layering of different materials and the wide range in particle size, shown in Fig. 1, indicate why each layer must be evaluated separately and why the taking of useful samples of practical size may be very difficult.

The horizontal extent of a placer is determined by digging test pits or drilling holes in rows (see *Augers, Augering*) that are at right angles to any expected elongation of the deposit. The shape of the minable deposit naturally reflects its origin and depositional environment. Residual placers may be broad and blanketlike, whereas stream and

FIGURE 1. Typical gold placer material showing complex interlayering of beds of gravel with wide range in particle size. Each layer must be sampled and washed separately during evaluation. The discontinuity of beds makes quantitative evaluation very difficult. (W. Lindgren, U.S. Geological Survey, Photo No. 310.)

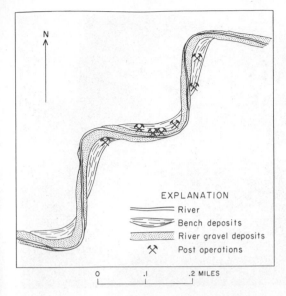

FIGURE 2. A sketch map of a placer mining area showing the position and orientation of rows of drill holes proposed to evaluate the river-gravel deposits and the position of productive terrace deposits on the concave sides of river bends. (After Prommel and Hopkins, N.D.)

beach placers tend to be elongate, but the valuable deposits are never as continuous as the gravel or sand deposit as a whole (Fig. 2).

Exploitation of Placers

The manner of exploitation is determined by local topographic conditions, the value per unit volume of gravel, the unit value of the mineral or metal (see Table 1), the shape and distribution of the valuable gravel bodies, the legal requirements of the country or local government, and the resources of the individual or group interested in the placer deposit. Deposits of exceptionally high value may be exploited by an individual using a pan, batea, or handmade washer. Lower-grade deposits must be larger and become workable only with larger processing capacity, which requires greater financial resources. The mining plants vary from small essentially mobile units, which are selective for the material to be mined, to large-capacity units—usually dredges.

The *dredges* float in a pond; dig, wash, and separate the valuable mineral or minerals from the waste; and simultaneously dispose of the waste by refilling the pit—all in one continuous operation (Fig. 3). Dredges are classified according to their method of digging and by their manner of movement around the deposit. Excavation may be by endless bucket lines, a single bucket on a rigid dipper stick, a shovel, a mobile or fixed dragline,

a cable-suspended clamshell or grab bucket, or suction pumps. Movement on ponds is accomplished by mounting the dredge on a boat or floats. Less commonly, a wheeled or tracked mount permits movement on land. The excavator may be combined with the mineral recovery plant in an integrated assemblage as on the large dredges, or where dredges are not practical, separate units for excavating, mineral recovery, and disposal of tailings may be connected by feed or conveyor mechanisms.

Hydraulicking is a form of placer mining by which the whole deposit is washed by high-pressure water sprays downslope through mineral-saving sluices, then letting the tailings accumulate below as the ground permits (Fig. 4). Other forms of placer mining include carrying the gravel to either ground sluices or built sluices where the desired minerals are recovered. In *ground sluicing* the tailings may be restacked to approach the original configuration of the terrain. Ground sluices are dug into the bedrock and fixed, whereas built sluices are manufactured, usually on the location, and may be semipermanent or portable. These sluices can be moved around the deposit as different areas of the placer are mined, and land-surface reconstruction may be made a simultaneous part of the operation.

The deposits discussed previously are sources of minerals of very high value per unit weight that make up less than 1% of the volume of the gravel. A second type is widely exploited for commodities of low unit value, especially titanium minerals,

FIGURE 3. Floating dredge with bucketline excavator on the left and stacker disposing of washed gravel on the right. The heavy minerals are washed from the gravel in the building, which contains the motors, pumps, sluices, and other machinery needed. This dredge in Idaho recovers euxenite, an ore of niobium and rare earths. Very similar dredges recover gold, platinum, tin and titanium ores, and other valuable heavy minerals. (D. L. Schmidt, U.S. Geological Survey, Photo No. 19.)

FIGURE 4. Small hydraulic mining operation in Alaska. The stream of water carrying sand and gravel is channeled into the sluice, where the gold is trapped by cross-ridges, or riffles. (Photo by R. E. Wallace, U.S. Geological Survey, Photo No. 68.)

most of which are obtained from placers. Deposits of these commodities tend to be larger and more uniform than placers of high unit value commodities and contain a much higher proportion of valuable minerals. Their composition, grade, and size can be determined more accurately by drilling and pitting because the desired mineral or commodity constitutes a rather large part of the total volume of the deposit. Examples include the placer iron deposits mined in New Zealand, the Philippines, and Japan; the glass-sand and feldspar deposit mined in California; and the sand and gravel deposits that yield construction material in many parts of the world. The last two groups, glass-sand and feldspar and sand and gravel, are sufficiently different from other placer deposits in the geologic and economic factors used in their evaluation, the means of exploitation, ultimate use of the product, and problems of land reclamation that they are usually not grouped with placer deposits; accordingly, they are not discussed here. Valuable heavy minerals may be by-products of their exploitation.

The minerals of low unit value are separated from waste by physical methods, as was true of the high-value commodities. Most mining operations for commodities of low unit value are on a larger scale than those for commodities of low unit value.

Marketing

Most products of the placer mining industry are marketed through well-established trade chan-

nels. In some countries the trade in precious metals, radioactive minerals, gemstones, or minerals thought to be especially important in the local economy is controlled by the government.

Reclamation of Mined Land

Return of mined land to a pleasing or useful condition has been of concern to major mining companies for many years (see *Mining Preplanning*). Marked success is indicated by the former ilmenite mines in Australia and the United States, and in the 1930s, gold placers underlying rice fields in Korea were dredged, the alluvium was restacked in its original order, and the ground was out of cultivation for only two years.

The methods used for land reclamation must be adapted to the type of mine and the nature of the terrain. Restoration of the original land surface may be difficult and prohibitively costly after hydraulic mining of steep slopes. Conversely, the surface may be self-healing after mining of beach or flood-gold deposits. Dredging, unlike hydraulic mining, makes only minor changes in the surface configuration but may disturb the original position of the subsurface materials. After mining by any method, the land can usually be left in a stable and useful condition.

Placer mining does not remove any part of the deposit except the mineral or metal of value. Hence, with metals or minerals of high value and high specific gravity, the volume removed is very small as compared to the whole deposit. Only in the mining of bulk minerals of low unit value and rather low specific gravity is there any appreciable shrinkage in the volume returned to the tailings. Chemical or metallurgical reagents are not used to recover the values—only water or, in a limited application, air. As such, this mining method is possibly the cleanest from the ecological view of all mining methods; no foreign material is introduced into the system, although silt and clay may be stirred up and remain suspended in the water. The natural silt load in rivers is commonly similar in composition to that stirred up into the water during placer mining and in large streams substantially exceeds it in amount. Where mining is by dredges floating on ponds that are detached from streams, the mining can add nothing to the streams.

PAUL M. HOPKINS
WALLACE R. GRIFFITHS

References

Boericke, W. F., 1936, *Prospecting and Operating Small Gold Placers,* 2nd. ed. New York: Wiley, 145p.

Lang, A. H., 1970, Placer and small-scale lode mining, in *Prospecting in Canada.* Ottawa, Ont.: Geological Survey of Canada (Econ. Geology Rept. 7), 250–262.

New South Wales Department of Mines, 1970, *Prospectors Guide, New South Wales,* 10th ed. Sydney, 262p.

Prommel, H. W. C., and P. M. Hopkins, n.d., *Preliminary Evaluation and Sampling of Gold Placer Deposits.* Denver, Colo.: Denver Equipment Co., (Bull. No. M4-B12).

Stone, D., and D. Stone, 1970, *Gold Prospecting.* Melbourne: Lansdowne Press Pty. Ltd.

Tolansky, S., 1962, *The History and Use of Diamond.* London: Methuen and Co., 166p.

Wells, J. H., 1973, Placer examination, *U.S. Bur. Land Management Tech. Rept. 4,* 209p.

Wolff, E. 1969, *Handbook for Alaskan Prospectors.* Fairbanks: University of Alaska, 460p.

Cross-references: *Exploration Geology; Hydrochemical Prospecting; Marine Mining; Mining Preplanning.* Vol. XIII: *Coastal Engineering; Coastal Zone Management.*

PLANE TABLE MAPPING

Plane table mapping involves the preparation of a map in the field by direct observation. Details can also be added to existing maps or aerial photographs by this process. In the *classical method,* a single operator uses simple equipment that can be hand-made whereas *stadia mapping* involves a minimum team of two and more expensive equipment. The basis of either method is to secure a sheet of paper or other material to a table and to locate points with rays drawn by means of an alidade. The operation involves two stages: (1) construct an accurate triangulation network (unless using existing maps), and (2) fill in detail as required; this may be topographical, geological, botanical, geomorphological, etc. Both stages may be carried out simultaneously.

History

Classical plane tabling dates from at least as early as the sixteenth century when the method was described by Digges (1571). An illustration dated 1590 (Fig. 1*a*) illustrates a plane table complete with alidade (and clinometer), and another of 1660 (Fig. 1*b*) shows the principles. The telescopic alidade was in use early in the twentieth century, but electronic distance measurement has not been applied to plane tabling (Allan, 1975).

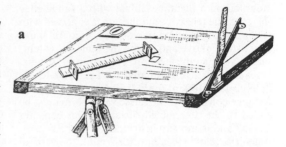

a

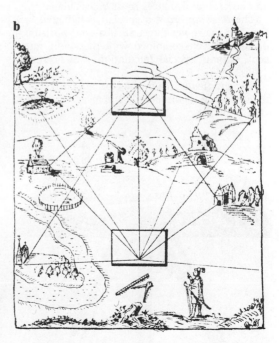

b

FIGURE 1. Historical plane tabling. (*a*) Plane table, 1590; (*b*) plane table map. (Sketch (*b*) after Schott, 1660; both figures from Debenham, 1940; reproduced by kind permission of The Blackie Publishing Group)

Equipment

A *plane table* is a rigid board, often 55 × 40 cm (22 × 15 in) but sometimes larger or smaller, mounted on a tripod by means of a ring clamp, ball-and-socket joint, or otherwise (Fig. 2). The mounting should be sturdy enough to withstand plotting activities.

The *mapping surface* is usually medium to heavy-weight drawing paper, sometimes mounted on a thin metal plate that is screwed to the table. Plastic or metal sheets may be used in moist conditions. Existing maps or aerial photographs may be used as a base for detailed mapping.

An *open-sight alidade* is a rule fitted with sights, usually two hinged sight vanes (see Fig. 2). The rule has a scaled straight edge. The vanes permit sighting to about 15° above or below horizontal.

A *telescopic alidade* is a transit telescope mounted on a metal baseplate with a ruling edge (Fig. 3). The telescope diaphragm is engraved with *stadia lines*. These are two short horizontal lines, equidistant above and below the main cross-hair, such that when the telescope is horizontal, the distance to a vertical graduated staff is a fixed multiple, usually 100, of the staff intercept. The focal distance constant of the telescope ($f + c$) must be added to the determined distance. This constant is small and in modern instruments is usually zero. When the telescope is not horizontal, the vertical and horizontal distances must be calculated.

The *Beaman arc* is a specially graduated vertical arc that gives direct vertical and horizontal distances when combined with the stadia intercept, without special calculations (see Forrester, 1946).

The *gradienter drum* may be attached to the tangent screw of the alidade and measures vertical

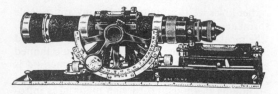

FIGURE 3. Telescopic alidade. (From Forrester, 1946; courtesy John Wiley & Sons, Inc.)

angles in linear units. Special tables are needed to calculate the vertical and horizontal distances (see Mather, 1919).

Stadia rods are folding or telescopic rods, usually 4 m (13 ft) in length, marked by painted scales that indicate feet and tenths of a foot (imperial rods) or meters, decimeters, and centimeters (metric rods) (Fig. 4). Special colored symbols may be used.

A *trough compass* is a long compass needle in a narrow box. A line is drawn on the mapping surface, and the compass sets against it to orient the table approximately at each new setup. Any compass with a reasonable straightedge—e.g., Brunton—can serve the purpose.

The *Indian clinometer* is a bar, fitted with a tall vane sight at the front and a peep sight at the rear (Fig. 5). The front vane carries scales of degrees and natural tangents. The bar is fitted with a leveling bubble and is pivoted at the front end on a baseplate. Leveling is by means of a screw at the rear.

Classical Plane Tabling

This method utilizes a plane table, open-sight alidade, and sometimes a trough compass and Indian clinometer. Rays are drawn to other identifiable points. Short distances are measured by taping, pacing, or estimating. Geological or other detail is sketched in by the tabler around each setup. There are four methods of operation, which may be combined as appropriate.

Radial Method   The radial method is suitable for small areas.

1. Set table up at a central point and mark position on paper.
2. Draw rays to each point whose location is desired.
3. Measure distances and mark points along relative ray to desired scale.
4. Sketch in required detail around table and located points.

Traverse Method   The traverse method is suitable for narrow areas.

1. Position table at first selected position, and mark on paper in correct relative location.
2. Draw a ray to second station.

FIGURE 2. Plane table and open-sight, or traverse table, alidade. (From Forrester, 1946; courtesy John Wiley & Sons, Inc.)

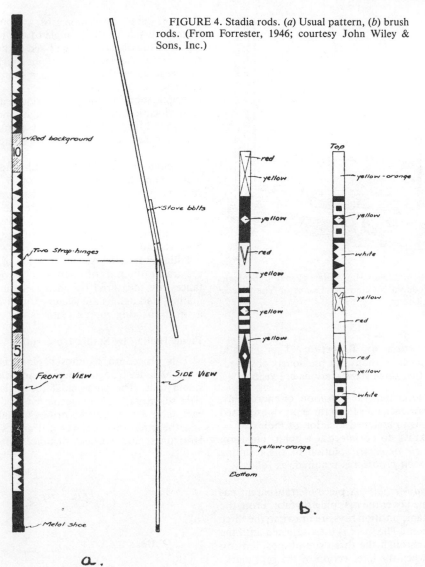

FIGURE 4. Stadia rods. (*a*) Usual pattern, (*b*) brush rods. (From Forrester, 1946; courtesy John Wiley & Sons, Inc.)

3. Measure distance and mark second station on paper.
4. After sketching detail around first station, move table to second station.
5. Orient table by sighting back to first station.
6. Draw ray to third station, measure distance, and sketch in detail around second station.
7. Repeat as necessary.

Triangulation by Intersection This is the most usual method. It involves the use of a baseline and intersecting rays.

1. At the first station, draw rays to three or more selected stations.

2. Measure carefully the distance to the second station. This gives the baseline. Plot to scale on the ray. Sketch detail around the first station.
3. Set table up on second station, and orient by sighting back to first station.
4. Draw rays to all visible stations sighted from first station and to any further stations in sight.
5. After sketching detail, move to one of the stations fixed by intersecting rays.
6. Orient table by sighting back along relative ray, and draw rays to further stations.
7. Continue as required to cover area. If possible, draw three rays to each station.

659

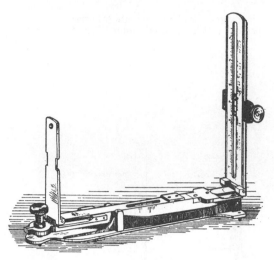

FIGURE 5. Indian clinometer. (From Debenham, 1946; reproduced by kind permission of The Blackie Publishing Group)

Triangulation by Resection This method serves to locate stations not previously observed, provided that three fixed stations are visible.

1. Set table on unmarked position, orient roughly, and draw back rays from three previously fixed positions to assumed location of table.
2. If backrays do not meet at a point, but form a *triangle of error,* solution of the resulting *three-point problem* is required, as follows.

Mechanical Method A piece of transparent paper is pinned over the mapping surface. From the assumed table position rays are drawn to the three fixed stations. The sheet is then adjusted until the rays pass through the three station positions on the map when the intersection of the rays marks the true table position.

Graphical Method Consider the *great triangle* whose apices are the map positions of the three fixed stations and the *great circle* on which those stations lie (Fig. 6).

Case 1. The triangle of error falls within the great triangle. The point sought lies within the triangle of error.
Case 2. The triangle of error lies outside the great triangle but inside the great circle. The ray drawn from the middle fixed point lies between the point sought and the intersection of the other two rays.
Case 3. The triangle of error falls outside the great circle. The point sought is on the same side of the ray from the most distant fixed point as the intersection of the other two rays.

In all cases the distance of the point sought from each side of the triangle of error is proportional to the distance to the fixed point along the ray forming that side. If the point sought lies near the great circle, the problem is insoluble and another station must be sought.

Application of the appropriate rule gives a closer location of the point sought and new rays are drawn. If a triangle of error persists, the process is repeated until a satisfactory intersection is achieved.

Calculation of Heights With Indian clinometer, hand level, Brunton, or other tool, measure the angle of elevation or depression. When the new station is fixed, measure the horizontal distance and calculate the vertical distance from the tangent. Allow for heights of eye and point observed above ground level.

Filling in Detail At each position of table, rays are drawn to points of interest in the vicinity; distances are measured by pacing, taping, or estimating, and details are sketched in. Tangent rays assist in locating curved features.

Plane Tabling by Stadia Observation

This procedure requires a plane table, telescopic alidade, trough compass, and one or more stadia rods. The party chief (geologist or other field scientist) holds or accompanies the rod. At each table setup rays are drawn to rod positions and the intercepts and vertical angles recorded. Horizontal and vertical distances are obtained

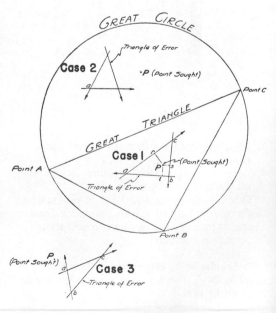

FIGURE 6. The three-point problem. (From Forrester, 1946; courtesy John Wiley & Sons, Inc.)

from stadia tables, a special slide rule, or a pocket calculator, using the following equations:

$$HD = RI \times F \times \cos^2 \alpha + (f + c)\cos \alpha \quad (1)$$

$$VI = RI \times F \times \sin \alpha \cos \alpha + (f + c) \sin \alpha + HI - MW \quad (2)$$

where

HD = Horizontal distance;
VI = Vertical distance (difference in elevation);
RI = Rod intercept (distance on rod between stadia lines);
F = Diaphragm factor, usually 100;
α = Angle of elevation or depression;
$(f + c)$ = Focal constant of telescope (see discussion of telescopic alidades);
HI = Height of instrument (telescope trunnion) above ground level;
MW = Rod intercept of central cross-wire—may be arranged to equal HI.

Methods of Operation Usually a network of first-class stations is established, either by electronic distance measurement, by using an existing map, or by a series of rays and stadia observations from the original station. Intersecting rays are not essential but give a better fix. No baseline is required. Each station is occupied by the plane table, and rays are drawn to the rod position selected by the geologist, who then fills in nearby detail by compass, pacing, or estimation. If necessary, the rod is sited on future table positions. If required, previously unlocated positions can be occupied by the table and located by resection (see "Triangulation by Resection"), solving the three-point problem, if necessary. The traverse method can also be used.

Calculation of Heights Since HI, MW, and α are recorded for each observation, the ground elevations of every rod position can be obtained from Eq. 2. Additional rod positions are occupied to obtain spot heights for contours if required.

Standard of Accuracy

A pencil or scribed mark is accurate to about 0.25 mm (0.01 in), so on a scale of 1:2,500, measurement to nearer than 0.6 m (2 ft) is unnecessary. At smaller or larger scales the required standard varies proportionately.

P. F. F. LANCASTER-JONES

References

Allan, A. L., 1975, New hybrid systems for the ground surveying of details at large scales. *Chartered Surveyor* (LHM Quarterly), **2**(3), 50–54.
Debenham, F., 1940, *Map Making*. London: Blackie & Son, 127–186.
Digges, L. (The Elder), 1571, *Pantometria*. London: Henrie Bynneman.
Forrester, J. D., 1946, *Principles of Field and Mining Geology*. New York: Wiley, 273–314.
Mather, K. F., 1919, The manipulation of the telescopic alidade in geologic mapping, *Denison Univ. Sci. Lab. Bull.* **19**.

Cross-references: *Compass Traverse; Field Notes, Notebooks; Geological Survey and Mapping; Maps, Logic of; Surveying, General.*

PLATE TECTONICS AND CONTINENTAL DRIFT

Plate tectonics is a unifying dynamic model that seeks to explain patterns of deformation and seismicity in the outer shell of the Earth. It is based on extensive geophysical data gathering done in the 1950s and 1960s. Two basic assumptions underlie the theory of plate tectonics: (1) The outermost shell of the Earth, called the *lithosphere*, rests on and is stronger than the immediately underlying layer, a layer of weakness called the *asthenosphere*. (2) The lithosphere is broken into a number of horizontally rigid shell segments, or *plates* (Fig. 1), that are in constant relative motion with respect to one another and the surface area of each of which is also constantly changing. Most energy-intensive tectonic processes take place along the boundaries between the plates.

Although the thickness of the lithosphere cannot be measured with any great degree of accuracy, researchers agree that it varies, in the interiors of plates, from about 70–80 km under the oceans to several hundred km under some parts of continents. The asthenosphere appears to extend below the lithosphere to a depth of about 700 km. It increases in strength downward, and some seismologists put its lower limit higher, at a minor discontinuity about 400 km deep.

Development of Evidence

Wegener (1912) showed how matching outlines of continents and geological features across oceans suggests that continents have moved apart since Permian time. He proposed a model according to which the lighter sialic continents had drifted through a denser simatic substratum.

Physicists were unable to accept the idea of strong continental rafts drifting through weak oceanic crust, as proposed by Wegener. Holmes (1931, 1945), however, pointed out that the continents could suffer mutual displacement without drifting through oceanic crust—e.g., they might ride passively on a convecting lower crust and mantle. He showed how the Mid-Atlantic Ridge might be the origin of continental displacement in

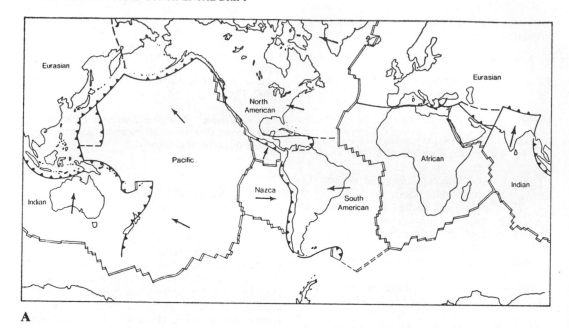

A

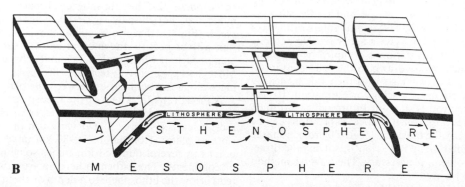

B

FIGURE 1. (*A*) Major plates of the present time. Diverging boundaries (spreading axes) are shown as double lines. Converging boundaries (subduction zones) are marked by toothed lines, with the teeth on the overriding plate. Transform boundaries are shown as single lines. Dashed lines indicate weak or uncertain boundaries. Arrows show direction of absolute plate motion, if the African Plate is assumed to be fixed. (From Petersen at al., 1980) (*B*) Block diagram illustrating schematically the configurations and roles of the lithosphere, asthenosphere, and mesosphere in plate tectonics. Arrows on lithosphere indicate relative movements of adjoining plates. Arrows in asthenosphere represent possible compensating flow in response to downward movement of lithosphere. One arc-arc transform fault appears to left between oppositely facing subduction zones (trenches); note that there is no relative movement between any trench and that plate bordering it that is not being consumed. Therefore, there must be relative movement between the two trenches at left, and this movement is taken up along the transform fault. Two ridge-ridge transform faults appear along ocean ridge at center. (After Isacks et al., 1968)

the Atlantic region, marking the emergence of an upward branch of a mantle convection system.

Later Hess (1962) and Dietz (1961) showed that many oceanic features suggest spreading of the ocean floors away from active oceanic rifts, with the continents riding passively along. Growth of ocean basins in this model could be accommodated either by an expanding Earth or by con-

sumption of oceanic crust away from rifts, presumably along deep-sea trenches. Vine and Matthews (1963) showed that the rate of seafloor spreading could be calculated by dating magnetic anomaly stripes on the ocean floors. Rates so measured turned out to be too high to be compatible with an Earth expanding at the rate of seafloor spreading. Soon afterward, Plafker (1965) showed from earthquake seismology that at the southern margin of Alaska the continent was actively overriding the ocean floor, and Oliver and Isacks (1967) detected lithosphere plunging into asthenosphere under Tonga by using an indirect method of measuring strength. Descent of the lithosphere into the asthenosphere at trenches thus seemed plausible; this process is now called *subduction*. In the meantime Wilson (1965) had recognized a then new type of strike-slip fault that links segments of trenches and/or segments of oceanic rifts, which he called *transform faults*. He recognized that transform faults might be links in a network of boundaries that divides the outermost shell of the Earth into a number of relatively rigid plates. McKenzie and Parker (1967) successfully applied this concept to the Pacific. These pieces of evidence together were sufficiently compelling to permit a synthesis. The resulting hypothesis, now called plate tectonics, was first proposed in its present basic form by Morgan in 1967 (Morgan, 1968), Le Pichon (1968), Isacks et al. (1968), and Heirtzler et al. (1968).

Plate Motion

All lithosphere plates are in continual relative motion. Certain geometric considerations make it possible to describe these motions and to set up something like laws of motion that reflect the natural constraints on plate movements.

We assume that the total area of all plates remains constant over a significant time span. Horizontal distortion within plates is insignificant away from plate margins. We may thus consider them rigid. Since displacement along transform faults is strike slip, it follows that plate motion must be parallel to contemporary transform faults. Because all this takes place on the surface of a sphere, each plate segment describes a path that is equivalent to a rotation on the spherical surface of the Earth. (Fig. 2).

It is possible to determine pivots or poles of rotation for each plate, with respect to motion at any instant. Spreading rates increase away from these poles (up to an "equator" at 90° from poles), as would be expected, but the instantaneous angular velocity referred to poles is the same everywhere for each pair of plates, even for pairs that do not touch. Again, these poles of rotation are geometrically unique to any pair of plates and bear no relation to the poles of rotation of planet Earth.

It is now a comparatively simple matter to determine the rates of spreading. Magnetic anomaly stripes can be identified and dated, and the distances of an anomaly from its spreading axis is then the only additional parameter needed to obtain the average linear spreading rate at any point along a rift. Spreading rates at spreading axes are

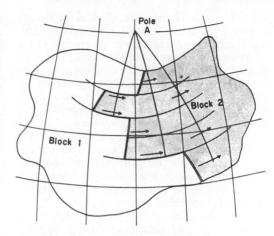

FIGURE 2. On a sphere, the motion of plate 2 relative to plate 1 must be a rotation about some pole. All faults on the boundary between 1 and 2 must be small circles concentric about pole A. (From Morgan, 1968)

usually given as half-rates—that is, the rate at which one lithosphere plate moves away from the boundary. The full spreading rate is the velocity differential between the two plates that diverge. Plate velocities at trenches are full rates, since one lithosphere plate does not move with respect to the trench. All rates are given relative to some plate boundary. In the pattern of motion of plates and plate boundaries, nothing is fixed, all velocities are relative. Spreading rates vary from less than 1 cm/yr in the Arctic Ocean to about 10 cm/yr in the equatorial Pacific Ocean. Where they cannot be measured directly, they can be extrapolated from known rates by finding the pole of rotation of the plate concerned from the trend of active transform faults.

Rates of convergence between plates at trenches and orogenic belts can be computed by vector addition of known plate rotations (see Le Pichon et al., 1976). Rates of slip along transform faults are easily obtained once rates of plate rotation are known.

Plate Boundaries

Plate boundaries may be defined as zones of seismic and tectonic activity (see Vol. XIII: *Seismiological Methods*) where two lithosphere plates adjoin. The boundaries are of three types: (1) diverging, (2) converging, and (3) transform. Along a diverging plate boundary oceanic lithosphere is created and spreads outward from the boundary. Along a converging plate boundary one of the two converging lithosphere plates sinks into the asthenosphere and is consumed by it. Along a transform boundary two lithosphere plates slip by each other horizontally, so lithosphere is neither created nor destroyed.

Plate boundaries are manifested, at the Earth's surface, by traces of great faults or fault systems and alignments of earthquake epicenters. Characteristic fault systems are rifts along diverging boundaries; thrusts along most converging plate boundaries, with their actual traces commonly obscured; and strike-slip faults along transform boundaries, relatively easy to trace in most oceanic crust but complex in continental crust.

The location of plate boundaries is not fixed, in either space or time. They are in constant relative motion, just as the plates are. Only accidentally would a movement vector with respect to a fixed frame be zero at any given time. Present boundaries are shown by type in Fig. 1.

In the ideal plate tectonic model, plates do not deform in any horizontal direction, except along their boundaries. Some of the most important geological processes are direct or indirect consequences of the evolution of plate boundaries.

Present plate boundaries have been traced over the entire Earth's surface by plotting characteristic seismic and tectonic activity. Some present boundaries between major plates are actually zones of mosaics of much smaller *microplates*. Each of these microplates is in relative motion with respect to all the others, including the major plates, making detailed analysis of relative plate motion there a very complicated matter. Examples include the Mediterranean region between the African and Eurasian plates and the Scotia Sea. These belts of microplates constitute a region of adjustment between major plates where continen-

tal lithosphere interferes with an orderly alignment of the standard plate boundary types. Similarly, plate boundaries within continents tend to be more diffuse than those in oceanic regions, as indicated by diffuse spreading out of seismic zones. Examples include the western margin of North America and eastern Asia.

Convergent Plate Boundaries Along convergent plate boundaries the plates on either side have appreciable components of motion toward one another. Such convergence could be accommodated by two possible mechanisms: (1) one or both plates might thicken and deform internally to take up the converging motion, or (2) one plate could escape downward into the asthenosphere, letting the other ride over it. The second mechanism, known as *subduction,* operates along all active convergent plate boundaries, but where continents are involved, the first mechanism may contribute to some extent.

Kinematically, subduction seems a fairly simple process. The geological consequences, however, are far-reaching and complex (Fig. 3), and not all of them are fully understood. In normal plate convergence the surface trace of the plate boundary is deeply depressed topographically, forming the well-known deep-sea trenches, some as deep as 10 km. Some trenches are filled with young sediments, which tend to obscure the structural depression. Where the downgoing slab enters the asthenosphere, magma, usually andesitic in composition, accumulates over it and forms a reservoir that feeds igneous intrusion and extrusion

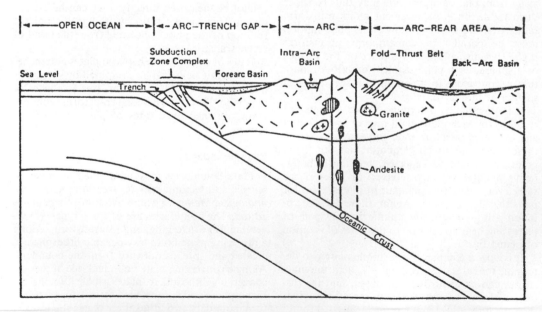

FIGURE 3. Diagrammatic cross-section of an arc system, showing major components and sedimentary basins. (From Condie, 1982)

immediately above. The mechanism of magma generation in this site is still controversial. The results are conspicuous, however: they include volcanic island arcs in oceanic crust and volcanoes, large batholiths, and mountain ranges in continental crust, usually a few hundred kilometers landward of the associated trenches.

Convergent plate boundaries, like all active plate boundaries, are seismically active. Earthquake foci cluster in a shallow zone near the trench and along the downgoing slab (Fig. 4). The zone of earthquake foci in the slab, the *Wadati–Benioff Zone,* extends as deep as 700 km in some places but has never been found deeper.

It seems that most convergent plate boundaries are initiated close to continental margins in previously unbroken oceanic lithosphere. This situation is suggested by the distribution of most present-day active island arcs close to continental margins. Some, which now are separated from the nearest continent by small ocean basins, seem to have been formed closer to the continental margin than they are now: the floor of the basin between

them and the continent, known as a *marginal,* or *back-arc, basin,* is much younger than the open ocean floor on the far side of the trench, indicating comparatively recent opening of the back-arc basin that now separates the arc from the continent. Thus, back-arc basins form by a kind of ocean floor spreading (plate growth) that entails oceanward migration of an arc-trench system.

Collision If a subducting plate carries a continent, or even a small portion of continental crust, such continental crust cannot be subducted when it reaches the trench, because its low density does not permit it to sink into asthenosphere. Hence, further convergence is impeded. If the overriding plate margin is oceanic, the sense of subduction may now be reversed (it may "flip"), and the previously overriding plate begins to be subducted. The abandoned slab becomes absorbed in the asthenosphere. If, however, the overriding margin is also continental (or carries only vestiges of oceanic crust), the arrival of continental lithosphere, however small, at the subducting margin causes *collision* when the two seg-

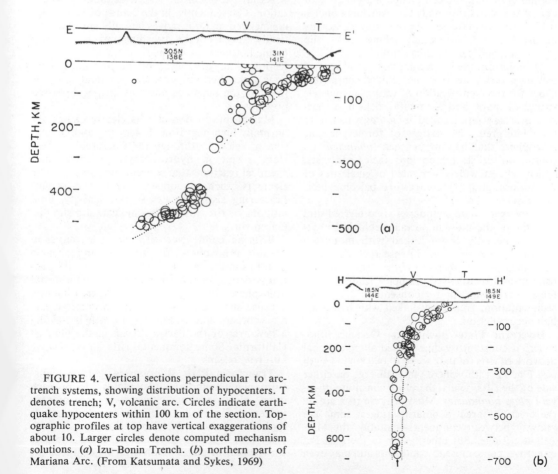

FIGURE 4. Vertical sections perpendicular to arc-trench systems, showing distribution of hypocenters. T denotes trench; V, volcanic arc. Circles indicate earthquake hypocenters within 100 km of the section. Topographic profiles at top have vertical exaggerations of about 10. Larger circles denote computed mechanism solutions. (*a*) Izu-Bonin Trench. (*b*) northern part of Mariana Arc. (From Katsumata and Sykes, 1969)

ments of continental lithosphere meet because low-density continental lithosphere cannot escape downward. Where only a comparatively small continental fragment collides, such as a microcontinent or an arc, subduction may be relayed oceanward, behind the newly arrived piece of continental lithosphere: The continental fragment has docked, and subduction continues behind it. No overall change in plate motions results. If both colliding continents are sizable, however, subduction will eventually cease. So the convergence vector across the boundary becomes zero, and the motion must be taken up elsewhere to preserve the zero sum of all plate motion vectors on the Earth's surface. Rearrangement of the global spreading and subduction pattern results.

Nevertheless, plates can continue to converge for a limited time after continents first collide. The Indian and Eurasian plates have converged for several hundred kilometers since the initial collision of the Indian and Asian continents. Two possible explanations are that the subducting plate may carry the continent along the lithosphere-asthenosphere boundary, underplating the overriding plate, or that the continental margin zone of one of the plates (normally the overriding one) is sufficiently yielding to take up convergence by distortion, both horizontal (along strike slip faults) and vertical (by thickening).

On collision, the intervening ocean will vanish, leaving a *suture* between the collided continents. Clues for the identification of sutures include (1) strings of mafic and ultramafic rocks in a characteristic association called *ophiolites,* which have been interpreted as vestiges of former oceanic lithosphere that became beached (*obducted*) on continental crust; (2) obvious facies contrasts, which suggest widely separated original sites of deposition; and (3) contrasting paleomagnetic histories (see above).

Convergent plate boundaries are preferred sites of some of the most important geological processes, especially those linked with mountain building (*orogeny*): e.g., intense rock deformation; magmatism, metamorphism, and ore deposition; and relatively rapid uplift leading to intense erosion and consequent copious sedimentation, frequently linked with hydrocarbon accumulation.

Divergent Plate Boundaries Oceanic lithosphere is constantly being created along a global network of rifts totaling some 70,000 km in length (see Fig. 1). Lithosphere plates diverge on either side of the rifts, which therefore constitute *divergent plate boundaries*. Most of these rifts are on the crests of great submarine ridges; some are midway between continents—notably, the Mid-Atlantic Ridge. But others, such as the East Pacific Rise, come close to continents and may even

enter them. Therefore, the commonly used name *mid-ocean ridge* is misleading.

Ridge elevation above adjacent ocean bottoms (abyssal plains) is normally between 1,000 m and 3,000 m; a representative width is of the order of 1,000 km. The associated rift is in a valley whose central axis may be as much as 2,000 m below the crest of the ridge. The rifts are over 10 km wide, as much as 30–32 km in the central Atlantic. The fast spreading East Pacific Rise carries no morphological rift. In a few places, divergent plate boundaries are more diffuse; they may rise above sea level, as in Iceland, the Azores, and the Afar Triangle.

No divergent plate boundary has a smooth, continuous trace; all are offset by transform faults (see Fig. 1). Occasionally a portion of a spreading axis will "jump" to a new location, abandoning its old trace. Once initiated, spreading along a divergent plate boundary may change the rate or direction of spreading, usually in response to global changes in plate motions. As far as is known, spreading along any rift can be stopped only by subduction of that rift.

As lava cools in the rift, it acquires magnetization. Consequently, in the course of spreading, magnetized seafloor moves laterally and symmetrically away from the spreading axis. Each newly formed element of oceanic crust acquires the direction of magnetization of the prevailing magnetic field. Periodic geomagnetic field referrals result in the magnetic anomaly stripes discussed previously.

Magma production at rifts clearly causes abnormally high heat flow. Fracturing and circulation of seawater through the fractured hot seafloor result in hydrothermal processes and chemical reactions that concentrate some metallic elements, such as copper, nickel, and cobalt. Fracturing also generates seismic activity. First motions of rift earthquakes indicate dip-slip displacement.

Rifts on Land New rifts may start either in oceanic lithosphere, dividing existing oceanic plates as in the case of the Cocos Ridge off Central America, or more commonly, in continental lithosphere, dividing continents and creating new oceans. The Red Sea Rift has just begun to create a new ocean, and the East Pacific Rise is creating a new arm of the Pacific Ocean in the Gulf of California. Some continental rifts do not evolve into true oceans.

Transform Plate Boundaries Along transform plate boundaries two plates slip by each other, ideally without either divergence or convergence. Since plate motions are about pivots (poles), as illustrated in Fig. 2, transform boundaries should follow small circles about the pole for the pair of plates involved. This is indeed the

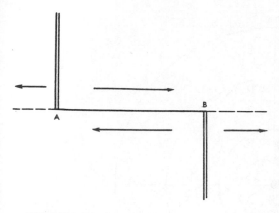

FIGURE 5. Displacement along a ridge-ridge transform fault. (From Dennis, 1967)

case for transform boundaries in oceanic lithosphere, where such boundaries are relatively sharp, narrow fault zones, the transform faults. Displacement along transform faults is as shown in Fig. 5.

All oceanic rifts are segmented by relatively short transform faults. Since plates cannot distort horizontally, there is no other way for divergent plate boundaries to curve and still maintain a constant spreading direction. The whole system of rifts and transform faults forms in essence one spreading system: In some segments, divergence predominates, as in most of the Atlantic Ocean; in some, transform motion predominates, as in the Gulf of California.

In continental lithosphere transform boundaries are not so clean and uncomplicated as in oceanic crust. They do tend to follow small circles about the poles of two adjoining plates, more or less. But displacement does not seem to propagate to the top of the crust in an orderly manner. Above transform boundaries in continental crust, transform motion is taken up by strike-slip faults that are not entirely in the theoretical location of small circle transform boundaries. Changes in fault strike result in converging or diverging components of motion, and this in turn results in extension and contraction of crustal rocks.

Former Plate Boundaries Since present-day plate boundaries have been traced by their present seismic and tectonic activity, it is difficult to reconstruct plate boundaries of the past: Seismic records of the geological past do not exist. Vestiges of former plate boundaries must be found by geological methods, and this involves recognition of geological features that characteristically and unambiguously mark plate boundaries.

Oceanic lithosphere is destroyed after a comparatively short life cycle; hence, pre-Jurassic plate boundaries can be detected only on conti-

nents, greatly displaced with respect to one another since their time of last activity. Spreading rifts, even those originating within continents, eventually become purely oceanic features and are thus ephemeral in long-term geological history. Only aborted rifts within continents may be preserved in the geological record, and their recognition is a matter of interpretation. Some major ancient transcurrent faults may mark former transform boundaries, but their identification as such is uncertain and controversial. Orogenic belts have marked converging plate boundaries since at least early Proterozoic time, but their association with plate boundaries in early Archean time is also controversial. Evidence of major andesitic volcanic activity may be a fairly good indication of contemporaneous plate convergence. Zonal alignment of ophiolites is considered a good indicator of past plate convergence. A promising tool in reconstructing past positions and motions of continents, and hence, by inference and within wide limits, in the location of past plate boundaries, is the plotting of *apparent polar wander paths*.

Paleomagnetic data indicate that there have been shifts in the position of the Earth's magnetic poles relative to sites of measurement over geologic time. These shifts could be explained in two ways: (1) the poles have moved, or wandered, over geologic time; and/or (2) the continents have moved with respect to the poles. Runcorn (1956) showed that poles computed for North America and for Europe had each described a different path (see Fig. 2). This meant that in fact there had been relative displacement between the two continents since Triassic time. The paths plotted in Fig. 6 are apparent polar wander paths; polar wander is apparent, but continental displacement is real, since the discrepancy in pole positions for the two continents is systematic, not random.

This is not a perfect method, but it does show which segments of continental lithosphere have had independent histories, and what changes in latitude and orientation they have undergone. Another clue is provided by comparing stratigraphic sequences. Some former continental and quasicontinental fragments, now docked along present continental margins, have been identified as exotic by these means. The west coast of North America is lined with such *accreted terranes,* which had migrated varying distances until colliding with, and being incorporated into, the western margin of the North American plate (see review in Nur, 1983).

Mechanism

It has been suggested that plate boundaries are surface expressions of convection in the asthen-

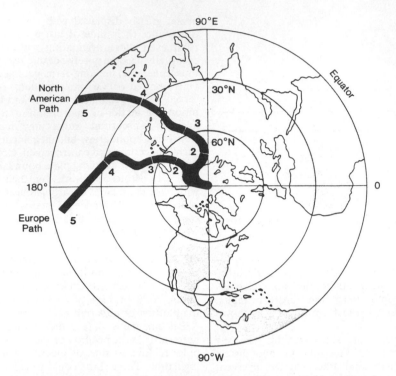

FIGURE 6. Apparent polar wander paths for North America and Europe. Bold numbers show radiometric ages of rocks used, in hundreds of millions of years. If the Atlantic Ocean had not opened about 200 my ago, the two paths should have nearly coincided. (From Peterson et al., 1980)

osphere. While this may be true to a limited extent, the picture is not that simple. The mass transfer budget does not balance within the realm of individual plates. However, there is certainly a relationship between plate motions and convective motion in the asthenosphere, and plate boundaries must mark some kind of discontinuity in asthenosphere currents. It may also be that the gravitational energy of the plates contributes to plate motion.

JOHN G. DENNIS

References

Condie, K. C., 1982, *Plate Tectonics and Crustal Evolution.* New York: Pergamon Press.

Dennis, J. G., 1967, International tectonic dictionary, English terminology, *Am. Assoc. Petroleum Geologists Mem. 7,* 196p.

Dickinson, W. R., and D. R. Seely, 1979, Stratigraphy and structure of forearc regions, *Am. Assoc. Petroleum Geologists Bull.* **63,** 2–31.

Dietz, R., 1961, Continent and ocean basin evolution by spreading of the sea floor, *Nature* **190,** 854–857.

Heirtzler, J. R., G. O. Dickson, E. M. Herron, W. C. Pitman, III, and X. Le Pichon, 1968, Marine magnetic anomalies, geomagnetic field reversals, and motions of ocean floor and continents, *Jour. Geophys. Research* **73,** 2119–2136.

Hess, H. H., 1962, History of the ocean basins, in *Petrologic Studies,* Buddington Vol. Boulder, Colo: Geological Society of America, 599–620.

Holmes, A., 1931, Radioactivity and earth movements, *Geol. Soc. Glasgow Trans.* **18,** 559–606.

Holmes, A., 1945, *Principles of Physical Geology.* New York: Ronald, 532p.

Isacks, B., J. Oliver, and L. R. Sykes, 1968, Seismology and the new global tectonics, *Jour. Geophys. Research* **73,** 5855–5899.

Katsumata, M. and L. R. Sykes, 1969, Seismicity and tectonics of the western Pacific: Isu-Mariana-Caroline and Ryukyu-Taiwan, *Jour. Geophys. Research* **74,** 5923–5948.

Le Pichon, X., 1968, Sea floor spreading and continental drift, *Jour. Geophys. Research* **73,** 3661–3697.

Le Pichon, X., J. Francheteau, and J. Bonin, 1976, *Plate Tectonics.* Amsterdam: Elsevier.

McKenzie, D. P., and R. L. Parker, 1967, The North Pacific: An example of tectonics on a sphere, *Nature* **216,** 1276–1280.

Morgan, W. J., 1968, Rises, trenches, great faults, and crustal blocks, *Jour. Geophys. Research* **73,** 1959–1982.

Nur, A., 1983, Accreted terranes, *Rev. Geophys. Space Phys.* **21,** 1779–1785.

Oliver, J. and B. Isacks, 1967, Deep earthquake zones, anomalous structures in the upper mantle, and the lithosphere, *Jour. Geophys. Research* **72,** 4259–4275.

Petersen, M. S., J. K. Rigby, and L. F. Hintzse, 1980,

Historical Geology of North America. Dubuque, Iowa: Wm. C. Brown, 232p.

Plafker, G., 1965, Tectonic deformation associated with the 1964 Alaska earthquake, *Science* **148**, 1675-1687.

Runcorn, S. K., 1956, Palaeomagnetic comparisons between Europe and North America, *Geol. Assoc. Canada Proc.* **8**, 77-85.

Vine, F. J., 1966, Spreading of the ocean floor: New evidence, *Science* **154**, 1405-1415.

Vine F. J. and D. H. Matthews, 1963, Magnetic anomalies over oceanic ridges, *Nature* **199**, 947-949.

Wegener, A., 1912, Die Entstehung der Kontinente, *Geol. Rundschau* **3**, 276-292.

Wilson, J. T., 1965, A new class of faults and their bearing on continental drift, *Nature* **207**, 343-347.

Cross-references: *Exploration Geophysics; Geomagnetism; Plate Tectonics, Mineral Exploration.*

PLATE TECTONICS AND MINERAL EXPLORATION

The advent of *plate tectonic theory* has provided exploration geologists with a new tectonic framework to study the temporal and spatial distributions of ore deposits. Studies of ore deposits within the earlier "geosynclinal" concept, initiated largely by Russian workers (e.g., Bilibin, 1955; Smirnov, 1968), were concentrated on the sequence of ore formation during the life of the geosyncline, and relatively little attention had been given to spatial relationships, apart from empirical definition of metal provinces. The inclusion of metallic ore deposits within the time-space models set up in plate tectonic terms for various tectonic regimes has thus considerably increased the scope of the search for new metal deposits. It is doubtful, however, that plate tectonic models can be applied to Archaean and perhaps Proterozoic rocks despite the fact that in most cases the relationships between plate tectonic and mineralization processes are obscure. These changes in approach have called for more precise geological data to identify features such as ancient oceanic crust, volcanic arcs, metamorphic facies, sedimentary environments, and other key factors that assist in the reconstruction of past tectonic regimes (e.g., Bird and Dewey, 1970; Mitchell and McKerrow, 1975).

Time-Space Relationships in Orogenic Belts

Early papers on plate tectonics and mineralization (e.g., Guild, 1972; Mitchell and Garson, 1972; Sawkins, 1972) highlighted the systematic relationships between the distribution patterns of deposits such as the porphyry coppers and tin-wolfram ores and those of *ocean trenches* and inferred *subduction zones.*

Porphyry-Copper Deposits These deposits occur in high-level granodiorite/diorite plutons commonly associated with volcanic rocks. They may be divided into gold-bearing and molybdenite-bearing types, each with characteristic hydrothermal alteration facies (Hollister, 1975). They are concentrated around the Pacific margin and appear to overlie subduction zones at distances up to several hundred kilometers from the trench (e.g., Sillitoe, 1972a). In the Andes, the porphyry-copper ores represent only part of a suite of mineralization typically occurring in, and derived from, the plutons and associated volcanic rocks, including copper-bearing breccia pipes, polymetallic veins in volcanics, and copper-bearing skarns in host rocks such as limestones (Sillitoe, 1973). The Andean deposits lie in discontinuous belts roughly parallel to the western coast of South America, and Sillitoe (1974a) has recognized characteristic longitudinal zones that he suggests may reflect the nature and behavior of segments within the underlying subduction zone. In several areas of the Andes the plutons young eastward, supposedly reflecting the gradual descent beneath the Andes of subducted ocean crust.

These relationships are clearly powerful tools for the exploration geologist, though there are important anomalies and complications that are not well understood. One of these is the small number of porphyry-copper deposits discovered so far in areas such as New Zealand and Japan, despite intensive exploration. Another is the great transverse width of the Mesozoic–Tertiary porphyry-copper province in southwest North America and northern Mexico (Guild, 1977).

Difficulties also arise in extrapolating the relationships to older orogenic belts. Sawkins (1972) pointed out that erosion of the Andes down to present sea level would eliminate the entire suite of porphyry-copper, vein, skarn, and breccia-pipe mineralization associated with Andean magmatism. Similar remarks must apply to many of the islands of the southwest Pacific where erosion of porphyry-copper deposits only 3-4 m.y. old is already well advanced. It is thus not surprising that porphyry coppers are relatively rare in pre-Mesozoic terrains, and deposits that have survived, like those of Devonian age in the Gaspe Peninsula (northern Appalachians), must require special conditions for their preservation. Plutons carrying low-grade disseminated pyrite-chalcopyrite mineralization, such as the Devonian examples in the Tasman Orogenic Belt of eastern Queensland (Australia), probably represent the eroded remnants of porphyry-copper systems.

Tin-Wolfram Deposits Another group of deposits that apparently show systematic relationships to tectonic features in orogenic belts are tin or tin-wolfram deposits (e.g., Mitchell and Garson, 1972). They occur in a variety of forms, including disseminations in altered, high-level, felsic plutons; in veins within and around the plutons; and as massive replacement bodies in country rock. A circum-Pacific tin belt can be outlined, though major discontinuities exist—e.g., in southern Chile. Along most of this belt it is clear that the tin deposit and related granitoids occur on the foreland side of the continental margin at considerable distance from the trench and commonly in shelf- or near-shelf-rock types where the sialic crust appears to be relatively thick—e.g., the Bolivian and north Queensland tin provinces. Because plutonic activity in a continental margin appears to young away from the trench, tin-wolfram deposits form late in the orogenic cycle, as noted by Bilibin (1955).

Sillitoe et al. (1975) have drawn attention to the similarities between the modes of occurrence of Bolivian tin deposits and porphyry coppers, particularly noting the relatively shallow depth of deposition (1–5 km). Hence, it is not surprising that the frequency of tin deposits declines rapidly with increasing age (Pereira and Dixon, 1965). However, the geology of tin deposits in other regions indicates a thicker overburden than most porphyry coppers, and confirmation of this situation is indicated by the existence of several Proterozoic tin deposits. Their ability to survive erosion is also shown by the presence of deposits of widely differing age within complex orogenic belts such as the North American Cordillera (Sainsbury et al., 1969) and the Tasman Orogenic Zones of eastern Australia (Solomon and Griffiths, 1974).

In some areas the tin provinces have been disrupted by subsequent plate movement and fracturing—e.g., the apparent separation of Malaysia and north Queensland, possibly by Tethyan shearing (Carey, 1975), and the separation of the Chinese and Japanese tin provinces by the formation of a marginal sea (the Sea of Japan).

Deposits Other Than Porphyry-Copper and Tin-Wolfram The Peruvian Andes display pronounced metal zonation across the orogen (see references of Sillitoe, 1972b). From the Pacific coast eastward there is a zonation from iron oxide (mainly skarn-type) deposits to porphyry coppers with gold and molybdenum to lead-zinc, silver-copper veins to silver, lead, copper and tin deposits that link up longitudinally with the Bolivian tin-wolfram province. These zones are not continuous along the length of the Andes but are clearly of great importance as indicators of zonation in other orogenic belts. Part of this pattern (without the tin province) can be seen in British Columbia (Brown, 1969) and other areas, giving a generalized trench-to-foreland zonation in the Americas of iron, copper (with gold), silver-lead-zinc, and tin-molybdenum (Sillitoe, 1972b).

This pattern has proved difficult to recognize in older terrains partly because of erosion and partly due to complexities introduced by deformation. Of particular importance is the thrusting of slices of oceanic crust into the rocks of the continental margin because the basalts of such material commonly contain massive sulfides of pyrite-chalcopyrite (-sphalerite) type (Sillitoe, 1972c), and the ultramafic units may contain chromite, osmiridium, etc. The Ordovician–Devonian cycle of the Tasman Orogenic Zone in eastern Australia is notable for its lack of porphyry coppers (though uneconomic root zones probably exist), the widespread occurrence of various types of massive sulfide deposits in volcanic and sedimentary rocks, and the presence of numerous gold and copper veins, particularly in thick flysch-type sediments. Slivers of ultramafic complexes contain uneconomic massive sulfide deposits, podiform chromites, and nickel sulfides that probably developed during serpentinization. The tin and tin-wolfram deposits near the old foreland provide the only close link with the Andean pattern. The massive sulfide ores are thought to be the product of shallow circulation of heated seawater (Solomon, 1976), and the gold deposits probably formed during cleavage deformation of graywacke-mudstone assemblages (V. Wall, Monash University, pers. comm.). As these factors are of minor importance in the near-surface Andes, it is predictable that the similarities in metal zoning are minimal.

Local variations due to the presence of massive sulfide, gold, and other deposits appear to complicate metal distribution in the northern Appalachian orogenic belt (Strong, 1973), though the distribution of porphyry coppers and tin deposits follows the Andean pattern. Increasing erosion of these Paleozoic belts would presumably tend to remove most of the tin-wolfram and remaining porphyry-copper mineralization, leaving the massive sulfide (and oxide) ores, gold veins, deposits in ultramafic rocks, etc. as the dominant styles of mineralization.

Another victim of erosion, and probably one of the earliest to be removed, is mercury. Small deposits are common throughout the Pacific coast, particularly in active Cenozoic volcanic terrains, but are rare in older rocks. Further complexity in zonation is introduced by galena-sphalerite (-barite-fluorite) deposits in shelf sediments near the foreland margins—e.g., those of the Carboniferous limestone successions in the Appalachians, Ireland, and northern England.

Review Andean-type metal zonation is important as a guide to mineral exploration of Mesozoic and younger orogenic belts but is of less significance in Paleozoic terrains because of erosion and deformation. Tin-wolfram deposits tend to form at depths great enough for many to survive throughout the Phanerozoic, and there are examples of Lower Paleozoic porphyry coppers.

However, zonation in the older Phanerozoic belts is more a reflection of the distribution and nature of massive sulfides, lead-zinc shelf deposits, and locally derived veins. Variations in the nature of the rocks of the continental margin and the depth of erosion thus allow for considerable local variation within and between orogenic belts, and detailed knowledge of the local geology is essential to the exploration geologist.

Time-Space Relationships in Intracontinental Regions

Tensional fracturing of continental plates appears to be associated with particular types of mineralization, important types involving tin (Sillitoe, 1974b), copper, and lead-zinc (Guild, 1974; Sawkins, 1976).

The early stages of rifting are accompanied by doming and generally alkaline volcanism and small-scale plutonism. Triple-junction rifting patterns with three arms at 120°, commonly with one arm failing to develop, supposedly mark the site of rising mantle plumes (Burke and Dewey, 1973). Further rifting and associated magmatism may lead to a separation of continental masses (e.g., the Red Sea) and eventually to development of an ocean.

Examples of the deposits accompanying initial rifting are the tin ores associated with linear belts of alkaline granitoids in Africa and South America, which Sillitoe (1974b) has related to mantle hot spots or plumes. The Messina Fault in the northern Transvaal is associated with a number of hydrothermal copper sulfide deposits and appears to be part of a triple-junction system. Copper mineralization in breccia pipes of the Tribag district of Ontario, Canada, appears to be part of a similar but older system (Sawkins, 1976). The base metal deposits of the Red Sea deeps are a style of mineralization associated with more advanced continental separation with accompanying basaltic volcanism.

The galena-sphalerite-barite-fluorite ores of the Mississippi Valley region of central North America and the Pine Point galena-sphalerite deposits in Canada occur in thick, carbonate-rich successions of intracontinental basins. Guild (1972) has pointed out the close relationships between the ore districts in these regions and the existence of major fractures, some with related alkaline magmatism, and has suggested that mantle hot spots may be the fundamental cause of the magmatism and mineralization.

Burke and Dewey (1973) suggested that the Zambian copper belt occupies a rifted arm. There are other deposits of major importance in basins that may also be related to rifting and associated magmatism of intracontinental regions.

Conclusion

Some metallic ore deposits of intracontinental regions are clearly associated with plate rifting and associated magmatism, and rifting may be important in the development of the host rocks in a considerable number of major occurrences (e.g., the Zambian Copper Belt). However, postulated relationships between mantle-plume activity and ore genesis are in many cases tenuous, and it is possible that many deposits are formed from migrating waters in large, stable continental basins.

MICHAEL SOLOMON

References

Bilibin, Yu. A. (E. A. Alexandrov, trans.), 1955, *Metallogenic Provinces and Metallogenic Epochs. Gosgeolhizdat, Moscow.* New York: Queens College Press.

Bird, J. M., and J. F. Dewey, 1970, Lithosphere plate-continental margin tectonics and the evolution of the Appalachian Orogen, *Geol. Soc. America Bull.* **81,** 1031–1060.

Brown, A. S., 1969, Mineralization in British Columbia and the copper and molybdenum deposits, *Canadian Inst. Mining and Metallurgy Trans.* **72,** 1–15.

Burke, K. C. A., and J. F. Dewey, 1973, Plume-generated triple junctions: Key indicators in applying plate tectonics to old rocks, *Jour. Geology* **86,** 406–433.

Carey, S. W., 1975, Tectonic evolution of Southeast Asia, *Proc. 4th Indonesian Petroleum Congr., Jakarta,* 1–31.

Guild, P. W., 1972, Massive sulfides vs. porphyry deposits in their global tectonic settings, *Proc. Joint Meeting MMIJ-AIME,* Tokyo.

Guild, P. W., 1977, Application of global tectonic theory to metallogenic studies, *Proc. 4th Symp. EAGOD Meeting* **2.**

Hollister, V. F., 1975, An appraisal of the nature and source of porphyry copper deposits, *Minerals Sci. Eng.* **7,** 225–233.

Mitchell, A. H. G., and M. S. Garson, 1972, Relationship of porphyry copper and circum-Pacific tin deposits to palaeo-Benioff zones, *Inst. Mining and Metallurgy Trans.* **81,** B10–25.

Mitchell, A. H. G., and W. S. McKerrow, 1975, Analogous evolution of the Burma Orogen and the Scottish Caledonides, *Geol. Soc. America Bull.* **86,** 305–315.

Pereira, J., and C. J. Dixon, 1965, Evolutionary trends in ore deposition, *Inst. Mining and Metallurgy Trans.* **74,** 505–527.

Sainsbury, C. L., R. R. Mulligan, and W. C. Smith, 1969, The Circum-Pacific "tin belt" in North America, *Proc. 2nd Tech. Conf. on Tin* **1,** 125–147.

Sawkins, F. J., 1972, Sulfide ore deposits in relation to plate tectonics, *Jour. Geology* **80,** 377–397.

Sawkins, F. J., 1976, Metal deposits related to intracontinental hotspot and rifting environments, *Jour. Geology* **84,** 653–671.

Sillitoe, R. H., 1972a, A plate tectonic model for the origin of porphyry copper deposits, *Econ. Geology* **67,** 184–197.

Sillitoe, R. H., 1972*b,* Relation of metal provinces in Western America to subduction of oceanic lithosphere, *Geol. Soc. America Bull.* **83,** 813–818.

Sillitoe, R. H., 1972*c,* Formation of certain massive sulphide deposits at sites of sea-floor spreading, *Inst. Mining and Metallurgy* **81,** B141–148.

Sillitoe, R. H., 1973, The tops and bottoms of porphyry copper deposits, *Econ. Geology* **68,** 799–815.

Sillitoe, R. H., 1974*a,* Tectonic segmentation in the Andes: Implications for magmatism and metallogeny, *Nature* **250,** 542–545.

Sillitoe, R. H., 1974*b,* Tin mineralization above mantle hot spots, *Nature* **248,** 497–499.

Sillitoe, R. H., C. Halls, and J. N. Grant, 1975, Porphyry tin deposits in Bolivia, *Econ. Geology* **70,** 913–927.

Smirnov, V. I., 1968, The sources of ore-forming material, *Econ. Geology* **63,** 380–389.

Solomon, M., 1976, The host rocks to "volcanic" massive sulphide deposits—A review and an explanation, in K. Wolf, ed., *Handbook of Strata-bound and Stratiform Ore Deposits.* Amsterdam: Elsevier, 21–54.

Solomon, M., and J. R. Griffiths, 1974, Aspects of the early history of the southern Tasman Orogenic Zone, in A. K. Denmead, G. W. Tweedale, and A. F. Wilson, eds., *The Tasman Geosyncline—A Symposium.* Brisbane: Geological Society of Australia, Queensland Division, 19–46.

Strong, D. F., 1973, Plate tectonic setting of Appalachian-Caledonian mineral deposits as indicated by Newfoundland examples, *American Society of Mining and Engineering Meeting, Pittsburgh (Preprint 73-1-3ZQ),* Pittsburgh.

Cross-references: *Indicator Elements.* Vol. XIII: *Geology, Applied.*

POPULAR GEOLOGY

The history of the public relations aspects of geology reveals that in the United States, as also in the United Kingdom, France, Germany, and several other countries, during the 1830s and 1840s, this science held a high position in public esteem. Later, especially in the early twentieth century, geology lost that position by default. At the present time, however, it appears that the pendulum is swinging back, re-establishing geology as a subject worthy of strong public support and interest. An effective and abundant popular literature plays a large role in such a renaissance.

During the Colonial and early Federal periods in the United States, the study of geology was largely limited to the educated aristocracy. Public figures such as Jefferson and Gallatin were considerable students of geology, and their interests were often translated into worthy public enterprises such as the explorations of Lewis and Clark, Pike, and Long.

It was in the 1830s and 1840s, however, that geology really came into its own. This is the great age of the lyceum, or public lecture series, that so impressed Agassiz when he arrived in Boston in 1846. Geology rode high on the wave of self-culture and interest in natural history that got under way about 1820, propelled by the ardent proselytizing activities of a sizable percentage of the most prominent geologists, including Eaton, Silliman, H. D. Rogers, Lyell, and Agassiz. Amos Eaton alone made 3,000 lectures outside the classroom, and it was in part due to the boundless enthusiasm of such men that a number of state legislatures were inspired to finance early surveys.

The lay literature of the time was largely that used by the geologists. Concepts were still simple, and the specialized geological terminology of 1840 was limited to a few hundred words. The textbooks of Eaton (1830), J. D. Dana (1837), Hitchcock (1841), and Lyell (1839) sold well among those hungry for learning, and Mantell's *Medals of Creation* (1839) and Miller's *Old Red Sandstone* (1841) were vastly popular informal books. Not only was the youthful science but also the poetry and beauty of nature were so appealing. The volumes of the early state surveys were also well received by laypersons. Yet even this enlightened age had its share of potboilers by nongeologists, and we must admit that the works of S. G. Goodrich (Peter Parley) are horrible examples of the genre.

Public interest in geology declined during the period from the Civil War to the turn of the century, despite interest in the great bonanzas and in Western exploration. Part of the decline may be blamed on the fact that during this period geologists began to develop their special and often unnecessary technical terms, a trend which continues to the present day and renders unacceptable, to the layperson, far too much of the professional literature. Equally significant was the sharp decline in the proportion of geologists interested in popularizing their science.

Some of the most famous exceptions were, of course, John Wesley Powell and Clarence King. *Exploration of the Colorado* (1875) and *Mountaineering in the Sierra Nevada* (1872) went through many editions and helped to keep alive the tradition of Western scientific exploration. Perhaps the biggest selling book was Darwin's *Origin of Species* (1860), which caused almost as much uproar and intellectual ferment in the United States as it did in England. Other geologist-writers included J. D. Dana, whose easier texts, like *The Geological Story Briefly Told* (1850), were fine for high school and lay use, and N. S. Shaler, who turned out a number of nontechnical introductions to physiography, such as *Aspects of the Earth* (1889). Another gifted popularizer was Alexander Wichell, whose vividly written *Walks and Talks in the Geological Field* (1886) was, for many years, on the reading list of the Chautauqua Association, an organization whose subscribers numbered 5 million in 1900. Among the nongeologist writers, the best was probably H. N. Hutchinson, a liberal English clergyman whose book *Extinct Monsters* (1892) was free of the theological controversy of earlier times.

From 1900 to World War II, it is no coincidence that geology and physiography all but disappeared from the high schools at the same time that interest in popularization all but disappeared within the field of geology. On the credit side, between 1910 and 1930 the U.S. Geological Survey undertook a modest public education program, publishing guidebooks (1915) covering some of the Western railroad routes and some booklets on the national parks, by men such as Arnold Hague, M. R. Campbell, and W. T. Lee. François Matthes, in his "Geologic history of the Yosemite Valley" (1930), proved that a scientific classic could be written in plain English. The National Park Service became a potent educational force in the 1920s as its public education program began to develop. Some state surveys—notably, those of New York, Pennsylvania, and Kentucky—began to issue popular publications, as did the American Museum of Natural History. Popular interest in mineral collecting, which would become so strong in the 1950s, began to revive with the founding of the journals *Rocks and Minerals,* in 1926, and *Mineralogist,* in 1933.

Among the few lay books that sold well were F. B. Loomis's *Field Book of Common Rocks and Minerals* (1923) and astronomer W. M. Reed's *The Earth for Sam* (1930). Two other authors, who are still turning out big sellers, are Roy Chapman Andrews, whose books on dinosaur hunting (1943, 1953) were well received, and Carroll Lane Fenton, who was establishing his reputation as a popularizer with works like *Our Amazing Earth* and *The Rock Book* (1940).

Since World War II, the outlook for popularizing geology has become brighter. Geology and earth science courses are being reintroduced into a number of high schools, and school libraries feature the better known geological texts and popular works, as well as fine journals such as *Natural History* and *Scientific American.* Some states have stepped up their educational and publishing programs, including California, Kansas, Oklahoma, and Illinois, and the U.S. Geological Survey (1948) now prints geological descriptions on the backs of its national parks maps.

The National Park Service remains preeminent as the great popularizer of geology in the United States. Perhaps 10 million young people of school age see geological wonders first hand every year and take home scores of thousands of earth science books from park concessions.

Interest in caves has burgeoned since 1941, when the National Speleological Society was formed; now its 1,700 members and perhaps 5,000 other young spelunkers spend their weekends underground. Even more spectacular has been the spread of interest in mineral collecting and cutting; some estimates suggest the minimum number of devotees near 250,000.

With the publication of the successful Merit Badge pamphlet on geology (1956), the Boy Scouts have become an important agent for promoting the earth sciences. Over 50,000 copies of this book have been sold since its introduction in 1953, and geology badges earned have risen from 300 to 3,000 a year. Geology month among the scouts, October 1957, featured "the biggest show on earth," during which 4,000 geologists assisted 38,000 scout units to make it by far the largest public relations program put on by the geology profession.

Geology now has an overall public relations organization in the American Geological Institute (AGI), founded in 1948. Its publications include a career booklet, of which a modest 18,000 copies were given away, as well as a guide to the nontechnical literature (1957). The AGI's operations, however, are still on a small scale compared with those of the American Petroleum Institute and of certain oil companies, which distribute free career materials and teachers' pamphlets by the millions.

In the 1980s there has been a noticeable increase in the sales of popular geology books, both absolute and in relation to other technical subjects; only aviation and space science attract more young readers now than do the earth sciences. The Fentons's *Rock Book,* e.g., sold twice as well after 1945 as it did when published in 1940, and Zim and Shaffer's *Rocks and Minerals* (1957) was the runaway leader in the highly successful Golden Guide series.

In addition to established names like the Fentons (1940, 1958), a new group of outstanding popularizers emerged. Among them were journalist Ruth Moore (1956); physicist George Gamow (1948); geologists O. P. Jenkins (1948), J. A. Shimer (1959), J. L. Dyson (1962), and F. J. Shepard (1959), paleontologists E. H. Colbert (1945, 1961, 1965), W. H. Matthews (1962), and F. H. T. Rhodes (1962); and mineralogists R. M. Pearl (1955), F. H. Pough (1953), and J. Sinkankas (1959).

Although a lack of objective data makes it difficult to contrast the current status of popular geology with its status in the 1840s, it is clear that in 1840 a high percentage of the most prominent geologists participated in public lectures and similar services; by 1940 surely no more than 1% or 2% of geologists lectured, counseled, or wrote for the public.

It is good to observe, however, that since World War II an increasing number of geologists have become aware that geology must be brought to the layperson if public interest and support for geological enterprises are to grow and if a supply of future geologists is to be assured. Let us hope that the geologists fill the gaps in the popular literature. New illustrating and publishing techniques promise more attractive books, and new marketing methods are providing far wider audiences than in the past. If geologists can engender the proselytizing enthusiasm of Amos Eaton and Agassiz and are willing to grant their popularizations the same care that they put into their professional work, then geology will indeed hold the American public's interest and approval.

Acknowledgment

This entry is largely based on an article published by Mark W. Pangborn, Jr., in the *Journal of the Washington Academy of Sciences,* 1959, 49(7).

RHODES W. FAIRBRIDGE

References

Andrews, R. C., 1943, *Under a Lucky Star; A Lifetime of Adventure.* New York: Viking Press, 300p.

Andrews, R. C., 1953, *All About Dinosaurs.* New York: Random House, 146p.

Boy Scouts of America, 1956, *Merit Badge Series: Geology,* 2nd ed. New York, 35p.

Colbert, E. H., 1945, *The Dinosaur Book.* New York: American Museum of Natural History.

Colbert, E. H., 1961, *Dinosaurs, Their Discovery and Their World.* New York: Dutton, 300p.

Colbert, E. H., 1965, *The Age of Reptiles.* New York: W. W. Norton, 228p.

Dana, J. D., 1837, *A System of Mineralogy.* New Haven, Conn.: Durrie & Peck, 144p.

Dana, J. D., 1895, *The Geological Story Briefly Told.* New York: American Book Co., 302p.

Darwin, C. R., 1860, *Origin of Species,* American ed. New York: Appleton, 432p.

Dyson, J. L., 1962, *The World of Ice.* New York: Alfred A. Knopf, 292p.

Eaton, A., 1830, *Geological Text Book, Prepared for Public Lectures of North American Geology.* Albany, New York: Webster and Skinners, 63p.

Fenton, C. L., and M. A. Fenton, 1940, *The Rock Book.* New York: Doubleday & Co., 357p.

Fenton, C. L., and M. A. Fenton, 1958, *The Fossil Book.* New York: Doubleday & Co., 496p.

Gamow, G., 1948, *Biography of the Earth.* New York: Viking Press. 194p.

Hitchcock, E., 1841, *Elementary Geology,* 2nd ed. New York: Dayton & Saxon, 346p. (30 editions in 20 years).

Hutchinson, H. N., 1892, *Extinct Monsters.* New York: Appleton, 254p.

Jenkins, O. P., 1948, *Geologic guidebook along highway 49—Sierran gold belt,* San Francisco, *California Div. Mines and Geology Bull. 141,* 164p.

King, C. R., 1872, *Mountaineering in the Sierra Nevada.* Boston: J. R. Osgood.

Lee, W. T., 1915, Guidebook of the Western United States, Part B, The Overland route, *U.S. Geol. Survey Bull. 612,* 244p.

Loomis, F. B., 1923, *Field Book of Common Rocks and Minerals.* New York: G. P. Putnam Sons, 352p.

Lyell, C., 1839, *Elements of Geology.* American ed., Philadelphia: James Kay (Lyell's numerous textbooks went through numerous editions).

Mantell, G. A., 1839, *The Wonders of Geology.* 2 vols. New Haven, Conn.: A. H. Maltby. Also *Medals of Creation.*

Matthes, F. E., 1930, Geologic history of the Yosemite valley. *U.S. Geol. Survey Prof. Paper 160,* 137p.

Matthews, W. H., 1962, *Fossils, An Introduction to Prehistoric Life.* New York: Barnes & Noble, 337p.

Miller, H., 1841, *The Old Red Sandstone.* Edinburgh.

Moore, R., 1956, *The Earth We Live On.* New York: Alfred A. Knopf, 416p.

Pangborn, M. W., Jr., 1957, *Earth for the Layman; A List of Nearly 1490 Good Books,* 2nd ed. Alexandria, Va.: American Geological Institute (Rept. No. 2), 68p.

Pearl, R. M., 1955, *How to Know the Minerals and Rocks.* New York: McGraw-Hill, 192p.

Pough, F. H., 1953, *A Field Guide to Rocks and Minerals.* Boston: Houghton Mifflin, 333p.

Powell, J. W., 1875, *The Exploration of the Colorado River of the West.* Washington, D.C.: U.S. Government Printing Office, 291p.

Reed, W. M., 1930, *The Earth for Sam.* New York: Harcourt, Brace & Co., 390p. (52,000 copies in print).

Rhodes, F. H. T., 1962, *Fossils, A Guide to Prehistoric Life.* New York: Golden Press, 160p.

Rhodes, F. H. T., 1962, *The Evolution of Life.* Baltimore, Md.: Penguin Books, 304p.

Shaler, N. S., 1889, *Aspects of the Earth.* New York: Charles Scribner's Sons, 344p.

Shepard, F. P., 1959, *The Earth beneath the Sea.* Baltimore, Md.: Johns Hopkins University Press, 275p.

Shimer, J. A., 1959, *This Sculptured Earth, The Landscape of America.* New York: Columbia University Press, 255p.

Sinkankas, J., 1959, *Gemstones of North America.* Princeton, N.J.: D. Van Nostrand Co., 675p.

U.S. Geological Survey, 1948, *Topographic Map of the Denver Mountain Area,* scale 1:190,680. Washington, D.C.

Winchell, A., 1886, *Walks and Talks in the Geological Field.* New York: Chautauqua Press, 329p.

Zim, H. S., and P. Shaffer, 1957, *Rocks and Minerals.* New York: Simon & Schuster, 160p.

Cross-references: *Geohistory, American Founding Fathers.* Vol. XIII: *Geological Communication; Geological Information, Marketing.*

PROFESSIONAL GEOLOGISTS' ASSOCIATIONS

During the 1800s most professional geologists were content to join one or more scientific societies and manage their professional or ethical problems on a more or less personal, in-house basis. In the 1900s, especially since the 1950s, with an increasing number of geological consultants dealing with geotechnical problems (see Vol. XIII: *Geotechnical Engineering*) of public concern, there has been an almost worldwide movement toward professional organizations, a trend most energetically pursued in the United States, Australia, Canada, and the United Kingdom.

Geotechnical problems of public concern focus on issues such as development of energy and water resources (see *Groundwater Exploration*), power station and dam-site location (see Vol. XIII: *Dams, Engineering Geology*) river control (see Vol. XIII: *River Engineering*), coastal protection (see Vol. XIII: *Coastal Engineering*), and urban planning (see Vol. XIII: *Urban Engineering Geology*). Consulting geologists are often involved not only in baseline surveys but also in project reviews, benefit/cost analyses, environmental impact assessment, and sometimes, litigation (see *Legal Affairs*). While in science most geologists are well equipped to marshal support for their statements in a truthful manner (see *Geological Methodology*), it is quite a different experience to stand up in a court of law and make sworn testimony. Value judgments are often required and relatively few geologists are adequately prepared for courtroom procedures or public relations in general (see Vol. XIII: *Geological Communication; Geological Information, Marketing*).

Not the least of professional geologists' worries are problems associated with contentious ethical issues. Medical practitioners, lawyers, accountants, engineers, and other specialists are certified by professional organizations that, to some extent, police their profession and attempt to suspend or remove individuals who are in violation of ethical codes. Although most geological associations have codes of ethics (see *Geological Fieldwork, Codes*), many geologists have tended to rely on good judgment and moral sense as a guide. When it comes to actual cases, individual problems and specific local issues are often less than clear. Conflicting interests also tread on ethical grounds when a geologist is invited to give expert witness evidence in an area where the geoscientist is liable to gain or lose financially. It is, e.g., well recognized that the expert witness whose consulting time is financially supported by the litigant tends to find irrefutable scientific evidence to support the argument of his or her employer. This procedure is no more or less the ethical approach of the lawyer, who is expected to present the case as well as possible. It is up to the other party to furnish a comparable expert to balance the case. Because these and many other problems arise, professional organizations are indicated.

One such group is the American Institute of Professional Geologists (AIPG), with general headquarters in Arvada, Colorado. This association works vigorously toward certification, quality education, maintenance of competence, and public recognition. It is the only nationwide organization that certifies the competence and eth-

ical conduct of American geological scientists in all branches of the profession. In addition to contributing to aspects of geologic professionalism, the association seeks to improve the professional application of geology for the good of the general public, assure proper geologic input to federal and state lawmaking and rule-making processes, enhance and preserve the standing of the profession, maintain adherence to an uncompromising code of ethics, establish and maintain the highest professional standards, and continuously evaluate the professional qualifications of geologists. A major activity of AIPG is certification of members as a certified professional geological scientist (CPGS). Such certification provides a means by which the public can recognize those geologists who are judged by their peers to be worthy of public trust. The designation of CPGS following a name identifies members of the association who hold proper and necessary qualifications.

The association has established sections to maintain and enhance professional activities at state and regional levels. It is quite probable that the most effective actions are the result of section activities. As a member society of the American Geological Institute (AGI), AIPG is making a concerted effort to coordinate the professional activities of all geological organizations. Efforts are being made to include more geochemists, geophysicists, and other earth scientists in AIPG membership.

Organizational Structures

Professional geologists' associations are governed by conventions that are variously referred to as charters, bylaws, constitutions, or rules. All have shades of difference but usually reflect the overall purpose of the society, which might be broadly instituted to investigate the mineral structure of the earth (Geological Society of London, 1974), to advance knowledge in the geological sciences (Geological Society of America, 1974; Geological Society of Australia, 1975), or regionally "to advance the science of geology and closely related fields of study and to promote a better understanding thereof throughout Canada" (Geological Association of Canada, 1973). Whatever the rules are called, they typically refer to policies and procedures related to membership (e.g., eligibility, sponsorship, dues, categories), gifts and bequests, research grants, awards, and management. Financial investment, legal council, liability, and bonding of officers, councillors, and committee members are now complex issues for societies supporting geoscientists in business. Most professional geologists' associations are incorporated as nonprofit enterprises and are managed with a keen sense of business acumen.

In addition to other mandates, the professional

associations facilitate various corporate matters, national scientific meetings, committee meetings, and regional meetings with division sections. The Geological Society of America, e.g., incorporates six sections, including Cordilleran, Rocky Mountain, Southeastern, Northeastern, North-Central, and South-Central, and five divisions of coal geology, engineering geology, geophysics, hydrogeology, and quaternary geology and geomorphology. The Geological Society of London, in a similar way, recognizes and sponsors constituted specialists groups—i.e., the Engineering Group, the Geological Information Group, the Marine Studies Group, and the Volcanic Studies Group. Affiliated groups include the Earth Sciences Education Methods Group, the British Sedimentological Research Group, and the Tectonic Studies Group. Such groups hold meetings and organize visits to the field to promote the exchange of information and encourage in-depth study of certain aspects of the science.

Most professional geologists' associations maintain relations or liaison with other national or international societies with similar goals or purposes. Cordial relations with other learned societies and groups are accomplished through joint meetings, but much of the continuity is maintained by delegates of the respective groups. The Geological Society of America, e.g, receives and, if necessary, pays for travel and subsistence of representatives from affiliated groups to participate in national meetings. Affiliated groups include the American Association for the Advancement of Science (AAAS), the American Commission on Stratigraphic Nomenclature (ACSN), the American Geological Institute (AGI), the American Society of Civil Engineers (ASCE), the International Geological Congress (IGC), and several committees of the National Research Council (NRC) such as the U.S. National Committee on Geochemistry, the U.S. National Committee on Geology, the U.S. National Committee on Rock Mechanics, the U.S. National Committee on Tunneling Technology, and the U.S. National Committee of the International Union for Quaternary Research.

In addition to affiliations between branches of professional societies, there are so-called supergroups composed of associations of associations. An example of national affiliations is highlighted by the eighteen associations that belong to the AGI as member societies: American Association of Petroleum Geologists (AAPG), American Geophysical Union (AGU), Association of Earth Science Editors (AESE), Association of Engineering Geologists (AEG), Association of Professional Geological Scientists (APGS), Geochemical Society (GS), Geological Society of America (GSA), Geoscience Information Society (GIS), Mineralogical Society of America (MSA), National Association of Geology Teachers (NAGT), Paleontological Society (PS), Seismological Society of America (SSA), Society of Economic Geologists (SEG), Society of Economic Paleontologists and Mineralogists (SEPM), Society of Exploration Geophysicists (SEG), Society of Mining Engineers of AIME (SME), and the Society of Vertebrate Paleontology (SVP).

The International Union of Geological Sciences (IUGS) is one of the largest and most active international scientific associations in the world. IUGS is a voluntary professional organization that is nongovernmental, nonpolitical, and nonprofit-making. Membership in IUGS is open to countries or defined geographical areas through a designated national organization or appointed committee for geological sciences, a national academy, geological survey, or similar national agency. Special scientific interests are represented in the union through affiliated organizations that currently include the Association of Arab Geologists (AGA), Association of Geoscientists for International Development (AGID), Association Internationale pour L'Etude des Argiles (AIPEA), Association des Services Géologiques Africains (ASGA), Carpathian Balkan Geological Association (CBGA), Commission for the Geological Map of the World (CGMW), European Association of Earth Science Editors (EDITERRA), Geological Society of Africa (GSA), International Association of Engineering Geology (IAEG), International Association of Geochemistry and Cosmochemistry (IAGC), International Association on the Genesis of Ore Deposits (IAGOD), International Association of Hydrogeologists (IAH), International Association for Mathematical Geology (IAMG), International Association of Planetology (IAP), International Association of Sedimentologists (IAS), International Commission for Palynology (ICP), International Mineralogical Association (IAM), International Palaeontological Association (IPA), International Union for Quaternary Research (INQUA), Society for Economic Geologists (EG), and the Society for Geology Applied to Mineral Deposits (SGA). Other multidisciplinary international joint programs of note include the International Geological Correlation Program (IGCP), a successful collaborative program operating since 1972, and the Inter-Union Commission on the Lithosphere (ICL), the latter co-sponsored by the International Union of Geodesy and Geophysics (IUGG), the International Council of Scientific Unions (ICSU), and the IUGS.

National and International Associations

The proliferation and diversification of professional associations and societies in the past several decades has prompted the production of directories for different branches or activities in the earth sciences. The *Directory of Geoscience Departments,* e.g., is published annually by the AGI. The twentieth edition of the directory (AGI Staff, 1983) provides an alphabetical list of earth science departments in colleges and universities in the United States and Canada and provides information pertaining to faculties and degrees, field courses and camps, and faculty specialties. The AGI also publishes a *Directory of the Geologic Division, U.S. Geological Survey* (AGI Staff, 1981), which contains staff listings and branch office rosters. The *U.S. Directory of Marine Scientists* (National Academy of Sciences, 1975) contains useful listings of organizations, foundations, and laboratories in addition to names of individual marine scientists. Directories of professional associations are more difficult to assemble. The *AAPG Bulletin* normally provides a forum where professional geologists may advertise consulting services in the "Professional Directory." The same journal also produces a worldwide "Directory of Affiliated Geological Societies," but the list is far from complete. *Geotimes,* published by AGI, periodically updates "A Directory of Societies in Earth Science." The August 1985 issue contains a rather comprehensive listing of local, regional, and national geoscience associations. A list of national and international professional geologists' associations is given in Table 1. Regional sections and specialists' groups have been omitted for brevity, and the list has been streamlined to primary service organizations. For additional listings, see *Geological Surveys, State and Federal; State Geological Surveys* (Vol. VIII, Pt. I); and *Information Centers* (Vol. XIII).

CHARLES W. FINKL, JNR.
RHODES W. FAIRBRIDGE

References

American Geological Institute, 1981, *Directory of the Geologic Division, U.S. Geological Survey.* Falls Church, Va., 160p.

American Geological Institute, 1983, *Directory of Geoscience Departments.* Falls Church, Va., 207p.

Geological Association of Canada, 1973, *Constitution and By-Laws.* Waterloo, Ontario: Department of Earth Sciences, University of Waterloo, 20p.

Geological Society of America, 1974, *Council Rules, Policies, and Procedures.* Boulder, Colo., 116p.

Geological Society of Australia, 1975, *Rules.* Sydney, 7p.

Geological Society of London, 1974, *The Charter and By-laws of the Geological Society of London.* London, 39p.

National Academy of Sciences, 1975, *U.S. Directory of Marine Scientists.* Washington, D.C., 325p.

Cross-references: *Geological Fieldwork, Codes; Geological Methodology; Geology, Scope and Classification.* Vol. XIII: *Geological Communication; Geological Information, Marketing.*

TABLE 1. List of Professional Geologists' Associations

Abilene Geological Society Box 974, Abilene, Tex., 79604	American Association of Stratigraphic Palynologists Inc. Kenneth M. Piel, Union Oil Company Research Center, Box 76, Brea, Calif., 92621	American Geological Institute 4220 King St., Alexandria, Va., 22302
Alabama Geological Society Box 6184, University, Ala., 35486		American Geophysical Union 2000 Florida Ave. NW, Washington, D.C., 20009
Alaska Geological Society Box 1288, Anchorage, Alaska, 99510	American Astronomical Society, Division for Planetary Sciences Institute for Astronomy, 2680 Woodlawn Drive, Honolulu, Hawaii, 96822	American Institute of Hydrology Alexander Zaporozec, Box 14251, St. Paul, Minn., 55114
Albuquerque Geological Society Box 26884, Albuquerque, N.M., 87125	American Congress on Surveying & Mapping 210 Little Falls St., Falls Church, Va., 22046	American Institute of Mining, Metallurgical & Petroleum Engineers 345 East 47th St., New York, N.Y., 10017
Alfred Wegener Foundation Ahrstrasse 45 (Wissenschaftszentrum), 5300 Bonn 2, West Germany	American Crystallographic Association 335 East 45th St., New York, N.Y., 10017	American Institute of Professional Geologists 7828 Vance Dr., Suite 103, Arvada, Colo., 80003
American Association for Crystal Growth Thomas Surek, Solar Energy Research Institute, 1617 Cole Blvd., Golden, Colo., 80401	American Gas Association 1515 Wilson Blvd., Arlington, Va., 22209	American Meteorological Society 45 Beacon St., Boston, Mass., 02108
American Association of Petroleum Geologists Box 979, Tulsa, Okla., 74101	American Geographical Society Collection University of Wisconsin, Milwaukee, Wis., 53201	American Mining Congress 1920 N St. NW, Washington, D.C., 20036

(continued)

TABLE 1. (*continued*)

American Petroleum Institute
1220 L St. NW, Washington,
D.C., 20005

American Quaternary Association
W.R. Farrand, Department of
Geological Sciences, University
of Michigan, Ann Arbor, Mich.,
48109

American Society of Agronomy
667 South Segoe Road, Madison,
Wis., 53711

American Society of Civil Engineers
Geotechnical Engineering Division, 345 East 47th St., New
York, N.Y., 10017

American Society of Limnology
& Oceanography
Great Lakes Research Division,
University of Michigan, Ann
Arbor, Mich. 48109

American Society of Photogrammetry
210 Little Falls St., Falls Church,
Va., 22046

American Water Resources Association
5410 Grosvenor Lane, Suite 220,
Bethesda, Md., 20814

Appalachian Geological Society
Box 2605, Charleston, W.Va.,
25329

Arab Geologists' Association
Box 1247, Central Post Office,
Baghdad, Iraq

Ardmore Geological Society
Box 1552, Ardmore, Okla., 73402

Arizona Geological Society
Box 40952, Tucson, Ariz., 85717

Asociación Colombiana de Geó-
logos y Geoffísicos del Petróleo
Apartado Aéreo 13726, Bogotá,
Colombia

Asociación de Geólogos de Costa
Rica
Apartado 6153, San José, Costa
Rica

Asociación de Geólogos Españ-
oles
Museo Nacional de Ciencias Naturales, Madrid 6, Spain

Asociación de Ingenieros de
Minas, Metalurgistas y Geólogos
de México
Apartado Postal 1260, Mexico,
D.F., Mexico

Asociación Geológica Argentina
Maipú 645, 1° Piso, 1006 Buenos
Aires, Argentina

Asociación Latinoamericana de
Editores en Geociencias
Escuela de Geología y Minas,
Apartado 50926, Caracas 105,
Venezuela

Asociación Mexicana de Geofisci-
cos de Exploración
Apartado Postal 53-077, Mexico
17, D.F., Mexico

Asociació Mexicana de Geólogos
Petroleros
Apartado Postal 53097, C.P.
06400, Mexico 17, D.F., Mexico

Asociación Paleontológica Argen-
tina
Maipu 645, 1° Piso, Buenos
Aires, Argentina

Asociación Venezolana de Geología, Minería y Petróleo
Apartado 60400, Caracas, Venezuela

Association des Géologues du
Bassin de Paris
Tour 14-15, 4ème étage, 4, place
Jussieu, 75230 Paris Cedex 05,
France

Association des Services Géologiques Africains
103, rue de Lille, 75007 Paris,
France

Association for Women Geoscientists
Box 1005, Menlo Park, Calif.,
94026

Association Française pour l'A-
vancement des Sciences
250, rue St. Jacques, 75005 Paris,
France

Association Internationale pour
l'Etude des Argiles
1, Place Croix du Sud, B-1348
Louvain-la-Neuve, Belgium

Association of American Geographers
1710 16th St. NW, Washington,
D.C., 20009

Association of American State
Geologists
R. Thomas Segall, Geological
Survey Division, Michigan De-
partment of Natural Resources,
Box 30028, Lansing, Mich.,
48909

Association of Australasian Paleontologists of the Geological
Society of Australia
10 Martin Place, Sydney 2000,
N.S.W., Australia

Association of Earth Science Editors
H.L. James, Montana Bureau of
Mines & Geology, Butte, Mont.,
59701

Association of Engineering Geologists
5313 Williamsburg Road, Brent-
wood, Tenn., 37027

Association of Exploration Geochemists
Box 523, Rexdale, Ont., M9W
5L4 Canada

Association of Geomorphologists
of Turkey
PK 652 Kızılay-Ankara, Turkey

Association of Geoscientists for
International Development
Asian Institute of Technology,
Box 2754, Bangkok 10501, Thai-
land

Association of Hungarian Geophysicists
1368 Budapest VI. Anker köz 1.
POB 240, Hungary

Association of Missouri Geologists
Box 250, Rolla, Mo., 65401

Association of Professional Engineers, Geologists & Geophysicists of Alberta
1010 One Thornton Court, Edmonton, Alberta, T5J 2E7, Canada

Association of Teachers of Geol-
ogy
Department of Earth Sciences,
The Open University, Milton
Keynes, MK7 6AA, England

Association Paléontologique
Française
Institut des Sciences de la Terre,
l'Université de Dijon, 6, Boule-
vard-2100, Dijon, France

Association Québécoise pour
l'Etude du Quaternaire
Département de Géographie,
Université de Sherbrooke, Sherbrooke, Quebec, J1K 2R1, Canada

Association Sénégalaise pour
l'Etude du Quaternaire de
l'Ouest Africain
Laboratoire de Géologie, Faculté
des Sciences, Dakar-Fann, Séné-
gal

Atlantic Coastal Plain Geological
Association
Delaware Geological Survey, Uni-
versity of Delaware, Newark,
Del., 19716

TABLE 1. (*continued*)

Atlantic Geoscience Center Bedford Institute of Oceanography, Box 1006, Dartmouth, N.S., B2Y 4A2, Canada

Atlantic Geoscience Society Department of Geology, Dalhousie University, Halifax, N.S., B3H 3J5, Canada

Austin Geological Society Box 1302, Austin, Tex., 78767

Australasian Institute of Mining & Metallurgy Box 310, Carlton South, Victoria 3053, Australia

Australia & New Zealand Association for the Advancement of Science Inc. 10 Martin Place, Sydney 2000, N.S.W., Australia

Australian Clay Minerals Society CSIRO Division of Soils, Black Mountain, Canberra, A.C.T., Australia

Australian Institute of Geoscientists Geological Society of Australia, 10 Martin Place, Sydney 2000, N.S.W., Australia

Australian Petroleum Exploration Association 12th Floor, 60 Margaret St., Sydney 2000, N.S.W., Australia

Australian Society of Exploration Geophysicists Science Centre, 35-43 Clarence St., Sydney 2000, N.S.W., Australia

Austrian Geological Society Geologische Bundesanstalt, Rasumofskygasse 23, A 1031 Vienna, Austria

Austrial Mineralogical Society Naturhistorisches Museum, Burgring 7, A 1010 Vienna, Austria

Austrian Paleontological Society Institut f. Paläontologie d. Universität, Universitätsstrasse 7/11, A 1010 Vienna, Austria

Bangladesh Association for the Advancement of Science Institute of Nutrition, University of Dacca, Dacca, Bangladesh

Bangladesh Geological Society Geological Survey of Bangladesh, Pioneer Road, Segunbagicha, Dacca 2, Bangladesh

Baton Rouge Geological Society Box 19151, University Station, Baton Rouge, La., 70893

Bay Area Geophysical Society #417, 625 Post St., San Francisco, Calif, 94109

Bay Area Mineralogists Menlo College, 1000 El Camino Real, Atherton, Calif., 94025

Baylor Geological Society CSB 367, Baylor University, Waco, Tex., 76798

Belgian Contact Group on Clays 1, Place Croix du Sud, B 1348 Louvain-la-Neuve, Belgium

Bemidji State Geological Association Sattgast Hall, Bemidji State University, Bemidji, Minn., 56601

Big Rivers Area Geological Society James V. Debroeck, Cominco American Inc., 1 Westbury Square, St. Charles, Mo., 63301

Billings Geological Society Box 844, Billings, Mont., 59101

Black Country Geological Society 569 Parkfield Road, Wolverhampton, WV4 6EL, England

Blackpool Archaeological & Geological Society 32 Beach Road, Cleveleys, Lancashire, England

Botanical Society-Paleobotany Section C.C. Baskin, School of Biological Sciences, University of Kentucky, Lexington, Ky., 40506

British Geomorphological Research Group Department of Geography, Amory Building, The University, Exeter, Devon, England

British Micropalaeontological Society P.P.E. Weaver, Institute of Oceanographic Sciences, Wormley, Godalming, Surrey, GU8 5UB, England

British Society of Soil Science School of Agriculture, University of Nottingham, Sutton Bonington, Loughborough, LE12 5RD, England

Bulgarian Geological Society 1000 Sofia, Box 228, Bulgaria

California Earthquake Society Box 686, San Marcos, Calif., 92069

Canadian Exploration Geophysical Society WES Urquhart Associates Ltd, 101 Glenvale Blvd., Toronto, Ont., M4G 2V8, Canada

Canadian Geoscience Council Department of Earth Sciences, University of Waterloo, Waterloo, Ont., N2L 3G1, Canada

Canadian Geotechnical Society Division of Building Research, National Research Council, Ottawa, Ont., K1A 0R6, Canada

Canadian Geothermal Resources Association Box 5059, Vancouver, B.C., V6B 4A9, Canada

Canadian Institute of Mining & Metallurgy 400-1130 Sherbrooke St., W, Montreal, Quebec, H3A 2M8, Canada

Canadian Society of Exploration Geophysicists 229, 640-5th Ave. SW, Calgary, Alberta, T2P OM6, Canada

Canadian Society of Petroleum Geologists 505, 206-7th Ave. SW, Calgary, Alberta, T2P OW7, Canada

Canadian Society of Soil Science National Research Council, Ottawa, Ont., K1A 0R6, Canada

Canadian Well Logging Society 229, 640-5th Ave. SW, Calgary, Alberta, T2P OM6, Canada

Carolina Geological Society Department of Geology, Duke University, Durham, N.C., 27708

Carpatho-Balkan Geological Association Nepstadion ut 14, Budapest, XIV, Hungary

Casper Geophysical Society Union Oil Co., Box 2620, Casper, Wyo., 82602

Centre International pour la Formation et les Echanges Géologiques 103, rue de Lille, 75007 Paris, France

Chamber of Geological Engineers PK 507 Kızılay-Ankara, Turkey

Cheltenham Mineral & Geological Society Mrs. M.J. Francis, 2a Hales Road, Cheltenham, Gloucestershire, England

(*continued*)

TABLE 1. (*continued*)

Chigaku Dantai Kenkyukai (Association for Geological Collaboration)
1-8-7, Minami-Ikebukuro, Toshima-ku, Tokyo 171, Japan

Circum-Pacific Council for Energy & Mineral Resources
Michel T. Halbouty, 5100 Westheimer Road, Houston, Tex., 77056

Circum-Pacific Jurassic Research Group (IGCP Project 171)
G.E.G. Westermann, Department of Geology, McMaster University, Hamilton, Ont., L8S 4M1, Canada

Clay Minerals Group of Jordan
University of Jordan, Amman, Jordan

Clay Minerals Group of the Mineralogical Society of Great Britain and Ireland
Department of Mineral Soils, Macaulay Institute for Soil Research, Craigiebuckler, Aberdeen, AB9 2QJ, Scotland

Clay Minerals Society
Box 2295, Bloomington, Ind., 47402

Clay Minerals Society of India
Division of Agricultural Physics, Indian Agricultural Research Institute, New Delhi 110 012, India

Clay Minerals Society of Japan
Hyogo University, Teacher Education, 37 17 Kgym 5 Suginami-Ku, Tokyo, Japan 168

Coastal Education and Research Foundation
P.O. Box 8068, Charlottesville, Va., 22906

Coast Geological Society
Box 3014, Ventura, Calif., 93003

Coastal Bend Geophysical Society
Continental Oil Co., Box 2226, Corpus Christi, Tex., 78403

Coastal Society
3426 North Washington Blvd., Arlington, Va., 22201

Colegio de Ingenieros de Minas, Metallurgistas, Petroleros y Geólogos de México
Avenue Juarez 97, Desp. 307, Mexico 1, D.F., Mexico

Colombian Society of Petroleum Geologists & Geophysicists
Apartado Aéreo 13726, Bogotá, Colombia

Colorado River Association
417 South Hill St., Los Angeles, Calif, 90013

Comisión Nacional de Geología
Rios Rosas, 23, Madrid 3, Spain

Commission for the Geological Map of the World
51, Boulevard de Montmorency, 75016 Paris, France

Committee for Coördination of Joint Prospecting for Mineral Resources in Southwest Pacific Areas
Mineral Resources Division, Private Mail Bag, G.P.O., Suva, Fiji

Corpus Christi Geological Society
Box 1068, Corpus Christi, Tex., 78403

Cumberland Geologists Society
D. Leviston, 74 Glenridding Drive, Barrow-in-Furness, Cumbria, LA14 4PB, England

Czechoslovakian Group for Clay Mineralogy & Petrology
Department of Petrology, Charles University, Albertov 6, 12843 Prague 2, Czechoslovakia

Dallas Geological Society
Suite 100, One Energy Square, Dallas, Tex., 75206

Delaware Mineralogical Society
Box 533, Newark, Del., 19715

Denver Region Exploration Geologists Society
Suite 508, 5025 Ward Road, Wheat Ridge, Colo., 80033

Deutsche Bodenkundliche Gesellschaft e.V.
K.H. Hartge, Institut fŕ Bodenkunde der Universität, Herrenhäuserstr. 2, 3000 Hannover 21, West Germany

Deutsche Geodätische Kommission bei der Bayerischen Akademie d. Wissenschaften Ständ
R. Sigl, Institut fŕ Astronomische und Physikalische Geodäsie der TU, Arcisstr. 21, 8000 München 2, West Germany

Deutsche Geologische Gesellschaft e.V.
Alfred-Bentz-Haus, Stilleweg 2, D-3000 Hannover 51, West Germany

Deutsche Geophysikalische Gesellschaft
M. Schmucker, Institut für Geophysik der Universität Herzberger, Landstr. 18, 3400 Göttingen, West Germany

Deutsche Gesellschaft für Kartographie
K.H. Meine, Rhöndorferstr. 40, 5340 Bad Honnnef 5, West Germany

Deutsche Gesellschaft für Polarforschung
D. Möller, Institut für Vermessungskunde der Universität, Pockelsstr. 4, 3300 Braunschweig, West Germany

Deutsche Mineralogische Gesellschaft
G. Troll, Institut für Mineralogie und Petrographie, Theresienstr. 41, 8000 München 2, West Germany

Deutsche Quartärvereinigung
3000 Hanover 51, Stilleweg 2, Postface 510153, West Germany

Deutscher Verein für Vermessungswesen
E. Plum, Im Höstert 17, 5350 Euskirchen, West Germany

Earthquake Engineering Research Institute
2620 Telegraph Ave., Berkeley, Calif., 94704

East African Academy
E.A. Community Building, Box 30756, Nairobi, Kenya

East Midlands Geological Society
Mrs. W.M. Wright, 311 Mansfield Road, Redhill, Nottingham, NG5 8JL, England

East Texas Geological Society
Box 216, Tyler, Tex., 75701

East Texas State University Geological Society
Department of Earth Sciences, East Texas State University, Commerce, Tex., 75428

Eastern Nevada Geological Society
24 First St., Ruth, Nev., 89319

Ecuadorian Geological & Geophysical Society
Casilla 371A, Quito, Ecuador

Edinburgh Geological Society
British Geological Survey, Murchison House, West Mains Road, Edinburgh, EH9 3LA, Scotland

Edmonton Geological Society
1010 One Thornton Court, Edmonton, Alberta, T5J 2E7, Canada

El Paso Geological Society
Department of Geological Sciences, University of Texas, El Paso, Tex., 79968

TABLE 1. (*continued*)

European Association of Exploration Geophysicists
Wassenaarseweg 22, The Hague, The Netherlands

European Geophysical Society
G.M. Brown Department of Physics, University College of Wales, Aberystwyth, SY23 3BZ, U.K.

Farnham Geological Society
7 Copse Way, Wrecklesham, Farnham, Surrey, England

Fine Particle Society
J.K. Beddow, University of Iowa, Ames, Iowa, 50010

Florida Paleontological Society
Florida State Museum, University of Florida, Gainesville, Fla., 32611

Forschungskollegium Physik des Erdkörpers
H.-P. Harjes, Institut für Geophysik der Universität, Postfach 102148, 4630 Bochum 1, West Germany

Four Corners Geological Society
Box 1501, Durango, Colo., 81301

Friends of Mineralogy
Raymond Lasmanis, Department of Natural Resources, Division of Geology & Earth Resources, Olympia, Wash., 98504

Ft. Worth Geological Society
900 Oil & Gas Building, Ft. Worth, Tex., 76102

Gemological Institute of America
1660 Stewart St., Santa Monica, Calif., 90404

Geochemical Society
Bryan Gregor, Department of Geological Sciences, Wright State University, Dayton, Ohio, 45435

Geochemical Society of Japan
Nihon Chikyu Kagakkai, c/o Business Center for Academic Societies of Japan, 2-4-16 Yayoi, Bunkyo-ku, Tokyo 113, Japan

Geological Association of Canada
Department of Earth Sciences, Memorial University of Newfoundland, St John's, Newfoundland, A1B 3X5, Canada

Geological Society of Africa
Department of Geology, University of Khartoum, Khartoum, Sudan

Geological Society of America
Box 9140, 3300 Penrose Place, Boulder, Colo., 80301

Geological Society of Australia Inc.,
10 Martin Place, Sydney 2000, N.S.W., Australia

Geological Society of Denmark
Oster Voldgade 5-7, DK-1350 Copenhagen K, Denmark

Geological Society of Finland
Geologian Tutkimuskeskus, SF-02150 Espoo, Finland

Geological Society of Glasgow
Department of Geology, University of Glasgow, Glasgow, G12 8QQ, Scotland

Geological Society of Greece
84 Veranzeoon Str, Athens 102, Greece

Geological Society of Iceland
H. Kristmannsdottir, National Energy Authority, Grensasveg 9, 108 Reykjavik, Iceland

Geological Society of India
16/1, Ali Asar Road, Bangalore 560 052, India

Geological Society of Iowa
Iowa Geological Survey, 123 North Capitol, Iowa City, Iowa, 52242

Geological Society of Iraq
Box 547, Baghdad, Iraq

Geological Society of Jamaica
Department of Geology, University of West Indies, Mona Campus, Kingston 7, Jamaica

Geological Society of Kentucky
811 Breckinridge Hall, University of Kentucky, Lexington, Ky., 40506

Geological Society of London
Burlington House, Piccadilly, London, W1V 0JU, England

Geological Society of Maine
Robert G. Gerber, Box 270, South Freeport, Maine, 04078

Geological Society of Malaysia
Department of Geology, University of Malaya, Kuala Lumpur 22-11, Malaysia

Geological Society of Minnesota
1711 Marshall Ave., St. Paul, Minn., 55104

Geological Society of Nevada
Box 12021, Reno, Nev., 89510

Geological Society of New Zealand
New Zealand Geological Survey, Box 80368, Lower Hutt, New Zealand

Geological Society of Norfolk
Natural History Department, Castle Museum, Norwich, Norfolk, NR1 3JU, England

Geological Society of Norway
Sars Gate 1, 0562 Oslo 5, Norway

Geological Society of the Oregon Country
Box 8579, Portland, Ore., 97207

Geological Society of the Philippines
Bureau of Mines Building, Pedro Gil St., Manila, Philippines

Geological Society of Poland
ul. Oleandry 2a, 30-063 Cracow, Poland

Geological Society of Puerto Rico
Box 40575, Minillas Station, San Juan, Puerto Rico, 00940

Geological Society of Sacramento
Department of Geology, California State University, 6000 J St., Sacramento, Calif., 95819

Geological Society of South Africa
Kelvin House, Holland St., Johannesburg, South Africa

Geological Society of Sweden
Postfack, S-104 05 Stockholm, Sweden

Geological Society of Thailand
Economic Geology Division, Department of Mineral Resources, Rama VI Road, Bangkok 4, Thailand

Geological Society of Trinidad & Tobago
Box 771, Port of Spain, Trinidad, West Indies

Geological Society of Turkey
P. K. 464, Kızılay-Ankara, Turkey

Geological Society of Washington
U.S. Geological Survey, MS 911, Reston, Va., 22092

Geological Society of Zimbabwe
A. Martin, Box 9427, Causeway Harare, Zimbabwe

Geologische Vereinigung e.V.
Geschäftsstelle, Postfach 249, D-5442 Mendig, West Germany

Geologists' Association
Burlington House, Piccadilly, London, W1V 9AG, England

Geophysical Society of Alaska
Geophysical Service Inc., 540 West International Airport Road, Anchorage, Alaska, 99502

(*continued*)

TABLE 1. (*continued*)

Geophysical Society of Edmonton
Imperial Oil Enterprises Ltd,
10025 Jasper Ave., Edmonton,
Alberta, T5J 0J0, Canada

Geophysical Society of Houston
Dresser Olympic, Box 1407,
Houston, Tex., 77001

Geophysical Society of Oklahoma
City
Data Finders, Box 19307, Oklahoma City, Okla., 73119

Geophysical Society of Tulsa
Amoco Production Co., Box 591,
Tulsa, Okla, 74102

Georgia Geological Society
Department of Geology, Room
340, Georgia State University, 1
University Plaza, Atlanta, Ga.,
30301

Geoscience Information Society
American Geological Institute,
4220 King St., Alexandria, Va.,
22302

Geothermal Resources Council
Box 1350, Davis, Calif., 95617

Gesellschaft Deutscher Metallhütten- und Bergleute
Fachsektion Lagerstättenforschung, Postfach 210, D-3392
Clausthal-Zellerfeld, West Germany

Glasgow Geological Society
M.C. Keen, Department of Geology, University of Glasgow,
Glasgow, G12 8QQ, Scotland

Grand Canyon Natural History
Association
Grand Canyon, Ariz., 86023

Grand Junction Geological Society
William L. Chenoweth, 707 Brassie Drive, Grand Junction,
Colo., 81501

Groupe Français des Argiles
Laboratoire de Cristallographie,
rue de Chartres, F-45046 Orléans Cedex, France

Guatemalan Geological Society
Instituto Centroamericano de Investigación y Tecnología Industrial, Avenida la Reforma 4-47,
Zona 10, Guatemala City, Guatemala

Gulf Coast Association of Geological Societies
Earth Enterprises Inc., Box 672,
Austin, Tex., 78767

Harrisburg Area Geological Society
Pennsylvania Bureau of-Topographic & Geologic Survey,
Box 2357, Harrisburg, Pa.,
17120

Herdman Geological Society
Jane Herdman Laboratories of
Geology, Box 147, University of
Liverpool, L59 3BX, England

Herrick Society
Department of Geology & Geography, Denison University,
Granville, Ohio, 42023

Himpunan Ahli Geofisika Indonesia (Association of Indonesian
Geophysicists)
Jalan Diponegoro 57, Bandung,
Indonesia

History of the Earth Sciences Society
Ellis L. Yochelson, Room E-501,
Museum of Natural History,
Washington, D.C., 20560

Hobbs Geological Society
(see Roswell Geological Society)

Houston Geological Society
6919 Ashcroft, Houston, Tex.,
77081

Hull Geological Society
K. Fenton, 17 Mill Walk, Cottingham, North Humberside,
HU16 4RP, England

Hungarian Geological Society
1061 Budapest, Anker köz 1-3,
Hungary

Hungarian Mining & Metallurgical Society
1061 Budapest, VI., Anker köz I.
em. 101-105, Hungary

Huntsville Association of Technical Societies
Box 1266, Huntsville, Ala., 35807

Idaho Association of Professional
Geologists
Box 7584, Boise, Idaho, 83707

Ikatan Ahli Geologi Indonesia
Jalan Diponegoro 57, Bandung,
Indonesia

Illinois Geological Society
104 McDudley Lane, Robinson,
Ill., 62454

Independent Petroleum Association of America
1101 16th St. NW, Washington,
D.C., 20036

Indian Geologists Association
Department of Geology, Panjab
University, Chandigarh 160 014,
India

Indian Mineralogical Association
M.N. Viswanathiah, Mineralogical Institute, University of Mysore, Mysore, Karnataka, India

Indian Science Congress Association
14, Dr Biresh Guha St., Calcutta
700 017, India

Indian Society of Earth Sciences
Department of Geology, Presidency College, Calcutta, 700
073, India

Indonesian Petroleum Association
Jalan Menteng Raya 3, Jakarta,
Indonesia

Institute of British Geographers
1 Kensington Gore, London,
SW7, England

Institute of Environmental Sciences
940 East Northwest Highway, Mt
Prospect, Ill., 60056

Institute of Petroleum
61 New Cavendish St., London,
W1M 8AR, England

Institute of Polar Studies
125 South Oval Mall, Ohio State
University, Columbus, Ohio
43210

Institute of Terrestrial Ecology
Furzabrook Research Station,
Wareham, Dorset, BH20 JAS,
England

Institute of the Expanding Earth
Box 144, Kingston 5, Jamaica,
W.I.

Institute on Lake Superior Geology
Department of Geology & Geological Engineering, Michigan
Technological University,
Houghton, Mich., 49931

Institution of Geologists
Burlington House, Piccadilly,
London, W1V 9HG, England

Institution of Mining & Metallurgy
44 Portland Place, London, W1N
4BR, England

Instituto Geológico y Minero de
España
Rios Rosas, 23, Madrid 3, Spain

Instrument Society of America
Instrument Division, Bourns Inc.,
6135 Magnolia Ave., Riverside,
Calif., 92506

TABLE 1. (*continued*)

International Association for Mathematical Geology
Kansas Geological Survey, 1930 Constant Ave., Campus West, Lawrence, Kan., 66044

International Association of Engineering Geology
Laboratoire Central des Ponts et Chaussées, 58, boulevard Lefebvre, 75732 Paris Cedex 15, France

International Association of Geochemistry & Cosmochemistry
Laboratoires de Minéralogie et de Pétrologie, Université Libre de Bruxelles, 50, Ave. F.D. Roosevelt, 1050 Brussels, Belgium

International Association of Hydrogeologists
U.S. National Committee, 2302 Fox Fire Court, Reston, Va., 22091

International Association of Planetology
Geological Survey, Hradebni 9, 110 15 Praha 1, Czechoslovakia

International Association of Sedimentologists
Laboratoire de Paléontologie Animale, Université de Liège, 7, Place du Vingt Août, B-4000 Liège, Belgium

International Association on the Genesis of Ore Deposits
G.B. Leech, 1113 Greenlawn Crescent, Ottawa, Ont., K2C 1Z4, Canada

International Commission for Palynology
B.P. 150, 33321 Bègles Cedex, France

International Council of Scientific Unions
51, rue de Montmorency, 75016 Paris, France

International Federation of Societies of Economic Geologists
Comandante Fortea 7-3, Madrid 8, Spain

International Geological Congress
Institute of the Lithosphere, 22, Staromonetny, Moscow, 109180, U.S.S.R.

International Geological Correlation Programme
Unesco, 7, place de Fontenoy, 75700 Paris, France

International Glaciological Society
Lensfield Road, Cambridge, CB2, 1ER, England

International Mineralogical Association
Institute of Mineralogy, University of Marburg, 3550 Marburg, West Germany

International Mountain Society
Box 3128, Boulder, Colo., 80307

International Palaeontological Association
Otto Walliser, Geologisch-Paläontologisches Institut, Goldschmidtstrasse 3, D-3400 Göttingen, West Germany

International Stop Continental Drift Society
Star Route 21, Winthrop, Wash., 98862

International Union for Quaternary Research (INQUA)
Ch. Schlüchter, Institute of Foundation Engineering, ETH-Hönggerberg, CH-8093 Zürich, Switzerland

International Union of Geodesy & Geophysics
(comprising the International Association of Geodesy; of Seismology & Geophysics of the Earth's Interior; of Meteorology & Atmospheric Physics; of Geomagnetism & Aeronomy; of Physical Sciences of the Oceans; of Hydrological Sciences; of Volcanology & Chemistry of the Earth's Interior) American Geophysical Union, 2000 Florida Ave. NW, Washington, D.C., 20009

International Union of Geological Sciences
Maison de la Géologie, 77, rue Claude Bernard, 75005 Paris, France

Inter-Union Commission on Lithosphere (ICL)
E.A. Flinn, National Aeronautics & Space Administration, Mail Code ERG-2, Washington, D.C., 20546

Interstate Natural Gas Association of America
1660 L St. NW, Washington, D.C., 20036

Irish Association for Economic Geology
Geology Department, Tara Mines Ltd, Knockumber, Navan, Co. Meath, Ireland

Irish Geological Association
14 Hume St., Dublin 2, Ireland

Israel Geological Society
Box 1237, Jerusalem, Israel

Israel Society for Clay Research
Geology Department, The Hebrew University of Jerusalem, Israel

Isituto di Geologia Applicata
Via Eudossiana 18, 00184 Rome, Italy

Italian Group of AIPEA
F. Veniale, Dipartimento di Scienze della Terra, Sezione Mineralogico-Petrografica Università, via Bassi 4, 27100 Pavia, Italy

Jackson Geophysical Society
Box 12304, Jackson, Miss., 39211

Kansas Geological Society
Suite 100 Landmark Square, 212 N Market, Wichita, Kan., 67202

Kentucky Geological Society
Department of Geology, University of Kentucky, Lexington, Ky., 40506

Kentucky Geological Society, Lexington Branch
Nicholas Rast, Department of Geology, University of Kentucky, Lexington, Ky., 40506

Kenya National Academy of Sciences
Ex. E.A. Community Building, Box 47288, Nairobi, Kenya

Korea Institute of Energy & Resources
219-5, Garibong-Dong, Guro-Gu, Seoul, Korea 150-06

Korean Federation of Scientific & Technological Societies
76-561 Yoksam-Dong, Kangnam-ku, Seoul, Korea

Lafayette Geological Society
Box 51896, Oil Center Station, Lafayette, La., 70505

Lamar University Geological Society
Box 10031, L.U. Station, Beaumont, Tex., 77710

Leeds Geological Association
Department of Earth Sciences, The University, Leeds, LS2 9JT, England

(*continued*)

TABLE 1. (continued)

Liberal Geological Society
Box 351, Liberal, Kan., 67901

Liverpool Geological Society
J.D. Crossley, C.F. Mott College of Education, The Hazels, Prescott, Lancashire, L34 1NP, England

Los Angeles Basin Geological Society
Box 1072, Bakersfield, Calif., 93302

Manchester Geological Association
D.D. Brumhead, 3 Falcon Close, New Mills via Stockport, SK12 4JQ, England

Marine Technology Society
2000 Florida Ave. NW, Suite 500, Washington, D.C., 20009

Meteoritical Society
5N4, NASA Johnson Space Center, Houston, Tex., 77058

Mexican Geological Society
Museo de Geología, Calle Jaime Torres Bodet 176, Col. Santa María La Ribera, Mexico 06400, D.F. Mexico

Miami Geological Society
Box 344156, Coral Gables, Fla., 33114

Michigan Basin Geological Society
Richard E. Kimmel, Geological Survey Division, Michigan Department of Natural Resources, Box 30028, Lansing, Mich., 48909

Microbeam Analysis Society
Institute for Materials Research, National Bureau of Standards, Washington, D.C., 20234

Mid-American Paleontology Society
2623 34th Ave. Court, Rock Island, Ill., 61201

Mineralogical Association of Canada
Department of Mineralogy & Geology, Royal Ontario Museum, 100 Queen's Park, Toronto, Ont., M5S 2C6, Canada

Mineralogical Association of South Africa
Box 61019, Marshalltown 2107, Republic of South Africa

Mineralogical Society
41 Queen's Gate, London, SW7 5HR, England

Mineralogical Society of America
2000 Florida Ave. NW, Washington, D.C., 20009

Minerals Professional Association
Harold York, 1656 South Kline Court, Lakewood, Colo., 80226

Mining & Metallurgical Society of America
275 Madison Ave., Room 2301, New York, N.Y., 10016

Mississippi Geological Society
Dora M. Devery, Box 422, Jackson, Miss., 39205

Montana Geological Society
Box 844, Billings, Mont., 59103

Montana Geophysical Society
Glacier Bit Service, 2501 Arnold Lane, Billings, Mont., 59102

Natchez Geological Society
Marvin E. Markley, 15 Covington Road, Natchez, Miss., 39120

National Association of Black Geologists & Geophysicists
Aminoil USA, West Memorial Park, 8588 Katy Freeway, Suite 131, Houston, Tex., 77204

National Association of Geology Teachers
Box 368, Lawrence, Kan., 66044

National Earth Science Teachers Association
Department of Geological Sciences, Michigan State University, East Lansing, Mich., 48824

National Petroleum Council
1625 K St. NW, Washington, D.C., 20006

National Speleological Society
2813 Cave Ave., Huntsville, Ala., 35810

National Water Well Association
500 West Wilson Bridge Road, Worthington, Ohio, 43085

Nebraska Geological Society
113 Nebraska Hall, Lincoln, Neb., 68588

Nepal Geological Society
Box 231, Kathmandu, Nepal

New England Intercollegiate Geological Conference
Department of Geology, Boston University, 725 Commonwealth Ave., Boston, Mass., 02215

New Mexico Geological Society
Campus Station, Socorro, N.M., 87801

New Orleans Geological Society
Box 52172, New Orleans, La., 70152

New York State Geological Association
Department of Geology, Hofstra University, Hempstead, N.Y., 11550

New Zealand Geophysical Society
Geophysics Division, DSIR Box 1320, Wellington, New Zealand

Nordic Clay Group
E. Roadset, Norsk Hydro Research Center, Box 4313-N-5013, Nygårdstangen, Norway

North American Cartographic Information Society
143 Science Hall, University of Wisconsin, Madison, Wis., 53706

North Dakota Geological Society
Box 82, Bismarck, N.D., 58501

North Texas Geological Society
Box 1671, Wichita Falls, Tex., 76307

Northern California Geological Society
Ed Horan, Natural Gas Corporation of California, 4 Embarcadero Center, #1400, San Francisco, Calif., 94111

Northern Ohio Geological Society
Department of Geological Sciences, Case Western Reserve University, Cleveland, Ohio, 44106

Norwegian Petroleum Society
Kronprinsens Gate 9, 0251 Oslo 2, Norway

Northwest Mining Association
633 Peyton Bldg., Spokane, Wash., 99201

Nottingham Geological Society
43 Eugene Gardens, Nottingham, NG2 3LF, England

Offshore Technology Conference
Box 833868, Richardson, Tex., 75083

Ohio Geological Society
Box 14322, Beechwold Station, Columbus, Ohio, 43214

Oklahoma City Geological Society
1020 Cravens Bldg., Oklahoma City, Okla., 73102

Palaeontographical Society
British Geological Survey, Exhibition Road, South Kensington, London, SW7 2DE, England

TABLE 1. (*continued*)

Palaeontological Association
P.W. Skelton, Department of
Earth Sciences, The Open University, Walton Hall, Milton
Keynes, MK7 6AA, England

Palaeontological Society of India
Geology Department, Lucknow
University, Lucknow-226007, India

Palaeontological Society of Japan
Department of Geology, Kyushu
University, Fukuoka (Hakata)
812, Japan

Paleontological Research Institution
1259 Trumansburg Road, Ithaca,
N.Y., 14850

Paleontological Society
John Pojeta Jr., U.S. Geological
Survey, E-501 U.S. National
Museum, Smithsonian Institution, Washington, D.C., 20560

Paleontological Society of China
Nanking Institute of Geology &
Paleontology, Nanking, People's
Republic of China

Paläontologische Gesellschaft
Paläontologisches Institut der
Universtät, Nussallee 8, 53
Bonn, West Germany

Pan American Institute of Geography & History
Comisión de Geoffsica, Apartado
11363, Lima 14, Peru

Pander Society
Department of Geology & Mineralogy, Ohio State University, 125
South Oval Mall, Columbus,
Ohio, 43210

Panhandle Geological Society
Box 2473, Amarillo, Tex, 79105

Peak District Mines Historical
Society
Peak District Mining Museum,
Matlock Bath, Derbyshire, England

Peninsula Geological Society
Division of Mines & Geology,
Ferry Building, San Francisco,
Calif., 94111

Permian Basin Geophysical Society
Box 361, Midland, Tex., 79702

Petroleum Exploration Society of
Australia
ANZ Banking Group Ltd, 1293
South Road, St Marys 5042,
S.A., Australia

Petroleum Exploration Society of
Great Britain
6 St James's Square, London,
SW1, England

Petroleum Exploration Society of
Libya
Box 820, Tripoli, Libya

Petroleum Exploration Society of
New York
Box 3443, Grand Central Station,
New York, N.Y., 10017

Petroleum Philatelic Society International
(with Les Amis du Pétrole) 2808
Baylor St., Bakersfield, Calif.,
93305

Philadelphia Geological Society
Juliet C. Reed, Department of
Geology, Bryn Mawr College,
Bryn Mawr, Pa., 19010

Pittsburgh Geological Society
Box 3432, Pittsburgh, Pa., 15230

Planetary Society
Box 91687, Pasadena, Calif.,
91109

Polish Geophysical Society
ul. Smolenskiego 16, 01-698 Warsaw, Poland

Polish Mineralogical Society
Instytut Geologii i Surowców Mineralnych AGH, ul. Mickiewicza
30, 30-059 Cracow, Poland

Polish Society of Friends of
Earth Sciences
ul. Mickiewicza 16/23, 01-517
Warsaw, Poland

Potomac Geophysical Society
2000 Florida Ave. NW, Washington, D.C., 20009

Quaternary Research Association
P.L. Gibbard, Botany School,
University of Cambridge, Downing St., Cambridge, CB2 3EA,
England

Real Sociedad Española de Historia Natural
Museo Nacional de Ciencias Naturales, Paseo de la Castellana
84, Madrid 6, Spain

Remote Sensing Society
Department of Geography, University of Reading, 2 Earley
Gate, Reading, RG6 2AU, England

Rio Grande Rift Consortium
Los Alamos National Laboratory,
Box 1663, MS D446, Los Alamos, N.M., 87545

Rocky Mountain Association of
Geologists
1220 University Building, 910
16th St., Denver, Colo., 80202

Rocky Mountain Mineral Law
Foundation
Fleming Law Building, Room 44,
University of Colorado, Boulder, Colo., 80309

Rocky Mountain Oil & Gas Association
345 Petroleum Building, Denver,
Colo., 80202

Roswell Geological Society
Box 1171, Roswell, N.M., 88201

Royal Astronomical Society
Burlington House, London, W1V
0NL, England

Royal Geological & Mining Society of the Netherlands
Box 157, 2000 AD Haarlem, The
Netherlands

Royal Geological Society of
Cornwall
West Wing, Public Buildings,
Penzance, England

Royal Society
G.M. Brown, Department of
Physics, University College of
Wales, Aberystwyth SY23 3BZ,
U.K.

Royal Society of Canada
344 Willington St., Ottawa, Ont.,
K14 0N4, Canada

Royal Society of Edinburgh
22-24 George St., Edinburgh,
EH2 2PQ, Scotland

Royal Society of New Zealand
Private Bag, Wellington 1, New
Zealand

Safford Centennial Society
Michael Bradley, U.S. Geological
Survey, A-413 Federal Building,
Nashville, Tenn., 37203

Salem Geological Society
1133 Chemeketa St., Salem, Ore.,
97301

San Diego Association of Geologists
8145 Ronson Road, Suite H, San
Diego, Calif., 92111

San Diego Society of Natural
History
Box 1390, San Diego, Calif.,
92112

San Joaquin Geological Society
Box 1056, Bakersfield, Calif.,
93302

(*continued*)

TABLE 1. (continued)

Saskatchewan Geological Society
Box 234, Regina, Sask., S4P 2Z6,
Canada

Schweizerische Geologische Gesellschaft
Institut de Géologie, 11, rue Em-
ile-Argand, CH-2000 Neuchâtel
7, Switzerland

Schweizerische Mineralogische
und Petrographische Gesellschaft
S. Graeser, Naturhistorisches Museum Basel, Augustinergasse 2,
CH-4051, Basel, Switzerland

Seismological Society of America
2620 Telegraph Ave., Berkeley,
Calif., 94704

Service Géologique National
B.P. 6009, 45060 Orléans Cedex,
France

Servicio Geológico de Bolivia
Ministerio de Minas y Petróleo,
Avenida 16 de Julio no. 1769,
La Paz, Bolivia

Shelf & Nearshore Dynamics &
Sedimentation
W.D. Grant, Ocean Engineering
Department, Woods Hole
Oceanographic Institution,
Woods Hole, Mass., 02543

Shreveport Geological Society
Box 750, Shreveport, La., 71162

Sigma Gamma Epsilon
Oklahoma Geological Survey,
University of Oklahoma, 830
Van Vleet Oval, Norman, Okla.,
73019

Singapore Association for the
Advancement of Science
Science Center, Singapore 2260

Sociedad Colombiana de Geolo-
gía
Apartado Aéreo 32214, Bogotá,
Colombia

Sociedad Geofísica del Perú
Geophysical Service Inc., Mayor
Armando Blondet 280, Lima,
Peru

Sociedad Geológica de Guatemala
Central American Institute for
Industry, Avenida La Reforma
4-47, Zone 10, Guatemala City,
Guatemala

Sociedad Geológica del Perú
Union Oil Co. of Peru, R. Rivera
Navarrete 791, 6° Piso, San Isi-
dro, Lima 1, Peru

Sociedad Venezolana de Espeleo-
logía
Apartado 47334, Caracas 1041A,
Venezuela

Sociedade Brasileira de Geologia
Instituto de Geosciências, Cidade
Universitária, Caixa Postal
20897, 01498 São Paulo, SP,
Brazil

Sociedade Geológica Portugal
Faculdade de Ciências, Universi-
dade de Lisboa, 1294 Lisboa
Codex, Portugal

Società Geologica Italiana
Istituto di Geologia, Citta Universitaria, 00100 Rome, Italy

Società Paleontologica Italiana
Istituto di Paleontologia, Via
Università n. 4, 41100 Modena,
Italy

Société Belge de Géologie
Rue Jenner, 13, B-1040 Brussels,
Belgium

Société Française de Minéralogie
& Cristallographie
Université P.&M. Curie, 4, place
Jussieu-Tour 16, 75230 Paris
Cedex 05, France

Société Géologique de Belgique
Université, Place du XX août, 7,
B-4000 Liège, Belgium

Société Géologique de France
77, rue Claude Bernard, 75005
Paris, France

Société Géologique du Nord
59655 Villeneuve d'Ascq, Cedex,
France

Societé Royale Belge d'Astron-
omie, de Météorologie et de
Physique du Globe
Avenue Circulaire, 3, B-1180
Brussels, Belgium

Society for Geology Applied to
Mineral Deposits
Francis Saupé, Centre de Re-
cherches Pétrographiques et
Géochimiques, B.P. 20, 54501
Vandoeuvre lès Nancy Cedex,
France

Society of Economic Geologists
185 Estes St., Lakewood, Colo.,
80226

Society of Economic Paleontologists & Mineralogists
Box 4756, Tulsa, Okla., 74159

Society of Exploration Geophysicists
Box 3098, Tulsa, Okla, 74101

Society of Independent Professional Earth Scientists
4925 Greenville Ave., Suite 170,
Dallas, Tex., 75206

Society of Mining Engineers
Caller D, Littleton, Colo., 80127

Society of Mining Engineers &
Technicians
ul. Powstanców 25, 40-952 Ka-
towice, P.B. 278, Poland

Society of Petroleum Engineers
Box 833836, Richardson, Tex.,
75083

Society of Vertebrate Paleontol-
ogy
Los Angeles County Museum of
Natural History, 900 Exposition
Blvd., Los Angeles, Calif.,
94720

Soil Science Society of America
677 South Segoe Road, Madison,
Wis., 53711

Sorby Natural History Society,
Geological Section
Sheffield City Museum, Western
Bank, Sheffield, England

South Texas Geological Society
B-100 Petroleum Center, San Antonio, Tex., 78209

Southeast Asia Petroleum Exploration Society
Box 423, Tanglin, Singapore 9124

Southeastern Geological Society
Box 1634, Tallahassee, Fla.,
32302

Southeastern Geophysical Society
Gemco, Box 61590, New Orleans,
La., 70160

Southern Geological Society Inc.
University of Southern Mississippi, Box 5044, Southern Station, Hattiesburg, Miss., 39406

Southwest Louisiana Geophysical
Society
Box 51463 OCS, Lafayette, La.,
70505

Spanish Clay Minerals Society
J.M. Serratosa, Grupo de Físico-
Química Mineral, CSIC, Mad-
rid, Spain

SPIE—The International Society
for Optical Engineering
Box 10, 1022 19th St., Bel-
lingham, Wash., 98227

Sri Lanka Association for the
Advancement of Science
120/10 Wijerama Mawatha, Colombo 7, Sri Lanka

TABLE 1. (*continued*)

Swinnerton Geological Society Department of Geology, The University, Nottingham, NG7 2RD, England	Union Geofisica Mexicana (Mexican Geophysical Union) Instituto de Geofísica, Ciudad Universitaria, Mexico 20, D.F., Mexico	Virginia Field Conference Box 3667, Charlottesville, Va., 22903
Swiss Association of Petroleum Geologists & Engineers U.P. Büchi, Bodenacherstr. 79, 8182 Benglen, Switzerland	Union of Iraqi Geologists Box 3092, Al Saadoon, Baghdad, Iraq	Vsesoyuznoe Paleontologicheskoe Obshchestvo Leningrad, U.S.S.R.
Texas A&M Geological Society College of Geosciences, Texas A&M University, College Station, Tex., 77843	U.S. National Committee on Geology Linn Hoover, U.S. Geological Survey, MS 917, Reston, Va., 22092	Wadia Institute of Himalayan Geology 15, Municipal Road, Dehra Dun 248 001, India
Tobacco Root Geological Society Box 470, Dillon, Mont., 59725	U.S. National Committee on History of Geology Kennard B. Bork, Department of Geology, Denison University, Granville, Ohio, 43023	Walker Mineralogical Club 100 Queen's Park, Toronot, Ont., M58 2O6, Canada
Toronto Geological Discussion Group Ray Goldie, Richardson Greenshields, 130 Adelaide St. West, Suite 1200, Toronto, Ont., M5H 1T8, Canada	Utah Geological Association Box 11334, Salt Lake City, utah, 84147	West African Science Association Box 7, Legon, Ghana West Texas Geological Society Box 1595, Midland, Tex., 79702
Tulsa Geological Society Inc 1150 S Victor, Tulsa, Okla., 74104	Utah Mineralogical Society 402 East State Highway, Bingham Canyon, Utah, 84006	Wyoming Association of Petroleum Landmen Box 1012, Casper, Wyo., 82602
Turkish Association of Geophysicists Bayindir sokak 7/14 Kızılay-ankara, Turkey	Vereinigung der Freunde der Mineralogie und Geologie B. Cruse, Moselweisserstr. 1, Postfach 411, 5400 Koblenz, West Germany	Wyoming Geological Association Box 545, Casper, Wyo., 82602 Wyoming Mining Association Box 886, Cheyenne, Wyo., 82001
Turkish Association of Petroleum Geologists Turkish Petroleum Corp., Mudafaa Cadesi 22, Ankara, Turkey	Vereinigung Schweizerischer Petroleumgeologen und Ingenieure (VSP/ASP) U.P. Búchi, Geologische Expertisen und Forschungen AG, Bodenacherstrasse 79, 8121 Benglen, Switzerland	Yellowstone-Bighorn Research Association Red Lodge, Mont., 59068 Yorkshire Geological Society J.M. Nunwick, 45, Boroughbridge Road, Northallerton, DL7 8BG, England
Twin City Geologists School of Mines & Metallurgy, University of Minnesota, Minneapolis, Minn., 55455		
Union Française des Géologues Maison de la Géologie, 77, rue Claude Bernard, 75005 Paris, France	Vermont Geological Society Box 304, Montpelier, Vt., 05602	Zentralverband der Deutschen Geographen H. Hagedorn, Geographisches Institut der Universität Am Hubland, 8700 Würzburg, West Germany

Source: Adapted from a directory of societies in earth science, *Geotimes* **29**(8), 15–24.

PROFILE CONSTRUCTION

The first recorded illustration of geological features in the third dimension—i.e., in vertical projection—seems to have been made by Niels Stensen (Nicolaus Steno) in *Prodomus* (1669). Today, three centuries later, Steno's method is still by far the most commonly used for such illustrations, but others include projections on inclined planes, models, stereographic and gnomonic projections, perspective drawings, and three-dimensional block diagrams.

The type of profile chosen depends, in part, on its purpose and on the amount of information available. This entry discusses only a few methods:

Two-dimensional sections
  Freehand projections
  Constant-thickness methods
  Right sections
Three-dimensional projections (block diagrams)
  Perspective block diagrams
  Axonometric block diagrams
    Isometric block diagrams
    Right-sectional block diagrams

Two-Dimensional Profiles

Freehand Projections A profile developed by freehand projection is illustrated in Fig. 1, modified from a map and section by J. E. Armstrong and E. F. Roots, of an area in north-central British Columbia. The field data are shown in the map, and the subsurface interpretation of these data is shown in profile on a vertical plane that intersects the surface along the line A-B. The country is mountainous, with local relief of 1,000 m or more (contours omitted).

Attitudes of bedding are shown by appropriate symbols at the points of observation. The line of profile runs, as nearly as possible, perpendicular to the regional trend.

To construct the profile A-B, the topographic relief is first plotted. Then attitudes of beds, as they would be seen in the section, are plotted along the surface trace. Obviously, if points of observation lie close to the profile, little or no projection along strike is necessary, but if the nearest attitude lies at some distance, such as that in the large outcrop due west of Tucha Lake, the attitude (if it is to be used) must be extrapolated horizontally to the profile plane. In such an ex-

trapolation, if the strike is not perpendicular to the profile plane, some curvature of the beds must be assumed to bring the observed bed into its correct stratigraphic position on the section line.

The dip shown on the section is, of course, the trace of the line of intersection of the observed bedding plane and the profile plane. The true dip will appear only if the strike of the bedding plane is perpendicular to the profile plane; if the bedding meets the profile plane at any angle other than a right angle, the dip plotted is an *apparent* dip. Apparent dips become progressively less than the true dip as the angular difference between the strike of the bedding and that of the profile plane decreases; if the two are parallel, the apparent dip is zero.

When the attitudes and geological boundaries (normal or faulted) are plotted in their correct relative positions, suitable extensions are sketched in. Extrapolations are usually restricted to a few hundred meters below or above the present surfaces. If subsurface data are available along or near the profile line, from drill holes, tunnels, or possible geophysical measurements, these may be incorporated in the drawing. Such profiles are imprecise, but they present a graphic interpretation

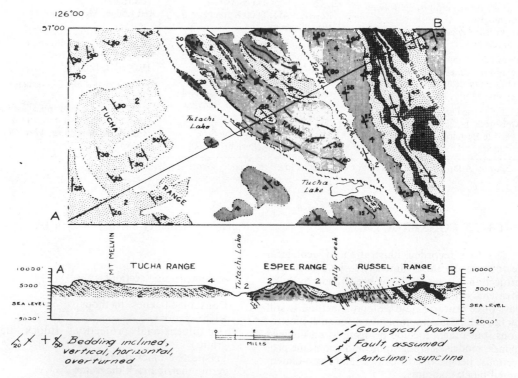

FIGURE 1. Part of Aiken Lake area, British Columbia, modified from Map 1030, Geological Survey of Canada (1954); geology by J. E. Armstrong and E. F. Roots. Rocks are Cambrian and (?) earlier: No. 2 (stippled), mainly quartzchlorite and other schists, some quartzite and quartzite conglomerate, minor limestone; No. 3 (crosshatched), white quartzite interbedded with 2; No. 4, limestone, in part micaceous, interbedded with 2.

of surface data and illustrate, to a first approximation, the disposition of the rocks in space.

To emphasize heights or depths in profile, the vertical scale may be exaggerated with respect to the horizontal. This is a common and useful practice in engineering drawings of road profiles, e.g., or in geological profiles across, say, continental areas. Geometrically, the result is to distort all angular relations and thus all the structural relations shown in the profile. Thus, what is gained in emphasis is lost in accuracy; for most geological profiles, the natural scale, equal for both horizontal and vertical measurements, is preferable.

Constant-Thickness Methods In 1945, J. Coates presented a useful summary of methods for the construction of geological profiles. Beds are assumed to maintain constant thickness; dips are plotted in the usual way on the profile plane, and the configuration is developed geometrically by arcs, tangents, or chords.

The arc method (Fig. 2), was first proposed by H. G. Busk (1929). Observed attitudes are drawn as before, on the profile plane. Normals (boundary rays) are drawn to each dip and produced to intersect in pairs. Thus, normals to dips A and B intersect at 01; normals to dips B and C intersect at 02, etc. With 01 as center and with the length to a marker bed, measured along a normal to A or B as radius, an arc is drawn across segment AB. Then with center 02, and the length to the intersection of the marker bed on the normal to B as radius, the bed is drawn as an arc across the segment to meet the normal to C.

Difficulties arise when the difference between successive dips is small, making the normals nearly parallel and thus placing their intersection at an inconveniently great distance below or above the surface. Other difficulties inherent in the method appear if dips are vertical or if the beds vary in thickness. Methods to meet these and other complications are discussed fully by Busk (1929).

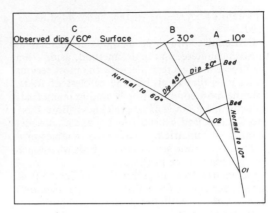

FIGURE 3. Method of chords. (After Coates, 1945)

The method of chords (Fig. 3) and the method of tangents (Fig. 4) are described by Coates (1945). In the *chord method,* beds are drawn across successive sectors at inclinations that are the averages of the pairs of dips defining each sector. In the *tangent method,* sectors are defined by the bisectors of the angles between any given pairs of dips; beds are drawn between these bisectors parallel to the surface dip observed within the sector. These methods are similar in principle to the method of arcs in that both assume constant thicknesses and both develop folds by extending beds across successive segments marked off by boundary rays (normals) to dips plotted as before on the profile plane. The forms developed, composed of straight-line segments, are perhaps less satisfying than the curves developed by the arc method, but both are independent of the points of intersection of the boundary rays. Coates recommends interpolation of dips at regular intervals as a first step in construction, and in the method of tangents, the angles between any two

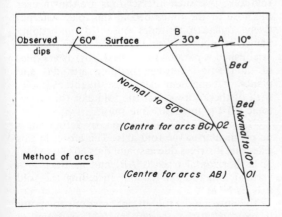

FIGURE 2. Method of arcs. (After Busk, 1929)

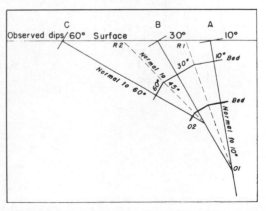

FIGURE 4. Method of tangents. (After Coates, 1945)

dips can, of course, be further subdivided as desired.

Difficulties with these methods are similar to those encountered in the arc method, aside from the problem of centers. Perhaps the most serious criticism is that in all but the gentlest of folds, only the most competent beds, and in many folds no beds, maintain constant thickness. Both Busk and Coates discuss this difficulty, and Coates developed a geometric procedure that enables progressive systematic attenuation of beds where such changes seem to be justified.

Geometric procedures for the development of continuous curves (called *involutes* and *evolutes*) from a series of attitudes observed along, or projected to, a profile plane were described by J. B. Mertie, Jr., in three papers (1940, 1947, 1948). The method enables development of elegant curves, but Rubey's remark (1926, 347) with reference to earlier rather elaborate geometrical procedures seems equally applicable here:

The writer's objections to these methods are that they are unwarranted theoretically and that they may lead to serious errors in the structure contour maps in areas where incompetent beds are steeply folded. The methods may possibly be theoretically applicable and they do not lead to such serious errors in areas of gentler dips, but the extra time required for their application remains a weighty objection to them.

Method of Right Sections   C. E. Wegmann described right sections (Fig. 5) in 1929, but they apparently remained unknown to the geological fraternity in North America until C. H. Stockwell and J. Hoover Mackin rediscovered them independently and described them almost simultaneously in 1950. Geometric procedures for drawing right sections were developed further by C. D. Dahlstrom (1954).

In this method, the projection plane chosen is the plane polar to the tectonic axis of the region—i.e., to the *b* axis of Sander. Thus, the attitude of the projection plane is fixed once the attitude of the tectonic axis is established. The projection plane will strike at right angles to the direction of plunge of the tectonic axis and will dip in the opposite direction an amount complementary to the dip of the tectonic axis.

The attitude of tectonic axis is established by a statistical study of the structural elements of the area—in particular, the orientations of axes of both minor and major folds—and of related lineations and fractures. Representative right sections can be drawn of any area in which the attitude of the tectonic axis is reasonably constant. Complications arise in areas that have been crossfolded, for in these a reasonably consistent attitude of *b* may not exist unless one reduces the area of observation to include only one wavelength of a given fold. Also one must note that in shear and flow folds, the tectonic *b* axis may not, indeed in the general case does not, coincide with the geometric *b* as defined by, say, bedding. The method is, in short, ideally suited for an examination of areas with orthorhombic or monoclinic structural symmetry, providing the tectonic *b* axis can be identified, but its use becomes progressively more difficult with increasing triclinicity of the symmetry.

The procedure in making a right section of a simple fold is illustrated in Fig. 5 (cf. Stockwell, 1950, Fig. 1). Attitudes are recorded for an idealized upright cylindrical syncline. They represent planes tangential to the cylinder, and therefore the line of intersection between them is parallel to the fold axis; two such intersections are shown near the nose of the outer bed in the fold. Alternatively, poles to attitudes observed on bedding planes lie in a plane, and the pole to the plane so established is parallel to the fold axis, as are lines of intersection between bedding attitudes and planes of congruent cleavages. In the example, each of these shows that the fold axis (b, or tectonic axis) plunges 30° due south.

The plane of the right section, by definition, is perpendicular to the fold axis. Therefore, in this example, it strikes due east and dips 60° N; in Fig. 5, its horizontal trace is AB, and its trace in a vertical longitudinal section, including the fold axis, is N'M'.

To construct the fold in right section, any point in plan, as at the nose of the fold, is projected along a ray parallel to the fold axis (in this in-

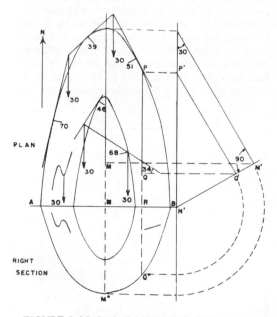

FIGURE 5. Method of determining plunge and construction of a right section of a cylindrical fold. (Cf. Stockwell, 1950, Fig. 1)

stance, along the fold axis); thus, in plan, it will follow the line from the nose of the fold through M, N, and appear in the right section somewhere along the line NM″. In longitudinal section, its trace lies along the line inclined at 30° to the horizontal, to its intersection with the trace of the plane of right section—i.e., with the line N′M′, which it meets at M′. The line from N′, perpendicular to and below AB is, of course, the trace of the longitudinal plans as seen in the plane of right section. The point M′ will lie on it at a distance below B equal to N′M′ and will appear in the right section at M″. Similarly, the point P projects in plan to R and in right sections lies along line RQ″. In longitudinal section, it projects along with ray P′Q′ and is carried around to appear in the plane of right section at Q″. Any other point in plan can be found in section in the same way, and so the form of the fold can be traced with any desired degree of precision.

The great advantage of a right section in regions of monoclinic or orthorhombic structural symmetry is that it is the only view that depicts the true form of the fold. Apparent dips are not calculated, and configurations at depth are not estimated or interpolated but are fixed by the geometry and the surface traces of the outcrops as observed. Again, although the attitude of the projection plane is fixed by the plunge, the point at which it crosses the fold is of no significance, for although the position of the horizon line AB will vary as the projection plane is moved one way or another, if the fold is truly cylindrical, the form will remain the same no matter where the trace of the section plane is placed.

The disadvantages of the method stem from the basic assumption that the fold is, in fact, cylindrical. Some folds of lower symmetry can be investigated by somewhat more complicated geometrical procedures, as described by Stockwell (1950). Faults or other planar or linear structures can be shown without difficulty in right section by ordinary drafting procedures.

An example of a right section of the Sheila Lake Fold, Manitoba, is shown in Fig. 6 (cf. Stockwell, 1950, Fig. 12). The trace of the axial plane strikes nearly due north, and a conventional vertical section would probably be drawn on a vertical east-west plane. However, the average plunge is 28°E, and since the several plunge lines as calculated tend, within reasonable limits, to be parallel to one another, the fold, to a good approximation, is cylindrical.

The right section is therefore drawn on a plane striking nearly north and dipping 62° westward. The Sheila Lake Fold appears recumbent, with the trace of the axial plane nearly horizontal. The difference in form between that which would probably be shown in a conventional east-west vertical section and that which appears in right section is

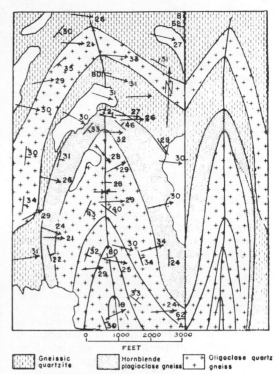

FIGURE 6. Plan and section of Sheila Lake Fold, Sherritt Gordon mine area, Manitoba. (Cf. Stockwell, 1950, Fig. 12)

obvious, as is the difference in interpretation of the movements that formed the fold. As seen in conventional section, shortening in an east-west direction or at least folding about a north-south axis would be indicated, whereas movements implied by the form in right section are oriented at nearly 90° to this. Furthermore, the form in right section enables examination of the type of fold—i.e., whether it is concentric, symmetrical, or some combination or modification of these types.

Three-Dimensional Projections (Block Diagrams)

If two intersecting profiles depicting geological features of an area are constructed and then combined with a surface plan in a single drawing, a three-dimensional representation of the geology of an area is achieved. Such three-dimensional representations are commonly called *block diagrams* (see Lobeck, 1924). Here, only two types of such diagrams are described: perspective block diagrams and axonometric block diagrams, including the special and popular isometric projections.

Perspective Projections Methods of *perspective drawing*—i.e., representation of objects by use of lines projected to one or two vanishing points—are well known and are fully described in many texts (see Lobeck, 1924, 1–118). In representing geological features, usually two vertical profiles are drawn at right angles to one another, the cross-sections being carried as before to some arbitrary depth. The two profiles then are assembled as the front and one side of a block drawn in one- or two-point perspective. In this projection, as indeed in any block diagram, surface features may not be sketched in; if they are omitted, the resultant drawing, often with profiles added to form the back and second side of a box, is commonly called a *fence diagram*. If surface features are added, the block becomes solid and the back and side profiles are hidden below surface.

The scale in any perspective drawing varies as the distance from the observer changes. Also, the effectiveness of the drawing in conveying an idea of solidity depends to a considerable extent on the skill of the artist. The perspective drawings of William Morris Davis, A. K. Lobeck, and other artists are beautiful illustrations, but such drawings require a high degree of skill. Moreover, the accurate measurement of distances or angles in such drawings is troublesome, and the precision of the results will depend on the accuracy of the original drawing. Thus, although perspective block diagrams, skillfully drawn, provide excellent illustrations, they are best done by an artist and are of limited use as working drawings.

Axonometric Projections Drawings that are constructed according to fixed, more or less mechanical, procedures and that require only modest artistic ability can provide simple yet effective three-dimensional representations of geological phenomena. These can be used as illustrations or as working drawings for the solution of structural problems, as aids in mine layouts, or in studies of stratigraphic relations and facies changes.

Simple isometric drawings are the most common. [The following description is paraphrased and condensed directly from Lobeck (1924, Part II).] The framework is a lozenge-shaped block, which may be turned at any angle to the observer (Fig. 7).

A *standard* isometric drawing is one in which the sides are drawn at 60° and 120° to the edges of the page, as in Block B, Fig. 7. This arrangement is popular because standard isometric cross-section paper is supplied in this pattern, or such blocks may be readily constructed with a 30–60° triangle.

Geological data such as attitudes, geological boundaries, etc., are transferred to the isometric drawing by corresponding coordinate lines, or the original orthogonal drawing may be converted to an isometric drawing by use of a special drafting instrument and then traced directly onto the appropriate face of the block. If relief within the area is to be represented, each contour line as seen

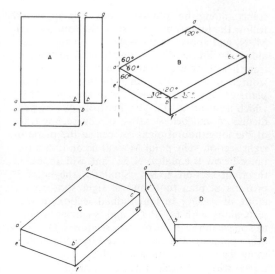

FIGURE 7. Isometric projection. (*A*) True dimensions of block in plan, front and side elevations; (*B*) isometric projection on same scale, with angles of 60° and 120° at corners; (*C*) isometric block B rotated 10° counterclockwise; (*D*) isometric block B rotated 30° clockwise. Line of sight in all blocks is inclined 36° to upper surface of block. (From Lobeck, 1924, Figs. 177 and 178; by permission)

in plan is redrawn in isometric projection and then is traced onto the drawing at its proper elevation with respect to a datum plane. For example, the base of the isometric block may be taken as sea level, and using an appropriate vertical scale, horizontal planes corresponding to each of the several contours can be marked off (Lobeck, 1924, 140–143).

In the standard isometric projection, the line of sight is inclined at 36° (nearly) to the upper surface of the block. Orientation of the sides of the block is arbitrary; the success of the drawing depends largely on the point of view, and several trials may be necessary to discover the best one.

If the lozenge-shaped block of a standard isometric projection is turned until the corners of the front face as projected are right angles (Fig. 7D), this face becomes a profile in true scale. Distances may be scaled and angles laid off directly on the front face; thus, ordinary vertical cross-sections may be represented without distortion. Drawings of this sort are useful in depicting three-dimensional views of structures in which features in one section—say, that perpendicular to the strike of a folded series—are complicated, and those in the section at right angles—say, that parallel to the strike—are relatively simple.

Right-Sectional Block Diagrams If the line-of-sight chosen for a block diagram is parallel to the tectonic, or *b*, axis, instead of the line of sight of fixed inclination characteristic of an isometric block diagram, and then if geological data are projected on lines parallel to the *b* axis, the developed drawing is a right section. This drawing can then be projected without change of pattern onto any surface, plane or curved, and of any orientation. Thus, it can be drawn without difficulty on the surface or sides of a block that is properly proportioned and oriented according to standard rules of axonometric projection. Topographic features likewise can be projected onto the surface of the block, following procedures outlined by Lobeck (1924).

The advantages of such drawings are twofold. First, the sectional profiles are constructed without recourse to apparent dips or to assumptions as to changes in thickness with depth. Second, since the right-sectional pattern appears unchanged on any random section, the true pattern of any geological features on any arbitrary section can be shown by simply rotating the chosen plane, on which the right-sectional pattern has been drawn, into the plane of the paper. The method fails if, within the area of interest, there is, in fact, no *b* axis of reasonably constant orientation. If there is a valid *b* axis, within reasonable limits the method is simple and informative.

J. W. AMBROSE

References

Ambrose, J. W., and H. T. Carswell, 1961, Right sectional block diagrams, *Geol. Assoc. Canada Proc.* **13**, 119–128.

Busk, H. G., 1929, *Earth Flexures.* Cambridge: Cambridge University Press, 13–33.

Coates, J., 1945, The construction of geological sections, *Mining, Geol. and Metall. Inst. Quart. Jour.* **17**(1), 1–11.

Dahlstrom, C. D. A., 1954, Statistical analysis of cylindrical folds, *Canadian Inst. Mining and Metallurgy Trans.* **57**, 140–145.

Gill, W. D., 1953, Construction of geological sections of folds with steep-limb attenuation, *Am. Assoc. Petroleum Geologists Bull.* **37**, 2389–2406.

Lobeck, A. K., 1924, *Block Diagrams.* New York: John Wiley & Sons, 206p.

Mackin, J. H., 1950, The down-structure method of viewing geologic maps, *Jour. Geology* **58**(1), 55–72.

Mertie, J. B., Jr., 1940, Stratigraphic measurements in parallel folds, *Geol. Soc. America Bull.* **51**, 1107–1134.

Mertie, J. B., Jr., 1947, Delineation of parallel folds, and measurement of stratigraphic dimensions, *Geol. Soc. America Bull.* **58**, 779–802.

Mertie, J. B., Jr., 1948, Application of Brianchon's theorem to construction of geologic profiles, *Geol. Soc. America Bull.* **59**, 767–786.

Rubey, W. W., 1926, Determination and use of thicknesses of incompetent beds in oil field mapping and general structural studies, *Econ. Geology* **21**, 333–351.

Stockwell, C. H., 1950, The use of plunge in the construction of cross-sections of folds, *Geol. Assoc. Canada Proc.* **3**, 97–121.

Wegmann, C. E., 1929, Beispiele tektonischer Analysen des Grundgebirges in Finnland, *Finlande Comm. Géol. Bull.* **13**(87), 98–127.

Cross-references: *Block Diagram; Cartography, General; Map Symbols.* Vol. XIII: *Geological Communication.*

PROSPECTING

Prospecting is the field search for mineral deposits, usually with the ultimate objective of economic development by mining operations. Oil and natural gas prospecting is a special category and involves specialized techniques; water search also is normally treated separately. Particular techniques of prospecting are likewise treated separately in this encyclopedia series (see prospecting topics in this volume and in Vol. XIII).

Prospectors may be divided into two general categories: those with academic, scientific training and those with only field training or in the purely amateur category. Sometimes the person with book learning is liable to look down his or her nose at this second category of prospector, which not only is gauche and sophism but also may be ignoring a wealth of local knowledge that may be of inestimable value. It may be wise to recall that Neolithic humans were prospecting and successfully mining nodular chert (for flint-implement manufacture) about 8,000 yr ago. Not long after, their descendants were effectively prospecting and developing deposits of gold, silver, copper, and lead. During the great Gold Rush epics of the 1800s, very few stakes were hammered by trained geologists. In the 1900s a great many mineral deposits have been located by people whose powers of observation and native intelligence more than make up for their lack of diplomas.

The academically trained prospector in the late twentieth century is an economic geologist with concentrated training in field and subsurface survey methods, photointerpretation (remote sensing), mineralogy, basic geochemistry, and elementary geophysics; some knowledge of modern sedimentology or marine geology and igneous petrology is also desirable.

Prospecting Methods

Field Survey and Sampling Basically this is field inspection, with sampling of interesting-looking outcrops. The trained prospector often uses a published geological survey map (see *Map Series and Scales*) as a general guide to the region and supplements it with air photos that disclose the topographic details as well as trend lines, fault traces, vegetational markers, etc. In this way much unpromising country can be eliminated, and one may make directly for the likeliest spots. Transportation is often a limiting factor, including travel on foot or by canoe, muleback, jeep, or helicopter, depending on terrain and funds available.

Samples are normally collected by eye or with no more than the aid of a X10 hand lens. In radioactive ore search a field Geiger counter or scintillometer can be used. Samples should normally be about 250–500 g, and care should be taken to identify them by number with a firm adhesive label and to protect them in a suitable bag or sack. Depending on the climatic conditions, the collecting bags may be brown paper, plastic, or calico.

The field notebook should be of high quality since it may get a lot of hard use and may have to be referred to for periods as long as a lifetime. The paper should be waterproof and so should the pen. There have been far too many reports of a season's work lost because a pack was saturated with water when a flood came down or a canoe overturned.

Field observations should be supplemented with photographs and in perilous situations the cameras and film should be kept in watertight plastic bags. In the tropics it is helpful to keep them cool in a thermal picnic bag or box; however, in humid tropics care should be taken to prevent condensation on lenses and film when they are brought out suddenly from a cool place. Many prospectors like the convenience of the Polaroid-type camera with its instant pictures on which notes can be directly inscribed while in the field. Some workers save further field time by making a running commentary on all their observations with the use of the cassette-type portable, battery-operated tape recorder. This procedure requires careful re-listening each evening with selective note taking. A typed field report can be maintained on a daily or weekly basis.

Some prospectors also keep a small butane-operated gas torch at their base camp so that quick field analyses can be obtained using the borax-bead methods and so on (see *Blowpipe Analysis*). A very small portable laboratory, containing pestle and mortar, test tubes, and the usual solvents or dye tests is not difficult to set up and can save transportation of many tons of doubtful specimens (see *Mineral Identification, Classical Field Methods*).

The standard field instrument is the *geologist's pick,* or *knapping hammer.* Care should be taken to strike the rock away from the body (and the eyes of assistants), since flying fragments are sometimes as dangerous as shrapnel.

When prospecting fluvial deposits and saprol-

ites for residual minerals, native gold, tin, and other heavy ores, the *prospector's pan* is very useful. It is a simple iron pan of arcuate section. A source of water is essential. The pan is half-filled with water; a shovelful of alluvial sand and gravel is dumped into the pan. Larger pebbles are picked out by hand, and the clay goes into suspension and must be decanted; then the pan is refilled with clear water and rotated in a swirling motion so that the light sands are gradually swept over the lip of the pan, leaving the heavy minerals and, one hopes, the gold nuggets as a residue.

The prospectors' field method in hilly country is either to move up the creeks or the ridge lines, depending on where the best outcrops are seen. Frequent systematic sampling is usual, although after close inspection most of the samples are thrown away. It should be remembered that, except in glaciated country, pebbles and boulders cannot move uphill. Therefore, as one goes upstream, one gradually narrows down the source of any interesting *float* (i.e., loose specimens). The *mother lode* should ultimately be located. This is usually a dike or vein or wall-rock contact zone where the valuable minerals are concentrated.

In areas where no geological map exists, many professional prospectors find it wise to construct a *reconnaissance geological map* as they go along. It may not involve very much more work, when a foot traverse of all outcrops is involved, and the preparation of a regional map will help to pinpoint favorable structures for closer study.

Pitting, Trenching, Shaft Sinking, and Bulldozing In country with thick soil and/or vegetation, outcrops are often far apart or even nonexistent. If cheap manual labor is available, it is often financially quite practicable to use pick and shovel techniques for opening pits or trenches or modest shafts. These are, however, time consuming but certainly cheaper than diamond drilling (see next section); their potential should be exploited before more expensive prospecting methods are applied.

The bulldozer is normally an expensive field aid, but in a mine area or civil engineering operation—e.g., a major dam construction—the geologist may have easy access to these valuable power tools. A fault zone or a mineral vein can be systematically scraped clean of soil and weathered overburden. It can then be followed, without fear of faulty correlation, as far as it goes. The bulldozer can also be used to clear off scrub and smaller forest trees.

Borehole Drilling This is not the place for engineering details of drilling techniques, but some notes on their application are appropriate. Auger drilling (q.v.) is the simplest procedure. The auger may be hand operated or mechanized (usually mounted on the back of a jeep or truck). The auger operates with a spiral flange or screw that collects the cuttings in chips or flakes that can then be washed and examined by hand lens. The auger is usually most effective in penetrating a soil and saprolite layer about 1–30 m in depth but will not penetrate solid rock. It also fails against boulders or corestones in the saprolite. Advantages include speed, cheapness, and easy transportability. Because of its ineffectiveness in hard rocks, the auger is of less use for metal ore prospecting (though it is useful in geochemical work on their saprolites) and is more applicable to prospecting for the soft sedimentary deposits, such as clay, coal, phosphate, and salts. It may also be used to evaluate mine tailings, with a view to re-exploitation for trace metals.

Cable-tool, churn, and rotary drilling are usually reserved for oil, natural gas, and deep artesian water drilling (see *Borehole Drilling*). However, they are also employed in serious prospecting for salts (halite and the potash salts), coal, limestone, phosphate, and similar sedimentary ores.

Core drilling, either by calyx or by diamond drilling, is often the most expensive form of drilling, but it also yields the most information to the geologist, who can examine and report on the rock core furnished by the drilling. The calyx drill is often used on dam sites or in coal mining operations where a very wide shaft is needed (up to 5 m diameter) to permit a geologist to go down the hole to examine the section at first hand. It can also be a life saver in mine rescue operations. Small-diameter diamond drilling is the best advanced prospecting technique for hard-rock mineral search. Both vertical and low-angle oblique holes can be put down. The narrow cores are easy to collect, examine, and store. The hard-rock samples, after the initial reporting, go on to the specialist. A detailed treatment of diamond drill work for the economic geologist is contained in Forrester (1946, 398).

Expensive drilling programs are rarely undertaken nowadays without prior application of some of the more sophisticated prospecting techniques that employ geochemical and geophysical methods. Cross-reference to entries on these subjects are given at the end of this one.

Development of Prospects

Once the geologist has made a discovery or marked out a number of hopeful-looking sites, a series of steps are necessary before any of them can be identified as a potentially exploitable deposit. The claim has to be staked and registered. At this stage it often changes hands, and before a more detailed scrutiny is made of the area, old mine sites in the neighborhood are reappraised and various economic matters (transportation costs, the labor situation, the state of the market) are reviewed. The economics of the subject are generally embraced under the heading of ore valuation.

In the early stages the geological prospector must often be something of an engineer, a businessperson, and a negotiator. He or she must become acquainted with state and federal legal matters (see *Legal Affairs*) (see Shamel, 1907). Such matters may seem to be far removed from the initial role of the prospector, as explorer, but unless this person takes the trouble to see his or her hard work through to the development stage, he or she must always be resigned to see the rewards fall into someone else's hands.

RHODES W. FAIRBRIDGE

References

Forrester, J. D., 1946, *Principles of Field and Mining Geology.* New York: John Wiley & Sons, 647p.

Jackson, C. F., and J. B. Knaebel, 1934, Sampling and estimation of ore deposits, *U.S. Bur. Mines Bull. 356.*

Keller, G. V., and F. C. Frischknecht, 1966, *Electrical Methods in Geophysical Prospecting.* Oxford: Pergamon Press, 517p.

Raeburn, C., and H. B. Milner, 1927, *Alluvial Prospecting.* London: T. Murby & Co., 478p.

Shamel, C. H., 1907, *Mining, Mineral and Geological Law.* New York: Hill Publ. Co., 680p.

Von Bernewitz, M. W., 1969, *Handbook for Prospectors,* 3rd ed. (rev. by H. C. Chellson). New York: McGraw-Hill, 529p.

Cross-references: *Augers, Augering; Biogeochemistry; Blowpipe Analysis; Borehole Drilling; Exploration Geochemistry; Exploration Geology; Exploration Geophysics; Hydrogeochemical Prospecting; Lake Sediment Geochemistry; Lithogeochemical Prospecting; Mineral Identification, Classical Field Methods; Pedogeochemical Prospecting; Photogeology; Photointerpretation; Plate Tectonics, Mineral Exploration.* Vol. XIII: *Economic Geology; Min eragraphy.*

PUNCH CARDS, GEOLOGIC REFERENCING

Most bibliographic referencing systems range in complexity from complex computer terminals to the simple index card file. The marginal *punch card referencing system* has features of both these systems. Unlike computer systems, which require a terminal for input, punch cards can be documented in the office, the laboratory, the library, at conferences, and even in the field. Data in the form of abstracts, diagrams, microfiche, and even stereo aerial photographs can be included directly on these cards (see *Geological Cataloging*). Suppliers of these unique reference cards are Royal McBee, Port Chester, New York; Professional

Aids, Chicago, Illinois; or Indecks, Arlington, Vermont.

This system realizes a maximum possible use with minimum expense, time, and effort. Complete data are printed directly on the reference card. Rapid, accurate coding and sorting are done by hand with uncomplicated equipment. Reference cards need not be filed or refiled in any particular order. The system also is highly selective; i.e., it allows extraction of all pertinent information on a subject with confidence that all possible references have been located.

Punch Card System

A bibliography compiled by the marginal punch card information file system is an inexpensive, effective nonconventional indexing, storage, and retrieval system that does not require a computer. The use of the punched card file system, whether computer referenced or noncomputer, permits easy access to file data for both general and specific information during all stages of compilation of a bibliography. In this system, indexing is by use of codes, edge notched in the standard marginally punched bibliographic card (Fig. 1). Bibliographic documentation is typed on the body of the card, which in addition to its use for indexing, is also the storage, search, and retrieval card. Search and retrieval is by a hand sorting rod or a simple manual sorting machine that retrieves the original indexed card. Only one card need be used for cross-referenced bibliographies such as for a subject, topic, subtopic geology bibliography. The system can search and sort on several categories at once or one at a time. It has the ability to locate any topic or combination of topics of up to 5,000 items of data. One can retrieve for complete data

FIGURE 1. Ease of sorting for the coded reference cards is the major feature of the marginal punch card system. A sorting rod inserted in the proper index hole allows all the notched reference cards simply to fall out.

on any category or only items that fit more than one category. In addition to bibliography cross-reference indexing, the system can be adapted for abstracts, statistical data, survey data, field data, and other records. This system can handle a file system of up to 10,000 cards with indexing, abstracts, or original articles at a sorting rate of 400 cards per minute (Casey, 1958).

Coding

To code a standard marginal punch card for compiling a bibliography, the following six basic classification requirements are necessary (Fisher, 1963):

1. Year of publication (including chronological index),
2. Author (alphabetically),
3. Subject classification (topics and subtopics),
4. State (and physiographic province where applicable),
5. Geologic age (where applicable),
6. Modifiers, or auxiliaries.

Coding is accomplished by assigning values to the holes around the edges of the cards for these classification requirements. A single factor may be assigned a single hole (direct index), which permits a direct sort or search of that factor. Several mutually exclusive topics can be coded by numerical coding. Groups of holes are assigned in so-called fields to permit topic and subtopic as well as sequence sorting or searching. Classification or coding is accomplished by notching the hole to the edge of the card, or when double-row punching is used, two holes are joined to form a slot. This can be done either by hand or machine punch. To sort, search, and retrieve an indexed card, a sorting needle or needles are inserted into the file of cards through the desired classification hole. The desired cards fall from the needle.

Year of Publication Three fields are provided on the standard bibliographic card for recording year of publication. The two fields on the extreme upper right of the card are for coding the last two digits of the year. The field to the left is for coding the century. References prior to the seventeenth century are not notched. Eighteenth-century references are notched in the hole indicating 17; nineteenth-century references are in the 18 position; and twentieth-century references are in the blank position on the left. Numeric values for the year digits as well as other numeric sequence classifications are assigned by coding the holes numbered 1-2-4-7 in a four-hole field as described in the following.

A field of four holes, numbered 1-2-4-7, can

be used to classify numeric value from 0 to 9 by notching as follows (Fig. 2):

0: No notch
1: Notch 1
2: Notch 2
3: Notch 1 and 2
4: Notch 4

5: Notch 1 and 4
6: Notch 2 and 4
7: Notch 7
8: Notch 1 and 7
9: Notch 2 and 7

This numeric coding can be expanded to permit additional classification to 14 as follows:

10: Notch 1-2-7
11: Notch 4-7
12: Notch 1-4-7

13: Notch 2-4-7
14: Notch 1-2-4-7

Therefore, one field of four holes will code 14 mutually exclusive numeric classifications; two fields, 149 numeric classifications, three fields, 1,499 classifications, etc.

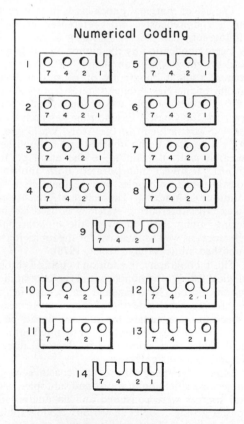

FIGURE 2. A unique numeric coding system has been designed so that a field of four holes, numbered 7, 4, 2, and 1, can indicate numeric values through 14. Two fields can indicate values to 149, three fields to 1,499, etc. This coding allows development of an extensive numeric subject index.

Author Coding Author coding occurs in the alphabetic field index along the right margin of the card. Letters of the alphabet are assigned numbers; A represented by 1, B by 2, etc. to M, which is represented by 13 using the preceding numeric coding. The second half of the alphabet is similarly coded, but the hole is also notched. Thus, N is represented by N–Z and 1, O by N–Z and 2, etc. to Z, which is represented by N–Z and 13. A card containing a reference listing on a particular topic is coded as 14 in the first field. The first three letters of an author's name are coded in their respective fields. Multiple authorship is indicated by notching the hole with the plus sign just to the left of the alphabetical field. In such cases, only the first letter of each author's name is coded in each of the three alphabetical fields. Another alphabet code uses the letters O-I-E-C-B that also appear in the alphabetical fields. It is primarily mnemonic and requires less notching, since the more common letters are represented by one or two notches. It is, however, a more complex system than the numerical alphabetical system.

Punch Card Bibliographies

The most common use of marginal punch card filing is the loan record system presently used in most large libraries. Cards are 3 × 5 in. and are usually marginally punched along two edges. Coding usually consists of borrower's status and the day and week of book return. Manual sorting by rod indicates which books are due for return when a certain day/week combination comes around.

A "Bibliography of New York Quaternary Geology" (Muller, 1965) was compiled using the marginal punch card system. The bibliography included a listing of published papers relating to New York Quaternary investigations, a geographic index, and a subject index. All data were tabulated on bibliographic marginal punch cards of the size and shape of the familiar IBM computer card with a single row of holes and indexing on a direct one-to-one basis.

Use of the marginal punch card system in maintaining a bibliography for radioactive carbon (^{14}C) dating studies was pointed out in 1965 (R. Fairbridge, pers. comm.). The present author suggested the possibility that the actual abstract of the original article could be printed within the body of the marginal punch card by xerography if the machine paper rack were loaded with blank marginal punch cards. Marginal punch card indexing for radioactive carbon dating or similar geologic correlation studies is especially useful. In addition to the usual author-date bibliographic information, the file can be indexed for the exact radioactive date as well as type radioactive carbon material (shell, wood, bone, etc.). Geological localities as well as environments (marine, nonmarine, etc.) could also be cross-indexed and sorted out by the radioactive dates.

Economic geology regional studies have found the punch card system useful for uranium studies (Finch, 1957) and general geochemical problems (Berger, 1958). A unique punch card system (Avery, 1967) includes aerial stereophotographs on the card together with identifying data. Microfiche film reference systems by which up to 500 pages of information can be recorded on a single 4 × 6 in. card have been combined with punch card systems (Kodak, 1971). This produces a very compact and efficient system that does not require that thousands of pages of information be typed into the system, which would be the case using computer retrieval systems.

Reference Retrieval Systems

Library reference systems such as the Dewey Decimal System and the Universal Decimal Classification System are often considered as numerical classification systems; however, the numbering system actually refers to an elaborate subject code. The Decimal Classification (Dewey Decimal System), first published in 1876, had 1,000 subject classifications divided into 9 main classes, with each of these divided into 9 subclasses. Each of these subclasses is divided further into 9 further subclasses, with general works constituting a tenth subclass in each category. This division into ten may be continued indefinitely—hence, the decimal classification.

The concept of "key-word" indexing has begun to appear in bibliographic reference coding systems in recent years. For some, it is simply a predetermined listing of subject headings (cf. Dolar-Mantuani, 1970), while for others it is an open-ended vocabulary of descriptive words and phrases, referred to as "Coordinate Indexing Systems" (Graves et al., 1967). Key-word categories are often generated by terms in the title or, less often, the abstract of the article under consideration. In general, key-word indexing is commonly used in computer-based retrieval systems where a large number of possible key words or combinations of key words can be searched rapidly.

Examples of machine retrieval geological or marine key-word indexing bibliographic systems are the American Geological Institute's GEO-REF system, the University of Tulsa Information Retrieval System (Graves et. al., 1967), and the Oceanic Index (Farmer, 1969). The Oceanic Index, as of January 1968, introduced a subject index designated "Keytalpha" for key-term alphabetical (Sinha, 1968). Significant key words for each article are arranged alphabetically, enabling retrieval by subject topics more rapidly and completely than an earlier index. Also added were numerical designators from the Universal Decimal Classification system. The American Geological Institute's GEO-REF system file of 2,000 earth science serials from 1969 uses both the Universal Decimal Classification system and a so-called natural language system for data retrieval. Twenty-one geologic categories, as well as descriptors and specifiers, are drawn from the cited document. The Smithsonian Science Information Exchange will search its data base on research topics in the life and physical sciences as supported by over 1,300 organizations (SSIE, 1977). Research topic key words as indicated by the researchers are the basis for the data retrieval. Computer systems have similar reference retrieval systems, such as Versafile, in which a question containing one or more key words will retrieve filed information.

Punch Card Referencing—A Case Study

The use of marginal punch cards in constructing a comprehensive cross-referenced geologic bibliography can best be shown by a case study. A recent application by the author of this punch card system was the compilation of a cross-referenced topic bibliography of coastal geologic studies of the New England states from 1800 to 1970. The bibliography was produced under contract for the Coastal Engineering Research Center, U.S. Department of the Army, Washington, D.C. It covered both published and unpublished research and, in addition to the usual bibliographic listings, incorporated both multiple coastal topics, subtopics, and regional state location cross-indexing. The marginal punch card index cross-referencing system was used for indexing, filing, and retrieving the bibliographic references as well as the basis for the topics in the published bibliography (Fisher, 1978).

This bibliography, in addition to being a chronological listing of regional coastal geologic studies for the various New England states, also incorporates a subject indexing of the various coastal aspects of these studies. Included are studies published in journals, federal and state publications, field guides, institution reports, doctoral dissertations, and masters theses.

Topic Referencing To set up a coastal geology subject classification of topics and subtopics, several sources were consulted and an analysis of coastal geologic topics in various bibliographies was conducted. Those bibliographies that are indexed by key words, such as the "Index to the Journal of Sedimentary Petrology" (Bloom, 1958) and "The Oceanic Index" (Sinha, 1968), are not suitable for developing a subject index because the key-word terms were developed from the content of the documents in the file and do not necessarily reflect future documents and future possible key words.

A subject index, rather than a report-generated key-word index, was considered to be the most useful for topic searching. The major subject categories were chosen by an analysis of other geologic and marine geology bibliographies as well as the type of coastal geologic studies in the New England area.

In general, coastal geologic subject categories were expanded by reference primarily to subtopics of these subjects as categorized in the U.S. Geological Survey's *Bibliography of North American Geology*. This bibliography has prepared indexes of geologic publications from 1785 to date.

Marginal Punch Card Coding Publication date and author are coded as discussed previously and are indexed along the upper right outer margin and along the right margin (Fig. 3). Subject coding by direct numerical index is along the entire lower outer margin and the lower left margin. States are coded along the lower inner margin with 1 = Maine, 2 = New Hampshire, etc. For all 50

states in the author's system, coding for the eastern states extends to number 25, with the remaining 25 western states indexed in the same holes but with the plus sign in the right corner notched. This technique doubles the use of a row of holes since few references would cover both an eastern and a western state. In cases where a state (inner row) and subject (outer row) coincide, a false drop will occur when sorting on this subject. Sorting on the state for this subject, however, will eliminate the inapplicable cards. The subject index includes the names of local beaches and estuaries. These names are coded by first letter, using the alphabetical system discussed under author coding. This coding is in the upper left margin. Geologic ages are coded by a numerical system in the upper center field. Most of the references of the present coastal bibliography were coded as "Recent" with some as "Pleistocene."

Modifiers (auxiliaries) are used to extend the range of a particular subject. For example, articles discussing postglacial submergence rates as determined by carbon dating are further indexed by using a numerical code for the numerical dates.

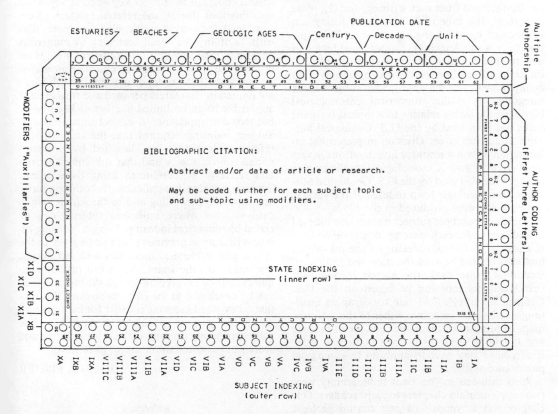

FIGURE 3. The following data were marginally coded by this punch card system: author, publication date, geologic age, state, several hundred coastal geology topics, together with their subtopics and topic modifiers (auxiliaries).

This code applies only to these subject cards; however, for a reference card dealing with beaches, modifiers further code for beach erosion, beach deposition, beach longshore transport, etc. The use of modifiers can therefore produce a very detailed marginal punch card reference file. Sorting is usually first done on various subtopic modifiers, and if sufficient references are not located, then the more general subject topic is sorted. This ability to sort on various levels of the subjects as well as the ability to incorporate both detailed and general information, even in numerical form, is a prime advantage of the system.

Final Bibliography The published bibliography proper (Fisher, 1978) was presented in the form of an extensive listing of topic and subtopic subjects together with a chronological author listing with sequential numbers for each of the New England states. After each topic and subtopic in the general subject outline, there are a series of letter-numbers that indicate the bibliographic author reference for that topic. A chronological author listing rather than the usual alphabetical listing was chosen for two reasons: (1) the historic development of coastal studies in a particular state can be discerned from such a listing, and (2), more important, this type of bibliographic listing can be updated by preparing, either annually or periodically, a supplemental chronological listing of new references. Each new reference can also indicate its subject matter content by simply indicating numerically the appropriate subject index categories. A similar numerical chronological bibliographic listing relative to a topical (subject) index has been used by the U.S. Geological Survey's Water Resources Division in presenting information on water resource investigations in various river basins. Chronological bibliographic listings were also used by the U.S. Geological Survey in their Geologic Map Indexes of the various states, which were produced in the 1950s.

Topic and subtopic subject matter was indexed in addition under each state by locational coding. If a further locational indexing of coastal subject matter is desired beyond the state and individual locational name level, the author recommends that locational indexing be based on the U.S. Geological Survey's 7 ½ min topographic quadrangle series with reference either to the different maps or portions of the maps. The use of latitude and longitude locations using a numerical index is of course very easy to apply in this marginal punch card system.

Also included in the final bibliography were two supplementary chapters to assist readers. One chapter was a synopsis of past coastal geologic studies of each New England state as indicated by the chronological bibliography. The other chapter was a synopsis of the coastal geology of each state, since it was anticipated that as many nongeologists as geologists would be using this bibliography in coastal planning and environmental impact studies.

Summary

In using any referencing system, computer or noncomputer, the major problem is retrieval assurance—the assurance that the bit of information on the topic or subjects, if within the data system, has been retrieved. First, when searching it is necessary to use the correct key word or key words that were used to index the original article. When topics are fitted into a subject classification such as the Universal Decimal Classification, searching a general subject classification often retrieves topics applicable to the search that might not appear if searching were limited to select key words.

There are advantages and disadvantages to the two basic retrieval systems, key-word index and subject index. Key-word indexing enables many general and/or specific terms to be indexed based on the data that are put into the file. The problem that often occurs in the retrieval is making the correct choice of the proper key word or key-word combinations during the reference search. Key-word indexing is most applicable to computer files with its ability for rapid searching of numerous terms and tape and disc storage where there is no limit to the terms that can be coded. Automated punch cards (IBM style) or marginal punch cards are limited in characters per card and thus are not applicable to an unlimited number of key words, but they are applicable to subject topic retrieval. Subject indexing requires that the content of a reference document be classified by predetermined subject topics and that subsequent search and retrieval be conducted using these terms. Subject indexing is applicable to both computer and noncomputer filing and to the simple punch card systems where unlimited subjects can be coded by numerical indexing. Subject indexing requires that some previous planning be given to the topics and subtopics, modifiers and auxiliaries, that make up the index. Key-word indexing requires little or no preplanning of the index, which can be developed as the filing progresses. In summary, key-word indexing is easier for indexing and search but presents problems in retrieval, while subject indexing is more difficult for indexing and search but is often more useful in retrieval.

JOHN J. FISHER

References

Avery, T. E., 1967, All sorts of stereograms, *Photogramm. Eng.* 33, 1397–1401.
Berger, I. A., 1958, Design of simple punched card sys-

tem with reference to geochemical problems, *Econ. Geology* **53**, 325–338.

Bloom, B. H., 1958, *Index to the Journal of Sedimentary Petrology,* Vols. 1–26 (1931–1956). Tulsa, Okla.: Society of Economic Paleontologists and Mineralogists (Spec. Pub. 6), 55p.

Casey, R. S., 1958, *Punched cards, Their application to service and industry.* New York: Reinhold Publishers, 200p.

Dolar-Mantuani, L., 1970, More on data retrieval, *Geotimes,* September, 8.

Eastman Kodak Co., 1971, *Micropublishing Systems and Services.* Rochester, N.Y.: Business Systems Markets Division, 8p.

Farmer, W. C., 1969, Oceanic Research Institute emphasizes up-to-date information, *NODC Newsletter,* April (No. 4-69), 5.

Finch, W. I., 1957, Application of punch cards for uranium data, *Econ. Geology* **52**, 180–191.

Fisher, J. J., 1963, Punch card coding for a selective geologic reference system, *World Oil* **156**, 89–93.

Fisher, J. J., 1978, *Bibliography of New England Marine Geology.* Kingston: New England Section, National Association of Geology Teachers, Deptartment of Geology, University of Rhode Island, 91p.

Graves, R. W., D. P. Helander, and S. J. Martinez, 1967, University of Tulsa Information Retrieval System, *Geol. Soc. America Mtg. Abs.,* 82.

Muller, E. H., 1965, Bibliography of New York Quaternary geology, *New York State Mus. and Sci. Service Bull. 398,* 116p.

Sinha, E., 1968, Oceanic index, *Citation Jour.* 5(1), 1.

Smithsonian Science Information Exchange (SSIE), 1977, *Newsletter* 7(2), 8p.

Cross-references: *Geological Cataloging; Map and Chart Depositories.* Vol. XIII: *Computerized Resources Information Bank; Earth Science, Information and Sources; Geological Communication.*

R

RADIOACTIVITY SURVEYS—See AERIAL SURVEYS; WELL LOGGING.

RADIOGEOLOGY—See EXPLORATION GEOPHYSICS. Vol. XIII: GEOCHRONOLOGY.

RECLAMATION PLANNING—See MINING PREPLANNING.

REFERENCE RETRIEVAL SYSTEMS—See COMPUTERS IN GEOLOGY; GEOLOGICAL CATALOGUING; PUNCH CARDS, GEOLOGIC REFERENCING. Vol. XIII: COMPUTERIZED RESOURCES INFORMATION BANK; EARTH SCIENCE, INFORMATION AND SOURCES.

REFLECTION PROFILING—See ACOUSTIC SURVEYS, MARINE.

REMOTE SENSING, GENERAL

In the geosciences, the term *remote sensing* refers to the procurement of images of the Earth's surface from airborne sensors or aerial or space platforms (Hunt, 1972). The sensors detect and then record, through photographic or optical electrical systems, various wavelengths of electromagnetic ground radiation. At present, the wavelengths of the electromagnetic spectrum being sensed for practical studies include visible light (0.4–0.1 μ), near infrared (0.7–1.0 μ), middle and far infrared (4–20 μ), and microwave radio (900 + μ).

There are various types of sensors in use, including cameras, multispectral scanners, microwave sensors, and force-field sensors (see *Photointerpretation*). The images are recorded on photographic film or magnetic tape, and they differ in scale, resolution, color, and procurability, depending on the sensor employed. Initially, from the early 1930s to the early 1970s, the sensors were

borne aloft in aircraft (Rabshevsky, 1970). Then, in 1972, the U.S. National Aeronautics and Space Administration (NASA) launched the first of its orbiting Earth Resources Technology Satellites (ERTS-1), the first satellite system dedicated solely to Earth resources sensing (Williams and Carter, 1976). This was followed in 1974 by the manned spacecraft Skylab, and in 1975 by a new unmanned spacecraft, LANDSAT. These platforms orbit the Earth at very high altitudes and carry sophisticated cameras and multispectral scanners. The state-of-the-art of remote-sensing technology is ascending rapidly in all phases—platforms, sensors, and imagery use and interpretation (Lillesand, 1979).

Camera Sensors

Camera and film systems are still the most important remote sensors in use. Prints or negatives can be used for three-dimensional viewing when obtained in sequence with overlap. Emulsions in use include films sensitive to visible light (black-and-white panchromatic film, and color film) and films sensitive to infrared (black and white and color).

Black-and-white films are still used with the greatest frequency because of their lower costs and higher resolutions than other films. The resolution capabilities of color films are being upgraded continuously, however, and costs are decreasing. Stereoscopic viewing and interpretation of color prints yields more data concerning geologic formations, surface soils, vegetation, and water than black-and-white prints and are favored by interpreters (see *Photogeology*).

Infrared films, producing black-and-white images, contain an emulsion sensitive to radiant energy extending beyond the visual range into the near-infrared portion (0.9 to 1.1 μ). Black-and-white infrared imagery is useful in areas perpetually covered by haze because of its penetrating capability. It also facilitates interpretation of water boundaries, tidal zones, and soil-moisture conditions. Color infrared (commonly called color IR) is often referred to as "false color" because the colors recorded are different than those appearing to the eye. Its primary uses are for crop or vegetation inventories and delineation of surface-water conditions. Healthy vegetation is presented in bright magenta red colors, whereas diseased vegetation is apparent in more somber hues.

Soil vegetation and water boundaries are enhanced by color IR, as are geologic features such as faults and joints. Water in rock fractures supports denser vegetative growth than in adjacent areas and appears as dark colors because it absorbs infrared.

A recent development is *multispectral photography*. The technique employs positive black-and-white transparencies, which transmit light filtered by various colors. The filters are used singly or are combined to achieve the most significant color enhancement to aid in interpretation. In addition to the applications described for other film-recorded imagery, the technique is extremely useful in detecting marine conditions such as pollution or shallow water depths.

Photographs of the Earth's surface are obtained in many scales. Low-altitude flights obtain imagery in scales ranging usually from 1:10,000 to 1:40,000. High-altitude aircraft such as the U-2 or RB-57 are used to procure imagery at scales on the order of 1:100,000. The Skylab color imagery, of very high resolution, was procured at a scale of 1:400,000.

Multispectral Scanners

Multispectral scanners (MSS) are electro-optical instruments mounted in aircraft and spacecraft. A scanning mirror senses the reflected and remitted energy of the ground; the energy beam is separated by an optical system into four discrete wavelengths (bands), and the signal from each band is recorded on magnetic tape. This was the system installed in the ERTS satellite. As the satellite moved around the Earth at an altitude of approximately 500 nautical mi (800 km), it scanned a swath 100 nautical mi wide, ultimately producing images covering 100 × 100 nautical mi. A given location on Earth was circumscribed every 18 days. The images were recorded in four discrete bands, where band 4 recorded in the 0.5–0.6 μ range (green), band 5 in the 0.6–0.7 μ range (red), band 6 in the 0.7–0.8 μ range (infrared), and band 7 in the 0.8–1.1 μ range (infrared). Bands 1 through 3 were designated television bands, for a three-television camera system on the satellite that malfunctioned from the outset of the flight.

The signals stored from each band on magnetic tapes are converted to four 70 mm film masters from which black-and-white renditions of each frame are made for examination and interpretation. Black-and-white positive transparencies are produced for multispectral viewing and printing using filtered light as described previously, or three of the bands can be combined (usually 4, 5, and 7) to produce a single-color infrared presentation that closely approximates the coloration of color IR film.

The ERTS imagery provides views of geomorphological features over large land areas, permitting interpretation and mapping of geologic formations on a scale not previously possible; large faults and other structural features extending for long distances are particularly evident. Because it was obtained periodically, the imagery is also useful to delineate limits of major floods and for making seasonal assessments of vegetation changes.

The scanner system is also used for thermal IR sensing to record heat radiation in the 8–14 μ portion of the infrared spectrum. With low-altitude, high-resolution imagery, temperature differences of less than 1°C have been detected. The imagery is used to detect drainage in areas with heavy vegetative cover and temperature differences in water bodies (thermal pollution), as well as underground fires and volcanic activity. The imagery is most useful when obtained at night or on cloudy days to avoid reflected sunlight radiation that obscures the thermal differences.

Microwave Sensors

At present, the primary microwave system in commercial use is sidelooking airborne radar (SLAR). Energy is directed at the ground from aircraft-borne equipment, reflected back to the aircraft, and sensed through a radar antenna. It is converted to an electronic signal and, through focusing optics, onto a film recorder. The results are black-and-white images of the Earth's surface, usually procured at scales of 1:250,000 to 1:500,000. These images are normally enlarged for interpretation to 1:125,000. Enlargement to larger scales is occasionally possible, but the poor resolution inherent with SLAR normally prohibits greater enlargement or the recording of objects greater than 15 m in width.

High terrain relief is apparent on SLAR imagery because of the shadow effect of the scanning radar waves on landforms. This characteristic makes the imagery useful in detecting significant geomorphological features and drainage patterns (Barr and Miles, 1970). Although resolution is poor compared to other sensors, SLAR can penetrate light clouds and even some ground vegetation to sense ground shape. Thus, it is particularly useful in areas of perennial or near-perennial cloud cover, areas where other types of imagery cannot be obtained.

Force-Field Sensors

Airborne magnetometers and gravity meters are force-field sensors that have long been used for geologic studies, particularly for mineral and oil exploration. Aerial magnetic surveys measure and record differences in magnetic properties of rocks.

These differences are plotted as contours and interpreted to identify anomalous zones that can be associated with geologic structures. The magnetic and gravitational methods are analogous in their fundamental relationships, as they are quantitative investigations made of a natural field of force.

Applications of Imagery

The applications of remote-sensing technology to terrestrial analyses are legion. The data generated from imagery interpretation are used for studies involving geology (structural, surficial, pedological, engineering soils, natural resources); hydrology (surface water, groundwater, drainage); pollution (air and water—physical, chemical, thermal); vegetation (agriculture, forestry, wetlands); and demography (land-use planning, transportation).

ROY E. HUNT

References

Barr, D. J., and R. D. Miles, 1970, SLAR imagery and site selection, *Photogramm. Eng.* **36**(2), 1155–1171.

Committee on Remote Sensing for Agricultural Purposes, 1970, *Remote Sensing.* Washington, D.C.: National Academy of Science, 424p.

Hunt, R. E., 1972, The geologic environment—Definition by remote sensing, *Internat. Conf. Microzonation* **2**, 577–592.

Lillesand, T. M., and R. W. Keifer, 1979, *Remote Sensing and Image Interpretation.* New York: Wiley, 612p.

Rabshevsky, G., 1970, Remote sensing of the Earth's surface, *Jour. Remote Sensing* **1**(4).

Rabshevsky, G., 1973, *Symposium on Significant Results Obtained From ERTS-1,* Vol. 1. Washington, D.C.: National Aeronautics and Space Administration (Spec. Pub. 327).

Siegal, B. S., and A. R. Gillespie, 1980, *Remote Sensing in Geology.* New York: Wiley, 702p.

Williams, R. S., Jr., and W. D. Carter, 1976, ERTS-1, A new window on our planet, *U.S. Geol. Survey Prof. Paper 929,* 362p.

Cross-references: *Aerial Surveys; Photogeology; Photointerpretation; Remote Sensing and Photogrammetry, Societies and Periodicals; Satellite Geodesy and Geodynamics; Sea Surveys; Surveying, General; Surveying, Electronic; VLF Electromagnetic Prospecting.* Vol. XIII: *Geomorphology, Applied; Oceanography, Applied; Remote Sensing, Engineering Geology.*

REMOTE SENSING AND PHOTOGRAMMETRY, SOCIETIES AND PERIODICALS

Societies

Of the scientific societies in existence today, photogrammetry-related organizations have shown the most interest in remote-sensing re-

search and development. The extent of their involvement is reflected in the number of technical sessions they sponsor as a part of their regular meetings and the number of relevant papers appearing in their journals. Only recently have remote sensing societies (Table 1) begun to be established. Fifty-nine national photogrammetry societies and agencies constitute the International Society of Photogrammetry (ISP) (Table 2). Many of them maintain committees devoted entirely to remote sensing. The Photogrammetric Society of England (Department of Photogrammetry and Surveying, University College of London, Gowen Street, London, WCIA 6BT, England) is also devoted to remote sensing/photogrammetry but is not a member of ISP. Several scientific organizations have established committees or working groups on remote sensing in recent years (Table 3), and a few organizations maintain a very close association with remote-sensing research and development (Table 4).

Periodicals

Scientific papers related to remote-sensing and photogrammetry research and development are published in a variety of journals. Those devoted entirely to remote sensing and photogrammetry are included in Table 5. Table 6 includes those periodicals that contain such papers on an irregular basis, and Table 7 includes the periodicals containing abstracts on remote sensing.

Symposia generate a large number of papers on remote sensing and photogrammetry. Proceedings of symposia sponsored by organizations included in Table 8 are an excellent source of readings on remote sensing and photogrammetry. Many societies and institutions also sponsor remote sensing symposia on an as-needed basis.

L. DAVID NEALEY

TABLE 1. Remote-Sensing Societies

Canadian Remote Sensing Society
Centre for Research in Experimental Space Science
York University, Downsview, 4700 Keele St.
Ontario, Canada M3J 2R2

Indian Society of Photointerpretation
Kaldis Road
Khra Dun, India

Remote Sensing Association of Australia
Commonwealth Forestry & Timber Bureau
Bank Street
Yakkalumla, A.C.T. 2600 Australia

Remote Sensing Society of England
c/o Department of Civil Engineering
Birmingham B47ET
England

TABLE 2. The International Society of Photogrammetry

Algeria	Institut National de Cartographie BP no. 32 Hussein-Dey	*Hungary*	Geodetic and Cartographic Society Ankerköz 1 1062 Budapest
Argentina	Instituto Geografico Militar Cabildo 381 Buenos Aires	*India*	Survey General of India P.O. Box 37 Dehra Dun (U.P.)
Australia	Australian Photogrammetric Society G.P.O. Box 1020 H Melbourne: VIC. 3001	*Indonesia*	Indonesian Surveyors Association Djalan Ganesja 10, Bandung
Austria	Oesterreichische Gesselschaft für Photogrammetrie Krotenthallergasse 3 A 1080 Wien	*Iran*	National Geographic Organiza- tion P.O. Box 1844 Tehran
Belgium	Societe Belge de photogrammetrie Boulevard Pacheco 34 (4ᵉ estage) B-1000 Bruxelles	*Iraq*	Director General of Surveys Baghdad
		Ireland	Irish Society of Surveying and Photogrammetry Engineering Department Trinity College Dublin 2
Denmark	Danish Society for Photogram- metry c/o Geodaetisk Institut Rigsdagsgarden 7 DK 1218 Copenhagen K		
		Israel	Photogrammetric Society of Israel Department of Surveys, Ministry of Labour 1 Lincoln Street P.O. Box 20090 Tel-Aviv
Egypt	Faculty of Engineering Alexandria University Hadara, Alexandria		
Finland	The Finnish Society of Photo- grammetry Institute of Photogrammetry Helsinki University of Technology 02150 Espoo 15	*Italy*	Societa Italiana di Fotogram- metria e Topografic (SIFFET) i. 20121 Milano Piazzale R. Morandi 2
France	Société Française de photogram- metrie 2, avenue Pasteur F. 94160 Saint Mande	*Japan*	Japan Society of Photogram- metry Faculty of Engineering University of Tokyo Hongo, 7-3-1, Bunkyo ku, Tokyo
German Demo- cratic Republic	Gesellschaft für Photogrammetrie in der Deutschen Demokra- tischen Republik 108, Berlin Clara-Zetkin-Str. 115/117		
		Libyan Arab Re- public	Department of Surveying Ministry of Planning P.O. Box 700 Tripoli
Germany (Fed- eral Republic of)	Deutsche Gesellschaft für Photo- grammetrie Institut für Photogrammetrie Technische Universitat D 3000 Hannover Nienburger Strasse, 1	*Luxembourg*	Ordre des Géometres du Grand- Duche de Luxembourg 54, avenue Gaston Diderich Luxembourg
Greece	Greek Society of Photogram- metry 4, Karageorgi Servias Athens (125)	*Malaya*	Jabatanarah Pemetaan Negara (Directorate of National Mapping) Jalan Gurney Kuala Lumpur

(continued)

TABLE 2. (*continued*)

Mexico	Sociedad Mexicana de Fotogram-metria Fotointerpretation y Geodesia San Antonio, Abad 124-5 piso "C" Mexico 8, D.F.	Rumania	Comite roumain de Photogram-metrie B-dul Expozitiei nr. 1A Sector 1 Bucuresti
Morocco	Conservation Fonciere, Service topographique Avenue Moulay Youssef Rabat	South Africa	Photogrammetric Society of South Africa Room 102, Lowliebenhof Building 193, Smit Street Braamfontein Johannesburg, 2001
Netherlands	Nederlandse Vereniging voor Fotogrammetrie c/o I.T.C. Box 6 Enschede	Spain	Real Sociedad Geografica Valverde 22 Madrid
Nigeria	Federal Ministry of Works and Housing Survey Division Tafawa Balewa Square P.M.B. 12596, Lagos	Sudan	Sudan Society of Photogrammetry Ministry of Defense Survey Department P.O. Box 306 Khartoum
Norway	Norges Kartekniske Forbund Gronprinsens Gatan, 17 Oslo 2	Sweden	Swedish Society for Photogram-metry The National Road Administra-tion
Pakistan	Survey of Pakistan P.O. Box 3906, Karachi		Fack S-10220 Stockholm 12
Papua and New Guinea	Photogrammetric Society of Papua and New Guinea Box 1036, Boroko Port Moresby	Switzerland	Societe suisse de photogrammetrie c/o Service topographique federal Seftigenstrasse 264 CH-3084 Wabern
Peru	Instituto de Topografia y Geodesia Universidad Nacional de los Ingenieros Faculdad de Ing. Civil Lima	Syrian Arab Re-public	Service Géographique de l'Armee Ministere de la Defense Damas
		Thailand	Royal Thai Survey Department Supreme Command Headquarters Bangkok 2
Philippines	Philippine Society of Photogram-metry Training Center for Applied Geo-desy and Photogrammetry College of Engineering University of the Philippines Quezon City	Tunisia	Direction de la Topographie et de la Cartographie 13, rue de Jordanie Tunis
Poland	Polish Society for Photogram-metry Polskie Towarzystwo Fotograme-tryczne Ul. Czackiego 3/5 Skr. poczt, 903, Warszawa 1	Turkey	Turkish National Society of Pho-togrammetry Ministry of National Defense T.C., M.S.B., Harita Genel Müdürlugü Cebeci, Ankara
Portugal	Associacao Portuguesa de Fotogrametria Instituto Geografico e Cadastral Praca de Estrela, Lisboa 2	United Kingdom	National Committee for Photo-grammetry c/o Hunting Surveys Ltd. Elstree Way, Boreham Wood Herts WD7 1SB

TABLE 2. (*continued*)

United States	American Society of Photogram- metry 105 N. Virginia Avenue Falls Church, Virginia 22046	Yugoslavia	Union of Geodesy Engineers of Yugoslavia Beograd Kneza Milosa 9/IV Postanski Fah 187
USSR	Comite national des Photogram- metres de l'URSS Glavnoe Upravlenie Geodezii i Kartografii Ulitsa Krzhizhanovskogo 14, korp. 2 Moscou, V. 218		

TABLE 3. Committees and Working Groups

International Union of Forestry Research Organizations
Norwegian Forest Research Institute
N-1432 AS-NLH
Norway

Association of American Geographers
1710 15th Street, N.W.
Washington, D.C. 20009

The International Astronautical Federation
250, rue Saint Jacques/Paris 75005, France

TABLE 4. Research and Development

American Congress on Surveying and Mapping
733 15th Street, N.W., Suite 430
Washington, D.C. 20005

Society of Photo-optical Instrumentation Engineers
P.O. Box 288
Redondo Beach, CA 90277

Geoscience Electronics Group
Institute of Electrical and Electronics Engineers
345 East 47th Street
New York, NY 10017

Highway Research Board
National Academy of Science
2101 Constitution Avenue, N.W.
Washington, D.C. 20418

Optical Society of America
2000 L Street, N.W.
Washington, D.C. 20036

American Institute of Aeronautics and Astronautics
1290 Avenue of the Americas
New York, NY 10019

Society of Photographic Scientists and Engineers
Suite 204
1330 Massachusetts Avenue, N.W.
Washington, D.C. 20005

TABLE 5. Primary Journals

Geodesy, Mapping and Photogrammetry (formerly *Geodesy and Aerophotography*)
Translations Board of the American Geophysical Union
Translated and edited by Scripta Technica, Inc.
American Geophysical Union
1909 K Street, N.W.
Washington, D.C. 20006
(Back issues available)

ITC Journal
Journal of the International Institute for Aerial Survey and Earth/Sciences
Enschede, The Netherlands
(Published five times a year)

Photogrammetria
International Society for Photogrammetry
P.O. Box 1345
Amsterdam, The Netherlands
(Bi-monthly)

Photogrammetric Engineering and Remote Sensing
Journal of the American Society of Photogrammetry
105 Virginia Avenue
Falls Church, VA 22046
(Published monthly; back issues available, information on request)

Photogrammetric Record
The Hon. Secretary, The Photogrammetric Society
Department of Photogrammetry and Surveying
University College of London
Gowen Street
London, WCIA 6BT, England

Remote Sensing of Environment—An Interdisciplinary Journal
Published by American Elsevier Publishing Company, Inc.
52 Vanderbilt Avenue
New York, NY 10017
(Published quarterly; back issues available, information on request)

Revue Photo Interpretation
Editions Technip
27, rue Ginoux
75737, Paris Cepex 15, France

Fotogrammetriska Meddelanden/Photogrammetric Information
Division of Photogrammetry
Royal Institute of Technology
S-100 44 Stockholm, Sweden
(Published irregularly)

Remote Sensing of the Electro Magnetic Spectrum
Newsletter of the Remote Sensing Committee of the Association of American Geographers
Center for Applied Urban Research
The University of Nebraska at Omaha
Box 668
Omaha, Nebraska 68101
(Published quarterly)

TABLE 6. Occasional Papers Included on an Irregular Basis

The Canadian Surveyor—Le Geometre Canadien
The Canadian Institute of Surveying
P.O. Box 5378, Station "F"
Ottawa, Ontario, Canada K2C 3J1
(Published five times a year)

IEEE Transactions on Geoscience Electronics
Journal of the Geoscience Electronics Group of the Institute of Electrical and Electronics Engineers
345 East 47th Street
New York, NY 10017
(Published quarterly)

Highway Research Record
Journal of the Highway Research Board
Transportation Research Board
National Research Council
2101 Constitution Avenue, N.W.
Washington, D.C. 20418

American Antiquity
Society for American Archaeology
1703 New Hampshire Avenue, N.W.
Washington, D.C. 20009
(Published quarterly)

Modern Geology
Gordon and Breach
Science Publisher Ltd.
42 William IV Street
London W.C.2, England

Surveying and Mapping
American Congress on Surveying and Mapping
430 Woodward Building
733 15th Street, N.W.
Washington, D.C. 20005

Fotointerpretacja W Geografii
15 Z1. Polskie Towarzystwo Geograficzne, Krakowskie
Przedmiescie 30, Warsaw,
Poland

Bildmessung and Luftbildwesen
Herbert Wichmann Verlag Gmbtt
Rheinstr. 122
7500 Karlsruhe 21, Germany
(Published six times a year)

TABLE 6. (*continued*)

Bulletin (Belge)
Societe Belge de Photogrammetrie
34, Boulevard Pacheco
1000 Bruxelles, Belgium
(Published three times a year)

*Schweizerische Zeitschrift für Vermessung, Photo-
grammetrie, and Kulturtechnik*
Schweizerische Gesellschaft für Photogrammetrie
Institut für Geogasie and Photogrammetrie ETH
Leonhardstr. 33
CH-8006 Zurich, Switzerland
(Published monthly)

Canadian Journal of Remote Sensing
Canadian Aeronautics and Space Institute
77 Metcalfe Street
Ottawa, Ontario, Canada KIP 5L6

*Rivisita de Fotogrammetria, Fotointerpretacion y Geo-
desia*
Apartado Postal 25447
Mexico
(Published four times a year)

Geodeticky a Kartograficky Obzor
Czechostavak Scientific and Technical Society
P.O. Box 20
Praha 1, Czechoslavakia

Annual Techniques
The Greek National Society of Photogrammetry
4 Karageergi
Servias, Athens (125) Greece

Photogrammetry
Polish Geodetic Association (Polish Society of
  Photogrammetry)
Czackiego 3/5
Warszawa, Poland
(Published quarterly)

Geodetical Review
Polish Geodetic Association (Polish Society of Photo-
  grammetry)
Czackiego 3/5
Warszawa, Poland

TABLE 7. Abstracts

Geo Abstracts G: Remote Sensing and Cartography
Geo Abstracts, Ltd.
University of East Anglia
Norwich NOR 88C, England
(Published six times a year)

*Quarterly Literature Review of the Remote Sensing of
  Natural Resources*
Technology Application Center
The University of New Mexico
Albuquerque, New Mexico 87131

TABLE 8. Proceedings of Symposia

The Environmental Research Institute of Michigan
P.O. Box 618
Ann Arbor, Michigan 48107

American Society of Photogrammetry
105 N. Virginia Avenue
Falls Church, Virginia 22046

University of Tennessee Space Institute
Tullahoma, Tennessee 37388

Canadian Aeronautics and Space Institute
77 Metcalfe Street
Ottawa, Ontario KIP 5L6 Canada

Alberta Remote Sensing Center
205, 100th Avenue Building
10405-100th Avenue
Edmonton, Alberta T5J 0A4 Canada

U.S. Army Corps of Engineers
Washington, D.C. 20314

Canada Centre for Remote Sensing
2464 Sheffield Road
Ottawa, Canada K1A 0E4 Canada

International Society of Photogrammetry
Contact its member agencies or societies.

National Aeronautics and Space Administration
Washington, D.C. 20546

Cross-references: *Photogeology; Photointerpretation; Professional Geologists' Associations; Remote Sensing, General.* Vol. XIII: *Conferences, Congresses, and Symposia; Earth Science, Information and Sources; Geological Communication; Remote Sensing, Engineering Geology.*

REVERSE CIRCULATION DRILLING— See BOREHOLE DRILLING.

ROCK CHIP GEOCHEMISTRY—See LITHOGEOCHEMICAL PROSPECTING.

ROCK-COLOR CHART

The Rock-Color Chart (issued by the Geological Society of America) is a system of 115 rectangular color chips (size: 1/2 × 5/8 in) arranged in an orderly scale of equal visual steps mutually relating the three attributes comprising color (hue, chroma, and value) of common rocks. The symbols for *hue* (red, yellow, green, blue, and purple), *value* (lightness of color), and *chroma* (relative strength or departure from a neutral color of the same lightness) are individually related quantitatively as shown in Fig. 1.

The color of a given sample is obtained by comparing it with the close-matching color chip of the chart. The results are numerically given following the order of hue/value/chroma, or symbolically, H/V/C.

As an example, a *light-brown rock* matching the color chip 7.5 YR 6/5 indicates that the rock has a *hue* (7.5 YR) 75% yellow and 25% red, a *value* (6/) of 60% of the way from black to white, and *chroma* saturation (/5) of 25% since the numerical *chroma* strength ranges from 0 to 20. Whenever a finer division is needed for any of the attributes of color, decimals may be used, such as 7.5 YR 6.5/5.2. This color notation, introduced early in the century by the Boston artist, A. H. Munsell (and adopted by the Optical Society of America), is a helpful system for the comparative study of rock and mineral colors in the laboratory, as well as in the field (especially for the color typification of recent sediments such as playas).

Rock charts (115 chips) are available with explanatory text from the Geological Society of America, P.O. Box 9140, Boulder, Colorado, 80301, and soil charts (196 chips) are available from the Kollmorgen Corporation, 2441 North Colvert Street, Baltimore, Maryland, 21218.

RAYMUNDO J. CHICO

References

Geological Society of America, 1964, *Rock-Color Chart*, 2nd ed. New York.

Judd, D. B., 1963, *Color in Business, Science, and Industry.* New York: John Wiley & Sons, 500p.

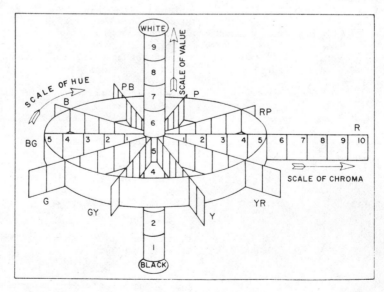

FIGURE 1. Hue, value, and chroma in relation to one another. The circular band represents the hues in their proper sequence, where R stands for red, YR for yellow red, G for green, B for blue, P for purple, etc. The paths pointing outward from the center show the steps of chroma, increasing in strength as indicated by the numerals. (Munsell Color Company)

Cross-references: *Soil Sampling; Soils and Weathered Materials, Field Methods and Survey.* Vol. XIII: *Engineering Geology Reports.*

ROCK DRILLING—See BOREHOLE DRILLING. Vol. XIII: RAPID EXCAVATION AND TUNNELING; SHAFT SINKING.

ROCK PARTICLES, FRAGMENTS

The breakdown of rocks by weathering produces fragments or detritus that is available for transport by geological agents to sites of deposition. Accumulations of such fragmental material are said to have clastic textures and can be described using particle characteristics such as shape and size. These textural properties are influenced by the type of parent material, the weathering processes, and the transport agents and processes.

Parent Material

The *parent material* from which the fragments are derived determines composition and the initial particle size and shape. Chemical makeup, cleavage, and hardness, among other factors, affect the material's response to weathering and transportation.

Weathering

Weathering is a complex of equilibrium processes and, as such, modifies the parent material to some extent to bring it into balance with near-surface conditions. Fragments are produced as the parent material breakdown by exfoliation, block separation, shattering, or granular disintegration. These breakdown mechanisms result primarily from physical processes that apply force to and create stress within the rock with little chemical alteration. Chemical reactions operating along microfractures or crystal boundaries, however, may contribute to the breakdown.

Stress may be created in rocks during weathering in many ways including temperature change, application and release of overburden pressure, shearing forces, crystal growth, water pressure or tension, and organic activities. Theoretically, differential thermal properties of separate mineral species composing rocks or the unequal expansion and contraction of exterior layers versus the interior should generate stress. Rocks are poor thermal conductors, so heat is not rapidly transferred, and as a result, the outer skin of a rock may behave differently from the deeper part. Cracking and spalling during forest fires seems to have been documented, but much of the effect of smaller diurnal temperature fluctuations remains problematic.

Reduction of overburden pressure either naturally by erosion, e.g., or artificially during mining or quarrying can result in the formation of fractures oriented perpendicular to the axis of pressure release. Rock bursts in mines and large exfoliation domes have been attributed to this process.

The formation of ice or mineral crystals within openings or along fractures in rocks produces very high pressures. Ice pressures develop because of the expansion that occurs as freezing takes place. Pressures from mineral crystals usually result as fluids evaporate from pore spaces, causing supersaturation and precipitation. The effectiveness of crystal growth depends on the type of mineral precipitating and the ionic concentration of the constituents in solution. Calcite is commonly associated with such breakdown as are other carbonates and sulfates or chlorides of calcium, sodium, potassium, and magnesium. Other salt-related processes such as thermal expansion and hydration are also recognized (Cooke and Smalley, 1968).

Increased pore pressure, and the formation or expansion of oriented layers of water along microfractures during wetting and drying cycles can also stress rock, causing fracturing. More complete discussions of weathering can be found in Carroll (1970), Ollier (1969), and Twidale (1968), and most recent geomorphology texts.

These disintegration processes produce coarse rock fragments, sand-sized material, and fine-grained materials that often have undergone some chemical alteration, such as clay minerals.

Transportation Agents and Processes

The fragments are shaped and reduced in size by mechanical processes during transport, becoming more modified from their original form with increased distance or time of movement. Transportation by streams, wind, waves, or nearshore marine currents results in graint-to-grain abrasion that rounds corners and edges (Dowdeswell et al., 1985), and grinding in glaciers produces angular and somewhat disk-shaped particles (Mazzullo et al., 1986).

Characterization of Fragments

Detrital particles, no matter what their source or manner of production, have properties that can be determined and quantified by field and laboratory techniques. Among the most important, or at least the most obvious and often measured, properties are particle type or composition, particle size, particle shape, particle surface texture, and the relationship of particles to each other, or the fabric of a deposit of fragments. Over the years numerous techniques have been devised and innumerable analysis carried out to characterize and understand sedimentary materials.

Composition The composition of fragments is rather easily determined by visual inspection with the aid of a hand lens or a microscope for finer sizes. Pebble or mineral counts can be conducted to determine the relative abundance of rock or mineral types. One hundred to 300 pebbles or grains are usually selected and identified. The choice of lithic groups or mineral classes must be meaningful and reasonable for the instruments available and experience of the investigator. The number of groups should not be too large but based on definitive properties. For example, a pebble count may distinguish clastic and carbonate sedimentary rocks and intrusive and extrusive igneous rocks, a total of four classes. Mineral counts may be as straightforward as distinguishing translucent and opaque grains, a total of two classes. More classes may be used, of course, if

necessary for the investigation, but overlapping boundaries must be avoided.

Grain Size Grain size is a much measured property that may in fact be overused in sedimentary studies (Table 1). The reasons for conducting size analysis are varied, but the purposes can be grouped generally into four categories. First, grain size may provide information about the energy characteristics of the transporting agent. Second, grain-size distributions may be diagnostic of sedimentary environments. Third, other physical properties such as compactability or permeability may be related to grain size (Fig. 1). Finally, grain-size distributions may be used simply to describe or characterize sediments (see Vol. XII, Pt. 1: *Particle-Size Distribution*).

Grain size is measured by many direct and indirect methods (Fig. 2). Large fragments can be measured directly with calipers or other devices. Three mutually perpendicular prime axes *a, b,* and *c* are measured and summed for each pebble. Di-

TABLE 1. Comparison of Grade Scales in Use

Udden-Wentworth	ϕ Values	German Scale[†] (after Atterberg)	USDA and Soil Sci. Soc. Amer.	U.S. Corps Eng., Dept. Army and Bur. Reclamation[‡]
Cobbles		(Blockwerk) ——200 mm——	Cobbles ——80 mm——	Boulders ——10 in.——
——64 mm——	−6			Cobbles ——3 in.——
Pebbles		Gravel (Kies)		
			Gravel	Gravel
——4 mm——	−2			——4 mesh——
Granules				Coarse sand
——2 mm——	−1	——2 mm——	——2 mm——	——10 mesh——
Very coarse sand			Very coarse sand	
——1 mm——	0		——1 mm——	
Coarse sand		Sand	Coarse sand	Medium sand
——0.5 mm——	1		——0.5 mm——	
Medium sand			Medium sand	——40 mesh——
——0.25 mm——	2		——0.25 mm——	
Fine sand			Fine sand	Fine sand
——0.125 mm——	3		——0.10 mm——	
Very fine sand			Very fine sand	——200 mesh——
——0.0625 mm——	4	——0.0625 mm——		
			——0.05 mm——	
Silt	Silt	Silt	Fines	
——0.0039 mm——	8			
		——0.002 mm——	——0.002 mm——	
Clay		Clay (Ton)	Clay	

Source: From Blatt et al., 1980, 57; copyright © 1980; reprinted by permission of Prentice-Hall, Inc., Englewood Cliffs, New Jersey.

[†]Subdivisions of sand sizes omitted.

[‡]Mesh numbers are for U.S. Standard sieves: 4 mesh = 4.76 mm, 10 mesh = 2.00 mm, 40 mesh = 0.42 mm, 200 mesh = 0.074 mm.

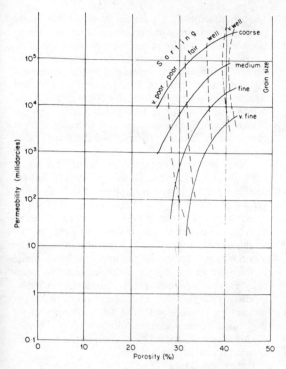

FIGURE 1. Relationship between physical properties and sediment texture. (From Selley, 1982)

viding by three then gives an approximation of the mean diameter. Smaller grains have been studied with sieves, sedimentation columns, hydrometers, pipettes, and other tools that are well described in many sources (Folk, 1968; Muller, 1967; Carver, 1971).

Shape

Particle shape is another property that supplies information about the type of source rock, the breakdown mechanisms, and the transporting agents. Both two- and three-dimensional shape indexes have been used to describe fragments. Roundness, a two-dimensional property, is often evaluated by visually comparing the sample grains to a standard chart. Three-dimensional shape such as form, sphericity, and flatness can be quantified by several methods. Zingg's classification of form is obtained by measuring the three prime axes of each particle and then using ratios to define spheres, discs, rods, and blades. A *sphericity index* (Krumbein, 1941) is obtained by measuring the three axes *a, b,* and *c* and entering the values in the formula:

$$\Psi = \sqrt[3]{\frac{bc}{a^2}}$$

Both indexes can be obtained from standard charts (Fig. 3).

More quantitative shape values can be obtained by Fourier series expansion as described by Ehrlich and Weinberg (1970). Their methods yield a mathematical model that can reproduce the grain shape. The method has been applied to a study of the St. Peter Sandstone (Mazzullo and Ehrlich, 1980, 1983) and may provide the basis for a valuable and attractive approach to grain-shape evaluation.

Surface Texture Particle surface textures are much more difficult to observe in the field and, as a result, are usually studied in the laboratory. Surface textures can supply useful information

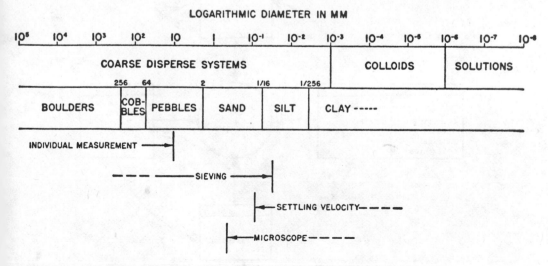

FIGURE 2. Measurement techniques applicable to different grain sizes. (From Krumbein and Sloss, 1963; copyright © 1951, 1963; reprinted with the permission of W. H. Freeman and Company)

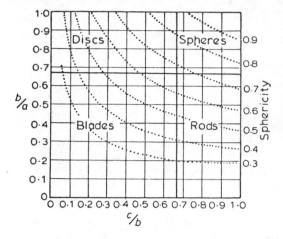

FIGURE 3. Krumbein's sphericity values (dotted lines) superimposed on a chart of Zingg's classification of particle shape. (From Briggs, 1977)

about the history of a sedimentary deposit; however, they are affected by grain origin, manner of transportation, the depositional environment, and diagenesis (Fig. 4) and must be interpreted with care. A great amount of literature exists on the theoretical and applied aspects of the interpretation of textures with the aid of scanning electron microscopes. A good introduction to the subject is found in Krinsley and Doornkamp (1973).

Fabric Fabric is the three-dimensional arrangement of the particles composing the deposit.

The orientation of a particle in space can be related to horizontal or compass direction alignment and the angle of inclination from the horizontal. Fabric studies commonly measure both. Preferred orientation developed by the transporting agent at the time of deposition can be altered significantly postdepositionally. Weathering, frost action, mass wasting, or animal traffic can all change particle orientation, and therefore, great care must be taken to sample and measure undisturbed material.

Fabric analysis is a time consuming and exacting task and, therefore, is usually carried out on pebble-sized or larger fragments, although some sand-sized particles have been measured. Particle orientation can be described either in reference to direction and inclination of the *a,* or long, axis or to the orientation of a place parallel to both the *a* and *b* axes. Measurements are most often made in the field at natural or artificial faces by carefully exposing or extracting individual particles. Undisturbed samples can also be collected for laboratory analysis with the aid of frames or by impregnation with various adhesive agents. Briggs (1977) presents a useful discussion of the goals and techniques of fabric analysis.

Presenting Data The quantitative data collected about the characteristics of fragments can be manipulated and analyzed by a variety of simple or sophisticated statistical techniques. The results of the measurements and analysis then must be presented or portrayed in some way. The results of compositional analyses can be simply and

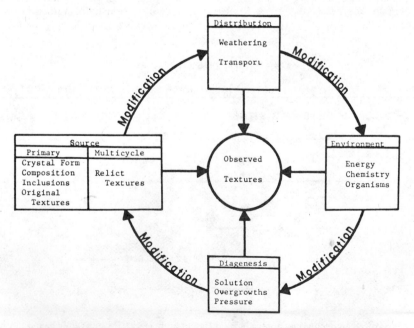

FIGURE 4. Interrelationship of factors affecting grain-surface microtextures. (From Stieglitz and Rothwell, 1978)

effectively presented on pie diagrams and histograms. Size distribution data can be displayed by plotting on triangular texture diagrams or on graph paper to produce frequency and cumulative frequency curves. Shape data lend themselves well to presentation on bar charts on which the subdivisions may be sphericity values or standard shapes such as rods or discs. Composition and grain-size data can be combined with shape by patterns placed on the bars of the chart. Surface textures are best illustrated by photographs or schematic diagrams. The data can be quantified in terms of relative numbers of textures or sequences of superimposed textures and may be combined with size, shape, and composition. Fabric information, being directional, can be portrayed on rose diagrams or polar scattergrams.

Significance of Fragment Characteristics

An extensive literature exists on the significance and the application of fragment characteristics to sedimentological problems. It is important to note that many grain characteristics are at least partly influenced by the specific genesis of the grains or the manner in which they are freed from the rock matrix during weathering. Attempts to use composition, size, shape, or surface textures for provenance, transportation, or environmental interpretations require an understanding of original grain properties. Examples of such studies are Blatt's (1967) investigation on the original characteristics of quartz grains and Boggs's (1968) paper on rock fragments.

Composition is frequently used to determine source rock contributions and to suggest transport direction. In coarse deposits such as conglomerates, identification is fairly straightforward; as particle size decreases, however, some confusion between plutonic and granoblastic metamorphic fragments may occur. Fine-grained igneous rocks such as rhyolite are optically indistinguishable from chert (Blatt et al., 1980). In sand-size deposits multiple crystal grains are rare, and determinations may be based on the optical extinction properties of the quartz grains (Blatt, 1967). Mack (1981) has studied the composition of modern stream sands derived from a complex fold-thrust belt in the Appalachians and was able to distinguish sand from different sources.

Grain-size distributions have been used to distinguish depositional environments (Friedman, 1961; Goldbery, 1980) and to decipher depositional and hydraulic properties of transporting agents (Visher, 1969; Middleton, 1976; Bridge, 1981). Statistical measures such as mean grain size, skewness, and kurtosis as well as patterns of log-probability curves have been expressed with the value of size analyses (Ehrlich, 1983).

Specific fragment shapes, shape indexes, or plots of combinations of shape indexes have been used in depositional and hydraulic studies. Beal and Shepard (1956) used the two-dimensional property of roundness as an environmental indicator. Spalletti (1976) used the ratio of the short axis (c) to the intermediate axis (b) and a geometricity ratio as indicators of shape-selective transport. The geometricity ratio is expressed as:

$$Gr = \frac{\text{rods \% + spheres \%}}{\text{discs \% + blades \%}}$$

Shape is related to particle size, and Ehrlich and others (1980) investigated the connection between shape frequency distributions and size frequency distributions. Those workers concluded that shape frequencies are multidimensional but that interpretation of such data can play an important role in sedimentological investigations. Using Fourier analysis techniques, Riester et al. (1982) investigated shape variations along a shoreline, and Mazzulo and Ehrlich (1983) documented vertical shape variations reflecting environmental alterations in the Ordovician St. Peter Sandstone.

Grain surface textures have been widely employed to distinguish transportational agents and depositional environments (Krinsley and Donahue, 1968). Because of the complexities of environmental conditions and influences of original characteristics, caution is needed when using surface textures as environmental indicators. Work done on grains from many recent environments, such as that of Al-Salen and Khalaf (1982), has improved understanding, and textures have been successfully used to characterize environments, e.g., by Higgs (1979).

Fabric analysis has been used to indicate transportation direction and in some cases conditions of deposition. Till deposits have been perhaps the most often investigated, but beach, stream, and slope materials have also been studied. Subglacial lodgement tills display a distinct orientation of elongate clasts parallel to the direction of basal ice flow (Evenson, 1971). Fabrics of coarse fluvial environments are affected by the individual particle's resistance to movement, competence reduction, or load increase and therefore must be interpreted with discretion (Rust, 1972). Stacked or imbricated particles exhibit a prominent upstream dip. Imbrication also develops on beaches where particles are aligned with their long axis parallel to the shore and dipping offshore (Black, 1967). Slope deposits typically have weaker but meaningful orientations.

In summary, even in this age of sophisticated techniques and instruments, a great amount of valuable data and information can be obtained from a field or relatively simple laboratory study of rock fragments.

RONALD D. STEIGLITZ

715

References

Al-Salen, S., and F. Khalaf, 1982, Surface textures of quartz grains from various recent sedimentary environments in Kuwait, *Jour. Sed. Petrology* **52**, 215–225.

Beal, M., and F. Shepard, 1956, A use of roundness to determine depositional environments, *Jour. Sed. Petrology* **26**, 49–60.

Black, B. J., 1967, Sedimentation of beach gravels: Examples from South Wales, *Jour. Sed. Petrology* **37**, 128–156.

Blatt, H., 1967, Original characteristics of clastic quartz grains, *Jour. Sed. Petrology* **37**, 401–424.

Blatt, H., G. Middleton, and R. Murray, 1980, *Origin of Sedimentary Rocks,* 2nd ed. Englewood Cliffs, N.J.: Prentice-Hall.

Boggs, S., Jr., 1968, Experimental study of rock fragments, *Jour. Sed. Petrology* **38**, 1326–1339.

Bridge, J. S., 1981, Hydraulic interpretation of grain-size distributions using a physical model for bedland transport, *Jour. Sed. Petrology* **51**, 1109–1124.

Briggs, S., 1977, *Sediments.* London: Butterworth & Co. Ltd., 192p.

Carroll, D., 1970, *Rock Weathering.* New York: Plenum Pub., 203p.

Carver, R., 1971, *Procedures in Sedimentary Petrology.* New York: Wiley-Interscience.

Cooke, R. U., and I. J. Smalley, 1968, Salt weathering in deserts, *Nature* **220**, 1226–1227.

Dowdeswell, J. A., M. J. Hambrey, and R. Wu, 1985, A comparison of clast fabric and shape in Late Precambrian and modern glacigenic sediments, *Jour. Sed. Petrology* **55**, 691–704.

Ehrlich, R., 1983, Size analysis wears no clothes or have moments come and gone? *Jour. Sed. Petrology* **53**, 1.

Ehrlich, R., P. Brown, J. Yarus, and R. Przygocki, 1980, The origin of shape frequency distributions and the relationship between size and shape, *Jour. Sed. Petrology* **50**, 475–484.

Ehrlich, R., and B. Weinberg, 1970, An exact method for characterization of grain shape, *Jour. Sed. Petrology* **40**, 205–212.

Evenson, E. B., 1971, The relationship of macro- and micro-fabric of till and the genesis of glacial landforms in Jefferson County, Wisconsin, in R. P. Goldthwait, ed., *Till: A Symposium.* Columbus: Ohio State University Press, 345–364.

Folk, R., 1968, *Petrology of the Sedimentary Rocks.* Austin, Tex.: Hemphills.

Friedman, G. M., 1961, Distinction between dune, beach and river sands from textural characteristics, *Jour. Sed. Petrology* **31**, 514–529.

Goldbery, R., 1980, Use of grain-size frequency data to interpret the depositional environments of the Pliocene Pleshet Formation, Beer Sheva, Israel, *Jour. Sed. Petrology* **50**, 843–856.

Higgs, R., 1979, Quartz-grain features of Mesozoic-Cenozoic sands from the Labrador and Western Greenland continental margins, *Jour. Sed. Petrology* **49**, 599–610.

Krinsley, D. H., and J. C. Donahue, 1968, Environmental interpretation of sand grain surface textures by electron microscopy, *Geol. Soc. America Bull.* **79**, 743–748.

Krinsley, D. H., and J. C. Doornkamp, 1973, *Atlas of Quartz Sand Surface Textures.* Cambridge: Cambridge University Press.

Krumbein, W., 1941, Measurement and geological significance of shape and roundness of sedimentary particles, *Jour. Sed. Petrology* **11**, 64–72.

Krumbein, W. C., and L. L. Sloss, 1963, *Stratigraphy and Sedimentation,* 2nd ed. San Francisco: W. H. Freeman and Company.

Mack, G. H., 1981, Composition of modern stream sand in a humid climate derived from a low-grade metamorphic and sedimentary foreland fold-thrust belt of north Georgia, *Jour. Sed. Petrology* **51**, 1247–1258.

Mazzullo, J., and R. Ehrlich, 1980, A vertical pattern of variation in the St. Peter Sandstone—Fourier grain shape analysis, *Jour. Sed. Petrology* **50**, 63–70.

Mazzullo, J., and R. Ehrlich, 1983, Grain shape variation in the St. Peter Sandstone: A record of eolian and fluvial sedimentation of an Early Paleozoic cratonic sheet sand, *Jour. Sed. Petrology* **53**, 105–119.

Mazzullo, J., and S. Magenheimer, 1937, The original shapes of quartz and grains, *Jour. Sed. Petrology* **57**, 479–487.

Mazzullo, J., D. Sims, and D. Cunningham, 1986, The effects of eolian sorting and abrasion upon the shapes of find quartz sand grains, *Jour. Sed. Petrology* **56**, 45–56.

Middleton, G., 1976, Hydraulic interpretation of sand size distributions, *Jour. Geology* **84**, 405–426.

Muller, G., 1967, *Methods in Sedimentary Petrography.* New York: Hafner Pub. Co.

Ollier, C. D., 1969, *Weathering.* London: Longman, 304p. (reprinted 1975).

Riester, D., R. Shipp, and R. Ehrlich, 1982, Patterns of quartz sand shape variation, Long Island littoral and shelf, *Jour. Sed. Petrology* **52**, 1307–1314.

Rust, B., 1972, Pebble orientation in fluvial sediments, *Jour. Sed. Petrology* **42**, 384–388.

Selley, R. C., 1982, *Introduction to Sedimentology.* London: Academic Press.

Spalletti, L., 1976, The axial ratio C/B as an indicator of shape selective transportation, *Jour. Sed. Petrology* **46**, 243–248.

Stieglitz, R., and B. Rothwell, 1978, Surface microtextures of freshwater heavy mineral grains, *Geosci. Wisconsin* **3**, 21–34.

Twidale, C. R., 1968, Weathering, in R. W. Fairbridge, ed., *Encyclopedia of Geomorphology.* New York: Reinhold Book Corp., 1228–1232.

Visher, G. S., 1969, Grain size distributions and depositional processes, *Jour. Sed. Petrology* **39**, 1074–1106.

Cross-references: *Samples, Sampling.* Vol. VI: *Grain-Size Studies.* Vol. XII, Pt. 1: *Particle-Size Distribution.*

ROCK SAMPLING—See EXPOSURES, EXAMINATION; LITHOGEOCHEMICAL PROSPECTING; PROSPECTING; SAMPLES, SAMPLING.

ROTARY DRILLING—See BOREHOLE DRILLING.

S

SAMPLES, SAMPLING

The geological *sample* is a representative unit of soil, rock, ore, fluid, or gas that is selected from a larger mass or volume to serve as an example of that larger body or to reflect some specific feature or variation within it. The simple rationale for *sampling* (Barnes, 1981) is that, one can take a specimen home, but not an outcrop. Sampling is undertaken to provide a type specimen for classification purposes and/or special-purpose (e.g., petrofabric) analysis, assay, or testing (including engineering-geological and geochemical testing).

Because the sample, once extracted, takes the place of the outcrop in some respect, the basic prerequisite of a sample is that it must be representative of whatever is sampled. Sample representativity is dependent, in large part, on the accuracy of geological observations that precede sampling (see *Field Notes, Notebooks; Exposures, Examination of*), and the procedures used for sample taking. Sampling procedures (and equipment, if appropriate) must, in turn, reflect the purpose of sampling, both with respect to sample analysis and/or testing and the subsequent decisions or designs that are to be based on the data derived.

Since sampling represents such an intricate and diverse practice, particularly in the area of applied geology, this entry can provide no more than a broad overview of the topic, with selected examples to offer some generalized insight into the philosophy and practice of sampling. The text is restricted to the sampling of soil and rock (see *Rock Particles, Fragments; Soil Sampling*). Therefore, no more than a passing mention is made to surface water (including seawater) or groundwater sampling, air quality (see *Atmogeochemical Prospecting*), liquid or gaseous petroleum (see *Petroleum Exploration Geochemistry*), fossil (see *Fossils and Fossilization*), or other sampling considered to fall within the broad realm of geological sampling. To contain further the length of this entry, all detailed discussion and examples quoted are restricted to the field of engineering geology.

Purpose of Sampling

Sampling is a means to an end rather than an end in itself. Sampling and subsequent testing and evaluation may provide critical input for, to mention only a few areas of applied geology (q.v. in Vol. XIII), the opening of new mining operations or petroleum fields (e.g., LeRoy et al., 1977), establishing the design of major civil engineering works (e.g., Hoek and Brown, 1980; Watson et al., 1975), or evaluating the potential impacts on groundwater of leachates from hazardous waste disposal sites (e.g., Watson, 1984a). Such differing projects call for very different sample types and sampling techniques.

Even within a single discipline, engineering geology, e.g., sampling procedures vary widely. For example, the appropriate sample(s) for the foundation design of a proposed civil engineering structure on a soils (unlithified sediment) foundation may well be some form of undisturbed sample(s) suitable for consolidation and triaxial testing to determine the compressibility, shear strength, and liquefaction potential of the soil under the static loading of the structure and the dynamic loading of the design earthquake (e.g., Watson et al., 1984).

In contrast, in evaluating the thermal properties of soils to support a buried (offshore) transmission cable(s) (Watson, 1984), more economical disturbed (rather than undisturbed) samples could be used. In this case, since a high-resistivity soil over even a portion of the alignment could initiate a local hot spot and cause a burnout of the cable, the critical consideration is establishing the detailed stratigraphy along the proposed cable route alignment. Thus, the key to sampling hinges on the selection of a cost-effective method, such as vibratory coring (Watson and Krupa, 1984), that will produce a long continuous core.

On other projects where soil or rock is used as a construction material, the taking of disturbed samples is not merely preferable but also essential, since on such projects the materials to be used are, by design, disturbed. In the case of a rock-rubble structure (a rockfill dam or breakwater), the critical design parameter is the angle of internal friction, a measure of how effectively the quarried rock fragments recombine to form the rockfill mass. This internal friction, in turn, is controlled to a large extent by the gradation of shot (blasted) rock. Sampling on such projects, therefore, requires the design of representative working blasts and the selection of representative samples of blasted rock for large-scale field-sizing analysis. Since no triaxial testing apparatus is sufficiently large to contain a sample of quarried rock, the engineering geologist must model the sample(s) for testing. Modeling is done by blending similar crushed stone to obtain a sample gradation curve that is identical (i.e., parallel) to the developed field curve.

A final example of the project objective's governing the sample type relates to the use of rock as a foundation material. In this case sampling is undertaken mainly for index and classification testing and, unlike in soils, is of only secondary importance in terms of foundation design. This is because the performance of most rock, en masse, is controlled not by the strength of the intact rock but by the defects, or discontinuities, within it (see also *Field Notes, Notebooks; and Exposures, Examination of*). Hence, in this case, the sampling program per se may be replaced by a two-phase site investigation program aimed at (1) structural mapping to determine the attitude, spacing, continuity, roughness, and infilling of all discontinuities and (2) in situ shear and/or plate-loading tests to help quantify the shearing resistance of potential failure wedges (see Vol. XIII: *Rock Mechanics*). Similar nonsampling procedures may also be appropriate for evaluating the stability of open cast mining slopes, rock cuts, and bridge piers (see Vol. XIII: *Rock Structure Monitoring*).

These represent only a few examples of the diversity of samples and sampling (within the field of engineering geology alone). They emphasize the fact that the purpose of sampling is the all-important variable in controlling the type of sampling (or nonsampling) program that is implemented for a particular task.

Sample Quality

In response to the question of how good a sample must be, Golder (1971) relates the story of Terzaghi, the father of soil mechanics, investigating the stability of cliffs at Folkstone Warren in southeast England. These cliffs are composed of a highly permeable chalk overlying a clay, a condition highly susceptible to active mass movement (Watson, 1984b). Terzaghi had asked for an undisturbed sample of the clay, a considerable request at that time (1938). After several days of intense efforts, a sample 15 in. long and 4 in. in diameter was proudly extruded from its thin-walled sampling tube. Terzaghi accepted it saying, "This is very good. You have done well. Now, let us see if it will break." He dropped the sample and it shattered, to the horror of the drilling contractor. By observing the way in which it broke, Terzaghi learned that the clay was stiff and fissured, and he was able to conclude that the fissures, like the discontinuities in a rock, would largely control the strength of the material. He could not have discovered this important fact from a disturbed sample.

Should all samples of clay be high-quality, undisturbed samples? Again, it depends on the project. For example, if the clay is a potential borrow source for the impervious core of an earth dam, then a disturbed sample would be not only less expensive but also more appropriate because the clay is remolded during working and placement in the dam, with a consequent change in engineering geologic behavior (e.g., permeability and strength).

Sampling Frequency

Sampling frequency is predominantly a function of the changing characteristics of the area being sampled and the nature of the project under investigation. Samples are usually taken either by the geologist in person from a natural or artificial exposure or are retrieved by remote means from subsurface soils and rock.

In the case of exposures, sampling is preceded by mapping aimed at delineating changes in the properties of the materials exposed (see *Exposures, Examination of*). Other than structural anomalies, variation is usually a function of lithological and weathering changes.

In subsurface investigations two levels of sampling are undertaken: (1) index sampling for classification purposes and (2) quality sampling for engineering-geologic testing. Test results on the former samples are combined with geophysical data to establish the basis for subsurface cross-sections.

In summary, the selection frequency of engineering samples is a function of the geological complexity of the project area and the type of project under consideration. As an example of the latter control, it would not be unusual in the foundation design for a nuclear power plant for more than 500 index samples to be taken and for close to 100 high-quality engineering-geologic samples to be selected for more sophisticated testing.

Sample Types and Sampling Techniques

The sample types considered consist either of soil or rock. Soil samples are normally referred to as either "disturbed" or "undisturbed." In this context, it should be remembered, however, that because all samples of granular and cohesive materials (no matter how carefully these are removed from in situ locations) show some degree of disturbance, the term *undisturbed sample* must be regarded as relative.

Undisturbed soil samples represent some of the most sophisticated samples taken by geologists. The undisturbed sample is mainly used on engineering-geologic projects to determine properties such as settlement characteristics, shear strength, and liquefaction potential. Because the tests to determine such properties may be significantly affected by any remolding or change of relative density in the test specimen, the desire in retrieving samples is to achieve the smallest practical degree of sample disturbance.

Sample disturbance is primarily a function of the thickness of the sampler (ratio of internal to external diameter) and the method of driving the sampler (e.g., hammer blows, hydraulic push). A thin-walled sampler pushed gently into the soil represents the ideal case. However, in materials such as very dense sands, the only suitable procedure may necessitate the use of a fairly substantial sampler such as the Dames & Moore Underwater Sampler (Fig. 1a). This very strong sampler must be driven by a relatively heavy hammer (300 lb, 136.1 kg). Conversely, in materials such as soft clays and peats, a thin-walled sampler (Fig. 1b) may be readily pushed or jacked into the soil.

Figure 1 also illustrates two additional widely used samplers, the Denison and the Pitcher. The Denison sampler (Fig. 1c) is frequently more suitable than the Dames & Moore underwater-type sampler in homogeneous, hard, and very stiff clays or in very dense sands. The Denison sampler, which works on the principle of the double-tube rock-core barrel, has a carbide or diamond bit attached to the outer barrel and a sharp cutting edge on the inner barrel. This double-cutting

action, together with the jetting assistance of drilling fluid that circulates between the inner and outer barrel, tends to minimize sample disturbance. The Pitcher sampler (Fig. 1d) is best suited to soils containing hard and soft layers since a spring-loaded sample barrel adjusts automatically to the relative density or consistency of the soil.

Disturbed soil samples are most often used by engineering geologists for laboratory index tests to confirm visual classification in the field. It is recommended that this classification of soils be based on the Unified Soil Classification System (Casagrande, 1948).

Disturbed samples may be cut and bagged in the field or retrieved from depth using the equipment for the Standard Penetration Test (SPT) (Terzaghi and Peck, 1967). It is worth mentioning that the SPT is the most useful test, since it yields not only a large undisturbed sample but also results that may be correlated with both static (e.g., relative density of cohesionless soils, consistency of cohesive soils) and dynamic engineering behavior (e.g., liquefaction potential under seismic loading). The correlations may be found in most textbooks on soil mechanics (e.g., Terzaghi et al., 1967).

Rock sampling, in the context of the present discussion, is relatively simpler than soil sampling. Because, as explained, rock samples are less often taken for design purposes, the main use of the rock sample is for index testing, microscopic analysis, and classification.

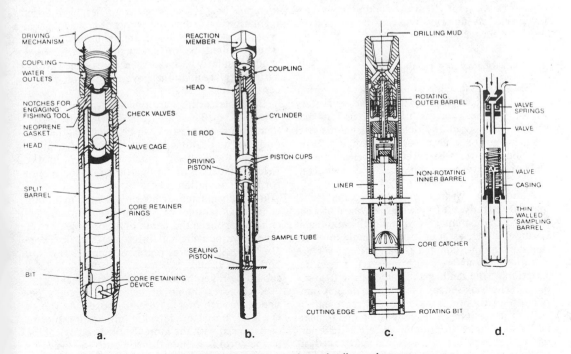

FIGURE 1. Some commonly used soil samplers.

It is good practice to mark all rock specimens taken to indicate their original orientation in situ. Normally, in sedimentary sections, an indication of the top of the bed is sufficient, but complete orientation data might be necessary if directional textural features such as cross-bedding or slump structures are present.

In structurally deformed or metamorphic rocks or where petrofabric analyses are planned, fully oriented samples are essential, and samples should be selected with particular care. In this way planes of weakness, such as cleavage, noted in thin section cut from the specimen, may be reoriented in space. Similar care with orientation is necessary on samples collected for paleomagnetic studies.

The size of the sample taken will vary according to project needs, but normally a hand specimen with a long dimension of about 8–10 cm is suitable. If detailed laboratory work is planned (multiple thin sections, insoluble residues, breakdown for microfossils, engineering testing, etc.), extra samples should be taken. Weight is a factor, especially in terms of the limitations that may be imposed by backpacking and/or air transportation.

Because the degree of weathering of rocks exerts an important influence on their strength and durability, weathering characteristics must be carefully noted and sampled for. Where weathering is not deep, the geologist may remember that the diggings from animal burrows may provide relatively unweathered samples, sometimes from surprising depths. Where weathering is deep and exposures are poor (frequent conditions in tropical areas), pitting or borings may be necessary to reach fresh rock.

Sample Labeling and Packing

The suitable labeling of all samples is of prime importance. Samples arriving in the laboratory without proper identification are useless and should be disposed of. Identifying such material from memory should never be attempted. Samples should always be carefully labeled immediately on retrieval to preclude any subsequent confusion.

Although the written content on a sample label may vary slightly, the following are typical of what is recorded: job number, owner identification, and boring number, as appropriate; sample number; sample location; sampled elevation (e.g., surface elevation, subsurface depth interval); a field description of the sample; the name of the field geologist who took the sample; and the date of sampling. Shortcuts on sample labeling represent, at best, false economy, especially when one remembers that samples from many different field parties or projects may be received and tested simultaneously in the laboratory, by personnel who have no direct connection with the field operations. In establishing quality-control procedures in the laboratory, the geologist should ensure that the original field label stays with the sample throughout testing and into storage.

The further details of sample labeling and packing in the field vary mainly as a function of the type of sample taken. The main sample types include the rock hand specimen, borehole cuttings and cores, and disturbed or undisturbed soil samples. The collection of the rock hand specimen probably represents the most common sampling practice. Rock specimens should be placed in suitable sample bags equipped with drawstrings and sewn-on waterproof labels.

Whenever practical, specimens should be neatly numbered with a waterproof felt-tip marker. Delicate samples should be wrapped (in newspaper) before bagging (some geologists wrap all samples). If distant or hazardous transshipping conditions are involved, a duplicate label is wrapped in with the sample as a precaution against damage to the external bag label. In any event, the number of the sample and label data should be logged separately in the field notebook.

Strong wooden boxes with steel strapping (e.g., nail kegs lined with straw) probably provide the most cost-effective method of shipping. Again, proper labeling is mandatory. Boxes or crates within a given shipment should be labeled sequentially (e.g., Box No. 3 of 6). If customs clearance is involved, the description *rock samples* is the preferred form to use. Avoid terms such as *minerals* or *fossils*. Some countries have strict regulations regarding the export of archaeological artifacts, skeletal material, etc., and *fossil*, in particular, might be misinterpreted. *Soil samples* should similarly be avoided because of the various agricultural regulations governing plant disease control.

Borehole cuttings are typically taken at intervals of 2–3 m. These are washed, dried, and placed in a dustproof (plastic) bag before being inserted into a regular sample bag. In the case of sampled cuttings, the geologist should record in his or her field book details of depth estimation (to account for mud circulation time).

Rock coring represents an attempt to obtain continuous samples of rock of selected diameter, either over the entire depth of a borehole or over predetermined critical intervals (see *Borehole Drilling*). Since cores represent samples that are both bulky and expensive, considerable care should be taken in their labeling and packing.

Rock cores should be placed in specially designed wooden (not cardboard) boxes. As shown in Fig. 2, basic core labeling information (not to be confused with the more detailed geological log of the core) is written directly onto the inside of the core box. As indicated, the depth of each core

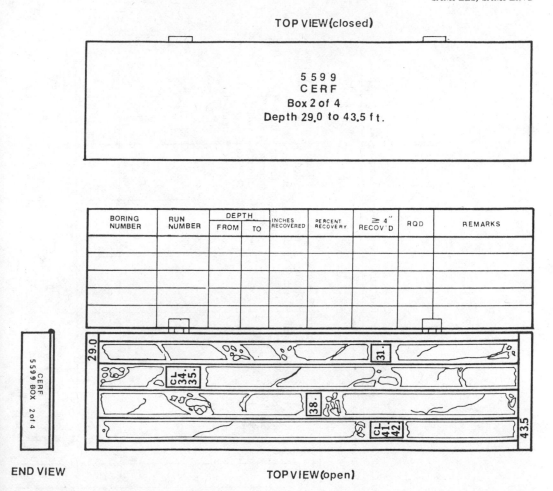

FIGURE 2. Specially designed wooden boxes are recommended for shipping and storing rock cores. Figure shows recommended format for labeling.

run is noted. The total length of the core recovered on each run is measured prior to removal of the core from the barrel (since once removed, highly jointed or fractured rock often appears to be longer than its true length). In the case of badly broken core, rigorous care should be exercised to ensure that pieces of core are placed in boxes exactly as they were in the core barrel. Marked spacers are placed in the core box to show the beginning and end of each core run. If samples are subsequently removed from core boxes for analysis or testing, additional spacers should be added to the box to show the depth intervals of samples taken.

Rock Quality Designation (RQD) is a useful index in evaluating the quality of rock for foundation support, rock bolting patterns, remedial grouting, etc. The RQD index, expressed as a percentage, is calculated by measuring the total length of all intact pieces of core of more than 4 in. (1.57

cm) and dividing this by the total length of the core run. The remarks column (see Fig. 2) may include any features of note (e.g., "elevation 32 m MSL, lost fluid circulation").

Disturbed soil samples are placed in plastic bags. Particular care should be taken with cohesive (e.g., clay) samples that will be used for moisture content determinations, to ensure that as much air as practical is expelled from the bag before sealing it. For further protection, plastic bags are placed in airtight jars or specially designed aluminum or plastic tubes and these, in turn, crated in sturdy sample containers for shipping.

Undisturbed samples are retrieved in reusable brass or aluminum liners (see Fig. 1) that, after sealing, are often carried to the laboratory by the geologist in person. Having taken the care to obtain a good undisturbed sample, further care should be exercised to avoid any postsampling disturbance that could equally alter the physical

TABLE 1. Some Environmental and Other Criteria that Influence the Choice of an Appropriate Offshore Sampling Rig

Drilling Rig (length of barge or ship or ship in meters)	Common Usage	Soil Sampling Capability: Undisturbed (U) Moderate-disturbance (M) Cuttings or dredged sample (D)	Rock-Coring Capability: Yes (Y) No (N)	Standard Penetration Test Capability: Yes (Y) No (N)	Environmental Criteria Impacted: Moderate Levels of: Yes (Y) No (N)					Operating Water Depth (m)	Practical Drilling Depth (m)	Man-power (drill and support vessel personnel)	Mobiliza-tion (moving or towing speed in knots)	Limiting Sea Condi-tions (waves/ swell height in m)	Support Vessels
					WAVE	WIND	CURRENT	SEDIMENT	ICE						
Barrel Float Wire-Line Sampler (5 m)	Interval soil samples in protected waters	U	N	Y	Y	Y	Y	Y	Y	3.5–6.0	20	3	2	0	Small tender (5 m)

722

Barge/Ship Vibratory Corer (30 m)	Rapid continuous soil sampling	M	N	Y	Y	N	Y	2–35	12	10 for live-ashore 16 for live-aboard	Ship 12 Barge 5	1–5	Small ship (30 m) or barge (30 m) and tug tender (18 m)
Small Jack-up Sampler/Corer	High-quality samples	U	Y	N	N	Y	Y	5–35	350	5 on rig (one shift) 5 on tug	4	5	Tug tender (30 m)
Small Drill Ship Dredge Sampler (50–80 m)	Bulk samples	D	N	N	Y	N	Y	5–35	15	15	12	1.5–2	None
Small Drill Ship Sampler/Corer (50–80 m)	High-quality samples	U	Y	Y	Y	N	Y	5–200	450	15	12	1.5–2	None
Large drill ship Sampler/Corer (100–180 m)	Interval or continuous samples	M	N	N	N	N	N	5–2,500	8,500	20	12	3	None

SAMPLES, SAMPLING

(engineering) properties of the sample. Undisturbed samples should most often be capped, taped, and sealed in wax in the field, transported in special containers, moved in an upright (in situ oriented) mode, and stored in temperature- and humidity-controlled rooms before being carefully extruded for laboratory testing. American Society for Testing and Materials (ASTM) specifications exist for the handling of samples for routine soil tests, and these should be followed rigorously. Samples high in organic content may have to be preserved by freezing.

Sample Support Equipment

The choice of drilling rigs to handle different types of sampling equipment for varying purposes

and under differing conditions is a most demanding task, the detailed discussion of which is beyond the scope of this entry. A simplified diagram is offered, however, by way of example, for the selection of an appropriate rig for offshore sampling (Fig. 3). Figure 4 illustrates some of the drilling and sampling rigs mentioned, while Table 1 shows some of the environmental and other criteria that influence the choice of an appropriate sampling rig.

The example offered is intended to demonstrate the relative complexity of sample-support selection. For further discussion see Anderson (1975), Hvorslev (1948), LeRoy et al. (1977), Le Tirant (1979), Terzaghi and Peck (1969), and Watson and Krupa (1984).

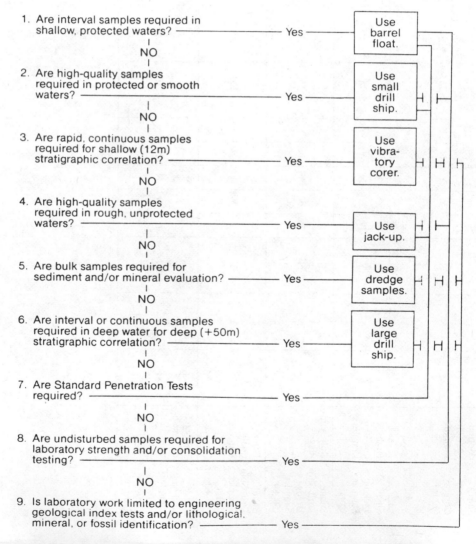

FIGURE 3. Example of a simplified decision-analysis diagram for selection of an appropriate offshore sampling rig. (After Watson and Krupa, 1984)

724

FIGURE 4. Some typical rigs for drilling and sampling offshore.

Conclusions

Samples must be, above all else, representative of the soil or rock horizons they are intended to represent. Sampling is a diverse practice governed in large part by the purpose of the sampling program, the conditions under which samples are taken (e.g., surface or subsurface, onshore or offshore), and the geological complexity of the area sampled.

IAN WATSON
ROY LEMON
STEVEN L. KRUPA

References

Anderson, G., 1975, *Coring and Core Analysis Handbook.* Tulsa, Okla.: Petroleum Publishing Co., 200p.

Barnes, J. W., 1981, *Basic Geological Mapping,* Geological Society of London Handbook Series. New York: Halsted Press, 112p.

Casagrande, A., 1948, Classification and identification of soils, *Am. Soc. Civil Engineers Trans.* **113**, 901–991.

Golder, H. Q., 1971, The practice of sampling soil and rock, *Am. Soc. Testing and Materials Spec. Tech. Pub. 483,* 3–8.

Hoek, E., and E. T. Brown, 1980, *Underground Excavations in Rock.* London: Institution of Mining and Metallurgy, 532p.

Hvorslev, J. M., 1948, *Subsurface Exploration and Sampling of Soils for Civil Engineering Purposes.* New York: American Society of Civil Engineers, 533p.

LeRoy, L. W., D. O. LeRoy, and J. W. Raese, 1977, *Subsurface Geology, Petroleum, Mining, Construction.* Golden: Colorado School of Mines Press, 942p.

Le Tirant, P., 1979, *Sea Bed Reconnaissance and Offshore Soil Mechanics for the Installation of Petroleum Structures.* Houston, Tex.: Gulf Publishing Co., 508p.

Merritt, A. H., 1974, Underground excavation: Geologic problems and exploration methods, in *Proceedings, Subsurface Exploration for Underground Excavations and Heavy Construction, ASCE,* 41–55.

Terzaghi, K., and R. B. Peck, 1967, *Soil Mechanics in Engineering Practice,* 2nd ed. (1st ed., 1951). New York: McGraw-Hill, 433–435.

Watson, I., 1984a, Contamination analysis—Flow nets and the mass transport equation, *Ground Water* **22**(1), 31–37.

Watson, I., 1984b, Hydrogeologic control and statistical prediction of active mass movement, *Assoc. Eng. Geologists Bull.* **21**(4), 479–494.

Watson, I., 1984c, Undersea transmission lines, engineering geology, in C. W. Finkl, Jnr., ed., *Encyclopedia of Applied Geology,* Vol. 13, Encyclopedia of Earth Science Series. Stroudsburg, Pa.: Van Nostrand Reinhold, 585–593.

Watson, I., J. A. Fischer, and D. P. Maniago, 1984, Nuclear plant siting, offshore, in C. W. Finkl, Jnr., ed., *Encyclopedia of Applied Geology,* Vol. 13, Encyclopedia of Earth Science Series. Stroudsburg, Pa.: Van Nostrand Reinhold, 371–377.

Watson, I., J. A. Fischer, and C. M. Urlich, 1975, Geo-

technical aspects of rock borrow for large breakwaters, *Offshore Technology Conf.* **2392**, 553–563.
Watson, I., and S. Krupa, 1984, Marine drilling exploration: Technical and environment criteria for rig selection, *Litoralia* **1**(1), 65–81.

Cross-references: *Augers, Augering; Environmental
Engineering; Exploration Geochemistry; Exposures,
Examination of; Field Geology; Fossils and Fossili-
zation; Prospecting; Soil Sampling.*

SANITARY LANDFILL—See
ARTIFICIAL DEPOSITS AND
MODIFIED LAND; CITIES,
GEOLOGIC EFFECTS.

SAPROLITE, REGOLITH AND SOIL

Most of the Earth's land surface (about 95%),
at least in the middle and lower latitudes, is covered by unconsolidated deposits of one sort or another (Hunt, 1986). In the higher latitudes, glacial
erosion has removed some of the surface cover,
exposing bare rock surfaces that account for
about 5% of the land area. In the unglaciated low-
lying tropical and peritropical regions, most of the
land surface is covered by soils and thick accumulations of detritus, largely derived from weathering mantles. Field geologists who have been directed to prepare maps of bedrock geology (see
Geological Survey and Mapping) for parts of
weathered cratons, shields, and stable platforms
appreciate the difficulties associated with mapping in strongly weathered terranes. Similar problems occur in other regions where there are extensive covers of sand, volcanic ash, loess,
alluvium, colluvium, wash deposits, vegetal accumulations, and soil. These kinds of surficial
covers also pose serious problems for economic
geologists who must attempt to elucidate bedrock
structure and composition at depth. By applying
indirect field techniques that involve geochemical
sampling (see *Exploration Geochemistry*) of cover
deposits and geophysical survey (see *Exploration
Geophysics*), it is often possible to interpret the
nature of bedrock under many meters of masking
cover. Whether attempting to discern the bedrock
geology below a surficial mantle or preparing
maps of surficial geology, it becomes necessary for
the field geologist to distinguish effectively the
major kinds of surface deposits—viz., saprolite,
regolith, and soil. The well-trained field geologist
thus employs the principles and practices of surficial geology and pedology.

Regolith

The terms *saprolite* and *regolith* are used to
designate a range of unconsolidated deposits that
occur at the surface of the Earth (see Table 1).
Regolith is the more general term of the two. It
applies to the layer or mantle of fragmental and
unconsolidated rock material, whether residual or
transported and of highly variable character, that
nearly everywhere forms the surface of the land
and overlies or covers the bedrock. This bane of
hard-rock geologists (see *Hard- versus Soft-Rock
Geology*) includes rock debris of all kinds—e.g.,
volcanic ash, glacial drift, alluvium, colluvium,
loess and other eolian deposits, organic accumulations, and soil (see Table 1). The term was originated by Merrill (1897) and is derived from the
Greek *rhegos,* blanket, combined with the suffix
lithos, stone. The term *regolith,* sometimes spelled
rhegolith, is frequently interchanged with *mantle
rock* (also spelled *mantlerock*), *rock mantle,* and
overburden, although the terms are not all synonymous. The term *overburden,* e.g., is mainly
applied by economic geologists to designate all
sorts of barren rock materials, either loose or
consolidated, that overlie mineral deposits. The
overburden, which must be removed prior to mining, thus might include *caprock,* which is a consolidated material. Sedimentologists, in contrast,
apply *overburden* in a similar context as *regolith*
in reference to the upper part of a sedimentary
deposit that compresses and consolidates the material below (see Vol. VI: *Compaction in Sediments*). This other usage applies mainly to the
loose soil, silt, sand, gravel, or other unconsolidated material that overlies bedrock, whether
transported or formed in place.

According to the original definition as proposed by Merrill (1897), *regolith* embraced the
following transported and in situ types (with additional comments by the writer):

(a). *Sedentary* or *in situ regolith:* (i) Residual
Deposits—lag gravels, loams, residual sands and
clays, grits, red clays, laterite, bauxite (for examples, see Table 1). These deposits therefore include *saprolite* (cf). (ii) Cumulative ("cumulose")
Deposits—peaty, organic soils and other organic-
rich accumulations (gytja, dy, etc.). The term
muck, as employed by Merrill, refers to highly decomposed organic material in which the plant
parts are not recognizable. Muck contains more
mineral matter and is usually darker in color than
peat. In some regions outside the United States—
Great Britain, e.g.—the term *muck* commonly refers to agricultural farmyard manure; this definition is quite different from the one employed
by Merrill.

(b). *Transported Regolith:* (i) Colluvial Deposits—scree, talus and cliff debris, avalanche, mud-
slide, rockslide, and landslide debris; *grèzes litées*
(bedded, partly cemented talus debris and soil);
appreciably displaced solifluction material (at the
foot of slopes). (ii) Alluvial Deposits—modern al-

TABLE 1. List of Surficial Materials with Examples of Naturally Occurring Transported
and Untransported Deposits from the Conterminous United States

Deposit	Comments
Untransported Surficial Materials	
Saprolitic Residuum	
Micaceous residuum	Without much quartz; clay, mostly kaolinite
Residuum with quartz	Mica content equal to clay
Red clay	Massive clay that is generally kaolinitic
Cherty red clay	Incorporates chert from parent rock
Residuum on Triassic formations	Shallow saprolite, reddish color
Sandy residuum	Weathered sandstone formations
Clay residuum	Swells when wet; derived from montmorillonitic shales
Loam	Texture ranges from sand to clay; contains nonswelling kaolinitic clays
Gravels	Intensively weathered upper Tertiary and Quaternary gravels; saprolite generally less than 10 m thick
Phosphatic clays	Poorly sorted clay and phosphate pebbles; 3–18 m thick; major source of phosphate fertilizer
Sandy, silty residuum	Includes some loess; generally less than 3 m thick
Thin Residuum	
Silt on limestone	Includes some loess blankets
Sandy ground	Mostly poorly consolidated sandstone formations
Shaley or sandy ground	On mixed sandstone and shale formations; where shaley, contains swelling clays
Sandy gypsiferous ground	Contains sinks, local dunes
Clayey ground	Mostly on weathered Permian or Triassic red beds
Organic Deposits	
Marshes, swamps	Includes peat deposits; may be locally thicker than 3 m
Sandy coastal ground	May include organic layers over a shallow water table
Transported Deposits	
Colluvial Deposits	
Sandy, stony colluvium	Derived mostly from sandstone and shale
Stoney colluvium on limestone	Contains admixed silt, loess
Stony colluvium on metamorphic rocks	Contains moderate amounts of silt and clay
Colluvium on volcanic rocks	Includes wide range of textural grades
Bouldery colluvium	Includes sandy colluvium on granitic rocks
Clayey and loamy colluvium	On poorly consolidated rocks
Shore Deposits	
Bayhead and bayside sandbars	Usually separated by rocky headlands
Sandy shores	Siliceous and calcareous sands
Sea islands	Irregularly shaped sandy islands
Coral	Includes patch reefs, knolls; sandy deposits from corals, corallines, shellfish fragments
Backshore deposits	Coastal organic-rich swamp, marshland deposits less than 8 m above sea level; have shallow groundwater tables
Glacial Deposits	
Pre-Wisconsin drift	Older glacial deposits mostly covered by loess and younger glacial materials

(continued)

TABLE 1. (*continued*)

Deposit	Comments
Terminal moraines	Mark equatorial boundaries of glacial drift and poleward limits of most residuum
End moraines	Hummocky ridges of poorly sorted weathered gravel, sand
Till, ground moraine	Poorly sorted bouldery, sandy deposits; generally less than 8 m thick
Ice-laid deposits	Mostly sand and silt forming smooth plains, sometimes with pitted ground
Deposits of mountain glaciers	Gravel and sand in U-shaped valleys; generally less than 8 m thick
Fluvioglacial deposits	Gravel, sand, silt, and clay deposited by glacial streams; commonly more than 18 m thick in some valleys
Stream Deposits	
Floodplain, alluvium gravel terrace	Well-bedded lenticular gravel, sand, and silt; fills may be up to 35 m thick
Fan gravels	Commonly exceed 350 m in thickness; rest on pediments; some have desert pavements
Fan sands	Lie between gravels upslope and silty beds of floodplains or playas in valleys; some extensively reworked into sand dunes
Ancient stream deposits	Broad fans of sand and gravel; generally less than 35 m thick
Lake Deposits	
Lacustrine sands, clays	From lakes formed by damming of rivers by continental ice sheets; overdeepening of river valleys by ice scour; infilling of closed basins
Eolian Deposits	
Sand sheets	Includes sand mounds and dunes; most have a basal layer of weathered, partly cemented, and stabilized older sand
Loess	Silty blanket deposits that frequently overlie Wisconsin and older glacial deposits or outwash

Source: Modified from Hunt, C. E., 1984, Surficial geology, *National Atlas of the United States of America.* Washington, D.C.: U.S. Geological Survey (Map No. NAC-P-0204-75M-O).

luvium (but not ancient alluvium, which constitutes distinct stratigraphic formations), including fluvioglacial deposits such as outwash sands (sandur) and piedmont gravels. (iii) Eolian Deposits—wind-blown accumulations (cf. Vols. III and VI) such as sand dunes, parna (clay dunes), and loess (Fig. 1). (iv) Glacial Deposits—moraional material, till or boulder clay (drift), drumlins, kames, eskers, etc. (for other examples, see Table 1).

Merrill (1897, 300) emphasizes that the upper part of the regolith incorporates the *soil* of pedologists (see Vol. XII, Pt. 1: *Soil*). It specifically includes that part of the unconsolidated mantle that supports plant life. The lower boundary of soil where it may merge into deep saprolite is ill defined, and the distinction between the two different kinds of deposits remains blurry. The lower limit of soil is usually taken as the depth to which biological activity extends. This concept of *pedochemical weathering* thus essentially embraces biochemical activity that takes place in the soil solum, the A and B horizons (Buol et al., 1980).

FIGURE 1. The Vjayovok exposure near Lubny (Ukranian SSR) showing a thick loess section dissected by the Boriskin and Voronov gullies. Thick loessial accumulations, intercalated with fluvioglacial materials, form part of the regolith in many areas marginal to Pleistocene glaciations.

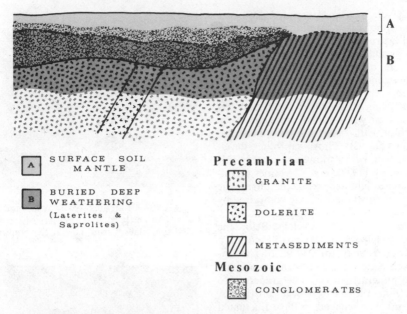

A ⟧ SURFACE SOIL MANTLE

B ⟧ BURIED DEEP WEATHERING (Laterites & Saprolites)

Precambrian

GRANITE

DOLERITE

METASEDIMENTS

Mesozoic

CONGLOMERATES

FIGURE 2. Deep tin mining quarries near Greenbushes (southwestern Australia) commonly show two distinct weathering mantles, a contemporary surface soil mantle, and buried laterite profiles, saprolites. Both weathering phases show continuous development in different kinds of parent (initial) materials.

Geochemical weathering (see *Rock Weathering*) takes place below the soil solum in the C horizon and lower zones of weathering. Figure 2 diagrammatically illustrates common field relations between soils, weathering mantles, and parent (initial) materials.

Other terms are commonly applied to regolith, and the field geologist should at least be aware of different usages. The term *head* is widely used in England and France for *regolith,* but as employed in these localities, it denotes both transported saprolite and solifluction debris. German workers apply the term *grus* to both rock-rotted debris and mechanical weathering detritus such as products of exfoliation, frost disintegration, and salt weathering (q.v. in Vol. III). Another term of Germanic origin that is sometimes used, *geest* (favored by McGee, 1891, 279), specifically refers to the gravelly or gritty varieties of regolith.

Saprolite

The term *saprolite,* which is based on the Greek *sapros,* "rotten," was proposed by Becker (1895) in reference to rock that was rotted in situ—i.e., rock that was chemically altered, but remaining coherent and not texturally disintegrated. The term *saprolith,* as sometimes used to designate sapropelic rocks, asphaltites, and related petroleum (see Pettijohn, 1957, 489), as well as the spelling *sathrolith* of Sederholm (1931), should be

avoided. *Saprolite,* as defined in the *Glossary of Geology* (Bates and Jackson, 1980), is a soft, earthy, typically clay-rich, thoroughly decomposed rock, formed in place by chemical weathering of igneous, sedimentary, and metamorphic rocks. Because saprolites are characterized by preservation of structures that were present in the unweathered rock, it is often possible to designate structured saprolites as "granitic saprolite," "basaltic saprolite," "doleritic saprolite," etc. when its origin can be clearly identified. Saprolites are commonly some shade of red or brown but may be white or gray. Although such colors are not always diagnostic of parent rocks, they do give some indication of the environments of formation. The zone of fluctuating groundwater, e.g., typically produces a mottled zone that is characteristic of alternating oxidizing and reducing conditions. Saprolitic zones occurring below the water table in a zone of permanent saturation are commonly pale colored, such as the pallid zones of laterite deep-weathering profiles.

In the tropics and peritropics, the thickness of saprolite can be quite remarkable: it frequently exceeds 100 m and may reach 200 m or more. Greater depths of weathering have been reported, but these may be related to shear zones where bedrock has been affected by hydrothermal alteration. This thick weathering mantle, together with derivative products from the reworking of the soft surface layers and also the formation of duricrusts (q.v. in Vol. XI), has frustrated geologists because they cannot easily map or gain access to the unaltered bedrock.

Saprolites assume a variety of morphologies that result from different influences of parent rock, drainage conditions, site (topographic position), and age. Profiles resulting from the deep chemical weathering of basic rocks such as dolerites and basalts on well-drained sites tend to be reddish in color with moderate morphological differentiation (Fig. 3). Those saprolites developed in ultramafic rocks show rather more differentiation. Smith (1977), e.g., describes a typical unmineralized laterite profile in ultramafic rocks in Western Australia as being comprised by a surface ferruginous (plinthite) zone that is immediately underlain by mottled clays. Below the mottled clays and lower in the profile occurs a silicified saprolite zone from 30 to 90 m. The saprolitic clays show the morphology of the original rock-forming minerals with a slow gradation to oxidized rock at the base of the profile. A groundwater table is believed to have fluctuated about the boundary of the ferruginous zone and mottled clay zone, based on observations of laterites presently forming in the tropics (Van Schuylenborgh, 1969). The classical laterite deep-weathering profile formed over metamorphic terrains displays a different but distinctive sequence of horizons with depth (see Figs. 4 and 5). Morphological zonation of laterite profiles developed

FIGURE 4. Laterite breakaway in southwestern Australia. This partly collapsed Fe-duricrust, developed over pallid zone clays, forms a distinctive landscape feature.

FIGURE 5. Lateritic regolith in southwestern Australia showing pisolitic (Fe gravel) veneer overlying mottled zone materials. The colluvial ironstone gravels are derived from upslope and are separated from the in situ mottled (kaolinitic) clays by a sharp boundary.

FIGURE 3. Saprolite profile, near Warrenton, developed in a basic dike on the piedmont in the northern neck of Virginia. Note the gradual transition from massive (structureless) saprolite in the upper part of the profile to structured saprolite at depth. Persistence of rock structure is evident near the base of the profile where corestones have resisted the weathering process.

from granitoid rocks, e.g., is commonly described in terms of an upper ferruginous zone (the ironstone crust being a cuirasse or carapace, depending on the degree of induration) (see Fig. 4), an underlying mottled zone (*flecken zone* of German workers) (see Fig. 5), and a still lower bleached or pallid zone (*bleich zone* of Walther, 1915) that grades into a soft rotten layer that retains essential rock structure (*zersatz zone*) before merging with the weathering front (Mabbutt, 1961). All three major zones—the mottled, pallid, and zersatz zones—may show degrees of silicification (see discussion in Finkl, 1984).

Economic Value of Saprolites

The *nickeliferous laterites,* also commonly referred to as *lateritic nickel deposits,* are associated with ultramafic rocks. The ultramafic terrains, greenstone belts, in which they occur on tropical cratons are easily distinguished from the surrounding reddish-colored, lateritized granitic terrains by their lighter yellowish-gray color. Although these surface colors are subtle, the trained observer with a keen eye will find them to be fairly reliable indicators of bedrock types. Lateritic nickel deposits such as those in New Caledonia are most commonly worked in open cuts. Ore-grade saprolitic materials are dug out by large shovels and then moved to refineries by truck, earth hauler, or overland conveyor belt.

The alumina-rich saprolite is *bauxite,* the principal ore of aluminum. These deposits, resulting from intense deep chemical weathering, show similar characteristics to nickel-rich laterites and are also mined from open pits and broad excavations. Bauxite (a hydrous aluminum oxide) is mined by these methods in the Darling Ranges of Western Australia, now a world leader in the extraction of this residual material. Other producers of bauxite include Surinam, the United States, the USSR, and southern Europe. The largest deposits of bauxitic saprolite in the United States occur in Arkansas. Some of the deposits were formed as saprolite on nepheline syenite, an igneous rock rich in aluminum (Gordon et al., 1958).

Gemstones (e.g., emeralds and rubies in Burma and Ceylon, diamonds in Africa) are often worked in old saprolites because they are not weathered and the enclosing weathered (soft) rock can be easily worked. Tin (in the form of cassiterite) is similarly contained in the granite saprolites of Malaya, Indonesia, and Australia. Often the saprolites have been eroded and the precious (heavy) minerals further concentrated by running water, alluvial placer deposits, e.g. Other saprolites containing heavy minerals are often worked by dredges or by powerful hydraulic jets (see *Placer Mining*).

Soils

The term *soil* is derived from the Latin *solum,* which at one time had the same general meaning as the modern term. According to *Webster* (Woolf, 1974) soil is "the upper layer of the earth that may be dug or plowed and in which plants grow." This definition is nearly as old as the word. Some geologists for a long time have considered soil to be an unconsolidated heterogeneous aggregate of disintegrated rock material that contains some organic matter (usually less than 5% by composition) near the ground surface (e.g., Geike, 1958). Many field geologists still find this

definition adequate for most purposes as they rarely attempt to identify specific kinds of soils.

There are four main technical perceptions of soil (see Vol. XII, Pt. 1: *Soil*) that may be conveniently referred to as the edaphic, geographic, pedologic, and engineering concepts. Edaphologists view soil as the natural medium for the growth of land plants, whereas the geographic concept of soil relates to the study of the areal distribution of soils (Butler, 1958). The pedologic concept (see Vol. XII, Pt. 2: *Pedology*) embraces the study of soils as independent natural bodies, each with a distinct morphology resulting from the integrated effect of climate and living matter, acting on parent materials, as conditioned by relief, over periods of time (Soil Survey Staff, 1951). Thus, according to this view soil is the unconsolidated mineral matter on the surface of the Earth that has been subjected to the influence of genetic and environmental factors—in short, a product of low-grade epidiagenesis. From the engineer's point of view, soil is a material with which and upon which one builds structures. It includes any unconsolidated material found between the ground surface and bedrock. Thus, soil in the engineer's view includes all regolith (q.v.) materials—e.g., unconsolidated accumulations of rock fragments and an organic matter that is intermixed—or the entire vertical section down to consolidated rock (Peck et al., 1974).

Although soil is frequently regarded by many field geologists as the annoying stuff that gets in the way of geological surveys and obscures the geology, appreciation of different expressions of the phenomenon can improve the quality and reliability of many field surveys (see *Geological Survey and Mapping*). When it becomes necessary to identify particular occurrences of distinct kinds of soils, field geologists are advised to contact local offices of the national soil survey (q.v. in Vol. XII, Pt. 2) of the country in which they are working. Some smaller or developing countries use the legend to the international *Soil Map of the World* (FAO/UNESCO Staff, 1961) or alternatively employ systems developed in other countries. The geographical application of modern soil classification systems has been summarized, on a worldwide basis, by Finkl (1982). The FAO/UNESCO legend is, e.g., followed in many parts of Central and South America and in parts of Africa. The main categories of soil units listed in the legend to the *Soil Map of the World* are summarized in Table 2. These modern terms, or others required by national soil surveys, should be used whenever practical. No matter what system of soil classification is used in a particular geographic area, it is clear that geologists must conform to accepted soil survey practices if they choose to deal specifically with soil in terms of mapping or classification units.

TABLE 2. Soil Map Units Listed in the Legend to the FAO/UNESCO Soil Map of the World

Fluvisols (water-deposited soils with little alteration)
  Dystric
  Eutric
  Calcaric
  Thionic

Regosols (thin soil over unconsolidated material)
  Dystric
  Eutric
  Calcaric
  Gelic

Arenosols (soils formed from sand)
  Ferralic
  Luvic
  Cambic
  Albic

Gleysols (mottled or reduced horizons due to wetness)
  Mollic
  Dystric
  Eutric
  Gelic
  Humic
  Calcaric
  Plinthic

Rendzinas (shallow soil over limestone)

Rankers (thin soil over siliceous material)

Andosols (volcanic ash with dark surfaces
  Ochric
  Mollic
  Humic
  Vitric

Vertisols (self-mulching, inverting soils, rich in mont-
    morillonitic clay)
  Pellic
  Chromic

Yermosols (desert soils)
  Haplic
  Calcic
  Gypsic
  Luvic
  Takyric

Xerosols (dry soils of semiarid regions)
  Haplic
  Calcic
  Gypsic
  Luvic

Solonetz (high sodium content)
  Orthic
  Mollic
  Gleyic

Planosols (abrupt A-B horizon contact)
  Eutric
  Dystric
  Mollic
  Gelic
  Humic
  Solodic

Kastanozems (chestnut surface color, steppe vegeta-
    tion)
  Haplic
  Calcic
  Luvic

Chernozems (black surface, high humus under prairie
    vegetation)
  Haplic
  Calcic
  Luvic
  Glossic

Phaeozems (dark surface, more leached than Kastan-
    ozem or Chernozem)
  Haplic
  Calcaric
  Luvic
  Gleyic

Greyzems (dark surface, bleached A2, and textural B)
  Orthic
  Gleyic

Cambisols (light color, structure, or consistence
    change due to weathering)
  Gelic
  Eutric
  Gleyic
  Dystric
  Ferralic
  Calcic
  Vertic
  Humic
  Chromic

Luvisols (medium to high base status soils with argillic
    horizons)
  Orthic
  Chromic
  Calcic
  Vertic
  Ferric
  Albic
  Plinthic
  Gleyic

Solonchaks (soluble salt accumulation)
  Orthic
  Mollic
  Takyric
  Gleyic

Podzoluvisols (leached horizons tonguing into argillic
    B horizons)
  Eutric
  Dystric
  Gleyic

Podzols (light-colored alluvial horizon and subsoil ac-
    cumulation of iron, alluminum, and humus)
  Humic
  Ochric
  Placic
  Gleyic
  Ferric
  Leptic

Acrisols (highly weathered soils with argillic horizons)
  Orthic
  Ferric
  Humic
  Plinthic
  Gleyic

TABLE 2. (*continued*)

Nitrosols (low CEC clay in argillic horizons)	Histosols (organic soils)
Dystric	Dystric
Eutric	Eutric
Humic	Gelic
Ferralsols (sesquioxide-rich clay)	Lithosols (shallow soils over hard rock)
Orthic	Arenosols (sandy soils)
Rhodic	Ferralic
Humic	Luvic
Acric	Cambic
Plinthic	Albic
Xanthic	

Source: Compiled from FAO/UNESCO Staff, 1964; Buol et al., 1980.

Field Relations of Regolithic Materials

Efforts to better understand the field relations of regolithic materials—i.e., the temporal and spatial interactions of weathering profiles (soils), sediments, and residuum—focus on basic tenets of stratigraphical geology and soil stratigraphy (q.v. in Vol. XII, Pt. 2). The principles of soil stratigraphy, perhaps less familiar to most field geologists than the basic laws of geologic chronology (see Vol. VI: *Top and Bottom Criteria*), have been summarized by Butler (1959), Ruhe (1969), Morrison (1967), Follmer (1978), and Finkl (1980), among others. Generalized examples of field relations among surficial materials are drawn here, for illustrative purposes, from the deeply weathered tropical cratons and from mid-latitude glacial landscapes.

The deeply weathered shield landscapes contain evidence of widespread stripping (erosion) of regoliths. The degree of stripping produces a wide range of distinct landscapes that characterize so much of Africa, Australia, Brazil, southern India, tropical South America, and even parts of Western Europe. The process and resulting surface form, detailed for parts of the peritropical African (Thomas, 1974) and Australian cratons (Finkl, 1979), are based on the concept of etchplains where there are "double planation surfaces" as explained by Budel (1982). The vertical succession of master horizons, including caps of ferruginous duricrust, in laterite profiles provides a measure of the degree of truncation. Regions with lateritic mantles that remain nearly intact—i.e., the oldest and stablest parts of the cratons—commonly display sandy ground surfaces that are underlain by ferricrete and a complete laterite profile (i.e., one that exhibits master mottled and pallid horizons). A moderate degree of stripping exposes the mottled zone, imparting a reddish-brown color to the contemporary soilscape. More extensive erosion that strips away the upper parts of the laterite profile typically exposes pallid zone materials. These strongly eroded landscapes are characterized by light-colored soils derived from the pale white to gray-colored kaolinitic clays of the lower laterite profile. Nearly complete removal of the deep weathering mantle exposes the irregular contour of the weathering front and leaves corestones behind on the ground surface. Inselberg landscapes (q.v. in Vol. III) are believed to form in this manner, by the stripping of weathered materials and subsequent exposure (exhumation) of bedrock highs, a process sometimes referred to as *etchplanation* (see discussion in Thomas, 1974).

Recognition of deeply weathered landscapes and especially their stripped counterparts is essential to competent interpretations of extant field conditions. The occurrence of different kinds of regoliths will determine the field methods that can be used to best advantage. Accurate identification of stripped surfaces and the nature of surficial materials, especially those derived from weathering mantles, provides a rational basis for the selection of sample media in geochemical surveys (see *Biogeochemistry; Exploration Geochemistry; Hydrochemical Prospecting; Pedogeochemical Prospecting; Samples, Sampling; Soil Sampling*). The same rationale applies to geophysical surveys where thick regoliths and saline groundwaters can affect survey results (see *Exploration Geophysics; VLF Electromagnetic Prospecting*).

Although completely different from the tropical and peritropical shield landscapes with their thick weathering mantles, other ectropical landscapes with extensive regoliths pose similar challenges to the field geologist—e.g., the glacial outwash and loess plains of northern latitudes. These glacial and periglacial landscapes with complex regoliths exhibit many different kinds, often a bewildering range, of surficial materials that frequently contain intercalated weathering zones (soils) (cf. Fig. 1).

Typical relations between soil mantles may be identified in two simple cases by what Butler (1959) calls overplaced and underplaced contacts. An *overplaced contact* is where two soil mantles

733

are locally separated by intervening deposits (Fig. 6). Horizons of the younger soil mantle (Fig. 6*A*) are replaced from the bottom upward by those of the older (lower), more strongly developed soil mantle (Fig. 6*B*). An *underplaced contact* is where an erosion surface intervenes between the development of two soil mantles (Fig. 7). Where the older soil mantle (Fig. 7*B*) is truncated, its horizons are replaced from the top downward by those of the younger, more weakly developed soil mantle (Fig. 7*A*). These basic field relations are complicated by the intensity (duration) of weathering phases and erosional-depositional cycles. Overplaced contacts, e.g., can be identified only where the younger soil mantle (cf. Fig. 6) is more weakly developed than the older; otherwise, the discon-

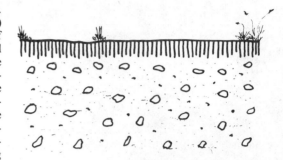

FIGURE 8. A single soil mantle formed in a single deposit of homogeneous parent materials.

tinuity is masked by pedogenic overprinting (Finkl, 1980). As shown in Fig. 6, the more strongly developed soil mantle (Fig. 6*B*) exists in both buried (covered by overburden) and relict (exposed at the ground surface) states.

Other basic field relations are depicted in Figs. 8 through 11. Figure 8 shows a single erosion-weathering cycle where deposition was immediately followed by soil development. The deposit formed during an unstable phase, and the weathering profile developed during a stable interval. Figure 9 shows two cycles where an early instability deposit is overlain by a weathering profile that was also subsequently weathered. Thus, there was deposition, then stability (weathering), a second instability involving erosion and deposition of materials similar to the underlying deposit, and then a second phase of weathering. Figure 10 records deposition followed by a different type of deposition, then there was a stable phase of soil development. In this case, there was no stable phase of sufficient duration for weathering to take place between depositions. An alternative field relation is presented in Fig. 11, which shows two

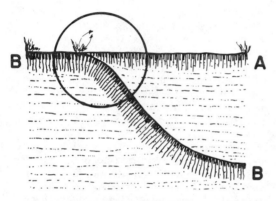

FIGURE 6. An overplaced contact between two soil mantles (circle). An intervening deposit similar to the original deposition separates the younger, more weakly developed mantle (A) from a fossil soil (B), which occurs at the ground surface and in buried situations.

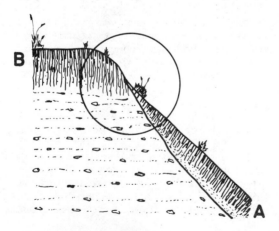

FIGURE 7. An underplaced contact (circle) where a younger, more weakly developed soil mantle (A) replaces an older truncated soil (B) downprofile from the eroded surface.

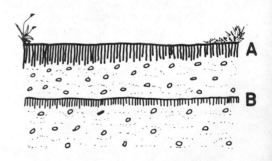

FIGURE 9. Two cycles of instability (erosion-deposition), each followed by a stable interval (soil development). The buried, weakly developed soil mantle (B) is overlain by a deposit similar to the original deposition. The second deposit was strongly weathered to form a surface soil mantle (A).

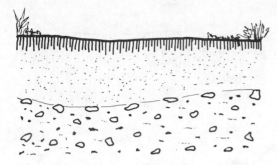

FIGURE 10. Two depositional sequences separated by an erosional unconformity, followed by stability (weak phase of soil development). No stable phase occurred between depositions.

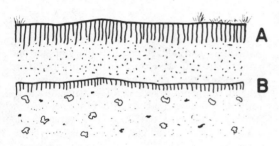

FIGURE 11. Two different depositions, each followed by a stable phase of weathering. The older buried soil mantle (B) is more weakly developed than the weathered ground surface (A). Unweathered parent materials separate the second deposit from the buried soil.

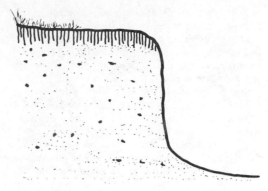

FIGURE 12. A simple soil developed in a single deposit, both subsequently truncated by an unstable erosive phase giving a deposition-weathering-erosion sequence.

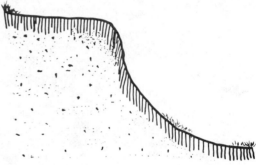

FIGURE 13. Intervention of erosion between deposition and weathering, truncation of deposit prior to soil development, to produce a deposition-erosion-weathering sequence.

different depositions, each followed by weathering during stable geomorphic conditions.

Relationships between erosion surfaces and weathering profiles, as described by Morrison (1967) and based on the K-cycle concept (erosion/ deposition/weathering models) of Butler (1959), are graphically summarized in Figs. 12 through 16. Figure 12, e.g., shows a deposit and its weathering mantle truncated by a subsequent unstable phase to produce a sequence of deposition, stability, then erosion. Figure 13, in contrast, shows erosion intervening between the depositional and stability phases, giving a sequence of deposition-erosion-stability. In Fig. 14, there are two successive phases of differing deposition, with erosion synchronous with the second deposition or intervening between the depositional phases, followed by weathering. The surface-weathering mantle here transgresses the lithologic discontinuity and the geomorphological unconformity. The sequence of events is thus deposition, then erosion followed by a second deposition, with subsequent weathering taking place across the contact.

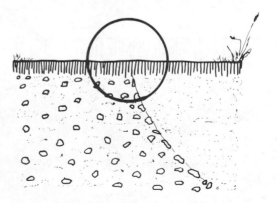

FIGURE 14. Two separate depositions followed by weathering across the contact (circle) to produce a sequence of deposition–erosion–second deposition–weathering.

Still other possibilities include situations where there is deposition, then erosion, followed by strong weathering and then a second deposition followed by a weak phase of weathering (Fig. 15). In this case, the second, weaker, phase of weathering does not mask the effects of the previous stronger phase of weathering. The strongly weathered profile occurs as a surface soil and as a buried soil associated with a paleogeomorphic surface. The contact between strongly and weakly developed weathering profiles coincides with the lithogeomorphic discontinuity at the ground surface. Figure 16 depicts an alternative to the pre-

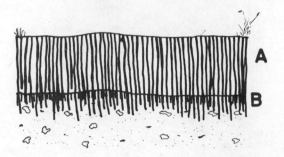

FIGURE 17. Two different depositions, each followed by stability (soil development). The intensely developed surface soil mantle (A) partly masks the paleogeomorphic unconformity, lithologic discontinuity, and weakly developed soil mantle (B) by pedogenic overprinting.

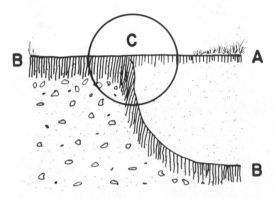

FIGURE 15. Two separate depositions, each followed by stable weathering intervals. The triple discontinuity (circle) is marked by a lithological contact, a paleogeomorphic unconformity, and pedogenic masking of a younger weak phase of weathering (A) by an older, strongly developed relict soil mantle (B). Surface expression of the discontinuities is placed by the soil boundary (C).

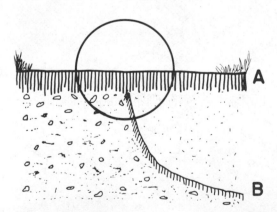

FIGURE 16. Two different depositions, each weathered. The triple discontinuity (circle) lacks surface expression due to pedogenic overprinting by younger strong weathering. The older, weakly developed soil mantle (B) is replaced downward by a younger phase of strong weathering (A).

vious sequence whereby the older weathering mantle is weakly developed and the younger one is more strongly developed. Consequently, there is no peodologic geomorphic discontinuity evident at the ground surface. The triple discontinuity is, however, marked at depth by breaks in lithology (parent materials), soil development, and subaerial erosion surfaces. Pedogenic overprinting (q.v. in Vol. XII, Pt. 2) by younger weathering mantles may also mask the relict occurrence of older weakly developed profiles (Fig. 17). The sequence may be summarized in terms of deposition, erosion, and weak weathering followed by a second deposition and strong weathering in both deposits.

C. W. FINKL, JNR.

References

Bates, R. L., and J. A. Jackson, eds., 1980, *Glossary of Geology.* Falls Church, Va.: American Geological Institute, 751p.
Becker, G. F., 1895, Reconnaissance of the gold fields of the southern Appalachians, *U.S. Geol. Survey, 16th Ann. Rept.,* Part III, 251–331.
Budel, J., 1982, *Climatic Geomorphology.* Princeton, N.J.: Princeton University Press, 443p (trans. L. Fischer and D. Busche).
Buol, S. W., F. D. Hole, and R. J. McCracken, 1980, *Soil Genesis and Classification.* Ames: Iowa State University Press, 404p.
Butler, B. E., 1958, The diversity of concepts about soils, *Jour. Australian Inst. Agric. Sci.* 24, 14–20.
Butler, B. E., 1959, Periodic phenomena in landscapes as a basis for soil studies, *CSIRO (Australia) Soil Publication No. 14.*
FAO/UNESCO Staff, 1961, *Soil Map of the World. Legend.* Paris: UNESCO, 59p.
Finkl, C. W., Jnr., 1979, Stripped (etched) landsurfaces in southern Western Australia, *Australian Geogr. Studies* 17(1), 33–52.

Finkl, C. W., Jnr., 1980, Stratigraphic principles and practices as related to soil mantles, *Catena* **7**(2/3), 169–194.

Finkl, C. W., Jnr., 1982, The geography of soil classification, *Quaestiones Geographicae* **8**, 55–59.

Finkl, C. W., Jnr., 1984, Chronology of weathered materials and soil age determination in pedostratigraphic sequences, *Chemical Geol.* **44**, 311–335.

Follmer, L., 1978, The Sangamon soil in its type area—A review, in W. C. Mahaney, ed., *Quaternary Soils*. Norwich, England: Geo Abstracts, 125–166.

Geike, J. [Campbell, R., and Craig, R. M. (eds.)], 1958, *Structural and Field Geology*. Edinburgh: Oliver and Boyd, 397p.

Gordon, M., J. I. Tracey, and M. W. Ellis, 1958, Geology of the Arkansas bauxite region, *U.S. Geol. Survey Prof. Paper 299*, 268p.

Hunt, C. E., 1986. *Surficial Deposits of the United States*. New York: Van Nostrand Reinhold.

Mabbutt, J. A., 1961, A stripped landsurface in Western Australia, *Trans. Inst. Br. Geogr.* **29**, 101–114.

McGee, W. J., 1891, The Pleistocene history of northeastern Iowa. *U.S. Geol. Survey, 11th Ann. Rept. (for 1889–1890), Part I*, 189–577.

Merrill, G. P., 1897, *A Treatise on Rocks, Rock-Weathering and Soils*. New York: Macmillan, 411p.

Morrison, R. B., 1967, Principles of Quaternary soil stratigraphy, in R. B. Morrison, and H. E. Wright, eds., *Quaternary Soils (Proc. VII INQUA Congr.)*. Vol. 9, 1–69.

Peck, R. B., W. E. Hanson, and T. H. Thornburn, 1974, *Foundation Engineering*. New York: Wiley, 514p.

Pettijohn, F. J., 1957, *Sedimentary Rocks*. New York: Harper, 781p.

Ruhe, R. V., 1969, Principles for dating pedogenic events in the Quaternary, *Soil Sci.* **107**(6), 398–402.

Sederholm, J. J., 1931, On the sub-Bothnian unconformity and on Archaean rocks formed by secular weathering, *Finlande Comm. Géol. Bull.* **95**, 1–81.

Smith, B. H., 1977, Some aspects of the use of geochemistry in the search for nickel sulphides in lateritic terrain in Western Australia, *Jour. Geochem. Research* **8**(1/2), 259–281.

Soil Survey Staff, 1951, Soil survey manual, *U.S. Department of Agriculture Handbook No. 18*, 503p.

Thomas, M. F., 1974, *Tropical Geomorphology*. New York: Wiley, 332p.

Van Schuylenborgh, J., 1969, Weathering and soil-forming processes in the tropics, in *Soils and Tropical Weathering*, Vol. 11, Natural Resources Research. Paris: UNESCO.

Walther, J., 1915, Laterit in West Australien, *Deutsch. Geol. Gesell. Zeitschr.* **67B**, 113–140.

SATELLITE GEODESY AND GEODYNAMICS

Geodesy is the discipline that deals with the measurements and representation of the Earth, including its external gravity field, in a three-dimensional, time-varying space. *Geodynamics* is the discipline that deals with the dynamics and dynamic history of the Earth. The artificial Earth satellite provides a tool to determine, among other things, present position of points on the surface of the Earth, their changes in time, and the broad geometry of the Earth's external gravity field. Satellite geodesy and geodynamics is thus a set of satellite techniques designed to solve some of the geodetic tasks, some of which are in common with geodynamics.

Tracking Modes

In the geodetic applications of satellites, we use directions and/or distances measured from points (stations) on the surface of the Earth to satellites. Taking these measurements is known as *tracking* the satellites, and the measurements are called *tracking data*. If the positions of the tracking stations are known in any coordinate system, the tracking data can be used to derive the satellite position variations with time, known as the *satellite orbit*, in the same coordinate system. The knowledge of satellite orbits is essential for most applications.

The tracking data can be obtained in a number of ways, using different instrumentation. Directions may be determined from photographs of the satellite taken against the star background. Knowing the directions to some of the stars from previous astronomic observations, directions to the satellite at specific instants of time can be derived from the coordinates of the satellite track on the photograph. This technique has been used, i.e., in establishing the worldwide *BC-4 network* (Schmid, 1974) or the *Baker–Nunn network* (Gaposhkin, 1973).

Instantaneous distances (ranges) can be observed with radar or laser equipment. An example of a systematic use of radar tracking is the *Secor equatorial network* (Mueller, 1974). Laser measurements are being used for selective tracking of most of the contemporary satellites (Lehr, 1969).

Radar altimetry—i.e., the determination of satellite altitude above a sea or large lake surface—can be considered another mode of satellite ranging. While the first two techniques require that the main instrumentation be ground-based, the radar altimeter is a satellite-borne device. It is becoming a standard equipment of the new generation of satellites.

Presently, the most widely used tracking mode is to observe range differences at station to satellite positions along the track at specified time intervals. These differences are observable via the Doppler shift in the satellite-emitted radio frequencies. This system was first conceived for the U.S. Navy Navigation Satellite System (Transit) and later refined for geodetic use (Krakiwsky et al., 1972; Anderle, 1974). More information about various tracking modes can be found in Veis (1965), King-Hele (1967), and Henriksen et al. (1972).

Computational Modes

Terrestrial geodetic information can be derived from tracking data in several ways. The most straightforward of these is the geometric mode with unknown orbit, where the satellite is used as a moving target to which observations are made simultaneously from a number of ground stations. Once the observations ($\mathscr{L}$) are made, the target satellite can be considered a stationary point in space, and relative position vectors ($\Delta \vec{r}_{ij}$) of the ground stations ($i, j = 1, 2, \ldots, n$) can be determined from the geometry of several such spatial configurations. The mathematical model for this determination can be written as

$$F(\Delta \vec{r}_{ij}, \mathscr{L}) = 0,$$

where F denotes the functional relationship between $\Delta \vec{r}_{ij}$ and $\mathscr{L}$. It should be noted here that if only direction observations to satellites are available, the scale must be supplied by terrestrial measurements.

The requirement for simultaneous observations to satellites may be replaced by the ability to interpolate or extrapolate (predict) the satellite orbit. The orbit may then serve as a target with known position in space, allowing us to solve for positions or position differences of ground stations in a fashion similar to that used in the geometric mode. This approach, the geometric mode with known orbit (orbital mode), is employed in the Transit positioning system where the orbit $[R(t)]$ is determined by a set of permanent tracking stations whose ground positions are assumed to be known. Position vectors of used stations from which observations to the satellite have been made are computed from

$$F[\vec{r}_i, \vec{R}(t), \mathscr{L}] = 0,$$

using the predetermined orbit.

In the course of predicting the orbit, the effects of various terrestrial as well as other perturbing phenomena have to be accounted for mathematically. This procedure can be used to improve the knowledge of the modeled phenomena, such as the Earth's external gravity field, tidal phenomena, air density distribution, and solar radiation pressure, by comparing the predicted with the observed orbits. This approach has become known as the *dynamic mode*. Denoting the parameters characterizing the pertinent terrestrial phenomena by $\mathscr{T}$ and those characterizing the other perturbing phenomena by $\mathscr{P}$, the mathematical model for the dynamic mode can be stated as

$$F(\vec{r}_i, \mathscr{T}, \mathscr{P}, \mathscr{L}) = 0.$$

Combinations of the described modes have been used by different investigators, taking some quantities as known and seeking the other. Details on mathematical models can be found in publications quoted here. It is also advantageous to combine in one mathematical model both satellite and terrestrial data of different kinds. As examples of such combined solutions, the works of Gaposhkin (1973), Seppelin (1974), or Mueller (1974) can be cited.

Geodetic Tasks

The geodetic tasks to which the satellite can be put may be divided into two categories: (1) determination of positions and their time variations and (2) investigation of the geometry of the Earth's external gravity field.

The general principles of point position determination have been described. In geodesy, however, it is customary to establish a complete network of points to obtain areal position control. Figure 1 shows one example of such a network, designated to serve as a global framework for further densification. As stated already, these positions were determined using directional (photographic) tracking and a geometric computational mode. The internal consistency (standard deviations) of the BC-4 points is of the order of 5 m.

Another example is given in Fig. 2. It represents a densification network determined from the Transit satellite system (Kouba, 1975). Internal consistency is about 1–2 m. I should mention here that Transit is the only satellite system available to geodetic organizations for positioning on a routine basis. A number of satellites are kept in orbit; the orbits are obtainable in real time; and the necessary instrumentation is not excessively expensive.

The three-dimensional coordinates of points on the surface of the Earth obtained by satellite methods can be employed for solving specific geodetic problems. Among these we may mention the determination of transformation parameters between different coordinate systems (e.g., Wells and Vaníček, 1975), investigation of terrestrial geodetic network distortions (e.g., Meade, 1974), and determination of two-dimensional horizontal positions and one-dimensional vertical positions for mapping and other uses.

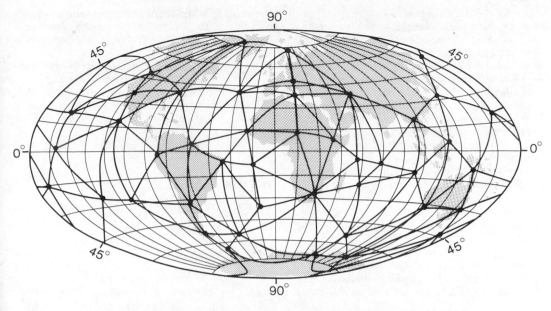

FIGURE 1. Worldwide BC-4 satellite triangulation network. (From Schmid, 1974)

FIGURE 2. Doppler satellite stations in Canada. (Courtesy of Geodetic Survey of Canada)

As a by-product of repeated or continuous position determination, we can expect to see the changes caused by tectonic plate movements and recent vertical crustal movements. Many years of observations are needed before the magnitudes of the position changes emerge above the level of noise inherent in the present satellite-positioning techniques. Several such projects are underway, but it is too early to report any concrete results. Ice movement tracking is another application of satellite positioning that stems from the same concepts.

Mapping of the sea level on the open seas has recently become possible with the advent of the satellite altimeter. The first global results, published by McCoogan et al. (1975), are encouraging, showing an accuracy of about 10 m, with the potential for an improvement by at least one order of magnitude. Tidal movements of land and sea are now also detectable through analysis of orbits—i.e., the dynamic mode (Lambeck et al., 1974). Direct detection may be possible in the near future.

A continuing program has been set up for monitoring polar motion as derived from observations made at the Transit tracking stations. The principles of this operation have been described by Anderle and Beuglass (1970), and a sample of the results is given in Fig. 3.

The geometry of the Earth's external gravity field is investigated using the dynamic computational mode to derive parameters describing the field. The mathematical model for the external gravitational potential (U) is usually set up by means of spherical harmonic functions:

$$U = \frac{GM}{R} \left[1 + \sum_{n=2}^{\infty} \left(\frac{R_o}{R} \right)^n \sum_{m=0}^{n} \right. \cdot$$

$$\left. (A_{nm}\cos m\lambda + B_{nm}\sin m\lambda) P_{nm}^* (\sin \phi) \right],$$

where

M	=	the mass of the Earth;
G	=	the Newton's gravitation constant;
R	=	the length of the satellite radius vector taken from the center of the Earth;

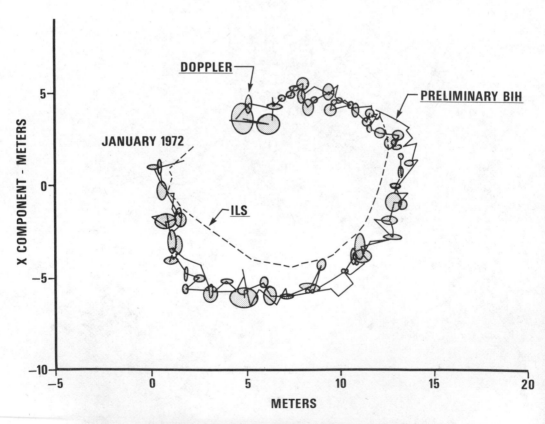

FIGURE 3. Polar motion in 1972. (After Anderle, 1973)

R_o = the radius of an arbitrarily adopted sphere, concentric with the Earth, with respect to which the potential can be considered external;

ϕ, λ = the geographical coordinates;

P^*_{nm} = the fully normalized associated Legendre's functions;

A_{nm}, B_{nm} = the parameters we seek.

There are other ways of setting up the mathematical model, for which the reader is referred to Balmino's and Koch's papers in Henriksen et al. (1972).

Once GM and the potential coefficients A_{nm}, B_{nm} are known, this equation fully describes the Earth's external gravity field (up to the centrifugal term). In principle, all these quantities could be derived; in practice, potential coefficients only up to the degree about 20 are sought. The shorter wavelength features at the altitudes of satellite orbits (above a few hundred kilometers) are too much attenuated to be normally sensed by the satellite.

To determine the mass of the Earth, or more specifically, the product GM, with sufficient accuracy, satellites orbiting at very high altitudes are needed. In fact, space probes are used for this purpose (Levallois and Kovalevsky, 1970), and the value of GM thus far determined is around 3.98601×10^{14} m^{3}s^{-2}.

Among the potential coefficients, the coefficient $A_{2,0}$ has a special place. It is directly related to the flattening of the Earth and was the first one derived from artificial satellite orbits. The value of flattening $f = 1:298.25$ currently adopted by the International Association of Geodesy resulted from satellite work (IAG, 1971).

Many determinations of the potential coefficients have been carried out in the past—e.g., Kaula (1966), Kozai (1966). From the potential coefficients other quantities characterizing the Earth's gravity field, such as the geoidal heights and deflections of the vertical and gravity anomalies, can be easily derived. For this purpose it is deemed preferable to take into the solution also terrestrial data, as we have already mentioned. Figure 4 shows the map of a global geoid, so produced, according to Rapp (1974). It should be noted that geoidal heights can also be computed from satellite-determined three-dimensional positions of points on the Earth's surface.

The deflections of the vertical can be taken as slopes of the geoid and as such are not normally computed separately. The relationship of gravity anomalies to the geoid is less straightforward, and it is hence usual also to derive them from the potential coefficients. A map of such global gravity anomalies is given in Fig. 5 (Gaposhkin, 1973).

Other geodetic parameters such as the semi-major axis of the best fitting mean Earth ellipsoid or the mean equatorial value of gravity can also be obtained from combined satellite solution. Examples for these are given in Seppelin (1974), Mueller (1974), and elsewhere.

For the sake of completeness, we have to mention that experiments are being carried out with satellite gradiometers (Forward, 1973). These instruments are designed to measure gravity gradients along the satellite.

Problems

Three main problem areas are recognized in the geodetic satellite applications where progress must be made before accuracy can be significantly improved. The most difficult theoretically are the problems connected with orbit modeling—i.e., orbit interpolation and extrapolation. Effects of the Earth's gravity field, air drag, solar radiation, etc. have to be modeled more precisely. This requires better data on these phenomena collected by the satellites; for details, the reader is referred to Escobal (1965), Kaula (1966), King-Hele (1967), Morando (1970), and Hagihara (1971). Drag-free satellites and satellite-to-satellite tracking (Henriksen et al., 1972) may help in solving or bypassing some of these problems in the future.

In all the tracking techniques, refraction, both tropospheric and ionospheric, plays an important role (Gaposhkin, 1973; Anderle, 1974; Schmid, 1974). Better understanding of the air temperature, vapor content, and ion distribution is needed to reduce errors from this source.

Instrumental problems such as noise in the instrumentation and inaccuracies of timing are probably the least serious. This is due to a fast progress in technology since the 1950s. Further improvements are being experienced almost daily.

P. VANICEK

References

Anderle, R. J., 1973, Pole position for 1972 based on Doppler satellite observations, *Naval Weapon Lab. Tech. Rept. 2952.*

Anderle, R. J., 1974, Transformation of terrestrial survey data to Doppler satellite datum, *Jour. Geophys. Research* 79, 35, 5319–5331.

Anderle, R. J., and L. K. Beuglass, 1970, Doppler satellite observations of polar motion, *Geod. Bull.* 96, 125–141.

Escobal, P. R., 1965, *Methods of Orbit Determination.* New York: Wiley.

Forward, R. L., 1973, Review of artificial satellite gravity gradiometer techniques for geodesy, *Hughes Research Labs. Rept. No. 469.*

Gaposhkin, E. M., 1973, 1973 Smithsonian standard earth III, *SAO Spec. Rept. 353.*

Hagihara, Y., 1971, *Celestial Mechanics,* Vol. 2, *Perturbation Theory.* Cambridge, Mass.: MIT Press.

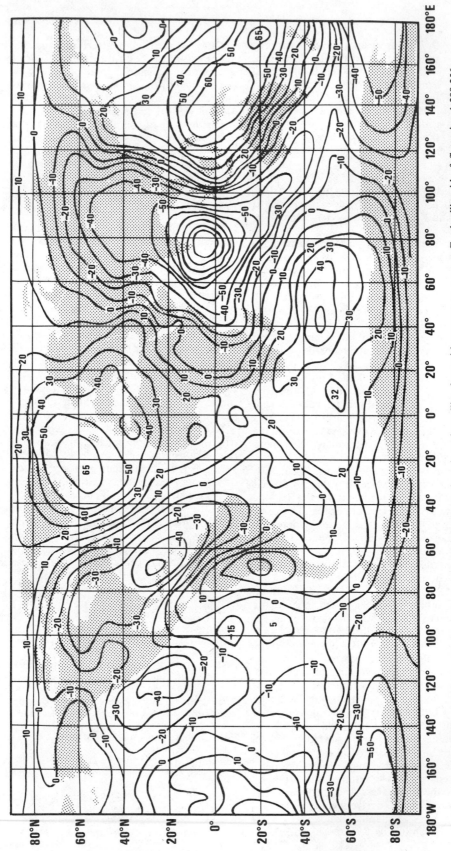

FIGURE 4. Global geoid, in meters, computed from gravimetric and satellite data with respect to a mean-Earth ellipsoid of flattening 1:298.256. (After Rapp, 1974)

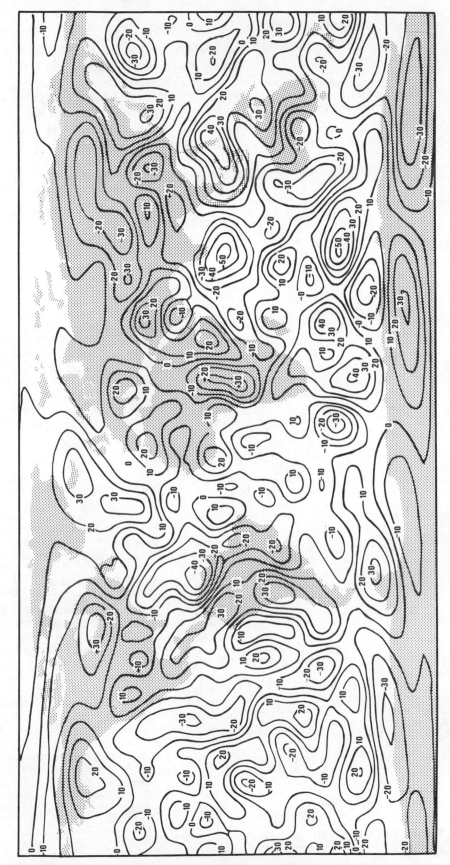

FIGURE 5. Global gravity anomalies, in milligals, computed from gravimetric and satellite data with respect to a mean-Earth ellipsoid of flattening 1:298.256. (Gaposhkin, 1973)

Henriksen, S. W., A. Mancini, and B. H. Chovitz, eds., 1972, The use of artificial satellites for geodesy, *Am. Geophys. Union Geophys. Mon. 15.*

International Association of Geodesy (IAG), 1971, *Geodetic Reference System 1967.* Paris: IAG.

Kaula, W. M., 1966, *Theory of Satellite Geodesy.* New York: Blaisdell.

King-Hele, D. G., ed., 1967, A discussion on orbital analysis, *Royal Soc. London Philos. Trans.* **262A.**

Kouba, J., 1975, Doppler satellite control in establishing geodetic control network, *Proc. Symp. Satellite Geodesy and Geodynamics, Contr. of Earth Physics Branch 45,* **3.**

Kozai, Y., 1966, The Earth's gravitational potential derived from satellite motion, *Space Sci. Rev.* **5,** 818–879.

Krakiwsky, E. J., D. E. Wells, and B. P. Kirkham, 1972, Geodetic control from Doppler satellite observations, *Canadian Surveyor* **26**(2), 146–162.

Lambeck, K., A. Cazenave, and G. Balmino, 1974, Solid Earth and ocean tides estimated from satellite orbit analysis, *Rev. Geophys. Space Phys.* **12**(3), 421–434.

Lehr, C. G., 1969, Geodetic and geophysical applications of laser satellite ranging, *IEEE Trans. Geosci. Electronics* **GE-7,** 261–267.

Levallois, J. J., and J. Kovalevsky, 1970, *Géodesie Générale,* Vol. 4. Paris: Eyrolles.

McCoogan, J. T., C. D. Leitao, and W. T. Wells, 1975, Summary of Skylab S-193 altimeter altitude results, *NASA Tech. Mem. X-69355.*

Meade, B. K., 1974, Doppler data versus results from high precision traverse, *Canadian Surveyor* **28**(5), 462–466.

Morando, B., ed., 1970, *Dynamics of Satellites.* New York: Springer.

Mueller, I. I., 1974, Global satellite triangulation and trilateration results, *Jour Geophys. Research* **79**(35), 5388–5348.

Rapp, R. H., 1974, The geoid: Definition and determination, *EOS* **55**(3), 118–126.

Schmid, H. H., 1974, Worldwide geometric satellite triangulation, *Jour. Geophys. Research* **79**(35), 5349–5376.

Seppelin, T. O., 1974, The Department of Defense world geodetic system 1972, *Canadian Surveyor* **28**(5), 496–506.

Veis, G., ed., 1965, *The Use of Artificial Satellites for Geodesy.* Athens, Greece: National Technical University.

Wells, D. E., and P. Vaníček, 1975, Alignment of geodetic and satellite coordinate systems to the average terrestrial system, *Geod. Bull.*

Cross-references: *Remote Sensing, General; Sea Surveys; Surveying, Electronic.* Vol. XIII: *Oceanography, Appied.*

SCANNING SYSTEMS—See ACOUSTIC SURVEYS, MARINE; AERIAL SURVEYS; ATMOGEOCHEMICAL PROSPECTING; BIOGEOCHEMISTRY; GEOBOTANICAL PROSPECTING; HYDROCHEMICAL PROSPECTING; LAKE SEDIMENT GEOCHEMISTRY; LITHOGEOCHEMICAL PROSPECTING; MARINE MAGNETIC SURVEYS; PEDOGEOCHEMICAL PROSPECTING; PETROLEUM EXPLORATION GEOCHEMISTRY; REMOTE SENSING, GENERAL; SATELLITE GEODESY AND GEODYNAMICS; SEA SURVEYS; VLF ELECTROMAGNETIC PROSPECTING; WELL LOGGING.

SCIENTIFIC METHOD (SCIENTIFIC METHODOLOGY)

Before entering into the discussion of what is meant by the term *scientific methodology,* I must first make clear what is to be understood, in the following text, by the word *science.* By this term we shall understand a particular mode of knowing, and a particular mode of coming-to-know, that is distinguished from ordinary or everyday knowledge by the following four discriminatory characters:

1. To be systematic—i.e., logically ordered;
2. To be progressive—i.e., always in the state of growing;
3. To be capable of explaining and even predicting things or events that have not been experienced;
4. To be communicated in a concise and logically ordered form either in verbal expressions (statements) or with appropriate symbols.

Scientific methodology, then, is a "science of science." For instance, geological methodology deals not directly with things geological but with the logical ways and procedures of arriving at and ordering geological knowledge. When speaking of scientific methodology, we think in the first place of that process of reasoning by means of which new knowledge in a particular scientific discipline is established. Another, less well-known, field of application of scientific methodology consists in logically ordering, coordinating, and critically comparing the various elements of established knowledge in a particular discipline. Such knowledge, which at a given time, is widely accepted and reasonably free of apparent inconsistencies, may be termed *encyclopedic knowledge,* forming the body of scientific knowledge in a particular discipline at that time. Correspondingly, the body of common knowledge in a certain field of experience, say, geology, may be termed prescientific geological knowledge. Common knowledge has vastly contributed to the development of applied geology by the discovery of tremendous mineral wealth, particularly in the newly developed continents.

This brief survey of scientific method in general necessarily centers on a selection of possible topics—i.e., those that have a direct bearing on the methodology of geological inquiry proper. Inasmuch as the latter involves the additional application of the exact and experimental sciences such as physics (in geophysics), or chemistry (as in geochemistry), or biology (as in paleontology), these may be considered auxiliary sciences to geological inquiry proper. Their particular methodological aspects may be studied in the pertinent entries of this encyclopedia and the special references given therein. In many instances, a consultation of the works by Popper (1962, 1963) is of advantage.

Scientific methodology of the nineteenth century made abundant use of induction as a method of reasoning from the particular to the general. Geological textbooks up to the present time preserve strong evidence of this now somewhat anachronistic method of reasoning, in mentioning and even stressing the significance and importance of certain so-called geological laws and principles.

Taken as a whole, scientific thinking since the early 1900s has been increasingly dominated by doubts about inductive reasoning as a valid method of arriving at reliable scientific knowledge. Instead, the method of proposing hypotheses and then testing and verifying them by means of observation and experiment has been developed as the proper scientific method of arriving at and establishing new scientific knowledge. The members of the positivistic school of philosophical thought and, among them, the members of the Vienna Circle of philosophers and logicians, have contributed much to the efforts of innumerable others toward this development.

Among the proponents of the method of hypotheses, Popper (1962, 1963) stands out by the rigor and consequential pursuit of certain ideas that he first proposed in 1935. Popper's ideas and basic procedures appear to be particularly suitable for the furthering of critical geological inquiry and may therefore be briefly characterized as follows.

Process of Growth of Scientific Knowledge

The basic views and methodological procedures underlying the process of growth of scientific knowledge, according to Popper (1962, 1963), can be summed up as follows: All empirical knowledge, including knowledge as "established" by the empirical sciences, is hypothetical knowledge—i.e., knowledge that is first presented in the form of a hypothesis or proposal of a trial solution of a problem. The statement (or statements) expressing a hypothesis, as well as the logical implications that can be deduced from that hypothesis (whence the name, *hypothetic deductive method*), must then be subjected to critical testing by means of observation and/or experiment. The invention and organization of a series of observations and experiments, designed either to support or falsify the hypothesis, are systematically guided by the assertions initially proposed by that hypothesis and the logical consequences of these assertions.

A hypothesis cannot be proved. Even the use of the phrase *verification of a hypothesis* is avoided in view of the implications or connotations of that phrase. A hypothesis can be corroborated to varying degrees by the variety and severity of the tests to which it is subjected. The degree of corroboration obtainable is great, depending on the degree of effort of the proposer or critic of the hypothesis in aiming at falsifying it by new observations and experiments expressly intended and carried out for that purpose. The scientific status of a hypothesis increases with the critical intention, but unsuccessful outcome, of the efforts to falsify it. But even so, the thus strongly corroborated hypothesis cannot be considered as the ultimate, final truth. On the contrary, the falsifiability of even the (currently) best corroborated hypothesis must be assumed in principle, even if it may take, as it did in the case of Isaac Newton's hypothesis on the gravity pull of masses, several centuries before its claim to universal validity was falsified, or at least modified, by Einstein's theory of relativity.

Hypotheses are either accidentally conceived or invented and are first proposed as trial solutions of problems. Problems are recognized, or come to the awareness of the researcher, as a consequence either of new observations or of the greater accuracy of measurements available by means of more powerful instruments. In general, one can say that a problem is recognized as such and identified with increasing conciseness when a researcher, through either new observations or critical search of the literature on "established" or encyclopedic science, fails to obtain a satisfactory solution of an apparent inconsistency despite numerous preliminary attempts to do so. The procedure of formulating hypotheses that promise a reasonable chance of solving a given problem improves with the increasing conciseness of the formulation of the problem at hand.

A single repeatable critical observation or experiment can be sufficient to falsify a hypothesis if it proves to be incompatible with or even contrary to the requirements or logical consequences of the hypothesis. In many cases, however, it is possible to modify, to change in some part or detail, the hypothesis first proposed (H_1) to the modified hypothesis (H_1'), which may then be corroborated in the usual way. In all other cases, further hypotheses (H_2, H_3, etc.) may be necessary to attempt the solution of the problem in question. In practically all cases, when an initial problem P_1 can be adequately solved by means of a suitable and well-corroborated hypothesis (say, H_1), a new problem or problems (P_2, P_3, etc.) emerge as a consequence of the wide range of observations or experiments undertaken to solve the problem P_1 first tackled. In principle, therefore, there appears to be no end to the progress of scientific knowledge and the search for it. This progress, according to Popper (1963), consists essentially in the progressive recognition and elimination of faults in the hypotheses proposed.

The writer wishes to acknowledge with sincere thanks permission kindly given by Professor Sir Karl Popper not only to communicate certain details from his current lectures and seminars but even to reproduce his following scheme representing the sequence in the procedure of problem solving by means of hypotheses and their critical testing:

$$P_1 \rightarrow H_1 \rightarrow T \rightarrow P_2,$$

where P_1 is an initially recognized problem, H_1 is a hypothesis proposed for its solution, T is (a full range of) critical tests by observation and experiments, and P_2 (or perhaps P_3, P_4, etc.) is one or more new problems emerging in consequence of the preceding critical procedure. This scheme may not at present be found as such in the publications by Popper (1962, 1963), but a wide range of pertinent discussions can be found in those works under the respective headings.

It may be added that Simpson (in Albritton, 1963, 40) proposes the term *strategy* for schemes representing the sequence of the various methodological steps planned and taken in a unit of scientific inquiry—e.g., in geological inquiry. The suitability of the earlier procedure and scheme by Popper to geological inquiry is discussed in *Geological Methodology.*

Probability of the Truth of a Hypothesis

In common scientific parlance, one often speaks of the probability that a hypothesis is true. Or one may speak of the degree of probability with which a hypothesis or theory approaches the correct or true explanation of a problem that it claims to solve. Popper (1962, 1963), in various places in the works cited, discusses this usage of the term *probability,* pointing out that the concept underlying this usage is markedly different from that underlying the probability calculus developed in mathematics and widely applied to modern natural science including geology. Popper (1962; 1963, 228) proposes the use of the term *verisimilitude* when expressing the approach of a hypothesis or theory to the truth or the degrees of that approach. This usage of the term, however, is nonmathematical, quite vague, but useful when comparing and roughly estimating the relative explanatory power of competing hypotheses in solving a given problem.

The concept of the probability, or rather, verisimilitude of a hypothesis is particularly useful in geological inquiry, where the exhaustive testing of hypotheses is commonly impossible because of incomplete conservation or accessibility of the necessary evidence or because of lack of the technical or financial means needed to carry out all the tests required by the hypotheses. The latter situation is typical for the exploratory phase of applied geology (cf. *Geological Methodology*).

Establishing Knowledge versus Ordering Knowledge

As indicated by the titles of his works and, in particular, the subtitle of the 1963 book, *The Growth of Scientific Knowledge,* Popper (1962, 1963) deals essentially with the methodology of critically establishing new scientific knowledge. In contradistinction to this, the textbooks, monographs, and encyclopedias of any scientific discipline, like this encyclopedia, deal essentially with the body of scientific knowledge as presently accepted and that should result, according to Popper, from "presently corroborated hypotheses." To this body of positive knowledge, as it is often called in loose scientific parlance, should apply the requirement of scientific knowledge—namely, of being systematic, logically ordered and logically consistent.

The authors of the encyclopedic works are not usually professional methodologists. There is no doubt, however, that the critical investigation and elaboration of the logical means of ordering corroborated scientific knowledge should be a task of the methodologist, no less important than that of establishing new knowledge. This is because the principles and methods of ordering the various elements of encyclopedic knowledge belong to the field of applied logic and require the methods and techniques of Boolean algebra, of the logic of classes and of relations, and others. In the field of biology, such examples of application of modern logic to the procedures of ordering are well known (cf. Gregg, 1954). In the field of geological science and its numerous constituent and auxiliary disciplines, such applications of modern symbolic logic seem to be practically unknown, although Whewell (1840) has termed geology a classificatory science. With the latter aspect of descriptive natural science in mind, Koch (1949) elaborated a system of categories, which could more explicitly be termed the natural taxonomy of the states of the natural physical world as discernible by means of sense perception.

When critically reexamining these categories, they can be considered as classes, and the definitions given for these categories (Koch, 10–14) are each enumerations of those characters, or sets of characters, of all natural phenomena that place them in the classes so defined. This system of categories can be applied to advantage to orderly and systematic procedure in the four basic operations of (descriptive) natural science—i.e., (scientific) observation, identification, classification, and (scientific, i.e., ordered) description including definitions.

In a given discipline of natural science, the more the state of things investigated, as well as the methods of investigation applied (e.g., sense perception and the use of natural language), correspond to the natural conditions under which the system of categories was derived, the more directly can the system be applied to these basic procedures (Koch, 1949, 20–31). The latter conditions apply in particular to objects of study of the geologist and mineralogist in the field and even in the laboratory and are briefly discussed in the entry on geological methodology.

Conclusion

In conclusion, it can be said that the rapid progress and expansion of modern scientific knowledge along the most diversified and often unconnected lines frequently necessitate a new ordering and coordinating of the body of knowledge so arrived at. Critical application of the means of modern logic to this process of ordering may reveal inconsistencies between certain bits of knowledge so gathered. Thus, new problems are exposed that in turn give rise to the method of hypothesis forming and hypothesis testing for the solution of these problems, as described earlier. Thus, the procedures of establishing new knowledge and of ordering that knowledge alternate with one another and are complementary and even necessary elements of scientific methodology.

L. E. KOCH

References

Albritton, C. C., Jr., 1963, *The Fabric of Geology.* Reading, Mass.: Addison-Wesley Publishing Co.
Gregg, J. R., 1954, *The Language of Taxonomy.* New York: Columbia University Press.
Koch, L. E., 1949, Tetraktys. The system of the categories of natural science and its application to the geological sciences, *Australian Jour. Sci.* 11(4) (Supp.), 1–31.
Popper, K. R., 1962, *The Logic of Discovery,* 3rd ed., London: Hutchinson & Co.
Popper, K. R., 1963, *Conjectures and Refutations. The Growth of Scientific Knowledge.* London: Routledge & Kegan Paul.
Whewell, W., 1840, *Philosophy of the Inductive Sciences,* 2 vols. London.

Cross-references: *Geological Methodology; Geology, Philosophy; Geology, Scope and Classification.* Vol. XIII: *Geological Communication; Geology, Applied.*

SEA SURVEYS

Surveying is the process of making measurements of the physical features of the oceans, marginal seas, and their adjoining coastal areas. It is the sort of work on which bathymetric charts are based, on which distributions of oceanic water properties are founded, and on which atlases of biological populations are compiled.

Data collection in support of research, however, does not (or should not) employ the same approach that surveys embody. It should be the collection of adequate data to provide an insight into some question that has been proposed—to answer the why and how rather than just the what and where. This requirement leads ordinarily into quite different sampling patterns and densities. Surveys, e.g., are usually laid out on a grid pattern designed to produce a certain number of data points per square kilometer, without regard to special areas of interest. Research data collection should emphasize the areas of critical interest for understanding—the areas of steepest gradients, or fastest rates of change, or broadest implications to other factors (Anonymous, 1963).

A further difference is the susceptibility to change. Surveys rarely stop or divert to investigate unusual phenomena. Research should; it is obligated to take into account the implications of the data as they are acquired and to modify the course of action to take advantage of what is being learned as the work goes forward.

Data Collection

The various data of interest to oceanographers and how they are collected have been studied and classified at various times by a number of agencies and organizations (Anonymous, 1960). One of the best, due to its thoroughness, has been compiled by the Naval oceanographic office (Anonymous, 1964). The table of measurement requirements proposed by the Oceanographic Office represents the opinion and judgment of numerous people and recognizes oceanographic requirements as neither absolute nor static and subject to change to meet continuously evolving needs.

A less rigorous classification of data parameters is presented in Table 1. It is reasonable to question whether the multiplicity of purposes or objectives of the research oceanographer and scientist are fully satisfied by this breakdown, but it may be complete enough to suffice for the present discussion. The variation of measurement technique and the scope of technology inherent in the several scientific disciplines included within oceanography present a complex, but challenging, problem. The selection of a proper approach, and eventually the optimum solution, requires a thorough analytical, yet practical, attack.

Selection and Standardization of Techniques

Techniques for doing things on board ship are laid out in a rigorous fashion for a simple reason. A prime responsibility of the sea-going individual is monitoring, collecting, and recording data that can be used with full confidence that it was collected in a standard, reproducible manner. Reducing a procedure to a routine of actions helps to guarantee consistent quality in data taking. In designing or choosing a technique, thought must be given to the following elements: a purpose for the measurement of observation (i.e., a use for the data to be obtained); which particular physical property to measure or observe; which equipment is most suitable for the measurement or observation in question, with special regard to the precision obtainable with them; the manner of taking the measurements; the units and forms for recording the measurements and/or observations; the place, time, and frequency of the measurements and/or observations; and the cost per measurement, in time or money, in relation to the expected results.

Oceanographic Platforms

Oceanographic platforms may be defined as the basic tools for placing oceanographers and/or their scientific equipment in a position to study systematically and interpret the physics, chemistry, biology, and geology of the marine environment, including water masses, interfaces, and boundaries.

Throughout the history of oceanography, surface ships have been the basic platform tool of the marine scientist. They will undoubtedly continue to be the prime oceanographic research and survey platforms in the future. For this reason, the discussion of techniques in this entry emphasizes surveys from ships. Since the 1970s, however, it has become increasingly evident that various types of platforms are required for specific areas of scientific investigations.

All types of platforms have their place and use, but the final choice is determined by the intended scientific objective and the limitations of the platform's capabilities. *Oceanographic platforms* have been divided into eight types for measuring 39 parameters (Elliott, 1964). Of these parameters, 28 are oceanographic and 11 are meteorological. The platforms and their oceanographic capabilities are listed in Table 2.

Types of Surveys

This review neither is inclusive nor does it imply that all these measurements are made on a single expedition.

Geodetic Surveys *Geodetic surveying* may be defined as that branch of surveying in which, due to the extent of the survey and the required precision of the results, it is necessary to consider the spheroidal form of the Earth's surface (see *Satellite Geodesy and Geodynamics*). The object of an extensive geodetic survey is usually twofold: (1) to locate certain points with great accuracy to connect different topographic or hydrographic maps and to furnish an accurate control of the whole survey and (2) to furnish data for perfecting our knowledge of the form and dimensions of the Earth.

Hydrographic Surveys Measurement of details of water areas and appropriate details of adjoining coastal areas is called a *hydrographic survey*. In addition to locating the points at their correct geographical positions, this practice results in the land and marine features' being in correct relationship to each other. This is an important consideration because marine navigators near a coast also locate themselves relative to the land—in many instances referencing the same landmarks used by the surveyor. Before a hydrographic survey can be conducted, a geodetic survey may be needed if available information does not provide adequate control of positions.

Limited Surveys *Limited surveys* are conducted to satisfy particular requirements. They are the most common type of survey. Limited surveys can be divided into several subcategories: exploratory (baseline) surveys, running surveys, beach surveys, bathymetric surveys, and photographic surveys (underwater and aerial) surveys.

Exploratory Survey When time or lack of equipment does not permit or where desired results do not justify carrying out a standard geodetic control or hydrographic survey, a limited exploratory survey may be conducted. This might be an advance investigation to determine the desirability of making a full detailed survey, an operation to make a preliminary chart of an anchorage (see *Harbor Surveys*), or an investigation of a reported shoal, e.g. The principles and techniques conform to those described earlier but are

TABLE 1. Classification of Data Parameters

Physical*	Chemical	Biological	Meteorological	Geophysical	Own Ship
Ocean temperature	Salinity	Plankton and nutrients	Atmospheric temperature	Bathymetry (bottom topography)	Date
Depth (pressure)*	Concentration of substances in suspension solution	Bottom samples	Barometric pressure	Gravity	Time
Conductivity	pH	Cores	Atmospheric pressures	Geothermal	Position
Sound velocity	Oxygen content	Sea life specimens	Relative humidity	Geomagnetism	Longitude
Sound attenuation	Plant nutrients	Genus and species	Cloud cover	Inclination	Latitude
Noise level	Geochemistry	Concentration	Wind vector	Declination	Heading
Current vector (surface and depth)		Migration	Rainfall	Earth tides	Speed
Waves, surface (sea state)		Underwater photography	Lightning	Structural profiles (geological)	Depth (for submersibles)
Internal waves		Deep scattering layers	Radiation	Heat flow	
Density		Background noise	Sea ice	Sediment profiles (sub-bottom, cores)	
Transparency (and visibility)		Fouling accumulation	Dew point	Bottom photography	
Radioactivity		Bioluminescence		Bottom samples (geological)	
Mass spectrography				Sound velocity in sea floor, attenuation	
Tide, tidal currents				Telluric currents	
				Seismicity	

*For many ocean parameters depth is the independent variable against which other variables are plotted.

TABLE 2. Types of Oceanographic Platforms and Their Application

Environmental Measurements	Airborne Systems	Bottom Mounted Systems	Expendable Systems	Research Vessels	Towed Systems	Submersibles	Satellite	Buoys Anchored	Buoys Freefloating
AIR									
1. Temperature	C			B				B	B
2. Pressure	C			B				B	B
3. Cloud Cover	C			D			D	D	D
4. Radiation, incident	C			B			B	B	B
5. Radiation, reflected	C			B			B	B	B
6. Relative Humidity	C			B				B	B
7. Wind Speed	C			B				B	
8. Wind Direction	C			B				B	
9. Precipitation				B				B	B
10. Lightning	D			D			D	D	D
11. Particle count	C			B				B	B
OCEAN									
Physical									
1. Ambient Light			C	C	B	C		B	
2. Bathymetry		B	B	D		D			
3. Current Speed and Direction		B	B	C				B	B
4. Density		B	C	C	B	C		B	
5. Gamma Radiation		B	C	C	B	C		B	
6. Geomagnetism	B	B		B	B	B	B	B	
7. Gravity	B	B		B		D		B	
8. Heat Flow, bottom				D		D			
9. Ice Cover	D						D		
10. Pressure		B	C	C	B	C		B	
11. Radio Activity		B	C	C	B	C		B	

	1	2	3	4	5	6	7	8	9
12. Sound Velocity		B		C	B	C		B	A
13. Temperature		B	C	C	B	C	D	B	
14. Transparency		B		C	B	C		B	
15. Turbulence		B		C				B	
16. Waves, height	D		A	A			D	A	A
17. Waves, length	D		A	A			D	A	A
18. Waves, period	D		A	A			D	A	A
19. Waves, direction	D		A	A			D	A	A
20. Waves, internal				C				B	
Chemical									
21. Nitrogen		B		C	B	C		B	
22. Oxygen		B		C	B	C		B	
23. Phosphate		B		C	B	C		B	
24. Salinity		B	C	C	B	C		B	
Geological									
25. Bottom Sediments				D					
26. Substrata				D					
27. Reflectivity				D		D			
Biological									
28. Biological sampling	D			B	B	C		B	

CAPABILITIES

		1	2	3	4	5	6	7	8	9
A	Surface	0	0	4	4	0	0	0	4	5
B	At instrument level	2	16	1	12	15	2	3	27	7
C	Versus depth or altitude	8	4	7	15	0	13	0	0	0
D	Remote sensing	8	0	0	7	0	3	8	2	2
	Total	18	20	12	38	15	18	11	33	14

adapted to meet contract requirements, instrument limitations, and training of personnel.

Running Survey A limited survey can be conducted as a ship steams along a coast. The position at the beginning of the run is determined as accurately as conditions permit. Natural ranges may be available from time to time to provide good course references. If charted shore objects are available, frequent fixes can be determined and the dead reckoning between them adjusted to avoid gaps in the plot. If accurately charted landmarks are not available, it is possible to use celestial navigation to establish good positioning if the vessel has the proper electronic instrumentation. As the survey progresses, the vessel steams at a safe distance from the shore, determining its courses and speeds as accurately as is practical. Continuous soundings are taken, preferably by a recording echo sounder. Bottom samples are taken at frequent intervals if conditions permit, and if required.

Beach Survey The most common purpose of a beach survey is to provide preliminary data for use in the planning or construction of piers, docks, or other harbor facilities. Another common purpose is to obtain data useful for landing military supplies, equipment, and personnel directly on the beach from landing craft or amphibious vehicles or simply for studying the problems associated with beach erosion (see vol. XIII: *Beach Replenishment, Artificial*). Since a beach survey seeks detailed information about a relatively small area, accurate positioning is essential.

Bathymetric Survey Sounding lines run at sea are of assistance in adding detail to existing charts or in constructing special charts to serve particular purposes. Most of the required information is obtained by ships proceeding between ports, either singly or in company with other ships.

The important factors are accurate depths and accurate positions. Inaccurate results may be worse than no information at all. Therefore, every effort should be made to obtain reliable data. Soundings that conflict with known or charted depths, in particular, should be carefully analyzed. Even when the equipment is operating correctly, false returns might be received due to sources external to the vessel: a shoal or phantom bottom may be due to marine life; there may be multiple echoes of interference; or a return may not be received because of aeration of the water or suspended matter in it. Unusual local conditions may be a source of error. If an error is believed probable, but no source is detected, full information should be submitted with the soundings so a charting agency may be able to interpret the results. This action is particularly important where the measured depths are less than those shown on the chart. If no error can be found, the charting agency may have no alternative but to enter the

shoal soundings on the charts affected and to take the first opportunity to send another survey vessel to verify or disapprove them.

Photographic Surveys One of the most significant contributions to modern map making has been the development of the precision aerial and underwater camera and the techniques for interpreting and utilizing the information appearing on the photographs made by it. Such photographs constitute a detailed and permanent record of all unobscured natural and artificial features of a given section of the Earth's surface or subsurface and, as such, furnish more completely than any other means the information required for making surveys. However, all photographs, whether aerial or underwater, are perspective views, and it is necessary to change these to orthographic views to obtain reliable map information. Although these photographs are often maplike in appearance, there are many errors, both systematic and random, that prevent the photograph from being a true map. The science of photogrammetry is used to eliminate or correct these errors and also to record properly all the photographed information into a true map presentation. Its development into a complex and exact science has made photogrammetry the most efficient, accurate, and economical method for surveying large areas (see *Photointerpretation; Remote Sensing, General*). This section is not intended to present the detailed theory or working procedure of photogrammetric instruments and methods but to acknowledge the fact that such methods and instruments do exist (see *Photogeology*).

Summary Surveys are usually conducted by personnel who have been given specialized training and are provided with complete equipment. Detailed information on conducting a survey can be found in the classic *Hydrographic and Geodetic Survey Manual*, H.O. Pub. No. 215, and in the U.S. Coast and Geodetic Survey Special Pub. No. 143, *Hydrographic Manual*.

Ocean Survey Instrumentation and Techniques

The requirements for equipment destined to invade the underwater world are stringent and not often understood by survey personnel. Because pressure builds up at 1.03 g/cm² (1/2 lb/in²/ft) of depth, construction of an instrument case to be lowered to 2,000 m requires careful layout and selection of materials.

Instruments used to profile the ocean, to measure its characteristics throughout its entire depth, are designed for dunking. They are built with sturdy metal casings and usually are surrounded by a frame known as a *bird cage*. Instruments used to observe ocean characteristics near the surface and over a wide area are often towed behind a research vessel. The instrument package is known as a *fish*. Instrument sensors may be attached at intervals along the wire or cable leading from the ship's winch to the fish. The hydrodynamic characteristics of a fish are important; the fish must glide smoothly through the water to avoid placing undue strain on the cable or umbilical connectors.

Even as late as the 1940s, oceanographic surveys were hampered by a lack of existing knowledge of ocean conditions and by the primitive nature of the instruments. Today, while some of the instruments remain essentially the same, a whole new suite of electronic tools and computer methodology have become available since World War II. Table 3 lists some of this instrumentation and the variables it is intended to measure.

The surveying techniques and equipment that land surveyors employ in their work usually involve optical observations and electromagnetic distance measurement (see *Surveying, Electronic*). Once moved under the surface of the sea, however, all this equipment becomes unusable—in the case of optical equipment, due to turbidity; and in the case of electromagnetic equipment, due to impossibly high attenuation.

Hydrographic surveyors thus depend on these systems to a lesser extent and make more use of medium- and long-range radio positioning equipment. Accepting that they must be able to locate the sensing devices and monitor their progress on the seabed and in the water column, these surveyors have to consider three very different methods to achieve the same results:

1. Seabed automated correlation: This system employs a relatively accurate computer model of the seabed topography, which is continuously compared against current bathymetric observations. The method can be very successful and has been used on an experimental basis, but it requires large and powerful computing equipment and is not commercially available at this writing.
2. Inertial navigation: Modern accelerometers are very sensitive and capable of continuously monitoring the relative motions of vehicles in any medium and integrating the resultant position. Their main disadvantage, particularly to the commercial surveyor, is that they require regular updating to correct for instrumental drift, and the equipment, which is very expensive, is used mainly by the survey service industry.
3. Hydro-acoustic ranging: Sound waves have been used for measurements and observations in the water column since the advent of echo sounders and passive and active military sonar devices (see *Acoustic Surveys, Marine*). The ability of water to transmit sound waves is well known, and thus, this is the area where most

commercial interest and development has been focused (Fig. 1).

Acoustic Positioning Methods

Acoustic positioning methods have now become the accepted norm for underwater position control in the offshore oil industry. Although many of the capabilities of acoustics have been familiar to scientists for some time, the development drive provided by the oil industry and the general advances in electronics have produced efficient and cost-effective equipment.

Practically all modern commercial acoustic positioning devices operate in three basic modes: (1) long baseline (Fig. 2), (2) short baseline (Fig. 3), and (3) supershort baseline. Equipment on the market today offers either one or a combination of these three types of operations.

Long Baseline This mode is the most accurate of the three and depends on a minimum of two seabed transceivers or transponders, which return the transmitted signal from a single hydrophone inserted in the water column. The two-way traveltime of the signal is then processed and a normal trilateration effected, thus producing the position of the hydrophone relative to the transponders. This mode is complicated by the need to calibrate initially the transponder positions in relation to a known datum, and this usually requires the use of a medium- or long-range continuous surface positioning system. It does, however, offer positioning over a relatively large area, particularly when more than two transponders are both laid and interrogated sequentially, the final position being computed from a least squares analysis of the resultant position lines.

Short Baseline This mode uses only one transponder but employs two hydrophones separated by a distance of about 5–20 m. Range to the single transponder is produced as in the long-baseline mode, but relative bearing is also available from the time difference at which the returning signal is received at the two hydrophones

Supershort Baseline This mode is very similar to short baseline but uses only a single hydrophone made up of a number of distinct units. The phase difference of signals received by various parts of the hydrophone array is used to determine relative bearing, while range is again achieved as in the long- and short-baseline modes.

Although it is still necessary to know the position of the single transponder in the latter two modes, the calibration may be more basic since a triangulation of multiple transponders is not involved; furthermore, the short- and supershort-baseline modes are most often used where short-range station keeping is required, and thus, the single transponder is usually located at a point of interest with a previously established position,

TABLE 3. Ocean Survey Instrumentation

Measurement Function	Devices	Characteristics and Limitations	Comments
Ocean temperature	Mechanical Bathythermograph (BT)	Depth: 0–900 ft Accuracy: ±10 ft Temperature: −2.2–32°C Accuracy: ±0.1°C Poor accuracy	In temperature measurement work the mechanical bathythermograph has long been the oceanographer's workhorse. The BT, together with those previously noted herein, and an airborne radiation thermometer (for surface depths, temperature range −2–35°C, accuracy ±0.2°C) comprise for the most part present-day temperature instrumentation. Various telemetering and self-containing recording systems, infrared devices, temperature probes, and expendable bathythermographs are also being used, developed, and evaluated. Specifications for improved systems have been documented as those requiring repeatability and accuracy ±0.1°C (for buoyancy problems the accuracy may be approximately ±0.05°C), quick reaction time sensors, durability, watertightness, linear output, and standardization of output signals compatible to computer and machine tabulation.
	Expendable Bathythermograph (XBT)	Depth: 0–1,380 ft Accuracy: ±15 ft Temperature: −2.2–35.5°C Accuracy: ±0.2°C	
	Deep-Sea Reversing Thermometer	Depth: 0–20,000 ft Accuracy: ±100 ft Temperature: −2 to 30°C Accuracy: ±0.01°C Not continuous; too delicate	
	Wire-Wound Resistance Thermometer	Depth: Surface Temperature: −5–30°C and −25–5°C Accuracy: ±0.1°C Poor accuracy except near surface	
	Thermocouple	Depth: 0–200 ft Temperature: −50–10°C Accuracy: ±0.5°C Requires laborious calibration: thermocouple reference junctions must be kept at a constant known temperature	
Depth	Edo 255 (Depth of bottom)	Accuracy: 0.1% of depth Experiences numerous breakdowns	Depth is most difficult to measure accurately. New depth measuring devices, e.g., require pressure transducers that are expendable, adaptable for use with various kinds of arrays, and reproducible. In this context, more transducers are coveted by the researcher. New developments in this field primarily include the strain gauge transducers for unlimited depth measurements with an accuracy of ±1 ft in 1,000 ft. The
	Unprotected Mercury Thermometer	Probable error of depths: ±15 ft for depths less than 3,000 ft; at greater depths to about 0.5% Too delicate; has limited application; is not continuous	

Category	Method/Instrument	Specifications	Comments
	Mechanical Pressure Transducer (Bourdon tube, bellows, helical coil, aneroid, etc.)	Depth: 0–1,500 ft Accuracy: ±15 ft Erratic	strain gauge principle has merit because it is more feasible to use a pressure-sensitive device (in an environment where the pressure varies from 0.44 psi to 0.46 psi per foot of depth) and then convert pressure to depth.
	Electronic Pressure Transducer ("Vibration," strain gauge, variable reluctance gauge, etc.)	Depth: 0–1,500 ft Accuracy: ± 4 ft (accuracy of transducer is generally limited by the recording apparatus) Erratic; transducer accuracy is limitation	
Conductivity/Salinity	Seawater Sampler: Nansen Bottle	Salinity accuracy: ±0.02 ppt	Although measuring techniques being used for salinity in situ and probes for conductivity are deemed by some as acceptable, many suggestions for improvement in this field have been put forth to provide accuracy for use in coastal waters—portability of the equipment, automated collection equipment, sensor capability of 20–40 ppt ocean salinity range, accuracy ±0.01 ppt (classical hydrography demands salinity measurements approximately ±0.01% relative), equipment capable of easy calibration and alignment, nonpolarized electrodes, etc.
	Volumetric Method (Titration)	Salinity accuracy: ±0.005 ppt	
	Conductivity Bridge Salinometer (University of Washington)	Repeatability: ±0.001 ppt Is not continuous, requires laborious analysis of samples	
	Electrical Method Conductivity Cell	Accuracy: ±0.1 ppt	Miniaturized transistor salinometers, conductivity-temperature-indicators, temperature-chlorinity titrators, neutron absorption, salinity-temperature-depth recorders, etc., are also being used to some extent for various activities.
	Foxboro Company Serfass Bridge	Accuracy: ±1.0 ppt (0.1 ppt if calibrated before and after use) Becomes unstable because of change in cell characteristic; actual measurements are largely empirical in nature, and a basically incorrect assumption is technically made that the ocean water composition is constant	
Ocean currents	Mechanical Current Meter (Ekman)	Speed: 0.15–2.5 knots Accuracy: ±0.1 knot Direction: 0–360° Accuracy: ±10° Not continuous or deck reading	Various methods are being considered for current measurement, such as a suspended-drop current meter (Lamont Geological Observatory), the Savonius Rotor, etc., to improve these instruments and their operating limits. Broad design criteria for further new developments have been documented as a need for a good solid-state type of current meter design with a speed range of 0.1–10 knots; accuracy: ±0.1 knot (lower thresholds and accuracies desired for special research projects); direction: ±10°; depth range: unlimited in some designs (but not necessarily for those with the speed ranges, accuracies, and directions mentioned above), simplicity in overall design and opera-
	Electromechanical Currents: Meter Types		
	Price Meter	Speed: 0.1–6.5 knots Accuracy: ±0.1 knot Direction: None Marginal accuracy; maintenance problems	

(continued)

TABLE 3. *(continued)*

Measurement Function	Devices	Characteristics and Limitations	Comments
	Roberts Meter Mod. 3	Speed: 0.2–7 knots Accuracy: ±0.1 knot Direction: 0–360° Accuracy: ±10° High threshold level; maintenance problems	tion, rugged, minimum maintenance, standardization of output signal range, and stability in three-dimensional space. Standardization of output signals should be compatible to present machine tabulation and/or computers. Drifters can be tracked remotely by high-precision fixing systems such as Loran C, Omega, etc. The trajectories and differential motions of the individual elements can yield much information about current dynamics.
	Low-velocity types		
	Hytech Crouse-Hinds	Speed: 0.1–7 knots Accuracy: ±0.1 knot Direction: 0–360° Accuracy: ±10°	
	Pruitt	Speed: 0.04–7 knots Accuracy: ±0.01 knot Direction: 0–360° Accuracy: ±10°	
	CM-3 (Japanese)	Speed: 0.2–5 knots Accuracy: ±0.1 knot Direction: 0–360° Accuracy: ±10°	
	Geomagnetic Electro Kinetograph (GEK)	Uncertain	
	Photographic Type (German paddle wheel)	Speed: 0.3–3 knots Accuracy: ±0.1 knot Direction: 0–360° Accuracy: ±10°	
	Parachute Drogues	(Speeds and accuracies determined from tests by the U.S. Hydrographic Office)	
	Drift Bottle	Crude, slow, and uncertain (for exceptions, see comments)	
Transparency and visibility determinations	Submarine Photometer	Depth: Approximately 500 ft	An example of a recording device design is the telerecording bathyphotometer (Boden, Kampa, & Snodgrass). Sophistications of all these instruments and new designs, which also assist in the determination of currents and biological conditions, should employ au-

	Hydrophotometer Mark 2	Depth: 200 ft	tomatic calibration, an ease of alignment, an ability to determine suspended particles in the ocean (by size, number, and type), an ability to measure attenuation of ambient light with depth; be capable of transparency measurement in relatively clear open ocean water; have a design simplicity; be capable of measurement of number of wavelengths; and be dependable.
	Secchi Disc	Depth: 20 ft (maximum)	
Density	No direct measuring instrument	No direct measuring device available for continuous measurements with the accuracy $\pm 10^{-6}$ desired	New instruments should not be sensitive to gravity-dissolved gases, acceleration, and organic matter in sample. Direct, accurate density-measuring systems are being sought.
Tide	Portable tide gauge	Not bottom-mounted, self-contained, or unattended	Nearly all the tide gauges in the United States are operated by the U.S. Coast and Geodetic Survey. Systems under consideration and in development include tide-buoy telemetering with output signals appropriate for computer work. Some advanced design criteria and the problems involved for improved gauges are accurate determination of a vertical datum, crustal movements, classification of tides affecting different coast lines, attain accuracy of ± 0.01 ft, determine parallax and declination cycles, tidal pulsations in deep water, and tidal buildup.
Bottom topography and sediment structure	Deep-Sea Multi Shot Camera (Type III, Navy Electronics Laboratory)	Depth: Greater than 20,000 ft Number of photographs per operation: Approx. 55	Designs ideally would provide high-power transducers, broad-band energy sources, high repetition rates, short pulse length for thin-bed resolution, a cored sediment more representative of in situ conditions, improved release mechanisms, deeper sediment penetration, nondistortion of sediment sample, adequate sediment samples for comprehensive laboratory analysis, optimum physical dimensions of piston corers, and portable shipboard analysis kits.
	Mechanical Bottom Signalling Device, "The Ball Breaker"	Depth: Unlimited	
	Substrata Acoustic Probe (Marine Sonoprobe)	Depth: 700 ft Sediment penetration: 200 ft	
	Precision Depth Recorder (Times Facsimile Corp.) and AN/UQN (Edo)	Depth: Up to 18,000 ft Sediment penetration: 120 ft (extreme maximum)	
	Fathometer-Echo Sounder (Model 255B-EDO Corp.)	Depth: 2.5–1,500 ft Accuracy: ± 1–6 ft, depending on depth scale in use Sediment penetration: Undetermined	

(continued)

TABLE 3. (*continued*)

Measurement Function	Devices	Characteristics and Limitations	Comments
	Corers: Gravity Type Phleger	Depth: Unlimited, determined by length of lowering cable Sediment penetration: 4 ft	
	Piston Type Kullenburg	Depth: Unlimited Sediment penetration: 6–12 ft	
	Ewing	Depth: Unlimited Sediment penetration; 20–60 ft	
	Grab Samplers: Clamshell Snapper	Depth: Unlimited Sediment penetration: Surface	
	Mud Sampler	Same as above Most now restricted in performance; more precision required; longer cores required	
Geothermal measurements	Thermistor chain	Range: 0–5°C Accuracy: ±0.005°C	Temperature measurements are taken at three or four varying depths (3–15 m) within the seafloor with observation times of 2–3 min intervals. In water depths to 10,000 m.
Gravity	Pendulums, Gyro-compensated Sweep Board Meters	For broad ocean areas: Capabilities: 977,000–984,000 mgal Accuracy: ±3 mgal	Data collection is with continuous analog readout with manual inputs and data reduction. Sampling interval is 4–8 km. Readings are averaged over a 4–6 min period. Observations are affected and limited by sea state. Reliability of data depends, to a large extent, on sea-keeping characteristics of the platform. Navigational and positioning accuracy limits accuracy of data.
Magnetism	Flux-gate Magnetometer, Search Coil Magnetometer, Nuclear Precession Magnetometer, Rubium Vapor Magnetometer	Total intensity range: 20,000–100,000 gammas Accuracy: ±15 gammas absolute	Sample rate is variable (from 5 samples/sec to 1 sample/10 min) using continual analog recording. Data are restricted to a zone from 7,500 m to 100 m above sea level during airborne operations and from ocean surface to 2,500 m below during slow marine operations. Station monitor data are collected only at selected points located on land for temporal variation measurements.

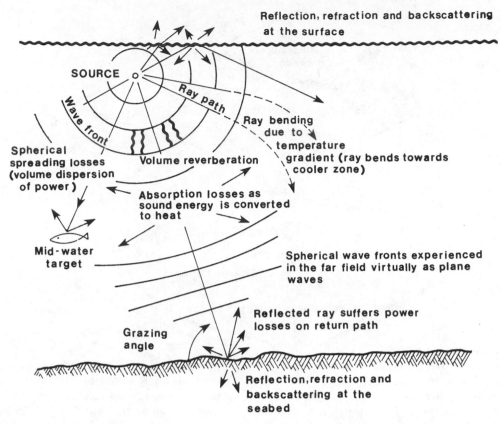

FIGURE 1. A summary of the characteristics of acoustic wave propagation.

such as a subsea wellhead. One drawback with the various short-baseline systems is that they need ship's heading and attitude information. Although many ships already have suitable equipment to monitor these parameters, this requirement presents an added expense, particularly for the short-term user.

Typical Combination A typical equipment combination will comprise a single or multiple array of seabed transponders, a hydrophone with one or more units, and a control and display unit at which the interrogated data will be made available in some usable form. Operating frequencies are normally in the 200 Hz to 1 MHz range. Sound wave attenuation in the water column is composed of spherical spreading and absorption losses and may be lessened by employing lower frequencies. In very simple terms, the lower the frequency the longer the range for a given acoustic level, but this gain must be set against the loss of some accuracy, the higher background noise, and the size, weight, and cost of transducers as frequency is decreased. Typical frequencies for a long baseline system would be in the range of 10

kHz to 20 kHz, with maximum ranges of 4 km to 5 km.

Commercial organizations have adopted various approaches to the important decision of which frequency to choose. The two solutions most commonly selected are to combine long- and short-baseline modes in one system and employ a compromise frequency of typically 40 kHz or to produce separate systems for each mode of operation and employ appropriate frequencies—e.g., approximately 16 kHz for long-baseline systems where range is more important than accuracy and 60 kHz for short-baseline systems where the opposite is the case. An important advantage of adopting the second solution to the frequency problem is that equipment can be less bulky and often more economical to potential clients.

Requirements Broadly speaking, the requirements for acoustic positioning during offshore operations fall into three areas: (1) positioning a surface vessel or structure over a particular point on the seabed (e.g., a drillship over a borehole), (2) positioning a surface vessel tracking within an area remote from any electromagnetic fixing sys-

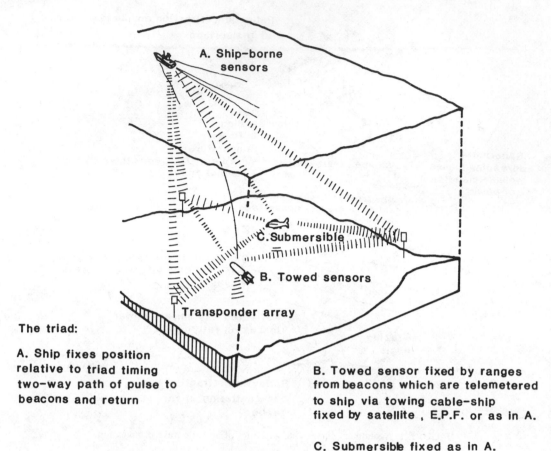

A. Ship-borne sensors

C. Submersible

B. Towed sensors

Transponder array

The triad:

A. Ship fixes position relative to triad timing two-way path of pulse to beacons and return

B. Towed sensor fixed by ranges from beacons which are telemetered to ship via towing cable-ship fixed by satellite , E.P.F. or as in A.

C. Submersible fixed as in A.

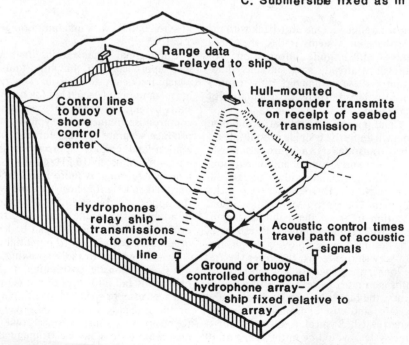

Range data relayed to ship

Control lines to buoy or shore control center

Hull-mounted transponder transmits on receipt of seabed transmission

Hydrophones relay ship-transmissions to control line

Acoustic control times travel path of acoustic signals

Ground or buoy controlled orthogonal hydrophone array-ship fixed relative to array

FIGURE 2. Long-baseline system configuration.

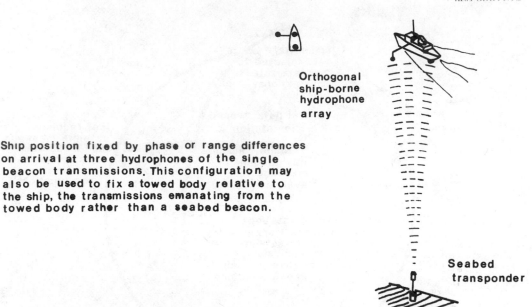

Orthogonal
ship-borne
hydrophone
array

Ship position fixed by phase or range differences
on arrival at three hydrophones of the single
beacon transmissions. This configuration may
also be used to fix a towed body relative to
the ship, the transmissions emanating from the
towed body rather than a seabed beacon.

Seabed
transponder

FIGURE 3. Shore-baseline system configuration.

tem (e.g., mid-ocean site investigations), and (3) positioning an underwater instrument package relative to the seabed or a mother ship (e.g., submersible control).

In the first requirement, short-baseline systems have proved useful for most wellhead recovery projects, and transponders are frequently left on wellheads prior to the drilling rig's departure. For *drillships,* which may sometimes be required to operate in considerable water depths, a more elaborate and dedicated acoustic system will be required. These ships use dynamic acoustic positioning that, when interfaced to suitable propulsion units, will maintain derrick position over a wellhead without the aid of anchors.

In the second requirement, long-baseline systems are typically employed since it will probably be necessary to cover a reasonable area to an acceptable order of accuracy. Groups or nets of *seabed transponders* will be laid in a way that provides maximum coverage of the area of interest. In some North Sea areas, these nets are used in preference to long-range surface positioning since they offer reliable 24 hr coverage and a high order of relative accuracy. In this case, the transponder units will be first calibrated against the long-range system. Farther afield, it will be necessary to use *Doppler satellite observations* for the calibration or use the net in a purely relative mode (Fig. 4).

Perhaps one of the most regular applications for acoustic systems of all types is in positioning the instrument package relative to the seabed or mother ship. Submersible inspections regularly integrate acoustic positioning into its sensing equipment and frequently that of the mother ship. In the case of submersible positioning, it is sometimes desirable to monitor continuously the locations of remote equipment at a separate control location. A relay mode is available on some systems, often operating concurrently with the normal or direct mode. In the relay mode, a discrete frequency is used to trigger the appropriate relay transponder, which in turn interrogates an existing fixed transponder array. The replies that are again received by the user may be used to obtain the relay transponder's position.

Final Result The final result of most acoustic systems is range and/or bearing information, usually available in both visual and digital form at the control unit. The preparation of operational software is, however, not as straightforward. In the case of long-baseline systems, the seabed array must initially be calibrated (even if it is only a relative calibration) before any useful work may be carried out. Much of the success of this calibration will depend on the ability of the software to handle a considerable quantity of acoustic raw data in the best possible way.

Most calibrations involve a relative geometry phase followed by an absolute positioning and orientation phase. During the relative geometry phase, the relationships of the various transponders comprising the complete array are established from an analysis of repeated data sets. Subsequently, the orientation of a chosen baseline and the absolute position of each transponder is de-

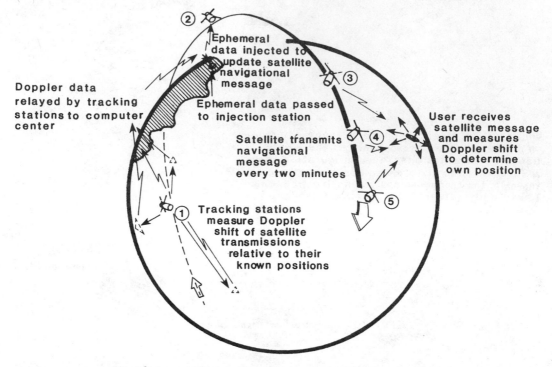

FIGURE 4. The Transit system of position fixing by artificial satellite.

termined by comparing surface positioning data and acoustic data while moving freely through the transponder net. With surface positioning systems possessing high data rates, such a calibration will take approximately 2–3 hr, but when calibrating against Doppler satellite observations, the long period between satellite passes (approximately 1½–2 hr in the North Sea area) and the corresponding slow data rate mean that calibration may take 2–3 day. Useful survey work can be undertaken before this period is completed by fixing, or freezing, the transponder net and then adjusting it in with the final observations after survey work is completed.

Complicating Factors Vertical displacement factors will be complicated during many operations by tidal effects. Depth alterations resulting from tidal movements must first be removed, and either predicted values or real observations will be used (Fig. 5).

Other problems not connected with the acoustic wave propagation are those of a more practical nature. The user vessel's directional movement during the sequential arrival of different signals will cause displacement errors. The data repetition rate may be increased but only at the expense of maximum range; therefore, unless a user's transducer is accidentally equidistant from all transponders, some corrections are necessary.

Another important problem affecting acoustic system capabilities is that of *ray bending*. Sound velocity increases with temperature, salinity, and pressure, and since these factors are not uniform, waves are refracted, or bent, as they travel between *hydrophones,* particularly if they are at different depths (Fig. 6). At distances over 2,000 m, errors due to an incorrect assumption of the velocity of sound wave propagation within the water

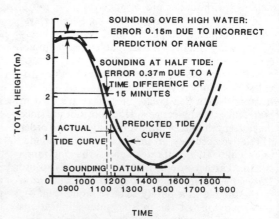

FIGURE 5. The effect of time and height differences on the tidal reductions.

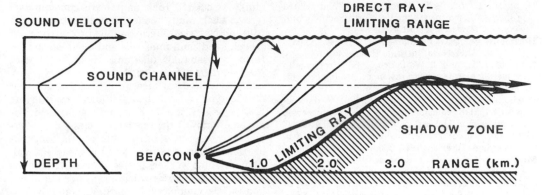

FIGURE 6. Characteristic raypaths in ocean depths.

column can cause considerable errors, while ray bending can result in the total loss of a return signal at the hydrophone unit.

Future Directions

Miniaturized responder units, which are convenient for divers to carry, provide these underwater workers with a powerful tool for both safety and underwater working. New miniresponder software developments will show the divers' progress and position on a video display and allow them to record their position on command as they locate points of interest on, e.g., a subsea pipeline. These small units work as relay transponders within a previously established acoustic array.

It is anticipated that acoustic devices might replace Doppler sonar in the future as an updating medium for inertial navigation. These packages are currently restricted by the high cost of inertial systems, but integrated with acoustics they may eventually provide a cost-effective positioning system with a high order of relative accuracy. Other developments in the field of low-frequency narrow-beam parametric sonar amplification offer interesting new areas for acoustic research.

It has not been possible to cover every aspect of the acoustic positioning field in this entry, and many important areas have no doubt been left uncovered. The future of acoustics in offshore projects seems assured, and when one remembers that most of the equipment capabilities outlined in this entry have been produced in approximately 5–6 yr, we can expect many exciting new developments in the future.

CHARLES P. GIAMMONA

References

Anonymous, 1960, *Oceanographic Instrumentation, Final Report of the Committee on Instrumentation.* Washington, D.C.: U.S. Navy Hydrographic Office (Spec. Pub. SP-41), 350p.

Anonymous, 1963, *Application of Adaptive Sampling Strategies to the Planning of Surveys.* Washington, D.C.: U.S. Navy (Ships Rept. No. 1340663, DDC Ad 415), 287p.

Anonymous, 1964, *Table of Proposed Oceanographic Measurement Requirements.* Washington, D.C.: Marine Science Department, U.S. Naval Oceanographic Office (Informal Manuscript Rept. No. 0-15-64).

Anonymous, 1969, *Marine Resources and Legal-Political Arrangements for their Development. Panel Reports of the Commission of Marine Science, Engineering and Resources,* Vol. 3. Washington, D.C.: U.S. Govt. Printing Office, 480p.

Anonymous, 1970, New concept for harnessing ocean waves, *Ocean Industry* **5,** 62–63.

Austin, C. F., 1966, *Manned Undersea Structures—The Rock Site Concept.* Annapolis, Md.: Naval Ordnance Test Station (Tech. Paper No. 4162), 230p.

Bascom, W., 1969, Technology and the ocean, *Sci. American* **221,** 199–217.

Bigler, A. B., and D. F. Winslow, 1969, Marine recreation: Problems, technologies and prospects to 1980, *Marine Technology Soc. Jour.* **1,** 43–60.

Brahtz, J. F., ed., 1968, *Ocean Engineering.* New York: Wiley, 720p.

Bretscheider, D. I., ed., 1969, *Topics in Ocean Engineering.* Houston: Gulf Publishing, 420p.

Cruickshank, M. J., C. M. Romanowitz, and M. P. Overall, 1968, Offshore mining—Present and future, *Eng. and Mining Jour.* **169,** 84–91.

Elliott, F. E., 1963, Oceanographic platforms and measurements, *U.S. Naval Inst. Proc.* **89**(12), 146.

Elliott, F. E., 1964, Choosing oceanographic platforms, *ISA Jour.* **2**(3), 57.

Ewing, M. A., A. P. Cary, and H. M. Rutherford, 1937, Geophysical investigations in the emerged and submerged Atlantic Plain: Part 1, Method and results, *Geol. Soc. America Bull.* **48**(6), 753–802.

Frosch, R. A., 1962, Underwater sound, *Internat. Sci. and Technology* **9,** 40.

Gillson, J. L., and others, 1960, *Industrial Minerals and Rocks,* Seeley W. Mudd Series. Minneapolis: American Institute of Mining, Metallurgy and Petroleum Engineering, 400p.

Groves, D., 1964, Challenger to challenge to chance, *U.S. Naval Proc.* **90**(3), 136.

Hamilton, E. L., 1970, Sound velocity and related properties of marine sediments, North Pacific, *Jour. Geophys. Research* **75**(23), 4423–4446.

Horton, J. W., 1957, *Fundamentals of Sonar.* Annapolis, Md.: U.S. Naval Institute Press, 246p.

Howard, T. E. and J. W. Paden, 1966, Problems in evaluating marine mineral resources, *Mining Eng.* **18**, 57–61.

Ketchum, D. D., and R. G. Stevens, 1961, A data acquisition and reduction system for oceanographic measurements, *Marine Sci. Instrumentation* **1**, 55.

Mero, J. L., 1965, *The Mineral Resources of the Sea.* Amsterdam: Elsevier Press, 200p.

Mourad, A. G., and M. J. Fubara, 1974, Requirements and applications of marine geodesy and satellite technology to operations in the ocean, in *Proceedings of the International Symposium on Applications of Marine Geodesy.* Columbus, Ohio: Battelle Memorial Institute, 123–131.

Nagy, B., and V. Columbo, eds., 1967, *Fundamental Aspects of Petroleum Geochemistry.* Amsterdam: Elsevier Press, 450p.

National Academy of Sciences, National Academy of Engineering, 1970, *Waste Management Concepts for the Coastal Zone.* Washington, D.C., 126p.

O'Hagan, R. M., 1964, Data processing at sea, *Geo-Marine Technology* **1**(2), 456–465.

Shepard, F. P., 1973, *Submarine Geology,* 3rd ed. New York: Harper & Row, 517p.

Strachan, R. A., 1962, Automatic data processing for numerical weather prediction, in *Proceedings of Computer and Data Processing.* Computer and Data Processing Society, 56–63.

Terry, R. D., ed., 1965, *Ocean Engineering,* Vol. 1. Washington, D.C.: National Security Industrial Association, 559p.

Wiegel, R. L., 1964, *Oceanographical Engineering.* Englewood Cliffs, N.J.: Prentice-Hall, 523p.

Cross-references: *Acoustic Surveys, Marine; Floating Structures in Waves; Harbor Surveys; Surveying, Electronic.* Vol. XIII: *Beach Replenishment, Artificial; Coastal Engineering; Oceanography, Applied.*

SEDIMENTARY ROCKS, FIELD RELATIONS

A detailed field description of a sedimentary rock is essential to an accurate interpretation of its geologic history (Lahee, 1952; Compton, 1985; Moseley, 1981). Much information concerning the rock-forming constituents can be determined in the field using a hand lens. Some aspects can be studied only in the field, including bedding characteristics and sedimentary structures, which are critical in the interpretation of sedimentary environments.

Rock Description

Grain Size Sediment is classified, on the basis of grain size, as gravel (>2 mm), sand (0.062–2 mm), silt (0.004–0.062 mm), and clay (<0.004 mm). Associated rock names are conglomerate (gravel size), sandstone (sand-size), and mudrock (silt and clay size). Terms applied to mixtures of gravel, sand, and mud (silt and clay) can be defined in triangular diagrams (Folk, 1954) (see *Rock Particles, Fragments*).

Field methods of measuring grain size depend on the type of rock being studied. In studying *conglomerates,* several strategies may be used. For regional analysis the ten largest clasts are determined by inspection of the outcrop; the long axis of each clast is measured and the mean size calculated. Some workers prefer to measure only the largest clast. In cases where grain-size variation, in a thick conglomeratic unit (e.g., a debris flow deposit) or a sequence of units, is being studied, a different method may be used. It involves measuring all clasts larger than some operational size (e.g., 16 mm) that cross a transect (a chalk line drawn on the outcrop or a tape laid on the outcrop) of unit length. The length (0.5–5 m) depends on the clast size and the desired spacing of transects. The average grain size of *sandstone* is estimated using a grain-size comparator with sand grains mounted on cardboard (Blatt, 1982, Fig. 4–1, 107) or with images printed on a plastic card (grain-size cards are available from American Stratigraphic, 6280 E. 39th Ave., Denver, CO 80207). Study of *mudrocks* is made difficult by the fine grain size. Where silt is the dominant component (>2/3 silt), it is visible with a hand lens. If silt and clay are present in subequal proportions (>1/3, <2/3 silt) the rock feels gritty when chewed. In mudrocks where clay is dominant (>2/3 clay) the rock feels smooth when chewed (Blatt et al., 1980, Table 11–1, 382).

Color The color of fresh and weathered surfaces is described using a rock-color chart (Goddard, 1948) (see *Rock-Color Chart*). Rock color is a function of the color and size of the detrital grains and/or the color of secondary minerals that coat them. Ferric oxides impart red and brown hues. Carbonaceous matter, and in some cases finely divided ferrous iron compounds, produce dark color through shades of gray to black, which are indicative of reducing conditions (Potter et al., 1980, Fig. 1.25, 55). Green phyllosilicates (e.g., illite and chlorite) produce the color of greenish mudrocks. Carbonate minerals are thought to be a factor in the bluish color of some calcareous mudrocks (Potter et al., 1980, 56).

Size Sorting and Roundness The uniformity of grain size, or sorting, can be estimated by visual comparison with images, illustrating various degrees of sorting (Fig. 1). A similar method for estimating grain roundness is sometimes used (Powers, 1953), but Blatt (1982, 111) states that "the differences in roundness among the grains are too small to be determined reliably with a visual estimate."

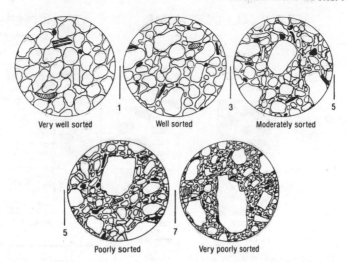

Very well sorted   Well sorted   Moderately sorted

Poorly sorted   Very poorly sorted

FIGURE 1. Terms for degrees of sorting. Drawings represent sandstones as seen with a hand lens. (From Compton, 1985, Fig. 4.1, 50)

Rock Names

Sedimentary rocks are composed of three constituents: (1) framework grains (sand- and gravel-sized detritus), (2) matrix, and (3) chemically precipitated cement. Most framework grains can be identified with a hand lens, and their abundance can be determined in the field using a comparison chart for visual percentage estimation (Terry and Chilingar, 1955). On this basis rock names are given, although detailed petrographic study is required for a more refined analysis.

Mudrock Terminology for mudrocks depends on the relative abundance of silt and clay. Fissile mudrocks are subdivided into silt shale (>2/3 silt), mud shale (subequal silt and clay), and clay shale (>2/3 clay); nonfissile mudrocks are subdivided into siltstone, mudstone, and claystone (Folk, 1954).

Sandstone There are many mineralogical sandstone classifications (Scholle, 1979, 94–96). Most are based on the relative abundance of three framework end members, which can be identified with a hand lens. They are quartz (glassy gray grains), feldspar (white or pink grains), and fine-grained rock fragments (dark grains). Typical rock names are quartzarenite, arkose, and litharenite. These names are commonly modified as to the grade scale that best reflects the average grain size (i.e., very coarse, 1–2 mm; coarse, 0.5–1 mm; medium, 0.25–0.5 mm; fine, 0.125–0.5 mm; and very fine, 0.062–0.125 mm). Examples of usage are fine-grained litharenite and very coarse-grained arkose.

Conglomerate Conglomerate is named on the basis of the average size of the gravel clasts (i.e., granule conglomerate, 2–4 mm; pebble conglomerate, 4–64 mm; cobble conglomerate, 64–256 mm; and boulder conglomerate, >256 mm).

Limestone A simple classification scheme that is very useful in the field utilizes two terms originally used by Grabau (1904): *calcarenite* (as commonly used it is analogous to sandstone and fine-grained conglomerate) and *calcilutite* (analogous to mudrock).

Many studies require a more detailed classification than simply calcarenite and calcilutite. The three basic components of limestone (micrite, sparry calcite, and allochems) are analogous to those in sandstone and mudstone (clay-silt, chemical cement, and sand grains) (Folk, 1980, 158). Of the four allochem types, three are easily recognized in the field: (1) ooids (generally spherical grains with radial and/or concentric structure), (2) intraclasts (round grains composed of micrite or other allochems), and (3) fossils (shell fragments recognizable by shape and composed of fibrous calcite or authigenic sparry calcite). Pellets are usually too small to be recognized with a hand lens. In hand specimen micrite appears dull in contrast to sparry calcite, which appears glassy. Identification of the various components is enhanced by putting a few drops of dilute HCl on the surface to be examined. Folk's (1962) classification (Fig. 2) combines a prefix based on the dominant allochem (i.e., bio-, intra-, or oo-) with a suffix determined by the relative abundance of micrite and sparry calcite (i.e., -micrite and -sparite). Examples of usage are *oomicrite* and *biosparite*. Another popular classification, proposed by Dunham (1962), incorporates the concept of grain support, a difficult property to evaluate in the field.

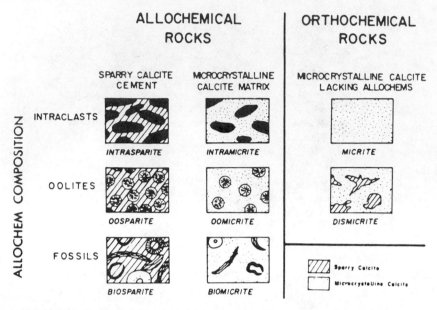

FIGURE 2. Folk's classification of limestones. (Modified from Scholle, 1978, 171)

Dolomite Although limestone is the dominant carbonate rock in most Phanerozoic sequences, dolomite is commonly a major constituent, especially in Paleozoic rocks (Garrels and Mackenzie, 1971). Some carbonate rocks are a mixture of calcite and dolomite, but most are relatively pure limestone or dolomite (Steidtmann, 1917). The standard test for dolomite involves the use of dilute HCl (powdered dolomite effervesces, and unpowdered dolomite does not). In addition, dolomite commonly weathers to a tan or buff color in contrast to limestone, which is usually gray or blue-gray. Coarse-grained dolomite rock has a granular surface imparted by well-formed rhombs.

Internal Organization and Structure

Strata Thickness and Contacts Sedimentary layers are called *beds* (thickness >1 cm) and *laminae* (thickness <1 cm). Bedding thickness is described as very thin (1–3 cm), thin (3–10 cm), medium (10–30 cm), thick (30–100 cm), and very thick (>1 m); laminae thickness is described as thin (<0.3 cm) and thick (0.3–1 cm) (Ingram, 1954). Examples of usage are thinly bedded and thickly laminated. Units with beds formed by splitting or parting are described as flaggy (1–10 cm), slabby (10–30 cm), blocky (30–100 cm), and massive (>1 m) (Collinson and Thompson, 1982, 9). Bedding thickness may be laterally uniform or variable, and beds may be continuous or discontinuous. In studying sedimentary rocks, special attention should be paid to the nature of bed contacts, which may be erosional, sharp, gradational, or interfingering. Beds that coalesce are said to amalgamate.

Internal Structures A group of strata of similar characteristics is called a *set;* a group of similar sets is called a *coset* (Fig. 3). Layers that lack angular discordance are referred to as *planar stratification* (Harms et al., 1982, 3-23). The depositional surface is probably close to being horizontal. Deposition on the lee side of ripples produces inclined layers called *foreset beds* (or laminae), collectively known as *cross-bedding*. Foresets are commonly curved with the concave side up. Cross-bedding occurs in sets described as tabular (formed by straight-crested ripples) (see Fig. 3). Set thickness ranges from 1 cm or 2 cm to many meters, although most sets are less than 1 m thick (Pettijohn, 1975, 107). Some small-scale cross-bedding is called *climbing-ripple* (ripple-drift) *cross-lamination* (Fig. 4). A distinctive medium-to-large-scale type, known as hummocky cross-bedding, has laminae and erosion surfaces (which bound the sets) that are concave- and convex upward (Fig. 5). This structure is significant because it is thought to reflect deposition resulting from the action of storm waves in lower shoreface and offshore marine environments (Harms et al., 1975, 88).

Penecontemporaneous deformation produces a great variety of structures. Soft-sediment folding is of two distinctive types. Convolute bedding and lamination involves folding into sharp anticlines, commonly overturned, and broad synclines (Fig. 6A). It is produced by plastic deformation of

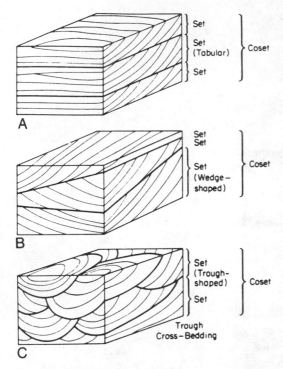

FIGURE 3. Terminology for cross-bedding. (From Blatt et al., 1980, Fig. 5-2, 131)

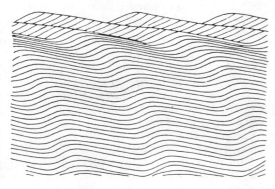

FIGURE 4. Patterns of climbing-ripple cross-lamination. (From Collinson and Thompson, 1982, Fig. 6.18, 69)

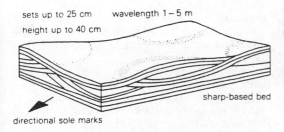

FIGURE 5. Hummocky cross-bedding. (From Collinson and Thompson, 1982, Fig. 6.63, 99)

partly liquefied sediment and is evidence of rapid deposition. Slump folds produced by lateral mass movement may involve units hundreds of meters thick. Soft-sediment folds usually occur in units that are bounded above and below by undisturbed sedimentary rock (Fig. 6B). This relationship distinguishes them from tectonically deformed beds. It is important to note the style of folding and to record orientations of fold axes and axial planes to distinguish further between tectonic and sedimentary deformation. The upward escape of water during consolidation of liquefied sediment produces dish and pillar structures. "Dishes are thin, subhorizontal, flat to concave-upwards clayey laminations . . . [and] . . . pillars are vertical to near-vertical cross-cutting columns and sheets of structureless sands" (Leeder, 1982, 112). Dish structures are distinguished from trough cross-lamination by the absence of foreset laminae. Sinking of sand into liquefied mud produces sand pillows (load casts). In some cases the sand bed breaks apart and collapses into the underlying mud, producing sandstone balls (ball-and-pillow structure, or pseudonodules) (Fig. 6C). Sandstone dikes form when liquefied sand is injected (usually from below) into an unlithified host sediment. Laminae parallel to the sides of the dike may be produced by shear as the sand moves through the fissure.

Animals moving through and on the sediment produce a complex suite of structures commonly known as *trace fossils* (Frey, 1975; Basan, 1978). These are very useful guides to interpreting sedimentary environments, sedimentation history (e.g., continuous versus discontinuous and rapid versus slow), and substrate consistency. Biogenic activity may obscure or destroy preexisting structures, in some cases producing massive and featureless units. Plants growing on aggradational surfaces leave traces of their roots concentrated along individual horizons in the sediment.

Algae (blue-green) produce distinctive laminated structures known as *stromatolites* (Walter, 1976). Although they commonly occur in intertidal deposits, they are by no means restricted to this environment. Stromatolites, which are found mainly in limestone, are usually columnar (Fig. 7) but may occur as planar laminae. It is important that stromatolitic limestone not be confused with tufa, travertine, caliche, or other nonalgal laminated limestone. The presence of unsupported voids (fenestrae), commonly filled with sparry calcite, characterizes many stromatolitic deposits.

Grain-size and bed-thickness variation within sedimentary units is an important aspect of the internal organization of sedimentary rocks. It may occur in sequences tens or hundreds of meters thick as well as within single beds or laminae. Thick sequences may coarsen upward (e.g., pro-

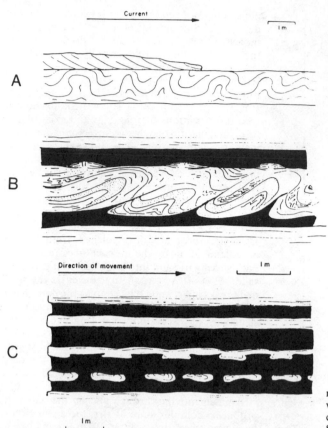

FIGURE 6. Structures produced by penecontemporaneous deformation. (*A*) Convolute bedding, (*B*) slumped beds, (*C*) load casts and sandstone balls. (Modified from Selley, 1976, Figs. 83, 85, 86)

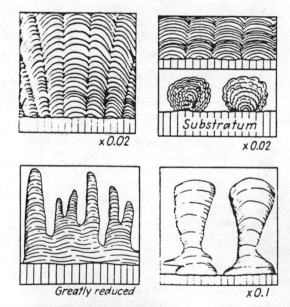

FIGURE 7. Typical columnar stromatolites. (Modified from Shrock, 1948, Fig. 245)

grading deltaic and barrier island deposits; see Fig. 16*C*) or fine upward (e.g., fluvial deposits; see Fig. 16*A* and *B*). Similarly, beds may show a progressive increase or decrease in thickness as one proceeds up-section. This is the case in many fan (subaerial and submarine) deposits (Fig. 8). Detailed field observations are essential so that these very important aspects are not overlooked.

Grain-size variation (grading) also occurs within beds a few centimeters to several meters thick. Sandstone beds commonly fine upward (normal grading) in response to deposition by waning currents (e.g., turbidity currents). Conglomerate beds may fine upward (normal grading) or coarsen upward (inverse or reverse grading). The latter is typical of some grain flow deposits.

Bedding-Plane Structures Ripples are the most commonly found bed forms in sedimentary rock. Larger bed forms (e.g., dunes and sandwaves) are rarely preserved on bedding surfaces and are manifest primarily as medium- and large-scale cross-bedding (see Fig. 15). In profile, ripples are asymmetric or symmetric (Fig. 9). The

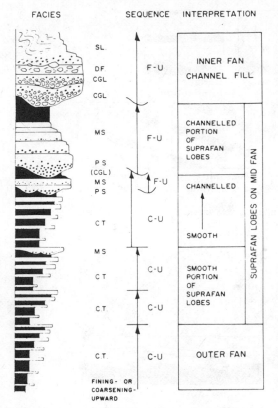

FIGURE 8. Hypothetical submarine fan stratigraphic sequence produced by fan progradation. CT, classical turbidite; MS, massive sandstone; PS, pebbly sandstone; CGL, conglomerate; DF, debris flow; SL, slumps. Arrows show thickening- and coarsening-upward sequences (C-U) and thinning- and fining-upward sequences (F-U). (From Walker, 1979, Fig. 14, 99)

difference is usually attributed to formation by unidirectional currents (asymmetric with steeper face sloping downcurrent) or wave action (symmetric). Although this is usually true, some "ripples with marked asymmetry may be caused by shoaling waves or by an interaction of waves and currents of similar direction" (Collinson and Thompson, 1982, 63). The shape and continuity of ripple crestlines, as seen in plan view, are equally as important as the ripple profile. Wave-formed ripples have relatively straight and continuous crestlines and some bifurcation. Where

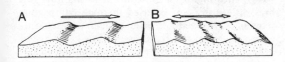

FIGURE 9. Ripples. *A:* asymmetric, *B:* symmetric.

wave motion is complex, interference ripples form. Asymmetric current-formed ripples show a continuum from straight-crested (nonbifurcating) through sinuous-crested to complex forms (e.g., lunate and linguoid) commonly with shallow scour pits. Some flat bedding planes show a lineation caused by minute ridges and hollows, known as parting *lineation*. It is parallel to the paleoflow direction.

Most erosional structures preserved on bedding surfaces are formed by erosion of cohesive mud, which is quickly buried by sand. They are seen typically in positive relief on the base (sole) of the overlying sandstone after removal of mudrock (Fig. 10). Scour around an obstacle lying on the mud substrate produces sole casts known as *current crescents*. They are horseshoe shaped and open in the downcurrent direction. Flute casts, also produced by scour, have an upcurrent end that is rounded or bulbous and a down-current end that flares out and merges with the bedding plane (Fig. 11). When objects (tools), carried by a current, impinge on the substrate, a shallow depression is formed. The names applied to casts of tool marks are descriptive of the inferred mode of origin; they include groove casts, skip casts, slide casts, and prod casts. Examples are illustrated by Pettijohn and Potter (1964).

Cracks caused by mud shrinkage are usually associated with subaerial exposure. Such desiccation mud cracks form polygons (from centimeters to meters in diameter) that are obvious on bedding surfaces but more difficult to recognize in vertical section.

Subaqueous shrinkage cracks (syneresis cracks) result from loss of pore water . . . because of a reorganization of originally highly porous clays, either due to flocculation, or because of salinity-induced changes of volume of certain clay minerals. In plan, syneresis cracks tend to have irregular or radiating patterns, sometimes crosscutting one another. Individual cracks are lenticular, pinching out rather than joining with other cracks. [Collinson and Thompson, 1982, 140]

Small circular pits, up to 1 cm in diameter and several millimeters deep, are formed by the impact of raindrops or hailstones on a cohesive mud surface. *Raindrop impressions* are distinguished with difficulty from trace fossils and gas-bubble escape features by an absence of disturbed laminae.

Grain Fabric Dimensional fabric in sedimentary rock usually involves preferred orientation of gravel. In one type the pebble long axes lie transverse to flow. This is the result of clasts rolling on the bed, such as in the case of many river gravels. Intraclasts in some limestones show similar orientation that may have been produced by wave action or tidal currents (Lindholm, 1980). In the

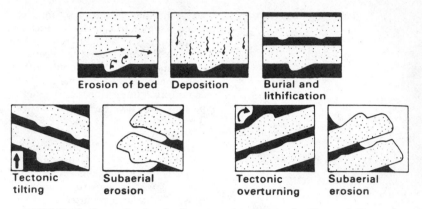

FIGURE 10. Stages in the development of a sole mark. (From Collinson and Thompson, 1982, Fig. 4.1, 37)

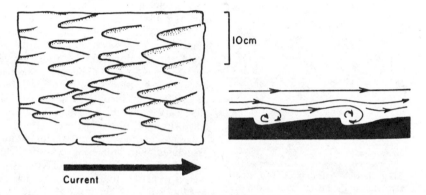

FIGURE 11. Flute casts. (From Selley, 1976, Fig. 73, 210)

other type the pebble long axes lie parallel to flow. It is thought to indicate "transport processes that maintain pebbles and cobbles in dispersion above the bed" (Harms et al., 1982, 6-4). In both types, clasts commonly are imbricated in the upcurrent direction (Lindsay, 1972).

On some bedding surfaces fossils show preferred orientation. Although this may represent a hydrodynamically stable rest position for transported shells, one must be aware that living organisms may respond to current action (Potter and Pettijohn, 1963, 37–40). Bivalve shells carried as traction load are usually deposited with the convex side up. In suspension transport they are deposited with the concave side up.

Paleocurrent Analysis The flow direction (paleocurrent) that was operative when ancient sediments were deposited can be determined by a variety of features including cross-bedding, asymmetric ripples, flute casts, and pebble imbrication. In addition, regional variation in grain size and facies can be used, provided that reasonable stratigraphic control exists. Some structures, although parallel to the current, indicate the line of movement but not the direction of flow. They include symmetric ripples, current lineations, groove casts, and clast orientation.

Paleocurrent measurements taken in beds with structural dip must be restored to their original position (Potter and Pettijohn, 1963, 259–262; Collinson and Thompson, 1982, Appendix A). Statistical analysis involves determination of the average direction (vector mean) and a measure of the central tendency (vector magnitude) (Potter and Pettijohn, 1963, 262–268; Curray, 1965; Briggs, 1977, 150–165).

Top-and-Bottom Criteria In areas where tectonic deformation has produced overturned folds, it is important to determine whether or not beds are in their original stratigraphic position (Shrock, 1948). Although superposition of formations and biostratigraphy can be used in thick continuous sequences, these are of little value in isolated outcrops. Many of the features previously discussed provide a valuable alternative, including crossbedding (only where foresets are concave up-

FIGURE 12. Geopetal structures in buried shells. (From Compton, 1985, Fig. 4.7C, 54)

ward), symmetric ripples, stromatolites, graded sandstone beds, sole marks, and mud cracks. Also useful are geopetal structures, which are cavities (e.g., bivalve fossils) partly infilled with sediment (Fig. 12). The upper part is commonly filled with sparry calcite.

Presentation of Data

Effective graphical presentation of field data is essential to analysis and interpretation as well as communication. The follwoing two types are used.

Graphic Logs There are many ways to present measured sections as graphic logs (Collinson and Thompson, 1982, 177–181). The method presented herein (Fig. 13) utilizes two columns. One is used to show lithology and is of constant width. The other is used to show sedimentary structures; grain size is indicated by column width. Gradual grain-size changes are reflected by a change in the width of the column. Fossil content (or other grains—e.g., ooids, glauconite) may be indicated by narrow columns next to the main columns. Paleocurrent measurements may be recorded opposite the units from which they were recorded. Azimuths are plotted, as on a map, with north toward the top of the page.

Maps Aerial variation of lithologic and structural aspects of sedimentary rock is clearly shown by plotting data on a map. The choice of maps depends on the type of data and the purpose of the study.

Although paleocurrent azimuths may be plotted for each outcrop where measurements were made, a map showing the moving average of these measurements is also useful (Fig. 14). "The final regional map should present the investigator's interpretation of paleocurrents, based on all the relevant evidence" (Potter and Pettijohn, 1963, 273).

Stratigraphic maps may be used to show the external geometry (e.g., isopach maps) of the rock body as well as its composition (e.g., facies maps) (Krumbein and Sloss, 1963, 432–500). Although many of these maps may be compiled by hand, some require (and most are facilitated by) statistical analysis using computers (Robinson, 1982).

Grain Size

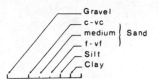

Lithology

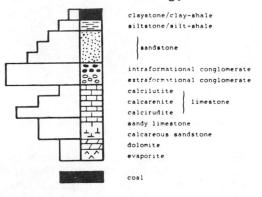

claystone/clay-shale
siltstone/silt-shale

sandstone

intraformational conglomerate
extraformational conglomerate
calcilutite
calcarenite } limestone
calcirudite
sandy limestone
calcareous sandstone
dolomite
evaporite

coal

Structures

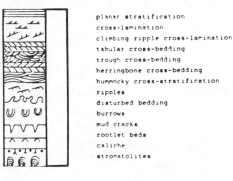

planar stratification
cross-lamination
climbing ripple cross-lamination
tabular cross-bedding
trough cross-bedding
herringbone cross-bedding
hummocky cross-stratification
ripples
disturbed bedding
burrows
mud cracks
rootlet beds
caliche
stromatolites

Contacts

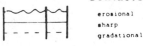

erosional
sharp
gradational

FIGURE 13. Graphic log of measured vertical section. See text for explanation.

Interpretation

Field study of sedimentary rock provides the basis to analyze its geologic history. This includes evaluation of flow regime and sedimentary environment, leading to reconstruction of basin paleogeography.

Flow Regime In subaqueous environments flow regime is defined by certain bed forms and their interaction with the water surface (Fig. 15).

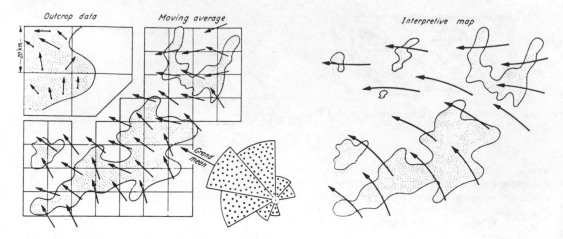

FIGURE 14. Maps showing paleocurrent data. (From Potter and Pettijohn, 1963, Fig. 10-12, 274)

The main variables are mean flow velocity, water depth, and grain size (Blatt et al., 1980, 136–146).

Sedimentary Environment A major role in the interpretation of sedimentary environments is played by mapping the rock body geometry with careful attention to sedimentary features. In ad-

dition to fossils, the most diagnostic features are sedimentary structures, particularly those with directional significance and implications to flow regime and sequences of sedimentary facies known to be characteristic of particular depositional environments (Fig. 16). Valuable summaries are

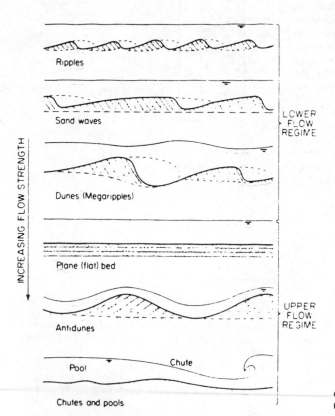

FIGURE 15. Sedimentary structures that characterize upper and lower flow regimes. (From Blatt et al., 1980, Fig. 5.3, 137)

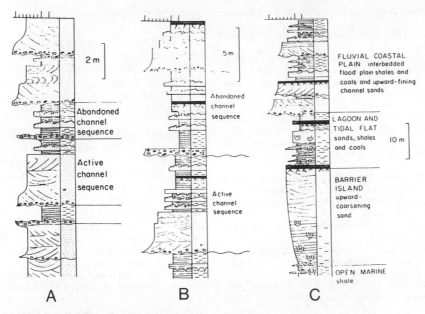

FIGURE 16. Sequences of sedimentary facies characteristic of (*A*) braided streams, (*B*) meandering streams, and (*C*) prograding barrier island complex. (Modified after Selley, 1976, Figs. 99, 101, 114)

given by Wilson (1975), Selley (1978), Ethridge and Flores (1981), Scholle and Spearing (1982), Davis (1983), Walker (1984), and Reading (1986).

ROY C. LINDHOLM

References

Basan, P. B., ed., 1978, Trace fossil concepts, *Soc. Econ. Paleontologists and Mineralogists Short Course Notes No. 5,* 201p.

Blatt, H., 1982, *Sedimentary Petrology.* San Francisco: W. H. Freeman, 564p.

Blatt, H., G. Middleton, and R. Murray, 1980, *Origin of Sedimentary Rocks.* Englewood Cliffs, N.J.: Prentice-Hall, 782p.

Briggs, D., 1977, *Sources and Methods in Geography: Sediments.* Boston: Butterworths, 192p.

Collinson, J. D., and D. B. Thompson, 1982, *Sedimentary Structures.* Boston: George Allen and Unwin, 194p.

Compton, R. R., 1985, *Geology in the Field.* New York: Wiley, 398p.

Curray, J. R., 1965, The analysis of two-dimensional orientation data, *Jour. Geology* **64,** 117–131.

Davis, R. A., Jr., 1983, *Depositional Systems: A Genetic Approach to Sedimentary Geology.* Englewood Cliffs, N.J.: Prentice-Hall, 669p.

Dunham, R. J., 1962, Classification of carbonate rocks according to depositional texture, in W. E. Ham, ed., *Classification of Carbonate Rocks.* Tulsa, Okla.: American Association of Petroleum Geologists (Mem. 1), 108–121.

Ethridge, F. G., and R. M. Flores, eds., 1981, Recent

and ancient nonmarine depositional environments: Models for exploration, *Soc. Econ. Paleontologists and Mineralogists Spec. Pub. No. 31,* 349p.

Folk, R. L., 1954, The distinction of grain size and mineral composition in sedimentary rock nomenclature, *Jour. Geology* **62,** 344–359.

Folk, R. L., 1962, Spectral subdivision of limestone types, in W. E. Ham, ed., *Classification of Carbonate Rocks.* Tulsa, Okla: American Association of Petroleum Geologists (Mem. 1), 62–84.

Folk, R. L., 1980, *Petrology of Sedimentary Rocks.* Austin, Tex.: Hemphill Pub. Co., 182p.

Frey, R. W., ed., 1975, *The Study of Trace Fossils.* New York: Springer-Verlag, 562p.

Garrels, R. M., and F. T. Mackenzie, 1971, *Evolution of Sedimentary Rocks.* New York: W. W. Norton, 397p.

Goddard, E. N., chair, 1948, *Rock-Color Chart.* Boulder, Colo.: Geological Society of America, 6p.

Grabau, A. W., 1904, On the classification of sedimentary rocks, *Am. Geologist* **33,** 228–247.

Harms, J. C., J. B. Southard, D. R. Spearing, and R. G. Walker, 1975, Depositional environments as interpreted from primary sedimentary structures and stratification sequences, *Soc. Econ. Paleontologists and Mineralogists Short Course No. 2,* 166p.

Harms, J. C., J. B. Southard, and R. G. Walker, 1982, Structure and sequences in clastic rocks, *Soc. Econ. Paleontologists and Mineralogists Short Course No. 9,* 249p.

Ingram, R. L., 1954, Terminology for the thickness of stratification and parting units in sedimentary rocks, *Geol. Soc. America Bull.* **65,** 937–938.

Krumbein, W. C., and L. L. Sloss, 1963, *Stratigraphy*

and Sedimentation. San Francisco: W. H. Freeman, 660p.

Lahee, F. H., 1952, *Field Geology.* New York: McGraw-Hill, 883p.

Leeder, M. R., 1982, *Sedimentology: Process and Product.* Boston: George Allen and Unwin, 344p.

Lindholm, R. C., 1980, Intraclast orientation in Cambro-Ordovician limestones in western Maryland, *Jour. Sed. Petrology* **50,** 1205–1212.

Lindsey, D. A., 1972, Sedimentary petrology and paleocurrents of the Harbell Formation, Pinyon Conglomerate, and associated coarse clastics, northwestern Wyoming, *U.S. Geol. Survey Prof. Paper 734-B,* 68p.

Miall, A. D., 1984, *Principles of Sedimentary Basin Analysis.* New York: Springer-Verlag, 490p.

Moseley, F., 1981, *Methods in Field Geology.* San Francisco: W. H. Freeman, 211p.

Pettijohn, F. J., 1975, *Sedimentary Rocks.* New York: Harper & Row, 628p.

Pettijohn, F. J., and P. E. Potter, 1964, *Atlas and Glossary of Primary Sedimentary Structures.* Berlin: Springer-Verlag, 370p.

Potter, P. E., J. B. Maynard, and W. A. Pryor, 1980, *Sedimentology of Shale.* New York: Springer-Verlag, 306p.

Potter, P. E., and F. J. Pettijohn, 1963, *Paleocurrents and Basin Analysis.* Berlin: Springer-Verlag, 296p.

Powers, M. C., 1953, A new roundness scale for sedimentary particles, *Jour. Sed. Petrology* **23,** 117–119.

Reading, G. H., ed., 1986, *Sedimentary Environments and Facies.* New York: Elsevier, 615p.

Robinson, J. E., 1982, *Computer Applications in Petroleum Geology.* New York: Hutchinson Ross Pub. Co., 154p.

Scholle, P. A., 1978, A color illustrated guide to constituents, textures, cements, and porosities of sandstone and associated rocks, *Am. Assoc. Petroleum Geologists Mem. 28,* 201p.

Scholle, P. A., and D. Spearing (eds.), 1982, Sandstone depositional environments, *Am. Assoc. Petroleum Geologists Mem. 31,* 410p.

Selley, R. C., 1976, *An Introduction to Sedimentology.* New York: Academic Press, 408p.

Selley, R. C., 1978, *Ancient Sedimentary Environments.* Ithaca, N.Y.: Cornell University Press, 287p.

Shrock, R. R., 1948, *Sequence in Layered Rocks.* New York: McGraw-Hill, 507p.

Steidtmann, E., 1917, Origin of dolomite as disclosed by stains and other methods, *Geol. Soc. America Bull.* **28,** 431–450.

Terry, R. D., and G. V. Chilingar, 1955, Summary of "Concerning some additional aids in studying sedimentary formations" by M. S. Shvetsov, *Jour. Sed. Petrology* **25,** 229–234.

Walker, R. G., ed., 1984, *Facies Models.* Toronto: Geological Association of Canada, 317p.

Walter, M. R., ed., 1976, *Stromatolites.* Amsterdam: Elsevier, 790p.

Wilson, J. L., 1975, *Carbonate Facies in Geologic History.* New York: Springer-Verlag, 471p.

Cross-references: *Hard- versus Soft-Rock Geology.* Vol. VI: *Sedimentology—Yesterday, Today, and Tomorrow; Sediment Parameters.*

SERENDIPITY

It is often said that the basis for scientific discovery, especially in the earth sciences, is serendipity—i.e., the discovery by accident. It should be stressed, however, that if one does not bother to climb the mountain, then there is no chance of discovering the rare fossil that may be found at the top.

Serendip is an old name for Ceylon. An early book of fairy tales called *The Three Princes of Serendip* came into the hands of Horace Walpole, Fourth Earl of Oxford, who noted in a letter (1754):

[A]s their Highnesses travelled, they were always making discoveries, by accidents and sagacity, of things they were not in quest of: for instance, one of them discovered that a mule, blind in the right eye had travelled the same road lately, because the grass was eaten only on the left side, where it was worse than on the right—now do you understand Serendipity? (Lonsdale, 1963)

Walpole's word has become a twentieth-century explanation of why research into pure science may have very practical and utilitarian consequences, whereas ad hoc research is not always very productive.

RHODES W. FAIRBRIDGE

Reference

Lonsdale, K., 1963, Origin of serendipity, *Science* **142,** 621 (letter).

Cross-reference: *Geological Methodology.*

SITE EXAMINATION—See EXPOSURES, EXAMINATION; FIELD NOTES, NOTEBOOKS; SURVEYING, ELECTRONIC. Vol. XIII: DAMS, ENGINEERING GEOLOGY; PIPELINE CORRIDOR EVALUATION; PUMPING STATIONS AND PIPELINES; ROCK STRUCTURE MONITORING.

SLOPE STABILITY ANALYSIS

The resistance of a sloping earth mass to failure or movement is termed its *stability.* The stability of a slope under varying conditions of stress involves vertical, horizontal, or oblique change in position of sections of the Earth's crust and of the engineering structures, such as highways or buildings, that may be connected with these sections. The earth materials involved are natural rock, soils, artificial fills, or combinations of these materials. Movement is in response to gravity or to seismic shock and, at least in unconsolidated materials, often is aided by the increased mobility of saturated materials.

Movements of earth slopes are examples of mass wasting. They include not only spectacular movements of material, which receive wide publicity, but also almost imperceptible creep, which receives scant notice. Mass movements of slopes can be roughly divided into landslides (or simply, slides), falls, topples, spreads, flows, and creep of earth masses (Varnes, 1978). The landslide is the type of mass movement in which failure is most apt to occur along fairly definite, visible, or inferred failure surfaces and thus is the type of slope failure most susceptible to mechanical analysis by means of the techniques currently available (see *Landslide Control*). Studies of other mass movements, however, indicate that appropriate methods of analysis are becoming available for them also (Johnson, 1970, 432–534; Saito, 1969).

In general, the stability of a slope subject to failure by sliding may be analyzed in three ways: empirical, stress-strain, and limit equilibrium. These methods can provide qualitative or precise quantitative assessment of stability. In recent years, there has been a trend toward the application of statistical theory to these methods of analysis to develop estimates of the probability of slope failure (Vanmarcke, 1980).

Empirical procedures provide a qualitative estimate of slope stability for reconnaissance purposes and for situations where accurate data on the physical nature and strength of the slope-forming materials and subsurface water conditions are lacking. Field measurement of attributes such as inclination of existing failed and unfailed slopes, nature of the materials, direction of slope exposure, and other parameters can be used as a basis for an empirical judgment of stability of a given slope (Lane, 1961).

Stresses and strains within a slope are calculated using relatively new techniques for stability analysis. The most widely used is called the *finite element method*. In this method, a cross section of the slope under consideration is divided into an array of triangular elements. The corners of the elements, called *nodal points,* are common to several triangular elements. Solutions are generated using a computer to calculate forces and displacements of the nodal points on the basis of geometry, boundary loading or unloading, and material properties. The determination of slope stability requires an additional judgment of the forces and/or displacements required to produce an unstable condition (Zienkiewicz and Cheung, 1967).

The *limit equilibrium method* is the most commonly used method of stability analysis. In this method, the forces resisting slope failure

(strength) are compared to the forces tending to cause failure, and a so-called factor of safety is calculated. This method can be used for rock slopes where the failure is principally along a preexisting fracture or zone of weakness (Pentz, 1971). The greatest application of the limit equilibrium method to date, however, has been in unconsolidated materials whose shear strengths can be ascertained relatively easily. Therefore, the remainder of this discussion treats the application of the limit equilibrium method to the analysis of stability of slopes in unconsolidated materials.

Two general classes of soils affect the stability of slopes in unconsolidated materials: (1) cohesive soils, such as clays, which have low permeability, and (2) cohesionless soils, such as sands and gravels, which have high permeability. The behavior of the two types in terms of slope stability is distinctly different.

Cohesive Soils

From the engineering point of view, the most common type of slide, and the one most amenable to quantitative investigation, is the *rotational shear slide,* which occurs in cohesive soils. An example of a rotational slide and its descriptive nomenclature are shown in Fig. 1.

In a rotational failure of cohesive soil, a large mass of the soil slides on a curved surface that intersects the slope. Collin (1846) was the first to note and analyze the curved nature of this slip surface. For an analytical solution, this surface may be approximated by assuming either a logarithmic spiral or a circular arc (Rendulic, 1935). Although the former assumption leads to an analytical solution if the soil is homogeneous, a much simpler approach is to assume that the failure surface approximates a circular arc (Fellenius, 1927; Taylor, 1948, 432).

The usual approach in the limit equilibrium method of stability analysis is to calculate the factor of safety of several trial circles. The circle having the lowest factor of safety is the surface of potential failure, and if the analysis for this circle indicates a factor of safety of less than 1, then failure is to be expected. Analyzing a sufficient number of trial slip surfaces to ensure locating the most dangerous one is laborious; however, the equations of equilibrium are commonly arranged into a form suitable for computer solution.

Method of Slices The form of the *Coulomb equation* (Coulomb, 1773) for the shearing resistance of a cohesive soil above the water table is

$$s = c + p \tan \phi,$$

where s is the shearing resistance of the soil per given area along a potential failure surface, c is the cohesion along a unit length of this surface,

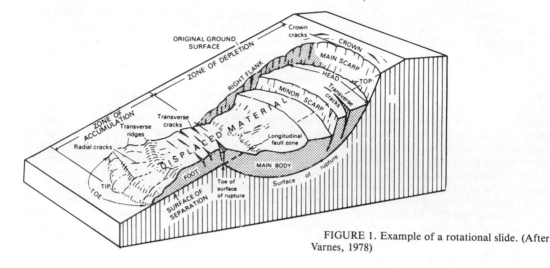

FIGURE 1. Example of a rotational slide. (After Varnes, 1978)

p is the compressive stress normal to the surface, and ϕ is the angle of shearing resistance (the angle of internal friction).

Stability along any assumed failure surface is dependent on the shearing resistance of the soil. Because shearing resistance is related to existing soil pressure, p, as shown in the Coulomb equation, and because p depends on depth, the shearing resistance will vary along the assumed circular failure surface. Thus, this variation must be taken into account by the method of analysis that is used.

Such a procedure is the *method of slices*, developed by the Swedish engineer Fellenius (1927) and more recently presented by Taylor (1948, 432–441). The method of slices is aptly described in most textbooks on elementary soil mechanics. This method requires a cross section, plotted to scale, of the slope being analyzed and the circular segment representing the sliding mass. The segment is divided into several vertical slices, and the shearing and resisting forces acting on each slice are evaluated. A typical slice and the forces acting on it are shown in Fig. 2.

The shearing or driving force along the assumed failure surface is the component of the weight of the slice acting in that direction ($W \sin \alpha$). The resisting forces (shearing resistance) are the cohesion force cl, which is the unit cohesion times the length of failure surface along the base of the slice, and the friction force $P \tan \phi$. P is the normal force ($P = W \cos \alpha$) acting along the base of the slice, and ϕ is the angle of internal friction. The unit cohesion c and the angle of internal friction ϕ are obtained from laboratory or field shear tests. For an approximate analysis, the side forces E and Y can be neglected because they effectively cancel out on summation of forces (Bishop, 1955).

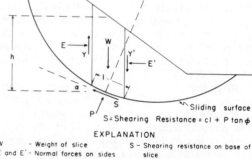

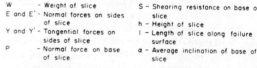

EXPLANATION

W	– Weight of slice	S –	Shearing resistance on base of slice
E and E'	– Normal forces on sides of slice	h –	Height of slice
Y and Y'	– Tangential forces on sides of slice	l –	Length of slice along failure surface
P	– Normal force on base of slice	α –	Average inclination of base of slice

FIGURE 2. Forces acting on a slice of a circular sliding mass.

By summing shearing and resisting forces parallel to the base of the slide, the following equation for factor of safety can be obtained for this slice:

$$\text{Factor of safety} = \frac{\text{shearing resistance}}{\text{shearing force}}$$

$$= \frac{cl + P \tan \phi}{W \sin \alpha}$$

When all the slices are considered, the factor of safety for the entire potential sliding mass becomes:

$$\text{Factor of safety} = \frac{\Sigma cl + \Sigma P \tan \phi}{W \sin \alpha},$$

in which Σcl and $\Sigma P \tan \phi$ represent the sum of the resisting forces for all slices, and $\Sigma W \sin \alpha$ is the total slide-inducing force.

Effect of Pore-Water Pressure When the assumed failure surface lies beneath the groundwater table, the resistance of the soil to sliding is reduced, owing to the buoyancy provided by the water. In this case, the factor of safety against sliding is given by

$$\text{Factor of safety} = \frac{\Sigma cl + \Sigma (P - U) \tan \phi}{\Sigma W \sin \alpha},$$

in which U represents the total buoyant force on the bottom of the slice. If the water table is above the assumed failure surface and no subsurface flow of water exists, the buoyant force at the base of the slice is

$$U = d \gamma_w l,$$

in which γ_w is the unit weight of the water, d is the vertical difference between the static water level and the assumed failure surface, and l is the length of the failure surface along the base of the slice.

Location of Critical Sliding Surface Circular failures in cohesive soils are typically categorized as slope failures or base failures, as shown in Fig. 3a and b. The location of the critical surface is determined by existing conditions. If materials of different shearing resistances are present, such as clay overburden on rock, the surface will typically be tangent to the firm base. If a layer of weak material exists, as much as possible of the surface will commonly be contained within this layer (Fig. 3c). Thus, consideration of existing field conditions will result in reduction of the number of trials necessary to locate the position of this surface.

Cohesionless Soils

For cohesive soils, Coulomb's equation predicts that the stability is a function of both height and inclination of the slope. For dry cohesionless soils, however, the shearing resistance or strength is expressed by Coulomb's equation as

$$s = p \tan \phi,$$

where all factors are as defined previously and stability is a function only of slope inclination. In cohesionless soils, failure ordinarily occurs along a surface of translation parallel to the slope, whereas in cohesive soils, rotational movement

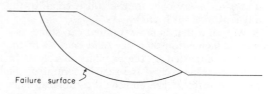

(a) Slope Failure

Failure surface

Firm base

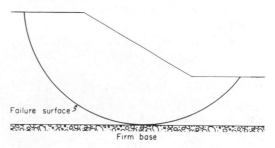

(b) Base Failure

Failure surface

Firm base

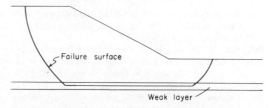

(c) Failure along Weak Layer

Failure surface

Weak layer

FIGURE 3. Examples of (a) slope failure and (b) base failure in a cohesive soil (modified from Fellenius, 1927; Terzaghi and Peck, 1967). (c) Illustrates failure along a weak layer.

along an approximately circular failure is more common. The critical inclination, which is known as the *angle of repose,* varies with the angle of internal friction of the cohesionless soil, and for dry cohesionless soil it is independent of the height of slope. As is the case for cohesive soils, shear failure occurs when the shearing force exceeds the shearing resistance of the soil. The factor of safety in regard to translatory failure is equal to the shearing resistance divided by the shearing force.

The factor of safety of the slope with respect to sliding may be expressed by the equation

$$\text{Factor of safety} = \frac{\tan \phi}{\tan \beta},$$

where β is the angle between the slope and the horizontal (Terzaghi and Peck, 1967, 232). Thus, the *critical inclination* (factor of safety equal to unity) for a dry cohesionless soil is equal to the angle of internal friction.

Wet sand that is not saturated can stand at a steeper inclination than dry sand because of the strength imparted by the capillary effect of the moisture. However, if the sand is saturated, the capillary effect is lost and the shearing resistance of the soil is lowered. The shearing resistance for saturated cohesionless soils can be obtained from the following form of the Coulomb equation:

$$s = (p - u)\tan \phi,$$

where u is the unit pore pressure at the failure surface.

ROBERT L. SCHUSTER
ROBERT W. FLEMING

References

Bishop, A. W., 1955, The use of the slip circle in the stability analysis of earth slopes, *Geotechnique* **5**(1), 7-17.

Collin, A., 1846, *Landslides in Clays* (trans. W. R. Schriever, 1956). Toronto: Toronto University Press, 160p.

Coulomb, C. A., 1773, Essai sur une application des règles de maximis et minimis à quelques problemes de statique, relatifs à l'architecture, *Mém. Acad. Sci. (Savants Etrang.)* **7**, 343-382.

Fellenius, W. K. A., 1927, *Erdstatische Berechnungen mit Reibung und Kohäsion (Adhäsion) und unter annahme kreiszylindtischer Gleitflächen*. Berlin: W. Ernst & Sohn, 40p.

Johnson, A. M., 1970, *Physical Processes in Geology*. San Francisco: Freeman, Cooper, 557p.

Lane, K. S., 1961, Field slope charts for stability studies, *5th Internat. Conf. Soil Mech. and Found. Eng. Proc.* **2**, 651-655.

Pentz, D. L., 1971, Methods of analysis of stability of rock slopes, in C. O. Brawner and V. Milligan, eds., *Stability in Open Pit Mining*. New York: Soceity of Mining Engineers, American Institute of Mining, Metallurgy, and Petroleum Engineers, 119-141.

Rendulic, L., 1935, Ein Beitrag zur Bestimmung der Gleitsicherheit, *Bauingenieur* **16**(19/20), 230-233.

Saito, M., 1969, Forecasting the time of slope failure by tertiary creep, *7th Internat. Conf. Soil Mechanics and Found. Eng. Proc.* **2**, 677-683.

Taylor, D. W., 1948, *Fundamentals of Soil Mechanics*. New York: John Wiley, 700p.

Terzaghi, K., and R. B. Peck, 1967, *Soil Mechanics in Engineering Practice,* 2nd ed. New York: Wiley, 729p.

Vanmarcke, E. H., 1980, Probabilistic stability analysis of earth slopes, *Engineering Geology* **16**(102), 29-50.

Varnes, D. J., 1978, Slope movement types and processes, in R. L. Schuster and R. J. Krizek, eds., *Landslides—Analysis and Control*. Washington, D.C.: National Academy of Sciences (Transportation Research Board Spec. Rept. 176), 11-33.

Zienkiewicz, O. C., and Y. K. Cheung, 1967, *The Finite Element Method in Structural and Continuum Mechanics*. London: McGraw-Hill, 272p.

Cross-references: *Environmental Engineering; Environmental Management; Landslide Control.* Vol. XIII: *Geomorphology, Appied; Pipeline Corridor Evaluation; Reinforced Earth; Rheology, Soil and Rock; Rock Structure Monitoring; Soil Mechanics.*

SOIL DYNAMICS—See SAPROLITE, REGOLITH, AND SOIL; SOILS AND WEATHERED MATERIAL, FIELD METHODS AND SURVEY. Vol. XIII: SOIL MECHANICS.

SOIL FABRIC

Soils are composed of discrete soil particles and particle groups with voids between that may be filled with liquid and/or air. The solid phase is not a continuous material, and the particles comprising the solid phase consist of minerals and some organic matter with varying sizes, from submicroscopic to visually discrete. On the basis of particle size, soils may be broadly categorized, in the order of increasing size, as clay, silt, sand, and gravel. The particles may encompass a wide range of shapes from nearly spherical, bulky grains to thin, flat plates or long, slender needles. The liquid phase of soils is a solution of various salts in water. The gaseous phase tends toward the composition of air with some water vapor. The variability of these phases, together with corresponding variation in specific interaction of the phases with each other, contributes to the nonuniformity in the soil characteristics and properties. Physical properties of the soil are much harder to alter than chemical properties, and they largely determine soil use (see Vol. XII, Pt. 1: *Microstructure, Manipulation*). The physical properties of importance in engineering are strength, compressibility, permeability, volume change, compactibility, and frost susceptibility.

Engineering structures may be designed with more confidence if one knows these properties and their change with environment and time. For an evaluation of stability of a soil mass, one is concerned with the load-deformation-time behavior of the soil. To analyze and predict soil behavior, it is necessary to understand how the individual particles or particle groups interact. The interaction between the particles is governed directly by the geometrical aspects of their arrangement, as well as the forces acting between them. The term *soil structure* (q.v. in Vol. XII, Pt. 1) is defined, then, as the arrangement of soil particles and the forces acting between adjacent particles. Therefore, soil structure is that property of soil that provides the stability of the soil system and that is responsible for response to externally applied and internally induced sets of forces. Thus, soil structure is the fundamental property that determines the physical properties of the soil for engineering purposes.

A major component of soil structure is the geometric aspects of the particle arrangement, and the term *fabric* denotes the geometric arrangement of particles, particle groups, and pore spaces in a soil. In granular soils (coarse-grained soils such as sand and gravel), the arrangement of individual particles is referred to as *packing* of particles. Consequently, the term *fabric* is used more exclusively for particle arrangement in fine-grained soils—i.e., clays and silts. The definitions are given herein in their most widely accepted meanings in engineering. Somewhat varying definitions are used by workers in other fields of soil studies as well as in engineering. For example, the term *structure* is used by some workers interchangeably with the term *fabric*. This is due partly to the relatively recent progress in soil structure research. The fabric of soils remained largely unknown until the development of suitable X-ray diffraction, optical, and electron microscope techniques made direct fabric observations possible starting in about the mid-1950s. In the late 1960s, research accelerated significantly, with a major contribution arising from the development of the scanning electron microscope.

Forces between Soil Particles

In a soil mass at equilibrium, there is a balance among the applied boundary and gravity forces, the pressure in the water, the electrostatic and electrodynamic forces, the bonds of organic and inorganic nature and the intergranular forces. Prior to external loading of soil mass, the *intergranular stresses* are generated by the difference between the gravity forces due to the weight of soil mass and the pressures in the fluid phase of the soil. This force governs between the particles in the case of coarse-grained soils such as clean sand and gravel with insignificant surface (electrical) forces. These soils are referred to as *cohesionless,* or *granular,* soils in soil mechanics literature, and the nature of their behavior is comparatively better known.

The *electrostatic* and *electrodynamic* forces are important in fine-grained soils such as clays, silts, and sands and gravels with a significant amount of fines (silts and clays). These soils are referred to as *cohesive* soils. The electrostatic forces arise primarily from the Coulomb repulsion between clay particles, which carry net negative charges. The electrodynamic forces are derived primarily from van der Wall's attractive forces. In addition to these forces, strong *bonds* may be established between soil particles by cementation, carbonate, and/or organic bonds in both cohesive and cohesionless soils.

Soil Fabric

Since fabric is more amenable than structure to direct measurement and quantification, and since much can be inferred about structure from its associated fabric, the study of fabric constitutes a most important step in understanding the engineering behavior of soils. Soil structure, in its broader definition, is very complicated and extremely difficult to measure and quantify, especially in clay soils due to the intricate and variable nature of these soils. Therefore, recent research has concentrated on the fabric aspect of soil structure (Mitchell, 1976). The levels and units of fabric organization are important. Fabric can be viewed at two general levels on the basis of degree of magnification required for a proper observation of fabric patterns: macrofabric and microfabric.

Macrofabric The macrofabric units and features are those aspects of soil fabric that can be seen with the unaided eye or a hand lens. The terms like *peds, crumbs,* and *aggregates* have been used to describe these macroscopic units. A *ped* is defined as a soil aggregate consisting of a cluster of particles separated from each other by surfaces of weakness such as fissures, voids, or skins of different composition material (Brewer, 1964). Some macrofabric features such as stratifications, fissuring, voids, and large-scale inhomogeneities are important in engineering applications, and the search for such features is of major concern in the soil investigation for any project (Rowe, 1972).

Microfabric The microfabric units are observable only under either a light or an electron microscope. At this level, fabric units may range from single particles acting independently to a group of particles acting as a unit. Fabric units consisting of two or more particles are termed *domains,* or *aggregates* (see Vol. XII, Pt. 1: *Aggregation; Microstructure, Manipulation*). The overall stability of a soil mass is defined in terms of the interactions that occur between various fabric units. Since the soil structure must reach a new equilibrium value under any added external load, fabric alterations will occur to provide the necessary changes in the internal constraints. These alterations probably take place as interactions between the macrofabric units. At times, one or the other may dominate the overall mechanical response.

Fabric Parameters

In fabric analysis, we are concerned with determining the geometrical aspects associated with particle arrangements. The parameters that define these aspects in the case of microfabric are particle size, particle shape, particle orientation, and porosity and pore size. These parameters are basically statistical quantities and are always associated with distribution of values in a natural soil. They may be characterized in terms of a mean value and other quantities defining their distributions.

Size of particles is an important parameter in defining and separating the two basic behavioral patterns—namely, those of cohesionless and cohesive soils. The former is usually associated with granular (coarse-grained soils) and the latter with predominantly fine-grained soils. The ordering of fabric parameters according to their importance in controlling the mechanical behavior also shows distinct differences in each of these groups of soils.

Packing in Granular Soils

Although in clays the smallest fabric element, in general, is the clay particle group or domain, the particles in sand are sufficiently large and bulky that they behave as independent units (see Fig. 4). In the case of cohesionless soils, porosity or density and particle shape are the most important fabric parameters. In *soil mechanics,* the relative distribution of the solid phase and the voids is expressed usually in terms of *void ratio* (see Vol. XII, Pt. 1: *Soil Mechanics*), which is defined as the ratio of the volume of voids to the volume of solids and is denoted as e. Depending on the relative positions of the grains, it is possible to have a wide range in void ratios. The highest void ratio possible for a given cohesionless soil, where each particle would still be touching its neighbors, is the maximum void ratio, e_{max}. The smallest void ratio is the minimum void ratio, e_{min}. The relationship between the actual void ratio of a soil and its limiting values, e_{max} and e_{min}, is expressed by relative density, D_r:

$$D_r = \frac{e_{max} - e}{e_{max} - e_{min}} \times 100\%.$$

A natural soil is said to be loose if its relative density is less than about 50%, dense if it is higher.

The packing behavior of a granular soil can be studied in terms of the maximum and minimum void ratios, and the state of packing can be expressed in terms of its relative density. The packing behavior depends to a large extent on geometric particle characteristics such as size and shape and their distributions, surface texture, as well as mineralogy. The geometrics and detailed surface features of any sand particle are a function of its mineral composition and structure and its environmental history.

The shape of the particles is important in affecting the packing characteristics of the granular soils. The shape characteristics affect the engineering behavior directly as well as through their influence on relative density. Three general classes of grain shapes have been defined: bulky grains, flakey or scalelike grains, and needlelike grains. The first two are far more important than the third. When the length, width, and thickness of the particles are of the same order of magnitude, the shape is bulky. Flakey grains are thin but not necessarily elongated. Small amounts of flakey mica can change the engineering behavior of a predominantly bulky grain soil.

In addition to particle shape, orientation also becomes important in granular soils especially in those with flakey grains. In general, characterization of the fabric of a cohesionless soil (packing) can be done in terms of grain shape factors, grain orientation, and interparticle contact orientations.

Two distinct parameters, sphericity and roundness, may be used to characterize the particle shape. *Sphericity* is defined as the ratio of the volume of the particle to the volume of the smallest circumscribing sphere, and *roundness* is understood to be the ratio of the curvature of the corners and edges of the particle to the average curvature of the particle. Perfect sphericity and perfect roundness in a two-dimensional view are both associated with a value of unity, and imperfect particles are characterized by values between zero and unity. Sphericity and roundness can be reliably measured in two dimensions by photographic techniques and estimated visually with consistent accuracy by use of standard charts. There are, however, a number of charts available with somewhat varying definitions of the shape factors. One of the charts is presented in Fig. 1 for determining visually the projection roundness of sands (Krumbein, 1941).

The size separation of granular soils can be achieved by mechanical sieving. Size and shape parameters (sphericity and roundness) are statistical quantities with associated distributions. They exert varying degrees of influence on packing characteristics. The maximum and minimum void ratios and the void ratio spread, Δe, which is the difference between the limiting void ratios, increase as mean particle roundness decreases (or angularity increases), and small changes in particle size distribution result in large changes in packing behavior, as shown in Fig. 2. Mean particle size and mineralogy are found to have little or no influence on the packing behavior of granular soils, whereas surface texture is found to have significant influence (Edil et al., 1975).

Microfabric of Cohesive Soils

Clay particles interact through the layers of adsorbed water and through mineral contact in cer-

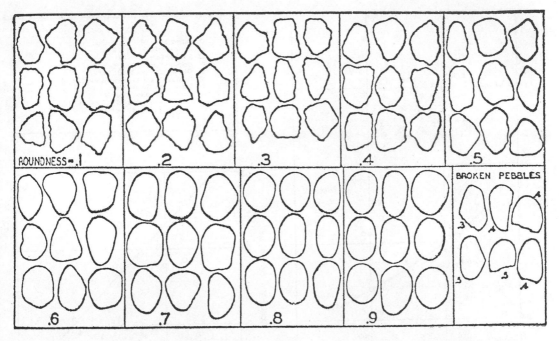

FIGURE 1. Chart for determining visually the projection roundness of sands. (After Krumbein, 1941)

tain particular cases. In placement or deposition of clay particles, interparticle forces of attraction and repulsion play a dominant part in particle arrangement. The presence and proportion of larger-sized particles in cohesive soils will also influence the final arrangement subsequent to deposition. As a result of both the large number of fabric studies in recent years and the variety and complexity of fabrics that have been observed, there has been a proliferation of terms for the description of fabrics and fabric features. In describing fabric features, both the form and the function of the units must be considered. Particle associations in clay suspensions are described by van Olphen (1963) by the following terms:

Dispersed: No face-to-face association of clay particles,
Aggregated: Face-to-face association of several clay particles,
Flocculated: Edge-to-edge or edge-to-face association of aggregates,
Deflocculated: No association between aggregates.

Various modes of association and the appropriate terminology are given in Fig. 3. Some experimental evidence supports the existence of some of these particle associations in clay suspensions and in some sedimentary soils (Schweitzer and Jennings, 1971; Rosenqvist, 1959). In most soils of engineering interest, however, individual particle associations are quite rare, and groups of several clay particles are the more usual fabric-forming units. Fabrics of sediments, residual soils, and compacted clays assume a variety of forms, but most of them are related to the configurations shown in Fig. 3, with the difference in density. The schematic representation of idealized fabrics for elementary particle arrangement are given in Fig. 4. Other variations undoubtedly exist and can be classified.

As can be inferred from the preceding discussion of the microfabric observed in natural soils, the interparticle spacing or porosity including pore size distribution, and the particle orientation characteristics, are the most important fabric parameters in the case of cohesive soils. Clay particles are typically plate shaped, and particle shape does not vary as much as in the case of granular soils. However, certain clay minerals such as attapulgite have rod-shaped particles. The mean particle size of various clay minerals varies significantly from each other. For example, kaolinite particles, often hexagonal plates, are 0.3–4 μ in diameter with a thickness of 0.05–2 μ; illite particles are plates with a thickness of 300 Å. Montmorillonite particles are thin plates typically around 30 Å thick and 0.1–1 μ in diameter. Particle size and shape are important in determining the specific surface area (surface area per unit weight) that, in turn, together with the net elec-

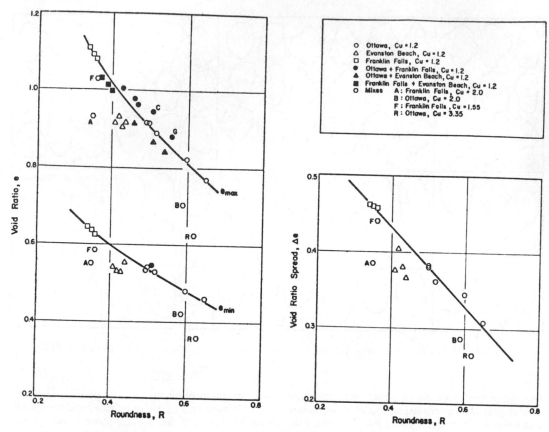

FIGURE 2. Void ratio versus roundness. (Note: C is the uniformity coefficient of grain-size distribution.) (Aftei Edil et al., 1975)

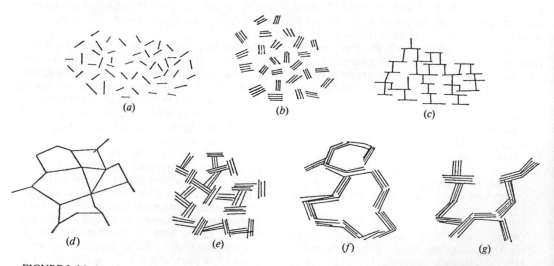

FIGURE 3. Modes of particle association in clay suspensions. (a) Dispersed and deflocculated, (b) aggregated but deflocculated, (c) edge-to-face flocculated but dispersed, (d) edge-to-edge flocculated but dispersed, (e) edge-to-face flocculated and aggregated, (f) edge-to-edge flocculated and aggregated, (g) edge-to-face and edge-to edge flocculated and aggregated. (After van Olphen, 1963)

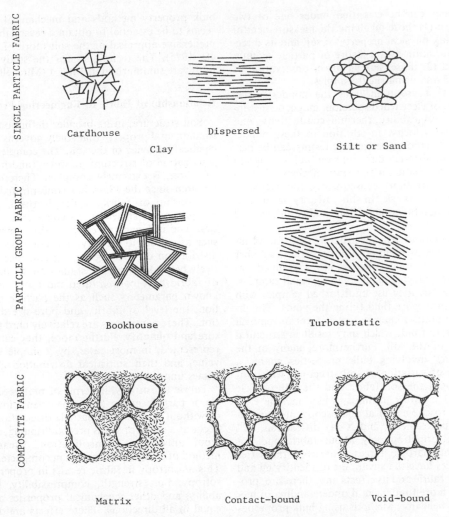

SINGLE PARTICLE FABRIC

Cardhouse Dispersed

Clay Silt or Sand

PARTICLE GROUP FABRIC

Bookhouse Turbostratic

COMPOSITE FABRIC

Matrix Contact-bound Void-bound

FIGURE 4. Schematic representation of idealized fabrics.

trical charge on the clay particle, determine the amount and nature of the surface forces. These forces are important in determining the initial fabric of a soil and the resistance of this fabric to change under external loading in engineering applications.

The particle orientation characteristics of clay comprise the directional parameter of fabric analysis, and this parameter is defined by the angular relationship between the linear or planar elements (single clay particles or domains) and set of chosen reference axis. The other geometric aspects (porosity, particle size, and shape) are basically scalar parameters. The terms *oriented* and *random* are very important in describing the orientation characteristics of a clay sample. When there is a parallel orientation between linear or planar elements, they are said to possess "preferred ori-

entation," or to be *oriented,* The term *oriented* implies a fabric involving essentially face-to-face arrangements of the particles. The term *random* implies that the particles are oriented equally in all directions and that a preferred direction of orientation is not recognizable. In this case, particle arrangements may be edge-to-edge or edge-to-face. Most clays have fabrics that are probably somewhere between perfectly oriented and perfectly random. To consider the intermediate cases, the concept of the "degree of orientation" must be introduced. The orientation in clays, like other fabric parameters, is a statistical parameter.

Methods of Fabric Observation and Analysis

Since the 1950s several methods have evolved for observing the fabric of soils. In general, such techniques can be classified under one of two headings: (1) those involving the measurement of some bulk physical property of soil and its directional variation as a measure of particle orientation and (2) those making direct observation of particles and using small samples.

The first group includes the nondestructive methods of electrical dispersion, magnetic susceptibility, permeability, thermal conductivity, and acoustical velocity. In addition to these nondestructive methods, destructive testing can be performed in different directions on bulk samples to assess the variation of a certain mechanical property with direction—consequently, the degree of orientation in fabric. In this category are permeability, strength, compression, and shrinkage and swelling tests.

Examinations of soil structure and fabric involving measurement of bulk properties, whether destructive or nondestructive, have certain advantages over direct techniques using small samples. In particular, they use undisturbed samples with their natural pore fluid filling the voids. The direct techniques require a replacement or removal of the pore fluid, which may result in structural damage of the soil. Furthermore, many of the techniques involving bulk properties enable a measure of variation of soil structure, which includes both the soil fabric and the interparticle forces. The direct methods, which include electron microscopy, optical microscopy, mercury intrusion porosimetry, and X-ray diffraction, will reveal information only about fabric and not about interparticle forces. However, the latter techniques have the advantage of identifying and isolating fabric and its effects and, therefore, providing a more fundamental understanding of mechanical behavior. Methods using bulk properties provide only average quantities, and the particular property measured may be affected by factors other than fabric only. Equally important considerations in direct methods are specimen preparation for fabric observations, methods of quantification of fabric, and statistical analysis and presentation of data. Methods for describing a complex phenomenon such as clay fabric are still in their early stages of development, and more exact quantitative methods of describing fabric are needed. The description of methods of fabric analysis and useful reviews and references of this topic can be found in the Proceedings of the International Symposium on Soil Structure (1973).

The techniques that are likely to give the most complete information include electron and light microscopy, X-ray diffraction, and pore size distribution (see Vol. IVB: *Soil Mineralogy*). Each of these methods has advantages and disadvantages. Consequently, the combined use of these three techniques, and the inclusion of information from

bulk property methods and mechanical testing, seems to be essential to obtain a reasonably comprehensive appraisal of the soil fabric (Krizek et al., 1975). The techniques for the study of soil fabric are summarized in Table 1 (Mitchell, 1976).

Relationship of Fabric to Engineering Property

Soil structure, in its broader definition, is the fundamental property that solely governs the mechanical behavior of the soil. The complete characterization of structure, however, and its quantification, is extremely complex. Therefore, the research since the 1950s has concentrated on the relationship of fabric, which is the important component of structure, to engineering properties. The influence of certain fabric parameters such as void ratio (density) and grain size and its distribution on the engineering response is relatively better known. The studies on the influence of fabric have concentrated more on the lesser known parameters such as the particle orientation, the level of fabric, and pore-size distribution. These parameters are relatively hard to measure and quantify. Furthermore, they cannot be represented, in most cases, by a simple average value, and their statistical distribution also becomes important.

Fabric anisotropy, in terms of preferred particle or particle group (domain) orientation, with or without intradomain particle orientation, may occur as a result of stress preconditioning, as e.g., from anisotropic consolidation, shear, and method of sample preparation in compacted soils. This anisotropy in fabric results in property anisotropy; i.e., strength, compressibility, permeability, and other mechanical properties are not equal in all directions. These effects are demonstrated on two laboratory-prepared samples of a kaolinitic clay. These samples were prepared by consolidating slurries of clays either anisotropically or isotropically, resulting in uniform oriented and random orientation samples respectively. These samples are designated as the anisotropic sample (AC) and the isotropic sample (IC). Specimens were trimmed in two mutually orthogonal directions from these samples to determine their fabric and various mechanical properties. In the case of the AC sample, specimens trimmed with their longitudinal axis parallel to the direction of the major principal consolidation stress are called V specimens, and those trimmed at 90° to this direction are referred to as H specimens. In the case of the IC sample, specimens with their longitudinal axis parallel to the general direction of drainage are referred to as V specimens, and specimens trimmed at 90° to the general direction of drainage are H specimens.

The overall fabric of the two samples can be

TABLE 1. Techniques for Study of Soil Fabric

Method	Basis	Scale of Observations and Features Discernible
Optical Microscope (Polarizing)	Direct observation of fracture surfaces or thin sections	Individual particles of silt size and larger, clay particle groups, preferred orientation of clay, homogeneity on a millimeter scale or larger, large pores, shear zones. Useful upper limit magnification about ×300
Electron Microscope	Direct observation of particles or fracture surfaces through soil sample (scanning electron microscope—SEM) observation of surface replicas (transmission electron microscope—TEM)	Resolution to about 100 Å; large depth of field with SEM; direct observation of particles; particle groups and pore space; details of microfabric
X-Ray Diffraction	Groups of parallel clay plates produce stronger diffraction than randomly oriented plates	Orientation in zones several square millimeters in area and several micrometers thick; best in single mineral clays
Pore Size Distribution	(1) Forced intrusion of nonwetting fluid (usually mercury) (2) Capillary condensation	(1) Pores in range from ~0.01 to >10 μm (2) 0.1 μm maximum
Acoustical Velocity	Particle alignment influences velocity	Anisotropy; measures microfabric averaged over a volume equal to sample size[a]
Dielectric Dispersion and Electrical Conductivity	Variation of dielectric constant and conductivity with frequency	Assessment of anisotropy; flocculation and deflocculation; measure microfabric averaged over a volume equal to sample size[a]
Thermal Conductivity	Particle orientations influence thermal conductivity	Anisotropy; measures microfabric averaged over a volume equal to sample size[a]
Magnetic Susceptibility	Variation in magnetic susceptibility with change of sample orientation relative to magnetic field	Anisotropy; measures microfabric averaged over a volume equal to sample size[a]
Mechanical Properties Strength Modulus Permeability Compressibility Shrinkage and swell	Properties reflect influences of fabric	Microfabric averaged over a volume equal to sample size[a]; anisotropy; macrofabric features in some cases

Source: After Mitchell, 1976.

[a]For a homogeneous sample. Discontinuities, stratification, and so on, on a macroscale can override effects of microfabric.

visually appreciated by examining the scanning electron micrographs given in Fig. 5. These micrographs indicate a rather definite orientation of the particles in the sample that was anisotropically consolidated, whereas the sample that was prepared by isotropic consolidation seems to possess a fabric consisting of randomly distributed groups of particles. The former, in general, exhibits a predominantly single-particle fabric, whereas the latter indicates a domain fabric. The effect of fabric on the stress-strain response of these two samples of extreme particle orientations is demonstrated in Fig. 6. The influence of the fabric parameters (void ratio and particle orientation) on the coefficient of permeability is depicted in Fig. 7. In a separate study (Edil and Krizek, 1975), the dependence of strength between the vertical and horizontal specimens (degree of

5 µ	5 µ
Horizontal plane normal to the major principal consolidation stress	Vertical plane parallel to the major principal consolidation stress

Anisotropically Consolidated from a Dispersed Slurry

5 µ	5 µ
Plane parallel to the general direction of drainage	Plane normal to the general direction of drainage

Isotropically Consolidated from a Flocculated Slurry

FIGURE 5. Scanning electron micrographs of typical fabrics. (From Edil and Krizek, 1977)

strength anisotropy) is plotted as a function of the difference in the particle orientation parameter in V and H directions as measured from the X-ray diffraction analysis of laboratory-prepared samples in Fig. 8. The degree of anisotropy in the strength response $(b_V - b_H)/b_V$ increased at an increasing rate as the degree of particle orienta-

tion $(Q_{max} - Q_{min})$ increases; within the limits of this experimental program, the strength of an H specimen is seen to be as much as double that of a V specimen from the same sample. Conversely, changes in void ratio may alter the strength by a factor of two to four for comparable samples.

The influence of fabric on other mechanical

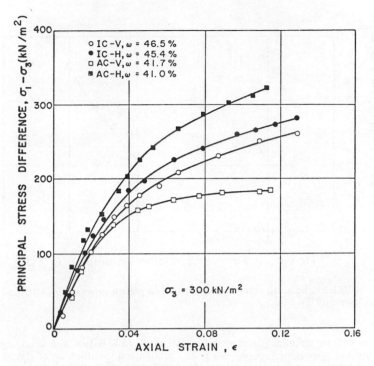

FIGURE 6. Effect of particle orientation on stress-strain response. (From Edil and Krizek, 1977)

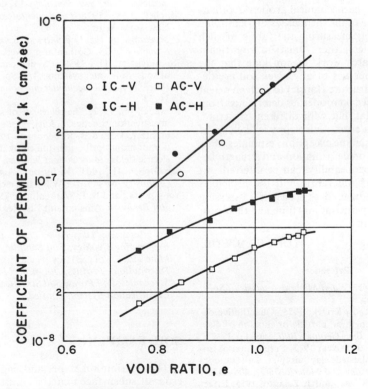

FIGURE 7. Effect of void ratio and particle orientation on permeability. (From Edil and Krizek, 1977)

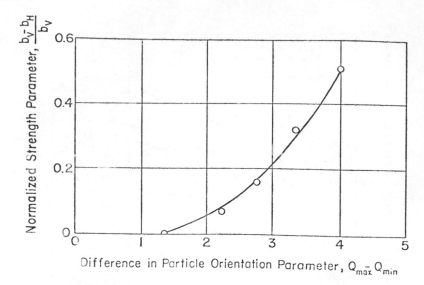

FIGURE 8. Undrained strength anisotropy versus particle orientation. (After Edil and Krizek, 1975)

properties such as volume change, compressibility, and sensitivity has also been demonstrated. Useful reviews and references can be found in the texts by Yong and Warkentin (1975) and Mitchell (1976). There are fewer quantitative studies of the fabric effects on the mechanical properties of natural soils because of the difficulties of identifying and quantitatively measuring the fabric of such soils. Nevertheless, there exists a significant amount of research work to conclude that the geotechnical properties of any given soil are dependent on the structure (fabric) to such an extent that analyses based on properties determined from the same material, but with different structure, may be totally in error. Anisotropy is likely to be the rule rather than the exception; estimates of its potential effects on deformation and strength behavior, and on permeability, can be inferred. An understanding of soil fabric and the principles governing it can be used as a basis for assessing both the soil properties at any time and their possible change in time.

TUNCER B. EDIL

References

Brewer, R., 1964, *Fabric and Mineral Analysis of Soils.* New York: Wiley.

Edil, T. B., and R. J., Krizek, 1975, Quantitative dependence of strength on particle orientation of clay, *Internat. Assoc. Eng. Geology Bull.,* no. 11, 19–22.

Edil, T. B., and R. J. Krizek, 1977, Preparation of isotropically consolidated clay samples with random fabrics, *Jour. Testing and Materials* 5(5), 406–412.

Edil, T. B., R. J. Krizek, and J. Zelasko, 1975, Effect of grain characteristics on packing of sands, *Istanbul Conf. Soil Mechanics and Found. Eng.* **1,** 46–54.

Krizek, R. J., T. B. Edil, and K. I. Ozaydin, 1975, Preparation and identification of clay samples with controlled fabric, *Eng. Geology,* no. 9, 13–38.

Krumbein, W. C., 1941, Measurement and geological significance of shape and roundness of sedimentary particles, *Jour. Sed. Petrology* **11,** 64–72.

Mitchell, J. K., 1976, *Fundamentals of Soil Behavior.* New York: Wiley.

Proceedings of the International Symposium on Soil Structure, 1973, Gothenburg, Sweden.

Rosenqvist, I. Th., 1959, Physico-chemical properties of soils: Soil water systems, *Am. Soc. Civil Engineers Proc., Jour. Soil Mechanics and Found. Div.* **85**(SM2), 31–53.

Rowe, P. W., 1972, The relevance of soil fabric to site investigation practice, *Géotechnique* **22**(2), 195–300.

Schweitzer, J., and B. R. Jennings, 1971, The association of montmorillonite studied by light scattering in electric fields, *Jour. Colloid Interface Sci.* **37,**443–457.

van Olphen, H., 1963, *An Introduction to Clay Colloid Chemistry.* New York: Wiley-Interscience, 301p.

Yong, R. N., and B. P. Warkentin, 1975, *Soil Properties and Behavior.* Amsterdam: Elsevier.

Cross-references: *Dispersive Clays; Expansive Soils, Engineering Geology; Soil Sampling.* Vol. IVB: *Soil Mineralogy.* Vol. XII, Pt. 1: *Aggregation; Pore-Size Distribution; Profiles, Physical Modification; Soil Mechanics; Soil Pores; Soil Structure.* Vol. XIII: *Soil Classification System, Unified.*

SOIL SAMPLING

The distribution, type, and engineering properties of subsurface materials must be known for the proper design of most engineering structures. The following typical steps, which lead to an engineering design, are intimately related, and changes in any one of the steps will affect each of the others:

Subsurface explorations and sampling,
Determination of engineering properties (laboratory and field testing),
Analysis and design.

The subsurface explorations and sampling constitute the first important step in *geotechnical engineering investigations*. A good understanding of the problems encountered in making subsurface explorations and of the various tools available for sampling is essential for the engineer to perform his or her design work properly. The objectives of subsurface exploration and sampling include the following:

Geology and stratigraphy of deposits (areal extent, depth, thickness, composition, and nature of each soil and rock stratum);
The location of groundwater and its variations;
The engineering properties of the soil and rock strata that affect the performance of engineering structures.

The results obtained in an engineering subsurface survey can be presented in terms of stratigraphic profiles and physical soil profiles. The former is a plot showing surface elevation and depths to the various strata along with their identity. The latter also shows variations in physical properties. Two-dimensional profiles are determined by interpolation between several one-dimensional profiles based on a single boring or probing.

A complete subsurface investigation for engineering purposes consists of the following stages:

Fact finding: This stage involves gathering available data on subsurface conditions and on the behavior of other structures in the vicinity of the proposed project. Much soils information can be obtained from the geological survey maps, soil survey maps, aerial photographs, and site inspections.
Exploratory investigations: The purpose of this stage is to obtain the stratigraphic profiles with some estimate of the engineering properties of the soils.
Detailed investigations: The purpose of these explorations is to obtain detailed soil profiles and proper samples to secure accurate information about critical strata from which design computations can be made.

The subsurface exploration and sampling for engineering purposes present similarities to the geological surveys during the first and second stages of investigation except perhaps for some differences in classification and terminology (see *Maps, Physical Properties*). These investigations, in general, draw on existing geological surveys. The main difference occurs during the last stage of investigation due to the special requirements of sampling for the determination of certain engineering properties to be used in a quantitative geotechnical design.

Classification of Samples

The samples obtained may be classified in accordance with condition or disturbance of the material. Since the engineering properties such as compressibility, shear strength, and permeability depend significantly on soil fabric, any disturbance of the original fabric prior to sampling will result in erroneous properties. Consequently, sample *disturbance* becomes a significant factor in soil sampling for engineering purposes. Soil samples may be classified roughly in three groups (Hvorslev, 1949).

Non-representative samples consist of a mixture of materials from various soil or rock layers or are samples from which some mineral constituents have been removed or exchanged by washing and sedimentations. These samples are also called *wash samples* because they often are washed or bailed out of a borehole. Such samples do not represent the material actually found at the bottom of the borehole and are unsuitable for positive identification of the material and for laboratory tests, but they often permit a preliminary classification and determinations of depths at which major changes in the subsurface strata occur.

Representative samples contain all the mineral constituents of the strata from which they are taken and have not been contaminated by material from other strata or by chemical changes, but the soil structure is seriously disturbed. These samples are suitable for general classification tests and positive identification of the material but not for major laboratory tests for determining the engineering properties.

Undisturbed samples may be defined as samples in which the material has been subjected to so little disturbance that it is suitable for all laboratory tests and for approximate determination of engineering properties in situ. The term is somewhat misleading since obtaining a truly undisturbed sample is impossible, but it is firmly established in engineering terminology. Soil samples may be subjected to disturbance before, during, and after the sampling. The disturbances to which soil samples may be subjected include changes in stress conditions, volume, water content, soil fabric, and chemistry and mixing and segregation of soil constituents. Although most of these disturbances can be avoided or reduced in extent by use of appropriate methods and equipment and careful work, the stresses acting on the sample in the ground will ultimately be reduced to atmospheric pressure, and the soil sample may be subject to some disturbance in terms of volume and fabric changes.

The required condition and the quantity or dimensions of samples to be obtained vary depending on the stages of the subsurface investigation and the various groups of laboratory tests. In general, for exploratory investigations representative samples with a volume of at least 100 cm^3, preferably 250 cm^3, should be obtained from each distinctive stratum; in thick and apparently uniform strata it is advisable to take such samples at intervals of not more than 1.5 m. In detailed investigations practically continuous samples should be obtained. Samples with a diameter of about 5 cm are normally satisfactory, but undisturbed samples of 7.5 cm are often obtained for more reliable test results. When undisturbed samples of very soft or cohesionless soils are too difficult to obtain, representative samples are taken.

Methods of subsurface exploration may be divided into three groups: indirect, semidirect, and direct. An outstanding review of various methods of soil exploration and a comprehensive bibliography are given by Hvorslev (1949). A more recent bibliography updates this work (Johnson and Gnaedinger, 1964). Indirect methods include geophysical methods and sounding methods, and samples are not obtained. Semidirect methods are common boring and drilling methods combined with intermittent sampling. Nonrepresentative samples such as wash samples are obtained in the course of the boring operations. Direct methods are boring and sampling methods that provide practically continuous representative or undisturbed samples and all accessible explorations such as test pits.

Borings

Borings comprise the most common procedure of subsurface exploration (see *Augers, Augering; Borehole Drilling*). Boreholes can be advanced in a number of ways, depending on the materials encountered and the exploration and sampling requirements. Boring methods, in general, include wash boring, percussion drilling, auger boring, rotary drilling, and coring. A borehole can also be advanced by continuous sampling. Boreholes are kept open by a variety of methods including the use of drilling mud, hollow-stem augers, and driven casings. Borings are made to determine stratigraphy, to take samples for laboratory testing, and to permit the access of equipment to test the characteristics of the soil in situ (field tests). Various methods of sampling may be used in con-

nection with a single method of boring and vice versa.

Sampling

Samples of soil close to the ground surface or in test pits and other accessible explorations may be obtained by excavating a chunk of soil by hand. *Chunk samples* usually are the best undisturbed samples obtainable. Samples are also obtained as a by-product of the boring process. Auger samples may be used to identify the soil strata; samples obtained by wash boring and bailing are nonrepresentative and of little value.

Many types of equipment have been designed to secure samples from deep boreholes, but not all types will be successful in every soil. The *split spoon*, or *split barrel*, is one of the most commonly used tools of soil sampling. It consists of a thick-walled steel tube split lengthwise. To the lower end is attached a cutting shoe; to the upper end is attached a connector to the drill rods and a check valve for venting. The standard size is 3.5–3.8 cm (1.4–1.5 in) ID (inside diameter) and 5 cm (2 in) OD (outside diameter), and the minimum length is 60 cm (24 in). Subsequent to drilling a borehole by augering, wash boring, or rotary drilling, split-barrel samplers attached to the end of the wash pipe or drill rod are driven into the bottom of the borehole with a 63.5 kg (140 lb) weight that is allowed 76 cm (30 in) free fall on a driving head. The sampler is first driven 15 cm (6 in) into the soil to ensure seating in virgin material. It is then driven an additional 30 cm (12 in) in 15 cm (6 in) increments, or until 100 blows have been applied. The number of blows for each 15 cm (6 in) is recorded. The standard penetration resistance, N, is the sum of the blows for the second and third increments. If it is not possible to obtain 30 cm (1 ft) of penetration, the log should indicate the number of blows and the fraction of the foot penetrated. In case more than 50 blows are required to achieve a penetration of less than 2.5 cm (1 in), refusal of the sampler has been achieved. After the sampler has penetrated to a depth of 45 cm (18 in) or to refusal, it is extracted and the sample is removed. Samples obtained by this method maintain the water content, composition, and stratification of the soil, although there may be appreciable disturbance of the structure. Therefore, the split-spoon samples are representative but disturbed samples.

The penetration resistance is an indication of the density of cohesionless soils and of the consistency of cohesive soils. Numerous empirical relations have been developed relating penetration resistance to engineering properties. However, these provide only indexes and are approximate. In the case of cohesionless soils, due to the difficulties involved in obtaining undisturbed samples, the standard penetration resistance becomes the most useful source of information in estimating the in situ density of such materials. The standard penetration test is the most widely used method and has a dual function of sampling and in situ testing that makes it possible to identify changes in the soil by two independent methods (Sowers and Sowers, 1970).

The simplest and most widely used deep undisturbed sampler is the *thin-wall,* or *shelby tube,* sampler. The sample disturbance has been found to be dependent on a number of factors such as soil displacement by the sampler, method of driving the sampler into the ground, friction on the inside of the sample tube, and handling and storing of samples. The displacement of the soil by the walls of the sampler is probably the most important source of disturbance. The area ratio, A_r, is a useful measure of the relative displacement and is expressed as

$$A_r = \frac{D_o^2 - D_s^2}{D_s^2} \times 100,$$

where D_o is the outside diameter of the sampler and D_s is the diameter of the sample. Good undisturbed sampling requires area ratios to be less than 10% (Hvorslev, 1949). The split-spoon samplers may have area ratios typically about 80–110% and will cause considerable disturbance in cohesionless and stiff and brittle soils. The thin-wall tube samplers are made of cold drawn steel tubing (the shelby tube); the most commonly used samplers are 5–7.5 cm (2–3 in) in diameter and 60–90 cm (24–36 in) in length. The area ratios for standard thin-wall samplers are about 10%. Tubes smaller than 5 cm (2 in) in diameter are usually considered unsatisfactory for obtaining undisturbed samples. The lower end of the thin-wall sampler is beveled to form a cutting edge; the upper end is fastened to a check valve. The sampling is accomplished by pushing the tube into the soil by a fast uninterrupted motion. On completion of the sampling operation, the sampler is withdrawn from the hole, and the tube is separated from the sampler head. The sample is sealed and shipped intact to the laboratory. The thin-wall sampler is used primarily for sampling soft and stiff cohesive soils.

In extremely soft soils, even thin-wall samplers may cause significant disturbance, and a large portion of the sample is frequently lost during extraction of the sampler. These difficulties can be reduced by closing the lower end of the thin-wall sampler with a piston that can be released or withdrawn when the actual sampling is to be started. The piston prevents shavings from the walls and bottom of the borehole from entering the sampler while it is being lowered. During sampling the tube is pushed ahead of the piston into the soil below.

The fixed piston prevents the soil from squeezing upward. If the sample tends to slide out of the tube, a slight vacuum is created between the sample and the piston, which helps hold it. There are *piston samplers* with retractable and freely floating pistons. The *Osterberg piston sampler* incorporates a hydraulic cylinder in the head of the sampler to actuate the advance of the thin-wall tube. Its advantages are that the sampler cannot be overpushed and that only one set of drill rods is required. Other variations of the piston sampler are in use (Lowe and Zacheo, 1975).

Core boring samplers or *core barrels* are advanced by the chopping action of the rotating barrel while it is being forced into the ground. Therefore, they combine drilling and sampling. The sample disturbance due to sampler displacement is minimized. The core barrels are primarily suited for use in stiff and dense soils and in rock.

Field Tests

A number of field tests have been developed to determine certain engineering properties of soils in place without taking a sample. These tests have the advantage of minimizing the disturbance caused by stress changes and other sampling distortions. The standard penetration test is one such field test. The others include the static cone penetrometer, field vane shear, borehole expansion (pressure meter), the dilatometer, borehold shear test, seismic surveys, California Bearing Ratio Test, field permeability (packer test), and large-scale tests such as plate bearing and pile load tests (In Situ '86, 1986).

Laboratory Testing and Evaluation

The objective of the last stage of investigation (detailed investigations) is to provide the quantitative data necessary for the engineering analysis and design. The appropriate engineering properties of the critical strata that were identified in exploratory investigations are evaluated using the results of laboratory tests conducted on undisturbed samples and the field test data. This information is then correlated stratum by stratum with the simpler identification tests, the penetration resistances, and the geophysical parameters. In any stratum there is ordinarily a range of values for any given property. It is not always safe to use the average values since the engineering behavior is usually controlled by the lowest values. If the range in test results is narrow, the average values can be used; otherwise, the lower values are given emphasis for design.

The subsurface investigation does not stop with the completion of the design; unforeseen conditions may exist and careful records of all soil, groundwater, and rock conditions should be kept during construction. If any conditions different than indicated from the earlier investigations are noted, the designer should be informed so that these deviations can be evaluated.

TUNCER B. EDIL

References

Hvorslev, M. J., 1949, *Subsurface Exploration and Sampling of Soils for Civil Engineering Purposes.* Vicksburg, Miss.: U.S. Army Engineers Waterways Experiment Station. (Reprinted by Engineering Foundation, New York.)

In Situ '86, 1986, Use of in situ tests in geotechnical engineering. *Am. Soc. Civil Engineers Geotechnical Spec. Publ. No. 6.*

Johnson, A. I., and J. P. Gnaedinger, 1964, Bibliography, soil exploration, *Am. Soc. Testing and Materials Spec. Tech. Pub. No. 351,* 137–155.

Lowe, J. III, and P. F. Zaccheo, 1975, Subsurface explorations and sampling, in H. F. Winterkorn and H. Y. Fang, eds., *Foundation Engineering Handbook.* New York: Van Nostrand Reinhold, 1–66.

Sowers, G. B., and G. F. Sowers, 1970, *Introductory Soil Mechanics and Foundations,* 3rd ed. New York: Macmillan.

Cross-references: *Augers, Augering; Borehole Drilling; Soil Fabric.* Vol. XIII: *Pipeline Corridor Evaluation; Soil Mechanics.*

SOIL SAMPLING, SURVEY—See PEDOGEOCHEMICAL PROSPECTING; SAMPLES, SAMPLING.

SOILS AND WEATHERED MATERIALS, FIELD METHODS AND SURVEY

The study of soils, particularly the investigation of soil morphology, genesis, and classification, belongs in the realm of pedology, a subdiscipline of the broader field of soil science (see Vol. XII, Pt. 1: *Soil Science*). Although soil scientists (pedologists) are specialists who study soils as natural objects at the surface of the Earth, geologists and other earth scientists also need sometimes to deal with soil systems. Geologists conducting field surveys, especially in the mineral exploration industries (see *Exploration Geochemistry; Exploration Geophysics; Soil Sampling*), often find soil covers to be a nuisance. Weathering mantles (see *Saprolite, Regolith, and Soil*) and soils can, however, provide useful information when properly identified and interpreted. Soils can, e.g., provide valuable sampling mediums as in pedogeochemical surveys where geochemists can often identify geochemical anomalies indic-

ative of mineralization at depth (Hawkes and Webb, 1962; Siegel, 1974). The residual (sedentary or in situ) parts of regoliths are frequently surface analogs of the underlying lithology and may facilitate the reconnaissance mapping of contrasting rock units. The gross lithologies (e.g., greenstone and whitestone belts) of many tropical and subtropical cratons are, e.g., frequently inferred from the colors of soil covers, the red hues being associated with the basic and ultramafic suites and yellow hues with acid rocks. Transported soil materials such as co-alluvium are also frequently indicative of their provenance and parent lithologies (e.g., Churchward, 1977), especially on the weathered cratons.

Concepts of Soil

Critical to any application of the soil imprint are the various concepts of soil. Because the term *soil* has different connotations, depending on the professional field in which it is being considered, it is advantageous to specify particular uses. Strzemski (1975), e.g., notes that there are two general ways to consider soils: (1) on the basis of the nature of soil properties and (2) on the basis of specified functions or uses of soil. According to Arnold (1983), the distinction of soil as a biochemically altered material involving pedogenesis versus a variable mixture of constituents is not widely appreciated. For convenience, Butler (1958) groups concepts of soil into three main categories: (1) the edaphic, (2) the pedologic, and (3) the geographic approaches.

In the *edaphic* approach, the main sphere of interest is restricted to soil conditions that are important to plant growth. The *pedologic* concept focuses on soils as independent natural bodies, each with a distinct morphology resulting from the integrated effect of climate and living matter, acting on parent materials, as conditioned by relief, over periods of time (Soil Survey Staff, 1951). According to this view, which is supported by the National Cooperative Soil Survey of the United States, soil is the unconsolidated mineral matter on the Earth's surface that differs from the material from which it has been derived in many physical, biological, chemical, and morphological properties and characteristics. This concept of soil is qualitative because it does not separate soil sensu stricto from nonsoil. Because it embraces pedogenically altered materials as well as those of minimal genetic expression, this concept of soil includes material bodies that were previously regarded as geological material (Soil Survey Staff, 1975).

The *geographic* approach, in contrast, considers the areal distribution of soils. One classic example of the geographic approach, as applied to the distribution of soils on a continental basis, is illustrated by Thorp's (1957) field study of Australian soils. A related but global approach is offered by Duchaufour (1978) in his ecological grouping of soils. Soil-mapping programs operate under the influence of the geographic (ecologic) concept and help provide simplified models of complex landscapes. Descriptions of soil distribution patterns and their interrelationships for specific geographic regions summarize valuable background information in the form of soil landscape units; see Hole (1976), Clayton et al. (1977), Duchaufour (1978); Aandahl (1982), CSIRO Staff (1983), and Hendricks (1985).

Still a different viewpoint is that of the engineer. In civil engineering, e.g., soil is any unconsolidated material found between the ground surface and consolidated bedrock. Thus, in the engineering sense, *soil* means deposited derivatives from the rock crust of the Earth in its natural undisturbed state or that which can be excavated and placed in an earthwork or that can be found in an earthwork already (Jumikis, 1962). To many engineers and some geologists, soil includes all regolith (see *Saprolite, Regolith, and Soil*) materials—e.g., unconsolidated accumulations of rock fragments and any organic matter that is intermixed or the entire vertical section down to consolidated rock (Peck et al., 1974).

Soil and the Field Geologist

Depending on the nature of a particular project, field geologists are often confronted with the dilemma of whether to attempt identification of specific kinds of soils or just to relegate the weathered occurrence to some general category such as residuum, alluvium, colluvium, soil cover, or some other descriptive term. If identification of specific kinds of soils is required, the field geologist might well be advised to seek the services of a soils specialist. In the United States the American Registry of Certified Professionals in Agronomy, Crops and Soils (ARCPACS) (677 South Segoe Road, Madison, WI 53711), provides national and regional lists of certified professional soil scientists (CPSS). The CPSS specializing in soil survey is qualified to make value judgments in the area of soil morphology, genesis, and classification. The field geologist is, however, by no means limited to guidance by a CPSS or other soils specialist because soil field manuals and classification systems (e.g., Macvicar, 1977; Mohr and van Baren, 1959; Northcote, 1960; Rozov and Ivanova, 1967; Soil Survey Staff, 1951, 1975; Stace et al., 1968) are available in most developed regions of the world. Useful summaries of the kinds of manuals and handbooks that are available for different environments may be found in Buol et al. (1980), Butler (1980), and Finkl (1982).

Identification of Soils and Soil Materials

Access to basic (national and international) soil classification systems and field manuals is essential to preparing for fieldwork. Familiarity with the potential range of possibilities for specific kinds of soils to occur within a particular geographic region sets the stage for field observation. The prepared observer, on the one hand, not only will understand and comprehend what he or she has been trained to look for but also will be able to extrapolate from one condition to another. Field geologists who are unfamiliar with theories and concepts of soils, on the other hand, will at the very least experience difficulty and frustration in their attempts to identify specific kinds of soils. Without proper training, many field workers will also become aggravated by their inability to articulate the nature of different kinds of soil materials. Because the latter case is a common field experience to be avoided, it is essential that concepts of soil, i.e., those that are to be applied in the survey—be clearly understood prior to the initiation of fieldwork.

Soils versus Sediments

The unique properties of soils, as opposed to sediments, bring to the field geologist new problems such as determining what technically constitutes a soil. The development of genetic horizons, which are distinct from depositional or inherited stratifications, is a basic criterion of soil formation. Soil-forming processes differentiate parent materials (rocks and loose sediments) into horizons, and the same kinds of horizons are produced repeatedly under the same genetic conditions (Valentine and Dalrymple, 1975). Soils, regardless of their age, are thus unique because they consist of horizons, i.e., pedogenetically altered materials. If this attribute of soils is regarded as normal, then the absence of soil horizons becomes significant because it signals the presence of weathered materials that are technically not soil.

Proof of Existence Proofs of existence are often taken for granted or ignored by many field workers. Failure to appraise critically those parts of sections that are called soils—i.e., reluctance to consider evidence for pedologic alteration as compared with geologic origin (Peterson, 1980)— is a distracting remission. *Soil* has unfortunately tended to become a catchall word for unknown features in geologic sections. Materials or layers believed to represent soil mantles or portions thereof require proof of existence, evidence of pedogenesis (Finkl, 1980b).

Discrete weathering units are usually detected by differentiating intra- and interlayer variations in soil properties. Proofs of the independent nature of weathering zones can be argued by reference to any of several concepts—i.e., the principle of random association (Walker, 1958, cited in Finkl, 1980b), principle of separate identity (Morrison, 1967), and principle of association (Butler, 1959). According to these principles, the separateness of a layer is established by showing that changes in the morphology or constitution of the substrate are not reflected in the overlay, and vice versa, and by demonstrating that vertical trends within superposed layers are independent of each other.

Lateral Continuity All sedimentary bodies have laterally bounding peripheries. Lateral termination may result from erosion so that the margin of a sedimentary layer no longer represents its original depositional limits, or it may pinch out, interdigitate (intercalate) with other deposits (Krumbein and Sloss, 1963). Lateral continuity is established in the field by following surface exposures or by tracing surface contacts between rock units. Lateral tracing of substrate sedimentary units requires analysis of samples from boreholes.

These same concepts, which provide a unified framework for correlation by stratigraphers and sedimentologists, can be applied to studies of soil mantles and weathering zones (Morrison, 1967; Retallack, 1977; Finkl, 1980b). Weathering is a pervasive process that affects all materials exposed at the Earth's surface. Biochemical weathering attacks fresh crystalline rock, sedimentary deposits, and even previously weathered materials (as in soil welding or pedogenic overprinting), transgressing both lithologic and pedologic boundaries (Finkl, 1980; Ruhe and Olson, 1980; Schaetzl and Sorensen, 1987). The imprints of soil development in transported (sedimentary) parent materials sometimes mask the lateral continuation of soil mantles. The same process in basement rocks becomes more apparent because the same weathering zone is seen passing indiscriminately from one rock type to the next. Changes in lithology, however, may alter the appearance of the weathering zone because the course of chemical alteration is modified by differences in mineralogy, texture, and structure. Such variations, including facies, must be taken into account when establishing lateral continuity.

Ascendancy and Descendancy According to the principle of ascendancy (Ruhe, 1969), a hillslope may be the same age or younger than the higher surface to which it ascends. The principle of descendancy states that a hillslope deposit is the same age as the alluvial fill to which it descends. Hillslopes are frequently covered by layers of sediments that have been subjected to pedogenesis. Thus, because hillslopes ascend to hill summits (or to higher-level surfaces), it is possible to determine the age of hillslope deposits and associated soils. Coupled with the axiom that a soil

may be no older than the material in which it developed, it is possible to date certain pedogenic events (at least in a relative sense) in stratified materials. One note of caution, however, pertains to landscapes where there is inversion of relief. Great care is required when ranking phases of weathering because the normal sequence may be reversed, with younger deposits and soils occurring on summits and broad divides while older deposits are associated with dissections (Finkl and Churchward, 1976; Finkl, 1984).

Description of Soils and Weathered Materials

Descriptions of soils and weathered materials should be as objective as possible—i.e., not supplemented by genetic interpretations. This objective, separation of genesis from morphology at the descriptive stage, is not easy to fulfill because the first breakdown of a soil profile requires identification of horizons and their genetic relationships (Taylor and Pohlen, 1970).

Field description of soil profiles requires locating boundaries between horizons and studying the profile as a whole before proceeding to the description of individual horizons by recording depth, color, texture, consistency, structure, and other relevant characteristics (see subsequent discussion). Although guidelines for the description of soils are prepared and followed by many national soil surveys, a large number of systems are based on the USDA's *Soil Survey Manual* (1951) or the FAO *Guidelines for Soil Description* (n.d.). Hodgson (1978) provides excellent summary reviews of various national and international systems for soil description, and readers should consult this work for access to the appropriate system in a particular region or for convenient comparison of different guidelines. Special-purpose soil surveys—i.e., those not prepared on a routine basis for use by agriculturalists—often require the application of proprietary guidelines (e.g., those prepared by mineral exploration companies; see *Exploration Geochemistry*) or other specialized procedures.

Soil Horizons It is not necessary to name the various horizons to describe a soil profile adequately. For many purposes, the field surveyor, in the case of an unknown or difficult profile, may label them simply as Horizon 1, 2, 3, etc. Profile descriptions are more useful, however, for pedogeochemical surveys and other uses as well if standard interpretive designations are employed. Soil profiles are made up of the master horizons O, A, B, C and others as summarized in Table 1. These master horizons are subdivided by various subordinate distinctions (Table 2) that feature important or unusual soil properties. The O horizon, e.g., is a master horizon of organic material above the surface of the mineral soil. The A ho-

TABLE 1. Designations for Master
Soil Horizons and Layers

Old System[a]	New System[b]
0	0
01	0i, 0e
02	0a, 0e
A	A
A1	A
A2	E
A3	AB or EB
AB	—
A and B	E/B
AC	AC
B	B
B1	BA or BE
B and A	B/E
B2	B or BW
B3	BC or CB
C	C
R	R

Source: After Gutherie and Witty, 1982.
[a]Based on Soil Survey Staff, 1962, *Supplement to Agriculture Handbook No. 18*. Washington, D.C.: U.S. Government Printing Office, 178–188.
[b]Derived from drafts issued by the Soil Conservation Service for the forthcoming revision of the *Soil Survey Manual* (1951).

TABLE 2. Subordinate Distinctions
within Master Soil Horizons

Old System	New System	Description
—	a	Highly decomposed organic matter
b	b	Buried soil horizon
cn	c	Concretions or nodules
—	e	Intermediately decomposed organic matter
f	f	Frozen soil
g	g	Strong gleying
h	h	Illuvial accumulation of organic matter
—	i	Slightly decomposed organic matter
ca	k	Accumulation of carbonates
m	m	Strong cementation
sa	n	Accumulation of sodium
—	o	Residual accumulation of sesquioxides
p	p	Plowing or other disturbance
si	q	Accumulation of silica
r	r	Weathered or soft rock
ir	s	Illuvial accumulation of sesquioxides
t	t	Accumulation of clay
—	v	Plinthite
—	w	Color or structural B
x	x	Fragipan character
cs	y	Accumulation of gypsum
sa	z	Accumulation of salts

Source: From Gutherie and Witty, 1982.

rizon is a master mineral horizon with maximum organic accumulation or a surface or subsurface horizon that is lighter in color than the underlying horizon. It has lost clay minerals, iron, and aluminum with consequent concentration of more resistant minerals. The B horizon is a master horizon of altered material that is more or less blocklike or prismlike in structure. It exhibits stronger colors that differ from those of the A horizon above or the C horizon below and is in addition characterized by an accumulation of clay, iron, and aluminum, with accessory organic matter. The master C horizon embraces the parent material and the parent rock.

Soil Color Color is one of the most obvious and easily determined soil characteristics. An indispensable characteristic of soil for its classification and study, soil color is significant because it is an indirect measure, a proxie, of other more important properties that are not so easily and accurately observed in the field. It is extremely useful for soil identification, especially when combined with other characteristics such as structure. The common coloring agent in the topsoil is organic matter; humus, the stablest fraction, is typically darker than less decomposed fractions. A gray color beneath the surface organic horizon, such as occurs in many Spodosols (Podzols) (see Buurman, 1984), usually indicates intense leaching—i.e., eluviation of soluble compounds (Fig. 1).

The color of skeleton quartz grains also provides useful information. In leached horizons from which iron, clay, and other materials have been removed, and particularly in the presence of acid humus, they are commonly bleached clean. By contrast, in illuvial horizons they are coated and dull.

Common coloring agents in the subsoil are iron oxides; in well-drained soils they range in color from red through brown to yellow, the differences being attributed in part to the degree of hydration. Some subsoil colors are not due to pedogenic iron compounds but reflect amounts inherited from the parent materials. Subsoil materials derived from basalts, e.g., are rich in iron and commonly red or strong brown (Fig. 2). Many youthful soils have the colors of the parent rock. These colors are often referred to as *lithochromic*.

Gray, reddish brown, and yellow mottles are nearly always present in imperfectly drained soils,

FIGURE 2. Duplex soil profile developed in basaltic parent materials near Balingup, southwestern Australia. Organic-rich brown-colored surface horizons are developed in transported materials that overlie older reddish-colored in situ basaltic residuum.

especially in horizons where the water table fluctuates. In the presence of organic matter, the proportion of gray increases with increasing wetness. In the low latitudes where laterites and lateritic soils are prevalent (McFarlane, 1976), thick mottled horizons marking the zone of fluctuating water table commonly overlie intensely bleached clays that lie in the zone of permanent water saturation (Fig. 3).

Soil profiles normally contain horizons that are uniform in color but they may also show distinct color patterns that reflect streaked, spotted, variegated, or mottled associations. Color patterns are three dimensional in soil horizons but are usually observed in two dimensions on exposed vertical surfaces. Mottling in soils is described by noting the color of the matrix and colors of the principal mottles and the pattern of mottling. Colors are best defined using standard Munsell notations and color names. Patterns of mottles are conveniently described according to contrast, size, and abundance (Fig. 4).

For uniform description of soil colors, it is necessary to compare them to standardized chip colors. Although several basic soil color charts are available, they all are based on the Munsell Color System. This system was devised by A. H. Munsell in 1905 for the systematic classification of colors. According to this system, all colors can be measured by three attributes: *hue*, which represents the dominant spectrum such as red, yellow,

FIGURE 1. Spodosol developed in Pamlico Sands on the Atlantic Coastal Ridge near Fort Lauderdale, Florida. Light-colored tongues of bleached (eluviated) materials extend down into the darker-colored B horizon, providing strong color contrasts in these sandy soils. (Rhodes W. Fairbridge pointing to deep eluviated tongues.)

FIGURE 3. Exposure of laterite profile in tin mine quarry near Greenbushes, southwestern Australia. Although developed in alluvium on the western margin of the Western Australian craton, the lateritic deep weathering displays prominent color zonation with a strongly mottled zone overlying a bleached basal kaolinite.

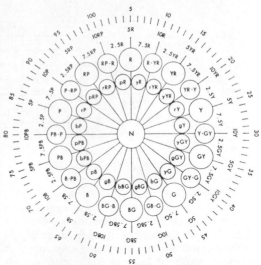

FIGURE 5. The partition of hue symbols in the Munsell Soil Color System and their relationship to one another. Hue is recorded by a combination of numerals 1 to 10 combined with initials of the major hues (red, yellow, blue, etc.).

green, blue; *value,* which represents the relative purity or strength of the spectral color. In the case of hue, five major colors (R, red; Y, yellow; G, green; B, blue; P, purple) and the respective complementary colors YR (yellow-red), GY (green-yellow), PB (purple-blue), and RP (red-purple) are arranged in a loop, and each of the colors is di-

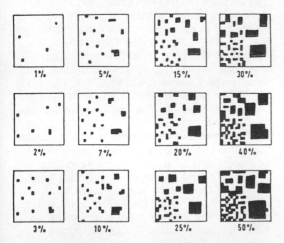

FIGURE 4. Chart for the visual estimation of percentage coverage by mottles, stones, nodules, etc. The abundance of mottles may be assessed in the field using this general chart, but in detailed examinations it may be necessary to include notes regarding the shape of mottles as well. *Note:* Each quarter of any one square has the same area of black. (After Folk, 1951)

vided by decimals from 0 to 10. Thus, the hues are divided into 100 units between R (red) and RP (red-purple) (Fig. 5). The notation of value consists of numbers from 0 for absolute white, with neutral, which has no depth of color. The colors between 0 and 10 are arranged so that they become successively lighter by visually equal steps and are expressed as 10/, 9/, etc. The chroma of each color gradually increases with increasing vividness. When the hue and value are systematically arranged, the chroma increases from 0 neutral gray to 1, 2, 3, etc. by equal steps and is noted as /1, /2, /3, etc. (Fig. 6).

When writing this notation of a color, the order is hue, value, chroma with a virgule (/) between the numbers for value and chroma. Thus, a color of 7.5R in hue, 6/ in value, and /2 in chroma is noted as 7.5R 6/2 (grayish red). The color notation using these three attributes, being a method in which color is expressed as a dot in the color space, may be referred to as a three-dimensional method, often called the *color solid* (Fig. 7).

Names of soil color are usually added to the Munsell notation in an effort to avoid careless mistakes in describing soil color. With brown as the standard, the more reddish color (5YR, 2.5 YR) is named "reddish brown," and those in which yellow is more dominant or that are lighter than brown are named "yellowish brown." If the value is higher than the one of brown, the color is "bright brown," or if it is slightly lower in chroma, it is called "dull brown." The color with

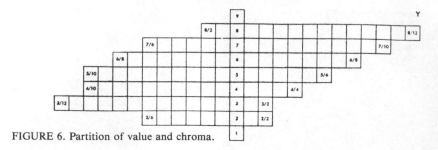

FIGURE 6. Partition of value and chroma.

a lower chroma is "gray brown"; the one with a lower value is "dark brown"; and one with an even lower value is called "blackish brown." Because there is an enormously wide range of soil colors the world over, it is often convenient to compare ranges of soil colors for different regions. One such comparison has been developed by the Japanese Research Council for Agriculture, Forestry, and Fisheries under the guidance of Oyama and Takehara (1970). This list of soil colors, as displayed in Table 3, provides comparisons of names for Munsell notations and is also now widely used in countries outside of Japan.

Soil Texture Texture, the particle size distribution of the solid inorganic constituents of the soil, refers specifically to particles less than 2 mm in diameter, the so-called fine earth fraction. The limits of the particle-size grades are not universal because many organizations employ different boundaries for particle-size classes (see Vol. XII, Pt. I: *Particle-Size Distribution*). The distribution of mineral particles of various sizes within the fine earth of a soil is probably the most important of its permanent properties and one that affects or

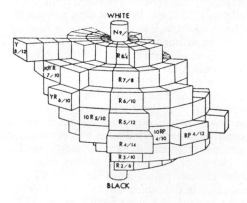

FIGURE 7. Color notation by the use of three attributes (hue, value, and chroma) in the Munsell Soil Color System. The circular band represents the hues in their proper sequence. The upright center axis is the scale for value. The paths radiating outward from the center show the steps of chroma that increase in strength from the center.

determines many others (Hodgson, 1978). The presence of larger-size particles or organic constituents is commonly recognized by qualifying the textural name. The particles are classified by diameter into four main fractions according to the International Scale as follows: (1) coarse sand (0.2–2 mm), (2) fine sand (0.02–0.2 mm), (3) silt (0.002–0.02 mm), and (4) clay (< 0.002 mm).

The term *soil texture* is adopted for descriptive classes based on relative proportions of the various particle-size grades within the fine earth. Because some workers use the term to include organic as well as mineral matter and because it is also used in the allied study of petrology in a different way (Brewer and Sleeman, 1960, 174), some researchers prefer to use the terms *particle-size distribution* and *particle-size class* rather than *texture*. For many users, the terms *texture* and *particle-size class* are synonymous, and both expressions are used interchangeably in this entry. There are various schemes of particle-size classification that attribute the fine earth fraction of the soil to one of a convenient number of groups, variously referred to as particle-size or textural classes (see Vol. XII, Pt. 1: *Particle-Size Distribution*). These are usually given names such as sand, sandy clay loam, silt loam, etc., as developed from traditional usage. Comparison of Figs. 8, 9, 10, 11, 12, and 13 (see pages 806–808), e.g., shows some ambiguity in the use of some textures among different organizations.

Because each of these fractions imparts particular physical qualities to the soil, experienced surveyors are able to assess the relative proportions of sand, silt, and clay to arrive at a close approximation of the texture as determined in the laboratory (see, e.g., Day, 1965). The sand fractions, e.g., are gritty whereas the silts are smooth, and most clays are plastic and sticky. Expertise in the estimation of field texture is best attained by handling soils where textures have already been determined by laboratory methods. In the field the soil is moistened to maximum plasticity and worked thoroughly with the fingers to eliminate

(*Text continues on page 805.*)

TABLE 3. Munsell Notations for Soil Colors and Color Names

Munsell No-tation		Munsell Color Name	Systematic Color Name
7.5R	7/1	—	grayish pink
7.5R	6/1	—	grayish brown
7.5R	6/2	—	grayish brown
7.5R	5/1	—	grayish brown
7.5R	5/2	—	grayish brown
7.5R	5/3	—	grayish brown
7.5R	4/1	—	grayish brown
7.5R	4/2	—	grayish brown
7.5R	4/3	—	grayish brown
7.5R	4/4	—	reddish brown
7.5R	4/6	—	reddish brown
7.5R	4/8	—	reddish brown
7.5R	3/1	—	dark grayish brown
7.5R	3/2	—	dark grayish brown
7.5R	3/3	—	dark grayish brown
7.5R	3/4	—	reddish brown
7.5R	3/6	—	reddish brown
7.5R	2/1	—	dark grayish brown
7.5R	2/2	—	dark grayish brown
7.5R	2/3	—	dark grayish brown
7.5R	1.7/1	—	dark grayish brown
10R	7/1	—	grayish pink
10R	6/1	reddish gray	grayish brown
10R	6/2	pale red	grayish brown
10R	6/3	pale red	grayish brown
10R	6/4	pale red	light reddish brown
10R	6/6	light red	light reddish brown
10R	6/8	light red	strong reddish orange
10R	5/1	reddish gray	grayish brown
10R	5/2	weak red	grayish brown
10R	5/3	weak red	grayish brown
10R	5/4	weak red	light reddish brown
10R	5/6	red	light reddish brown
10R	5/8	red	strong reddish orange
10R	4/1	dark reddish gray	grayish brown
10R	4/2	weak red	grayish brown
10R	4/3	weak red	grayish brown
10R	4/4	weak red	reddish brown
10R	4/6	red	reddish brown
10R	4/8	red	deep reddish orange
10R	3/1	dark reddish gray	dark grayish brown
10R	3/2	dusky red	dark grayish brown
10R	3/3	dusky red	dark grayish brown
10R	3/4	dusky red	reddish brown
10R	3/6	dark red	reddish brown
10R	2/1	reddish black	dark grayish brown
10R	2/2	very dusky red	dark grayish brown
10R	2/3	—	dark grayish brown
10R	1.7/1	—	brownish black
2.5YR	7/1	—	pinkish beige
2.5YR	7/2	—	pinkish beige
2.5YR	7/3	—	pinkish beige
2.5YR	7/4	—	dull orange
2.5YR	7/6	—	dull orange
2.5YR	7/8	—	light reddish orange
2.5YR	6/1	—	pinkish beige
2.5YR	6/2	pale red	pinkish beige
2.5YR	6/3	—	pinkish beige
2.5YR	6/4	light reddish brown	light brown

(continued)

TABLE 3. (*continued*)

Munsell Notation		Munsell Color Name	Systematic Color Name
2.5YR	6/6	light red	light brown
2.5YR	6/8	light red	strong reddish orange
2.5YR	5/1	—	grayish brown
2.5YR	5/2	weak red	grayish brown
2.5YR	5/3	—	grayish brown
2.5YR	5/4	reddish brown	light brown
2.5YR	5/6	red	light brown
2.5YR	5/8	red	strong reddish orange
2.5YR	4/1	—	grayish brown
2.5YR	4/2	weak red	grayish brown
2.5YR	4/3	—	grayish brown
2.5YR	4/4	reddish brown	brown
2.5YR	4/6	red	brown
2.5YR	4/8	red	deep reddish orange
2.5YR	3/1	—	dark grayish brown
2.5YR	3/2	dusky red	dark grayish brown
2.5YR	3/3	—	dark grayish brown
2.5YR	3/4	dark reddish brown	brown
2.5YR	3/6	dark red	brown
2.5YR	2/1	—	dark grayish brown
2.5YR	2/2	very dusky red	dark grayish brown
2.5YR	2/3	—	dark brown
2.5YR	2/4	dark reddish brown	dark brown
2.5YR	1.7/1	—	brownish black
5YR	8/1	white	pale pinkish beige
5YR	8/2	pinkish white	pale pinkish beige
5YR	8/3	pink	pale orange
5YR	8/4	pink	pale orange
5YR	7/1	light gray	pinkish beige
5YR	7/2	pinkish gray	pinkish beige
5YR	7/3	pink	pinkish beige
5YR	7/4	pink	dull orange
5YR	7/6	reddish yellow	dull orange
5YR	7/8	reddish yellow	light orange
5YR	6/1	gray	pinkish beige
5YR	6/2	pinkish gray	pinkish beige
5YR	6/3	light reddish brown	pinkish beige
5YR	6/4	light reddish brown	light brown
5YR	6/6	reddish yellow	light brown
5YR	6/8	reddish yellow	strong orange
5YR	5/1	gray	grayish brown
5YR	5/2	reddish gray	grayish brown
5YR	5/3	reddish brown	grayish brown
5YR	5/4	reddish brown	light brown
5YR	5/6	yellowish red	light brown
5YR	5/8	yellowish red	deep orange
5YR	4/1	dark gray	grayish brown
5YR	4/2	dark reddish gray	grayish brown
5YR	4/3	reddish brown	grayish brown
5YR	4/4	reddish brown	brown
5YR	4/6	yellowish red	brown
5YR	4/8	yellowish red	deep brown
5YR	3/1	very dark gray	dark grayish brown
5YR	3/2	dark reddish brown	dark grayish brown
5YR	3/3	dark reddish brown	dark grayish brown
5YR	3/4	dark reddish brown	brown
5YR	3/6	—	brown
5YR	2/1	black	dark grayish brown
5YR	2/2	dark reddish brown	dark grayish brown
5YR	2/3	—	dark brown

TABLE 3. (*continued*)

Munsell No-tation		Munsell Color Name	Systematic Color Name
5YR	2/4	—	dark brown
5YR	1.7/1	—	brownish black
7.5YR	8/1	—	pale beige
7.5YR	8/2	pinkish white	pale beige
7.5YR	8/3	—	pale orange
7.5YR	8/4	pink	pale orange
7.5YR	8/6	reddish yellow	pale orange
7.5YR	8/8	—	light orange
7.5YR	7/1	—	beige
7.5YR	7/2	pinkish gray	beige
7.5YR	7/3	—	beige
7.5YR	7/4	pink	dull orange
7.5YR	7/6	reddish yellow	dull orange
7.5YR	7/8	reddish yellow	light orange
7.5YR	6/1	—	beige
7.5YR	6/2	pinkish gray	beige
7.5YR	6/3	—	beige
7.5YR	6/4	light brown	light yellowish brown
7.5YR	6/6	reddish yellow	light yellowish brown
7.5YR	6/8	reddish yellow	strong orange
7.5YR	5/1	—	grayish brown
7.5YR	5/2	brown	grayish brown
7.5YR	5/3	—	grayish brown
7.5YR	5/4	brown	yellowish brown
7.5YR	5/6	strong brown	yellowish brown
7.5YR	5/8	strong brown	deep orange
7.5YR	4/1	—	grayish brown
7.5YR	4/2	dark brown	grayish brown
7.5YR	4/3	—	grayish brown
7.5YR	4/4	dark brown	yellowish brown
7.5YR	4/6	—	yellowish brown
7.5YR	3/1	—	dark grayish brown
7.5YR	3/2	dark brown	dark grayish brown
7.5YR	3/3	—	dark yellowish brown
7.5YR	3/4	—	dark yellowish brown
7.5YR	2/1	—	dark grayish brown
7.5YR	2/2	—	dark grayish brown
7.5YR	2/3	—	dark yellowish brown
7.5YR	1.7/1	—	brownish black
10YR	8/1	white	pale beige
10YR	8/2	white	pale beige
10YR	8/3	very pale brown	pale beige
10YR	8/4	very pale brown	dull orange yellow
10YR	8/6	yellow	dull orange yellow
10YR	8/8	yellow	strong orange yellow
10YR	7/1	light gray	beige
10YR	7/2	light gray	beige
10YR	7/3	very pale brown	beige
10YR	7/4	very pale brown	dull orange yellow
10YR	7/6	yellow	dull orange yellow
10YR	7/8	yellow	gold
10YR	6/1	light gray	beige
10YR	6/2	light brownish gray	beige
10YR	6/3	pale brown	beige
10YR	6/4	light yellowish brown	light yellowish brown
10YR	6/6	brownish yellow	light yellowish brown
10YR	6/8	brownish yellow	gold

(*continued*)

TABLE 3. (*continued*)

Munsell Notation		Munsell Color Name	Systematic Color Name
10YR	5/1	gray	grayish brown
10YR	5/2	grayish brown	grayish brown
10YR	5/3	brown	grayish brown
10YR	5/4	yellowish brown	yellowish brown
10YR	5/6	yellowish brown	yellowish brown
10YR	5/8	yellowish brown	brownish gold
10YR	4/1	dark gray	grayish brown
10YR	4/2	dark grayish brown	grayish brown
10YR	4/3	brown	grayish brown
10YR	4/4	dark yellowish brown	yellowish brown
10YR	4/6	—	yellowish brown
10YR	3/1	very dark gray	dark grayish brown
10YR	3/2	very dark grayish brown	dark grayish brown
10YR	3/3	dark brown	dark yellowish brown
10YR	3/4	dark yellowish brown	dark yellowish brown
10YR	2/1	black	dark grayish brown
10YR	2/2	very dark brown	dark grayish brown
10YR	2/3	—	dark yellowish brown
10YR	1.7/1	—	brownish black
2.5Y	8/1	—	grayish yellow
2.5Y	8/2	white	grayish yellow
2.5Y	8/3	—	grayish yellow
2.5Y	8/4	pale yellow	dull orange yellow
2.5Y	8/6	yellow	dull orange yellow
2.5Y	8/8	yellow	strong orange yellow
2.5Y	7/1	—	grayish yellow
2.5Y	7/2	light gray	grayish yellow
2.5Y	7/3	—	grayish yellow
2.5Y	7/4	pale yellow	dull orange yellow
2.5Y	7/6	yellow	dull orange yellow
2.5Y	7/8	yellow	gold
2.5Y	6/1	—	grayish yellow
2.5Y	6/2	light brownish gray	grayish yellow
2.5Y	6/3	—	grayish yellow
2.5Y	6/4	light yellowish brown	dark yellow
2.5Y	6/6	olive yellow	dark yellow
2.5Y	6/8	olive yellow	gold
2.5Y	5/1	—	grayish olive
2.5Y	5/2	grayish brown	grayish olive
2.5Y	5/3	—	grayish olive
2.5Y	5/4	light olive brown	brownish olive
2.5Y	5/6	light olive brown	brownish olive
2.5Y	4/1	—	grayish olive
2.5Y	4/2	dark grayish brown	grayish olive
2.5Y	4/3	—	grayish olive
2.5Y	4/4	olive brown	brownish olive
2.5Y	4/6	—	brownish olive
2.5Y	3/1	—	grayish olive
2.5Y	3/2	very dark grayish brown	grayish olive
2.5Y	3/3	—	dark brownish olive
2.5Y	2/1	—	dark grayish olive
5Y	8/1	white	grayish yellow
5Y	8/2	white	grayish yellow
5Y	8/3	pale yellow	grayish yellow
5Y	8/4	pale yellow	dull yellow
5Y	8/6	yellow	dull yellow
5Y	8/8	yellow	strong yellow
5Y	7/1	light gray	grayish yellow

TABLE 3. (*continued*)

Munsell No- tation		Munsell Color Name	Systematic Color Name
5Y	7/2	light gray	grayish yellow
5Y	7/3	pale yellow	grayish yellow
5Y	7/4	pale yellow	dull yellow
5Y	7/6	yellow	dull yellow
5Y	7/8	yellow	olive yellow
5Y	6/1	light gray	grayish yellow
5Y	6/2	light olive gray	grayish yellow
5Y	6/3	pale olive	grayish yellow
5Y	6/4	pale olive	dark yellow
5Y	6/6	olive yellow	dark yellow
5Y	6/8	olive yellow	olive yellow
5Y	5/1	gray	grayish olive
5Y	5/2	olive gray	grayish olive
5Y	5/3	olive	grayish olive
5Y	5/4	olive	olive
5Y	5/6	olive	olive
5Y	4/1	dark gray	grayish olive
5Y	4/2	olive gray	grayish olive
5Y	4/3	olive	grayish olive
5Y	4/4	olive	olive
5Y	3/1	very dark gray	grayish olive
5Y	3/2	dark olive gray	grayish olive
5Y	2/1	black	dark grayish olive
5Y	2/2	black	dark grayish olive
7.5Y	8/1	—	grayish yellow
7.5Y	8/2	—	grayish yellow
7.5Y	8/3	—	grayish yellow
7.5Y	7/1	—	grayish yellow
7.5Y	7/2	—	grayish yellow
7.5Y	7/3	—	grayish yellow
7.5Y	6/1	—	grayish yellow
7.5Y	6/2	—	grayish yellow
7.5Y	6/3	—	grayish yellow
7.5Y	5/1	—	grayish olive
7.5Y	5/2	—	grayish olive
7.5Y	5/3	—	grayish olive
7.5Y	4/1	—	grayish olive
7.5Y	4/2	—	grayish olive
7.5Y	4/3	—	grayish olive
7.5Y	3/1	—	grayish olive
7.5Y	3/2	—	grayish olive
7.5Y	2/1	—	dark grayish olive
7.5Y	2/2	—	grayish olive
10Y	8/1	—	grayish yellow
10Y	8/2	—	grayish yellow
10Y	7/1	—	grayish yellow
10Y	7/2	—	grayish yellow
10Y	6/1	—	grayish yellow
10Y	6/2	—	grayish yellow
10Y	5/1	—	grayish olive
10Y	5/2	—	grayish olive
10Y	4/1	—	grayish olive
10Y	4/2	—	grayish olive
10Y	3/1	—	grayish olive
10Y	3/2	—	grayish olive
10Y	2/1	—	dark grayish olive
N—8		white	light gray
N—7		light gray	light gray

(*continued*)

TABLE 3. (*continued*)

Munsell No-tation		Munsell Color Name	Systematic Color Name
N—6		gray	light medium gray
N—5		gray	medium gray
N—4		dark gray	dark medium gray
N—3		very dark gray	dark gray
N—2		black	grayish black
N—2		—	black
N—1.5		—	black
2.5GY	8/1	—	pale yellow green
2.5GY	7/1	—	grayish leaf
2.5GY	6/1	—	grayish leaf
2.5GY	5/1	—	grayish olive green
2.5GY	4/1	—	grayish olive green
2.5GY	3/1	—	grayish olive green
2.5GY	2/1	—	dark grayish olive green
5GY	8/1	—	pale yellow green
5GY	7/1	—	grayish leaf
5GY	6/1	—	grayish leaf
5GY	5/1	—	grayish olive green
5GY	4/1	—	grayish olive green
5GY	3/1	—	grayish olive green
5GY	2/1	—	dark grayish olive green
7.5GY	8/1	—	pale yellow green
7.5GY	7/1	—	grayish leaf
7.5GY	6/1	—	grayish leaf
7.5GY	5/1	—	grayish olive green
7.5GY	4/1	—	grayish olive green
7.5GY	3/1	—	grayish olive green
7.5GY	2/1	—	dark grayish olive green
10GY	8/1	—	pale yellow green
10GY	7/1	—	grayish leaf
10GY	6/1	—	grayish leaf
10GY	5/1	—	grayish green
10GY	4/1	—	grayish green
10GY	3/1	—	dark grayish green
10GY	2/1	—	dark grayish green
5G	7/1	—	light grayish green
5G	6/1	—	light grayish green
5G	5/1	—	grayish green
5G	4/1	—	grayish green
5G	3/1	—	dark grayish green
5G	2/1	—	dark grayish green
5G	1.7/1	—	greenish black
10G	7/1	—	light grayish green
10G	6/1	—	light grayish green
10G	5/1	—	grayish green
10G	4/1	—	grayish green
10G	3/1	—	dark grayish green
10G	2/1	—	dark grayish green
10G	1.7/1	—	greenish black
5BG	7/1	—	light grayish green
5BG	6/1	—	light grayish green
5BG	5/1	—	grayish green
5BG	4/1	—	grayish green
5BG	3/1	—	dark grayish green
5BG	2/1	—	dark grayish green
5BG	1.7/1	—	greenish black

TABLE 3. (*continued*)

Munsell Notation		Munsell Color Name	Systematic Color Name
10BG	7/1	—	light grayish green
10BG	6/1	—	light grayish green
10BG	5/1	—	grayish green
10BG	4/1	—	grayish green
10BG	3/1	—	dark grayish green
10BG	2/1	—	dark grayish green
10BG	1.7/1	—	greenish black
5B	7/1	—	grayish sky
5B	6/1	—	grayish blue
5B	5/1	—	grayish blue
5B	4/1	—	grayish blue
5B	3/1	—	dark grayish blue
5B	2/1	—	dark grayish blue
5B	1.7/1	—	bluish black
5PB	7/1	—	light bluish gray
5PB	6/1	—	bluish gray
5PB	5/1	—	bluish gray
5PB	4/1	—	dark bluish gray
5PB	3/1	—	dark bluish gray
5PB	2/1	—	bluish black
5PB	1.7/1	—	bluish black
5P	7/1	—	light purplish gray
5P	6/1	—	purplish gray
5P	5/1	—	purplish gray
5P	4/1	—	dark purplish gray
5P	3/1	—	dark purplish gray
5P	2/1	—	purplish black
5P	1.7/1	—	purplish black
5RP	7/1	—	light purplish gray
5RP	6/1	—	purplish gray
5RP	5/1	—	purplish gray
5RP	4/1	—	dark purplish gray
5RP	3/1	—	dark purplish gray
5RP	2/1	—	purplish black
5RP	1.7/1	—	purplish black
5R	7/1	—	grayish pink
5R	6/1	—	grayish red
5R	5/1	—	grayish red
5R	4/1	—	grayish red
5R	3/1	—	dark grayish red
5R	2/1	—	dark grayish red
5R	1.7/1	—	reddish black

Source: Based on soil color names from *Standard Soil Color Charts* (Oyama and Takehara, 1967)

effects due to aggregation. Field texture is estimated in two main ways. From a knowledge of the plasticity, stickiness, grittiness, and other properties of the textural classes, the texture is assessed directly. Estimates of the percentages of sand, silt, and clay may be obtained if required by reference to analyses of medial samples. The percentages of sand, silt, and clay may be alternatively estimated directly from feel and the tex-

tural class obtained by reference to textural diagrams (cf. Figs. 8 through 13).

Soil Consistency The terminology for evaluating soil consistency calls for separate terms at three moisture contents: dry, moist, and wet. The term *consistency* refers to the inherent qualities of soil materials that are expressed by the degree and kind of cohesion and adhesion or by resistance to deformation or rupture. As described in the *Soil*

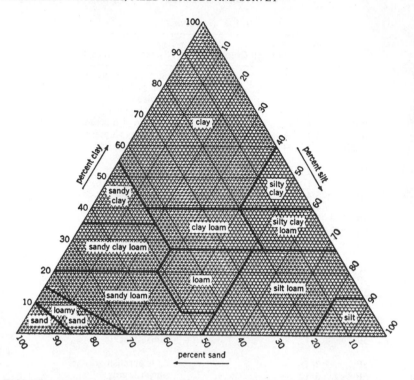

FIGURE 8. Texture triangle showing the basic particle-size classes of the U.S. Department of Agriculture for sand (0.05–2 mm), silt (0.002–0.05 mm), and clay (<0.002 mm). (From Soil Survey Staff, 1951)

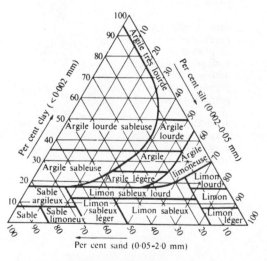

FIGURE 9. Particle-size classes used by the Belgian Soil Survey. Key: Sable (sand), Sable limoneux (loamy sand), sable argileux (clayey sand), limon sableux léger (light sandy loam), limon sableux lourd (heavy sandy loam), limon sableux (sandy silt), limon lourd (heavy silt), limon (silt), limon léger (light silt), argile limoneuse (silty clay), argile légère (light clay), argile sableuse (sandy clay), argile (clay), argile lourde (heavy clay), argile très lourde (very heavy clay), argile lourde sableuse (heavy sandy clay). (From Steffins, 1971)

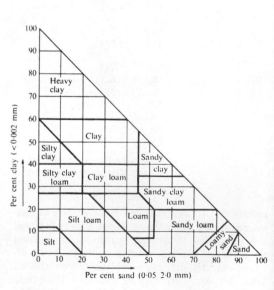

FIGURE 10. Particle-size classes used by the Canadian Soil Survey. (From Canada Department of Agriculture, 1970)

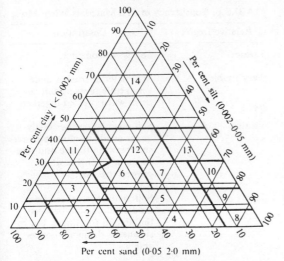

FIGURE 11. Particle-size classes used by L'Institut National de la Recherche Agronomique. Key (with standard abbreviations): (1) S, sable (sand); (2) SL, sable limoneux (silty sand); (3) SA, sable argileux (clayey sand); (4) LLS, limon léger sableux (light sandy silt); (5) LMS, limon moyen sableux (medium sandy silt); 6 LSA, limon sablo-argileux (sandy clayey silt); (7) LAS, limon argilo-sableux (clayey sandy silt); (8) LL, limon léger (light silt); (9) LM, limon moyen (medium silt); (10) LA limon argileux (clayey silt); (11) AS, argile sableuse (sandy clay); (12) A, argile (clay); (13) AL, argile limoneuse (silty clay); (14) ALO, argile lourde (heavy clay). (From Jamagne, 1967)

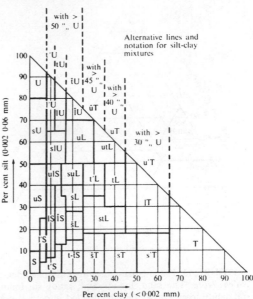

FIGURE 12. Particle-size classes used by the German Federal Republic. Key (with standard abbreviations): S, sand (sand); uS, schluffiger Sand (silty sand); lS, lehmiger Sand (loamy sand); tS, toniger bis stark tiniger Sand (clayey to very clayey sand); U, Schluff (silt); sU, sandiger Schluff (sandy silt); lU, lehmiger Schluff (loamy silt); sL, sandiger Lehm (sandy loam); uL, schluffiger Lehm (silty loam); tL, toniger Lehm (clayey loam); sT, sandiger Ton (sandy clay); lT, lehmiger Ton (loamy sand); T, Ton (clay); −, very; ', slightly. (From Kohl, 1971)

Survey Manual (Soil Survey Staff, 1951), the terms used in soil descriptions for consistency are as follows:

Consistency When Wet This is determined at or slightly above field capacity. Stickiness is the quality of adhesion to other objects. For field evaluation, the soil material is pressed between the thumb and finger and its adherence noted. Degrees of stickiness are described in Table 4.

Plasticity is the ability to change shape continuously under the influence of an applied stress and to retain the impressed shape on removal of the stress. For field determination of plasticity, roll the soil material between thumb and finger and observe whether a wire or thin rod of soil can be formed. The degree of resistance to deformation at or slightly above field capacity is expressed according to the usage provided in Table 5.

Consistency When Moist When determined at a moisture content approximately midway between air dry and field capacity, most soil materials exhibit a form of consistency characterized by a tendency to break into smaller masses rather than into powder, some deformation prior to rupture, an absence of brittleness, and an ability of the material after disturbance to cohere again

TABLE 4. Degrees of Stickiness for Wet Soil Materials

Relative Degree	Description
Nonsticky	After release of pressure, practically no soil material adheres to thumb or finger.
Slightly sticky	After pressure, soil material adheres to both thumb and finger but comes off one or the other rather cleanly. It is not appreciably stretched when the digits are separated.
Sticky	After pressure, soil material adheres to both thumb and finger and tends to stretch somewhat and pull apart rather than pulling free from either digit.
Very sticky	After pressure, soil material adheres strongly to both thumb and finger and is decidedly stretched when they are separated.

Source: After Soil Survey Staff, 1951.

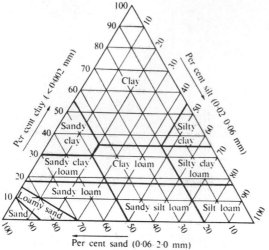

Per cent silt (<0·002 mm)

Per cent clay (<0·002 mm)

Per cent silt (0·02 0·06 mm)

Clay

Sandy clay

Silty clay

Sandy clay loam

Clay loam

Silty clay loam

Sandy loam

Loamy sand

Sand

Sandy silt loam

Silt loam

Per cent sand (0·06 2·0 mm)

FIGURE 13. Particle-size classes used in England and Wales. *Note:* The sand, loamy sand, sandy silt loam, and sandy loam classes may be divided into fine, medium, and coarse subclasses according to the size of the sand fraction, where fine = more than two-thirds of the sand fraction (0.6–2 mm) is between 0.06 mm and 0.2 mm; coarse = more than one-third of the sand fraction is larger than 0.6 mm; and medium = less than two-thirds of the sand fraction is between 0.06 mm and 2 mm and less than one-third of the sand fraction is larger than 0.6 mm. (After Hodgson, 1976)

TABLE 6. Consistency of Soil Materials When Moist

Relative Degree	Description
Loose	Soil is noncoherent.
Very friable	Soil material crushes under very gentle pressure but coheres when pressed together.
Friable	Soil material crushes easily under gentle to moderate pressure between thumb and forefinger and coheres when pressed together.
Firm	Soil material crushes under moderate pressure between thumb and forefinger, but resistance is distinctly noticeable.
Very firm	Soil material crushes under strong pressure but is barely crushable between thumb and forefinger.
Extremely firm	Soil material crushes only under very strong pressure, cannot be crushed between thumb and forefinger, and must be broken apart bit by bit.

Source: After Soil Survey Staff, 1951.

when pressed together. Because the resistance to rupture decreases with increasing moisture content, accuracy of field descriptions of this consistency is limited by the accuracy of estimating moisture content. To evaluate this consistency, select and attempt to crush in the hand a mass that appears slightly moist. Terms used to express the consistency of moist soil material are given in Table 6.

Consistency When Dry The consistency of dry soil materials is characterized by rigidity, brittle-

TABLE 5. Field Determination of Plasticity

Deformation under Applied Stress	Description
Nonplastic	No wire is formable.
Slightly plastic	Wire is formable; soil material is easily deformable.
Plastic	Wire is formable; moderate pressure is required for deformation of the soil material.
Very plastic	Wire is formable; much pressure is required for deformation of the soil material.

Source: After Soil Survey Staff, 1951.

ness, maximum resistance to pressure, more or less tendency to crush to a powder or to fragments with rather sharp edges, and inability of crushed material to cohere again when pressed together. To evaluate, select an air-dry mass and break in the hand. Terms used to describe consistency when dry are summarized in Table 7.

Soil Structure The structure of the soil is the manner in which primary soil particles are aggregated within it. Most methods for describing soil structure in the field follow procedures outlined in the *Soil Survey Manual* (Soil Survey Staff, 1951). This systematic method is based on the form and arrangement of the aggregates, the size of the aggregates, and the degree of development or grade of the structure. Even though natural structures such as peds (natural soil aggregates) may be extremely irregular, they can be referred to as four main types according to general form—viz., platelike, prismlike, blocklike, and spheroidal (Fig. 14). Platelike, prismlike, and blocklike are fitting structures; i.e., the sides of the peds are the casts of adjacent peds, indicating that when the soil is wet the swollen aggregates fit closely together. Platelike peds are usually arranged horizontally, whereas prismlike peds are arranged vertically. For unusual arrangements it

TABLE 7. Consistency of Dry Soil Materials

Relative Degree	Description
Loose	Soil material is noncoherent.
Soft	Soil mass is very weakly coherent and fragile, breaks to powder or individual grains under very slight pressure.
Slightly hard	Soil material is weakly resistant to pressure, easily broken between thumb and forefinger.
Hard	Soil material is moderately resistant to pressure; it can be broken in the hands without difficulty but is barely breakable between thumb and forefinger.
Very hard	Soil material is very resistant to pressure, can be broken in the hands only with difficulty, is not breakable between thumb and forefinger.
Extremely hard	Soil material is extremely resistant to pressure; it cannot be broken in the hands.

Source: After Soil Survey Staff, 1951.

is necessary to amplify the description— e.g., "vertical platelike structure." Spheroidal structures are nonfitting because their surfaces are not casts of those adjacent ones. They commonly consist of spheroids or polyhedrons having curved or plane surfaces. The primary types of soil structures and their size limits, which differ according to shape and arrangement, are given in Table 8.

The degree of development, or *grade,* of structure is the degree of aggregation and expresses the difference between the strength of the aggregate and the strength of the bonds between them. It varies with the moisture content of the soil and, unless otherwise stated, is described at the normal moisture content of the soil horizon. Grades of structure are given in Table 9.

The size of peds is determined in the field by reference to standard charts. These charts provide a range of sizes for structural types (peds) and are keyed to relative terms such as *fine, medium, coarse,* and *very coarse.* Ped size is determined by holding representative samples next to patterns on the chart. Examples of different size classes are provided here for angular and subangular blocky peds (Fig. 15), granular peds (Fig. 16), platy peds (Fig. 17), prismatic and columnar peds (Fig. 18), polyhedral peds (Fig. 19), and lenticular peds (Fig. 20).

Compound structures and mixed structures commonly occur in soils. Some soils, e.g., have a

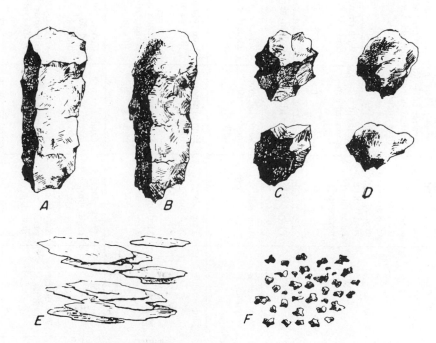

FIGURE 14. Line drawings illustrating some types of soil structures: (*A*) prismatic, (*B*) columnar, (*C*) angular blocky, (*D*) subangular blocky, (*E*) platy, and (*F*) granular. (From Soil Survey Staff, 1951).

TABLE 8. Types and Classes of Soil Structures

Class	TYPE (Shape and arrangement of peds)						
	Platelike with one dimension (the vertical) limited and greatly less than the other two; arranged around a horizontal plane; faces mostly horizontal.	Prismlike with two dimensions (the horizontal) limited and considerably less than the vertical; arranged around a vertical line; vertical faces well defined; vertices angular.		Blocklike; polyhedronlike, or spheroidal, with three dimensions of the same order of magnitude, arranged around a point.			
				Blocklike; blocks or polyhedrons having plane or curved surfaces that are casts of the molds formed by the faces of the surrounding peds.		Spheroids or polyhedrons having plane or curved surfaces which have slight or no accommodation to the faces of surrounding peds.	
		Without rounded caps	With rounded caps.	Faces flattened; most vertices sharply angular.	Mixed rounded and flattened faces with many rounded vertices.	Relatively nonporous peds.	Porous peds.
	Platy	Prismatic	Columnar	(Angular) Blocky[1]	Subangular blocky[2]	Granular	Crumb
Very fine or very thin.	Very thin platy; < 1 mm.	Very fine prismatic; <10 mm.	Very fine columnar; < 10 mm.	Very fine angular blocky; <5 mm.	Very fine subangular blocky; < 5 mm.	Very fine granular; <1 mm.	Very fine crumb; < 1 mm.
Fine or thin.	Thin platy; 1 to 2 mm.	Fine prismatic; 10 to 20 mm.	Fine columnar; 10 to 20 mm.	Fine angular blocky; 5 to 10 mm.	Fine subangular blocky; 5 to 10 mm.	Fine granular; 1 to 2 mm.	Fine crumb; 1 to 2 mm.
Medium	Medium platy; 2 to 5 mm.	Medium prismatic; 20 to 50 mm.	Medium columnar; 20 to 50 mm.	Medium angular blocky; 10 to 20 mm.	Medium subangular blocky; 10 to 20 mm.	Medium granular; 2 to 5 mm.	Medium crumb; 2 to 5 mm.
Coarse or thick.	Thick platy; 5 to 10 mm.	Coarse prismatic; 50 to 100 mm.	Coarse columnar; 50 to 100 mm.	Coarse angular blocky; 20 to 50 mm.	Coarse subangular blocky; 20 to 50 mm.	Coarse granular; 5 to 10 mm.	
Very coarse or very thick.	Very thick platy; >10 mm.	Very coarse prismatic; >100 mm.	Very coarse columnar; >100 mm.	Very coarse angular blocky; >50 mm.	Very coarse subangular blocky; > 50 mm.	Very coarse granular; > 10 mm.	

Source: From Soil Survey Staff, 1951.

[1](a) Sometimes called *nut.* (b) The word "angular" in the name can ordinarily be omitted.

[2]Sometimes called *nuciform, nut,* or *subangular nut.* Since the size connotation of these terms is a source of great confusion to many, they are not recommended.

TABLE 9. Grades of Soil Structures

Grade	Description
Structureless	Soil has no observable aggregation; is *massive* if coherent, *single grain* if noncoherent.
Weakly developed	Poorly formed indistinct peds are barely observable in place. When disturbed, soil material that has this grade of structure breaks into a mixture of few entire peds, many broken peds, and much unaggregated material.
Moderately developed	Many well-formed peds are evident, but not distinct, in undisturbed soil. Soil material of this grade, when disturbed, breaks down into a mixture of many distinct entire peds, some broken peds, and a little unaggregated material.
Strongly developed	Peds are quite evident in undisplaced soil, adhere only weakly to one another, and withstand displacement. When disturbed, soil material of this grade consists very largely of entire peds and includes few broken peds and little or no unaggregated material.

Source: After Soil Survey Staff, 1951.

Note: In field practice, grade of structure is determined mainly by noting the durability of aggregates and the proportions between aggregated and unaggregated material that result when the aggregates are displaced or gently crushed.

coarse prismatic structure, the prisms of which are composed of coarse, blocky aggregates. Such a compound structure may be recorded as "coarse prismatic breaking to coarse blocky." Morphological assessments of structure provide a good basis for the study of the genesis of structure, which involves processes that are complex and not well understood. Many fitting structures have their origin in the swelling and shrinking of the soil clay with alternate wetting and drying. Prismatic structure appears to be due to slow, prolonged shrinkage of the soil, with consequent formation of cracks perpendicular to the surface of drying. Some blocky structures may result from fairly rapid shrinkage of the soil clay on drying. Some platy structures may result from very rapid drying with consequent formation of cracks parallel to the surface of drying, from pressure, from freezing and thawing, and from both macro- and microsedimentation. Most granular and crumb

structures are a direct expression of the organic cycle, but some originate independently of organisms. Structureless horizons (massive or single grain) tend to form below the zone disturbed by widely fluctuating moisture conditions or by roots. Other structureless horizons are formed where the layer-silicate clays responsible for swelling and shrinkage have become disrupted and the residual colloidal particles are dispersed and do not form aggregates.

Soil Porosity Porosity is dependent on the shape, size, and abundance of the crevices, passages, and other soil cavities that are included under the general name of soil pores (see Vol. XII, Pt. 1: *Soil Pores*). Nikiforoff (1941) recognized three main kinds: textural, structural, and specific. *Textural porosity* refers to the voids between the primary particles such as those between grains of sand. *Structural porosity* refers to voids (cracks and fissures) between structural units. *Specific porosity* refers to voids produced by roots, insects, worms, and other animals and by bubbles

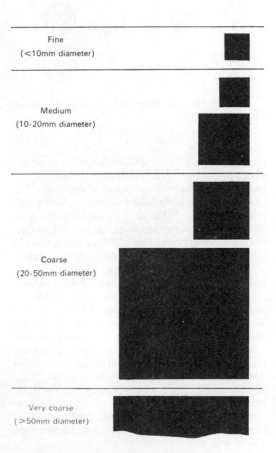

Fine
(<10mm diameter)

Medium
(10-20mm diameter)

Coarse
(20-50mm diameter)

Very coarse
(>50mm diameter)

FIGURE 15. Size classes of angular and subangular blocky peds. (After Hodgson, 1974)

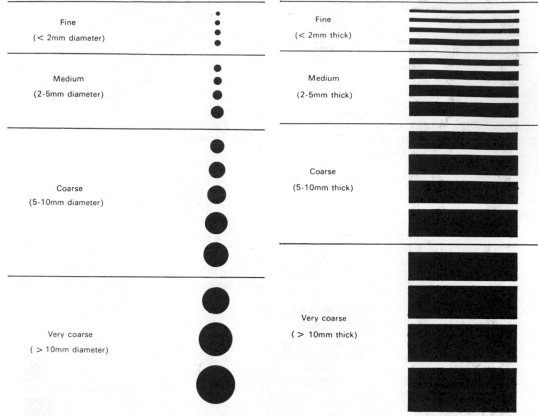

FIGURE 16. Size classes for granular peds. (After Hodgson, 1974)

Fine (< 2mm diameter)

Medium (2-5mm diameter)

Coarse (5-10mm diameter)

Very coarse (> 10mm diameter)

FIGURE 17. Size classes for platy peds. (From Hodgson, 1974)

Fine (< 2mm thick)

Medium (2-5mm thick)

Coarse (5-10mm thick)

Very coarse (> 10mm thick)

of air or other gases. Standard descriptions of soil pores employ six criteria: shape, distribution, orientation, continuity, size, and abundance. The shape of pores is classified according to the definitions given in Table 10. The size of pores is indicated by diameter classes as follows: micro (less than 0.075 mm), very fine (0.075-1 mm), fine (1-2 mm), medium (2-5 mm), coarse (over 5 mm).

A pore diameter of 0.075 mm theoretically corresponds with a moisture tension of 40 cm and is a useful boundary between pores that drain freely and those that do not (see Vol. XII, Pt. 1: *Soil Water*). Micropores are present in all soils but are difficult to observe without a microscope and normally are not mentioned in field descriptions. Pore diameter may be estimated by comparison with objects of known diameter; e.g., 40 gauge copper or nichrome wire is almost exactly 0.075 mm, 18 gauge wire (paper clip wire) and thin mechanical pencil leads are 1 mm, and lead of an ordinary wood pencil is about 2 mm in diameter. The size of macropores can also be conveniently estimated by comparison with standardized diagrams (see Fig. 19). The percentage of macro-

pores per unit area of the ped interior may be estimated using the accompanying diagrams shown in Fig. 19. Thus, in ordinary field descriptions, porosity may be expressed in terms of abundance,

TABLE 10. Classification of Soil Pores

Pore Shape	Description
Vesicular	Approximately spherical or ellipsoidal in shape, not appreciably elongated in any direction.
Interstitial	Irregular in shape, with faces that are curve inward, bounded by curved or angular surfaces of adjacent grains or peds or both.
Tubular	More or less cylindrical in shape—i.e., roughly circular in cross-section but greatly elongated along the third axis.

Source: After Soil Survey Staff, 1951.

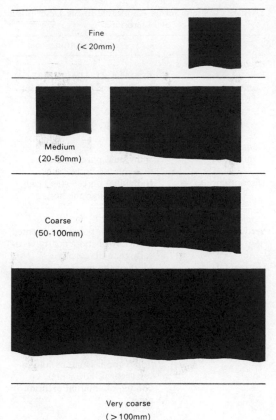

FIGURE 18. Size classes for prismatic and columnar peds. (From Hodgson, 1974)

size, and shape of the pores—e.g., "abundant fine vesicular pores" and "few coarse interstitial pores."

Because the volume of many interstitial soil pores changes with soil moisture, their significance is closely related to soil structure. Where peds fit together, the sides of each ped being molds of adjacent ones, the interstitial pores become smaller as the peds swell with increasing moisture and ultimately become virtually nonexistent. Where the peds do not fit together, it is evident that the size of interstitial pores is much less dependent on moisture content. Additional terms used in the field description of soil pores are briefly defined in Table 11.

Soil Organic Matter A large part of soil organic matter comes from the plant and animal material that decomposes at the surface and may ultimately become intimately incorporated with the soil. Another part comes from the plant and animal remains that decompose entirely within the lower horizons. In most soils, decomposition of organic matter takes place by a combination of aerobic and anaerobic microbiological activity,

with foci of each kind existing close together and the relative proportions changing with changing moisture conditions. As a rule, the wetter the site, the more anaerobic is the activity. Although there is much detailed information on the different kinds of organic matter distributed in the soil profile, there is a wide range of terminologies. Words with local European meanings have been redefined from time to time and may now have various meanings—viz., *mor, duff, moder, mull, raw humus, turf* (see Kubiena, 1953; Kononova, 1961).

The processes of soil formation from peat are not well understood, and in most places peaty soils are difficult to classify. Peats are divided broadly into sedentary peats and sedimentary or limnic peats (see Vol. VI: *Peat-Bog Deposits*). *Sedentary peat* is formed from plants growing in situ and is subdivided further according to the main kinds of vegetal remains. For the purposes of soil survey,

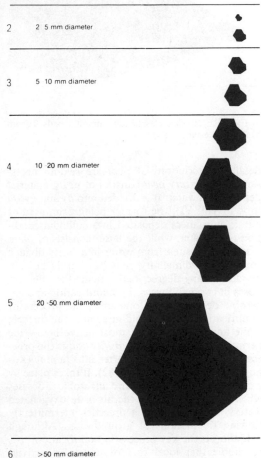

FIGURE 19. Size classes for polyhedral peds. (From McDonald et al., 1984)

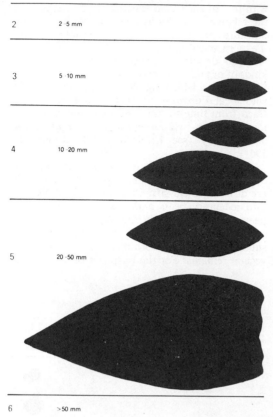

2   2-5 mm

3   5-10 mm

4   10-20 mm

5   20-50 mm

6   >50 mm

FIGURE 20. Size classes for lenticular peds. (From McDonald et al., 1984)

TABLE 11. Field Terms Applicable to the Description of Macropores

Term	Description
Abundance Classes	
Few	1–50/dm² (1–3/in² of surface)
Common	51–200/dm² (4–14/in² of surface)
Many	More than 200/dm² (more than 14/in² of surface)
Continuity Classes	
Continuous	Individual pores extend through the horizon.
Discontinuous	Individual pores extend only partly through the horizon.
Orientation Classes	
Vertical	Most of the pores are oriented vertically or more nearly vertically than diagonally.
Horizontal	Most of the pores are oriented horizontally or more nearly horizontally than diagonally.
Oblique	Most of the pores are oriented at the angle of 45° to the vertical or more nearly diagonally than horizontally or vertically.
Random	Pores are oriented in all directions, and it is impossible to say vertical, horizontal, or oblique orientation predominates.
Distribution within Horizons	
Inped	Most of the pores are within the ped.
Exped	Most of the pores are between ped surfaces—i.e., along the interface between adjacent peds.
Morphology of Individual Pores	
Vesicular	Approximately spherical or ellipsoidal in shape, not appreciably elongated in any direction.
Interstitial	Irregular in shape, with faces that are curved inward, bounded by curved or angular surfaces of adjacent mineral grains or peds or both.

Source: Modified from Johnson et al., 1960.

detailed descriptions of peat are seldom necessary. *Sedimentary peat* consists of peaty material deposited in water. It is divided into *dy* and *gyttja* (q.v. in Vol. VI), the former being composed of humic substances deposited in a colloidal condition from water, while the latter consists of plant material deposited from water in a finely divided condition. Sedimentary peat can hardly be subdivided to any degree in the field. Peat decomposes in two main ways, by humification and anaerobic decay. At any one time, the process of humification is most rapid at or near the surface, while anaerobic decay is most active below the permanent water table. *Humification* is the process of changing organic matter into humus (Kononova, 1961; Stevenson, 1982). It takes place by a combination of aerobic and anaerobic processes when peat is exposed to the air or to oxygenated waters. In many places it precedes intermittently flowing fluctuations of groundwater. Although oxidation is the key process in humification, it is inevitable that local foci of anaerobic activity should play a significant part in an environment so dominated by organic matter. *Anaerobic decay*

takes place in the absence of appreciable amounts of oxygen (Paul and McLaren, 1975). Where peat is submerged for substantial periods, it becomes soft and plastic without significant reduction in the carbon/nitrogen ratio. Hydrolysis and decomposition by anaerobic microorganisms cause the original cellulose to soften and the lignin to assume a yellow color (Alexander, 1977).

Sampling the Soil Profile

The soil profile is sampled in various ways according to the purpose for which the samples are required. Consequently, it is necessary to know the requirements of sampling before the samples are collected (see *Samples, Sampling*).

For rapid examinations, such as those to find soil boundaries during surveys, various types of soil augers are used. In ordinary use soil augers are 1–1.5 m long, with provisions for adding extra lengths for deeper boring. It is convenient to have a scale marked on the shaft of the auger to estimate the depth of sampling. The core or post-hole type of auger is more useful than the screw type. The core type is especially favored in dry compacted or loose sandy soils and the screw type for identifying differences in compaction between horizons and the presence of pans or gravelly horizons. The core type, which gives a larger and less modified sample, is not well suited for use in wet clay soils. Although soil augers are simple in design, considerable skill is required to use them effectively in making dependable observations of the soil profile. Consequently, their value depends to a considerable extent on the experience and skill of the surveyor who must record personal impressions of the samples.

Examinations of deep peat deposits are made with a special sampler such as the *Davis peat sampler*. This instrument consists of ten or more sections of steel rods, each about a meter long, and a cylinder of brass or duraluminum. The cylinder is provided with a plunger, cone shaped at the lower end and with a spring catch near the upper end. The sampler is pressed into the peat until the desired depth is reached for taking the sample; then the spring catch allows withdrawal of the plunger from its enclosing cylinder. With the plunger withdrawn and locked in that position, the cylinder may be filled with a solid core of organic material by a further downward movement.

Power augers, mounted on the rear of a truck or on a trailer, are used in special surveys, often to cut through cemented or very compact dry soils. Power augers of the core or screw type are especially useful in deeply weathered terrains when samples are required from depth. Many laterite profiles on the weathered cratons, e.g., commonly extend to depths of 30–100 m or more. Penetration of the thick duricrust (ferricrete cap) is usually achieved by locating the drill bit in a crevasse or old root cast. In truncated laterite profiles where the duricrust has been largely removed by erosion, deep sampling may be conducted on a regular grid pattern.

Exposures of various sorts provide excellent sites for sampling of soil and weathering profiles. Although fresh cuts are the ideal, older exposures can provide much useful information. Walls or faces of cuts must be cleaned to remove slough or overburden; the surveyor must take great care to assure that samples do not contain anthropogenic or disturbed materials. Railway and road cuts that expose complete profiles—i.e., the zone of weathering down to bedrock or initial material—may provide valuable information that can expedite tedious surveys. Natural exposures also provide useful sample sites, but they must be carefully examined to ensure only the sampling of in situ materials. Undistorbed samples for determinations of bulk density, porosity, percolation rate, soil strength, and other purposes may be collected with core samplers. For microbiological and biochemical studies, it is essential to take appropriate precautions against contamination and changes that may occur rapidly after sampling.

The collection of soil samples must be accompanied by appropriate written information that details the purpose for which the sample was taken and the kinds of analyses that are to be performed. Soil sample cards are commonly used by many organizations to ensure that the essential minimum information is supplied with the soil samples. The sampling location of each sample should be sufficiently detailed to enable the sampling spot to be found again, should the need arise. Most proper soil surveys require that the following features be noted in site descriptions: vegetation (native plant cover), topography (relief, slope, aspect, altitude, drainage), parent rock, land use, erosion, and other relevant information. This kind of information may be summarized in tabular form (Table 12), which encourages completion by the field party. A typical soil description card, as shown in Fig. 21, requires the recording of additional data to provide a complete record of extant field conditions for a particular site. Detailed surveys may require the compilation of more comprehensive records as provided, e.g., in the form of soil profile descriptions. Such descriptions, shown in Tables 13 and 14, provide detailed information on profile morphology and are the end products of the field description of soils. These representative profile descriptions also provide a basis for mapping and classification.

Soil Maps

Like other special-purpose maps, soil maps assume a wide range of forms. Soil maps may depict various sorts of soil units, at large and small scales, or specific attributes of soils as employed in engineering and environmental surveys (Simonson, 1974; see also *Maps, Environmental Geology*). Producing a soil map usually requires compiling information from many exposures, cores, and pits dug by the surveyor (Olsen, 1981). Data pertaining to soil color, texture, structure,

SOIL DESCRIPTION SHEET

Field No.		Rainfall:					
Lab. No.		Grid Ref.					
Landform	Location	Elevation	Slope and Terrain Class	Drainage/Water Table	Vegetation	Erosion	Parent Material

Horizons		Sample Depth & Number	Colour	Texture	Consistence	Structure	Channels, Roots, Casts Etc.	Clayskins (C) & Pores (P)	Identifiable Minerals	Mottles	Stones & Concretions	Boundary
Symbol	Thickness											

Special Features/Notes:

Tentative/Established Series		
Phase	Group	
Surveyor:	Date:	

FIGURE 21. A typical soil description sheet. (From Leamy and Panton, 1966, and based on data from the Soil Survey Staff, 1951)

TABLE 12. General Pro Forma for Recording Site Characteristics
Relevant to Soil Survey

Index Data and Soil Properties	Comments	
Index Data		
Profile Number:	Observer:	Date:
Grid reference:	Soil class:	
Latitude and longitude:		
Locality:		
Map or aerial photograph number:		
Site Characteristics		
Relief:		
Elevation		
Slope, aspect		
Regional relief		
Local relief		
Microtopography		
Climate and weather:		
Geomorphological development of site:		
Age of site:		
Soil erosion and deposition:		
Drainage:		
Flood hazard:		
Land use and vegetation:		
Geology of soil parent material:		
Rock outcrops:		
Fauna:		
Anthropogenic impacts on site:		
Soil Surface		
Stoniness:		
Surface form and condition, etc.:		
General Soil Information		
Soil moisture regime:		
Presence of salt or alkali:		
Pans, indurated layers, etc.:		
Unusual organic features (termiteria, burrows, holes, wallows, etc.):		

Source: Modified from Hodgson, 1978.

porosity, consistency, pH (soil reaction), organic matter, kind of clay, and other profile characteristics are combined with observations and knowledge of geology, geomorphology, topography, vegetation, land use, and other landscape attributes to map different soil areas on aerial photographs. Although the main base of the soil survey is the soil profile, most modern surveys use aerial photographs as the base map and as a means for rapidly extending soil units, by extrapolation, from one geographic area to another. Many soil survey reports, such as those prepared by the U.S. Soil Conservation Service, contain colored maps of small-scale soil units (usually soil associations) and planisiacs at a larger scale where boundaries of soil units are shown directly on the areal photograph. Such systems are cost effective since they do not require the preparation of many colored map sheets and, in addition, provide quantitative information that may be gleaned from the aerial photographs.

Soil maps show reference information in appropriate legends, as needed by the reader, and often include basal topographic information. For identification purposes, the soil map has a title and date, the name of the accompanying publication or the reference number, and suitable acknowledgments to the surveyors and others responsible for the information that is shown. For orientation and location, the map gives clearly the direction of north and the scale of the map (areal and linear) and contains small-scale diagrams showing the position of the map in relation to relevant political units and the position of each sheet in relation to others. Detailed, step-by-step instructions for the preparation of soil maps are provided by the Soil Survey Staff (1951) of the U.S. Department of Agriculture and more re-

TABLE 13. Soil Profile Description for a Polygenetic Soil Developed on Valley Side Slopes of Dissected Topography on the West Australian Craton, 135 km South of Perth

Sample	Morphology		Description
	Horizon	Depth (cm)	

Site Description

Waratan soil-profile section at Forests Department Grimwade 80 Sheet, FZ:58.2 F.D. Grid Reference, Lewana Plantation, Section C, 1 m west of Radiata Road in deep cutting. *Vegetation:* Pasture. *Parent Rock:* Aluminous gneiss, *Physiography:* Bridgetown Dissection, Taylor geomorphic surface. *Slope/Aspect:* 18°/50°. *Elevation:* 270 m. *Profile Drainage:* Somewhat poorly to moderately well drained. *Sampled and Described by:* Charles W. Finkl, Jnr., January 1970.

Sample	Horizon	Depth (cm)	Description
P578.1–2	A1	0–17	Black (10YR 2/1M), very dark gray brown (10YR 3/2D) sandy loam; moderate medium angular blocky breaking to fine and very fine granular structure; nonsticky and nonplastic, friable, loose; many roots; abundant macropores; few ironstone gravels; abrupt smooth boundary; 16–17 cm thick.
P578.3–4	B1	17–37	Dark reddish brown (5YR 3/3M), reddish brown (5YR 4/3D), dark reddish grey (5YR 4/2R) sandy loam; massive structure; sticky and plastic, very friable, firm to hard; many roots; abundant macropores; few stones; clear smooth boundary; 18–24 cm thick.
P578.5a	B2	37–67	Dark reddish brown (5YR 3/4M), reddish brown (5YR 4/4D) sandy clay loam; massive structure; sticky and plastic, firm, hard; stone line; many roots; many macropores; with abundant quartz and rock fragments; clear wavy boundary; 27–33 cm thick.
		Stone line	
P578.5b	IIB$_{2b}$	67–80	Yellowish red (5YR 4/8M), yellowish red (5YR 5/8D) mixed with dark reddish brown (5YR 3/4M), yellowish red (5YR 4/6D), yellowish red (5YR 5/6R) mottled clay loam; strong medium angular blocky structure; sticky and plastic, very firm, hard; many roots; common macropores; many pedotubules; gradual smooth boundary; 10–19 cm thick.
P578.5c	IIB$_{3b}$	80–90	Reddish brown (5YR 4/4M), yellowish red (5YR 4/6D) mixed with strong brown (7.5YR 5/8M), reddish yellow (7YR 6/8D), reddish yellow (5YR 6/8R) clay; strong medium angular blocky breaking to strong fine granular structure shattering to powder, sticky and plastic, very firm, hard; common roots; few macropores; gradual smooth boundary; 10–15 cm thick.
P578.6	IIC$_{1b}$	90–125	Yellowish brown (10YR 5/6M), brownish yellow (10YR 6/8D) mixed with reddish brown (5YR 4/4M,D) sandy clay loam; moderate medium subangular blocky structure; sticky and plastic, hard; few roots; few macropores; few pedotubules; few mica flakes, clear smooth boundary; 33–35 cm thick.
P578.7	IIC$_{2b}$	125–175	Yellowish brown (10YR 5/8M) mixed with reddish brown (5YR 4/4M,D) sandy loam; moderate thin platy or flaky structure; nonsticky and slightly plastic, hard; gradual smooth boundary; 48–52 cm thick.
P578.8	IIC$_{3b}$	175–225	Dark reddish brown (5YR 3/2M), yellowish red (5YR 5.6D) very fine sand; nonsticky and nonplastic, very firm, very hard; grading to weathered rock.

TABLE 14. Soil Profile Description for a Monogenetic Soil Developed on Interfluves
of the Appalachian Piedmont in Fauquier County, Virginia, near Warrenton

Morphology		
Horizon	Depth (in)	Description

Site Description
Bucks silt loam on Spring Hill Estate (7 mi SE of Warrenton) about 0.1 mi west of Rogues Road (SR602) about 70 ft north of Gupton Barn gravel road. *Vegetation:* Mixed hay pasture: white clover, timothy, orchardgrass. *Parent Rock:* Triassic red shale and diabase. *Physiography:* Piedmont divides, 2% slope. *Profile Drainage:* Runoff, rapid; internal drainage, slow to medium; permeability: moderately well drained. *Sampled and Described by:* Charles W. Finkl, Jnr., July 1965.

Horizon	Depth (in)	Description
Ap1	0–4	Brown (19YR 4/3) silt loam, yellowish brown (10YR 5/4) dry; moderate very fine subangular blocky breaking to weak very fine granular structure; slightly hard; friable, slightly plastic; abundant roots; few macropores; neutral (pH 6.8–7); few small rounded quartz pebbles throughout; clear smooth boundary; 3–6 in thick.
Ap2	4–8	Brown (10YR 4/3) silty clay loam, yellowish brown (10YR 5/4) dry; strong medium subangular blocky breaking to weak very fine granular; hard; friable, slightly sticky, slightly plastic; many roots; few pores; neutral (pH 6.8); few very fine strong brown concretions (7.5YR 5/8); abrupt smooth boundary; 4–5 in thick.
B₁₁	8–11	Dark brown (7.5YR 4/4) silty clay, light yellowish brown (10YR 6/4) dry; strong medium subangular blocky breaking to weak very fine granular structure; slightly hard; firm to friable, slightly sticky, slightly plastic; many roots; few pores; neutral (pH 6.6); purplish cast throughout; brown (7.5YR 5/4) concretions; mica particles; black concretions; abrupt smooth boundary; 3–4 in thick.
B₁₂	11–20	Dark yellowish brown (10YR 4/4) gritty silty clay, light yellowish brown (10YR 6/4) outside ped and strong brown (7.5YR 5/6) inside ped; strong medium subangular blocky; firm; friable, slightly sticky, slightly plastic; few roots; few pores; neutral (pH 6.6); black concretions; yellowish striations; mica particles throughout; clear smooth boundary; 9–10 in thick.
B₂₁	20–26	Two colors strongly intermixed: yellowish brown (10YR 5/6), brownish yellow (10YR 6/6), dry and strong brown when mixed (7.5YR 5/6) silt loam; strong coarse subangular blocky breaking to strong fine subangular blocky structure; firm; friable, nonsticky, nonplastic; mica particles; medium acid (pH 6); clear smooth boundary; 5–7 in.
B₂₂	26–32	Strong brown (7.5YR 5/6 to 5/8) silt loam; strong coarse subangular blocky breaking to medium subangular blocky and sometimes breaking to moderate medium platy structure; friable, nonsticky; strongly acid (pH 5.5); bedding planes evident with reddish and yellowish colors in alternating layers and sometimes black mottles between planes; purplish cast throughout; mica particles; well-decomposed shale in evidence; concretions of yellowish brown (10YR 5/6) and brownish yellow (10YR 6/6); strongly acid (pH 5.5); clear wavy boundary; 5–6 in thick.
B₃	32–50	Strong brown (7.5YR 5/8) silty clay loam; weak coarse platy breaking to weak medium and fine subangular blocky and sometimes breaking to weak medium platy structure (peds will break equally well with or against bedding planes, thus giving the two types of structures); friable, slightly sticky, slightly plastic; strongly acid (pH 5.2); purplish cast; mica particles; black mottling between structure planes; gradual smooth boundary; 17–19 in thick.
C₁	50–	Yellowish brown (10YR 5/6) silty clay loam; black mottling between layers in peds; mica particles; strongly acid (pH 5.2) (made with Oakfield Probe).

cently in a handbook prepared by the German Geological Survey (Kohl, 1971).

Although mapping conventions differ widely from one soil survey to another, most tend to follow traditional guidelines or those mandated by governmental departments. By way of one example, soil boundaries are variously shown as solid, dashed, or dotted lines. Boundaries of soil units that merge gradually into one another through broad transitions or that are not clearly defined are normally shown by dotted or dashed lines. This cartographic representation distinguishes transitional zones from ordinary soil boundaries that indicate narrow transitions shown by solid lines. All these symbols are shown in a map reference together with other relevant symbols. Soil units are commonly identified by symbols, either letters or numbers (see *Map Symbols*).

Separate colors are frequently allotted to general soil units that are closely similar; symbols provide the means for identifying individual soil units. The New Zealand Soil Survey, e.g., uses colors as a guide to the identification of soil areas—viz., green (brown-gray and yellow-gray earths), brown (yellow-brown soils), red (yellow-brown loams), purple (brown granular clays), blue (red-brown loams), gray-brown (Podzols), yellow (recent soils), and orange (gley and organic soils). Lighter shades of these colors are allocated to the less leached or less developed soils of each group and the darker shades to the more leached and better developed soils. In this very broad way sequences within main groups are indicated as well as the main group. Other soil surveys follow different conventions useful in coloring soil maps and maps in general (see *Map Colors, Coloring*), but the trend is toward the production of a single general small-scale color map and several planisiacs that accompany each report.

CHARLES W. FINKL, JNR.

References

Aandahl, A. R., 1982, *Soils of the Great Plains.* Lincoln: University of Nebraska Press, 282p.

Alexander, M., 1977, *Introduction to Soil Microbiology.* New York: Wiley, 467p.

Arnold, R. W., 1983, Concepts of soils and pedology, in L. P. Wilding, N. E. Smeck, and G. F. Hall, eds., *Pedogenesis and Soil Taxonomy,* Vol. 1. Amsterdam: Elsevier, 1–21.

Brewer, R., and J. R. Sleeman, 1960, Soil structure and fabric: Their definition and description, *Jour. Soil Sci.* **11,** 172–185.

Buol, S. W., F. D. Hole, and R. J. McCracken, 1980, *Soil Genesis and Classification.* Ames: Iowa State University Press, 404p.

Butler, B. E., 1958, The diversity of concepts about soils, *Australia Inst. Agric. Sci. Jour.* **24,** 14–20.

Butler, B. E., 1959, Periodic phenomena in landscapes as a basis for soil structures, *CSIRO (Australia) Soil Publ. No. 14.*

Butler, B. E., 1980, *Soil Classification for Soil Survey.* Oxford: Clarendon, 129p.

Buurman, P., 1984, *Podzols.* New York: Van Nostrand Reinhold, 450p.

Canada Department of Agriculture, 1970, *The System of Soil Classification.* Ottawa: Minister of Supply and Services.

Churchward, H. M., 1977, Landforms, regoliths and soils of the Sandstone–Mt. Keith area, Western Australia, *CSIRO (Australia) Land Resources Management Series No. 2.*

Clayton, J. S., W. A. Ehrlich, D. B. Cann, J. H. Day, and I. B. Marshall, 1977, *Soils of Canada,* Vol. 1: *Soil Report.* Ottawa: Canada Department of Agriculture, 243p.

CSIRO Staff, 1983, *Soils: An Australian Viewpoint.* Melbourne: Academic, 928p.

Day, P. R., 1965, Particle fractionation and particle-size analysis, in C. E. Black, ed., *Methods of Soil Analysis, Part 1* (Am. Soc. Agron. Mon. No. 9), 545–567.

Duchaufour, P., 1978, *Ecological Atlas of Soils of the World* (G. R. Mehuys, C. R. de Kimpe, and Y. A. Martel, trans.). New York: Masson, 178p.

FAO, n.d., *Guidelines for Soil Profile Description.* Rome: Multigrafica, 53p. (Soil Survey and Fertility Branch, Land and Water Development Division, FAO).

Finkl, C. W., Jnr., 1980a, The geography of soil classification, *Quaestiones Geographicae,* **8,** 55–59.

Finkl, C. W., Jnr., 1980b, Stratigraphic principles and practices as related to soil mantles, *Catena* **7,** 169–194.

Finkl, C. W., Jnr., ed., 1982, *Soil Classification.* New York: Van Nostrand Reinhold, 391p.

Finkl, C. W., Jnr., 1984. Chronology of weathered materials and soil age determination in pedostratigraphic sequences, *Chem. Geology* **44,** 311–335.

Finkl, C. W., Jnr., and H. M. Churchward, 1976, Soil stratigraphy in deeply weathered shield landscape in southwestern Australia, *Australian Jour. Soil Research* **14**(2), 109–120.

Folk, R. L., 1951, A comparison chart for visual percentage estimation, *Jour. Sed. Petrology* **21,** 32–33.

Gutherie, R. L., and J. E. Witty, 1982, New designation for soil horizons and layers in the new *Soil Survey Manual, Soil Sci. Soc. America Jour.* **46**(2), 443–444.

Hawkes, H. E., and J. S. Webb, 1962, *Geochemistry in Mineral Exploration.* New York: Harper & Row, 415p.

Hendricks, D. M., 1985, *Arizona Soils.* Tucson: University of Arizona Press, 244p.

Hodgson, J. M., ed., 1974, *Soil Survey Field Handbook.* Harpenden: Rothamsted Experimental Station, Lawes Agricultural Trust, 99p.

Hodgson, J. M., 1978, *Soil Sampling and Soil Description.* Oxford: Oxford University Press, 241p.

Hole, F. D., 1976, *Soils of Wisconsin.* Madison: University of Wisconsin Press, 223p.

Jamagne, M., 1967, Bases et techniques d'une cartographie des sols, *Annales Agronomy* **18,** 1–142.

Jumikis, A. R., 1962, *Soil Mechanics.* New York: Van Nostrand, 791p.

Kohl, F., 1971, *Kartieranleitung: Anleitung und Richtlinen zür Hertellung der Bodenkarte 1:25,000.* Hanover: Bundesanstalt fur Bodenforschung und den Geologischen Landesamten der Bundesrepublik Deutschland.

Kononova, M. M., 1961, *Soil Organic Matter* (T. Z. Nowakowski and G. A. Greenwood, trans.). New York: Pergamon, 450p.

Krumbein, W. C., and L. L. Sloss, 1963, *Stratigraphy and Sedimentation*. San Francisco: W. H. Freeman, 660p.

Kubiena, W. L., 1953, *The Soils of Europe*. London: Murby, 317p.

Leamy, M. L., and W. P. Panton, 1966, Soil survey manual for Malayan conditions, *Division of Agriculture Bull. No. 119* (Ministry of Agriculture and Co-operatives Malaysia).

Macvicar, C. N., 1977, *Soil Classification*. Pretoria: Department of Agricultural Technical Services, 150p.

Mcdonald, R. C., R. F. Isbell, J. G. Speight, J. Walker, and M. S. Hopkins, 1984, *Australian Soil and Land Survey: Field Handbook*. Melbourne: Inkata, 160p.

McFarlane, M. J., 1976, *Laterite and Landscape*. London: Academic, 151p.

Mohr, E. C. J., and F. A. van Baren, 1959, *Tropical Soils*. The Hague: Uitgeverij W. van Hoeve, 498p.

Morrison, R. B., 1967, Principles of Quaternary soil stratigraphy, R. B. Morrison and H. E. Wright, eds., *Quaternary Soils*. Reno, Nevada: Center for Water Resources Research, Desert Research Institute, 1–69.

Nikiforoff, C. C., 1941, Morphological classification of soil structure, *Soil Sci.* **52**, 193–207.

Northcote, K. H., 1960, A factual key for the recognition of Australian soils, *CSIRO Div. Soils, Div. Rept. 4/60*.

Olsen, G. W., 1981, *Soils and the Environment*. New York: Chapman and Hall, 178p.

Oyama, M., and H. Takehara, 1970, *Standard Soil Color Charts*. Tokyo: Ministry of Agriculture and Forestry, 13p. and color charts.

Paul, E. A., and A. D. McLaren, eds., 1975, *Soil Biochemistry*. New York: Marcel Dekker, 277p.

Peck, R. B., W. E. Hanson, and T. H. Thornburn, 1974, *Foundation Engineering*. New York: Wiley, 514p.

Peterson, F., 1980, Holocene desert soil formation under sodium salt influence in a playa-margin environment, *Quat. Research* **13**, 172–186.

Retallack, G. J., 1977, Triassic paleosols in the upper Narrabeen Group of New South Wales, Part II. Classification and reconstruction, *Geol. Soc. Australia Jour.* **24**(1), 19–36.

Rozov, N. N., and E. N. Ivanova, 1967, Classification of soils in the USSR, *Soviet Soil Sci.* **2**, 147–156.

Ruhe, R. V., 1969, Principles for dating pedogenic events in the Quaternary, *Soil Sci.* **107**(6), 398–402.

Ruhe, R. V., and C. G. Olsen, 1980, Soil welding, *Soil Sci.* **130**, 132–139.

Schaetzl, R. J., and C. J. Sorensen, 1987, The concept of "buried" versus "isolated" paleosols: Examples from northeastern Kansas, *Soil Sci.* **143**(6), 426–435.

Siegel, F. R., 1974, *Applied Geochemistry*. New York: Wiley, 353p.

Simonson, R. W., ed., 1974, *Non-Agricultural Applications of Soil Surveys*. Amsterdam: Elsevier, 178p.

Soil Survey Staff, 1951, *Soil Survey Manual*. Washington, D.C.: U.S. Department of Agriculture (Handbook No. 18), 503p.

Soil Survey Staff, 1975, *Soil Taxonomy: A Basic System of Soil Classification for Making and Interpreting Soil Surveys*. Washington, D.C.: U.S. Department of Agriculture (Handbook No. 436), 754p.

Stace, H. C. T., G. D. Hubble, R. Brewer, K. H. North-cote, J. R. Sleeman, M. J. Mulcahy, and E. G. Hallsworth, 1968, *A Handbook of Australian Soils*. Glenside, South Australia: Rellim, 435p.

Steffins, R., 1971, Le sols de la Lorraine Belge, *Pedologie Mém. No. 4*.

Stevenson, F. J., 1982, *Humus Chemistry*. New York: Wiley, 443p.

Strzemski, M., 1975, *Ideas Underlying Soil Systematics*. Warsaw: Foreign Scientific Publications Department of National Center for Scientific, Technical and Economic Information (Trans. of 1971 Polish ed.), 541p.

Taylor, N. H., and I. J. Pohlen, 1970, *Soil Survey Method: A New Zealand Handbook for the Field Study of Soils*. DSIR (New Zealand) Soil Bureau Handbook 25.

Thorp, J., 1957, Report on a field study of soils of Australia, *Earlham College Sci. Bull. No. 1*.

Valentine, K. W. G., and J. B. Dalrymple, 1975, The identification, lateral variation, and chronology of two buried paleocatenas at Woodhall Spa and West Runton, England, *Quat. Research* **5**, 551–590.

Cross-references: *Cat Clays; Dispersive Clays; Expansive Soils, Engineering Geology; Hard- versus Soft-Rock Geology; Saprolite, Regolith, and Soil; Soil Fabric; Soil Sampling.* Vol. XII, Pt. 1: *Soil Science.*

SPARKER SURVEYS—See ACOUSTIC SURVEYS, MARINE.

SUBSURFACE EXPLORATION, SURVEY—See ACOUSTIC SURVEYS, MARINE; AUGERS, AUGERING; BOREHOLE DRILLING; EXPLORATION GEOPHYSICS; GROUNDWATER EXPLORATION; VLF ELECTROMAGNETIC PROSPECTING. Vol. XIII: PIPELINE CORRIDOR EVALUATION; ROCK STRUCTURE MONITORING; SEISMOLOGICAL METHODS; TUNNELS, TUNNELING.

SURVEYING, ELECTRONIC DISTANCE MEASUREMENT

In today's sophisticated world, *speed* and *accuracy* are the key words that have caused members of surveying and navigation communities to seek the solution to their everyday undertakings from the ever-expanding fields of electronics and electro-optics. Construction surveyors, land surveyors, photogrammetrists, geodesists, geophysicists, hydrographers, and navigators need proper instruments for the execution of their practical or theoretical goals. Worldwide industry has entered into a race and competition not seen before to supply these instruments. The result is the utilization of a multitude of principles, designs, sizes, and shapes with an equal variety of prices.

Distance, either instantaneous, continuous, or distance difference, is the basic unit for all measurements described here. The great variety of possibilities offered and problems created by the modern state of the art of electronic surveying is best understood by exploring the following aspects: applications, electronics, environmental factors, geometry, reductions, and instrumentation.

Applications

In construction surveying, engineers in charge of building massive structures are also responsible for detecting any minute changes in these structures caused by strain (Laurila, 1976). The distances used to observe these changes are usually short, 500–1000 m; thus, the distance capacity of the instrument is of no particular importance. The accuracy must be of the order of 1 mm or better, and therefore light is used as the carrier.

Land surveyors, involved in cadastral surveying and land division, need speed and ease of operation. An accuracy of about ±10 cm is usually sufficient, and distances can vary from a few hundred meters to tens of kilometers; microwaves and light waves may be used equally as carriers. Photogrammetrists, while expanding the existing control network for stereo- or single-image photomapping, fall in the same category as land surveyors. In *geodetic surveying* where trilateration is to be established, the instrument must have a long distance capability of up to 100 km with an accuracy of about $\pm 2 \times 10^{-6}$ of the distance. In measuring the reference baselines, usually 10–20 km long, an instrument operating with light waves as the carrier is preferred. The accuracy obtained should be 10^{-6} or better (Parm, 1973). *Geophysical surveying* is the most demanding with respect to distance and accuracy. Measurements over faults and fracture zones, as well as long-term observations of lateral crustal movements, require instruments with the capability of measuring up to 60–100 km with an accuracy of about 10^{-7} (Carter et al., 1976; Laurila, 1976, 1983). Gaseous lasers with an output power of 10 mW are the only instruments commercially available today for that purpose. Multiwavelength instruments utilizing two laser light units with differing wavelengths simultaneously with a microwave unit have been developed and offer the most accurate results because of their capability to integrate the mean refractive index along the ray path (Huggett and Slater, 1974).

In the close-to-shore *hydrographic survey,* the utilization of medium long waves ($\lambda = 150$ m) was very popular in the 1940s and 1950s in the form of comparison continuous waves or hetero-

dyned signals. They are still widely used but the microwave modulation technique is gaining ground because of the higher obtainable accuracy. In general, distances up to 200 km can be measured with an accuracy of about ±2–3 m with the medium long wave technique, and distances to about 100 km can be reached with an accuracy of about ±0.5–1 m with the use of microwave modulation. The offshore hydrographic survey can be said to take place when distances from ground stations are of the order of 400 km or more. Utilization of long continuous waves ($\lambda = 3$ km; e.g., Decca) has long been a classical approach. A modified instrument called Digital Shoran uses a pulse-matching technique with a wavelength of $\lambda = 1$ m. A combination of pulse matching and phase comparison is adapted in Loran-C with a wavelength of $\lambda = 3$ km. All instruments used in hydrographic surveying are called *dynamic positioning devices,* which means that the ship's position can be continuously observed and recorded. Accuracies vary between a few meters and about 50 m depending on the type of instrumentation and the geometry of the pattern configuration.

Long-distance navigation can be classified as ground-based, aircraft-based, and satellite-based systems. *Ground-based systems* are similar to those used in offshore hydrographic surveys but are provided with greater output power. All the systems have finite distance capability and thus limited coverage. The exception is Omega, which operates with very long wavelength ($\lambda = 30$ km). The Omega signals propagate in the wave guide formed between the Earth and the ionospheric D-region, reaching distances so long that at present the existing eight ground stations offer a worldwide coverage (Laurila, 1976). Accuracies of the ground-based systems depend greatly on the geometry of the pattern configuration, and variations from some tens of meters to a few kilometers are possible.

The *aircraft-based system* is an application of the well-known dead reckoning principle. The aircraft carries a *Doppler radar* with two reflectors mounted at fixed angles with each other and with the vertical. Because of the Doppler effect, a frequency shift occurs between the transmitted and received signals. This shift is a function of the ground speed of the aircraft and the drift angle caused by the wind. By feeding this information to a small computer, the actual route of the aircraft to its destination can be programmed. This system is entirely self-contained, relatively inexpensive, and suitable to small craft use. The ground speed error is about 0.17%, and the drift error is about ±0.17° (McMahon, 1970).

The *satellite-based system* in the United States is the U.S. Navy Navigation Satellite System, or Transit. It provides intermittent position-fixing updates with a worldwide coverage. At present six satellites are in circular polar orbits about 1,075 km high, circling the Earth every 107 min. Each satellite is a self-contained beacon transmitting two very stable frequencies of about 150 MHz and 400 MHz to the ground-based tracking stations. Also, it transmits timing marks and a navigation message that describes the satellite's position as a function of time. By receiving this information, the user can calculate his or her position with reference to the global datum from which it can be transformed to any appropriate local datum. The system accuracy is about ±20–40 m (Laurila, 1976).

The newest satellite application, the Global Positioning System (GPS), has been under development since 1973 by the Department of Defense. A total of 18 satellites will be put in orbit, about 10,900 nautical miles above the earth by the end of 1987. They will cover the entire world providing an all-weather, 24-hour, three-dimensional navigation system with an accuracy of about 10 m (Laurila, 1983).

Land-to-air surveys serve two purposes. Trilaterations with long sides of about 400–900 km are established by crossing the sides at a certain altitude by an airplane carrying electronic surveying equipment and making continuous measurements to the terminals of the line. By summing together the partial distances and minimizing the sum, the point where the airplane is exactly above the line can be found. The minimum sum distance is then reduced onto the ellipsoid. For photogrammetric mapping (see *Photointerpretation*), the fix of the airplane at the instant of exposure can be found by simultaneous distance measurements from two known ground stations. By using stabilized verticality finders, the so-called nadir point of the photograph is found. Its location on the photograph coincides with the location of the airplane projected on the ground. Until the early 1960s the instrumentation used for both applications was a pulsed-type transceiver operating with a wavelength of about $\lambda = 1$ m; many phases of the operations were manually controlled. Today microwave modulation techniques are utilized with complete automation. The accuracy in measuring long lines varies between 1 m and 3 m, depending on the brand of instruments used and the length of the line (Cubic Corporation, 1963; Marshall, 1971).

Land-to-space surveys became a reality when Sputnik 1 was put into orbit on October 4, 1957. Early tracking was made by using specially designed cameras, such as the Baker-Nunn, PC-1000, and DC-4. In the early 1960s tracking satellite by electronic or electro-optical means became more prominent. Radio waves as carrier were used in the Secor system (*Se*quential *Co*llation of *R*ange) with the wavelength of about $\lambda =$ 0.7–1.3 m. In the ground stations the carrier is phase modulated, and distances to the satellite are made sequentially by comparing the phases of the transmitted and received modulation signals. From three ground stations with known locations, distances are measured to three satellite positions in space simultaneously with distances measured from a ground station with unknown location to the same three satellite positions. The location of the unknown point on the ground is then obtained as a spatial intersection from the established space locations of the satellite. The accuracy of the point location is about ±2–3 m (Prescott, 1965). In the late 1960s electro-optical tracking methods became popular because of the lesser effect of atmospheric refraction. Crystalline ruby lasers are used exclusively. These Pockels Q-switched lasers generate 0.5–7.5 joule of energy during pulse lengths of 15–30 nsec. The system accuracies of these lasers are about ±1 m (Lehr et al., 1969; U.S. Department of Defense, 1975). The special cameras, Secor, and lasers were used to establish a worldwide satellite triangulation network.

When the Apollo 11 astronauts Armstrong and Aldrin placed the Lunar Ranging Retroreflector (LRRR) on the lunar surface on July 21, 1969, a foundation stone was set for real extraterrestrial research of the Earth's physical characteristics. With three retroreflectors placed on the lunar surface, forming a triangle with sides over 1,000 km long, phenomena such as the libration of the moon can be observed (Bender et al., 1973). The main objective, however, at present is to investigate the secular changes in the geocentric positions of lunar ranging stations, including components due to local and regional crustal deformations and tectonic plate movements (Carter et al., 1976). This task is coordinated by the Lunar Ranging Experiment (LURE) team. There are two permanent lunar ranging stations in the United States. One is the McDonald Observatory at Mt. Livermore, Texas. There, a Q-switched ruby laser is used with a pulse length of 4 nsec containing about 2 joule of energy at the wavelength of 6,943 Å. The other station, the University of Hawaii LURE Observatory, is located at the summit of Mt. Haleakala, Maui. The laser used is the Q-switched YAG-Nd (yttrium-aluminum-garnet crystal doped with neodymium). The pulse length of 0.2 nsec contains 0.5 joule of energy at the infrared wavelength of 10,640 Å. This radiation is frequency doubled, yielding the final output radiation of 5,320 Å in the green region (Carter, 1973). The system accuracy of the McDonald laser is about ±8–15 cm, while that of the Haleakala laser is about ±3 cm.

One of the applications of electronic distance measurements is *satellite altimetry*. The first satellite altimeter, the S-193 altimeter, rode in the

Skylab, which was placed into Earth's orbit on May 14, 1973. The frequency used was 13.9 GHz that, with the 1.1 m diameter parabolic antenna, yielded the 3 dB beam width of 1°. The main objective was to investigate geodetic and geophysical phenomena such as geoid undulations, correlations with ocean floor topography, current detections, wave heights, and land topography. The system accuracy was about ±1 m. In 1978, SEASAT, a satellite carrying a radar altimeter, was launched into orbit. Its system accuracy is expected to be about ±10 cm, and its primary use is in oceanographic research (McGoogan, 1974).

A real breakthrough in aerial navigation occurred in the late 1970s with the development of the *Inertial Navigation System* (INS). In this system the position of a new point is obtained in the three-dimensional Cartesian coordinate system with the aid of three gyros and three accelerometers, initially set to corresond to the location of a known point. The instrument is truly self-contained and can be used anywhere in the world independent of the environment, such as terrain or atmosphere.

To investigate the mathematical structure of the Earth, together with its polar motion, very long baselines are needed. They are obtainable today through the use of Radio Interferometric Surveying (RIS), developed by the National Ocean Survey/National Geodetic Survey in Rockville, Maryland. In this surveying system, highly stable atomic clocks at each site are used to control the recordings of radio signals from the extragalactic radio sources. These sources are called the *quasars* or *quasi-stellar sources* and are immeasurably luminous celestial objects, which were already known by the mid-1960s. They lie far outside our own galaxy, billions of light years from Earth. The RIS observations serve geodetic, geophysical, and astronomical studies in measuring the length of the connecting line by determining the delay and the rate of change of the delay between the time that a wave front from the radio source arrives in the radio telescope at one end of the baseline, and the time it arrives in the radio telescope at the other end of the same line. In this system the separation of the stations can be extended up to the diameter of the earth (Laurila, 1983).

Electronics

To obtain an idea of the operation principle, capabilities, and limitations of instruments to be chosen for a particular task, an outline of electronics is presented. The spectrum of electromagnetic radiation used as the carrier in modern electronic and electro-optical instruments is large. Some samples are given in Table 1.

The way of generating the carrier waves depends on the wavelength and the type and pur-

TABLE 1. Electromagnetic Radiation

Radiation Source	Wavelength
Tungsten lamp	0.57 µm
He-Ne laser	0.73 µm
Ruby laser	0.69 µm
Ga-As diode	0.90 µm (infrared region)
YAG-Nd	1.06 µm
Microwave	0.008–0.100 m
Short wave	1 m (satellite tracking)
Medium long wave	150 m (hydrographic surveying)
Long wave	3,000 m (hydrographic surveying and navigation)
Very long wave	30,000 m (navigation)

pose of instruments used. In older model *geodimeters* the radiation source was a tungsten lamp. The light was noncoherent with large loss of illumination power; measurements could be made only at night. Many long- and some short-distance electro-optical instruments today employ He-Ne lasers with the output power varying between 1 mW and 10 mW. Measurements can be made during the day or at night. Most short-distance electro-optical instruments use gallium-arsenide (Ga-As) diodes as the radiation source. These cannot be used at long distances because of the relatively weak illumination power. All lasers used for satellite tracking and lunar ranging are of a solid type, such as the ruby and the YAG-Nd lasers. Microwave and short-wave instruments generate the carrier waves by utilizing the so-called velocity-modulated tubes; the most commonly used is the reflex klystron. Many instruments designed for hydrographic surveying and that employ medium or long waves generate these waves by crystal-controlled oscillators where vacuum tubes such as triodes or pentodes can be used. Long-distance navigation systems employing long or very long carrier waves have very sophisticated atomic oscillators (cesium, rubidium) to achieve the ultimate stability. These systems also use atomic clocks to synchronize stations together without the need of transmission between them.

In some applications carrier waves can be used as such for the distance (or distance difference) determination. In most cases the carrier waves are modulated to produce the desired signal forms. Three commonly used modulations are (1) *amplitude (intensity) modulation,* where the frequency and phase of the carrier wave do not change but the amplitude of the carrier wave alternates sinusoidally according to the amplitude of the modulation wave; (2) *frequency modulation,* where the amplitude of the carrier wave is kept constant but the frequency varies according to the amplitude and polarity of the modulation wave; and (3) *phase modulation,* where the amplitude of the carrier wave is kept constant but the phase of the carrier wave is varied according to the phase of the modulation wave. The following samples illustrate the process of modulations.

If light is used as the carrier, the desired modulation is amplitude, or intensity, modulation. In older model instruments using noncoherent light as the carrier, the *Kerr cell* filled with nitrobenzene was used as the modulator. Before the light beam enters the Kerr cell, it passes first through a polarization plate and will become plane polarized (e.g., horizontally). After leaving the Kerr cell, the light beam meets another polarization plate (vertical polarizer) and thus cannot pass through it. If RF (radio frequency)-controlled voltage is applied to the electrodes across the Kerr cell, it changes the refractivity characteristics of the Kerr cell in the form of a rotating ellipse with varying shape. This change causes the light intensity of the now emerging light beam from the second polarizer (analyzer) to vary sinusoidally. In newer model instruments, especially those using the He-Ne lasers, solid-type Pockels crystals (potassium dideuterium phosphate) have replaced the Kerr cell. The Pockels modulator is faster and more accurate than the Kerr cell, which is slightly affected by the impurities in nitrobenzene. There is a great advantage to using instruments utilizing a Ga-As diode as the light source since the intensity of light emission varies according to the applied voltage. No extra modulator is needed; intensity modulation is obtained directly from the Ga-As diode by RF-controlled voltage. In microwave instruments, which are widely used for long-distance surveying, the carrier wave is frequency modulated simply by feeding the modulation frequency to the reflector of the reflex klystron.

The detection or demodulation of RF signals (either received or locally generated) is done in the detector for the AM signals and in frequency discriminator for the FM and PM signals. In electro-optical instruments noncoherent light signals or coherent laser signals are detected and amplified in the photomultiplier. In the case of infrared radiation produced by Ga-As diodes, the signals are detected in junction diodes called the photodiodes.

The phase comparison between transmitted and received modulation signals can be made in several ways. In older model geodimeters the transmitted (reference) signal was delayed, or retarded, through a variable impedance line to match with the received signal in phase. The amount of delay indicated the fractional part of the unit length (usually one-half the modulation wavelength) included in the distance. It was later replaced by a resolver, which is a rotatable coil system. The resolver is more accurate and easier to read than the delay line. Resolvers are used in both microwave instruments and light-wave instruments. The lat-

est development is the digital, or pulse-counting system for phase measurement. It is used only in a few microwave instruments but is very common in light-wave instruments measuring short distances. When the transmitted modulation signal crosses the zero axis of the polarity, it opens the gate of a pulse-generating device. When the returned signal crosses the same zero axis, it closes the gate. The number of pulses formed during this process is counted, and by knowing the total number of pulses included in the whole unit length, its fractional part is obtained. This method of phase measurement is suitable in instruments designed for complete automation. In all cases of phase measurements, several additional modulation frequencies are needed to measure distances longer than the basic unit length. The phase measurement in instruments utilizing continuous waves or heterodyned beat signals is done by servo-driven resolvers.

The preceding paragraphs placed the emphasis on instruments measuring distances by counting the number of unit lengths and their fractions included in the distance. In *radar-type systems,* short pulses of electromagnetic energy are sent to the secondary instrument that relays them back to the primary instrument. Elapsed time is observed between the transmission and reception of such pulses, and the distance is obtained by the well-known echo principle.

In hydrographic survey and in navigation, the distance differences from ground stations to the vessel are usually observed. In this case the phases of signals arriving from pairs of stations are compared. To make the phase comparison possible, distinction between the signals is necessary. It can be achieved by several means. Ground stations may transmit with the same common frequency but on a time-sharing basis. This system requires very stable storage circuits (e.g., Decca Hi-Fix). The ground stations may transmit at different but harmonically so related frequencies that at the survey receiver they can be further multiplied by a harmonic set of numbers to produce pairs with the same comparison frequency (e.g., Survey Decca). The ground stations may transmit with frequencies differing slightly from each other (a few hundred hertz). These transmissions are received in the survey receiver and heterodyned down to a low-frequency beat signal. They are also received at a fixed ground reference station and heterodyned. The reference beat signal is then sent by way of an FM link to the survey receiver where the phases of the two beat signals are compared (e.g., Raydist N). The single-sideband (SSB) technique is gaining popularity in the process of sending information signals (beat notes) from the ground stations to the survey vessel. By making use of the same carrier wave, one beat note is sent in the form of the upper sideband and the other in the form of the lower sideband to the survey receiver (e.g., Raydist DR-S).

Environmental Factors

Factors independent of the functioning of the instrument are always present in distance measurements and, therefore, must be separately analyzed. Such factors are the velocity of electromagnetic radiation, curvature of the ray path of the radiation, soil conductivity affecting the velocity, and ionospheric and ground reflections.

The first reliable determination of velocity in vacuo (speed of light) was made by Simon Newcomb in 1882 and yielded the value of $c_o = 299,860 \pm 30$ km/sec. Based on measurements made by using microwave and light-wave techniques, a value of $c_o = 299,792 \pm 0.4$ km/sec was adopted as standard by the International Union of Geodesy and Geophysics in 1958. The modern technique provides the means to determine frequencies in the light spectrum with an accuracy of about 10^{-9} (Evenson et al., 1972). Barger and Hall (1973) determined the wavelength of 3.39 μm for a line of methane, which was measured with respect to the ^{86}Kr standard. From this wavelength, $\lambda = 3.392,231,404 \pm 1.2 \times 10^{-8}$ μm, and the measured frequency, $f = 88.376,181,627 \pm 5 \times 10^{-8}$ THz (Evenson et al., 1973), a new value for the speed of light was determined:

$$c_o = 299,792.4587 \pm 0.0011 \text{ km/sec.} \quad (1)$$

The Comité Consultatif pour la Définition du Mètre (Terrien, 1974), backed by the International Astronomical Union (1974), recommended that the value

$$c_o = 299,792,458 \text{ m/sec,} \quad (2)$$

with an uncertainty of $\pm 4 \times 10^{-9}$ be adopted as a constant for the propagation velocity of electromagnetic radiation in vacuo. It was also recommended that the meter be redefined by the value of c_o, thus making absolute distance measurements possible.

In actual measurements of distances, the average velocity of electromagnetic radiation is needed in ambient conditions. This value is obtained as follows:

$$c = \frac{c_o}{n}, \quad (3)$$

where n is the average refractive index along the path of propagation. To determine the instantaneous refractive index for light-wave propagation, first the so-called group refractive index must be computed for the particular wavelength utilized (Barrell and Sears, 1939):

$$(n_g - 1)10^7 = N_g \times 10$$
$$= 2,876.04 + \frac{3 \times 16.288}{\lambda^2} + \frac{5 \times 0.136}{\lambda^4}, \quad (4)$$

where λ is the effective wavelength of the radiation in micrometers. To compute the ambient instantaneous (point-to-point) refractive index, the Barrell and Sears (1939) formula is modified as follows (Laurila, 1969, 1976):

$$N = \frac{N_g P - 41.8e}{3.709T}, \quad (5)$$

where N is the refractive number, $N = (n-1)10^6$; T is the temperature in Kelvin units; P is the pressure in millibars; and e is the partial pressure of water vapor in millibars. In the radio spectrum the refractive index is practically independent of the wavelength and the *Essen formula,* where the parameters T, P, and e, given in the same units as those in Eq. (5), can be used as follows:

$$N = \frac{77.62}{T}P - \left(\frac{12.92}{T} - \frac{37.19 \times 10^4}{T^2}\right)e. \quad (6)$$

To obtain the average refractive index along the propagation path, a second-order polynomial has proved to be accurate enough even for high-precision survey. By taking the integrated mean of such a function, we obtain

$$\overline{N} = A + \frac{h}{2}B + \frac{h^2}{3}C, \quad (7)$$

where A, B, and C are constants to be determined for each line separately through the least squares principles, and h is the elevation at which the N-values, Eqs. (5) or (6), have been observed and computed. In actual ground-to-ground or ground-to-air surveying, the length of the line, the elevations of its terminals, and the curvatures of the Earth and the propagation path must be recognized. Jacobsen (1951) derived the following formula:

$$\overline{N} = A + \frac{B(H+K)}{2}$$
$$+ \frac{C}{3}[(H+K)^2 - HK] \quad (8)$$
$$- \frac{D_A^2}{12}\left(\frac{1}{R} - \overline{Q}\right)[B + C(H+K)],$$

where D_A is the length of the line, H and K are the elevations of its terminals, R is the curvature radius of the Earth, and Q is the average curvature of the propagated ray path. The average curvature of the ray path, in turn, is obtained as follows:

$$-\overline{Q} = B + C(H+K)$$
$$- \frac{CD_A^2}{6}\left[\frac{1}{R} + B + C(H+K)\right]. \qquad (9)$$

In space applications, the electromagnetic radiation propagates through the entire atmosphere. The influence of the atmosphere is considered as retardation in the velocity of the radiation and is given as a correction ΔS in the measured distance to artificial satellites or to the moon. Based on extensive meteorological observations within the continental United States, Laurila (1968, 1976) and Schnelzer (1972) found that this correction, when applied to light-wave propagation (lasers), is fairly constant, -239.7 ± 0.5 cm, and determined at sea level and at the zenith angle of $\psi = 0°$. At other altitudes and zenith angles, the correction will be

$$\Delta S = (\Delta S_h)\sec\psi, \qquad (10)$$

where ΔS_h is the correction at elevation h, and ψ is the zenith angle. Saastamoinen (1973) derived the following formulas for the range correction in light-wave and radio-wave propagations respectively:

$$\Delta S = 0.002357\sec\psi(P+0.06e - B\tan^2\psi) + \delta_L \qquad (11)$$

and

$$\Delta S = 0.002277\sec\psi$$
$$\left[P+\left(\frac{1225}{T}\right) + (0.05)e\right] - B\tan^2\psi + \delta_R, \qquad (12)$$

where T is the temperature in Kelvin units, P is the pressure in millibars, and e is the partial pressure of water vapor in millibars, all observed at station height; ψ is the zenith angle; B is a correction factor due to the curvature of the Earth and its atmosphere; and δ_L and δ_R are correction terms due to the curvature of the ray path of electromagnetic radiation. Corrections ΔS in Eqs. (11) and (12) must be subtracted from the measured distances.

The variation of soil conductivity affects greatly the velocity of long waves ($\lambda = 3000$ m). The velocity over seawater can be 470 km/sec larger than that over average land, and special correction charts must be used when combining propagations over land and water. Long-wave propagation is also affected by the sky-wave reflection from the ionosphere. While the ground-wave component is used for the phase comparison proper, the reflected wave component causes an error in the observed phase value. This error cannot be corrected, but its magnitude is predictable.

In the microwave technique used for ground-to-ground applications, part of the concentrated beam of the radiated energy, although very narrow, reflects from the ground or objects off the line, causing distortions in the distance readings. The behavior and magnitude of such reflections can be predicted and their influence reduced by slightly changing the carrier frequency step by step during the measuring sequence.

Geometry

The configuration of the geometric pattern is of either circular or hyperbolic mode. In the former, the location of points is determined by the intersection of two distance circles. In the latter, the location is made by the intersection of two or more hyperbolas obtained by measuring distance differences between a point concerned and two or more pairs of ground stations.

In the circular mode, the distance circles are concentric and equally spaced, thus providing a circular coordinate system with a constant scale. This grid is easy to construct and to use. In the hyperbolic mode, the unit of the coordinate is the space between two zero-phase hyperbolas. It is called the *lane,* and its value at any point is given by the following formula where phase differences are measured:

$$N = \frac{f}{c}(B + r_M - r_S), \qquad (13)$$

where f is the comparison (or transmission) frequency, c is the average velocity of electromagnetic propagation, B is the length of the base, r_M is the distance from the point to the master station (center station in the three-station formation), and r_S is the distance from the point to the slave station (side station). The width of a lane along the base is equal to one-half of the wavelength used, and the corresponding width at any point to be located is given by the following formula:

$$\ell = \ell' \times \frac{1}{\sin \beta/2}, \qquad (14)$$

where ℓ' is the lane width on the baseline and β is the subtended angle at which the base can be seen from the point. It can be shown that the direction of a hyperbola at that point coincides with the bisector of the angle β. Because of this fact, it is obvious that the angle γ between two intersecting hyperbolas, designated as R and G, is found to be the following:

$$\gamma = (\beta_R + \beta_G)/2. \qquad (15)$$

Thus, $\sin \beta R$, $\sin \beta G$, and $\sin(\beta R + \beta G)/2$ define the *geometric accuracy* of the location.

Reductions

In the process of distance measurement by electronic means, there are four main reductions to be made: (1) converting the observed arc distance to chord distance; (2) leveling the sloped chord distance to the horizontal; (3) reducing the horizontal chord distance to the sea level; and (4) converting this ellipsoidal chord distance to the ellipsoidal arc distance. In all these reductions the ray path is assumed to be an arc of a circle with the radius R', and the Earth is assumed to be a sphere with radius R.

The observed arc-to-chord correction is

$$\Delta D_1, = \frac{-k^2 D_A^3}{24R^2}, \qquad (16)$$

where D_A is the observed arc distance and $k = R/R'$.

The ellipsoidal-chord-to-ellipsoidal-arc correction is

$$\Delta D_2 = \frac{D_A^3}{24R^2}, \qquad (17)$$

and the combined correction, Eqs. (16) and (17), then will be

$$\Delta D_3 = \Delta D_1 + \Delta D_2$$

or

$$\Delta D_3 = \frac{1-k^2}{24R^2} D_A^3. \qquad (18)$$

These equations are used in cases where k is determined based on observations along the ray path. If k is determined only at the terminals of lines, the following combined correction is utilized (Saastamoinen, 1962):

$$\Delta D_4 = \frac{(1-k)^2}{24R^2} D_A^3 \qquad (19)$$

The horizontal and the sea level reductions can be made either as corrections to the sloped chord distance or the ellipsoidal chord distance, or even the ellipsoidal arc distance can be computed directly from the sloped chord distance. The sequence will be as follows:

Correction to horizontal:

$$C_L = - \left(\frac{\Delta H^2}{2D_C} + \frac{\Delta H^4}{8D_C^3} \right) \qquad (20)$$

Reduction to sea level:

$$C_E = - \left[\frac{D_C H_m}{R} - \frac{(D_C H_m)^2}{R^2} \right], \qquad (21)$$

where ΔH is the elevation difference between the terminals of the line, D_C is the sloped chord distance, and H_m is the mean elevation of the terminal points of the line.

The corrections C_L and C_E are used in conjunction with Eq. (18), and D_C in Eqs. (20) and (21) can be replaced with D_A without causing an error larger than 10^{-7} at any measurable distance (Laurila, 1976).

d_C as the function of D_C:

$$\left[\frac{(D_C^2 - (H-K)^2 R^2}{(R+K)(R+H)} \right]^{1/2}, \qquad (22)$$

where d_C is the ellipsoidal chord distance, and H and K are the elevations of the terminals of the line. This equation is used in conjunction with Eqs. (16) and (17) because in this case D_C cannot be replaced with D_A (Ewing, 1955; Laurila, 1960, 1976).

d_A as the function of D_C:

$$\left[\frac{12R^2 M}{12(R+H)(R+K) - M} \right]^{1/2}, \qquad (23)$$

where d_A is the ellipsoidal arc distance and $M = D_C^2 - (H - K)^2$. This equation is used in conjunction with Eq. (16) because D_C cannot be replaced by D_A (Wong, 1949) (see Laurila, 1960).

Instrumentation

Instruments available today are too numerous to be analyzed here. A summary is given according to the purpose and the usage of different devices in Tables 2 through 6.

SIMO H. LAURILA

TABLE 2. Land-to-Land Survey, Long Distances up to 150 km

Manufacturer	Model	Remarks
AGA, Sweden	Geodimeter NASM-series	Emission source tungsten or mercury vapor lamp
	M-series	Tungsten or mercury vapor lamp; M 6BL and M 8 He-Ne laser
Cubic Corporation, United States	Electrotape DM-20	Microwave
Anritsu, Japan	Autosurveyor ADM-4	Microwave
Electrim, Poland	Telemetre OG-2	Microwave
Ertel-Grundig, West Germany	Distameter III	Microwave
Fairchild, United States	Microchain MC8	Microwave
Funkmechanik, East Germany	PEM-2	Microwave
Laser Systems & Electronics, United States	Rangemaster and Ranger	He-Ne laser, built for Keuffel & Esser Co.
Metrimpex, Hungary	GET-series	Microwave
Spectra Physics, United States	Geodolite 3G	He-Ne laser
Tellurometer, Pty., South Africa	Tellurometer MRA-series and CA-1000	Microwave
USSR	Quartz	He-Ne laser
	SVV-1 and EOD-1	Mercury vapor lamp
	RDG and LUS	Microwave
Siemens-Albiswerke	SIAL MD 60	Microwave (modified Distomat DI-60)
Wild-Heerbrugg, Switzerland	Distomat DI-50	Microwave
	DI-60	
Zeiss, East Germany	EOS	Filament lamp

TABLE 3. Land-to-Land Survey, Short Distances up to 6–8 km

Manufacturer	Model	Remarks
AGA, Sweden	Geodimeter M 700	Emission source He-Ne laser in all three instruments
	M 710	
	M 76	
Askania, West Germany	Adisto S 2000	Ga-As diode
Askania-Franken, West Germany	Adisto S 1000	Ga-As diode
	ART	Ga-As diode
Cubic Industrial Corp., United States	Cubitape DM-60	Ga-As diode
Hewlett-Packard, United States	H-P 3800B and H-P 3805A	Ga-As diode
Kern, Switzerland	DM 500, 1000, and 2000	Ga-As diode
	Mekometer ME 3000	Xenon flash tube
Laser Systems & Electronics, United States	Microranger	Ga-As diode
	Ranger I and II	He-Ne laser, all built for Keuffell & Esser Co.
National Physical Lab., England	Mekometer I, II, and III	Xenon flash tube, experimental
Tellurometer, Pty., South Africa	MA-100 and CD-6	Ga-As diode
USSR	DST-2	Tungsten lamp, experimental
	KDG-3 and GD-13	Ga-As diode
Wild-Heerbrugg, Switzerland	DI-3 and DI-10	Ga-As diode
Zeiss, East Germany	EOK 2000	Ga-As diode
Zeiss, West Germany	SM-11 and RegElta 14	Ga-As diode

TABLE 4. Land-to-Sea Survey, Hydrographic Survey

Manufacturer	Model	Remarks
Cubic Industrial Corp., United States	Autotape DM-40A and DM-43	Microwave as carrier, circular or hyperbolic mode
Decca Survey, Ltd., England	Hi-Fix series	Medium long wave, circular or hyperbolic mode
	Lambda	Long wave, circular mode, lane identification
	Pulse/8	Long wave, operates with Loran-C
	Survey Decca	Long wave, hyperbolic mode
	Trisponder	Microwave, circular mode
Lorac Service Corp., United States	Lorac A and Lorac B	Medium long wave, heterodyne principle, hyperbolic mode
SERCEL, France	RANA	Four frequencies in the medium long waveband, hyperbolic mode
	Toran O	Medium long wave, utilizes cesium standard, circular mode
Teledyne Hastings-Raydist, United States	Raydist DR-S and DRS-H	Medium long waves, heterodyne principle, use SSB technique, circular or hyperbolic mode
	Raydist "T"	As above but emphasis on hyperbolic mode
Tellurometer, Pty., South Africa	MRB-201	Microwave, circular mode

TABLE 5. Land-to-Air and -Space Survey

Manufacturer	Model	Remarks
Cubic Industrial Corp., United States	CR-100	Microwave, land-to-air system
	Shiran	Microwave, land-to-air system
	Secor	Short-wave land-to-space system
Tellurometer, Pty., South Africa	MRB-301	Microwave land-to-air system
	AFCRL laser	Ruby laser for satellite tracking
	NASA LURE laser	Ruby and YAG-Nd laser for lunar ranging
	SAO laser	Ruby laser for satellite tracking

TABLE 6. Navigation

Model	Remarks
Consol-Consolan	Theta system (angular mode), inexpensive, suitable for small boat use
Decca Navigator System	Long-wave, hyperbolic mode, lane identification
Loran-C	Long-wave, hyperbolic mode, combined pulse-matching and phase-measuring technique
Omega	Very long-wave, worldwide navigation system
Differential Loran-C and Omega	Differential systems utilize fixed monitor station in the survey area to transmit corrections to the users
Aircraft Doppler Navigation	Airborne Doppler system to determine the ground speed and drift angle of the aircraft
Transit	Satellite Doppler worldwide navigation system
GEOLE	Satellite navigation system measuring range and range rate (French)
INS	Inertial navigation system, self-contained, operates with gyros and accelerometers

Note: For detailed analysis of instruments presented in tables 2 through 6, see S. H. Laurila, 1976, *Electronic Surveying and Navigation* (New York: Wiley-Interscience), 545p.

For new instruments, marketed after 1976, see S. H. Laurila, 1983, *Electronic Surveying in Practice* (New York: Wiley-Interscience), 388p.

References

Barger, R. L., and J. H. Hall, 1973, Wavelength of the 3.39 m laser-saturated absorbtion line of methane, *Appl. Physics Letters* **22**(4), 196–199.

Barrell, H., and J. E. Sears, 1939, The refraction and dispersion of air for the visible spectrum, *Royal Soc. London Philos. Trans.* (Series A) **238.**

Bender, P. L., et al., 1973, The lunar laser ranging experiment, *Science* **182,** 229–238.

Carter, W. E., 1973, University of Hawaii LURE Observatory, in R. S. Mather and P. V. Angus-Leppan, eds., *Proceedings of the Symposium on Earth's Gravitational Field and Secular Variations in Position.* Sydney: Australian Academy of Science, 433–441.

Carter, W. E., E. Berg, and S. Laurila, 1976, The University of Hawaii Lunar Ranging Experiment Geodetic-Geophysics Support Programme, *Hawaii Inst. Geophysics Contr. No. 744,* 6p.

Cubic Corporation, 1963, AN/USQ-32 Microwave, Geodetic Survey System (Shiran), *ASD Tech. Doc. Rept. ASD-TDR-62-872,* 245p.

Evenson, K. M., et al., 1972, [article title not given] *Phys. Rev. Letters* **29,** 1346.

Evenson, K. M., et al., 1973, [article title not given] *Appl. Physics Letters* **22,** 192.

Ewing, C., 1955, The parallel radius method of solving the inverse Shoran problem, Ph.D. dissertation, Ohio State University, 231p.

Huggett, G. R., and L. E. Slater, 1974, Precision electromagnetic distance-measuring instrument for determining secular strain and fault movements, paper presented at the International Symposium on Recent Crustal Movements, Eidgenossische Technische Hochschule, Zurich, August 26–31, 17p.

Jacobsen, C. E., 1951, High-precision Shoran test, Phase I. Dayton, Ohio: Wright Air Development Center (Air Force Tech. Rept. No. 6611).

Laurila, S. H., 1960, *Electronic Surveying and Mapping.* Columbus: Ohio State University Press, 294p.

Laurila, S. H., 1968, Refraction effect in satellite tracking, *Hawaii Inst. Geophysics Tech. Rept. No. 68-19,* 44p.

Laurila, S. H., 1969, Statistical analysis of refractive index through the troposphere and the stratosphere, *Bull. Géod., No. 92,* 139–153.

Laurila, S. H., 1976, *Electronic Surveying and Navigation.* New York: Wiley-Interscience, 545p.

Laurila, S. H., 1983, *Electronic Surveying in Practice.* New York, Wiley-Interscience, 388p.

Lehr, C. G., et al., 1969, The laser system at the Mount Hopkins Observatory, paper presented at the International Symposium on Electromagnetic Distance Measurement and Atmospheric Refraction, Boulder, Colo., 8p.

McGoogan, J. T., 1974, *Precision Satellite Altimetry.* Wallops Islands, Va.: NASA Wallops Station, 7p.

McMahon, F. A., 1970, AN/APN-187 Doppler Velocity Altimeter Radar Set, *Digest of U.S. Naval Aviation Weapons Systems* (Avionics ed., Navair 08-1-503), May, 49p.

Marshall, A. G., 1971, *Development of the MRB 201/ 301.* Farmingdale, N.Y.: Research and Development, Tellurometer Division of Plessey, 37p.

Parm, T., 1973, A determination of the velocity of light using laser geodimeter, Finnisch. *Geod. Inst. Veröffentl. No. 73/3,* 12p.

Prescott, N. J. D., 1965, Experiences with secor planning and data reduction, in *Symposium on Electromagnetic Distance Measurement,* Oxford, September. London: Hilger-Watts Ltd., 312–338.

Saastamoinen, J., 1962, Reduction of electronic length measurements, *Canadian Surveyor* **16,** 98–100.

Saastamoinen, J., 1973, Contributions to the theory of atmospheric refraction, *Bull. Géod.,* No. 107, 13–34.

Schnelzer, G. A., 1972, A comparison of atmospheric refractive index adjustments for laser satellite-ranging, *Hawaii Inst. Geophysics Pub. No. 72-14,* 88p.

Terrien, J., 1974, News from the Bureau International des Poids et Measures, *Meteorologia* **10,** 75–77.

U.S. Department of Defense, Defense Mapping Agency, 1975, *Department of Defense National Geodetic Satellite Program.* Washington, D.C., 215p.

Wong, R. E., 1949, Conversion of Shoran measurements to geodetic distances, *Mapping and Charting Research Laboratory Tech. Paper No. 62.*

Cross-references: *Satellite Geodesy and Geodynamics; Sea Surveys; Surveying, General.* Vol. XIII: *Remote Sensing, Engineering Geology; Rock Structure Monitoring.*

SURVEYING, GENERAL

Surveying is the art and science that determines the dimensional relationships between physical features, natural and artificial, on, below, or above the Earth's surface. This is done by measuring directions or angles and distance between points. Surveying is the first step in the making of plats and maps for location and of charts for air and marine navigation (Gagrow, 1964).

Four principal types of surveys include (1) land surveys, (2) engineering surveys, (3) geodetic surveys, and (4) cartographic surveys. *Cartographic surveys* include geodetic control surveys, topographic feature surveys, and hydrographic feature surveys, plus various specialty surveys, including air surveys, to compile any necessary thematic data for a specific map or chart. *Geodetic surveys* take into account the size and shape of the Earth and establish the basic horizontal and vertical control points, or stations and benchmarks, of known elevation above sea level (see *Satellite Geodesy and Geodynamics*). These stations then provide the framework of control for the mapping and charting of large areas, such as a continent, so that each place, or point, on each map is uniquely related to all other points on the Earth and so that each map can be fitted into a perfectly connected series of maps covering a large area, such as the United States (Bomford, 1952). *Topographic surveys* start from the basic geodetic control and determine, in great detail, the configuration of the terrain—i.e., the relative positions and elevations of all features to be shown on the topographic map. *Hydrographic surveys* also start from the basic geodetic control and determine the depths of water areas and the configuration of the bottom. These surveys also determine the positions of all submerged features of interest to navigation, such as submerged dangers and the positions of deep-water passages and channels for shipping (Shalowitz, 1962). Some hydrographic surveys also involve studies of the continental shelf and offshore oil and mineral exploration.

From the National Geodetic Surveys, geographic and state plane coordinate and other projection systems are extended to which all other surveys, including land boundary and engineering route and layout surveys and cartographic compilations, ultimately must relate to fix their position.

Land Surveys

Land surveying, one of the oldest of professions, dating back to several thousand years B.C., deals with the basis of capitalistic systems, the right to hold real estate. Reference to property includes boundaries, and this is where land surveying is involved (Dowson and Shepard, 1952; Clark, 1959; Brown and Eldridge, 1962). The individual as well as society as a whole is a jealous guardian of its rights and privileges, willing to do battle in defense of property rights. Society is unqualified to establish boundaries and so relies on the competence of certain of its members on whom it has conferred the powers and privileges to conduct its affairs. Of the four classes of surveying, land surveying is the one most subject to encroachment by engineers and technicians who think they are qualified to make land surveys. The legal practice of land surveying requires far more experience and judgment than the mere facility to drag a tape and manipulate the knobs on a theodolite.

In the United States, where individuals and corporations may own land, the professional *land* or *property-line* surveyor is in effect the judge's advocate who determines on the ground the rightful property lines of landowners as decreed by courts of law. Legislative acts in various states insist on special licensing for the surveyor of land-title or property lines or locations. Issuance of license is qualified by his or her philosophical knowledge of the legal line perhaps even more than technical knowledge of the physical line. On this point it must be emphasized that a lawyer without knowledge of survey technology would not qualify as a professional land surveyor. Neither would an engineer, fully equipped with the technology of the physical survey, unless he or she further qualified with the philosophy and experience concerned with the anomaly of the legal property bound. In the true sense of land sur-

veyor licensure, both qualifying elements are essential.

The art and science of land surveying is also involved in almost every civil engineering project, inasmuch as their construction frequently entails a transfer of ownership of either the property or some property right. The same general knowledge is required to clear title to land for a development project such as to locate and evaluate property or to perform a boundary survey for no other purpose than to advise an owner as to the location of corners.

Engineering Surveys

Engineering surveying includes the development of design data, such as horizontal and vertical control, the plotting of culture and topography, and the development of profiles and cross-sections for the study and selection of sites for engineering construction. It also includes the layout of structures, quantity and measurement determinations, so-called as-built and utility surveys, and mine surveys.

Engineering surveys are in effect *physical measurement surveys* to locate and control the alignment in construction. Such surveys run the gamut from the preliminary exploratory surveys of lesser accuracies and higher accuracies for laying out highways, railroads, transmission lines, pipelines, etc. to the utmost precision that may be required for aligning missile base operations, complex highway interchanges, tunnels, or cyclotrons or to true up a giant aircraft under construction.

On construction projects, besides the survey engineers, many technicians or subprofessional surveyors may be employed under supervision on routine repetitive production measurements. It might be said that perhaps more technicians or subprofessional surveyors are employed in the general field of engineering surveys than in all other types of surveying that, by their nature, do not tend toward repetitive routine production operations but require continuous creative sensing and professional judgment.

Geodetic Surveys

In the surveying and mapping of large areas, the exact curvature of the sea level surface of the Earth must be taken into account. For this reason, the basic surveys of a country are called *geodetic surveys* after the word *geodesy,* which is the science that treats mathematically of the figure and size of the Earth.

Geodetic surveys include highly accurate measurements of directions and distances over long lines, the measurement of differences in elevation, astronomic observations, and measurement of the Earth's gravity. These field measurements provide the geodesist with information for studies of the figure and size of the Earth as well as the information needed to compute the horizontal positions and elevations of monumented geodetic control stations. These stations are the basic framework for local surveys, mapping, and the planning and construction of lines of transport and communication, among many other purposes.

Distances are usually measured indirectly by means of *triangulation,* a method that consists of a system of connected triangles with all angles carefully and accurately measured but with only an occasional length actually measured on the ground. Each measured length is known as a base, or *baseline.* By use of the measured angles and the measured bases, the lengths of the sides of the connected triangles can be computed trigonometrically. If the latitude and longitude of one point are known, together with the azimuth or direction to one of the other stations in the scheme of triangulation, then the latitudes and longitudes of all of the other points, and the azimuths or directions of all other lines, can be computed. Baselines must be measured with very high accuracy. Such accuracy is achieved today by means of electronic distance measuring instruments such as the *geodimeter* (see *Surveying, Electronic Distance Measurement*).

The angles of triangulation are measured by *theodolites.* These instruments are similar to the *surveyor's transit* except that they are more precisely constructed and have microcomputers for reading the circles. Angle observations are made almost entirely at night by observing on lights to reduce errors caused by atmospheric refraction. To lengthen the line of sight, the points of observation are elevated. In mountainous country, the triangulation is laid out to extend from mountaintop to mountaintop. In low country or in country with trees, portable steel towers are used.

Notwithstanding the high precision with which the angles of triangulation are measured, there is a tendency for arcs of triangulation to swerve away from their true orientation. This swerving could readily be corrected by the simple expedient of making astronomical azimuth determinations at suitable intervals and adjusting the triangulation to these azimuths, if it were not for the effect of deflections of the verticle. A *plumb line* is commonly thought to point toward the center of the Earth, or at least toward the axis of the Earth, but this is not quite the case. Large mountain masses or other topographic features exert a sidewise attraction on the plumb line or on the levels of instruments that may affect the astronomical observations by several seconds of arc, by as much as 20 sec or 30 sec in extreme cases.

Uncorrected astronomical azimuths cannot be used, therefore, as *true azimuths.* Fortunately, there is a method for determining quite accurately the effect of the deflection of the vertical on astronomical azimuth observations at a triangulation station. If the astronomical longitude is determined at such a station and compared with the geodetic longitude—i.e., the longitude carried through the triangulation—the deflection of the vertical in an east-west direction becomes known within a small fraction of a second. This value of the *deflection of the vertical* can then be used to correct the astronomical azimuth observations and thus obtain a true azimuth to be held fixed in the adjustment of the triangulation (Roelofs, 1950). This corrected azimuth is known as a *Laplace azimuth* and the station as a *Laplace station.* Longitude and azimuth observations are therefore made at intervals along an arc of triangulation.

Horizontal control surveys can also be accomplished by the *transverse method.* Transverse differs from triangulation in that all lengths have to be measured directly in much the same manner as baselines, and it lacks the automatic checks on the angle observations that are obtained from the triangle closures when the triangulation method is used. Today the distance measurements for traverse are accomplished with electronic distance measuring instruments such as the *geodimeter* and the *gellurometer,* among others. Traverse is inferior to triangulation for the most accurate geodetic surveys, since the measurements are all along a single line and the results cannot be adjusted and computed with the certainty and rigidity of triangulation.

Geodetic control surveys and the *monumented stations* resulting from those surveys are classified as either horizontal or vertical. Positions for the horizontal control stations are given in terms of latitude and longitude and state coordinates. Elevations of the vertical control stations (*benchmarks*) are given in feet and meters above mean sea level (Kiely, 1947; Middleton and Chadwick, 1956; Rayner and Schmidt, 1963).

The use of interferometric receivers in conjunction with signal-sending satellites for geographic positioning and ellipsoidal heights makes survey control possible in isolated places.

Increased use of electronic distance measuring devices offers speed as well as accuracy to measurements in the field, improving production of geodetic networks. Use of the *active-type geodetic satellite,* which is photographed simultaneously from at least three points on the Earth, against known star backgrounds, permits accurate measurement of far greater distances on Earth than was possible with classical triangulation techniques. Computer operations speed up adjustment of geodetic networks. These rapid adjustments and satellite triangulations will eventually bring about more accurate measurements of the size and shape of the Earth. The existing National Geodetic Control Network based on the 1927 North American Datum has been undergoing readjustment by the National Geodetic Survey to result in the new 1983 North American Datum, made available to the public in 1985. Various states are presently effecting legislation to convert their state coordinate systems to the 1983 NAD.

Cartographic Surveys

Cartographic surveying includes, for topographic data, the establishment, computation, and adjustment of control surveys of the order required by the detail and scale of the map being made; the extension of control by aerotriangulation; the preparation of the manuscript by field and/or photogrammetric surveys; and the field inspection and field editing of the manuscript.

Control Surveys Control surveys are required to ensure accurate positions and elevations for the features shown on the map. These surveys are of two kinds: (1) geodetic (or basic control) surveys of third-order or higher accuracy, which determine positions or elevations for points that are marked on the ground with metal tablets, and (2) less accurate supplemental control surveys, which determine elevations or positions for points that are identified on aerial photographs but not marked with tablets.

The latitude and longitude of basic horizontal control points may be determined by triangulation or traverse or a combination of these two methods. The elevations of basic control points are established by precise leveling. From this basic network, additional elevations needed for photogrammetric compilation are determined by supplemental control surveys. The most common types are stadia traverse with plane table and alidade, barometric altimetry, and fly levels with spirit level or transit.

Topographic-Planimetric Surveys and Maps Most of the modern topographic maps are plotted from aerial photographs with instruments that operate on the principle of the stereoscope. So that the map data plotted from these photographs may be accurately positioned on the map manuscript, the positions of several image points on each stereoscopic pair of photographs must be known. *Aerotriangulation* is the photogrammetric process by which positions, and sometimes elevations, are determined for these image points (usually referred to as *pass points*). To extend control by aerotriangulation, the positions of the ground control stations must be accurately identified on the aerial photographs. The coordinates of the pass points are determined relative to those of the ground control stations by analytical,

graphic, or mechanical methods of adjustment or by a combination of these methods (Boldov and Selifanov, 1963).

A stereoscopic model of the terrain may be viewed in a plotting instrument when the photogrammetrist recovers the angular relationships of two overlapping photographs as they existed at the original exposure stations (see *Photointerpretation*). The models are oriented, both horizontally and vertically, with respect to the map manuscript by means of the pass points determined by aerotriangulation and the elevations determined by supplemental control surveys or aerotriangulation. Contour lines and planimetric details are plotted orthographically on the map as a floating mark is moved about on the apparent surface of the stereographic model. To produce contours, the floating mark is maintained at a fixed elevation as it traces the locus of ground points at that elevation. To produce planimetry, the floating mark is varied in elevation to match the variations in elevation of the feature being delineated.

Maps are occasionally constructed in the field by the *plane table and alidade method* or from measurements made with transit and steel tape. The latter procedure, however, is usually limited to the preparation of large-scale, highly detailed maps or for plotting contours in areas of very low relief (Kiely, 1947).

Field surveying methods are used for checking the photogrammetrically compiled map manuscript and for map revision. Additions and corrections of planimetric details and contours are usually made by the plane table and alidade method, whereas accuracy tests are accomplished by geodetic survey methods. Features such as names, civil boundaries, and public-land lines must be added to the map according to the best documentary records and evidence recovered in the field.

Hydrographic Surveys Hydrographic surveys are surveys of bodies of water and the bottom topography. They usually include mapping of the shoreline and foreshore, measurement of depths of the water and delineation of the bottom contours, finding and positioning of submerged dangers to navigation, positioning of natural and artificial channels of deep water that provide safe passage for ships, location of navigational aids, and measurements of the rise and fall of the tides and of the direction and velocity of tidal currents (Shalowitz, 1962).

Hydrographic surveys are used in the preparation of nautical charts for navigation, in harbor dredging and maintenance work, and in coastal and sanitary engineering projects involving, among other aspects, silting and water-circulation studies and offshore oil and mineral exploration. Hence, observations and samplings may include bottom composition and subbottom character, as well as relief, and physical and chemical properties of the water.

Before making hydrographic surveys, a system of visible control points or radio aids to navigation must be established on shore. Two principal operations are measuring water depths and determining positions of observations with respect to land and surrounding features. Soundings in modern hydrographic surveys are made by electronic sounding devices. Sound signals are transmitted from a transceiver in the hull of a ship to the ocean floor. The reflected signal is detected by the same unit, amplified, and graphically recorded. Depth of water is measured by the time interval between the transmitted and return signal. Corrections are applied for changes in sound velocity caused by differences in physical properties of the water medium. Consecutive soundings are recorded on a graph (*fathogram*) to provide a profile of the ocean floor along the line of soundings.

The soundings, which are vertical measurements of depth, must be positioned in latitude and longitude so that each such measurement is related horizontally to all others. When the hydrographic survey is close to land, this positioning is done by observing angles to signals, or targets, of known position that have been placed along the shore for this purpose. When surveys are beyond visibility of shore signals, this positioning is accomplished by any one of several types of electronic positioning instruments such as Shoran, Electronic Position Indicator (EPI), satellite positioning and navigational systems, and Radist.

Aerial Survey Services Airborne sensing devices provide data for many types of surveys. Of these devices, the widely used aerial camera provides data for surveys in the form of black-and-white, color, and infrared photography. Aerial photography may be utilized for almost every kind of survey, particularly for cartographic surveys, design and construction surveys, land or property surveys, and surveys that require interpretation of photographic images or terrain features, such as forest inventory surveys and geologic surveys.

Imagery of the Earth's surface may be recorded by sophisticated satellites in Earth orbit and by airborne scanning devices that make use of infrared radiation or radar or other image sensing systems. Infrared sensors detect differences in the energy radiated from the Earth's surface due to differences in temperature and physical characteristics. Radar devices detect the reflected energy from radio waves transmitted from the aircraft. Some satellites televise images back to Earth.

Airborne distance-measuring systems are used to obtain position fixes and to measure distances between ground stations. In Hiran and Shoran, electromagnetic waves are transmitted from the airborne equipment to ground stations where they are amplified and transmitted back to the airborne receiver. The time required for these waves to travel to the ground stations and back is measured and recorded by equipment in the aircraft. Supplemented by data from meteorological observations, these time measurements are used to obtain accurate distances between the ground stations.

Shiran is an airborne electronic ranging system that employs phase comparison of electromagnetic waves for measuring air to ground ranges up to 450 nautical mi. An airborne interrogator transmits a wave signal that is received and transmitted back to the aircraft by transponders on two to four ground stations. Since the system has a four-station ranging capability, it can be used for distance measurement between ground stations or for controlled photography missions in which the position of the nadir point for each photograph can be determined. Secor is an electronic distance measuring system that makes use of an orbiting satellite for measuring distances between widely separated ground stations, particularly for extending control from continents to remote islands.

Airborne magnetometer surveys offer an inexpensive, quick means of obtaining magnetic data for the study of subsurface geology. Various electronic navigational aids and positioning systems are used to guide aircraft from these surveys. Geologists and oil and mineral prospectors use the magnetic data in digital form or in the form of magnetic maps. Information can be extracted from these data more readily when aerial photographs of the area are used to aid in the interpretation of the magnetic data.

Airborne profile recording systems make use of a narrow-beam radar altimeter and a pressure-sensing device for obtaining a profile of the terrain along the path of the survey aircraft. When photographs and profile data are obtained simultaneously, elevations can be completed for points along each flight strip of a photographed area. This system is becoming more widely accepted as a tool for mapping remote areas because it can be used without supplemental ground control.

Organization of Surveying

In the United States, the surveying profession is organized into the *American Congress of Surveying and Mapping* (210 Little Falls, Falls Church, Virginia, 22046). There are numerous state and regional societies. On the federal government level, surveying and mapping are handled by the agencies indicated in Table 1.

On the initiative of French surveyors, an International Congress of Surveyors was held in Paris in 1878 with the purpose of founding a per-

TABLE 1. Federal Surveying and Mapping Agencies

Agency	Address
Aeronautical Chart and Information Center, DMA	2nd and Arsenal St. Louis, MO 63118
Agricultural Stabilization and Conservation Service	Asheville, NC 28801
Topographic Center, DMA	6500 Brooks Lane Washington, D.C. 20305
Bureau of Land Management	Washington, D.C. 20240
National Geodetic Survey, NOS, NOAA	6001 Executive Blvd. Rockville, MD 20852
Forest Service, USDA	Alexandria, VA 22313
U.S. Geological Survey	National Center Reston, VA 22092
Hydrographic Office	U.S. Navy Oceanographic Office Suitland, MD 20390
Lake Survey Center, NOAA	630 Federal Bldg. Detroit, MI 48226
Mississippi River Commission (Corps of Engineers)	P. O. Box 80 Vicksburg, MS 39180
Soil Conservation Service, USDA	Beltsville, MD 20705
Tennessee Valley Authority	Maps and Surveys Branch Chattanooga, TN 37401
Weather Bureau, NOAA	Washington, D.C. 20235

Note: World surveying is united into the *Fédération Internationale des Géomètres* (International Federation of Surveyors).

Key: DMA, Defense Mapping Agency; NOS, National Ocean Survey; NOAA, National Oceanic and Atmospheric Administration; USDA, U.S. Department of Agriculture.

manent International Federation of Surveyors. A second congress was held in Brussels in 1910, and in 1926, the International Federation of Surveyors (IFS; *Fédération Internationale des Géomètres,* FIG) was founded in Paris. During more than thirty years, the IFS has done important work in the scientific, technical, economic, and social fields. This work culminated in the congresses of Zurich (1930), London (1934), Rome (1938), Lausanne (1949), Paris (1953), the Netherlands (1958), Vienna (1962), and Rome (1965). The proceedings of these congresses have been published. Organizations from eighteen countries are affiliated with the IFS.

Spectacular advances in geodesy and photogrammetry were apparent by the tenth Congress in Vienna in 1962. The IFS is today one of the largest international technical associations in the world with many associate organizations; it includes leaders from scientific and scholastic fields, directors and officials of cadastre, technical operators in the geotopocartographic sphere as well as those representing the technique of real estate, of which town planning and regional development are a part. The groups and technical commissions of the IFS, as of 1967, include

Group A, Professional Organization and activities: First Commission, Professional practice; Second Commission, Professional Education; Third Commission, Professional Literature;

Group B, Surveying and Mapping: Fourth Commission, Cadastre and Rural Land Management; Fifth Commission, Survey Instruments and Methods; Sixth Commission, Engineering and Functional Surveys;

Group C, Land Administration: Seventh Commission, Legal and Social Studies of Urban Land Systems; Eighth Commission, Town Planning and Land Development; and Ninth Commission, Valuation and Management of Real Estate.

The following IFS offices or committees exist: Comité du Dictionnaire Multilingue de la F.I.G. (at the Instut für Angewandte Geodäsie, Frankfurt am Main), Office International du Cadastre et du Régime Foncier (an important central data assembly office), and Surveying Engineering.

WALTER S. DIX

References

American Congress on Surveying and Mapping Staff, 1961, *Index to Surveying and Mapping, 20 vols.* (1941-1960). Falls Church, Va.

Joint American Congress on Surveying and Mapping, American Society of Civil Engineers Staff, 1972, *Definitions of Surveying and Associated Terms.* Falls Church, Va.: American Congress on Surveying and Mapping, 210p.

American Society of Civil Engineers, *Manuals of Engineering Practice* (e.g., *Technical Procedure for City Survey,* Manual 10; *Definition of Surveying Terms,* Manual 15; *Land Subdivision,* Manual 16; *Horizontal Control Surveys to Supplement the Fundamental Net,* Manual 20). New York: American Society of Civil Engineers.

Boldov, V. G., and V. P. Selifanov, 1963, Scientific and technical information in geodesy and cartography, *Geodesy and Aerophotography* 5, 305-307 (trans. from Russian).

Bomford, B. G., 1952, *Geodesy.* Oxford: Oxford University Press, 405p.

Bowditch, N., 1958, *American Practical Navigator.*

Washington, D.C.: U.S. Navy Hydrographic Office, 1,524p.

Breed, C. B., and G. L. Hosmer, 1966, *Surveying*. New York: Wiley, 1,331p.

Brown, C. M., and W. H. Eldridge, 1962, *Evidence and Procedure for Boundary Location*. New York: Wiley, 480p.

Clark, F. E., 1959, *Clark on Surveying and Boundaries*. Indianapolis: Bobbs-Merrill, 1,029p.

Dowson, E., and V. L. O. Shepard, 1952, *Land Registration*. London: Her Majesty's Stationery Office, 211p.

Eldridge, W. H., 1962, Bibliographies on surveying in the U.S.A., *Surveying and Mapping* **22**, 2-6.

Eldridge, W. H., 1965. Current activities in America in surveying literature, *Surveying and Mapping* **25**, 242-243.

Eller, R. C., ed., 1963, *Bibliography of Property Surveying Literature*. Falls Church, Va.: American Congress on Surveying and Mapping, 142p.

Fédération Internationale Géomètre (International Federation of Surveyors) Staff, 1963, *Dictionaire Multilingue*. Amsterdam: Argus, 500p.

Gagrow, L., 1964, *History of Cartography*. London: Watts, 312p.

Griffith, S. V., 1962, Surveying and mapping in the U.S.A., *Surveying and Mapping* **22**, 10-21.

Karo, H. A., 1964, Progress in international cartography, *Surveying and Mapping* **24**, 401-410.

Kiely, E. R., 1947, *Surveying Instruments—History and Classroom Use*. New York: Columbia University Press, 400p.

McNair, A. J., 1964, Professional status of surveying Europe and the United States, *Surveying and Mapping* **24**, 435-442.

Middleton, R. E., and O. Chadwick, 1956, *Treatise on Surveying*, 2 vols. London: Philosophical Library.

Rayner, W. H., and M. O. Schmidt, 1963, *Surveying*. New York: Van Nostrand, 530p.

Roelofs, R., 1950, *Astronomy Applied to Land Surveying*. Amsterdam: Ahrend and Zoon, 259p.

Shalowitz, A. L., 1962, *Shore and Sea Boundaries*, 2 vols. Washington, D.C.: U.S. Govt. Printing Office, 1,168p.

Skelton, R. H., 1930, *Boundaries and Adjacent Properties*. Indianapolis: Bobbs-Merrill, 580p.

Tennessee Valley Authority Staff, 1951, Surveying, mapping, and related engineering, *Tennessee Valley Authority Tech. Rept. 23*, 415p.

Authored by experts of the National Geodetic Survey, The new adjustment of the North American Horizontal Datum (a series of twenty-five articles) *ACSM Bull.*, August 1975 (No. 50) through November 1981 (No. 75).

Cross-references: *Acoustic Surveys, Marine; Aerial Surveys; Geological Surveys, State and Federal; Harbor Surveys; Marine Magnetic Surveys; Remote Sensing, General; Satellite Geodesy and Geodynamics; Sea Surveys; Surveying, Electronic.*

SURVEYS, MUSEUMS—See GEOLOGICAL SURVEYS, STATE AND FEDERAL. Vol. XIII: EARTH SCIENCE, INFORMATION AND SOURCES; INFORMATION CENTERS.

T

TEPHROCHRONOLOGY

Tephrochronology comprises the identification, measurement, correlation, and dating of volcanic ash layers to establish sequences of geologic and archaeologic events. The term derives from the Greek word *tephra* ($\tau\epsilon\phi\rho\alpha$), originally used by Aristotle in his *Meteorologica* to denote volcanic ash and adopted by Thorarinsson (1954, 1) for "all the clastic volcanic material which during an eruption is transported from the crater through the air, corresponding to the term lava to signify all the molten material flowing from the crater." The clastic volcanics typically utilized in tephrochronological studies are pumice, ranging in composition from rhyolite to andesite, and less commonly, basaltic ash. The size descriptive terms and those of the corresponding sedimentary rocks are given in Table 1. (See also Fisher and Schmincke, 1984.)

A widespread mantle of tephra—i.e., volcanic ash or lapilli—deposited rapidly and rapidly buried, forms an excellent stratigraphic marker and has been used as such in many parts of the world. Air-fall deposits are normally used but the general principles are also applicable to pyroclastic flow deposits such as *nuées ardentes*. The ease with which such material is removed by subaerial erosion means, however, that tephrochronology can be applied only to deposits of Holocene, Pleistocene, and much less accurately, Pliocene age. Tropical weathering rapidly obliterates unconsolidated volcanic ash, and material deposited in temperate lakes, bogs, and valley bottoms stands the best chance of preservation. Due to deflation, a wet, cold climate is more favorable than an arid one. Local redeposition by water is a common feature.

The earliest tephrochronological studies were made in Iceland (Thorarinsson, 1944, 1954, 1970) and were followed by successful work in the western United States (summarized in Wilcox, 1965), New Zealand (Healy et al., 1964; Vucetich and Pullar, 1969; Neall, 1972), and Japan (Tsuya, 1955; Nakamura, 1964). More recently, detailed studies of the volcanoes of the Azores and the Canary Islands (Walker and Croasdale, 1971; Booth, 1973), of Somma-Vesuvius (Lirer et al., 1973), and

Published by permission of the Director, British Geological Survey.

in Mexico (Mooser, 1967; Bloomfield and Valastro, 1974, 1977) have been made. A *World Bibliography and Index of Quaternary Tephrochronology* was published by the International Union for Quaternary Research (Westgate and Gold, 1974), and an up-to-date review of tephra studies has been given by Self and Sparks (1981).

At first, tephrochronology was restricted to postglacial deposits and was intended as an adjunct to, and to some extent a replacement for, varve chronology, dendrochronology, and palynology, all of which are of limited value in almost treeless volcanic regions. Later, the scope widened considerably and the subject is now of great interest to the archaeologist, stratigrapher, geomorphologist, pedologist, and oceanographer. The methods used to identify, correlate, and differentiate tephra layers were initially based mainly on their field relationships and physical characteristics, but modern mineralogical and chemical techniques are being increasingly utilized.

Field Relationships

The position of one tephra layer relative to others and to soil horizons and its thickness, color, and degree of consolidation, together with the nature of internal stratification and grading, are all diagnostic. Shower, or mantle, bedding, giving distinctive fall units, is common in an air-fall pumice, and reverse grading is usual (Fig. 1). There is often thinning over local topographic highs and corresponding thickening on lee slopes, and other forms of thickness variation are shown in Fig. 2. Intraformational contacts are planar and conformable on lowland surfaces but wavy, with

TABLE 1. Descriptive Particle-Size Terms for Volcaniclastics and Sedimentary Rocks

Volcaniclastic	Size	Sedimentary
Block	256 mm	Boulder
	64 mm	Cobble
Lapilli	4 mm	Pebble
	2 mm	Granule
Coarse ash	1/16 mm	Very coarse to fine sand
Fine ash	>1/16 mm	Silt and clay

FIGURE 1. The air-fall Lower Toluca Pumice (L.P.) resting on a strong 24,500 yr. B.P. paleosol and separated from the thick Upper Toluca Pumice (U.P.) by an unconformity. Both pumices are from the Nevado de Toluca volcano, central Mexico. (From Bloomfield and Valastro, 1974, 903)

local angular unconformities, on more dissected uplands.

Layers of volcanic ash thin progressively when traced away from a volcano, and isopach maps give valuable information on the eruptive conditions, such as the axis of dispersal, determined mainly by the prevailing wind, and the total volume of tephra produced—an indication of the magnitude of the eruption. Figure 3 shows a typical isopach map of the A.D. 1104 eruption of Hekla in Iceland, and Fig. 4 shows the directions in which tephra spread during the initial phase of each of Hekla's historic eruptions (the widths of the arrows are roughly proportional to the volumes of the layers). Isochrones may also be drawn to represent the precise beginning of tephra fall in favorable cases, such as the eruption of Askja in Iceland on March 29, 1875. Tephra from this volcano reached Stockholm, 1,900 km away, the next day (Thorarinsson, 1970).

Vegetation killed off by a hot tephra cloud will normally be found as charcoal fragments at the base of the deposit, and radiocarbon dating of this material will give the true age of the eruption. However, if the ash layer is thin, trees and other vegetation can continue to grow after an eruption and thus form roots that are younger than the overlying tephra deposits. This possibility must be

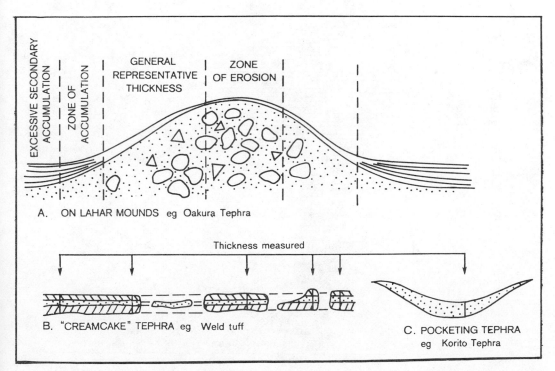

FIGURE 2. Thickness variations in the tephras of western Taranaki, New Zealand, showing so-called creamcake and pocketing features. (From Neall, 1972, 515)

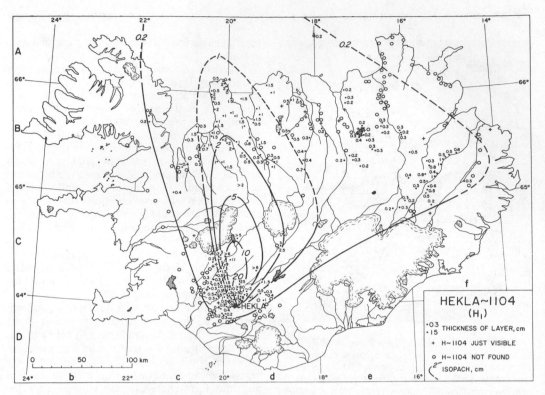

FIGURE 3. Isopach map of the A.D. 1104 tephra layer from the Hekla volcano, Iceland. (From Thorarinsson, 1970, Fig. 9, 306; copyright © 1970 by The Regents of the University of California)

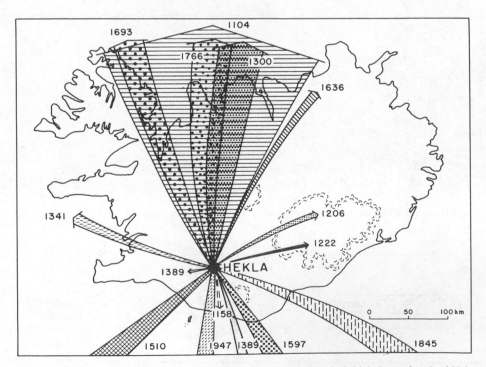

FIGURE 4. Map showing the directions of tephra spread during the initial phase of each of Hekla's historical eruptions. (From Thorarinsson, 1970, Fig. 8, 305; copyright © 1970 by The Regents of the University of California)

borne in mind when selecting suitable material for radiocarbon dating. On the one hand, a thin ash layer, producing little vegetational disturbance, forms an accretion to the existing soil, and it is difficult to identify individual accretions in a pumice soil solum only 20–30 cm thick. On the other hand, thick tephra deposits completely bury existing vegetation, and new soil forms on the sterile surface, well separated from the buried old soil. These buried soils, or paleosols, are best recognized by their dark color and well-developed blocky structure; they contain an accumulation of organic matter and an increased abundance of root channels. Weakly developed paleosols show only a blocky structure.

A clearly defined paleosol, rich in organic matter and underlying a tephra layer, can be used with care to date eruptions. Radiocarbon analyses give the mean residence time of formation of the soil carbon rather than the true age—i.e., the time that has elapsed since the beginning of humus formation in the regolith—but for Pleistocene material, the ages obtained approximate very closely to the period of soil formation. In younger Holocene deposits, the discrepancy between radiocarbon and calendar years must also be taken into account. By such methods it has been possible to show that, e.g., in the Central Volcanic Region of New Zealand, there have been 25 major eruptions in the past 40,000 yr.

Drift pumice occurs on several marine postglacial strandlines in northern Europe and the western Arctic and is a valuable correlation aid. Its distribution, combined with the known oceanic current circulation, points to Iceland as the probable source. Pumice found in many archaeological sites in northern Scotland was probably obtained from such strandlines. Widespread Quaternary tephras have been recognized in deep-sea sediments as much as 1,000 km east of New Zealand and in the Pacific to the west of the United States.

Physical Characteristics

The size, shape, and density of tephra are valuable aids to identification and correlation. The size decreases and the density increases away from the source volcano, and isograde maps, similar in form to isopach maps, can give indications of the eruptive conditions.

The three main constituents of pumiceous tephra deposits are pumice, lithics (rock fragments), and crystals, and their relative proportions serve to differentiate individual layers. Triangular diagrams showing the relative weight percentages are used and, e.g., show the differences between the Pompei and Avellino pumices produced by Somma-Vesuvius, Italy (Lirer et al., 1973).

Granulometric studies of tephra, using mechanical analyses, give useful diagnostic information. From them the median diameter Md_ϕ and the deviation σ_ϕ (a measure of the degree of size sorting) are the parameters most usually derived. The phi notation of Krumbein (1936) is used, and the set of sieves covers the range 32 mm to $\frac{1}{32}$ mm. The aperture size of one sieve is half that of the preceding one, and there is thus a size range of from phi -5 to $+5$, with a one phi interval between each. Both Md_ϕ and σ_ϕ characterize distinctive fall units, and from them, it is also possible to distinguish between pyroclastic fall and pyroclastic flow deposits (Walker, 1971). Grain-size data also enable the computation of the median terminal fall velocity of the tephra, leading to the derivation of limiting values for the height of the eruptive cloud and the wind strength (Lirer et al., 1973; Walker et al., 1971; Wilson, 1972).

A combination of granulometry and isopach data has been used by Walker (1973) to classify explosive volcanic eruptions, the significant parameters being the area of dispersal and the degree of fragmentation of the material. The three main types are hawaiian/strombolian, surtseyan, and plinian. The last, the powerful plinian gas-blast eruptions, produce the widespread pumiceous tephra blanket most useful in tephrochronological studies.

Heiken (1972) related ash morphology to magma composition and type of eruption. High-viscosity magma produces ashes whose morphology is governed mainly by vesicle density and shape, whereas low-viscosity magmas produce mostly droplets and spheres. Phreatomagmatic eruptions cause fragmentation and result in blocky or pyramidal glass ash particles. In most cases the shapes of the lithic fragments are equant and angular to subangular—a reflection of the fabric of the rock. A combination of petrography and scanning electron microscopy may be used to characterize and differentiate ash types (Heiken and Wohletz, 1985).

So-called soil monoliths that show critical stratigraphic relationships may be collected from suitable tephra sections for storage and reference. These are usually 2 m long sections, stabilized using polybutyl methacrylate, vinylite resin, and cellulose acetate.

Mineralogical Characteristics

The pumice, lithics, and crystals making up most tephra layers all have a distinctive *mineralogy* that can be used for correlative purposes, but the minerals are the most diagnostic. Separated phenocrysts of feldspars, pyroxenes, amphiboles, micas, and olivine have been used, and the study of single crystals has been greatly aided by the use of the spindle stage, developed by Wilcox and used by co-workers at the U.S. Geological Survey to determine variations in refractive index and other optical properties (Fig. 5).

Another method is to measure the *refractive index* of the glass in pumiceous tephra. This index increases with the degree of hydration, and although secondary alteration processes and initial inhomogeneity are possible additional variables, ash from one particular deposit normally has the same refractive index. Steen-McIntyre (1975) has found that the *rate* of hydration is dependent primarily on the chemical composition of the glass and the postdepositional environment but that the *extent* of hydration depends on the degree of crystallization, the specific surface of the fragment, and time. Hydration is completed in about 25,000 yr, but older glass is superhydrated and excess water collects in vesicles in amounts that appear

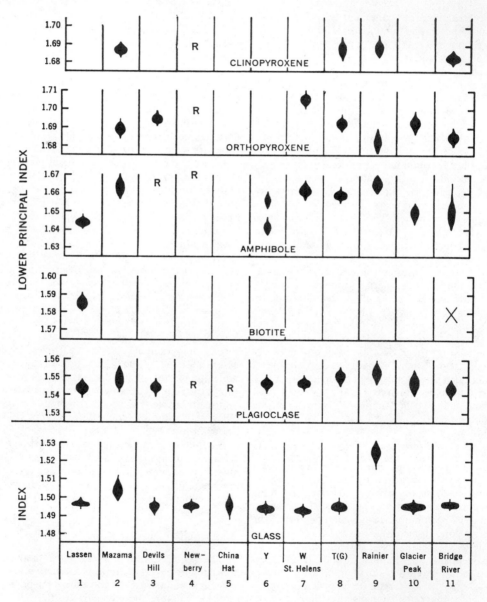

FIGURE 5. Refractive indexes of constituents of pumices and ash of postglacial or very late glacial age from vents in the Cascade Range, western United States. The widest part of the symbol indicates the predominant index, and the vertical height indicates the total range of the index. R, rare phenocrysts; X, possibly xenocrystic; blank, not found. (From Wilcox, 1965, Fig. 2, 810)

to be related to the age of the material. Pumiceous tephra as old as 0.7–2 m.y. may be dated by this method.

Chemical Characteristics

Chemical analyses of the major elements in pumices and other tephra serve to classify the parent volcanic rock but are usually insufficiently distinctive to be of value in correlation, due mainly to compositional variations within the same deposit (Heiken and Wholetz, 1985). The parental magma often changes composition during the course of a single powerful eruption. In general, the amount and nature of the phenocryst (or crystal) fraction is more important than the composition of the glass when considering differences in bulk chemical composition.

X-ray fluorescence, classical analysis, and electron beam microprobe techniques have all been used. Smith and Westgate (1969) found that Ca, K, and Fe are enough to characterize a pumice sample, but in central Mexico, principal component analysis suggests that a particular deposit could be defined by Si, Mg, Ca, and Fe. However, minor and trace element analyses are more significant. La, Th, Co, and the La/Yb ratio have proved good discriminants in ash samples from the Cascade Range volcanoes of the Pacific Northwest (Randle et al., 1971), and K/Rb ratios have been found useful in Mexico. The abundance of relatively immobile Ti has also been used to characterize Cascade pyroclastics (Czamanske and Porter, 1965).

The chemical composition of the individual minerals within a particular tephra is a much better discriminant than the bulk rock composition. This is particularly true of the easily separable titanomagnetite, which has been used extensively, e.g., in New Zealand. Here, specific abundances of Cr, V, Ni, Co, and Mn can be used to identify particular tephras (Kohn and Neall, 1973).

Direct Dating Methods

The most obvious direct dating method involves a careful search through ancient documents, and this method has been used with great success in, e.g., Iceland to date the eruptions of Hekla in historical times (Thorarinsson, 1967). Here the earliest recorded eruption to produce a distinctive tephra layer occurred in A.D. 1104. Another example is the volcano Popocatepetl in central Mexico, some eruptions being recorded in the Aztec codices.

Pumice fragments are occasionally found in the forks of living trees, but before recorded history, recourse must usually be made to radiocarbon dating of included charcoal or of paleosols, referred to earlier. The age limit for this method is about 35,000 yr B.P., but under very favorable conditions, it may be extended back as far as 50,000 yr B.P. In some cases direct dating, utilizing the rate of peat accumulation from the top of an ash layer to the present ground surface, is possible.

Older tephra may be dated by analyzing potassium-rich minerals such as micas by the K/Ar or ^{40}Ar/^{39}Ar methods, but these are usually only applicable to samples older than about 500,000 yr. There is thus a time gap of some 450,000 yr that cannot be spanned by normal radiometric methods, although with caution the uranium series method could be used. This method involves the measurement of uranium isotopes and their long-lived decay products ^{230}Th and ^{231}Pa in bone samples but is subject to possible anomalies (Szabo et al., 1969).

Measurements of the demagnetized natural remnant magnetism (NRM) of tephra layers and associated sediments can be used to determine the direction of the Earth's magnetic field during Quaternary time. Marked changes in this direction, known as geomagnetic field excursions and lasting a few hundred years, may provide a means for dating and correlating such deposits in restricted areas.

K. BLOOMFIELD

References

Bloomfield, K., and S. Valastro, 1974, Late Pleistocene eruptive history of Nevado de Toluca volcano, central Mexico, *Geol. Soc. America Bull.* **85**, 901–906.

Bloomfield, K., and S. Valastro, 1977, Late Quaternary tephrochronology of Nevado de Toluca volcano, central Mexico, *Overseas Geol. Miner. Resour., London,* No. 46, 15p.

Booth, B., 1973, The Granadilla Pumice Deposit of southern Tenerife, Canary Islands, *Geol. Soc. London Proc.* **84**(3), 353–370.

Czamanske, G. K., and Porter, S. C., 1965, Titanium dioxide in pyroclastic layers from volcanoes in the Cascade Range, *Science* **150**, 1022–1025.

Fisher, R. V., and H.-U. Schmincke, 1984, *Pyroclastic Rocks.* New York: Springer-Verlag, 472p.

Healy, J., C. G. Vucetich, and W. A. Pullar, 1964, Stratigraphy and chronology of late Quaternary volcanic ash in Taupo, Rotorua, and Gisborne districts, *New Zealand Geol. Survey Bull,* n.s., 73.

Heiken, G., 1972, Morphology and petrography of volcanic ash, *Geol. Soc. America Bull.* **83**, 1961–1988.

Heiken, G., and K. Wohletz, 1985, *Volcanic Ash.* Berkeley, Calif.: University of California Press, 246p.

Kohn, B. P., and V. E. Neall, 1973, Identification of late Quaternary tephras for dating Taranaki lahar deposits, *N.Z. Jour. Geology and Geophysics* **16**, 781–792.

Krumbein, W. C., 1936, Application of logarithmic moments to size frequency distribution of sediments, *Jour. Sed. Petrology* **6**, 35–47.

Lirer, L., T. Pescatore, B. Booth, and G. P. L. Walker, 1973, Two Plinian pumice-fall deposits from Somma-Vesuvius, Italy, *Geol. Soc. America Bull.* **84,** 759–772.

Mooser, F., 1967, Tefracronologia de la Cuenca de Mexico para los ultimos treinta mil anos, *México Inst. Nac. Antropol. History Bol.* **30,** 12–15.

Nakamura, K., 1964, Volcano-stratigraphic study of Oshima Volcano, Izu, *Tokyo Univ. Earthquake Research Inst. Bull.* **42,** 649–728.

Neall, V. E., 1972, Tephrochronology and tephrostratigraphy of western Taranaki (N108-109), New Zealand, *New Zealand Jour. Geology Geophysics* **15,** 507–557.

Randle, K., G. G. Goles, and L. R. Kittleman, 1971, Geochemical and petrological characterization of ash samples from Cascade Range volcanoes, *Quat. Research* **1,** 261–282.

Self, S., and R. Sparks (eds.), 1981, *Tephra Studies.* Dordrecht, Holland: Reidel, 481p.

Smith, D. G. W., and J. A. Westgate, 1969, Electron probe technique for characterizing pyroclastic deposits, *Earth and Planetary Sci. Letters* **5,** 313–319.

Steen-McIntyre, V., 1975, Holocene and Pleistocene volcanic ash beds—extent of glass hydration may give a rough estimate of age (abs.), Southwestern Anthropological Association/Sociedad Mexicana de Antropologia Meeting, Santa Fe and Los Alamos, New Mexico, March 27–29.

Szabo, B. J., H. E. Malde, and C. Irwin-Williams, 1969, Dilemma posed by uranium-series dates on archeologically significant bones from Valsequillo, Puebla, Mexico, *Earth and Planetary Sci. Letters* **6,** 237–244.

Thorarinsson, S., 1944, Tefrakronologiska studier pa Island (Tephrochronological studies in Iceland), *Geog. Annaler* **26,** 1–217.

Thorarinsson, S., 1954, The tephra fall from Hekla on March 29th 1947, in *The Eruption of Hekla 1947–1948,* vol. 2. Reykjevik: H. F. Leiftur (Societas Scientiarium Islandica), 1–68.

Thorarinsson, S., 1967, The eruptions of Hekla in historical times, in *The Eruption of Hekla 1947–1948,* vol. 1. Reykjevik: H. F. Leiftur (Societas Scientiarium Islandica), 1–170.

Thorarinsson, S., 1970, Tephrochronology and medieval Iceland, in R. Beyer, ed., *Scientific Methods in Medieval Archaeology.* Berkeley: University of California Press, 295–328.

Tsuya, H., 1955, Geological and petrological studies of volcano Fuji, 5. On the 1707 eruption of volcano Fuji, *Tokyo Univ. Earthquake Research Inst. Bull.* **33,** 341–383.

Vucetich, C. G., and W. A. Pullar, 1969, Stratigraphy and chronology of late Pleistocene ash beds of central North Island, New Zealand, *New Zealand Jour. Geology and Geophysics* **10,** 1109–1135.

Walker, G. P. L., 1971, Grain-size characteristics of pyroclastic deposits, *Jour. Geology* **79,** 696–714.

Walker, G. P. L., 1973, Explosive volcanic eruptions—A new classification scheme, *Geol. Rundschau* **62,** 431–446.

Walker, G. P. L., and R. Croasdale, 1971, Two Plinian-type eruptions in the Azores, *Geol. Soc. London Quart. Jour.* **127,** 17–55.

Walker, G. P. L., L. Wilson, and E. L. G. Bowell, 1971, Explosive volcanic eruptions. I: The rate of fall of pyroclasts, *Royal Astron. Soc. London Geophys. Jour.* **22,** 377–383.

Westgate, J. A., and C. M. Gold, 1974, *World Bibliography and Index of Quaternary Tephrochronology.* Edmonton, Canada: University of Alberta, 528p.

Wilcox, R. E., 1965, Volcanic ash chronology, in H. E. Wright, Jr., and D. G. Frey, eds., *The Quaternary of the United States.* Princeton, N.J.: Princeton University Press, 807–816.

Wilson, L., 1972, Explosive volcanic eruptions. II: The atmospheric trajectories of pyroclasts, *Royal Astron. Soc. London Geophys. Jour.* **30,** 381–392.

Cross-references: *Lichenometry.* Vol. XIII: *Geochronology; International Geochronological Time Scale.*

TERRAIN EVALUATION SYSTEMS

The term *terrain evaluation* is used to denote the assessment of properties of terrain as a whole, whether over large or small areas, that are applicable to any given purpose. The term is used mostly when areas are being assessed for engineering or urban and regional planning purposes, but it is not restricted to these purposes. Forms of terrain evaluation have been applied to similar assessments for rural purposes (see *Land Capability Analysis; Open Space*) including agriculture and pasture, but in general, the term has not been used for this type of work.

The concept of terrain evaluation originally stemmed from the desire to make predictions about the properties of unknown terrain from known properties of terrain in other areas. The necessity for this prediction technique arose during World War II (1939–1945) when, in the course of the fighting, there was neither available time nor, indeed, opportunity to survey and determine the properties of terrain over which the opposing forces had to move. During and after the war, the idea was taken up in the civilian field, first, with the wartime lack of labor, to acquire the knowledge about the terrain and do the civil engineering necessary for the prosecution of the war and, second, in the postwar years, to open up large tracts of land that had hitherto not been developed.

The earliest published work in the field of terrain evaluation was that of Christian and Stewart (1953), who developed a system for terrain evaluation, although it was not called that, as an aid to the development of northern Australia, and Belcher (1943), who at Purdue University, became interested in the spatial distribution of soils and other natural attributes that have a bearing on the engineering characteristics of terrain. Belcher (1942) was also an early advocate of the use of aerial photographs to determine these distributions. A synthesis of these two approaches was achieved later independently by Aitchison and Grant (1967), in Australia, who developed a system for terrain evaluation for general engineering and urban and regional planning purposes. Beckett and Webster (1965), in England, were interested in the extrapolation of terrain properties relevant to military use from known to unknown areas. Brink and Partridge (1967), in South Africa, considered aspects of terrain for road building and engineering geological purposes. Grabau (1963), in the United States, attempted to make predictions of ground mobility for military purposes. Many others among whom are Dowling (1963) and Clare and Beavan (1962) in England and Hittle (1949), Lueder (1951), and Miles (1962) in South Africa, used some of the principles of terrain evaluation without, however, systematizing their approaches.

Principles of Terrain Evaluation

Terrain evaluation depends on the ability to classify terrain and to evaluate for each member of each class the properties of the terrain that are significant for the desired purpose. Inherent in the process is the ability to predict terrain properties from known to unknown areas. Many systems for classifying terrain for various purposes have been proposed. These range from landscape approaches—i.e., classifying the terrain in terms of its inherent characteristics—to parametric approaches in which the terrain is classified in terms of parameters significant for some given purpose (Grabau, 1963). The inherent difficulty in using the parametric approach is that each purpose often requires parameters that are different, at least, in some respect from those required for any other purpose; hence, each purpose requires not only its own terrain classification but also that classification to be based on the parameters collected in a statistically significant number in a random manner. The predictive capability of such a system based, as it is, on measured terrain properties rather than naturalistic terrain attributes is low.

Conversely, the landscape approach assumes that parameters significant for any given purpose will be related to the inherent characteristics of the terrain; hence, the terrain classification is completed at least in outline before the collection of parameters is attempted. The predictive capability of such a system is high. Experience has shown that the assumption, although difficult to prove, appears reasonable.

Most systems for terrain evaluation use the landscape approach for the terrain classifications stage of the process because of the inherent difficulties of collecting the sufficient samples necessary for the parametric approach and the higher predictive capability of the system. Some systems, however, combine the landscape and parametric approaches, mostly using parameters that relate only to the physiography of the terrain in the clas-

sification stage. In general, this type of system has had the most widespread application.

The various types of terrain evaluation systems available for use have been reviewed by Mitchell (1973). The best documented and most widely used system for engineering and urban and regional planning use is due to Grant (1975a, 1975b). This system, known as the P.U.C.E. (Pattern, Unit, Component Evaluation) system for Terrain Evaluation and revised by Grant and Finlayson (1978), uses a combination of the landscape and parametric approaches for terrain classification. The parameters used are physical and relate only to the physiography of the terrain. The system then develops methods for evaluation and makes provision for storing, processing, and retrieving of information gathered both at the classification and evaluation stages.

Summary of the P.U.C.E. System

Terrain Classification The P.U.C.E. System divides all land into four basic classes.

Terrain Component The terrain component is defined explicitly in terms of its three-dimensional slope characteristics using two orthogonal slope axes and one axis that may not be orthogonal to the other two; its soil, in which each layer can be classified into one class of the Unified Engineering Soil Classification (U.S. Department of Interior, 1974) and the whole profile into one class of its subdivided primary profile form (Northcote, 1974); and its vegetation association. Terrain components can be mapped at a scale of 1:10,000 or greater.

Terrain Units Terrain units are physiographic features of the landscape (landforms) with consistent soil associations and vegetation formations that consist of a limited repetitive association of a small number of terrain components. Terrain units are limited in number, can be listed, and can be mapped at a scale of 1:50,000 or greater.

Terrain Patterns Landscapes, which are usually coincident with a consistent pattern or an aerial photograph, have a consistent local relief amplitude and drainage pattern. They consist of a limited repetitive association of a small number of terrain units. Terrain patterns can be mapped at a scale of 1:250,000 or greater.

Provinces These units exhibit consistent geology at the group level. They consist of a limited repetitive association of a small number of terrain patterns.

Description of System The P.U.C.E. system is based on the premise that all terrain is produced by the interaction of climate and geology, with time as the operator. The classification is basically geological, physiographical, pedological, and botanical, with the effects of climate reflected in the nature of the landforms, soils, and vegetation. The nomenclature used in the P.U.C.E. system is essentially numerical and is compatible with a digital computer for storing, processing, and retrieving information. *Terrain components* are allocated eight digits: the types of slopes along the orthogonal axes (i.e., the combinations of convex, concave, and planar slopes) (1), the maximum angles of slope along each of the orthogonal axes (2, 3), the soil (serial) (4, 5), land-use (in developed areas) or surface cover (in undeveloped areas) (standardized by listing) (6), and the vegetation association (serial) (7, 8). *Terrain units* are allocated four units: the landform (standardized by listing) (1, 2), the soil (standardized by listing) (3), and the vegetation formation (standardized by listing) (4). *Terrain patterns* are allocated three digits: the maximum local relief amplitude (classified using significant class intervals) (1), the stream density (classified using significant class intervals) (2), and serial (3). *Provinces* are allocated five digits: the geological area (1), the geological period within each era (2), and serial (3, 4, 5).

The terrain classification of any area is established by a process of aerial photograph interpretation using stereopairs of photographs of suitable scale followed by a short period of fieldwork to establish the soil and vegetation characteristics and to validate the interpretation of the members of each class and their boundaries. An example of terrain classification for the Albury–Wodonga (Australia) area is given in Grant et al. (1981).

Evaluation The P.U.C.E. terrain evaluation system is capable of working at four levels of detail consistent with the terrain classes, terrain pattern, terrain unit, and terrain component. Thus, for initial broad-scale feasibility studies, terrain patterns and terrain units are recognized and evaluated for the purpose proposed; terrain patterns may be mapped. Over the more limited area defined by the feasibility study, terrain units may be mapped and evaluated in more detail for planning purposes. Finally, over a more limited area defined by the planning process, terrain components may be recognized; mapped, if necessary; and evaluated to the extent that construction can immediately proceed. An example of such an evaluation is included in Grant et al. (1981).

The technique of terrain evaluation is equally applicable to the planning of new developments in the open countryside or of redevelopment of city areas that are no longer viable in their present context within the city scene; each type of development, of course, imposes its own constraints on the purposes of and methods used for evaluation. In areas of new development, the number of degrees of freedom for location is usually large; in areas of redevelopment the number of degrees of freedom is most often small.

The evaluation stage consists of specifying the properties of the terrain and its attributes that are critical for the desired purpose and, for those properties, the level of detail required to be measured or estimated for each phase in the feasibility, planning, or construction process. The evaluation continues by measuring or estimating and recording the values of these necessary parameters in a sufficient number of terrain samples in terms of the level in the terrain classification appropriate to them and extrapolating to all occurrences of all members of all classes, the properties of each member of each class when these properties have been established by testing a significant number of terrain samples.

While the terrain classification stage in the process of terrain evaluation can be done in a standardized ordered form, evaluation is always tied to a particular objective, whether large or small, that requires the collection of a set of information different from the set required for any other objective. This is true whether broad-scale purposes or more detailed projects are under consideration. In the engineering or urban and regional planning sense, the required information can be considered as existing in three sets. One set of information is required for feasibility studies, another more detailed set for planning, and another yet more detailed set for construction purposes. These sets of information may overlap and contain some items in common. For all purposes, there is virtually an unlimited number of sets of information required; such a situation defies any attempt at overall standardization of procedures, although the procedures for the collection of some items may be standardized.

Storage, Processing, and Retrieval of Terrain Information One of the advantages of using a terrain evaluation system, such as the P.U.C.E. system, for collecting information about the terrain is that, provided the system is properly constructed, the information obtained may be stored, processed, and retrieved for future use. The actual system for information processing must necessarily be of an incremental type in which it is possible to add information continually as it is collected and to retrieve data as they are required.

A key to a successful system for information processing is a method for encoding the information presented to the system. Encoding ensures that information is always presented to the system in a uniform manner. Because terminology is standardized, there is no duplication of information and no difficulties are experienced in interpreting the precise meaning of every item placed in the system.

As has already been pointed out, each objective of the terrain evaluation process requires the collection of a set of information that is essentially different from the set required for any other objective. A collected set of information can be added to the system as an increment. Equally, a set of information once placed in the system can be withdrawn as a dividend at any time. Also, if a set of required information contains some items that have already been collected on any other set, that information can be retrieved from the system as particular items are added to the required set; that is, there is no duplication in the collecting of information.

K. GRANT

References

Aitchison, G. D., and K. Grant, 1967, The P.U.C.E. Programme of terrain description, evaluation and interpretation for engineering purposes, *Proc. 4th Regional Conf. for Africa on Soil Mechanics and Found. Eng.* **1**, 1–8.

Beckett, P. H. T., and R. Webster, 1965, *A Classification System for Terrain*. Christchurch, England: Military Engineering Experimental Establishment (Rept. No. 987), 28p.

Belcher, D. J., 1942, Use of aerial photographs in war time engineering, *Roads and Streets* **85**(7), 35.

Belcher, D. J., 1943, The engineering significance of soil patterns, *Proc. 23rd Ann. Mtg. of the Highway Research Board* **23**, 569.

Brink, A. B. A., and T. C. Partridge, 1967, Kyalami land system: An example of physiographic classification for the storage of terrain data, *Proc. 4th Regional Conf. for Africa on Soil Mechanics and Found. Eng.* **1**, 9.

Christian, C. S., and G. A. Stewart, 1953, General report on survey of the Katharine-Darwin region, 1946, *CSIRO Australia Land Research Ser. No. 1*, 156p.

Clare, K. E., and P. J. Beavan, 1962, Roadmaking materials in British Borneo: I. Geology, roadstone and soil classification. *Harmondsworth Road Research Lab. Note No. LN/86;* II. Properties of soils, *Harmondsworth Road Research Lab. Note No. LN/87;* III. Soil location and stabilization, general conclusions, *Harmondsworth Road Research Lab. Note No. LN/88.*

Dowling, J. W. F., 1963, The use of aerial photography in evaluating engineering soil conditions in a selected area of northern Nigeria, *Harmondsworth Road Research Lab. Note No. LN/379.*

Grabau, W. E., 1963, A comparison of quantitative versus descriptive systems for mobility analysis, *U.S. Army Corps Engineers Waterways Expt. Sta. Tech. Rept.*

Grant, K., 1975a, The P.U.C.E. programme for terrain evaluation—I. Principles, *CSIRO Australia Div. Appl. Geomech. Tech. Paper. No. 15,* 2nd ed., 32p.

Grant, K., 1975b, The P.U.C.E. programme for terrain evaluation—II. Procedures for terrain classification, *CSIRO Australia Div. Appl. Geomech. Tech. Paper No. 19,* 2nd ed., 68p.

Grant, K., and A. A. Finlayson, 1978, The assessment and evaluation of geotechnical resources in an urban or regional environment, *Eng. Geology—Internat. Jour.* **12**, 219–293.

Grant, K., T. G. Ferguson, A. A. Finlayson, and B. G. Richards, 1981, Terrain analysis, classification, assessment and evaluation for regional development purposes of the Albury-Wodonga Area, New South Wales and Victoria. Vol. 1: Terrain analysis and classification; Vol. 2: Terrain assessment and evaluation, *CSIRO Div. of Appl. Geomechanics Tech. Paper No. 30,* 244p., 3 maps.

Hittle, J. E., 1949, Airphoto interpretation of engineering sites and materials, *Photogramm. Eng.* **15**, 589.

Kantey, B. A., and A. A. B. William, 1962, The use of soil engineering maps for road projects, *Civil Eng. South Africa* **4**, 149; **5**, 41.

Lueder, D. R., 1951, The preparation of an engineering soil map of New Jersey, *Am. Soc. Testing and Materials Spec. Tech. Pub. No. 122,* 73.

Miles, R. D., 1962, A concept of land forms, parent materials and soils in airphoto interpretation studies for engineering purposes, in *Transactions of the Symposium on Photo Interpretation*. Delft: International Society for Photogrammetry, 462–476.

Mitchell, W. C., 1973, *Terrain Evaluation—The World's Landscapes*. London: Longman, 221p.

Northcote, K. H., 1974, *A Factual Key for the Recognition of Australian Soils,* 4th ed. Adelaide: Rellim, 123p.

Townsend, J. R. G., 1981, *Terrain Analysis and Remote Sensing*. Winchester, Mass.: Allen & Unwin, 240p.

U.S. Department of the Interior, Bureau of Reclamation, 1974, *Earth Manual,* 2nd ed. 810p.

Cross-references: *Environmental Engineering; Environmental Management; Land Capability Analysis.* Vol. III: *Terrain, Terrane.* Vol. XIII: *Geology, Applied; Military Geoscience; Terrain Evaluation, Military Purposes.*

THERMAL SURVEYS—See AERIAL SURVEYS; PHOTOGEOLOGY. Vol. XIII: REMOTE SENSING, ENGINEERING GEOLOGY.

U

UNIFORMITARIANISM

Uniformitarianism is the doctrine that the geological processes operating in the past were similar to those that can be demonstrated today. It is also known as the *principle of uniformity* and is often summarized concisely in the phrase "the present is the key to the past" (see Read, 1949). Without making an assumption like this, it would indeed be difficult to interpret the tangle of strata and associated landforms that represent the result of processes now no longer active. Leonardo da Vinci appears to have been the first person to realize that the occurrence of marine fossils in rocks a long way from the present-day sea must imply that the seas once covered much of the land areas of today. As da Vinci pointed out, cockle shells move extremely slowly, and they could not possibly have traveled the 250 mi from the present sea coast to Lombardy during a flood lasting only 40 days! Yet the fossil shells showed the same muscle scars and growth lines of their present-day marine counterparts and would presumably have been secreted by the same types of organisms.

The next step in the application of the principle of uniformity was made by Johann Gottlob Lehmann (1759), a mining geologist, and Christian Fuchsel, a physician, when they independently applied what is called the concept of *super-position of strata,* now regarded as one of the basic laws of stratigraphy. According to this concept, if one observes a water-laid bed overlying another bed or formation, the one on top will be younger than the one underneath. Where the attitude of the two beds is different, material forming the younger stratum may have been partly derived by erosion of the underlying layers. Again, independently, the Russian scientist, M. V. Lomonosov (1711-1765) had also developed a logical theory of an evolving Earth.

It was Fuchsel (1761), writing in Latin, who introduced the word *stratum* for a layer of bed. He also showed that many layers of rock were characterized by their own particular kinds of fossil organisms, realizing that strata could be matched or differentiated by the inspection of the organic remains found in them. He followed the example of da Vinci in applying the study of present-day processes of sedimentation to the interpretation of the older rocks on the assumption that they constituted records of similar processes operating in the past. Here, then, are examples of pioneer scientists successfully applying the doctrine of uniformitarianism.

Almost to the end of the eighteenth century the concept had not been formalized or widely publicized. Instead it was overlooked, due partly to the competing influence and teachings of Abraham Gottlob Werner (1749-1817), professor at the School of Mines in Freiberg, Saxony. An able teacher, Werner developed what became known as the *Neptunist theory.* This theory presumed that at an early stage in its history, the Earth possessed an irregular nucleus completely covered by an ocean whose waters contained in solution or suspension all the materials required to make the crust. First, this ocean deposited granite, and the granite was followed by a variety of other crystalline rocks. Werner called these two groups the *primitive rocks* since they were massive and devoid of fossils. Then came the *transition rocks,* consisting partly of similar crystalline chemical precipitates (so he believed) and partly of debris derived from the older rocks. During this time, the ocean was believed to be shrinking in volume. Then came rocks consisting entirely of debris eroded off the emerging land surface—i.e., clastic sediments, together with marine fossils—and last came alluvial deposits to cloak the lower parts of the land areas.

Numerous problems arose with this hypothesis, but it should be noted that Werner accepted the orderly succession of rocks as indicating the development of Earth history, which itself is an application of the principle of uniformity. Conversely, he disregarded the very same principle in his interpretation of the significance of many individual deposits. As a result, it was left to Nicholas Demarrest to show that basalts like those Werner found in the geologic sequence in many places are actually produced by volcanoes today. By the principle of uniformity, rocks must be igneous (formerly molten), not chemical precipitates from the ocean.

As the number of geologists multiplied and fieldwork progressed, the various interpretations of the formation of crystalline rocks according to the Neptunist theory were largely disproved, in favor of the *Vulcanist theory,* under the leadership of Hutton, Hall, Pallas, and de Saussure.

In the meantime, Hutton (1788) reverted to the strict view of uniformitarianism. He wrote:

No powers are to be employed that are not natural to the globe, no action to be admitted of except those of which we know the principle, and no extraordinary events to be alleged in order to explain a common appearance. The powers of nature are not to be employed in order to destroy the very object of those powers; we are not to make nature act in violation to that order which we actually observe, and in subversion of that end which is to be perceived in the system of created things. . . . Chaos and confusion are not to be introduced into the order of nature, because certain things appear to our practical views as being in some disorder. Nor are we to proceed in feigning causes when those seem insufficient which occur in our experience.

Hutton's views, when fully expounded by Sir Charles Lyell (1830), have become the basis of modern geology and, in fact, the entire structure of logical science. Indeed, Lyell's book, *Principles of Geology,* with its many editions, was probably to become one of the three most influential books in the English-speaking world of the 1800s. The others, of course, were the *Bible* and the *Origin of Species* by Charles Darwin (1859), one of Lyell's closest friends and companion on field trips. The best students of Werner eventually recognized the weaknesses in their master's teachings, so the doctrine of uniformitarianism became the basic premise of geological studies. Combined with careful and continued fieldwork, Lyell's great masterpiece remains a fundamental building block in the reading of the history of the Earth as written in its rocks and landforms. In Germany, the work of von Hoff (1771–1837) complemented that of Lyell.

At least two distinct interpretations have developed as to the implications of uniformitarianism, and these have led to much discussion. These interpretive philosophies may be generalized as follows:

Logical rationalism employs the spirit and letter of Hutton's argument—i.e., that one interprets the evidence of past events on a basis of logical reasoning, a basis of induction, avoiding supernatural or ad hoc hypotheses. Intuitive insight will not be accepted unless it is supported by demonstrable evidence. That evidence will be based on experimental procedures or field observations of rock relationships or of present-day dynamic phenomena. This part of uniformitarianism was needed particularly at the beginning of the last century as a counter to the *theory of catastrophism,* the presumption of unheralded disasters or cataclysms that might terminate one cycle (physical or biological) and initiate another. Today, logical rationalism is accepted as a standard procedure. Gould (1965) calls it "methodological uniformitarianism."

Another philosophy, *paleo-conservatism,* presumes that in the past the natural processes were always just as they are today—thus, a conservative view implying a continuity of the status quo through time. This aspect is called "substantive uniformitarianism" by Gould (1965) and is "false and stifling to hypothesis formation." Yet, unfortunately, it still pervades much geological thinking. In its defense one may comment that for simplicity in any approach to a problem of the past, employing the procedure of multiple hypotheses (see *Geology, Philosophy*), it is appropriate to assume as hypothesis No. 1 that conditions of geography, climate, oceanography, etc. would be closely comparable to those of today before embarking on any lines of reasoning that call for fundamental changes in distribution or quantity—e.g., polar shift, continental spreading, displacement of climatic belts, alteration of ocean salinity, etc. Such concepts call for limited geological catastrophes (Miller, 1940), a mild version of *neocatastrophism* (Schindewolf, 1962), consisting of periodic revolutionary or threshold events initiating new evolutionary developments (Fairbridge, 1964).

Other interpretations and classifications can be found in the literature, e.g., Coates and Vitek (1980) suggested that critical thresholds that can be demonstrated for certain geomorphic processes such as landslides form an intermediate position between catastrophism and uniformitarianism. However, careful examination of the argument suggests that thresholds are consistent with uniformitarianism since they occur today. They may not have been allowed for in all the teachings of all writers, such as W. M. Davis (1884), but this does not mean that they are inconsistent with the original concept of uniformitarianism.

If one rigidly adopts the status quo approach as an inflexible rule, then it is true that the principle of uniformity does not always work. It has therefore become the practice to use the conservative procedure until the evidence indicates that this approach is too simple. The evidence for the former existence of ice sheets covering such widely separated areas today as parts of Australia, Antarctica, South Africa, South America, and India in the Permo-Carboniferous periods defies a conservative uniformitarian interpretation unless some drastic revolutionary event is accepted.

Several explanations have been put forward to explain the Permo-Carboniferous glaciation. Today its outer boundaries extend over an entire hemisphere, 20,000 km in diameter. Possibilities considered include the following:

1. A polar shift, to the middle of the Indian Ocean, has been considered, but this is not enough; the glacial limits would still reach up to 90° of arc from the south pole—i.e., to the equator of that time.

2. Some workers have suggested a rapidly changing polar shift, beginning in Brazil, then

Africa, India, Australia to Antarctica, so that only a limited area not exceeding 40° of arc was glaciated at any one moment; the stratigraphy and paleontology are not perfectly preserved, but the apparent correlations would seem to rule out this explanation.

3. The rate of the Earth's axial rotation may have been slowed down to coincide with that of its orbital revolution like that of the moon. [It used to be thought that Mercury was like this, but modern observations indicate an axial rotation of almost two times per sidereal year.] In this way, one side of the planet Earth would be constantly cold and the other side extremely hot. Apart from difficulties in celestial mechanics, requiring the aperiodic deceleration and acceleration of the globe, the stratigraphic-paleontologic evidence of northern hemisphere conditions at this time suggests a normal climate.

4. The so-called glacial deposits are not really glacial at all. They are pebbly mudstones (diamictite) produced by mudslides and other products of gravitative sedimentation that lead to pseudoglacial striations. Impoverished faunas might be related to high salinity rather than excessive cold. Objections to this approach are that the deposits in question are distributed over all sorts of basement terrains, including low relief shield areas, unsuitable for pebbly mudstone (geosynclinal) facies. Besides, the timing is unique: the deposits claimed to be glacial occur over the Southern Hemisphere at approximately the same horizon, while in the Triassic there is no trace of such formations.

5. Accepting that they are really glacial and formed on a continent or group of closely related land areas never more than 10,000 km across, a theory of continental drifting was proposed by Wegener, Taylor, and others, which would carry the floating continents apart until they reached their present distribution. Modern seismic evidence shows that continents are closely related to their underlying (and modified) mantle; iceberg-type floating thus seems to be impossible.

6. A theory of continental spreading was proposed by Gutenberg, followed up by a concept of ocean enlargement with growth of suboceanic crust (from deep mantle sources) developed variously by Egyed, Carey, Heezen, and others, meets the need for a splitting apart of continents but avoids the free floating of continents. The mid-oceanic ridge is taken to be the locus of new splitting and ocean growth, being fed by expanding mantle material brought up from high-pressure levels near the core and thus expanding under phase change and with geochemical modification—e.g., the hydration of olivine-type magma to form much less dense serpentine (theory of Hess)—possible with convection currents (theories of Vening-Meinesz and Dietz). Studies of *paleomagnetism* (q.v. in Vol. V) have lent strong support to the idea of continental spreading, and this has been confirmed by determining the age of the various parts of the ocean floor by paleontology and potassium argon dating. The youngest parts of the ocean floor lie along the mid-oceanic ridges, and the seafloor becomes progressively older as it is traced away to the margins of the oceans. All the ocean floors appear to be relatively young.

The doctrine of paleoconservative uniformity cannot always be applied to the formation of rocks. Certain rock types—e.g., some of the sedimentary iron ores—are not being formed today. We therefore must make use of associated features to deduce the conditions under which they were formed. Thus, in the case of the Northamptonshire iron ores of England, the ore consists of *ooliths* of chamosite (iron silicate), in a siderite (iron carbonate) matrix. They must be sedimentary in origin because they appear to be bedded, they contain occasional marine fossils, and they can be traced southward into beds of ooliths of limonite in a calcareous matrix and then into oolitic limestones. We know that oolitic limestones form today on the Bahama Banks in shallow water where seawater rich in calcium carbonate becomes warmed and evaporated until aragonite and calcite precipitates (partly biogenic) are produced with accumulate on the shallow seafloor, but we have no analogues for the gradual change to ferruginous ooliths in Northamptonshire. The conditions were clearly marine, but how this part of the sea came to produce thick sedimentary deposits containing over 30% iron remains a mystery. There are no signs of barriers, and the modern seas offer no comparative model. In fact, iron salts appear in neither the normal sequence of compounds deposited by evaporation of seawater nor in most fossil evaporite deposits—e.g., the Stassfurt Beds in Germany. The limited iron salts merely give a reddish color due to their presence as an impurity in the last salts to form sylvite and carnallite.

Re-examination of mountain ranges is showing that they can be explained by the theory that the Earth's surface consists of a series of floating plates moving away from the oceanic ridges and interacting along their margins. These plates may be under tension (divergent plates), resulting in rift valleys on land or in the ocean, or they may be under compression (convergent plates). The latter can produce mountain ranges where two plates collide—e.g., the Himalayas between India and Asia. Movement of plates over more slowly moving magma sources can produce chains of volcanoes such as the Hawaiian chain, with the oldest islands occurring farthest from the magma source. A combination of erosion and isostatic adjustment to the sudden local overloading of the plate produces a sequence from volcanic islands with fringing reefs, to atolls, with increasing age. The Hawaiian, Pitcairn, and MacDonald chains in the Pacific Ocean are all dog-legged, demonstrating a change in direction of movement of the Pacific Plate about 40 m.y. ago. Extending this idea of changes in the direction of plate movement with time, together with changing plate boundaries, it is possible to explain many, if not most, former mountain ranges. This concept is called *plate tectonics,* and much geological research in the late 1960s and 1970s was concerned with exploring the applicability of these ideas.

Similar difficulties occur with plutonic rocks. We can see a volcano in action. We can trace a lava flow back to a pipe. We can trace dikes and sills laterally into larger igneous bodies such as laccoliths and batholiths, but we can never watch a magma crystallizing in a batholith. Then again there are special kinds of magmas such as *carbonatite intrusions* (q.v. in Vol. V) and those diatreme pipes in which diamonds occur. We do not seem to have any active volcanism of such types today, and so, once again, the origin must be inferred.

In the science of paleoclimatology, it is clear that glacial deposits are now recognized and accepted as occurring in areas where no ice is found today. Similarly, red, highly leached soils and weathering zones in the Sudan and East Africa, occurring in regions where the present-day climate is semiarid, are nevertheless recognized as products of humid monsoon conditions. Evidently, we are dealing here with the effects of climatic changes; these can be proved by studies of pollen in the associated sediments and by archaeological studies—e.g., humans inhabited much of the Sahara Desert during certain stages of the Pleistocene period. Indeed, the data of paleomagnetism have now swelled to statistically valid proportions, and even if there were no glacial remnants, the paleomagnetics would imply the same conclusion.

The modern view of the Permo-Carboniferous problem calls for a revolutionary spreading of the continents during Mesozoic time accompanied by relative shift of the poles (but retaining the same axial tilt to the circum-solar orbit). This is indeed a far cry from paleoconservative uniformitarianism, for it postulates an evolving, aperiodic, noncyclic, non-steady-state growth of the planet Earth, for which we know of no comparative model elsewhere in our solar system or galaxy. Furthermore, global expansion introduces problems in physics. Current concepts of the iron core and phase changes of great magnitude present serious difficulties that remain to be resolved.

How is this impasse compatible with uniformitarianism? Gould (1965) has pointed out that the logical rationalistic interpretation of the doc-trine is merely the universally accepted basis for modern science—namely, the principle of induction. From the historical point of view, however, it should be appreciated that somewhat over a century ago this procedure was not by any means always accepted, and divine interference was quite commonly a basic presumption. The rejection of the miraculous in common science is really only an outgrowth of the teachings of Lyell and Darwin.

The second interpretation of uniformitarianism, the paleoconservative approach, can always be the first of the multiple hypotheses. It involves essentially an application of the *simplicity principle* but with an absolutely vital corollary: if the status quo assumption does not meet the observed data, then further hypotheses must be proposed within the framework of the logical rationalistic principle. Simplicity implies maximum economy of hypotheses; thus, one must avoid the so-called electronic mousetrap approach. Nevertheless, modern *systems analysis* (see Vol. III: *Geomorphology*) in earth sciences show that in this field closed systems are the exception and open systems are the rule, so that simplicity per se is less likely to meet the full presentation of data on a given problem than a presumption of complex interactions.

One must be careful to avoid biasing the reported data by paleoconservative assumptions. For this reason and others (mentioned earlier) a third approach to uniformitarianism has been the growing tendency to refer to it as *actualism*. This word is almost a synonym, but it has come to imply a rejection of paleoconservatism and to approach the reading of the past in terms of possibilities that are observable or experimentally demonstrable.

STUART A. HARRIS
RHODES W. FAIRBRIDGE

References

Berry, E. W., 1929, Shall we return to cataclysmal geology? *Am. Jour. Sci.* (Ser. 5) **17**, 1–12.

Bubnoff, S. von, 1963, *Fundamentals of Geology.* Edinburgh: Oliver & Boyd, 287p. (trans. W. T. Harry, from German 3rd ed.).

Coates, D. R., and J. D. Vitek, 1980, Perspectives on geomorphic thresholds, in D. R. Coates and J. D. Vitek, eds., *Thresholds and Geomorphology.* London: George Allen and Unwin, 3–23.

Darwin, C., 1859, *The Origin of Species.* London: John Murray, 490p.

Davis, W. M., 1889, Geographic classification, illustrated by a study of plains, plateaus and their derivitives, *Am. Assoc. Adv. Sci. Proc.* **33**, 428–432.

Fairbridge, R. W., 1964, The importance of limestone and its Ca/Mg content to paleoclimatology, in A. E. M. Nairn, ed., *Problems of Palaeoclimatology.* London: Interscience Publishers, 431–530.

Fairbridge, R. W., 1965, The Indian Ocean and the status of Gondwanaland, *Problems in Oceanography* 3, 83–136.

Gould, S. J., 1965, Is uniformitarianism necessary? *Am. Jour. Sci.* 263, 223–228.

Hutton, J., 1788, The theory of the earth; or an investigation of the laws observable in the composition, dissolution and restoration of land upon the globe, *Royal Soc. Edinburgh Trans.* 1, 209–304.

Lehmann, J. G., 1759, *Essai d'une Histoire Naturelle de la Terre.* Paris.

Lyell, C., 1830, *The Principles of Geology.* London: J. Murray.

Miller, R. L., 1940, Revision of the concept of uniformitarianism, *Pennsylvania Acad. Sci. Proc.* 14, 68–77.

Read, H. H., 1949, *Geology.* London: Oxford University Press, 248p.

Schindewolf, O. H., 1962, Neokatastrophismus? *Deutsch. Geol. Gesell. Zeitschr.* 114, 430–445.

Cross-references: *Geohistory, American Founding Fathers; Geological Methodology; Geology, Philosophy; Geology, Scope and Classification; Geomythology; Plate Tectonics, and Mineral Exploration.*

V

VEGETATION MAPPING

Vegetation

Vegetation may be defined as the mosaic of plant communities in the landscape. Its significance as a resource and as an indicator of site qualities requires that the most accurate information be available on the character of the individual plant communities, their location, and their extent. This information permits correlating vegetation and its seasonal variations with various features of the environment. When the location and extent of the plant communities are shown on a topographic (contour) map, the geographical distribution pattern of vegetation in the landscape is revealed, and the ecological significance of the plant communities can be inferred (Küchler, 1967).

Criteria

One of the first steps in mapping vegetation is selecting an appropriate method to analyze the character of the plant communities. The criteria employed in analyzing and classifying vegetation are numerous and varied and are summarized in Table 1.

All plant communities consist of *growth forms* (trees, shrubs, grasslike plants, forbs, etc.) and *taxa* (genera, species, etc.). Any analysis of vegetation should describe the *structure* (the spatial distribution pattern of growth forms) of each plant community as well as its *floristic composition* (the particular combination of taxa compos-

TABLE 1. Criteria for Analyzing and
Classifying Vegetation

Physiognomic Criteria
  Growth forms
  Structure
  Seasonal Aspects

Floristic Criteria
  Taxa
  Combinations of taxa

Evolutionary Criteria
  Developmental (seral) phases
  Final state of vegetation evolution (climax)

ing the community). Structure and floristic composition are the basic characteristics of plant communities everywhere.

Physiognomy and Structure

A structural analysis may be based on either a strictly hierarchical physiognomic classification such as Fosberg's (1961) or a flexible approach, allowing the use of symbols and formulas, such as Küchler's (1967). The latest international classification of vegetation physiognomy and structure was presented by UNESCO (1973). The result is always a description of the general appearance of the vegetation (its physiognomy) and of as many details of the structure as the scale of the proposed vegetation map permits. Such *physiognomic methods* depend exclusively on growth forms and not on plant names. They are therefore especially valuable in all regions where the compilation of floristic details has not been completed. These methods are particularly useful because they can be employed anywhere on the globe with a minimum of training.

The structure of vegetation is equally revealing in all latitudes. In all plant communities, growth forms can be grouped into one or more layers or strata (e.g., tree layer, shrub layer, moss layer, etc.), with each layer consisting of one or several related growth forms growing under similar environmental conditions. The basic analysis of the structure usually emphasizes, therefore, the recording of all growth forms in each individual stratum, and its height and density. The stratification is not limited to forests; it is common even in grasslands. The recognition and distinction of structural details is of fundamental importance in analyzing a plant community, especially from an ecological point of view.

Floristic Composition

Floristic analysis of vegetation is concerned with recording the taxa of which the individual plant communities are composed. It usually permits a considerable refinement over the physiognomic approach. However, whereas the various physiognomic methods always present the physiognomy of the vegetation, the various *floristic methods* may produce very different results because the criteria employed are not always comparable. Floristic vegetation maps of a given area may therefore differ appreciably, depending on the particular floristic method employed in their preparation.

The recording of floristic data is usually based on sample plots (quadrats) of appropriate size. Each quadrat should be large enough to encompass most of the more common species. Thus, sample plots must be large in mature forests but may be much smaller in relatively uniform grasslands. The number of quadrats varies with the communities and with circumstances, but it should always be as high as possible. A single sample plot per community is inadequate.

Within a sample plot, every species is recorded by name and coverage. The latter represents the area covered by a species if it were projected vertically downward to the ground. Coverage is estimated and expressed in percentages of the total area of the quadrat. A variety of other criteria may be used singly or in combination. The more common ones, recorded directly, include abundance, density, and vitality. Indirectly—i.e., by comparison—the quadrats reveal the percentage of plots in which a given species occurs; this is termed the *frequency* of a species. A comparison of different floristically analyzed plant communities establishes a species' fidelity—i.e., the extent to which it is limited to one or more community types. In the field, a species with a high fidelity reveals therefore the type and extent of the community with which it is affiliated and thus becomes a valuable aid for vegetation mappers.

While the selection of quadrats may be systematic or random, it remains somewhat subjective. Once the floristic features of the plant communities have been recorded, however, they lend themselves to statistical manipulation. It is then feasible to distinguish floristically not only the larger plant communities but also their divisions and subdivisions in the finest detail. A vegetation map based on such a floristic analysis can therefore be an instrument of extraordinary accuracy. Excellent discussions on establishing the floristic composition of plant communities may be found in Cain and Castro (1959), Ellenberg and Mueller-Dombois (1974), and Long (1974).

Dynamism of Vegetation

Vegetation is dynamic because it is engaged in change and evolution more or less continuously. When vegetation is undisturbed by humans, and has time to evolve, it will eventually develop into plant communities that are in harmony with their environment. The communities are then in a relatively stable condition termed *climax*. As humanity's disturbing influence spreads more widely over the globe, climax vegetation becomes rare and is replaced by seminatural, or cultural, types of vegetation, which are unstable. Human activ-

ities result in an extraordinary variety of cultural vegetation types including wheat fields, vineyards, coffee plantations, overgrazed pastures, second-growth forests, rice paddies, highway rights-of-way, apple orchards, etc. None of these can maintain itself without human aid and vanishes when abandoned, because as soon as humans cease to influence the vegetation, it begins to evolve toward climax conditions. Only the climax vegetation can regenerate itself continuously and survive as long as the environment does not change. Vegetation maps can reveal the developmental phase of each plant community and thereby make the interpretation of the vegetation more meaningful. They also serve as valuable historical records and standards of reference. As time passes and the vegetation evolves, the vegetation map shows what changes took place, as well as their direction and rate of change.

Environmental Relations

Undoubtedly, vegetation maps are particularly useful because of their ecological implications, relating the various plant communities to their respective sites. All taxa can tolerate only a limited range of environmental conditions. The environment, however, is exceedingly complex, including items such as high and low temperature extremes, length of growing season, amount, reliability and seasonal distribution of precipitation, depth and duration of snow cover, depth and fluctuations of the water table (groundwater), strength and persistence of winds, innumerable physical and chemical features of the soil, amount and quality of humus, and activities of soil microbes. Every taxon in a plant community must tolerate every individual feature of its environment or it will perish. The various qualities of any given site represent therefore a set of conditions, all of which must be within the range of tolerance of every member of the plant community. If this is not so, the floristic composition of the plant community will change.

The limited range of tolerance of each taxon is further restricted by competition among the plants for light, water, and nutrients. As a result, a plant community is an intricate and highly refined expression of all environmental qualities prevailing at a particular site. A vegetation map is therefore a basic tool in establishing *ecological regions* (Küchler, 1973).

Scale

Vegetation maps can successfully integrate the complex environment only if their scale is large enough to reveal all needed details. A large scale, however, implies a relatively small area. Vegetation maps of large areas, even the whole world, are therefore generalizations of the more detailed maps. They are less informative and less precise. They are, however, preferable in many aspects of regional planning and of education, especially in classroom use because they permit a quick survey of whole countries and continents.

Aerial Photography

The efforts to refine mapping methods have benefited greatly from advances in aerial photography. A set of high-quality aerial photographs permits a rapid and accurate tracing of the boundaries of the individual vegetation units, thereby assuring the greatest economy in time, energy, and funds and at the same time substantially increasing the accuracy and reliability of the vegetation map.

Photographs taken vertically permit a direct transfer of the vegetation boundaries to a contour map, fitting the various plant communities into the landscape. Good aerial photographs reveal fine details and facilitate measurements and checking for sources of error. They also permit the preparation of a key, which allows an extension of limited ground observations far beyond the range of an individual photograph. More recently, the advances in the use of colors, infrared and other wavelengths, satellite imagery, radar, and other forms of remote sensing have opened new avenues to basic and applied research and are already being exploited in studying and mapping vegetation (Zonneveld et al., 1979; Van Dijk et al., 1977; Radambrasil, 1974).

Trained in photointerpretation (q.v.) and equipped with a mirror stereoscope for three-dimensional viewing of the photographs, a vegetation mapper is in a position to derive the greatest benefits. However, aerial photographs are not vegetation maps. The mapper is still obliged to work in the field, to establish the structure and floristic composition of the plant communities, and frequently to check on the ground what the camera recorded from above. It takes much time to become an accomplished photointerpreter, whereas the photographs eventually become obsolete.

Application

Persons trained in the interpretation of vegetation can derive a vast store of information from a vegetation map. This information can then be applied in many ways, especially in all forms of land use such as agriculture, range management, and forestry as well as in planning. A vegetation map thus becomes a basic tool in maintaining and increasing the productivity of the land. It is also valuable for geologists and pedologists and a great variety of commercial interests and is used increasingly in military matters. Finally, the vegetation map is a versatile teaching device in fields

such as geography, botany, zoology, plant and animal ecology, soil science, and geology, among others. For some years, two international journals have been devoted to vegetation mapping—e.g., *Cartographic Geobotany* (Ozenda, 1963 ff.) and *Geobotanical Mapping* (Sochava, 1969 ff.). At this writing (1981), the whole field of vegetation mapping is undergoing a rapid development, evolving with the technology of our era.

A. W. KÜCHLER

References

Cain, S. A., and G. M. de Oliveira Castro, 1959, *Manual of Vegetation Analysis.* New York: Harper & Bros., 325p.

Ellenberg, H., and D. Mueller-Dombois, 1973, *Aims and Methods of Vegetation Ecology.* New York: Wiley, 547p.

Fosberg, F. R., 1961, A classification of vegetation for general purposes, *Tropical Ecology* 2, 1–28.

Küchler, A. W., 1967, *Vegetation Mapping.* New York: Ronald Press Co., 472p.

Küchler, A. W., 1973, Problems in classifying and mapping vegetation for ecological regionalization, *Ecology* 54, 512–523.

Long, G., 1974, *Diagnostic phytoécologique et aménagement du terroir. I: Principes généraux et méthodes.* Paris: Masson & Cie, 252p.

Ozenda, P., ed., 1963ff, *Documents de cartographie écologique.* Grenoble: Laboratory of Vegetation Biology, Domaine Universitaire.

Radambrasil, 1974ff, *Projeto Radambrasil.* Brasilia, Brazil: Ministry of Mining and Energy, National Department of Mineral Production.

Sochava, V. B., ed., 1969ff, *Cartographic Geobotany.* Leningrad: Komarov Institute, Academy of Sciences of the USSR.

UNESCO, 1973, International classification and mapping of vegetation, *Ecology and Conservation, No. 6.* Paris: United Nations Educational, Scientific and Cultural Organization.

Van Dijk, A., D. C. P. Thalen, and I. S. Zonneveld, 1977, *The Application of Satellite-Multi-Spectral Scanning in Land Survey: Preliminary Experiences with Optical and Digital Image (Pre)Processing of an Area near Kahama, Tanzania.* Enschede: International Institute for Aerial Survey and Earth Sciences (ITC).

Zonneveld, I. S., H. A. M. J. Van Gils, and D. C. P. Thalen, 1979, Aspects of the I.T.C. approach to vegetation survey, *Documents phytosociologiques,* n.s. 4, 1029–1061.

Cross-references: *Geobotanical Prospecting; Photointerpretation.*

VLF ELECTROMAGNETIC PROSPECTING

The fundamental principle involved in *electromagnetic prospecting* can best be described as follows. A primary electromagnetic field induces currents in a conductive region in the Earth that, in turn, produces a secondary field on the surface. The secondary field adds to the primary field and produces a change in the propagation characteristics of the original electromagnetic field. The shape of the field anomaly curve plotted over the extent of the conductive body gives an indication of its depth below the surface of the Earth and also its conductivity in relation to its surrounding media. VLF, or very low frequency electromagnetic prospecting (VLF-EM), refers to the frequency of the primary field. The VLF radio band ranges from 10 kHz to 30 kHz in frequency with corresponding wavelengths of 30 km to 10 km. While a number of electromagnetic prospecting techniques utilize this band of frequencies, *VLF-EM prospecting* is widely accepted as referring only to passive systems—i.e., systems that do not undertake to provide the primary field. This allows the system to operate in the so-called far field of the transmitter (i.e., many wavelengths from the transmitter) where propagation characteristics should be essentially uniform.

The principle of VLF-EM prospecting was first observed in Sweden by Paal (1965) in 1963–1974. He noticed strong VLF signals as unwanted noise in a *Slingram loop frame prospecting system*. The source of the primary field proved to be one of a number of VLF transmitting stations built to provide a worldwide navigation system. Commercial prospecting systems available at present include Geonics, Scopas, Kem, TSIM, Radiophase, and Ephase. Some commercial systems now offer to provide a VLF transmitter also.

E Mode and H Mode VLF Prospecting

To understand the physical mechanism of VLF-EM prospecting techniques, the propagation characteristics of the signal across the surface of the earth must be examined. This propagation has been the subject of many detailed investigations since the 1960s, however, and the fine details are relatively unimportant in VLF prospecting; consequently, details given here are brief. All measurements taken in VLF systems are relative, so changes in the absolute phase or field strength are unimportant unless they preclude the recording of reliable information. The extremely long wavelength ensures very little attenuation along the path length and high phase stability. If the surface of the Earth is considered to be a perfect conductor, then the signal at the Earth's surface will be vertically oriented. An imperfectly conducting Earth allows the penetration of part of the signal, and so a horizontal component to the E field (E_h) of different phase ϕ is created at the Earth's surface. The resultant E field is elliptically polarized (Fig. 1). A change in the surface impedance of the Earth will produce a change in the magnitude (E_h)

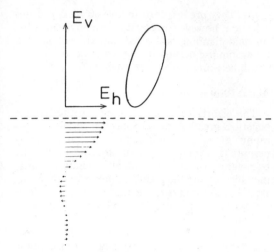

FIGURE 1. A conductive Earth allows the formation of a horizontal component of the electric field (E_h), and the resultant signal is elliptically polarized. The horizontal field penetrates beneath the surface with severe attenuation.

and phase (ϕ) of this horizontal component and a change in the polarization ellipse (Jordan and Balmain, 1968).

A surface layer of high conductivity beneath a layer of lesser conductivity will change the effective surface impedance of the Earth, and this will vary as the depth to the conductive layer increases and as the electrical characteristics of the layer become increasingly different. The currents induced in the conductive layer will radiate a secondary magnetic field that adds to the primary magnetic field to produce an elliptically polarized magnetic field on the surface. The strength of the reradiation depends on the orientation of the conductor and its proximity to the surface (Fig. 2).

A distinction has been drawn by Parasnis (1972) between VLF prospecting methods monitoring the electric field component of the electromagnetic wave (E mode) and those monitoring the magnetic component (H mode). While the H mode method depends on the detection of a vertical magnetic field (H_v) and the corresponding anomaly shape, quantitative measurements can be made using the E mode method, and the next section outlines the mathematical basis for this method.

E Mode Theory

The propagation of an electromagnetic field above the surface of an idealized conductor will not result in signal penetration beneath the surface of the conductor. If, however, the surface is an imperfect conductor, some energy penetrates the surface, producing a small horizontal electric field at the surface. The electric field component

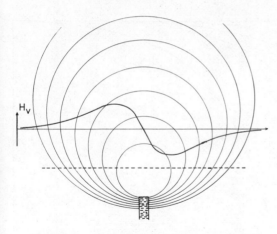

FIGURE 2. An electromagnetic wave propagating into the plane of the page induces currents in the uppermost portion of the subsurface conductor. (This current reradiates a magnetic field, thus contributing a vertical component to the magnetic field H_v.) As the surface is traversed (perpendicular to the direction of propagation), the change in H_v is shown.

is no longer vertical since, in fact, elliptical polarization exists with the parameters of the ellipse being dependent on the electrical characteristics of the medium. The horizontal electric field (E_h) will no longer be in phase with the original vertical electric field (E_v) but will differ by some amount ϕ. A theoretical investigation of the phenomenon was commenced by Norton (1937a, 1937b) who defined the wave tilt function W as the complex ratio of the horizontal to the vertical electric field at the Earth-air interface, and for grazing incidence this is given by

$$W = \frac{E_h}{E_v} = \left(\frac{\mu_1}{\mu_0}\right)\frac{\gamma_0}{\gamma_1}\left(1 - \frac{\gamma_0^2}{\gamma_1^2}\right)^{1/2},$$

where μ_0 and μ_1 are the permeabilities of the air and the Earth and γ_0 and γ_1 are the propagation coefficients in air and an infinite Earth respectively—i.e.,

$$\gamma = (j\sigma\mu\omega - \epsilon\mu\omega^2)^{1/2} = \left[-\frac{4\pi^2}{\lambda}(\epsilon - jX)\right]^{1/2},$$

where σ is the conductivity of the medium, ϵ is the dielectric constant of the medium, j is $\sqrt{-1}$, X is the imaginary part of the dielectic constant and is related to the conductivity, and ω is the angular frequency of transmission ($\omega = 2\pi f$). The modulus and argument of W are the magnitude ratio and the phase difference respectively of the two electric field components. A change in the wave tilt parameter W will result from a change in the magnitude (E_h) and phase difference ϕ of this

horizontal component. The amplitude and phase of the vertical electric field (E_v) are used as references for the E_h measurement.

In recent years a more practical E mode technique has been developed that allows a direct measurement of the "apparent resistivity" and phase of an Earth plane (Collett and Jensen, 1982). This technique is based on the measurement of the VLF surface impedance Z_s and is defined by

$$Z_s = E_h/H_h = (j\omega\mu\rho)^{1/2},$$

where E_h is the horizontal electric field in line with the transmitter, H_h is the horizontal magnetic field perpendicular to E_h, and ρ is the resistivity of the Earth ($= 1/\sigma$). The apparent resistivity ρ_a is given by the inverse equation

$$\rho_a = |Z_s|^2/\omega\mu$$

A number of small, highly portable units are available to make these measurements (e.g., TSIM, EM16-R, VLF-3). Apparent resistivity and phase data can be interpreted in terms of horizontal layers (Thiel, 1986; Poddar and Rathor, 1983), and also two-dimensional effects (Fischer et al., 1984; Thiel, 1984, 1985).

Above a uniform Earth, the phase is virtually independent of the conductivity for values above 0.1 Sm^{-1} and for the normal range of dielectric constant for the overburden (i.e., $1.0\epsilon_0$ to $20\epsilon_0$) (Fig. 3). The amplitude ratio varies linearly with the square root of the conductivity above 0.001 Sm^{-1} and to a first order is independent of the dielectric constant for these values (Fig. 4). It is interesting to note that an inphase and phase quadrature measurement relative to the primary field E_v will not be independent because the phase difference is of the order of 45° above an infinitely conducting half-space.

The depth of penetration of the electromagnetic wave in the Earth is related to the conductivity of the Earth. A measure of this penetration depth is the skin depth defined as the depth where the incident amplitude drops to $1/e$ of this value. The mathematical relationship between the skin depth δ and the conductivity σ is

$$\delta = \sqrt{\frac{1}{\pi f \mu \sigma}}\ \text{m},$$

and skin depths in the VLF radio band are indicated in Fig. 5.

A subsurface layer of high conductivity beneath a layer of lesser conductivity will change the surface impedance of the Earth, and this effect will vary with depth and conductivity. If a two-layer Earth is considered (Fig. 6), the surface

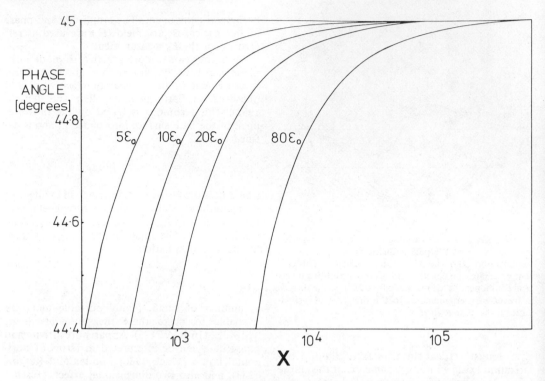

FIGURE 3. Phase angle between the vertical and horizontal electric field components as a function of conductivity and dielectric constant (uniform infinite earth). X is defined in text.

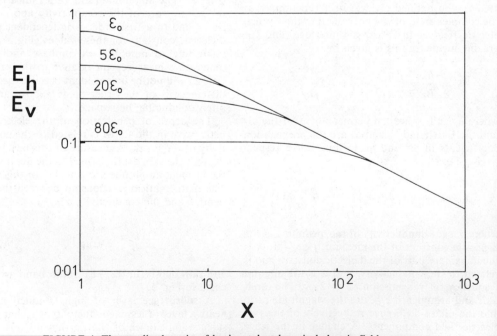

FIGURE 4. The amplitude ratio of horizontal and vertical electric field components as a function of conductivity and dielectric constant (uniform infinite Earth).

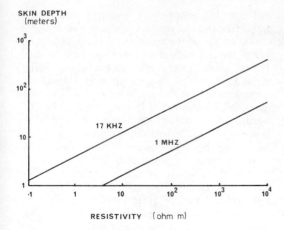

FIGURE 5. Plot of skin depth versus resistivity.

impedance is modified by the factor Q defined by Wait (1971):

$$Z_s = QZ_1$$

where Z_1 is the surface impedance of a single layer of infinite extent as calculated in the above equation. Q is given by

$$Q = \frac{Z_1/Z_2 + \tanh(\gamma_1 h)}{1 + (Z_1/Z_2)\tanh(\gamma_1 H)},$$

where h is the depth to the next layer and γ_2 is the propagation coefficient of the upper layer. Figure 7 shows the phase change in the horizontal E field caused by a second layer of various conductivities as the depth of overburden increases. This plot postulates an overburden of conductivity of 0.001 Sm^{-1}, and a dielectric constant of 10 ϵ_0, and considers a second layer (of infinite depth) of dielectric constant 80.0 ϵ_0, as would be the case for regions of high water content. A change in the dielectric constant of the second layer will have only a minor effect on the numerical values given in Fig. 7. Notice also that the maximum phase difference induced by the first layer is 45° so phase differences approaching 90° are possible. The rel-

ative change in amplitude induced by the second layer is indicated in Fig. 8. In Figs. 7 and 8, for a conductivity of 0.001 Sm^{-1}, the effect of a change in the dielectric constant of 10 ϵ_0 to 80 ϵ_0 can be seen as the line deviates from the straight line. Should the second layer be less conducting, then the response will be the mirror image of both figures: Fig. 7 would be reflected through the line $\phi = 0$, and Fig. 8 would be reflected through the line $E_h/E_v = 1$. This has been shown by Cagniard (1953). The overburden conductivity determines the depth of penetration, and the ratio of conductivities of the upper and lower layers determines the magnitude of the maxima on the two plots. A number of authors have published mathematical derivations of the effect of electromagnetic waves incident on a multilayered Earth (Wait, 1962; Dosso, 1964, 1965), and the theory is a logical extension to the two-layer problem.

Conductive orebodies are usually electrically short; i.e., their dimension in the direction of the inducing electric field is much less than one wavelength. For this reason and also because many orebodies are stratified, the reradiation from the orebody has an effective radiation pattern based

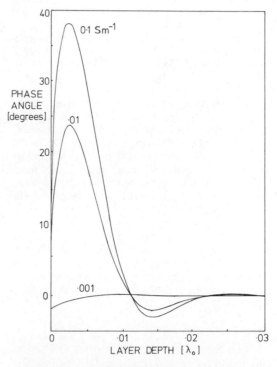

FIGURE 7. Phase change of the horizontal electric field above a two-layer Earth (upper layer has conductivity 0.001 Sm^{-1}, dielectric constant 10 ϵ_0, depth as shown; lower layer, conductivity shown, dielectric constant 80 ϵ_0, infinite depth).

FIGURE 6. Two-layer Earth model.

859

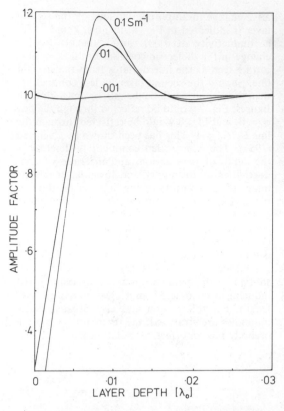

FIGURE 8. Relative change in amplitude ratio (multiplying factor) of horizontal to vertical electric field component above a two-layer Earth described in Fig. 6.

on the Hertzian dipole, and this radiation pattern provides the limits on the surface of the Earth in which a perturbation in the field can be detected. This means that a maximum perturbation in the horizontal electric field will be evident directly above the orebody. The maximum vertical magnetic field will be found some distance from the orebody, a distance related to the depth of the conductor. If the orebody lies parallel to the direction of propagation, then the reradiation will be maximized. Horizontal electric field traverses parallel to the direction of propagation will provide a larger detectable anomaly distance than a traverse perpendicular to the direction of propagation, although the anomaly field strength maximum will be identical.

Real Earth Parameters

Two books that contain considerable information in references, tables, and discussion have been written particularly for geophysical investigations by Keller and Frischknecht (1966) and Heiland (1946). The conductivity of a material is dependent on a number of physical parameters

including the temperature, pressure, and mineral water content and concentration as well as the intrinsic nature of the material. A conductivity survey of the United States (Keller and Frischknecht, 1966), using local radio station frequencies, yielded a conductivity contour map; similar conductivity maps of the world at VLF, LF, and MF have been suggested (Morgan, 1967). The U.S. conductivity contour map shows that most of the country has a conductivity of greater than 0.02 Sm^{-1} and that up to about 40% of the area has a conductivity exceeding 0.01 Sm^{-1}. In Canada (Ontario), overburden conductivities of about 0.01 Sm^{-1} have been given as part of VLF-EM results (Barringer, 1971). In Western Australia, the overburden has been described as having conductivities of 0.02–0.01 Sm^{-1} with weathered depths of 35–70 m. Most conductive ores lie in the conductivity range of 1 Sm^{-1} to 10^4 Sm^{-1} in their pure form.

The only common polar material in rocks is water, so the dielectric constant is usually a measure of the water content of rocks. Since the dielectric constant is of secondary importance in the EM prospecting method, large discontinuities are required before noticeable changes in the propagation of the wave at the surface occur. This can be seen from Figs. 7 and 8. Most minerals have a dielectric constant in the range 5 ϵ_0 to 20 ϵ_0 (Keller and Frischknecht, 1966).

Effective changes in the magnetic permeability are rarely significant, although high concentrations of magnetic materials may cause noticeable anomalies. Few minerals possess a magnetic permeability significantly different from the free space value, and only minerals like magnetite and pyrrhotite will contribute significantly to influence the polarization ellipse.

VLF-EM Measurements

VLF-EM prospecting methods gain information from the perturbation in the magnetic or electric field in the presence of a body of differing conductivity. Consequently, field surveys record parameters of the electromagnetic field either continuously or in grid formation. The primary field is used as a basis for comparison in both amplitude- and phase-based measurements. The EM-16 (Geonics, 1964), e.g., one of the first H mode VLF instruments available, uses two perpendicular coils mounted in a handset. The operator locates the primary field direction by locating the magnetic field intensity minimum in the horizontal plane when rotating the instrument in the horizontal plane. The residual signal from this pickup coil is fed back in phase quadrature (i.e., 90° out of phase) to the second coil. The instrument is then rotated in the vertical plane and, by varying the strength of the primary signal required and the angle of elevation, effectively measures the relative strength and direction of the secondary magnetic field from a subsurface conductor.

VLF surface impedance meters are placed on the ground for measurement. The electric field is monitored either by staked earth contact probes separated by 10 m, or an insulated wire antenna lying on the surface. The more recent instruments allow direct readings of apparent resistivity and phase from a digital display on the instrument.

An airborne E mode system, Radiophase (Barringer and McNeill, 1968), records the horizontal electric field with a trailing whip aerial in addition to the two components of the magnetic field. All field strengths are measured in phase quadrature with the primary field. To ensure that all available information is obtained, all parameters of all fields are monitored in some systems, and perturbations in more than one of the channels are used to indicate areas needing further study. Airborne techniques do not measure the phase directly but rely on the measurement of the amplitude of a quadrature phase component to gain a parameter that, in the ideal situation, is proportional to the phase but that in practice is never the case.

The mathematical solutions yielding the field components in practice are extremely difficult to compute, so the derivation of useful information from an anomaly curve is usually done by comparison with anomaly curves obtained from model experiments. An example of the mathematical method involved is given in Keller and Frischknecht (1966) and the solution for a subsurface conducting sphere derived. This mathematical method, however, is of little practical use, and geophysical companies prefer to use model-generated curves and high-speed computers to perform the matching task. Interpretation methods in VLF-EM methods therefore include pattern recognition to determine the shape and orientation of the conductor, and having determined these factors, simple mathematical formulas are used to find its depth and size (Barringer, 1971; Geonics, 1968).

In an H mode VLF anomaly pattern of a spherical conductor, the depth h is given by

$$h = \Delta x,$$

where Δx is the horizontal distance between the maximum points of the vertical field. The radius of the sphere a is given by

$$a = 1.3h \sqrt[3]{T \text{ (max)}},$$

where T is proportional to the vertical magnetic field and $T = 1$ at an angle of 45° to the horizontal, based on the assumption that the primary magnetic field remains constant. The equations for a cylindrical body are

$$h = 0.86\Delta x \qquad a = 1.22h \sqrt{T \text{ (max)}}.$$

The depth indication h is more reliable than the dimension of the conductor a.

The effect of topology on H mode contours is marked and can make pattern recognition impossible. The first derivative (slope) of the profile is often more diagnostic than the simple crossover because it has the effect of minimizing topological interference (Whittles, 1969). This is because the anomaly slope at crossover is usually much greater when a conductor is present. The width of the first derivative profile at zero slope is an indication of depth to a spherical conductor. This result follows directly from the equation above. A similar technique called Fraser filtering is commonly used (Fraser, 1969).

VLF-EM System Effectiveness

A review article by Paterson and Ronka (1971) concluded that the method is capable of moderate depth of exploration in nonconductive rocks but is severely limited in conductive ground. Relatively small conductive orebodies have produced VLF-EM anomalies. Conductivity changes in the overburden or at the overburden-bedrock interface may generate anomalies. Poor conductors also produce VLF anomalies (e.g., sheared contacts, breccia zones, narrow faults, etc.) so the method is suitable for geological mapping.

In Australia ground-based surface impedance measurements have been used to delineate conductive ore veins (Thiel, 1983), high silica content "reefs," coal seams, fault lines, basalt intrusions, and other geological structures, including glacial ice mapping (Thiel, 1986). The H mode technique has proved less successful in Australia (Langron, 1972) primarily because of the depth of weathering.

Other applications include bedrock mapping in India (Podder and Rather, 1983) and in Canada (Collett and Jensen, 1982).

One well-documented VLF-EM survey is a Radiophase survey over Lake D'Alembert, Quebec, Canada (Paterson, 1970). This survey demonstrated the effectiveness of the system in detecting fault lines and regions of high subsurface conductivity. Conductivity contrasts due to subsurface gravel deposits have been successfully detected (Peden et al., 1972). Such instruments have been effective in many situations; however, the method becomes limited in the detection of small highly conducting zones.

In conclusion, VLF-EM prospecting systems can be used to detect subsurface conductivity changes of large areas. Its efficiency is lowered with increasing surface conductivity, decreasing size of target, and decreasing conductivity contrast. The systems (in some cases) have the advantage of simplicity of operation, minor costs, and in other cases, airborne operation. It is usual, as with most geophysical prospecting techniques, to use the system simultaneously with other methods of prospecting.

DAVID V. THIEL

References

Barringer, A. R., and J. D. McNeill, 1968, Radiophase—a new system of conductivity mapping, in *5th Symposium on Remote Sensing of Environment.* Ann Arbor, Mich.: University of Michigan, 157.

Barringer Research, 1971, *E. Phase,* Case History Series. Toronto.

Cagniard, L., 1953, Basic theory of the magneto-telluric method of geophysical prospecting, *Geophysics* **18**(3), 605–635.

Collett, L. S., and O. G. Jensen, (eds.), 1982, Geophysical applications of surface wave impedance measurements, *Geophysical Survey of Canada, Paper 81-15.*

Dosso, H. W., 1964, The electric and magnetic fields in a stratified flat conductor for incident plane waves, *Canadian Jour. Physics* **43**, 898–909.

Dosso, H. W., 1965, A multilayer conducting earth in the field of plane waves, *Canadian Jour. Physics* **44**, 81–89.

Fischer, G., B. V. Le Quang, and I. Muller, 1984, VLF ground surveys, a powerful tool for the study of shallow two-dimensional structures, *Geophys. Prospecting* **31**, 977–991.

Fraser, D. C., 1969, Contouring of VLF-EM data, *Geophysics* **34**(6), 958–967.

Geonics, Ltd., 1964, *Geonics EM-16,* Operator's Manual. Toronto.

Geonics, Ltd., 1968, *Deep-Penetrating Electromagnetic Detector, Ronka EM-17,* Operating Manual. Toronto.

Heiland, C. A., 1946, *Geophysical Exploration.* Englewood Cliffs, N.J.: Prentice-Hall, 1,013p.

Jordan, E. C., and Balmain, K. G., 1968, *Electromagnetic Waves and Radiating Systems,* 2nd ed. Englewood Cliffs, N.J.: Prentice-Hall, 753p.

Keller, G. V., and F. C. Frischknecht, 1966, *Electrical Methods in Geophysical Prospecting,* International Series of Monographs in Electromagnetic Waves. Oxford: Pergamon Press, 517p.

Langron, W. J., 1972, A study of results of the VLF-EM methods of prospecting in Australia and Papua, *Australasian Inst. Mining and Metallurgy Proc.* **241,** 27–38.

Morgan, R. R., 1967, Omega navigational system conductivity map, *IEE Conf. Pub. No. 36,* 111–137.

Norton, K. A., 1937a, The propagation of radio waves over the surface of the earth and in the upper atmosphere, *Proc. IRE* **24**(Pt. I), 1367–1387.

Norton, K. A., 1937b, The propagation of radio waves over the surface of the earth and in the upper atmosphere, *Proc. IRE* **25** 1203–1236.

Paal, G., 1965, Ore prospecting based on VLF radio signals, *Geoexploration* **3**(3), 139–147.

Parasnis, D. S., 1972, *Principles of Applied Geophysics,* 2nd ed. London: Chapman and Hall, 214p.

Paterson, N. R., 1970, Airborne VLF-EM test, *Canadian Mining Jour.* November 1–4, 1–4.

Paterson, N. R., and V. Ronka, 1971, Five years of surveying with the VLF-EM method, *Geoexploration* **9**(1), 7–26.

Peden, I. C., G. E. Webber, and A. S. Chandler, 1972, Complex permittivity of the Antarctic ice sheet in the VLF band, *Radio Sci.* **7**(6), 645–650.

Poddar, M., and B. S. Rathor, 1983, VLF survey of the weathered layer in southern India, *Geophys. Prospecting* **31**, 524–537.

Schlak, G. A., and J. R. Wait, 1968, Attenuation function of propagation over a nonparallel stratified ground, *Canadian Jour. Physics* **46**, 1135–1136.

Thiel, D. V., 1984, One dimensional surface impedance measurements above an anisotropic earth, *Explor. Geophysics* **15**(1), 43–46.

Thiel, D. V., 1985, VLF surface impedance measurements at Zeehan, Tasmania, *Explor. Geophysics* **16**(4), 387–390.

Thiel, D. V., 1986, A preliminary assessment of glacial ice profiling using VLF surface impedance measurements, *Jour. Glaciology* **32**(112), 376–382.

Wait, J. R., 1962, *Electromagnetic Waves in a Stratified Media,* International Series of Monographs on Electromagnetic Waves, vol. 3. New York: Pergamon Press, 372p.

Wait, J. R., 1971, *Electromagnetic Probing in Geophysics.* Boulder, Colo.: Golem Press, 391p.

Whittles, A. B. L., 1969, Prospecting with radio frequency EM-16 in mountainous regions, *Western Miner,* February, 51–56.

Cross-references: *Exploration Geophysics; Marine Magnetic Surveys; Prospecting; Remote Sensing, General.* Vol. XIII: *Magnetic Susceptibility, Earth Materials.*

W

WATER ENGINEERING, SURVEY—See ALLUVIAL SYSTEMS MODELING; CANALS AND WATERWAYS, SEDIMENT CONTROL. Vol. XIII: CHANNELIZATION: COASTAL INLETS, ENGINEERING GEOLOGY; RIVER ENGINEERING; WELLS, WATER.

WATER GEOCHEMISTRY—See HYDROCHEMICAL PROSPECTING

WELL LOGGING

Well logging is a continuous qualitative and quantitative analysis of underground formations and their interstitial fluids. The purpose of the analysis is to forecast the outcome of well production tests, to correlate physical properties of rocks and fluids encountered in different wells in the search for fluids, and to define lithologic bodies.

Surface Methods

Surface well logging always requires wellbore samples in some form as a basis for logging. The accuracy and reliability of surface methods, therefore, depend on sampling methods. Surface samples may be in the form of bit cuttings and/or cores. *Bit cuttings* are circulated to the surface with the drilling fluid, whereas *cores* are unaltered samples of rocks penetrated by the core bit, on using specialized drilling assemblies.

Bit cuttings and cores are used for *qualitative* evaluation of lithology and fluid saturation while the hole is being drilled. Cores are also used for detailed *quantitative* petrophysical evaluation by engineers and geologists. The logs obtained using surface methods are as follows.

Lithology Log Bit cuttings from wells are examined by geologists and described as to texture, color, taste, smell, and type and distribution of lithology and saturating fluids (oil, gas, or water). The compilation of the examinations and descrip-

tions, plotted versus the depth of the borehole, is called a lithologic, cutting, or strip log.

Mud Log A mud log is often compiled in conjunction with a lithologic log. The mud log is a record (plotted versus depth) of the analysis and description of the properties of the drilling fluid, which is often called *mud*. The mud log portrays the quantitative analysis of the mud's physical properties such as density, viscosity, percent of solids, fluid loss (the ability of a mud to retain its makeup fluid under differential pressures of around 7.031 kg/cm^2, or 100 psi), and the relative amount of hydrocarbons and water.

Drillers' Logs A drillers' log, similar to a mud log, describes the mechanical variations encountered while drilling the well. Measurements such as the drilling rate (drilling-time logging), number of bits used, amount of weight used on the bit, size of bits, rate at which the bit is rotated, and hole angle and direction are recorded versus depth and time. Frequently, drillers' recordings of formation tops (and his or her lithologic calls) are exceptionally accurate. A computer program has been recently developed to record these parameters while drilling the well.

Subsurface Methods

A second general class of well logs involves measurements made in the borehole beneath the ground surface. The subsurface measurements are made using a variety of instruments that are lowered into the wellbore by means of a cable or wireline. Recent developments in well logging allow the recording of subsurface measurements while drilling. At the present time, however, most subsurface well logging is performed after the drilling mechanism has been removed from the wellbore (see *Borehole Drilling*).

The petrophysical properties that are measured in subsurface well logging include (1) electric and dielectric properties when saturated or partly saturated with fresh and/or saline water; (2) natural radioactivity (GR/Spectralog); (3) stimulated emission of radioactivity due to neutron bombardment (density logs; neutron logs, three types; pulsed neutron logs; and activation logging); (4) acoustic properties (sidewall, short-, long-spaced); (5) thermal conductivity; and (6) nuclear magnetism. Subsurface methods can be classed as electrical and nonelectrical.

Electrical Well Logging

Methods that utilize electrical recording instruments in a wellbore are termed *electrical well logs.* These logging methods record variations in electrical potential due to the electrical properties of the subsurface formations and their reaction to applied electrical forces. Resistivity (the reciprocal of conductivity) indicates changes in the lithology, porosity, and fluid content.

All resistivity-measuring devices basically seek to delineate the relationship between the resistivity of a rock 100% saturated with formation water, R_o, and the resistivity of the same rock partly saturated with hydrocarbons, R_t. The R_o/R_t ratio is an exponential function of the water saturation, S_w, and $(S_w)^n = R_o/R_t$, where n is the saturation exponent.

Self- or Spontaneous Potential Logs A circuit, as shown in Fig. 1, is connected between a moving electrode, M, and a fixed surface electrode, N. The potential deflection encountered is registered by recorder, R, by means of a *galvanometer* that responds to variations of minute currents flowing through the circuit (C, SR, R, P, N, earth, and M) as electrode M is hoisted up the hole by means of the winch, W. The electrodes, which are made of lead or iron, are stable and in electrochemical equilibrium with the well fluid. The formation potentials recorded are a result of an *electrochemical self-potential,* which consists of a diffusion potential and a shale potential (Nerntz potential) and a streaming (or electrokinetic) potential. Formation resistivity (R_w), bed boundaries, and bed thickness could be obtained from an SP log.

Normal Logs Normal logs are resistivity measurements for which four electrodes are always required. Two are current electrodes, A and B, and two are potential electrodes, M and N (Fig. 2).

The potential V from electrode A to electrode M may be expressed as $V = iR/4\pi$ (AM), where i is the current intensity issuing from A, R is the resistivity of the homogeneous media, and AM is the electrode spacing. As shown in Fig. 2, A and M are closely spaced, whereas B and N are far apart. Thus, the only potential of measurable significance is the one transmitted from A to M. The

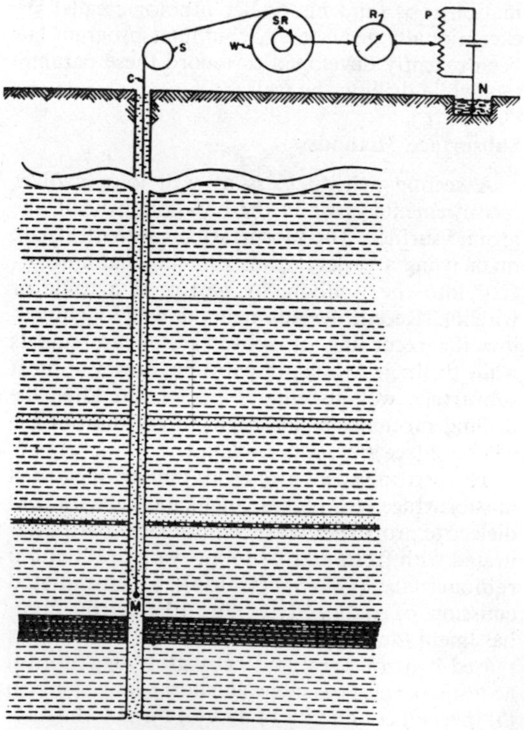

FIGURE 1. Diagram of circuit for recording SP logs. (From Schlumberger Well Surveying Corporation)

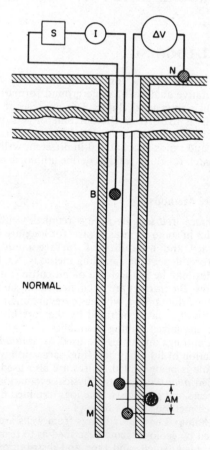

NORMAL

FIGURE 2. Electrode arrangement for normal resistivity curve. (From Schlumberger Well Surveying Corporation; after L. G. Chombart, pers. comm.)

potential is recorded halfway between A and M as the electrode assembly is raised up the wellbore on the logging cable.

In 1934, the long normal curve was introduced by the Schlumberger brothers. This was preceded by the short normal curve, which was also introduced by them in 1931. Various sizes of electrode spacings have been used in the development of resistivity logging, which are being standardized to the 16 in. short normal (AM = 16 in.) and the 64 in. long normal (AM = 64 in.). These normal electrode spacings were used almost universally but are now almost totally out of service.

A short normal resistivity log (AM = 16 in., versus spherical focused 16 in.) measures the contaminated or invaded zone resistivity, R_i, and consequently the porosity of the formation. This technique, however, is no longer common. The longer (AM = 64 in.) seeks to measure an intermediate resistivity value. True resistivity of a formation, R_t, can be obtained providing filtrate invasion is not deep.

Lateral Logs The first wellbore resistivity measurement was introduced by Schlumberger using a 2 m lateral spacing. It was reintroduced in 1936, using a much longer spacing. Various electrode spacings have been used in successive development of lateral resistivity logging to be eventually standardized to the 18 ft 18 in. lateral; however, this log is hardly used anymore. It is described here for historical purposes and because of the need to interpret log data from many old oil wells.

Lateral logs are known as a three-electrode curve because only three electrodes, out of the four usually used, are effective in the measurements. Lateral curves are resistivity measurements for which four electrodes are always required. The two current electrodes A and B and the two potential electrodes M and N are shown in Fig. 3. The electrodes M and N are 32 in. apart, and the electrode A is 18 ft 18 in. from point O, which is the center of MN and also the reference level for measurements. Inasmuch as electrode B is located far away from the electrode assembly AMN, it has no significant effect on the assembly. The voltage difference, V, transmitted to MN is equal to $[(iR/4\pi) \times (1/AM - 1/AN)]$ for a constant current i. The voltage recorded at O is proportional to the resistivity R, providing the medium is sufficiently thick and homogeneous and the borehole effects are negligible.

The purpose of the lateral log was to measure R_t, the true resistivity, of the formation undisturbed by contamination or invasion effects. This is accomplished by choosing a spacing long enough so that the radius of investigation exceeds the radius of the invaded zone. Thus, the lateral log is affected only to a very small extent by contamination of the invaded zone.

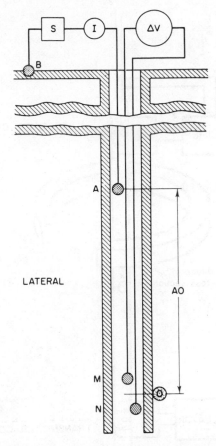

FIGURE 3. Arrangement of electrodes for lateral resistivity curve. (From Schlumberger Well Surveying Corporation; after L. G. Chombart, pers. comm.)

Induction Logging The *induction log* was developed by H. Doll of Schlumberger to overcome difficulties found in determining true resistivity, R_t, in formations when the hole was filled with oil-base drilling fluid or was drilled with air. Figure 4 is a schematic diagram of an induction logging device. Alternating current, which is applied to the transmitter coils, sets up an alternating magnetic field around the coil, and the magnetic field induces electric eddy currents in conductive formation. The latter currents, which are proportional to the conductivity of the formation, also develop alternating magnetic fields. These, in turn, induce a current in the receiver coils. The latter is amplified, rectified, and sent to the surface to be recorded.

In addition to these two coils, auxiliary coils are included to minimize the effects of borehole drilling fluid in adjacent formations. Auxiliary coils focus the generated magnetic field on the bed opposite the device, instead of permitting the field to fan out radially in all directions. Basically, this

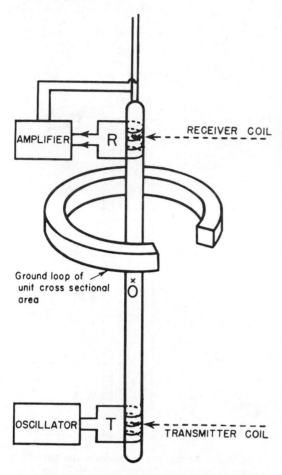

RECEIVER COIL

AMPLIFIER   R

Ground loop of
unit cross sectional
area

OSCILLATOR   T

TRANSMITTER COIL

FIGURE 4. Induction logging device. (From Schlumberger Well Surveying Corporation)

is done to increase the depth of investigation of the tool, as far out as possible from the invaded zone (i.e., a zone altered by the invasion of drilling fluid filtrate).

The distance between the two main coils is the *spacing*. Depth measurement is taken at a point midway between the main coils. Inasmuch as there is no need for an electrical contact, induction logs may also be run in empty boreholes or those filled with oil-base *drilling fluids*.

The signal of induction logs is proportional to the conductivity of the medium surrounding the device and is recorded as conductivity at the surface. Conductivity and its reciprocal, resistivity, are recorded for comparison with other logs.

The apparent conductivity recorded by the instrument is the result of the combined effects of the adjacent formations, the invaded zone, and the uncontaminated zone. The geometrical locations and conductivities of each of these regions determine the reading obtained. Inasmuch as re-

sistivity is a reciprocal of conductivity, these factors may be related by the following equation:

$$\frac{1}{R_a} = \frac{G_m}{R_m} + \frac{G_i}{R_i} + \frac{G_t}{R_t} + \frac{G_s}{R_s},$$

where R_a = average resistivity, G_m = geometrical factor for drilling fluid column, G_i = geometrical factor for invaded zone, G_t = geometrical factor for the uncontaminated zone, G_s = geometrical factor for adjacent formations, and $G_m + G_i + G_t + G_s = 1$.

Geometrical factors are the result of two special considerations: (1) a set of concentric cylinders around the borehole and (2) sets of essentially horizontal beds of formations. The concentric cylinders comprise the uncontaminated zone, the invaded zone, and the drilling fluid column. Geometrical factors show that the effect of the borehole is negligible if it is less than 10 in. in diameter. In practice, however, this is true only if the drilling fluid resistivity is greater than 0.50 ohm-meters.

Three resistivity curves, with different depths of investigation, can be recorded simultaneously. The *deep induction log* measures the resistivity in the deep zone; the *medium induction log* represents the resistivity in the transitional (medium) zone; and the *Laterolog 8* records the resistivity of the invaded zone. The major advantage of this log is the fact that the three resistivity curves permit an estimation of the invasion diameter and, consequently, result in an improved R_t value (Tixier et al., 1963).

An important logging instrument is the Dresser Atlas Dielectric downhole induction-type (high-frequency) tool designed to investigate formation characteristics in areas of brackish or low-formation-water salinity.

Focused-Current Logs The focused-current log was one of the early logs run by Conrad Schlumberger. Its use was temporarily abandoned but was revived in 1949 by Birdwell and reintroduced by Schlumberger. Different types of tools have been developed, such as Laterolog, Guard Log, and LL3. The logging is achieved by focusing a current beam into a thin sheet (or disk) that penetrates the formation parallel to the bedding plane.

Laterolog 7, which is obsolete now, uses a measuring current electrode and two bucking current electrodes, adjusted so that the potential difference of two pairs of electrodes M1M1' and M2M2' is maintained at zero. The resistivity is computed by measuring the potential difference between the M electrodes and the infinity in the relation with the current of the measuring electrode A (Fig. 5).

The *Dual Laterolog* (Fig. 5) has been developed to provide resistivity measurements at dif-

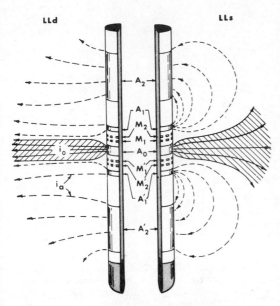

FIGURE 5. Schematic diagram of Dual Laterolog electrodes. (From Schlumberger, 1972, Fig. 4-3, 21)

ferent depths of investigation. As pointed out by Schlumberger (see Vol. 1: *Log Interpretatioy*), the *deep Laterolog* has a depth of investigation deeper than the original Laterolog, by using larger bucking current electrodes. The beam of current of the *shallow* Laterolog penetrates a shorter distance into the formation, through a different manner of using the bucking current electrodes. It is important to decide when to select the *DIL* (Dual Induction Log) versus the *DLL* (Dual Laterology) (induction vs. Laterolog).

Microresistivity Logs   The goal of the microresistivity tools is to provide a measurement of the resistivity in the infiltrated zone R_{xo}. Measurement of R_{xo} improves the accuracy of the measurement of the true resistivity, R_t, and the saturation. Microresistivity tools use a hydraulically operated pad, which is pressed against the formation to ensure a close contact with the borehole wall.

Micrologs   In 1949, the Schlumberger Well Surveying Corporation introduced the Microlog primarily for the purpose of providing a well log that would exhibit a fine lithologic definition that is free of adjacent bed effects. It was developed to measure the formation factors of the rocks in place by means of closely spaced electrodes embedded in an insulated pad. Although this objective was partly achieved, the microlog proved to be primarily a mud cake detector.

The Microlog used two curves, *micro-normal* and *micro-inverse,* the measuring devices of which are similar to the normal and the lateral logs, ex-

cept that the electrode spacings are of the order of a few inches only. In a permeable zone, the resistivity read by the Micro-inverse log, with the lowest depth of investigation, is mainly that of the mud cake and is lower than the resistivity computed from the Micro-normal curve. The Microlog is used mainly in the determination of effective pay thickness. If formation is sufficiently porous and permeable to admit mud filtrate and develop a mud cake, then it may be permeable enough to produce gas and oil when they are present at sufficient saturations.

Microlaterologs and Proximity Logs   The advantages of the Microlaterolog and the proximity log over the Microlog are due to the same effects that improve the Laterologs over the normal and lateral logs (Fig. 6). A very thin focused beam of current, which is sent into the formation, opens up after traversing a few inches, but the tool reading basically reflects the resistivity of the formation before the opening. Charts were developed for correcting for the mud cake effect. Proximity logs and Microlaterologs differ in their electrode configuration. The proximity log has a somewhat deeper depth of investigation than the Microlaterolog; thick mud cake and rough hole conditions have little effect on its reading. The Microlaterolog is preferred when mud cake is up to ⅜ in. (about 1.0 cm) in thickness and there is a shallow invasion by drilling fluid filtrate.

Microspherical Focused Logs   The microspherical focused log, which represents one of the later developments of the microresistivity tools by Schlumberger, improves the response of the tool in the case of shallow invasion and thin mud cakes. It is also compatible with different tools and can be run together with induction log, and Laterolog.

Nonelectrical Logs   *Gamma-Ray Logs*   The gamma-ray log was introduced in 1939 for the purpose of measuring continuously the natural radioactivity of rocks. This log enabled determination of the lithology of cased wells that have not been logged before. It made it possible to investigate old wells and to determine tops and bottoms of formations for which records were not available.

The gamma-ray log measures and records natural gamma-ray emission of rocks using the schematic circuit presented in Fig. 7. Gamma-ray energy given off by rocks is on the average equivalent to 1 MEV. The distance that can be investigated by this log is about 30 cm in water-saturated sediments. The gamma-ray intensity is reduced by about 30% in the presence of casing. Detectors are actuated, however, even though the gamma rays are degraded as a result of energy loss. Most of the high energy count and most of the counting rate recorded are due to the gamma rays that originate close to the well walls.

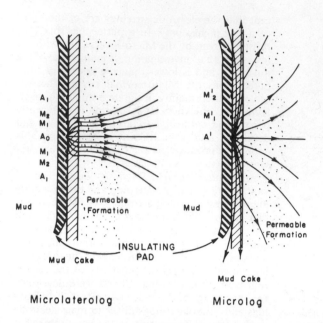

A₁
M₂
M₁
A₀
M₁
M₂
A₁

Mud

Permeable Formation

Mud Cake

Microlaterolog

M'₂
M'₁
A'

Mud

Permeable Formation

Mud Cake

INSULATING PAD

Microlog

FIGURE 6. Comparative distribution of current lines for Microlaterolog and Microlog. (From Schlumberger, 1972, Fig. 6-3, 35)

Gamma rays are detected directly as a result of their interaction with matter by the ionization process—i.e., by releasing one or more electrons from neutral atoms.

Three different types of gamma-ray detectors used for natural radioactivity logging are shown in Fig. 8: (1) the scintillation unit, (2) the Geiger–Mueller counter, and (3) the ionization chamber. The main advantages of the *ionization chamber* were simple construction and low-voltage operation. The difficulty of measuring currents of the order of 10^{-13} amp is a distinct disadvantage, however.

The *Geiger–Mueller counter* was advantageous because it produced large pulses that are easily transmitted. It is difficult to build, however, and is inefficient, like the ionization chamber. The ionization chamber and the Geiger–Mueller counter require long length to give adequate counting rates.

High efficiency (50–60%) and short length (a few centimeters) are the main advantages of a *scintillation detector.* As a result, this log gives a maximum of formation detail and a minimum of statistical fluctuation. Its primary disadvantage was the temperature sensitivity of the photomultiplier; this problem, however, has been overcome by insulation.

The gamma ray measures the radioactivity of the rocks and, consequently, delineates sands, shales, and carbonate rocks. Radioactivity, or gamma-ray, logs are sometimes used as *tracer logs.* The tracers may be gaseous, liquid, or solid isotopes, which are often tritium, radio-iridium, radio-iodine, or radio-cobalt.

The latter application of the gamma-ray log is in *gamma-ray spectral logging.* According to Holtz and Fertl (1979), in addition to the total gamma-ray counts, the spectralog measures gamma rays emitted by (1) potassium (K^{40}) at 1.46 MEV, (2) the uranium series nuclide bismuth (Bi^{214}) at 1.764 MEV, and (3) the thorium series nuclide thallium (Tl^{208}) at 2.614 MEV (see Fig. 8).

Gamma-Gamma Logs The gamma-gamma log was introduced in 1953 under the name of *density log.* Rock density in place is determined when using this technique for the purpose of aiding geophysicists and making allowance for density variations with depth in *gravimetric surveys.*

It became apparent, however, that the log had more promise for measuring *porosity,* and most of the developments have been directed toward the improvement of the porosity resolution of the density log. The main objective of the gamma-gamma log, from the formation evaluation standpoint, is the determination of the reservoir rock porosity by measurement of the rock's bulk density.

A collimated source of gamma radiation (usually Cs^{137} embedded in a wall skid) irradiates the formation. The gamma rays, which are emitted, interact with the electrons of matter around the wellbore. Two gamma-ray detectors about the source respond to the resulting back-scattered rays. The intensity of the back-scattered rays is dependent on the electron density of the matter penetrated. The intensity varies according to the density of the surrounding rock in much the same manner as light is absorbed and back-scattered in a fog of variable density.

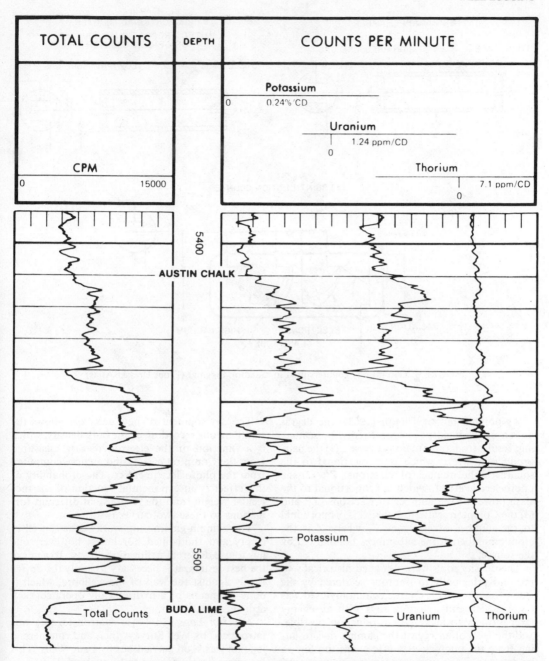

FIGURE 7. Spectralog response in a carbonate formation. High potassium content is characteristic of shales and is not related to the shaliness as read from the total counts curve (from Fertl, 1979, Fig. 11). This Spectralog is shown over the Eagle Ford Shale Formation in the Pearsall area, Frio County, Texas. The total counts (gamma ray) response appears to be completely unrelated to the shaliness of the formation. The K curve indicates a relatively clean, calcareous Eagle Ford Formation, particularly between 5,450 ft and 5,500 ft. Note the frequent correlation between decreasing K and increasing U concentration. Selective perforating in such zones often yields commercial production without any massive well stimulation.

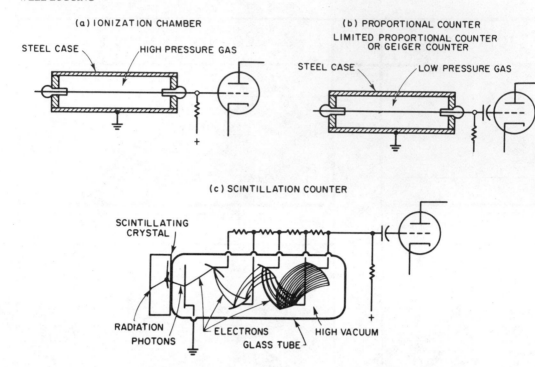

FIGURE 8. Gamma-ray counters used in radiation logging. (From Dresser Atlas)

As pointed out by Pirson (1963), the degradation of the incident photon energy, by scattering, results from three main processes: (1) the pair-production effect, which is proportional to the square of the number of electrons, Z^2; (2) the photoelectric effect, which is proportional to the sixth power of the number of electrons, Z^6; and (3) the Compton scattering, which is proportional to the number of electrons Z, where Z is the atomic number of the substances bombarded by the source.

The density tools were designed historically to respond principally to photons scattered by the Compton effect. The source commonly used was Cs^{137}, which emits gamma rays with an energy level too low to make pair production possible, and the level of energy of the gamma rays resulting from the photoelectric effect was too low to be detected. The count rate for gamma rays scattered by the Compton effect decreases exponentially with the electron density. Inasmuch as the sum of the (electrons)/(molecular weight of a compound) ratio is about constant, there is a direct relationship between the electron density and the bulk density. Two detectors are used to compensate for the borehole effect, especially the mud cake, and the computation is made in the surface equipment. The bulk density is recorded, as well as the correction applied by the system. The depth of investigation of the density tool is about 7 cm.

The development of new detectors allows the simultaneous recording of the bulk density, which is a function of the electron density measured from the Compton effect, and a curve computed from the photoelectric effect. The probability of this effect is an exponential function of the molecular weight, and the response of different formations to these two curves is different.

The gamma-gamma log works equally well in empty and fluid-filled boreholes; however, one has to use different calibration charts. To achieve the best results, it is necessary to apply the device firmly against the wall of the wellbore, which is done by means of a hydraulically operated back arm (Fig. 9).

Neutron Logs In 1941, neutron logging was introduced by Well Surveys Inc., and commercial logging was soon carried out by Lane Wells. This log immediately proved to be successful.

During logging operations, a neutron source bombards the formation with energetic neutrons at the rate of several million per second. The radius of a spherical neutron cloud that surrounds the source varies with the type of formation. Inasmuch as the neutrons are emitted from the source at high speed, they collide with nuclei of matter (mainly hydrogen) in the borehole and in the formation. Fast neutrons become epithermal neutrons and, finally, thermal neutrons before they are ultimately captured by chlorine or boron

far-detector to the near-detector counts. The tool detects thermal neutrons. The depth of investigation of this tool is about 40 cm in the case of low porosity and lower in the case of high porosity.

As the count rate decreases at the detector, the porosity increases. The neutron tools are mainly sensitive to the presence of hydrogen in the matrix or in the formation fluid. Shales show a high porosity because the neutrons do not distinguish between the free water and bound water of clays. Liquid hydrocarbons and water-bearing formations give about the same response because their hydrogen index (hydrogen density related to hydrogen density of fresh water at 75°F) is almost identical. Gas, however, has a much lower hydrogen index. The neutron log is useful for identifying gas-bearing formations, which show a very low porosity.

Acoustic (Sonic) Logs In 1948, the idea of using elastic waves to investigate boreholes was proposed by Humble Oil and Refining Co. After various stages of development, it was found in 1954 that the velocity of sound in rocks is related to their porosity. Thus, it became apparent that the acoustic log, or sonic log, is a valuable tool in formation evaluation.

The main objective of the acoustic log in formation evaluation is the determination of reservoir fluids and resistivity of interstitial water. The elastic log has also been used widely as a well completion tool for establishing the degree of cement bond to casing.

The determination of reservoir rock porosity is accomplished by measuring the transit time of elastic waves through the rocks over a short distance, usually about 0.6 m. A relationship among this transit time, lithology of the rocks, and the porosity, ϕ, has been established by M. R. J. Wyllie:

$$\phi = \frac{\Delta t - \Delta t_m}{\Delta t_m - \Delta t_f},$$

where Δt is the transit time through a unit length of formation, Δt_m is the transit time through a unit length of matrix, and Δt_f is the transit time through a unit length of fluid. Although this relation is fairly accurate for consolidated rocks, that is not the case for semiconsolidated shaly rocks. Thus, it is necessary to introduce corrections for disseminated clays and compaction to obtain reasonable porosity values. According to Schlumberger Wireline Services (see Vol. 1: *Log Interpretation*), the decrease in transit time is proportional to the value of the secondary porosity; i.e., the amount of secondary porosity can be estimated by subtracting porosity obtained by using density log from that obtained by using sonic log.

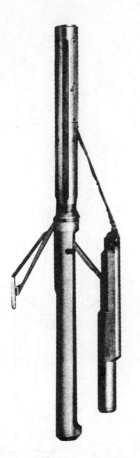

FIGURE 9. Dresser Atlas Densilog. (From Dresser Atlas, 1974, Fig. 7)

present in the formation or its interstitial waters. Their capture energies are emitted in the form of gamma rays. The neutron source, normally americium-beryllium or americium-plutonium, should be visualized as a spherical cloud, the radius of which is modulated by the hydrogen richness of the substance surrounding the source and the outer boundary of which shows a prevalence of slow neutrons.

The first type of neutron tool was used to detect emission gamma rays. The correlation between counts and porosity was normally matched with available cores or known information from nearby wells. Only one detector was used. Later, the sidewall neutron log was developed, whereby the tool is pressed against the formation. Inasmuch as only epithermal neutrons are detected, the log is not sensitive to the presence of strong thermal absorbers in the formation. Subsequently, the compensated neutron tool was developed, which is eccentralized by a bow spring. It has two detectors to minimize borehole effects, and the porosity is computed from the ratio of the

Compressional waves, having the fastest velocity, arrive first, followed by shear waves, the Rayleigh waves, and finally, the mud waves (signal going through the mud column only), which have very slow velocities (Fig. 10).

The sonic log reading is accomplished the following way: Two transmitters, in which electrical energy is converted into vibrational energy, are fired in sequence. They generate a short wave train of constant amplitude. Two or four receivers, depending on the tool design, receive the waves after they have traveled through the drilling fluid and the formation. The receivers are located 30 cm and 100 cm away from the transmitters, and the transit time per unit of length is computed from the difference between two traveltimes divided by the distance (span) between the receivers. The purpose of having two transmitters and four receivers is to compensate for the borehole effect (Fig. 11). The transit time should be measured between the corresponding points of the wave form at each receiver, and these points should be chosen as close to the interception as possible. A sufficiently high threshold level of noise is chosen before these pulses trigger the circuit. This step eliminates the effect of background noises. Inasmuch as the differential arrival time at the threshold amplitude of the received pulses is not the same as that of the points of interception, an error is introduced, which may be compensated by amplification of the wave receiver R_2 slightly more than R_1. The ratio of the wave attenuation for a 1 m average interval is about 1.4.

An acoustic log should be run in a fluid-filled open hole. The presence of drilling fluid in the hole is required to provide acoustic coupling between the tool and the formation. The absence of drilling fluid would prevent an effective energy transmission to the formation. Similarly, a gas-cut drilling fluid will reduce the coupling, and logging a gas- or air-filled hole is impossible. The cycle-skipping feature in elastic logging is char-

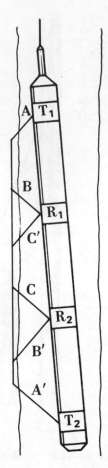

FIGURE 11. Acoustic tool design, showing dual transmitters and dual receivers. (From Dresser Atlas, 1974, Fig. 5)

acteristic of a gas-cut drilling fluid and of fractures. It also happens in large or caved holes, where the amplitude of the signal at the receiver is too low.

Sonic tools include special transmitters polarizing the vibrational energy. A series of tools has been designed where the distance between receivers and transmitters has been increased up to 8 ft, 10 ft, and 12 ft and longer (long-spaced sonic logs). This new design enables a greater depth of penetration and allows a more accurate transit-time computation in shales, especially in the shallow formations. In some cases it also makes possible sonic logging behind casing. The amplitude of the signal can also be recorded, and in open-hole logging, the pattern of the wave trains can help identify fractures.

The *cement bond log* uses the same principle to estimate the compressive strength of the cement. Only one receiver is used to record the amplitude.

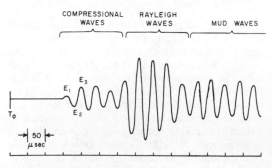

FIGURE 10. Schematic diagram of sonic wave train in a hard formation. (From Schlumberger, 1972, Fig. 7-4)

The tool must be kept centered in the well because even a very slight eccentralization will reduce considerably the amplitude of the signal. The minimum amplitude of the reflected acoustic wave indicates good bond, whereas free casing exhibits a high amplitude, which is a function of the type of drilling fluid and its properties, the casing size, and the casing weight.

Temperature Logs Well temperature is measured as a function of depth for establishing the presence of anomalies due to lithology and fluid flow, especially gas flow. The log is used to locate the point of gas entry into casing, the lost circulation zone, water entry, casing leaks, tubing leaks, the top of the cement column behind the casing, gas-oil contacts, and oil and gas productive zones by measuring the volumetric gas-flow rates from individual formations producing simultaneously into a well. It also helps determine the presence of gas channeling and the results of well stimulation in gas wells. The measurement of thermal potential is accomplished with a resistance thermometer (or more conveniently a thermistor), which is run into the well where the temperature is recorded at the surface by the degree of unbalance of a Wheatstone bridge. The temperature indicated is an exponential function of time, even though such thermometer elements require very little heat to reach the temperature equilibrium, and each thermistor has its own time constant. The speed of logging must be adjusted to the time constant of the thermometer so that measurements can be obtained within a certain degree of accuracy.

Nuclear Magnetic Logs The objective of nuclear magnetic logs is to delineate fluid-bearing zones, under all possible borehole conditions, and to distinguish between water and oil zones. A more specific objective of the log is to determine the amount of free fluids (fluids, the physical or chemical bonding of which allows proton precession) in the reservoir rock (both oil and water) by making direct measurements of the free-hydrogen content of rocks. Furthermore, this log is intended to identify the nature of the free fluids and the amount of each as a percentage of pore space. Thus, the nuclear magnetic log should give some idea of the free oil and water, the amount of each that may be produced (relative permeability), and the well productivity.

Nuclear magnetism is responsive only to hydrogenous fluids in the reservoir during operating borehole conditions. Hydrogen is the only common nucleus present in the reservoir rock fluids that has a sufficiently large magnetic moment and nuclear spin to be identified under wellbore conditions. A few other elements are present as traces in sediments and have the required properties— e.g., aluminum, sodium, iron, potassium, calcium, silicon, sulfur, and oxygen. They have too weak a magnetic moment and spin, however, to make detection possible.

According to Pirson (1963), the magnetic moment and nuclear spin of the hydrogen nucleus cause it to behave like a bar magnet and gyroscope combination in the magnetic field of the Earth. This is analogous to a spinning top in the Earth's gravitational field. If the top is disturbed by a push, the axis of rotation precesses around the direction of the gravitational field. The spinning bar magnets, constituted of protons, tend to be oriented in the direction of the magnetic lines of force on application of a strong magnetic field to a hydrogenous substance. Thus, magnetic polarization results, which is proportional to the applied field strength. Sudden removal of the magnetic field causes the proton polarization to disappear under the scrambling effect of temperature, but the new equilibrium is not reached until a certain time after removal of the polarizing magnetic field. The relaxation time is dependent on the nature of the hydrogenous fluid and of the enclosing rock. In the scrambling process, an alternating magnetic field that generates a voltage in a suitable receiver coil is created by the freely precessing protons. The initial voltage received at time zero is directly proportional to the number of hydrogen nuclei that have been affected by the magnetizing coil.

The hydrogenous substance has to be in a fluid state of matter to respond (free fluid); therefore, there is no response from solid hydrocarbons, such as tar, and from asphalt and other polar hydrocarbon compounds that are adsorbed on the rock surfaces and render them oil-wet. The same is true of water of hydration in clays.

Pulsed Neutron Capture Logs With the pulsed neutron capture logs, the measurement of the lifetime of a neutron and, consequently, the capture cross-section of the formation is made possible. The new tools include Neutron Life Time (Atlas), TDT (Schlumberger), C/O (Atlas), and GS7 (Schlumberger).

Chlorine is one of the elements that have the highest capture cross-section. According to Dresser Atlas (1974), the capture cross-section of a formation is going to increase with increasing porosity and formation water salinity and a decreasing amount of hydrocarbons. No chemical source is used for running this log, but an electrical source is turned on for a very short period. Then, for different time windows after the neutron pulse, the gamma rays emitted by absorbed thermal neutrons are counted. From the exponential gamma-ray decay, the lifetime of neutrons can be computed. Knowing the cross-section capture of the rock and of the formation fluids, the water saturation can be determined.

This log can be run in an open or cased borehole and through tubing in a producing well. One of the main applications of the pulsed neutron capture log is to monitor the location of the water table during different stages of reservoir development. The log run before starting production will look very similar to a resistivity survey because the capture cross-section decreases when the resistivity increases in a hydrocarbon-bearing formation. In addition, it can be possible to find oil in an old reservoir, drilled before sophisticated logging techniques were developed. Dresser Atlas (1974) estimated that the volume investigated by this log is about 1 m³.

Dipmeter Logs The purpose of a dipmeter log is to record the dip angle and the dip direction of a formation. In the field, four microresistivity curves are recorded, which are 90° apart but in a common plane. The deviation and the azimuth of the well, the position of each pad with respect to north, and two caliper curves are recorded as well. As the tool is opened hydraulically, each pad is pressed against the formation.

At the end of the survey, the curves are correlated with each other and a continuous plot of the dips is made. Other applications of this tool are the identification of structures, stratigraphic traps, faults, and unconformities. By an overlap of the microresistivity curves, through another computer program, density, width, and direction of fractures can be identified. Excellent dip log computer programs are available from, e.g., Dresser Atlas.

Logging Programs for Different Types of Formations

The choice of a resistivity log depends on drilling fluid resistivity, formation water resistivity, depth of filtrate invasions, and estimated resistivity of the formation. For fresh-water muds, the induction log will give the most accurate formation resistivity values; if the muds are very salty, however, the focused resistivity logs will give better results. Often, both an induction log and a laterolog will be run, especially if some zones have low resistivity whereas some others have high resistivity or if a vertical resolution better than the one obtained with the induction log is required.

Microresistivity devices can be run in combination with either the induction or the laterolog tool. In some special cases—e.g., in the case of presence of very thick mud cake and very large mud filtrate invasion—a proximity log can be run in combination with a microlog.

In empty holes or with oil-base drilling fluids, only the induction log can be run, and the true resistivity and the saturations are found from two curves recorded—i.e., deep-induction and medium-induction logs.

The porosity estimation may depend largely on the exact knowledge of the lithology of rocks. In many fields already developed, only the sonic log is run, and the porosity is computed from the actual transit time, knowledge of the matrix transit time, and compaction factor. During the exploration phase or if the lithology is complex, the gamma-gamma log and the neutron log are run together, and porosity (and lithology) can be found from the crossplots. A crossplot of formation density versus neutron porosity gives a good resolution between sandstone, limestone, and dolomite. In addition, the most common evaporites—i.e., rock salt, anhydrite, and gypsum—are easily identified (see Fig. 12).

Evaporites are easily identified on a crossplot of formation density versus sonic transit time; however, resolution between sandstone, limestone, and dolomite is poor. In contrast, the sonic-versus-neutron crossplot shows good resolution for sandstone, limestone, dolomite, salt, and gypsum. Anhydrite, however, is not easily differentiated from low-porosity limestone (Fig. 13). By comparing density vs. neutron porosity and sonic porosity vs. neutron porosity crossplots, the secondary porosity index can be computed. All three porosity logs can also be combined into one plot that enables determination of lithology, shaliness, and gas effect, independent of porosity. Numerous crossplot concepts for well-logging parameters have been presented by Fertl (1978). Raw and/or calculated logging parameters can be crossplotted on linear, semilogarithmic, and other exponential scales. Input data include resistivity, acoustic, and nuclear logging measurements and, wherever available, core, test, and production data.

The following logging program is presented here as an example: First, an induction log is run in combination with both an elasticity and a natural gamma-ray log. The resistivity at two or three different depths of investigation is recorded. The transit time, natural radioactivity, and spontaneous potential are also recorded. Then a formation density log is run in combination with a neutron log. The bulk density, neutron porosity, borehole size, and sometimes, the natural radioactivity (for correlation purposes) are recorded. In case of need, a laterolog is run in combination with a microresistivity log. A dipmeter log is very often run last. Spectral gamma ray can be combined with the neutron and density logs.

Logs are recorded in analog and digital forms at the wellsite. Prolog, Cyberlook, and Laserlog interpretations are also performed at the wellsite. Oil saturation, porosity, and shaliness are calcu-

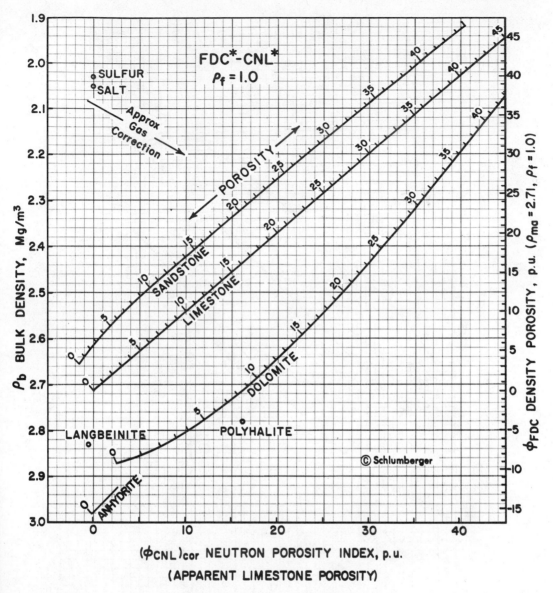

FIGURE 12. Cross-plot of formation density log versus neutron log, showing responses for various lithologies and porosities (water-filled holes). (From Schlumberger Technology Corporation)

lated. Dip of the formation can also be computed. The wellsite computer can also produce different crossplots and histograms. It can play back the logs with other than standard parameters. Playback can also achieve a better depth accuracy. Finally, printouts can be obtained at the wellsite.

As shown in Fig. 14, e.g., the Cyberlook log presents porosity and water saturation data computed from the resistivity and porosity logs. It is normally produced within 1–2 hr after completing log acquisition. This log enables rapid computation of reservoir parameters and helps with important decisions, such as setting casing, drilling further, or plugging and abandoning. The dual

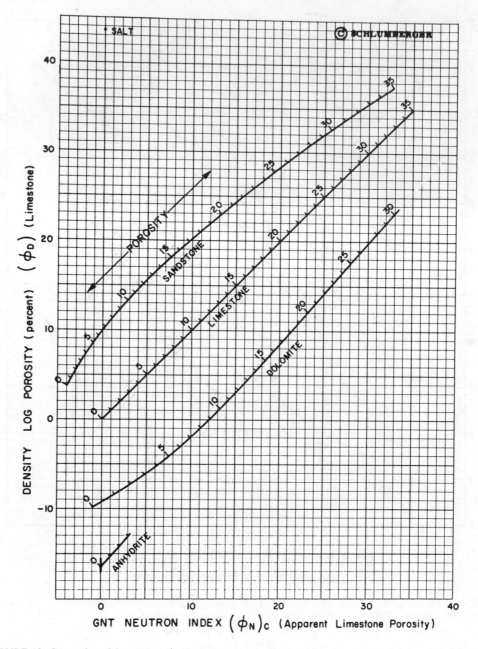

FIGURE 13. Cross-plot of formation density log versus neutron log, showing responses for various lithologies and porosities (water-filled holes). (From Schlumberger Technology Corporation)

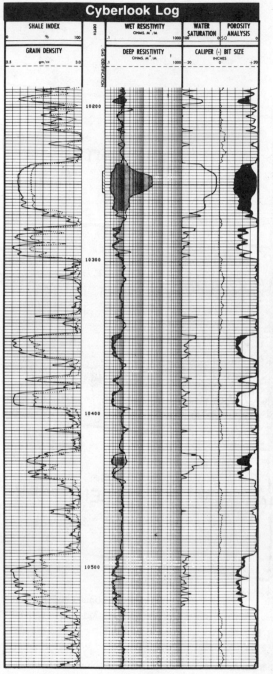

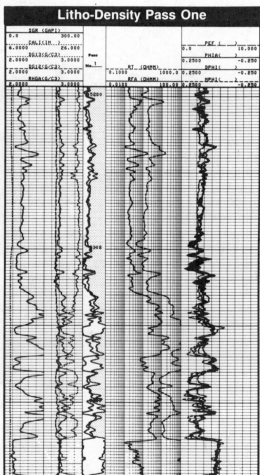

FIGURE 14. Cyberlook logs and Litho-Density log (LDQL). (From Schlumberger Well Services, 1985, 5)

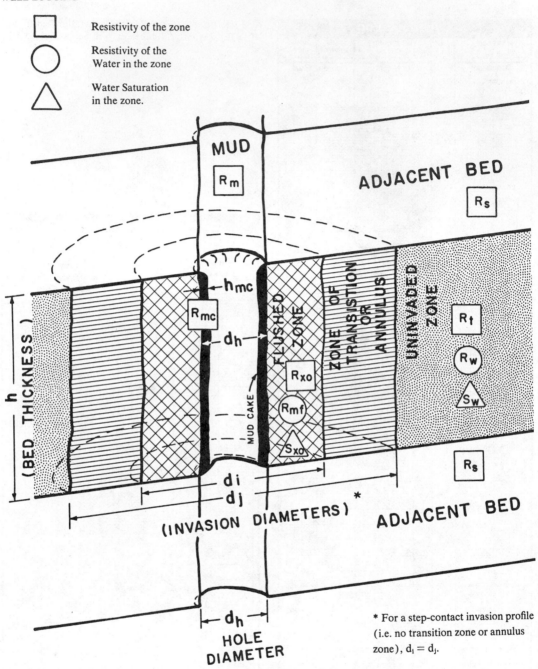

□ Resistivity of the zone

○ Resistivity of the
Water in the zone

△ Water Saturation
in the zone.

FIGURE 15. Symbols used in log interpretation (schematic). (From Schlumberger Technology Corporation)

water concept deals systematically with the effects of shaliness on logs. Litho-Density, EPT electromagnetic propagation, and NGS natural gamma ray spectrometry logs are used in this Schlimberger program for enhanced results.

It is of utmost importance to state here, however, that although computer processing allows quick and easy data handling, hand plotting still provides an effective check for an experienced log analyst.

Symbols used in log interpretation (schematic) are presented in Fig. 15.

GEORGE V. CHILINGAR
FRANÇOIS L. MARGOT
LUCAS G. ADAMSON
ROBERT A. ARMSTRONG
WALTER H. FERTL
IRAJ ERSHAGHI
MOHAMED N. HASHEM

References

Alger, R. P., 1972, *Modern Logging Programs and Interpretation Methods.* Houston, Texas: Schlumberger Well Surveying Corporation, II-59–II-81.

Brock, J., 1985, Applied open-hole log analysis, in G. V. Chilingar, ed., *Contributions in Petroleum Geology and Engineering.* Houston, Texas: Gulf Publishing Co.

Chilingar, G. V., and C. M. Beeson, 1969, *Surface Operations in Petroleum Production.* New York: Elsevier Publishing Co., 397p.

Dakhnov, V. N., 1959, *Geophysical Well Logging*, G. V. Keller, trans. Moscow: Gostoptekhizdat.

Desbrandes, R., 1968, *Theorie et Interpretation des Diagraphies.* Paris: Edition Technip, 545p.

Desbrandes, R., 1985, *Encyclopedia of Well Logging.* Houston, Texas: Gulf Publishing Co., 584p.

Dresser Atlas, 1974, *Log Review,* vol. 1. Houston, Texas, 177p.

Fertl, W. H., 1978, Open hole crossplot concepts—A powerful technique in well log analysis, paper presented at the European Offshore Petroleum Conference and Exhibition, London, October 24–27.

Fertl, W. H., 1979, Gamma ray spectral data assists in complex formation evaluation, *Soc. Petrol. Well Log Analysts 6th European Eval. Symp. Trans.,* March, 11p.

Fertl, W. H., and G. V. Chilingar, 1978, Formation evaluation of tar sand using geophysical well-logging techniques, in G. V. Chilingar and T. F. Yen, eds., *Bitumens, Asphalts and Tar Sands.* Amsterdam: Elsevier, 259–276.

Guyod, H., 1944–1945, Electrical well logging fundamentals, *Oil Weekly* **114–115**, Aug. 7–Dec. 24.

Hamilton, R. G., 1960, Application of the proximity log, paper presented at the Society of Well Logging Analysts Annual Symposium.

Haun, J. D., and L. W. LeRoy, 1958, *Subsurface Geology in Petroleum Exploration.* Golden, Colo.: Colorado School of Mines, 887p.

Hotz, R. F., and W. H. Fertl, 1979, Spectralog applications in complex formation evaluation, *Drilling-DCW,* August.

Lynch, E. J., 1962, *Formation Evaluation.* New York: Harper & Row, 422p.

Pardue, G. H., R. L. Norris, L. H. Gollwitzer, and J. H. Moran, 1962, Cement bond log—A study of cement and casing variables, *SPE Paper No. 453.* Los Angeles, Calif.: Soc. Petroleum Engineers Fall Meeting, 15p.

Pirson, S. J., 1963, *Handbook of Well Log Analysis.* Englewood Cliffs, N.J.: Prentice-Hall, 326p.

Pirson, S. J., 1968, Redox log interprets reservoir potential, *Oil and Gas Jour.* **29**, 69–75.

Schlumberger Educational Services, 1987, *Log Interpretation Principles/Applications.* Houston, Texas, 198p.

Schlumberger Well Services, 1985, *Cyber Service Unit. Wellsite Products and Calibration Guide,* CP28. Houston, Texas, 77p.

Schlumberger Well Surveying Corp., 1972, *Log Interpretation,* vol. 1. Houston, Texas, 113p.

Serra, O., 1987, *Fundamentals of Well-Log Interpretation. 1. The Acquisition of Logging Data,* Developments in Petroleum Science, 15A. Amsterdam: Elsevier, 424p.

Tixier, M. P., R. P. Alger, W. P. Biggs, and B. N. Carpenter, 1963, Dual induction Laterlog, a new tool for resistivity analysis, *SPE Paper No. 713.* Soc. Petroleum Engineers Fall Meeting, Oct. 6-9.

Walker, T., 1967, Utility of micro-seismogram bond log, *SPE Paper No. 1751.* Fort Worth, Texas: Soc. Petroleum Engineers Regional Meeting, 5p.

Youmans, A. H., E. C. Hopkinson, and H. I. Oshry, 1964, Neutron lifetime, a new nuclear log, *Trans. AIME* **231**, 319–328.

Cross-references: *Borehole Drilling; Exploration Geophysics.* Vol. XIII: *Magnetic Susceptibility, Earth Materials; Well Data Systems.*

INDEXES

AUTHOR CITATION INDEX

SUBJECT INDEX

899

Van der Wall's attractive forces, 779

VCG. *See* Vertical center of gravity

VEGETATION MAPPING, 8, **853–855**

Vehicle and foot survey, 253

Veins, map symbols, 551

Velometer, distance measurement, 86

Verification of a hypothesis, 745

Verisimilitude, truth of a hypothesis, 746

Vertebrate paleontology, 206

Vertical air photos, 650

Vertical center of gravity (VCG), 201

Vertical illuminations, maps, 458

Very low frequencies (VLF), electromagnetic field, 10

Vesicles, 335

Vibratory suction dredge, marine mining, 570

VLF anomaly pattern, 861

VLF ELECTROMAGNETIC PROSPECTING, **855–862**

VLF-EM measurements, 860

VLF-EM prospecting, 856

VLF-EM survey, 861

Void ratio, 780

Volcanoes, map symbols, 522

Vortex tube, sediment ejector, 62

Vulcanist theory, 848

Wadati-Benioff Zone, plate tectonics, 665

Waste producers, cities, 70

Water
bores and wells, map symbols, 519
quality, allowable impurities for human consumption, 195
resistance, explosive materials, 37
resources, 112
samples, acquiring and preserving, 194
sampling, 122
supply, to cities, 73
table, 115
tanks and dams, map symbols, 521

Watershed Protection and Flood Prevention Act, 391

Waxes, 587

Weathered materials, description of, 794

Weathering, 290, 711
mantles, 729

Weeks Act of 1911, 392

Welded tuff (ignimbrite), 338

Wellbore resistivity, 865

WELL LOGGING, **863–879**

Well drillers, 318

Well logs
borehole geophysics, 132
induction log, 866
density logging, 51
driller's logs, 863
dual laterlog, 867
gamma-gamma log, 868, 870
laterlog, 866
laterlog 7, 866
laterlog 8, 866
microlaterlog, 867

microlog, 867
neutron log, 870
nonelectrical logs, 867
normal logs, 864
resistivity log, 320
sonic log, 872
spectralog, 868, 869
pontaneous logs, 864
temperature logs, 320

Westward drift, magnetic field, 297

Westward precession, magnetic field, 297

Wet ashing, 31

Wettability, mineral property, 591

Wheeler Survey, 268

Wide-beam sonar, 2

Wiedespuren, 209

Wilderness Act of 1964, 392

Wild and Scenic Rivers Act of 1968, 392

Wohnbauten, 209

Word compounds, use on maps, 416

World maps collections, 427

Xanthoria elegans, 398

X-ray fluorescence, 31, 560

Yellowstone Act of 1872, 392

Young's modulus, 38

Yttrium, element, 352

Zeeman effect, 26

Zersatz zone, laterite deep weathering profile, 730

Zinc, element, 349

Zirconium, element, 357

Zoom stereoscopes, 651